Handbook of the Marine Fauna of North-West Europe

Edited by

P. J. HAYWARD AND J. S. RYLAND

Marine and Environmental Research Group,
School of Biological Sciences,
University of Wales, Swansea

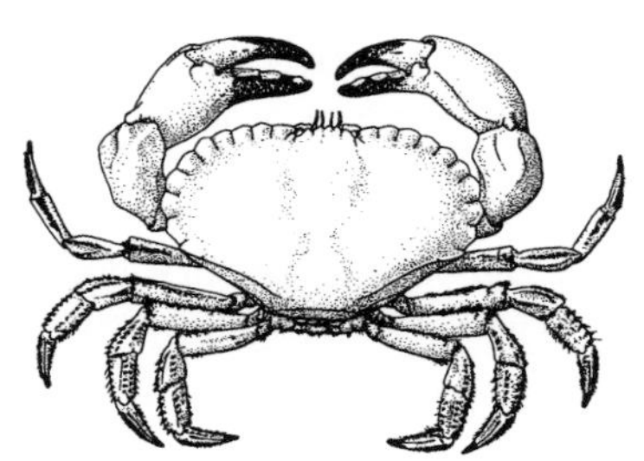

OXFORD
UNIVERSITY PRESS
Great Clarendon Street, Oxford OX2 6DP
Oxford University Press is a department of the University of Oxford.
It furthers the University's objective of excellence in research, scholarship,
and education by publishing worldwide in
Oxford New York
Athens Auckland Bangkok Bogotá Buenos Aires Calcutta
Cape Town Chennai Dar es Salaam Delhi Florence Hong Kong Istanbul
Karachi Kuala Lumpur Madrid Melbourne Mexico City Mumbai
Nairobi Paris São Paulo Shanghai Singapore Taipei Tokyo Toronto Warsaw
with associated companies in Berlin Ibadan
Oxford is a registered trade mark of Oxford University Press
in the UK and in certain other countries
Published in the United States
by Oxford University Press Inc., New York

First published 1995
Reprinted (with corrections) 1996, 1998, 2000

A catalogue record for this book is available from the British Library
Library of Congress Cataloging in Publication Data
Handbook of the marine fauna of north-west Europe / edited by P. J. Hayward and J. S. Ryland
1. Marine fauna—Europe. I. Hayward, P. J. II. Ryland, J. S. (John Stanley)
QL253.H36 1995 591.94—dc20 94-36396
ISBN 0 19 854054 X (Hbk)
ISBN 0 19 854055 8 (Pbk)
Printed and bound in Great Britian by
Antony Rowe Ltd, Chippenham, Wiltshire

Handbook of the Marine Fauna of North-West Europe

PREFACE

This book is based on the two-volume *The marine fauna of the British Isles and north-west Europe* (Clarendon Press, 1990). It is a more compact work, with fewer species and rather less detail than was included in its progenitor. It is not, however, simply an abridgement. While the total number of species treated has been reduced, a small number of additional species has been included, and recent taxonomic revisions have resulted in occasional name changes. Most of the illustrations have been reprinted, but a few new figures have been added and others replaced. A number of plates have been reorganized to aid clarity or provide more convenient comparisons.

It has been difficult to make the selection of animals to be covered in a concise edition, and in doing so we have given considerable thought to the purposes of this book. These are to provide swift and simple identifications of the animal species likely to be encountered during a shore-based marine field course, or during inshore coastal studies using scuba or the Agassiz trawl. We consider we have been generous in the case of molluscs and decapods, which tend to attract the eye more readily than hydroids, small worms, or copepods. Thus, the coverage of these groups is still extensive. Some groups requiring specialist techniques for collection, examination, and identification have been excluded entirely, for example the Protozoa, Pogonophora, and Tardigrada. The selection of fish is limited to intertidal and coastal species, as are those of sponges, amphipods, and mysids. Despite these exclusions, we are confident that this book will serve its purpose, allowing student, amateur, and professional biologists to identify with accuracy the preponderance of animal species encountered between the tide-marks and in the shallow subtidal around the British Isles and neighbouring coasts of mainland Europe. With the exception of the ctenophores and large pelagic scyphozoans, which frequently beach on sandy shores, all of the invertebrate animals included belong to the benthos. The plankton and the meiofauna, although interesting and abundant, and certainly deserving of attention, are too numerous and diverse for even a representative selection to be included here. Together with sea birds and marine mammals they have an extensive literature of their own.

The format established for the larger predecessor has proved successful and is followed here. Introductory accounts to each group are necessarily more brief and the keys, designed for the identification of smaller suites of species, have had to be modified. Wherever possible these have been simplified in comparison to the originals, and have occasionally been wholly rewritten; we have attempted to improve the labelling of structures in the plates, and in general to make this book easy to use. Inevitably, rare animals are occasionally encountered. If an animal does not appear to key out or fails to fit any of the descriptions, recourse must be had to the larger work, or to a specialist text.

The task of preparing these concise accounts and modified keys has been the responsibility of the contributors following the directions, admittedly sometimes vague, of the editors. Each contributor has had a slightly different view as to how compaction should be approached. While the editors have ensured that each conforms generally to the outlines provided, none the less each chapter or section has a certain individuality which we hope will add to the interest of the book.

Swansea
1994

P. J. H.
J. S. R.

ACKNOWLEDGEMENTS

The editors wish to thank all of the contributors for responding so readily to the call for new scripts, so soon after their Sisyphean labours with the original Fauna. We much appreciated the positive and constructive comments of colleagues and students, which assisted us in making improvements to certain areas of the text, keys, and illustrations. We would also express our most sincere thanks to Jan Greengo for her dedicated, and greatly valued, help throughout the preparation of this book.

Full acknowledgements for the illustrations can be found in the original *Fauna*: Hayward, P. J. and Ryland, J. S. (1990), *The marine fauna of the British Isles and north-west Europe* (2 vols), Oxford: Clarendon Press.

CONTENTS

CONTRIBUTORS

P. F. S. CORNELIUS
Dept. of Zoology, The Natural History Museum, London
Chapter 4, Hydrozoa (with J. S. Ryland).

P. E. J. DYRYNDA AND ELISABETH A. DYRYNDA
School of Biological Sciences, University of Wales, Swansea
Chapter 3.

R. GIBSON
Dept. of Biology, Liverpool John Moores University
Chapter 5, Nemertea (with E. W. Knight-Jones).

P. J. HAYWARD
School of Biological Sciences, University of Wales, Swansea
Chapters 1 and 2; Chapter 8, Amphipoda; Chapter 10, Bivalvia, Cephalopoda Scaphopoda, and Polyplacophora.

M. J. ISAAC
Dept. of Adult and Continuing Education, University of Wales, Swansea
Chapter 8, Ostracoda—Copepoda, Cumacea, Tanaidacea

P. E. KING
School of Biological Sciences, University of Wales, Swansea
Chapter 9 (with P. J. A. Pugh).

E. W. KNIGHT-JONES
School of Biological Sciences, University of Wales, Swansea
Chapter 5, Nemertea (with R. Gibson); Chapter 6, Hirudinea;
Chapter 7 (excluding Entoprocta); Chapter 13, Hemichordata.

PHYLLIS KNIGHT-JONES
School of Biological Sciences, University of Wales, Swansea
Chapter 6, Sabellariidae, Sabellidae, Serpulidae, and Spirorbidae.

P. MAKINGS
School of Biological Sciences, University of Wales, Swansea
Chapter 8, Mysidacea.

R. L. MANUEL
Dept. of Zoology, University of Oxford
Chapter 4, Anthozoa.

J. MOYSE
School of Biological Sciences, University of Wales, Swansea
Chapter 8, Cirripedia, Reptantia; Chapter 12 (with P. A. Tyler).

CONTRIBUTORS

E. NAYLOR
School of Ocean Sciences, University of Wales, Bagnor
Chapter 8, Isopoda.

A. NELSON-SMITH
School of Biological Sciences, University of Wales, Swansea
Chapter 6 (excluding Sabellariidae—Spirorbidae).

P. J. A. PUGH
British Antarctic Survey, Cambridge
Chapter 9 (with P. E. King).

J. S. RYLAND
School of Biological Sciences, University of Wales, Swansea
Chapter 4, Scyphozoa, Hydrozoa (with P. F. S. Cornelius), Ctenophora; Chapter 11; Chapter 13; Urochordata; Chapter 14.

G. SMALDON
formerly: School of Biological Sciences, University of Wales, Swansea
Chapter 8, Caprellidae, Natantia.

P. A. TYLER
Department of Oceanography, University of Southampton
Chapter 12 (with J. Moyse).

G. D. WIGHAM
School of Biological Sciences, University of Plymouth
Chapter 10, Prosobranchia.

NATHALIE YONOW
School of Biological Sciences, University of Wales, Swansea
Chapter 10, Opisthobranchia.

ILLUSTRATORS

SIMON CHEW
formerly: Blackpool and the Fylde College

DANIEL COLE
formerly: Blackpool and the Fylde College

NIGEL GERKE
formerly: Carmarthen School of Art

R. GIBSON
Dept. of Biology, Liverpool John Moores University

TONI HARGREAVES
formerly: Blackpool and the Fylde College

P. J. HAYWARD
School of Biological Sciences, University of Wales, Swansea

PHYLLIS KNIGHT-JONES
School of Biological Sciences, University of Wales, Swansea

PAUL J. LLEWELLYN
School of Biological Sciences, University of Wales, Swansea

P. J. A. PUGH
British Antarctic Survey, Cambridge

J. S. RYLAND
School of Biological Sciences, University of Wales, Swansea

NATHALIE YONOW
School of Biological Sciences, University of Wales, Swansea

1 INTRODUCTION

From Shetland to Scilly, the British Isles extend through 11 degrees of latitude, a distance of about 1250 km. Between eastern England and western Ireland they range through 13 degrees of longitude, about 900 km. The northern and western boundaries of this region face the open Atlantic; oceanic influences, locally modified by geographical and topographical factors, largely modulate the character of the marine environment. The eastern side of the British Isles bounds a shallow epicontinental sea; oceanic influences in the North Sea decrease towards the south, where the proximity of the European landmass has a different modifying effect. In the south, the complex hydrography of the Channel results in rapidly changing physical characteristics along its length, again with discernible environmental consequences. The Irish Sea, and major embayments such as the Bristol Channel, Liverpool Bay, the Scottish firths, and the Wash, are also characterized by particular physical conditions and, if geography is stretched to include the Channel Isles, the Gulf of St. Malo exemplifies further modifications of the marine environment. The effects of major physical factors are everywhere mediated by local geographical and topographical influences. Thus, throughout the region there exist marked gradients in the marine environment which are reflected in profuse variation in the characteristics of different marine habitats, with consequent effects on the nature of the marine fauna.

The marine fauna of the British Isles is rich and varied. The coastline encompasses almost all types of temperate intertidal habitat, from hypersaline and brackish lagoons, estuaries, and coastal marsh and mudflats, to sandy and rocky shores with every degree of exposure and widely varying profile. Subtidal habitats are equally diverse. Each local habitat reflects prevailing environmental factors and is further characterized by its biota. Thus the marine fauna itself demonstrates gradients of change throughout the British Isles.

The sea has been of economic importance throughout the histories of the nations inhabiting the British Isles, and marine science has an increasing significance today. Long-established fishing traditions have developed into major industries which now demand careful international management, based on continuous research and monitoring. Industrialization and urban development, particularly of coastal wetlands and estuaries, have had a profound impact on both inshore and offshore habitats. Improved ecological understanding is necessary if the effects of current levels of development are to be assessed accurately. This is needed with increasing urgency; coastal barrage schemes, for example, almost certainly leading to further industrialization, will have profound effects on marine habitats over very wide areas. Environmental quality is accorded increasing importance for both economic and aesthetic reasons, and marine biological research is important in this context, not only in view of the need to measure and control the effects of urban, industrial, and radioactive effluents, but also because some marine ecosystems are useful as early indicators of environmental contamination.

One field of marine research with a long history is that concerned with fouling organisms. Rapid, dense, and persistent growth of sessile invertebrates, leading to extensive fouling of harbours, docks, the hulls of ships, and power station water intake systems, continues to be a problem. Toxic substances, such as organo-tins, used to control fouling may also adversely affect the naturally occurring fauna, especially in enclosed waters. The development of offshore structures for oil drilling, mineral extraction, and power generation is likely to increase during the rest of the twentieth century, and marine fouling research will continue to be important. The impact of such future developments on marine habitats and their animal communities will require continuous monitoring. A final consideration, perhaps likely to loom larger in the future, is marine biology as a leisure pursuit. Academic research is long established, and will certainly continue, but there is a rapidly expanding amateur interest, at least partly fuelled by the popularity of scuba diving.

Study or use of marine habitats needs to be informed and considerate. Increased access to our marine heritage must be accompanied by habitat awareness and responsibility. The adverse effect of visitors on the marine biota is not a new phenomenon, and the deterioration of shores in Torbay in the Victorian era was recorded by Edmund Gosse (1907, *Father and son*), but the number of people able to visit the coast is hugely greater now. Marine biology will also continue to expand steadily as both an academic and applied discipline, and will contribute increasingly to other, more specialized fields of biological science. Specialization must be founded upon a basic background in marine biology, which undoubtedly needs to be more widely taught.

The remaining sections of this introductory chapter deal briefly with those features of the marine environment most important to an understanding of the nature of local marine habitats, together with a discussion of distribution patterns.

HYDROGRAPHY

The primary distinction between the sea areas to the west and to the east of the British Isles (Fig. 1.1) is related to the proximity in the west of the edge of the European continental shelf, generally accepted to coincide with the 200 m isobath. From the Celtic Sea the edge of the shelf trends north-eastwards close to the western coasts of Ireland and Scotland, turning abruptly south-east north of the Shetlands as the edge of a deep trough, the Rinne, which flanks the western and southern coast of Norway. The 100 m isobath follows a similar route, but lies closer inshore in the south-west, extends into the southern Minch between the Hebrides and mainland Scotland, and subsequently continues south-east from Shetland to Aberdeen before turning east, marking the edge of a broad area of deep shelf situated between north-east Scotland and the Rinne. Beyond 200 m depth, topographic features of the sea floor influence the deep water circulation of the north-east Atlantic. The deep channel between the Shetland–Hebrides shelf and The Faeroes is interrupted by a north-west trending ridge, the Wyville Thomson ridge, which forms a barrier affecting the free interchange between cold water of the deep Norwegian Sea and the warmer water of the north-east Atlantic. However, shallow water circulation patterns are more regularly related to the 200 m or 100 m isobaths. Much of the sea floor of the central and southern Irish Sea lies below the 50 m isobath, as do the central and western areas of the English Channel. To the east, the Southern Bight of the North Sea is mostly shallower than 50 m while the central region lies between 50 m and 100 m depth. Thus, the west of the British Isles is characterized by a great and relatively rapid shallowing towards the coast, while the North Sea shoals gradually and gently towards the south.

There are no marked areas of strong upwelling around the British Isles. The region is characterized by fluctuating gradients of changing physical factors related to the residual flow of the North Atlantic drift. The rate of flow, volume, and course of the North Atlantic drift vary from year to year, and surface current patterns may similarly be modified by seasonal effects such as wind and temperature. However, the gross features of the circulation are relatively constant. The North Atlantic drift passes north-eastwards along the western coasts of Ireland and Scotland. A portion is deflected into the Celtic Sea, and from there both northwards into the Irish Sea and eastwards along the Channel. These two elements of the circulation are particularly variable. North of Scotland the flow divides; the larger part continues north-eastwards but a portion turns southwards into the North Sea. It is thought that from the Hebrides to Shetland and into the

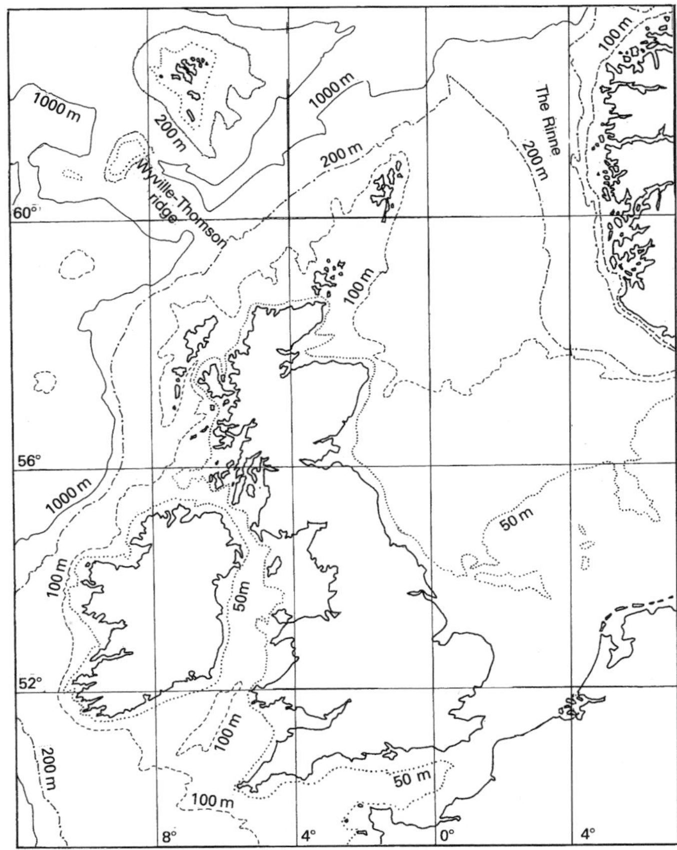

Fig. 1.1 Bathymetric map of the British sea area.

North Sea much of the water of the North Atlantic drift flows along the 200 m isobath, and as it flows southwards in the North Sea is partly influenced by the 100 m isobath. Circulation within the North Sea is variable, but important features are a northern, clockwise flow, which incorporates northward-flowing water originating from the Kattegat, and a smaller anticlockwise gyre in the south, which varies considerably according to the volume of water flowing eastwards through the Dover Strait and the southward-flowing coastal current which passes down the eastern side of Britain. This eastern coastal current represents a portion of the North Atlantic drift which passes between Orkney and Shetland, and between Orkney and mainland Scotland. The volume of this flow, and its degree of deflection towards the eastern coast, vary greatly from year to year. It should be noted that the circulation patterns described above all relate to surface currents, and all may be influenced and modified by water movements at greater depth.

Surface waters are characterized by their plankton species, and the distribution and abundance of indicator species in a particular year give a rough guide to the origins of the different water masses. Most simply, through late spring to autumn, this allows distinction to be made between resident plankton communities in the North Sea, English Channel, and certain large embayments such as the Bristol Channel and Morecambe Bay, and oceanic species marking influxes of fresh oceanic water. In regions of mixing, a plankton community characterized by the chaetognath *Sagitta elegans* is found, and the distribution and extent of '*elegans* water' each year gives an approximate indication of the direction and volume of residual currents of the North Atlantic drift.

TEMPERATURE AND SALINITY

Temperature is probably the most important of the major physical environmental factors which mediate the life histories of marine organisms. Its effects are expressed in different distribution patterns, rates of growth, and in the timing of reproductive cycles. Sea surface temperatures around the British Isles vary considerably through the year (Fig. 1.2), and both the range and the pattern of isotherms differ between the western shores and the shallow North Sea.

During the winter months the surface temperature gradient for the whole region trends roughly east–west. In the eastern North Sea, along the German and Danish coasts, mean winter surface temperature is typically less than 3 °C. North–south isotherms extend almost to the middle of the North Sea, but near the east coasts of England and Scotland they are sharply deflected to run east–west, influenced by warmer Atlantic water in the northern North Sea and warmer flows through the Dover Strait. However, from the Thames estuary to Caithness, inshore waters have a mean winter temperature of only 5 or 6 °C. Winter minima increase steadily westwards along the Channel, and in the Western Approaches the winter mean reaches 10 °C. Off the west coast, winter isotherms again tend to align north–south, and most of the west coast of England has a mean winter low of 7–8 °C. In the Bristol Channel and Morecambe Bay it is often lower than this. In Morecambe Bay, in particular, the winter low may match that for east coasts and, under certain meteorological conditions, approaches the very low values obtained for the Danish coast. The western coast of Ireland is the warmest in the region and from Donegal southwards the mean winter low ranges from 9 to 10 °C.

During the summer the temperature gradient runs approximately north–south on both sides of the British Isles, with highest values, of 18 °C and upwards, obtained along the south-east coast of the North Sea, and a summer mean declining from 17 to 13 °C northwards along eastern Britain. On the west coast the highest mean summer temperatures, up to 16 °C, are found in relatively enclosed areas, such as Morecambe Bay, while for western Ireland 14–15 °C is normal for the whole coast. Thus, the annual range in sea surface temperature may be as much as 12 °C off south-eastern England, or as little as 5 °C along the west coast of Ireland.

For intertidal species air temperature may be as significant a factor in their distribution as sea temperature. The seasonal pattern of air temperature over the British Isles is very similar to, though less complex than, that for sea surface temperature (Fig. 1.2). In winter there is a gradient from a mean of 3.9 °C in the east to 7.2 °C in south-west Cornwall. In summer it extends north–south, from 13 °C in Orkney to 17 °C off the south-east coast of England.

Mean bottom seawater temperatures for the whole of the North Sea and the western shelf are only a little lower, in winter, than those of the surface. The distribution of isotherms is not dissimilar to that for surface temperature, and a vertical range of 1–2 °C occurs over the whole region. During summer, vertical ranges in temperature are naturally greater, and differ between west and east coasts. On the west coast the pattern of isotherms is similar to that for surface temperature, with a vertical range of 2 °C in the central Irish Sea and up to 6 °C in the southwest Celtic Sea. In the North Sea a body of very cold water lies on the bottom over much of the central and northern part of the region, giving a steep gradient in mean bottom temperature from 17 °C off the coast of The Netherlands to less than 7 °C off southern Norway where a vertical range of as much as 13 °C has been recorded.

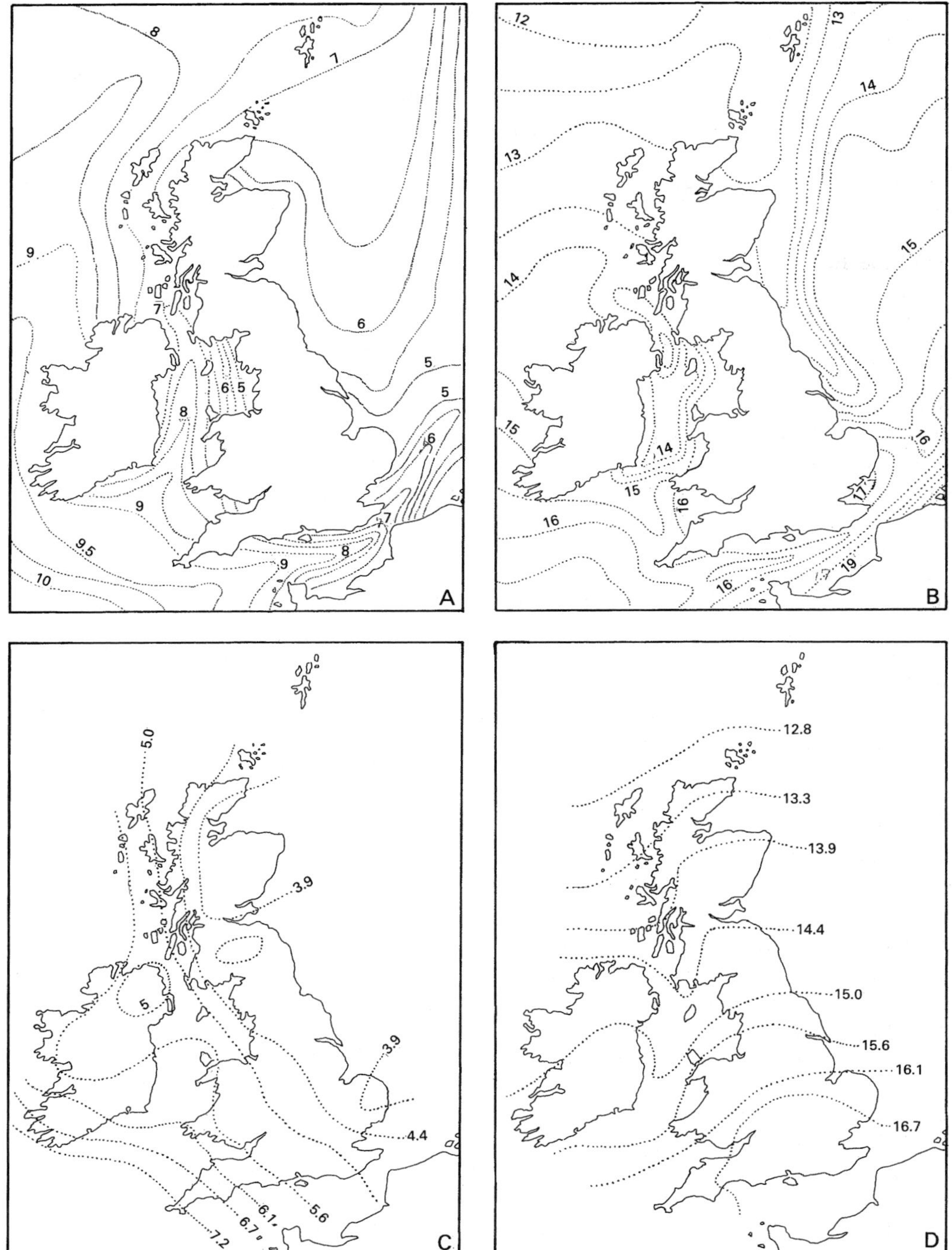

Fig. 1.2 Temperature, degrees centigrade. (A) Mean winter sea surface temperature; (B) mean summer sea surface temperature; (C) mean winter air temperature; (D) mean summer air temperature.

Salinity is regarded as the second most important physical characteristic of the marine environment. Throughout the British sea area the total concentration of dissolved salts has a low range of about 31–35‰ (g/kg) (Fig. 1.3). The 35‰ isohaline marks the boundary of Atlantic oceanic water; values lower than this reflect the mixing of oceanic water with varying proportions of water from other sources. Surface and bottom salinity differ little; for both, values of 34–35‰ pertain throughout most of the Irish Sea and the North Sea. In major embayments, in the German Bight, and along the southern shores of the North Sea a mean summer surface salinity as low as 30‰ may be further reduced to 29‰ by winter river outflows, while bottom salinities of as little as 27‰ may similarly be depressed to 23‰. However, over much of the British sea area, seasonal variation in both surface and bottom salinity is related to the penetration of oceanic water, having a salt concentration in excess of 35‰, into the English Channel and North Sea during the winter months.

Temperature/salinity profiles are used together to define the limits of particular water bodies, which may be further characterized by other properties, such as nutrient content and plankton communities. Temperature/salinity profiles of the water column during the summer months sometimes indicate a sharp boundary, or thermocline, between warm surface waters, often of slightly lower salinity, and cold deeper waters. The thermocline has biological significance in both preventing the interchange of nutrients between upper and lower water layers and in limiting the vertical migration of organisms. Regular vertical mixing of water resists the establishment of a thermocline, and it has been shown that the likelihood of mixing can be predicted by the equation: $\log_{10}(H/U^3)$, where H is water depth at a particular point, and U the maximum tidal stream velocity at that point. Values of less than 2 indicate a regime of continual mixing and no

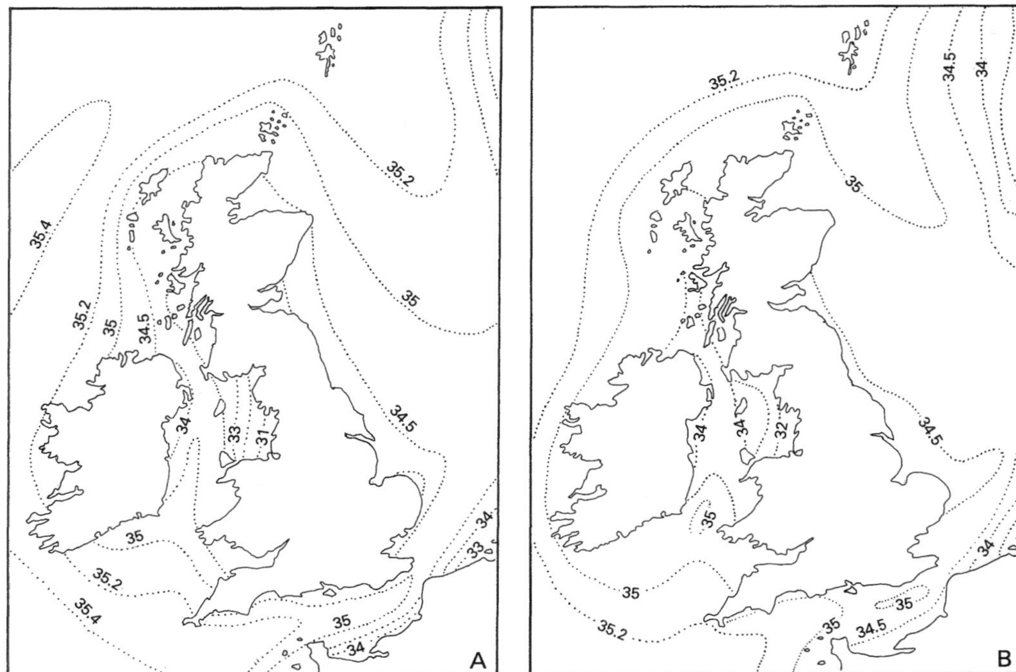

Fig. 1.3 Mean surface salinity (‰). (A) Winter; (B) summer.

thermocline, whereas sea areas with values of 2 or more tend to be layered during the summer months. The transition boundary between layered and non-layered waters tends to be sharp, and such boundaries constitute what are referred to as oceanic fronts. In the British sea area this boundary [$\log^{10}(H/U^3) \geqslant 2$] coincides with the limits of unmixed Atlantic oceanic water, except in the southern North Sea. Here, the boundary extends well south of the unmixed Atlantic waters, perhaps influenced by a deep-lying body of cold water at the bottom of the north and central North Sea. Thus, the southern part of the North Sea, the coastal waters of eastern England, the Channel, Bristol Channel, and much of the Irish Sea are unlayered, mixed, and often turbid throughout the year. Where these warm, shallow waters are high in dissolved organic nutrients, as off south-east England and The Netherlands, high marine productivity results, and such areas tend to be important nursery grounds for benthic fish.

MARINE PRODUCTIVITY

Plankton productivity, measured in terms of the average daily production of organic carbon over the sea surface, is the most important biological factor determining the distribution and biomass of marine organisms. Generally, productivity is highest in shallow seas, off large estuaries or in regions of upwelling, and lowest in polar waters and over the great ocean basins. The primary source of organic carbon is through photosynthetic fixation by phytoplankton, and the rate of production is dependent on light, temperature, and the availability of nutrient salts, principally of nitrate and phosphate, for phytoplankton growth. Light available for photosynthesis penetrates to a far greater depth in clear oceanic water than in the often turbid waters of shallow seas. However, high productivity demands a high level of nutrients, and in oceanic waters there is a continual loss of nutrients as organisms die and sink to depths beyond which photosynthesis is impossible.

Each sea area is characterized by a particular type of production cycle, dependent upon prevailing physical environmental factors and biological factors, especially the period of delay between the initiation of phytoplankton growth and the appearance of planktonic herbivores. In the temperate north-east Atlantic, plankton productivity displays characteristic seasonal peaks. As daylength and sea temperature rise in early spring phytoplankton growth increases rapidly, developing a 'bloom', which around Britain peaks between late April and May. This growth is fuelled by nutrient levels enhanced by winter stirring—the overturn of coastal waters by wind and storm, bringing dissolved organic materials to the sea surface. The phytoplankton bloom is followed by a bloom of grazing zooplankton, the peak of which coincides with the decline of the phytoplankton. The effects of intensive grazing are enhanced by a decrease in nutrients, and the phytoplankton bloom declines swiftly through June. An important factor in reducing nutrient levels is the summer thermocline, which effectively prevents mixing of surface and deeper waters, and there is a consequent net loss of nutrients from the upper layers. In late autumn the thermocline breaks down, vertical mixing begins again, and there is a second, smaller productivity peak, which eventually declines as both daylength and temperature drop with the onset of winter.

In the British sea area, winter nutrient levels are highest in the southern North Sea, off the Thames Estuary and the Wash, and in the northern Irish Sea. Values for both phosphate and nitrate are lowest in the central and northern North Sea, in the Celtic Sea, and off western Britain beyond the 100 m isobath. In these latter areas, also, a summer thermocline is characteristic. The shallow coastal waters of the British Isles show little summer layering of the water column and here productivity is highest, achieving an average daily rate of at least 500 mg C/m^2, and often very much more. In these areas light penetration is perhaps an important limiting factor, with photosynthesis occurring only in the top few metres of water.

TIDES

Tides are a significant factor in the ecology of marine organisms, as well as an important consideration for marine biologists. All coastlines are subjected to regular vertical changes in sea level, although in some parts of the world these are so small at times as to be scarcely discernible. Around the British Isles tidal ranges are variable, sometimes complexly so, and some coastlines experience often spectacular tidal amplitudes.

Tidal movements of the oceans result from the mutual gravitational attraction of the earth, moon, and sun. Tidal predictions are based on the known, regular movements of the three bodies, but tidal forces are subject to modification by meteorological influences, amongst others, and actual tidal effects may vary considerably from prediction. The tide-producing, or tractive, force is the resultant of two opposing agencies: the gravitational forces exerted by the sun and the moon, and a centrifugal force generated by the rotation of the earth. The tractive force varies in tune with the earth's yearly orbit of the sun, the 28 day orbit of the moon, and the 24 hour rotation of the earth itself. Analysis of tidal rhythms shows that the components attributable to the sun and to the moon are distinct and separable; when both are aligned with the earth their separate gravitational forces combine to give maximum tidal effect, resulting in the fortnightly spring tides. At the first and last quarter of the moon, the two forces are to some extent opposed; tidal effects are then dampened, resulting in the low-amplitude neap tides. Overall, however, the moon is the dominating influence on tidal rhythms, and is responsible for the mean tidal period of around 12 hours, giving two high and two low tide periods in each day, a tidal pattern referred to as semi-diurnal. The two high tides of a single day may not be of equivalent height and, similarly, the two low tides may be unequal; this phenomenon of diurnal inequality is related to the declination of the moon relative to different points on the earth's surface. Generally, around the British Isles the degree of diurnal inequality is slight compared with that in certain other parts of the world.

The tide-producing tractive force gives rise to tidal streams in the oceans, characterized by wave forms whose velocity is dependent upon water depth and gravitational force. Where these tidal streams impinge on continental land masses and are subject to the modifying effects of shelf and coastal topography, reflection of the wave sets up the oscillating system which gives rise to the tides. In enclosed basins, such as the Bristol Channel, these oscillations are particularly marked, and their period is dependent entirely on water depth and the length of the basin. When the period of oscillation is close to the mean tidal period, very large tidal amplitudes may result; at points in the Bristol Channel, for example, spring tide ranges of 10 m or more are usual.

The tidal cycles at different points around the British Isles can be correlated by reference to a further modifying effect of the centrifugal force generated by the earth's rotation, the Coriolis force. In the northern hemisphere the Coriolis force results in moving bodies, such as tidal streams, being deflected by a measurable amount to the right of their path. Thus, as the Atlantic flood tide flows into the northern part of the North Sea, the tidal stream is deflected towards the coasts of Scotland and England; the deflection is reversed as the ebb flows northwards again. The resulting complex movement of water in the North Sea, through a tidal cycle of 12 hours, can best be described by reference to three nodal, or amphidromic, points around which tidal streams and successive changes in tidal elevation appear to move in an anticlockwise direction. These amphidromic points may be fixed by plotting co-tidal lines, viz., lines joining points on the surface of the North Sea where high tide occurs at precisely the same time. The co-tidal lines are seen to radiate from three points (Fig. 1.4): one situated midway between East Anglia and the coast of The Netherlands, a second in the central North Sea, and a third close to the south-west

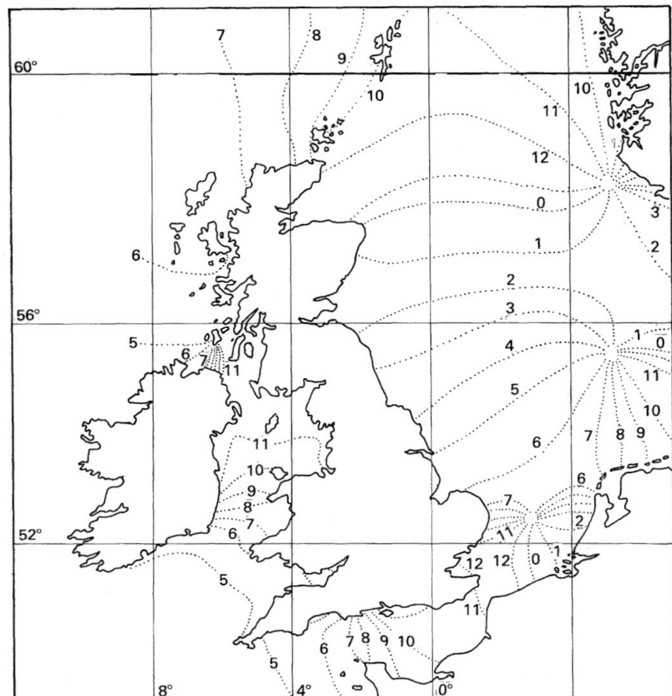

Fig. 1.4 Co-tidal lines and amphidromic points.

coast of Norway. At each amphidromic point there is no tidal fluctuation in sea level, and tidal range increases progressively away from the point. The anticlockwise flow of the three North Sea amphidromic systems is synchronized to the extent that from Caithness to Sheppey, high water occurs successively later down the whole of the east coast. Off the western British Isles co-tidal lines tend to be correlated with amphidromic systems of the north Atlantic; there is no simple progression of high tide times, tidal range is highly variable and often rather low. In the Channel, the combined effects of the Coriolis force and the frictional drag on the tidal stream as it passes eastwards along a shallowing and narrowing trough, result in inequalities in tidal range on each side of the Channel. On both ebb and flow, vertical movement of the tide tends to be higher on the south than on the north shore of the Channel. The amphidromic system is poorly developed, but can be plotted with reference to an inland point in southern England.

Tidal phenomena have been extensively studied around the British Isles, particularly with regard to their significance for intertidal marine ecology. The tidal range for a given locality refers to the vertical variation in sea-level throughout a tidal period. The 28 day cycle of neap and spring tides oscillates about a mean tidal level (MTL), and four important standard levels may be recognized: mean high water of spring (MHWS) and neap (MHWN) tides, and mean low water of spring (MLWS) and neap (MLWN) tides. These levels refer to the mean heights in any year, while 'extreme' levels (e.g. ELWS) refer to the lowest tides of the year. Two important features of tidal cycles are related to the astronomical and gravitational factors discussed above. First, the tidal range changes progressively, with each tidal period, throughout the spring/neap cycle, with the highest spring tides of each cycle occurring about 1.5 days after the full and new moon. Secondly, spring and neap tidal amplitudes, also, change progressively through the year, with the greatest spring amplitude occurring at the time of the equinoctial tides, in late March and late September, when the sun passes the earth's equator. The fact that the tidal period, between two high tides, is on average greater than 12 hours results in tides becoming progressively later each day. This cycle coincides with the neap/spring cycle of 14.5 days, and thus at any locality spring tides occur at approximately the same time of day on each cycle, this time advancing by 1 or 2 hours in the course of a complete year.

The tidal range, MHWS–MLWS, varies most commonly between 3 and 7 m; in the Bristol Channel and the Gulf of St. Malo it may be greater than 10 m, while along the Channel coast it may be only 1.5–3 m. Smallest tidal ranges, from 0.5 to 1.5 m, are found along the south-west shores of Scotland. The effect of the tidal cycle on a particular shore is expressed as successive periods of emersion and immersion, of varying duration, the consequences of which for the intertidal fauna and flora are mediated by topographic and climatic factors. The latter—rain, freshwater runoff, wind, sunshine, and air temperature—will vary in intensity according to latitude and the time of day at which low tides, especially low springs, occur. Barometric pressure and wind strength may have profound effects on tides. A combination of low pressure and onshore winds can result in low tides not reaching predicted levels, while unusually high pressure may depress the tide below its expected minimum. The tide has an influence on the broad distribution of marine organisms, both vertically and horizontally, and also a profound effect on the zoned distribution of the intertidal fauna and flora.

GEOLOGY

Sedimentary and igneous rocks of all geological ages contribute to the structure of the British Isles, are represented in coastal topographies, and provide sources of material for marine deposits. Generally, the youngest and least resistant rocks outcrop along the south-eastern coasts; here, excluding man-made structures such as groynes and port installations, intertidal hard substrata tend to be relatively impermanent, and typically with low and rounded profiles. The geological age of rocks increases steadily northwards and westwards; coastal outcrops are correspondingly harder, more resistant to erosion and, depending on the geomorphology of each region, more or less precipitous. The igneous intrusions of south-western England, and the Precambrian metamorphic series of north-west Scotland and the Western Isles, are exceptionally hard. These rocks weather less readily than sedimentary rocks and the effect of coastal erosion is to produce smooth, rounded profiles which provide few opportunities for habitat diversity. Generally, the limestone shores of north-east and south-west England, South Wales, and southern and western Ireland support the richest intertidal faunas. The physical and chemical erosion of limestone, particularly when well-bedded, leads to the development of maximum diversity of intertidal rocky habitats. However, shores of slate, interbedded sandstones and marls, or fissile shales, when protected from high levels of marine erosion, often support faunas almost as rich as those of limestone shores.

The nature of the marine fauna developed along a certain coastline, including both intertidal and subtidal elements, is also dependent on coastal topography, which is itself related both to regional geomorphology and hydrography. Thus, shallowing, retrenched coastlines tend to be areas of marine deposition, characterized by sandy beaches and offshore bars. When sheltered from the prevailing swell such areas may accumulate the finest deposits, giving rich faunas of soft-bottom invertebrates and high populations of demersal fish. Rocky promontories may develop a hard-substratum fauna, but this is often impoverished through the scouring effect of a high sand-table. High-energy depositing coastlines are characterized by coarse gravel deposits, frequently with a high proportion of shell which may support a diverse and dense community of encrusting organisms. Salient coastlines of promontories, frequently associated with offshore islets, stacks, or reefs, are erosional coastlines typically fully exposed to the Atlantic swell; the magnitude, and hence erosional effect, of a swell is related to its 'fetch', the distance of uninterrupted sea surface over which it develops. Deposition on such coasts is localized, and limited to short, coarse-grained, high-energy beaches between promontories. Offshore, the bottom is typically of bedrock, boulders and cobbles, and patches of coarse gravel. Both intertidal and subtidal habitats are richly populated; seaweed cover is typically dense and animal diversity high. Estuarine and lagoonal habitats, finally, develop where coastal topography and regional drainage patterns interact to give conditions of extreme shelter and permit continual deposition of land-derived erosional products. Three exceptions to this may be noted, however. Drowned ria coastlines, such as those found in south-west England, typically have extensive, sheltered, fully marine inlets in which both the rate of freshwater runoff and of sedimentation are too low to encourage the development of estuarine conditions. Fjords are drowned glacial valleys, often very deep and typically with a shallow barrier, or 'threshold', close to the seaward mouth, which may to some extent result in different physical conditions between the benthic environment of the fjord and that of the adjacent sea. The deep sea lochs of western Scotland may be regarded as fjords. Finally, sheltered lagoons may develop, as on the south coast of Devon, where storm beaches on a high-energy coast isolate narrow stretches of coastline, along which lenses of seawater are trapped, freshening to brackish conditions through slow, persistent freshwater seepage.

The varying rock types of Britain and Ireland thus give rise to differing coastal forms and provide different opportunities for the development of marine habitats. Similarly, the rock and soil type of a particular region, together with its coastal topography and drainage pattern, will influence the nature of the offshore, subtidal environment, with consequent effects on the marine fauna. These influences are modified by other factors of the physical marine environment, and strong tidal streams, exposure to heavy swell, and various climatic features may result in erosional products being deposited at sites remote from their origin. However, in very few areas is the nature of the offshore environment predominantly determined by proximate, presently acting influences. Instead, the most important factor determining the nature of the marine benthic environment around the British Isles is the Pleistocene glaciation of north-western Europe. Much of the North Sea, the Channel, and the Irish Sea is floored with thick deposits of mixed clay, sand, and gravel which have been continuously reworked and redistributed by hydrographical factors since the retreat of the ice sheets. Much of the eastern half of the North Sea is covered with sand and gravel deposits predominantly of Pleistocene age, sorted by tidal streams and currents, with its present-day distribution related to prevailing current patterns, and augmented to a varying degree by post-Pleistocene erosion. Central areas of the North Sea have extensive muddy bottoms, probably consisting predominantly of fine particles transported from other areas of the sea, although again this will be augmented to some extent by post-Pleistocene sediments. However, muddy areas along the southern coast of the North Sea are in all likelihood entirely derived from post-glaciation river-borne sediments.

Westwards along the Channel sediments steadily coarsen, as tidal streams increase in velocity. The western half of the Channel has extensive areas of coarse gravel and large deposits of dead shell, their distribution again related to prevailing current regimes. The exposed western coasts of Ireland and Scotland have offshore bottoms of boulders, gravel and coarse sand, with patches of bare bedrock, and are not influenced to any degree by glacial deposits. Sands and gravels, on the western coasts of the British Isles, have a high content of shell debris and other organic carbonates, while eastern benthic sand deposits are predominantly of silica.

ANIMAL DISTRIBUTIONS

The marine faunas of the British Isles are not everywhere the same. A survey of a moderately sheltered rocky shore on the coast of Northumberland will reveal a slightly different suite of species from a similar survey on the coast of south Devon. Similarly, dredge hauls from the Firth of Clyde and the western English Channel may be equally rich in species, but different groups of species will occur in each haul. Certain familiar species occur commonly on all British coasts; certain others may have very limited distributions, being restricted, for example, to north-east or south-west coasts. Many species may simply be more common at one geographical extreme than the other, occurring with diminishing frequency along a north–south or east–west gradient. No marine species is truly ubiquitous, and even the commonest and most widely distributed species do not occur at constant frequency or density over the whole of their geographical range. Similarly, widely distributed species rarely display the same population structure, or the same reproductive cycle, over the whole of their range. Distribution pattern is part of the array of biological features characterizing each individual species, and patchiness is characteristic of all species distribution patterns.

The distribution of a particular marine organism can be defined at different levels, and discontinuities—patchiness—may occur on more than one scale. At the broadest level, each species occupies a particular geographical range, which may be very extensive or very limited. Some large pelagic fish, planktonic cnidarians, or benthic molluscs with very protracted larval stages may be truly pan-oceanic in distribution; abyssal molluscs and echinoderms may be widely distributed over entire ocean basins; many intertidal or shallow sublittoral organisms are distributed along the whole length of one or more continental margins. Conversely, for example, perhaps a majority of bryozoans, with only short swimming larval stages, have relatively restricted geographical distributions; viviparous intertidal prosobranchs may occur as groups of polymorphic or allopatric species, distributed successively along a continental margin or through a chain of islands; shallow-water or intertidal animals of oceanic islands are sometimes specifically distinct from those of the next island or the nearest continental land mass. Within the limits of its geographical range, each species is further restricted by a series of physical, biological, and ecological factors, which generally result in a concentration of populations in habitats particular to each species. These factors are frequently complexly interrelated, but it is often possible to isolate them and thus to define successive scales of patchiness. For example, within the north-east Atlantic region a particular species may be limited by lowest winter sea surface temperatures to parts of western Norway, the Western Isles of Britain, and the west coast of Ireland. It may then be further restricted to moderately exposed rocky shorelines free of estuarine or other freshwater influence; resource competition may confine the same species to a particular vertical zone, from parts of which it may be excluded by predation, parasitism, habitat disturbance, or other ecological factors.

The study of animal and plant distributions is formally defined as biogeography. At its broadest, biogeography plots the geographical distribution of species, describes and analyses types of pattern, and attempts to explain determined patterns in terms of the biology, ecology, and evolutionary history of the organisms. At this level biogeography is largely an academic and intellectual pursuit. However, by shifting its emphasis towards the study of local distributions, within the defined geographical boundaries of particular species, and of the ecological requirements of each species, biogeography becomes a necessary discipline for the marine biologist. Marine biogeography has a long history in British marine biology, the first attempt to describe patterns of marine distributions being Forbes and Godwin-Austen (1859, *The natural history of the European seas*). Early interest was doubtless stimulated by certain features of Britain's marine faunas. The latitudinal extent of the British Isles, and the major hydrographic differences between eastern and western coasts result in rather different faunal assemblages in the west and in the east, and observable faunal changes from south to north. Discontinuities in the distribution of particular species, or assemblages of species, may be further emphasized by changes in substratum or habitat. It has long been known that, while certain elements of the British marine fauna comprise species which range far to the north, others are widespread in the western Mediterranean and along the coasts of north-west Africa instead; each of these elements displays different distributions around the British coasts (Fig. 1.5).

Biogeographers divide the marine realm into 'provinces', each characterized by a certain minimum proportion of autochthonous, or endemic, species, whose evolutionary origins, or the largest part of whose evolutionary history, are inferred to lie within that province. Such provinces may also be defined by physical barriers, such as land masses, deep sea trenches, or divergent current systems, and may display clear boundaries in the form of abrupt changes in fauna. In the north-east Atlantic region, influenced for much of its extent by the Gulf Stream drift, there appear to be no major biogeographical boundaries, although a few abrupt hydrographic boundaries, such as the Wyville Thomson Ridge, do correlate with equally abrupt changes in bottom fauna. However, the fauna of the western Mediterranean and north-western Africa is clearly different from that of northern and western Norway, and the particular interest of the faunas of the British Isles derives from the fact that they are largely a mixture of northern cold water (or boreal) species, and southern warm water (or lusitanian) species. There are very few marine animals endemic to the British Isles, although for some, such as the dog whelk *Nucella lapillus*, the British Isles is the centre of their distributional range. However, many southern species reach their northernmost limits in British waters, and perhaps as many northern species are close to their southern limits here. Probably for few species can an actual furthest limit be demonstrated; the geographical ranges of most species probably contract and expand in concert with long-term climatic cycles, and distant population outliers may persist or disappear according to purely local environmental factors. Instead, most species will show a gradual range attenuation, to north and south, which is modified by local factors and influenced by the physiological and ecological characteristics of each species. The resulting overlay of distribution patterns provides a rich testing ground for all aspects of biogeographical research. Particularly interesting examples are provided by intertidal organisms. Many lusitanian species reach their northern limits around south-western and western Britain. The extent of their distribution and the persistence or relative abundance of their populations vary widely between species, suggesting different tolerances to different environmental factors. The top shell *Gibbula umbilicalis*, for example, has an almost continuous distribution from the Isle of Wight westwards and northwards to Orkney, including all coasts of Ireland, while *Monodonta lineata* occurs on the west coasts of Ireland and Wales and around the Cornubian peninsula, but occurs only sporadically on the east coast of Ireland. *Balanus perforatus*, a lowshore barnacle, is found on the Pembrokeshire coast, around the south-west peninsula of England, and eastward to the Isle of Wight, but is entirely absent from Ireland. Conversely, the echinoid *Paracentrotus lividus* is abundant along the western coasts of Ireland, but, apart from sporadic records from Devon and the Channel Isles, occurs nowhere on the coasts of England and Wales.

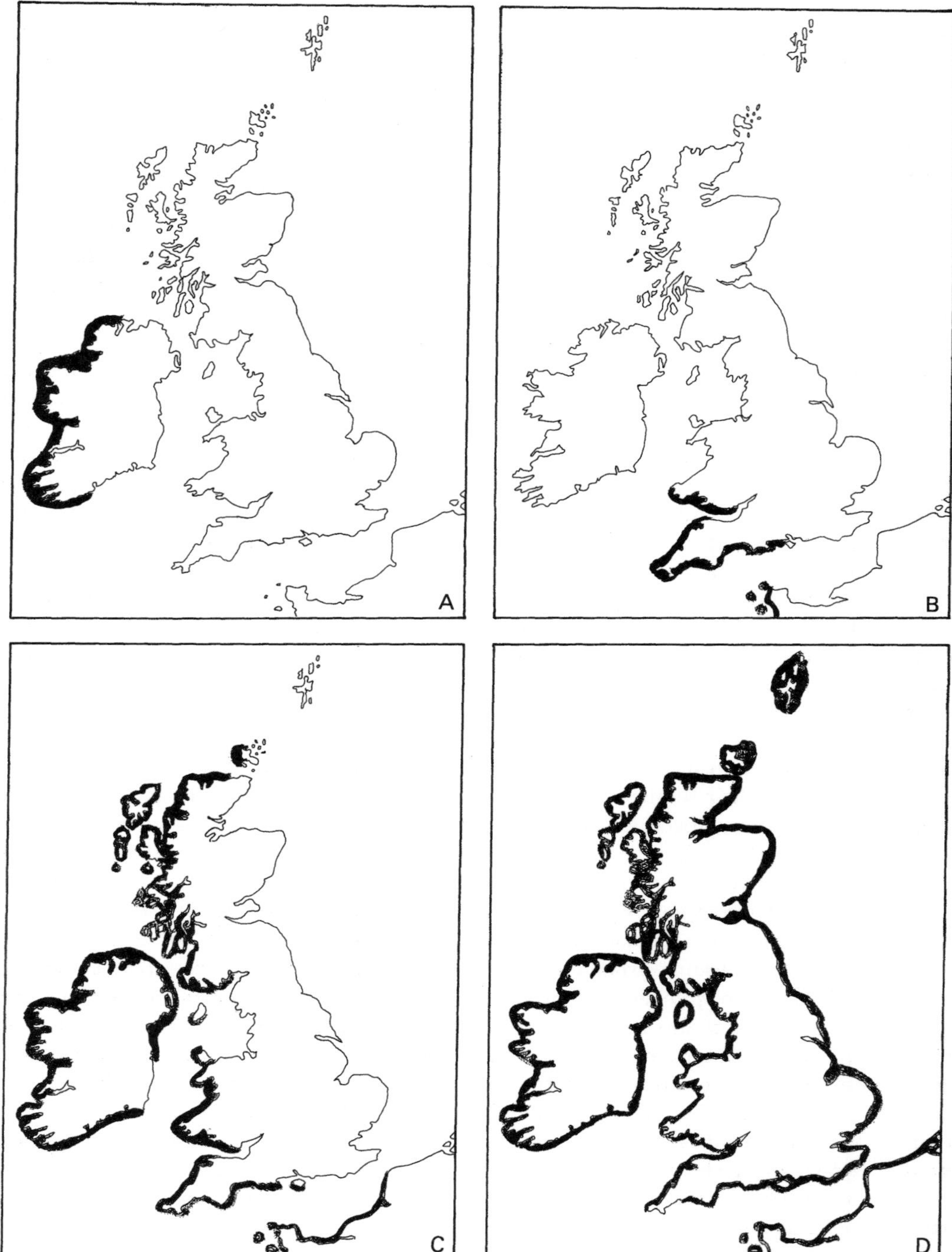

Fig. 1.5 Contrasting distributions in four intertidal animals: (A) *Paracentrotus lividus*; (B) *Balanus perforatus*; (C) *Gibbula umbilicalis*; (D) *Semibalanus balanoides* (after Lewis 1964).

The geographical distribution of a particular species reflects both its evolutionary history and its present-day ecological tolerances. The ultimate barrier to the expansion of any species is its own physiology, and on the broadest scale the distribution of each species tends to be limited by physical environmental factors, the most important of which is probably temperature. Most marine animals are able to live within a fairly broad temperature range. So-called stenothermal organisms may tolerate an annual temperature range of only 4 or 5 °C; such narrow ranges of temperature are not found close inshore anywhere around the British Isles. Most shallow-water temperate organisms are, rather, eurythermal, and are able to tolerate annual temperature ranges of 10 or even 15 °C. Temperature affects all physical and biological processes, which tend to proceed more quickly as temperature rises; metabolic rates in marine animals may be doubled by a 10 °C temperature rise. Thus growth, reproduction, and metabolic rate of a particular species may vary across its geographical range, affecting both its adaptive and competitive potential in different ways, with consequent effects on its relative abundance, reproductive output, and overall biomass at different places. All biological processes function most efficiently within an often narrow optimum temperature range, and all reproductive stages of marine organisms are mediated by temperature, with critical or optimal temperatures for all stages, from oogenesis, to maturation, spawning, fertilization, and embryogeny. Thus, certain apparently widely distributed species may achieve maximum population level over only a small part of their geographical range, and may breed successfully only in certain optimum habitats. Elsewhere, populations may be sustained by larval drift or adult dispersion, with the population structure related to the reproductive output of populations in neighbouring, or remote, habitats, and to the survival of dispersive stages. Some widely distributed species exist as a series of physiological races, with each race adapted to annual temperature regimes, or other climatic conditions, existing in its immediate vicinity. Local physiological adaptation is particularly likely to occur when a suitable habitat for the species is not present over the whole of its geographical range, but is distributed as discrete areas, each separated by areas of unsuitable habitat.

The continental shelf edge is a significant barrier to expansion for many benthic species; but, for certain offshore species, adapted to a narrow range of fairly low temperatures, the outer continental shelf is a migration pathway. Consequently, certain decapods, molluscs, and echinoderms, for example, occurring offshore around the British Isles, may be distributed in similar habitats as far south as the coasts of West Africa. Major river outflows, or deltas, may be important barriers to similar migration of coastal species. Current systems may act as barriers to geographical expansion, particularly when they are associated with upwellings, when abrupt temperature changes may enhance the effect. The anticlockwise, southern North Sea gyre may be responsible for limiting the northwards expansion of the immigrant prosobranch *Crepidula fornicata* along the eastern coasts of England, simply by preventing the northward movement of planktonic larvae. Other good examples may be found in the Isles of Scilly. Both the southern, warm-water barnacle *Balanus perforatus*, and the northern, cold-water *Semibalanus balanoides* are rare around the Lizard peninsula, and the latter is absent from the Scilly Isles. The southern limpet, *Patella depressa*, does not occur in the Scilly Isles, and the northern winkle *Littorina littorea* occurs only rarely. These curious distributions are explained by a clockwise current eddy off the Lizard which ensures that dispersing larvae, whatever their origin, are swept away from the Scilly Isles. For the most part, however, current systems are most important as pathways for migration and, although some intertidal organisms show clear range attenuation, or even marked geographical limits, in British waters, the hydrographic conditions experienced by the British Isles generally favour broad distributions of marine benthic organisms in the British sea area.

The evolutionary history of each species is often reflected in its geographical distribution, as is the history of its environment. Climate and environment change through time. Species adapt to change, or migrate, or become extinct; environmental change may lead to the fragmentation of continuous distributions, leaving isolated populations surviving in reduced areas of optimum habitat. Evidence of formerly wide distributions for particular species may be preserved in sediments and fossil deposits, indicating long-term changes in the species' geographical ranges. However, such change may also be observed on far shorter time scales. In recent time, the severe winter of 1962–63 caused local extinctions of many intertidal or shallow sublittoral species. In some instances, the affected areas were soon recolonized by immigrant populations, and through larval dispersion from adjacent areas. In others, the immediate result was a marked contraction in geographical range, which has been regained subsequently by some organisms, while others seem to have suffered a permanent range contraction. Periodic change in climatic and hydrographic factors is reflected in periodic change and fluctuation in marine benthic faunas. For example, in the early 1920s the western English Channel experienced substantial influxes of oceanic water and was characterized by high levels of dissolved phosphorus in winter, a rich spring macroplankton, the chaetognath *Sagitta elegans*, and large herring populations. A marked change in the 1930s, perhaps related to rising sea temperature, gave rise to a regime of low winter phosphorus and poor spring macroplankton, characterized by *Sagitta setosa* and pilchards, with total disappearance of the herring. A number of benthic bivalve species, perhaps only established during the 1920s, disappeared from the western Channel during the 1930s, and a number of northern species, such as *Antalis entalis*, became locally extinct along the south-west coasts of England.

Continuous adaptation to environmental change may lead ultimately to the reproductive isolation of populations of a particular species over its geographical range, and eventually to allopatric speciation, the appearance of distinctly different, locally adapted species derived from a common ancestor. Possible allopatric species groups can be seen in certain intertidal littorinids.

Within the broad limits of its geographical range, each species occurs in a particular habitat, defined by the particular, optimum conditions for its growth and reproduction. Bordering each area of optimum habitat may be a zone of suboptimal habitat in which the species is able to exist, but in which it does not achieve maximum growth or longevity, or is unable to achieve maximum reproductive output, or is at a competitive disadvantage to other species. The optimum habitat may be limited by physical and/or biological factors. The most simple demonstration of this is with reference to substratum. Level-bottom communities around the British Isles change in composition as the substratum changes. As fine muds intergrade with silty sand, which then trends to progressively coarser sand, both the species composition of the benthic fauna and the relative abundance of species change. Certain species of infaunal polychaetes, bivalve molluscs, and echinoderms decline in relative abundance, perhaps at different rates, eventually to be replaced by others. Such patterns are ultimately related to current and tidal effects, and change in response to these around the whole of the British Isles.

Substratum has an important influence on the distribution of sessile organisms. For example, as the limestones and shales of southern England give way eastwards to soft chalk and clay, patterns of faunal change occur, perhaps reinforced by gradients of change in other factors, such as temperature. Hard-substratum sessile species, crevice-dwelling species, and rock borers may be discontinuously distributed along a coastline, with breaks occurring where the rock type changes or gives way to unconsolidated substrata. Limestones are the most richly colonized rocks, being sufficiently resistant to erosion to enable populations to persist, yet eroding sufficiently along bedding or jointing planes to provide crevices, gullies, and pools, thus increasing the diversity of local habitat. Softer limestones, shales, and marls will have a different fauna of boring organisms from crystalline limestones, but fewer encrusting species. The hard igneous and metamorphic rocks of western Scotland provide the least diversity of habitat, and sessile faunas will flourish only where the smoothly eroded rock surfaces are provided with local shelter.

The vertical zonation of intertidal rocky shore organisms is a further expression of local distribution patterns varying in response to environmental factors (Fig. 1.6). The barnacle belts and seaweed zones of rocky shores, with their attendant fauna, are sensitive indicators of local environmental conditions. Gently sloping shores with maximum shelter from wave and swell are characterized by a narrow supralittoral barnacle belt, and a broad midshore region dominated by the brown algae *Fucus serratus* and *Ascophyllum nodosum*. Such shores have perhaps the richest intertidal faunas. As exposure increases and the shore profile steepens, the fucoid algae give way to shrubby red algae and a different, less diverse fauna. The extreme is seen in storm-beaten western promontories where the barnacle belt extends some tens of metres above high-water mark, and the algal zones are reduced to a narrow fringe of exposure-resistant species, such as *Alaria esculenta*. While the geographical distribution of intertidal species may be defined by gross environmental factors, particularly sea temperature, local distribution patterns are determined by the availability of habitat, mediated by the effects of exposure, shore aspect, tidal factors, and climate. Air temperature, rainfall, and sunshine, in relation to the magnitude, duration, and time of daily low tide periods, have important effects on the distribution of intertidal species, which differ between northern and southern, and eastern and western coasts of Britain.

A final aspect of local distribution patterns may be termed 'ecological distribution'. The occurrence of parasitic and symbiotic species is naturally dependent on the distribution of their hosts. Similarly, epiphytic and epizootic species occur only where their preferred plant or animal substratum occurs. The persistence of such species may depend on the biological characteristics of their substratum. For example, the brown seaweed, *Fucus serratus*, supports a rich and characteristic fauna of bryozoans, hydroids, tubeworms, sponges, and sea squirts, with their associated

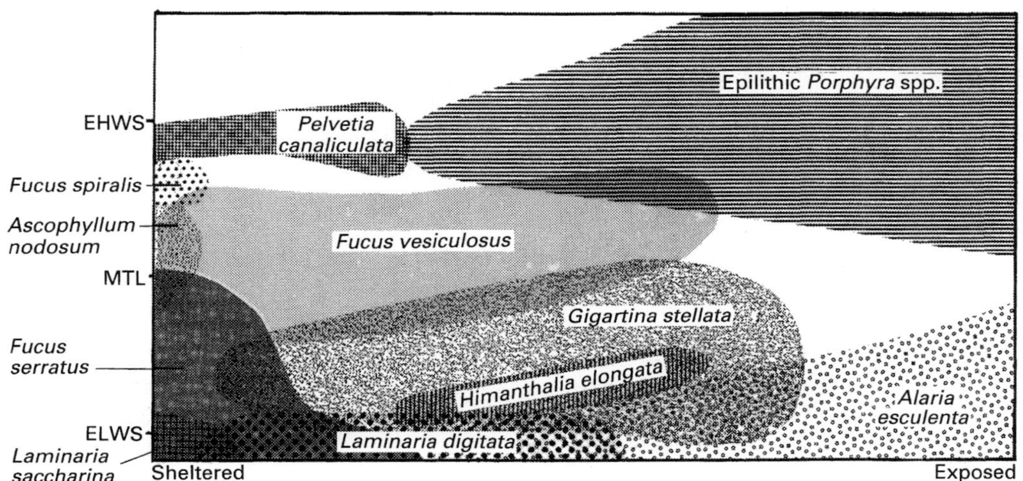

Fig. 1.6 Zonation of intertidal rocky shore seaweeds in relation to tidal level and exposure (after Jones and Demetropoulos 1968).

predators. However, this community develops only where sheltered conditions allow the *Fucus* plant to achieve maximum size and bushiness, and where the cover of seaweed is also at a maximum. In turbulent conditions, plants tend to be small, spindly, and sparsely distributed; the epiphytic fauna may be only minimally developed, or quite absent, not simply as a result of harsher environmental characteristics, but also through the narrower range of microhabitats provided by the *Fucus* population.

2 USING THIS BOOK

This book is a manual for the identification of approximately 1500 species of the north-west European marine fauna. The process of identification depends on the use of simple dichotomous keys, by which means a species' identity is determined after choosing correctly from a sequence of paired contrasting diagnostic characters. Despite the obvious attractions of pictorial guides, the dichotomous key remains the most efficient and precise means of identifying organisms and, although keys to large taxonomic groups may appear intimidating, in practice it may not take more than half a dozen steps to arrive at a name for the specimen to hand.

The selection of marine organisms included here is arranged in the accepted systematic sequence, from sponges (Chapter 3) to fishes (Chapter 14). For each phylum there is a brief introduction to the morphology of the group, in which structures important in identification are described and the minimum necessary terminology is defined. These introductory accounts are accompanied by labelled diagrams, which should be referred to before commencing identification of a specimen. For the larger phyla, such as Crustacea and Mollusca, there are introductory accounts, also accompanied by labelled diagrams, to each class or order included. For smaller phyla, classes, or orders, or for those groups for which only a small selection of species or genera is given, only a single key may be provided. For most, however, identification should begin with the key to families. These follow in numbered taxonomic (or alphabetic) sequence, the running head on each page allowing easy location. In most cases the family entry is followed by a brief morphological diagnosis, with notes on occurrence and the known diversity of the family, when appropriate. Having arrived at the correct family, the next stage in the process of identification may require a second key, to the species included. Each species keyed out has a text entry and an illustration, and for most some short confirmatory characteristics are given, together with notes on occurrence and distribution. For some groups the amount of detail given here may be extensive; for others, as in the case of prosobranchs when all necessary taxonomic characters can usually be portrayed in a good illustration, very little is required beyond notes on colour. Diagnostic characters used in the keys are not usually repeated in the species entries.

Every effort has been made to present the keys, on which accurate identification depends, in as simple a manner as possible. Unnecessary jargon has been pruned or excluded. However, in some instances the precise morphological term must be employed, as for example in describing the constituent articles of crustacean limbs. In these cases, please refer to the diagrams. References to appropriate, labelled figures have been inserted in the keys where it seemed useful, but if in doubt over the terminology, read the introduction to the group and refer to the figures. Each term is defined in the text and may be located using the *Index of technical terms*.

Experienced biologists will know to which phylum, class, order, or even family the specimen they are identifying belongs, and should use the index to find the appropriate key, description, or illustration. Beginners often experience the greatest difficulty in deciding to which group the organism they are regarding belongs. If unsure, the following Guide may help.

GUIDE TO ANIMAL GROUPS

---

1. Fish: vertebrate animals with fins, gills, eyes, mouth, anus (if you can't recognize a fish give up marine biology and take up bird watching) **Chapter 14**

   Not fish **Go to 2**

---

2. Worms and worm-like animals: choose from A–F, below

   Not worms **Go to 3**

   A. Smooth bodied, with neither segments nor annulations. Body long and thin, short and cylindrical, or flat and leaf-like; usually soft, occasionally stiff. Sometimes with tiny eyes, but without distinct body regions, and completely lacking appendages
      i. Thin, flat, leaf-like worms. Pastel-coloured or striped. Free living (Fig. 2.1) **Turbellaria** (p. 136)
      ii. Thin, flat, often with suckers or hooks. Parasitic, attached to or in fish or macroinvertebrates (Fig. 2.2) **Flukes (Digenea, Monogenea)**, not treated further here
      iii. Slender, soft, elongated worms, 5 mm to 10 m long. Often slimy. Head end may bear minute eyes; often brightly coloured (Fig. 2.3) **Nemerteans** (p. 143)

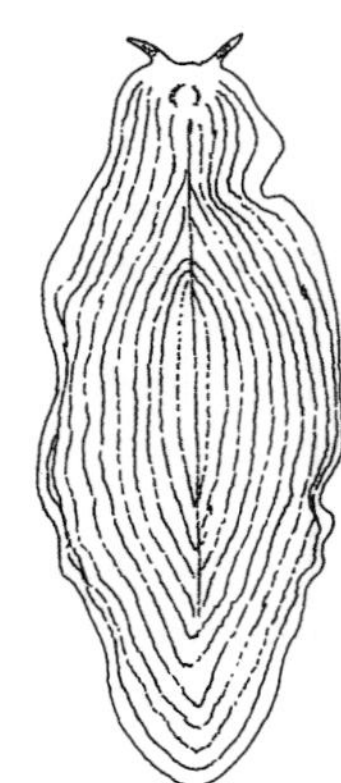

Fig. 2.1 A turbellarian, *Prostheceraeus vittatus* (Montagu).

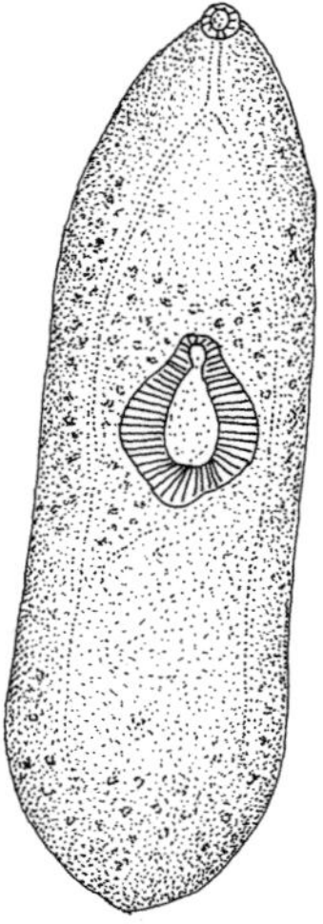

Fig. 2.2 A digenean fluke, *Cryptocotyle*.

Fig. 2.3 A nemertean worm, *Lineus longissimus* (Gunnerus).

iv. Small, white, stiffly coiled worms. Often with a terminal bristle. Free-living or parasitic (Fig. 2.4) **Nematodes**, not treated further here

B. Body smooth, with neither segments nor annulations, but clearly divided into a knob-like or finger-like anterior region (proboscis), separated by a short collar from an elongate posterior trunk. No appendages. Burrowing in sand or mud (Fig. 2.5) **Enteropneusts** (p. 687)

C. Body smooth, unsegmented. No appendages. Head end with a horseshoe-shaped crown of long, slender tentacles. Small worms boring in limestone, or burrowing in subtidal sands **Phoronids**, not treated further here

D. Body stout, unsegmented; skin smooth, sticky, or coarsely papillate. Often with rows of extensible 'tube feet'. Mouth terminal, surrounded by a ring of branched, often bushy, retractile tentacles (Fig. 2.6) **Holothurians (sea cucumbers)** (p. 681)

E. Body unsegmented, but clearly annulated (i.e. with many fine rings). Body regionated, the anterior proboscis contracile or retractable. Some

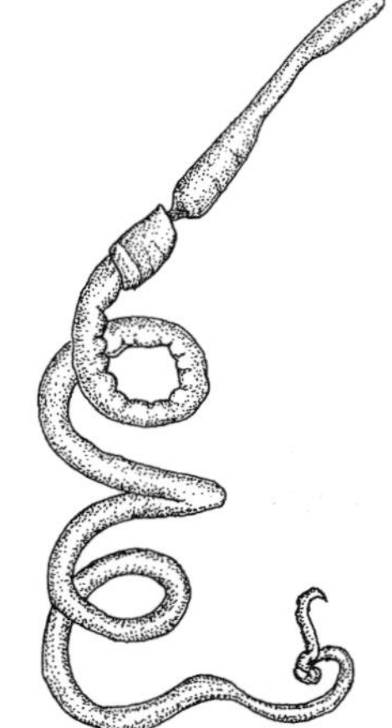

Fig. 2.5 The enteropneust, *Saccoglossus ruber* (Tattersall).

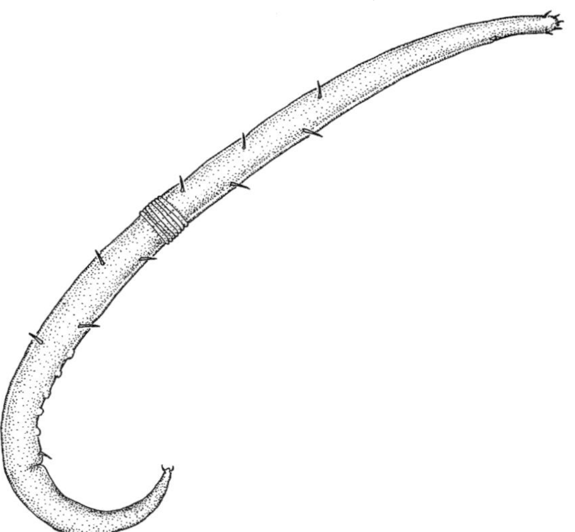

Fig. 2.4 Generalized marine nematode (after Platt and Warwick 1983).

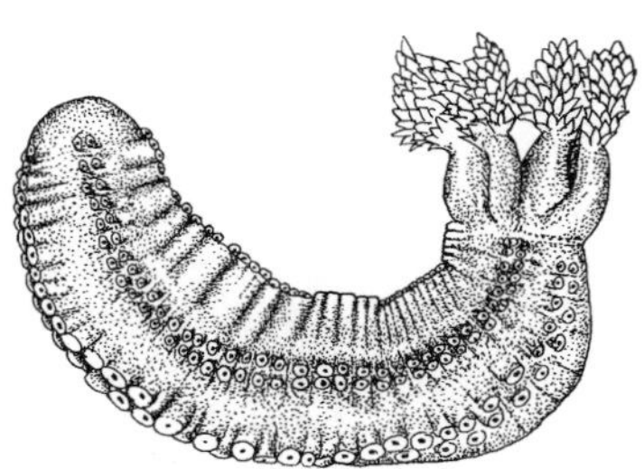

Fig. 2.6 A sea cucumber, *Ocnus lactea* (Forbes and Goodsir).

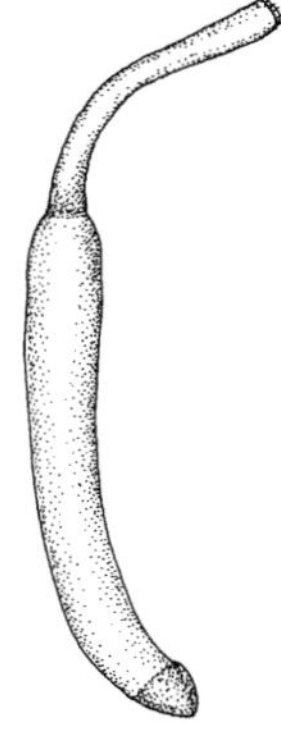

Fig. 2.7 A sipunculan, *Golfingia vulgaris* (de Blainville).

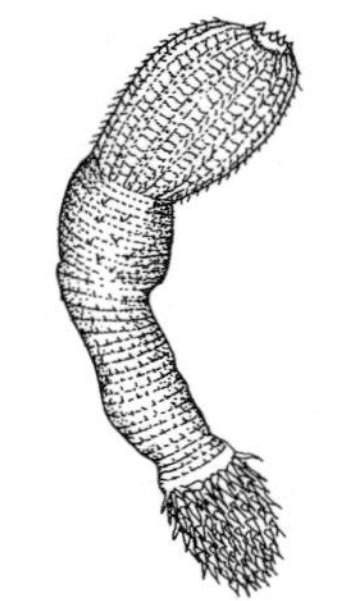

Fig. 2.8 The priapulid, *Priapulus caudatus* Lamk.

species bear one or two pairs of chitinous hooks, but none has serially repeated appendages

i. Body wall thick, finely ringed, often papillate. Proboscis slender, retractable, with a fringe of short tentacles surrounding the mouth (Fig. 2.7) **Sipunculans** (p. 278)

ii. Body wall thick, coarsely ringed. Proboscis retractable, inflated, longitudinally ridged. Posterior end with many small tentacular processes (Fig. 2.8) **Priapulids** (p. 278)

iii. Body wall soft, finely ringed. Proboscis contractile, forked in some species, with a deep groove along one face (Fig. 2.9) **Echiurans** (p. 282)

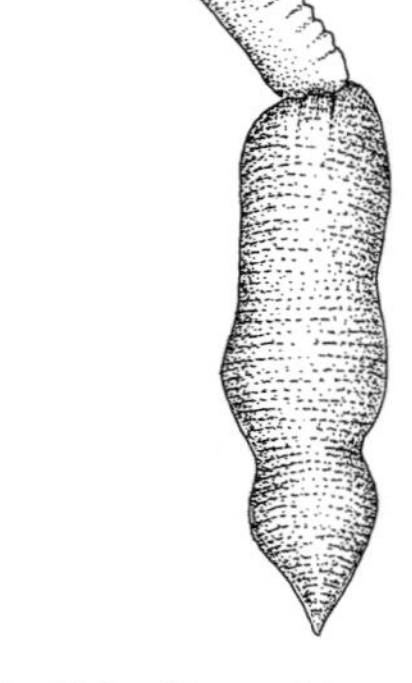

Fig. 2.9 The echiuran, *Thalassema thalassemum* (Pallas).

F. Body clearly segmented, the segmentation sometimes obscured by fine annulation. Segments all alike; or dissimilar, with a distinct head region, and in some species with the rest of the body also regionated. Appendages present or absent

i. Flat bodied, finely ringed, with a sucker at each end. No appendages. Attached to fish or macroinvertebrates (Fig. 2.10) **Hirudinea (leeches)** (p. 266)

ii. Body clearly segmented, with or without superimposed annulation; each segment bearing paired bunches of chaetae, often on lobed appendages (parapodia), and with or without additional paired processes—leaf-like, thread-like, branched, or as short tentacles. Anterior end with few to many tentacles, or forming a distinct head, with a variety of often complex appendages. Free living, or in tubes in soft substrata, or attached to a range of other substrata (Fig. 2.11) **Polychaetes** (p. 165)

iii. Body slender, cylindrical, finely ringed. Paired chaetae on most segments, but often indistinct; without any other appendages. Body not regionated, no defined head. Mostly upper shore, or in estuarine muds (Fig. 2.12) **Oligochaetes** (p. 268)

iv. Body cylindrical, segmented, less than 10 mm long. Short appendages at head and tail ends only, other segments with few bristles; head with distinctive chitinous case. Among middle- and upper-shore seaweeds, and amongst drift weed (Fig. 2.13) **Chironomidae (midge larvae)**, not treated further here

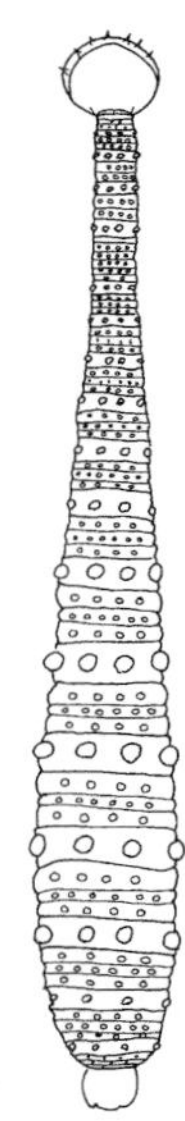

Fig. 2.10 A marine leech, *Pontobdella muricata* (Linn.).

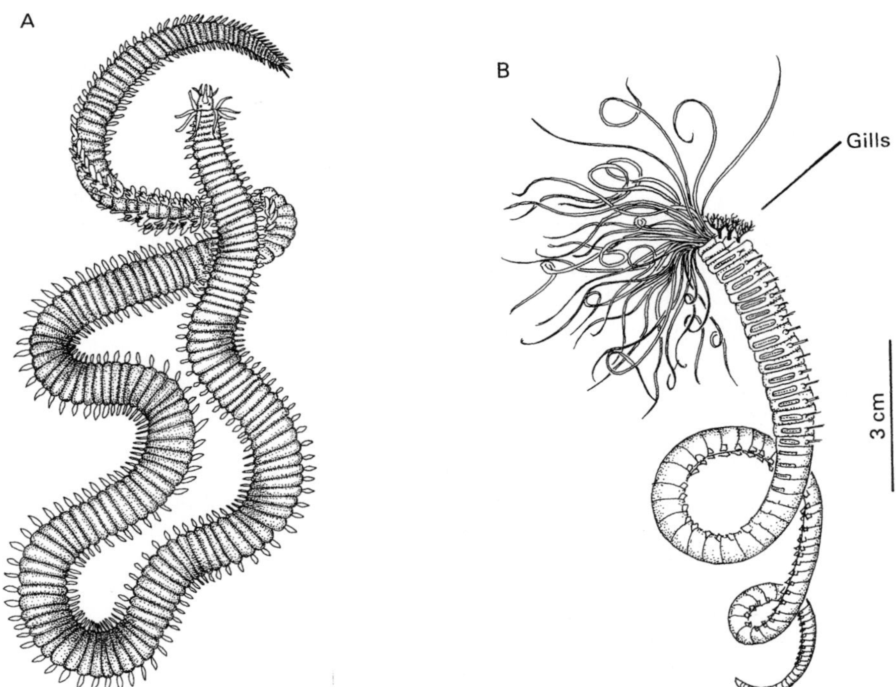

Fig. 2.11 (A) An errant polychaete, *Eulalia viridis* (Müller). (B) A sedentary polychaete, *Amphitrite edwardsi* (Quatrefages).

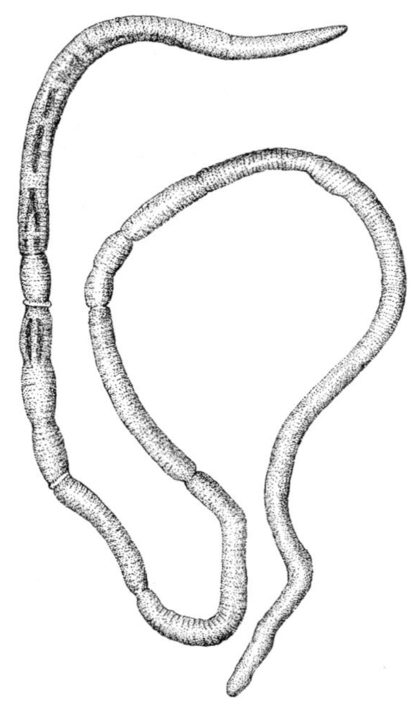

Fig. 2.12 An oligochaete, *Tubificoides benedeni* (Udekem).

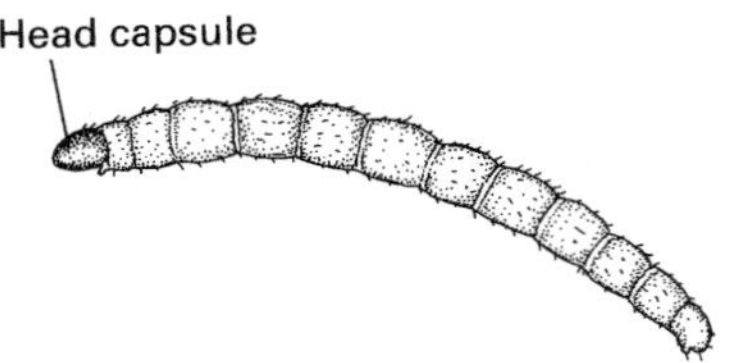

Fig. 2.13 A larval midge, *Clunio* (Chironomidae).

3. Animals with jointed legs (arthropods): choose from A–I, below

   Without legs **Go to 4**

   A. Legs few to many; tapered, cylindrical, of similar segments. All basically alike, two branched and appearing feathery
      i. Small crustaceans with oval to cylindrical bodies; head and thoracic region broad, with prominent antennae, abdominal region tapered, with paired terminal processes (furcal rami) (Fig. 2.14) **Copepods** (p. 294)
      ii. Tadpole-shaped crustaceans, with a broad anterior carapace, and a slender abdomen bearing paired terminal uropods (Fig. 2.15) **Cumaceans** (p. 319), **see also *Nebalia*** (p. 319)

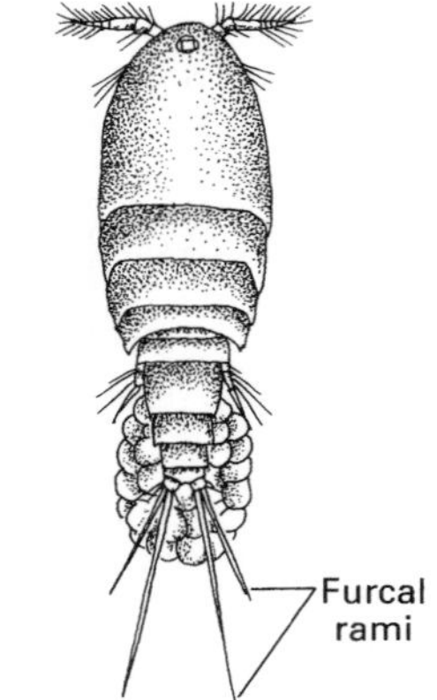

Fig. 2.14 A copepod, *Westwoodia nobilis* (Baird).

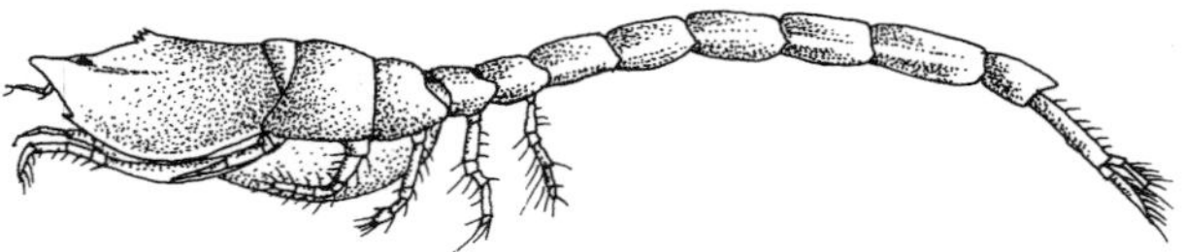

Fig. 2.15 A cumacean, *Iphinoe trispinosa* (Goodsir).

iii. Transparent, shrimp-like animals with prominent black eyes, especially feathery legs, and a broad tail fan with a conspicuous pair of statocysts at its base (Fig. 2.16) **Mysids** (p. 328)

B. Tiny crustaceans with few, simple legs. Body enclosed within an oval, bivalved shell, the tips of the appendages protruding along its margin (Fig. 2.17) **Ostracods** (p. 291)

C. Insects: animals with six legs, a well-defined head, with or without wings. Representatives of several orders occur on the middle to upper shore (p. 291)

D. Myriapods: arthropods with stiff annulated bodies, well-defined heads, bearing jaws, and numerous short, identical legs. Upper shore carnivores **Centipedes and millipedes** (p. 470)

E. Arthropods with eight long, identical legs.
 i. Body short and flat, wholly or partly segmented, not regionated. Anterior end with a short, tubular cephalon (Fig. 2.18) **Pycnogonids (sea spiders)** (p. 477)
 ii. Body regionated, not clearly segmented; anterior portion a well-defined cephalothorax (or gnathosoma), usually bearing paired, chelate palps (Fig. 2.19) **Arachnids (spiders and mites)** (p. 463)

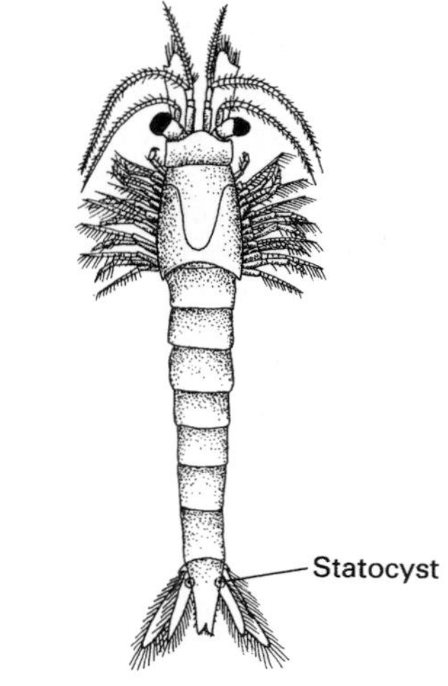

Fig. 2.16 A mysid, *Praunus inermis* (Rathke).

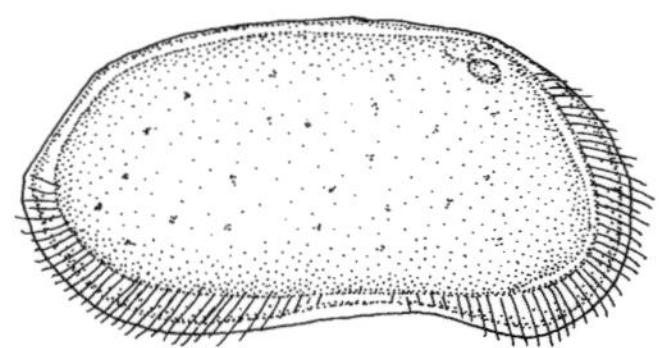

Fig. 2.17 An ostracod, *Cythere albomaculata* Baird.

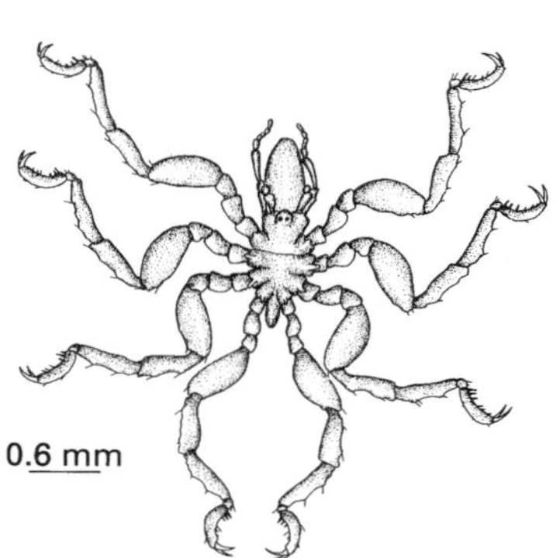

Fig. 2.18 A pycnogonid, *Achelia longipes* Hodge.

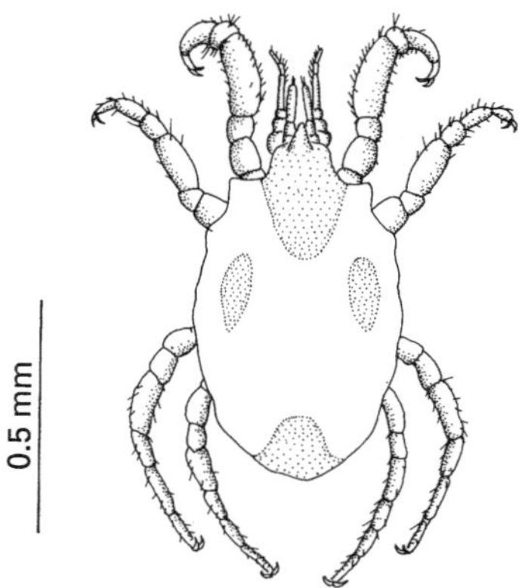

Fig. 2.19 A marine mite, *Halacarellus basteri* Johnston.

F. Small crustaceans. The body typically dorso-ventrally flattened, the thorax (pereon) bearing five to seven pairs of more or less identical walking legs (Fig. 2.20) **Isopods** (p. 340)

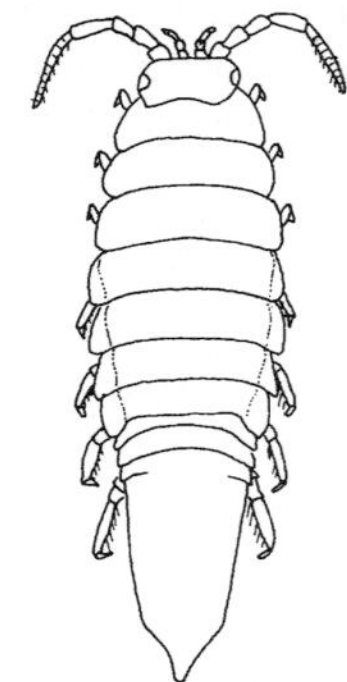

Fig. 2.20 An isopod, *Idotea granulosa* Rathke.

G. Small crustaceans; typically with laterally flattened bodies, distinctly segmented and without an anterior carapace. Thorax (pereon) with seven pairs of variously modified legs, the first two pairs—termed gnathopods—usually conspicuously different from the rest (Fig. 2.21) **Amphipods** (p. 361)

H. Small crustaceans; the body elongate, cylindrical or dorso-ventrally flattened. Anterior end with a carapace. Thorax (pereon) with seven pairs of legs, the first pair bearing claws (chelipeds), the rest essentially identical (Fig. 2.22) **Tanaids** (p. 324)

I. Large crustaceans. Thorax enclosed by a well-developed carapace and bearing 10 pairs of legs, at least the first pair with large claws (chelae). Abdomen with or without pleopods, visible or reduced and reflected beneath the thorax **Decapods (shrimps, crabs, etc.)** (p. 406)

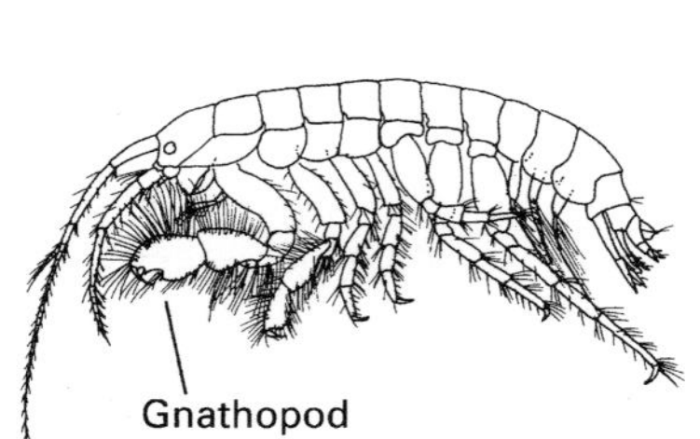

Fig. 2.21 An amphipod, *Lembos websteri* Bate.

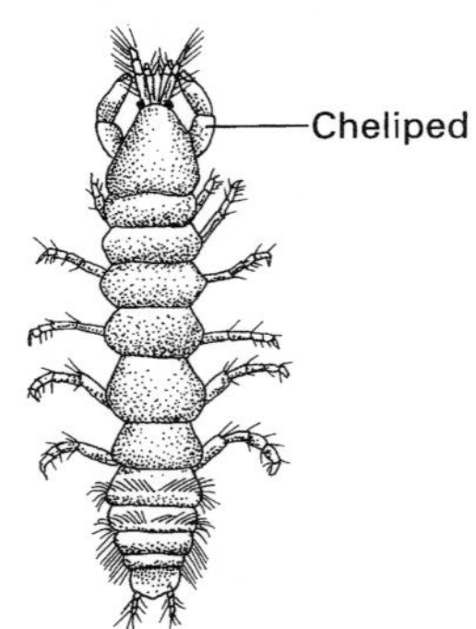

Fig. 2.22 A tanaid, *Tanais dulongii* (Audouin).

---

4. Shelled animals: choose from A–D, below

Body hard or soft, but without a distinct shell **Go to 5**

A. Sessile, i.e. permanently attached to a surface, either directly, or by means of a flexible stalk

i. Shell a short cone of calcareous plates directly attached to the substratum, closed by additional hinged plates (Fig. 2.23) **Acorn barnacles** (p. 307)

ii. Shell a series of paired, flat plates linked by a stiff skin, enclosing the animal on each side. Attached to fixed or floating substratum by a flexible stalk (Fig. 2.24) **Goose barnacles** (p. 305)

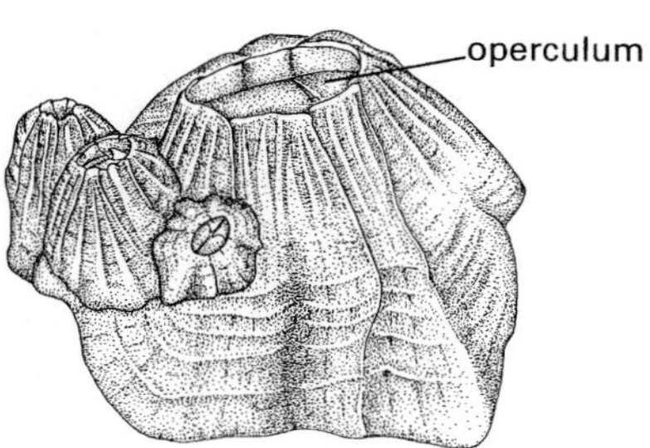

Fig. 2.23 *Balanus perforatus* Bruguière, an acorn barnacle.

B. Shell bivalved, the two valves hinged along one edge. Usually free living, but rarely cemented immovably to substratum (Fig. 2.25) **Bivalvia (cockles, clams, etc.)** (p. 573)

C. Shell a single structure, not bivalved

i. Shell cylindrical, straight or curved, open at both ends (Fig. 2.26) **Scaphopods** (p. 490)

ii. Shell spiralled, with a single well-defined aperture; or conical to oval, with or without one or more small openings at the top (Fig. 2.27) **Prosobranchs (limpets, winkles, etc.)** (p. 491)

D. Shell a transverse series of interlocking plates, encircled by a soft, uncalcified skirt (Fig. 2.28) **Chitons** (p. 484)

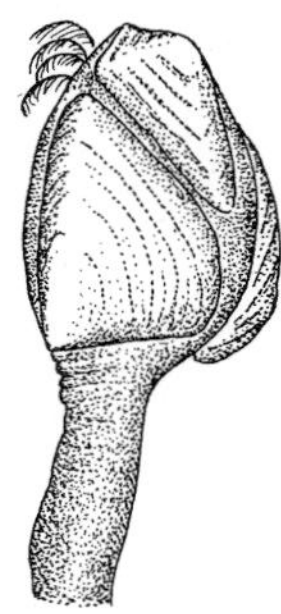

Fig. 2.24 A goose barnacle, *Lepas anatifera* (Linn.).

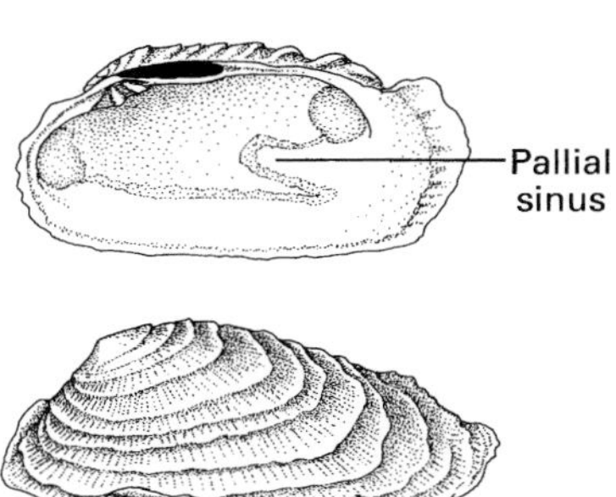

Fig. 2.25 A bivalve, *Irus irus* (Linn.).

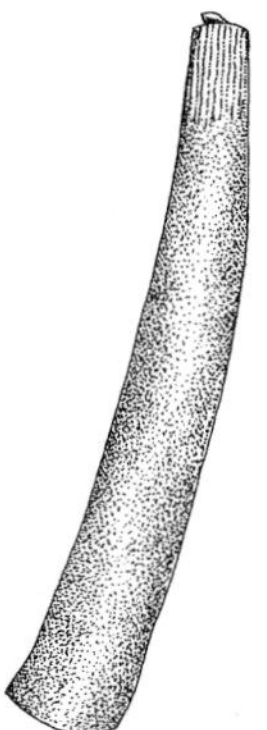

Fig. 2.26 A scaphopod, *Antalis vulgare* (da Costa).

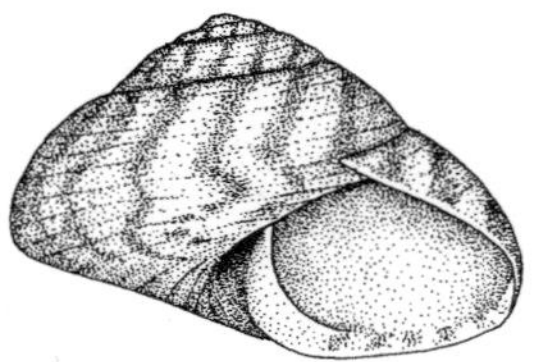

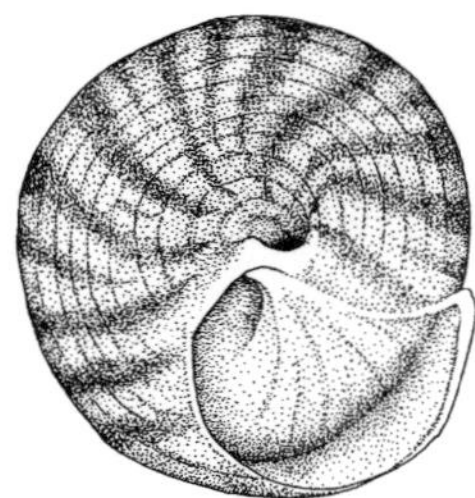

Fig. 2.27 A prosobranch gastropod, *Gibbula umbilicalis* (da Costa).

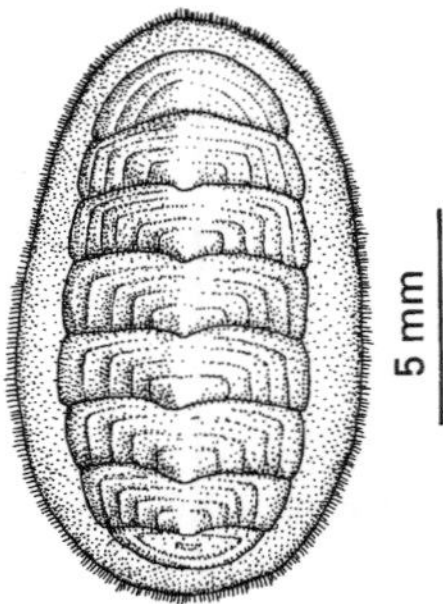

Fig. 2.28 A chiton, *Tonicella rubra* (Linn.).

5. Radially symmetrical animals, i.e. without clearly appreciable left and right sides. Body soft or hard: choose from A–B below

   Not obviously radially symmetrical: with left and right sides, or without apparent symmetry **Go to 6**

   A. Soft-bodied, transparent animals. Pelagic, or attached to a substratum
      i. Ovoid, solid, with five rows of minute, synchronously beating swimming plates. Pelagic (Fig. 2.29) **Ctenophores (sea gooseberries)** (p. 133)
      ii. Discoid, or umbrella-shaped. With or without peripheral tentacles, and central feeding tentacles. Pelagic, but often stranded on the shore **Scyphozoa (jellyfish)** (p. 64)
      iii. Cup-shaped, attached to the substratum by a short stalk. Rim of cup lobed (Fig. 2.30) **Stauromedusae** (p. 67)

   B. Body stiff and hard, rigid or flexible, often spiny. Star-shaped, ovoid, or sausage-shaped
      i. Spheroidal to ovoid, without arms. With many stout or slender spines **Echinoids (sea urchins)** (p. 676)

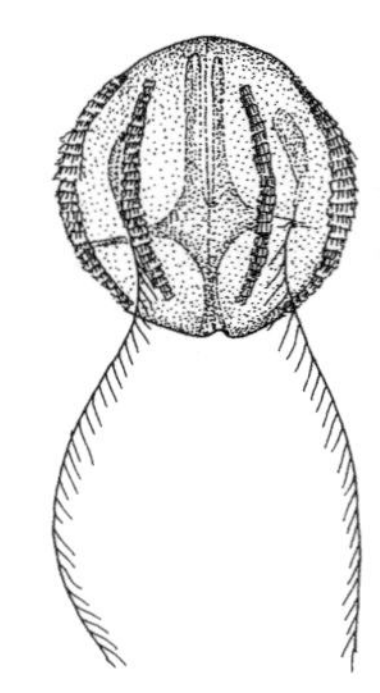

Fig. 2.29 A ctenophore, *Pleurobrachia pileus* (Müller).

ii. Star-shaped, with five or more broad, tapered, triangular arms **Asteroids (starfish)** (p. 664)

iii. Body a round, flat, central disc, with five long, thin, cylindrical arms **Ophiuroids (brittlestars)** (p. 671)

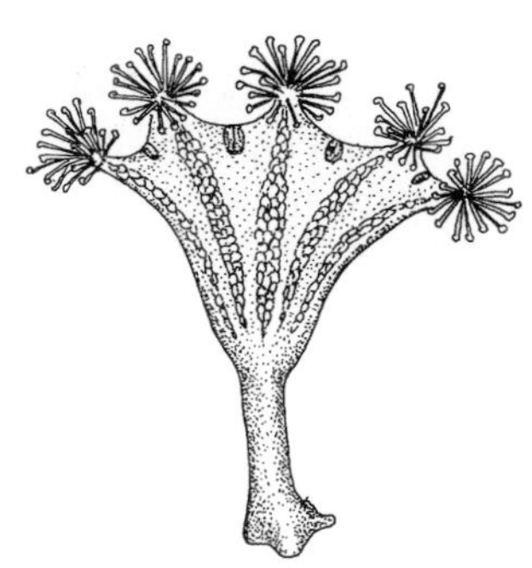

Fig. 2.30 A stalked jellyfish, *Haliclystus auricula* (Rathke).

---

6. Soft-bodied, non-colonial animals. Slug-like, with well-defined head, often bearing eyes and tentacles, and a creeping sole. Dorsal surface often with papillae, lobes, tufted gills, or other structures **Opisthobranchs (sea slugs)** (p. 539)

   Body form and texture various. Colonial or non-colonial, but without a defined head and usually sedentary or sessile **Go to 7**

---

7. Sac-like structures, attached to the tissues of other invertebrates and obviously parasitic try **copepods** (p. 294), **Rhizocephala** (p. 311), **isopods** (p. 340)

   Free-living or epibiotic, for example encrusting the exoskeleton of a crab, but not parasitic **Go to 8**

---

8. Pelagic animals sporadically stranded on the shore; body more or less transparent, consisting of a float, a 'sail', and beneath, bunches of long or short fishing tentacles ***Velella*** (p. 78) or ***Physalia*** (p. 112)

   Sessile, not pelagic **Go to 9**

---

9. Unitary animals (i.e. not colonial); sometimes in dense clumps, but the individuals none the less

separate, with no common tissue: choose from A–B, below

Colonial animals; the constituent members of the colony often clearly distinct, but obviously permanently fused to their neighbours **Go to 10**

A. On hard substrata, occasionally on seaweed, or burrowed in sand. With single or multiple whorls of tentacles bordering a terminal disc, with central mouth, surmounting a contractile, muscular column (Fig. 2.31) **Anthozoa (sea anemones)** (p. 117)**(see also the hydroids *Candelabrum* and *Corymorpha*).**

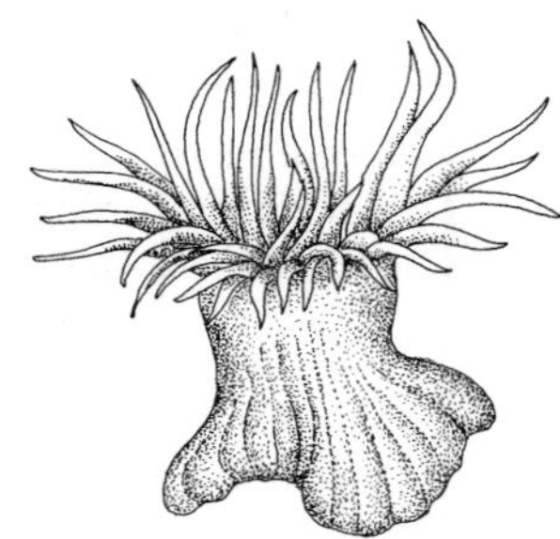

Fig. 2.31 A sea anemone, *Actinothoe sphyrodeta* (Gosse).

B. Attached to various substrata; erect or repent, the body oval, sac-like, firm or gelatinous, with two distinct, sphinctered siphons (Fig. 2.32) **Urochordata (sea squirts)** (p. 689)

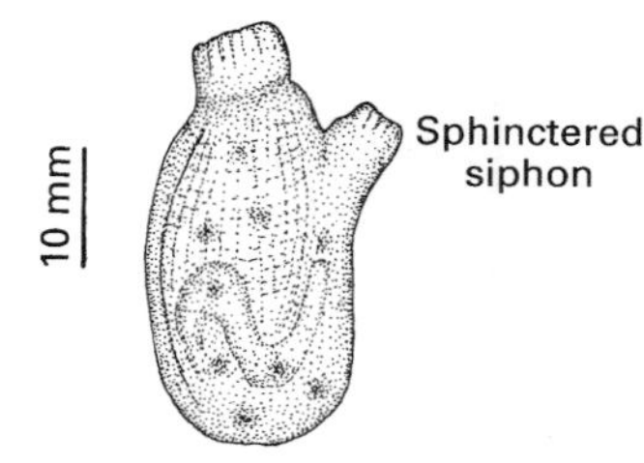

Fig. 2.32 A solitary sea squirt, *Ascidiella scabra* (Müller).

---

10. Colony a sheet or mound, sometimes bushy or tree-like, with regular branching, but more often rather formless. Surface texture smooth or spongy, often rather amorphous, with few or many conspicuous openings (oscula), but without recognizable 'individuals' (zooids) **Porifera (sponges)** (p. 35)

Colony clearly composed of serially repeated units or modules (zooids), monomorphic or polymorphic but always easily defined **Go to 11**

---

11. Individual feeding polyps always with tentacles; few or many, long and lax or short and stiff, contractile or retractile: choose from A–D, below

Tentacles absent **Go to 12**

A. Colony consisting of stalked modules linked by a stolon, each stalk with a swollen distal end (capitulum), bearing short, ciliated tentacles which may contract but do not withdraw into the capitulum (Fig. 2.33) **Entoprocts** (p. 284)

B. Colony form various; often calcified, sometimes gelatinous. Zooids box-shaped, flask-shaped, or irregular (if stalked, or attached to a stolon, the tentacles clearly retractile through an opening which may be closed by a single hinged operculum). Zooids with discrete boundaries, without visible tissue connection between. Tentacles with cilia—visible as a flickering iridescence under low magnification (Fig. 2.34) **Bryozoans** (p. 629)

C. Colony form various; never calcified. Tentacles contractile, sometimes withdrawn into a surrounding chitinous cup, which may be closable, but never by a single, hinged operculum; always without cilia. In transparent, branching forms body tissue is seen to be continuous through the colony (Fig. 2.35) **Hydrozoans (hydroids and allies)** (p. 70)

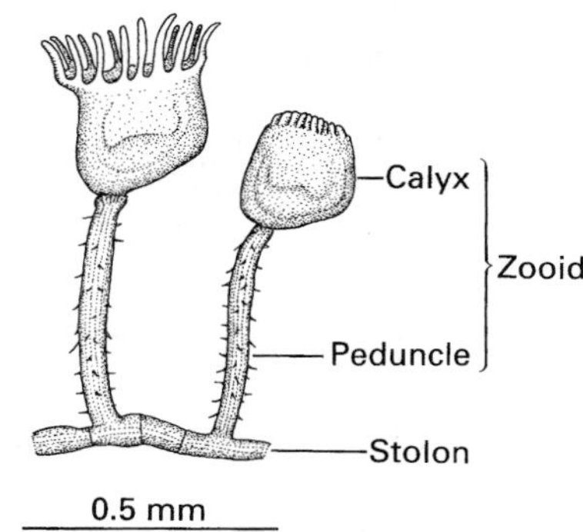

Fig. 2.33 The entoproct *Pedicellina cernua* (Pallas).

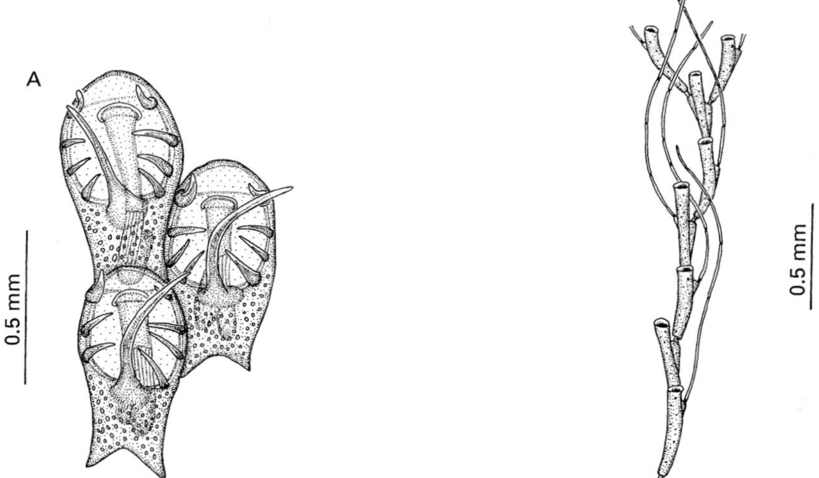

Fig. 2.34 Zooids of (A) the encrusting bryozoan *Electra pilosa* (Linn.); and (B) the erect bryozoan *Crisidia cornuta* (Linn.).

D. Colony a series of diffuse, annulated chitinous tubes encrusting dead shells. Zooids each with a few branched, ciliated tentacles
Pterobranchs (p. 687)

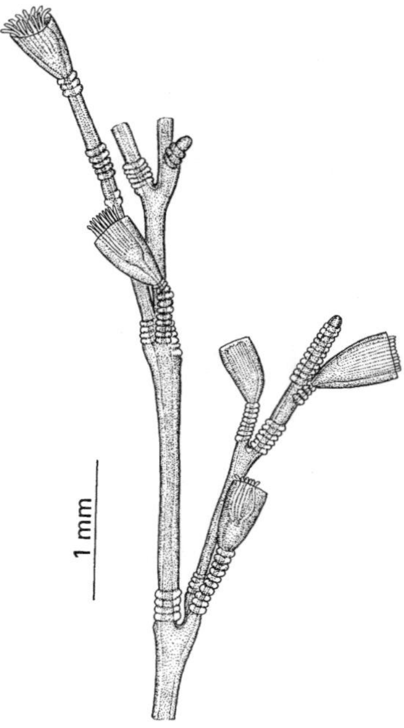

Fig. 2.35 Part of a colony of the hydroid *Obelia dichotoma* (Linn.).

---

12. Colony a flat sheet, a mound, or a fleshy, often sandy, stalked cushion, gelatinous or rather stiff. Zooids clearly visible (cf. sponges), sometimes grouped in regular or irregular series around common cloacal openings, but each with its own incurrent opening (Fig. 2.36) **Urochordates (sea squirts)** (p. 689)

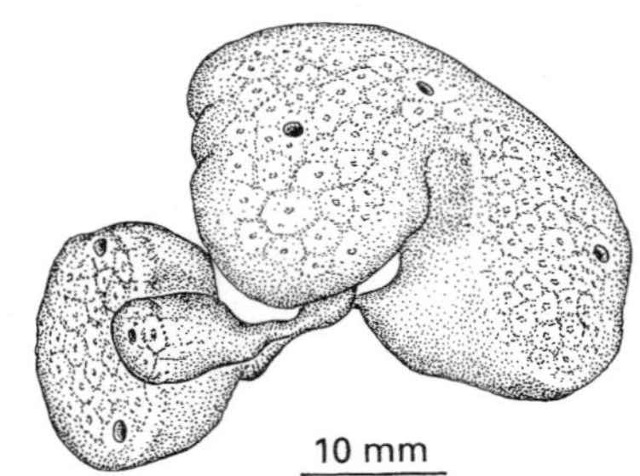

Fig. 2.36 A colonial sea squirt, *Polyclinum aurantium* Milne Edwards.

COLLECTION AND CONSERVATION

Field studies in marine biology usually involve the collection of specimens. Many fish and some macroinvertebrates are easily identified in the field and need not be removed from their habitats, but accurate identification of most invertebrates requires low-power microscopic examination. Collection and removal of specimens should be practised with restraint, and every care should be taken to avoid damaging disturbance to marine habitats. Rocks and boulders need to be lifted to reveal the communities living beneath them or on their undersides, but must always be replaced in their original positions. Constant overturning, as too frequently happens on shores where fishing bait is collected, results in the death of plants and animals living on both surfaces of, and the substratum below, the rocks, and soon leads to a sterile shore. Most large brown algae, and many of the shrubby red algae, are long-lived perennials that may take years to achieve maximum size. It may be necessary to remove foliage to examine encrusting organisms, but while wholesale removal of large plants opens up space for new sporelings, it also destroys entire communities of sessile and sedentary organisms associated with the plants. Similarly, indiscriminate splitting open of crevices on soft rock shores destroys microhabitats which may have taken several years to develop. Digging and sieving sand for infaunal animals probably causes less disturbance than results from storms and tidal surges, but the comparatively stable mudflat communities are severely disturbed by excessive digging (as for bait), and it may take several seasons without digging before the substratum and its animal populations are restored.

Intertidal and shallow subtidal animals may be remarkably hardy, and will withstand storage for some hours in clean seawater, kept out of doors and in the shade. It is thus perfectly feasible to identify, sex, measure or weigh specimens, and return them to the sea without their having suffered any damage. It is always best to examine specimens as soon as possible after collection. For some groups—especially anemones, sea slugs, and many polychaetes—identification depends on features such as colour, or the length, shape or number of contractile tentacles, which can only be appreciated in healthy living specimens. However, for some animals it is necessary to narcotize and kill the specimen before it can be identified to species. In most cases only a single specimen need be sacrificed. This necessity arises most often with the smaller crustacea, but is not as damaging to natural populations as may appear. Populations of amphipods, for example, consist of many, short-lived, overlapping generations which are replaced several times in the course of a single year. Techniques for narcotizing and preserving specimens are given where appropriate, and repeated wastage of specimens can be avoided to some extent by retaining a small reference collection.

Marine and coastal conservation are rightly attracting an increasing priority on the most heavily populated, or most frequently visited, coastlines. A number of national marine reserves now exist or are planned, and in some areas of Britain further protection of marine biological resources is afforded by either statutory regulations or local by-laws. Thus, before venturing on to the sea-shore it is always important to be sure that you are not disturbing a protected area.

TAXONOMY

The number and variety of living organisms have generally defeated the name-coining abilities of generations of naturalists, and most animal species do not have a common, or vernacular, name. Common names for species are included here where, as with fish, they are long established and clearly unambiguous; this latter point especially needs to be stressed as an accepted vernacular name may have been used for different organisms in different parts of Britain. By contrast, each recognized species does have a unique, two-word Latin name, which is no more difficult to pronounce or remember than a name in, for example, English, French, or Dutch. Although the pronunciation of latinized words is subject to some variation, it should be noted that a terminal -e is always sounded: thus *Belone* (bel-on-ee) or *Polynoe* (poly-no-ee). This binomial system of taxonomic nomenclature was established by the Swedish biologist Carolus Linnaeus in the tenth edition of his great work of classification, *Systema Naturae* (1758), and is now accepted worldwide. The procedures for naming plants and animals are now governed by broadly similar codes, administered by their respective international commissions. The name of any animal consists of a unique species (or trivial) name, formally with its originating author following. Thus, the cod is *morhua* Linnaeus. Closely related species are grouped into genera (sing., genus); the cod is classified in *Gadus*, to give the binomial *Gadus morhua* Linnaeus. In cases where subsequent taxonomic research indicates that a species has been placed in an inappropriate genus, it may be relocated: the name of the author is then enclosed within parentheses, as in the shore rockling *Gaidropsarus mediterraneus* (Linnaeus), to indicate that it has been reclassified.

The animal kingdom is classified in an ascending hierarchy of taxa, with genera grouped into families, families into orders, orders into classes, and classes into phyla (sing., phylum). These major categories may be further divided (or grouped) into suborders, subfamilies (superorders, superfamilies), etc. The finer levels of classification are of limited importance to field biologists, and in this guide, for the most part, only families, orders, and classes are used and defined, principally to emphasize observable distinctions between these major groupings. Each species included has been given its current Latin name; this may occasionally differ from that used in earlier faunas, although illustrations clearly depict the same species. Names in use do change, or appear to change, for a number of reasons: two or more names given to a single species are said to be 'synonyms'. Space does not permit synonymies to be given here, but the literature cited at the end of the book includes systematic and monographic treatments, which give the sources of the classification and taxonomy employed here.

REFERENCES AND FURTHER READING

This concise fauna naturally includes only a selection of the marine animals occurring around the north-west European coasts. Text descriptions in some cases will mention additional, similar species; and notes on the total taxonomic diversity, in most cases, are given following the family diagnoses. Species not included here may be identified using *The marine fauna of the British Isles and north-west Europe* (Hayward and Ryland 1990), or one of the specialist texts listed in Chapter 15. These are grouped in chapter sequence and consist of the most recent, or most useful, monographic treatments of each group, together with any important taxonomic research papers, and a range of textbook sources which detail the biology, ecology, morphology and classification of each animal group. Most of these works will be found in university libraries, but only few in municipal public libraries. All, however, may be borrowed from the British Library, by application through a public lending library.

3 SPONGES

The sponges (Phylum Porifera) are sedentary, aquatic, fundamentally radially symmetrical invertebrates. They possess a minimal diversity of cells and the simplest grade of metazoan organization. They are filter feeders characterized by an internal monolayer of flagellated *choanocytes* collectively circulating one or more unidirectional water streams through the *aquiferous system*, a network of pores, canals, and chambers perforating the body.

The 5000 or so Recent species are classified in four classes, principally according to the composition and structure of the skeletal elements:

The exclusively marine Calcarea possess calcium carbonate spicules. The group presents considerable taxonomic difficulties at all levels, and only a selection of the better understood species is included in this guide.

The Hexactinellida possess six-rayed spicules. They typically occur in deep marine waters, and thus are excluded here.

The Demospongiae represent some 95 per cent of the world sponge fauna and include the only freshwater species. Many have a skeleton of siliceous spicules, which are either *monaxons* (single-rayed) or *tetraxons* (four-rayed) (Fig. 3.1). This is sometimes supplemented or entirely replaced by an organic, *collagenous* skeleton of flexible *spongin* fibres. A few species have no skeleton.

The Sclerospongiae is a small marine group characterized by a massive calcareous skeleton. There are no known British representatives.

Sponge organization is graded in complexity (Fig. 3.1). The simplest scheme is the *asconoid* arrangement. The body consists of one or more simple, thin-walled, tubular units, each with a central cavity opening apically as a single, exhalant *osculum*. Inhalant canals lined by specialized *pinacocytes* bridge the *mesohyl* between the *pinacoderm* and the *choanoderm*, which lines the cavity as a simple layer (Fig. 3.1). The asconoid arrangement occurs in the *olynthus*, the post-settlement phase of calcareous sponges, and in adults of the Leucosoleniidae.

The *syconoid* organization entails a folding of the sponge wall, with the choanoderm localized to side chambers off a central cavity. These interdigitate with cavities lined by pinacoderm, extending from the outside of the sponge. The result is that the mesohyl is thicker than in asconoid sponges. This condition is best known within the Calcarea, e.g. *Sycon*.

The most complex organization, the *leuconoid* arrangement, entails further annexation of the choanoderm into *chambers*. Each is fed from *incurrent canals*, passing from incurrent pores, the *ostia*, and entering the chamber through *prosopyles*. Water exits via a single *apopyle* leading to an *excurrent canal*. Tributary canals coalesce to converge on excurrent apertures or *oscula*. The aquiferous system of leuconoid sponges can be complex, and is often accompanied by a considerable thickening of the mesohyl, and the loss of radial symmetry. The leuconoid condition occurs in many Calcarea, e.g. *Leuconia*, and almost all other sponges.

Growth occurs at the edges of encrusting sponges, at terminal points in arborescent forms, and more generally across the surface of massive, mound forms. Growth is often indeterminate, ultimate size being governed by external agents such as water motion or predation. It often involves the repetitive addition of units of a standard template, but these are not so rigidly defined as the polyps and zooids of other colonial phyla. Such a *modular* growth scheme provides a high capacity to survive injury.

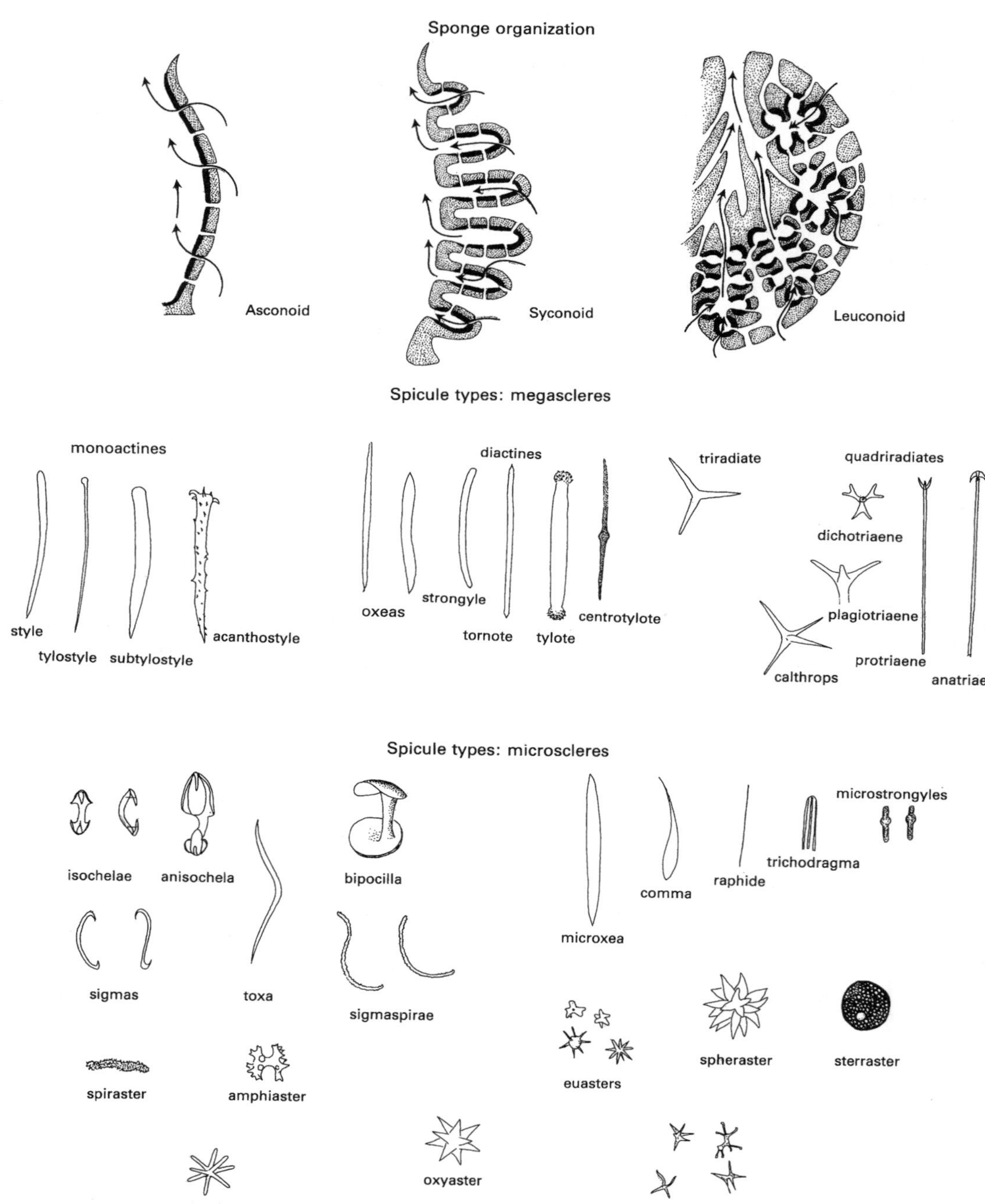
Sponge organization
Asconoid
Syconoid
Leuconoid
Spicule types: megascleres
monoactines
style
tylostyle
subtylostyle
acanthostyle
diactines
oxeas
strongyle
tornote
tylote
centrotylote
triradiate
quadriradiates
dichotriaene
plagiotriaene
calthrops
protriaene
anatriaene
Spicule types: microscleres
isochelae
anisochela
toxa
bipocilla
sigmas
sigmaspirae
spiraster
amphiaster
strongylaster
oxyaster
microxea
comma
raphide
trichodragma
microstrongyles
euasters
spheraster
sterraster
streptasters

Fig. 3.1

Individual sponges are usually hermaphroditic. Amoeboid cells of the mesohyl develop into gametes. Spermatozoa are released from, and enter, the sponge via the aquiferous system. Fertilization and embryogenesis are internal within the mesohyl. Short-lived, lecithotrophic, ciliated larvae are released via the aquiferous system, and metamorphosis follows settlement. Growth and reproduction are seasonal. As with freshwater sponges, some marine species produce overwintering gemmules.

In total, 67 species, representing 29 families of the Calcarea and Demospongiae are included in this guide, constituting around one-third of the British continental-shelf sponge fauna. Common species, together with some particularly distinctive rarer ones, are included.

CRITERIA USED IN IDENTIFICATION

Sponge taxonomy is based on the presence and form of spicule types. While most species can be identified according to spicules alone, others require an integrated approach. Problems are frequently exacerbated by intraspecific variability in characteristics. Some species have distinctive forms, e.g. spheres, funnels, or arborescent patterns. Others share very common forms, e.g. encrusting sheets, or irregularly massive mounds, or can be highly variable in form. By virtue of indeterminate growth, maximum size values cannot be rigidly defined. Surface morphology can be more consistent than gross form; while some species are smooth, many are cratered, or bear papillae or projecting spicules. Inhalant pores are usually invisible to the naked eye, but oscula are often prominent. They may be single or numerous; scattered or grouped; simple or armoured with spines; flush or raised upon papillae. Species with dense-packed spicules can have a characteristically firm consistency, and those with a high content of spongin fibres may be exceptionally delicate. Many sponges contract when disturbed, causing a marked reduction in size, firmer consistency, and the closure of oscula. Some sponges exude mucus when disturbed or damaged.

Sponges can be very colourful in life; browns, oranges, reds, and yellows are predominant; white, blues, or black ones are less frequent. Sometimes the colour is sufficiently unusual and consistent to serve as a diagnostic character. However, a wide range of colour variants is more typical. Endosymbiotic algae can tint or transform surface colour, by imparting green, blue, yellow or red hues.

Many sponges accumulate aromatic terpenoids which may serve defensive roles against infection, overgrowth, or predation. Such volatile metabolites may produce distinctive odours which can be difficult to describe in the absence of more familiar equivalents, but which may be species specific. The most important character for identification is still the skeleton. In some species the spicules are arranged randomly, in others they are localized and ordered, e.g. radially; and/or may be differentiated into an *ectosome* or skin, and an internal *choanosome*. Spongin fibres can also be distinctively arranged.

In most Demospongiae, spicules are categorized by size into *megascleres* and *microscleres* (Fig. 3.1). There is considerable variation in form, and some variants from particular families have specific names. However, most spicules fall into one of several general categories (Fig. 3.1). Those of the Calcarea, and the megascleres of Demospongiae, are classified by the number of axes and points, e.g. monaxons can be either *monoactines* or *diactines*. Thereafter, classification relies on the form of the ends, whether any swellings occur, or whether the surface is thorny or otherwise. Within each spicule category there can be considerable inter- and intraspecific variation in form, e.g. shape and curvature of the shaft, the point, any swellings, and the thorn distribution. Greater diversity is encountered in the microscleres. The types of spicules known for each species are described in this guide, but size ranges, which can be variable, are only included in the absence of reliable alternative criteria in the keys.

Marine sponges are most widespread within shallow sublittoral rocky areas. A few species regularly occur on the shore, but none is confined to that zone. Essentially sublittoral species can occur intertidally in well-shaded crevices and caves on the lowest shore. Definitive vertical distributions cannot be specified for most species; the littoral and shallow sublittoral distribution of many species is fairly well known, but beyond regular diving depths data are far more limited. Since few species are definitively depth zoned within the 0–40 m range, and the distinction between truly sublittoral and littoral species is not absolute, depth range is not considered particularly useful for identification purposes. Geographical distributions and abundances are also imperfectly known and cannot be relied upon as aids to identification.

Most sponges are typically epilithic, although many also foul artifical hard substrata. Some species are characteristic of biogenic materials, such as macroalgae or mollusc shells. In many localities, heavy silting precludes sponges from upwardly orientated facets. However, certain species are morphologically adapted to tolerate silting, and commonly occur on such surfaces. Sponges flourish most prolifically under conditions of strong water flow. Many species thrive in wave-induced water motion, as occurs in the surge gullies of the shallow sublittoral zone. Within tidal rapids, intrinsic active filtration is sometimes augmented by extrinsically induced passive filtration. Massive forms characteristic of strong currents often shrink if transferred to more sheltered areas. Although most species are exclusively marine, many extend into estuaries and lagoons.

IDENTIFICATION

Ideally, the true form, dimensions, surface detail and colour of the live sponge should be recorded by photography. Many species substantially and irreversibly alter their appearance once collected. Odour or the presence of mucus should also be recorded. A sufficiently large sample should be taken to portray surface characteristics, skeletal structure, and to provide adequate quantities of spicules and/or spongin for analysis, i.e. the whole sponge in the case of small forms. Removal with the substratum is advised for very thin encrusting forms. Preservation should be undertaken soon after collection. Seawater formalin is a good short-term preservative for all sponges, but because of its tendency to decalcify, it is unsuitable for Calcarea beyond a few days. A safe long-term preservative for all sponges is 70% ethanol, but it usually leaches or otherwise alters the natural colour.

Skeletal structure may be examined in thick sections of the sponge made along the appropriate axis with a razor blade. The disposition of spicules or spongin fibres can be seen with a low-power compound microscope (× 100). To examine spicules or spongin fibres, the sponge tissue must be removed. Small cubes of sponge representing all sections from the surface to the interior should be submerged in a digesting agent. If it is unclear whether the specimen is a calcareous sponge or not, this can be established by immersing a subsample in dilute hydrochloric acid. The best agent for spicule extraction is sodium hypochlorite or any household bleach containing this substance. Whereas some species digest quickly, some may take up to 12 h, and others will not break down without specialist techniques. Note that spongin fibres may be damaged by digesting agents. Once digested, the dissolved materials should be rinsed with several changes of water, allowing sufficient time for the smallest microscleres to settle before decanting.

For examination, magnifications of × 100 to × 1000 are required. The presence and detailed structure of spicules should be sufficient for determining most species; however, spicule dimensions are relevant in a few cases. Usually the dominant spicule types are immediately apparent, but rarer types may require a lengthy search. Some sponges incorporate spicules released from their neighbours, and this can cause problems with identification.

KEY TO FAMILIES

1.	Spicules wholly calcareous (test with weak hydrochloric acid)	**Class Calcarea** (p. 42)
	Spicules siliceous or absent	**2**
2.	Spicules absent	**3**
	Spicules numerous	**6**
3.	No skeleton. Encrusting, pigmented sponges	**4**
	Spongin fibre network present, tough and elastic (spongy)	**5**
4.	Surface lobed and velvety	**5. Oscarellidae**
	Surface very smooth	**29. Halisarcidae**
5.	Spongin without inclusions	**28. Aplysillidae**
	Spongin full of sand grains	**27. Dysideidae**
6.	Megascleres include tetraxons, microscleres are asters	**7**
	Spicules not as above	**8**
7.	Microscleres include sterrasters	**6. Geodiidae**
	Sterrasters absent	**7. Pachastrellidae**
8.	Megascleres are triaenes and oxeas, microscleres are sigmaspirae	**8. Tetillidae**
	Spicules not as above	**9**
9.	Megascleres are calthrops, microscleres include toxas	**7. Pachastrellidae**
	Spicules not as above	**10**

10. Megascleres are tylostyles or subtylostyles, usually with at least superficial radial organization **11**

Spicules form and arrangement not as above **14**

---

11. Sponge spherical, showing strongly radially symmetrical spicule arrangement throughout, megascleres very long subtylostyles, microscleres are asters **12. Tethyidae**

Not as above **12**

---

12. Sponge forms a cushion from which arise numerous long papillae, microscleres absent **10. Polymastiidae**

Not as above **13**

---

13. Sponge massive or encrusting; microscleres usually absent, but microstrongyles can occur **9. Suberitidae**

Sponge massive or boring, microscleres include spirasters **11. Clionidae**

---

14. Without cheloid microscleres **15**

With cheloid microscleres or, if not, with at least three other microsclere types **22**

---

15. Skeleton contains axial tracts of substantial spongin fibres, and megascleres which may be monaxons, oxeas, styles, or strongyles; these may be twisted in form. Sponge erect with flexible consistency, surface rough with projecting spicules (Axinellida). **16**

Not as above **18**

---

16. Acanthostyles occur, sponge branching **15. Raspaillidae**

Acanthostyles absent, sponge branching or cup-shaped **17**

---

17. Very long styles present (some >1.5 mm long), sponge branching **14. Hemiasterellidae**

Styles not exceptionally long (usually <1.5 mm), sponge branching or cup-shaped **13. Axinellidae**

---

18. Acanthostyles occur, or styles with toxas. Sponge usually thin, encrusting **24. Hymedesmiidae** *(Hymedesmia brondstedi)* **26. Clathriidae**

    Acanthostyles absent **19**

---

19. Oxeas, if present, not stout. Spongin does not occur as substantial fibres. Spicules without regular arrangement except at the surface in some species **20**

    Skeleton dominated by stout oxeas with or without substantial quantities of spongin fibre **21**

---

20. Skeleton principally of slender oxeas, sometimes with styles **16. Halichondriidae**

    Skeleton principally of styles, although oxeas may occur **17. Hymeniacidonidae** **21. Biemnidae** *(Hemimycale columella)*

---

21. Tangential dermal spicule skeleton occurs, spongin occurs in very small quantities at the spicule joints, imparting a very delicate consistency to the sponge (easily confused with the Haliclonidae) **19. Adociidae**

    Spongin typically evident as heavy fibres imparting a fairly tough but flexible consistency (easily confused with the Adociidae) **18. Haliclonidae**

---

22. Cheloid microscleres absent, sigmas, commas, and other forms occur **21. Biemnidae**

    Cheloid microscleres present **23**

---

23. Without acanthostyles **24**

    With acanthostyles **25**

---

24. With anisochelae and often a number of other microsclere types **20. Mycalidae**

    Without anisochelae **22. Desmacidonidae**

---

25. With toxas **26. Clathriidae**

    Without toxas **26**

26. Sponge mound, massive, or branched in form, often exceeding 20 mm in thickness, with a variety of microscleres **23. Myxillidae**

    Sponge forming thin (<20 mm) patches or sheets **27**

27. Endosomal skeleton includes an array of acanthostyles orientated vertically away from attachment surface and embedded basally in a layer of spongin fibre **24. Hymedesmiidae**

    Endosomal skeleton of acanthostyles not arranged as above, spongin not present in substantial quantities **25. Anchinoidae**

Class Calcarea

Skeleton composed solely of calcareous spicules, not differentiated by size into megascleres and microscleres. Exclusively marine.

KEY TO SPECIES

1. Sponge composed of delicate, thin-walled tubes (<1 mm diameter) **2**

   Sponge more substantial and solid **4**

2. Sponge consisting of a mass of tubes forming a tight anastomosing meshwork. Triactines with rays equal in length ***Clathrina coriacea***

   Sponge consisting of a loose array of mainly erect tubes. Triactines with rays unequal in length **3**

3. Erect tubes little branched ***Leucosolenia botryoides***

   Erect tubes frequently branched ***Leucosolenia complicata***

4. Gross shape a flattened sac ***Scypha compressa***

   Not as above **5**

5. Gross shape cylindrical or vase-shaped with a single terminal osculum ***Sycon ciliatum***

   Gross shape encrusting or massive, usually several oscula **6**

6. Surface smooth to touch ***Leuconia nivea***

   Surface distinctly rough to touch ***Leuconia gossei***

Subclass Calcinea

Triactines, if present, have rays set approximately at equal angles, and equal in length.

Order Clathrinida

Asconoid organization retained at the adult stage. The sponge consists of delicate, thin-walled tubes.

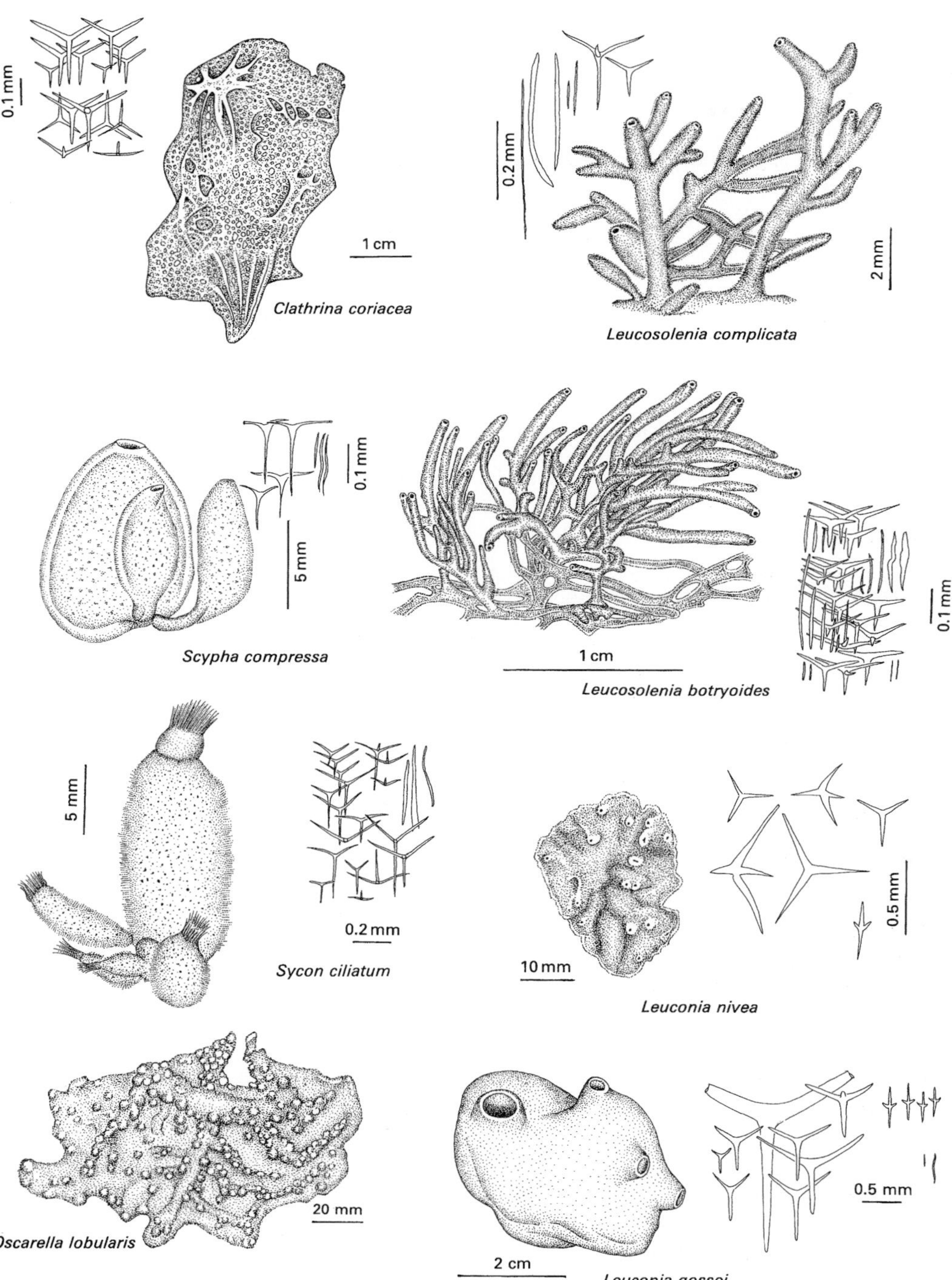
0.1 mm
1 cm
Clathrina coriacea
0.2 mm
2 mm
Leucosolenia complicata
0.1 mm
5 mm
Scypha compressa
1 cm
Leucosolenia botryoides
0.1 mm
5 mm
0.2 mm
Sycon ciliatum
10 mm
0.5 mm
Leuconia nivea
20 mm
Oscarella lobularis
2 cm
Leuconia gossei
0.5 mm

Fig. 3.2

1. CLATHRINIDAE

Clathrina coriacea (Montagu) Fig. 3.2
Characteristically clathrate (mesh-like), anastomosing mass of delicate tubes. Diameter to 25 mm, thickness to 10 mm. Oscula not distinct. White. Skeleton a mixture of triradiates, quadriradiates and sometimes oxeas.

Epilithic, low-littoral, and shallow sublittoral. Common, all coasts.

Subclass Calcaronea

Triactines with rays of unequal length, set at unequal angles.

Order Leucosolenida

Asconoid structure occurs throughout life. Sponge composed of very thin tubes.

2. LEUCOSOLENIIDAE

Leucosolenia botryoides (Ellis and Solander) Fig. 3.2
Encrusting holdfast of thin tubes, from which arise erect, little-branched tubes, each bearing a single, terminal osculum. Often about 10 mm in diameter and height. Surface minutely hispid. Consistency very delicate. Translucent white. Skeleton includes tri-radiates and quadriradiates.

Epilithic and epibiotic on algae. Common, all coasts.

Leucosolenia complicata (Montagu) Fig. 3.2
Extensive encrusting holdfast of reticulating ascon tubes giving rise to erect tubes bearing lateral branches and terminal oscula. Sponge often diminutive to the naked eye. Surface of tubes minutely hispid. Consistency very delicate. Translucent white. Skeleton of triradiates, quadriradiates, and oxeas.

Epilithic and epibiotic on algae and sessile invertebrates. Cryptic low-littoral, sublittoral. Common.

Order Sycettida

Syconoid or leuconoid structure, choanocytes localized to choanocytic chambers.

3. SYCETTIDAE

Sycon ciliatum (Fabricius) Fig. 3.2
Characteristically cylindrical, more or less tapering as a vase. Bears a single, terminal osculum, often fringed by a circlet of spines. Up to 50 mm in length and 7.5 mm wide. Delicate to firm, flexible. Pure or off-white. Skeleton of triradiates, quadriradiates, and oxeas.

Occurs singly or in clusters on rock or seaweeds, cryptic low-littoral and sublittoral. Very common, all coasts.

4. GRANTIIDAE

Scypha (Grantia) compressa (Fabricius)
Purse sponge Fig. 3.2
Laterally compressed sac, tapering at both ends, attached basally, and with a single, terminal osculum. Typically 10–20 mm high, 5–10 mm wide and 2.5 mm thick. Surface lightly hispid. Consistency firm but flexible. White or off-white. Skeleton of triradiates, quadriradiates, and oxeas.

Occurs individually or collectively upon rock or seaweeds, in the sublittoral or the lower littoral where it may feature on open surfaces as well as in crevices. Very common, all coasts.

Leuconia gossei (Bowerbank) Fig. 3.2
Encrusting or irregularly massive form, with undulating ridges that can impart a near cerebriform appearance. Surface slightly rough to touch. Oscula naked. Consistency firm. White or off-white. Skeleton of triradiates, quadriradiates, and oxeas.

Epilithic, lower littoral, and sublittoral.

Leuconia nivea Grant Fig. 3.2
Thick, often undulating encrusting sheets or cushions featuring a scattering of small, naked oscula. Surface smooth. Consistency firm and friable. Pure white. Skeleton consists of triradiates, quadriradiates, and microxeas.

Epilithic, sublittoral.

Class Demospongiae

Sponges with siliceous spicules, in many species divisible into megascleres and microscleres. Megascleres are usually monaxons or tetraxons, but are triaxons in one subclass. Hexaxons absent. In some cases the skeleton is supplemented, dominated, or may be entirely substituted by spongin. In a few species, the skeleton is absent.

Subclass Homoscleromorpha

Spicules often triactines, frequently supplemented by diactines and tetractines. No differentiation into megascleres and microscleres, all spicules being small (less than 100 μm), and without regional differentiation. One genus has neither spicules nor spongin.

Order Homosclerophorida

Definition as for the subclass.

5. OSCARELLIDAE

Oscarella lobularis (Schmidt) Fig. 3.2
Encrusting sheet with rounded lobes, sponge to 100 mm across. Surface smooth, oscula sometimes visible. Gelatinous but firm. Colour strongest in the lobes: often yellowish or brownish; sometimes entirely red, violet, blue, or green. No spicules or spongin skeleton.

Epilithic, or epibiotic on algae, cryptic low-littoral, or sublittoral.

Subclass Tetractinomorpha

Where present, megascleres are tetraxons or monaxons. Microscleres usually asters, but sigmas, raphides, microxeas, and other types can occur.

Order Choristida

Microscleres are asters and sometimes microxeas. Megascleres are tetractines and oxeas. In some genera, tetractines or microscleres, or both, are absent, only oxeas remaining.

6. GEODIIDAE

Megascleres are triaenes and oxeas, microscleres include sterrasters.

Pachymatisma johnstonia (Bowerbank in Johnston). Fig. 3.3
Massive, rounded mounds or thick plates. Smooth surface, oscula distinct and often in lines. Diameter to 150 mm, height to 100 mm. Surface colour usually bluish to violet-grey; always whitish within. Distinct skin, 1 mm thick. Megascleres are strongyles, and also orthotriaenes, usually with straight rays. Microscleres are microstrongyles, oxyasters with a few elongate rays, and characteristic, near spherical sterrasters, 90–200 μm long and 70–160 μm broad. Microrhabds and centrotylotes often occur.

Epilithic, sublittoral.

7. PACHASTRELLIDAE

Microscleres include calthrops and also streptasters. Euasters absent.

Dercitus bucklandi (Bowerbank) Fig. 3.3
Encrusting to massive, to 500 mm in diameter and 50 mm in thickness. Surface smooth and even with scattered oscula of variable size. Consistency fairly firm. Dark grey to black, colour characteristic when combined with gross form. Megascleres are calthrops. Microscleres include microxeas and toxas.

Epilithic, sublittoral.

Order Spirophorida

Radial skeleton results in a spheroid form. Megascleres are triaenes and oxeas. Sigmaspirae are microscleres that are characteristic of the order.

8. TETILLIDAE

Definition as for the order.

Tetilla cranium (Müller) Fig. 3.3
Spherical or ovoid. Diameter to 50 mm. Surface finely hispid and with conuli. Oscula single, grouped, or not visible. Off-white or yellowish. Radially symmetrical skeleton. Megascleres are anatriaenes, protriaenes and oxeas. Microscleres are sigmaspirae.

Epilithic, cryptic lower littoral and sublittoral; sometimes on *Axinella infundibuliformis*.

Order Hadromerida

Megascleres are tylostyles or subtylostyles, with at least superficial radial organization. If present, microscleres are asters or microxeas. Spongin often occurs but not as fibres.

9. SUBERITIDAE

Megascleres are tylostyles, subtylostyles, or sometimes styles. Microscleres usually absent, but can occur.

---

1. Thin, encrusting **2**
   Usually, but not always, massive **3**

---

2. Surface hispid ***Prosuberites epiphytum***
   Surface smooth; very soft consistency, gelatinous internally, endophytes often cause bright blue-green or orange-yellow colour ***Terpios fugax***

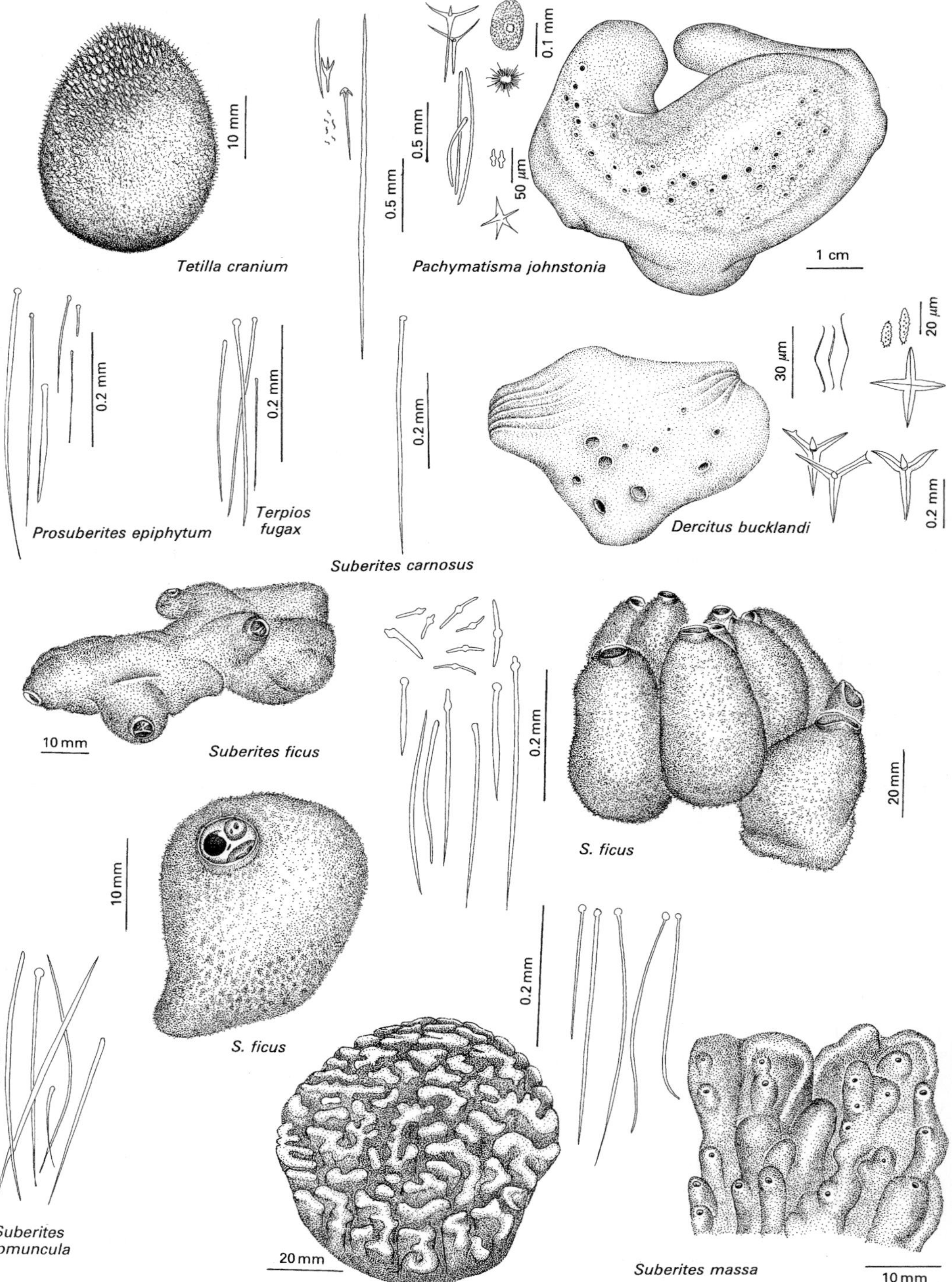
10 mm
Tetilla cranium
0.5 mm
0.5 mm
0.1 mm
50 µm
Pachymatisma johnstonia
1 cm
0.2 mm
Prosuberites epiphytum
0.2 mm
Terpios fugax
0.2 mm
Suberites carnosus
30 µm
20 µm
0.2 mm
Dercitus bucklandi
10 mm
Suberites ficus
0.2 mm
20 mm
S. ficus
10 mm
S. ficus
0.2 mm
0.2 mm
Suberites domuncula
20 mm
Suberites massa
10 mm

Fig. 3.3

3. Sponge mound composed of anastomosing plates imparting a brain-like appearance **_Suberites massa_**

   Sponge not as described above **4**

4. Spicules consist of tylostyles alone **_Suberites carnosus_**

   Other spicule types occur also **5**

5. Spicules include centrotylote strongyles **_Suberites ficus_**

   Centrotylote strongyles do not occur **_Suberites domuncula_**

Prosuberites epiphytum (Lamarck) Fig. 3.3
Thin encrusting patches, diameter to 150 mm. Surface finely hispid and uneven, oscula not visible. Firm consistency. Megascleres only tylostyles with discrete spheroid heads, and narrower, gradually tapering and curved shafts. Microscleres absent.

Epilithic, cryptic low-littoral, and sublittoral.

Suberites carnosus (Johnston) Fig. 3.3
Similar to *S. ficus.* Typically spheroid and anchored by a short stalk, sometimes encrusting or irregularly massive, to 150 mm in height. Yellow to brown. In the spheroid form, usually a single, apical osculum. Surface even, almost smooth. Contraction substantial. Consistency soft when dilated, firm when contracted. Megascleres exclusively tylostyles with egg-shaped basal swellings and spindle-shaped shafts.

On stones and rock. Common.

Suberites domuncula (Olivi) Fig. 3.3
Similar to *S. ficus.* Usually massive, rounded spheroid or lobed form, to 200 mm wide and 100 mm high, sometimes encrusting. Typically orange, but can be whitish, blue or red, or a mixture. Surface even but slightly rough. One or more distinct oscula. Consistency firm yet elastic. Megascleres include curved tylostyles, the terminal basal swelling appearing three-lobed in optical section. A second swelling may occur below this on the shaft. Styles and oxeas also occur.

Epilithic, also on stones and shells. Usually sublittoral. Common, all coasts.

Suberites ficus (Esper) Fig. 3.3
Massive, highly variable form. Often spheroid or lobed, sometimes clubbed, fig-shaped, or elongate (hanging down), or encrusting. Size can exceed 300 mm. Grey or orange-red, and yellow within the core. Endophytic chlorophytes may cause dark-green patches. Surface even, but slightly rough. One or more large oscula. Consistency firm, yet elastic. Out of water, contraction substantial and the sponge becomes firm. Megascleres include tylostyles and styles, both usually curved. In the tylostyles, the single basal swelling can occur terminally, or subterminally. Oxeas also occur. Microscleres, sometimes absent, can be centrotylote microstrongyles.

Epilithic, also on stones and shell, often exceeding the size of the initial substrate, so becoming free-standing. Usually sublittoral. Occasionally in the cryptic low-littoral. Common, all coasts.

Suberites massa Nardo Fig. 3.3
Massive, hemispheroid, or irregular, composed of anastomosing ridges bearing small oscula. Surface slightly rough. Diameter to 250 mm, height to 150 mm. Bright orange to whitish yellow; in illuminated situations significantly darker. Consistency fairly firm. Contraction significant out of water; the ridged gross structure becomes more apparent and the oscula close. Megascleres are tylostyles, each with a spheroid basal swelling, characteristically much wider than the slender, tapering shaft which may be straight or curved.

Epilithic, on stones and shells, occasionally in the lower littoral, more widely in the sublittoral. Locally abundant in turbid, brackish waters of some southern inlets, otherwise uncommon.

Terpios fugax Duchassaing and Michelotti Fig. 3.3
Thin encrusting patches, to 50 mm across. Smooth surface, oscula not visible. Fundamentally ochre-yellow or bright brown, but symbiotic algae can impart a deep blue-green or bright orange-yellow cast. Very soft in consistency. Megascleres all tylostyles with a slender shaft, usually smoothly curved, gradually tapering to a point. Basal swelling spheroidal but variable.

Epilithic, typically in cryptic low-littoral situations, also sublittoral. Common.

10. POLYMASTIIDAE

Megascleres tylostyles or subtylostyles, always in two or three size categories; styles can also occur. Microscleres usually absent. Characteristic cushion-formed massive body covered with large erect papillae, some bearing an osculum, others ostia.

---

1. Erect papillae solid, broad-based and fairly steeply tapering, usually opaque seen underwater; surface of base and papillae smooth
Polymastia boletiformis

Erect papillae delicate with a gradual taper, often translucent seen underwater, surface of base rough, and of papillae smooth
Polymastia mammillaris

Polymastia boletiformis (Lamarck) Fig. 3.4

Cushion-formed or massive, upper surface bearing many robust and steeply tapering, cylindrical papillae. Diameter of sponge to 100 mm, papillae 20–120 mm long. Orange, ochre-yellow, dark grey or green. Surface of base and papillae smooth. Oscula confined to papillae; often visible in live specimens. Firm consistency, papillae flexible. Papillae not translucent, nor significantly paler when seen underwater. Megascleres are tylostyles, mostly with distinct but irregular basal swellings, and spindle-formed shafts. No microscleres.

Epilithic, often upward facets of rock, sublittoral. Common.

Polymastia mammillaris (Müller) Fig. 3.4

Cushion formed with many subcylindrical, fairly slender, gradually tapering papillae. Diameter of base to 120 mm, thickness 10–20 mm; height of papillae to 120 mm. Orange, yellow, pink, or grey, papillae paler and sometimes translucent. Interior of base orange. Surface of base rough, whereas papillae are smooth. Papillae bear ostia or oscula, the latter often visible in live specimens. Base firm in consistency, papillae are flexible. Megascleres include stout and spindle-formed tylostyles, most with subterminal basal swellings. Styles also occur. Microscleres are absent.

Sediment-tolerant, often colonizes upper facets of rock, sublittoral. Common.

11. CLIONIDAE

Sponges that excavate extensive burrows within calcareous substrates, or form massive colonies. Megascleres are tylostyles or subtylostyles, and microscleres are spirasters, amphiasters, and thorny microxeas. Inhalant and exhalant papillae always feature at the surface, in the case of burrowing sponges, protruding through circular holes at the surface of the substrate.

Cliona celata Grant Fig. 3.4

Penetrant form burrows in limestone or mollusc shells, excavating a network of galleries. Communication maintained via protruding circular papillae, to 3 mm across, each bearing ostia or an osculum. This form may outgrow its substrate and progress to the massive form. Mound forms also occur on non-calcareous rocks, in extreme cases, growing as thick, rounded plates, up to 1 m across, 250 mm thick, and 500 mm high. Surface of mound sponges uniformly covered by flat-topped papillae bearing ostia, and by rows of large, rimmed oscula, arranged along the crest in plate forms. Bright to deep yellow. Consistency very firm and inflexible. Sponge contracts significantly when out of water, papillae being withdrawn. Megascleres include tylostyles with distinct, subterminal basal swellings, the shafts somewhat spindle-formed, with the lower half curved. Oxeas are fine, mostly in bundles, but often absent. Microscleres are spirasters, and occur solely in young sponges.

Sublittoral, found at wave-exposed, open coast sites and in silty estuaries. Very common.

12. TETHYIDAE

Massive, spherical sponges with a distinctly radially symmetrical skeleton, and a distinct cortical region. Megascleres include very long subtylostyles, also styles. Microscleres are asters.

Tethya aurantium (Pallas) Fig. 3.4

Massive, spherical, with a holdfast of root-like projections. Diameter to 60 mm. Surface with tubercles, polygonal in shape. Osculum often single and apical, ostia submicroscopic. Yellow or orange. Consistency very firm when sponge is contracted. Megascleres are styles and subtylostyles, the latter being straight, spindle-formed, and sometimes very

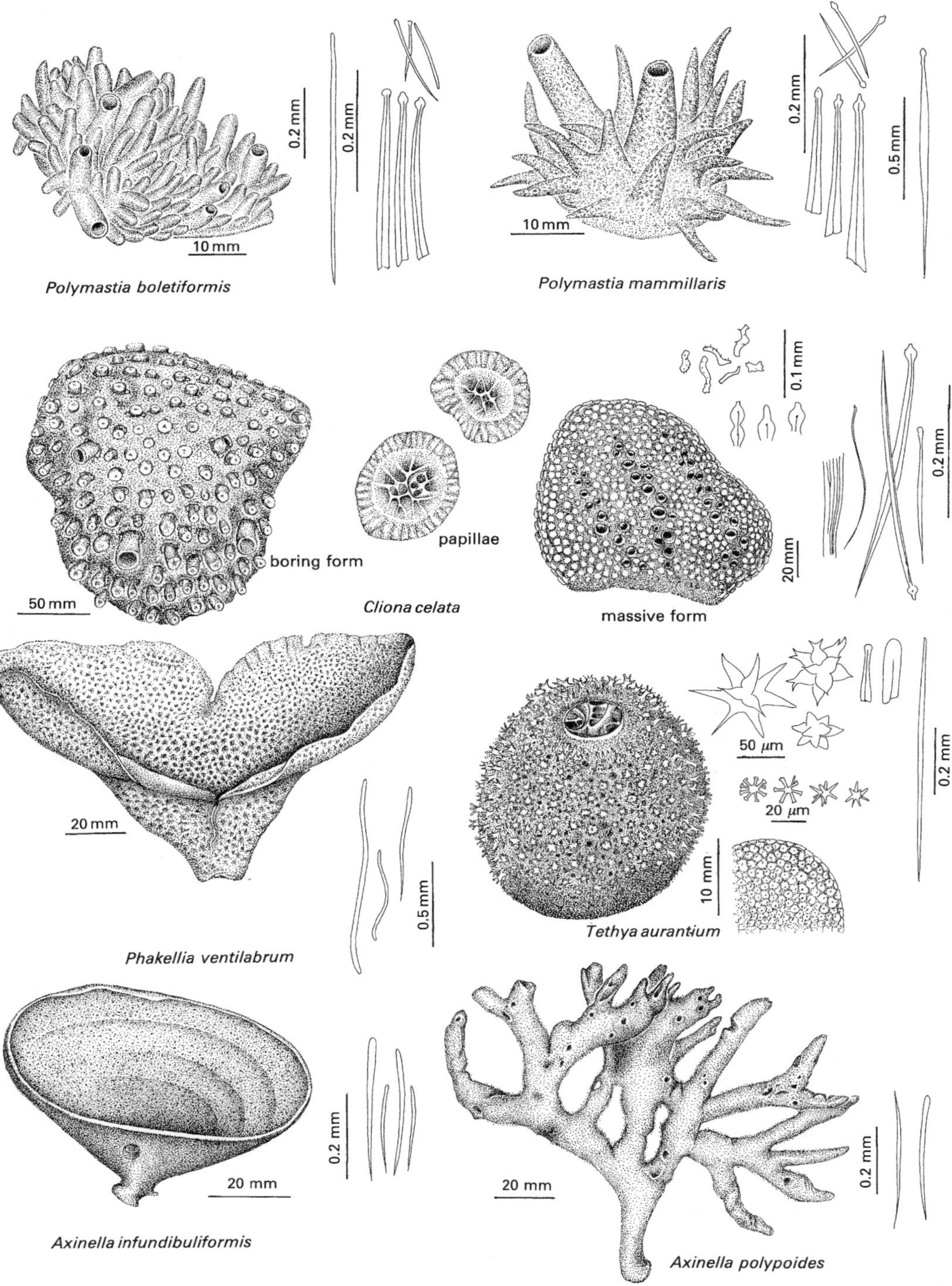
10 mm
0.2 mm
0.2 mm
Polymastia boletiformis
10 mm
0.2 mm
0.5 mm
Polymastia mammillaris
0.1 mm
0.2 mm
papillae
boring form
50 mm
Cliona celata
20 mm
massive form
20 mm
50 μm
20 μm
0.2 mm
0.5 mm
10 mm
Phakellia ventilabrum
Tethya aurantium
0.2 mm
20 mm
20 mm
0.2 mm
Axinella infundibuliformis
Axinella polypoides

Fig. 3.4

long (1.5–2.7 mm). Microscleres include spirasters and strongylasters, the latter having 9-12 rays.

Epilithic, occasionally cryptic low-littoral, usually sublittoral. Common.

Order Axinellida

Skeleton of spicules and a substantial proportion of spongin fibres, subdivisible into two components: a more concentrated and central, axial skeleton of spicules and fibres, and a more superficial extra-axial skeleton in which the spicules are held by spongin in plumose or plumoreticulate arrays. The two are often discernible physically in that the latter is noticeably more flexible than the former. Megascleres are some combination of oxeas, styles, tylostyles, and strongyles. Spicules are often sinuous. Microscleres can be trichodragmas or asters. The sponge surface usually is rough with projecting spicules.

13. AXINELLIDAE

The main skeletal tracts are of spongin fibres enclosing megascleres. These are partly condensed into an axial skeleton, and consist of extra-axial plumose or plumoreticulate elements leading to the surface. Megascleres are monoactines, diactines, or both.

---

1. Erect, funnel or fan-shaped **2**
   Erect, ramifying and branched ***Axinella polypoides***

---

2. Edge of funnel or fan, thick or rounded ***Axinella infundibuliformis***

   Edge of funnel or fan, tapered and sharp ***Phakellia ventilabrum***

Axinella infundibuliformis (Linnaeus) Fig. 3.4
Funnel- or sometimes fan-shaped, rim thick and rounded. Short stalk. Diameter of funnel to 250 mm. Surface even, and finely hispid, oscula numerous and scattered over entire surface. Ochre-yellow with a brownish tint. Megascleres include styles, most with an abrupt curvature near the base, and a point finely drawn out but somewhat short. Oxeas are slightly curved. Microscleres are trichodragmas.

Epilithic, sublittoral.

Axinella polypoides Schmidt Fig. 3.4
Erect and ramifying, branches flattened, partly anastomosing, ends rounded. Height to 150 mm. Firm but flexible. Pale brown. Surface smooth, oscula small, scattered. Megascleres include curved styles, and oxeas, also curved but more slender.

Epilithic, sublittoral.

Phakellia ventilabrum (Johnston) Fig. 3.4
Funnel- or fan-shaped, edge tapered and much sharper than in *Axinella infundibuliformis*, to which it is similar. Height to 450 mm, thickness to 5 mm. Surface usually smooth, sometimes rough and finely hispid with small scattered oscula. Pale yellow (in alcohol). Megascleres are strongyles and styles, both twisted and irregular in form.

Epilithic, sublittoral.

14. HEMIASTERELLIDAE

Megascleres are monoactines, diactines, or both, these being enclosed in spongin fibres that radiate upwards in a plumose fashion. Microscleres can include asters and microxeas.

Stelligera stuposa (Montagu) Fig. 3.5
Erect, tree form, branches bifurcating, rounded in section. Height to 180 mm. Surface spiny with small oscula. Consistency firm but flexible. Sponge charged with mucus. Pale brown or yellow. Megascleres include very long, solid styles, to 1.87 mm in length, thickness at base to 0.25 mm. Strongyles are unequal ended, one end being less rounded and more drawn out than the other; oxeas are uniformly slender, and with steeply tapering points. Microscleres are spherasters.

Epilithic, sublittoral.

15. RASPAILIIDAE

Axial skeleton of spongin fibres enclosing monoactines, diactines, or both. Extra-axial skeleton of radial or plumose fibres extending from the axis to the surface where long terminal styles project as spines. Very small acanthostyles occur. True microscleres are absent.

---

1. Tree form, fairly slender branches, colour pale yellow-brown ***Raspailia hispida***

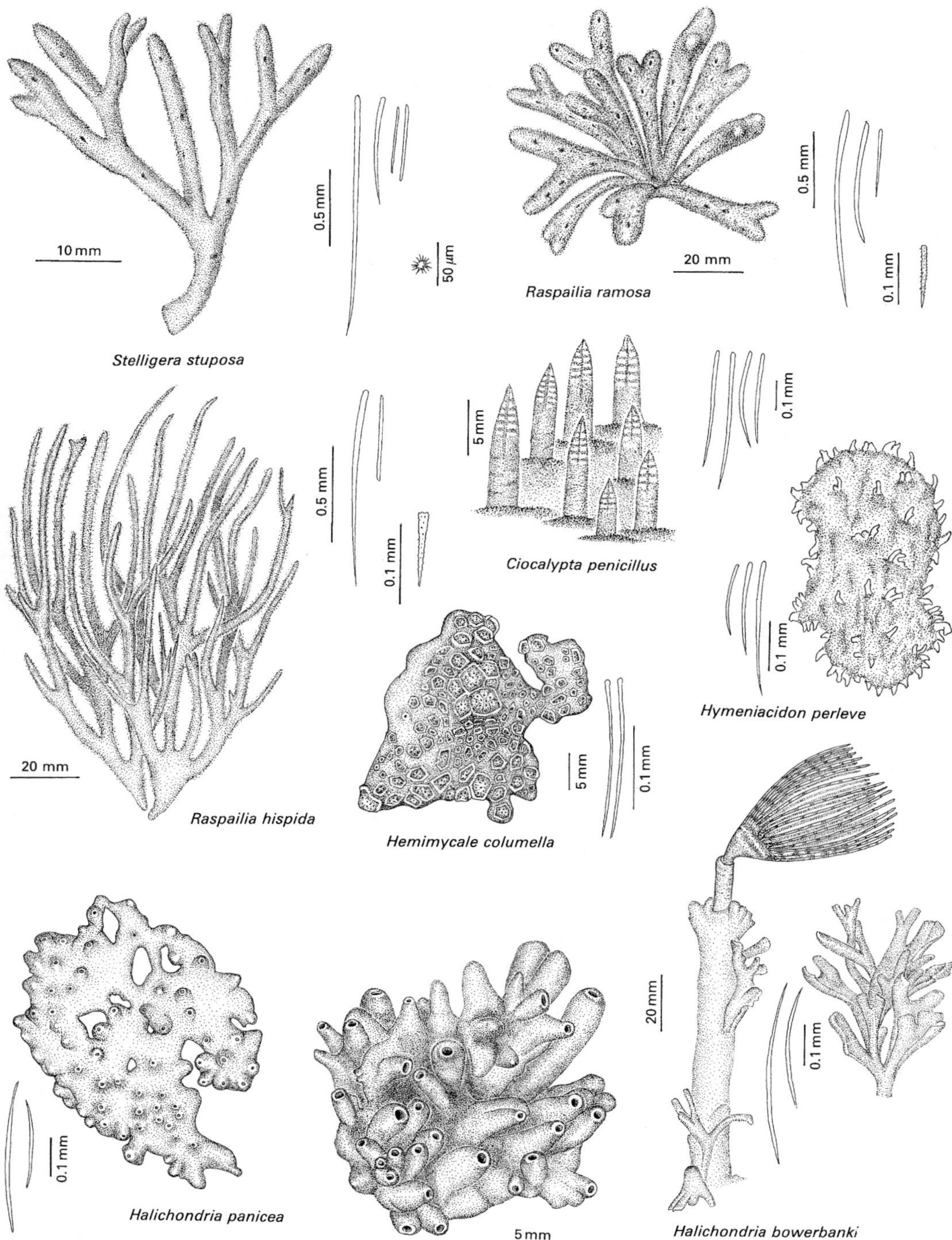
10 mm
0.5 mm
50 μm
Stelligera stuposa
20 mm
Raspailia ramosa
0.5 mm
0.1 mm
5 mm
Ciocalypta penicillus
0.1 mm
0.5 mm
0.1 mm
20 mm
Raspailia hispida
0.1 mm
Hymeniacidon perleve
5 mm
0.1 mm
Hemimycale columella
20 mm
0.1 mm
0.1 mm
Halichondria panicea
5 mm
Halichondria bowerbanki

Fig. 3.5

Tree form, stout branches, colour dark reddish-brown
Raspailia ramosa

Raspailia hispida (Montagu) Fig. 3.5
Branches long and slender, mostly forked, sometimes a little flattened. Whole sponge to 350 mm high, branches to 130 mm. Surface even, but somewhat hispid and with small oscula. Consistency fairly soft, and elastic. Megascleres include curved styles, acanthostyles that are straight and wholly thorned, and oxeas.

Epilithic, sublittoral, often with *R. ramosa*. Common.

Raspailia ramosa (Montagu) Fig. 3.5
Branches digitate, fan-shaped, or irregular but with well-rounded ends. Sponge to 150 mm, length of branches to 25 mm. Surface even but hispid, oscula small but distinct. Consistency flexible. Megascleres are styles, somewhat curved, and occasionally with a basal swelling. Strongyles are lightly curved. Oxeas occur. Acanthostyles are entirely thorny, and straight.

Epilithic, sublittoral. Often with *R. hispida*. Common.

Order Halichondrida

Megascleres are oxeas, styles, or strongyles. No microscleres. Spicule skeleton structured at the surface—a layer of tangential dermal spicules can be evident. The endosomal skeleton with spicules in disarray is termed 'halichondroid'.

16. HALICHONDRIIDAE

Principal megascleres are oxeas, although styles may occur. Subdermal spaces separate the often reticulated dermal skeleton from the randomly arranged endosome.

---

1. A cluster of tall conical projections arising from a cushion-like base often hidden by sediment
Ciocalypta penicillus

Sheet encrusting, massive or erect and branching **2**

---

2. Often with slender digits (diameter 5 mm or less), or flat, thin fronds; soft and pliable, not readily fractured by twisting. Odour often weak or absent
Halichondria bowerbanki

Not as above, although sometimes with stout anastomosing digits (10 mm or more); sponge firm and rubbery, quite readily fractured by twisting. Odour strong **_Halichondria panicea_**

Ciocalypta penicillus (Bowerbank) Fig. 3.5
Encrusting cushions giving rise to hollow, slender and tapering projections of characteristic form. Diameter of cushion to 100 mm, projections typically 50 mm high, basally 5 mm thick. Upper surface of cushion rough. Consistency of the projections is fairly firm but flexible. Brownish- or whitish-yellow, digits may be translucent. Megascleres are styles or oxeas, either may predominate and one may occur without the other. They are usually curved. Some spongin present.

Typically on upward-facing rock surfaces, sometimes with only the digits showing, the base being covered by sand or silt. Common.

Halichondria bowerbanki Burton Fig. 3.5
Gross form highly variable. Can be sheet-encrusting, in fairly thin patches (often less than 5 mm thick, diameter to 100 mm); otherwise massive, usually mound form (to 250 mm in thickness, 1 m across). Sheet and mound forms often give rise to elongate processes, slender and digitate, or flattened as expanded fronds. Surface smooth to slightly rough, oscula large and distinct in massive forms, not so in sheet forms. Consistency soft and pliable. Odour absent, or as for *H. panicea* but much weaker. Pale yellow or yellow-white. Megascleres fairly slender, curved oxeas. Microscleres absent.

Mainly on rock, but often epibiotic on seaweeds and sessile animals, e.g. the tubes of *Sabella pavonina* (Polychaeta). Usually sublittoral, but sometimes in well-shaded cryptic situations in the lowest littoral zone. Open coast, and outer and mid zones of silty estuaries, alongside *Halichondria panicea* with which it can be readily confused.

Halichondria panicea (Pallas)
Breadcrumb sponge Fig. 3.5
Gross form highly variable. Encrusting to various thicknesses, or irregularly massive, sometimes spherical, or mound-formed, or occurring as upright masses, anastomosing lobes, or digits. Thickness of encrusting sheets from approximately 2.5 mm, and the horizontal extent from a few centimetres to a metre or more; erect plates can be to 350 mm in height; anastomosing digits from 10 mm in diameter. Surface smooth to slightly rough, oscula are large and prominent, often as raised craters, arranged as lines along the crests of some erect forms. Consistency firm and flexible or rubbery, but sponge can be fractured if twisted significantly. Characteristic strong pungent odour. Megascleres only oxeas, elongated, slender, spindle-formed, curved.

On rocks and some larger seaweeds, e.g. kelp stipes. Encrusting sheet forms typical of the lower littoral zone, where it may be abundant in cryptic and semi-cryptic situations. More massive forms are confined to the sublittoral. Open coasts and outer zones of estuaries. Very common, all coasts (easily confused with *H. bowerbanki*, with which it often occurs, particularly in the outer zones of estuaries).

17. HYMENIACIDONIDAE

Megascleres are styles, sometimes with oxeas, microscleres being absent. The ectosomal skeleton is often structured, with spicule brushes protruding beyond the sponge surface. Significant subdermal cavities may separate the ectosome from the endosome, the skeleton of which is not organized. Spongin may occur. Characteristically, the surface of the sponge bears short erect processes.

Hymeniacidon perleve (Montagu) Fig. 3.5
Encrusting or cushion-formed, occasionally rounded. To 150 mm across, and 20–50 mm thick. Surface uneven, usually covered with small erect processes. Oscula sparse and to 2 mm diameter. Consistency fairly firm but springy. Yellow, orange-yellow, orange-red, blood red, or superficially green. Megascleres are all lightly curved styles with long drawn-out points. Sometimes two size categories occur.

Epilithic. Tolerant to desiccation, can occur on open surfaces in the mid-littoral. Most common in cryptic low-littoral and shallow sublittoral. Can occur on upward-facing rocks in silty estuaries such that the main body of the sponge is smothered, the bright orange papillae alone protruding through the silt layer. Most common on the open coast.

Order Haplosclerida

Skeleton reticulated, the pattern isodictyal with rectangular or triangular meshes, sided by single spicules, spicule bundles, or constructed entirely of spongin fibre without spicules. Principal spicules usually fairly stout oxeas or strongyles, in both cases relatively uniform in length within a species. Microscleres occur in some cases.

18. HALICLONIDAE

Megascleres typically oxeas, stout in appearance and of uniformly small size. These are arranged within an isodictyal skeleton consolidated by spongin, either in small quantities at the joints of spicule meshes or as more substantial fibres. In some species, spongin is the predominant skeletal component. Microscleres are usually absent.

NB. Several species occur in British waters, but the identity of most is not clearly defined, a prominent exception being the species given below.

Haliclona oculata (Pallas) Fig. 3.6
The limited encrusting holdfast gives rise to a stalked tree form with digitate branches, mostly round in section and with rounded ends; branches typically numerous, but under sheltered conditions occasionally flattened, antler-shaped, and few in number. Height to 300 mm, diameter of branches about 7 mm. Oscula small but distinct, often slightly raised, frequently ordered into lines along branches. Consistency robust, but easily compressible and very flexible. Pale brown, sometimes yellowish, greenish, rose, or purple. Megascleres are spindle-formed, mostly curved, sturdy oxeas.

Epilithic, sublittoral but occasionally cryptic low-littoral. Open coasts and outer reaches of estuaries. Very common.

19. ADOCIIDAE

Megascleres are often in multispicular tracts with an isodictyal arrangement. Oxeas and strongyles can occur. Microstyles can be sigmas and toxas. All species have a tangential dermal spicule skeleton. Spongin occurs, but to a lesser degree than in the Haliclonidae. See note for the Haliclonidae.

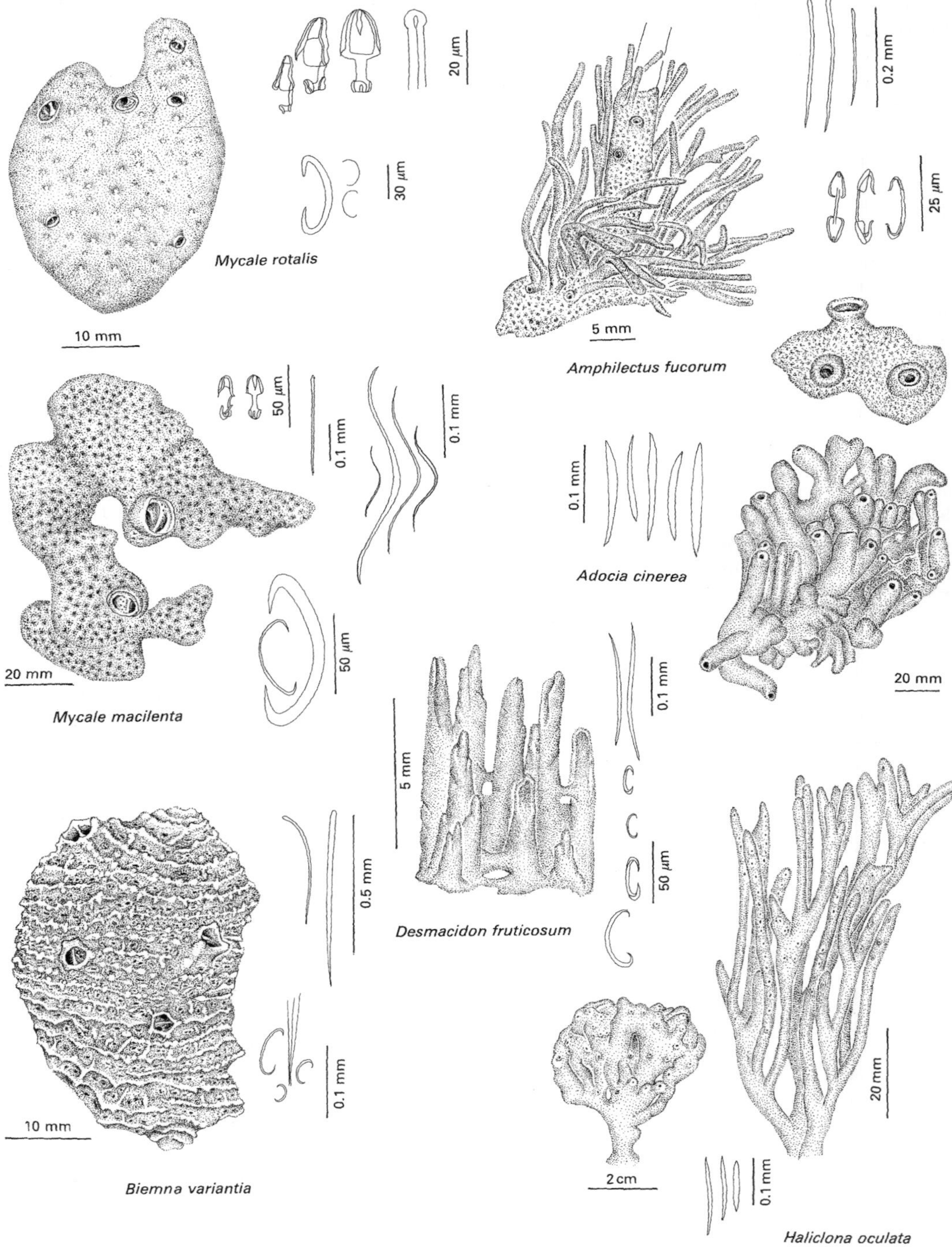
20 μm
30 μm
Mycale rotalis
10 mm
0.2 mm
25 μm
5 mm
Amphilectus fucorum
50 μm
0.1 mm
0.1 mm
0.1 mm
Adocia cinerea
50 μm
20 mm
Mycale macilenta
20 mm
5 mm
0.1 mm
50 μm
0.5 mm
Desmacidon fruticosum
0.1 mm
10 mm
Biemna variantia
2 cm
20 mm
0.1 mm
Haliclona oculata

Fig. 3.6

Adocia cinerea (Grant) Fig. 3.6
Encrusting, cushion form with chimney-formed conical elevations or short branches. Height to 35 mm, horizontal extent to 200 mm. Surface even but slightly rough. Oscula scattered, located upon the chimneys, diameter 1–2 mm. Consistency very soft, delicate, and compressible, but form readily springs back after squeezing lightly. Whitish violet, blood red, dark purple, reddish brown, orange, brown, ash-grey, or bright yellow. Skeleton a three- or four-sided meshwork of individual oxeas joined at their ends by small quantities of spongin, hence the soft consistency. Oxeas slender (for a Haplosclerid), curved, seldom straight, cylindrical with short points, and smooth. No microscleres.

On rock and stable stones. Sublittoral and cryptic low-littoral. On open coasts and in outer sections of estuaries.

Order Poecilosclerida

Skeleton always composed of both spicules and spongin fibre. Considerable diversity in the form of spicule assemblages. Megascleres a variety of monoactines or diactines, often spinose. The many microsclere types include chelae, sigmas, and toxas.

20. MYCALIDAE

Diffuse, radially arranged, plumoreticulate and fibre skeleton in which megascleres are monoactines, i.e. styles or subtylostyles. Microscleres always include anisochelae, and often some combination of sigmas, toxas, raphides, isochelae, and other cheloid variants. Up to seven microsclere types may occur in a single species.

---

1. Encrusting patch form, surface porous in appearance, but smooth. Colour yellow or deep red **Mycale macilenta**

   Mound form, surface warty and solid in appearance. Colour scarlet red, or pink **Mycale rotalis**

Mycale macilenta (Bowerbank) Fig. 3.6
Encrusting, diameter to 50 mm. Surface even but porous. Soft consistency. Megascleres include subtylostyles, mostly straight and a little spindle-formed, with slightly formed elliptical basal swellings. Microscleres include anisochelae of three size categories. Also sigmas of two sorts, and toxas with a gentle curve and broad crook in their middle. Trichodragmas are absent.

On rock and shells, sublittoral. Common, all coasts.

Mycale rotalis (Bowerbank) Fig. 3.6
Sheet to mound form. Diameter to 50 mm. Surface rounded and warty. Oscula sometimes evident. Consistency soft. Megascleres are subtylostyles with elliptical basal swellings. Shaft spindle-formed, point somewhat short. Microscleres include anisochelae of three size categories; sigmas of two sizes, both slender. Toxas and trichodragmas are absent.

Epilithic, sublittoral.

21. BIEMNIDAE

Plumoreticulate skeleton of smooth styles with some spongin. Microscleres abundant and diverse, including sigmas, microxeas, toxas, commas, and other varieties.

---

1. Sponge a uniform light-greyish colour; surface rough with conical elevations. A diversity of microscleres **Biemna variantia**

   Sponge surface pink or yellowish with whitish rimmed craters. No microscleres **Hemimycale columella**

Biemna variantia (Bowerbank) Fig. 3.6
Mature sponges fan- or cup-shaped. Diameter to 80 mm. Surface rough with conical elevations, oscula sparse and scattered. Consistency soft and flexible. Light grey, sometimes yellowish. Skeleton is of anastomosing spongin fibres and numerous spicules. Megascleres are styles. Microscleres are sigmas, raphides, trichodragmas, and commas.

Rock and stones, sublittoral.

Hemimycale columella (Bowerbank) Fig. 3.5
Thick encrusting or irregularly massive. Thickness from 10 mm, diameter to 300 mm. Surface characteristically cratered, each crater enclosing one or more pores. Ridges sometimes give honeycombed effect. Oscula small (1 mm in diameter) and often indistinct. Megascleres are styles, slightly curved,

with slightly expanded bases; mixed with a few strongyles.

Epilithic, sublittoral.

22. DESMACIDONIDAE

Reticulate or plumoreticulate skeleton, and diactinal megascleres of a uniform type throughout the sponge. Microscleres can be sigmas and chelae.

---

1. Colour orange-pink, sponge grooved, broad extensions orientated towards the water surface. Megascleres tornotes, microscleres anchorate isochelae
   Desmacidon fruticosum

   Colour usually bright-orange. Microscleres palmate isochelae, megascleres are styles.
   Amphilectus fucorum

Amphilectus fucorum (Esper) Fig. 3.6
Form variable: encrusting patches, 5 mm thick; when growing on substrates of limited area, e.g. tubes of *Sabella* (Polychaeta), can produce numerous slender erect digits, 3 mm in diameter and to 100 mm long. Occasionally massive. Surface even and finely rough. Oscula scattered in encrusting and massive forms, sometimes raised upon flat-topped chimneys in encrusting forms, and in approximate rows along branches of erect forms. Consistency delicate and very flexible. Usually bright orange, sometimes reddish, yellow-brown or grey-brown. Megascleres are slightly curved styles, 135–480 μm long, 3–19 μm thick. Microscleres are palmate isochelae, 14–28 μm long.

On rock, algae, or *Sabella* tubes. Occasionally cryptic low-littoral, but typically shallow sublittoral. Common.

Desmacidon fruticosum (Montagu) Fig. 3.6
Encrusting to massive, mound form, with vertical extensions, often grooved, usually orientated towards the water surface, irrespective of the axis of the substrate. Surface finely hispid. Oscula small and scattered. Orange-pink. Substantial mucus content. Megascleres are tornotes. Microscleres are anchorate isochelae and sigmas.

Epilithic, sublittoral. Common.

23. MYXILLIDAE

Endosomal skeleton is a regular reticulum of monoactinal megascleres, styles, acanthostyles, perhaps with echinating acanthostyles. Ectosomal spicules are diactinal. The diversity of possible microscleres include arcuate isochelae, sigmas, anisochelae, bipocilli, forceps, and others. Toxas and palmate isochelae do not occur.

---

1. Tornotes end in a simple, steeply tapering point, no thorns
   Myxilla fimbriata

   Tornote ends are covered with thorns
   Myxilla incrustans

   Tornote ends each have three thorns
   Myxilla rosacea

Myxilla fimbriata (Bowerbank) Fig. 3.7
Cushion-formed. Size to 80 mm. Surface even, and a little rough. Oscula scattered. Consistency variable in firmness, somewhat elastic. Orange in life, brown to black in alcohol. Megascleres joined by small amounts of spongin; include acanthostyles and tornotes with simple ends. Microscleres are spatulate isochelae.

Epilithic, sublittoral, all coasts.

Myxilla incrustans (Johnston) Fig. 3.7
Encrusting, irregularly massive or lobed. Height to 90 mm, diameter to 115 mm. Scattered oscula, often upon conical humps. Moderately firm consistency. Much mucus. Yellow to orange, or brownish. Megascleres joined by small quantities of spongin; include acanthostyles with sparse scattered thorns, and tornotes, most with sparsely thorny ends. Microscleres are spatulate isochelae and sigmas, typically with a quarter spiral.

Epilithic, often in cryptic low-littoral situations, but mainly sublittoral. Common.

Myxilla rosacea (Lieberkuhn) Fig. 3.7
Encrusting, sometimes branched. Size to 60 mm. Surface overall pitted and somewhat rough. Oscula sometimes visible. Consistency firm but somewhat flexible. Typically dull rose-red, otherwise orange or brownish yellow. Megascleres linked by small

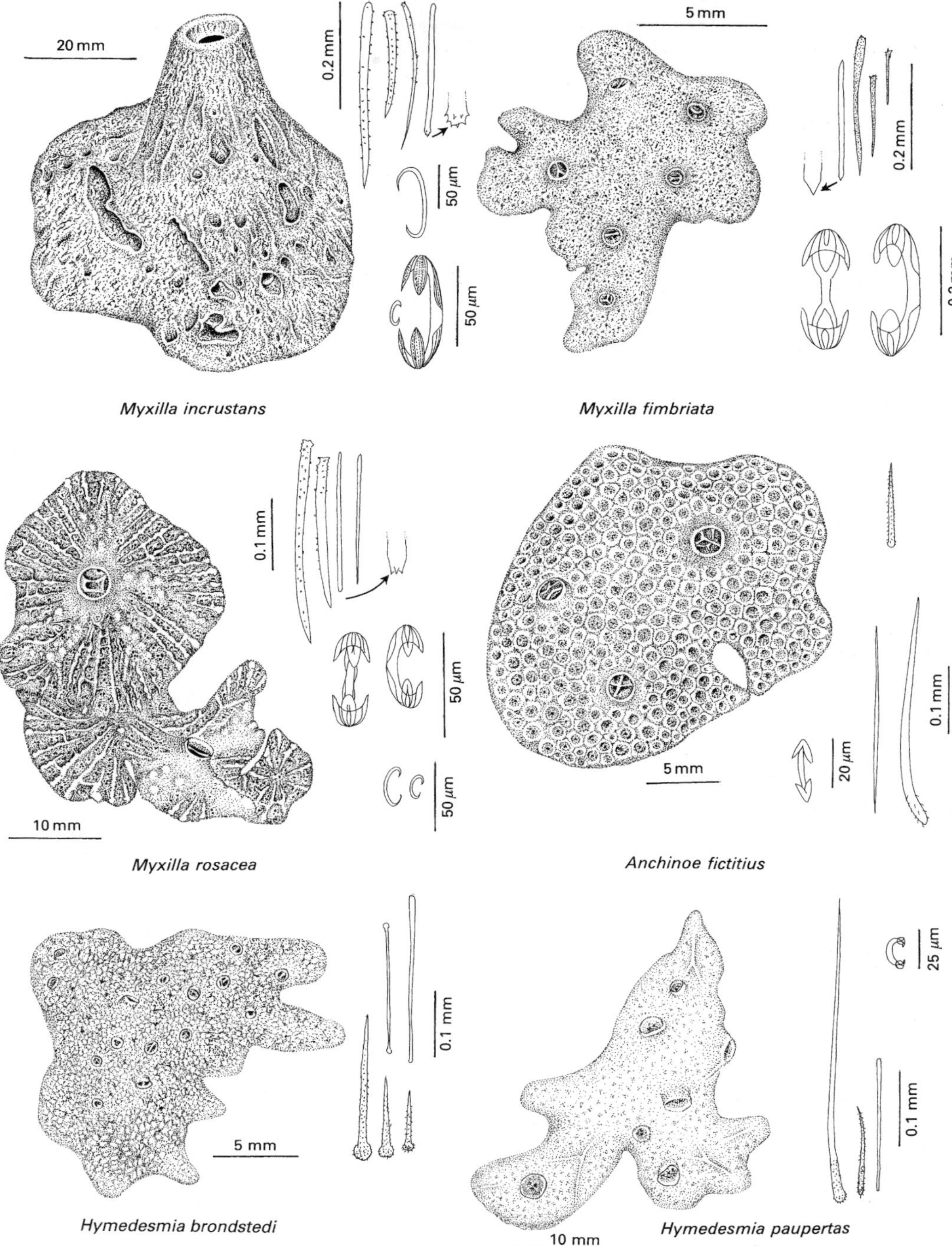
20 mm
0.2 mm
50 µm
50 µm
Myxilla incrustans
5 mm
0.2 mm
0.2 mm
Myxilla fimbriata
0.1 mm
50 µm
50 µm
10 mm
Myxilla rosacea
5 mm
20 µm
0.1 mm
Anchinoe fictitius
0.1 mm
5 mm
Hymedesmia brondstedi
25 µm
0.1 mm
10 mm
Hymedesmia paupertas

Fig. 3.7

quantities of spongin at the joints, include acanthostyles with scattered thorns; tornotes, mostly with triple-thorned ends. Microscleres include spatulate isochelae and sigmas.

Epilithic, sublittoral. All coasts.

24. HYMEDESMIIDAE

All have a thin, sheet-encrusting growth form. Endosomal megascleres are acanthostyles, orientated vertically and attached by the base to the substratum with spongin, i.e. the characteristic 'hymedesmoid' arrangement. Smaller accessory acanthostyles stand among principal ones. Ectosomal spicules are smooth, often diactines or monoactines, occurring individually or grouped, and orientated vertically or strewn without organization throughout the thin body. Usually more slender than the principal spicules. The substantial diversity of microscleres can include a variety of isochelae, sigmas, forceps, as well as many peculiar forms that characterize the genera.

---

1. Sponge light brown and very thin, slimy to touch out of water. No microscleres
Hymedesmia brondstedi

Sponge vivid turquoise-blue in life. Isochelae present with strongly curved shafts and rounded teeth
Hymedesmia paupertas

Hymedesmia brondstedi Burton Fig. 3.7
Very thin encrusting patches, diameter to 25 mm, thickness to 0.5 mm. Yellowish or brownish. Surface very smooth, feels slimy to touch out of the water. Oscula sometimes not visible, or small and scattered. Skeleton of acanthostyles with spherical heads, overall thorny, 83–220 μm in length. Tornotes and strongyles also occur, one end often being larger than the other. Microscleres absent.

On rock, stones, also shells; sublittoral. All coasts, common.

Hymedesmia paupertas (Bowerbank) Fig. 3.7
Thin encrusting patches. Diameter to 150 mm, thickness 3 mm. Surface hispid, sometimes smooth; cratered effect evident in live specimens *in situ*. Deep turquoise-blue. Megascleres include two kinds of acanthostyles. The larger ones are 300–550 μm long, with a pronounced basal swelling, 12 μm in diameter, a slightly curved shaft, bearing sparsely distributed, small, robust thorns. The smaller acanthostyles are 120–160 μm long, with weakly pronounced basal swelling, 5 μm in diameter, and covered with densely packed, long thorns. Strongyles are straight, slender, 220–320 μm long and 3–5 μm thick. One end is thicker than the other. Microscleres are isochelae with strong curved shafts.

Epilithic, sublittoral.

25. ANCHINOIDAE

Principal endosomal skeleton of plumose to plumoreticulate columns of diactines echinated by styles or acanthostyles. Ectosomal spicules diactinal, often as in the endosome. Microscleres are isochelae and sigmas. There is no dense ectosomal layer of spicules, and spongin fibre is not a major component of the skeleton. These negative characters distinguish the group from the Crellidae, with which it can be confused.

Anchinoe (Phorbas) fictitius (Bowerbank) Fig. 3.7
Encrusting sheet form, can be extensive, to 500 mm in diameter, to 15 mm in thickness. Surface with a cratered effect in life (similar to *Hemimycale columella*). Oscula small and scattered. Vivid light-red or pink. Megascleres include two size categories of acanthostyles. In the larger, the thorns are confined to the basal region, and those of the smaller category are wholly thorny. Tornotes also occur. Microscleres are arcuate isochelae.

Epilithic, sublittoral.

26. CLATHRIIDAE

Megascleres are dominated by styles or acanthostyles. These are organized into unispicular or multispicular tracts, also containing spongin which can occur in substantial quantities in some species. The tracts also support echinating acanthostyles. Styles also occur as accessory ectosomal spicules, these usually being finer than endosomal ones. There are no diactinal megascleres. Microscleres occur as palmate isochelae and toxas. All species covered here form encrusting sheets.

---

1. Fairly thick (can exceed 10 mm), deep red encrusting sheets with rounded edges, containing stout styles and also toxas
Ophlitaspongia seriata

Thin encrusting sheets, orange or red **2**

2. Colour orange or orange-red **3**

Colour deep red **4**

3. Colour orange-red, often conspicuously wrinkled, sheet form. Stout acanthostyles, toxas devoid of thorns

Antho involvens

Colour orange, usually smooth. More slender acanthostyles. Toxas with thorny ends occur

Microciona spinarcus

4. Many toxas exceptionally slender with curvature confined to the central section

Microciona strepsitoxa

All toxas exhibit a typically gradual curvature

Microciona atrasanguinea

Antho involvens (Schmidt) Fig. 3.8
Thin encrusting patches or sheets. Smooth, sometimes wrinkled, slightly hispid. Scattered small oscula, not easily seen. Orange-red. Megascleres are acanthostyles, sparsely covered with thorns. Also styles and subtylostyles. Microscleres are toxas and palmate isochelae.

Epilithic, sublittoral.

Microciona atrasanguinea Bowerbank Fig. 3.8
Thin encrusting patches or sheets, diameter can be 200 mm, thickness less than 2.5 mm. Surface finely hispid. Oscula tiny, scattered. Sponges *in situ* often exhibit a web-like pattern of raised exhalant channels, converging on each osculum. Dark red. Megascleres include curved styles; acanthostyles usually curved and with thorns confined to the basal swelling, occasionally the thorns are absent. Smaller acanthostyles are straight and thorny over their entire length. Microscleres are toxas and two-toothed isochelae.

Epilithic, cryptic littoral, sublittoral. Common.

Microciona spinarcus Carter and Hope Fig. 3.8
Thin encrusting patches, diameter to 25 mm. Surface even, only hispid in places. Orange. Megascleres include long styles that are slightly curved, sometimes with slight basal swellings, which can be slightly thorny. Acanthostyles are slightly to strongly curved, and almost entirely, but sparsely, thorny. Microstyles include very slender toxas which are characteristically terminally thorny, and also isochelae.

Sublittoral.

Microciona strepsitoxa Hope Fig. 3.8
Thin encrusting patches, to 1 mm. Surface smooth when live, spiny when dry. Oscula numerous. Scarlet red. Megascleres include subtylostyles with slight basal swellings, sometimes thorny and lightly curved. Tylostyles have distinct basal swelling which is also slightly thorny, and a slender shaft. Acanthostyles or acanthotylostyles are thorny overall. Many toxas are very slender, with the curvature confined to the central section. Isochelae occur.

Epilithic, sublittoral.

Ophlitaspongia seriata (Grant) Fig. 3.8
Encrusting patches with rounded edges. Diameter to 75 mm, thickness to 12.5 mm. Surface even, partly hispid. Oscula small but distinct, numerous and scattered. Dark, dull red. Megascleres are rather broad styles. Microscleres are toxas.

Epilithic, sublittoral.

Order Dictyoceratida

No spicules present, but there is a well-developed organic skeleton of spongin fibres. This always forms an anastomosing network of differentiable primary and secondary fibres. The order includes the commercially exploited *Spongia* (Fam. Spongiidae, not found in the British Isles).

27. DYSIDEIDAE

Choanocyte chambers are very large (40 μm), and often oval with a wide exhalant opening, a form of chamber termed 'eurypylous'. Fibres of spongin skeleton reticulate and laminated. Foreign materials often incorporated in quantity into the fibres.

Dysidea fragilis (Montagu) Fig. 3.8
Irregularly encrusting with rounded edges, 50 mm across and from 7.5 mm in thickness; otherwise a lobed mass, to 300 mm across. Surface covered with conuli, 1.5 mm in height, 2–2.5 mm across the base.

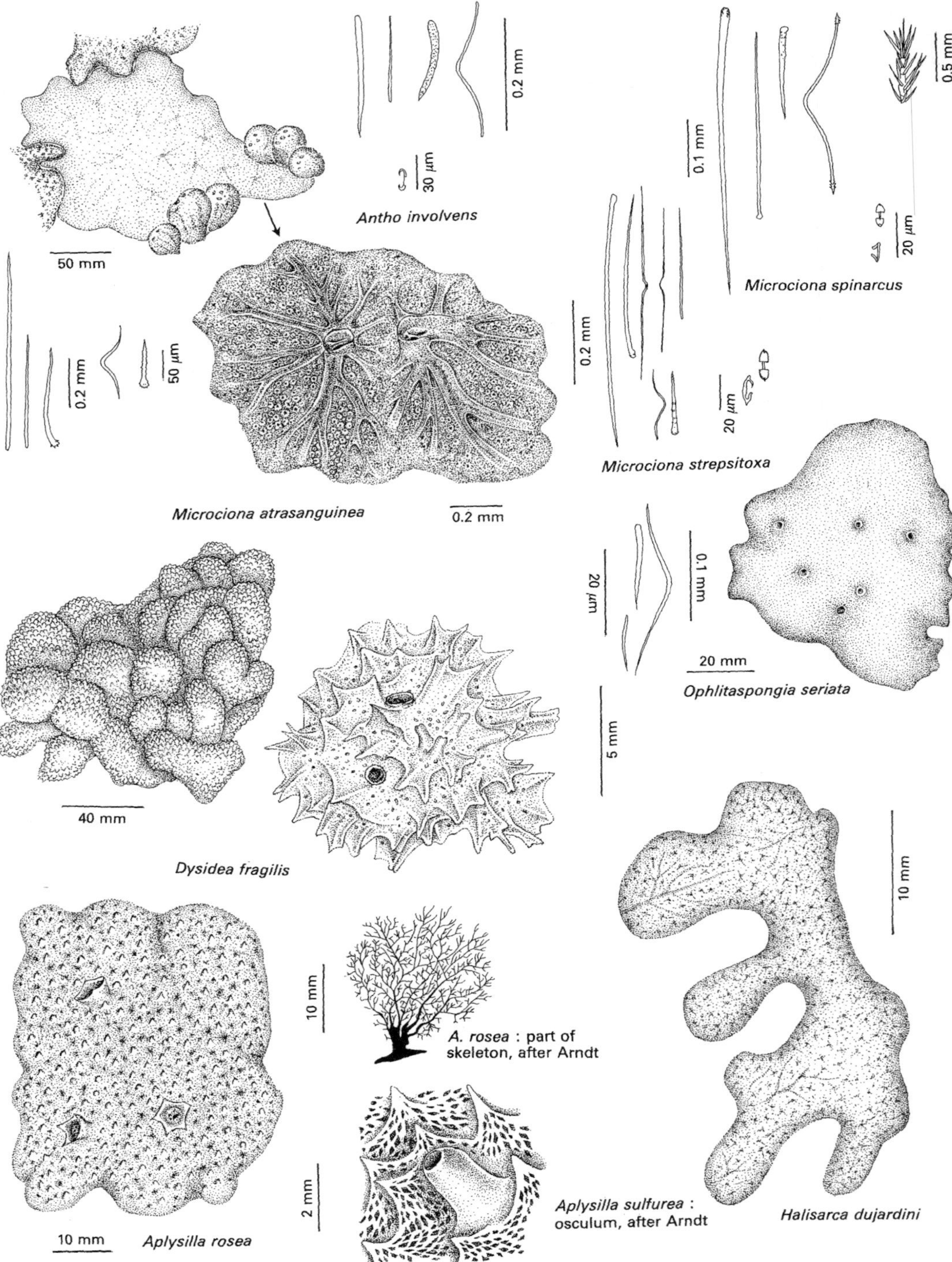
0.2 mm
30 µm
Antho involvens
0.1 mm
0.5 mm
20 µm
Microciona spinarcus
50 mm
0.2 mm
50 µm
0.2 mm
20 µm
Microciona strepsitoxa
Microciona atrasanguinea
0.2 mm
20 µm
0.1 mm
20 mm
Ophlitaspongia seriata
5 mm
40 mm
Dysidea fragilis
10 mm
10 mm
A. rosea : part of
skeleton, after Arndt
2 mm
Aplysilla sulfurea :
osculum, after Arndt
Halisarca dujardini
10 mm
Aplysilla rosea

Fig. 3.8

Oscula distinct, scattered, and sometimes raised, 2–5 mm wide. Typically brownish or greyish white, sometimes other colours. Characteristic subtle odour. Spicules are absent. The skeleton is a meshwork of spongin fibres, 100–200 μm in diameter, and characteristically incorporates foreign hard material, sand grains, spicules of other sponges, foraminifera, etc.

Epilithic, cryptic low-littoral, and sublittoral.

Order Dendroceratida

No mineral spicules, skeleton usually entirely composed of spongin fibres characteristically arranged in a dendritic pattern, rarely anastomosing. Skeleton absent in one family.

28. APLYSILLIDAE

Spongin skeleton present.

---

1. Colour red. Surface conuli well-separated (towards 5 mm), oscula not raised — ***Aplysilla rosea***

   Colour yellow. Surface conuli dense packed (1 mm), oscula on raised chimneys — ***Aplysilla sulphurea***

Aplysilla rosea Schulze — Fig. 3.8

Small, flat encrusting patches. Diameter to 40 mm, thickness to 5 mm. Surface with thin conulae, 2–3 mm high and 5 mm apart. Skeleton composed of tree-like systems of branched spongin fibres.

Epilithic, littoral, or sublittoral.

Aplysilla sulfurea Schulze — Fig. 3.8

Small, flat encrusting patches. Thickness 3–6 mm. Surface covered with conuli, 0.5–1 mm in diameter and about 1 mm apart. Oscula distinct, 1–2 mm wide, each located upon a raised chimney. Pale to dark sulphur-yellow in life, gradually changes to a patchy blue in alcohol. Skeleton consists of smooth, tree-formed system of branching fibres.

Epilithic, cryptic low-littoral, sublittoral.

29. HALISARCIDAE

No spongin skeleton.

Halisarca dujardini Johnston — Fig. 3.8

Encrusting, diameter to 400 mm, thickness to 5 mm. Surface smooth and slimy. Oscula sparse, sometimes raised. Dull yellowish brown or whitish. No spicules or distinct spongin fibres, although very fine fibrils do occur.

Epilithic, epibiotic on algae. Cryptic low-littoral, sublittoral.

4 HYDROIDS, SEA ANEMONES, JELLYFISH, AND COMB JELLIES

PHYLUM CNIDARIA

The Cnidaria, formerly combined with the Ctenophora (p. 133) and known as Coelenterata, are most obviously characterized by their *radial*, or sometimes more strictly *biradial*, *symmetry*. The basic structure is sac-like with a single terminal opening, the mouth, which also functions as an anus. The internal space is the *coelenteron* or *gastrovascular cavity*, which may be subdivided by radially arranged partitions, the *mesenteries* or *septa*. The mouth is surrounded by a circle or circles of *tentacles*. The body wall consists of only two layers of cells, ectoderm and endoderm, between which is the *mesogloea*. The mesogloea forms a non-cellular membrane in hydroids, a thick fibrous layer in anemones, and a jelly-like filling in medusae. The musculature, formed from epitheliomuscular cells, may be well developed. All cnidarians are characterized by the presence, especially on the tentacles, of stinging capsules, *cnidae* or *nematocysts*, of various types. These capsules are unique to the phylum. Nematocysts consist basically of a capsule with an armed or naked thread coiled inside it. When stimulated the thread is fired by eversion.

Two basic structural types of cnidarian are recognized: a *polyp* is sessile, more or less cylindrical, has the mouth at the free, distal end, and thin mesogloea; a *medusa* is free swimming, bell- to saucer-shaped, with the convex surface upward and the mouth and tentacles on the under surface. Larger forms are referred to as jellyfish. Cnidarian species may be exclusively medusoid, exclusively polypoid, or pass through both polypoid and medusoid phases. The medusae are often reproductive, producing gametes; on fertilization the zygote develops into a *planula* larva. Polyps may propagate asexually by fission (longitudinal or transverse) or by budding new polyps from themselves. Such buds in the freshwater *Hydra*, for example, may separate from the parent polyp; but in marine cnidarians the buds usually do not separate, leading to the formation of a colony. Medusae rarely bud. Many colonies of polyps build a common skeleton which may be chitinous or calcareous, internal or external.

Class I Scyphozoa

Marine. The pelagic medusoid stage predominates in most orders and the sessile polyp (*scyphistoma*) is usually very small. Marginal tentacles in a single (or variously modified) whorl surround the mouth, which is generally situated at the end of a tubular *manubrium*. Four endodermal interradial *septa* partly divide the coelenteron into a central stomach and four *perradial pouches*; these pouches may give rise to *radial canals* which join a marginal *ring canal*. Symmetry is tetramerous and defined in terms of the disposition of the four stomach pouches in the *perradii* and their separating septa in the *interradii*. Between these main axes lie eight *adradii*. The mouth is tetramerous with the angles lying in the perradii. Endodermal gonads are borne on the septa or pouches. The margin usually bears some reduced tentacles (*rhopalia*) enclosing statoliths. The upper surface of the medusa is *exumbrellar* or *aboral*; the under surface is *subumbrellar* or *oral*. The subumbrellar surface usually contains four invaginated *interradial funnels*. The mouth hangs at the end of a quadrangular, tubular manubrium, and the corners of the mouth may be prolonged as oral arms. There is no velum (cf. Hydrozoa).

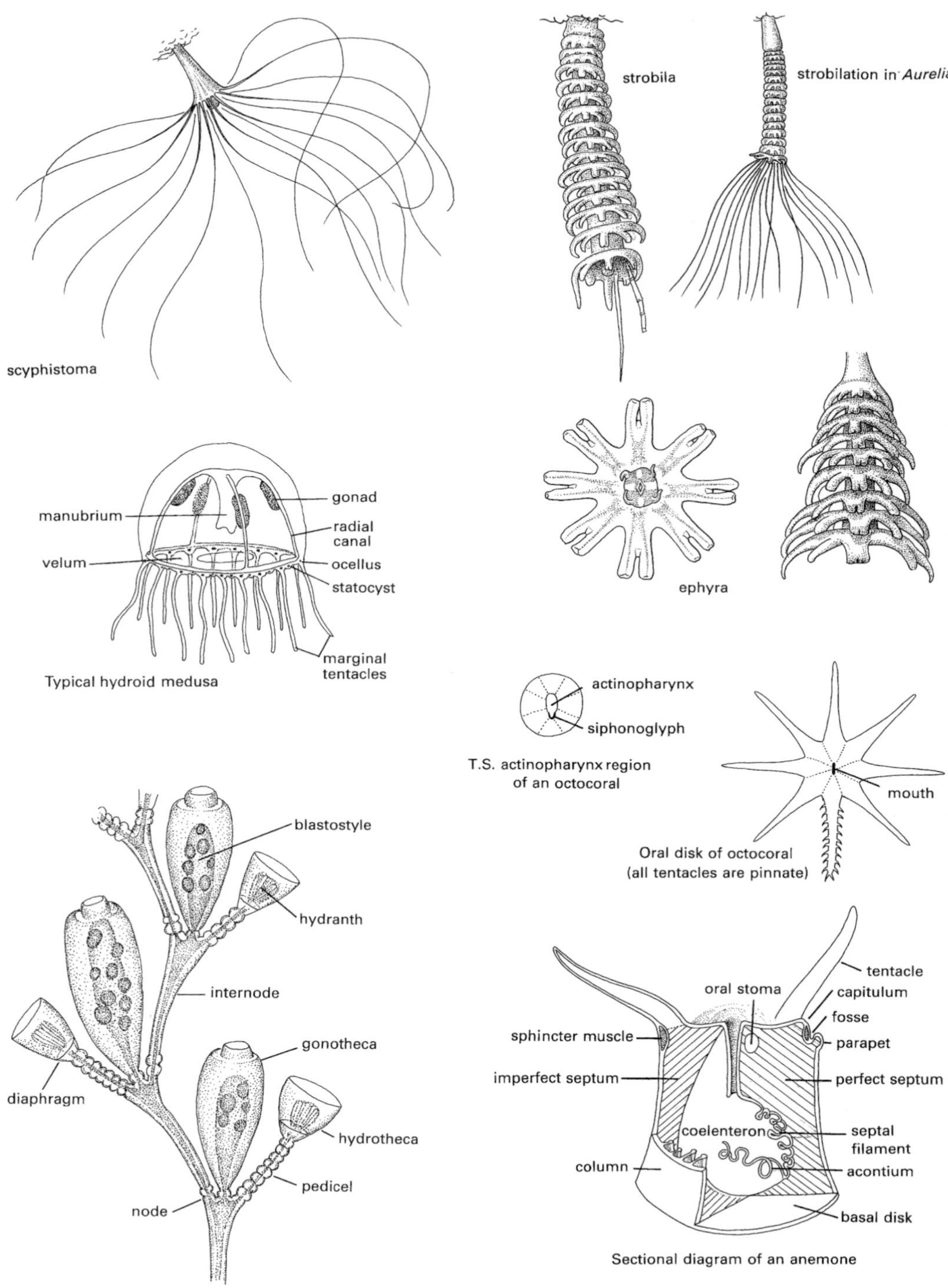
scyphistoma
strobila
strobilation in Aurelia
ephyra
gonad
manubrium
radial canal
velum
ocellus
statocyst
marginal tentacles
Typical hydroid medusa
actinopharynx
siphonoglyph
T.S. actinopharynx region of an octocoral
mouth
Oral disk of octocoral (all tentacles are pinnate)
blastostyle
hydranth
internode
gonotheca
diaphragm
hydrotheca
pedicel
node
Part of a typical thecate hydroid colony
tentacle
oral stoma
capitulum
fosse
sphincter muscle
parapet
imperfect septum
perfect septum
coelenteron
septal filament
column
acontium
basal disk
Sectional diagram of an anemone

Fig. 4.1

Few scyphistoma stages have been described (Fig. 4.1). Scyphistomas may produce more scyphistomas from stolons by budding, and produce young medusae (ephyrae) by transverse fission (strobilation), either singly or in simultaneous groups.

There are four orders. Three of these (Coronatae, Semaeostomeae, and Rhizostomeae) are jellyfish; the fourth, Stauromedusae, are sessile.

Class II Cubozoa

Box jellyfish (single order, Cubomedusae). Medusae are often roughly square in section. One or more interradial tentacles hang from pedalia at each corner, and perradial sense organs are situated on each side of the bell.

Class III Hydrozoa

Characterized by the alternation of attached, asexually replicating, typically colonial *polyps* and pelagic, sexually reproducing *medusae*, the latter perhaps best being regarded as the adult. However, the medusa is generally small and may be suppressed, while the polyp tends to propagate into conspicuous colonies. In neither polyp nor medusa is the coelenteron septate. Free medusae are most commonly budded from polyps, are radially (or tetramerously) symmetrical, and have the subumbrellar space partially enclosed by a thin shelf or *velum*. The mouth is at the end of a tubular *manubrium*. Four *perradial canals* (and sometimes others) connect the gastric cavity with a *ring canal* in the bell margin. No rhopalia. The polyps are radially symmetrical and are generally colonial, the coelenterons being continuous from polyp to polyp. There is generally a chitinous (or sometimes calcareous) *exoskeleton*. New polyps are budded from adnate stolons or from other polyps. Many are polymorphic.

The principal order is the Hydroida (p. 70); a pleustonic (floating on the sea's surface) representative of another order (Siphonophora, p. 110) and an aberrant athecate hydroid (p. 78) may be stranded on Atlantic beaches. These are the Portuguese man-o'-war *Physalia*, and by-the-wind sailor *Velella*, respectively.

Class IV Anthozoa

Marine. Exclusively polypoid; solitary or colonial. Gastrovascular cavity divided by six, eight, or many longitudinal, radially arranged *septa*; terminated distally by the *oral disc* from which, between the septa, arise the hollow tentacles. Symmetry is secondarily *biradial*; the mouth and *actinopharynx* descending from it, elongate in cross-section, one or both ends modified as ciliated grooves or *siphonoglyphs*. Septa are thickened and convoluted at their free edge as *septal filaments* and may include well-formed retractor muscles; gonadal cells develop within them. Fertilization is usually external. Dispersal is by pelagic larvae, usually *planulae*, but asexual propagation by external budding, fission, fragmentation, and (apparently) some form of internal budding or parthenogenesis is widespread. Many contain internal skeletal material of various kinds or secrete a calcareous exoskeleton.

Divided into two subclasses:

1. Octocorallia or Alcyonaria. Septa eight; tentacles pinnate. Classification into orders is dubious but includes Alcyonacea (soft corals), Gorgonacea (sea fans), and Pennatulacea (sea pens).
2. Hexacorallia or Zoantharia. Septa six or multiples of six; tentacles simple. Orders include Actiniaria (anemones), Corallimorpharia (jewel anemone), Scleractinia (corals), Zoanthidea (zoanthids), and Ceriantharia (tube anemones).

Class Scyphozoa

KEY TO SCYPHOMEDUSAE

1. Pelagic medusa (jellyfish), often large, swimming with aboral surface upwards **2**
   Fixed medusa, small (<4 cm diameter), attached with oral surface upwards **Order Stauromedusae** (p. 67)

2. Jellyfish with thick, dome-shaped, 'solid' umbrella; no marginal tentacles; manubrium tips branched to form eight arms with many mouth openings arranged in two series of elongate clumps
Order Rhizostomeae, 4. Rhizostomatidae (*Rhizostoma octopus*)

Jellyfish with saucer-shaped umbrella, floppy out of water; marginal tentacles usually present; manubrium with four simple or frilly arms
(Order Semaeostomeae) 3

3. Marginal tentacles short, very numerous; gastrovascular system in form of numerous, branching radial canals; mature gonads appear as four horseshoe-shaped ribbons **3. Ulmaridae (*Aurelia aurita*)**

Characters not as above. (All the remaining jellyfish can sting: handle with care) **4**

4. Gastrovascular pouches unbranched; tentacles on margin of umbrella
(1. Pelagiidae) 5

Pouches branched; tentacles arising from subumbrellar surface, some distance from margin **(2. Cyaniidae) 6**

5. Marginal tentacles in eight groups of three, the series separated by sense organs; umbrella patterned with red <-shaped markings
Chrysaora hysoscella

Eight marginal tentacles alternating with eight sense organs; umbrella with brownish nematocyst warts ***Pelagia noctiluca***

6. Jellyfish yellowish to reddish, large (may greatly exceed 30 cm diameter); >65 tentacles per group; North and Irish Seas northwards
Cyanea capillata

Jellyfish pale yellow to blue, not exceeding 30 cm diameter; <65 tentacles per group; Bay of Biscay northwards ***Cyanea lamarckii***

Order Semaeostomeae

1. PELAGIIDAE

Scyphomedusae with tentacles arising from umbrella margin; gastrovascular system comprising unbranched pouches; no ring canal; manubrium with elongate oral arms with frilled lips. Two British species.

Chrysaora hysoscella (Linnaeus) Fig. 4.2
Umbrella with 32 marginal lappets; 24 marginal tentacles in eight groups of three; generally with distinctive <-shaped brown-red markings pointing towards an apical circle or spot; to about 30 cm diameter.

All coastal waters from Biscay to Norway; also Mediterranean and West Africa. Strobilation occurs during winter and large adults are present in British waters from July to late September.

Pelagia noctiluca (Forskål) Fig. 4.2
Umbrella with 16 marginal lappets, eight marginal tentacles; to about 10 cm diameter.

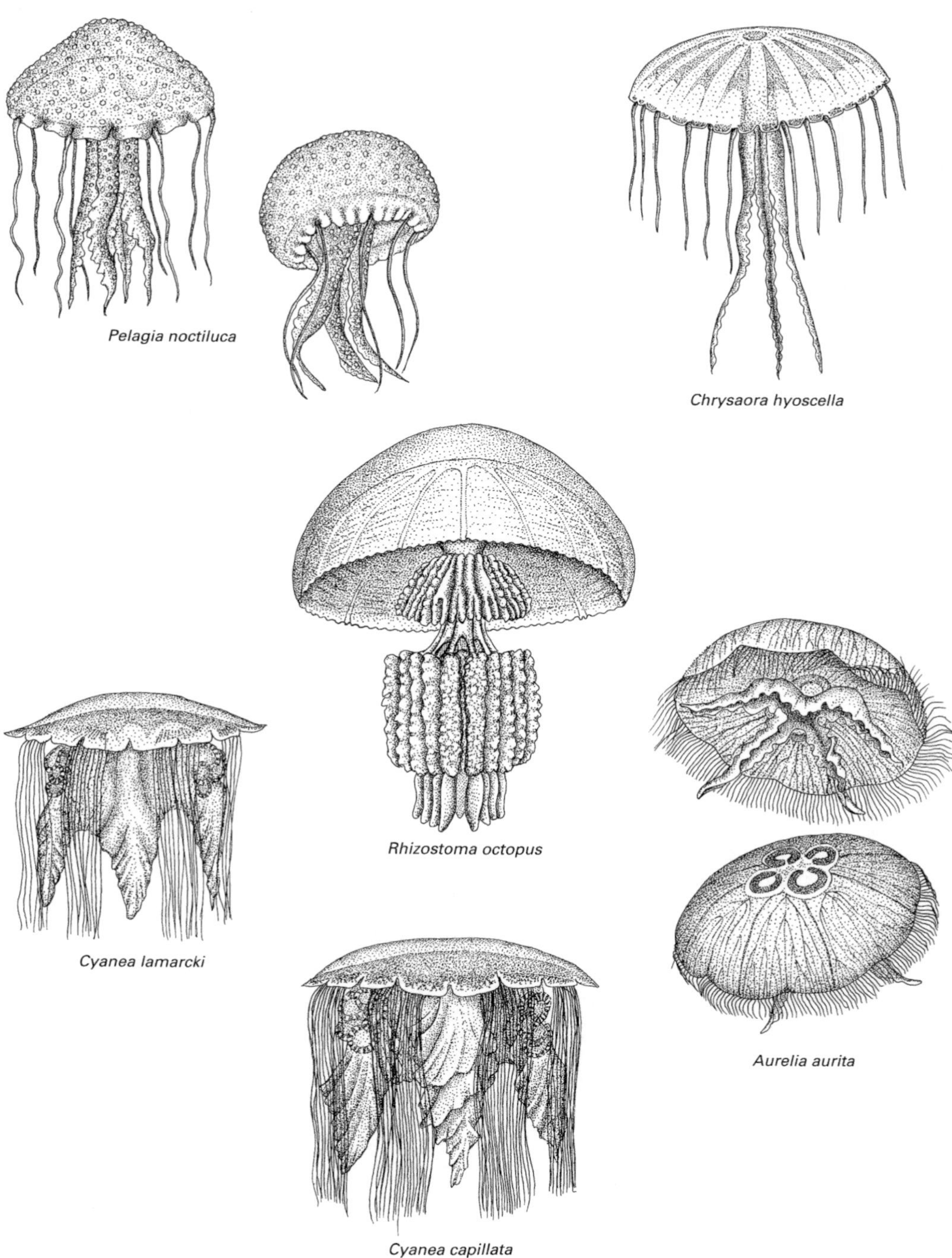
Pelagia noctiluca
Chrysaora hyoscella
Rhizostoma octopus
Cyanea lamarcki
Aurelia aurita
Cyanea capillata

Fig. 4.2

Widely distributed oceanic species; off Atlantic coasts from Biscay to Norway, sometimes carried around Scotland into northern North Sea. Development is direct, with no scyphistoma; small medusae occur in autumn and size increases until the following late summer.

2. CYANIIDAE

Scyphomedusae with marginal tentacles arising from subumbrellar surface; gastrovascular system comprising branched pouches; no ring canal; oral arms of manubrium with wide, much-folded tips. Two British species.

Cyanea capillata (Linnaeus) Fig. 4.2
A northern boreal medusa with a circumpolar distribution. Around the British Isles it occurs in the North Sea, the Irish Sea, and off Northern Ireland; into the western Baltic. Strobilation of scyphistomas takes place during winter and, in British waters, small medusae appear by April and May, reaching up to 0.5 m diameter by early autumn. Diameters up to 2 m occur in north boreal waters. This species stings severely and the stinging powers are long retained by stranded medusae.

Cyanea lamarckii Péron and Lesueur Fig. 4.2
Much smaller than *C. capillata* and without the severe sting. Off all British coasts from April to July but uncommon; Biscay to Kattegat, western Norway and Iceland.

3. ULMARIDAE

Scyphomedusae with branched radial canals and a ring canal. Two species recorded for the British sea area, only the following occurs commonly.

Aurelia aurita (Linnaeus) Fig. 4.2
A cosmopolitan species found all around British coasts. Strobilation takes place in winter; ephyrae occur in the plankton from January to April; increasingly large adults, up to about 40 cm diameter, are present from late April to about August or, less usually, September. Often very abundant.

Order Rhizostomeae

4. RHIZOSTOMATIDAE

Medusae lacking marginal tentacles; tips of manubrium forming eight arms bearing many, clustered mouths; each arm ending in a terminal club-shaped appendage. One British species.

Rhizostoma octopus (Linnaeus) Fig. 4.2
Thick dome-shaped umbrella filled with tough mesogloea; up to 90 cm diameter; milky colour with bluish (male) or brown (female) gonads and marginal lappets. Scyphistoma unknown.

A southern species occurring commonly off (and stranded on) Channel and western shores in the British Isles as far as the Clyde; the southern North Sea; more rarely to the Kattegat and western Norway. Present in all seasons, with large specimens especially from midsummer to autumn and major strandings possible during autumn gales.

Order Stauromedusae

Attached Scyphozoa which develop directly from the scyphistoma. Each consist of a *bell* (umbrella or *calyx*) and a more or less distinct aboral stalk which terminates in an *adhesive disk*. The septa of the bell are rounded in section and effectively hollow, the oral surface being indented as four tapering, *interradial funnels* (pits or infundibula). The bell typically has eight adradial marginal *arms* (lobes), each supporting a cluster of knobbed hollow *tentacles* (secondary tentacles). In some genera eight per- and interradial marginal *primary tentacles* (anchors) are present, used for temporary attachment when the animal moves in a looper caterpillar fashion, transferring the stalk disk to a new position. No rhopalia. Gonads are borne on the septal walls as eight separate half gonads, or pairs may be partially or completely fused in the perradii. Gametes are shed into the water; zygotes develop into unciliated vermiform planula larvae from which the scyphistoma develops after a period of attached encystment.

Most species occur on the lower shore attached to algae, *Zostera*, and stones. The sexes are separate and all evidence points to their being annuals.

KEY TO SPECIES

---

1. Arms absent; >20 tentacles in each octant of bell margin, in one to three rows; gonads horseshoe-shaped; bell <10 mm in diameter
   Depastrum cyathiforme

Arms present; arm tentacles numerous, in eight bunches; gonads various; mature bell >10 mm in diameter **2**

---

2. Ams cruciformly arranged in four pairs; deep and shallow sinuses alternating ***Lucernaria quadricornis***

   Eight equal, radially-arranged arms; inter-arm sinuses equal **3**

---

3. Well-developed marginal primary tentacles (anchors) present in each sinus (*Haliclystus*) **4**

   Primary tentacles typically absent in adult **5**

---

4. Marginal primary tentacles large, trumpet-shaped, with thickened rim and a central rudimentary tentacle ***Haliclystus salpinx***

   Marginal primary tentacles kidney-shaped, shortly stalked, aboral to the margin ***Haliclystus auricula***

---

5. Peduncle short; arms short and wide, with up to 80 tentacles; gonad halves united (presenting in oral view a perradial, cruciform arrangement) ***Craterolophus convolvulus***

   Peduncle about as long as height of bell (although it may be involuted within the base of the bell), arms distinct, with up to 35 tentacles; gonad halves separate or partially separate, extending adradially to arm tips (*Lucernariopsis*) **6**

---

6. Bell to 35 mm diameter; colour variable but uniform, with 1–4 bright turquoise spots seen in interradii midway down bell; aboral surface papillate; a globular swelling at the base of each tentacle of the outermost tentacle row; gonad halves linear, separate ***Lucernariopsis campanulata***

   Bell to 18 mm diameter; colour maroon with conspicuous white spots on the disk in the pattern of a Maltese cross; white flecks in V-formation seen in aboral interradii; a swollen band at base of outermost tentacle row encircling outer half of arm; gonad halves united perradially at base, Y-shaped in oral view; characteristically displays 'stalk-less' posture with stalk involuted within the bell ***Lucernariopsis cruxmelitensis***

---

5. CLEISTOCARPIDAE

Gonads united by a transverse, circumferential membrane which divides each of the stomach pouches into outer and inner chambers. Two British species.

Craterolophus convolvulus (Johnston) Fig. 4.3
Bell- or urn-shaped, to 25 mm high and 25 mm diameter with short stalk, 5 (–8) mm, and wide basal disk; adradial arms short, equidistant, with 60–80 tentacles in clusters. Gonad halves united, evenly lobed along each side, extending perradially to bell margin, seen as four 'feathers' in oral view. Non-translucent; grey, green, brown, or red, sometimes with whitish spots along mid-line of gonad.

Low intertidal and shallow sublittoral, on algae and *Zostera*; all coasts; rarely numerous. North Atlantic: New England, Iceland, Denmark, Brittany.

Depastrum cyathiforme (M. Sars) Fig. 4.3
Bell goblet-shaped, to 10 mm high and 8 mm diameter; stalk about as long as bell, 12(–14) mm, contractile. Arms absent; 8–15 tentacles in 1–3 rows in each adradial octant of the bell margin; bell constricted below tentaculate rim. Grey-brown, ripe gonads whitish to pinkish.

Low intertidal and shallow sublittoral, usually under or among stones and boulders (never on

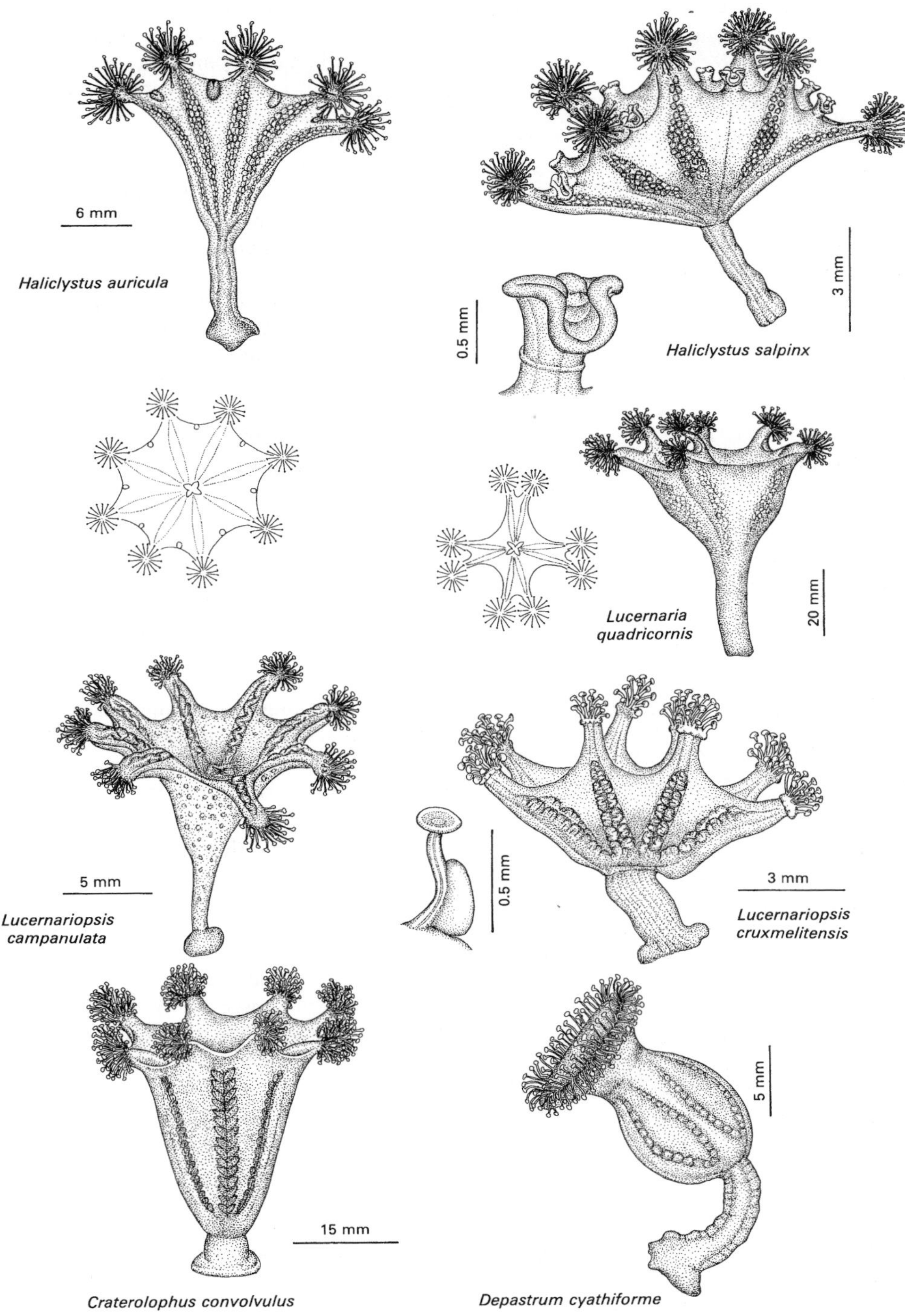
6 mm
Haliclystus auricula
0.5 mm
3 mm
Haliclystus salpinx
Lucernaria quadricornis
20 mm
5 mm
Lucernariopsis campanulata
0.5 mm
3 mm
Lucernariopsis cruxmelitensis
5 mm
15 mm
Craterolophus convolvulus
Depastrum cyathiforme

Fig. 4.3

algae); local. North Atlantic: Norway, British Isles south to Scilly, and Brittany.

6. ELEUTHEROCARPIDAE

Perradial stomach pouches simple. Six species recorded for the British sea area.

Haliclystus auricula (Rathke) Fig. 4.3
Bell funnel-shaped, to 15 mm high and 25 mm diameter; stalk as long as bell. Arms adradial, with 30–60(–100) tentacles in each tuft; per- and interradial primary tentacles ovoid. Half gonads separate, extending to arm tips; seen as eight linear-lanceolate 'leaves' in oral view. Body translucent, pale to deep grey, green, yellow, brown, red, or purple; a few whitish spots (nematocyst storage vesicles) near perradial margins.

Low intertidal and shallow sublittoral; on algae and *Zostera*; all coasts; widespread, the most numerous British species. North Atlantic and North Pacific.

Haliclystus salpinx Clark Fig. 4.3
Bell broadly funnel-shaped, generally 20–25 mm high and 20–25 mm in diameter, stalk as long as bell height; adradial arms moderately prominent, with 60–70(–100) slender tentacles in each cluster; primary tentacles large, prominent and columnar. Gonad halves separate, meeting at vertex of funnel, almost reaching margin adradially. Pink-red in colour.

Low intertidal and sublittoral; to about 5 m, on *Laminaria*, other algae, and *Zostera*. North Atlantic and North Pacific. Not yet recorded in British waters although known from the vicinity of Bergen, Norway.

Lucernaria quadricornis Müller Fig. 4.3
Bell funnel-shaped, to 60 mm high and 70 mm diameter; stalk a little longer than bell height but very contractile. Arms arranged in four pairs; wide perradial sinuses alternating with narrow, interradial sinuses; up to 140 tentacles in each cluster, thin stalks, small heads. Specimens very flaccid; symmetrical oral view seldom presented. Non-translucent, green-black to brown-black, gonads cream-orange in aboral view.

Low intertidal to 550 m; under stones and boulders at British localities, also on algae further north; local but may be numerous in subarctic waters; mainly northern in British Isles but south to Isles of Scilly.

Lucernariopsis campanulata (Lamouroux) Fig. 4.3
Bell funnel-shaped, to 20 mm high and 30 mm diameter, stalk 20(–25) mm, ending in a disk; entire aboral surface papillated. Adradial arms long, the per- and interradial sinuses of equal size; up to 45 tentacles, all except short inner ones with discoid ends, and each of outermost with a basal, glandular swelling. Gonad halves separate, linear, extending to arm tips; in oral view appearing coiled, as adradial 'ropes'. Generally translucent, uniformly green, brown, or red; 1–4 bright turquoise spots situated deeply at the base of each of the four interradial funnels of the oral surface, most clearly seen in oral or lateral view.

Low intertidal and shallow sublittoral; on algae and *Zostera*; all coasts but scarce bordering North Sea; north-west France.

Lucernariopsis cruxmelitensis Corbin Fig. 4.3
Bell widely funnel-shaped, 8 mm high and 12 mm diameter; stalk 8 mm, but base of bell characteristically involuted around stalk, presenting in life a 'stalkless' appearance; basal disk broad. Arms well developed, equidistant; up to 35 tentacles per cluster, each with a rounded head. Gonad halves thick, linear, united perradially at base, extending adradially to arms; with swellings along their length, in oral view appearing as thick 'ropes'. Translucent, maroon in colour, orally with a conspicuous Maltese cross of bright white spots (nematocyst storage organs) in the perradii and delimited by the gonads.

Low intertidal and shallow sublittoral; usually on *Chondrus* and *Gigartina*, rarely *Zostera*; south-west England from Swanage to north Devon, Atlantic coasts of Ireland.

Class Hydrozoa

Order Hydroida

Hydrozoans are typified by sessile polyp and free-swimming medusa generations in the life cycle. Almost all are marine. The hydroid stage is either solitary or colonial, and either buds free medusae or has sessile structures representing vestigial medusae. Gametes liberated from the medusae or from their sessile counterparts fuse to form zygotes which develop into *planula* larvae. The planula settles and develops into a *primary polyp* from which a new colony grows, completing the life cycle.

The skeleton is nearly always external to the tissues, but in the Hydractiniidae it secondarily forms an underlying mat. Over most of the colony the skeleton comprises a branching tube of chitinous material termed the *perisarc*. The tube of tissue inside is the *coenosarc* and consists of the usual three cnidarian layers. The colonies formed by most hydroids comprise (1) a basal portion, the *hydrorhiza*, which might be *stolonal*, mat-like, or a fibrous mass of aggregated tubes, and usually (2) an erect stem which in many species is branched (Fig. 4.4). Usually a colony comprises several to many erect shoots of well-defined growth habit, joined basally by a stolon lacking polyps. In a few hydroids, however, colonies start growing stolonally, subsequently sending up erect shoots, and having feeding and sometimes also reproductive polyps on both components of the colony.

The main stem, or *hydrocaulus*, may fork, the side branches are called *hydrocladia*. The branching pattern may be regular or irregular, bushy, *pinnate* (feather-like), or have branches arising all round the stem; or it may be a combination of these patterns. The first level of branching is termed *first order*, the next *second order*, and so on. Many hydrocauli and hydrocladia are *flexuose* (zigzag). Stems and branches comprising a single tube are termed *monosiphonic* or simple, and those of two or more fused together are called *polysiphonic* or compound.

At intervals most hydrocauli and hydrocladia have annuli or rings, which may be *transverse*, if at right angles to the axis of the skeletal tube, or *oblique*. Annuli are roughly circular grooves in the exoskeleton which permit limited bending of the mostly stiff outer tube. The positions of the annuli on the stems and branches are termed *nodes* and the intervening portions *internodes*. Annuli occur also in the *pedicels* or stalks of many polyps.

The polyps originate from the coenosarc and their *enterons* are continuous with it. There are several kinds of polyps, the feeding polyps, or *hydranths*, being usually the largest and most conspicuous. In polymorphic forms the feeding polyps, which may lack tentacles, are termed *gastrozooids*. Hydranths are borne on the hydrocladia, and often on hydrocauli, stolons, and other hydrorhizae, depending on the species. In the suborder Thecata the hydranths can, in most families, contract within a robust protective perisarcal cup, the *hydrotheca*, although in some thecates this has degenerated to a mere collar. In the suborder Athecata there are no true hydrothecae, there being at most a frail, often deciduous, extension of the stem perisarc, termed a *pseudohydrotheca*. These occur in just a few species and are clearly secondary in origin. In thecates the margin or rim of the hydrotheca may be *smooth* or *even*, *undulating*, *crenulate*, *indented*, *cusped*, *denticulate*, or *castellate* (Fig. 4.4). Some cusps are themselves deeply indented at the apex so that the rim has alternating deep and shallow *embayments* between the points, the *bimucronate* condition. In some families the hydrothecal *aperture* is closed by a lid, or *operculum*, of one to several flaps. In many genera new hydrothecae grow inside old ones so that a chain or nest of *renovated hydrothecae* or *renovated hydrothecal rims* becomes built up.

Some groups of thecates have the hydrotheca free and on a short *pedicel*, which is often *annulated*. The hydrothecae in others can be considered as a series showing progressive incorporation of the theca into the perisarc tube. Hydrothecae lacking a stalk are termed *sessile*, those in which part or all of the adjacent or *adcauline* wall is fused with the perisarc of the stem or branch are termed *adnate* or *fused*, and those in which the outer or *abcauline* wall also becomes incorporated to a varied extent are described as *sunken*. Hydrothecal pedicels in some genera of Campanulariidae have two unusually deep annuli at the top, delimiting a *subhydrothecal spherule* and permitting great lateral movement. In some genera the hydrotheca has internally near the base a *diaphragm* which may be *transverse*, if at right angles to the hydrothecal axis, or *oblique*. A small *hydropore* in the centre allows continuity with the coenosarc below. *True diaphragms* are continuous with the hydrothecal wall, *false diaphragms* arise later as outgrowths from it. Hydrothecae of some thecates have *internal septa* of irregular structure which partition them into chambers of varied shape. They are most prevalent in the Sertulariidae and Aglaopheniidae.

The hydranth comprises a short, but often very extensile, body or *column*. The mouth, often on a raised *hypostome*, is terminal. There are almost always numerous tentacles. They surround the mouth in one or two circlets or *whorls*, or in an *oral* (upper) whorl around the mouth and an *aboral* (lower) one more basally on the column. In some genera they are scattered over much of the column, or may be disposed in several more or less loosely defined whorls, or partly in whorls and partly scattered. Hydranths of thecates usually have a single whorl in which alternate tentacles are directed upwards and downwards, an arrangement termed

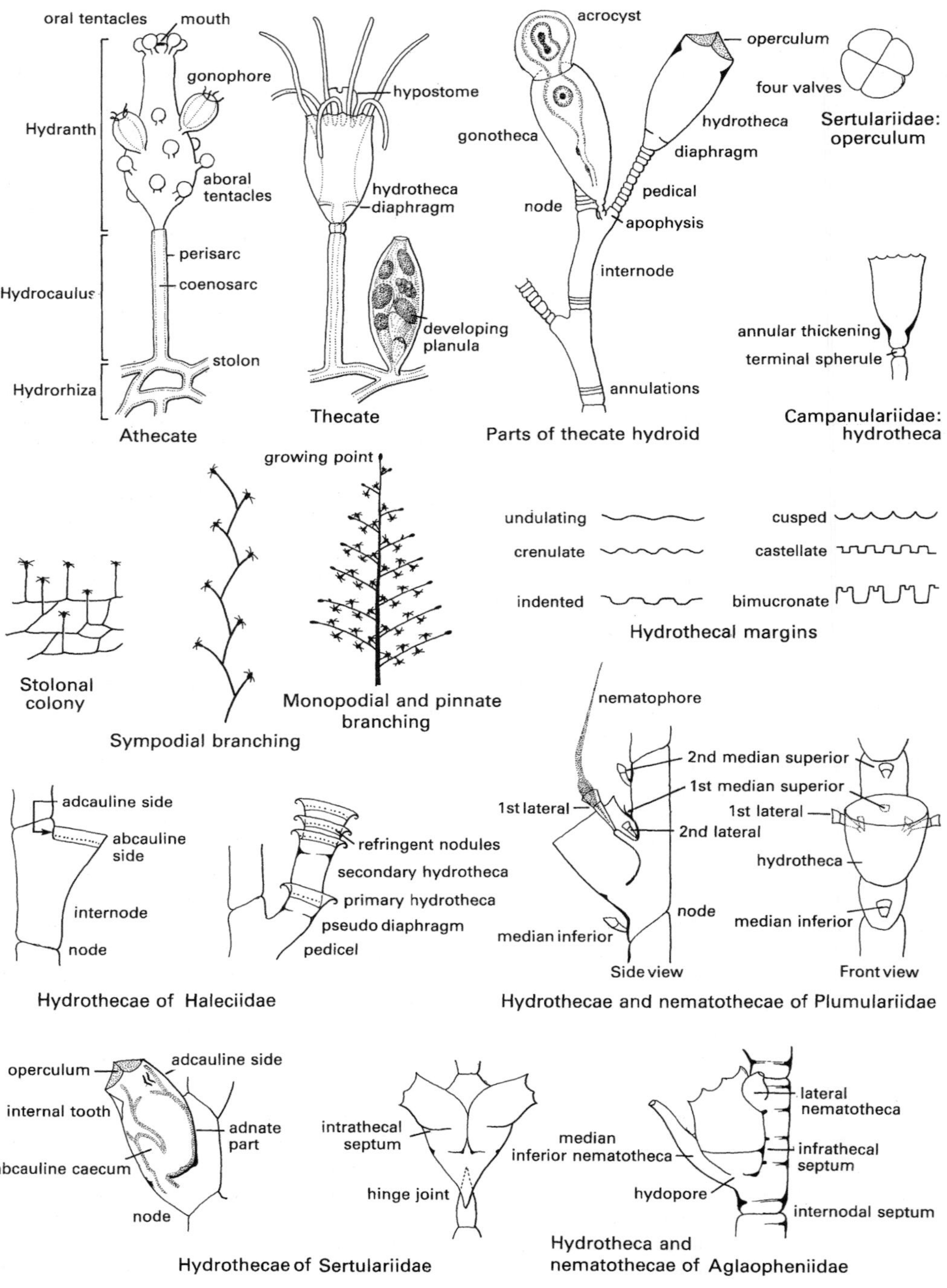
oral tentacles
mouth
gonophore
Hydranth
aboral tentacles
perisarc
coenosarc
Hydrocaulus
stolon
Hydrorhiza
Athecate
hypostome
hydrotheca
diaphragm
developing planula
Thecate
acrocyst
operculum
gonotheca
hydrotheca
diaphragm
pedical
node
apophysis
internode
annulations
Parts of thecate hydroid
four valves
Sertulariidae: operculum
annular thickening
terminal spherule
Campanulariidae: hydrotheca
growing point
Stolonal colony
Sympodial branching
Monopodial and pinnate branching
undulating
crenulate
indented
cusped
castellate
bimucronate
Hydrothecal margins
nematophore
2nd median superior
1st median superior
1st lateral
2nd lateral
1st lateral
hydrotheca
node
median inferior
median inferior
Side view
Front view
Hydrothecae and nematothecae of Plumulariidae
adcauline side
abcauline side
internode
node
Hydrothecae of Haleciidae
refringent nodules
secondary hydrotheca
primary hydrotheca
pseudo diaphragm
pedicel
operculum
adcauline side
internal tooth
adnate part
abcauline caecum
node
intrathecal septum
hinge joint
Hydrothecae of Sertulariidae
lateral nematotheca
median inferior nematotheca
infrathecal septum
hydopore
internodal septum
Hydrotheca and nematothecae of Aglaopheniidae

Fig. 4.4

amphicoronate, but in some species all point the same way. Short ectodermal *webs* join the tentacles basally in some thecates. *Capitate* tentacles have knobbed tips and occur in some athecate families, but most, termed *linear*, are either parallel-sided almost throughout their length or taper barely perceptibly to the pointed tip. *Moniliform* tentacles have the nematocysts in groups, giving a banded appearance; *cateniform* tentacles have them in a spiral, and in *filiform* tentacles they are scattered irregularly. Other arrangements occur. In some species of thecates all three are to be found and they have little use in identification.

Nematophores occur on the stems and branches of some athecates and in many species of the thecate families Plumulariidae, Aglaopheniidae, Campanulinidae, and Haleciidae. These are minute, extensile polyps, which usually lack mouths and tentacles but are armed with numerous nematocysts. They are thought to defend the colony against predation and settlement by larvae. Those of most thecate genera have a surrounding perisarc cup, the *nematotheca*, which may be sessile and immovable, termed *fixed*, or on a short, movable pedicel and *mobile*, but in some genera the nematophores are *naked*.

In the Hydractiniidae, long, atentaculate polyps called *dactylozooids* occur, comprising *spiral zooids*, which coil up, and *tentaculozooids*, which remain roughly straight.

Increase in size by hydroid colonies is by growth, in which new branches and polyps, the 'individuals' or modules, are added to the colony. The life cycle of most hydroids involves a succession or 'alternation' of generations, in which one or other of the generations includes a sexual phase. Species retaining the basic life cycle produce medusae by asexual budding. In the Hydrozoa the medusae are called *hydromedusae*. Most are less than 20 mm across. Most species of hydromedusae were first described separately from, and in ignorance of, their hydroid stages and the two were only later linked. Even in some British species the hydroid and medusa have still to be connected. The medusae, when present, are the sexual phase of the life cycle. Each medusa is either male or female, bearing gonads from which spermatozoa or ova are typically discharged into the sea. The zygotes resulting from fertilization develop in the plankton into minute planulae which settle to grow into new hydroid colonies and so complete the life cycle. However, in many species the medusa is reduced and remains attached to the hydroid in a variety of vestigial states such as *eumedusoids* and *sporosacs*. The latter are secondarily released in some species. *Gonomedusoids*, a kind of eumedusoid found in *Gonothyraea*, have vestigial radial canals and tentacles, but in the more widespread sporosacs these are absent and the medusa is reduced to a mere sac of gonadal tissue. A sporosac held external to the gonothecal aperture and lacking medusoid features is termed an *acrocyst*.

In many athecates medusae are budded from unmodified feeding hydranths which may, in some species, regress as the medusae develop. In other athecates, and in all thecates, medusae and their vestigial derivatives are borne on specialized polyps, lacking mouth or tentacles, termed *blastostyles*. In the thecates these are enclosed in corresponding *gonothecae* but in athecates they are naked. A blastostyle or other kind of polyp bearing gonophores and sporosacs or medusae was termed a *gonozooid* in older literature. In thecates the blastostyle and gonotheca together are sometimes referred to as a *gonangium*.

The gonothecae usually differ between species and, in species not releasing medusae, often between male and female. Female gonothecae tend to have wider apertures to allow the egress of mature planulae, whereas male gonothecae, from which spermatozoa are released, usually have narrower apertures. A typical planula might be 1 mm long and 0.2 mm wide. In species that liberate medusae there is no such distinction between male and female gonothecae. In some genera of thecate hydroids the gonangia are protected in modified branches, such as the *coppiniae* (muff-like tangles of elongate reduced hydrothecae) of some Lafoeidae, and the *corbulae* and *phylactocarps* of some Aglaopheniidae.

Species not releasing medusae are usually *dioecious*. The spermatozoa of species having sessile gonophores are released into the sea and presumably find their way to the ova which are fertilized and subsequently develop *in situ*, typically into planulae. In some species of Tubulariidae development continues further, resulting in a modified swimming polyp, the *actinula*. Most gonothecae (restricted to thecates) terminate in a simple circular opening, through which the planulae eventually escape. However, in many species of *Diphasia*, for example, the embryos develop in a protective *brood chamber* formed from a chitinous extension from the end of the gonotheca. In other thecates, for example *Orthopyxis* and several of the Sertulariidae, planulae mature in a gelatinous sac or *acrocyst* (see above).

The two suborders which are both well represented around British coasts can be distinguished thus:

Suborder ATHECATA (ANTHOMEDUSAE; formerly Gymnoblastea), releasing so-called *anthomedusae*. Hydroida in which the hydranths are not protected by true hydrothecae, or the gonangia by gonothecae; and in which the gonads of the medusae are usually borne on the manubrium.

Suborder THECATA (LEPTOMEDUSAE; formerly Calyptoblastea), releasing so-called *leptomedusae*. Hydroida in which hydrotheca and gonotheca are fundamentally present, although the hydrotheca is reduced in some families; and in which the gonads of the medusae are usually borne on the radial canals.

There are exceptions in both these groupings. Thus in the thecate families Haleciidae and Plumulariidae the hydrotheca is usually too small to accommodate the hydranth, while several species of athecates have an extension of the perisarc enveloping much of the hydranth and termed a *pseudohydrotheca*.

Most hydroids occur sublittorally. Only a few are characteristic of the shore, most of the species found there being stragglers from sublittoral strongholds. In the first 0–15m hydroids grow abundantly on *Laminaria* and other algae. They continue below this, often in a thick turf in association with Bryozoa, to a depth of 30–40 m around British coasts. Many occur much deeper but only in shallow waters do hydroids normally form dense communities.

Intertidally, hydroids occur most commonly in rock pools, particularly just below the surface, and in moist situations such as in wide crevices beneath overhangs and algae. Others grow on the blades of brown and red algae, and on eel grass. Most thrive best in places of moderate water movement where feeding opportunities are greatest, and the richest dredge hauls are often made in turbulent waters off headlands or in narrow straits.

Hydroids seem particularly sensitive to human pollution, and along the south-east coast of England and around some other heavily populated and industrialized areas they may be less common than formerly. Many species are tolerant of reduced salinity, silty conditions, and temperature fluctuation, but most have rather narrow limits of tolerance.

Hydroids are best collected by removing them from their substrata with forceps and placing them in collecting vessels. Many will survive well in aquaria but cleanliness and temperature control are crucial. Anaesthetization can be achieved with menthol crystals, 15% $MgSO_4{\cdot}7H_2O$ or 7.5% $MgCl_2{\cdot}6H_2O$, or with very dilute formalin or ethanol. Colonies for preservation should be fixed first in 8% formaldehyde solution (i.e. a 20% solution of commercial formalin) for 24 h and transferred to 70% methylated ethanol for permanent storage. Formaldehyde is harmful and should be handled with care. The perisarc of most hydroids can be stained effectively with borax carmine and the tissues with haematoxylin. Gurr's hydromount or equivalent is an adequate mountant and has the great advantage that specimens can be transferred into it from water, ethanol, or formalin. Care should be taken not to squash the hydrothecae and hence distort their shape: it is advisable to support the coverslip, and short lengths of thicker monofilament nylon fishing-line are ideal for this purpose.

KEY TO SUBORDERS

(medusa characters omitted)

---

1. Hydranth with no well-defined hydrotheca; sometimes a weak pseudohydrotheca (delicate extension of the stem perisarc); gonotheca usually lacking, gonophores borne on unspecialized hydranths, on reduced hydranths (blastostyles) or directly on stolon, stem, or pedicel
   Athecata

Hydranth with usually large, robust hydrotheca of definite shape, although in some groups too small to accommodate hydranth; gonophores borne on blastostyles and protected by a gonotheca **Thecata** (p. 85)

Suborder Athecata (Anthomedusae)

Hydroida fundamentally lacking well-defined hydrotheca and gonotheca. Tentacles capitate or linear, scattered over hydranth, in oral and aboral whorls, or inserted in one whorl.

The pelagic species *Velella velella* (Fig. 4.5) is not included in the key: see **3. Velellidae**.

KEY TO FAMILIES

---

1. Some or all tentacles capitate (club-tipped), very short in the worm-shaped *Candelabrum* **2**

   Tentacles all linear (tips not clubbed) **3**

---

2. Hydranth large, solitary, worm-shaped **4. Candelabridae**

   Colony a normal hydroid, erect or stolonal **5. Corynidae**

---

3. Tentacles scattered over much of hydranth **6. Clavidae**

   Tentacles clearly in one or two whorls **4**

---

4. Tentacles in one whorl, or in two closely adjacent whorls seeming to be one **5**

   Tentacles in two clearly separate whorls **8**

---

5. Hypostome narrowing below and widening above, obconical to spherical; never a medusa stage **7. Eudendriidae**

   Hypostome conical to domed; with or without medusa **6**

---

6. Hydranths borne on definite, perisarc-covered stem, erect, branched in some species; rarely stolonal **8. Bougainvilliidae**

   Hydranths on stolonal network or sheet-like hydrorhiza; rarely, on short erect stems **7**

---

7. Hydrorhiza from a closely anastomosing mat to completely sheet-like; erect stems never present; usually with polyp-types additional to hydranths and gonophores **9. Hydractiniidae**

Hydrorhiza an open network of stolons; sometimes with short erect stems of up to about four hydranths; only hydranths and gonophores represented **10. Pandeidae**

---

8. Hydranths solitary, massive, up to about 100 mm thick, anemone-like, floppy, lacking normal hydroid perisarc sheath **1. Corymorphidae**

Usually colonial or aggregated, hydranths atop rigid perisarc-covered stems; more typically hydroid-like in appearance **2. Tubulariidae**

---

1. CORYMORPHIDAE

Large, solitary hydroids with cylindrical hydrocaulus ensheathed in delicate, non-supportive perisarc; usually with basal anchoring filaments; hypostome conical; tentacles in oral and aboral whorls; gonophores attached above aboral tentacles, usually on branched blastostyles. Three species recorded from Britain.

Corymorpha nutans M. Sars Fig. 4.5
Hydranth white to pale red, floppy, to about 10 cm; elongate, in life often with upper end recurved and bent, basalmost part in substratum; terminal region red with lower whorl of about 32 long, thick, white, non-contractile tentacles; upper group of about 80 finer, shorter ones. Developing medusae attached just above lower group, on 15–20 branched stalks.

Anchored in silt to gravel; coastal, to at least 100 m; throughout British Isles; Mediterranean to northern Norway.

2. TUBULARIIDAE

Solitary or colonial; perisarc of stems extending to base of hydranth. Hydranth large, tentacles in oral and aboral whorls; gonophores often in branched clusters attached between tentacle whorls. Six species recorded from Britain, in the genera *Tubularia, Ectopleura* and *Hybocodon.*

---

1. Many more oral tentacles than aboral; stems not fused ***Tubularia larynx***

Oral and aboral tentacles roughly similar in number; stems sometimes fused **2**

---

2. Hydranths as high as wide, or taller; stems erect, often fused basally; up to about 45 mm ***Tubularia indivisa***

Hydranth wider than high; lower parts of stems often procumbent, not fused basally; up to about 25 mm ***Tubularia bellis***

Tubularia bellis Allman Fig. 4.5
Colony up to about 25 mm. Erect stems sparsely branched, often partly procumbent, coenosarc orange to red-orange, grading darker basally; widely ringed for much of basal part; with creeping stolon. Hydranth wider than high; tentacles about 15–20 in oral whorl, 20 in aboral; spread of aboral whorl in life about 10 mm. Gonophores in short clusters of 4–5 above aboral tentacles; ♂ and ♀ with four conspicuous distal conical processes; central column scarlet. Dispersive stage an actinula.

In rock pools and on rocks, mid to lower shore; recorded Shetlands, south-west Ireland, North Sea oil rigs; probably overlooked.

From *T. larynx* by shorter gonophore clusters, proportionate width of hydranth, procumbent tendency of stem bases, and their widely spaced annuli.

Tubularia indivisa Linnaeus Fig. 4.5
Stems clustered, stiff, pale horn-coloured, sinuous; annuli sparse; often fused together basally in intertwined groups, usually 30–150 (exceptionally to 400) mm. Hydranth pale pink to scarlet, large, drooping, about 40 white oral tentacles, 20–30 aboral. Gonophores in 3–4 pendulous branches with up to about 20 retained medusae each releasing a single actinula.

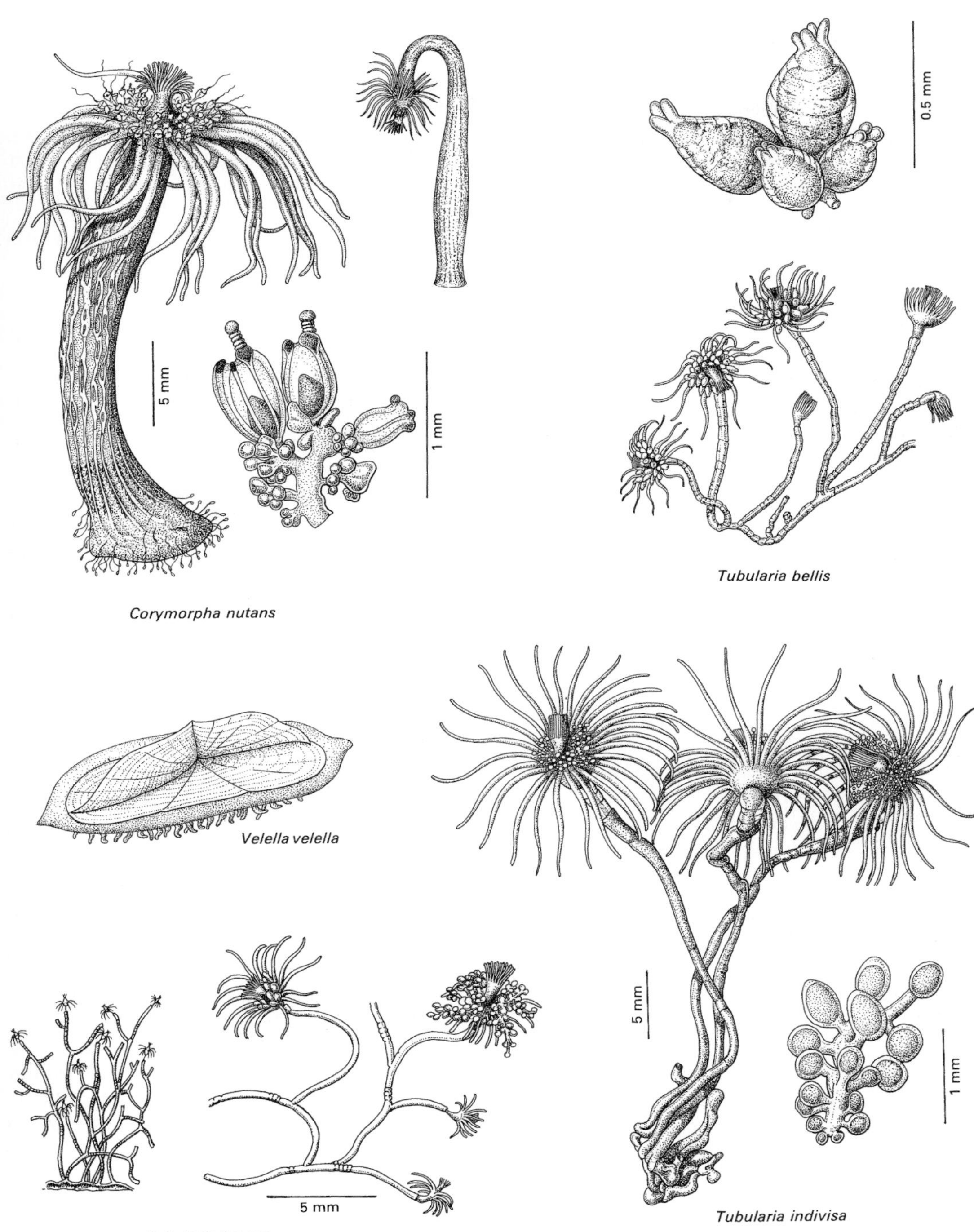
0.5 mm
5 mm
1 mm
Corymorpha nutans
Tubularia bellis
Velella velella
5 mm
1 mm
5 mm
Tubularia larynx
Tubularia indivisa

Fig. 4.5

On various solid substrata; lower shore pools to at least 280 m; common throughout British Isles.

Tubularia larynx Ellis and Solander Fig. 4.5
Colony a loose collection of occasionally forked stems up to about 45 mm. Each hydranth rose-red, 14–20 white oral and about 20 aboral tentacles. Stems straw-coloured, with numerous spaced annuli. Gonophores in long clusters, supporting about 20 retained medusae, each with four low tubercles distally.

On other hydroids and solid substrata; lower shore to at least 100 m; common throughout British Isles; tolerant of slightly reduced salinity.

3. VELELLIDAE

Hydranth floats at the sea surface. Float flattish, chambered, bearing a triangular fin; the blastostyles and tentacles below it, concentric around the mouth. A single North Atlantic species.

Velella velella (Linnaeus)
By-the-wind sailor Fig. 4.5
Float flat, oval, raised diagonally as a flange-like fin or 'sail'; up to 100 mm in length; deep blue.

May occur stranded in large numbers on south-westerly coasts, in all seasons, and occasionally as far north as the Hebrides during summer, following persistent southerly or south-westerly winds. Occasionally in Faeroe Islands.

4. CANDELABRIDAE

Large, solitary, hydranths with thin, flexible perisarc closely ensheathing basalmost region; tentacles capitate, numerous, scattered over hydranth; attachment by stumpy basal organs; gonophores on special blastostyles. No medusa stage. One British species.

Candelabrum cocksi (Cocks) Fig. 4.6
Worm-shaped, up to 50 mm when contracted but extending up to 100 mm; upper part covered in 200 or more minute club-tipped tentacles of a variety of reds, pinks, and white; lower down with many pinkish-horn branched processes, each supporting several pink gonophores. Liberates an actinula.

Habits little known. Recorded under stones and algae from ELWS to about 3000 m; in Britain most records south-west England, but sparingly reported north to Arctic Ocean.

5. CORYNIDAE

Colonial, with erect stems and firm perisarc reaching at least to hydranth base; hypostome conical; tentacles capitate, some species with small linear aboral tentacles; gonophores borne on hydranth amongst or below tentacles; reproduction by sporosacs or free medusae. More than a dozen species listed for Britain, only the following are at all frequent.

---

1. Perisarc closely ringed throughout; gonophores producing sporosacs (*Coryne*) **2**

   Perisarc mostly smooth; gonophores producing medusae (*Sarsia*) **3**

---

2. Colony elongate, up to about 150 mm, hydranth usually with thin deciduous perisarc sheath basally, diagnostic if present ***Coryne muscoides***

   Colony short, 25 mm or less, hydranth always naked ***Coryne pusilla***

---

3. Perisarc nearly smooth throughout; slightly rugose in places but not usually ringed ***Sarsia tubulosa***

   Perisarc ringed in places **4**

---

4. Colony tall, with erect, frequently branched stem up to about 100 mm; stems pale brown to dark brown; medusae borne amongst tentacles, released ***Sarsia eximia***

   Colony stolonal or with short erect shoots up to about 40 mm, usually branched only two or three times; stems light coloured to pale brown; medusoids in whorl below capitate tentacles, not released ***Sarsia loveni***

Coryne muscoides (Linnaeus) Fig. 4.6
Colony long, branched, up to about 150 mm. Stem brown, closely ringed throughout; with many short side-branches. Hydranths terminal, long, pink,

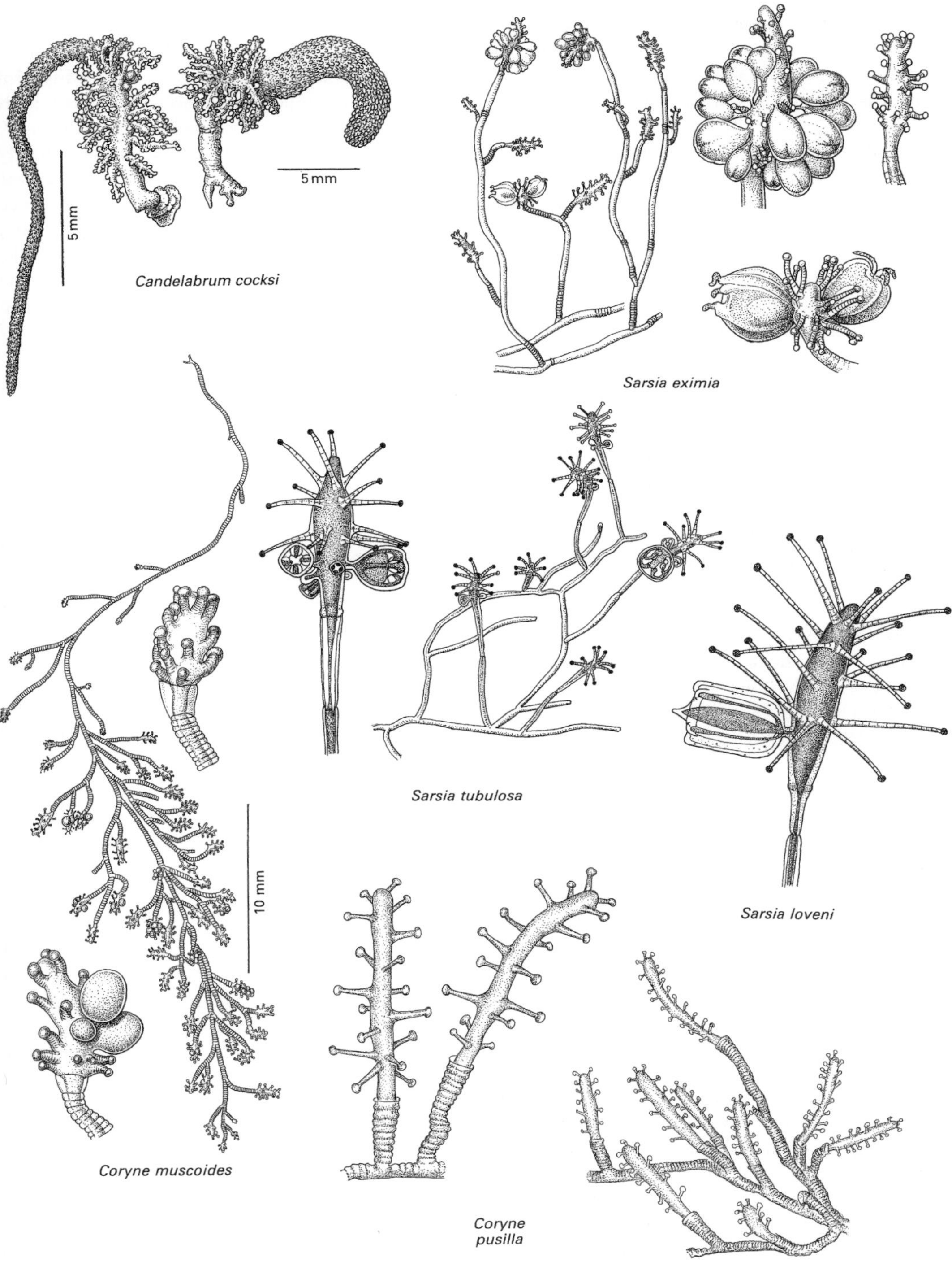
5 mm
5 mm
Candelabrum cocksi
Sarsia eximia
Sarsia tubulosa
Sarsia loveni
10 mm
Coryne muscoides
Coryne pusilla

Fig. 4.6

tapered basally; up to about 30 capitate tentacles sometimes in distinct whorls; tip of supporting branch with diagnostic (but deciduous) membranous perisarc sheath. Gonophores spherical, attached to hydranth amongst tentacles. Dispersive stage a planula.

On algae and other substrata; lower shore to about 2 m below LWST; records north to Isle of Man, Northern Ireland, Norfolk, Orkneys, but status around mainland Scotland and north-east England unclear.

Coryne pusilla Pallas Fig. 4.6

Colony a loose tangle, up to about 25 mm. Stem sinuous, forked 3–4 times; perisarc pale to mid-brown, ringed throughout, lacking a pseudohydrotheca. Hydranth long, hardly tapering; up to about 30 long, thick, capitate tentacles scattered over length, sometimes in loose to distinct whorls. Gonophores globular, scattered among tentacles in lower half of hydranth. Dispersive stage a planula.

Typically on brown algae; mid to lower shore; all around British Isles; at least from northern Norway to Cape Verde Islands.

Sarsia eximia (Allman) Fig. 4.6

Colony of erect, narrow, straw to brown, branched stems, up to about 100 mm; sometimes creeping. Main stem sometimes forked, ringed at base, otherwise smooth; branches tending to arise on one side of stem, with numerous rings basally, smooth above, sometimes themselves branching. Hydranths long, pink; up to about 30 straight, capitate tentacles, four just below mouth and rest scattered. Gonophores on short stalks in axils of tentacles over most of hydranth, each producing a single medusa.

On rocks, floating structures, *Laminaria*, and other algae; lower shore, probably to shallow coastal; all around British Isles.

Sarsia loveni (M. Sars) Fig. 4.6

Colonies erect, irregularly branched, up to about 40 mm, or limited to an anastomosing stolon; stems colourless to pale yellow, corrugated irregularly, annulated above and below origins of branches and near stem base, smooth elsewhere. Hydranths pink with 16–30 slender, capitate tentacles, scattered or in up to about four whorls. Gonophores in form of retained, reduced medusae; dispersive stage a planula.

On rocks, stones, and weed; around LWST and offshore to 200 m; recorded Plymouth, north-east England, west Scotland, Norway, and Kattegat; tolerant of reduced salinity; other records problematic.

Sarsia tubulosa (M. Sars) Fig. 4.6

Colony of hydranths on pedicels arising singly from an anastomosing stolon, to about 10 mm; or erect, slightly branched, up to about 30 mm; stems mainly smooth. Pedicels unringed, slightly rugose; perisarc thin, translucent horn-coloured. Hydranths orange to pink; narrowing below; up to about 20 long, white, capitate tentacles in ill-defined distal whorls. Sometimes a loose, delicate perisarc sheath over the basal third of the hydranth. Gonophores attached singly on short stalks in one irregular whorl below lowermost tentacles. Medusa released.

On rocks, stones, and weeds; rock pools on lower shore, occasionally sublittoral; tolerant of reduced salinity; around British Isles; circumpolar boreo-arctic.

6. CLAVIDAE

Solitary or colonial; hydranths with conical hypostome, tentacles filiform, scattered; firm perisarc on hydrorhiza only, or on erect stems also; fixed sporosacs or free medusae. Eight species recorded from British sea area.

---

1. Colony stolonal or encrusting; hydranths attached directly to basal structure **Clava multicornis**

   Colony tall, erect, with long branches ensheathed in perisarc; hydranths terminal **Cordylophora caspia**

Clava multicornis (Forskål) Fig. 4.7

Colony of naked hydranths in small groups; up to about 30 mm when fully extended, white through pink to red, brown, or blue; up to about 30 white tentacles scattered over upper part. Gonophores clustered on short branches below tentacles. Dispersive stage a planula, one to two from each gonophore.

Often on *Ascophyllum* and other algae; intertidal, MTL to ELWS; also on and under stones and in rock pools; tolerant of brackish water; all around British Isles.

Cordylophora caspia (Pallas) Fig. 4.7

Stems to about 100 mm, occasionally branched from alternate sides; each branch ringed basally.

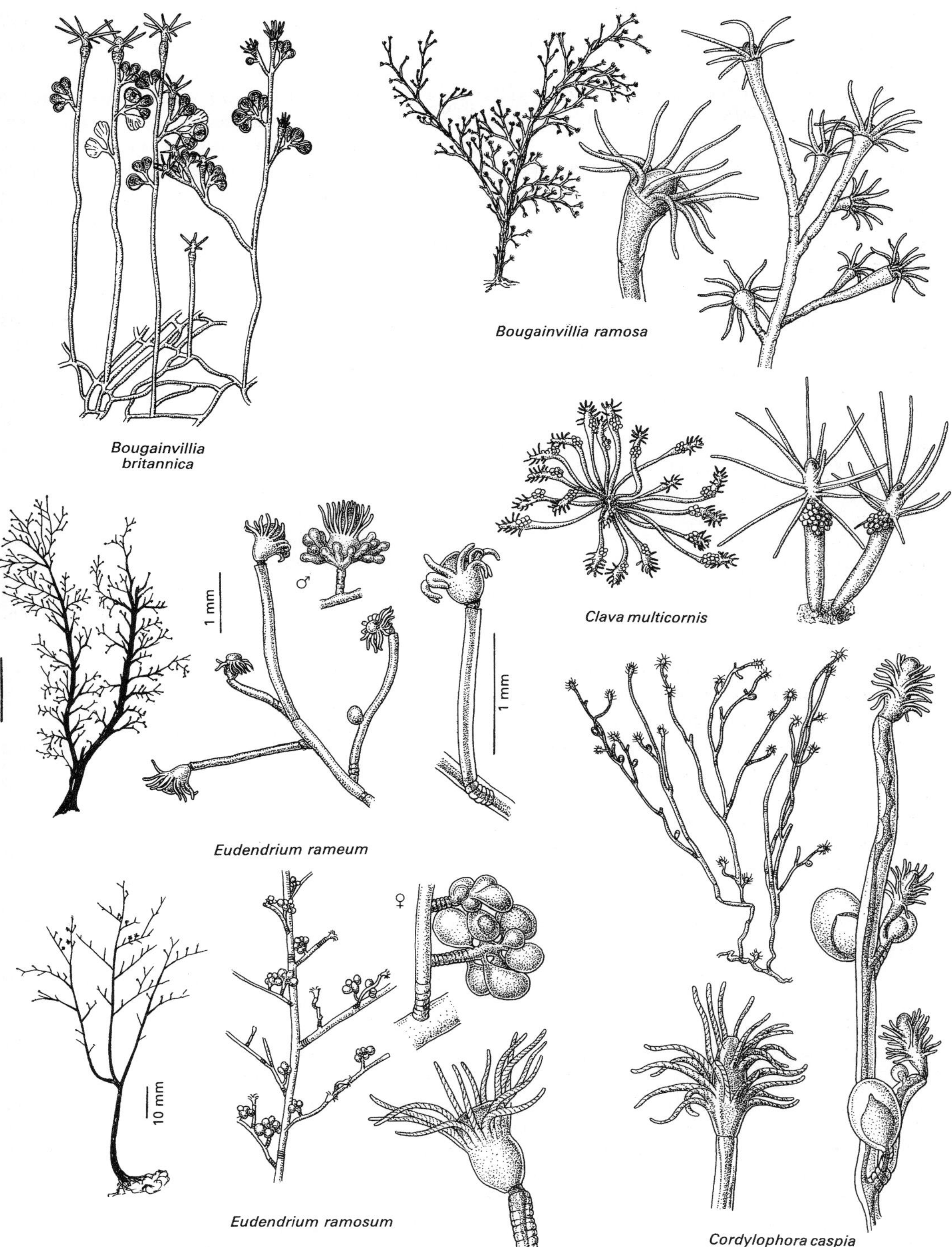
Bougainvillia britannica
Bougainvillia ramosa
Clava multicornis
2 cm
1 mm
♂
1 mm
Eudendrium rameum
♀
10 mm
Eudendrium ramosum
Cordylophora caspia

Fig. 4.7

Hydranths terminal, white to pale pink, long, tapered below; hypostome conical but truncate above; 12–16 long, colourless, rather straight, linear, tentacles. Gonophores pear-shaped, on short stalks, up to about three on each final branch. Dispersive stage a planula.

On various substrata, often in shade; brackish to nearly fresh water; shallow depths; widespread, likely to occur all around British Isles.

7. EUDENDRIIDAE

Colonial, stems erect, usually branched; perisarc extending to hydranth base. Hydranth with bulbous hypostome atop short constriction; one or more whorls of tentacles beneath it; fixed sporosacs, often on a blastostyle; ♂ gonophore often a short chain of sporosacs, ♀ usually single; no medusa. About 10 species of *Eudendrium* have been reported from British waters, but the taxonomic status of most is uncertain.

---

1. Most branches monosiphonic, colony polysiphonic only basally; even main stems slender, branches often long and tending upwards; limp out of water **_Eudendrium ramosum_**

   All but distal branches polysiphonic; stems thick, trunk-like, branches short, pointing in all directions; rigid out of water; ♂ gonophores single spheres in clusters on and below hydranth, ♀ inserted singly along each pedicel **_Eudendrium rameum_**

Eudendrium rameum (Pallas) Fig. 4.7

Colony large, bushy; trunk and major forks thick, to 100 mm, occasionally 250 mm, rigid out of water. Main stems brown, repeatedly branched, monosiphonic only distally. Hydranths roughly alternate on finer branchlets, on long, sinuous, tightly ringed pedicels; red, tapering below; tentacles 20–24, in one whorl. Gonophores ovoid, yellow, short-stalked, on and below hydranths in clusters of up to about 15 (male) or several inserted singly along each pedicel (female). Male gonophores two-chambered. Dispersive stage a planula.

On various solid substrata; about 5–100 m, sometimes deeper; common around British Isles; near-cosmopolitan.

Eudendrium ramosum (Linnaeus) Fig. 4.7

Colony tall, alternately branched, with long monosiphonic shoots arising from lower region; polysiphonic only near base; limp out of water. Stems to about 100 mm, occasionally 200 mm, brown, narrow, branches arising acutely and often turning upwards to lie roughly parallel with primary branch. Hydranth large, vermilion to pink, tapered below; hypostome spherical, white; tentacles up to about 24, occasionally to 30, in one whorl. Male gonophores on hydranths, in chains of 2–3 spherical sporosacs, smallest proximally; female in circlet of about 3–4 at base of hydranth, sometimes another circlet below, pear-shaped on short stalk; supporting hydranths atrophy. Dispersive stage a planula.

On variety of substrata; about 5–200 m, sometimes deeper; common around British Isles; widespread in Atlantic.

8. BOUGAINVILLIIDAE

Colonies erect and branched, sometimes stolonal; perisarc often extending part-way up hydranth; hydranth with conical hypostome and one whorl of linear tentacles; fixed sporosacs or medusae. Eight species recorded in the British sea area.

---

1. Parts of stem polysiphonic **_Bougainvillia ramosa_**

   Stem monosiphonic throughout **2**

---

2. Colony arising from a closely anastomosing stolonal mat **_Dicoryne conferta_**

   Branches of stolon widely spaced **_Bougainvillia britannica_**

Bougainvillia britannica Forbes Fig. 4.7

Colony an anastomosing stolon with rather distant, erect shoots to about 15 mm, usually arising from stolon junctions; some branches at acute angle, each ending in a hydranth. Hydranth club shaped with short conical hypostome, white; tentacles 6–16. Gonophores spherical to pear-shaped, 2–3 together near top of shoot or in branched clusters of up to about six; each in persistent perisarc sheath. Dispersive stage a medusa.

On gastropod and bivalve shells; coastal; medusa recorded all around British Isles and southern Norway.

Bougainvillia ramosa (Van Beneden) Fig. 4.7
Colony tall, erect; from short, wholly monosiphonic, little-branched stems to much-branched, polysiphonic, arborescent colonies with thick trunks; to about 100 mm. Branches numerous, somewhat regularly arranged, often alternate and in loose spiral; secondary and even tertiary branching. Hydranths terminal, tapered, pale pink; tentacles up to about 20. Gonophores pear-shaped to ovoid, on stalks one-third to one-half their length, attached below hydranths singly or up to about three in small branched cluster. Dispersive stage a medusa.

On variety of substrata; lower shore probably to about 50 m; common all around British Isles.

Dicoryne conferta (Alder) Fig. 4.8
Colony moss-like, stolon anastomosing; with single, stalked hydranths or erect, monosiphonic, branched stems, to about 20 mm. Branching acute; perisarc smooth, brown. Hydranths pale brown, cylindrical with conical hypostome; tentacles up to 16, rarely 20, linear. Gonophores stalked, in clusters of up to about 15, on blastostyles which lack tentacles but have large extensile hypostome; arising from stolons or stems, releasing distinctive sporosacs with two ciliated processes.

On shells of living gastropods and empty ones with or without hermit crabs; 5–300 m; scattered records around British Isles, mostly northerly; Barents Sea to southern Africa.

9. HYDRACTINIIDAE

Colonial, stolonal to hydrorhizal; polyps polymorphic; hydranths naked, tentacles linear, in one or more whorls; hydrorhiza often including chitinous spines; reproduction by fixed sporosacs or free medusae. Seven species listed from Britain.

---

1. Tentacles in two whorls of about eight, those in lower whorl shorter than those in upper; reproduction by sporosacs **Hydractinia echinata**

   Tentacles in one whorl of about 16, often alternately short and long; reproduction by medusae 2

2. Hydranth 10–15 mm **Podocoryne borealis**

   Hydranth up to about 5.5 mm **Podocoryne carnea**

Hydractinia echinata (Fleming) Fig. 4.8
Colony a white to pale-pink mat. Hydranths up to about 13 mm, with upper circlet of about eight long tentacles and lower circlet of about eight shorter ones. Gonozooids shorter than hydranths, with few, stubby, tentacles and ring of yellow to white (♂) or pink (♀) gonophores. Also long, slender, coiling dactylozooids. Basal crust with many blunt conical spines, about 2 mm high with jagged edges, mingled amongst polyps. Dispersive stage a planula.

On gastropod shells inhabited by hermit crabs, sometimes also on other solid substrata; MTL probably to about 30 m; common throughout British Isles; recorded Arctic to Morocco.

Podocoryne borealis (Mayer) Fig. 4.8
Colony encrusting. Hydranths 10–15 mm, cylindrical to club-shaped, perisarc collar at very base only; one whorl of up to 16 long tentacles. Perisarc spines smooth, amongst hydranths. Gonophores pear-shaped, on some hydranths, which are often smaller than infertile ones; to about 20 in a close cluster, about one-third from top of hydranth; hydranth regresses as gonophores mature. Most colonies include tentaculozooids, resembling small, isolated tentacles arising direct from stolon, with or without minute basal perisarc collar. No spiral zooids recorded. Medusa released.

On shells of gastropods, barnacles, and larger crustaceans; also shells occupied by hermit crabs, and even on hydroids and *Aphrodita*. Coastal, to at least 100 m; all British coasts; southern Norway and Iceland to Belgium.

Podocoryne carnea (M. Sars) Fig. 4.8
Colony encrusting. Hydranths up to 5.5 mm, cylindrical to club-shaped, no basal perisarc collar; hypostome conical to rounded; tentacles up to 16, sometimes 20, in one whorl. Perisarc spines smooth, amongst hydranths. Gonophores variable in size, on shorter hydranths having fewer tentacles, in clusters of up to about 10, about one-third from top of hydranth; hydranth regresses as gonophores mature.

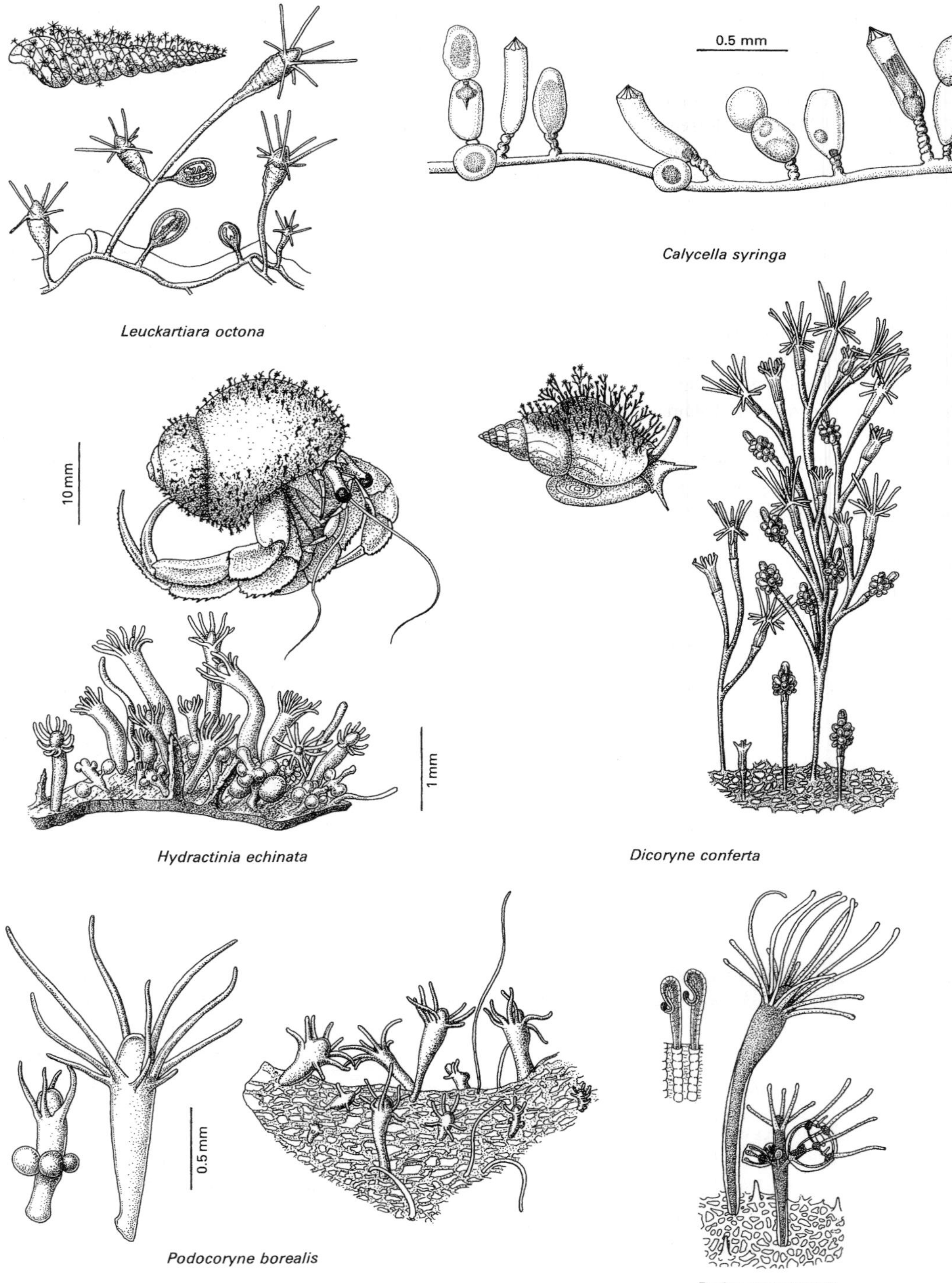
0.5 mm
Calycella syringa
Leuckartiara octona
10 mm
1 mm
Hydractinia echinata
Dicoryne conferta
0.5 mm
Podocoryne borealis
Podocoryne carnea

Fig. 4.8

Spiral zooids in some colonies, especially around apertures of gastropod shells occupied by hermit crabs; short, hydranth-like, curling actively, lacking tentacles and defined hypostome. Tentaculozooids may be present. Medusa released.

On shells of several living gastropods, especially *Hinia reticulata*, and on those with hermit crabs; also on bivalves, barnacles, larger crustaceans, and possibly stones; intertidal pools to about 50 m; around British Isles, north to mid-Norway.

10. PANDEIDAE

Colonies stolonal to shortly erect; hydranths with one whorl of filiform tentacles; perisarc varied in development, sometimes absent; medusae released. Four species recorded for Britain.

Leuckartiara octona (Fleming) Fig. 4.8
Colony creeping, erect stems short to about 3 mm; up to four hydranths on each; hydranths terminal or at each branching; perisarc thin, extending over hydranth to base of tentacles as a wrinkled pseudohydrotheca. Hydranths small, cylindrical to tapered, white, hypostome conical; tentacles 4–10, in one circlet, straight, stiff, amphicoronate, lowermost the shorter. Gonophores subsessile to stalked, on stolon and/or up to five per erect stalk. Medusa released.

On various substrata, often *Turritella* and *Corystes*, other hydroids, *Aphrodita*, *Agonus*, spider crabs, *Antalis*, *Scaphander*, and other gastropods; 0–200 m; common around British Isles and to north-west Europe; near-cosmopolitan.

Suborder Thecata (Leptomedusae)

Hydroida with well-defined hydrotheca (reduced in some genera) and gonotheca; tentacles always linear, inserted in one whorl and not increasing in number after rupture of deciduous hydrothecal cap. Medusa typically with gonads overlying radial canals; usually having statocysts, sometimes ocelli.

KEY TO FAMILIES

1. Hydrotheca bilaterally symmetrical, sometimes only slightly so — **2**

   Hydrotheca fundamentally radially symmetrical — **4**

2. Hydrothecae on both sides of stem (in *Hydrallmania* the two series are so close as to appear to form a single row), or colony stolonal — **6. Sertulariidae**

   Hydrothecae on one side of stem; colony not stolonal — **3**

3. Hydrotheca always large enough to accommodate hydranth; hydrotheca usually with conspicuous adherent nematotheca in mid-line; rim with about eight deep clefts and cusps; gonotheca in protective basket or unprotected — **8. Aglaopheniidae**

   Hydrotheca usually too small to accommodate hydranth; no nematothecae adherent to hydrotheca in mid-line; rim usually smooth; gonotheca always unprotected — **7. Plumulariidae**

4. Hydrotheca closed by conical operculum of several roughly triangular flaps (may be deciduous) — **5**

   Hydrotheca lacking operculum — **6**

5. Operculum delimited basally by creased hinge-line **1. Campanulinidae**

Operculum not delimited basally by creased hinge-line **3. Phialellidae**, also *Opercularella* (this page)

6. Hydrotheca pedicellate, more or less goblet-shaped **2. Campanulariidae**

Hydrotheca sessile, or tubular and narrow **7**

7. Hydrotheca short, far too small to accommodate the large hydranth **5. Haleciidae**

Hydrotheca deep, tubular, easily accommodating the slender hydranth **4. Lafoeidae**

1. CAMPANULINIDAE

Small colonies, stolonal or with short, erect stems; hydrotheca deep, cylindrical; with conical operculum of several triangular flaps, with or without hydrothecal diaphragm; hydranth slender, fully contracting within hydrotheca; hypostome conical; some species with nematophores; reproduction by sporosacs or medusae. Six species recorded for Britain.

1. Colony mostly erect ***Opercularella lacerata***

Colony stolonal ***Calycella syringa***

Calycella syringa (Linnaeus) Fig. 4.8

Hydrocaulus a creeping stolon. Hydrothecae arising individually, up to about 1 mm including pedicel; cylindrical; rim scalloped into an operculum of 8–9 slender projections, operculum deciduous. Gonotheca smooth, ovoid, on short pedicel attached to stolon; developing gonophores contained in acrocysts projecting singly from gonothecal apertures.

On other hydroids and algae; coastal; near-cosmopolitan; common throughout British Isles; often on stranded material.

Opercularella lacerata (Johnston) Fig. 4.9

Main stem erect, to about 10 mm, slightly flexuose, repeatedly branched; perisarc annulated to sinuous throughout; giving rise at nodes to short pedicels. Hydrotheca widest in middle, with conical operculum of 9–12 long, pointed flaps not demarcated from hydrotheca. Gonothecae axillary or on stolon; male subcylindrical; female obconical, with wide aperture. Dispersive stage a planula.

On various substrata; 0–50 m; widespread in British Isles; Mediterranean to Arctic.

2. CAMPANULARIIDAE

Colonies stolonal or erect; hydrotheca pedicellate, bell-shaped, fundamentally radially symmetrical, without operculum; hydranths with pear-shaped to spherical hypostome; tentacles linear, in one whorl, amphicoronate; reproduction by sporosacs, gonomedusoids or free medusae. A large family represented in the British sea area by about 20 species.

1. Colony with erect stems each supporting several to many hydrothecae **2**

Colony mainly stolonal, each stem or pedicel supporting one or few hydrothecae **14**

2. Hydrothecal rim even, sinuous, or crenulate **3**

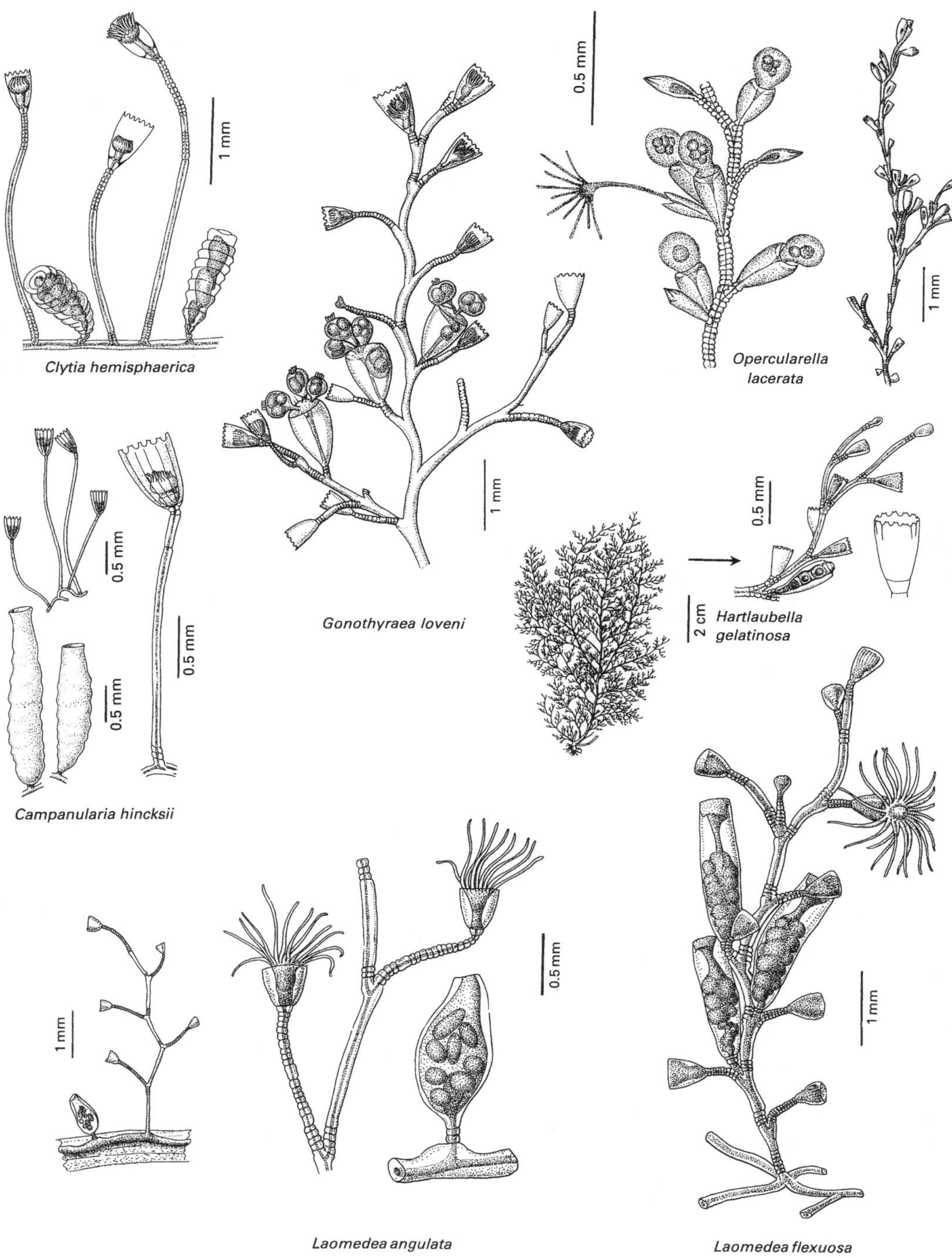
1 mm
Clytia hemisphaerica
1 mm
Gonothyraea loveni
0.5 mm
Opercularella lacerata
1 mm
0.5 mm
0.5 mm
0.5 mm
Campanularia hincksii
0.5 mm
2 cm
Hartlaubella gelatinosa
1 mm
0.5 mm
Laomedea angulata
1 mm
Laomedea flexuosa

Fig. 4.9

Hydrothecal rim definitely cusped (but often abrading smooth in *Gonothyraea loveni, Hartlaubella gelatinosa*, and *Obelia longissima*) **8**

---

3. Internodes curving **4**

   Internodes quite straight, or only slightly curving **6**

---

4. Hydrotheca thickened; internode usually asymmetrically thickened ***Obelia geniculata***

   Neither hydrotheca nor internode thickened **5**

---

5. Hydrotheca usually at least one and a half times long as broad, rim sinuous to crenulate; aperture of mature gonotheca raised; releasing medusae ***Obelia dichotoma*** (but see also ***O. longissima***)

   Hydrotheca not much longer than broad; rim even; end of mature gonotheca flatly truncate; releasing planulae ***Laomedea flexuosa***

---

6. Internodes usually rigidly straight, rather long; terminal tendrils present in autumn; gonotheca on stolon; releasing planulae (found only on eel grasses) ***Laomedea angulata***

   Internodes at least slightly curving; terminal tendrils unusual; gonothecae usually axillary; releasing medusa 7

---

7. Colony very long, stem brown to black, monosiphonic except sometimes towards base in older colonies, side-branches ceasing growth at roughly uniform length but this gradually decreasing distalwards ***Obelia longissima***

   Colony loosely fan-shaped, stem polysiphonic basally, pale to mid-brown, never black; side-branches not uniform in length, growth of many continuing indefinitely ***Obelia dichotoma***

---

8. Hydrothecal cusps sharp (Fig. 4.4) **9**

   Hydrothecal cusps blunt; square, notched, or rounded **10**

---

9. Hydrothecal rim with embayments between cusps all roughly the same depth ***Clytia hemisphaerica***

   Hydrothecal rim usually bimucronate. Mature colony tall and bushy, with scores or hundreds of hydranths; main stem polysiphonic ***Obelia bidentata***

---

10. Hydrothecal rim with rounded cusps **11**

    Hydrothecal rim with square cusps, often notched (abrade easily) **12**

---

11. Mature colony large, main stem polysiphonic; hydrothecae in whorls, most with sub-hydrothecal spherule; no medusa stage ***Rhizocaulus verticillatus***

    Mature colony short, always monosiphonic; no spherule; medusae released ***Clytia hemisphaerica***

---

12. Central stem dark brown to black ***Obelia longissima***

    Central stem pale **13**

---

13. Small, slender colony, stem monosiphonic; main stem forked, but no true secondary branching; medusae retained as sessile medusoids, when mature held external to gonotheca ***Gonothyraea loveni***

    Large, bushy colony, stem polysiphonic; with some second- and

third-order branching; large ova, developing into planulae within gonotheca; no medusa
Hartlaubella gelatinosa

---

14. Sub-hydrothecal spherule absent **Clytia hemisphaerica**

Sub-hydrothecal spherule present (Fig. 4.4) **15**

---

15. Hydrothecal rim even; hydrotheca smooth **Orthopyxis integra**

Hydrothecal rim cusped or undulating; hydrotheca typically with lines running down from rim
Campanularia hincksii

Campanularia hincksii Alder Fig. 4.9
Unbranched, often long pedicels topped by hydrothecae rising from a tortuous stolon; to about 15 mm. Hydrothecae large, bell-shaped; rim castellate, gaps deep, with line from centre of each nearly to base of hydrotheca. Pedicel long; a spherule below hydrotheca; shaft mostly smooth to sinuous, usually ringed basally and sometimes in middle. Male and female gonothecae uniform, on stolon; loosely tubular, sides crimped in a succession of shallow rings; end truncate, sometimes flared; aperture wide, terminal. Dispersive stage a planula.

On hydroids and hard substrata; about 10–200 m, perhaps deeper; fairly common off all British coasts; widespread; north to about Lofoten Islands and Iceland.

Clytia hemisphaerica (Linnaeus) Fig. 4.9
Stems unbranched or slightly branched, up to about 10 mm, arising from creeping stolon. Pedicels long, variously annulated. Hydrothecae large, bell-shaped; rim with 10–14 rounded, symmetrical, triangular cusps. Gonothecae borne on hydrorhiza or, less often, on hydrocaulus, long-ovoid to cylindrical, smooth or corrugated; truncate at top. Medusa released.

On plant and animal substrata, intertidal to deep sublittoral; on all British and Irish coasts, one of our commonest hydroids; nearly cosmopolitan.

Gonothyraea loveni (Allman) Fig. 4.9
Stems erect, arising from creeping stolon; irregularly branched; to about 30 mm. Colourless to pale horn, annulated at base and above origins of branches. Hydrothecae alternate, bell-shaped, margin with about 14 blunt cusps; pedicels usually ringed throughout. Male and female gonothecae similar; axillary, on short pedicels; obconical, truncate distally; containing several vestigial medusae which project from the gonotheca in twos or threes.

On stones and algae, particularly browns; often in pools; tolerant of brackish water; MTL to 200 m; common around British Isles; all coasts of Europe.

Hartlaubella gelatinosa (Pallas) Fig. 4.9
Colony erect, bushy, to about 350 mm; main stems rather thick; polysiphonic. Side-branches numerous, tending to be in pairs, successively on opposite sides of stem (as in *Obelia bidentata*); branches straight, polysiphonic basally, monosiphonic and flexuose distally; with short branchlets at each flexure. Hydrothecae bell-shaped, small; rim castellate with 12–14 blunt cusps, often abraded smooth; pedicels varied in length, ringed throughout or with smooth central portions. Male and female gonothecae similar, axillary, long-ovoid, flat-topped with raised terminal aperture. Dispersive stage a planula.

Intertidal to 15 m or deeper, often in gentle current; tolerant of silt and of brackish water; southern Scotland, Oslo Fjord, and Denmark south to the Mediterranean and Black Sea.

Laomedea angulata Hincks Fig. 4.9
Colony a series of short shoots arising from a rather straight stolon. Erect shoots rigidly zigzag, to about 15 mm; internodes almost straight. Hydrotheca on long, ringed pedicel; thin-walled; rim even. Gonothecae on stolon; long-ovoid; aperture terminal, narrow. Dispersive stage a planula.

Recorded only on eel grasses; ELWS to about 8 m; known today from south coast of England southwards, but in 19th century records ranged north to the Isle of Man; status in Ireland unclear.

Laomedea flexuosa Alder Fig. 4.9
Stems flexuose, monosiphonic, usually unbranched, arising from creeping stolon; to about 35 mm; internodes curving, annulated at base and above origin of each pedicel. Hydrothecae alternate, large, cup-shaped, rims even; on annulated pedicels. Female gonotheca subcylindrical, truncate distally;

male shorter, slightly tapered distally towards narrow truncate aperture; on ringed axillary pedicels. Dispersive stage a planula.

On stones and weeds; mid-shore to about 40 m, occasionally to 90 m; one of the commonest British hydroids; all European coasts.

Obelia bidentata Clarke Fig. 4.10

Colony comprising erect, slightly zigzag, polysiphonic but slender main stems, to about 150 mm; lateral branches tending to be in pairs, successively on opposite sides of stem. Minor branches delicate, with downward curvature in life, pinnate; branchlets alternate, monosiphonic, flexuose, bearing hydrothecal pedicels at nodes. Hydrotheca elongate, bell-shaped; rim with 10–20 cusps, usually bimucronate. Gonotheca long-obconical with slightly raised aperture. Dispersive stage a medusa.

On inert substrata such as wood, shells, wrecks, and on sandy bottoms, sometimes algae; tolerant of brackish water; sublittoral to at least 200 m, rarely intertidal in pools; also stranded. Around British Isles known from the Wash to near Portsmouth; also warmer parts of northern France to The Netherlands; temperate- to warm-water global.

Obelia dichotoma (Linnaeus) Fig. 4.10

Colonies varying from unbranched, through branched, monosiphonic stems up to 50 mm high, to loosely fan-shaped, to elongate with stem polysiphonic basally; these reaching 350 mm in height, perhaps especially in calm habitats. Branches shallowly zigzag. Internodes nearly straight, with many short branches arising at nodes; solitary, monosiphonic stems similar, lacking branches; hydrothecal pedicels arising at nodes, ringed. Hydrotheca wide-obconical to long bell-shaped, often slightly polyhedral in section; rim crenulate to smooth; often with about 15 fine striae running basally from indentations on rim, best seen with side-lighting. Gonotheca obconical; truncate distally, with short terminal collar. Medusae released in summer.

On animal, plant, and inert substrata; intertidal to about 100 m; throughout British Isles; near-cosmopolitan.

Obelia geniculata (Linnaeus) Fig. 4.10

Erect stems to about 50 mm arising from long, rather straight stolon; usually zigzag, sometimes branched; annulated at each flexure. Perisarc thickened asymmetrically. Hydrothecae alternate, cup-shaped, often thickened, sometimes asymmetrically; rim even; on short, annulated pedicels. Gonothecae long, obconical, with aperture atop short terminal collar; axillary, on short pedicels. Displays bioluminescence. Medusa released in summer in British waters.

Usually on algae, especially *Laminaria* and *Fucus*; MTL to sublittoral; common; all British coasts; nearly cosmopolitan.

Obelia longissima (Pallas) Fig. 4.10

Colony typically single-stemmed, exceptionally to about 350 mm, occasionally with main stem forked; monosiphonic except in older colonies when main stem polysiphonic basally. Internodes of both stem and branches long; branches arising singly or in twos or threes. Hydrotheca elongate, obconical to long bell-shaped; rim shallow-castellate, 11–17 cusps but often abraded even; not polygonal in section; pedicels slightly tapered, ringed, arising at nodes. Gonotheca obconical; end truncate, usually with short collar. Medusae released in spring in British waters.

On plant and inert substrata, including rock and sand; intertidal pools to 30 m or more; all British coasts; probably near-cosmopolitan. Fully grown colonies are among the most graceful of all British hydroids.

Orthopyxis integra (Macgillivray) Fig. 4.11

Tortuous, anastomosing stolon with many erect pedicels, up to 10 mm. Pedicel corrugated, with a spherule below hydrotheca. Hydrotheca bell-shaped, rim even; wall often greatly thickened; a shelf projecting inwards at base. Gonotheca borne on stolon; shortly stalked, irregularly ovoid with a wavy profile; top truncate, aperture wide. Planulae in acrocyst when well developed.

On algae, especially reds, and on inert substrata; LWST to offshore; rarely in deep intertidal pools; widespread in British Isles, but seldom recorded around Ireland, in the Irish Sea, or off western Scotland; nearly cosmopolitan.

Rhizocaulus verticillatus (Linnaeus) Fig. 4.11

Stem erect, polysiphonic, irregularly forked; to about 10 cm. Hydrotheca bell-shaped, rim with

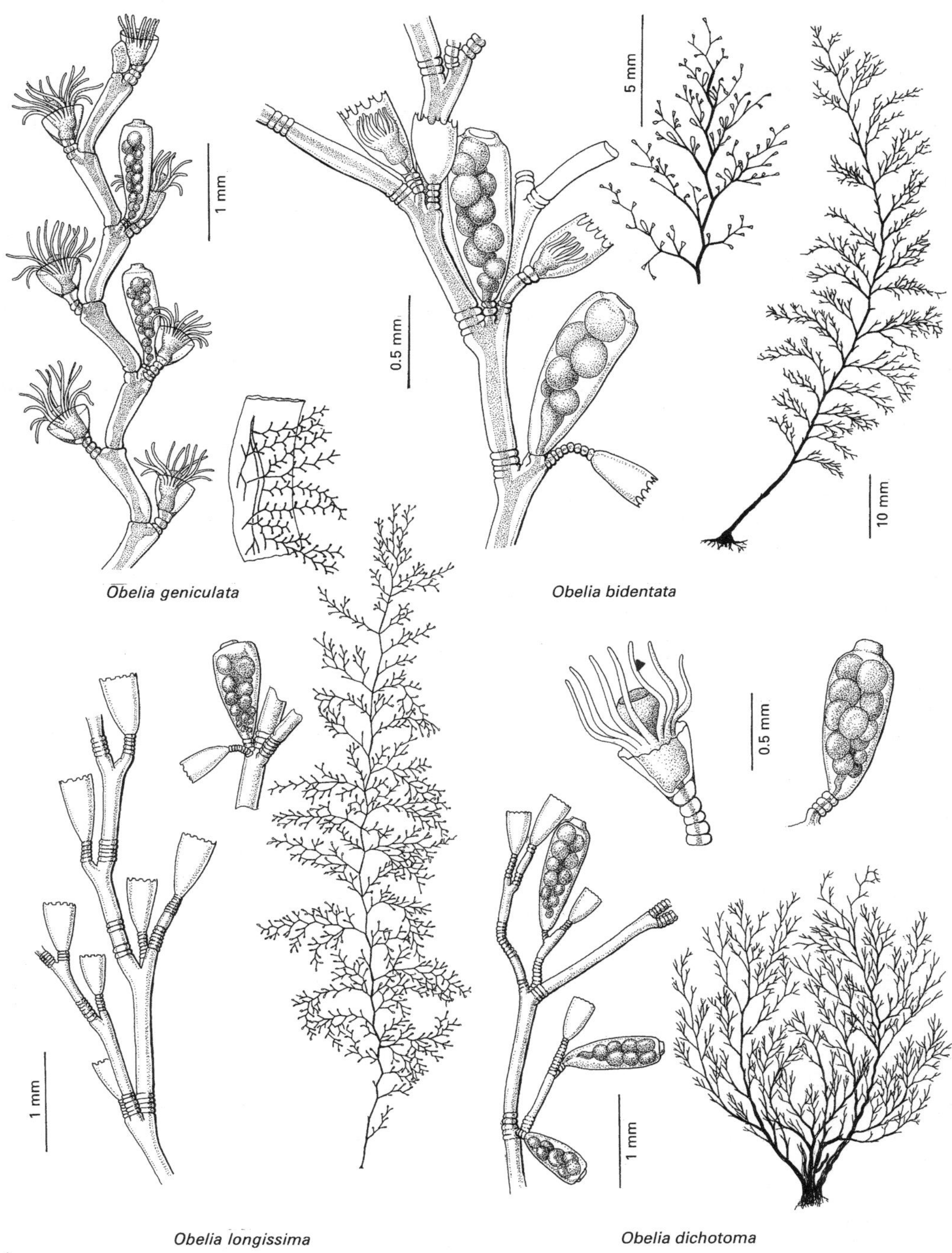
1 mm
0.5 mm
5 mm
10 mm
Obelia geniculata
Obelia bidentata
0.5 mm
1 mm
1 mm
Obelia longissima
Obelia dichotoma

Fig. 4.10

10–16 blunt cusps; on long pedicels, arranged in whorls of 4–6, reminiscent of horsetails (*Equisetum*). Male and female gonothecae alike, flask-shaped with distal neck of varied length, narrowing abruptly to short stalk below; on stem. Dispersive stage a planula.

On shells and stones; 20–600 m; all British coasts, Arctic south to Brittany and perhaps Biscay.

3. PHIALELLIDAE

Colonies erect, short, branched; hydrothecae radially symmetrical, with conical operculum of many triangular flaps not delimited basally; hydranth retracting within hydrotheca, with single whorl of linear tentacles and conical hypostome; gonotheca cylindrical, borne on stem. Medusa released. A single British species.

Phialella quadrata (Forbes) Fig. 4.11
Stems erect, rising from a creeping stolon, to about 5 mm; simple or forked. Perisarc annulated. Hydrotheca obconical, with operculum of long, triangular segments. Gonotheca large, shaped like inverted bottle, usually borne on stolon.

Coastal; on various substrata; common. Medusa recorded off all coasts except eastern and south-eastern England, but distribution of hydroid stage is less well known.

4. LAFOEIDAE

Colony stolonal or erect; hydrotheca bell-shaped to tubular, deep, radially symmetrical, with or without diaphragm, pedicellate or sessile to adnate, rim even, lacking operculum; hydranth contracting wholly within hydrotheca, with conical hypostome and single whorl of filiform tentacles; gonothecae usually aggregated beneath modified protective hydranths in a coppinia; reproduction usually by fixed sporosacs, sometimes by released medusae. Four species in the British fauna.

---

1. Colony mainly stolonal **2**

   Colony mainly erect **3**

---

2. Hydrotheca with occasional rings, tubular, adherent, never with defined pedicel, upward-curving from stolon, diameter up to about 175 μm; stolon tortuous; minute, typically on another hydroid, often *Abietinaria abietina* ***Filellum serpens***

   Hydrotheca not ringed; arising at right angles from stolon, diameter about 350 μm upwards; stolon straighter; larger, on wide variety of substrata ***Lafoea dumosa***

---

3. Hydrothecae tubular, in 4–8 regular and parallel rows, partly adnate and outward-turning distally; sessile ***Grammaria abietina***

   Hydrothecae typically tapered below, in two or more rows, entirely free, usually pedicellate ***Lafoea dumosa***

Filellum serpens (Hassall) Fig. 4.11
Colony a tortuous stolon, from which tubular upward-curving hydrothecae arise to a height of about 1.5 mm. Hydrotheca slightly wider than stolon, basal quarter to one-third adnate; rim even, usually slightly flared; often with renovations; no operculum or diaphragm. Gonothecae protected in hermaphrodite coppiniae. Dispersive stage a planula.

Epizoic, especially on *Abietinaria* spp. and other sertulariid hydroids, shells, and other inert substrata; sublittoral to edge of continental shelf; common throughout British Isles; cosmopolitan.

Grammaria abietina (M. Sars) Fig. 4.11
Colony erect, branching irregular to nearly pinnate; stem and nearly all branches polysiphonic; up to about 100 mm. Hydrotheca long, tubular, merging with stem; outward-curving; rim even, often renovated. Hydrothecae in 4–8 longitudinal rows, alternate in adjacent rows. Gonothecae protected in hermaphrodite coppiniae resembling muffs. Dispersive stage a planula.

On both silty and rocky substrata; 10–250 m; a cold-water species, Shetland south to Northumberland but apparently not the west coast of Scotland southwards; also Denmark north to Arctic Ocean.

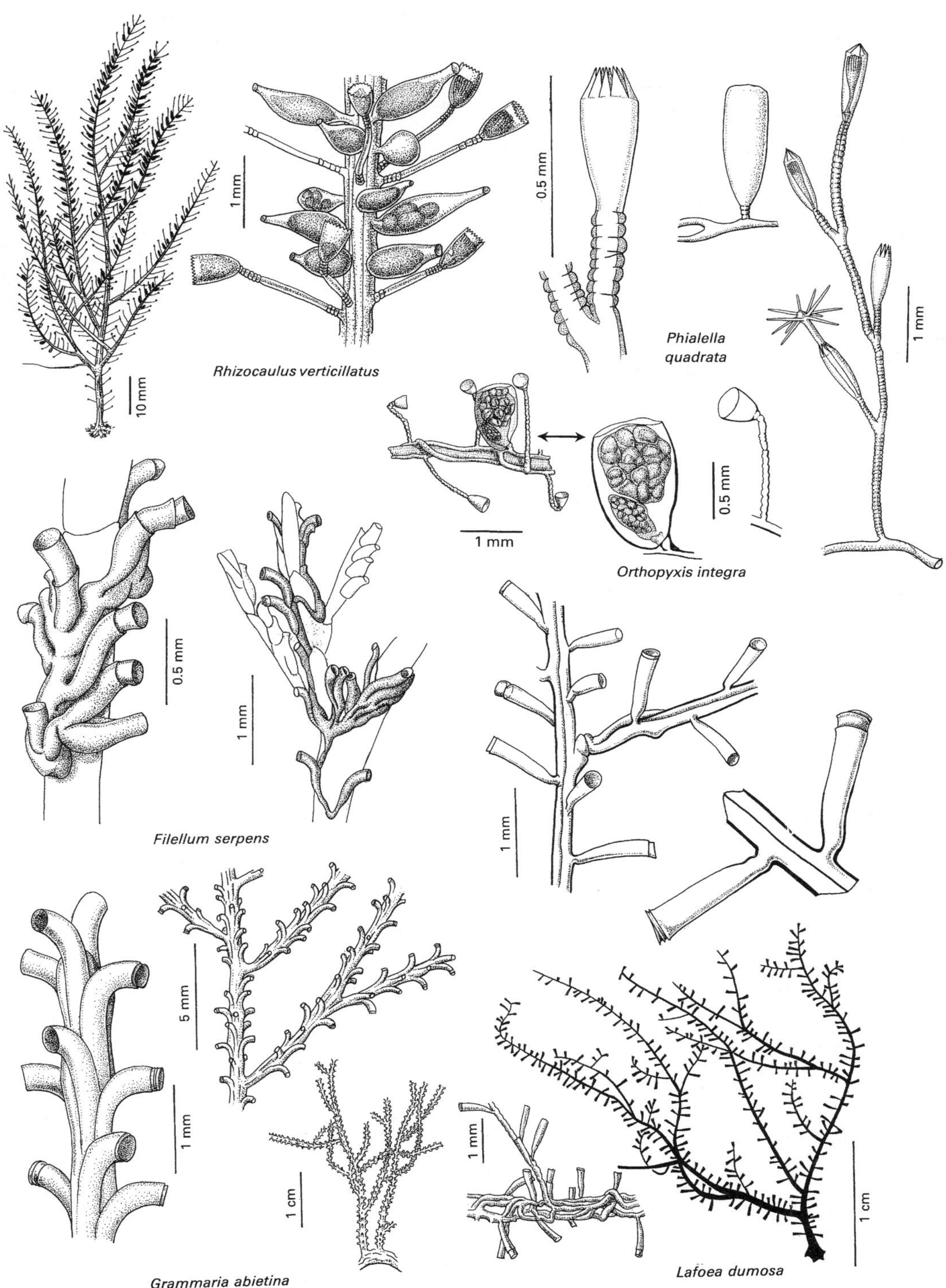
10 mm
1 mm
Rhizocaulus verticillatus
0.5 mm
Phialella
quadrata
1 mm
1 mm
0.5 mm
Orthopyxis integra
0.5 mm
1 mm
Filellum serpens
1 mm
5 mm
1 mm
1 cm
Grammaria abietina
1 mm
1 cm
Lafoea dumosa

Fig. 4.11

Lafoea dumosa (Fleming) Fig. 4.11
Colony either stolonal, or erect with stems irregularly forked and polysiphonic, sometimes both; to about 100 mm. Hydrotheca long, up to about 1 mm, smooth; narrowing slightly to form a pedicel, no diaphragm; rim plain, without operculum, occasionally renovated. Gonothecae borne in somewhat rare coppiniae. Dispersive stage a planula.

On various substrata; intertidal to offshore; all coasts, but scarce in eastern English Channel and southern North Sea; not recorded off Belgium.

5. HALECIIDAE

Colony erect, procumbent or stolonal; hydrotheca usually shorter than wide, too small to accommodate hydranth, cylindrical; rims even, often renovated; lacking operculum; usually no pedicel; hydranths larger than hydrothecae, hypostome conical, with single whorl of filiform tentacles; male and female gonothecae usually dissimilar, often with associated hydranths; reproduction usually by fixed sporosacs, sometimes by eumedusoids and free-swimming medusae. The British fauna includes about 12 species, mostly in the genus *Halecium*.

---

1. Hydrothecal rim flared **2**

   Hydrothecal rim straight **3**

---

2. Colony minute, up to 20 mm, comprising one or a few erect shoots; monosiphonic throughout
   Halecium tenellum

   Colony large, usually over 20 mm, arborescent; polysiphonic in older parts
   Halecium muricatum

---

3. Colony regularly pinnate, in one plane, branches neatly parallel; never stolonal
   Halecium halecinum

   Colony irregular, branches not precisely parallel; or not in one plane; or stolonal
   4

---

4. Colony polysiphonic basally
   Halecium beanii

   Colony monosiphonic throughout
   Halecium lankesteri

Halecium beanii (Johnston) Fig. 4.12
Colony erect, irregularly branched, finer branches delicate; to about 200 mm. Hydrothecae alternate, one below each joint, shortly tubular, widening a little towards aperture, rim plain; often renovated. Gonothecae borne on hydrocladia on short pedicel; male elongate-ovoid, female kidney-shaped, with short tube on concave (axial) side. Dispersive stage a planula.

On other hydroids and shells; 5–100 m depth; common throughout British Isles; nearly cosmopolitan.

Halecium halecinum (Linnaeus). Herring-bone hydroid Fig. 4.12
Colony stiffly erect, regularly pinnate; to 250 mm. Stems and branches polysiphonic basally, tapering and monosiphonic distally. Branches alternate, parallel, regularly spaced. Hydrothecae alternate, often renovated; rim plain. Gonothecae borne on upper sides of branches; male ovoid; female oblong, tapering below, broad and truncate above, with short tubular upward-pointing aperture at one upper corner. Dispersive stage a planula.

On stones and shells; coastal; common off all coasts; widely distributed in Atlantic.

Halecium lankesteri (Bourne) Fig. 4.12
Colony a collection of erect, monosiphonic shoots, limp out of water, up to about 80 mm. Internodes unequal in length, usually a hydrotheca on each. Hydrothecae alternate, short, walls divergent; rims even, not flared; often renovated. Gonothecae on short pedicels; male cylindrical, aperture terminal; female kidney-shaped with aperture in centre of concave side. Dispersive stage a planula.

On stones, other hydroids, crabs, bryozoans, and algae; LWST to about 50 m; a southern species reaching south-west Wales and Norfolk; not reported from Ireland; probably overlooked, may occur more widely.

Halecium muricatum
(Ellis and Solander) Fig. 4.12
Colony erect, up to 200 mm, imperfectly pinnate; main stem and larger branches polysiphonic;

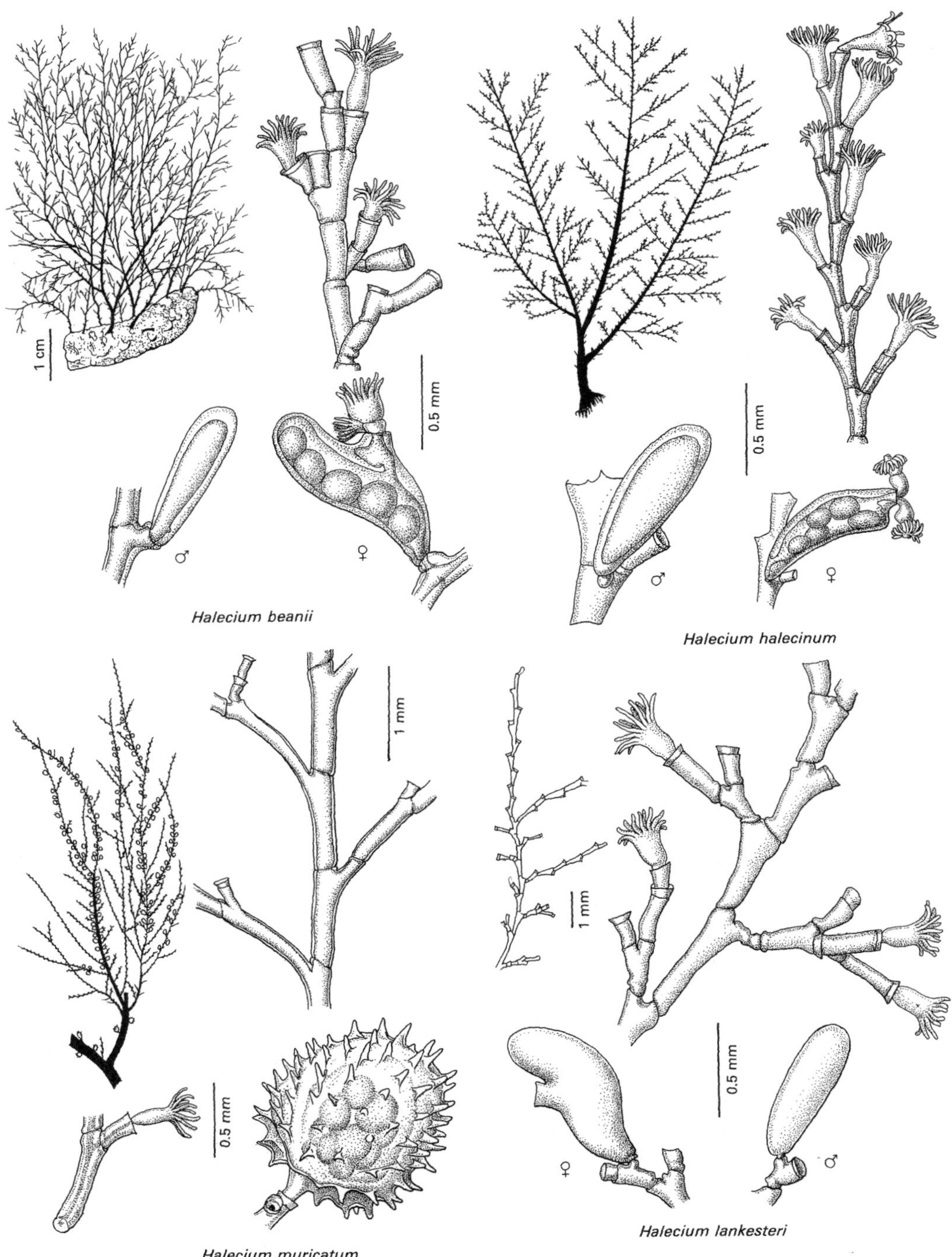
1 cm
0.5 mm
♂
♀
Halecium beanii
0.5 mm
♂
♀
Halecium halecinum
1 mm
1 mm
0.5 mm
Halecium muricatum
0.5 mm
♀
♂
Halecium lankesteri

Fig. 4.12

branches alternate. Internodes equal. Hydrothecae alternate, on final, monosiphonic, branches; short, rims even, usually flared; often renovated. Male and female gonothecae similar, flattened-ovoid with rows of spines. Dispersive stage a planula.

Mostly on rocks and shells; 10–1350 m; Arctic to boreal, reaching south-west Scotland, Isle of Man, Norfolk, central North Sea, Denmark, and west Sweden; no Irish or English Channel records in the 20th century.

Halecium tenellum (Hincks) Fig. 4.13

Small; erect shoots up to 20 mm, delicate, monosiphonic, irregularly branched. Stem zigzag; internodes usually straight, long, and narrow. Hydrothecae one per internode, each on a prominent process; short, rim flared. Male gonotheca ovoid, flattened, aperture narrow, terminal; female similar, proportionately broader. Dispersive stage a planula.

On other hydroids and on bryozoans; LWST to 500 m; nearly cosmopolitan; probably throughout British Isles, but sparsely recorded.

6. SERTULARIIDAE

Colonies erect; hydrothecae in two or more rows, usually sessile to adnate or wholly sunk within perisarc, bilaterally symmetrical, rim often cusped, usually with operculum of 1–4 flat flaps (never conical); hydranth tubular, with conical hypostome and single circlet of linear tentacles, completely retracting within hydrotheca, in some genera folding on retraction to form a side-pocket or *abcauline caecum* (Fig. 4.4); gonophores usually fixed sporosacs, rarely releasing a medusoid. A large family with more than 20 species recorded from the British sea area.

---

1. Hydrothecal rim even or notched, never cusped **2**

   Hydrothecal rim cusped, not notched **11**

---

2. Hydrotheca totally contained within perisarc or nearly so **3**

   Terminal quarter or more of hydrotheca projecting from perisarc **5**

---

3. Branches all round stem ***Thuiaria thuja***

   Branches in one plane **4**

---

4. Branches mostly opposite ***Thuiaria articulata***

   Branches mostly alternate; hydrothecal aperture directed outward (young, pinnate colony prior to acquiring adult growth habit) ***Thuiaria thuja***

---

5. Hydrothecae in two closely set rows, but appearing uniseriate (in one row) ***Hydrallmania falcata***

   Hydrothecae clearly biseriate or tetraseriate (triseriate in occasional specimens of *Diphasia fallax*) **6**

---

6. With axillary hydrothecae (see text for distinctions) ***Abietinaria abietina*** and ***A. filicula***

   Without axillary hydrothecae **7**

---

7. Side-branches narrower than main stem; colony regularly pinnate **8**

   Side-branches same width as main stem; colony irregularly pinnate, or not pinnate at all **10**

---

8. Inner wall of hydrotheca almost or completely adnate ***Diphasia nigra***

   Inner wall of hydrotheca at most three-quarters adnate **9**

---

9. Sides of hydrocladium approximately parallel; inner wall of hydrotheca about three-quarters adnate ***Diphasia alata***

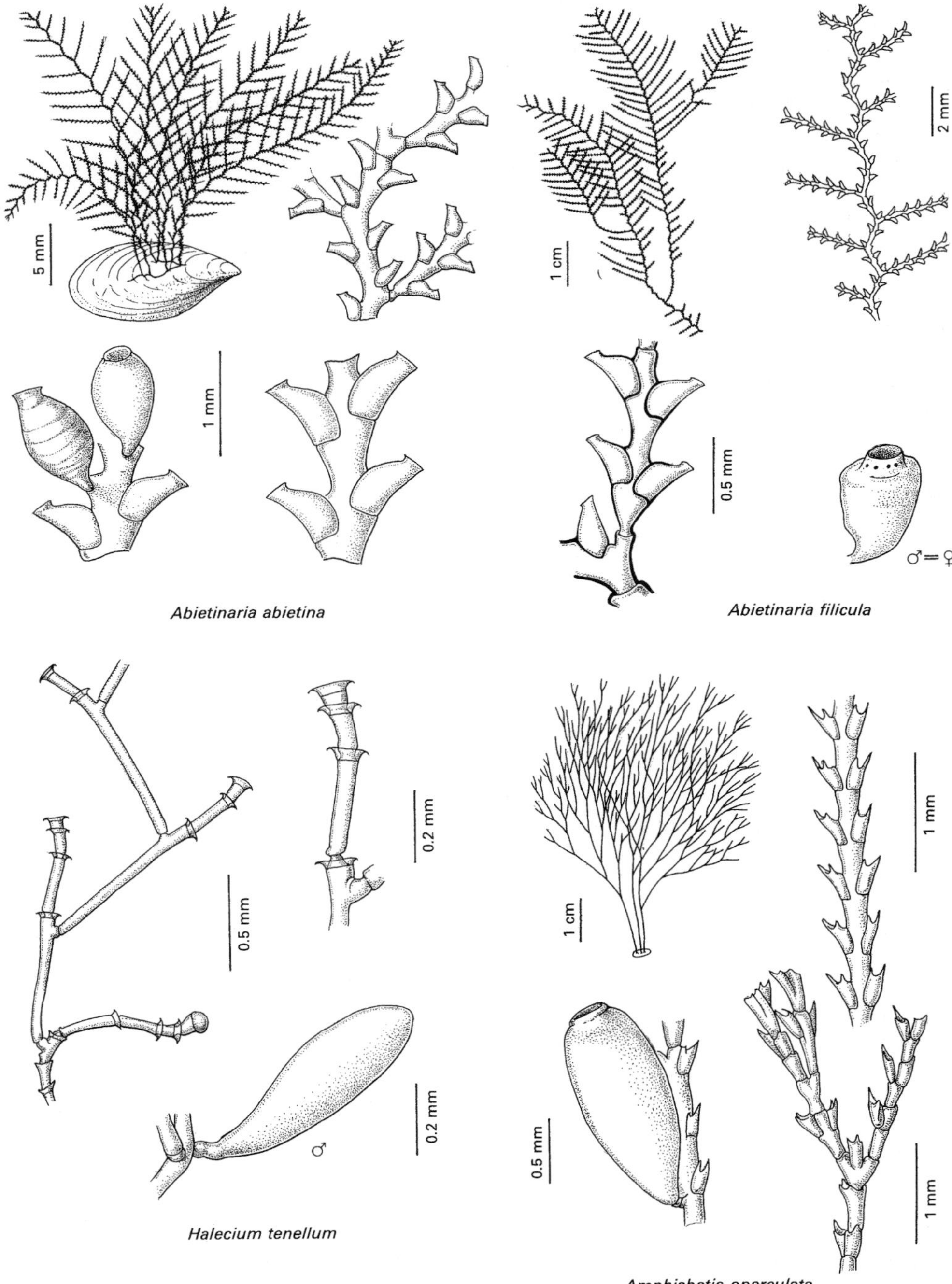
5 mm
1 mm
1 cm
2 mm
0.5 mm
♂=♀
Abietinaria abietina
Abietinaria filicula
0.2 mm
0.5 mm
0.2 mm
♂
1 cm
1 mm
0.5 mm
1 mm
Halecium tenellum
Amphisbetia operculata

Fig. 4.13

Sides of hydrocladium narrowing below each pair of hydrothecae; inner wall of hydrotheca about one-half adnate **Diphasia pinaster**

---

10. Hydrotheca half adnate; flexure abrupt, about 45° (see text for distinctions) **Diphasia attenuata** and ***D. rosacea***

Hydrotheca two-thirds or more adnate; flexure gradual, <45° **Diphasia fallax**

---

11. Two or three cusps on hydrothecal rim, one sometimes minute **12**

Four cusps on hydrothecal rim **17**

---

12. Three hydrothecal cusps **Tamarisca tamarisca**

Two hydrothecal cusps **13**

---

13. Hydrothecae in (sub)opposite pairs (also sometimes *Sertularia cupressina*) **14**

Hydrothecae alternate **16**

---

14. Hydrothecal cusps markedly unequal **Amphisbetia operculata**

Hydrothecal cusps approximately equal **15**

---

15. Nodal annuli all transverse, except sometimes basal-most **Dynamena pumila**

Transverse and oblique nodal annuli approximately alternating **Tridentata distans**

---

16. One (rarely both) cusps on hydrothecal rim long (see also young *Amphisbetia operculata*) **Sertularia cupressina**

Both cusps on hydrothecal rim short (young colonies, or rather common aberrant branches on mature colonies) **Hydrallmania falcata**

---

17. Hydrothecae rugose (for distinctions see text) **Sertularella rugosa** and ***S. tenella***

Hydrothecae smooth **18**

---

18. Three (rarely, one, two, or four) low, rounded projections subdistally on inside of hydrothecal wall; internodal perisarc and hydrothecal wall usually smooth **Sertularella gaudichaudi**

No projections on inside of hydrothecal wall; internodal perisarc and hydrothecal wall usually undulating to rugose **19**

---

19. Stems 50 mm or less, monosiphonic **Sertularella gayi**

Stems to 250 mm, polysiphonic **Sertularella polyzonias**

Abietinaria abietina (Linnaeus) Fig. 4.13
Stems thick, to 300 mm; flexuose, sometimes forked; side-branches alternate, regularly pinnate. Hydrothecae alternate, bulbous below, narrowing above into a short tubular neck; aperture oblique, rim even, shallow notch just below, on inner side (cf. *A. filicula*). Male and female gonothecae similar, shortly stalked, ovoid, with a plain aperture. Embryos yellow, developing in an external acrocyst.

On shells and stones; 10 m to offshore, often found on strandline; all coasts, common; Arctic to Mediterranean.

Abietinaria filicula
(Ellis and Solander) Fig. 4.13
Colony erect, branched, pinnate, smaller than *A. abietina*; up to about 100 mm. Hydrothecae sub-

alternate to alternate, on stems and branches and in axils; flask-shaped, bulbous below; one- to two-thirds adnate; aperture circular, even-rimmed; notch below aperture on inner side deeper than in *A. abietina*. Male and female gonothecae similar, ovoid, aperture terminal, on short collar. Dispersive stage assumed to be a planula.

Substrata not recorded; 5–550 m but mostly shallower than 50 m; formerly recorded all around British Isles but all recent records are from the Isle of Man and Norfolk northwards.

Amphisbetia operculata (Linnaeus) Fig. 4.13
Colony recalling tuft of fine, wavy hair; to 100 mm. Hydrothecae in subopposite pairs; aperture sloping inwards towards stem, its outer angle pointed. Gonotheca ovoid, smooth, with terminal aperture; operculum a simple flap. Reduced medusoids develop within gonotheca.

On brown algae, often *Laminaria digitata* stipes; LWST, often in pools in south-west, and sublittoral; widely distributed, north to Shetlands.

Diphasia alata (Hincks) Fig. 4.14
Colony robust, up to about 150 mm, regularly pinnate with closely set, parallel branches; main stem thicker than branches, usually monosiphonic throughout. Hydrothecae opposite or nearly so; long, S-shaped, two-thirds adnate, distal third sharply out-turned at about 90° with immediate upward flexure of about 45°; aperture broad, rim even to sinuous. Male and female gonothecae alike, obovoid with quadrangular section; aperture terminal, surrounded by four perisarc ridges ending in rounded points. Dispersive stage a planula.

Substrata unknown; 7–140 m; scattered British Isles records: Shetlands to south-west Ireland, also north-west France; mid-Norway to Azores.

Diphasia attenuata (Hincks) Fig. 4.14
Colony erect, to about 50 mm, bending, pinnate with branches spaced apart; monosiphonic throughout; stem and branches similar in width, both sometimes ending in tendrils; axils about 65°. Hydrothecae in opposite pairs, tubular, half to two-thirds adnate, gradually out-turned; aperture roughly circular, but rim with deep notch on inner side. Male gonotheca cylindrical, with six longitudinal ridges ending in points; females hexagonal, with whorls of conical spines distally; aperture terminal. Dispersive stage a planula.

Usually on other hydroids; sublittoral at least to edge of continental shelf; all around British Isles but reportedly commonest in south and west.

Diphasia fallax (Johnston) Fig. 4.14
Colony erect, pinnate, monosiphonic, to about 100 mm; some secondary branching; terminal tendrils frequent. Hydrothecae (sub)opposite; short, tubular, at least three-quarters adnate, slightly out-turned above; aperture circular, rim even. Nodal annulus below each pair of hydrothecae. Gonothecae on branches; male elongate, widest above, with four erect spines surrounding terminal aperture; female similar but longer, four long conical processes joining above aperture to form brood-chamber protecting an acrocyst. Sometimes monoecious. Dispersive stage a planula larva.

Recorded on other sertulariid hydroids; 20–250 m; in British waters today recorded only north of a line passing from the Clyde to Heligoland. Arctic to southern North Sea.

Diphasia nigra (Pallas) Fig. 4.14
Colony large, stiffly erect, to about 200 mm, robust and rigidly pinnate, main stem thicker than branches. Hydrothecae alternate, tubular, wholly adnate; rim even. Gonothecae borne on the branches; male obovoid, aperture terminal, surrounded by four blunt spines; female similar but larger, lacking spines. Dispersive stage a planula.

On mussels and probably other substrata; about 80 m and deeper; a warm-water Atlantic species known from south-west England, north-west France, and western English Channel.

Diphasia pinaster sensu Hincks Fig. 4.15
Colony erect, untidily pinnate, up to about 150 mm; branches long; some secondary branching. Hydrothecae opposite to subopposite; each sharply out-turned in middle, half to three-quarters adnate; aperture oblique, circular, even-rimmed, usually with inner notch. Male gonotheca ovoid, tetrangular, with four spines; female larger, long-ovoid, tetrangular, with row of 1–3 spines on each corner, aperture terminal. Dispersive stage a planula.

75–900 m; Bergen southwards; all around British Isles, reportedly most common in south.

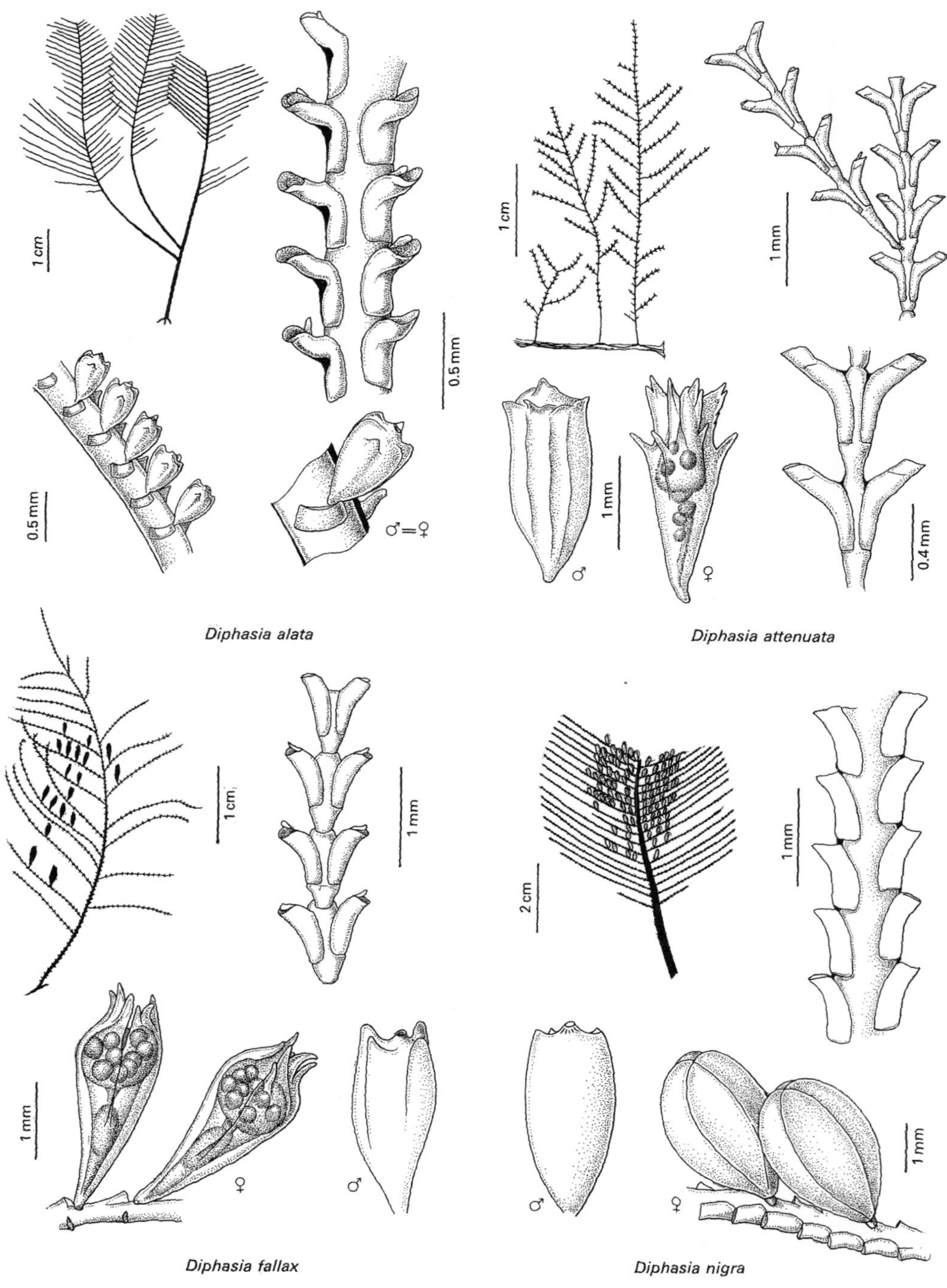
1 cm
0.5 mm
0.5 mm
♂=♀
Diphasia alata
1 cm
1 mm
1 mm
♂
♀
0.4 mm
Diphasia attenuata
1 cm
1 mm
1 mm
♀
♂
Diphasia fallax
2 cm
1 mm
♂
♀
1 mm
Diphasia nigra

Fig. 4.14

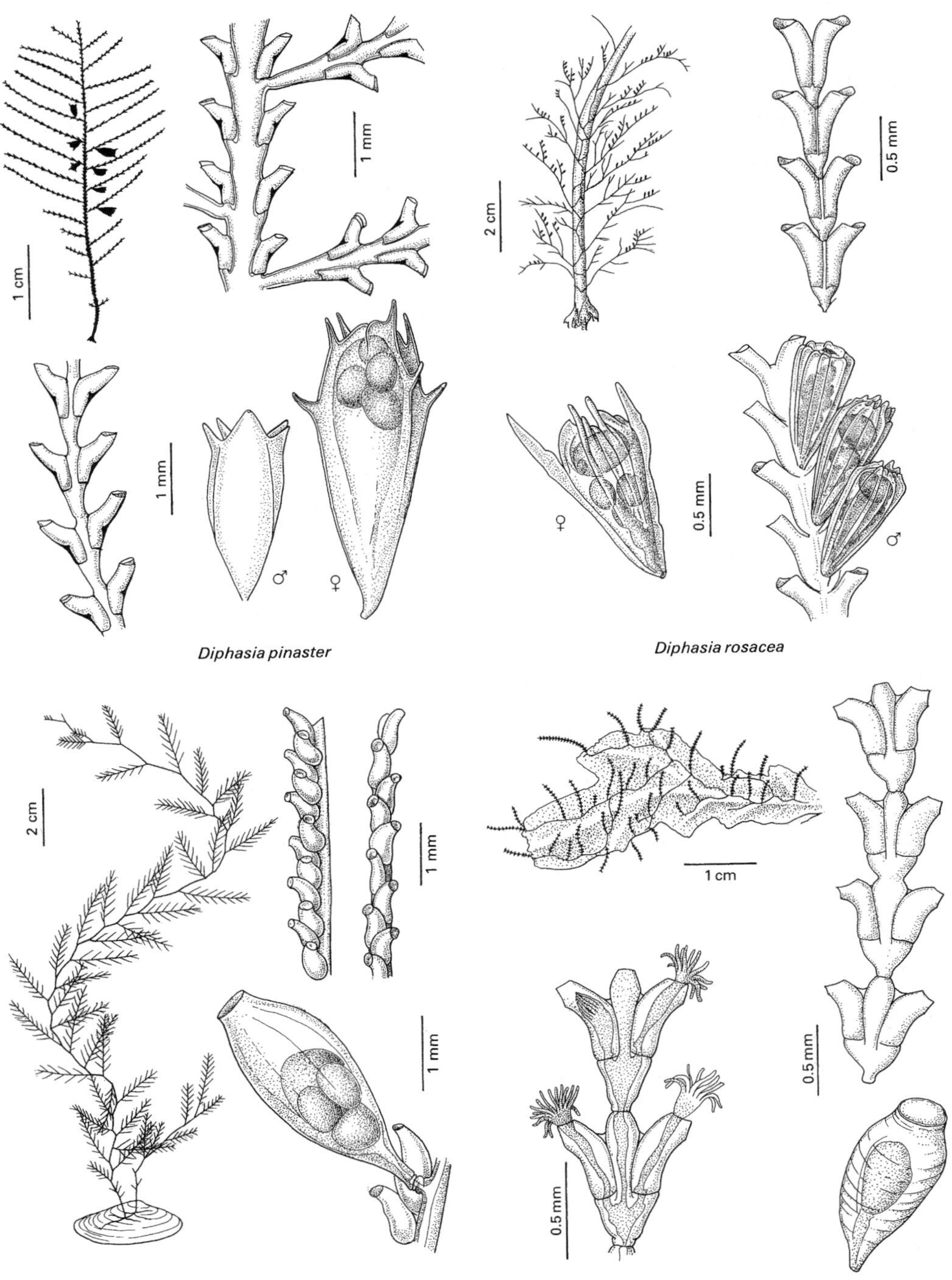
1 cm
1 mm
1 mm
♂
♀
Diphasia pinaster
2 cm
0.5 mm
♀
0.5 mm
♂
Diphasia rosacea
2 cm
1 mm
1 mm
Hydrallmania falcata
1 cm
0.5 mm
0.5 mm
Dynamena pumila

Fig. 4.15

Diphasia rosacea (Linnaeus) Fig. 4.15
Stems gently curving, delicate, irregularly and unequally branched; to about 25 mm. Hydrothecae in pairs, long, tubular, upper portion free and out-turned; aperture oblique, rim even, with slight notch on inner side. Male gonotheca club-shaped, with eight longitudinal ridges which end in spinous points. Female gonotheca pear-shaped, shortly stalked, long, with eight longitudinal ridges terminating in spines. When mature the processes form a distal brood chamber in which planulae develop in an acrocyst.

On hydroids and shells; lower intertidal to about 80 m; common around British Isles; Arctic to South Atlantic.

Dynamena pumila (Linnaeus) Fig. 4.15
Stems short, erect, up to about 30 mm; usually unbranched. Internodes short, hydrothecae in pairs, which in profile form inverted triangles. Hydrotheca short, half adnate, free and out-turned in distal half; aperture two-cusped. Perisarc horn-coloured. Gonotheca irregularly ovoid, aperture wide. Dispersive stage a planula brooded in an acrocyst.

On brown algae and rocks, sheltered to moderately exposed shores; MTL to shallow sublittoral. Common on all rocky coasts around British Isles; Arctic to western France. Tolerant of brackish conditions.

Hydrallmania falcata (Linnaeus) Fig. 4.15
Main stems very tall, in open spiral, to at least 500 mm; monosiphonic; lateral branches regularly pinnate, hydrocladia alternate. Hydrothecae tubular, contiguous in groups of 4–10; in two closely set rows. Male and female gonothecae alike, ovoid, shortly stalked. Dispersive stage a planula.

On shells and stones, particularly in sandy areas; 20–100 m; common off all British coasts; Arctic to south-west Europe.

Sertularella gaudichaudi (Lamouroux) Fig. 4.16
Stem short, erect, unbranched, to about 35 mm. Hydrotheca one per internode, tubular, tapering or flask-shaped; with subterminal constriction, one-third to half adnate; aperture four-cusped; 2–4 low, rounded projections on inside of hydrothecal wall near aperture. Gonothecae ovoid, annulated; aperture terminal. Planula released.

On inert substrata, intertidal and coastal; probably all around British Isles but no Irish records. Taxonomic status unclear.

Sertularella gayi (Lamouroux) Fig. 4.16
Stems erect, loosely pinnate; to 250 mm; branching irregular; stems and larger branches polysiphonic; smaller branches monosiphonic. Hydrothecae distant, alternate, one to each internode; out-turned, walls frequently sinuous proximally, narrowing above to a four-cusped aperture. Tissues often sulphur-yellow. Gonotheca elongate-ovoid, aperture small with two or more cusps. Dispersive planulae develop in an acrocyst.

Sublittoral, possibly sometimes intertidal; widespread in north Atlantic and common off all coasts of British Isles to edge of continental shelf. Possibly conspecific with *S. polyzonias.*

Sertularella polyzonias (Linnaeus) Fig. 4.16
Stems erect, irregularly branched; to 50 mm; main stem and branches monosiphonic. Hydrothecae alternate, one to each internode, out-turned, bulging below, narrowing slightly to four-cusped aperture. Tissues often sulphur-yellow. Male and female gonothecae alike, ovoid, wrinkled; aperture small with four cusps. Dispersive-stage planulae developing in an acrocyst.

On shells and algae; intertidal to 50 m, sparsely to 300 m; all British coasts; widely distributed in North Atlantic. Possibly conspecific with *S. gayi.*

Sertularella rugosa (Linnaeus) Fig. 4.16
Stems arising from basal stolon, to 20 mm, un- or sparingly branched; or hydrothecae attached to stolon. Hydrothecae alternate, barrel-shaped, deeply ridged, with notch near top on outer side; aperture with four minute cusps. Male and female gonothecae alike, large, ovoid, strongly annulated; aperture four-cusped. Dispersive stage a planula.

On bryozoans, particularly *Flustra foliacea*, also algae and other substrata; LWST to about 50 m; all around British Isles, widespread on European coasts.

Sertularella tenella (Alder) Fig. 4.16
Colonies usually comprising upright stems to 20 mm connected by tortuous basal stolons, but hydrothecae sometimes borne directly on stolon. Hydrothecae alternate, quarter adnate; walls with

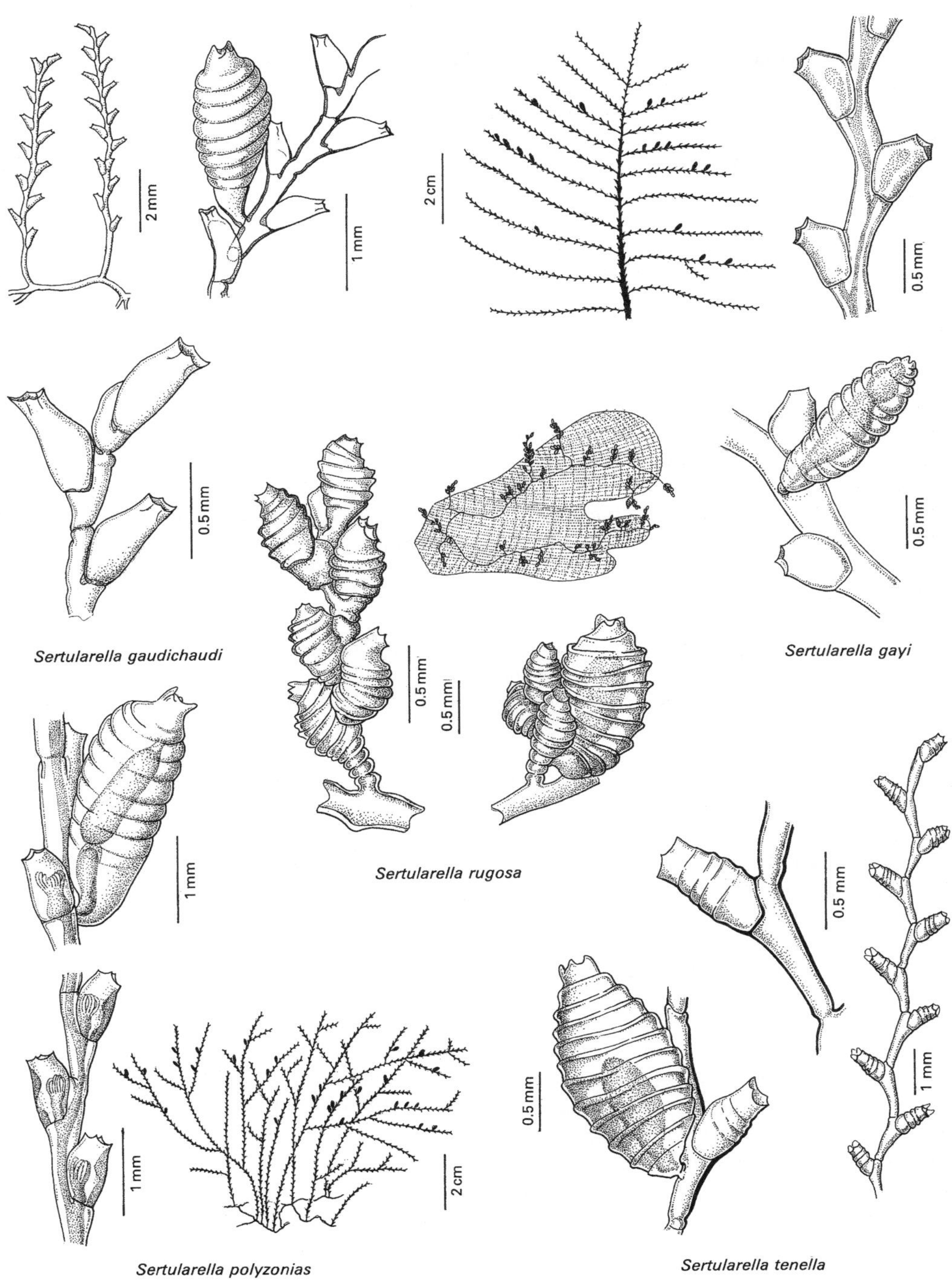

2 mm
1 mm
2 cm
0.5 mm
0.5 mm
Sertularella gaudichaudi
0.5 mm
0.5 mm
0.5 mm
Sertularella gayi
1 mm
Sertularella rugosa
0.5 mm
0.5 mm
1 mm
1 mm
2 cm
Sertularella polyzonias
Sertularella tenella

Fig. 4.16

3–6 shallow ridges; aperture transverse, four-cusped, no notch on outer side. Gonothecae large, ovoid rugose. Dispersive stage a planula.

On other hydroids; 10–150 m, perhaps deeper; reported all around British Isles, widespread in north Atlantic.

Sertularia cupressina Linnaeus.
'Whiteweed' Fig. 4.17

Colony long and bushy, to 500 mm. Young colonies pinnate. Main stems straight, monosiphonic, side-branches alternate but giving impression of spiral arrangement. Hydrothecae alternate, tubular; rim typically with one or two long, sharp cusps. Male and female gonothecae alike, on hydrocladia, short-stalked, elongate, rectangular, tapering below. Planulae developing in an acrocyst. Includes *S. argentea* L.

A common and often abundant sublittoral species, characteristically on sand, sometimes at ELWS and in low-shore pools on flat shores; to at least 100 m; found all around British Isles, widespread in Europe.

Tamarisca tamarisca (Linnaeus) Fig. 4.17

Colony 150 mm or more, straggly; monosiphonic throughout; branches alternate, widely spaced. Hydrothecae on both stem and branches, (sub)opposite; large, up to 1.6 mm long; tubular, half adnate, variously outward-curving. Male gonotheca kite-shaped, flattened, tapering to short pedicel below. Female gonotheca conical, distal end hidden by two large, ragged flaps cut into several fingers. Sometimes monoecious. Dispersive stage a planula.

Substratum apparently unrecorded; 10–250 m; local throughout British Isles; Brittany to Arctic Ocean.

Thuiaria articulata (Pallas) Fig. 4.17

Colony tall, rigid, regularly pinnate, up to 250 mm. Stem flat and wide. Branches usually opposite. Hydrothecae in two rows, apertures pointing successively left and right; sunk within perisarc; aperture directed outwards. Male and female gonothecae alike, in 1–2 rows on branches; slender ovoid; aperture circular, wide; planulae brooded in an acrocyst.

On stones and shells; 18–300 m; frequent north of a line roughly passing through Dublin and London; Arctic, circumpolar, south to Brittany.

Thuiaria thuja (Linnaeus).
'Bottle-brush hydroid' Fig. 4.17

Colony 50–250 mm, in form of a bottle-brush; branches all around stem. Young colony pinnate, abrupt change to bottle-brush arrangement of branches when about 20 mm high. Hydrothecae on branches only, alternate, biseriate (rarely, triseriate); entirely sunk within branch. Male and female gonothecae alike, ovoid to obconical; aperture circular; acrocyst in female. Dispersive stage a planula.

Young, still pinnate, colonies have zigzag stems and alternate branches, whereas similar-sized colonies of *T. articulata* have straight stems and opposite branches.

On shells and similar substrata; 2–800 m; northern, reported south to Clyde and Farne Islands; often cast up on strandline from Yorkshire northwards, occasionally drifting south as far as Norfolk.

Tridentata distans (Lamouroux) Fig. 4.17

Resembles *Dynamena pumila* but delicate, with longer cusps on hydrothecal rim. Hydrothecae in opposite pairs, sharply to gradually out-curved; about half adnate, less in longer hydrothecae; aperture two-cusped, occasionally with minute third cusp on outer margin. Nodes of two kinds, oblique (or 'hinge') and transverse in side view (see Fig. 4.4; *D. pumila* has only transverse nodes). Hydranth with abcauline caecum (cf. *D. pumila*). Gonothecae ovoid; walls smooth; aperture terminal.

On various plant, animal, and inert substrata; intertidal to 60 m; warmer-water species, regular south coast England to north-west Wales, Norfolk, and probably as far as Yorkshire.

7. PLUMULARIIDAE

Colonies erect, often regularly pinnate; hydrothecae in one row, usually too small to accommodate hydranth; hypostome of hydranth conical, tentacles linear, inserted in a single whorl; nematothecae usually present and movable; gonophores nearly always fixed. Fourteen species recorded from Britain.

---

1. Main stem monosiphonic throughout **2**

   Main stem polysiphonic or apparently so, at least at base **6**

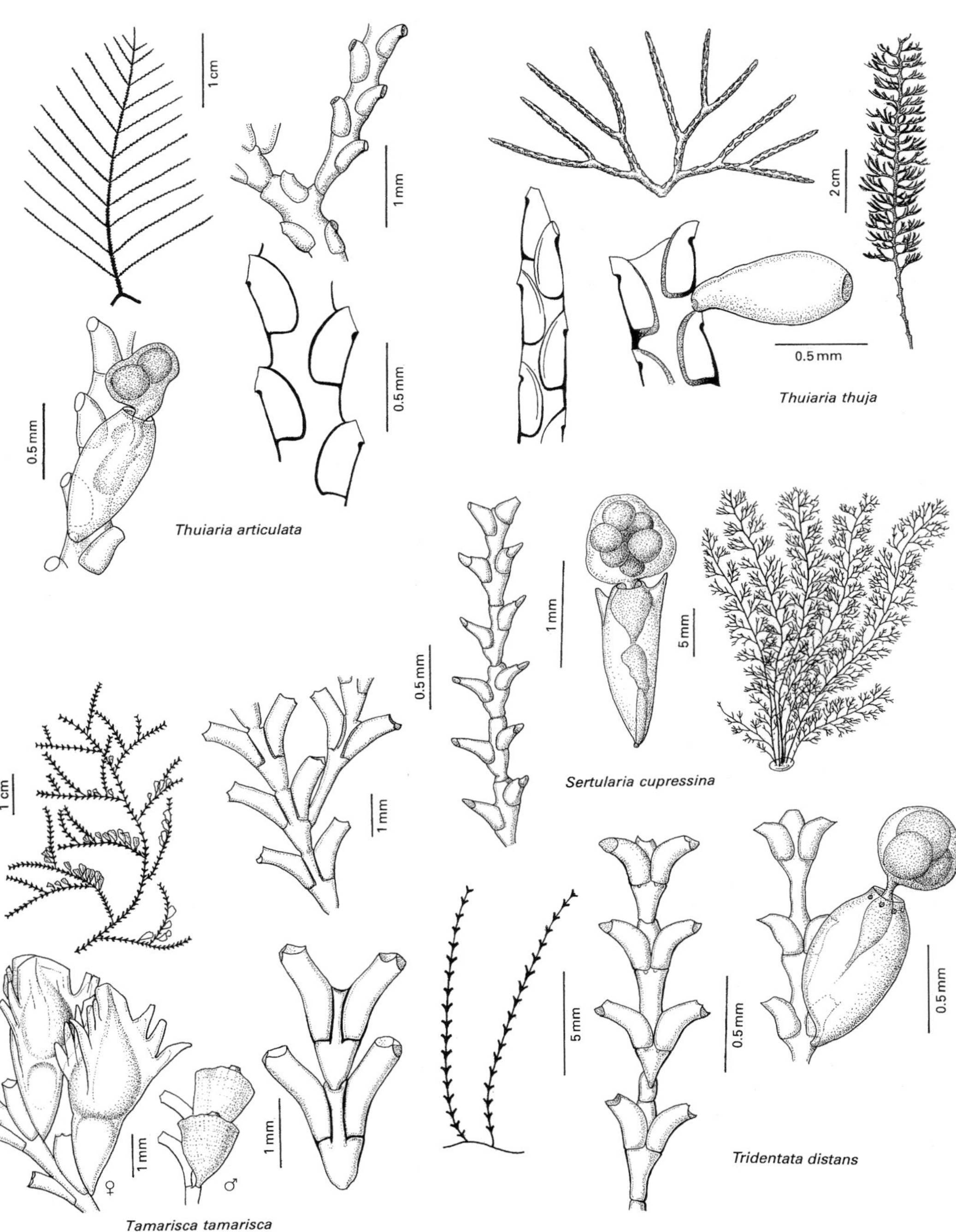
1 cm
1 mm
2 cm
0.5 mm
0.5 mm
0.5 mm
Thuiaria thuja
Thuiaria articulata
1 mm
5 mm
0.5 mm
Sertularia cupressina
1 cm
1 mm
5 mm
0.5 mm
0.5 mm
1 mm
1 mm
♀
♂
Tridentata distans
Tamarisca tamarisca

Fig. 4.17

2. Colony comprising plumes with opposite branchlets **_Halopteris catharina_**

   Colony comprising plumes with alternately arranged branchlets **3**

3. Usually more than one hydrocladium arising per stem internode; one annulus between successive hydrothecae **_Kirchenpaueria pinnata_**

   One hydrocladium arising from each stem internode; two or three annuli between successive hydrothecae **4**

4. Nematothecae absent; one naked nematophore below each hydrotheca, none above a hydrotheca or in the hydrocladial axil **_Kirchenpaueria similis_**

   Nematothecae present, occurring above and below a hydrotheca and in the hydrocladial axil **5**

5. A pair of nematothecae above the hydrotheca, one below it, and one on the short internode; female gonotheca flask-shaped, with a slender neck **_Plumularia setacea_**

   A single nematotheca above the hydrotheca, one below it, none on the short internode; female gonotheca barrel-shaped (young colony, yet to acquire polysiphonic stem) **_Ventromma halecioides_**

6. Mature colony not exceeding 50 mm **_Ventromma halecioides_**

   Mature colony more than 50 mm **7**

7. Hydrocladia arising all round main stem **8**

   Hydrocladia arranged in two rows, occasionally three **9**

8. Main stem forked **_Nemertesia ramosa_**

   Main stem not forked **_Nemertesia antennina_**

9. Colony rigid; main stems almost perfectly straight; forking and sub-branching uniform at about 45° **_Polyplumaria flabellata_**

   At least final branches limp; main stems not perfectly straight, forking and sub-branching dissimilar in angle **_Schizotricha frutescens_**

Halopteris catharina (Johnston) Fig. 4.18
Colony delicate, pinnate; hydrocladia widely spaced, usually opposite; to 40(–100) mm. Hydrothecae on hydrocladia separated by one transverse and one oblique constriction; cup-shaped, one-third to half adnate; hydrothecae on stem also. Male gonotheca ninepin-shaped, female much broader; aperture wider in female; inserted below hydrotheca.

On variety of substrata; in large intertidal pools and sublittoral to at least 150 m; widespread, all British coasts.

Kirchenpaueria pinnata (Linnaeus) Fig. 4.18
Pinnate stems clustered on branched basal stolon; commonly 30(–100) mm; hydrocaulus straight to slightly zigzag; hydrocladia alternate, several to each stem internode, bearing hydrothecae on the stem side; hydrothecae wide and short with plain rim, separated by a single annulus. No nematothecae. Gonothecae pear-shaped, contiguous, in double row, attached to main stem or on stolon; usually with several spinous projections of varied length at the top.

On stones and algae; in pools, MLW to sublittoral; common off all British coasts; Iceland to Mediterranean.

Kirchenpaueria similis (Hincks) Fig. 4.18
Pinnate stems clustered on branched basal stolon; 30–60 mm; hydrocladia alternate, one per stem

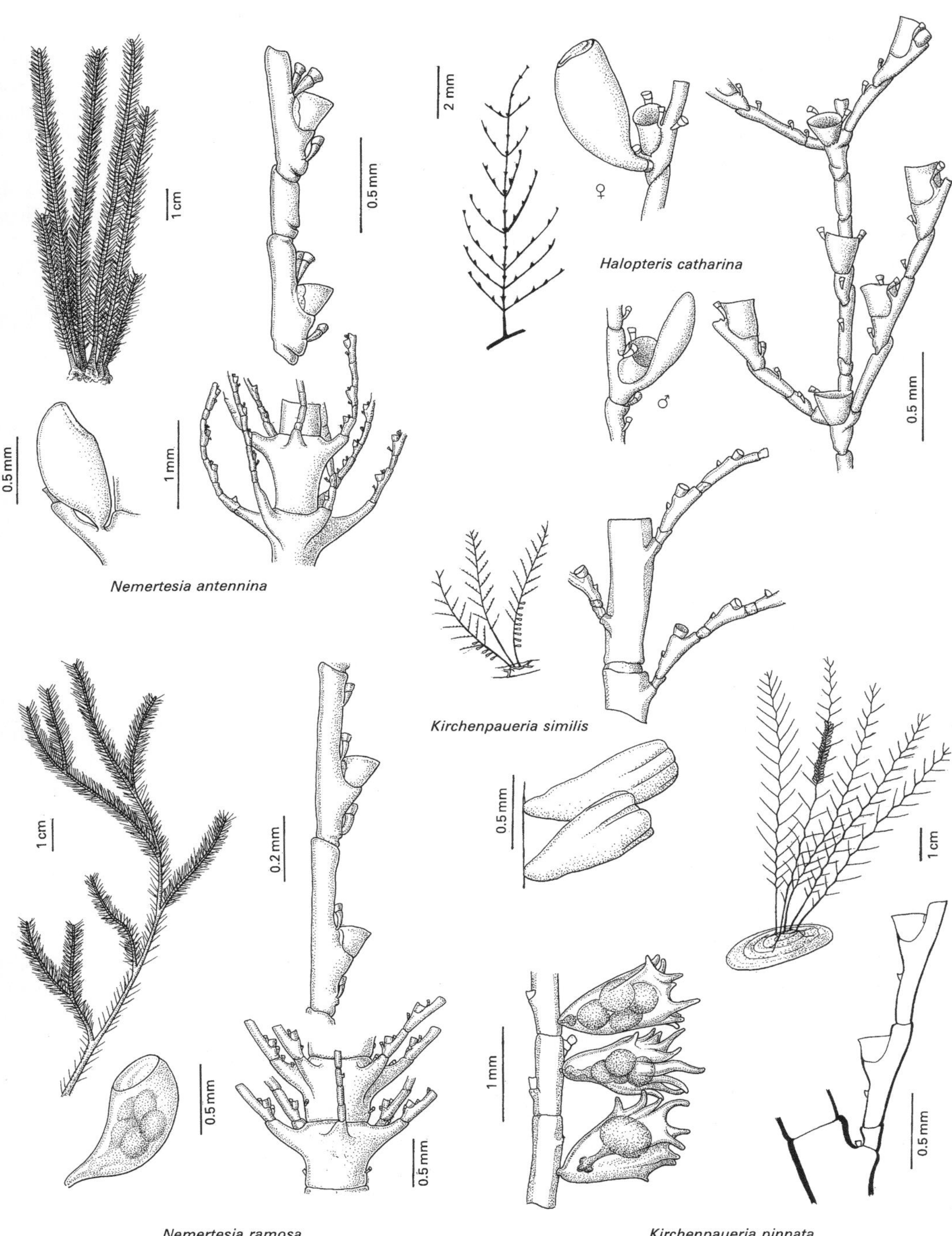
1 cm
0.5 mm
2 mm
♀
Halopteris catharina
♂
0.5 mm
0.5 mm
1 mm
Nemertesia antennina
Kirchenpaueria similis
1 cm
0.2 mm
0.5 mm
1 cm
0.5 mm
0.5 mm
1 mm
0.5 mm
Nemertesia ramosa
Kirchenpaueria pinnata

Fig. 4.18

internode; hydrothecae wide and short, rim even; two annuli between successive hydrothecae. No nematothecae. Gonothecae inverted pear-shaped, separated, in double row attached to main stem or on stolon; said sometimes to have spinous projections.

Little known; recorded widely but similarity to *K. pinnata* makes separation dubious and the two might prove to be conspecific.

Nemertesia antennina (Linnaeus). 'Sea beard' Fig. 4.18

Forms clumps of up to about 50 stems, orange-buff, stiffly erect, up to 250 mm; almost never branched. Hydrocladia in whorls of 6–10, straight to slightly incurved; with alternating short, athecate, and longer, thecate, internodes. Hydrothecae short and wide, rims even. Gonothecae borne singly, in axils of hydrocladia, male and female alike but distinguishable by colour of contents (male whitish, female orange), curved ovoid. Usually one planula per female gonotheca.

On shells and sandy bottoms; inshore to deeper water; common in British waters; in North Atlantic found at least from Iceland to north-west Africa.

Nemertesia ramosa (Lamouroux) Fig. 4.18

Stems yellow-buff, erect, up to 150 mm; irregularly branched and sub-branched. Hydrocladia in whorls of 6–10; jointed, with straight internodes of uniform length, a hydrotheca on each; no athecate internodes. Gonothecae in axils of hydrocladia, curved-ovoid, smooth; aperture facing main stem.

Similar in habitat to *N. antennina*; inshore to deeper water; common throughout British Isles; in North Atlantic from Iceland at least to north-west Africa.

Plumularia setacea (Linnaeus) Fig. 4.19

Colony pinnate, stems 25–40 mm; hydrocladia alternate; branchlets with thecate internodes separated by one athecate internode. Hydrotheca short, rim plain. Monoecious. Gonothecae axillary; female ampullate, male narrow and smaller, lower on stem than female gonothecae.

On algae, pool sides, and rock faces on the shore, and on variety of substrata (especially *Nemertesia ramosa*) sublittorally; common off all British coasts; nearly cosmopolitan, north to Iceland.

Polyplumaria flabellata G. O. Sars Fig. 4.19

Colony regularly pinnate; main stem and branches forking always at about 45°; up to about 350 mm. Hydrotheca short-tubular, tapering slightly basally; length about 1.5 times breadth; rim even, inclined at about 45°. Male and female gonothecae similar, inverted pear-shaped, obliquely truncate above; four small nematothecae near base.

30–800 m; in British Isles recorded Shetland, western Scotland, North Sea, Plymouth, Scillies; also Roscoff; north to Iceland.

Schizotricha frutescens (Ellis and Solander) Fig. 4.19

Colony erect, pinnately branched, up to about 150 mm. Main stem polysiphonic, forked in some specimens; hydrocladia limp, often subequally divided near base; insertion both alternate and opposite. Hydrothecae with annulus between every one to three; tubular, entirely adnate; aperture broad, rim even or dipping down slightly near stem. Male and female gonothecae similar; inverted pear-shaped, truncate above; two small basal nematothecae.

Usually on rocks and stones, but reported once on a sponge; 30–1000 m; known throughout British Isles but few records from eastern English Channel and southern North Sea; Iceland to Biscay.

Ventromma halecioides (Alder) Fig. 4.19

Colony a group of small, pinnate shoots up to about 50 mm, sometimes longer in still waters. Hydrocladia alternate, rather short, bearing 3–7 hydrothecae, terminal one directed outwards. Hydrothecae with 2–3 annuli between them; short, walls divergent; rim even; entirely adnate. Male gonotheca cylindrical, developing late. Female gonotheca barrel-shaped, walls with 9–12 grooves; attached near base of lower hydrocladia, sometimes on stolon.

Often on other hydroids, also algae, eel grasses, and stones; intertidal pools and lagoons, and LWST to shallow sublittoral, probably 20 m maximum; scattered around British Isles; nearly cosmopolitan in boreal to tropical waters.

8. AGLAOPHENIIDAE

Colonies erect; feather-shaped, at least in side-shoots; hydrothecae usually completely adnate, often having an adnate nematotheca in midline; all nematothecae immovable; hydranth where known uniform, small,

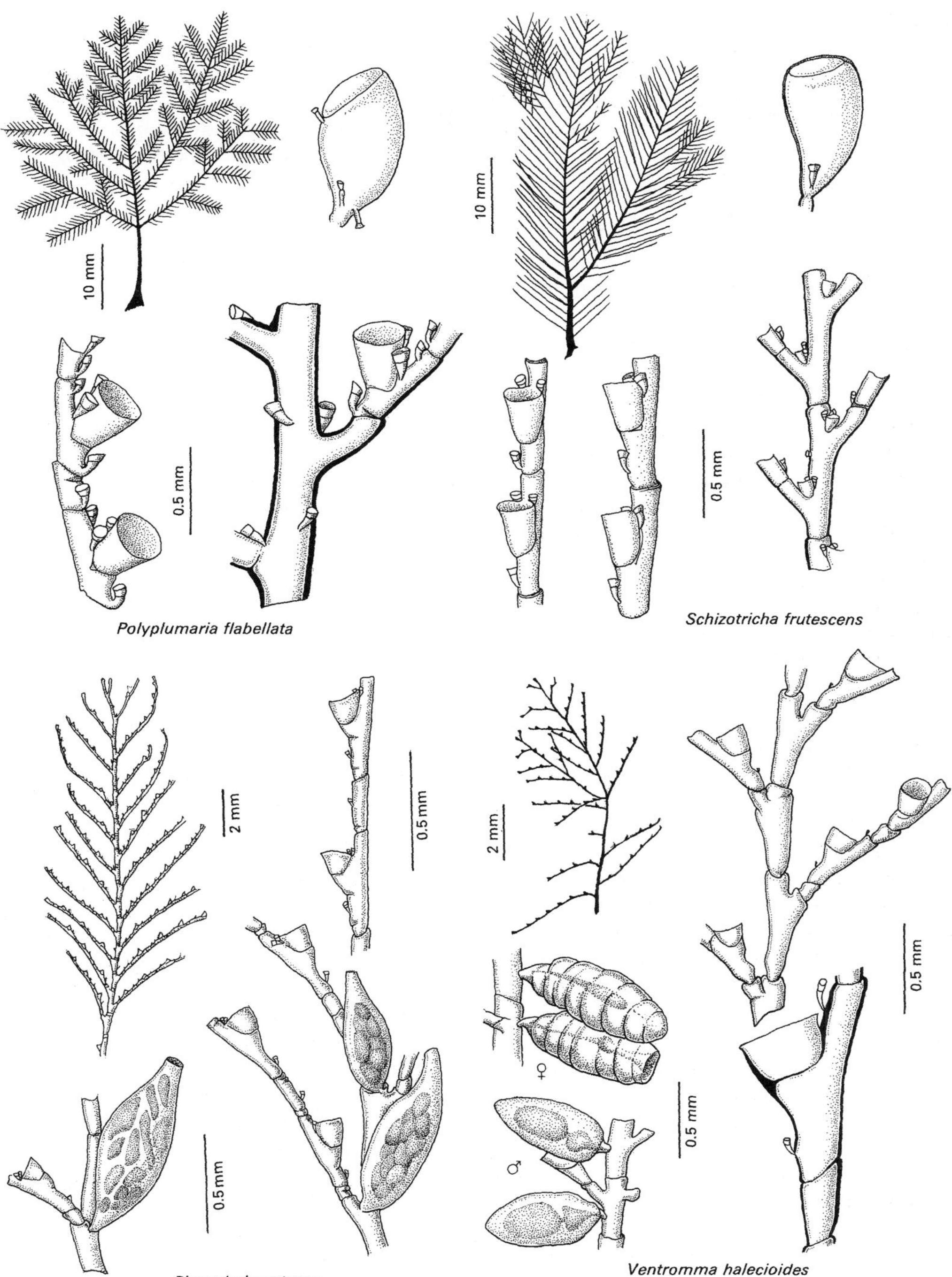
10 mm
0.5 mm
Polyplumaria flabellata
10 mm
0.5 mm
Schizotricha frutescens
2 mm
0.5mm
0.5mm
Plumularia setacea
2 mm
♀
♂
0.5 mm
0.5 mm
Ventromma halecioides

Fig. 4.19

with conical hypostome and 10 tentacles, withdrawing totally into hydrotheca; gonothecae unprotected, or protected in loose to closed baskets formed from modified branchlets, called phylactocarps and corbulae; no medusae in European species. Seven species recorded around the British Isles.

---

1. Main stem polysiphonic; both series of hydrocladia arising from same side of stem **_Lytocarpia myriophyllum_**

   Main stem monosiphonic; hydrocladia in more or less opposed series **2**

---

2. Median nematotheca on outer wall of hydrotheca long, pointed, free for most of length, projecting well above hydrothecal rim; gonothecae borne on stem, not enclosed **_Gymnangium montagui_**

   Median nematotheca adnate for much of length, not pointed; projecting at most only slightly above hydrothecal rim; gonangia borne in hydrocladial 'pods' (corbulae) **3**

---

3. Median nematotheca free for about half of length and reaching nearly to hydrothecal rim, tubular **_Aglaophenia tubulifera_**

   Median nematotheca much shorter, adnate most of length, never reaching hydrothecal rim, usually gutter-shaped **_Aglaophenia pluma_**

Aglaophenia pluma (Linnaeus) Fig. 4.20

Colony comprising erect plumes, to about 80 mm, joined by branched stolon; main stem monosiphonic. Hydrocladia slightly curving, alternate. Hydrothecae not deep; rim with nine cusps of varied length. Median nematotheca not reaching hydrothecal rim. Male corbula with gaps between the 5–8 ribs, female with ribs fused.

Often on *Halidrys siliquosa* but also reported on rock, gravel, *Sargassum*, and *Laminaria*; LWS and deep intertidal pools to about 20 m, locally abundant; from western Scotland southwards; east to Isle of Wight, no reports from east coast of Britain in past 100 years; south at least to South Africa, possibly cosmopolitan. For other British and European species see Svoboda and Cornelius (1991).

Aglaophenia tubulifera (Hincks) Fig. 4.20

Colony comprising erect, monosiphonic plumes joined by an infrequently branched stolon; to about 60 mm. Hydrocladia straight to slightly curving, alternate, absent basally. Hydrothecal rim nine-cusped; median (outer) cusp longest. Up to 12 corbulae per erect plume; corbula with 8–13 ribs.

On rocks, boulders, and pebbles; from 10 m to at least 80 m; Irish Sea north to western Scotland; south to Cape Verde Is.

Gymnangium montagui (Billard) Fig. 4.20

Colony of tall, erect plumes, to about 150 mm; stem monosiphonic. Branches in two rows but both directed to some extent forwards; alternate, closely set, slightly curved, parallel. Hydrotheca small, cup-shaped, margin with one cusp on each side; a long, hollow spine (nematotheca) on outer side, curving upward, rising well above level of aperture. Gonothecae borne on stem; not enclosed in corbula.

On algae, shells, and rock; coastal; in north-east Atlantic recorded from Morocco to Galway and Isle of Arran.

Lytocarpia myriophyllum (Linnaeus) Fig. 4.20

Stems erect, solitary; up to 300 mm. Stem polysiphonic, forked in some colonies; hydrocladia alternate, close together, in two rows on same face of stem. Hydrotheca large, cylindrical; rim wavy. Gonothecae in pairs on modified hydrocladia, protected by long, unfused ribs which form a loose, open basket (phylactocarp).

Sublittoral on sandy bottoms; widely distributed in British Isles; northern Norway to Mediterranean.

Order Siphonophora

Floating or swimming hydrozoan colonies in which medusae are not released. The polyps (zooids) exhibit a high degree of polymorphism. Zooids are budded from a stem (siphosome) derived from the primary polyp. The only genus included, *Physalia*, belongs to the suborder **Cystonectae** in

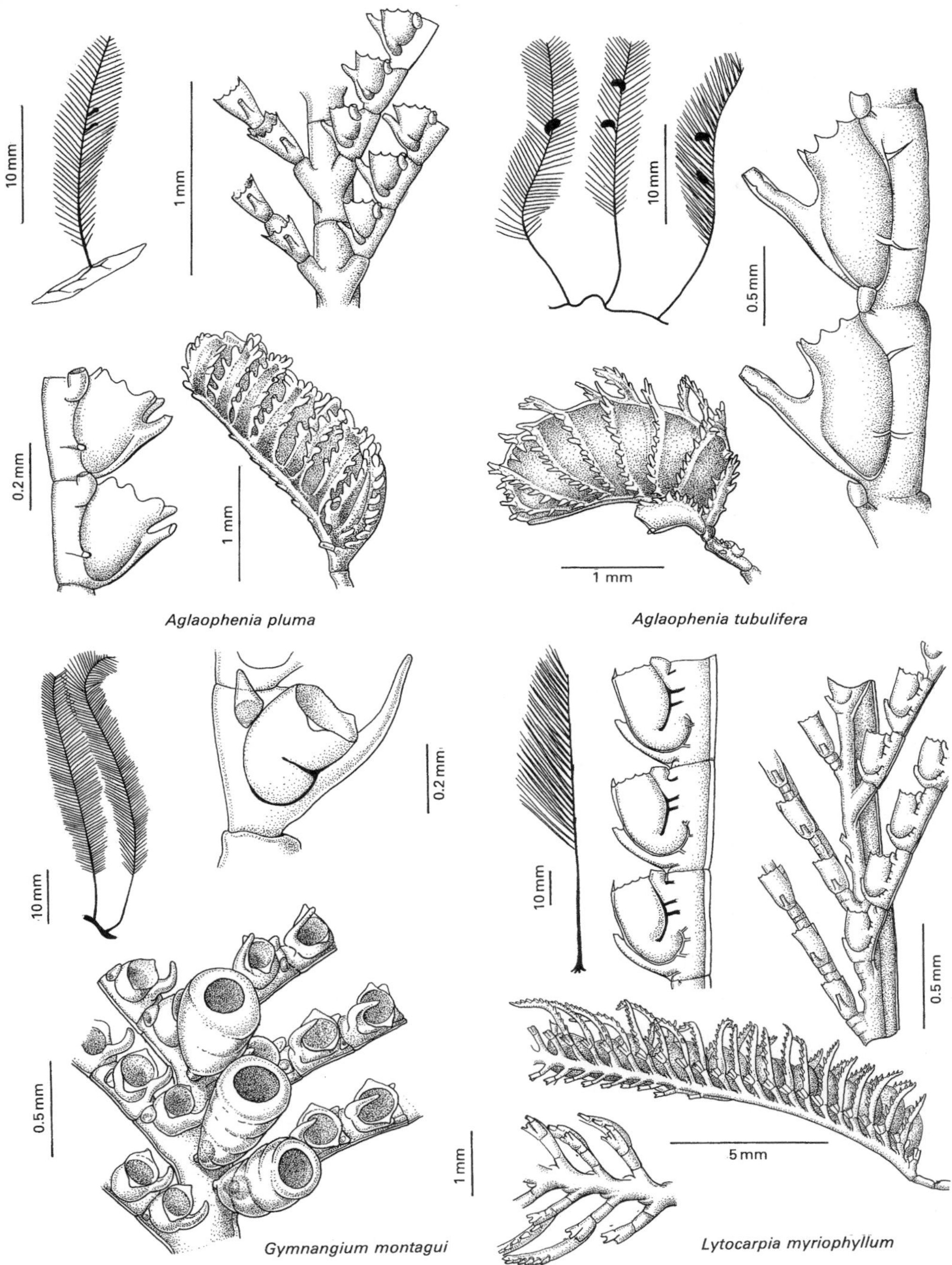
10 mm
1 mm
0.2 mm
1 mm
Aglaophenia pluma
10 mm
0.5 mm
1 mm
Aglaophenia tubulifera
10 mm
0.2 mm
0.5 mm
1 mm
Gymnangium montagui
10 mm
0.5 mm
5 mm
Lytocarpia myriophyllum

Fig. 4.20

which the zooids are crowded below a large float (pneumatophore).

1. PHYSALIIDAE

Physalia physalis (Linnaeus).
Portuguese man o'war Fig. 4.21

Pneumatophore elongate, thin-walled, crested, up to about 300 mm in length; blue, suffused pink when living. Various types of polyps hang below the float. In the sea the tentacles trail behind the float, often for tens of metres. Viewed against the light the tentacles appear beaded, the 'beads' being batteries of nematocysts, extremely virulent to humans.

May appear stranded on south-westerly beaches during summer following prolonged southerly or south-westerly winds; intermittent. The tentacles should not be touched.

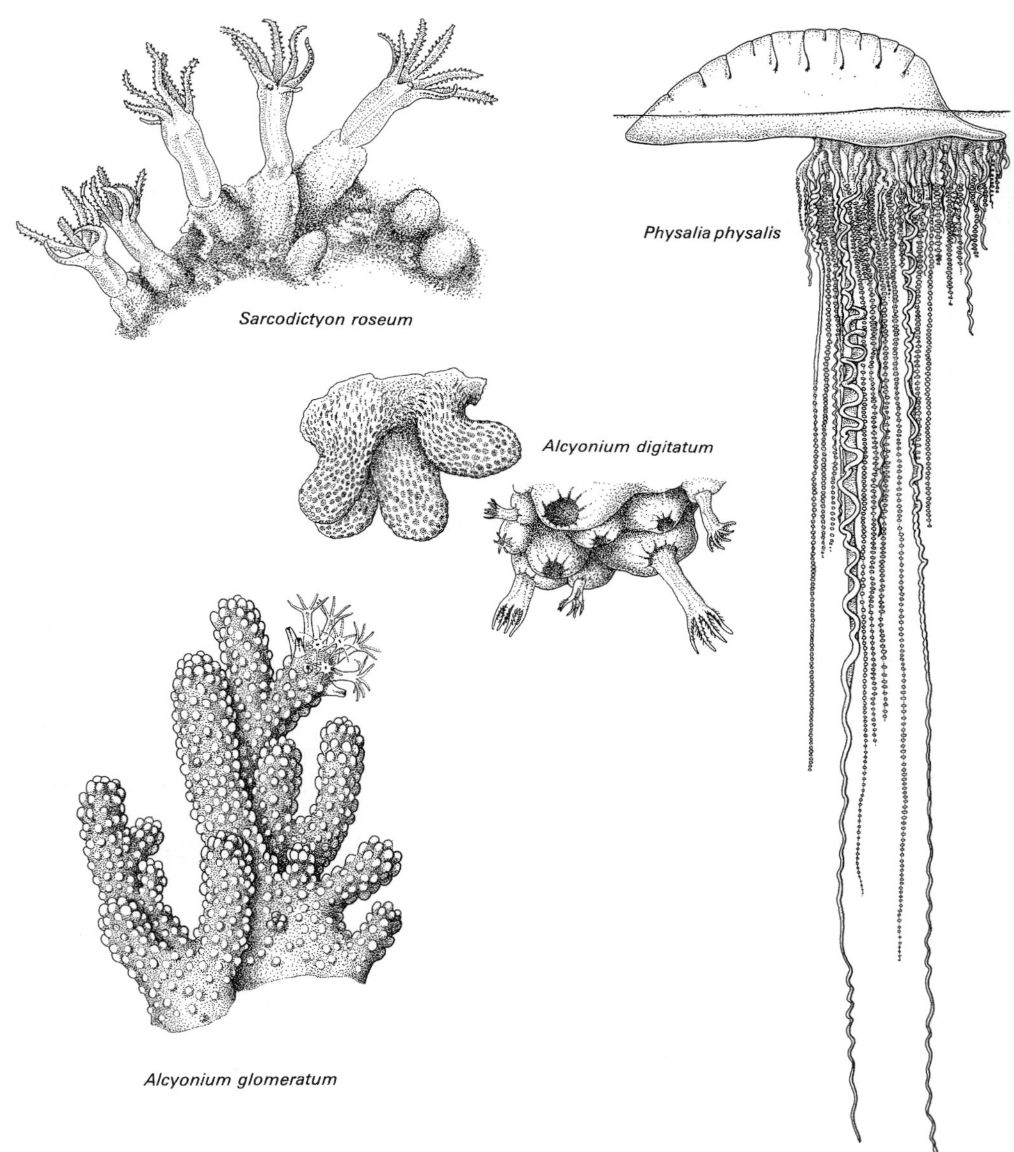

Fig. 4.21

Class Anthozoa

Polypoid coelenterates with no medusoid stage (Fig. 4.1). Anthozoan polyps consist of a cylindrical body or *column* closed at the distal end by the transverse *oral disc*. The hollow tentacles arise from the disc, usually being arranged in concentric rings or *cycles*. At the centre of the disc is the mouth which leads to the throat or *actinopharynx*, a laterally compressed tube, with one or both angles modified as ciliated gutters (*siphonoglyphs*), projecting into the body cavity (*coelenteron*). This cavity is divided into alcoves by radially arranged *septa* (mesenteries), which are attached to the internal surfaces of the disc, body wall, and base. Some septa reach the actinopharynx and are termed *perfect*; those that do not are *imperfect*. Septa commonly arise in pairs of equal size. The free, inner edge of a septum is formed into a convoluted *septal filament*. The base of a polyp may take the form of an adherent disc, as in most anemones, or it may be confluent with a common basal mass of tissue, *coenenchyme*, as in many colonial forms.

Anthozoan polyps are very contractile, and most of them can also retract their tentacles and oral disc into the body when disturbed. They are thus capable of changing their size and shape. Except in the colonial octocorals, where the form of the colony is more important than that of the polyps, the keys and descriptions apply to 'open' polyps with expanded tentacles. Measurements also apply to specimens in this condition but are only a rough guide to relative size: differences of 50% either way are quite possible. Tentacle length is a useful recognition feature and is expressed in relation to the disc diameter of a healthy, well-expanded specimen: *long* tentacles are longer than disc diameter; *moderate* tentacles are half to equalling disc diameter; and *short* tentacles are less than half diameter.

Colour is, unfortunately, rarely an accurate guide to identification. Many anemones are so variable in this respect that it is best to ignore this aspect of their appearance and concentrate on other features when trying to identify them.

KEY TO SUBCLASSES

1. Always colonial; polyps with eight pinnate tentacles arranged in a single cycle (Fig. 4.1); most colonies contain calcareous particles (*sclerites* or spicules) and some have a rod-like axial skeleton **Octocorallia**

   Solitary or colonial; polyps with at least 12 tentacles which are arranged in at least two cycles and never pinnate; sclerites absent but stony corals have a limestone skeleton **Hexacorallia**

SUBCLASS OCTOCORALLIA (ALCYONARIA)

Colonial; polyps with eight pinnate tentacles arranged in a single cycle. Polyps comprise retractile distal *anthocodium* and proximal, more skeletalized *anthostele* into which the anthocodium retracts. A short, wart-like anthostele is termed a *calyx*. One siphonoglyph. Most octocorals contain internal calcareous *sclerites* or *spicules*; some have a rod-like skeletal *axis*. Gastrovascular cavities linked by endoderm-lined channels (*solenia*).

KEY TO ORDERS AND FAMILIES

1. Colony free, living upright with its base in mud or sand; overall shape a spike or feather-like, consisting of a very long, modified axial polyp forming the stem and supporting the secondary polyps (sea pens: PENNATULACEA) **2**

   Colony attached to a substratum; polyps arising singly **4**

2. Colony a spike. Polyps not in fused lateral series (leaves) but clustered around the stem **5. Funiculinidae**

   Colony pinnate (even if long and thin); polyps in fused lateral series (leaves) **3**

3. Colony red; leaves long **6. Pennatulidae**

   Colony whitish or yellowish; leaves short **7. Virgulariidae**

4. Colony tree-like, with branching axial skeleton (sea fans: GORGONACEA) **5**

   Colony massive or encrusting, not tree-like, no axial skeleton **6**

5. Colony small, to 150 mm tall, with few branches; always white; spicules of polyp in eight arrowheads **3. Paramuriceidae**

   Colony usually large, to 300 × 400 mm, with many branches; typically pink but occasionally white; no spicules in retractile portion of polyp **4. Plexauridae**

6. Polyps arising from a creeping stolon of coenenchyme (STOLONIFERA) **1. Clavulariidae**

   Polyps arising from a massive, fleshy, often lobed mass of coenenchyme (soft corals: ALCYONACEA) **2. Alcyoniidae**

Order Stolonifera

Cylindrical polyps arising from a ribbon-like stolon or sheet spreading over solid surfaces. Tentacular portion of polyp retractile within a stiffened, proximal portion (anthostele).

1. CLAVULARIIDAE

Skeleton of usually separate spicules; a cuticle on the stolon.

Sarcodictyon roseum (Philippi) Fig. 4.21
Stolon forming a narrow creeping band, often branched; usually red, sometimes yellowish or colourless. Polyps up to 10 mm; white, arising singly at close intervals.

On rocks or shells; LWST to offshore; common on all coasts, but easily overlooked; south to Mediterranean.

Order Alcyonacea

Soft corals. Colonies encrusting or lobate. Coenenchyme filled with sclerites (spicules). Basal portion of colony often lacking polyps.

2. ALCYONIIDAE

Colony fleshy, with spindle-shaped sclerites.

Three British species. *Parerythropodium coralloides* (Pallas) occurs rarely on western coasts.

1. Thick, stubby lobes; polyps similar in colour to rest of colony ***Alcyonium digitatum***

   Longer, slender lobes; white polyps contrast with pale

orange-yellow to deep red colony mass
Alcyonium glomeratum

Alcyonium digitatum Linnaeus. Dead men's fingers Fig. 4.21
Colonies massive; usually in form of thick, fleshy, finger-like lobes up to about 200 mm; young colonies form thin encrustation. Translucent polyps evenly distributed over colony surface. Colonies typically white or dull orange, also yellowish or even brownish.

Attached to rocks, shells, and stones; LWST to about 50 m; common on all coasts; Iceland to western Europe.

Alcyonium glomeratum (Hassall) Fig. 4.21
General colony shape as in *A. digitatum* but the lobes slender, not more than 20 mm wide, often branched, and long (to 300 mm). Colour from deep red to pale orange-yellow, with contrasting white polyps.

In gullies or caves sheltered from strong wave action; below about 10–15 m; south and west coasts as far north as Scotland; south to Biscay.

Order Gorgonacea

Sea fans. Arborescent colony attached (usually) to a solid substratum by a basal plate. Peripheral axial epithelium secretes a skeletal axis of gorgonin (horny proteinaceous material) with or without non-spicular calcareous matter, or of fused sclerites surrounded by gorgonin. Thin coenenchyme (rind) contains spicules and, sometimes, strands of gorgonin.

3. PARAMURICEIDAE

Anthocodial sclerites arrayed in chevrons; neck between anthocodium and anthostele more or less devoid of spicules; polyp retracts into protruding anthostele. One British species.

Swiftia rosea Madsen Fig. 4.22
A small sea fan up to about 150 mm; always white or pale greyish in colour. Eight chevrons of spicules on the anthocodium.

On steep slopes or cliff faces below about 10 m; widespread but local, west Scotland and Ireland; also Biscay and Mediterranean.

4. PLEXAURIDAE

Sclerites include warty spindles and short clubs. Polyps either completely rectractile, or anthocodia withdraw into anthosteles. A single British species.

Eunicella verrucosa (Pallas) Fig. 4.22
Colony erect, fan-like, branching profusely; up to 300 mm. Usually pink but white or yellowish specimens occur occasionally. Anthocodia without chevrons of sclerites.

Attached to rocks at depths from about 10 m (less in Channel Isles); south and west coasts, locally common; to northern Ireland, (?) west of Scotland; south to north-west Africa and the Mediterranean.

Order Pennatulacea

Sea pens. Only octocoral order adapted for life in soft substrata. Secondary, dimorphic polyps arise from the rachis of the elongate, primary axial polyp, which proximally is dilated as a terminal bulb or peduncle . Gastrovascular cavity of primary polyp lacks mesenteries but contains the horny, calcified skeletal axis. There are two suborders: Sessiliflorae, in which the polyps are borne on the rachis itself; and Subsessiliflorae, in which polyps are fused in pinnately arranged leaves.

Suborder Sessiliflorae

5. FUNICULINIDAE

Elongate sea pens. Polyps laterally and ventrally on the rachis. Axis quadrangular. One species recorded for Britain.

Funiculina quadrangularis (Pallas) Fig. 4.22
Large sea pen, can exceed 2 m. Autozooids irregularly arranged on rachis or in short oblique rows. Axis white, diagnostically quadrangular in section.

In the muddy bottom of sea lochs and elsewhere; 20 m to deep water; west and north coasts of Ireland and Scotland; north Atlantic, perhaps world-wide.

Suborder Subsessiliflorae

6. PENNATULIDAE

Pinnate sea pens. Leaves bilaterally symmetrical along rachis; each leaf containing a single row of fused anthocodia, the outermost being the oldest. One species in the British sea area.

Pennatula phosphorea Linnaeus Fig. 4.22
Axial polyp rather stout and fleshy, divided into bulbous peduncle, anchoring the colony, and distal, polyp-bearing rachis; to about 250 mm. Polyps elongated and graduated in size, fused together in bunches to form triangular leaves in two opposing lateral rows on rachis. Colour deep red.

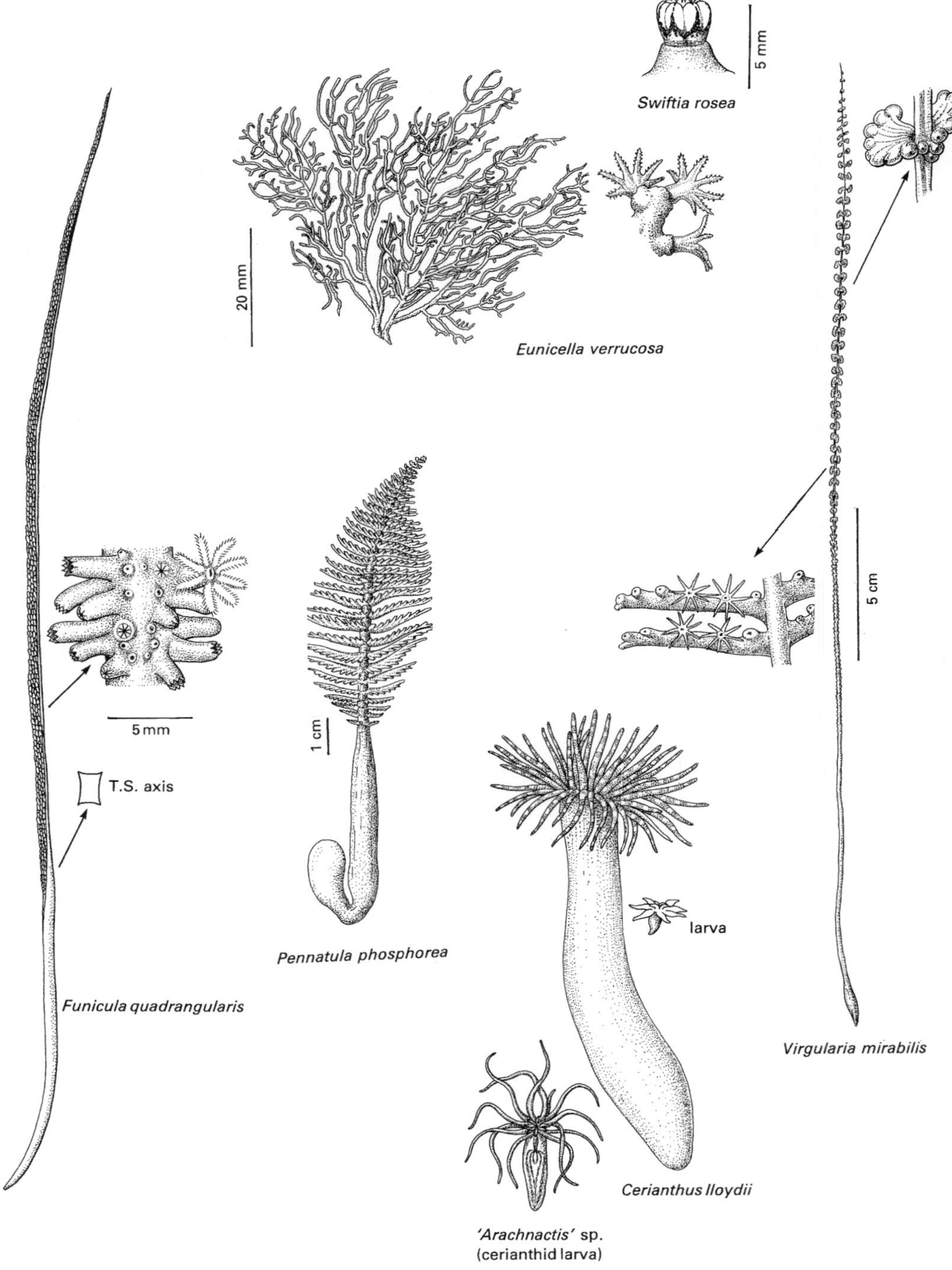

5 mm
Swiftia rosea
20 mm
Eunicella verrucosa
5 cm
5mm
T.S. axis
1 cm
Pennatula phosphorea
Funicula quadrangularis
larva
Virgularia mirabilis
Cerianthus lloydii
'Arachnactis' sp.
(cerianthid larva)

Fig. 4.22

Living erect in mud or sand below about 15 m; local, on all coasts except south; widespread in North Atlantic.

7. VIRGULARIIDAE

Slender pinnate sea pens. Leaves small, bilaterally symmetrical along rachis. Axis rounded in section. A single British species.

Virgularia mirabilis (Müller) Fig. 4.22

Axial polyp very long and slender, up to 0.6 m. Polyps in small clusters of 3–8 in lateral rows on rachis; spicules absent; colour off-white to creamy yellow.

Standing erect in soft mud into which the whole colony can withdraw when disturbed; in sheltered localities, i.e. harbours and sea lochs, below about 10 m, or in deeper open water; locally common on all coasts; North Atlantic and Mediterranean.

Subclass Hexacorallia (Zoantharia)

Solitary, aggregated, or colonial anthozoans. Some (Scleractinia) secrete calcareous exoskeleton. Cylindrical polyps usually with septa in multiples of six, disposed in *pairs* (complementary adjacent septa) and *couples* (mirror-image septa on opposite sides of the axis of symmetry). The space between members of a pair is the *endocoel*, that between pairs the *exocoel*. The *directive septa* are at opposing ends of the actinopharynx. Longitudinal muscles typically endocoelic (i.e. facing one another). Often with an annular *sphincter muscle* at the top of the column which can close the polyp with the disk and tentacles withdrawn. Endodermal zooxanthellae sometimes present. Hexacorals found around the British Isles are classified into five orders: Actiniaria (anemones), Corallimorpharia (jewel anemone), Scleractinia (corals), Zoanthidea (zoanthids), and Ceriantharia (tube anemones).

KEY TO ORDERS

1. Polyp solitary, living in soft tube buried in bottom sediment; tentacles of two types: long, marginal in 3–4 cycles; and short, labial from top of actinopharynx **Ceriantharia 8. Cerianthidae**

   Polyps solitary or colonial, living attached to rocks or buried in soft sediments but never inhabiting a tube; tentacles all of one type, arising from the oral disk **2**

2. Polyps with calcareous skeleton fixed to substratum **Scleractinia** (p. 118)

   Polyps with no calcareous skeleton **3**

3. Polyps colonial, arising from an encrusting basal mass of coenenchyme (not always easily observed); tentacles in two cycles (easily observed) **Zoanthidea** (p. 120)

   Polyps not colonial (not permanently joined at bases) but sometimes living in dense aggregations **4**

4. Tentacles not arranged in distinct cycles but in radial rows, each tentacle with a prominent spherical knob on the tip; polyps typically gregarious **Corallimorpharia 9. Corallimorphidae**

Tentacles usually arranged in three or more distinct cycles, not knobbed (except in rare species not included here); polyps gregarious or solitary
Actiniaria (p. 120)

Order Ceriantharia

Tube anemones. Solitary anthozoans, with an elongate column, that live in a felt-like tube of mucus, cnidae, and mud, buried to the level of the oral disc in soft sediment. Single siphonoglyph. Aboral end contains a pore (no basal disc). Tentacles of two kinds: short labial tentacles encircling mouth; long marginal tentacles around periphery of disc. Arrangement of septa distinctive, coupled but not in pairs, lacking retractor muscles. Nematocysts include a distinctive type known as ptychocysts. Pelagic arachnactis larvae develop tentacles and septa before settling.

8. CERIANTHIDAE

Characters as for order. A single species around Britain.

Cerianthus lloydii Gosse Fig. 4.22
Column with rounded basal end; up to 150 mm. Disc markedly concave, with short labial tentacles. Marginal tentacles long and slender, to about 70; up to 50 mm in length.

In soft felt-like tubes in mud, sand, or gravel. The cerianthid is not attached to a tube and can retract very rapidly when disturbed. Common on all British coasts where suitable substrata occur; LWST to offshore; Greenland and Spitzbergen to (?) Mediterranean.

Order Corallimorpharia

A small order of anemone-like anthozoans, most easily distinguished by the radial arrangement of tentacles which are often knobbed (capitate).

9. CORALLIMORPHIDAE

Tentacles simple, each provided with an acrosphere (terminal knob). Two to eight tentacles in radial lines arising from each endocoel, alternating with a single marginal tentacle arising from the exocoel. One species recorded for Britain.

Corynactis viridis Allman.
Jewel anemone Fig. 4.23
Column usually short and squat, usually overlapped by oral disc. Tentacles short to moderate, arranged in radial rows increasing in size towards the margin; each with a spherical knob (acrosphere) at its tip. Disc about 10 mm diameter; span of tentacles to 25 mm. Colours varied, with disc, tentacles, and acrospheres often in contrasting colours.

Uncommon in shaded places near LWST; common in depths below 5 m on south and west coasts, often occurring in dense aggregations beneath rocky overhangs, in caves, and similar places; northern Scotland to south-west Europe and the Mediterranean.

Order Scleractinia

Stony corals. Solitary or colonial. Polyp morphology similar to actinians but siphonoglyphs lacking. Most important characteristic is the secretion of a calcareous (aragonitic) exoskeleton (*corallum*). Radial sclerosepta are deposited as ridges between the septa of the polyp, and the living tissues of the polyp can be retracted between the sclerosepta.

Only solitary corals are found intertidally or in shallow waters around the British Isles.

---

1. Tentacles yellow, without distinct terminal knobs; disc without a contrasting pattern ***Balanophyllia regia***

   Tentacles rarely yellow, with small but distinct terminal knobs; disc usually with a contrasting pattern
   Caryophyllia smithii

---

10. DENDROPHYLLIIDAE

Corallum with porous walls. Free margin of sclerosepta granular or dentate. Three species recorded for Britain.

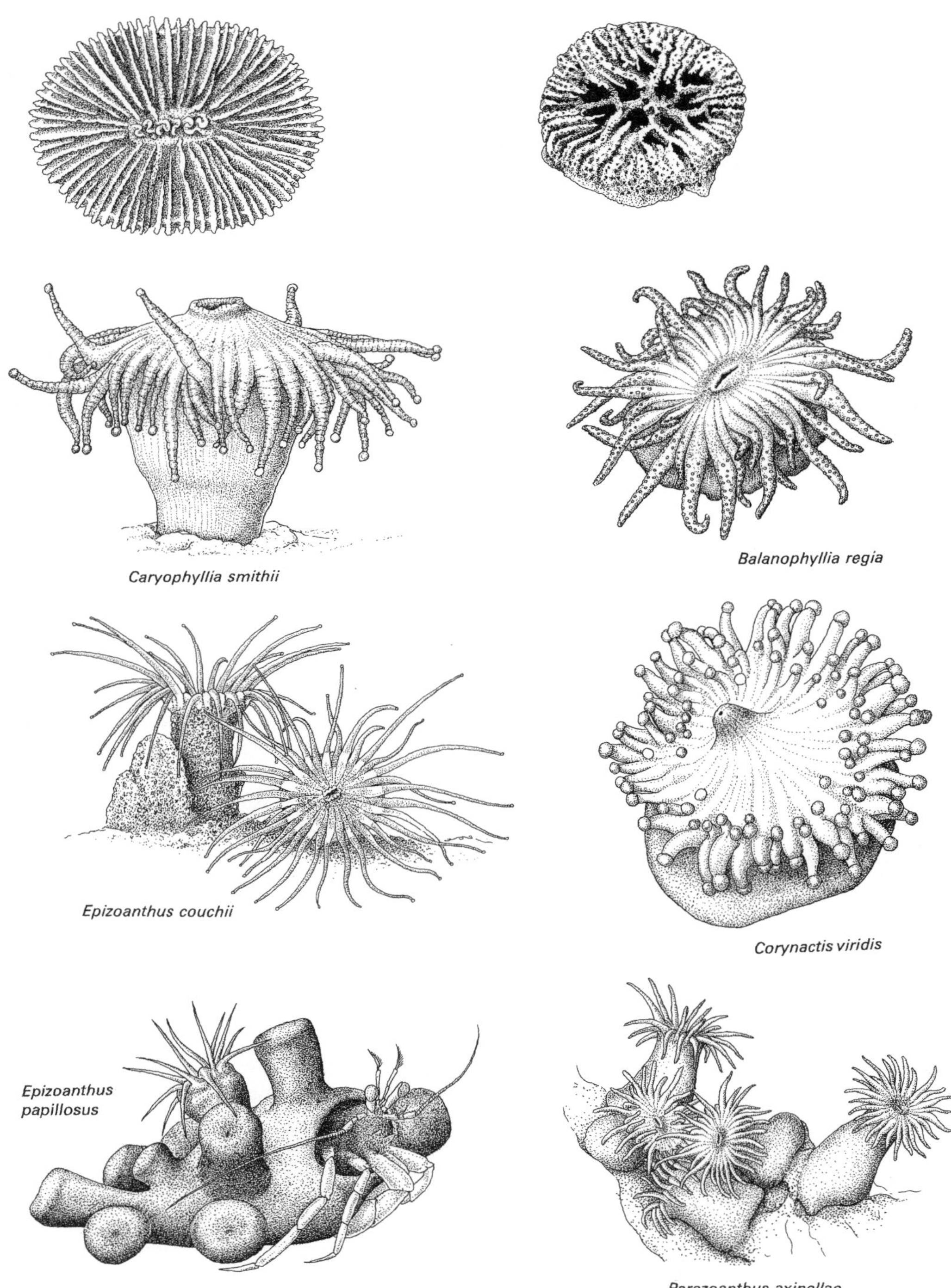
Caryophyllia smithii
Balanophyllia regia
Epizoanthus couchii
Corynactis viridis
Epizoanthus papillosus
Parazoanthus axinellae

Fig. 4.23

Balanophyllia regia Gosse Fig. 4.23

Polyp with up to about 48 rather short tentacles; span about 25 mm. Corallum broad and low, texture porous, spongy, and brittle. Overall colour rich translucent golden yellow, often shading to orange or scarlet around the mouth.

Attached to rocks in caves or surge-gullies around LWST to 25 m; south-west England and Pembroke; southward to Morocco, Canary Isles, and Mediterranean.

11. CARYOPHYLLIIDAE

Corallum with solid walls; sclerosepta raised (exsert), margins smooth. Four species recorded for Britain.

Caryophyllia smithii Stokes and Broderip. Devonshire cup-coral Fig. 4.23

Polyp with up to 80 moderate or long, tapering tentacles, each ending in a prominent spherical knob. Corallum hard, dense, not perforated as in *Balanophyllia*. Colour variable, usually translucent.

An epizoic barnacle, *Megatrema anglicum*, is commonly attached to the corallum.

On rocks and shells; LWST in south and west, down to about 100 m, often in abundance; all coasts except eastern England; to south-west Europe and Mediterranean.

Order Zoanthidea

Colonial, arising from flat stolons or basal mat containing canals lined with endoderm; or immersed in coenenchyme. No skeleton but many incorporate sand particles into the body wall. Actinopharynx with a single siphonoglyph. Differ structurally from anemones in the arrangement of septa; in most, apart from the directives, pairs comprise one perfect and one imperfect septum; dorsal directives imperfect, ventrals perfect. New septa arise laterally to the ventral directives. Tentacles in two cycles, corresponding to endocoels and exocoels.

---

1. Colony in association with a hermit crab, or free **Epizoanthus papillosus**

   Colony in association with sponges or gorgonians, or on shell, rocks, etc. **2**

---

2. Tentacles colourless, minutely capitate **Epizoanthus couchii**

   Tentacles yellowish (rarely white), tapering to fine points **Parazoanthus axinellae**

---

12. EPIZOANTHIDAE

Oral sphincter muscle in the mesogloea; polyps sand encrusted. Three species recorded for Britain.

Epizoanthus couchii (Johnston) Fig. 4.23

Polyps arising from a thin encrustation of coenenchyme, which forms an irregular network or sheet that is usually hidden by silt or overgrown by other organisms. Polyps up to 15 mm tall and 5 mm diameter; up to 32 long tentacles, transparent except for a tiny white knob at the tip.

Found on rocks and shells, to at least 100 m, very occasionally on the shore; common and widespread off all coasts except North Sea; inconspicuous and easily overlooked; also recorded France.

Epizoanthus papillosus (Johnston) Fig. 4.23

Colonies on shells occupied by hermit crabs, where they form carcinoecia of rather few polyps directed upwards and outwards; or form small polyp clusters lying free on the bottom. Polyps to about 15 × 6 mm (more in free form), with up to 48 tentacles.

Recorded on all coasts but few recent records; mostly in deep water; North Atlantic.

13. PARAZOANTHIDAE

Oral sphincter endodermal; polyps sand encrusted. Three species recorded for Britain.

Parazoanthus axinellae (Schmidt) Fig. 4.23

Colonies of bright yellow polyps arising from encrusting mass of coenenchyme. Polyps to 20 × about 7 mm, with 26–34 moderately long tentacles arranged in two distinct cycles. Most specimens yellow, shading to orange around the mouth, occasionally white.

Attached to rocks, gorgonians, sponges, shells, etc.; 6–100 m; south and west coasts, locally abundant; south to Mediterranean.

P. anguicomus (Norman) is a similar but northern species. It is whitish, slightly larger, with 34–44 tentacles, and found in deeper water.

Order Actiniaria

The anemones form the largest order of anthozoans and the most important in British waters. Before using the key it is important that the terms described below are understood.

The cylindrical *column* of an anemone is often divided into distinct regions. The major (proximal), thicker-walled part is the *scapus*; in many genera the top edge of the scapus folds inwards to form a distinct rim, the *parapet*, which encloses a groove or *fosse*. Between the fosse and the oral disc is a delicate, thin-walled zone, the *capitulum*. This arrangement can be observed on the beadlet anemone *Actinia*. Less common is for the scapus to lead directly to a retractile distal region of smooth texture, the *scapulus*. This usually occurs in genera which have the scapus covered with *periderm*, a cuticle-like material. Other anemones show no differentiation of the column.

Various features may be present on the scapus: hollow warts or *verrucae*; flattened adhesive *suckers*, differentiated by their pale colour; or solid *tubercles*. A few species possess a ring of specialized warts, *acrorhagi*, on the parapet or in the fosse. *Cinclides* are tiny apertures in the column wall that are often visible as raised mounds or tiny dark dots. The periderm, when present, may be tough and closely adherent, or thin, paper-like, and only loosely attached; it often incorporates foreign material. Other species may secrete a thin, non-adherent sheath of soft mucus.

The proximal end of the anemone may be an adherent *basal disc* forming a distinct angle (*limbus*) at its junction with the column wall. Alternatively, the base may be rounded, with no sign of a limbus. This usually occurs in burrowing forms which use the rounded base (*physa*) as a digging organ.

The tentacles are arranged on the disc in concentric cycles, the innermost being primary, the next secondary, and so on. The number of tentacles in each cycle increases progressively. Commonly there are six tentacles in cycle 1, six in cycle 2, 12 in cycle 3, 24 in cycle 4, and so on. This is a *hexamerous* arrangement; but other numerical arrangements can occur. The distinction between the perfect and imperfect septa is very marked in some anemones, there being none of intermediate size. Such clearly differentiated large and tiny septa are respectively *macrocnemes* and *microcnemes*.

Some anemones possess thread-like organs, *acontia*, originating on the septa, which can be discharged through the cinclides (or sometimes the mouth) when the animal is disturbed. Care must be taken to avoid confusing acontia with septal filaments (thickened and convoluted free margins of septa), which may protrude through wounds.

Asexual division is common in some species, occurring in one of three ways: *transverse fission* (across the column) is rare; *longitudinal fission*, which normally results in two anemones of roughly equal size; and *pedal laceration*, in which small portions of the column tear off at the limbus, eventually growing a disc, tentacles, and other parts. Recently it has been shown in several anemones that the brooded young found in the coelenteron are genetically identical to the parent. This implies that they are not sexual progeny but the product of either internal budding or parthenogenesis. Specimens resulting from asexual division typically have an irregular arrangement of tentacles and other parts, and this can be a useful feature for identification.

KEY TO SPECIES OF ACTINIARIA

---

1. Anemones with a distinct adherent basal disc which forms a limbus. Tentacles 48 or more in adults. Typical anemones which attach to firm substrata **2**

   Anemones with no basal disc but sometimes capable of firm basal adhesion; proximal end rounded when free, never with a limbus. Tentacles 36 or fewer. Mostly burrowing forms with more or less elongate columns, capable of fast retraction when disturbed **25**

---

2. Column divisible into scapus and capitulum between which are parapet and fosse (may disappear in extreme extension) **3**

   Column divisible into scapus and scapulus, or undivided. Never with parapet and fosse **14**

---

3. Acontia absent; verrucae present on column, or acrorhagi in parapet region, or both **4**

   Acontia present (fairly readily discharged through cinclides on scapus); no verrucae or acrorhagi **11**

---

4. Never with verrucae on column; acrorhagi present in parapet region **5**

   Verrucae present on column; acrorhagi present or absent **7**

5. Acrorhagi in fosse, usually of contrasting colour to column (nearly always blue); tentacles readily retractile (*Actinia*) **6**

   Acrorhagi on parapet, same colour as column; tentacles not readily retractile ***Anemonia viridis*** (p. 125)

6. To 50 mm diameter; column and tentacles variously coloured but not spotted ***Actinia equina*** (p. 125)

   To 100 mm diameter, column red with regular greenish spots; tentacles red or purplish ***Actinia fragacea*** (p. 125)

7. Tentacles rather stout, decamerously (i.e. 10 + 10 + 20, etc.) arranged (*Urticina*) **8**

   Tentacles not very stout, hexamerously or irregularly arranged **9**

8. Scapus with numerous verrucae usually covered with debris; tentacle span up to 200 mm; colour variable with many combinations ***Urticina felina*** (p. 125)

   Verrucae smaller and rarely with adherent debris; tentacle span up to 300 mm; colour usually pale ***Urticina eques*** (p. 125)

9. A distinct small red spot on each verruca ***Anthopleura ballii*** (p. 125)

   No red spots on verrucae **10**

10. Tentacles regularly hexamerous, in four or five cycles. Acrorhagi absent; often verrucae on the parapet ***Bunodactis verrucosa*** (p. 125)

    Tentacles irregular, apparently in only two or three cycles. Spherical acrorhagi present in the fosse ***Anthopleura thallia*** (p. 125)

11. Tentacles very numerous, presenting a fluffy appearance; mouth with prominent ribbed lips extending a little on to disc; may be large (up to 250 mm high) ***Metridium senile*** (p. 128)

    Tentacles never more than about 200. Small, column not exceeding about 20 mm diameter **12**

12. Column brownish or greenish, usually with vertical stripes of orange, yellow, or white. Mouth inconspicuous, without prominent lips ***Haliplanella lineata*** (p. 128)

    Column without vertical stripes (apart from tonal differences at insertion of septa) **13**

13. Column greatly elongated in extension; in this state the parapet and fosse disappear and the capitulum is very long, often equal to scapus ***Diadumene cincta*** (p. 128)

    Column never greatly elongate; capitulum relatively short; parapet and fosse permanent. Mouth with prominent ribbed lips extending on to disk ***Metridium senile*** (p. 128)

14. Acontia absent; column smooth, no suckers or tubercles. Offshore **15**

    Acontia present (not always readily discharged); column with or without suckers or tubercles. If offshore, then tubercles present or acontia fairly readily discharged **16**

---

15. Tentacles hexamerously (i.e. 6 + 6 + 12 + 24, etc.) arranged; can be shed by pinching off at base; span of tentacles up to 300 mm or more
Bolocera tuediae (p. 125)

Tentacles distinctively arranged, the first two cycles different in number (i.e. 6 + 10 or 12 + 16 or 18, etc.); cannot be shed; tentacle span not exceeding 60–70 mm
Stomphia coccinea (p. 128)

---

16. Base broadly expanded into two lobes which embrace hermit crab. Acontia usually magenta, sometimes white; freely discharged
Adamsia carciniopados (p. 131)

Base normal, often broad but never expanded into lobes. Acontia always white **17**

---

17. Column not divisible into regions; tubercles absent but scapus often rough or wrinkled, sometimes with periderm **18**

Column divisible into scapus and scapulus; scapus with tubercles and typically covered with periderm **31**

---

18. Body wall rather tough and thick; cinclides only on proximal part of column. Anemone usually associated with hermit crab
Calliactis parasitica (p. 131)

Body wall soft, smooth, and not very thick; cinclides not restricted to lower column. Anemone not associated with hermit crab **19**

---

19. Column with suckers that appear as distinct pale spots, often with debris stuck to them **20**

Suckers absent **22**

---

20. Tentacles nearly always irregularly arranged. Suckers rarely have adherent debris
Sagartia elegans (p. 129)

Tentacles regularly hexamerous. Suckers usually with adherent debris **21**

---

21. Column often trumpet-shaped in extension (oral disc much wider than column). Suckers usually large and prominent; up to 500 or more small tentacles
Cereus pedunculatus (p. 129)

Column never trumpet-shaped (oral disc not much wider than column). Suckers usually rather inconspicuous; up to about 200 tentacles **32**

---

22. Tentacles long and stout, not readily retracted into column. Anemone predominantly brown
Aiptasia mutabilis (p. 128)

Tentacles of more 'normal' proportions, readily retracted into column. Anemones not brown **23**

---

23. Acontia readily discharged; tentacles never very long, always plain white
Actinothoe sphyrodeta (p. 129)

Acontia not very readily discharged; tentacles often long, always patterned (*Sagartiogeton*) **24**

---

24. Extended column pillar-like, to 120 mm high; up to 200 tentacles arranged

hexamerously; column grey or buff with vertical stripes **Sagartiogeton undatus** (p. 131)

Rarely exceeding 30 mm when extended; irregular arrangement of tentacles owing to occurrence of pedal laceration; always with some orange coloration on column and disc **Sagartiogeton laceratus** (p. 131)

---

25. Column stout, typically pear-shaped. Tentacles up to 36 with seven in inner cycle **Mesacmaea mitchellii** (p. 133)

Column slender, more or less cylindrical. Tentacles of first cycle four, six, or eight **26**

---

26. Tentacles very long, at least half column length. Anemone very small, rarely more than 15 × 2 mm. In isolated lagoons or tidal creeks **Nematostella vectensis** (p. 133)

Tentacles less than half column length; column (when full grown) substantially larger than above. Found on open coasts or in the sea **27**

---

27. Tentacles 12 (6 + 6), with 'V' or 'W' markings on them. Burrowing anemones **28**

Tentacles 16 or more, never with 'V' or 'W' markings. Burrowing or crevice-dwelling anemones **29**

---

28. Anemones up to 300 × 30 mm; tentacles much longer than disc diameter when expanded; a small multi-lobed organ (*conchula*) projecting from one corner of mouth **Peachia cylindrica** (p. 133)

Anemone small, up to 50 × 5–6 mm; tentacles always shorter than disc diameter: no conchula **Halcampa chrysanthellum** (p. 133)

---

29. Anemone inhabiting holes in rocks **Edwardsiella carnea** (p. 133)

Anemone burrowing in mud or sand (*Edwardsia*) **30**

---

30. Scapus with eight longitudinal rows of tubercles; tentacles 16 (8 + 8), colourless with spots and bands of buff and red-brown **Edwardsia claparedii** (p. 133)

Scapus lacking rows of tubercles; tentacles 32 (in three cycles), translucent buff or pink with white markings **Edwardsia timida** (p. 133)

---

31. Cinclides present near the limbus; lining of actinopharynx blackish. May occur on lower shore **Cataphellia brodricii** (p. 131)

Cinclides absent; lining of actinopharynx reddish or brownish. Not known to occur on shore **Hormathia coronata** (p. 131)

---

32. Pedal disk up to 50 mm diameter; up to six cycles of tentacles; reproduces only sexually ***Sagartia troglodytes** (p. 129)

Pedal disc not exceeding about 15 mm diameter; not more than five cycles of tentacles; produces viviparous asexual young ***Sagartia ornata** (p. 129)

(*These two species may occur together.)

Suborder Nynantheae

Septal filaments with ciliated tracts; retractor muscles facing into endocoels.

14. ACTINIIDAE

Tentacles simple, one arising from each endocoel and each exocoel. Septa not divided into macrocnemes and microcnemes. Sphincter muscle endodermal. Acontia absent. At least nine British species.

Actinia equina (Linnaeus).
Beadlet anemone Fig. 4.24
Column smooth, its height and diameter about equal, up to about 50 mm. Prominent acrorhagi in the fosse. Tentacles of moderate length, to about 200. Colour red, green, or brown; acrorhagi typically blue. Reproduction by viviparity common.

A typical shore form extending from HWN to shallow water offshore, on rocks, in pools, etc; common on all coasts; Arctic to West Africa and Mediterranean.

Actinia fragacea Tugwell.
Strawberry anemone Fig. 4.24
Compared with *A. equina* this species is larger, up to 100 mm column diameter. Column red, with regular greenish flecks, like the pips on a strawberry; tentacles red or purplish.

Occurs on the lower shore; widespread but local.

Anemonia viridis (Forskål).
Opelet or snakelocks anemone Fig. 4.24
Column smooth, usually wider than high, to about 50 mm diameter. Disc wide, with up to 200 long and sinuous tentacles; the tentacles rarely retracted. Column and acrorhagi brownish, tentacles brown, grey or bright grass green, usually with purple tips. Contains zooxanthellae.

On the shore in pools and other places open to the light, and down to about 20 m; south and west, north to Scotland; reaching south-west Norway; south-west Europe and Mediterranean.

Anthopleura ballii (Cocks) Fig. 4.24
Column from tall and trumpet-shaped to squat; to about 70 mm high; small verrucae arranged in neat vertical rows, each terminating at the parapet in an acrorhagus. Disc wide; tentacles moderate to long, to about 96; arranged hexamerously; tentacle span up to 100 mm. Column variably coloured; acrorhagi and verrucae paler, each verruca tipped with a tiny but distinct red spot. Contains zooxanthellae.

In holes or crevices, buried in gravel, etc.; midshore pools and lower shore to about 25 m; south and west coasts as far north as Isle of Man, rather local; south-west Europe to Mediterranean.

Anthopleura thallia (Gosse) Fig. 4.24
Column tall, up to 50 mm. Verrucae prominent, in irregular rows and increasing in size towards the parapet, usually with debris stuck to them; small acrorhagi present in fosse. Tentacles moderate in length, up to about 100. Column olive green, brown, grey, or reddish; verrucae darker, acrorhagi white or pink; disc and tentacles almost any colour. Reproduces by longitudinal fission.

In pools and crevices on shores exposed to strong wave action; MTL to LWST; extreme south-west, local; also Atlantic coast of France.

Bolocera tuediae (Johnston) Fig. 4.24
Column smooth and featureless. Tentacles moderate to long, thick; to 200; hexamerously arranged. Anemone dull pink, or brownish; up to 300 mm or more across the tentacles.

Sublittoral, 20 to at least 2000 m; all coasts but rare in the south; North Atlantic and Arctic.

Bunodactis verrucosa (Pennant) Fig. 4.24
Column typically about 40 mm. Verrucae prominent and arranged in neat vertical rows. Tentacles moderate to long, hexamerously arranged, up to 48. Column grey or pink, verrucae darker with the six principal rows usually white. Reproduction by viviparity common.

A typical shore form found in crevices, pools (especially *Corallina* pools) or attached to bedrock beneath a layer of sand or gravel; rarely found below the tidal zone; south and west, locally common; to south-west Europe and Mediterranean.

Urticina felina (Linnaeus).
Dahlia anemone Fig. 4.25
Column typically squat, with a firmly adherent base, to about 100 mm diameter; verrucae numerous on scapus, not in rows; with adhering debris. Tentacles stout, short or moderate in length, arranged in multiples of 10, up to about 160; their span up to about 200 mm. Colour variable; column plain or with irregular patches of different colours, verrucae usually grey; disc usually patterned with red lines; tentacles banded or plain.

In pools and crevices, usually around low water mark, occasionally midshore in deep, shaded pools, on rocky shores, common; offshore to at least 100 m; Arctic to Biscay.

Urticina eques (Gosse) Fig. 4.25
Similar to *U. felina* but tentacles (also arranged in multiples of 10) relatively longer, verrucae reduced in size and rarely with adherent debris. Colour usually pale: red, yellowish, or cream, the pale

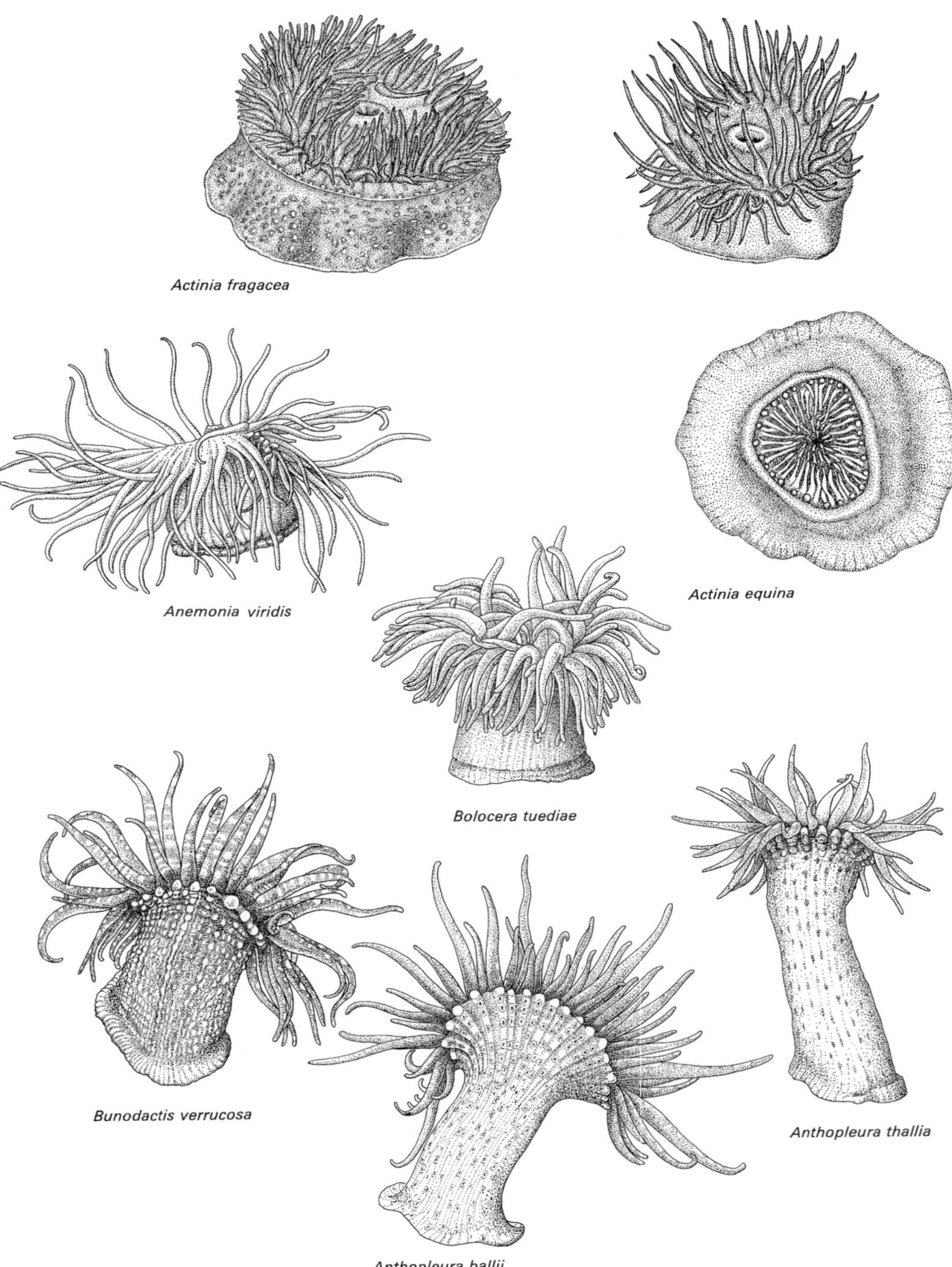
Actinia fragacea
Anemonia viridis
Actinia equina
Bolocera tuediae
Bunodactis verrucosa
Anthopleura thallia
Anthopleura ballii

Fig. 4.24

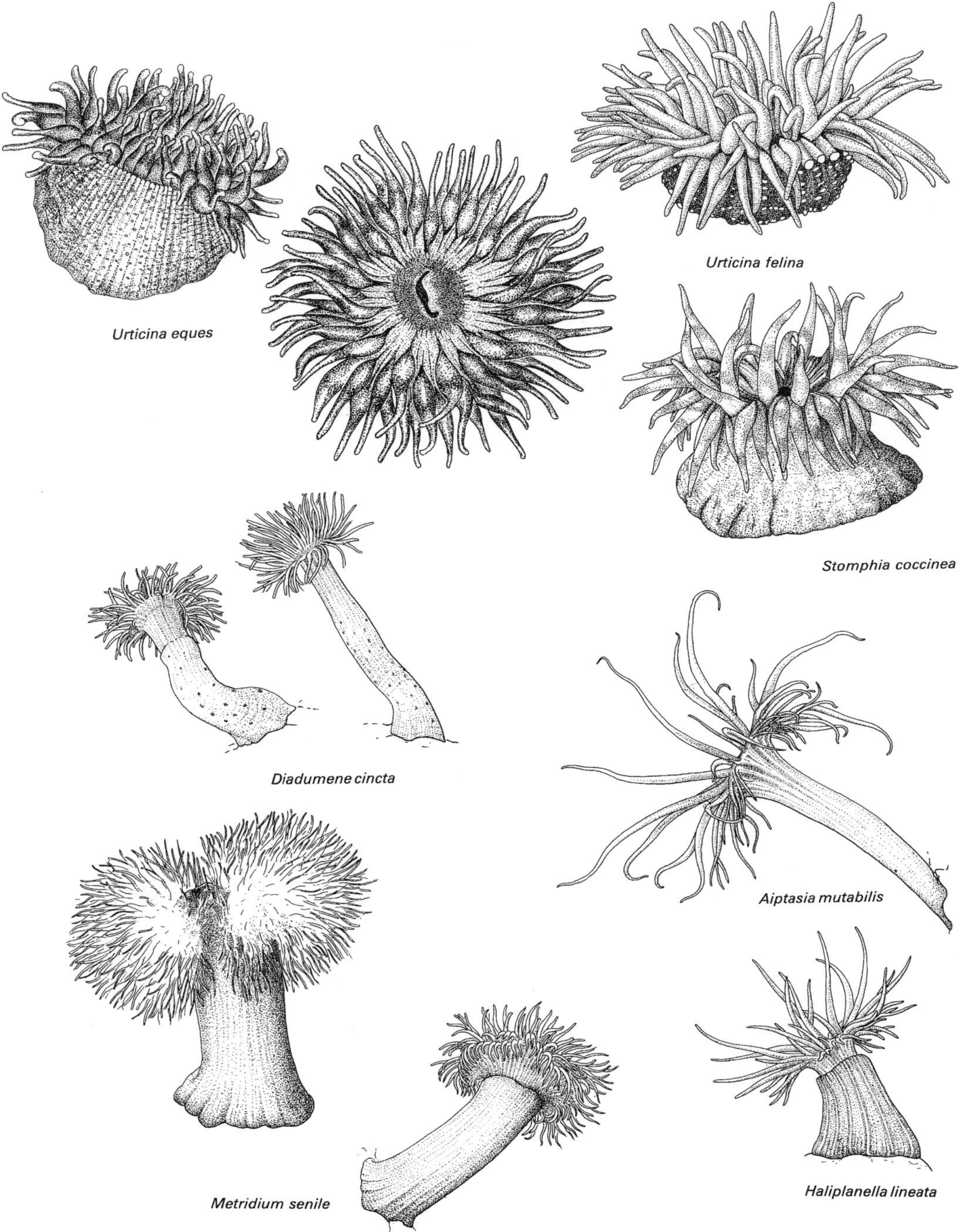
Urticina eques
Urticina felina
Stomphia coccinea
Diadumene cincta
Aiptasia mutabilis
Metridium senile
Haliplanella lineata

Fig. 4.25

forms usually blotched with red or orange. Often larger than *U. felina*, with a tentacle span of 300 mm or more.

Typically offshore, LWST to at least 400 m; all coasts (doubtful English Channel); to Arctic.

15. ACTINOSTOLIDAE

Septa not divided into macrocnemes and microcnemes. Sphincter muscle mesogloeal. Acontia absent. A single British species.

Stomphia coccinea (Müller) Fig. 4.25
Column smooth, to about 70–80 mm diameter. Tentacles moderate in length, and arranged 6 + 10 or 12 + 16 or 18, etc. Colour variable, white, red, orange, buff; column irregularly blotched, tentacles usually banded.

On stones and shells (especially *Modiolus*); 10–400 m; all coasts except south; Arctic-boreal.

16. AIPTASIIDAE

Sphincter muscle mesogloeal but weak, or absent. Acontia present. Septa not divided into macrocnemes and microcnemes. Two British species.

Aiptasia mutabilis (Gravenhorst) Fig. 4.25
Column typically tall and trumpet-shaped, flaring out to the wide oral disc which bears up to 100 long tentacles; height about 100–120 mm, span 150 mm. Cinclides fairly prominent when column well-extended; acontia emitted rather reluctantly. Rarely retracts its tentacles. Colour brown, usually with opaque white lines on the disc. Contains zooxanthellae.

Under stones, beneath overhangs, in lower shore pools, but more abundant in the shallow sublittoral amongst *Laminaria* and *Saccorhiza* holdfasts; to about 30 m. Restricted to extreme south-west; also Channel Isles and south to Mediterranean.

17. DIADUMENIDAE

Sphincter muscle endodermal and weak, or absent. Acontia present. Septa not divided into macrocnemes and microcnemes. Two species around the British Isles.

Diadumene cincta Stephenson Fig. 4.25
Column smooth, long and slender; in contraction the parapet, fosse, and capitulum become distinct. Oral disc about same diameter as column, with up to about 200 long, slender tentacles. Column typically to about 60–70 mm. Reproduction by pedal laceration. Anemone dull orange, the deep orange actinopharynx usually showing through; cinclides visible as dark dots.

On stones, mussels, growing through sponges, etc.; sometimes in variable salinity; on the shore and sublittoral down to about 40 m; reported from all coasts but local, occurring in aggregations; also Netherlands and France.

Haliplanella lineata (Verrill) Fig. 4.25
Column smooth, divided into scapus and capitulum, with parapet and fosse; usually about 10–20 mm diameter. Tentacles long, arranged irregularly; up to about 100. Reproduction by longitudinal fission. Column typically olive green with prominent orange or yellowish longitudinal stripes.

On stones, shells, pier piles, etc., usually intertidal; all British coasts, very local; often in harbours and in water of variable salinity; widely distributed through Europe and rest of world.

18. METRIDIIDAE

Sphincter muscle mesogloeal. Acontia present. Septa not divided into macrocnemes and microcnemes. Column divided into scapus and capitulum, with parapet and fosse. A single British species.

Metridium senile (Linnaeus).
Plumose anemone Fig. 4.25
An anemone of very variable form. Base wide, often irregular (result of pedal laceration); column smooth; in full expansion the parapet may be visible as a prominent collar. Tentacles variable in length and number. Colour plain; commonly white or orange. Two forms or morphs can be recognized. Var. *dianthus* (the plumose form): column up to 500 mm; tentacles rather short, very numerous, creating a fluffy effect; var. *pallidum*: a small shore form, often found in brackish water; column usually no taller than wide; up to about 200 long tentacles.

In overhangs, caves, and beneath boulders on the lower shore, and on pier piles and rock faces to at least 100 m; common and widespread on all coasts; Scandinavia to Biscay.

19. SAGARTIIDAE

Sphincter muscle mesogloeal. Acontia present. Septa not divisible into macrocnemes and microcnemes. Column not divided into scapus and capitulum; never with parapet and fosse. Seven species recorded for Britain.

Actinothoe sphyrodeta (Gosse) Fig. 4.26
Column typically short and squat but can become taller than wide, to 20 × 15 mm. Up to about 120 moderate to long tentacles irregularly arranged. No suckers. Cinclides visible as dark dots on the upper column; acontia readily emitted. White to greyish or orange. Distinguished from white specimens of *Sagartia* spp. by the lack of suckers and its habit of not inserting its base into a hole or crevice. Reproduces by longitudinal fission.

On open rock faces exposed to wave or tidal action; sometimes on *Laminaria* or *Himanthalia* buttons; occasional on the shore at LWST in caves, etc.; commoner in sublittoral down to at least 50 m; south and west, north to Shetland; also south to Biscay.

Cereus pedunculatus (Pennant).
Daisy anemone Fig. 4.26
Column taller than wide, typically narrowest just above basal disc, flaring. Oral disc wide, usually 40–50 mm. Tentacles short, small, numerous, 500–700 in an average specimen; arranged hexamerously. Cinclides present high on the column; acontia freely emitted. Column dark, suckers forming prominent pale spots, usually with adhering debris. Contains zooxanthellae. Viviparous, often producing many young when collected.

In holes and crevices, typically in pools, also buried in sand or mud with the disc expanded at the surface; from MTL down to at least 50 m; south and west coasts north to Scotland; locally abundant; south-west Europe and Mediterranean.

Sagartia elegans (Dalyell) Fig. 4.26
Base up to 30 mm diameter, wider than column; column taller than wide, flaring to the oral disc. Suckers prominent; only rarely with adherent debris. Cinclides just visible on upper part of column; acontia emitted freely. Up to about 200 tentacles, moderate in length, arranged irregularly owing to frequent occurrence of pedal laceration.

Colour of oral disc and tentacles variable, five varieties being recognized:

1. disc and tentacles all white; no pattern: var. *nivea*;
2. disc orange, tentacles white; no pattern: var. *venusta*;
3. disc variable, tentacles rose-pink or magenta: var. *rosea*;
4. disc variable, tentacles dull orange: var. *aurantiaca*;
5. disc and tentacles patterned; var. *miniata*.

Varieties 1, 2, and 5 are common; 4 is rare.

A common species, in pools or under stones, beneath rocky overhangs, typically with its base inserted into a hole or crevice; on the shore, and extending to at least 50 m; all coasts, locally abundant; Scandinavia to Mediterranean.

Sagartia ornata (Holdsworth) Fig. 4.26
Column widest basally; diameter of base usually less than 15 mm. Tentacles long, up to four to five cycles. Suckers inconspicuous, often without adherent debris; cinclides few, but large. Lower part of column may be covered by loose investment of mucus and detritus. Column and oral disc translucent shades of green or brown. Pattern at base of tentacles rather similar to that in *S. troglodytes* and some specimens may be difficult to place in their correct species. Viviparous young are produced asexually, building up aggregations of identical anemones.

In holes, crevices, and under stones, and in *Laminaria* holdfasts; also found on stones, shells, and *Zostera* over muddy sand flats, including brackish localities; all British coasts.

Sagartia troglodytes
(Price in Johnston) Fig. 4.26
Column typically broadest at base; usually a tall pillar up to 120 mm. Tentacles moderate, to about 200, neatly arranged in the usual hexamerous pattern (cf. *S. elegans*). Suckers small and inconspicuous but often with adherent debris; cinclides on upper part of column, acontia not readily emitted. Column typically dull in colour; disc and tentacles exceedingly variable; usually with a speckled pattern, or trace of one; usually with conspicuous, or at least discernible, B-shaped markings at the base of tentacles. Unlike *S. elegans* this species never occurs in reddish shades. Reproduces sexually; populations therefore more variable than those of *S. ornata*; the two may occur together.

Typically buried in sand or silty mud, attached to a stone or shell up to 150 mm below, in full

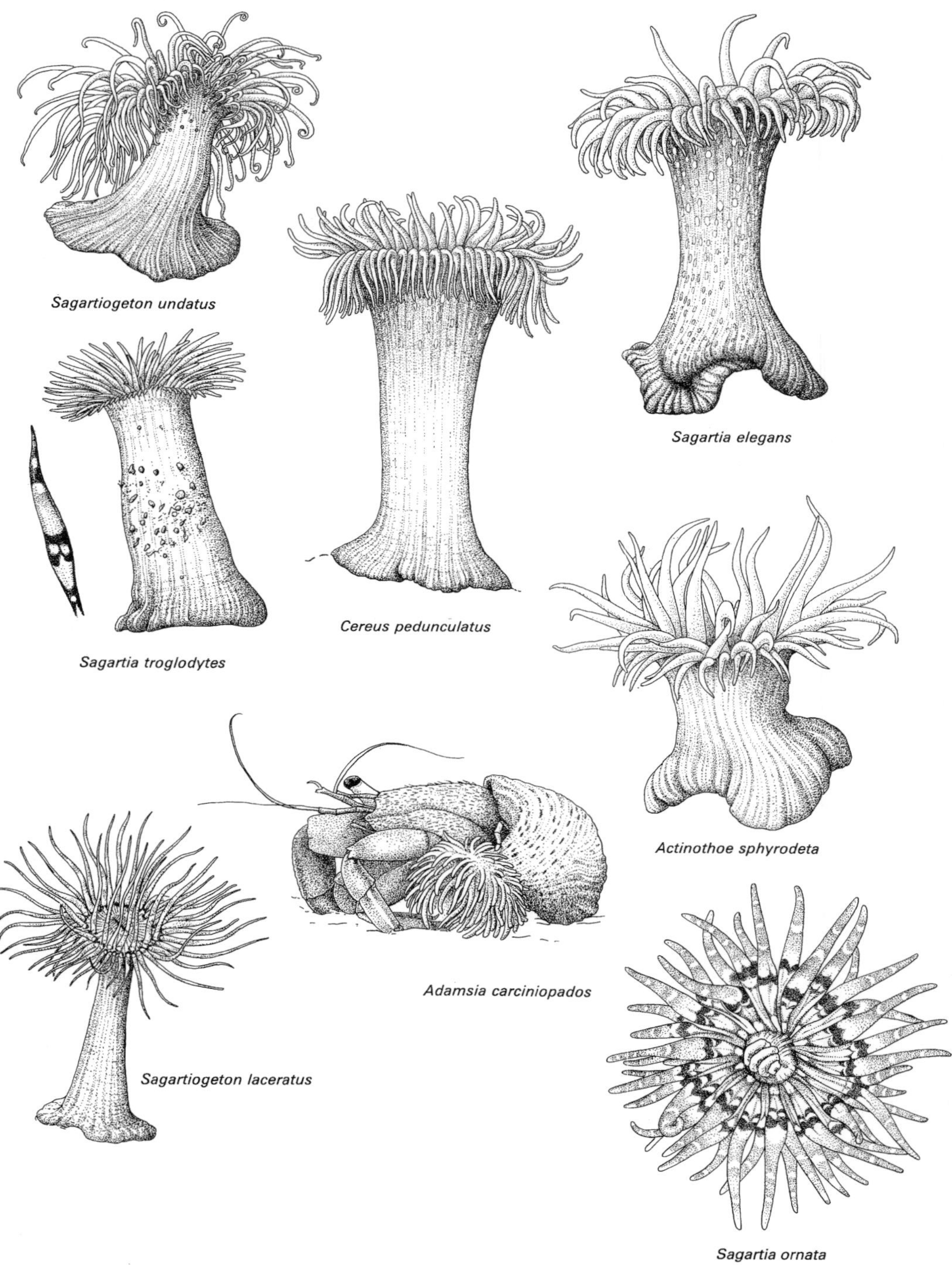
Sagartiogeton undatus
Sagartia elegans
Cereus pedunculatus
Sagartia troglodytes
Actinothoe sphyrodeta
Adamsia carciniopados
Sagartiogeton laceratus
Sagartia ornata

Fig. 4.26

or reduced salinity; also in pools, crevices, kelp holdfasts on rocky shores; from MTL to 50 m; all coasts; Scandinavia to Mediterranean.

Sagartiogeton laceratus (Dalyell) Fig. 4.26
General form similar to *S. undatus* but smaller, rarely more than 30 mm high, 15 mm base diameter; tentacles arranged irregularly, owing to habitual pedal laceration. No suckers. Column patterned as *S. undatus*; acontia not readily emitted. This species, unlike *S. undatus*, has some orange coloration in the column and disc, and pale chevrons on a dark ground near the tentacle bases. Tends to occur in small groups.

On hard substrata; sublittoral, to at least 100 m; all coasts; Scandinavia to Biscay.

Sagartiogeton undatus (Müller) Fig. 4.26
Column up to 120 mm high and 20 mm diameter, becoming very low in contraction. Oral disc wider than column, with an unusually wide mouth; tentacles up to 200, long, arranged hexamerously. No suckers. Column grey or buff, with regular vertical stripes of brown flecks of variable intensity; acontia not readily emitted. Disc and tentacles translucent grey or buff, with a pattern of dark markings and cream spots; often with irregular dark radial wedges; tentacles with two dark longitudinal stripes. Easily distinguished from *Sagartia troglodytes* by its lack of suckers and the long striped tentacles. Pedal laceration not recorded.

Typically buried in sand or gravel, attached to a stone or shell, or in crevices or holes in rocks, lower shore and sublittoral to at least 100 m; all coasts, frequent but rarely abundant; Scandinavia to Mediterranean.

20. HORMATHIIDAE

Sphincter strong, mesogloeal. Acontia present. Septa not divisible into macrocnemes and microcnemes; only 6–12 pairs perfect. Column tough, tuberculate, cuticular. Often on detached substrata, crustaceans, or pagurid-inhabited shells. Eight species recorded for Britain.

Adamsia carciniopados (Otto) Fig. 4.26
Base broadly expanded, forming two lobes which embrace and overlap a shell inhabited by the hermit crab *Pagurus prideauxi*. Oral disc elliptical, tentacles short and slender, up to 500. Colour mainly white, mostly with numerous pink or magenta spots on column; cinclides present on lower column, through which pink, rarely white, acontia are freely emitted.

Occasional LWST, common sublittorally at all depths on sandy or gravelly substrata; all coasts; Norway to Mediterranean.

Calliactis parasitica (Couch) Fig. 4.27
Column taller than wide, tentacles moderate in length, rather slender, up to about 700; to 80 × 50 mm. Acontia readily emitted from cinclides just above limbus. Ground colour cream or buff, blotched and streaked with reddish or greyish brown, tending to form vertical stripes.

Typically found on whelk (*Buccinum undatum*) shells inhabited by hermit crab *Pagurus bernhardus* but occasionally on stones or empty shells; rarely intertidal; locally common offshore at depths to about 60 m; southern coasts, north to Bristol Channel and west of Ireland; south-west Europe and Mediterranean.

Cataphellia brodricii (Gosse) Fig. 4.27
Column with a broad base; up to 60 × 40 mm; scapus with small dark tubercles, usually concealed by a wrinkled layer of brown periderm. Oral disc small, with about 100 short to moderate tentacles arranged in a hexamerous pattern. Lining of actinopharynx blackish. Cinclides visible as tiny dark spots just above the limbus. Acontia emitted reluctantly. Viviparous.

In crevices or attached to bedrock beneath sediments, under stones, etc.; from LWST to 20 m; Devon, Cornwall, south-west Ireland, France.

Hormathia coronata (Gosse) Fig. 4.27
General form and size very similar to *Cataphellia brodricii*, but lacks cinclides. Actinopharynx lining red or buff.

Usually found on shells, worm tubes, or other organic substrata; sublittoral only; south and west, rather local; Ireland and western Scotland to south-west Europe and Mediterranean.

21. HALOCLAVIDAE

Basilar muscles lacking; column generally elongate. Sphincter weak or lacking. Single siphonoglyph. No acontia. Septa not differentiated into macrocnemes and microcnemes. Three British species.

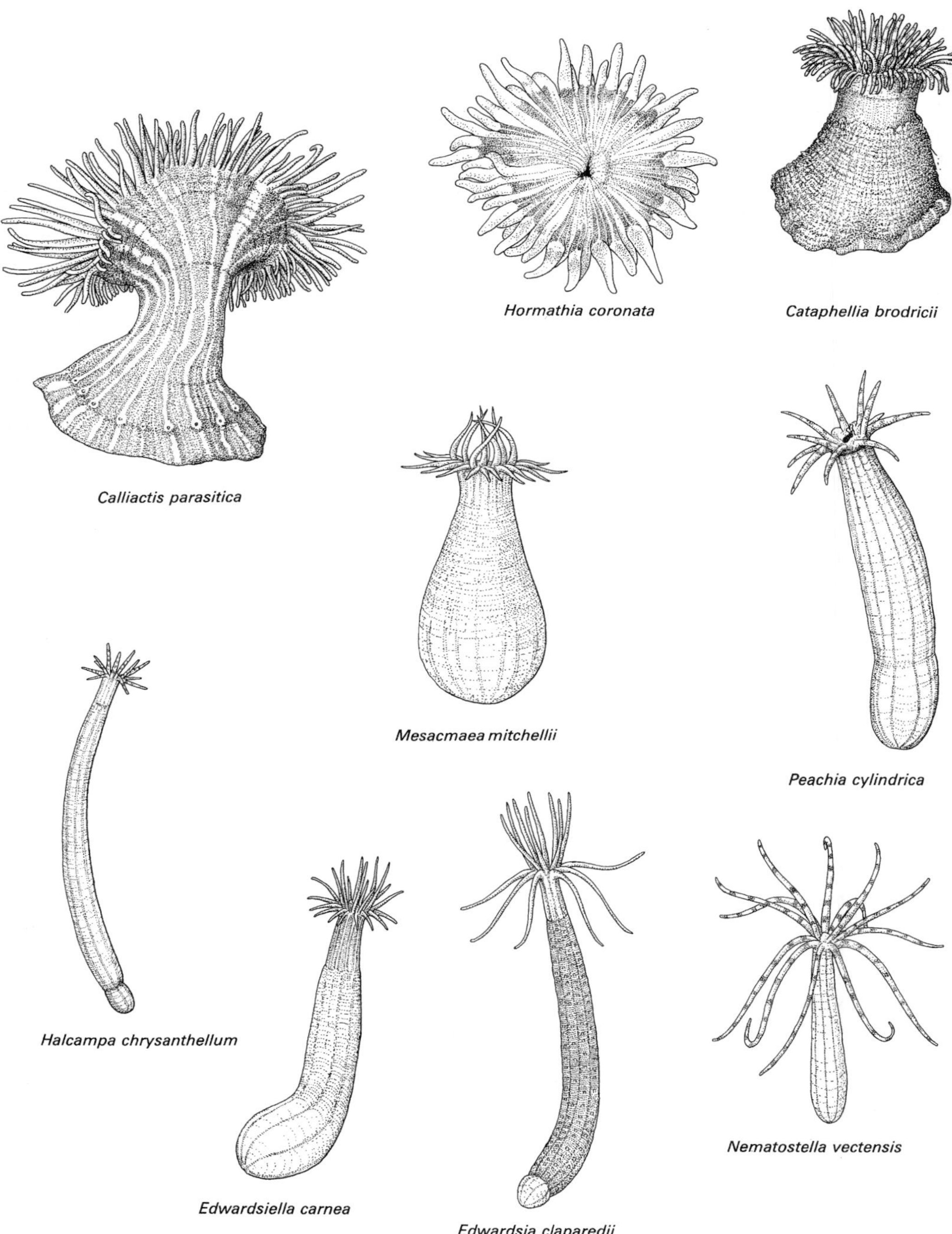
Hormathia coronata
Cataphellia brodricii
Calliactis parasitica
Mesacmaea mitchellii
Peachia cylindrica
Halcampa chrysanthellum
Nematostella vectensis
Edwardsiella carnea
Edwardsia claparedii

Fig. 4.27

Mesacmaea mitchellii (Gosse) Fig. 4.27

Column turnip-shaped, divided into scapulus and scapus; base rounded, used for digging, capable of becoming a disc without a limbus and able to adhere firmly to a solid surface; up to 70 × 35 mm. Tentacles long, up to 36, with seven in inner cycle.

Buried in sand or gravel with just the disc and tentacles protruding; always sublittoral, 15–100 m south and west coasts; to mid-Scotland; south-west Europe and Mediterranean.

Peachia cylindrica (Reid) Fig. 4.27

Column elongated when buried but contracting to short sausage-shape when dug up, to about 60 × 20 mm; indistinctly differentiated into physa, scapus, and capitulum. A small lobed organ, the *conchula*, protruding from one corner of the mouth; tentacles 12 in two cycles.

Burrowing in sand or gravel, occasionally on the lower shore but mainly sublittoral; all coasts; western Europe and Mediterranean.

22. HALCAMPIDAE

Column lacking a base; typically an elongate cylinder divisible into physa, scapus, and capitulum. Scapus may be cuticular and bear adhesive papillae (*tenaculi*). Sphincter mesogloeal. Septa divisible into macrocnemes and microcnemes. Tentacles few. A single British species.

Halcampa chrysanthellum (Peach) Fig. 4.27

Column elongated and worm-like, usually 30–40 × 4–5 mm. Disc small, tentacles 12, always short or moderate; fewer than the septa.

Burrows in sand or gravel; occasional at LWST (e.g. in *Zostera*) but mainly sublittoral; all coasts; further distribution beyond France uncertain.

23. EDWARDSIIDAE

Column vermiform, differentiated into two or more regions. Basal end rounded. Sphincter, acontia, basilar muscles absent; cinclides may be present. Eight macrocnemes and at least four microcnemes present; tentacles few. Some genera characterized by *nemathybomes*, pockets in the column wall containing batteries of large nematocysts. The condition of the ciliated tracts (on the septal filaments) provides generic characters. Eight species recorded in the British sea area.

Edwardsia claparedii (Panceri) Fig. 4.27

Column elongated and worm-like, up to 60–70 × 5 mm; divisible into rounded basal physa; long cylindrical scapus, bearing eight longitudinal rows of tubercles (underlying nemathybomes); and short, naked scapulus. Disc small, 16 tentacles, arranged 8 + 8.

Burrowing in mud or muddy sand; LWST to sublittoral, often in or around *Zostera* beds; south and west coasts, local.

Edwardsia timida Quatrefages is similar but more slender, and the column lacks the rows of tubercles; physa with cinclides. Tentacles up to 32 (usually 20–28) arranged in three cycles.

Recorded only from lower shore in muddy coarse sand or gravel; south and west only, very local.

Edwardsiella carnea (Gosse) Fig. 4.27

Column cylindrical, up to 20 × 3–4 mm, divided into scapus, which lacks nemathybomes and bears a thick, rough layer of brownish periderm, and naked scapulus; basal end of scapus adherent to the substratum. Up to about 34 moderate to long, slender tentacles. Ciliated tracts short. Whole anemone translucent orange.

In holes and crevices in rocks, often in large local aggregations; always in shaded places, caves, beneath overhangs, etc., MTL to shallow water; all coasts, but mostly in south and west; Scandinavia to France, perhaps Mediterranean.

Nematostella vectensis Stephenson Fig. 4.27

Column moderately elongated, up to 15 × 1.5 mm differentiated into physa, scapus, and capitulum. Disc small, 10–18 tentacles arranged irregularly owing to habitual reproduction by transverse fission. No nemathybomes. Ciliated tracts long but broken into discrete sections. Whole anemone transparent, colourless except for occasional marks.

Burrowing in fine mud or attached to plant life in brackish lagoons and tidal creeks (not in the open sea); several localities on south and east coasts; very local but abundant where found.

PHYLUM CTENOPHORA

Biradially symmetrical, usually transparent, gelatinous planktonic species that lack cnidae, and swim by means of eight meridional rows of ciliary plates ('comb rows'). Digestive system comprising stomach and biradial canal system, with a single opening (mouth). Many are provided with a pair of long retractile tentacles which catch animal prey by means of sticky colloblasts . 'Comb jellies'.

Class Tentaculata

Tentacles present.

Order Cydippida

Ovoid body; tentacles branched and retractable into pouches.

Pleurobranchia pileus Müller.
Sea gooseberry Fig. 4.28
Gooseberry-shaped body, 17–20 mm high; comb rows of equal length, extending from the aboral pole almost to the mouth; retractile tentacles extending to 15–20 times body length, one side of each bearing filaments.

All British waters; from south-west Norway and Baltic to Atlantic south and west of the British Isles. Present from about March to November with the greatest numbers in summer.

Order Lobata

Body subovoid with two large oral lobes; tentacles small, not in pouches.

Bolinopsis infundibulum Müller Fig. 4.28
Body pear-shaped in sagittal plane, somewhat flattened in tentacular plane, up to 15 cm high; large oral lobes comprising one-third of body height; comb rows of unequal lengths; tentacles short, not retractile into pouches. Very fragile.

Boreal distribution from the Arctic south to northern North Sea, Irish Sea and Atlantic waters

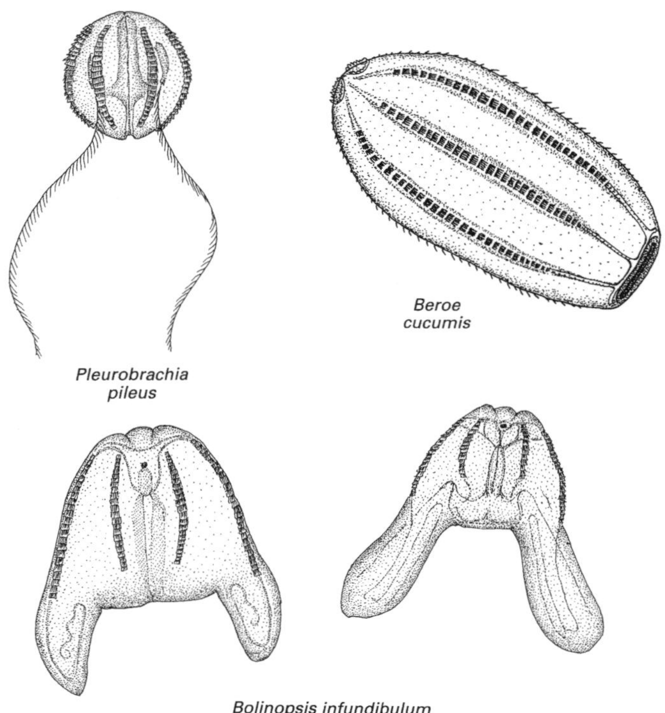

Fig. 4.28

to the south and west of Ireland; rarely western Channel. Most common between May and August; variable between years.

Class Nuda

Without tentacles.

Order Beroida

Conical-cylindrical body form with wide mouth and pharynx; capture food with rim of mouth.

Beroe cucumis Fabricius Fig. 4.28

Thimble-shaped, with the oral end somewhat flattened, up to 15 cm high; comb rows of equal length, extending from the aboral pole for about three-quarters of length. Mature specimens may be pink in colour. (Genus name is pronounced be-row-ee.)

Boreal distribution from Arctic to Kattegat, North Sea, English Channel, Irish Sea, and Atlantic waters to the south and west of Ireland. Most common between May and September; variable between years.

5. FLATWORMS AND RIBBON WORMS

PHYLUM PLATYHELMINTHES

These 'flatworms' are bilaterally symmetrical, with marked cephalization, mostly with a mouth, gut, and protonephridia, and typically with hermaphrodite reproductive systems, but without a special body cavity, anus, or circulatory system. Five classes may be recognized. The Aspidogastraea, Digenea, Monogenea, and Cestoda are entirely parasitic and their hosts include many marine species, particularly among fish, molluscs, and macrocrustaceans. These classes are remarkable for storing yolk, not in oocytes which may develop, but in special 'yolk cells' (modified oocytes?). These are formed in *yolk glands*, which appear to be specialized branches of the ovary. Early embryos ingest the yolk cells thus provided.

The remaining class, Turbellaria, contains the free-living platyhelminths.

Class Turbellaria

All but two of the orders and suborders of Turbellaria include common marine species. Some are predominantly marine. Most creep more often than they swim, and rely for locomotion on sticky epidermal cilia, helped by body-wall muscles. Many, ranging from the smallest to the largest, are like most animals in storing yolk directly within the eggs which are to be fertilized. Many others, however, have yolk glands and yolk cells.

Only a few Turbellaria can be identified in the field. The bold pattern of *Prostheceraeus* (Fig. 5.1) is readily seen by SCUBA divers, or found, with some other polyclads and the 'stony' capsules of *Fecampia*, by turning over shore rocks at low water. The triclad, *Procerodes*, is easily visible on smooth undersides of small boulders, pulled out from gravel or sandy beds of freshwater streams running out over mid-tide beaches. Some species are parasitic or commensal, or have specific prey, so may be collected by searching the larger animals on which they depend. Many live interstitially in sandy beaches, and are obtained by the ice-water technique, for example. Small forms dealt with in this brief introduction were collected on rocky shores near Swansea (Mumbles Point, Bracelet Bay, Oxwich, and Port Eynon) by violently rinsing successive clumps of various algal species in a plastic sandwich box containing enough sea water to fill a jar. Many detached organisms were thus brought back in jars, and the Turbellaria were pipetted away from these in shallow dishes under a binocular microscope, and examined in solid watch-glasses or under coverslips supported by peripheral vaseline. Some internal structures were seen in slightly compressed live worms, but many species are rather opaque or pigmented. Most taxonomically important details shown in figures of turbellarian anatomy were evidently derived by reconstruction from serial sections, which are essential for serious study of this class.

The following are the main subdivisions of marine Turbellaria:

Subclass I Archoophora

Without yolk glands, but storing yolk in their eggs.

Order 1 Acoelida

Lacking gut cavity and protonephridia. Mouth leads to phagocytic central area where organisms are digested in food vacuoles. Eight families.

Order 2 Macrostomida

Gut a simple sac. Body width little more than dorsoventral thickness. Two families.

Order 3 Polycladida

Central gut gives off many ramifying branches on each side, within very flattened body. Twenty-four families.

Subclass II Neophora

With yolk glands connected to oviducts and producing yolk cells.

Order 1 Lecithophorida

With only one or two yolk glands. Gut a simple sac.

Suborder 1 Dalyellina

Pharynx anterior, thick-walled and barrel-shaped. Six families, including the Umagillidae (parasites of echinoderms not included here, although common at Plymouth).

Suborder 2 Plagiostomina

Pharynx less thick-walled. Mouth may be anterior or posterior. Five families.

Suborder 3 Mesostomina

Pharynx in mid-body, usually short, thick-walled, with radial muscles like rosette in axial view. Five families.

Suborder 4 Kalyptorhynchina

With a proboscis, extrusible anteriorly. Mouth in mid-body, with bulbous thick-walled pharynx. Six families.

Order 2 Proseriatida

With a series of yolk glands on each side, alternating with small unbranched lobes from gut. Two families.

Order 3 Tricladida

Gut with three major branches (one anterior, two posterior), each ramifying amongst a series of yolk glands on each side. Three marine families.

KEY TO COMMON REPRESENTATIVES OF SELECTED FAMILIES

The families are numbered and the appropriate number precedes the name of each species shown in Fig. 5.1 overleaf.

---

1. Body either coloured green by symbionts or with lateral margins curving together ventrally — **1. Convolutidae**

   Body not green. Margins not thus convoluted — **2**

---

2. Forming chains of zooids by fission — **3. Microstomidae**

   Not reproducing by fission — **3**

---

3. Body width often >3 mm, always 3 × dorso-ventral thickness, usually >0.3 × length (Polycladida) — **4**

   Width of uncompressed body <1.5 mm, either <2 × dorso-ventral thickness or (Tricladida) <0.3 × length — **5**

---

4. Without marginal tentacles — **4. Leptoplanidae**

   With a pair of anterior marginal tentacles — **5. Euryleptidae**

---

5. Commensal in viscera of molluscs — **6. Graffillidae**

   Not associated with molluscs or echinoderms — **6**

---

6. Attaching hard bottle-shaped capsules to undersides of shore rocks **7. Fecampidae**

Not forming such seemingly stony capsules **7**

---

7. Body white, with broad dark-brown transverse bands **8. Plagiostomidae**

Body without transverse bands **8**

---

8. With four eyes **9. Pseudostomidae**

With no more than two eyes **9**

---

9. Length <2 mm **10**

Length >2 mm **14**

---

10. Head with anterolateral lobes or tentacles (*Vorticeros*) **2. Macrostomidae**

Head without paired lobes or tentacles **11**

---

11. Head pointed. Tail bilobed Astrotorhynchinae in **10. Trigonostomidae**

Without these characters **12**

---

12. With extrusible proboscis anteriorly **11. Gyratricidae**

Without proboscis **13**

---

13. Pharynx anterior, inconspicuous. Tail capable of spatulate expansion, with adhesive papillae **2. Macrostomidae**

Pharynx in mid-body, looks like rosette under compression. Tail adhesive but small and tapering **10. Trigonostomidae**

---

14. Thread-like, leech-like, with two eyes touching each other **12. Monocelidae**

Flattened 'planarians', with two eyes well separated **13. Procerodidae**

Subclass Archoophora

Order Acoelida

Eight families, distinguished by positions of mouth and genital pore(s), and presence or absence of pharynx, spermatheca, penis, etc.

1. CONVOLUTIDAE

Mouth antero-ventral; spermatheca and penis present. Seven genera.

Convoluta convoluta (Abildgaard) Fig. 5.1
Length 0.5–3 mm, pear-shaped, speckled brown, with lateral margins curving ventrally. Two eyes. Spermatheca in mid-body near female pore. Male pore (with penis) near posterior end.

One from algae, LWST, Port Eynon. Recorded from Barents Sea to Black Sea.

Convoluta roscoffensis Graff
Length up to 4 mm, body more uniformly slender, coloured green by symbiotic Prasinophyceae. When tide is out, aggregated worms form green streaks >10 cm long, at about HWNT on sandy gravel beaches near Aberthaw, South Wales, most northern locality of species.

Order Macrostomida

Two families.

2. MACROSTOMIDAE

Without asexual reproduction. Ten genera.

Macrostomum appendiculatum (Fabricius) Fig. 5.1
Length 0.5–2 mm, colourless. Mouth an antero-ventral longitudinal slit, behind a pair of small eyes. Pharynx inconspicuous, gut simple. Female pore and atrium may resemble a rosette. Hind end can be spread out for adhesion and bears peripheral *rhabdites* (secreted rods). Male pore, with penis, is mid-ventral nearby.

On *Enteromorpha* in splash-zone pools at Port Erin and Oxwich.

Vorticeros auriculatum (Müller) Fig. 5.1
Length 0.5–1.5 mm or more. Head bilobed, frequently raised while crawling. Two eyes. Tail adhesive, but always tapering. Epidermis colourless, but internal organs obscured by mesenchyme pigment, dark red-brown or pale yellow. Extension of head lobes to form tentacles, as usually figured, was not seen at Swansea.

Norway to Mediterranean. Both colour morphs occur at Swansea. The yellow one seems to be *Vorticeros luteum* Hallez, from Plymouth, Boulogne, Naples, and Adriatic.

3. MICROSTOMIDAE

Multiplying by transverse fission.

Microstomum rubromaculatum Graff Fig. 5.1
Chains 2–3 mm long, of several zooids, producing smaller chains by median fission (arrowed). Head marked by reddish-brown pigment, tail by adhesive papillae. Mouth antero-ventral.

In shallow water with drifting algae, Scandinavia to Mediterranean.

Order Polycladida

Suborder Acotylina

Lacking ventral sucker behind female pore.

4. LEPTOPLANIDAE

With ventral mouth in mid-body; without tentacles on anterior margin; three easily confused British genera are recorded.

---

1. Bearing a pair of cerebral tentacles antero-dorsally (*Stylochoplana*) **2**

   Without any tentacles **3**

---

2. Male and female pores separate ***Stylochoplana maculata***

   Male and female pores united ***Stylochoplana agilis***

---

3. Faintly spotted with brown. Behind pharynx lie a seminal vesicle and a prostatic organ, both bulbous. Latter has a penis papilla bearing a long stylet, which sometimes protrudes through male pore ***Notoplana atomata***

Tinged brown, not spotted. Male organs less conspicuous, without a penis stylet **_Leptoplana tremellaris_**

Leptoplana tremellaris (Müller)
Body delicate, 12–25 mm long, whitish or tinged brown. Two clusters of 20–25 small cerebral eyes extend forward from brain; 6–12 larger 'tentacular' eyes form a dense rounded group on each side of brain. Resembles *Stylochoplana* (see Fig. 5.1) but has neither spots (apart from eyespots), tentacles, nor conspicuous receptaculum.

Scandinavia to Mediterranean.

Notoplana atomata (Müller)
Dorsally whitish brown or grey-brown, with darker spots, ventrally grey-white. Eyes as in *Leptoplana*. An unusually large seminal receptacle may be distinguished posteriorly, as in *Stylochoplana*.

Scandinavia to Channel Islands.

Stylochoplana agilis Lang
This may or may not be separate from the following species.

Stylochoplana maculata (Quatrefages) Fig. 5.1
Body 3–16 mm long, brownish with darker spots. Tentacles are best seen in side view (lower figure, of worm on vertical side of glass dish) for they are transparent, the eyes being sub-basal. Eyes as in *Leptoplana*. The most posterior of the mid-line chain of pharynx and reproductive vesicles is the receptaculum seminis.

One from algae, LWST, Port Eynon. Scandinavia to Mediterranean.

Suborder Cotylina

With a ventral sucker in mid-body, behind female pore.

5. EURYLEPTIDAE

With a pair of tentacles on anterior margin. Ventral mouth in cerebral region, in front of bell-shaped pharynx. Five British genera.

---

1. Length of marginal tentacles much greater than their basal width **3**

   Length of tentacles similar to, or less than, their basal width **2**

---

2. Ultimate branches of gut end in peripheral vesicles, each of which opens dorsally forming a ring of pores around body. Male genital pore distinct from mouth **_Cycloporus papillosus_**

   Without a circle of pores around body. Male pore with penis stylet opens into mouth **_Stylostomum ellipse_**

---

3. Dorsal surface yellowish-white with many dark longitudinal lines **_Prostheceraeus vittatus_**

   Without conspicuous dark longitudinal stripes **4**

---

4. Mouth posterior to brain **_Eurylepta cornuta_**

   Mouth anterior to brain **_Oligocladus sanguinolentus_**

Cycloporus papillosus (Sars) Fig. 5.1
Body elliptical, up to 16 × 9 mm, yellowish-white with many small brownish-red spots. Cerebral eyes separate from those on marginal tentacles. Feeds on botryllid ascidians, to which it clings by mid-ventral sucker. Scandinavia to Mediterranean.

Cycloporus maculatus Hallez
As above, but dorsal surface yellowish-orange, with regularly distributed white spots. Cerebral eye group merges with tentacular group on each side.

Recorded only from Pas-de-Calais?

Eurylepta cornuta (Müller) Fig. 5.1
Body oval, up to 30 × 10 mm, translucent or whitish, tinged with red and dotted white. Eyes 60–70 on each marginal tentacle (extending to tips) and about 200 in each cerebral group, the latter being mostly anterior to pharynx, which is indicated by dotted line. Mouth (arrowed) ventral to cerebral groups.

Norway to Mediterranean.

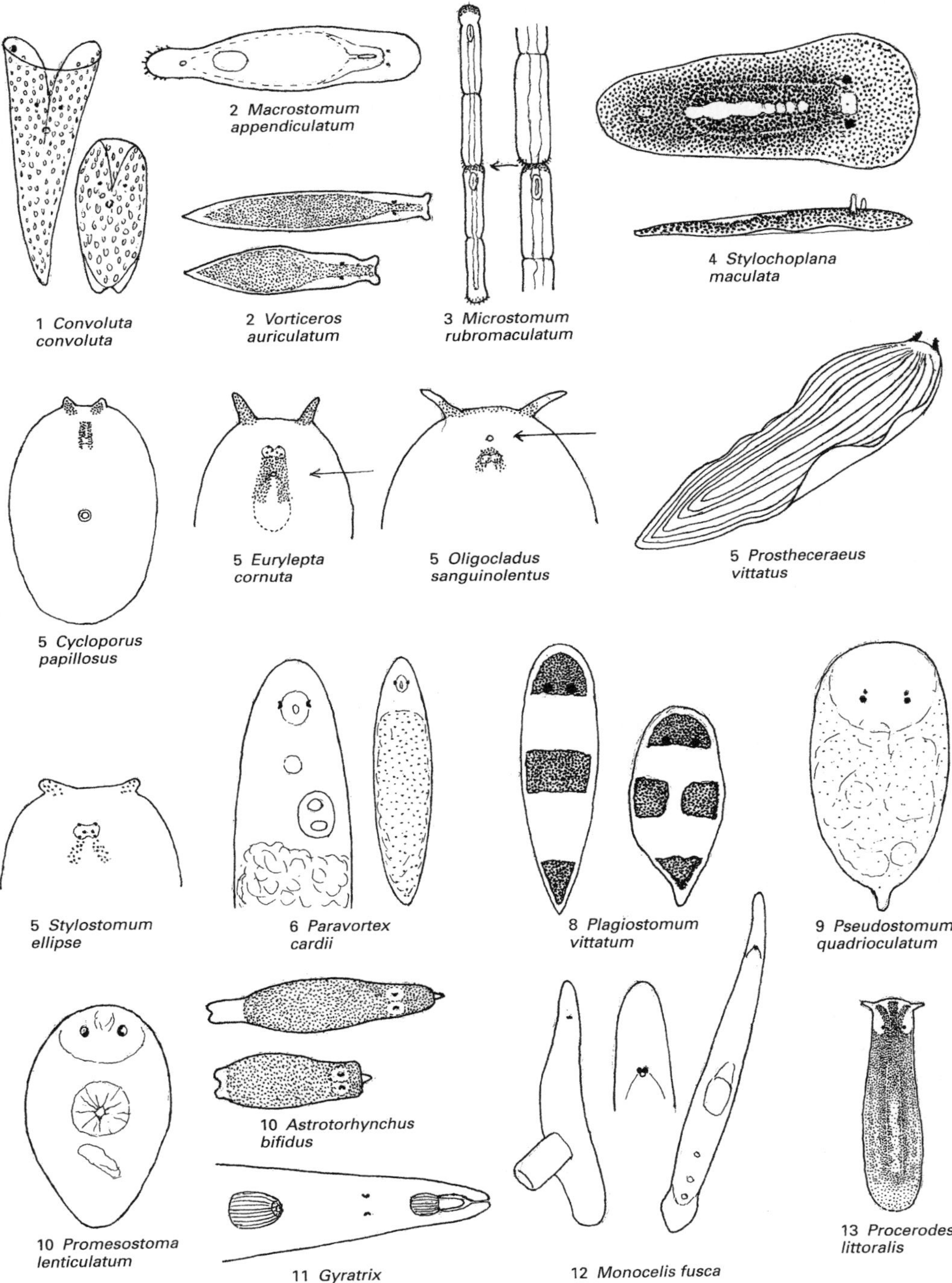

1 Convoluta convoluta
2 Macrostomum appendiculatum
2 Vorticeros auriculatum
3 Microstomum rubromaculatum
4 Stylochoplana maculata
5 Cycloporus papillosus
5 Eurylepta cornuta
5 Oligocladus sanguinolentus
5 Prostheceraeus vittatus
5 Stylostomum ellipse
6 Paravortex cardii
8 Plagiostomum vittatum
9 Pseudostomum quadrioculatum
10 Promesostoma lenticulatum
10 Astrotorhynchus bifidus
11 Gyratrix hermaphroditus
12 Monocelis fusca
13 Procerodes littoralis

Fig. 5.1

Oligocladus sanguinolentus (Quatrefages) Fig. 5.1
Body translucent, up to 22 × 13 mm, whitish, yellowish or brown with marginal band of blue-grey, mid-dorsal line and lateral offshoots brownish or red. Eyes: about 50 with each tentacle and 20–30 in each cerebral group, the latter being dorsal to pharynx. Mouth (arrowed) in front of cerebral groups. Distal halves of tentacles lack eyes.

Under stones, red algae and *Laminaria* intertidally, and dredged to 60 m depth. Scandinavia to Mediterranean.

Prostheceraeus vittatus (Montagu) Fig. 5.1
Body up to 50 × 25 mm, very mobile, yellowish white with many thin, dark, longitudinal stripes.

Under rocks at low tide, or crawling more conspicuously at 5–10 m depth. Scandinavia to Mediterranean.

Stylostomum ellipse (Dalyell) Fig. 5.1
Body up to 10 × 4 mm, translucent but with pigmented food in gut dorsal to a whitish lattice-work of ovaries. Eyes few, grouped as usual, but including three pairs characteristically associated with the brain.

Common intertidally, among algae, *Mytilus*, *Balanus*, and *Ciona*, and at depths to about 200 m. Scandinavia to Mediterranean. Southern Ocean.

Subclass Neoophora

Order Lecithophorida

Suborder Dalyellina

6. GRAFFILLIDAE

About three genera free living and two parasitic or inquiline in molluscs.

Paravortex cardii Hallez Fig. 5.1
Length up to about 1 mm. Mouth and pharynx inconspicuous, ventral to pair of eyes. Common gonopore in front of mid-body, flanked by testes and ovaries, with branching yolk gland filling body posteriorly. Larger individuals contain one or more egg capsules, each with several yolk cells and (usually) two eggs.

In gut of *Cardium edule* from Swansea Bay.

7. FECAMPIDAE

Fecampia erythrocephala Giard
The hard bottle-shaped cysts, about 10 mm long, are common attached to undersides of stones at low tide on shores of the Menai Straits and Plymouth Sound. The reddish parasites, up to 12 mm long, are said to be visible through the thin carapaces of young *Carcinus* (less than 5 mm diameter).

Irish Sea and English Channel.

Suborder Plagiostomina

8. PLAGIOSTOMIDAE

Mouth anterior, separate from genital pore. Eyes two, sometimes four. At least nine marine genera.

Plagiostomum vittatum Leuckart Fig. 5.1
Length 1–2 mm, streamlined when swimming (as on left), more rounded when crawling. Body white, usually with three transverse bands of dark brown, one across head, one across tail and one central. The latter may be interrupted mid-dorsally, and other variations are common. Eyes two, rounded.

From algae in mid-shore pools, Bracelet Bay. Bergen to Boulogne.

9. PSEUDOSTOMIDAE

Mouth anterior, combined with genital pore. Eyes four.

Pseudostomum quadrioculatum (Leuckart) Fig. 5.1
Length 0.5–0.9 mm. Body colourless, rounded, tapering to short adhesive tail. Four rounded eyes, anterior pair smaller and ventral to the other. Pharynx and gut inconspicuous.

Numerous on algae, LWST, Port Eynon. Barents Sea to Black Sea.

Suborder Mesostomina

10. TRIGONOSTOMIDAE

Promesostoma lenticulatum (Schmidt) Fig. 5.1
Length 0.5–1.5 mm, usually oval, tapering posteriorly, may become bell-shaped anteriorly. Eyes two. Mouth midventral. Pharynx like rosette.

One from LWST at Port Eynon resembled Gamble's description from Port Erin. Described from Faeroe Islands.

Subfamily Astrotorhynchinae

One genus, one or two species

Astrotorhynchus bifidus (McIntosh) Fig. 5.1
Body 0.3–2 mm long, yellowish or reddish brown, colourless at conical front and bifid hind ends. Eyes

two. Penis (seen by sectioning) like a dexiotropic and barbed corkscrew in *bifidus*, but like a laeotropic screw without barbs in 'variety *regulatus*' Graff (from Port Erin and Denmark).

Greenland and White Sea to Plymouth and Boulogne. Several in mid-shore pools near Swansea in July.

Suborder Kalyptorhynchina

11. GYRATRICIDAE

Gyratrix hermaphroditus Ehrenberg Fig. 5.1

Length 1 or 2 mm, tapering anteriorly, with proboscis protrusible through anterior pore. Two eyes. Pharynx in mid-body. Penis, with stylet, can be protruded through postero-ventral pore.

Recorded in shore pools at Plymouth, and said to thrive equally in fresh and salt water, but different species may be involved.

Order Proseriatida

12. MONOCELIDAE

Monocelis fusca Oersted Fig. 5.1

Up to 3 mm long, thread-like, colourless at 1 mm long, reddish at 2 mm and brown or greenish when adult. Two eyes touching each other, associated with brain and otolith. Hind end can widen to form an adhesive disc. Pharynx just behind mid-body, followed by three mid-ventral pores (seen under slight compression), the first to receive sperm, the second to extrude a penis, and the third to lay egg capsules. Worm on left has extruded pharynx.

Abundant at Oxwich, with *Enteromorpha* and *Tigriopus* in high shore pools. Bergen to Boulogne.

Order Tricladida

13. PROCERODIDAE

Procerodes littoralis (Strom) Fig. 5.1

Body about 5–7 × 1 mm, pale grey-brown dorsally, widest in posterior half, around pharynx. Head bears pointed tentacular lobe on each side, and three diverging stripes in front of a pair of small eyes. When disturbed this species hastens along like a leech (and like many Turbellaria), whereas other littoral triclads observed by Hartog moved smoothly.

Found most easily at mid-tide, where fresh water runs out over gravel and sand, under rocks to which it attaches spherical brown egg capsules, 1 mm in diameter. Murmansk to Brittany.

PHYLUM NEMERTEA

Unsegmented, bilaterally symmetrical, ciliated acoelomate worms, with separate mouth and anus, a closed blood system, and an eversible muscular proboscis contained within a dorsal fluid-filled chamber, the rhynchocoel. Classification uses internal characters.

Nemerteans occur under stones, in sand, mud, and rock crevices, amongst turf algae, kelp holdfasts, trawled hydroids, and other dense populations of sedentary animals. Ideally they should be studied alive, noting colours, pigment patterns, head shape, positions of *mouth* and *proboscis pore*, numbers and distributions of *eyes* and *cephalic grooves*, and the presence or absence of a *caudal cirrus*. Some internal organs (e.g., the *cerebral ganglia* or *proboscis stylets*) can often be distinguished by transmitted light, particularly in pale or translucent species which have been gently compressed under a supported glass slide. $MgCl_2$(0.3N) can be used as a narcotic during this process and washed away afterwards to allow recovery. Although the broad classification of the phylum also uses characters revealed only by histological sectioning, the study of living material is sufficient for the identification of many species. A considerable number of species have indeed been described from external features alone, with no details of internal anatomy; this has led to several such forms being taxonomically misplaced. Proper study needs microscopy and drawings from life, followed by histological examination. Species identification of preserved material is particularly difficult, and often scarcely practicable.

KEY TO FAMILIES

1.	Pigmented body with conspicuous rectangular pattern of white lines	**22**
	Without such a pattern	**2**
2.	With longitudinal grooves on opposite sides of head	**17**
	Without longitudinal grooves on head	**3**
3.	With four eyes forming a rectangle	**9**
	Eyes two, none, or more than four	**4**
4.	Elongated body densely pigmented dorsally but pale ventrally	**12**
	Without such a marked difference between dark and pale surfaces	**5**
5.	Flattened, leech-like body with posterior ventral sucker (Fig. 5.5Q). In mantle cavity of bivalve molluscs	**16. Malacobdellidae**
	Body much longer than wide, without sucker. Not with bivalves	**6**
6.	With two small eyes. Parasitic on gills or eggs of crabs. Proboscis very short and reduced (Fig. 5.4G)	**9. Carcinonemertidae**
	Eyes rarely two, usually four, many or none. Free living. Proboscis at least one-third length of body	**7**
7.	Mouth indistinct, in front of brain, usually opening with proboscis pore at or near anterior tip. Proboscis armed with one or more stylets. TS shows lateral nerves internal to body wall muscle layers (Enopla, Hoplonemertea)	**8**
	Mouth usually distinct, below or behind brain, well separated from proboscis pore. Proboscis not armed with stylets. Lateral nerves embedded in or external to body wall muscle layers (Anopla)	**16**
8.	With four eyes arranged to form corners of a square or rectangle. Mostly small worms (less than 20–30 mm long), with or without distinctive colour patterns or markings on dorsal surface; markings, when present, often restricted to head and may partially mask the eyes	**9**
	Eyes usually 0 or more than 4, often numerous. Body often long and slender, frequently much coiled. Rarely with distinct colour pattern, but often with dorsal surface more deeply pigmented than ventral	**12**

9. Body soft, slender, colour pattern when present restricted to head or consisting of one or two dark but narrow longitudinal stripes (Fig. 5.5A–N) **13. Tetrastemmatidae**

Without colour pattern or with pattern extending full body length; if striped, stripes are pale on darker background **10**

10. Body soft, delicate, without colour pattern **11**

Body firm, stout, with distinct colour pattern of transverse bands, marbled or pale longitudinal bands (Fig. 5.4M) **12. Prosorhochmidae**

11. Head rounded or diamond shaped, without distinct terminal notch, eyes small. Mid- to lower shore, typically associated with algae (Fig. 5.5A–N) **13. Tetrastemmatidae**

Head somewhat bilobed, with terminal notch, front pair of eyes significantly larger than rear pair. Upper shore, rarely found other than in summer months when mature individuals give birth to live young (Fig. 5.4N) **12. Prosorhochmidae**

12. Live body length more than 40 × width. Proboscis sheath not reaching beyond middle of body. Head distinct with many eyes, or very slender, indistinct, with four minute eyes (Fig. 5.4I-L) **11. Emplectonematidae**

Live body length usually less than 20 × width. Proboscis sheath reaches to, or almost to, posterior tip of body. Eyes two or numerous **13**

13. Eyes numerous, arranged in four longitudinal rows on head. Proboscis armed with pad bearing several stylets (Polystilifera) (Fig. 5.5P) **14**

Eyes not as above, proboscis with one central stylet (Fig. 5.4L) **15**

14. Dorsally marked with colour pattern of pale longitudinal stripes (Fig. 5.5O) **14. Drepanophoridae**

Dorsal body surface without pattern of longitudinal stripes (Fig. 5.5P) **15. Paradrepanophoridae**

15. Body stout, thick, somewhat dorso-ventrally flattened, very contractile. Head with median dorsal longitudinal ridge or swelling. Proboscis large relative to size of body (Fig. 5.4H) **10. Cratenemertidae**

Body soft, delicate, not strongly contractile. Head without median dorsal longitudinal ridge or swelling. Proboscis comparatively small relative to size of body (Fig. 5.4A–G) **8. Amphiporidae**

16. With distinct longitudinal grooves on opposite sides of head **17**

Without longitudinal grooves on head, either entirely without grooves or with single transverse furrow encircling rear of head **20**

17. With longitudinal median groove on upper and lower surfaces of head (Fig. 5.3S) **7. Valenciniidae**

With lateral longitudinal grooves on each side of head **18**

18. With caudal cirrus **19**

Without caudal cirrus (Fig. 5.3F–M) **6. Lineidae**

19. Head more or less pointed, body distinctly flattened dorso-ventrally, broad with sharp lateral margins (Fig. 5.3A–D) **5. Cerebratulidae**

Head rounded or pointed, body not distinctly flattened dorso-ventrally, more or less slender with rounded lateral margins (Fig. 5.3N–Q) **6. Lineidae**

20. With numerous eyes along borders of rounded head. Body long, slender, marked by narrow interrupted dark longitudinal stripes (Fig. 5.3R) **7. Valenciniidae**

Without eyes. Either without obvious colour pattern or with well-marked pattern of longitudinal and/or transverse white bands **21**

21. With transverse furrow encircling rear of head. Head broad, bluntly pointed (Fig. 5.3E) **5. Cerebratulidae**

Without transverse furrow at rear of head. Head rounded or, if pointed, extremely slender **22**

22. Head rounded, usually forming distinct anterior lobe. With strikingly marked colour pattern of white transverse and/or longitudinal bands (Fig. 5.2J–R) **4. Tubulanidae**

Head slender, pointed, or rounded. Without colour pattern of transverse and/or longitudinal bands **23**

23. Mouth far behind brain, four or five body widths from anterior end (Fig. 5.2B–E) **2. Cephalothricidae**

Mouth just behind brain, less than four and usually less than two body widths from anterior end **24**

24. Head long, slender, tapering to point, body thickest in posterior half (Fig. 5.3T) **7. Valenciniidae**

Head rounded, body either thickest in anterior half or of more or less uniform width throughout **25**

25. Head not delimited from trunk. Body long, slender, tapering gradually from thicker anterior regions (Fig. 5.2A) **1. Carinomidae**

Head usually forming distinct lobe. Body of more or less uniform thickness for most of its length **26**

26. Head with rhynchodaeum visible internally as a triangular shape at its obtusely blunt tip (Fig. 5.2F) **3. Hubrechtidae**

Rhynchodaeum not distinguishable or not forming triangular shape (Fig. 5.2J–R) **4. Tubulanidae**

Class Anopla

Mouth below or posterior to cerebral ganglia. Longitudinal nerve cords embedded in body wall. Proboscis without stylets. Two orders.

Order Palaeonemertea

Outermost muscles of body wall circular. Nerve cords either external to these muscle layers or embedded in the longitudinal musculature.

1. CARINOMIDAE

Remarkable in possessing muscle fibres (circular and longitudinal) within the epidermis. Anteriorly the nerve cords lie just below the epidermal basement membrane, but posteriorly they move deeper into the inner longitudinal muscle layer. Only one genus.

Carinoma armandi (McIntosh) Fig. 5.2A

Up to 20 cm long, thread-like when extended. Head rounded in front, flattened and slightly enlarged behind the tip. No grooves or eyes. Body mostly cylindrical, anterior part whitish, middle pale buff, posterior end flattened, translucent and somewhat pointed. Superficially resembles *Tubulanus linearis*, but lacks cerebral organs.

In sand near LW, amongst tubes of *Lanice*. Known only from Southport and St. Andrews.

2. CEPHALOTHRICIDAE

Head somewhat pointed, with mouth four or five body widths distant from the anterior tip. Also with no cerebral organs, and nerve cords situated in the inner longitudinal muscle layer throughout. Most cephalothricids lack an inner circular muscle layer. Species separation difficult.

1. On disturbance contracts into a tight spiral (*Callinera* of the Tubulanidae also does this) **2**

On disturbance may contract into a tight knot, but does not coil spirally **3**

2. Up to 15 cm long and 1 mm or more wide, with an incomplete layer of inner circular muscle fibres in the foregut region ***Procephalothrix filiformis***

Up to 1 cm long and 0.2 mm wide, without inner circular muscles ***Cephalothrix arenaria***

3. Up to 30 cm long, 0.5-1 mm wide, translucent white, yellowish, or greyish ***Cephalothrix linearis***

Up to 6 cm long, 0.5 mm wide, whitish or translucent tinged with orange, red, or purple on tip of head **_Cephalothrix rufifrons_**

Cephalothrix arenaria Hylbom Fig. 5.2B

Length 6–10 mm. Translucent with yellowish tinge. Characterized mainly by its small size.

Common in fine, clean sand, 4 m depth, near Kristineberg, Sweden.

Cephalothrix linearis (Rathke) Fig. 5.2C

Lower shore and sublittoral, burrowed in coarse sand or mud, or among *Laminaria* holdfasts, *Corallina*, on hydroids or beneath stones. Britain, Sweden, Mediterranean, Greenland, eastern North America, Japan.

Cephalothrix rufifrons (Johnston) Fig. 5.2D

Gregarious, common intertidally beneath boulders, usually in clean coarse sand, and amongst *Corallina* in pools, occasionally in brackish water. Dredged to depths of 39 m. Britain, Sweden, Mediterranean.

Procephalothrix filiformis (Johnston) Fig. 5.2E

Head bluntly pointed, white or translucent, body yellowish white or orange, often darker posteriorly.

Under stones or in muddy gravel from midshore to 40 m or more depth. St. Andrews, western Ireland, Channel Islands, France.

3. HUBRECHTIDAE

With a sensory cerebral organ on each side of head. Otherwise typically palaeonemertean except in having a mid-dorsal blood vessel. *Hubrechtia* (from Naples) has an inner circular muscle layer, but *Hubrechtella* lacks this.

Hubrechtella dubia Bergendal Fig. 5.2F

Body up to 1.5 cm long, 0.5 mm wide, white, translucent anteriorly but not in the tapering extensible posterior part. Head not wider than body, rounded at front but flattened when creeping. Slight constriction in mouth region. The short rhynchodaeum can usually be seen as a triangular shape at the tip of the head. The proboscis and rhynchocoel are half the length of the body.

In surface of sublittoral muds, at depths from 8–100 m or more. Sweden, Norway, Denmark.

4. TUBULANIDAE

Some are brown or red and may bear conspicuous white stripes. Others are small and translucent enough to show the cerebral organs by transmitted light. These organs are embedded one on each side in the constriction separating the rounded, somewhat flattened head from the more or less cylindrical body. Cerebral organs are lacking, however, from *Callinera* and *Carinesta*. The nerve cords usually lie just below the basement membrane of the epidermis, as in most palaeonemerteans, but *Carinina* (alone amongst nemerteans) has the nerve cords within the epidermis, outside the thin basement membrane.

---

1. Brown or reddish with transverse white rings and usually also longitudinal white stripes **2**

   Without such conspicuous white stripes **6**

---

2. Head colour and pattern like that of body **3**

   Head white **4**

---

3. With four longitudinal stripes (mid-dorsal, mid-ventral, and two lateral) **_Tubulanus superbus_**

   With three such stripes (mid-dorsal and two lateral) **_Tubulanus annulatus_**

---

4. White head lacks markings and is followed by a dark unconstricted collar of brownish red. Rest of body yellowish **_Tubulanus albocapitatus_**

   White head bears two black pigment patches **5**

---

5. Up to 1.5 cm long, with up to 18 transverse rings. Indistinct mid-dorsal longitudinal stripe as a series of white flecks present or absent **_Tubulanus banyulensis_**

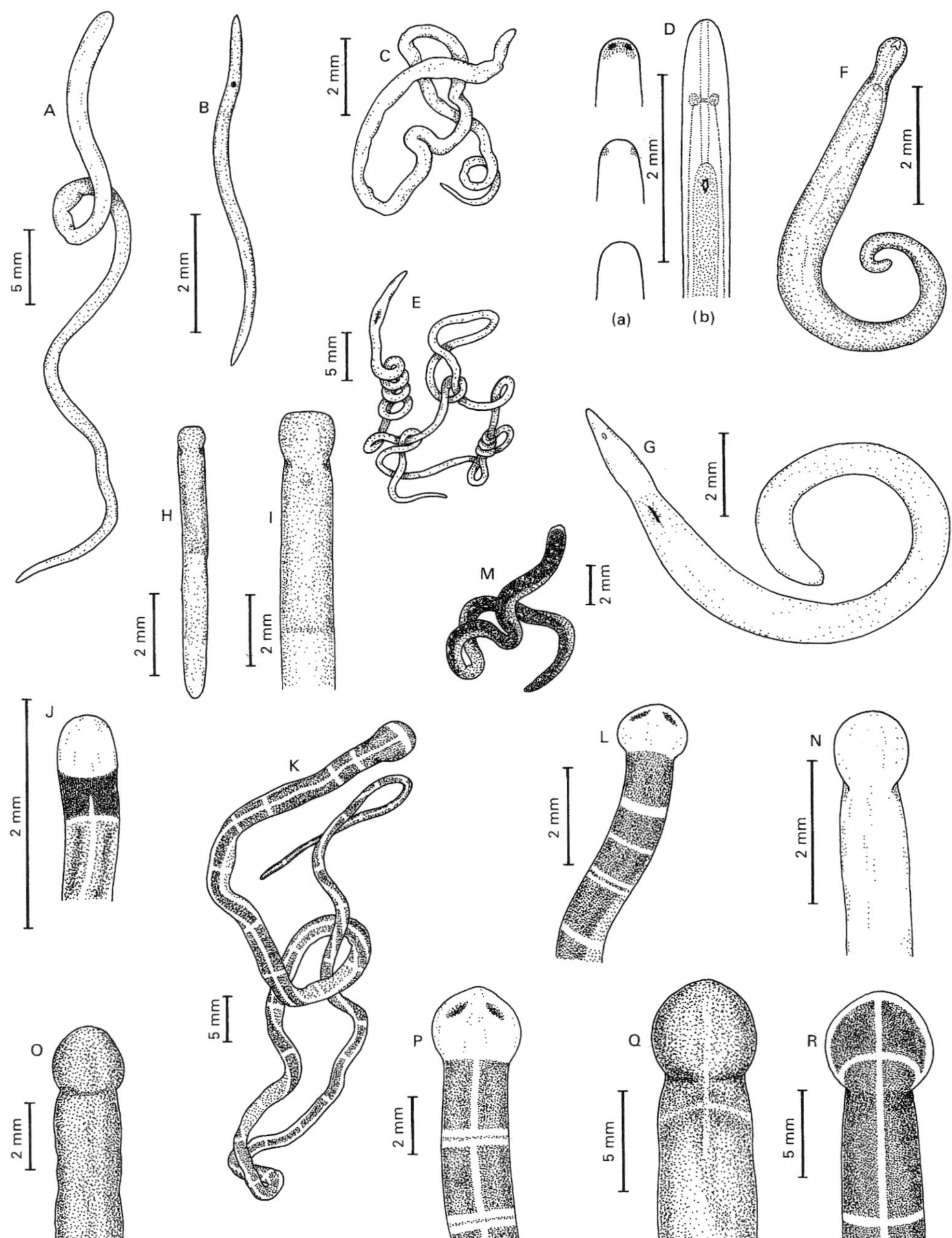

Fig. 5.2 Palaeonemertea. 1, CARINOMIDAE: A. *Carinoma armandi*. 2, CEPHALOTHRICIDAE: B. *Cephalothrix arenaria*, C. C. *linearis*, D. C. *rufifrons*, E. *Procephalothrix filiformis*. 3, HUBRECHTIDAE: F, *Hubrechtella dubia*. 4, TUBULANIDAE: G. *Callinera buergeri*, H. *Carinina arenaria*, I. C. *coei*, J. *Tubulanus albocapitatus*, K. *T. annulatus*, L. *T. banyulensis*, M. *T. inexpectatus*, N. *T. linearis*, O. *T. miniatus*, P. *T. nothus*, Q. T. *polymorphus*, R. *T. superbus*.

Up to 10 cm long, with 40 or more rings and three or four longitudinal stripes **Tubulanus nothus**

---

6. Richly pigmented, brown, orange, or red **7**

   Pale yellow, pink, grey, milky, or translucent **10**

---

7. Up to 30 cm or more long and 5 mm wide. Dark reddish brown **8**

   5 cm or less in length, 2 mm wide. Red-orange or red-brown **9**

---

8. Head wider than body **Tubulanus polymorphus**

   Head about the same width as body **Tubulanus theeli**

---

9. Vermilion or reddish brown, with minute specks of white **Tubulanus inexpectatus**

   Uniformly orange, or paler ventrally **Tubulanus miniatus**

---

10. Head rounded and flattened, broader than body or marked off by a slight constriction in which a cerebral organ opens on each side **11**

    Head pointed, narrow, and not marked off from body, without cerebral organs **13**

---

11. Length up to 15 cm **Tubulanus linearis**

    Length up to 3 cm, width 0.3–2 mm, translucent yellowish with a sudden change to a lighter tinge about halfway along body **12**

---

12. Thin, 0.3–0.5 mm wide, lives interstitially in sand **Carinina arenaria**

    Stout, usually about 1 cm long, 1 mm wide, lives in mud **Carinina coei**

---

13. Posterior part of proboscis sheath partly surrounded (ventrally and laterally) by a mass of transverse muscles (distinguishable through skin, close behind head) **Callinera buergeri**

    Not as above **Carinesta anglica**

Callinera buergeri Bergendal Fig. 5.2G
Length 9–30 mm, width 0.3–1.5 mm. Head translucent, pointed, not broadened or marked off from body. Proboscis pore ventral, about one body width from tip of head. Mouth two or three body widths further back. Body whitish, rounded anteriorly, somewhat flattened posteriorly, hind end (partly missing in figure) often coiled spirally.

Mud at depths of 6–50 m, especially in polluted situations. Sweden, Norway, Denmark.

Carinesta anglica Wijnhoff
Head elongated and pointed, becoming wrinkled when contracted. Body translucent, whitish anteriorly, tinged rosy-brown posteriorly.

Sand or muddy sand at low water, near Plymouth.

Carinina arenaria Hylbom Fig. 5.2H
Length 7–15 mm, width 0.3–0.5 mm. Body yellowish, translucent, posterior part with lighter tinge.

Clean coarse sand, 4 m depth, West Sweden.

Carinina coei Hylbom Fig. 5.2I
Length about 10 mm, width 1–2 mm. Otherwise like *C. arenaria*.

Mud, depths 20–30 m. Sweden, Denmark.

Tubulanus albocapitatus Wijnhoff Fig. 5.2J
Length about 1.5 cm, width 0.5 mm. Head small and white. Body yellowish, but reddish brown dorsally, with white lines mid-dorsally and laterally, and forming 11 transverse rings.

Muddy sand or gravel at 50–60 m depth. Plymouth.

Tubulanus annulatus (Montagu) Fig. 5.2K
Length up to 75 cm or more, width 3–4 mm. Head broad and rounded.

Occasionally intertidal under rocks or in sand or mud at LW, more commonly sublittoral to depths of 40 m or more. With a wide geographical range in the northern hemisphere, from the Pacific coast of North America, North Sea, and Mediterranean coasts of Europe.

Tubulanus banyulensis (Joubin) Fig. 5.2L
Length 1–1.5 cm, width 1–1.5 mm. Head rounded with two black markings anteriorly. Body tapering, reddish or greenish, with ventral surface sharply contrasting pink or yellow. Up to 18 white rings, sometimes with indistinct mid-dorsal longitudinal stripe formed as a series of flecks.

Dredged from 4–16 m depth. West Ireland, Mediterranean.

Tubulanus inexpectatus (Hubrecht) Fig. 5.2M
Length up to 3.5 cm, width 1 mm. Tip of head colourless, body red or red-brown, paler ventrally, with minute specks of white.

Clean gravel at 90 m depth. West Ireland, Capri.

Tubulanus linearis (McIntosh) Fig. 5.2N
Length up to 15 cm, width 0.5–1 mm. Body translucent white, often ringed orange or yellow-brown posteriorly.

Sand at LW and sublittorally. Britain, Mediterranean.

Tubulanus miniatus (Bürger) Fig. 5.2O
Length 3–4.5 cm, width 2 mm. Body orange-red.

Dredged from 40-70 m depth. Naples, Plymouth.

Tubulanus nothus (Bürger) Fig. 5.2P
Length up to 10 cm, width 2.5 mm. Resembles *T. annulatus* but with the head white and bearing two curved black patches anteriorly, and sometimes with an inconspicuous mid-ventral line.

Dredged. Naples, Plymouth.

Tubulanus polymorphus Renier Fig. 5.2Q
Length up to 50 cm or more, width 5 mm. Head rounded, much wider than body. Body uniformly reddish or orange-brown. In alcohol or formalin loses colour except for a dark band in the region of the foregut and nephridia. There is a lateral sensory organ on each side, near the posterior margin of the band. In this region too there is a dorsal muscle-cross between the outer and inner circular muscle layers of the body wall, but no ventral muscle-cross.

Sand and gravel down to 50 m depth or more. Britain, Scandinavia, Mediterranean, and westwards to Pacific coast of North America.

Tubulanus superbus (Kölliker) Fig. 5.2R
Length up to 75 cm or more, width 5 mm. Similar to *T. annulatus* but with up to 200 white rings and a distinct mid-ventral white line in addition to the other three longitudinal stripes.

Sand or gravel down to 80 m or more depth, occasionally intertidal. Sweden, Britain, Mediterranean.

Tubulanus theeli (Bergendal)
Length 3 cm or more, width 1–1.5 mm. Reddish brown and like *T. polymorphus*, but with head usually not broader than body and with both dorsal and ventral muscle-crosses in the foregut region.

Sand or gravel, 20–35 m depth. West Sweden, Denmark.

Order Heteronemertea

Body wall muscles in three layers, outer and inner longitudinal, middle circular. Nerve cords lie just outside the middle layer.

5. CEREBRATULIDAE

Proboscis when everted has in its wall an arrangement of muscle layers like that of the body wall (i.e. a middle layer of circular muscles between outer and inner layers of longitudinal fibres). Most have body somewhat flattened, with lateral edges often sharp and fin-like. Some swim readily, with undulations in the sagittal plane. *Cerebratulus* has a horizontal slit along each side of the head, ending posteriorly in the aperture of a cerebral organ (figured for *C. marginatus*).

---

1. With a longitudinal groove on each side of head and a caudal cirrus posteriorly **2**

Without longitudinal cephalic grooves or a caudal cirrus
Oxypolia beaumontiana

---

2. Posterior end wider than anterior, not tapering at its rear
Cerebratulus fuscus

Posterior end narrower than anterior, body tapering gradually hindwards **3**

---

3. Mottled with dark and pale markings, especially near antero-dorsal margins
Cerebratulus pantherinus

Not mottled **4**

---

4. Greyish brown, greyish green, slate blue, or brown, with whitish lateral margins ***Cerebratulus marginatus***

Pink or yellowish **5**

---

5. Head yellowish, narrower than trunk
Cerebratulus roseus

Head white-tipped, swollen
Cerebratulus alleni

Cerebratulus alleni Wijnhoff
Poorly described, without measurements recorded. Head swollen with white tip, body otherwise pink.

Sand at LW, Plymouth.

Cerebratulus fuscus (McIntosh) Fig. 5.3A
Length up to 15 cm, width 2–5 mm. Yellowish, pinkish, or greyish brown. Tapers anteriorly to a bluntly pointed head, with 4–13 eyes on each side. Caudal cirrus originates from a broad, almost square posterior end. Readily autotomizes posteriorly. Swims readily. Sketch of head shows reddish cerebral ganglia by transparency.

Sand or shelly gravel at LW and to 50 m or more depth; as deep as 1590 m off Portugal. Fairly common, Alaska, Florida, Greenland, Britain, Mediterranean, South Africa.

Cerebratulus marginatus Renier Fig. 5.3B
Length up to 100 cm, width to 25 mm or more, tapering towards both ends. Often recorded in the literature as eyeless, for eyes are small and inconspicuous. Otherwise, similar to *C. fuscus*. Swims actively. Head illustrated in side view shows canal of cerebral organ opening into cephalic slit.

Sand or mud sublittorally at depths of 20–150 m or more, rarely at LW. Fairly common, widely distributed in northern hemisphere from Japan eastwards to Europe.

Cerebratulus pantherinus Hubrecht Fig. 5.3C
Length 4–7 cm or more. Not well described. Marked antero-dorsally by a mottled pattern of brown, green, yellow, and white.

Sand at 50 m depth or deeper. Plymouth, Roscoff, Naples.

Cerebratulus roseus (Delle Chiaje) Fig. 5.3D
Length up to 50 cm, width 5–6 mm. Caudal cirrus may be 2 cm long. No eyes. Pink in foregut region, otherwise yellowish, with colourless lateral margins.

In limestone burrows and sand, LW to 30 m depth. Plymouth, Mediterranean.

Oxypolia beaumontiana Punnett Fig. 5.3E
Length about 12 cm, width 5 mm. Head flat and pointed, bounded by a transverse furrow in front of mouth. Body white anteriorly, tinged with brown or red posteriorly.

Intertidally and to 50 m depth, on soft rocks well covered by colonial invertebrates. Plymouth.

6. LINEIDAE

Proboscis with only two muscle layers; in everted position these are outer circular and inner longitudinal muscles. All members of this family have a horizontal slit along each side of head, as in *Cerebratulus*.

---

1. With a caudal cirrus **2**

Without a caudal cirrus **7**

---

2. Body up to 20 cm long, bright red, with head tinged yellow near tip. Lateral sensory organs present. Rhynchocoel has lateral pouches in foregut region
Micrella rufa

Mostly smaller, not red tipped with yellow anteriorly. Without lateral organs or lateral pouches to rhynchocoel (Genus *Micrura*) **3**

---

3. Reddish or brownish with transverse white bands ***Micrura fasciolata***

   Without transverse white bands **4**

---

4. Body milky white ***Micrura lactea***

   Body pigmented **5**

---

5. Head with two lateral rows of eyes, five to eight in each ***Micrura scotica***

   Head mostly white or white-tipped, without eyes **6**

---

6. Bright red dorsally, white or pink ventrally. Head bears a reddish patch anteriorly ***Micrura aurantiaca***

   Dark purplish-brown dorsally, somewhat paler ventrally. Otherwise white head bears a transverse band of yellow ***Micrura purpurea***

---

7. Body thick, wrinkled, more or less cylindrical, speckled with brown or green ***Euborlasia elizabethae***

   Body thin, smooth, thread-like or ribbon-like, not speckled **8**

---

8. With two pale stripes dorsally, extending the length of the body ***Lineus bilineatus***

   Without conspicuous longitudinal stripes **9**

---

9. Body mostly white or pale coloured **10**

   Body darkly pigmented **11**

---

10. Head sharply pointed with deep cephalic slits. Without eyes ***Lineus acutifrons***

    Head bluntly rounded with shallow cephalic slits. With eyes ***Lineus lacteus***

---

11. Body dark brown, black, or green **12**

    Body reddish or mid-brown **13**

---

12. Black or dark brown, except for younger specimens. Usually over a metre long and may reach 20–25 m or more ***Lineus longissimus***

    Green to greenish black, less than 10 cm long ***Lineus viridis***

---

13. If disturbed contracts without coiling ***Lineus ruber***

    If disturbed contracts into tight spiral coils ***Lineus sanguineus***

Euborlasia elizabethae (McIntosh) Fig. 5.3F
Length up to 30 cm, width 5–6 mm. Body rounded, but becomes flattened posteriorly when contracted. Head tapering, bluntly pointed, white or pale yellow, bearing on each side a short horizontal cephalic slit, which continues internally as a canal. Body brown, speckled with paler shades and regularly marked with pale transverse belts.

Sand or mud, LW to 50 m depth. Herm (Channel Islands), Mediterranean.

Lineus acutifrons Southern Fig. 5.3G
Length (estimated) up to 15–17 cm, width 5–7 mm, depending upon degree of contraction. White anteriorly, gradually becoming pink, red, or brown posteriorly, or red in front and pink behind. Head acutely pointed, without eyes.

Sand near LW, western Ireland.

Lineus bilineatus (Renier) Fig. 5.3H
Length up to about 70 cm, width 6 mm. Body reddish brown, purplish, or chocolate, paler in juveniles, marked with two white or yellowish longitudi-

nal dorsal stripes adjoining the mid-dorsal line. On the head these diverge, outlining an elongate triangular shape. No eyes.

Lower middle shore to sublittoral. Fairly common on gravel, sand, mud, amongst coralline algae, under stones, or between mussels and oysters. Scandinavia, Britain, North America, South Africa.

Lineus lacteus (Rathke) Fig. 5.3I

Length to 60 cm, width 1–2 mm. Head bears 6–15 eyes in a dorso-lateral row on each side. Body pink anteriorly, fading to white or pale yellow posteriorly. Mouth unusually far behind cerebral ganglia and not indicated in illustration, which shows cerebral organs behind ganglia, with canals opening into lateral slits, and rhynchodaeum running to end of snout.

Low water and sublittoral on gravel, sand, or stones. Britain, Mediterranean, Black Sea.

Lineus longissimus (Gunnerus) Fig. 5.3J

Length commonly 5–15 m, occasionally up to 30 m or more, width about 5 mm. Head bears 10–20 eyes in a row on each side, but these are usually obscured by the dark pigmentation of the body. Colour olive brown to black, often with an iridescent sheen.

Fairly common under rocks or on mud, sand, or shell sediments; LW and sublittorally. Iceland and the Atlantic, North Sea, and Baltic coasts of Europe.

Lineus ruber (Müller) Fig. 5.3K

Length up to about 8 cm, rarely more, width 2–3 mm. Head with 2–8 eyes in an irregular row on each side. Body reddish brown, paler ventrally. When irritated contracts by becoming shorter and broader without coiling. Larvae 10–15 per egg string, weakly photopositive for only a short time after emergence.

Very common intertidally on muddy sand under stones, but also amongst barnacles and mussels, on rock pool algae, in laminarian holdfasts, etc. Extends sublittorally and into brackish water. Circumpolar in northern hemisphere.

Lineus sanguineus (Rathke) Fig. 5.3L

Length up to 20 cm, width 2–3 mm. Very similar to *L. ruber* but with 4–6 eyes on each side, arranged further back and more regularly, and when irritated contracts into tight spiral coils.

Intertidal under rocks embedded in black mud. Scandinavia, western Europe, Britain.

Lineus viridis (Müller) Fig. 5.3M

Like *L. ruber* but dark olive green to greenish black, or occasionally pale green and sometimes lighter ventrally. Larvae small, 400–500 per egg string, strongly photopositive for 2–3 weeks after emergence.

Locally abundant on mud below midshore boulders. Britain, Mediterranean, and widespread in the northern hemisphere.

Micrella rufa Punnett

Length up to 20 cm, width 2–3 mm. Head pointed, with shallow cephalic slits but no eyes. Body uniformly bright red, shading into a yellowish tinge near tip of head.

Mud at LW, near Plymouth.

Micrura aurantiaca (Grube) Fig. 5.3N

Length up to 8 cm, width 1.5–2 mm. Head white, bluntly tapered, without eyes, usually with an antero-dorsal spot of red, violet, or brown. Body rounded and red dorsally, flattened and white or pink ventrally. Caudal cirrus small and translucent, often indistinct.

Beneath stones in rock pools and sublittorally from coralline grounds. Plymouth, Herm, Mediterranean.

Micrura fasciolata Ehrenberg Fig. 5.3O

Length 10–15 cm, width 1–4 mm. Usually reddish brown, with white transverse bars, sometimes yellowish, greenish brown, or purple, with the white bars indistinct or lozenge-shaped. Head white, or coloured but white-tipped, with 3–12 eyes in a row on each side.

Sand, gravel, mud, rocks, with *Pomatoceros* tubes, or in laminarian holdfasts, LW to 80 m depth. Scandinavia, Britain, Mediterranean.

Micrura lactea (Hubrecht) Fig. 5.3P

Length up to 9 cm, width 1–1.5 mm. Eyes absent. Milky-white overall, sometimes tinged pink, yellow, or brown.

Dredged from depths of about 30 m. Port Erin, Plymouth, Mediterranean.

Micrura purpurea (Dalyell) Fig. 5.3Q

Length up to 20 cm, width 2–3 mm. Head with transverse band of yellow. Snout whitish or

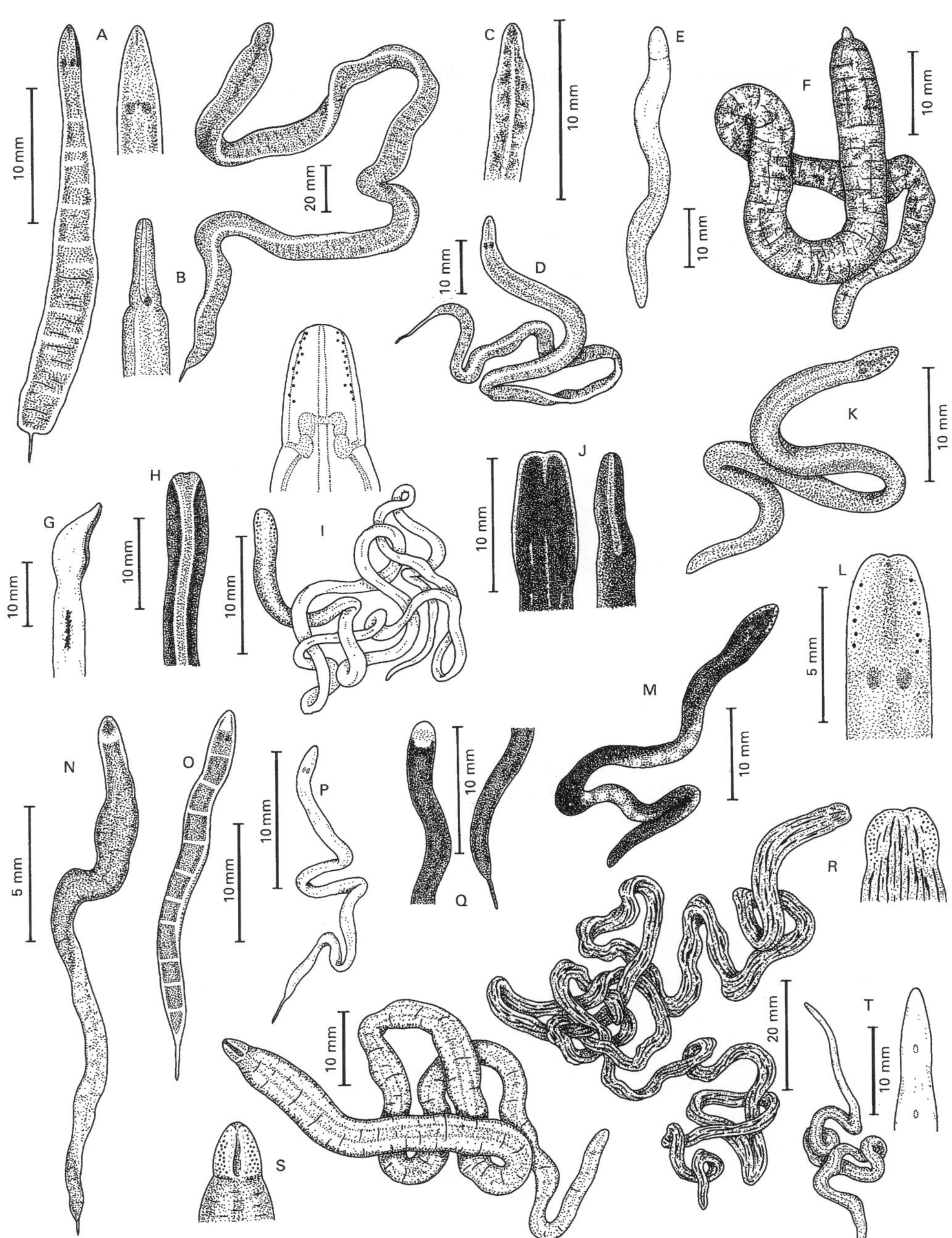

Fig. 5.3 Heteronemertea. 5, CEREBRATULIDAE: A. *Cerebratulus fuscus*, B. *C. marginatus*, C. *C. pantherinus*, D. *C. roseus*, E. *Oxypolia beaumontiana*. 6, LINEIDAE: F. *Euborlasia elizabethae*, G. *Lineus acutifrons*, H. *L. bilineatus*, I. *L. lacteus*, J. *L. longissimus*, K. *L. ruber*, L. *L. sanguineus*, M. *L. viridis*, N. *Micrura aurantiaca*, O. *M. fasciolata*, P. *M. lactea*, Q. *M. purpurea*. 7, VALENCINIIDAE: R. *Baseodiscus delineatus*, S. *Poliopsis lacazei*, T. *Valencinia longirostris*.

translucent. Eyes absent. Body dark purplish brown, often appearing iridescent, similar but sometimes paler ventrally.

Sand, gravel, mud, or rocks, LW to 40 m depth. Scandinavia, Britain, Mediterranean.

Micrura scotica Stephenson
Length 6.5 cm, width 2.5 mm. Head tapering, red-tipped, with a row of 5–8 eyes on each side. Anterior body light brown, tinged purple. Posteriorly only the gut and its caeca are similarly pigmented, the body otherwise appearing pale, especially at margins.

Dredged from about 30 m depth, Firth of Clyde.

7. VALENCINIIDAE

Proboscis with only two muscle layers, the outer of longitudinal and the inner of circular fibres when everted. Horizontal lateral cephalic grooves absent. NB The minute palaeonemertean *Callinera* may key out here, for its tapering head resembles *Valencinia* (see 4. TUBULANIDAE).

---

1. With longitudinal stripes **Baseodiscus delineatus**

   Without longitudinal stripes 2

---

2. With a mid-dorsal and a mid-ventral cephalic furrow **Poliopsis lacazei**

   Without cephalic furrows **Valencinia longirostris**

Baseodiscus delineatus (Delle Chiaje) Fig. 5.3R
Length up to 100 cm or more, width 2–4 mm. Head slightly bilobed, with many eyespots. Body light brown marked with interrupted reddish-brown longitudinal stripes which are occasionally fused, 5–12 stripes dorsally, fewer and paler ventrally.

Sublittoral on shells or gravel with sand or mud, Plymouth; world-wide in tropical and subtropical latitudes.

Poliopsis lacazei Joubin *Fig. 5.3S*
Length up to 50 cm, width 5–8 mm. Head with about 40 small eyes on each side of the mid-dorsal furrow. Body bulky, cylindrical, wrinkled when contracted, pink to greyish red anteriorly, yellowish posteriorly.

Uncommon but widespread. Mostly dredged from 40–50 m depth on sand or shelly substrata, Plymouth, Calais, Mediterranean, Mauritius, rarely intertidal Chile.

Valencinia longirostris Quatrefages Fig. 5.3T
Length up to 15 cm, width 2–3 mm. Proboscis pore (see figure) about halfway between mouth and tip of head, which is slender, whitish, without eyes or grooves. Body thickest posteriorly, mostly pink, yellowish grey, or brown.

In sand or among roots of sea grasses, 1–10 m depth. Jersey, Mediterranean.

Class Enopla

Mouth anterior to cerebral ganglia and usually united with proboscis pore. Nerve cords internal to body wall musculature. Proboscis usually armed with stylets. Two orders.

Order Hoplonemertea

Body worm-like. Gut straight. No posterior sucker. Two suborders.

8. AMPHIPORIDAE

Mostly plump species with many eyes and two pairs of cephalic grooves. Colours usually drab, variable, and without distinctive patterns. Rhynchocoel extends almost whole length of body and has wall of circular and longitudinal muscles forming separate layers.

---

1. Head bears mid-dorsal longitudinal ridge **Amphiporus hastatus**

   Head without longitudinal ridge 2

---

2. Eyes form four longitudinal rows **Amphiporus allucens**

   Eyes not arranged in four longitudinal rows 3

---

3. Less than six eyes 4

   More than six eyes, usually more than 20 5

---

4. Only two large eyes near tip of head. Body length less than 20 × width **_Amphiporus bioculatus_**

   With two anterior eyes, plus a few more further back. Body length more than 30 × width **_Amphiporus elongatus_**

5. Eyes form separate anterior and posterior groups on each side of head. Posterior cephalic furrows meet dorsally to form backwards pointing V, which is located behind the cerebral ganglia **_Amphiporus lactifloreus_**

   Eyes form single continuous row on each side. Posterior cephalic furrows form V located above cerebral ganglia **_Amphiporus dissimulans_**

Amphiporus allucens Bürger Fig. 5.4A
Length 4–4.5 cm, width 2.5–3 mm. Pale yellow anteriorly with pink ganglia. Most of body red. Eyes form double row on each side. Only one pair of transverse cephalic grooves, which meet mid-ventrally.

Dredged from 20–30 m depth. Plymouth, Naples.

Amphiporus bioculatus McIntosh Fig. 5.4B
Length 8–10 cm, width 5–6 mm. One pair of large eyes near tip of head. Originally described as orange or brownish, paler ventrally, head reddish. Other reports of widely differing colour forms indicate that the species needs much more investigation.

Sand or *Laminaria* holdfasts, LW to 30 m depth. Britain, Roscoff, possibly eastern North America.

Amphiporus dissimulans (Riches) Fig. 5.4C
Length 5–7.5 cm, width 1–2 mm. Head oval with many eyes around dorsolateral margins. Posterior furrows meet dorsally over cerebral ganglia, the latter visibly pink by transparency. Body pink, orange, yellowish, or brownish, often darker posteriorly.

Shells, mud, sand, or gravel, LW to 40 m depth. Britain, Scandinavia.

Amphiporus elongatus Stephenson Fig. 5.4D
Length 7.5 cm, width less than 1 mm, tapering to bluntly pointed head. Two eyes anteriorly, three others between the cephalic furrows. Body yellow, with whitish lateral margins. Poorly described.

Sand, Firth of Clyde.

Amphiporus hastatus McIntosh Fig. 5.4E
Length 9–10 cm, width 3–20 mm. Head with many eyes, a mid-dorsal pale longitudinal ridge and only one pair of cephalic grooves meeting dorsally and ventrally as forward pointing Vs, anterior to cerebral ganglia. Colour pinkish, yellowish, reddish, or brownish, the ventral surface and head being paler.

Sand, mid-tide to 35 m depth. Atlantic North America, Greenland, Britain, Mediterranean, Scandinavia.

Amphiporus lactifloreus (Johnston) Fig. 5.4F
Length up to 10 cm, usually less, width about 2 mm. Head broad but bluntly pointed, with many eyes in two groups on each side. Posterior cephalic furrows meet mid-dorsally as backward pointing V behind cerebral ganglia. Colour variable depending upon reproductive state and gut contents; typically pinkish with paler head, tail, and lateral margins, posteriorly green, brown, grey, or black.

Locally abundant in sand or gravel, from just below *Pelvetia* zone to 250 m or more depth. Britain, North Atlantic, Mediterranean.

9. CARCINONEMERTIDAE

On gills or egg masses of crabs. With two eyes and no cephalic furrows. Rhynchocoel wall lacks muscles and is remarkably short, extending only a little way behind the cerebral ganglia. Proboscis reduced, weak, without reserve stylets.

Carcinonemertes carcinophila (Kölliker) Fig. 5.4G
Juveniles on gills 1.5 cm long, maturing on eggs 2-7 cm long. Body yellowish, orange, or reddish.

On *Carcinus*, *Liocarcinus*, Xanthidae, and Galatheidae. Britain, north Atlantic.

10. CRATENEMERTIDAE

Generally like amphiporids, but with wall of rhynchocoel composed of single layer of interwoven longitudinal and circular muscle fibres. NB *Amphiporus hastatus* may also key out here because of its mid-dorsal longitudinal cephalic ridge (see 8. AMPHIPORIDAE).

Nipponnemertes pulcher (Johnston) Fig. 5.4H
Length up to 9 cm, width 1–5 mm. Head distinct, with many eyes and a mid-dorsal longitudinal ridge. Two pairs of transverse furrows, the anterior fused ventrally and possessing many short, forwardly directed secondary ridges; the posterior furrows fuse dorsally to form a backward pointing V. Brown or reddish, especially dorsally.

Sublittorally on sand, gravel, or shells down to 569 m depth, occasionally beneath stones at extreme LW. Greenland, Britain, Roscoff, Chile, Antarctica.

11. EMPLECTONEMATIDAE

Elongate forms, with rhynchocoel restricted to anterior half of body.

---

1. Head narrow, with two pairs of eyes and two pairs of cephalic furrows. Length up to 4 cm **_Nemertopsis flavida_**

   Head broader than body, with many small eyes and indistinct transverse cephalic furrows. Length often 10–40 cm (Genus *Emplectonema*) **2**

---

2. Up to 20 cm long, with numerous sickle-shaped spicules scattered in the skin. Reddish, paler posteriorly **_Emplectonema echinoderma_**

   Up to 50 cm long or more, without spicules. Densely pigmented dorsally, much paler ventrally **3**

---

3. Speckled dorsally, with short longitudinal streaks of brown. Central stylet straight **_Emplectonema neesii_**

   Greyish-green or blue-green dorsally. Central stylet curved **_Emplectonema gracile_**

Emplectonema echinoderma (Marion) Fig. 5.4I
Length up to 20 cm, width 1–2.5 mm. Head a rounded diamond shape, with about 20 small eyes on each side. Body yellowish or orange-red, paler posteriorly, sometimes with dorsal longitudinal streaks of pigment on head. Skin contains many transparent refractive spicules (a in figure). Sketch (b), on the right of the figure, in which the neck is widened by compression, indicates how rhynchocoel and mouth open together, ventral to cerebral ganglia and just behind cerebral organs.

Lower shore and shallow water in sand, with stones or *Zostera*. Madeira, Mediterranean, uncommon in Britain.

Emplectonema gracile (Johnston) Fig. 5.4J
Up to 50 cm long, 3–4 mm wide. Head rounded and rather flattened, with 20–30 eyes on each side. Usually greenish dorsally, pale ventrally. Produces much mucus when irritated. Central stylet curved, with a long slender basis.

On shore amongst boulders, on silt, gravel or sand, in crevices, laminarian holdfasts, or *Mytilus* beds. Also dredged to 100 m depth. Britain, North America, Chile, northern Europe, Mediterranean, Russia, Japan.

Emplectonema neesii (Oersted) Fig. 5.4K
Length up to 50 cm, occasionally 1 m or more, width 5–6 mm. Head rounded with many eyes. Body flattened but fairly bulky, pale yellowish-brown finely streaked with darker brown dorsally, much paler ventrally. Sometimes more uniformly reddish. Central stylet straight, basis short.

On the shore in crevices, under boulders, amongst *Mytilus*, and on sand, silt, gravel or shingle, also dredged to 30 m or more depth. Iceland, Britain, Mediterranean.

Nemertopsis flavida (McIntosh) Fig. 5.4L
About 4 cm long and 0.5 mm wide, tapering towards both ends. Head somewhat like *Amphiporus elongatus*, with four eyes and the posterior pair of cephalic furrows forming a V dorsally. White, yellowish pink, or reddish brown, with pale margins and translucent snout.

Intertidal, under stones and in rock pools, in holdfasts, and dredged to 300 m depth. Denmark to Mediterranean, possibly also from South Africa.

12. PROSORHOCHMIDAE

Ill-defined assemblage of generally small species, including the terrestrial *Argonemertes dendyi* (Dakin) which has the circular and longitudinal muscles of its rhynchocoel wall interwoven, as in the Cratenemertidae. In *Oerstedia* and *Prosorhochmus* the rhynchocoel muscles form separate layers. Cephalic

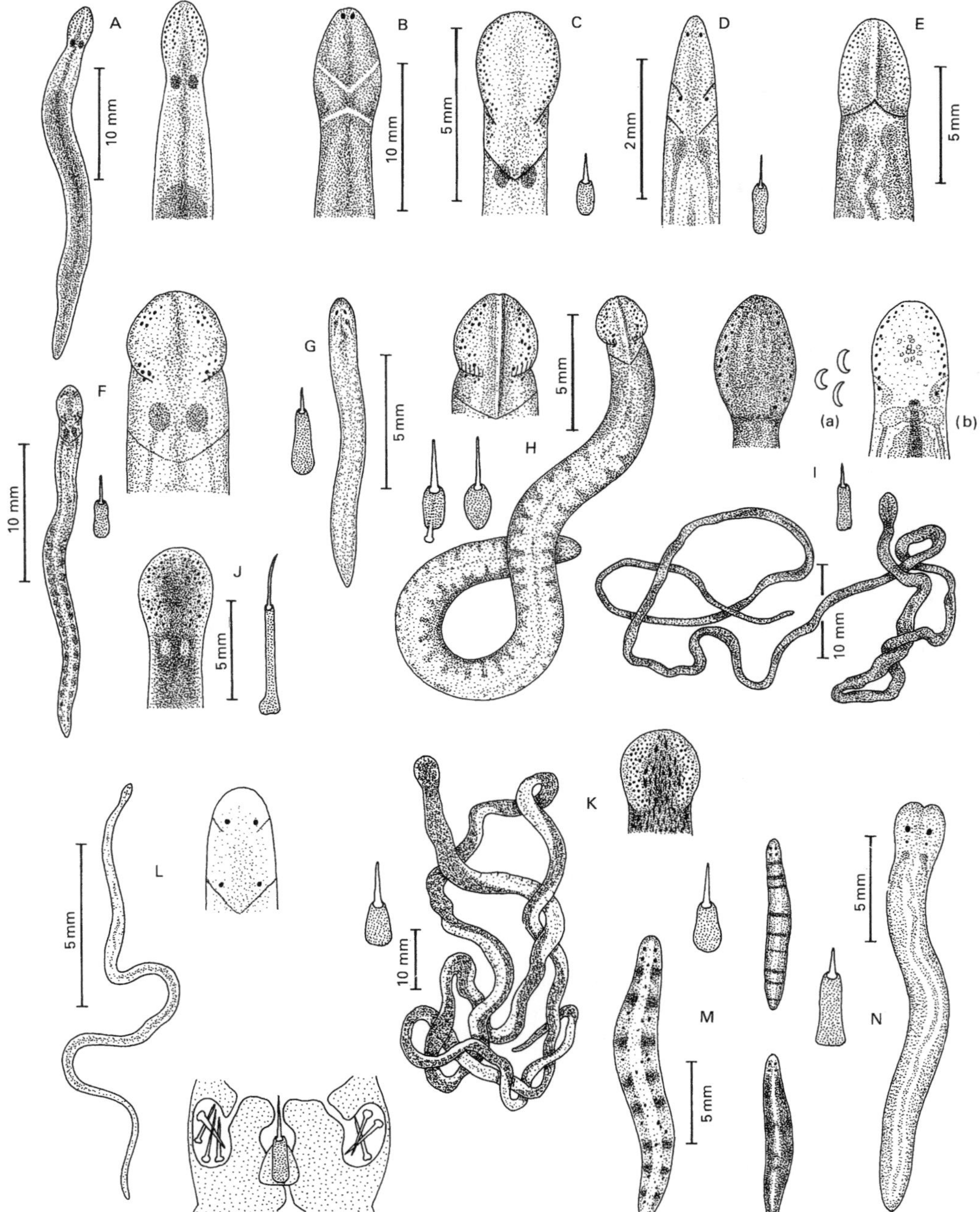

Fig. 5.4 Hoplonemertea. 8, A*mphiporidae*: A. *Amphiporus allucens*, B. *A. bioculatus*, C. *A. dissimulans*, D. *A. elongatus*, E. *A. hastatus*, F. *A. lactifloreus*. 9. CARCINONEMERTIDAE: G. *Carcinonemertes carcinophila*. 10. CRATENEMERTIDAE: H. *Nipponnemertes pulcher* 11. EMPLECTONEMATIDAE: I. *Emplectonema echinoderma*, J. *E. gracile*, K. *E. neesii*, L. *Nermertopsis flavida*. 12. PROSORHOCHMIDAE: M. *Oerstedia dorsalis*, N. *Prosorhochmus claparedii*.

grooves usually shallow or indistinguishable in marine forms.

---

1. Body somewhat flattened, pale without patterns of pigment, tapering posteriorly, with head as wide as or wider than body. Ovoviviparous, juveniles emerging through anus **_Prosorhochmus claparedii_**

   Body cylindrical, bearing patterns of pigment, tapering towards both ends. Not ovoviviparous **_Oerstedia dorsalis_**

Oerstedia dorsalis (Abildgaard) Fig. 5.4M

Length usually 1–1.5 cm, occasionally 3 cm, width 1–2 mm. Head rounded but narrow, not demarcated from body. Eyes four, in a square. Body cylindrical, marbled, or transversely banded with shades of brown, often speckled with white or yellow, sometimes with a pale mid-dorsal stripe. Usually paler ventrally. Colour forms *O. immutabilis* (Riches) and *O. nigra* (Riches) may be distinct species.

Common, often abundant amongst rock-pool algae or dredged from mud, gravel, sand, stones, or shelly sediments, to 80 m or more depth. Widespread in northern hemisphere from Pacific North America eastwards to Europe.

Prosorhochmus claparedii (Keferstein) Fig. 5.4N

Length up to 3.5–4 cm. Head may appear bilobed, with a terminal notch. Eyes four, in a transversely elongated rectangle, anterior pair larger than the others. Body pale yellow or orange. Rhynchocoel and brooded young may be seen through body wall.

Under stones and in rock crevices, upper and lower shore and sublittorally. Britain, France, Spain, Mediterranean, Adriatic.

13. TETRASTEMMATIDAE

Small slender nemerteans, which include the freshwater species of *Prostoma*. Rhynchocoel extends throughout body, its wall having two separate muscle layers.

---

1. Dark red stripe extends mid-dorsally throughout body **_Tetrastemma herouardi_**

   No dark mid-dorsal longitudinal stripe **2**

---

2. Head with mid-dorsal patch of pigment **3**

   Head without mid-dorsal patch of pigment **6**

---

3. Cephalic pigment patch subquadrangular, covering some eyes **4**

   Cephalic pigment patch crescentic, between eyes **5**

---

4. Patch reddish, covering only anterior half of area between eyes **_Tetrastemma longissimum_**

   Patch black or dark brown, covering most of area between eyes **_Tetrastemma melanocephalum_**

---

5. Patch a simple crescent, its ends pointing anteriorly. Body slender, 1.5 cm long and 1 mm wide, greenish **_Tetrastemma coronatum_**

   Crescent usually bears a third median point. Body bulky, 4 cm long and 2 mm wide, brownish **_Tetrastemma peltatum_**

---

6. With two longitudinal brown streaks joining anterior and posterior eyes on each side of head **_Tetrastemma vermiculus_**

   Without longitudinal streaks joining eyes **7**

---

7. With a transverse band of pigment just behind head **_Tetrastemma robertianae_**

   Without such a transverse band of pigment **8**

---

8. Head diamond shaped and wider than most of body **9**

Head not diamond shaped, not wider than body **10**

9. Anterior eyes much bigger than others ***Tetrastemma ambiguum***

Eyes all of similar size ***Tetrastemma cephalophorum***

10. Body stout, cylindrical, broadest near posterior end ***Tetrastemma beaumonti***

Body somewhat flattened, of uniform width or tapering posteriorly **11**

11. Shining whitish gland cells present between anterior eyes ***Tetrastemma helvolum***

Without such a conspicuous mass of gland cells **12**

12. Proboscis with four accessory stylet pouches visible when compressed in transmitted light ***Tetrastemma quatrefagesi***

Only two accessory stylet pouches **13**

13. Eyes distinct, moderately large, body tapering posteriorly ***Tetrastemma candidum***

Eyes minute, body not tapering posteriorly ***Tetrastemma flavidum***

Tetrastemma ambiguum Riches Fig. 5.5A
Length 1-1.5 cm. Pale yellow, usually tinged reddish brown dorsally.

Sublittoral to 60 m depth, on sand, mud, limestone fragments, and stones. Plymouth.

Tetrastemma beaumonti (Southern) Fig. 5.5B
Length 0.3–0.6 cm. Remarkably stout posteriorly, tapering towards head. Anterior eyes usually larger than others. Whitish, occasionally tinged pink.

Dredged from 3–20 m depth on gravel and sand. Atlantic coast of Ireland, possibly Isle of Man.

Tetrastemma candidum (Müller) Fig. 5.5C
Length 1–3 cm, width 1 mm. Four eyes equally distinct. Lacks clearly diagnostic characters. Colours seem extremely variable (yellowish, orange, reddish, greenish, whitish), perhaps including unrecognized species. Apart from being more restless and actively moving, easily confused with *T. flavidum*.

Mid-shore to depths of 55 m or more; from rock-pool algae, in *Ascophyllum* bladders, under rocks on shells or gravel, with colonial invertebrates. Apparently circumpolar in northern hemisphere.

Tetrastemma cephalophorum Bürger Fig. 5.5D
Length 1.5 cm, width 1.5–2 mm. Head diamond-shaped, with four large eyes. Reddish brown dorsally, pale yellowish laterally and ventrally.

Shell gravel or stones at 10–15 m depth. Britain, Mediterranean.

Tetrastemma coronatum (Quatrefages) Fig. 5.5E
Length up to 1.5 cm, width 0.5–1 mm. Head blunt, with shallow cephalic furrows posteriorly and a dark crescent between eyes. Otherwise pale, usually greenish.

With intertidal algae, tubicolous polychaetes, stones and sandy detritus to 40 m depth. Scandinavia, Britain, Mediterranean, Black Sea.

Tetrastemma flavidum Ehrenberg Fig. 5.5F
Length up to 1.5 cm, width 0.5–0.75 mm. Body rather flattened, head rounded or slightly tapered with four small eyes. Typically bright pink, sometimes tinged yellowish or reddish, or translucent. May be confused with *T. candidum*.

LW to depths of 100 m in mud, sand, gravel, or amongst laminarian holdfasts. Britain, Scandinavia, Mediterranean, Red Sea (NB British records of this species are confused).

Tetrastemma helvolum Bürger Fig. 5.5G
Length up to 2 cm, width less than 1 mm. Like *T. candidum* but generally longer and more slender. Light honey-yellow, head paler.

Dredged 4–80 m depth from coralline, muddy, shelly, or sandy sediments, or amongst algae. Britain, Mediterranean.

Tetrastemma herouardi (Oxner) Fig. 5.5H
Length up to 0.6 cm, width 0.75 mm. Pale transparent pink, with single mid-dorsal stripe of dark red.

With algae and sedentary invertebrates. Britain, Roscoff.

Tetrastemma longissimum Bürger Fig. 5.5I
Length may exceed 2 cm, width 1 mm. Like *T. coronatum* but brownish yellow. Head generally distinct from body, almost colourless but with transverse band of red between the eyes.

With algae, shells, or sand; intertidal and sublittoral to 20 m depth. Britain, Mediterranean.

Tetrastemma melanocephalum (Johnston) Fig. 5.5J
Length 3–6 cm, width 2–2.5 mm, with dark subquadrangular pigment patch on head. Body yellowish, sometimes reddish brown.

Common in various intertidal habitats, with rock pool algae, in crevices, on sand, and to 40 m depth. Scandinavia, Britain, Mediterranean, Madeira, Canary Islands, Black Sea.

Tetrastemma peltatum Bürger Fig. 5.5K
Up to 4–5 cm long, 2 mm wide, with dark three-pronged crescent of pigment on head, and anterior eyes much larger than others. Body brownish, sometimes greenish posteriorly.

Plymouth, Adriatic, Mediterranean, Chile.

Tetrastemma quatrefagesi Bürger Fig. 5.5L
Up to 0.6 cm long. Transparent yellowish, not clearly distinguishable from *T. flavidum*.

Tetrastemma robertianae McIntosh Fig. 5.5M
Up to 3–3.5 cm long, 0.7–1 mm wide. Anterior eyes larger than others. Transverse brown band around body behind head, may be incomplete ventrally. From it, two dorso-lateral brown stripes run backwards longitudinally.

Shallow sublittoral to 70 m depth on mud, shelly gravel, and stones. Britain, Scandinavia.

Tetrastemma vermiculus (Quatrefages) Fig. 5.5N
Up to 2 cm long, 0.8 mm wide. Head oval, with two pairs of cephalic furrows. Body yellowish, pinkish, or pale orange. Brown longitudinal streak joins anterior and posterior eyes on each side of head.

Lower shore and to 40 m depth in rocky places. Scandinavia, Britain, Mediterranean, Madeira, Atlantic North America, Gulf of Mexico.

14. DREPANOPHORIDAE

Cerebral sense organs each with two sensory canals and a glandular appendage. Mouth and proboscis pore adjacent but separate. Cephalic furrows subdivided into secondary slits.

Punnettia splendida (Keferstein) Fig. 5.5O
Up to 5 cm long and 5 mm wide. Body flattened, reddish brown, dorsally with five longitudinal whitish stripes and thin lateral margins which are also whitish. Head spatulate, demarcated by a transverse cephalic furrow on each side, each containing 7–8 secondary slits which are longitudinal (i.e. transverse to the furrow). Eyes about 70, in longitudinal rows. Swims actively when irritated.

Sublittorally to 40 m depth, with algae, gravel, or stones. Plymouth, Channel Islands.

15. PARADREPANOPHORIDAE

Cerebral sense organs each with one sensory canal and no glandular appendage. Otherwise similar to Drepanophoridae except that mouth and proboscis pore may share a common opening, while cephalic furrows may or may not include secondary slits. NB From arrangement of eyes *Amphiporus allucens* may also key out here (see 8. AMPHIPORIDAE).

Paradrepanophorus crassus (Quatrefages) Fig. 5.5P
Up to 16 cm long and 9 mm wide. Body flattened but bulky, tapering to both ends. Brownish, paler ventrally. Head demarcated by white cephalic furrows with brown secondary slits. Mouth and proboscis adjoining but separate. Eyes in four longitudinal rows.

Under stones just below LW and to 5 m depth, between worm tubes and rocks. Secretes a thin

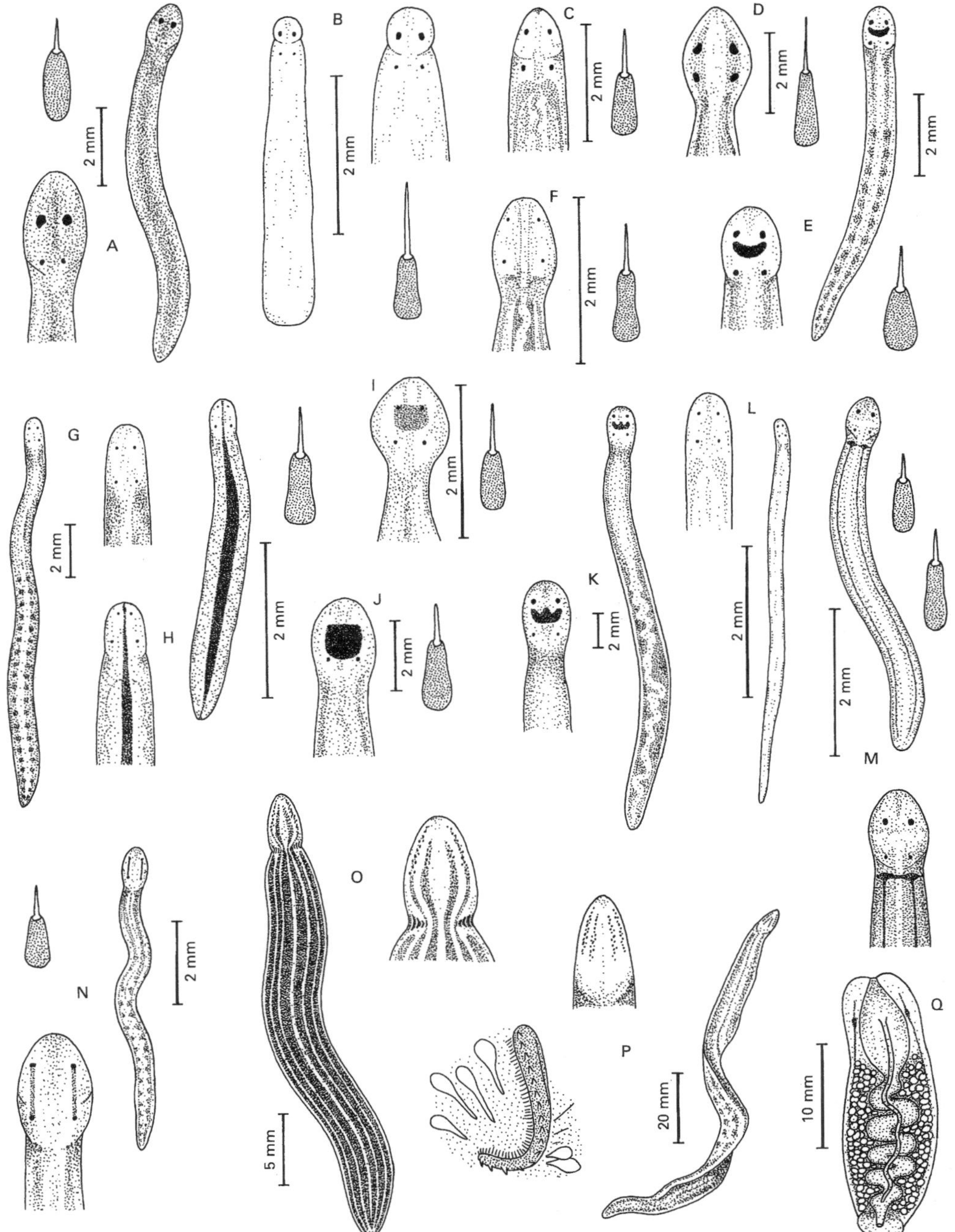

Fig. 5.5 Hoplonemertea and Bdellonemertea. 13, TETRASTEMMATIDAE: A. *Tetrastemma ambiguum*, B. *T. beaumonti*, C. *T. candidum*, D. *T. cephalophorum*, E. *T. coronatum*, F. *T. flavidum*, G. *T. helvolum*, H. *T. herouardi*, I. *T. longissimum*, J. *T. melanocephalum*, K. *T. peltatum*, L. *T. quatrefagesi*, M. *T. robertiana*, N. *T. vermiculus*. 14, DREPANOPHORIDAE: O. *Punnettia splendida*. 15, PARADREPANOPHORIDAE: P. *Paradrepanophorus crassus*. 16, MALACOBDELLIDAE: Q. *Malacobdella grossa*.

parchment tube, fixed to underside of boulders. Lough Hyne (Ireland), Mediterranean.

Order Bdellonemertea

Body leech-like, with a posterior sucker. Gut sinuous. Proboscis without stylets.

16. MALACOBDELLIDAE

Body broad, flattish and short, with a posterior ventral sucker. Rhynchocoel opens into foregut, which is barrel-shaped. One genus, entocommensal in bivalves.

Malacobdella grossa (Müller) Fig. 5.5Q

Up to 4 cm long and 15 mm wide. Rhynchocoel as long as body and straighter than gut. Immature specimens whitish, mature females with greenish grey ovaries, mature males with pinkish testes.

Particularly common in *Mya truncata* and *Zirfaea crispata*, but also occurs in many other host bivalve species. Usually only a single worm per host, typically on or between gill lamellae, exceptionally up to five individuals in a single host. Widely distributed on the coasts of Europe, Atlantic, and Pacific North America.

6 ANNELIDS

These, the 'true' worms (in Phylum Annelida), typically have a slender cylindrical body characteristically made up of a series of ring-like segments, seen externally as annular constrictions, with an anterior subterminal mouth and a posterior terminal anus. In less specialized forms, the segments between those of the head and the last one more or less resemble each other, each containing a similar set of organs. This may be apparent externally as a segmental series of similar pores or protruding structures, particularly those used in locomotion. Worms normally move using peristaltic waves of contraction or sigmoid rippling of the body. In polychaetes, which are entirely aquatic, a typical segment carries, on each side, well-developed fleshy lobes bearing a large number of chitinous bristles. These function like short legs, and may aid swimming as the body flexes more vigorously in a horizontal plane. In the oligochaetes, fleshy outgrowths are lacking and the bristles are fewer; they are almost invisible in the familiar earthworms, although quite long in some aquatic species. Leeches lack them altogether; using terminal suckers, they may 'loop' along a solid surface and, when swimming, the dorso-ventrally flattened body flexes in a vertical plane.

Although the head end of oligochaetes and leeches carries sense organs, it appears simple and smooth. Most polychaetes have a variety of appendages on their head which, amongst those living in a burrow or tube, may form a conspicuous fan- or brush-like structure; this may be all that can be seen of the animal.

Class Polychaeta: Bristle worms

The largest group, all aquatic and almost entirely marine. They are mainly free living; some are commensal and very few are parasitic. The body form varies widely, reflecting a range of habit from pelagic, through crawling with occasional swimming, to active burrowing or tube-dwelling. There are many well-defined families but no widely accepted or consistent scheme of higher taxa so, for ease of reference, families are included here in alphabetical order.

The basic anatomy is illustrated in Fig. 6.1. Of the segments forming the head, the *prostomium* (which is normally a flattened sphere projecting forwards above the mouth) has *antennae* (often one mid-dorsally and one or more pairs antero-laterally), which are essentially tactile, and a pair of chemosensory *palps* ventrally, in front of, or flanking, the mouth. One or more pairs of *eyes*, often large and frequently with a lens, are mounted dorsally. Ciliated *nuchal organs* of various types may pass back dorsally from the rearward edge of the prostomium. The *peristomium*, the segment surrounding the mouth, usually incorporates several anterior body segments. Their appendages may have several origins but are referred to, collectively, as *tentacles*. These are often similar to the antennae, although they may be longer and more mobile for food gathering. Amongst the more specialized tube-worms, the tentacles are stiff and feathery, forming a funnel-like *branchial crown* the cilia of which generate a current for respiration as well as suspension feeding. The anterior part of the gut can often be everted through the mouth as a *proboscis* whose surface frequently bears papillae, hard ridges, or chitinous teeth. In some species there may be one or more pairs of dark, chitinous *jaws* operated by strong muscles. At the posterior end, the last segment (*pygidium*) often has a pair of *anal cirri*.

The locomotory structure (*parapod*) on each side of a typical body segment is usually divided into dorsal (*notopod*) and ventral (*neuropod*) lobes, each containing a bundle of bristles (*chaetae*); those in the dorsal bundle are called *notochaetae*, with *neurochaetae* in the ventral one. Each lobe is normally supported internally by a strong chitinous rod (*aciculum*) whose outer end sometimes protrudes like a short, stout bristle. One or both lobes may be subdivided, with more than one bundle of chaetae; alternatively, one (most often, the notopod) may be entirely lacking. Each lobe typically has a *cirrus*

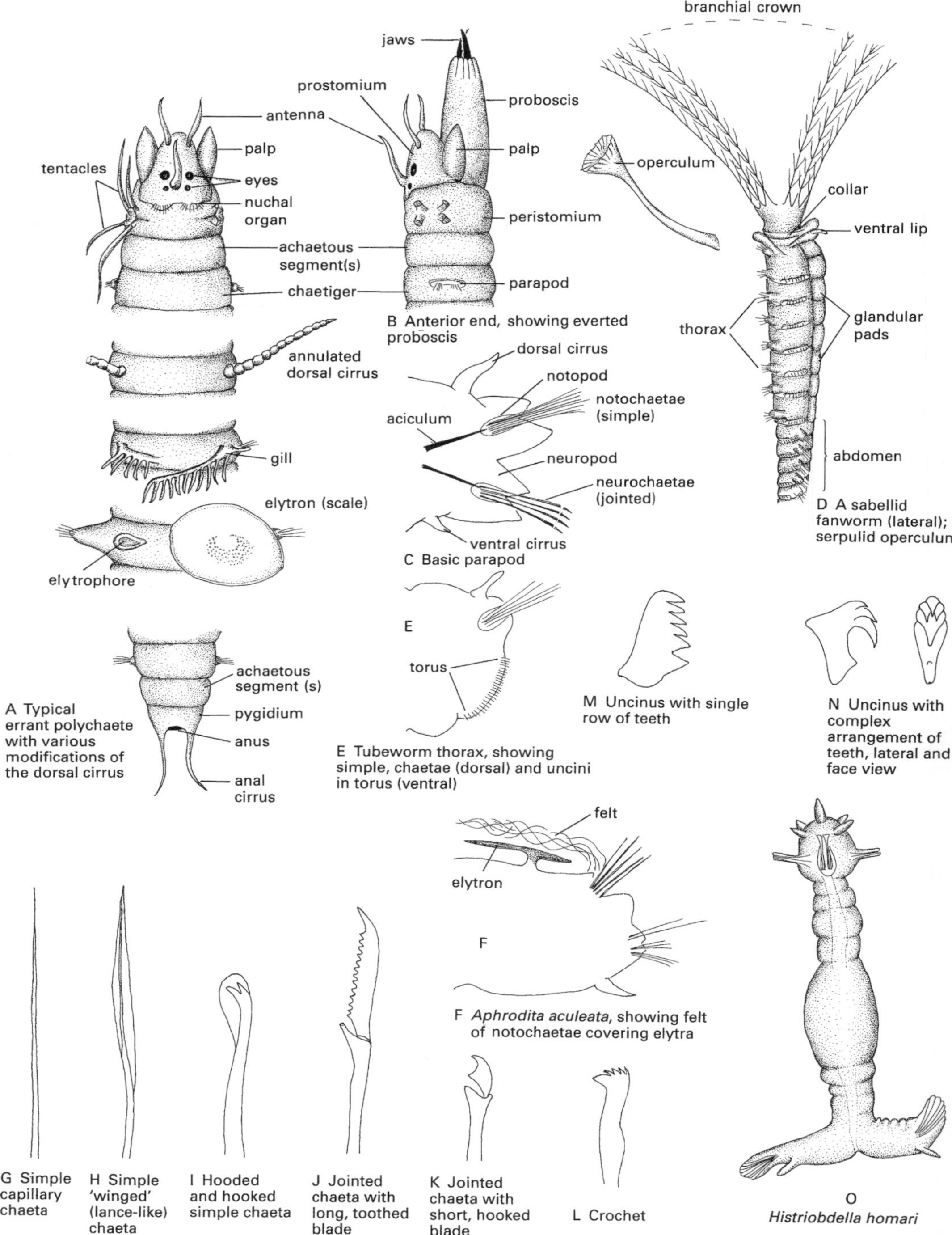
jaws
prostomium
antenna
proboscis
palp
tentacles
eyes
nuchal organ
peristomium
achaetous segment(s)
chaetiger
parapod
B Anterior end, showing everted proboscis
annulated dorsal cirrus
gill
elytron (scale)
elytrophore
achaetous segment (s)
pygidium
anus
anal cirrus
A Typical errant polychaete with various modifications of the dorsal cirrus
dorsal cirrus
notopod
notochaetae (simple)
aciculum
neuropod
neurochaetae (jointed)
ventral cirrus
C Basic parapod
E
torus
E Tubeworm thorax, showing simple, chaetae (dorsal) and uncini in torus (ventral)
branchial crown
operculum
collar
ventral lip
thorax
glandular pads
abdomen
D A sabellid fanworm (lateral); serpulid operculum
M Uncinus with single row of teeth
N Uncinus with complex arrangement of teeth, lateral and face view
felt
elytron
F
F *Aphrodita aculeata*, showing felt of notochaetae covering elytra
G Simple capillary chaeta
H Simple 'winged' (lance-like) chaeta
I Hooded and hooked simple chaeta
J Jointed chaeta with long, toothed blade
K Jointed chaeta with short, hooked blade
L Crochet
O
Histriobdella homari

Fig. 6.1

associated with it. These are usually tactile and shaped like small antennae, although the dorsal one may be modified into a paddle blade, a branching gill, or a protective scale.

Chaetae occur in various types. They may be *simple* (all in one piece) or *jointed* (also called compound), usually with a short terminal part (often hooked) hinged to a longer shaft. The simplest are fine and hair-like (*capillary* chaetae); other simple forms may be stout and serrated or hooked at the free end. In species that swim well, they may be flattened like oars. Burrowing species often have strong, hooked chaetae (*crochets*) by which they anchor themselves to the wall of the burrow. For this purpose, tubeworms use *uncini*, which are small plates whose outer edge is serrated. They are stacked in a row within the groove of a parapod modified in the form of parallel swollen lips. The arrangement is known as a *torus* and resembles a short, closed zip-fastener. Crochets or uncini usually occur in neuropods, whereas notopods have a bundle of normal bristles.

A *chaetiger* is a segment that bears chaetae (and, usually, other components of a parapod). The term is used here particularly when the exact location of a characteristic feature is needed; because the number of segments incorporated with the peristomium is not readily apparent, it would be imprecise to refer to the *n*th segment. Where the number of readily visible segments is small and may be diagnostically important, information is also given in terms of chaetigers. Where a worm may have 100 or more segments, those involved in its head make little difference and reference is then made to the number of segments. Almost invariably, it is necessary to have the head end of a broken specimen before an identification can be made; in some groups, the anus is surrounded by characteristic papillae or lobes and, in these, the posterior end is also necessary.

Details of the habitat in which a polychaete was found, together with the nature of its burrow or tube if it makes one, may be helpful in confirming its identity. Whenever possible, work with specimens while they are still alive, noting coloration and any characteristic movements. It may then be possible to release the animal unharmed. If identification of a living worm proves to be impossible, or if a collection is to be made for future reference, it is best to relax specimens with a narcotizing agent before killing them. This is particularly useful when it is desirable to have the proboscis everted. A good general purpose reagent is 7.5% magnesium chloride in sea water; individual animals can also be relaxed in a small dish by slowly adding 70% ethanol to the water. As soon as the worms cease to respond to touch, they should be fixed in 10% seawater formalin for at least 24 hours. It is inadvisable to leave them in formalin indefinitely: after fixing, wash with fresh water and transfer to 70% ethanol for storage (but note that ethanol alone is not an adequate fixative). Colours are rapidly lost in ethanol so, where they are particularly important, the worm could be transferred to 1–2% aqueous propylene phenoxytol instead.

KEY TO FAMILIES

---

1. Dorsal surface covered with overlapping scales, a dense coat of felted or fur-like chaetae (which conceals such scales) or transverse ridges of bristles **2**

   Dorsal surface not so covered (NB tufts of gills in various forms, or stiff chaetae, may project dorsally from the lateral parapodia) **3**

---

2. Large, broad worms covered dorsally by a dense, greyish-brown felt; **or** medium-sized, with overlapping dorsal scales fringed and partly covered by a tangled chaetal fur; **or** various sizes but more elongated, most covered by overlapping scales and bearing typical chaetae in lateral bundles (N.B. such scales are readily shed by many species) **3. Aphroditidae**

Small, oval, flattened body in which each segment bears dorsally a range of bristles and branching gills extending from each side almost to mid-line; a crest extends along dorsal mid-line for a few segments behind head
2. Amphinomidae

---

3. Small, with elongated lateral processes at one end giving an anchor- or pick-like body form, lacking chaetae **4**

   Lacking such body form **5**

---

4. Extremely small (2 mm or less) with indistinct segmentation, lacking parapods. Body bifurcates posteriorly into thick lateral processes each terminating in a sucker; head bears a number of short appendages. Found amongst the eggs or in the branchial cavity of lobsters
Histriobdella homari (p. 197, Fig. 6.10)

   Small (up to 20 mm) with segmentation emphasized by elongated paddle-like parapods. Prostomium extended to each side, resembling a pick-head; second segment bears a whip-like appendage on each side, stiffened by a long aciculum and usually swept back along the body, the length of which it nearly equals. Of delicate, transparent appearance, swimming in the plankton *Tomopteris helgolandica* (**26. Tomopteridae**)

---

5. Segmentation of body indistinct, indicated largely by rows of spherical capsules. Parapods simple, uniramous, and with few chaetae
22. Sphaerodoridae

   Segmentation of body usually indicated distinctly by transverse constrictions, prominent parapods, or both. Spherical capsules lacking **6**

---

6. Anteriorly, antennae short or absent; tentacles and palps short or absent (NB spionids that have lost their palps may key out here) (Fig. 6.2A–E, H–J) **7**

   Anteriorly, antennae, tentacles, or palps well developed (Fig. 6.2K–U) **17**

---

7. Stout but soft and fragile worm with a smooth, more or less cylindrical mid-region lacking typical parapods but bearing specialized appendages dorsally (anteriorly, a pair of curved wings or fins and posteriorly three large fans or paddles). In life, always occupies a leathery tube buried in clean sand, with each end opening at the surface
Chaetopterus variopedatus (**6. Chaetopteridae**)

   Lacking such specialized appendages **8**

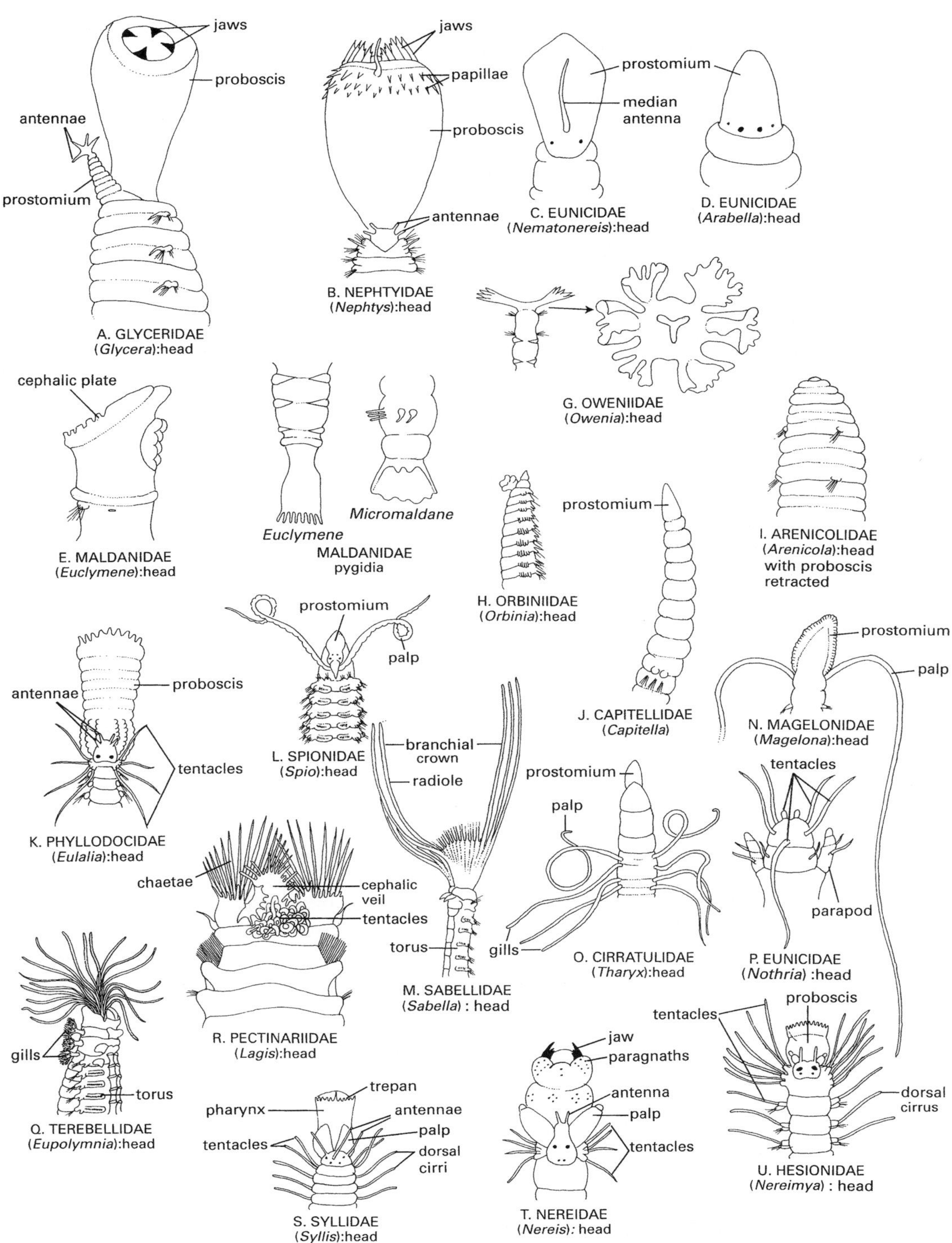
jaws
proboscis
antennae
prostomium
A. GLYCERIDAE
(Glycera):head
jaws
papillae
proboscis
antennae
B. NEPHTYIDAE
(Nephtys):head
prostomium
median antenna
C. EUNICIDAE
(Nematonereis):head
D. EUNICIDAE
(Arabella):head
G. OWENIIDAE
(Owenia):head
cephalic plate
E. MALDANIDAE
(Euclymene):head
Euclymene
Micromaldane
MALDANIDAE
pygidia
H. ORBINIIDAE
(Orbinia):head
prostomium
J. CAPITELLIDAE
(Capitella)
I. ARENICOLIDAE
(Arenicola):head
with proboscis
retracted
prostomium
palp
L. SPIONIDAE
(Spio):head
proboscis
antennae
tentacles
K. PHYLLODOCIDAE
(Eulalia):head
branchial crown
radiole
torus
M. SABELLIDAE
(Sabella) : head
prostomium
palp
N. MAGELONIDAE
(Magelona):head
prostomium
palp
gills
O. CIRRATULIDAE
(Tharyx):head
tentacles
parapod
P. EUNICIDAE
(Nothria) :head
chaetae
cephalic veil
tentacles
R. PECTINARIIDAE
(Lagis):head
gills
torus
Q. TEREBELLIDAE
(Eupolymnia):head
trepan
pharynx
antennae
palp
tentacles
dorsal cirri
S. SYLLIDAE
(Syllis):head
jaw
paragnaths
antenna
palp
tentacles
T. NEREIDAE
(Nereis): head
tentacles
proboscis
dorsal cirrus
U. HESIONIDAE
(Nereimya) : head

Fig. 6.2

8. Prostomium either conical, with four minute terminal antennae (Fig. 6.2A) or more or less rectangular with four small antennae, one at each anterior corner (Fig. 6.2B) **9**

Prostomium not so **10**

9. Body cylindrical, at least anteriorly; parapods small. Prostomium conical, annulated, with four antennae forming a cross at its tip. Often reddish **10. Glyceridae**

Body flattened; parapods well developed, biramous with well-separated lobes. Prostomium not annulated, more or less rectangular with four small antennae, one at each anterior corner. Usually grey or white **14. Nephtyidae**

10. Body not divided into distinct regions; segments not markedly longer than wide; prostomium conical, without appendages or with a short median antenna only (Fig. 6.2C, D) **8. Eunicidae I**

Body usually divided into distinct regions, **or** broad anteriorly with a narrow, ventrally concave 'tail' **or** some body segments much longer than wide; prostomium variable **11**

11. Some body segments much longer than wide; anterior end obliquely flattened or hooded, lacking appendages or bearing a frilled membranous crown (Fig. 6.2E, G) **12**

Segments may be multi-annulate, mostly shorter than wide and often with wrinkled, reticulated epidermis in anterior region; posteriorly, they may be more elongated and narrower. Anterior end more or less conical (Fig. 6.2H–J), lacking appendages, reminiscent of an earthworm **13**

12. Posterior end in the form of a funnel, a more or less oblique plate, or spoon-like (Fig. 6.2F). The elongated segments are usually in posterior part of body **13. Maldanidae**

Posterior end simple, rounded. Segments elongated in anterior mid-region but becoming progressively shorter posteriorly **18. Oweniidae**

13. Numerous (more than 11) pairs of tufted gills dorsally along middle or posterior parts of body. Neuropod consists of a simple torus containing a single row of stout crochets not hooded at tip. All but the first few segments 5-annulated **4. Arenicolidae**

Gills, if present, usually numerous simple lobes, paddle-shaped or cirriform structures; if tufted, no more than four pairs anteriorly. Crochets may be lacking, inserted in notopodial bundles, or in multiple rows on a papillated torus; if in a simple torus, their tips are hooded. Segments simple or with fewer than five annuli **14**

---

14. Dorsal and/or ventral tori present, containing crochets in rows **15**

Parapods biramous as small lobes or reduced to bundles of fine chaetae; no tori, crochets absent **16**

---

15. Anterior region broad and flattened, with lateral parapods; posteriorly, body narrower and more cylindrical with dorsal parapods. Anterior neuropod in the form of a torus bearing several rows of crochets and a fringe of post-chaetal papillae. A number of segments at junction of the two body regions may also bear a row of papillae ventrally. Gills, chaetae, and parapodial appendages may be sufficiently numerous to give a brush-like appearance to the dorsal surface posteriorly **17. Orbiniidae**

Anterior region usually with markedly wrinkled epidermis, posterior region smooth with rather elongated segments, body cylindrical and of much the same diameter throughout. Simple tori present, crochets small. Parapods not prominent; superficial resemblance to an earthworm. Two or more pairs of prominent genital spines may occur mid-dorsally in anterior region **5. Capitellidae**

---

16. Long cirriform gills on many of body segments, inserted on ventral parapods but sweeping upwards and backwards. Segments may be multi-annulate but rarely with reticulated epidermis. Conical prostomium may be prolonged anteriorly in a short finger-like process. Anus surrounded by papillae, may be enclosed in a short tube or hood **16. Opheliidae**

Gills either absent, or tufted and restricted to no more than four anterior segments. Parapods lateral, occasionally bearing short cirri. All segments multi-annulate, with markedly reticulated epidermis. Prostomium blunt and T-shaped or bilobed. Pygidium usually with 4–5 elongated cirri **21. Scalibregmidae**

---

17. Elongated, typically worm-like. Segments more or less alike throughout, body not divided into distinct regions (except in epitokes). Prostomium usually well developed, with sensory appendages but none markedly long or specialized (Fig. 6.2K, S, T, U). Parapods usually well developed, adapted for active movement **28**

Often short and relatively broad-bodied; frequently divided into regions that are distinctly different. Prostomium reduced, with appendages that are either numerous and specialized or very long (or both) (Fig. 6.2L–R). Parapods often reduced; mainly burrowing or tubicolous forms **18**

---

18. Head with two long palps; a few filamentous gills may be present, but anterior end lacks numerous filamentous appendages (Fig. 6.2L, N) **19**

Head with numerous bristles, tentacles, or feathery appendages (Fig. 6.2M, O, Q, R) **21**

19. Body not divided into distinct regions (although some appendages, e.g. gills, may occur on only one part of it) **23. Spionidae**

Body divided into distinct regions **20**

20. Body in two regions distinctly different from each other; anterior end like a duck's bill (Fig. 6.2N); occupies burrows but no permanent tube **12. Magelonidae**

Body in three regions distinctly different from each other; anterior end blunt, often with a funnel-like terminal mouth; occupies a tough, membranous permanent tube **6. Chaetopteridae**

21. Anterior end with feathery appendages, normally stiff and arranged in two semicircles, forming a funnel (Fig. 6.2M) **22**

Anterior appendages not feathery **23**

22. Occupying a straight, sinuous, or coiled calcareous tube; one appendage often modified as a stalked plug (operculum) sealing tube when worm is retracted into it **32**

Usually occupying a more or less erect tube of cemented fine sediment (in one species, of thick mucus; some small species are free-living); no operculum **28. Sabellidae**

23. Anterior appendages tentacle-like (Fig. 6.2O, Q) **24**

Anterior appendages stiff bristles (Fig. 6.2R) **26**

24. Tentacles usually originate from behind head (Fig. 6.2O). Body not divided into distinct regions. Gills simple, elongate, may be present along much of body **7. Cirratulidae**

Tentacles inserted on head (Fig. 6.2Q). Body clearly divided into a swollen anterior and more slender posterior region. Often one or more pairs of branched gills behind the tentacles **25**

25. Tentacles retractile **1. Ampharetidae**

Tentacles not retractile **25. Terebellidae**

26. Body covered with papillae and more or less encased in mucus, often encrusted with sediment particles which do not form a definite tube. Prostomium and buccal segment cylindrical and retractile, not markedly truncated. Anterior chaetae usually directed forwards, making a cage around head **9. Flabelligeridae**

Body not papillated, in life always occupying a rigid tube of cemented sand grains. Anterior end obliquely truncated with a crown of stout flattened chaetae, sealing tube when worm is retracted into it (Fig. 6.2R) **27**

27. Anterior chaetae arranged in a comb-like series on either side, crossing in the mid-line. Most posterior part of body short, spoon-shaped, segmented but mostly lacking chaetae. Tube free, solitary, a smooth elongated cone open at each end **19. Pectinariidae**

Anterior chaetae arranged in three concentric semicircles on each side. Most posterior part of body a long, unsegmented anal tube lacking chaetae and reflected forwards along ventral side of worm. Tubes attached to solid substrata, with a rough surface and open at only one end, either forming part of a massive reef or more or less solitary on stones and shells **27. Sabellariidae**

28. Parapods uniramous, with leaf-like or globular cirri; four antennae (at front of head) and sometimes a median one (which may be further back); no palps; 2–4 pairs of tentacles; proboscis jawless but bearing papillae (Fig. 6.2K) **20. Phyllodocidae**

Parapods uni- or biramous but cirri not leaf-like or globular when present; usually two antennae (may be none or three); two palps present, usually short and stout; tentacles absent or various (Fig. 6.2S-U); proboscis often bears dark chitinous jaws, spines, or teeth **29**

29. Parapods clearly biramous, with dorsal and ventral lobes more or less equal. Four pairs of tentacles, may be quite long but not divided into sections, distinct from dorsal cirri of adjacent segments. Proboscis with a pair of jaws and small scattered teeth **15. Nereidae**

Parapods appear uniramous (some elements of missing lobe may still be present); tentacles absent, or if present, not as four pairs **30**

30. First three or four body segments lack chaetae, compressed into a more or less distinct short region behind the head, bearing 6–8 pairs of relatively long tentacles. Palps moderately long and divided into two sections (Fig. 6.2U). Proboscis with or without jaws, or with simple horny lips **11. Hesionidae**

Tentacles absent or, if present, as one or two pairs only. Palps not divided into two sections and usually squat. Proboscis with at least one large tooth, usually with complex jaws **31**

31. Head with three relatively long antennae (Fig. 6.2S). Two pairs of tentacles followed by dorsal cirri down the rest of the body, all usually long and divided into many sections **24. Syllidae**

Head with two short antennae or none, five long tentacles or none (Fig. 6.2P). Feathery gills may be present on part of the body but dorsal cirri not prominent **8. Eunicidae II**

32. Tube straight or sinuous. Thorax with more than four chaetigers **29. Serpulidae**

Tube helical, usually coiled flat against substratum. Thorax with three or four chaetigers **30. Spirorbidae**

1. AMPHARETIDAE

Body divided into two regions: anterior broad and short with biramous parapods (notopod a conical boss with a bundle of long capillary chaetae neuropod with a row of simple spear-like chaetae, or uncini); posterior with neuropodial flaps bearing a row of uncini, the notopod vestigial or lacking. Generally resemble terebellids but can retract their tentacles completely and have simple gills (a group of four on each side in the species included here). A large family with more than 30 species, in 17 genera, recorded for the British sea area. Many of these are known only from deep, offshore dredgings.

1. Posterior region (lacking bundles of long chaetae but with well-developed neuropodial flaps) not much longer than broader anterior region, with about 12–15 segments. A prominent bundle of stiff golden 'collar chaetae' in front of gills on each side of head region. Gills free more or less to the base. At least two anal cirri **2**

Posterior region long and narrow, with 50 or more segments. No 'collar chaetae'. A pair of stout nuchal hooks protruding from thoracic membrane dorsally, just behind gills (Fig. 6.3B). Gills united by a web-like membrane for more or less half their length. No anal cirri **3**

2. Twelve posterior chaetigers, mostly bearing only neuropodial flaps with uncini. Anus surrounded by numerous cirri (one on each side may be larger than the others). Tentacles bear small papillae, giving them a slightly feathery appearance. Prostomium pointed, without prominent ridges. Uncini with 8–10 teeth in two rows (Fig. 6.3D) ***Ampharete acutifrons***

Fifteen posterior chaetigers. Notopodial boss of anterior region bears a small club-like dorsal cirrus which remains as a vestige posteriorly. Two anal cirri, of variable length. Tentacles smooth or lightly ringed. Prostomium rectangular, with a prominent glandular ridge on each side. Uncini with 5–7 teeth in a single row (sim. Fig. 6.3C) ***Amphicteis gunneri***

3. Four anterior segments have a row of fine spear-like chaetae as the neuropod. Posterior margin of thoracic membrane cut into 10–20 well-defined teeth (Fig. 6.3B). Web joining gills poorly developed. Body of medium length (30–60 mm) ***Melinna cristata***

Three anterior segments with a row of spear-like chaetae; these are lacking

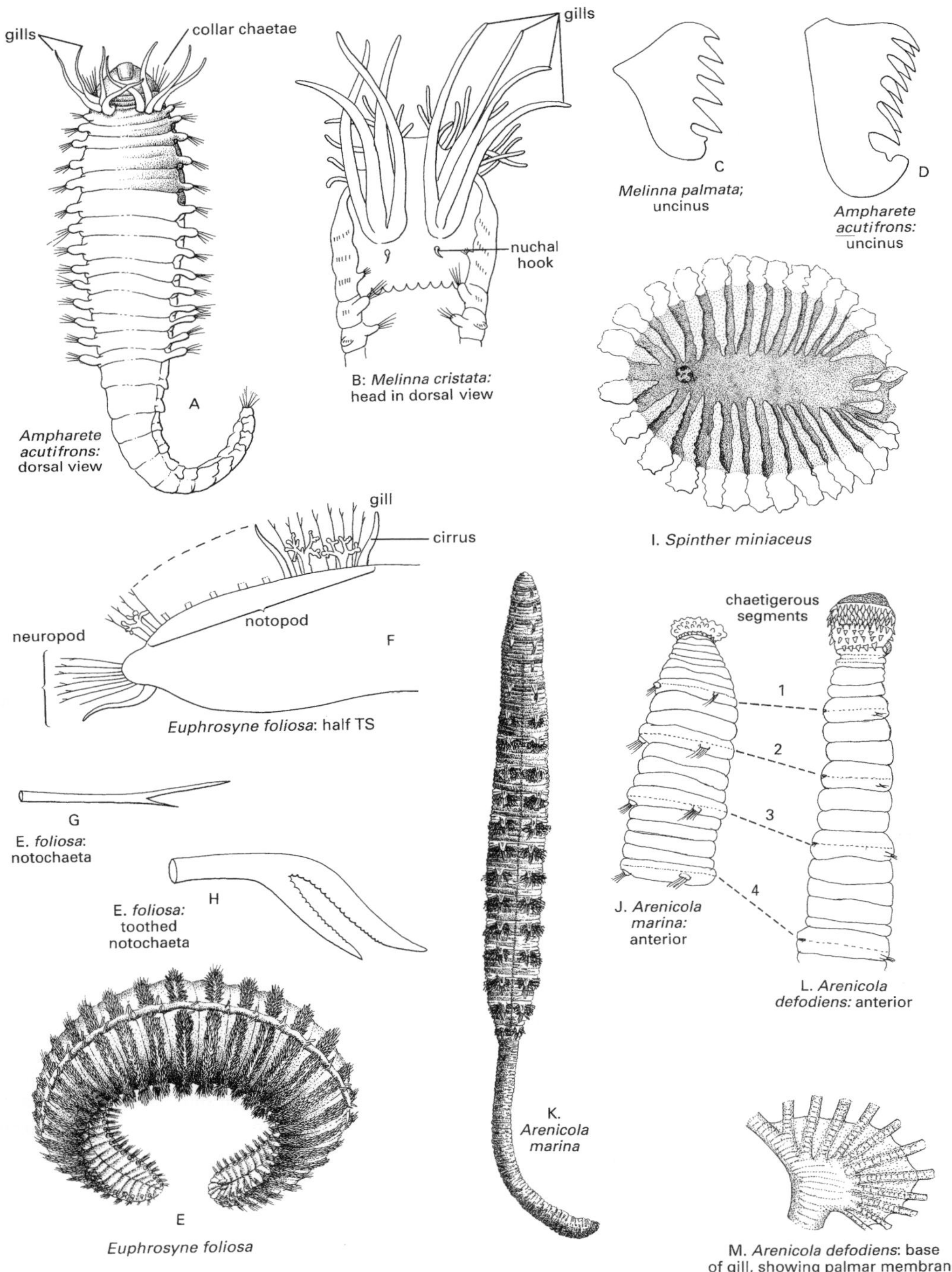
gills
collar chaetae
A
Ampharete acutifrons: dorsal view
gills
nuchal hook
B: *Melinna cristata:* head in dorsal view
C
Melinna palmata; uncinus
D
Ampharete acutifrons: uncinus
I. *Spinther miniaceus*
gill
cirrus
notopod
neuropod
F
Euphrosyne foliosa: half TS
G
E. *foliosa*: notochaeta
H
E. *foliosa:* toothed notochaeta
E
Euphrosyne foliosa
K. *Arenicola marina*
chaetigerous segments
1
2
3
4
J. *Arenicola marina:* anterior
L. *Arenicola defodiens:* anterior
M. *Arenicola defodiens*: base of gill, showing palmar membrane

Fig. 6.3

from one segment before uncinigerous tori appear. Posterior margin of thoracic membrane either entire or cut into 4–8 indistinct teeth. Web joining gills may extend as far as two-thirds their length. Small (15–20 mm) ***Melinna palmata***

Ampharete acutifrons Grube Fig. 6.3

Fourteen anterior and 12 posterior chaetigers, 15–35 mm long (A). Pointed prostomium with 2 small ridges outlining an irregular pentagon. Two eyespots. Peristomium narrow and can be partially retracted. Buccal tentacles bear small papillae, appearing pinnate. Ten to 15 long, golden 'collar chaetae' on each side in front of gills, in a stiff forward-pointing fan. Four cirriform gills, free to the base, in a group on each side. First parapod a tiny achaetous tubercle behind gills: subsequently, the notopod is a simple conical boss with spear-like bristles and the neuropod (from chaetiger 2) a flattened, stalked button bearing a transverse row of uncini with teeth in two parallel rows (D). In the posterior region, the neuropod is a palette shaped flap with a cirriform prolongation at the top. On first 4–5 posterior segments, a reduced notopod without bristles. Pygidium with numerous anal cirri, of which one on each side is usually long. Males greenish-white, females pale salmon-pink; blood greenish. Tube membranous, covered with sand grains.

At low water, in muddy sand amongst sea grasses, and sublittorally. Circum-Arctic, European coasts, Baltic; south to Madeira, Morocco, and into the Mediterranean.

Amphicteis gunneri (M. Sars)

Similar to *Ampharete acutifrons* but with 17 anterior and 15 posterior chaetigers, 20–40 mm long. Prostomium made rectangular by two prominent glandular crests. Two groups of eyespots. Peristomium wide ventrally. Buccal tentacles smooth or lightly ringed. Eight to ten 'collar chaetae'. Gills more or less fused at base. Uncini, with a single row of teeth, from chaetiger 4. Notopod with a little club-shaped dorsal cirrus; posteriorly, it is represented only by this cirrus. Pygidium with two anal cirri. Yellowish-white or pink, with white spots and brown blotches. Gills green with bands of brown, white, and yellow.

At low water and in the sublittoral to considerable depths, on mud or muddy sand. Circum-Arctic, most European coasts south to Morocco, and in the Mediterranean; also mid-Atlantic, Gulf of Mexico, southern Africa, East Indies, Antarctic.

Melinna cristata (M. Sars) Fig. 6.3

Sixteen anterior chaetigers, first rudimentary; about 50 posterior ones; 30–60 mm long, very thin posteriorly. Prostomium trilobed, without glandular crests. Several eyespots on each side. Peristomium mostly covered by a collar-like extension of the next segment ventrally. Few smooth buccal tentacles. Each group of four long, filiform gills united by a palmar membrane for up to half their length. Chaetigers 2–5 have a deep dorsal depression covered by a thoracic membrane with a posterior border cut into 10–20 well-defined teeth (B). Behind gills on each side of chaetiger 4, a chitinous nuchal hook (B). Laterally on chaetigers 2,3,5 and 6 the neuropod is represented by a transverse row of fine spear-like bristles. From the sixth, notopods are typical but lack a cirrus; from the seventh, neuropodial tori with triangular uncini whose teeth are in a single row (as C). Posteriorly, the neuropod is a rectangular flap without a cirriform extension; notopod is vestigial. Pygidium lacks cirri; anus within a small funnel. Yellowish or pinkish-white, gills blotched with olive-green; blood red. Tube membranous, encrusted with mud.

On the shore and sublittorally on muddy bottoms, often amongst sea grasses. Most North-West European coasts; Arctic; also North Pacific, sub-Antarctic.

Melinna palmata Grube Fig. 6.3

Smaller than *M. cristata* (15–20 mm long), with annulated gills bound together by a web for up to two-thirds their length. Thoracic membrane behind them has a posterior border which is either entire and concave or cut into 4–8 indistinct rounded flaps. Spear-like bristles lacking from chaetiger 6. Pale or brownish pink, blotched with red. Gills greenish with brown bands; blood red.

At low water in mud amongst sea grasses, also sublittorally on muddy bottoms. Western Scotland, East Anglia, and Channel coasts. Northern Spain, scattered locations in Mediterranean; also Morocco, Black Sea, Arabian Gulf.

2. AMPHINOMIDAE

Body often short, flattened, and oval. Prostomium reduced; proboscis eversible, without jaws or papillae. A nuchal organ forming a crest (the *caruncle*) extends back mid-dorsally for several segments.

Dorsal surface covered to a varying extent by rows of simple bristles, branched gills, and cirri. Most are carnivorous, sucking the tissue fluids of sessile invertebrates, and may be brightly coloured to match them. Some are more typically worm-like: these include the tropical 'fire-worms' *Hermodice*, with hollow poison-chaetae which break off after penetrating the skin of anyone handling them roughly, causing intense irritation. A small family with about 10 species recorded for Britain. The genus *Spinther* (Fig. 6.3I) includes small, flat, almost circular worms which live on sponges.

Euphrosine foliosa Audouin and Milne Edwards — Fig. 6.3

30–36 chaetigers; 10–30 mm long, up to 10 mm wide (E). Prostomium extends a little rearwards ventrally, with two large dorsal and two smaller ventral eyes. A thick, median antenna between dorsal eyes and two lateral ones, half its size, ventrally. Mouth ventral, with a pair of reduced palps anterior to it. Caruncle extends back to chaetiger 5. Notopods spread across dorsal surface, leaving only a small bare region in the midline (F). On each side, 7–9 dichotomously branching gills with enlarged tips, and two cirri, one at inner end of gill row, the other between 2nd and 3rd gill from that end. Notochaetae, much longer than gills, form an irregular row anterior to them; all bifurcated, some with smooth branches of very unequal length (G), others equal and toothed internally (H). Neurochaetae around edge of body, all with two smooth unequal branches; one cirrus below each bundle. Two short, thick anal cirri. Colour orange, brick-red, or vermilion.

Under stones, amongst shells and algae or in crevices, from low water to moderate depths. Cosmopolitan.

3. APHRODITIDAE

A large family characterized by the elytra (scales) which cover much of the dorsal surface. Representatives of three main subgroups (often listed as separate families) are included here. The Aphroditinae have few segments and an oval body, often quite large. They are slow moving, burrowing or creeping on soft bottoms where they feed on living animals, carrion, or detritus. Some notochaetae are fine and extremely long, intertwining to form a felt which completely covers the back in *Aphrodita*; in others this is less well developed, or may be lacking. Four species, in three genera, are presently recognized from Britain. The Polynoinae are the most numerous and typical scale-worms. The British fauna includes more than 40 species, in 14 genera. At least 20 species are referred to the genus *Harmothoe*; these are often difficult to tell apart and only the most common species are included here. Polynoid scale-worms may be oval and flattened, or cylindrical and elongated; they are generally quite small. Elytra usually occur on every second or third segment, the others having a long dorsal cirrus. Elongated worms may lack scales posteriorly. Polynoids are carnivorous, with four large chitinous teeth in pairs above and below the opening of the proboscis. They are often commensal with animals living in tubes or burrows, when their elytra are usually smooth; those with strongly ornamented elytra are more likely to be free living. The Sigalioninae have an elongated body with very many segments, appearing tetragonal in cross-section. Elytra occur on alternate segments in the anterior part of the body but then on every segment, right to the tail end. There are no dorsal cirri, although all but the first parapods have a cirriform gill inserted dorsally, on the stalk of the elytra when present. Parapods also typically carry a dorsal row of three *ctenidia* (ciliated cup-shaped or leaf-like protuberances). Some or all neurochaetae are jointed. Like polynoids, sigalionids have four chitinous hooks in the proboscis; they are carnivores, usually moving fast on sandy bottoms. The six genera represented in the British fauna include perhaps a dozen species.

NB: Elytra are readily shed by these worms; their position is shown by the mushroom shaped 'stalk' (*elytrophore*) (Fig. 6.4B-F), which is the basal portion of the modified dorsal cirrus.

---

1. Large, broad, and flattened oval-bodied worm whose dorsal surface is covered with a greyish-brown felt and fringed by stout iridescent bristles — ***Aphrodita aculeata***

   Dorsal scales more or less visible — **2**

---

2. Medium size, flattened, and oval-bodied. Dorsal scales visible but partially covered by chaetal 'fur'; body fringed laterally by tangled chaetae — ***Laetmonice filicornis***

Size varies, but generally shorter and more slender; dorsal scales not at all covered by chaetae, body fringed laterally only by straight chaetae of normal length and, often, by elongated dorsal cirri **3**

---

3. Elytra (scales) inserted on alternate segments, **or** one segment in three, **or** lacking posteriorly; segments not bearing elytra have dorsal cirri, elongated and protruding dorso-laterally. Prostomium always bears three antennae, one median (anterior) and two lateral, usually smaller (Fig. 6.4A–C). Body mostly short, flattened and oval, though not very broad; 12, 15, or 18 pairs of elytra (Polynoinae) **4**

   Elytra inserted on alternate segments anteriorly and on every segment posteriorly; segments not bearing elytra lack dorsal cirri, but almost all segments bear dorsally a ciliated cirriform gill on each side, beneath the elytron if present (except *Pholoë*). Prostomium bears vestigial or 1–3 visible antennae (Fig. 6.4D–F). Body usually long and slender, with up to *c.* 150 pairs of elytra (Sigalioninae) **17**

---

4. Elytra (15 pairs) leave an appreciable number of posterior body segments (*c.* 10–50) uncovered **5**

   Elytra cover substantially whole length of body **6**

---

5. Posterior quarter body length (8–15 segments) uncovered. Length 30–40mm ***Lagisca extenuata***

   Posterior half body length (*c.* 50 segments) uncovered. Length 30-120mm. Typically shares tube of terebellid *Eupolymnia* ***Polynoë scolopendrina***

---

6. Twelve pairs of elytra **7**

   More than 12 pairs of elytra **8**

---

7. Elytra scarcely overlap and leave dorsal surface uncovered in mid-line; mostly smooth and unfringed ***Lepidonotus clava***

   Elytra overlap considerably, covering entire dorsal surface; with many prominent papillae, and fringed along external border ***Lepidonotus squamatus***

---

8. Fifteen pairs of elytra **9**

   More than 15 pairs of elytra **16**

---

9. Elytra transparent, not overlapping markedly and lacking prominent papillae or pigmentation, not strongly fringed. Fragile worms typically found on echinoids **10**

   Elytra opaque, often with brownish markings, prominent papillae or a fringed external border. Sturdy appearance **11**

---

10. Antennae and cirri bear short spines ***Adyte assimilis***

   Antennae and cirri smooth ***Adyte pellucida***

---

11. Lateral antennae inserted subterminally. Antennae, palps, and cirri smooth. Elytra smooth and unfringed, banded brown with darker spots. Normally found on spatangoids or asteroids ***Malmgrenia castanea***

   Lateral antennae inserted ventrally. Antennae, palps, and cirri hairy or spiny **12**

---

12. No elytra on last four segments; their external border finely fringed, surface covered with small papillae, greyish-white with a central black spot. Shares tube of terebellid or *Chaetopterus* ***Gattyana cirrosa***

Body completely covered with elytra, variously decorated **13**

13. Elytra not fringed; papillae absent or numerous but so small as to be inconspicuous **14**

Elytra fringed with only few hairs or weak spines; papillae generally small but may become large near free edge **15**

Elytra well fringed with strong spines, variously covered with papillae which may be large near free edge; generally brownish; worm fairly small (12–25 mm) ***Harmothoë impar***

14. Elytra brown with black circular, semi-circular, or chevron pattern; worm medium size (10-35 mm), often shares tube or burrow ***Harmothoë lunulata***

Elytra pale yellowish with broad brown band around posterior edge: dorsal cirri and ventral chaetae very long, giving marked lateral fringe to body; fairly large (30–60 mm), often shares tube or burrow ***Harmothoë glabra***

15. First pair of elytra almost white, often with central black spot; most dark brown with greenish sheen, becoming paler and marbled with lighter spots towards posterior end; worm fairly small (*c.* 15 mm) ***Harmothoë spinifera***

Elytra greyish to dark brown with purplish sheen, sometimes with white outer edge; worm fairly large (30–50 mm), may share tube or burrow ***Harmothoë imbricata***

Elytra greyish to violet brown, either lighter or darker in centre; worm large (60–80 mm) and broad ***Eunoë nodosa***

16. Eighteen pairs of translucent off-white elytra; worm large (80–90 mm) but fairly broad ***Alentia gelatinosa***

Approximately 20–70 pairs of small elytra; worm large (35–220 mm) and slender; shares terebellid tubes ***Lepidasthenia argus***

17. No median antenna, lateral antennae vestigial. Elytra (150 pairs or more) fringed with 10–20 feathery papillae (Fig. 6.4I) ***Sigalion mathildae***

Median antenna present, elytra lack feathery fringe **18**

18. Median antenna present but short (as are other anterior processes). Elytra (40–60 pairs) fringed with stout, simple, or moniliform papillae. Small (10–20 mm) ***Pholoë inornata***

Median antenna long, short lateral antennae (or processes resembling lateral antennae) present. Large (50–200 mm), *c.* 150 pairs of elytra **19**

19. Body rectangular in section, elytra fail to meet in mid-line or only barely do so. Some jointed ventral chaetae have banded tips, but none are bifid ***Neoleanira tetragona***

Body oval in section, elytra overlap well and completely cover the dorsal surface. Some jointed ventral chaetae have bifid tips, but none are banded **20**

20. Elytra have a fringe of simple spines and are opaque, covered with small conical tubercles (Fig. 6.4G) ***Sthenelais boa***

Elyctra are smooth, translucent, and mostly lack a fringe, but are notched in the external edge (Fig. 6.4H) ***Sthenelais limicola***

APHRODITINAE

Aphrodita aculeata (Linnaeus)
Sea mouse Fig. 6.1
Body oval, more pointed to rear; dorsal surface convex, completely covered by a close 'felt' of fine chaetae, ventral surface forms a flattened sole. About 40 segments; 100–200 mm long, 30–70 mm wide. Prostomium spherical, with two sessile eyes and a single tiny antenna between. Two long awl-shaped palps; two pairs of subequal tentacles, shorter than palps. Fifteen pairs of smooth elytra, tightly overlapping, hidden beneath felt (F). In addition to the very long fine chaetae forming this covering, the notopod bears short, thick, dark-coloured bristles which poke up through it, and fine, silky, iridescent ones forming a fringe around its edge. Long cirri project through fringe from segments not bearing elytra. Neuropod with few short, stout bristles extending laterally and a short ventral cirrus. Dorsal felt dark grey; flanks iridescent from blue through green and yellow to bronze; sole brownish yellow.

Sublittoral in sand or muddy sand; rarely, stranded at low tide. Most north-west European coasts, Mediterranean.

Laetmonice filicornis Kinberg Fig. 6.4
Body broad, oval with 34–36 chaetigers; length 20–35 mm. Dorsal felt loose, variably developed. Prostomium rounded, divided into three lobes. Two eyes on short stalks. Antenna long; two palps nearly as long, covered with fine papillae. Two pairs of filamentous tentacles, shorter than palps. Fifteen pairs of smooth interlocking elytra, visible through felt. Chaetigers not bearing elytra with long, thin dorsal cirri, notopod with felt-forming chaetae; elytron-bearing chaetigers with stout multi-barbed, harpoon-like chaetae (M), rarely enclosed between two long valves. Neuropod has a few strong chaetae with a spur near tip, from which runs a terminal fringe of fine, stiff filaments (N). First few chaetigers also with fine pinnate chaetae. Dorsal surface often bluish or violet, seen through the grey felt.

On muddy bottoms to considerable depths. Most north-west European coasts; Pacific.

Hermione hystrix (Savigny) is similar but 50–60 mm long and without a dorsal felt. Its elytra are smooth, brownish, with a slight iridescence. It occurs on shell and gravel bottoms along western coasts (also the Mediterranean and Indian Ocean).

POLYNOINAE

Adyte assimilis (McIntosh)
Body elongated, narrow, and fragile, about 40 segments, 18–20 mm long. Prostomium in two rounded lobes; four eyes, anterior pair larger and with a lens. Median antenna long, lateral ones shorter. Palps thick, brown; tentacles short. Antennae, palps, and tentacles smooth. Dorsal cirri relatively short. Antennae, tentacles, and dorsal cirri terminate in an elongated point; ventral cirri short and awl-shaped. Elytra (15 pairs) rounded or oval, transparent; finely dotted, with fine papillae around free edge. Notochaetae short, slightly curved, with a row of fine spines. Neurochaetae longer and finer; a moderate swelling near tip with a webbed spur, beyond which a row of fine spines runs to slightly hooked tip. Pale coloured, with a longitudinal brownish or greenish band dorsally.

In shallow sublittoral, on *Echinus*. Western European coasts.

Adyte pellucida (Ehlers)
Body narrow, very fragile, with about 40 segments; about 30 mm long. Four eyes, anterior pair largest. Median antenna long, with an elongated tip; lateral antennae and tentacles a little shorter. These and similar dorsal cirri with short knobbed processes. Palps tapering and smooth. First ventral cirrus very long, rest short. The anterior of 15 pairs of elytra overlap in the middle; further back they leave mid-dorsal surface bare. Elytra transparent, free surface dotted with small processes and, occasionally, a few larger, rounded tubercles; readily shed. Notochaetae short but thick; neurochaetae longer, with fine spines. Two long anal cirri. Body translucent, back yellowish with light brown transverse bands, occasionally with whitish blotches. Elytra finely dotted with yellow, white, pink, brown, or purple.

Shallow sublittoral, on echinoderms. Western coasts of Europe, Mediterranean.

Alentia gelatinosa (M. Sars)
Body straight-sided, narrowing at each end, flattened, with 43 chaetigers; 80–90 mm long.

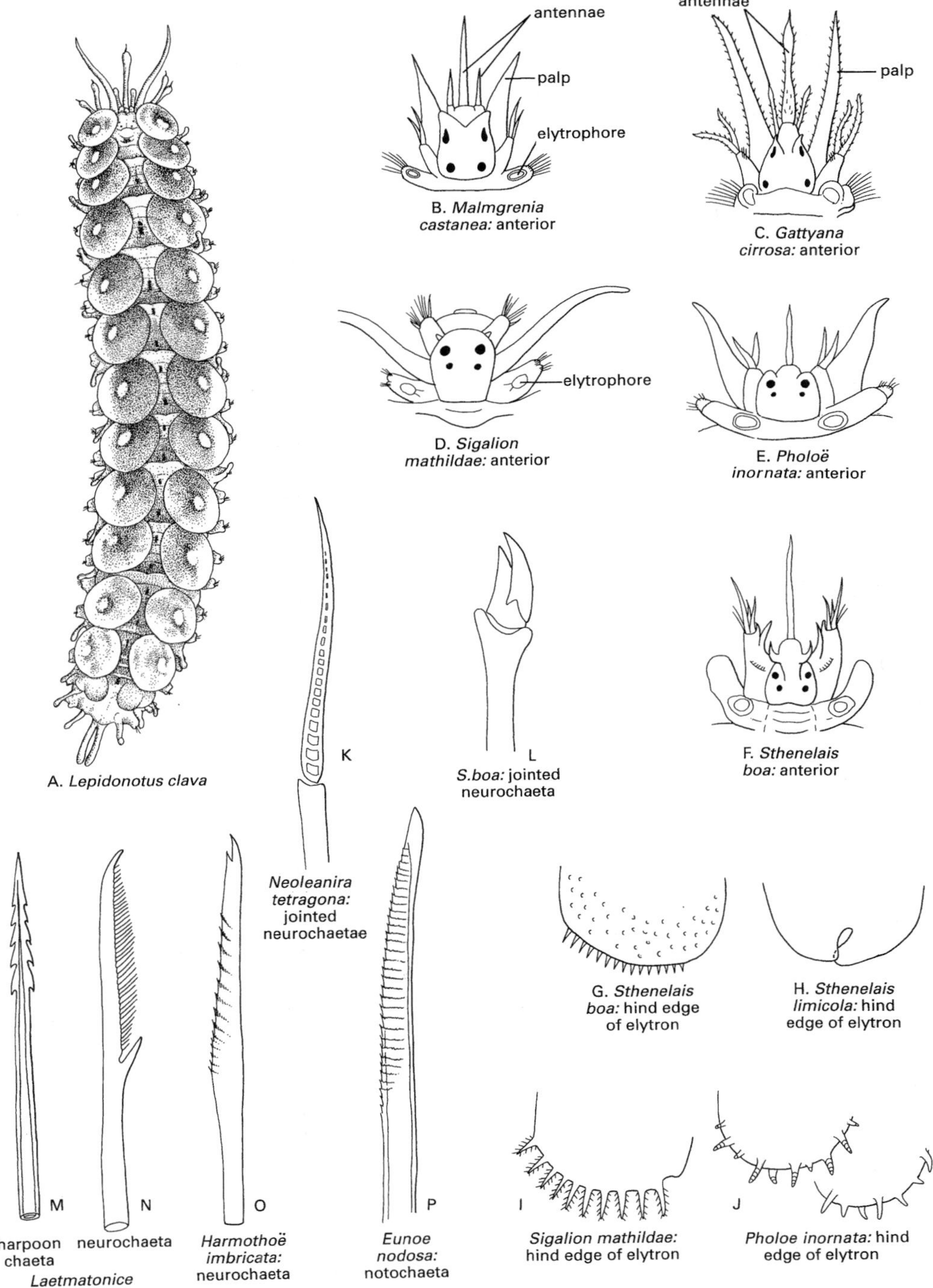
antennae
palp
elytrophore
B. Malmgrenia castanea: anterior
antennae
palp
C. Gattyana cirrosa: anterior
elytrophore
D. Sigalion mathildae: anterior
E. Pholoë inornata: anterior
A. Lepidonotus clava
K
Neoleanira tetragona: jointed neurochaetae
L
S.boa: jointed neurochaeta
F. Sthenelais boa: anterior
G. Sthenelais boa: hind edge of elytron
H. Sthenelais limicola: hind edge of elytron
M
N
harpoon chaeta
neurochaeta
Laetmatonice filicornis
O
Harmothoë imbricata: neurochaeta
P
Eunoe nodosa: notochaeta
I
Sigalion mathildae: hind edge of elytron
J
Pholoe inornata: hind edge of elytron

Fig. 6.4

Prostomium divided into two rounded lobes, each bearing two large eyes on outer side; a semicircular nuchal fold covers it posteriorly like a high collar. Median and lateral antennae subequal, shorter than palps which are smooth and tapering. Tentacles almost as long as palps; antennae and tentacles also smooth, with a subterminal swelling and pointed tip. Dorsal cirri similar, extending beyond chaetae; ventral cirri short and tapering. Elytra (18 pairs) large, soft, with brown reticulated pattern; free surface covered with many small, truncated brownish papillae. Notochaetae few and delicate, very finely toothed towards tip. Neurochaetae numerous and long, with a broader finely toothed region towards tip, which is lightly bifid in those more ventrally inserted. Ventral side orange, dorsal side marked with transverse brown and white bands; elytra a translucent dirty white or brownish.

Under stones, sometimes in *Nerine* burrows, on muddy bottoms in the shallow sublittoral. Around most British coasts; north-east Atlantic.

Gattyana cirrosa (Pallas) Fig. 6.4

Body broad, not narrowing markedly at ends, with 34–36 chaetigers; 25–50 mm long, 5–10 mm wide. Prostomium with well-marked frontal horns; four little black eyes, first pair looking downwards. Median antenna as long as palps, lateral ones half as long. Tentacles a little shorter than palps, decorated with papillae like the antennae (C). Dorsal cirri extend about as far as chaetae; ventral cirri shorter. Fifteen pairs of elytra, none on last four chaetigers but covering whole body; free edge fringed with short hair-like processes, surface appears smooth but covered with many little conical processes, some with two or four cusps. Notopod with many long, delicate, finely spined chaetae; neurochaetae longer and thicker, with a spiny blade. Two long anal cirri. Colour milky or greyish white, sometimes broken by a black spot in the middle of each elytron.

In the tube of terebellid or chaetopterid worms. On most British coasts; Arctic, north Atlantic, north Pacific.

Harmothoë species Fig. 6.4

General characters are as follows. Only noteworthy variations from these will be given for the individual species listed, although it should be appreciated that the relative lengths of contractile structures such as antennae may be deceptive.

Body short, straight-sided or narrowing at rear end, with 35–40 chaetigers. Frontal horns of prostomium variously developed. Four eyes. One median antenna and two shorter lateral antennae. Two stout palps, two pairs of long tentacles with small chaetae around bases. Dorsal cirri more or less long, ventral cirri short. Antenna, tentacles, and cirri usually with pointed tip and subterminal swelling, and covered with small, elongated or knobbed processes; palps smooth or bear finer papillae in longitudinal rows. Notochaetae short and stout, with many rows of spines distally; neurochaetae longer, finer, spiny, and often slightly bifid at tip (O). Fifteen pairs of overlapping elytra cover whole body dorsally, inserted from 2nd to 32nd chaetiger, more or less alternately. Two long anal cirri.

Harmothoë imbricata (Linnaeus)

Body 30–50 mm long. Dorsal cirri extend well beyond chaetae. Elytra appear smooth but covered with small conical tubercles; fringed along free edge with short hairs, and in large worms with a row of large, roughly globular, stalked nodules on hind margin. Mottled grey-black, through bluish grey to brown or brownish purple with metallic glints. Elytra may have dark centre with lighter spots around edge; some specimens have a broad chestnut median longitudinal band, with white on each side.

At low water in kelp holdfasts or tubes of terebellid and chaetopterid worms; on stony or shelly bottoms to considerable depths. European coasts from the Arctic to the Mediterranean; Indian Ocean, North Pacific.

Harmothoë impar Johnston

Body 12–25 mm long. Frontal horns prominent. Dorsal cirri extend beyond chaetae. Elytra easily shed, strongly fringed on free edge, mostly covered with chitinous tubercles and spiny outgrowths of varied shape; these may be enmeshed in a reticular colour-pattern. Dorsal coloration greenish brown, head end darker, with a complex pattern of stripes and spots; elytra often have a yellowish median spot; reticular pattern orange-brownish. Ventrally, pale and iridescent. Bioluminescent.

At low water under stones, amongst shells or in kelp holdfasts; offshore to some depth. North-western European coasts, Mediterranean.

Harmothoë glabra Malmgren

Body 30–60 mm long. Frontal horns blunt or absent. Dorsal cirri extend beyond chaetae. Elytra easily shed, delicate, appearing smooth but with very many small conical processes, and a few very short hairs along the free edge. Ventral coloration silvery white; back pale, perhaps with transverse brown bands. Head appendages and cirri may be brown. Elytra pale yellow with a chestnut band around posterior edge, or colourless. Bioluminescent.

At low water and in shallow sublittoral, under stones and particularly in tubes of chaetopterids, terebellids, and even *Arenicola*. North-western European coasts, Mediterranean.

British records of this species were formerly referred to *H. longisetis* (Grube), a southern species which possibly does not occur in Britain.

Harmothoë lunulata (Delle Chiaje)
Body 10–35 mm long. Frontal horns not well marked. Dorsal cirri extend only as far as chaetae. Elytra smooth and unfringed, decorated with very variable patterns. Coloration very variable: ventral side with a median dark-red band and transverse brownish stripes posteriorly; elytra with brownish markings in circular, semicircular, or V shapes, sometimes broken, with a central black spot. Head reddish, antennae and cirri with brown markings. Bioluminescent.

At low water, under stones or in crevices; most frequently commensal with various annelids or holothurians. Western coasts of Europe (northern records need checking) to Mediterranean.

Harmothoë spinifera (Ehlers)
Body about 15 mm long. Dorsal cirri extend only as far as chaetae. Surface of elytra with small, rounded tubercles, free edge with a few short hairs. Ventral side colourless anteriorly, greenish behind. Dorsal side with transverse dark stripes. First pair of elytra very pale with a black spot, contrasting strongly with rest, which are dark brown or black with a metallic greenish glint; colouring becomes more pale and mottled posteriorly. Antennae, palps, and cirri brownish.

Under stones, in crevices and in kelp holdfasts at low water, also in shallow sublittoral. Western coasts of Europe to Mediterranean.

Lagisca extenuata (Grube)
Body of about 40–50 segments, 30–40 mm long, 8–10 mm wide. Prostomium with sharp frontal horns; four eyes. Median antenna long, lateral ones half that length. Tentacles similar to median antenna in length and form, as are dorsal cirri (which extend beyond chaetae). Ventral cirri short. Antennae, tentacles, and cirri ornamented with long papillae. Elytra with little conical chitinous tubercles over entire surface, a posterior row of large globular warts and a fringe of long processes. Last 8–15 chaetigers uncovered, forming a tapering tail. Notochaetae numerous, directed dorsally; some neurochaetae bifid. Two papillated anal cirri. Coloration very variable: underside usually pale and iridescent, the back with brownish bands forming a complex design. Elytra marbled brown and grey or reddish, usually with a lighter median spot. Head reddish with a pale transverse band. Antennae and cirri ringed with brown.

Under stones and amongst rocks throughout the tidal zone, also offshore to some depth. European coasts from the Arctic to the Mediterranean.

Eunoë nodosa (M. Sars)
May reach length 60–80 mm and width 25–30 mm. Slight frontal horns. Dorsal cirri extend well beyond chaetae. Free edge of elytra fringed or not; free surface covered with spiny chitinous papillae and often, also, large conical or globular nodules. Elytra grey or purple-brown, often with a lighter or darker median spot; antennae and cirri reddish with a brown band and white tip.

On bottoms of gravel and muddy sand to considerable depths. Arctic and northern Europe.

Lepidasthenia argus Hodgson
Body narrow and straight-sided, with more than 200 segments; 35–215 mm long. Four eyes. Median antenna long, fairly thick, with subterminal swelling and pointed tip; lateral ones more slender. Palps stout. Tentacles and dorsal cirri similar to antennae; appendages smooth. Dorsal cirri extend as far as chaetae, ventral cirri much smaller. Elytra (22–27 pairs) smooth, translucent, and unfringed, covering sides of body to the rear end but leaving uncovered a large part of the back. Notopod reduced to a small protuberance with four or five long, smooth chaetae, often absent. Many neurochaetae, all with bifid tips. Two short anal cirri. Fawn coloured, with a brown band across each segment; a red mid-line on underside. Antennae and cirri with a brown ring below tip. Elytra with a dark median spot and a white posterior crescent.

South-western coasts of Britain, commensal in the tube of *Amphitrite*.

Lepidonotus clava (Montagu) Fig. 6.4
Body of uniform width, with 26 chaetigers; 25–30 mm long (A). Four eyes. Median antenna

shorter than palps, lateral ones half its length; tentacles and dorsal cirri about as long as median antenna. All smooth, with a marked subterminal swelling and pointed tip; palps stout, tapering, with longitudinal rows of small papillae. Ventral cirri short, thick, with a pointed tip; first one larger, with a swollen tip, directed forwards. Elytra (12 pairs) overlap only in young specimens; extend to rear end but leave part of back uncovered; first four pairs with widely scattered chitinous tubercles, large and small, rest almost smooth, none fringed. Notochaetae fine and spiny; neurochaetae stout, with small spines below tip. Two anal cirri similar to dorsal cirri. Back dark, with white blotches; elytra marbled brown, chestnut, and yellowish-white, with a large white median spot. Appendages with a dark ring below subterminal swelling and sometimes another near base.

Under stones in the tidal zone. Western Europe to Mediterranean; Indian Ocean, North Pacific.

Lepidonotus squamatus (Linnaeus)
Similar to *L. clava* but elytra overlap considerably, covering entire dorsal surface; all bear many chitinous tubercles of various types, and strongly fringed around free edge. Coloration variable but often pale yellowish with a dark median spot on elytra; larger tubercles brownish, making a mosaic pattern.

Under stones on the shore; amongst old shells or serpulid tubes in the shallow sublittoral and to considerable depths. North-western European coasts; North Pacific.

Malmgrenia castanea (McIntosh) Fig. 6.4
Body broad, narrowing posteriorly, with 36–41 chaetigers; 18–20 mm long. Four eyes, front pair almost invisible from above. Median antenna a little longer than palps, lateral ones half as long (B). Palps smooth, stout, short. Tentacles shorter than palps, dorsal cirri similar (extending about as far as chaetae); ventral cirri slender and short. Elytra (15 pairs) smooth except for a small group of rounded tubercles on anterior (hidden) part; not fringed. Notopod small, with a few short blunt chaetae; neurochaetae longer, with a short spiny blade. Two long anal cirri. Coloration very variable; underside pink, brownish, or purple, back light brown. Elytra with transverse chestnut bands and dark spots.

Found around the mouth and in the ambulacra of *Spatangus* or, less often, on sea stars. European coasts from the Arctic to the Mediterranean.

Polynoë scolopendrina Savigny
Body narrow, with 80–100 segments; 30–120 mm long. Four eyes, forward pair larger but scarcely visible from above. Median antenna fairly long, covered with short knobbed processes. Tentacles and dorsal cirri similar, lateral antennae and ventral cirri shorter. Palps of variable length, with rows of fine papillae. Elytra (15 pairs) cover only front half of body and do not all overlap across mid-line; each with a broad band of little tubercles across anterior (partly hidden) portion and a few short knobbed processes around free edge. Segments not bearing elytra have a row of three tubercles across dorsal surface. Notochaetae small, spiny, often blunt-ended; neuropod with one large, bluntly pointed bristle in more anterior segments and many smaller ones. Two short papillated anal cirri. Underside usually light brown, becoming darker towards rear, with a white or reddish mid-ventral line; back mottled brown. Elytra mottled brown with a central dark spot surrounded by a paler area.

At low water under stones, in crevices and, particularly, commensal with *Eupolymnia* or *Lysidice*. North-western Europe, Mediterranean.

SIGALIONINAE

Neoleanira tetragona (Oersted) Fig. 6.4
Body narrow, tetragonal in section, with up to 300 segments; up to 200 mm long and 7 mm wide. No visible eyes. Median antenna long, with two small finger-like ctenidia at its base. Slender, shorter lateral antennae; a long dorsal and short ventral tentacle on each side. Two very long tapering palps. No dorsal cirri. Elytra (about 150 pairs) covering entire body, but only barely meeting in mid-line; transparent, fringed with long papillae along their free edge. One cirriform gill inserted on elytrophore or above each parapod except the foremost; three cup-shaped ctenidia and several short finger-like processes along dorsal edge of each parapod. Notochaetae long, fine, and spiny; neurochaetae jointed, the tapering terminal segment with a ladder-like internal structure (K). Short awl-like ventral cirri. Colourless.

In deep water on muddy bottoms. Off north-western British coasts; Arctic, North Sea, North Atlantic, Mediterranean.

Pholoë inornata Johnston Fig. 6.4
Body oblong; dorsal surface convex, ventral surface flat, with a longitudinal gutter; 45–70 segments, 10–20 mm long. Four small eyes. Median antenna

short, papillated; palps short and stout, smooth (E). Two pairs of tentacles, much shorter than palps, with scattered papillae. Elytra (40–60) with concentric rings, and fringed by long, more or less moniliform processes (J); only more posterior meet at mid-line. Parapods covered with short papillae. Notochaetae fine, curved; neurochaetae fewer, much longer and thicker, jointed, the terminal segment a short hook. Two thread-like anal cirri. Back pale pink, elytra speckled with rusty brown.

At low water under stones; in the shallow sublittoral amongst old shells. North-western Europe; Pacific.

Sigalion mathildae Audouin and Milne Edwards Fig. 6.4

Body cylindrical, rolling into a spiral when disturbed. About 200 segments, 100–150 mm long and 4–5 mm wide. Four very small eyes, often not visible. No median antenna, lateral antennae very small; palps long and smooth. Chaetiger 1 directed forwards beneath long prostomium, with a small dorsal and a rudimentary ventral tentacle. Elytra rectangular, smooth, and colourless, fringed with 10–20 pinnate papillae (I). A cirriform gill inserted on elytrophore or above each parapod. Three ciliated mushroom-like ctenidia on dorsal edge of each parapod. Notopod with one finger-like process dorsally above a bundle of numerous long, fine, curved chaetae. Neurochaetae mostly jointed, terminal region of several segments, often strongly recurved. Greyish white; underside iridescent, with a prominent red mid-ventral blood vessel.

At low water, 150–200 mm beneath the surface of fine sand. North-western Europe, Mediterranean.

Sthenelais boa (Johnston) Fig. 6.4

Body narrow, with more than 150–200 segments; 100–200 mm or more long. Four black eyes; two button-like nuchal organs at rear of prostomium. One long, smooth median antenna, with swollen base also bearing two finger-like ctenidia resembling short lateral antennae (F). Two long, smooth palps. Chaetiger 1 directed forwards beneath rounded prostomium: with one short lateral antenna on each side, a ctenidium dorsally at its base, and one long dorsal, one short ventral tentacle. Elytra cover whole body, overlapping on mid-line and down each side, giving the worm a snake-like appearance; covered with small, flattened tubercles and with a fine fringe around free edge (G). A long cirriform gill on all but the most anterior few segments; dorsal edge of each parapod has a row of three mushroom-shaped ctenidia. All but the foremost chaetiger have a notopod with several finger-like processes and a bundle of numerous long, fine chaetae; most neurochaetae jointed with a hooked tip (L). Two long anal cirri. Pale grey to yellow or brownish, sometimes with transverse rusty or dark brown bands; occasionally bright orange.

At low water in muddy sand amongst sea grasses or under stones. On western coasts of Europe to the Mediterranean.

Sthenelais limicola (Ehlers) Fig. 6.4

Similar to *S. boa*, with about 120–200 segments, but 50–100 mm long. Elytra unfringed but have a deep notch in outer edge (H). Colourless and transparent, with a large brown blotch on each elytron forming, towards the front, an inverted 'V'-shape across the back.

Sublittoral, on sandy or muddy bottoms. Off coasts around the southern half of Britain; Atlantic, Mediterranean.

4. ARENICOLIDAE

Lugworms are relatively stout worms with a cylindrical body made up of short segments, each (except the first three or four) superficially divided into five annuli. The notopod is a conical swelling bearing a bundle of relatively large simple chaetae; the neuropod, a simple torus containing a single row of crotchets with one large and 1–3 subsidiary terminal teeth. Body divided into an anterior region lacking gills, a mid-region with branching, tufted gills, each inserted dorsal to the notopod; in the best-known species, this is followed by a narrower 'tail' region lacking both gills and chaetae. Lugworms occupy a fairly deep burrow in sand or muddy sand; the head end is usually indicated at the surface by a shallow conical depression; the characteristic coiled worm cast appears at the top of the tail shaft.

---

1. Nineteen chaetigerous segments anteriorly, the last 13 bearing tufted gills, followed by a narrower posterior region lacking gills and chaetae **2**

   30–60 chaetigerous segments, 20–40 bearing gills; no achaetous 'tail' **3**

---

2. Chaetiger 1 and 2 separated by two annulations, chaetiger 2 and 3 by three annulations. Gill filaments short and shrubby, but not linked by a membrane basally ***Arenicola marina***

   Chaetiger 1 and 2 separated by two annulations, chaetiger 2 and 3 by two annulations. Gill filaments long and pinnate, linked basally by a membrane ***Arenicola defodiens***

3. First 11–12 chaetigers lack gills; 20–30 pairs of gills posteriorly ***Arenicolides branchialis***

   First 15–16 chaetigers lack gills; 30–40 pairs of gills posteriorly ***Arenicolides ecaudata***

Arenicola marina (Linnaeus)
Blow Lug Fig. 6.3
Thick anterior region of 19 chaetigers, last 13 bearing gills (J); tail region narrower and more fragile, lacking chaetae and gills (K). Body 150–200 mm long. Flesh pink when young, becoming greenish-yellow; stains yellow when handled.

On lower shore in clean to muddy sand, exposed or sheltered. An important bait species. North-west European coasts from the Arctic to the Mediterranean.

Arenicola defodiens Cadman and Nelson-Smith
Black Lug Fig. 6.3
A larger, stouter worm than *A. marina*, characteristically black or dark yellow, readily distinguished by its gills and the annulation of the thoracic region. *A. defodiens* occupies a J-shaped burrow with a defined faecal cast but no feeding depression. It occurs low in the intertidal, and is perhaps more common on exposed beaches than *A. marina*. Where they occur together, the two species occupy distinctly different zones on the beach. The geographical distribution of this species is still uncharted, but it probably occurs widely around Britain.

Arenicolides branchialis Audouin and Milne Edwards
Body moderately thick, narrowing at each end, 30–45 chaetigers, 150–200 mm long. First 11–12 chaetigers without gills, then 20–30 pairs; no 'tail'. Very dark green or black with a metallic glint, occasionally reddish-brown.

In sediments between rocks or stones on lower shore; casts much finer than *A. marina*. Western Europe and Mediterranean.

Arenicolides ecaudata Johnston
Similar to *A. branchialis*, with 40–60 chaetigers, 120–250 mm long. First 15–16 chaetigers without gills, then 30–40 pairs.

In rich mud between stones or in rock-crevices at low water; casts much finer than *A. marina*. Western Europe.

5. CAPITELLIDAE

Very common, widespread worms with long, slender bodies. Head end conical, without appendages. Their cylindrical shape, reddish-brown colour, and rather insignificant parapods make them the most earthworm-like of polychaetes. A short, relatively broad anterior region has short segments, most of which bear simple hair-like chaetae only; the long posterior region has more elongated segments with small hooded crochets in simple tori, which may be visible only when they catch the light. The British fauna includes 10 species, in seven genera.

1. A single genital pore dorsally between chaetigers 8 and 9, at the centre of four stout chaetae (or pairs of chaetae) arranged in a diagonal cross (*except* in female *C. capitata*, which lacks these spines). Anterior region of nine or 10 chaetigerous segments **2**

   Four or more paired genital pores dorsally on last segments of anterior region or early segments of posterior region, without associated spines. Anterior region of 12 chaetigerous segments **3**

2. Long, slender worm (up to 100 mm, 90 or more segments). Blood-red, in muddy sand or rich mud ***Capitella capitata***

   Short (about 10 mm, 35–45 segments). Almost colourless, in relatively clean sediments amongst stones ***Capitellides giardi***

3. Long crochets in last six segments of anterior region. Genital pores on last four anterior segments. Segments of posterior region at first long and cylindrical, then shorter and bead-like, finally bell shaped with parapodial swellings increasing their width posteriorly. Single long, stout anal process ***Heteromastus filiformis***

   Chaetigers of anterior region bear bundles of capillary chaetae only; no parapodial lobes or swellings. Genital pores on 7–20 anterior segments of posterior region. Posterior segments more or less uniform. No anal process ***Notomastus latericeus***

Capitella capitata (Fabricius) Figs 6.5, 6.6
Ninety or more segments, 20–100 mm long. Body rather fragile, narrowing at each end, capable of considerable expansion or contraction. Anterior region usually of nine chaetigers, first six with capillary chaetae only; last two with only hooded crotchets, number 7 with capillary chaetae, crochets or both. Posterior segments bear crochets only, twice as long in the first few as in the rest. A single genital pore opens mid-dorsally between chaetigers 8 and 9, in the female on an oval swelling, in the male in the middle of four stout chaetae (or bundles of chaetae)—'genital spines'—arranged in a diagonal cross. No gills or branchial lobes. Blood-red colour.

At low water and offshore, in muddy sand or rich mud and under stones; often indicates polluted conditions. European coasts from Arctic to the Mediterranean, widespread elsewhere around Atlantic and Pacific coasts.

Capitellides giardi (Mesnil)
From 35 to 45 segments, 10 mm long. Two body regions very distinct: anterior usually of nine chaetigers, first six with capillary chaetae only, from 7th with long, slender hooded crochets like those of posterior region, where segments are markedly longer than in anterior part. Genital pore single, opening mid-dorsally between chaetigers 8 and 9, surrounded in both sexes by a pair of genital spines on each segment. No gills or branchial lobes. Almost colourless, tinged pale pink.

At low water, amongst stones. Irish Sea, Western Approaches, English Channel.

Heteromastus filiformis (Claparède) Fig. 6.5
About 140 segments, 100 mm long. Body quite thin. Anterior region of 12 chaetigers, 2–6 with only short, stout capillary chaetae, 7–12 with long hooded crochets. Posterior segments at first long, cylindrical and biannulate, then shorter and rounded, finally bell-shaped, with swollen parapodial ridges at posterior end (B). Abdominal crochets are shorter, stouter, and more strongly toothed. Branchial lobes occur on parapodia from about segment 80. Four pairs of genital pores, on chaetigers 9–12. Pygidium has a single long, stout terminal process. Anterior region dull red, posterior reddish-green or yellow.

Lower shore to shallow sublittoral, in muddy sand. North Sea, English Channel, north-east Atlantic, Mediterranean.

Notomastus latericeus M. Sars Fig. 6.6
From 100 to 150 segments, 150–300 mm long. A very fragile worm, difficult to remove intact from its very contorted gallery in the sediment. Anterior region of 12 chaetigers with reticulated appearance; all bear capillary chaetae. Posterior region bears hooded crochets; between dorsal and ventral rami on each side there is a button-like lateral organ. Simple gills formed by extensions of the rami along most of posterior region (C). Paired genital pores dorsally on many (7–20) anterior segments of posterior region (B). Pygidium terminates in a membranous flap. Anterior region purple or dark red, becoming brighter or yellowish posteriorly; posterior region colourless and transparent.

Lower shore to deep sublittoral, in clean or muddy sand, often amongst sea grasses. European coasts from the Arctic to the Mediterranean.

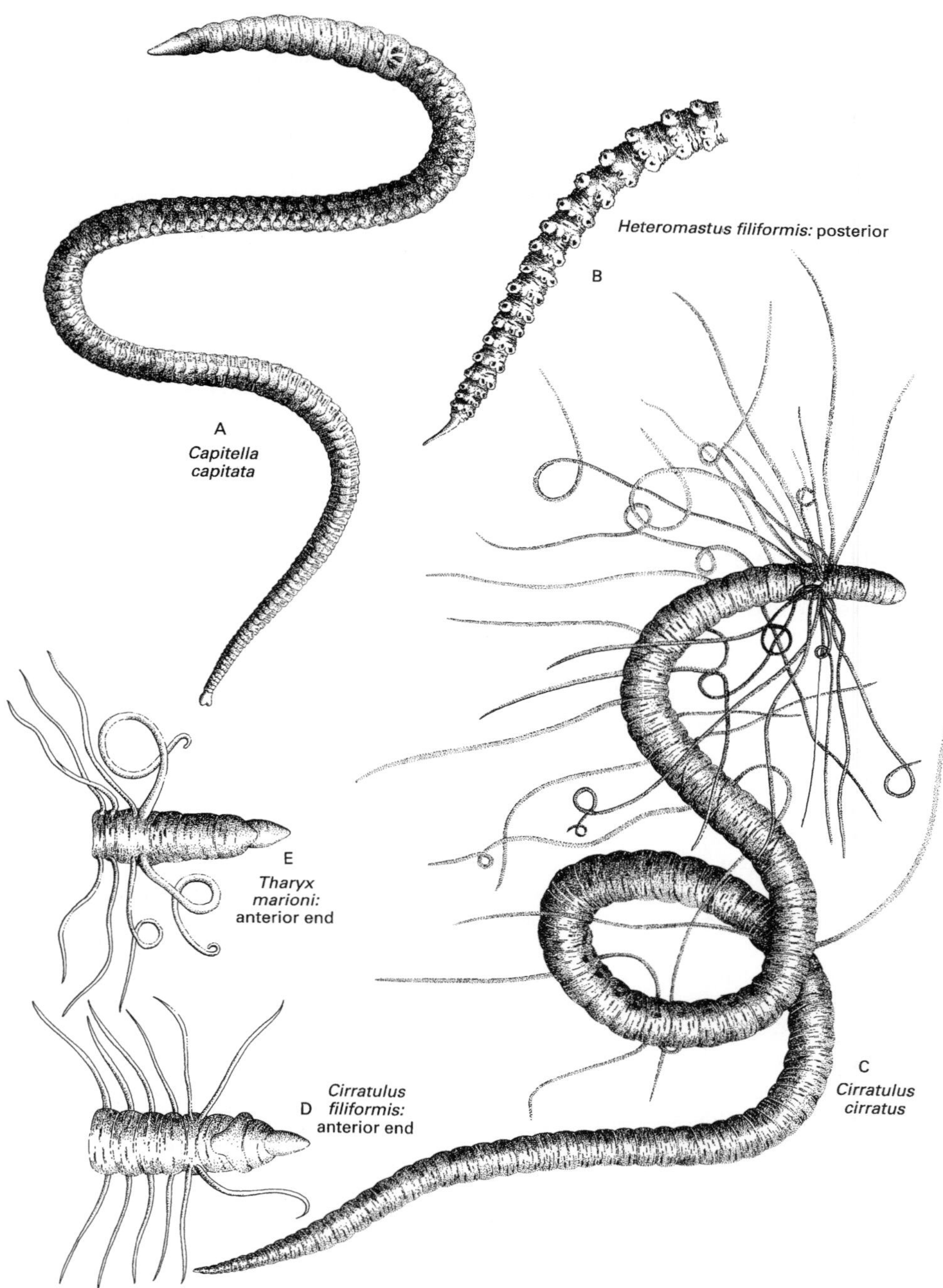
A
Capitella capitata
B
Heteromastus filiformis: posterior
C
Cirratulus cirratus
D
Cirratulus filiformis: anterior end
E
Tharyx marioni: anterior end

Fig. 6.5

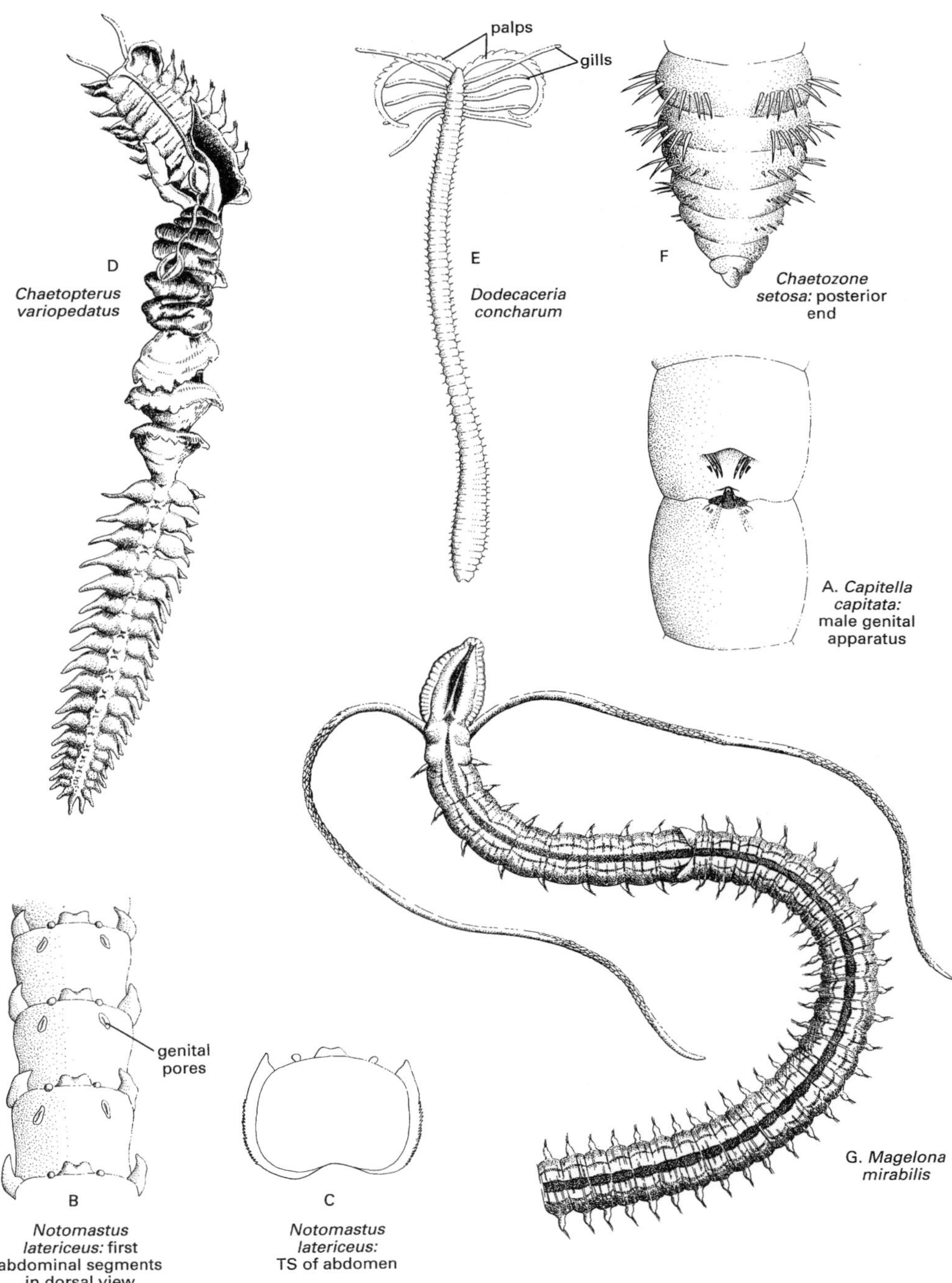

D *Chaetopterus variopedatus*

E *Dodecaceria concharum*

F *Chaetozone setosa:* posterior end

A. *Capitella capitata:* male genital apparatus

G. *Magelona mirabilis*

B *Notomastus latericeus:* first abdominal segments in dorsal view

C *Notomastus latericeus:* TS of abdomen

Fig. 6.6

6. CHAETOPTERIDAE

Soft and fragile worms with body divided into three distinct regions. Prostomium small, often indistinct; mouth terminal, lacking a proboscis. Two conspicuous palps. Anterior region of a few segments bearing notopods only; middle and posterior regions have biramous parapods. Segments of mid-region few, elongated, and may bear much-modified structures; posterior region of many segments, all alike. Chaetiger 4 carries extra large spines. Neuropods carry uncini. The worms occupy tough permanent tubes. Six species recorded for Britain, in three genera.

---

1. Large and thick (150–250 mm long, 10–25 mm wide) with relatively small palps but various large, highly specialized structures dorsally in mid-region. Tube opaque, whitish, not annulated **_Chaetopterus variopedatus_**

   Medium length (20–60 mm), slender (up to 2 mm wide). Palps relatively large, mid-region without large specialized appendages. Tube translucent, yellowish, often annulated **2**

---

2. Chaetigerous segments 60–70, less than 40 mm long. Anterior region has more than 10 chaetigers, mid-region usually has more than 10 chaetigers **_Phyllochaetopterus anglicus_**

   Chaetigerous segments 140–150, undamaged adult more than 40 mm long. Anterior region has 9–10 chaetigers, mid-region usually fewer than 10 chaetigers **_Spiochaetopterus typicus_**

Chaetopterus variopedatus (Renier) Fig. 6.6
Large, stout, fragile worm (150–250 mm long, up to 25 mm wide) (D). Mouth funnel-shaped with lower edge protruding forwards. Dorsally, two relatively short thread-like palps. Anterior region usually of nine (8–12) chaetigers, flattened, with triangular notopods only, each with a small fan of spear-like chaetae (chaetiger 9 has rudimentary neuropods). Chaetiger 4 has several greatly enlarged chaetae. Mid-region of five segments; first very elongated and bears dorsally a pair of curved fins supported by an embedded bundle of fine chaetae; second has a feeding cup on dorsal mid-line. Three remaining segments each have dorsally a large rounded paddle. All five segments have sucker-like ventral structures fringed with uncini. Posterior region has 20–70 chaetigers in which notopods form slender finger-like structures pointing upwards and supported by fine embedded chaetae; neuropods consist of a latero-ventral fin with a small lateral cirrus, and a more ventral fin partly fused with that of the other side. Each bears three rows of pectinate uncini. Yellowish or greenish white; mature females pinkish. Certain parts of the worm, and the mucus that it secretes abundantly, give off a bluish luminescence. Tube circular in section, made up of many layers of flexible whitish membrane. The central part forms a flattened U-shape in clean or muddy sand; each end is narrower but open, and protrudes slightly above the surface.

Around or below low water. Cosmopolitan.

Phyllochaetopterus anglicus Potts
Thin worm of medium length (20–40 mm long, 1–2 mm wide), 60–70 chaetigers. Mouth a semicircular or straight slit; two very short tentacles and a pair of large palps, often spiralled or rolled up. Anterior region usually of 13 (10–18) chaetigers, rather flattened, with spear-like chaetae dorsally. Chaetiger 4 has a single greatly enlarged chaeta on each side. Ventrally, a glandular cushion extends over last half of the region. Mid-region of 7–24 segments. Notopods have bilobed leaf-like dorsal flaps, supported by embedded fine chaetae, and gill-like crescents laterally; neuropods have a tiny lateral lobe and a large ventral one partly fusing with that of the other side, each bearing triangular uncini. Posterior region of many short segments; notopods small, erect, and cylindrical, with only one or two lanceolate chaetae, neuropods simple flanges bearing pectinate uncini. Yellowish, with red markings anteriorly. The tube is of a tough translucent yellow plastic-like material, often annulated. As the worms often break and regenerate, several can be found in one tube, or side-branches are made.

At low water on rocks; dredged from the sublittoral. Sometimes tubes form an extensive 'turf'. Irish Sea, south-west Britain, north-east Atlantic, Mediterranean.

Spiochaetopterus typicus M. Sars
Long, thin worm (50–60 mm long, up to 2 mm wide), 140–150 chaetigers. Mouth funnel-like, with a fleshy lower lip. Two very long palps, often spiralled. Anterior region of 9–10 chaetigers, rather flattened, with long spear-like dorsal chaetae. Chaetiger 4 has a single greatly enlarged chaeta on each side. Ventrally, two glandular cushions occupy posterior half of this region. Mid-region of 2–10 very elongated segments with leaf-like notopods supported by capillary chaetae and simple neuropods containing fine-toothed uncini. Posterior region of many short segments with small cylindrical notopods each with a few lanceolate chaetae; neuropods contain triangular uncini. Yellowish, with brown and white glandular patches. Tube of a tough translucent yellow plastic-like material with regular annulations.

In both shallow and deep water on bottoms of sand, mud, and gravel. Northern waters.

7. CIRRATULIDAE

Worms generally living in mud or rock crevices, with a more or less cylindrical body usually tapering at each end. First three segments lack parapods and chaetae. On one of chaetigers 1–4 either a pair of long, coiled palps or a number of finer tentacular filaments (some species not included here have a single long median tentacle). Long, contractile gill-filaments laterally on a number of segments; these, protruding from the mud or crevice like a number of smaller worms, are often all that can be seen of the animal. Palps and some gills may be shed on collection, but a few gills usually remain. The anus tends to be dorsal, leaving the pygidium as a terminal button or cone. The British fauna includes 15 species in six genera.

---

1. One of the first few segments bears a pair of palps markedly larger than the thread-like gills **2**

   One of the first few segments bears few to numerous tentacles of similar size to the gills (which occur throughout most of the body) **4**

---

2. Stout chaetal spines present, particularly posteriorly. Fewer than 100 segments **3**

   Fine hair-like chaetae only. More than 200 segments ***Tharyx marioni***

---

3. Spines form almost a complete ring around posterior segments, each tapering to a simple point. Gills on numerous (at least 20) segments. Body relatively thin, narrowing posteriorly ***Chaetozone setosa***

   Spines inserted laterally, with either a spoon-shaped tip or a short, toothed blade. Gills on only 4–6 anterior segments. Body relatively thick, broad and flattened posteriorly ***Dodecaceria concharum***

---

4. Eight or fewer tentacles in a cluster on each side of first chaetiger. Segments relatively long and distinct **5**

   Many more than eight tentacles on each side of first chaetiger. Segments short and tightly packed together. Gills and tentacles roll up tightly when animal is disturbed ***Cirriformia tentaculata***

---

5. Prostomium blunt, with 4–8 large black eyes on each side; following three segments (which lack chaetae) similar to each other, with normal straight margins. Large, blunt chaetae present in middle and posterior segments in addition to hair-like ones ***Cirratulus cirratus***

   Prostomium pointed, without eyes; second achaetous segment behind it is heart-shaped dorsally, with a lobe extending back in the mid-line over the third. Chaetigerous segments bear capillary chaetae only ***Cirratulus filiformis***

Chaetozone setosa Malmgren Fig. 6.6
Body with 70–90 segments, 20–25 mm long. Prostomium conical, pointed, without eyes. Two large palps, grooved, very long and fragile, inserted

on anterior edge of chaetiger 1. Gills on a large number of segments throughout about first half of body. Chaetigers throughout body bear capillary chaetae; in posterior region, simple, pointed, stout chaetae are inserted in a transverse series which, towards rear end, almost encircles each segment (F). Grey, brownish- or bluish-black.

On muddy bottoms in the shallow sublittoral. European waters from the Arctic to the Mediterranean.

Cirratulus cirratus (O. F. Müller) Fig. 6.5
Body with 75–130 segments, 30–120 mm long. Prostomium a blunt cone with a row of 4–8 large, black eyes on each side, sometimes meeting in mid-line. Two to eight grooved tentacles on anterior edge of chaetiger 1 on each side, sometimes meeting in mid-line (C). Gills similar size to tentacles, from chaetiger 1 almost to tail end. Stout, bluntly pointed chaetae ventrally from chaetiger 10 to 12 and dorsally from 20 to 23. Orange, pinkish- or brownish-red; tentacles and gills red or yellow.

Lower shore, in mud or muddy sand beneath or between rocks. Most north-west European coasts, also in South Atlantic.

Cirratulus filiformis Keferstein Fig. 6.5
Body with 150–200 segments, 30–42 mm long. Prostomium pointed, without eyes; second of three achaetous segments behind it roughly heart-shaped dorsally, with a lobe overlapping the third in the mid-line (D). One or two tentacles on anterior edge of chaetiger 1 on each side. Gills of similar size to tentacles, from chaetiger 1 to tail end. Throughout body, only capillary chaetae. Brownish- or greenish-yellow.

Lower shore, amongst lithothamnia or in rock-crevices. Irish Sea, Channel, and North Sea.

Cirriformia tentaculata (Montagu)
Body with 300 or more very short, tightly packed segments; 150–200 mm long, flattened or concave ventrally. Prostomium a blunt cone without well-developed eyes but bearing several pigment spots. On chaetigers 6–7 (rarely 4–5), numerous tentacular threads in two groups which may meet in mid-line. Gills from chaetiger 1 almost to tail end. Stout, bluntly pointed chaetae on all but first few chaetigers, ventrally sooner than dorsally. Orange, reddish-brown, or greenish-bronze; tentacles and gills red.

Lower shore, in rich mud under stones, in muddy sand or amongst sea grasses. Most north-west European coasts.

Dodecaceria concharum Oersted Fig. 6.6
Body with 45–80 segments, 20–60 mm long; cylindrical but becoming flattened and broader posteriorly (E). Prostomium small, without eyes in adult; next (buccal) segment long, achaetous and triannulate ventrally; with a pair of large, wrinkled, grooved and coiling palps, and a pair of slender gill filaments on its posterior edge. Chaetigers 1–4 (rarely 5) also carry a pair of gills (E). Chaetigers 6–7 with capillary chaetae only, but subsequently there are also crochets with tips hollowed like small spoon-bowls. Brownish- or blackish-green; eggs green.

At low water, in galleries excavated amongst lithothamnia, in soft rock or old shells. Most north-west European coasts from the Arctic to the Mediterranean.

Tharyx marioni (de Saint-Joseph) Fig. 6.5
Body with 200 segments or more, 35–100 mm long, very thin. Prostomium a blunt cone without eyes; three following, achaetous segments relatively long. Two large, grooved, and coiling palps in anterior edge of chaetiger 1 (E). Gills from chaetiger 1 on about the first half of body. Chaetigers throughout body bear only capillary chaetae. Reddish-brown, eggs greenish.

At low water, amongst lithothamnia and in rock-crevices. Irish Sea, Channel, North Sea.

8. EUNICIDAE

All members of this family have a strong, muscular eversible proboscis with black chitinous jaws, often numerous and elaborate but arranged in the same basic plan (Fig. 6.7U). The body is usually elongated and typically worm-like, with distinct although small parapods in which dorsal elements are reduced and ventral ones well developed. However, the presence and arrangement of appendages, including gills, is variable. The British fauna includes more than 60 species in almost 30 genera. Some specialists regard the subfamilies referred to here as separate families within the Order Eunicida. Here, purely for practical purposes, the Eunicidae has been divided into two groups which emerge from the general key separately. *Histriobdella homari*, a parasite of lobsters, also keys out separately on the basis of its bizarre external appearance; it is included in this family because of its jaw structure.

'EUNICIDAE I'

This group includes the subfamilies Arabellinae and Lumbrinereinae, plus *Nematonereis* from the Eunicinae. All have a conical prostomium with either no appendages or only a single median antenna. There are never gills; the notopod is either completely absent or represented only by a small button-like dorsal cirrus and, sometimes, an aciculum. The first body segment (following the peristomium) has neither parapods nor chaetae. Pectinate chaetae (like a flattened brush) are found only in *Nematonereis*.

---

1. Short median antenna; short dorsal cirri and pectinate chaetae present **Nematonereis unicornis**

   No prostomial appendages; dorsal cirri rudimentary or absent, no pectinate chaetae 2

---

2. Chaetae all simple (mostly with a long, thin, curved blade); no crochets 3

   Simple or jointed hooded crochets present (Fig. 6.7M, N) 4

---

3. One very large, plain, blunt chaeta in lower bundle of each parapod **Drilonereis filum**

   Chaetae all much the same size **Arabella iricolor**

---

4. Jointed crochets present in chaetigers 20–30, simple crochets in remainder; usually well under 300 mm long **Lumbrineris latreilli**

   No jointed crochets 5

---

5. Simple crochets lacking from first 1–5 chaetigers, present in remainder; acicula yellow; may be large **Lumbrineris tetraura**

   Simple crochets lacking from first 22–35 chaetigers, present in remainder; acicula black; usually under 30 mm long **Lumbrineris fragilis**

Arabella iricolor (Montagu) Fig. 6.7

Body narrowing to a point at each end, with about 400 segments; up to 500–600 mm long. Four eyes in a transverse row at rear edge of prostomium (E). Parapods with two lobes, postero-ventral one an elongated conical shape (H). Dorsal cirri small tubercles, no ventral cirri. Several protruding yellow acicula to each parapod. Chaetae all simple, short, and strong (F, G). Four short anal cirri. Pinkish-grey, iridescent; sometimes four longitudinal rows of dark spots on fore segments (lost in alcohol).

At low water and in the shallow sublittoral, in muddy sand. Coasts of south-west Britain, Europe to the Mediterranean; Red Sea, Arabian Gulf, Indo-Pacific.

Drilonereis filum (Claparède) Fig. 6.7

Slender body with very many segments, 80–160 mm long. Prostomium usually lacking eyes, often depressed along dorsal mid-line; two dark nuchal organs dorsally at rear edge (I). Parapods simple, low swellings in first few chaetigers, becoming two-lobed and longer further back (J). Dorsal cirri reduced to little buttons. Many protruding yellow acicula to each parapod. Chaetae all simple, with one very large, straight, blunt-ended one ventrally in each parapod. Four anal cirri. Yellowish, pink, or dirty green, strongly iridescent.

At low water and below in muddy sand, often associated with *Cirriformia tentaculata*. On southern and western coasts of Britain, Europe to the Mediterranean; Red Sea, Arabian Gulf.

Lumbrineris fragilis (O. F. Müller)

Body with about 300 segments, length 150–350 mm, constant width throughout. Prostomium without eyes (*sim.* Fig. 6.7K); two small nuchal organs (may be withdrawn from view). Parapods with two lobes, postero-ventral one an elongated conical shape (*sim.* Fig. 6.7H). Dorsal cirri indistinct or absent. A few black acicula dorsally. Chaetae all simple: in first 60–100 chaetigers with curved blades and, from chaetiger 22–35, also hooded crochets. Posterior parapods bear only crochets. Pygidium a four-lobed

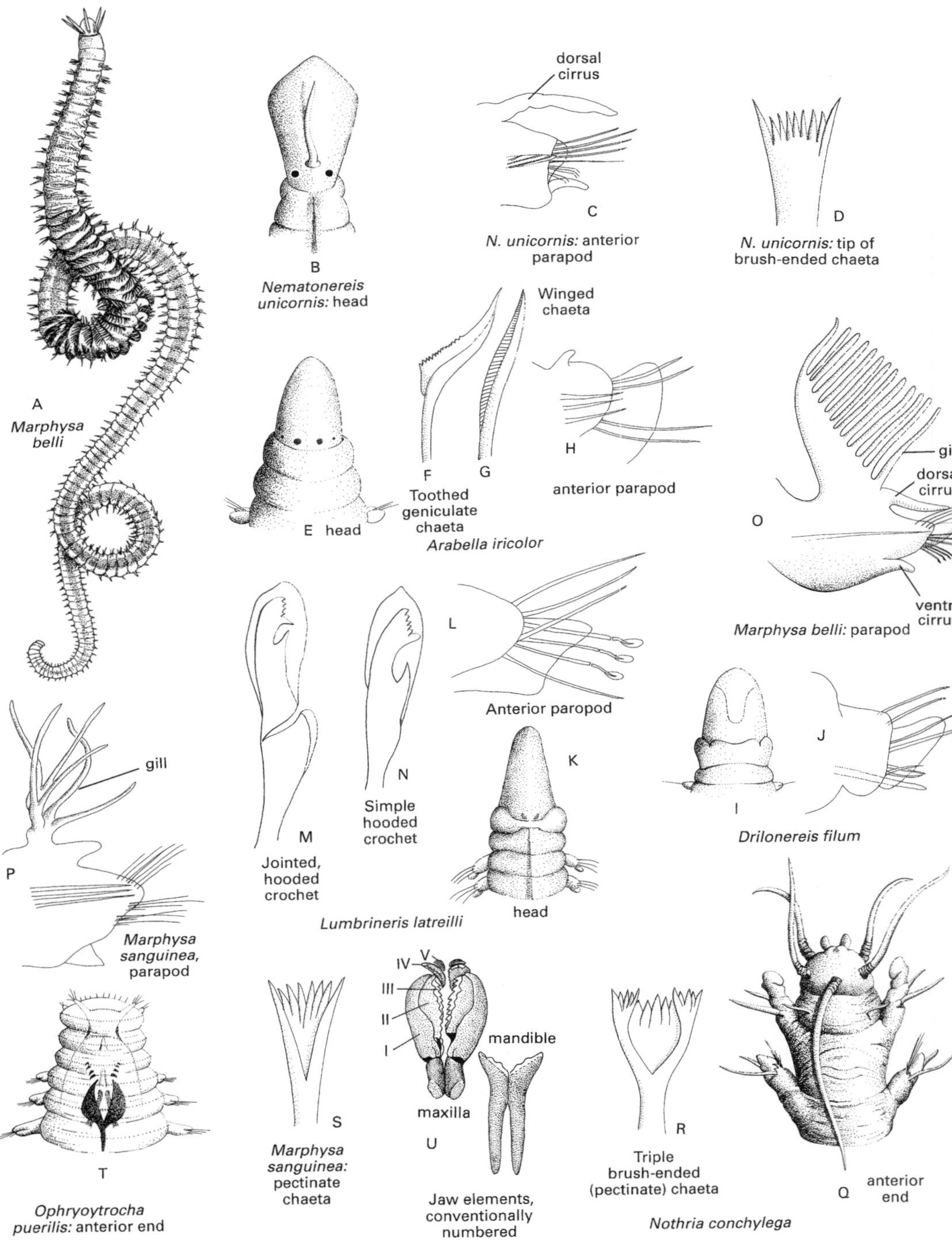
A
Marphysa belli
B
Nematonereis unicornis: head
dorsal cirrus
C
N. unicornis: anterior parapod
D
N. unicornis: tip of brush-ended chaeta
Winged chaeta
F
G
Toothed geniculate chaeta
H
anterior parapod
E head
Arabella iricolor
gill
dorsal cirrus
O
ventral cirrus
Marphysa belli: parapod
L
Anterior paropod
M
Jointed, hooded crochet
N
Simple hooded crochet
K
head
Lumbrineris latreilli
I
J
Drilonereis filum
gill
P
Marphysa sanguinea, parapod
T
Ophryoytrocha puerilis: anterior end
S
Marphysa sanguinea: pectinate chaeta
IV
V
III
II
I
mandible
maxilla
U
Jaw elements, conventionally numbered
R
Triple brush-ended (pectinate) chaeta
Q
anterior end
Nothria conchylega

Fig. 6.7

button. Orange, yellow, or brown (becoming grey in alcohol), iridescent.

Sublittoral, on bottoms of muddy sand or gravel. West coast of Europe from the Arctic to the Mediterranean; North Pacific.

Lumbrineris latreilli Audouin and Milne Edwards Fig. 6.7

As *L. fragilis* but smaller (50–300 mm long)(K). Acicula yellow. Curved, bladed, simple chaetae occur in first 40–60 chaetigers and, in first 20–30, jointed chaetae with a short end-section in the form of a hooded hook (M); in remaining segments, simple hooded crochets (N). Four short, unequal anal cirri. Pink, orange, or brown, iridescent.

At low water and below, amongst sea grasses in muddy sand, in gravel or black mud under stones. On most European coasts from Iceland to the Mediterranean; North Atlantic, Red Sea, Indo-Pacific.

NB: *L. gracilis* Ehlers also has jointed crochets (in the first 10–15 chaetigers); it is smaller (30–50 mm long) and very slender, occurring along western shores.

Lumbrineris tetraura (Schmarda)

As *L. fragilis* but large (300–400 mm long, with more than 500 segments). Acicula yellow. Curved, bladed, simple chaetae occur in first 40–80 chaetigers; no jointed crochets, simple ones from chaetiger 1 to 5. Four short anal cirri. Pale pink or red, iridescent.

At low water, in clean or only slightly muddy sand, shell gravel or algal holdfasts, to considerable depths. Cosmopolitan.

Nematonereis unicornis (Grube) Fig. 6.7

Body thread-like, often thrown into many bends, with 300–400 segments, about 150–200 mm long. Prostomium with a single, short dorsal antenna flanked by a pair of large eyes (B). Dorsal cirri finger-like, extending laterally a little beyond parapods (C); ventral cirri short, pear-shaped. Acicula dark; those protruding like chaetae occur from about chaetiger 20 and are tipped with a small, hooded hook. Chaetae in upper bundle of each parapod simple and brush-ended (D); in lower bundle, jointed, with a short end section. Anal cirri slender, two long and two short. Body pale pink in front, orange or greenish behind, iridescent.

At low water and in the shallow sublittoral, under stones on muddy sand or in rock-crevices. On south-western coasts in Britain; north-east Atlantic, Mediterranean; Indian Ocean.

'EUNICIDAE II'

Includes the subfamilies Dorvilleinae, Eunicinae except *Nematonereis*, and Onuphinae. The prostomium has many appendages, usually two antennae, two palps, and 0–5 tentacles. The notopod is reduced but may contain a few chaetae; gills may also be present dorsally.

---

1. Gills (and usually tentacles) present **2**

   Gills and tentacles absent (short, stout dorsal cirri may be present but are not readily confused with gills) **5**

---

2. Gills a pair of single finger-like filaments on most segments, more or less resembling dorsal cirri; worm may be occupying a tube which it carries around **3**

   Most or all gills have a number of filaments, arranged in a tuft or comb-like; worm does not occupy a free tube **4**

---

3. First body segment lacks parapods but with a pair of short tentacles dorsally. Tube membranous, encrusted with shell fragments or coarse sand grains ***Nothria conchylega***

   First body segment lacks both parapods and tentacles; tube transparent, horny ***Hyalinoecia tubicola***

---

4. Gills comb-like, on about 20 segments of anterior half of body; dorsal cirri long. Large (up to 200 mm long), fairly sturdy ***Marphysa belli***

   Gills as a tuft, on most segments of body; dorsal cirri short. Very large (may be well over 300 mm long) but readily breaks up ***Marphysa sanguinea***

---

5. Palps, antennae, and dorsal cirri reduced to short unjointed papillae; translucent, whitish; black jaws visible within body, resembling a closed pair of tongs ***Ophryotrocha puerilis***

Palps relatively prominent; antennae with three or four sections, dorsal cirri with two sections. Brightly coloured in life, details of jaws not visible within body ***Dorvillea rubrovittata***

Dorvillea rubrovittata (Grube)
Body narrowing at each end, 15–30 mm long, flattened ventrally, dorsally convex; 40–50 very distinct segments. Two short, broad palps, curling backwards; two short antennae with three or four joints in adults. Buccal segment twice as long as next, both achaetous. Chaetiger 1 without dorsal cirrus; on following chaetigers, a cylindrical cirrus, with a long basal section strengthened by an aciculum and a shorter, conical end section. Ventral cirri finger-like. Parapods small but long, bluntly conical, with chaetae in two bundles: upper ones simple, stout, flattened and blunt ended; lower, finer, jointed, with a hooded bidentate blade. Yellow, banded with greenish-yellow in juveniles, reddish-orange in adults (colourless in alcohol).

On the lower shore, under stones and amongst small algae; in the shallow sublittoral amongst old shells. South and west coasts of Britain, northern seas, north-east and tropical Atlantic, Mediterranean.

Hyalinoecia tubicola (O. F. Müller)
Body rather square in section, 60–120 mm long, with about 80–120 segments. Two large globular palps; two ovoid frontal antennae. Five occipital antennae on a short annulated basal section; the two most anterior short, lateral ones about one-third longer (reaching back to chaetiger 9–10) and middle one longer again. Buccal segment almost same size as next. Gills all simple, finger-like; starting on chaetiger 22–26, at first very small, then longer than dorsal cirri (which become much smaller around chaetiger 30). Ventral cirri stout but short, on chaetigers 1–3. Ventral element of each parapod resembles a flattened cirrus until chaetiger 13–16, then much smaller as far as 20. On chaetigers 1–2, parapods contain fine bristles and crochets which may appear jointed; from 3, with short lance-like bristles and brush-ended chaetae, together with two stout chaetae each having two hooded terminal teeth. Acicula yellow. Body yellowish in alcohol, iridescent.

The worm occupies an unattached horny tube resembling the quill of a feather, open at each end but with internal mitre valves. Offshore on sandy or muddy bottoms. Around most of the British coast; more or less cosmopolitan.

Marphysa belli (Audouin and Milne Edwards) Fig. 6.7
Body slender, 100–200 mm long, 200–300 segments (A). Five antennae, indistinctly annulated, subequal, and scarcely extending beyond head. Buccal segment twice as long as next. Dorsal cirri long, finger-like; ventral cirri similar but shorter. Gills prominent, like a coarse comb with 10–18 teeth (O), from chaetiger 12–15 to about 35. Conical parapods with two bundles: upper with fine lance-like and brush-ended chaetae; below, jointed chaetae with a hooded bidentate hook, and some (in anterior third of body) with a long, pointed outer element. Dark stout chaetae from about chaetiger 35. Acicula black. Body pinkish or purplish grey, iridescent, with bright red gills.

On the lower shore in muddy sand, often amongst sea grasses, frequently associated with the capitellid *Notomastus latericeus*. Western coasts of Britain, Europe to the Mediterranean.

Marphysa sanguinea (Montagu) Fig. 6.7
Body broad, flattened, 300–600 mm long, breaks up very readily; 300 segments or more. Details as for *M. belli* but dorsal cirri smooth, short, conical. Gills first simple, then as a tuft of 4–7 filaments (P). Some brush-ended notochaetae with much larger, fewer teeth (S); all jointed neurochaetae with a long, pointed outer element.

Occupies a mucus-lined gallery in muddy sand under stones and in rock-crevices on the lower shore, also amongst old shells sublittorally. Valued as angling bait in the Channel Isles ('verm') and can bite painfully when harassed. Western coasts of Britain; more or less cosmopolitan.

Nothria conchylega (M. Sars) Fig. 6.7
Body cylindrical, 100–150 mm long, with more than 150 segments. Two ovoid frontal antennae, five

occipital antennae on a short annulated basal section; posterior lateral of these reach back to chaetiger 5–6, median one a little longer. Buccal segment shorter than next, with two slender tentacles. Chaetigers 1–2 large, with parapods extending forwards to each side of head (Q). Dorsal cirri slender and finger-like on anterior chaetigers, same length as gills back to 18–20, then much shorter, disappearing at about chaetiger 30. Gills simple (or, very rarely, bifid) from chaetiger ll–13 to end of body. Ventral cirri finger-like on chaetigers 1–2 and directed forwards; globular from chaetiger 3 back. On chaetigers 1–2, flattened simple bristles and large crochets; from 3, chaetae apparently jointed. From chaetiger 4, parapods contain lance-like chaetae, chaetae with a triple brush-end (R), and also, from 9–12, stout chaetae with a hooded bidentate tip. Colour very variable: anterior segments often banded with violet/brown on a yellowish-brown background. Rust-coloured blotches at base of parapods.

Worm occupies a flattened, membranous tube covered with shell fragments embedded in silt, which it drags around with it. On sandy or muddy bottoms in the shallow sublittoral and offshore. Around most of the British coast; cosmopolitan.

Ophryotrocha puerilis Claparède and Mecznikow Fig. 6.7

Body 3–10 mm long; 20–30 segments, each with a ciliated ring around it. Antennae and palps as small, elongated papillae to each side of rounded prostomium. Buccal segment similar size to next, both achaetous. Parapods bluntly conical. Dorsal and ventral cirri reduced to small, rounded lobes. One to three stout, simple chaetae dorsally; two to four jointed chaetae inserted lower down, with a single, simple, pointed one ventrally. One straight aciculum. Colourless or whitish; jaw apparatus visible through body wall as a black shape resembling a closed pair of tongs (T). The whole worm is reminiscent of a small maggot.

Amongst sedentary animals and algae in the intertidal of sheltered shores; sometimes in the body cavity of holothurians. All around the British coast; cosmopolitan, often appears in marine aquaria.

Histriobdella homari van Beneden Fig. 6.1

Very small (0.25–0.5 mm long) with a slender body of five unequal segments. Head and 4th (genital) segment swollen, globular; posterior end broadly bifurcated, giving the worm the appearance of a pick-axe or anchor (O). Head bears five tentacular appendages: at its rear, a pair of stouter appendages each bear a terminal sucker. No parapods nor chaetae. Paired posterior processes each bear a small cirriform tubercle midway along rear edge and a terminal sucker; anus lies dorsally between them. The sexes are separate, with complicated genital organs visible within the body.

Parasitic amongst the eggs or in the branchial cavity of lobsters; surprisingly agile. South and west coasts of Britain, north-west Europe.

9. FLABELLIGERIDAE

Body short and spindle shaped, covered with small papillae and characteristically encrusted with sand or mud grains incorporated into mucus secreted by the papillae, or completely sheathed in clear mucus. Some chaetae of anterior segments long and directed forwards, often forming a cage around head. Two thick palps and a number of slimmer simple gills are borne on a buccal tube formed from prostomium and peristomium, and can be retracted with it. Flabelligerids typically feed on particles on the sand or mud surface, living under stones or in shallow burrows and creeping around only slowly: *Flabelligera* species are commensals of sea urchins, feeding on faecal material. Seven British species, in four genera.

---

1. Cephalic cage of numerous long chaetae more or less encloses anterior end. Stout, curved and hooked chaetae may be prominent ventrally on posterior segments. Buccal tube short, not very extensible **2**

   Cephalic cage represented by only 1–3 collar chaetae on each side. Chaetae throughout body are hair-like. Buccal tube long and mobile ***Diplocirrus glaucus***

---

2. Body completely encased in a thick, clear mucus sheath. In posterior half of body, each segment has one or two prominent jointed chaetae with a darker, curved blunt-ended tip. Anteriorly, two groups of 20–30 ciliated gill filaments. May be found amongst spines of echinoids ***Flabelligera affinis***

Body encrusted with grains of sand or mud. All but the most anterior segments have a row of stout, curved spines on each side ventrally. Anteriorly, a row of four finger-like gills on each side
Pherusa plumosa

Diplocirrus glaucus (Malmgren) Fig. 6.8
Body with 20–30 (rarely 50) segments, 20–25 mm long; swollen and fairly smooth anteriorly, posterior region like a string of beads, with a marked constriction between each segment. Four flattened gills on anterior edge of long buccal tube and four much more slender cirriform ones behind them. Four eyes; two long palps, broader than first gills. One to three long, forward directed collar chaetae on chaetiger 1, clearly divided into short sections like telescope segments, iridescent. Chaetae of subsequent segments (D) progressively shorter; all are simple and annulated. Pearly grey or silvery white.

Sublittoral, on muddy or sandy bottoms. Northern European coasts.

Flabelligera affinis M. Sars Fig. 6.8
Body with 30–50 segments, 20–60 mm long (A). Two pairs of eyes, almost fused. Two groups of 20–30 delicate, ciliated cirriform gills. Two broad palps with thick, wrinkled edges. Four bundles of long chaetae, 60–120 altogether, project forwards from chaetiger 1 to form a cephalic cage. Notopods bear fine hair-like chaetae; from chaetiger 25, each neuropod contains one or two stout, jointed chaetae with an annulated stalk and a short, curved, blunt-ended tip of darker colour (B). Body greenish, gills green. Palps yellow or orange, gut bright red (visible by transparency). Body encased in mucus which is clear unless stained with mud.

At low water, under stones and amongst algae; young specimens often found amongst the spines of echinoids. Also sublittoral, on muddy bottoms. Northern European coasts, including the western Baltic and Western Approaches.

Pherusa plumosa (O.F. Müller) Fig. 6.8
Body with 25–70 segments, 50–60 mm. Four brown eyes, anterior pair larger than posterior. Eight finger-like gills in a semicircular row, more or less divided into two groups. Two broad palps with thick wrinkled edges. Cephalic cage formed from long chaetae of chaetigers 1–3, directed forwards; annulated and weakly iridescent. Parapodial lobes scarcely developed. From chaetiger 4, hair-like chaetae in a bundle dorsally and, ventrally, a transverse row of sigmoid spines (C) of variable shape, together with finer chaetae. Deep orange or rusty yellow when young; adults greenish, dark brown or steel-grey. Encrusted with sand or mud grains, giving an appearance of sandpaper; cephalic chaetae may be festooned with mucus, like cobwebs.

At low water in muddy rock-crevices, amongst mussels or under stones; also sublittoral on muddy bottoms. European coasts from the Arctic possibly to the Mediterranean.

10. GLYCERIDAE

Long, slender worms with numerous segments, further subdivided by annular constrictions; often red, pink, or flushed with pink anteriorly. Anterior end bears a small, elongated, tapering and multiannulate prostomium with four minute antennae at the end; other anterior appendages and parapods are insignificant. A living specimen will soon evert a long, thick proboscis from beneath the base of the prostomium. Some species deliver a bite which is said to resemble a bee sting. Dorsal surface of parapods may or may not bear gills; in some species, these are retractile and thus may not be visible, especially in preserved specimens. Dorsal cirrus small and globular; ventral cirrus elongated but nevertheless small. Glycerids are carnivores burrowing actively in clean or muddy sand. When swimming, the body is thrown into spiral undulations; they may coil up tightly when disturbed. The British fauna includes 16 species in four genera. Some authorities recognize a second family, Goniadidae, retaining only *Glycera* in the Glyceridae.

---

1. Body divided into two distinct regions, posterior part broader and more flattened. Parapods uniramous anteriorly, biramous posteriorly. Mouth (at end of a large eversible proboscis) armed with two large and numerous small jaw pieces in a circle (Fig. 6.8F). Segments divided into two by an annular constriction
Goniada maculata

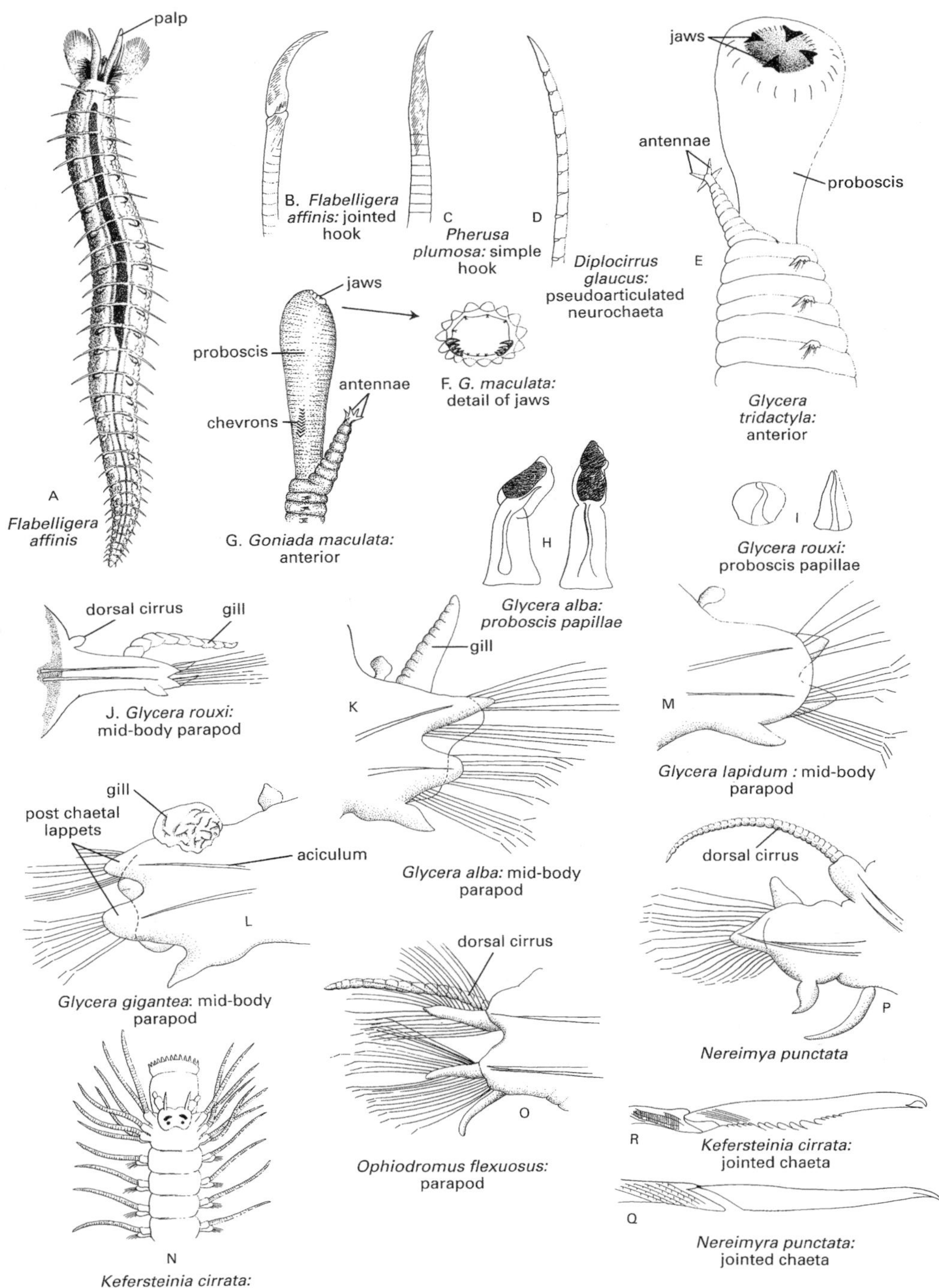
palp
A
Flabelligera affinis
B. Flabelligera affinis: jointed hook
C
D
Pherusa plumosa: simple hook
Diplocirrus glaucus: pseudoarticulated neurochaeta
jaws
antennae
proboscis
E
Glycera tridactyla: anterior
jaws
proboscis
antennae
chevrons
G. Goniada maculata: anterior
F. G. maculata: detail of jaws
H
Glycera alba: proboscis papillae
I
Glycera rouxi: proboscis papillae
dorsal cirrus
gill
J. Glycera rouxi: mid-body parapod
gill
K
Glycera alba: mid-body parapod
M
Glycera lapidum : mid-body parapod
gill
post chaetal lappets
aciculum
L
Glycera gigantea: mid-body parapod
dorsal cirrus
O
Ophiodromus flexuosus: parapod
dorsal cirrus
P
Nereimya punctata
N
Kefersteinia cirrata: anterior region
R
Kefersteinia cirrata: jointed chaeta
Q
Nereimyra punctata: jointed chaeta

Fig. 6.8

Body uniform throughout, although tapering posteriorly. All parapods more or less obviously biramous. Proboscis with four large fang-like jaws arranged in a cross. Segments divided by two or three annular constrictions **2**

---

2. Gills visible, inserted dorsally on parapods over most of body **3**

   Gills either absent or retracted (especially in preserved specimens) **4**

---

3. Gills fairly long, finger-like; worm small to moderately large (under 200mm long) **6**

   Gills short and globular; worm very large (over 200 mm long) ***Glycera gigantea***

---

4. Parapods have two postchaetal lappets, one to each ramus (Fig.6.8L). Body segments biannulate. Prostomium with 10 or more annulations **5**

   Parapods have single large postchaetal lappet, extending across rami. Body segments triannulate. Prostomium with eight (rarely up to 11) annulations ***Glycera lapidum***

---

5. Prostomium with 12–14 annulations, each secondarily divided into two; worm large (over 200 mm), greyish. Gills globular when protruded ***Glycera gigantea***

   Prostomium with 10–12 annulations; worm moderately large (up to 200 mm), reddish. Gills finger-like when protruded ***Glycera rouxi***

---

6. Papillae on proboscis resemble a fingertip, having a bent upper part bearing a chitinous 'nail' **7**

   Papillae on proboscis elongated or globular, but do not bear a chitinous plate ***Glycera rouxi***

---

7. Prostomium with 8–10 annulations. Worm small (30–75 mm), milky white ***Glycera alba***

   Prostomium with 14–18 annulations. Worm of medium size (60–100 mm) pinkish ***Glycera tridactyla***

Glycera alba (O. F. Müller) Fig. 6.8

Body with 100–150 biannulate segments, 30–75 mm long. Prostomium with 8–10 annulations. Papillae on proboscis may be simple, globular or conical, but include many resembling a fingertip with an oblique chitinous 'nail' (H). Parapods with two postchaetal lappets, upper fairly long and triangular, lower short and rounded (K). Gills non-retractile, finger-like (K). Milky-white.

At low water and in the shallow sublittoral. West of Britain, North Sea, Channel, North Atlantic.

Glycera lapidum Quatrefages Fig. 6.8

Body with 140–170 triannulate segments, 30–75 mm long. Prostomium with 8–11 annulations. Papillae on proboscis simple, mostly elongated but some globular. Parapods with single large postchaetal lappet (M). No gills. Greyish, flushed with pink anteriorly.

Shallow sublittoral. Arctic, north-west Europe, Atlantic.

Glycera gigantea Quatrefages Fig. 6.8

Large: 300–400 biannulate segments, 200–350 mm long. Prostomium with 12–14 annulations, each with a secondary constriction. Papillae on proboscis simple, mostly elongated but some globular. Parapods with two postchaetal lappets, both short and rounded (L). Gills retractile, globular (L). Greyish, flushed with pink anteriorly.

At low water and in the shallow sublittoral, often amongst rocks. South and west of Britain, north-east Atlantic.

Glycera rouxi Audouin and Milne Edwards Fig. 6.8

Body with 200–250 biannulate segments, 100–200 mm long. Prostomium with 10–12 annulations.

Papillae on proboscis simple, conical, and globular (I). Parapods with two postchaetal lappets, upper fairly long and triangular, lower short and rounded (J). Gills retractile, finger-like (J). Reddish.

Most British coasts, north-west Europe to the Mediterranean.

Glycera tridactyla Schmarda Fig. 6.8

Body with 120–180 biannulate segments, 60–100 mm long. Prostomium with 14–18 annulations (E). Papillae on proboscis resemble a fingertip with an oblique chitinous 'nail'; a few are simple, short, and rounded. Parapods with two postchaetal lappets, upper fairly long and triangular, lower short and rounded. Gills non-retractile, finger-like. Pink, iridescent.

Intertidal and shallow sublittoral; rolls up at the slightest touch. South and west of Britain, Atlantic coast to the Mediterranean; Red Sea.

Goniada maculata Oersted Fig. 6.8

About 200 biannulate segments, 50–100 mm long. Prostomium with 10 annulations. Proboscis with two large multi-toothed jaws, three smaller ones ventrally and four yet smaller ones dorsally (F); a row of 7–11 chitinous chevrons on each side at base (G); various small papillae anteriorly, none bearing a chitinous plate. Parapods in anterior half of body uniramous (but with 2–3 lappets); in flattened posterior half, biramous with rami well separated. No gills (but dorsal cirrus is fairly large). Pale green or yellowish, becoming orange posteriorly, often flecked with brown.

Low water to deep sublittoral. Arctic, north-west Europe including most British coasts; north Pacific, Arabian Gulf.

11. HESIONIDAE

Worms resembling centipedes. Prostomium with four eyes, two or three antennae, and two palps, each divided into two sections. Eversible proboscis with or without chitinous jaws or fringing papillae. First three or four segments compressed into a short, more or less distinct region bearing 6–8 pairs of tentacles, usually annulated. Dorsal lobe of parapods less well developed than ventral, mostly very reduced. Dorsal cirri mostly long and annulated; ventral cirri shorter. Dorsal chaetae, if present, simple and slender; ventrally, stouter jointed chaetae. Two anal cirri. Active but fragile worms, some carnivorous but others commensal with burrowing worms, asteroids, or shrimps. Approximately 20 species are recorded for the British Isles.

---

1. Dorsal lobe of parapod scarcely developed except for cirrus; tentacles and dorsal cirri both long and annulated **2**

   Dorsal lobe only slightly less well developed than ventral tentacles; dorsal cirri short and only feebly annulated, if at all ***Ophiodromus flexuosus***

---

2. Eight pairs of tentacles; jointed chaetae of neuropod with saw-toothed blade coarsely bifurcated at tip ***Kefersteinia cirrata***

   Six pairs of tentacles; jointed chaetae with straight edged blade delicately bifurcated at tip ***Nereimyra punctata***

Kefersteinia cirrata (Keferstein) Fig. 6.8

Body slender, 20–75 mm long; 36–65 chaetigers. Prostomium short and square. Two short antennae; proboscis cylindrical with a large circular terminal opening bordered by a fringe of about 40 fine papillae (N). Eight pairs of long annulated tentacles. No dorsal chaetae. Ventral chaetae jointed with a broad saw-like blade having a bifurcated tip (R). Coloration varies according to age and sex—yellow, brownish, purple or scarlet, transparent anteriorly, with whitish cirri.

At low water and in the shallow sublittoral in rock-crevices, under stones, amongst shells and worm tubes, or in kelp holdfasts. On most British coasts; north-west Europe to the Mediterranean.

Nereimyra punctata (O. F. Müller) Fig. 6.8

Body thick anteriorly, slender towards rear; 10–25 mm long, 3–4 mm wide; 40–50 chaetigers. Prostomium broad, bilobed. Two short antennae; proboscis rather globular, with a ventral opening fringed by about a dozen well-spaced pointed papillae—horny, dorsal and ventral folds resemble jaws. Six pairs of long, annulated tentacles. Dorsal lobe of parapod with only 2–4 small, hair-like chaetae and a long, annulated cirrus (P). Ventral chaetae jointed, the long blade with straight or only very fine-toothed edge and slightly bifurcated tip (P, Q). Dorsally, yellowish with darker transverse bands; ventrally, salmon-pink with a brown band down mid-line.

At low tide and in deeper water in rock-crevices, kelp holdfasts or amongst worm tubes and shells. All around Britain; Atlantic coast of Europe.

Ophiodromus flexuosus Delle Chiaje Fig. 6.8

Body fairly stout but fragile; 35–70 mm long, 3–4 mm wide; 55–60 chaetigers. Prostomium broad, oval, with three antennae, median one very short. Proboscis globular, constricted around its opening, no papillae or jaws. Six pairs of tentacles, short and not annulated. Parapods nearly biramous (O); dorsal lobe with numerous long, fine hair-like chaetae, dorsal cirri not much longer than these and feebly articulated. Ventral lobe a little larger, bearing a fan of stouter, long, jointed chaetae with a long, thin, fine-pointed blade. Dorsal surface dark brown with transverse iridescent turquoise bands; ventral surface yellowish or brownish with a paler longitudinal line. Head white with red eyes, cirri banded brown and white. The colours rapidly disappear in alcohol.

In burrows of other worms, the ambulacra of *Astropecten*, or on the surface of muddy sand below low water mark. All around Britain; north-west Europe to the Mediterranean.

12. MAGELONIDAE

Body long, slender, divided into two distinct regions by an intervening segment different from the others. Prostomium flattened, like a duck's bill, often wider than rest of body; no antennae, no eyes. A pair of long, thick palps fringed with papillae down one side inserted ventrally at the base of the prostomium. (NB These may be missing, damaged, or regenerating atypically, as they form an important part of the diet of bottom-feeding fish.) Mouth ventral, with a very large proboscis. Parapods biramous with well-developed lappets. Probably only one species occurs in northern European waters.

Magelona mirabilis (Johnston) Fig. 6.6

Body thread-like; 50–170 mm long, up to 2.5 mm wide, rather square in section; about 150 chaetigers. Anterior region of eight chaetigers bears only hair-like chaetae. Chaetiger 9 has greatly developed dorsal lappets (G) which almost meet in mid-line; special chaetae, paddle-shaped and terminating in a fine, pointed process, form a broad fan with longer, spear-shaped ones in each half of parapods. Remaining chaetigers with smaller, incurved parapodial lappets and relatively few, short, hooded chaetae with double hooks. Two small anal cirri. Palps and anterior region pink, posterior region greenish-grey with white blotches.

Burrowing in clean sand at low water and sublittorally, apparently without a tube. All around Britain, north-west Europe to the Mediterranean.

13. MALDANIDAE

Polychaetes with long, cylindrical bodies usually truncated at one or both ends, and very elongated segments swollen at one end by parapodial lobes. Well described as bamboo worms.

Prostomium lacks appendages, head consists of a median keel lying between a pair of nuchal slits. Keel may be strongly convex, and hood-like, or flattened and incorporated into an oblique cephalic plate whose membranous rim is notched or cut into many teeth. Mouth ventral; proboscis small, globular, eversible. Parapods biramous with a bundle of hair-like chaetae dorsally; ventrally there are single, simple stout chaetae (anterior few segments) or many-toothed crochets in a single or double row, set in a swollen transverse torus. Tori occur at anterior end of chaetigers 8–9, and towards posterior ends of the rest. A few pre-anal segments lack chaetae but retain parapodial swellings. Glandular bands or patches are scattered over anterior and middle part of body. Pygidium funnel-shaped, fringed with cirri (the mid-ventral is usually the largest), anus opening terminally in the centre of the funnel, often on a conical protuberance; or an oblique plate similar to the cephalic plate, with the anus dorsal, opening outside the plate; or spoon-like. Maldanids live in mucus tubes encrusted with particles of sediment, usually head-down in sandy or muddy bottoms. Thirty species are recorded for Britain, in 15 genera.

Identification is difficult if the specimen is incomplete. The relative length and shape of segments, and the degree to which the parapodial lobes protrude, depend on the way in which non-living specimens died or were preserved. Segments may swell at one end to form a sheath partly enclosing adjacent ones; this should not be confused with the relatively thin, delicate collars characteristic of *Rhodine*.

1. Head a more or less oblique, flattened cephalic plate with a notched or toothed rim **2**

   Head rounded, with no encircling rim **4**

2. Anus terminal, within a pygidial funnel fringed with cirri **3**

   Anus dorsal, inserted outside a flattened pygidial plate not fringed with cirri ***Maldane sarsi***

3. Rim of cephalic plate cut into 7–14 teeth posteriorly. Pygidial funnel long, striated longitudinally, fringed by numerous cirri either more or less of equal size or alternating large and small fairly regularly. Parapods of three pre-anal segments lack chaetae ***Euclymene lumbricoides***

   Rim of cephalic plate continuous except for a notch on each side and one posteriorly. Pygidial funnel short, smooth, fringed by a few large cirri amongst more numerous small ones. Parapods of five pre-anal segments lack chaetae ***Heteroclymene robusta***

4. Posterior edge of last few segments extended as a collar. Crochets of mid-region parapods inserted in a double row ***Rhodine loveni***

   No collars. Crochets inserted in single rows **5**

5. Chaetigers 1–3 bear ventrally a single simple, stout bristle on each side. More than 20 segments, medium size (over 50 mm long) ***Nicomache lumbricalis***

   All chaetigers bear ventrally two or more comparatively large crochets. Fewer than 20 segments, very small (5 mm long or less) ***Micromaldane ornithochaeta***

Euclymene lumbricoides (Quatrefages) Fig. 6.9
Nineteen chaetigers plus three achaetigerous pre-anal segments (C), up to 150 mm long. Rim of cephalic plate cut into 7–14 teeth posteriorly (A, B). Nuchal slits straight, extending about half-way back across plate. Chaetigers 1–3 bear ventrally on each side only a single simple, stout bristle; following ones, crochets in a single row. Pygidium a long, longitudinally striated funnel fringed by 30–40 cirri, either all much the same size or fairly regularly alternating between larger and smaller ones (C, D). Pinkish or pale brown with red and white glandular bands.

At low water, amongst sea grasses and stones on sand. Western coasts of Britain and Europe to the Mediterranean.

Heteroclymene robusta Arwidsson Fig. 6.9
Similar to *E. lumbricoides* but with five achaetigerous pre-anal segments (E). Rim of cephalic plate with one notch on each side and a less marked one posteriorly (F). Nuchal slits extending back across most of plate. Pygidial funnel short, smooth, fringed by a few long cirri amongst many short ones (G). Yellow-brown with various red bands and patches.

At low water and in shallow sublittoral. West of Britain, north-west Europe. (*Euclymene oerstedii* (Claparède) is similar but variable, apparently because of the frequency with which atypical regenerating specimens are found.)

Maldane sarsi Malmgren Fig. 6.9
Nineteen chaetigers plus two achaetigerous pre-anal segments (P), 50–100 mm long. Cephalic plate with a markedly convex keel, rim with a notch on each side (O). Nuchal slits short and divergent. Chaetiger 1 with neither a single, stout bristle nor crochets ventrally; subsequent segments have crochets in a single row. Pygidium a plate with a rim notched on each side, sometimes wavy or weakly toothed ventrally. Anus opens on a wrinkled tubercle above dorsal edge of plate (P). Anterior region brown with darker patches, posterior paler.

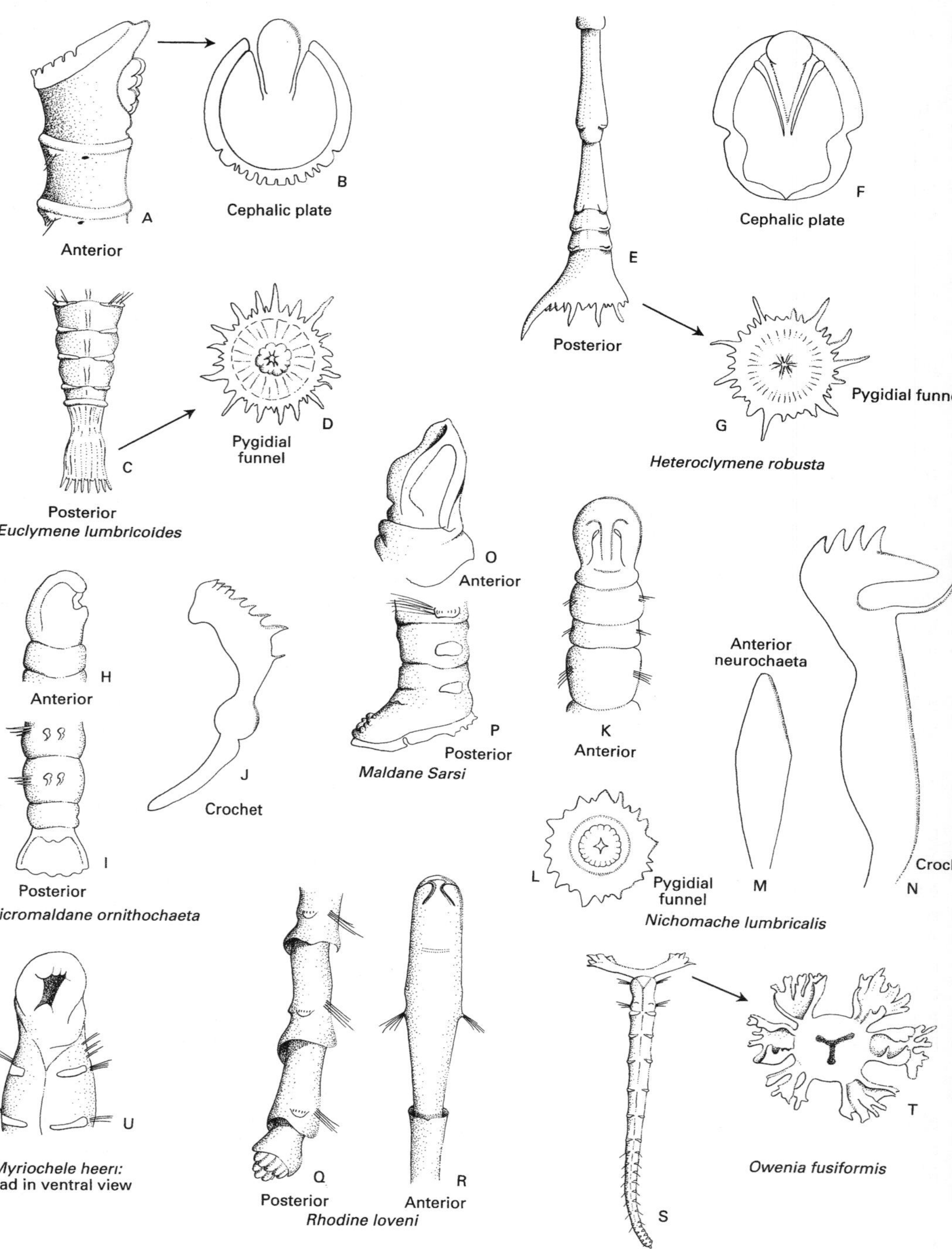
A
Anterior
B
Cephalic plate
C
Posterior
D
Pygidial
funnel
Euclymene lumbricoides
E
Posterior
F
Cephalic plate
G
Pygidial funn
Heteroclymene robusta
H
Anterior
I
Posterior
J
Crochet
Micromaldane ornithochaeta
O
Anterior
P
Posterior
Maldane Sarsi
K
Anterior
L
Pygidial
funnel
Anterior
neurochaeta
M
N
Croc
Nichomache lumbricalis
U
Myriochele heeri:
head in ventral view
Q
Posterior
R
Anterior
Rhodine loveni
S
T
Owenia fusiformis

Fig. 6.9

At low water and in shallow sublittoral on shell gravel, sandy or muddy bottoms. North-west Britain, North Sea, north-west Europe.

Micromaldane ornithochaeta Mesnil Fig. 6.9
Thirteen to 17 chaetigers plus one achaetigerous pre-anal segment (I); very small (3–5 mm). Head rounded like a hood (H), without a membranous rim. Nuchal slits fairly long, curved. All chaetigerous segments bear ventrally a few relatively large crochets (J) in a single row. Pygidium a short funnel with an irregularly fringed or wavy rim (I). Colourless.

At low water amongst algal holdfasts and lobes of lithothamnia. South-west Britain, Irish Sea, Atlantic coast of Europe.

Nicomache lumbricalis (Fabricius) Fig. 6.9
Body with 22–23 chaetigers plus two achaetigerous pre-anal segments, 60–160 mm long. Head rounded like a hood (K). Nuchal slits short and curved. Chaetigers 1–3 bear ventrally on each side a single simple, stout bristle (M); following ones, crochets (N) in a single row. Pygidium a short funnel fringed by 15–25 short cirri, all about the same size (L). Reddish-brown along back and sides of anterior end.

At low water and in shallow sublittoral. Around most of Britain, Irish Sea, Arctic, North Sea, north-west Europe.

Rhodine loveni Malmgren Fig. 6.9
Forty chaetigers, up to 100 mm long. Head rounded like a hood (R). Nuchal slits short, horseshoe-shaped. Anterior segments (including buccal segment) bear hair-like chaetae dorsally; crochets ventrally in a single row from chaetiger 5 and in a double row in the mid-region. Anterior edge of first few chaetigers and, more conspicuously, posterior edge of last few is extended as a collar encircling the next segment (Q). Pygidium small, not forming a funnel or plate, terminating in small papillae surrounding anus (Q). Yellowish- or greyish-brown with red patches.

Shallow sublittoral to medium depths, often in brackish water. North-west Britain, Arctic, North Sea, west Baltic, Atlantic coast of Europe to Mediterranean.

14. NEPHTYIDAE

Medium or large, active worms with smooth, pearly white dorsal and ventral surfaces. Well-bristled yellowish parapods extend to each side, giving a flattened appearance. Their swimming is characteristic, with a rapid lateral wriggling, starting from the rear and increasing in amplitude towards the head. Prostomium with four very small antennae at its anterior corners; proboscis large, eversible, bearing usually prominent papillae. The worm burrows rapidly in muddy sand. On all but the first few segments there is a reddish, sickle-shaped gill between notopod and neuropod. Dorsal cirrus usually vestigial, ventral cirrus small. There is a single long tail-filament. About 10 species are recorded for Britain, in the genera *Nephtys* and *Aglaophamus*. Large specimens are used for bait and are called 'catworms' by anglers.

---

1. Dorsal and ventral cirri of chaetiger 1 equally long (about equal to posterior antennae) **4**

   Dorsal cirrus of chaetiger 1 smaller than ventral cirrus (usually reduced to a mere tubercle) **2**

---

2. In posterior parapods, the branchial cirrus is as long as the gill. Some posterior chaetae sharply bent. Body more or less oval in section **_Nephtys cirrosa_**

   Branchial cirrus markedly shorter than gill. Some chaetae curved but not sharply bent. Body markedly rectangular in section **3**

---

3. In anterior parapods, a small globular process (acicular cap) at the point where aciculum reaches surface of each lobe, most obvious in notopod. Papillae of proboscis prominent, especially at anterior end. Worm large (up to 200 segments, 200 mm long) **_Nephtys hombergi_**

   No acicular cap. Papillae of proboscis are very small, except for a single very large one dorsally. Worm of medium size (up to 70 segments, 65 mm long) ***Nephtys incisa***

---

4. A single very large papilla present mid-dorsally on everted proboscis. Ventral cirrus of parapod relatively large (approx. same size as gill) ***Nephtys longosetosa***

   The everted proboscis has many papillae but not a single very large one mid-dorsally. Ventral cirrus of parapod shorter than gill ***Nephtys caeca***

Nephtys caeca Fabricius Fig. 6.10
Body with 90–150 segments, 150–250 mm long (A). Proboscis with numerous papillae but no large mid-dorsal one. Dorsal and ventral cirri of chaetiger 1 are of more or less equal size, as long as posterior antennae. Chaetae in posterior parapods strongly toothed, some curved (F), but none sharply bent. Branchial cirri shorter than gills (*sim* D).

Intertidal and at low water. Common and widespread (all around Britain; Arctic and North Atlantic).

Nephtys cirrosa Ehlers Fig. 6.10
Body with 90–95 segments; 60–100 mm long, relatively thin. Proboscis with many papillae, including a large mid-dorsal one (B). Dorsal cirrus of first chaetiger reduced to a small tubercle; ventral cirrus rather longer than posterior antenna. Chaetae in posterior parapods very finely toothed, some sharply bent (E). In posterior segments, branchial cirrus is as long as the rather slender gill (C).

Intertidal and at low water. All around Britain; Atlantic coast of Europe.

Nephtys hombergi Savigny Fig. 6.10
Similar to *N. cirrosa* but with 90–200 segments, 100–200 mm long; body markedly rectangular in section. Amongst anterior parapods, point where the aciculum reaches surface is marked by a small globular fleshy cap, most obvious in notopod (D). No posterior chaetae sharply bent. Branchial cirri shorter than gills (D).

Intertidal and at low water. Common and widespread (all around Britain, north-west Europe, Mediterranean).

Nephtys incisa Malmgren
Body with 60–70 segments; 25–65 mm long; rectangular in section. Proboscis with numerous but very small papillae, except for very large mid-dorsal one. Ventral cirrus of chaetiger 1 as long as posterior antenna, dorsal cirrus smaller. Notopod very well separated from neuropod. Chaetae in posterior parapods finely toothed, none sharply bent. Branchial cirri short.

Sublittoral, in shallow to deep water. Around much of Britain (? rare in south-west); north-west Europe, Mediterranean.

Nephtys longosetosa Oersted
Body with 90–120 segments, 50–150 mm long. Anterior papillae of proboscis quite elongated, one in mid-dorsal particularly long. Dorsal and ventral cirri of chaetiger 1 of more or less equal size, as long as posterior antennae. Posterior chaetae strongly toothed, none sharply bent. Branchial cirri short and slim. Ventral cirri relatively large (about as long as gills).

Sublittoral, in shallow water. West coast of Britain, Atlantic coast of Europe.

15. NEREIDAE

Nereids are typical errant polychaetes, having a long slender body with very many similar segments. The prostomium bears two ovoid palps, each with a small terminal button, at its anterior corners; between are two small antennae and four dorsal eyes arranged in a rectangle. A large eversible proboscis has two curved chitinous jaws at its anterior end and, usually, small horny teeth (*paragnaths*) scattered over its surface. These are arranged in groups on the dorsal surface; those used in identification lie laterally in the proximal part, just anterior to the palps. Ventrally, the paragnaths usually form a continuous transverse band. Pressure applied behind the head will cause the worm to evert its proboscis. Figures 6.12C–E illustrate the way in which the regions of an everted proboscis are numbered conventionally. There are four pairs of tentacles lateral to, and slightly behind, the prostomium. Parapods are generally biramous, with dorsal and ventral cirri; chaetae occur in a single bundle dorsally and two, superior and inferior, ventrally. Simple bristles may be present but the jointed ones are more important: the socket at the end of the shaft may be symmetrical (*homogomph*, Fig. 6.11I,J) or asymmetrical, with the tooth on one side larger than the other (*heterogomph*, Fig. 6.11K); the blade may be short, like a

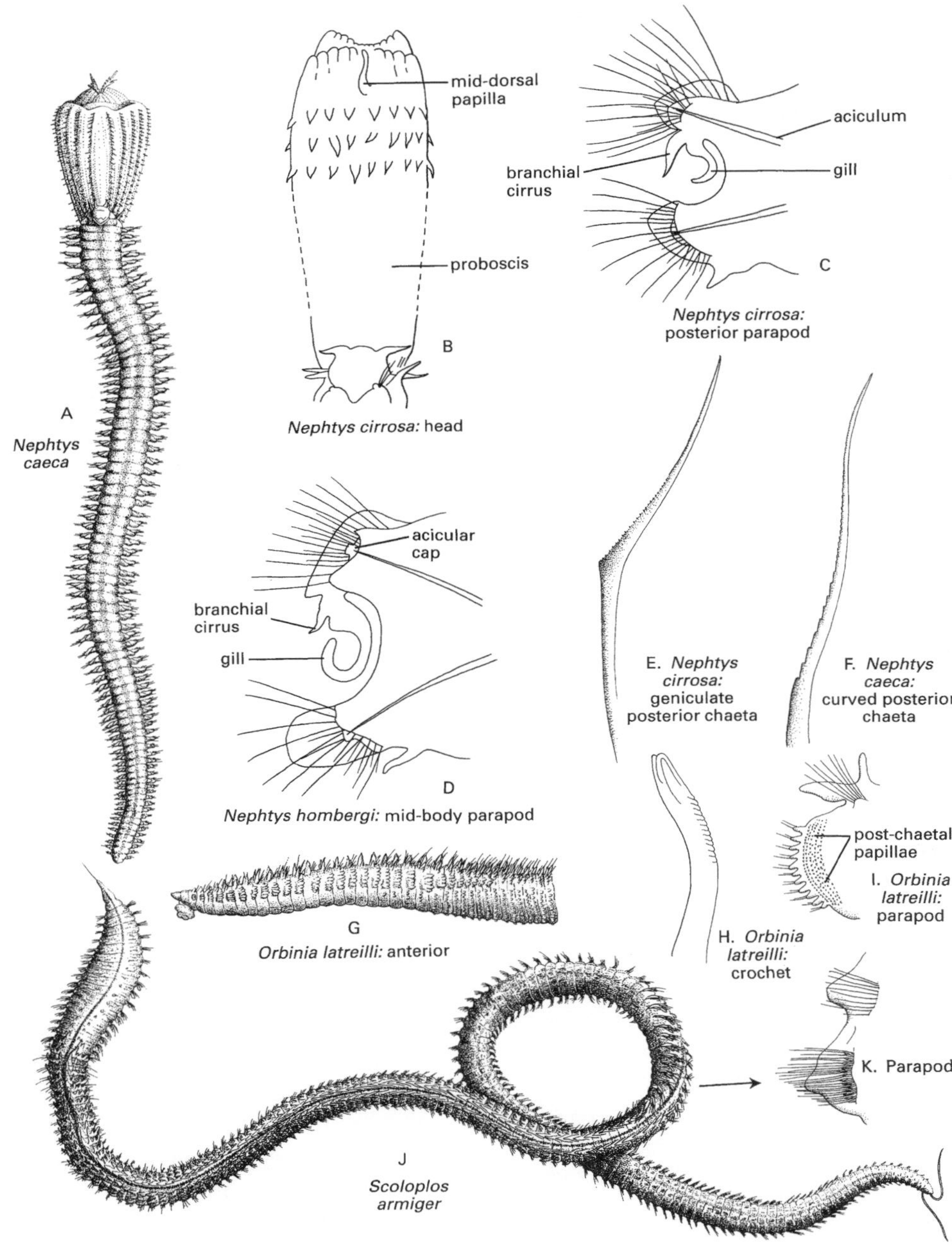

Fig. 6.10

bill-hook or pruning-saw, or long like a lance. Stout chaetae with a paddle-like blade which seems fused with its shaft, rather than articulating on it, occur on epitokes (see below) but also occasionally in the more familiar forms. Usually the notopod contains homogomph jointed chaetae with a lance-like blade; each bundle in the neuropod contains both bill-hook and lance-like chaetae, but in the superior bundle, the former are heterogomph and the latter homogomph, while in the inferior one both are heterogomph. There are two anal cirri. Most nereids produce a tenuous mucous tube in which sand grains become embedded. At maturity, they usually become transformed into a pelagic *epitoke* (*heteronereis*) which shows sexual dimorphism; generally, anterior sense organs are enlarged, posterior parapods are modified for more efficient swimming, and the anus becomes surrounded by papillae.

Nineteen species of ragworm have been recorded on British coasts, in nine genera.

---

1. Dorsal lappets of parapod markedly longer than ventral lappets, especially in posterior part of body (Fig. 6.11G, H) **2**

   Dorsal lappets more or less same size as ventral (Fig. 6.11F) **4**

---

2. Dorsal lappets broad and leaf-like, at least in mid-region (Fig. 6.11G); tentacles short (all but rearmost pair no longer than body width) (Fig. 6.11B) **3**

   Dorsal lappets long but narrow, not leaf-like (Fig. 6.11H); tentacles all longer than body width, rearmost may be one-quarter body length (Fig. 6.11E); small worm (60 mm long or less) ***Platynereis dumerili***

---

3. Dorsal cirri long, extending beyond lappets (which are broadest in mid-region); large worm (up to 200 mm long) ***Neanthes fucata***

   Dorsal cirri insignificant (Fig. 6.11G); lappets broad throughout, giving a frilly edge to body; very large (may be well over 200 mm long) ***Neanthes virens***

---

4. Dorsal cirri extend beyond lappets **5**

   Dorsal cirri shorter than lappets **7**

---

5. Antennae markedly shorter than palps; on proximal dorsal surface of everted proboscis, two patches of large paragnaths (Fig. 6.11D); a medium-sized cylindrical worm (up to 120 mm long) ***Nereis pelagica***

   Antennae same length as, or only slightly shorter than, palps **6**

---

6. Tentacles all the same length (about equal to body width); on proximal dorsal surface of everted proboscis, two patches of small paragnaths (Fig. 6.12B); a small, delicate, flattened worm (less than 80 mm long) ***Nereis zonata***

   Posterior pair of tentacles longer than others, extending back to about chaetiger 6; on proximal dorsal surface of everted proboscis, two transverse chitinous ridges resembling eyebrows (Fig. 6.12A); a large, stout worm (over 100 mm long) with a markedly rounded back ***Perinereis cultrifera***

---

7. Most tentacles longer than body width (Fig. 6.11C); antennae more or less as long as palps; large worm (200–300 mm long) with a 'broad-shouldered' appearance ***Neanthes irrorata***

   Most tentacles only as long as body width or shorter (posterior pair may be an exception); antennae markedly shorter than palps **8**

---

8. A medium-sized worm (60–100 mm long) of normal tapering shape; jointed chaetae of notopod heterogomph (Fig. 6.11K) ***Hediste diversicolor***

   A long worm (200–500 mm) of more or less constant width; jointed chaetae of notopod homogomph (Fig. 6.11I) ***Nereis longissima***

Hediste diversicolor (O. F. Müller)

Ragworm Fig. 6.11

Body with a prominent dorsal blood vessel; 60–120 mm long, 90–120 chaetigers. Antennae much

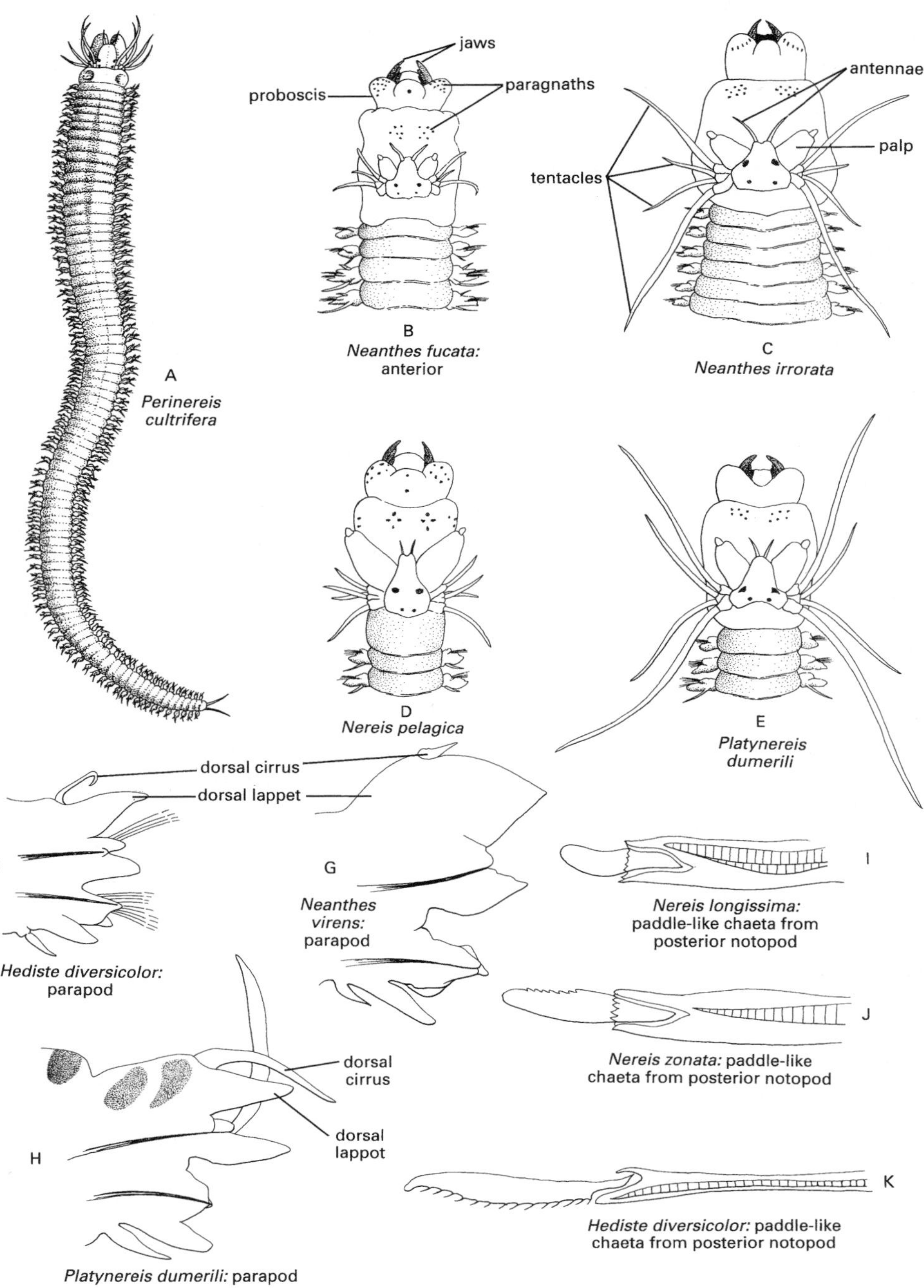
jaws
proboscis
paragnaths
antennae
palp
tentacles
A
Perinereis cultrifera
B
Neanthes fucata: anterior
C
Neanthes irrorata
D
Nereis pelagica
E
Platynereis dumerili
dorsal cirrus
dorsal lappet
F
Hediste diversicolor: parapod
G
Neanthes virens: parapod
I
Nereis longissima: paddle-like chaeta from posterior notopod
J
Nereis zonata: paddle-like chaeta from posterior notopod
dorsal cirrus
dorsal lappot
H
Platynereis dumerili: parapod
K
Hediste diversicolor: paddle-like chaeta from posterior notopod

Fig. 6.11

shorter than palps; tentacles about as long as body width. Parapods with short, thick lappets about equal in length dorsally and ventrally (F); dorsal cirri shorter than lappets. Posteriorly, one large paddle-like chaeta in each neuropod (K). Colour variable—greenish, yellowish, or orange.

Intertidal, making burrows in black muddy sand, often under brackish conditions. Commonly used as bait by anglers. Around most British coasts; north-west Europe to the Mediterranean.

Neanthes fucata (Savigny) Fig. 6.11
Long parapods give it a broad, flattened appearance; 100–200 mm long, 90–120 chaetigers. Antennae and palps of same length (B). Tentacles slightly shorter than body width, rearward pair slightly longer. Dorsal lappets markedly larger than ventral, broad and leaf-like in mid-region; dorsal cirri extend beyond lappets. Bright pink or yellowish, with a white line passing longitudinally down each side of dorsal blood vessel; parapods whitish.

Commensal with hermit crabs in whelk shells, also free living. On most British coasts; north-west Europe to the Mediterranean.

Neanthes irrorata (Malmgren) Fig. 6.11
Body broad at first, then slender and flattened; 200–300 mm long, 130–140 chaetigers. Antennae and palps about same (medium) length. Length of tentacles slightly greater than body width, rearward pair longer, reaching back to about chaetiger 15 (C). Dorsal lappets as long as ventral anteriorly, becoming progressively longer posteriorly from chaetiger 25. Dorsal cirri about as long as lappets. Brick-red with scattered grey and white spots.

Occupies a delicate tube of mucus in rock-crevices, under stones or amongst eel grass. South-west coasts of Britain; Atlantic and Mediterranean, coasts of Europe.

Neanthes virens (Sars) King rag Fig. 6.11
Body 200–300 mm or more long; 100–175 chaetigers or more, dorsal surface with a flabby, wrinkled appearance. Antennae shorter than palps. Tentacles about as long as body width except rearward pair, which reach back to about chaetiger 8. Dorsal lappet very large and leaf-like (G), giving a frilly edge to body. Dorsal cirri insignificant. Jointed chaetae all long and thin (none with a blade like a paddle or bill-hook), also many fine, simple chaetae. Dark green with a bluish iridescence; parapodial lappets fringed in yellow.

Occupies a mucus-lined burrow in black muddy sand; swims vigorously, with a graceful sinusoidal motion. Taken commercially and much valued by anglers as bait. On most British shores, particularly in the south and west; Atlantic coast of Europe.

Nereis longissima Johnston Fig. 6.11
Body flattened, with very regular segmentation, 200–500 mm long, more than 200 chaetigers. Antennae much shorter than palps. Tentacles shorter than body width, except rearward pair, which are rather longer. Dorsal and ventral lappets of similar length; dorsal cirri as long as lappets. No neurochaetae with bill-hook blades in posterior segments but, from chaetiger 65–70 rearwards, each notopod has one or two paddle-shaped chaetae with a stout shaft and short, narrow blade (I). Everted proboscis bears paragnaths only as two small patches on proximal dorsal surface; they are pale and hard to see. Blue-grey or pinkish.

At low water and in the shallow sublittoral in muddy sand or rich mud amongst eel grass. Around most British coasts; Atlantic coast of Europe.

Nereis pelagica Linnaeus Fig. 6.11
Body smooth, cylindrical, 60–210 mm long; 30–100 chaetigers. Peristomium noticeably wider than next segment; prominent dorsal blood vessel. Palps unusually large, antennae markedly shorter (D). Tentacles about as long as body width. Dorsal lappets slightly longer than ventral; dorsal cirrus quite long, especially towards rear when it is three or four times as long as lappets. Golden or bronze, often with a greenish metallic sheen.

At low water amongst algae, in kelp holdfasts and mussel beds; very active. On most British coasts; Arctic, north-west Europe to the Mediterranean: South Pacific.

Nereis zonata Malmgren Figs 6.11, 6.12
Body slender, slightly flattened, 30–80 mm long; 80–100 chaetigers. Antennae and palps rather short. Tentacles about as long as body width. Dorsal lappets slightly longer than ventral; dorsal cirri quite long, about twice as long as lappets. Around chaetiger 25–30, one or several large paddle-like chaetae in the notopod. Fawn, pink, or yellow, occasionally with whitish transverse bands.

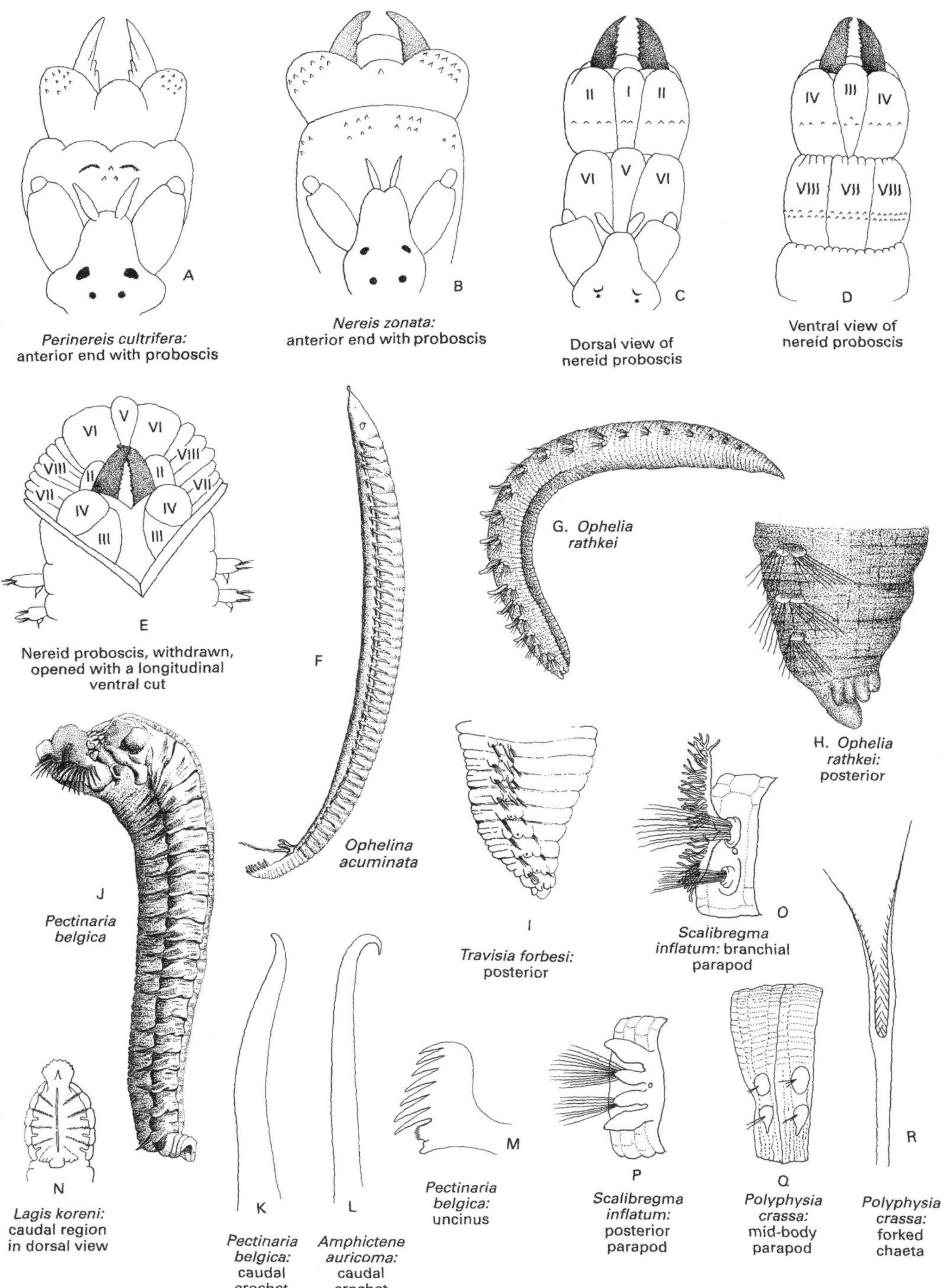
A
Perinereis cultrifera: anterior end with proboscis
B
Nereis zonata: anterior end with proboscis
II
I
II
VI
V
VI
C
Dorsal view of nereid proboscis
IV
III
IV
VIII
VII
VIII
D
Ventral view of nereid proboscis
VI
V
VI
VIII
II
II
VIII
VII
VII
IV
IV
III
III
E
Nereid proboscis, withdrawn, opened with a longitudinal ventral cut
F
Ophelina acuminata
G. Ophelia rathkei
H. Ophelia rathkei: posterior
J
Pectinaria belgica
I
Travisia forbesi: posterior
O
Scalibregma inflatum: branchial parapod
R
N
Lagis koreni: caudal region in dorsal view
K
Pectinaria belgica: caudal crochet
L
Amphictene auricoma: caudal crochet
M
Pectinaria belgica: uncinus
P
Scalibregma inflatum: posterior parapod
Q
Polyphysia crassa: mid-body parapod
Polyphysia crassa: forked chaeta

Fig. 6.12

In the shallow sublittoral amongst shells, stones, and worm tubes. Around most British coasts; Arctic, north-west Europe to the Mediterranean; Adriatic, Indo-Pacific.

Perinereis cultrifera (Grube)
Ragworm Figs 6.11, 6.12
Body flattened ventrally, rounded dorsally; 100–250 mm long; 100–125 chaetigers. Antennae shorter than palps. Tentacles about as long as body width. Dorsal lappets slightly longer than ventral, dorsal cirrus slightly longer again. Everted proboscis with two transverse dark chitinous ridges on proximal dorsal surface. Bronze-green with bright red dorsal vessel and parapods.

Makes galleries in the mud filling rock-crevices, under stones and amongst eel grass; swims clumsily by throwing its body into S-shapes. Commonly used as bait by anglers. All around British coasts; north-west Europe to the Mediterranean; Indian Ocean.

Platynereis dumerilii (Audouin and Milne Edwards) Fig. 6.11
Body very slender at rear; 20–60 mm long; 70–90 chaetigers. Antennae not quite as long as palps. Tentacles very long (E): rearmost pair may reach one-quarter of body length. Dorsal lappets markedly larger than ventral, with dark glandular patches (H), becoming longer and narrower towards rear. Dorsal cirri longer than lappets. Posteriorly, one or two jointed notochaetae with a bill-hook blade (not with a long lance-like blade). Paragnaths very small, widely scattered, and hard to see (E). Coloration variable—greenish, yellowish, pink, reddish.

At low water and in the shallow sublittoral, occupying a mucous tube in rock-crevices or kelp holdfasts. All round Britain; cosmopolitan.

16. OPHELIIDAE

Small worms with three distinct body shapes: short and maggot-like, slender and torpedo-shaped, or with an anterior region appreciably broader than the posterior; with relatively few segments and often a deep ventral gutter. Parapods ill-developed, all chaetae simple, hair-like bristles; finger-like or paddle-shaped gills obvious along each side of body for at least half its length. The British fauna includes 17 species in six genera; several are common in sandy beaches.

---

1. Anterior region of body broad; posterior relatively slender, with a marked ventral gutter. Anus surrounded by a prominent hood or tube; ventrally, two long, stout anal papillae **2**

   Body stout, maggot-like, broad throughout; no ventral gutter. Anal tube and papillae insignificant. May smell strongly of garlic when first collected ***Travisia forbesi***

---

2. Gills (46–48) laterally on all but first and last two or three chaetigers (Fig. 6.12F). Prostomium terminating in a single finger-like median process. Anus surmounted by a dorsal hood with a toothed or fringed posterior edge ***Ophelina acuminata***

   Eight to 10 pairs of gills in mid-region. No prostomial process. Posterior end with one large ventral papilla and several thinner dorsal papillae (Fig. 6.12H) ***Ophelia rathkei***

Ophelia rathkei McIntosh Fig. 6.12
Body up to 15 mm long. Anterior region broad with 10 or 11 abranchiate chaetigers. Narrower posterior region bears 8–10 pairs of long, narrow, paddle-shaped gills (G). The last two or three chaetigers appear telescoped. Normally pinkish, with iridescent glint; gills bright red. When breeding, males may appear milky-white, females greenish-grey.

Lower shore and shallow sublittoral, in mixed medium to coarse sand. Probably all British coasts, Atlantic coasts of Europe.

Ophelina acuminata Oersted Fig. 6.12
About 50 segments, multi-annulated; 25–60 mm long. Conical prostomium extended in a finger-like median process slightly swollen at its tip. Long cirriform gills on all chaetigers but first and last two or three (F). Anus surmounted by a spoon-shaped hood, open ventrally, whose margin is divided into 8–10 little teeth. Yellowish or pearly grey with bright red gills.

In the shallow sublittoral, on sand. European coasts from the Arctic to the Atlantic.

Travisia forbesi Johnston Fig. 6.12
Body with 23–26 chaetigers, triannulated; 20–30 mm long. Integument reticulated. Prostomium small, oval, with an anterior conical process. Cirriform gills laterally in about 15 obvious pairs, from chaetiger 2 but small posteriorly and missing from last 2–5 segments. Pygidium like an oval button amongst small anal papillae (I). Whitish or flesh-pink, may be encrusted with sand; a strong garlic-like smell when first collected.

At low water, in fine, clean sand. European coasts from the Arctic to the Atlantic.

17. ORBINIIDAE

Long worms with a slightly flattened and broader anterior region bearing lateral parapods, and a rounded, more fragile posterior region bearing dorsal parapods (Fig. 6.10J). Broad, lanceolate, simple gills present dorsally on all but the first few chaetigers. The main genus, *Orbinia*, has anterior neuropods as tori with crochets in several rows, fringed by a row of post-chaetal papillae (Fig. 6.10I); at the junction of the anterior and posterior regions several chaetigers also have a row of subpodial papillae on each side, which may meet in the ventral mid-line. There are never fewer than five subpodial papillae in each segment which bears them. In several less common species, one or more large, dark, spear-like bristles projects from each neuropod of the last few anterior chaetigers. In *Scoloplos*, there are markedly fewer post-chaetal and subpodial papillae; crochets may be present on all, a few, or none of the anterior segments. A few members of the family completely lack subpodial papillae. In all, tori and crochets are lacking from all but the first few posterior segments. Eleven species recorded for Britain.

---

1. Gills along dorsal surface from chaetiger 5. Towards rear of anterior region, more than 10 post-chaetal papillae form a fringe down neuropodial tori, which contain large crochets in several rows. Numerous (17–33) segments in this mid-region also have a row of subpodial papillae, which encircle body ventrally in several (7–11) segments (Fig. 6.10G) **2**

   Gills occur from chaetiger 9 to 17. Few (maximum three) or no post-chaetal papillae. No more than two subpodial papillae occur on only 1–4 segments of mid-region ***Scoloplos armiger***

---

2. Subpodial papillae occur on chaetigers 22–42/55, encircling body ventrally on chaetigers 25–36; 12–25 post-chaetal papillae on best-developed tori ***Orbinia latreilli***

   Subpodial papillae occur on chaetigers 17–34, encircling body ventrally on chaetigers 20–27; 10–15 post-chaetal papillae on best-developed tori ***Orbinia sertulata***

Orbinia latreilli (Audouin and Milne Edwards) Fig. 6.10
Anterior region of 30–40 chaetigers (G); altogether 300–400 segments, 300–400 mm long. Anterior neuropods are swollen tori, typically with three or four rows of large, yellow, curved crochets (H), fringed behind by a row of 12–25 post-chaetal papillae (I) back to chaetiger 40/43. Anterior flesh-pink, posterior yellowish.

On lower shore in clean sand. West of Britain, North Sea, Channel, and Atlantic coasts of Europe, Mediterranean.

Orbinia sertulata (Savigny)
As ***O. latreillii***, with differences indicated in key above. Anterior region of 22–24 chaetigers; altogether up to 400 segments, up to 300 mm long. Anterior dark red, posterior greyish-brown.

At low water and in shallow sublittoral, on bottoms of sand or muddy sand. West of Britain, North Sea, Channel, and Atlantic coasts of Europe, Mediterranean.

Scoloplos armiger (O. F. Müller) Fig. 6.10
Anterior region of 12–20 chaetigers; altogether more than 200 segments. 50–20 mm long (J). Anterior

neuropods are swollen tori with curved crochets in all, few, or none (but, where absent, stout curved capillary chaetae may resemble them if worn down or broken). Few or no post-chaetal papillae (K, *cf.* I); 1–2 subpodial papillae occur on two or three segments only (never more than four per segment). Bright orange-pink, with main blood vessels well visible.

At low water or in shallow sublittoral, in fine muddy sand, often amongst sea grasses. West and north of Britain, Arctic, north-west Europe; Indian Ocean, Pacific, Antarctic.

18. OWENIIDAE

Head either without appendages or bearing a frilled membranous crown. Thoracic region made up of a chaetigerous buccal segment and three short segments bearing only dorsal hair-like chaetae. Abdominal region of long segments, progressively shorter posteriorly, with ventral rows of very many small crochets inserted in a long torus on each side. Pygidium small, rounded, slightly bilobed. Tube open but narrowing at each end. Three British species, only two of which are common.

---

1. Head with a frilled membranous crown surrounding mouth (Fig. 6.9T) ***Owenia fusiformis***

   Head end rounded and obliquely truncated, lacking appendages of any kind (Fig. 6.9U) ***Myriochele heeri***

Owenia fusiformis Delle Chiaje Fig. 6.9

Body with 20–30 segments, 30–100 mm long (S). Head bears a crown of six short, membranous processes cut into numerous frilly lobes, with a terminal three-lobed mouth at its centre (T). Tube encrusted with sand grains or shell fragments, each overlapping the more posterior one like a crudely tiled roof. The worm is able to move around while still in its tube. The tubes may be so numerous as to bind the sediment, but many are usually empty. Body greenish or yellowish, with paler glandular bands; crown reddish.

At low water and in the shallow sublittoral. All around Britain, north-west Europe, Mediterranean, Indian Ocean, Pacific.

Myriochele heeri Malmgren Fig. 6.9

About 27 segments, 20–30 mm long. Anterior end obliquely truncated; head rounded, without crown or other appendages. Mouth subterminal (U). Tube membranous, encrusted with small sand grains not usually overlapping; may be very numerous in sediment. Yellowish.

Sublittoral, at moderate depths. North-west Britain, Arctic, north-west Europe, Mediterranean, Black Sea, Pacific, Antarctic.

19. PECTINARIIDAE

Body is relatively short and thick, divided into three regions. The blunt anterior bears two comb-like bundles of stout, golden spines (used for digging) crossing in the mid-line, and numerous prehensile tentacles surrounded by a cephalic veil; two pairs of tentacles and two pairs of feathery gills. It includes three segments bearing only bundles of hair-like bristles dorsally. The posterior (abdominal) region has 12–13 segments which, in addition, bear ventral uncini with hooked teeth in many rows; it is about as thick as the anterior region. A narrower, much shorter caudal region (*scaphe*) is divided from it by a marked constriction and is segmented, but lacks chaetae except a few crochets on each side around its base. The worm occupies a characteristic smooth, unattached tube, open at each end, constructed of a single layer of sand grains or shell fragments, cemented together so that their edges meet closely. It lives in sediments at low water and below, usually on fairly exposed sandy beaches, head-down, and with the smaller end of the tube protruding from the surface. Six British species.

---

1. Tube distinctly curved (17 segments bear bundles of hair-like chaetae) ***Amphictene auricoma***

   Tube straight or nearly straight **2**

---

2. Fifteen segments bear bundles of hair-like chaetae; first three segments of caudal region carry a club-shaped papilla on each side (Fig. 6.12N) ***Lagis koreni***

Seventeen segments bear bundles of hair-like chaetae; caudal region has no papillae **Pectinaria belgica**

Amphictene auricoma (O. F. Müller) Fig. 6.12

Length 20–40 mm. Chaetigers 4–16 bear uncini; no achaetous segments anterior to caudal region. Cephalic veil cut into long tongues, its dorsal edge crenellated. 'Comb' of 10–15 spines on each side. Uncini bear five or six rows of hooks above a series of very fine teeth. Crochets at base of caudal region are fine with a hooked point, 8–10 on each side (L). This region oval, with a membranous edge cut into lappets curving dorsally. Pinkish-white, gills bright red. Tube clearly curved, about 70–80 mm long.

At low water in sand and sublittorally on more muddy bottoms. European coasts from the Arctic to the Mediterranean; north-east Pacific.

Pectinaria belgica (Pallas) Fig. 6.12

Length 30–70 mm (J). As ***A. auricoma*** but cephalic veil cut into a few long tongues, its dorsal edge entire. 'Comb' of 8–15 spines on each side. Uncini bear seven or eight rows of hooks above numerous very small ones (M). Crochets at base of caudal region stout, only slightly hooked, 6–12 on each side (K). This region an elongated oval, first few segments with a wavy margin, rest more or less smooth-edged. Pinkish-white, gills carmine red. Tube almost straight, about 90 mm long.

In muddy sand at low water and, more often, sublittorally. North-west Europe from Scandinavia to the Mediterranean; north-east Pacific (earlier reports more likely to be of ***L. koreni***).

Lagis koreni Malmgren Fig. 6.12

Up to 50 mm long; as ***A. auricoma*** but 15 segments bear hair-like chaetae, and chaetigers 4–15 bear uncini. Two achaetous segments anterior to caudal region. Uncini with a massive base and 6–8 rows of thick hooks above four rows of little, indistinct teeth. Crochets at base of caudal region short, stout, and with a blunt, slightly hooked tip, 3–7 on each side. This region an elongated oval with edges raised dorsally; on first three segments bearing a little club-shaped papilla on each side (N). White with pink iridescence, gills carmine red, ventral vessel red, visible by transparency. Tube slightly curved, about 80 mm long, always with a little mucous extension.

At very low water on long, open beaches, usually in very great numbers. Arctic and north-east Europe; Adriatic.

20. PHYLLODOCIDAE

A large family of long, thin, often quite large errant polychaetes, in which the dorsal cirri of all but the first few segments are in a flattened leaf-like shape and held out vertically as a long fringe down each side of the animal (Fig. 6.13A) (hence the popular name 'paddleworms'). Prostomium more or less pear-shaped, tapering anteriorly; with two pairs of short antennae in front, normally two small, dark eyes dorsally with the fifth, median antenna (if present) usually placed just in front of them. At the back of the prostomium, dorsally or in an indentation of its rear edge, there may be a nuchal papilla like a domed button, or rarely a pair, resembling ear flaps. Proboscis eversible, lacking jaws but typically with large papillae around its anterior margin and smaller ones on its outer surface. In some species, the proboscis is mostly smooth but may appear ridged or papillated if incompletely everted. The first two or three segments lack paddles (and often also chaetae), and dorsal and some ventral cirri are tentacle-like. They may be as short as the antennae, but posteriorly are often moderately long, usually extending back fairly close to the body. Segment 2 is often fused with the first, but they can be distinguished as separate in lateral or ventral view; less often, the first may be partly fused with the prostomium or appear to surround it like a collar. There are many body segments (often several hundred), which are closely similar to each other. Ventral cirri are generally similar in shape to the dorsal paddles, but smaller. Between them lies a relatively simple parapodial mound, bearing a single bundle containing a few jointed chaetae, consisting of a long cylindrical shaft swollen at its outer end, with a short, flattened blade. The swollen joint and one edge of the blade are toothed. Simple chaetae may be present on segments bearing tentacles. A pair of anal cirri vary in form from short, thick paddles to cylindrical or thread-like tentacles. Phyllodocids are very active worms, primarily carnivorous, and they secrete large quantities of mucus. They are often brightly coloured and may bear patterns of dots and lines across their dorsal surface. More than 30 species have been recorded from Britain, in 14 different genera.

---

1. Head bears only four antennae (Fig. 6.13E) **2**

   Head bears five antennae (Fig. 6.14H) **13**

---

2. Behind the head, two pairs of short, conical tentacles on the first of two segments not bearing paddle-like dorsal cirri (Fig. 6.13B) **3**

   There are three or four pairs of tentacles, which may be quite long (Figs. 6.14H, L); three segments do not bear paddles, although the division between the first two may be difficult to distinguish **6**

---

3. Proboscis has numerous hard papillae, many bearing spines, around its proximal half (not visible if only partially everted). Shaft of jointed chaetae bears two large spines at its joint with blade (Fig. 6.14B). Dorsal surface shows a pattern of brownish blotches ***Eteone picta***

   Proboscis more or less ridged and folded to produce soft papillae, but hard spiny ones lacking. Shaft of jointed chaetae bears only one large spine (Fig. 6.13K) or, if several, only one is prominent. Dorsal surface may have blotches but they do not form well-marked pattern **4**

---

4. Second body segment bears a ventral cirrus on each side but lacks chaetae ***Eteone foliosa***

   Simple chaetae as well as ventral cirri present on second body segment **5**

---

5. Paddles in mid-region broader than long; prostomium also broader than long, narrowing only near front ***Eteone flava***

   Paddles in mid-region longer than broad (Fig. 6.13H); prostomium about as long as wide, narrowing well back towards eyes ***Eteone longa***

---

6. Three pairs of tentacles on first two of three distinctly separate segments lacking paddle-like dorsal cirri; ventral pair of second segment flattened (Fig. 6.14G) ***Pseudomystides limbata***

   Four pairs of tentacles, all cylindrical but of variable length, on three segments lacking paddles; segment 1 more or less fused with 2, so there may appear to be only two such segments, especially from above **7**

---

7. Tentacles long and slender: last pair reaches back beyond body segment 6. Nuchal papilla set unobtrusively in an indentation at rear of prostomium (Fig. 6.14D) **8**

   Tentacles short and stout: last pair reach no further than body segment 4 if deflected rearwards. Nuchal papilla either set on prostomium or absent **11**

---

8. Worm quite long (up to 300 mm), with paddles subrectangular in mid-body (Fig. 6.13J); prostomium heart-shaped, concave at sides; proximal part of everted proboscis bears small papillae in distinct rows down each side; in life, pale yellow, green or brown dorsally **9**

   Worm long or very long (up to 750 mm) with leaf-like paddles (Fig. 6.13A); prostomium pear-shaped, straight or convex at sides; proximal part of everted proboscis bears small papillae diffusely arranged; in life, dark green with a metallic sheen dorsally **10**

---

9. Over 100 mm long, with more than 250 segments; in life, dorsal surface greenish-yellow, with green paddles each having brown markings ***Phyllodoce groenlandica***

   Under 100 mm long, 250 segments or fewer; in life, dorsal surface whitish or yellow, with brown paddles. Three brown blotches across each segment create a characteristic pattern ***Phyllodoce maculata***

---

10. Worm long (up to 600 mm). Prostomium longer than broad, with more or less straight sides (Fig. 6.13A). In life, dorsal surface dark green with a bluish sheen; paddles brownish or yellowish ***Phyllodoce lamelligera***

    Worm very long (up to 750 mm). Prostomium broader than long, with convex sides (Fig. 6.14C). In life, dorsal surface very dark green-brown with a steel blue iridescence; paddles green with a red-brown blotch ***Phyllodoce laminosa***

---

11. Nuchal papilla prominent, set on a narrow rearward projection of prostomium behind eyes ***Paranaitis kosteriensis***

    Nuchal papilla absent **12**

---

12. Worm moderately long (up to 100 mm), with fewer than 200 segments; in life, yellowish-green with rust coloured paddles ***Nereiphylla rubiginosa***

    Worm long (up to 300 mm), with several hundred segments; in life, an iridescent blue or slate grey; paddles green, edged with yellow ***Nereiphylla paretti***

---

13. Two nuchal organs present on posterior dorsal margin of prostomium, resembling ear-flaps ***Notophyllum foliosum***

    Ear-like nuchal organs absent; nuchal papilla absent or indistinct **14**

---

14. First three segments lack paddle-like dorsal cirri (Fig. 6.13C); distinctly separate from each other, although first (bearing a pair of tentacles) may not otherwise be distinct from prostomium. The first obvious segment may thus appear to bear two pairs of short tentacles **15**

    First three segments lack paddles but division between first two may be difficult to distinguish. The first obvious segment may thus appear to bear two pairs of short tentacles and one pair of moderate length (Fig. 6.14N) **17**

---

15. Worm small (12 mm long or less) ***Eulalia expusilla***

    Worm of moderate size (30–150 mm long) **16**

---

16. Last pair of tentacles long, reaching back to body segment 10 or 12 (Fig. 6.13D), paddles elongated and pointed (Fig. 6.13I). In life, a uniform bright-green colour ***Eulalia viridis***

    Tentacles short and stout (last pair reach no further than body segment 4 if deflected to rear) (Fig. 6.13C); paddles oval and blunt-ended (Fig. 6.13G); in life, yellowish or greenish-yellow ***Eulalia bilineata***

---

17. Ventral cirrus of segment 2 differs from other tentacles in having a flattened blade along one side, resembling a cleaver (Fig. 6.14O) ***Pterocirrus macroceros***

    Tentacles all of tapering cylindrical shape **18**

18. Worm of moderate length (60–150 mm), with several hundred segments; body marked with transverse lines of dark spots; proboscis has about 50 papillae around margin ***Pirakia punctifera***

    Worm short (20–60 mm), with fewer than 150 segments; spots may be present but do not form a distinctive pattern; proboscis has about 20 papillae around margin **19**

19. Proboscis covered with numerous small papillae ***Pirakia fucescens***

    Proboscis smooth when fully extended (Fig. 6.14M) ***Eumida sanguinea***

Phyllodoce groenlandica Oersted — Figs 6.13, 6.14

Body with 250–700 segments, 150–300 mm long. Prostomium heart-shaped, a nuchal papilla in crevice between its posterior lobes. Four small antennae. Proboscis with rows of large flattened papillae distally, small rounded papillae proximally; 17 papillae surround its opening. First and second tentacular segments indistinguishable dorsally but third clearly separate, bearing simple chaetae; four pairs of tentacles, longest reaching back to about body segment 10. Paddles of mid-region subrectangular, fairly elongated. Anal cirri cylindrical. Greenish-yellow with transverse bands of brown and blue; paddles greenish, heavily blotched with brown. In alcohol, yellow-grey with brown markings and bluish iridescence.

In sand and rock-crevices at low water; all around British Isles and north-west Europe, also Arctic; Sea of Japan.

Phyllodoce maculata (Linnaeus) — Fig. 6.13

As *P. groenlandica*, but with about 250 segments; 30–100 mm long. Proboscis transversely and longitudinally ridged distally to give appearance of large irregular papillae; flattened oval papillae proximally (E). Longest tentacles reaching back to body segment 6–10. Paddles of mid-region broad but short. Whitish or yellowish with transverse dark-brown bands; paddles brown. In alcohol, yellowish or brownish. Eggs dark orange or green.

In muddy sand, under stones, and in mussel beds. All around British Isles and north-west Europe, also Arctic.

Phyllodoce lamelligera Johnston — Fig. 6.13

Body with very many segments, 60–600 mm long (A). Prostomium an elongated, truncated cone. Four short antennae. Proboscis ridged longitudinally (appears hexagonal in section) and transversely, opening surrounded by 16 papillae. Three tentacular segments, first indistinguishable from second dorsally. Four pairs of moderately long tentacles. Posterior two tentacular segments bear simple chaetae. Paddles large, in an asymmetrical elongated leaf-shape. Dark green, with a blue metallic sheen. Paddles olive-green or brownish-yellow. In alcohol, iridescent with brownish paddles.

Under large stones on the lower shore; in kelp holdfasts and offshore. Around the south and west coasts of Britain; Channel and Atlantic coasts of Europe, also Mediterranean; Pacific.

Phyllodoce laminosa Savigny — Fig. 6.14

Similar to *P. lamelligera* but 150–750 mm long. Prostomium a rounded heart-shape, with nuchal papilla visible in posterior indentation (C). Opening of proboscis surrounded by 16–20 rounded papillae, basal part covered with tightly packed small conical papillae not arranged in rows (C). Rearward dorsal pair of tentacles long (C).

Under stones at low water or in damp rock-crevices. Around most of the British Isles; North Sea, Channel, Atlantic, also Mediterranean.

Eteone flava (Fabricius)

Body flattened, with about 300 segments; 40–120 mm long. Prostomium short, with concave sides; nuchal papilla indistinct. Four small antennae. Proboscis with more or less strongly marked transverse ridges;

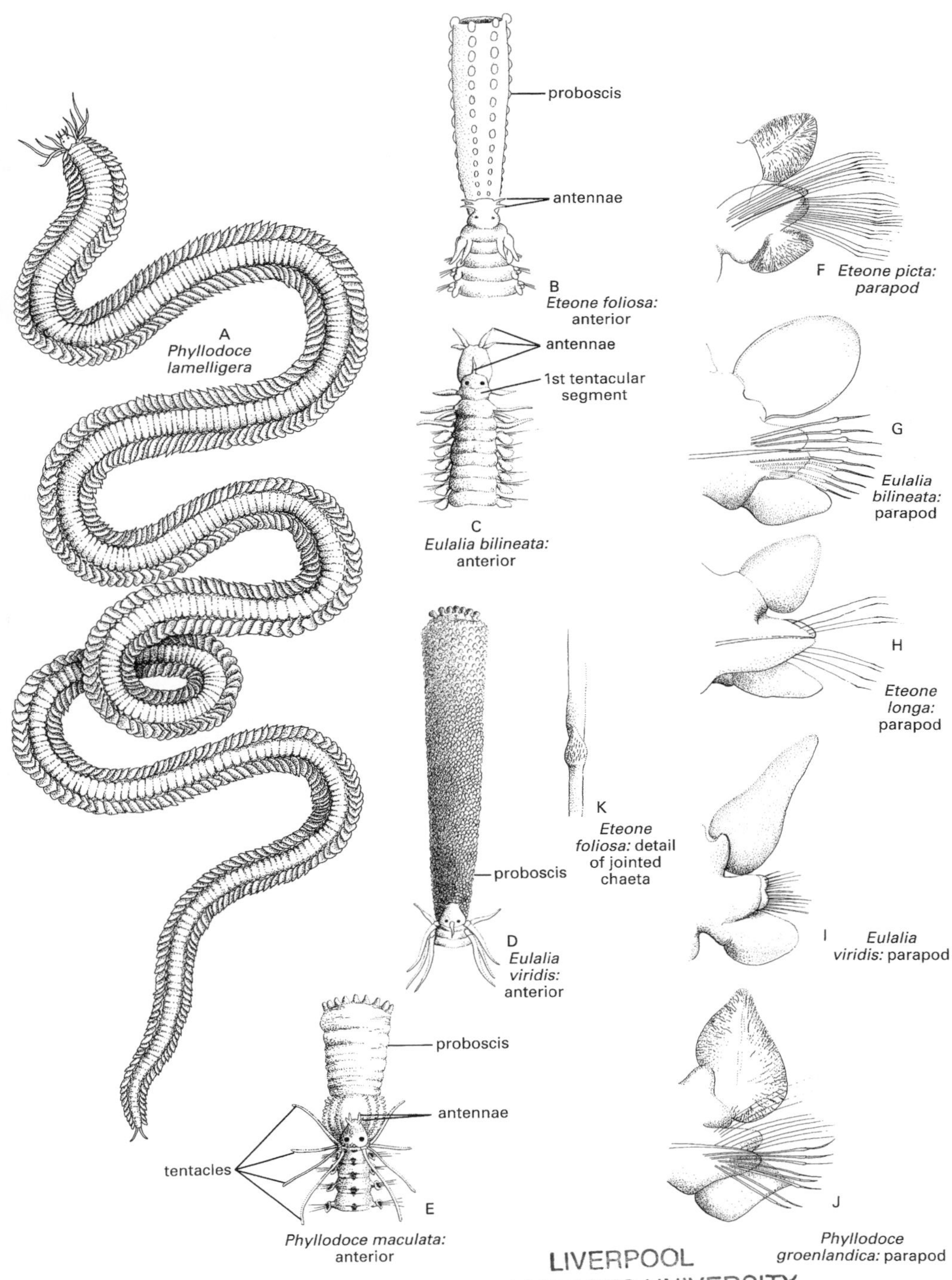
proboscis
antennae
B
Eteone foliosa: anterior
A
Phyllodoce lamelligera
antennae
1st tentacular segment
C
Eulalia bilineata: anterior
F Eteone picta: parapod
G
Eulalia bilineata: parapod
H
Eteone longa: parapod
K
Eteone foliosa: detail of jointed chaeta
proboscis
D
Eulalia viridis: anterior
I Eulalia viridis: parapod
proboscis
antennae
tentacles
E
Phyllodoce maculata: anterior
J
Phyllodoce groenlandica: parapod

Fig. 6.13

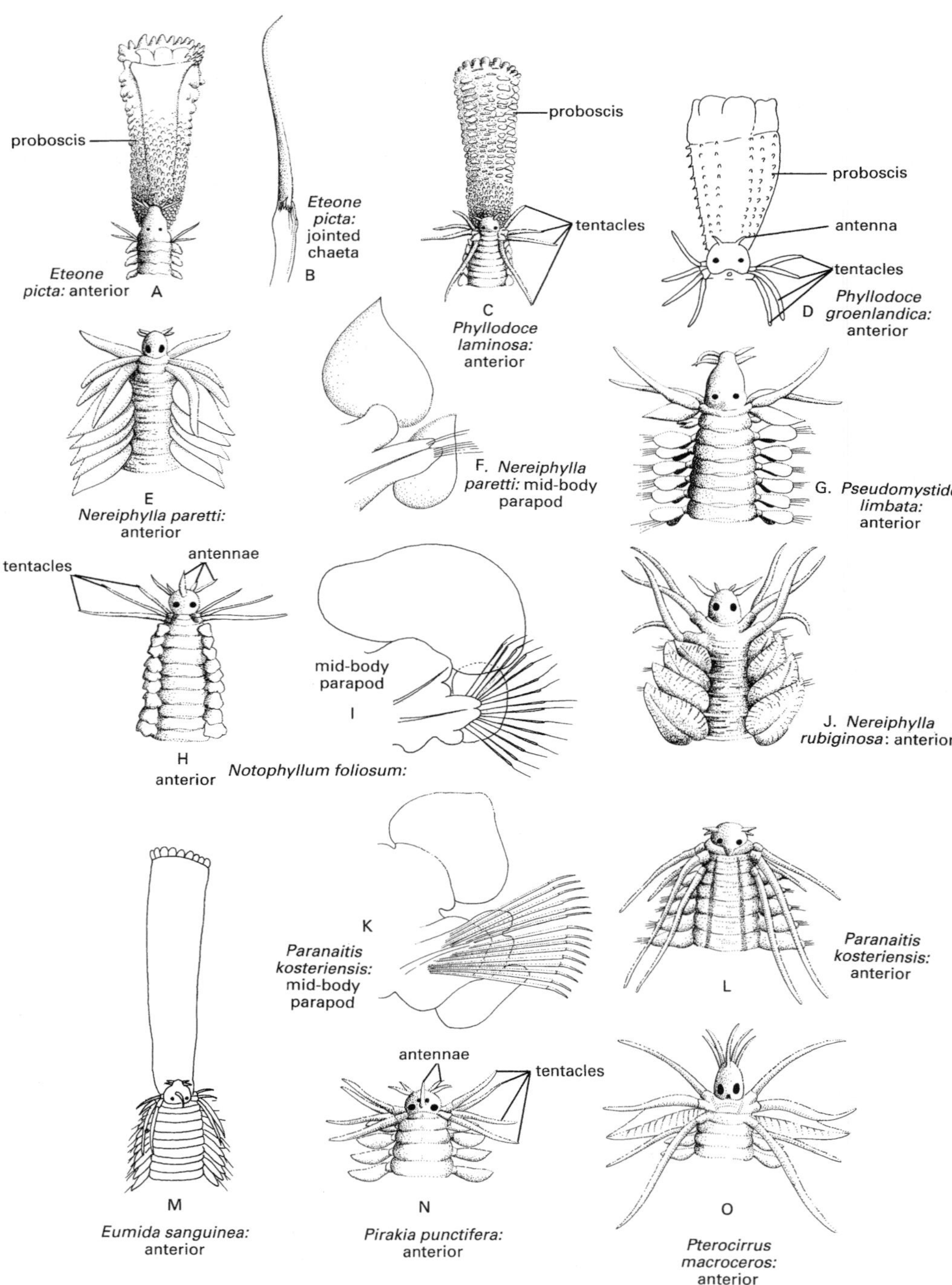
proboscis
Eteone picta: anterior A
Eteone picta: jointed chaeta
B
proboscis
tentacles
C
Phyllodoce laminosa: anterior
proboscis
antenna
tentacles
D Phyllodoce groenlandica: anterior
E
Nereiphylla paretti: anterior
F. Nereiphylla paretti: mid-body parapod
G. Pseudomystides limbata: anterior
tentacles
antennae
H
anterior
mid-body parapod
I
Notophyllum foliosum:
J. Nereiphylla rubiginosa: anterior
K
Paranaitis kosteriensis: mid-body parapod
L
Paranaitis kosteriensis: anterior
antennae
tentacles
M
Eumida sanguinea: anterior
N
Pirakia punctifera: anterior
O
Pterocirrus macroceros: anterior

Fig. 6.14

flaring out towards its opening, which is fringed with many medium-sized papillae. Two distinct tentacular segments; first has two pairs of short tentacles, second a pair of small ventral cirri and simple chaetae. Paddles broad, coming to a blunt point. Joint of compound chaetae finely toothed, terminating in a single, large spine (as Fig. 6.13K). Anal cirri broad, oval. Yellow or brick-red. In alcohol, lemon or brown.

In sand at low water and in the shallow sublittoral; around most of British Isles; North Atlantic to Arctic.

Eteone foliosa Quatrefages Fig. 6.13
As *E. flava* but rounded dorsally, with about 350 segments; 120–300 mm long. Prostomium a truncated cone. Proboscis with ridges forming flattened papillae; short, widening towards its opening which has two fleshy lobes resembling lips, separated at each side by smaller papillae (B). Second tentacular segment has no chaetae. Paddles not pointed, their stalk extended across blade like a short finger. Yellowish-white with little violet spots at base of parapods; brownish gut visible within. In alcohol, dirty white with iridescent sheen, or brownish.

In clean sand or rock-crevices at low water; south-west Britain, Irish Sea, Atlantic coast of Europe.

Eteone longa (Fabricius) Fig. 6.13
As *E. flava* but very slender, with about 200 segments; 25–60 mm long. Proboscis smooth or with small transverse ridges, flaring towards its opening which is surrounded by about 15 conical papillae. Paddles rounded, oval (H). Whitish or pale grey, with brown blotches or broad transverse bands of brownish-green. In alcohol, dark yellow with brownish paddles.

Offshore, around most of British Isles; North Atlantic to Arctic.

Eteone picta Quatrefages Figs 6.13, 6.14
Body flattened, with 60–150 segments; 30–60 mm long. Four short stout antennae. Proboscis with two rows of large, soft papillae down each side, flaring outwards towards its opening which bears a row of about 20 pointed papillae around ventral edge. Distal half otherwise smooth; basal half (not visible if proboscis is only partly everted) covered with numerous hard papillae, some flattened and others stalked, many with chitinous spines or teeth (6.14A). Two distinct tentacular segments; first has two pairs of short, stout tentacles, second a pair of small ventral cirri and simple chaetae. Paddles broad, oval or triangular in shape, with a blunt point and thick stalk (Fig. 6.13F). Joint of compound chaetae bears two large hooked spines (Fig. 6.14B). Colour very variable: pink or yellowish-green with, on each segment, blotches in purple or rust which form a pattern of longitudinal and transverse lines. Each paddle has a brownish spot.

At low water amongst stones or algae; also offshore. Around most of the British Isles; Channel and Atlantic coasts of Europe, also Mediterranean.

Eulalia bilineata (Johnston) Fig. 6.13
Body with 100–150 segments; 30–90 mm long. Prostomium broad and rounded. Five short antennae (C). Proboscis uniformly covered with scattered small papillae. Three tentacular segments: first separated from prostomium dorsally only by an ill-marked transverse groove, with one pair of short tentacles; segments 2 and 3 bear simple chaetae and (respectively) two pairs and one pair of tentacles, also short (C). Paddles oval, bluntly pointed (G). Joint of compound chaetae covered with medium-sized teeth. Two short, cylindrical anal cirri. Mature female mid-green with a faint and broken median double line; male yellowish-white.

At low water in crevices, under stones, and in kelp holdfasts; offshore amongst old shells. Mostly on south and west coasts of Britain; North Sea, Channel, Atlantic and Mediterranean coasts of Europe.

Eulalia expusilla Pleijel
As *E. bilineata* but with 65–80 segments; 8–12 mm long. Opening of proboscis surrounded by 24 large, blunt papillae. Paddles leaf-like, rounded. Jointed chaetae with a very finely toothed shaft. Two oval anal cirri. Brownish, with a longitudinal green line; paddles brown or green. Eggs green.

Offshore, from around south and west Britain; North Sea, Channel, North Atlantic coast of Europe.

Eulalia viridis (Linnaeus) Fig. 6.13
As *E. bilineata* but with 60–200 segments; 50–150 mm long. Long cylindrical proboscis covered by very many small, rounded papillae, opening

surrounded by 14–30 large ones, base sometimes smooth or longitudinally striated (D). Posterior tentacles reaching back to body segment 10–12. Paddles elongated, narrowing to a point (I). Shaft of jointed chaetae with a coarsely toothed joint. Two anal cirri, stout or flattened. Bright mid- to deep-green, sometimes with black spots or transverse lines dorsally.

Very frequently recorded, perhaps because it is so obvious moving about in the open on intertidal rocks; also in kelp holdfasts, rarely in sediments or offshore. All around British Isles and north-west Europe; Mediterranean.

Pirakia fucescens (de Saint-Joseph)

Body slender, with 95–100 segments; 20–30 mm long. Five short antennae. Proboscis covered with many small, rounded papillae, opening surrounded by 20 similar ones. Three tentacular segments, first indistinguishable from second dorsally. Four pairs of short tentacles. The two posterior tentacular segments also bear simple chaetae. Paddles heart-shaped, pointed distally. Light brown with regular darker spots; paddles brown or greenish. Eggs violet, mature males whitish.

Offshore; Irish Sea, Channel, Atlantic coast of Europe. The taxonomic status of this species is uncertain and in need of review.

Pirakia punctifera (Grube) Fig. 6.14

As *E. fucescens* but body stout, with about 350 segments; 60–150 mm long. Five short antennae (N), median one with a dark blotch around its base. Proboscis very long, covered with very many small, conical papillae, opening surrounded by about 50 large, rounded ones. Tentacles moderately long (N). Greenish or yellowish, with one or two lines of brown or dark-green spots dorsally across each segment; each paddle has a mass of brown spots at its centre. Mature females orange, eggs pink; mature males whitish.

At low water, under stones, and in muddy gravel, also offshore. Mainly around south and west coasts of Britain; North Sea and Atlantic coasts of Europe, also Mediterranean.

Pterocirrus macroceros (Grube) Fig. 6.14

Similar to *E. fucescens* but body broad, with about 170 segments; 30–80 mm long. Five relatively long antennae, all inserted anteriorly (O). Proboscis covered with many small, conical papillae, opening surrounded by 48 larger, rounded ones. First tentacular segment distinct except when worm is tightly contracted. Ventral pair of tentacles on second segment each flattened, resembling a cleaver (O); last pair long. Paddles elongated, narrowing to an acutely pointed tip. Blade of jointed chaetae very long. Pale green, straw-yellow, or pale brown, with a darker blotch between eyes. Eggs green.

On the lower shore amongst bryozoans and hydroids; offshore, amongst old shells. Mainly on south and west coasts of Britain; North Sea and Atlantic coasts of Europe, also Mediterranean.

Eumida sanguinea (Oersted) Fig. 6.14

As *E. fucescens* but body stout, with 60–140 segments; 30–60 mm long. Proboscis smooth when fully everted (M) but otherwise wrinkled; opening surrounded by about 20 papillae. First tentacular segment appears fused with prostomium dorsally; last pair of tentacles quite long. Paddles broad and bluntly pointed. Blade of jointed chaetae long. Colour variable, from greyish-white with brown spots through yellowish-green to reddish-brown with white transverse bands. In alcohol, olive- or rust-coloured. Eggs green or reddish.

At low water, under stones or in kelp holdfasts; offshore amongst old shells. All around Britain and north-west Europe; Mediterranean, Arabian Gulf, Australasia.

Pseudomystides limbata (de Saint-Joseph) Fig. 6.14

Body slender, with 40–90 segments; 6–10 mm long. Prostomium an elongated cone, prolonged anteriorly to give the appearance of a snout (G). Four small, stout antennae. Proboscis covered with large conical papillae, opening surrounded by 8–10 papillae. Three distinct tentacular segments; first and second each bear dorsally a pair of cylindrical tentacles of moderate length. The second also has simple chaetae and a pair of flattened ventral cirri, intermediate in form between tentacles and paddles (G). The third has some jointed chaetae and a pair of normal ventral cirri (like dorsal paddles, but much smaller) but no paddles dorsally. Paddles elongated, oval, with a blunt tip. Jointed chaetae with a broad blade markedly toothed down one edge. Light brown; eggs green.

Offshore from shelly or coralline bottoms. South and west coasts of Britain; Channel and Atlantic coasts of Europe.

Nereiphylla paretti Blainville Fig. 6.14
Body slender, with several hundred segments; 150–300 mm long. Four small flattened antennae (E). Distal part of proboscis transversely ridged, opening surrounded by papillae; basal part covered with fine papillae. Three tentacular segments, first indistinguishable from second dorsally and scarcely more so ventrally. Four pairs of stout tentacles, last two flattened. The two posterior tentacular segments also bear simple chaetae. Paddles large: broad and heart-shaped in mid-body (F), narrower but longer anteriorly (E). Colour variable—back bright blue with an iridescent sheen, slate colour or grey, or yellow and blue; ventral surface yellow or pink, sometimes with a blue iridescence. Paddles green edged with yellow, tentacles yellow.

At low water under stones, also offshore. South-west Britain; North Sea, Channel and Atlantic coasts of Europe, also Mediterranean.

Nereiphylla rubiginosa
(de Saint-Joseph) Fig. 6.14
Body slender, with 50–60 segments; 10–100 mm long. Prostomium small, appearing to be mounted within first tentacular segment (J). Four stout antennae. Base of proboscis covered with many papillae, its opening surrounded by eight large ones. Three tentacular segments, first indistinguishable from second dorsally and scarcely more so ventrally. Four pairs of short, stout tentacles; two posterior tentacular segments also bear simple chaetae. Paddles heart-shaped, with an acutely pointed tip (J). Yellowish-green with two longitudinal lines, darker green or blue, down dorsal mid-line. Paddles dark yellow or rust-red with darker spots. In alcohol, body cinnamon with rust-red paddles.

Under stones or amongst algae along the sublittoral fringe, also offshore. Around the south and west coasts of Britain; Channel and Atlantic coasts of Europe.

Notophyllum foliosum (M. Sars) Fig. 6.14
Body broad, dorsal surface almost entirely hidden by overarching paddles; 80–110 segments, 15–55 mm long. Five antennae (H). Two ciliated nuchal organs extend backwards behind eyes, resembling ear-flaps. Proboscis has many papillae around basal part, opening surrounded by large, rounded ones. Three tentacular segments, first difficult to distinguish from second dorsally. The two rearward dorsal tentacles are long, reaching back to body segment 9–10. The last two tentacular segments bear simple chaetae, also present in notopod of normal segments. Paddles kidney-shaped and very wide (I). Greenish-grey, having green paddles edged with brown. In alcohol, brownish or greenish. Eggs greyish.

Offshore, amongst old shells, stones, and serpulid tubes. Around most of the British Isles; North Sea, Channel and Atlantic coasts of Europe, also Mediterranean.

Paranaitis kosteriensis (Malmgren) Fig. 6.14
Body wide, dorsal surface rounded. About 155 segments, 60–80 mm long. Prostomium narrowing abruptly to a rearward extension bearing nuchal papilla. Four short antennae (L). Proboscis with two longitudinal rows of papillae on each side at base. Three tentacular segments, first enclosing posterior part of prostomium to each side in front and indistinguishable from second, dorsally, behind. Four pairs of tentacles (L). Third tentacular segment bears simple chaetae. Paddles kidney-shaped and very wide, with a thick stalk (K). Back striped with transverse red or purple bands; paddles pale, with a spot of similar colour at their base.

Offshore, coastal and deep. Off the south and west coasts of Britain; North Sea, North Atlantic.

21. SCALIBREGMIDAE

Body short and stocky or long and slender, often swollen anteriorly. Prostomium bilobed or two-horned. Segments annulated, integument usually wrinkled or reticulated. Parapods small but may have fairly prominent finger-like cirri posteriorly; bearing hair-like and pitchfork (Fig. 6.12R) bristles but no crochets. Branched gills may be present anteriorly. Four species in the British fauna.

---

1. Body elongated, with 50–60 segments, all but first few 4-annulate. Finger-like cirri, at least ventrally, on parapods of posterior half of body. Pygidium bears four or five medium–long cirri **2**

   Short, thickset, maggot-like, with about 30 triannulate segments. Parapods simple mounds, each with a bundle of chaetae but no appendages (Fig. 6.12Q). Anus surrounded by short papillae **_Polyphysia crassa_**

---

2. Body long (50–100 mm), swollen anteriorly then slender. Four pairs of branching gills dorsally on chaetigers 2–5 (Fig. 6.12O). Chaetae of first chaetiger similar to others. Parapods of posterior region bear finger-like cirri dorsally and ventrally. Proboscis smooth, globular
Scalibregma inflatum

Body short (5–20 mm), slender, spindle shaped. No gills. Notopod of chaetiger 1 bears a number of stout, blunt spines. In posterior half of body, finger-like cirri are ventral only. Proboscis cylindrical and fringed with papillae
Sclerocheilus minutus

Scalibregma inflatum Rathke Fig. 6.12
About 50–60 segments, first few triannulate, then 4-annulate; 50–100 mm long. Prostomium rectangular, with two short frontal horns; proboscis smooth and globular. Four pairs of branching gills dorsally on chaetigers 2–5 (O). From about chaetiger 16, finger-like dorsal and ventral cirri (P). Pygidium bears four or five thread-like or finger-like cirri. Purple-red with yellow blotches (eggs yellow).

At low water and in the shallow sublittoral, deep in sand or mud. From the Arctic all around Britain to Atlantic coast of Europe.

Sclerocheilus minutus Grube
About 54 segments, chaetigers 2–4 biannulate, then each with four annuli; 5–20 mm long. Prostomium T-shaped; proboscis cylindrical, fringed with papillae. Gills absent. In posterior half of body, finger-like ventral cirri are present and there is a button-like lateral organ between parapodial lobes. Pygidium bears four or five finger-like or clubbed cirri. Reddish-brown; mature females may be greyish-white.

Shallow sublittoral, in the mud between stones and shells. English Channel, Atlantic and Mediterranean coasts of Europe.

Polyphysia crassa (Oersted) Fig. 6.12
About 30 segments, triannulate; 20–25 mm long. Prostomium small and bilobed. Gills and parapodial appendages are absent (Q). Anus surrounded by short papillae. Coloration when alive flesh-pink.

Shallow sublittoral, in a thick tube of mud cemented with mucus, on muddy bottoms. Irish Sea, Firth of Clyde, North Sea, Atlantic coast of Europe.

22. SPHAERODORIDAE

Medium-sized worms, long and thin or very small, short and grub-like. Integument covered by papillae, segmentation not indicated externally by grooves or constrictions. Simple uniramous parapods with few chaetae are inconspicuous except for their dorsal cirri in the form of large, spherical capsules. Amongst the small, stout forms, several more spherical tubercles lie between these cirri, aligned in longitudinal rows. Prostomium indistinct, often more or less retracted, and bears only a few short tentacular papillae. Eight species recorded from British waters.

1. Elongated (10–60 mm), thin, with a single series of spherical capsules down each side of body (dorsal surface covered with smaller papillae) **2**

Short (2–4 mm), stout, dorsal surface bears many series of spherical capsules **3**

2. About 120 chaetigers (indicated by spheres); chaetae simple
Sphaerodorum flavum

About 50 chaetigers; chaetae jointed
Ephesiella abyssorum

3. Eight to 16 chaetigers, each bearing a row of six spheres dorsally
Sphaerodoridium claparedi

Seventeen to 22 chaetigers, each bearing 10–12 spheres dorsally
Sphaerodoropsis minutum

Sphaerodorum flavum Oersted Fig. 6.19
Body elongated and narrow, 10–60 mm long; about 120 chaetigers. Four dark, comma-shaped eyes dorsally beneath skin at anterior end (N). Mouth ventral; proboscis cylindrical, closely covered with fine papillae. Integument bears scattered small

papillae dorsally and ventrally. Parapods short and conical (P), each containing a few simple, stout, pointed chaetae (R). Dorsal cirri in form of large spherical capsules bearing a small, pointed papilla (P); ventral cirri cylindro-conical. Pygidium with two spheres a little larger than the more anterior ones, and a single anal cirrus. Pale yellow with a brown or pinkish tinge in the gut region; spheres whitish.

Amongst stones, shell gravel, or coralline fragments at low water and sublittorally. All around Britain; north-east Atlantic, Mediterranean.

Ephesiella abyssorum (Hansen)
Similar to *S. flavum* but with about 50 chaetigers. Ventral cirri squatter and less pointed. Chaetae jointed, with a hooked terminal section. Yellowish-white.

In kelp holdfasts. Southern Britain, north-east Atlantic, Mediterranean.

Sphaerodoridium claparedi (Greeff) Fig. 6.19
Body short and stout, about 2 mm long; 8–16 chaetigers. Two eyes. Mouth ventral, proboscis globular. Striated gizzard visible externally (O). Dorsally, each segment bears a row of six large spherical capsules, the outermost at each end being the dorsal cirrus of a parapod; these spheres also aligned longitudinally. Globular papillae smaller ventrally, arranged irregularly or in four longitudinal rows. Each parapod a ribbed elongated cone bearing two club-shaped papillae between which the chaetae emerge (Q); chaetae jointed, with a long, hooked end (S). Pygidium bears small papillae and a single anal cirrus. Yellowish-white.

Amongst shells and algae at low water and sublittorally; may swim at the surface at night. South and west coasts of Britain, north-east Atlantic.

Sphaerodoropsis minutum (Webster and Benedict)
Similar to *S. claparedi* but 2–4 mm long, with 17–22 chaetigers. Ten to 12 spherical capsules in each dorsal row, but only small, scattered papillae ventrally. The pygidium bears two large globular papillae and a long cirrus.

Arctic waters, north-east and north-west Atlantic.

23. SPIONIDAE

Worms with a relatively elongated body not divided into distinct regions. Prostomium small, without antennae; anterior end of variable shape even within a single species, may have a pair of short frontal horns, also a small mid-dorsal occipital tentacle at rear. Two very long, often sturdy palps are very mobile in life and usually roll up spirally when the animal is disturbed. Parapods biramous with leaf-like dorsal and ventral lobes. Dorsal gills, usually simple but rarely feathery, on a variable number of segments. These, and sometimes the extended dorsal lobes, arch over the dorsal surface. In addition to simple hair-like chaetae, posterior segments have shorter, stouter ones with a hooded tip, usually hooked, with 1–3 teeth. Pygidium terminates in an anal funnel or numerous anal cirri. There may be one or more longitudinal, ciliated sensory grooves down the back.

The palps and gills are readily lost. The palps are almost invariably shed when the worm is killed, thus removing the main character which identifies it as a spionid.

A large family with more than 50 species recorded from Britain, representing perhaps 16 different genera.

---

1. Chaetiger 5 obviously different from all others, lacking gills and parapodial lobes but bearing a few extra large chaetal spines **_Polydora_**

   No single anterior segment markedly different from its neighbours **2**

---

2. Gills elongated, finger-like, more or less arching over dorsal surface and usually with bright-red blood vessels **3**

   Gills absent. Tufts of long, silky threads laterally on chaetigers 5–15 **_Spiophanes bombyx_**

---

3. Gills commence from chaetigers 1–2 and are present on at least front third of body (often more) **4**

   A large pair of gills may be present on chaetiger 2, but remainder commence between 11 and 20, and are present on no more than middle third of body (or less); anus surrounded by four stout papillated processes (Fig. 6.15I) **_Pygospio elegans_**

---

4. Anus surrounded by a sucker-like or membranous funnel lacking cirri (Fig. 6.15H); posterior half of body with a few short, stout chaetae, both dorsally (together with long delicate hair-like chaetae) and ventrally **5**

   Anus surrounded by thread-like or petal-like cirri; posterior half of body with short, stout chaetae ventrally only **6**

5. Gills present for almost whole length of body; prostomium extends rearwards in a blunt point but has no median tentacle (Fig. 6.15C); posteriorly at least 20 segments have short, stout chaetae ventrally before the dorsal ones appear ***Scolelepis squamata***

   Gills absent from about last third of body; prostomium has a short but distinct median tentacle (Fig. 6.15D): stout chaetae ventrally but not dorsally ***Scolelepis foliosa***

6. Gills present for almost whole length of body **7**

   Gills absent from about last half of body; prostomium rounded in front, lacking horns (Fig. 6.15E) ***Laonice cirrata***

7. Prostomium rounded or indistinctly lobed at front, ending behind in a small, stout process ***Spio filicornis***

   Prostomium with well developed frontal horns (Fig. 6.15F) **8**

8. Medium size (up to 60 mm long, 2 mm wide); 6–8 short anal cirri; stout chaetae of posterior segments terminate in two hooded teeth (Fig. 6.15N), four or five in each neuropod ***Malacoceros fuliginosus***

   Large (100–160 mm long, 6 mm wide); 15–30 long anal cirri; stout chaetae of posterior segments terminate in three hooded teeth (Fig. 6.15O), 20–25 in each neuropod ***Malacoceros vulgaris***

Laonice cirrata (M. Sars) Fig. 6.15
Body flattened, 90–120 mm long, 3–5 mm wide; 160 chaetigers. Prostomium rounded anteriorly, with an occipital tentacle but no frontal horns (E). Sensory crest runs dorsally along first 28–30 chaetigers; 35–44 pairs of long cirriform gills from chaetiger 2. In gill-bearing segments, dorsal lobe of each parapod is large, extending upwards like a pointed ear (L), independent of gill. Stouter chaetae with two hooded teeth occur ventrally from chaetiger 40–50 rearwards, absent from notopods. Anus surrounded by about 12 cirri. Yellowish in front, brownish towards the rear.

Sublittorally to considerable depths on bottom of muddy sand, shell gravel, or shingle. All around Britain; cosmopolitan.

Malacoceros fuliginosus (Claparède) Fig. 6.15
Body 50–60 mm long, up to 2 mm wide; 100–150 chaetigers. Prostomium with a pair of frontal horns and four eyes, extending rearwards in a crest but lacking an occipital tentacle (F). Palps banded. Finger-like gills from chaetiger 1 (although very small there), partly fused with lobe of notopod for first third of body (after which that lobe becomes small). Stouter chaetae with two hooded teeth (N) occur ventrally from chaetiger 30–45 rearwards, 4–5 in each neuropod, never dorsally. Pygidium with 6–8 petal-like anal cirri. Salmon-pink, darker in front; when undisturbed, covered with sandy mucus.

At low water, in galleries in muddy sand or under stones; may form dense groups in rich mud. Very active when disturbed, swimming in loose spirals. All round Britain; north-east Atlantic, Mediterranean.

Malacoceros vulgaris (Johnston) Fig. 6.15
As *M. fuliginosus* but body stout (100–160 mm long, 6 mm wide) though fragile; 200–350 chaetigers. No eyes in adult. Stouter chaetae with three hooded teeth (O), 20–25 in each neuropod. Anus surrounded by 15–30 thread-like cirri. Pink or orange in front, greenish-brown towards the rear; gills bright red.

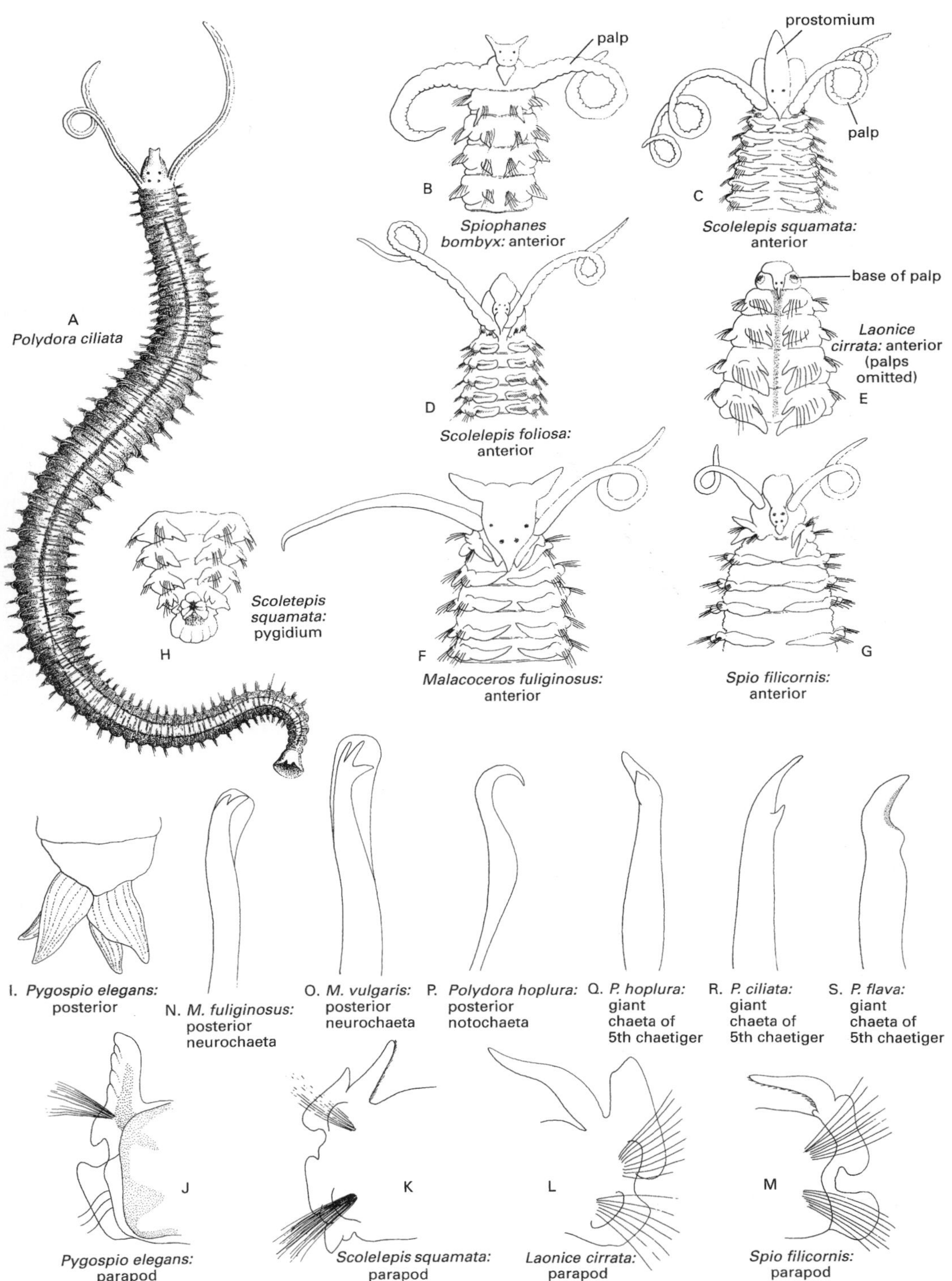

A
Polydora ciliata
palp
B
Spiophanes bombyx: anterior
prostomium
palp
C
Scolelepis squamata: anterior
D
Scolelepis foliosa: anterior
base of palp
Laonice cirrata: anterior (palps omitted)
E
H
Scoletepis squamata: pygidium
F
Malacoceros fuliginosus: anterior
G
Spio filicornis: anterior
I. Pygospio elegans: posterior
N. M. fuliginosus: posterior neurochaeta
O. M. vulgaris: posterior neurochaeta
P. Polydora hoplura: posterior notochaeta
Q. P. hoplura: giant chaeta of 5th chaetiger
R. P. ciliata: giant chaeta of 5th chaetiger
S. P. flava: giant chaeta of 5th chaetiger
J
Pygospio elegans: parapod
K
Scolelepis squamata: parapod
L
Laonice cirrata: parapod
M
Spio filicornis: parapod

Fig. 6.15

At low water, in muddy sand under stones and amongst eel grass; always well spaced, does not occur in large numbers. Commonest in the south and west of Britain.

Polydora

Prostomium blunt in front, extending rearwards in a low crest. Palps long, slender, ciliated, very mobile. Longitudinal sensory grooves mid-dorsally. Dorsal lobes of parapods do not usually contribute to the gills; number of gills variable, beginning from chaetiger 7–9 (rarely from chaetiger 2). Chaetiger 5 markedly different from all others, lacking gills and parapodial lobes but with characteristic, extra-large chaetal spines dorsally together with long lance-like chaetae. Short, stout chaetae ventrally from chaetiger 7–8, tipped with two small, hooded teeth; this type of chaeta absent dorsally, although large chaetae terminating in a hooked point may be present there (*P. hoplura*). Anus surrounded by a membranous funnel, complete or four-lobed. All species make a U-shaped tube from small particles (usually of mud, but may be whitish and calcareous if excavating in lithothamnia or other encrusting coralline algae): much of this tube may be embedded in a burrow excavated in limestone rock or shells. In life, the palps protrude, waving vigorously.

---

1. Dorsally on the 10–20 rearmost segments, large yellow chaetae with a pointed tip bent over like a meat-hook (Fig. 6.15P); of medium size (length 50 mm or more) ***Polydora hoplura***

   No such chaetae posteriorly; small (length 40 mm or less) **2**

---

2. Dorsally from chaetiger 8 rearwards, tufts of numerous parallel fine chaetae which scarcely protrude from the surface ***Polydora flava***

   No such tufts of chaetae **3**

---

3. Large chaetae on chaetiger 5 have a lateral tooth at the tip (Fig. 6.15R) **4**

   Large chaetae on chaetiger 5 have no lateral tooth; tip hollowed on one side like a spoon (*sim.* Fig. 6.15S). Gills on most body segments ***Polydora caeca***

---

4. Gills from chaetiger 8 rearwards for more than half the remaining number of segments; over 20 mm long ***Polydora ciliata***

   Gills from chaetiger 10 (rarely 9) rearwards for fewer than half the remaining segments; 10 mm long or less ***Polydora giardi***

Polydora caeca (Oersted)

Body 20–40 mm long, 1 mm wide; 70–130 chaetigers. Chaetiger 5 has on each side dorsally four or five spoon-ended giant chaetae, lacking a lateral tooth. Gills from chaetiger 8 to one-half or two-thirds down the body. Yellowish or pale pink.

In lithothamnia and kelp holdfasts. Most common in south-west Britain; Atlantic, Mediterranean, Indian Ocean.

Polydora ciliata (Johnston) Fig. 6.15

Body 20–30 mm long, 0.7–1 mm wide; 60–180 chaetigers (A). Chaetiger 5 has on each side dorsally six or seven giant chaetae with a lateral tooth (R). Gills from chaetiger 7 to all but the 10 rearmost ones. Yellowish-brown.

In limestone rock and stones, old shells or lithothamnia. All around Britain; Arctic, Baltic, Atlantic, Mediterranean, Indo-Pacific.

Polydora flava Claparède Fig. 6.15

Body 20–45 mm long, slender; 100–150 chaetigers. Chaetiger 5 has on each side dorsally several spoon-ended chaetae, lacking a lateral tooth (S). Gills from chaetiger 8 (rarely chaetiger 7 or 9), often lacking from last third of body. Dorsally from chaetiger 8 rearwards, tufts of numerous parallel fine chaetae which scarcely protrude from the surface.

In coarse sand, rock-crevices, lithothamnia, and old shells. Most common in south-west Britain, north-east Atlantic, Mediterranean, Indo-Pacific.

Polydora giardi Mesnil

Body 8–10 mm long, 0.5 mm wide, 50–100 chaetigers. Chaetiger 5 has on each side dorsally four or five giant chaetae with a lateral tooth. Gills from chaetiger 10 (rarely chaetiger 9) to about 25, the first two or three pairs small. Yellowish or pink.

In lithothamnia and kelp holdfasts. South-west Britain, Channel; perhaps Mediterranean, north Pacific.

Polydora hoplura Claparède Fig. 6.15

Body 50–60 mm long, 1–2 mm wide, 200 or more chaetigers. Chaetiger 5 has on each side dorsally several giant chaetae with a lateral tooth (Q). Dorsally on the 10–20 rearmost segments, large yellow chaetae with a pointed tip bent over like a meat-hook (P). Reddish or yellowish.

Amongst serpulids and particularly oysters. Mostly in south-west Britain; north-east Atlantic, Mediterranean, southern Africa.

Pygospio elegans Claparède Fig. 6.15

Body slender, 10–15 mm long; 50–60 chaetigers. Prostomium extends posteriorly in a pointed process as far as chaetiger 2; 2–8 small eyes, scattered in no clear pattern. Gills fringed for entire length by an extension of dorsal parapodial lobe (J) from chaetiger 11–20, only 7–9 pairs in females but 20–28 pairs in males. In males, also a pair of large erect gills on chaetiger 2, which stand separate from the notopod. Behind the gilled region, parapodial lobes are much reduced. Short, stout chaetae with two hooded teeth ventrally from chaetiger 8, 4–5 in each neuropod. Pygidium with four, stout papillated processes (I). Yellowish or greenish with a brown gut.

Occupies long, flexible tube of fine sand grains embedded in mucus; in mud collected in rock-crevices or between old shells, from the mid-shore to the sublittoral. All around Britain, north-east Atlantic, Mediterranean, north Pacific.

Scolelepis foliosa (Audouin and Milne Edwards) Fig. 6.15

Body 100–160 mm long, 6–9 mm wide, fragile; 200–250 chaetigers. Prostomium small, with a little occipital tentacle having a swollen base (D). Gills from chaetiger 2, well developed in the first 50–60 segments, each fringed almost to top by extended dorsal parapodial lobe; rearwards, lobe does not contribute to long, slender gills. Gills absent from last third of body. Short, stout chaetae with a hooded tip occur from chaetiger 58–60 ventrally (about 20 per neuropod) and chaetiger 60–65 dorsally (about 10 per notopod). Short, oblique anal funnel. Red in front, grey-green and almost transparent behind; gills bright red.

At low water, in clean or only slightly muddy sand, usually solitary; does not swim when disturbed. On most British coasts; north-east Atlantic, Mediterranean.

Scolelepis squamata (O. F. Müller) Fig. 6.15

Body of medium length, slender; 50–80 mm long, 2–3 mm wide; 150–200 chaetigers. Prostomium elongated, extending rearwards in a stout, pointed process which overhangs chaetigers 2 and 3. Palps particularly long. Gills from chaetiger 2 as far as last seven or eight; extended dorsal lobe fringes each gill for much of its length at front end but less and less developed posteriorly (K). Stout chaetae from about chaetiger 40 ventrally (10–12 per neuropod) and chaetiger 60–65 dorsally (2–5 per notopod). Delicate scalloped anal funnel. Bluish-green with contrasting red blood vessels in palps and gills; mature males whitish, females brighter green; when undisturbed, covered with sandy mucus.

Mid- to lower shore in clean or only slightly muddy sand on exposed beaches; swims in spirals when disturbed. On most British coasts, north-eastern Atlantic, Mediterranean.

Spio filicornis (O. F. Müller) Fig. 6.15

Body about 30 mm long, 2 mm wide, fragile; 80–90 chaetigers. Prostomium ending behind in a short, stout pointed process (G). Palps relatively short. Gills from first to all but last few chaetigers. Dorsal lobe of parapod extends only a short way up anterior gills and is separate from more posterior ones (M). Short stout chaetae with two or three hooded teeth ventrally only, from chaetiger 10–15 rearwards (6–9 per neuropod). Anus dorsal, between four petal-like anal cirri. Pink, with a brown gut and cream flecks laterally.

Low water and shallow sublittoral, in clean sand. On most British coasts, Arctic, Baltic, north-east Atlantic, Mediterranean; north Pacific.

Spiophanes bombyx (Claparède) Fig. 6.15

Body flattened, 50–60 mm long, 1.5 mm wide; about 180 chaetigers. Prostomium with two long frontal horns and a stout, pointed rearward process (B). Palps relatively short. Two longitudinal ciliated sensory grooves run down dorsal mid-line; a ciliated ridge crosses these transversely on each segment. Gills absent. Dorsal lobe of each parapod extended upwards in a blade, folded over across body, which becomes longer and more narrow on more posterior segments. Chaetigers 5–15 bear laterally, between lobes of each parapod, a gland producing a tuft of long, twisted silky threads. From chaetiger 15 rearwards, each neuropod contains one large, curved chaeta. Short, stout chaetae with two hooded teeth occur ventrally from chaetiger 15 but are absent dorsally. Pygidium has two finger-like anal cirri. Bright pink in front, then darker red or greenish-brown.

At low water, in a slender but stiff sandy tube which protrudes slightly from the surface. On most British coasts; Atlantic, Mediterranean, north Pacific.

24. SYLLIDAE

Small, active, carnivorous worms found mostly amongst colonial sedentary animals and seaweeds, occasionally in a mucus tube; a few are interstitial in muddy sand. Typically with long, finger-like and often annulated dorsal cirri. Pharynx eversible, usually with a single piercing tooth or a *trepan* (circlet of small teeth), connected to a barrel-shaped muscular proventriculus (gizzard) which may be visible through the body wall. Head rounded, normally with four large eyes and three antennae, followed by two pairs of tentacles; a pair of flattened ventral palps may be separate, joined at the base or wholly fused together in the mid-line. Parapods uniramous, mostly with only a few jointed chaetae. Reproduction may be by stolons budded from an asexual stock, which may form a chain behind the asexual individual (Figs 6.16L,M; 6.17Q). More than 70 species are known from Britain. Major divisions (subfamilies) of this large group and their main characteristics are:

Syllinae—palps free for their entire length, antennae and cirri distinctly annulated (moniliform), ventral cirri present (*Langerhansia*, *Haplosyllis*, *Syllis*, *Trypanosyllis*, *Typosyllis*);

Eusyllinae—palps joined only at the base, antennae and cirri sometimes show superficial constrictions, ventral cirri present (*Amblyosyllis*, *Eusyllis*, *Odontosyllis*, *Pionosyllis*);

Exogoninae—palps fused for their entire length, antennae and cirri short, not annulated; ventral cirri present but may be indistinct; often very small (*Brania*, *Eurysyllis*, *Exogone*, *Sphaerosyllis*);

Autolytinae—poorly developed palps fused and folded back ventrally (thus not readily visible), antennae and dorsal cirri (which may be absent except on the first chaetiger) not annulated, no ventral cirri; number and arrangement of teeth may be an important feature, often requiring preserved specimens to be cleared (*Autolytus*, *Myrianida*, *Proceraea*).

---

1. Ventral cirri lacking **2**

   Ventral cirri present **7**

---

2. Antennae, tentacles, and dorsal cirri more or less elongated; anterior end of pharynx armed with a circle (trepan) of 10 or more teeth (Figs 6.16H, I; 6.17D–H) **3**

   Antennae, tentacles, and dorsal cirri globular (Fig. 6.17I); further spherical tubercles across back; pharynx armed with a single tooth ***Eurysyllis tuberculata***

---

3. Dorsal cirri (also, to some extent, antennae and tentacles) flattened, tongue-shaped; antennae short, median one up to twice as long as those on either side; trepan with 50–60 small teeth. May trail a long chain of stolons (Fig. 6.17Q) ***Myrianida pinnigera***

   Dorsal cirri, antennae, and tentacles cylindrical **4**

---

4. Ciliary band across each segment dorsally; shaft of jointed chaetae thin **5**

   Ciliary bands absent; shaft of jointed chaetae thick. Antennae and tentacles quite long; median antenna up to twice length of lateral ones. Trepan of 10 large, alternating with 10 small, teeth (Fig. 6.17E) ***Proceraea picta***

---

5. Antennae short (about twice head length), median one about same length

as those on either side (Fig. 6.17A). All but most anterior dorsal cirri of uniform length, about equal to body width. Trepan of 10 blunt teeth, large but of varying size **_Autolytus prolifer_**

Lateral antennae of short to medium length, median one 2–3 times as long (Fig. 6.17C) **6**

---

6. All but most anterior dorsal cirri of uniform length, less than body width. Trepan of 16–20 pointed teeth of equal size (Fig. 6.17G) **_Autolytus aurantiacus_**

After the most anterior ones, length of dorsal cirri alternates between half body width and about equal to body width. Trepan of 10 large, blunt teeth, each separated by two or three small, pointed ones (Fig. 6.17H) **_Autolytus longeferiens_**

---

7. Palps well developed, fused together for most or all of length and clearly visible from above (Fig. 6.17J–L). Small (usually less than 6 mm and not more than 10 mm long) **8**

Palps fused together only at base, if at all (Fig. 6.16B); usually, but not necessarily, visible from above. Of medium size (usually more than 6 mm and often more than 10 mm long) **14**

---

8. Antennae, tentacles, and dorsal cirri of medium length (about equal to body width) and only moderately swollen near base; two pairs of tentacles. Very small (body length 1–3 mm) **9**

Antennae, tentacles, and dorsal cirri of varying length but often very short and swollen; only one tentacle on each side. Small (body length 2–10 mm) **10**

---

9. Dorsal cirri with a square-cut lip (Fig. 6.17M); some jointed chaetae with a short blade, others long with a larger single tooth at tip **_Brania pusilla_**

Dorsal cirri with a long, pointed tip (Fig. 6.17N); jointed chaetae all have a fairly long blade, bearing twin larger teeth at tip **_Brania clavata_**

---

10. Antennae, tentacles, and dorsal cirri all of much the same length (up to half body width), with a swollen base and pointed tip. Body may be encrusted with sand or mud **11**

Antennae, tentacles, and dorsal cirri more or less club-shaped, lacking a pointed tip; median antenna slightly or markedly longer, tentacles rudimentary, cirri short (much less than half body width) **12**

---

11. Body covered with papillae, often encrusted with sand or mud. Antennae, tentacles, and dorsal cirri about half body width. Jointed chaetae with slender blades, some long, some of medium length (Fig. 6.17O) **_Sphaerosyllis hystrix_**

Papillae only on parapods and anal segment. Antennae, tentacles, and dorsal cirri very short. Jointed chaetae with short, stout blade (Fig. 6.17P) **_Sphaerosyllis bulbosa_**

---

12. Antennae all of medium length (at least half head width). Jointed chaetae with a short double hook or a long delicate spine **_Exogone naidina_**

Lateral antennae very short **13**

---

13. Median antenna of medium length. Jointed chaetae with short, broad blades **_Exogone hebes_**

Median antenna also very short. Jointed chaetae with a short blade or a long spine **_Exogone verugera_**

---

14. Palps separate for their whole length; dorsal cirri (and usually antennae also) distinctly segmented (Fig. 6.16A, C) **20**

Palps fused at base; dorsal cirri and antennae not segmented (but may be superficially annulated) (Fig. 6.16J,K) **15**

15. Nuchal organ present, resembling elongated ear-flaps folded back across tentacular segment (Fig. 6.16J); dorsal cirri annulated, length about four times body width and coiled in a loose spiral. Body with 16 segments ***Amblyosyllis formosa***

Nuchal flaps absent; dorsal cirri may not be annulated, and less than twice body width in length. Body with 40 or more segments **16**

16. Occipital flap present as a disc forming half or more of a circle, extending forwards dorsally across prostomium from tentacular segment (Fig. 6.16K). Front of pharynx bears small teeth in a row or ring, but no large tooth **17**

Occipital flap, if present, represented only by a narrow fold; a single large tooth present at front of pharynx (Fig. 6.16I) **19**

17. Antennae at least as long as head width; longest tentacle about twice body width. Dorsal cirri cylindrical, blunt-ended, alternately long and short. Body with up to 100 segments. Jointed chaetae with a short, smooth edged hook (Fig. 6.16F, G) **18**

Antennae short (less than head width), tentacles short (longest about equal to body width). Dorsal cirri fusiform, with pointed tip, not alternating in length. Body with no more than 40 segments. Jointed chaetae with a long, finely toothed blade (*sim.* Fig. 6.16E) ***Odontosyllis gibba***

18. Antennae, tentacles, and anterior dorsal cirri slightly annulated. Each segment has a transverse granulated band dorsally. Hook of jointed chaetae single ***Odontosyllis ctenostoma***

Antennae, tentacles, and dorsal cirri smooth. No obvious body markings. Hook of jointed chaetae double ***Odontosyllis fulgurans***

19. Antennae and anterior cirri annulated; front of pharynx bears a small, dark, toothed chitinous ring; jointed chaetae with a short blade (Fig. 6.16D) ***Eusyllis blomstrandi***

Antennae, tentacles, and cirri smooth; pharynx lacks a dark, toothed ring; some jointed chaetae with a long, narrow blade (Fig. 6.16E) ***Pionosyllis lamelligera***

20. Body flattened; pharynx with a single large tooth and a circlet of 10–12 small teeth (Fig. 6.16H) **21**

Body slender; pharynx with a single large tooth and sometimes also a circle of soft papillae (Fig. 6.16B) **22**

21. Body of medium length (8–12 mm); antennae of equal length; dorsal cirri uniformly short (less than body width) ***Trypanosyllis coeliaca***

Body long (30–60 mm); median antenna longer than lateral ones; dorsal cirri at least as long as body width, alternating with longer ones ***Trypanosyllis zebra***

22. Jointed chaetae present **23**

Jointed chaetae lacking ***Haplosyllis spongicola***

23. All jointed chaetae have a short, stout blade **24**

Many jointed chaetae have a very long, thin blade ***Langerhansia cornuta***

24. In the mid-region and posteriorly, stout chaetae present with a bifid end resembling a bird's open beak ***Syllis gracilis***

Stout chaetae with a simple blunt or pointed end (rarely, only delicately bifid) **25**

25. Alternate dorsal cirri thicker at the tip, curling back over dorsal surface and about as long as body width; tooth at anterior end of pharynx ***Typosyllis krohni***

Dorsal cirri narrowing towards tip **26**

26. Dorsal cirri not markedly alternating, all shorter than body width; tooth near anterior end of pharynx ***Typosyllis armillaris***

Dorsal cirri alternating in length, all longer than body width (Fig. 6.16A); tooth inserted about one-third of length of pharynx from its anterior end ***Typosyllis prolifera***

Amblyosyllis formosa (Claparède) Figs 6.16, 6.17
Body 10–15 mm long, with 16 trapezoidal segments (6.17R). Two long nuchal organs project behind head like ciliated ear-flaps (6.16J). Median antenna longer than lateral ones. Palps folded against ventral surface, well separated but scarcely visible from above. Tentacles long, in unequal pairs. Pharynx very long and slender, coiled several times when retracted, crowned with a circle of six or seven two- or three-cusped teeth. Dorsal cirri very long, usually rolled in a loose spiral and annulated, as are the antennae. Usually a creamy white, with two or more brown/violet stripes across back of each segment. In mature individuals, eggs green or brown; sperm white.

On the lower shore on green algae; also in shallow sublittoral amongst old shells. Around most of the British coast, Europe to the Mediterranean; Japan.

Autolytus aurantiacus (Claparède) Fig. 6.17
Body 8–10 mm long; 60–100 segments. Median antenna longer than lateral ones (as C). Fused palps clearly visible, extending beyond head. Pharynx long and slender, coiled in one complete turn when retracted, crowned with 16–20 almost equal-sized teeth (G). Dorsal tentacles about as long as lateral antennae, ventral ones shorter. Dorsal cirri of chaetiger 1 at least as long as median antenna, of chaetiger 2 about one-third or one-quarter this length, thereafter much shorter (less than body width). Colourless or yellowish, often with a red spot below each parapod. Appendages of head and anterior segments orange-tipped. Eggs reddish-purple.

On lower shore and in shallow sublittoral, under stones and amongst algae. South and west coasts of Britain, Europe to the Mediterranean.

Autolytus longeferiens de Saint-Joseph Fig. 6.17
As *A. aurantiacus* but 10–20 mm long, with 50–90 segments. Pharynx as long as body when extended, coiled in numerous turns when retracted, crowned with 10 large blunt teeth, each separated by two or three very pointed small ones (H). Dorsal cirri of chaetiger 1 as long as lateral antennae, following ones alternating long and short (longest not attaining length of first). Orange; on anterior segments, three longitudinal red bands. Eggs greyish.

In kelp holdfasts. South-west Britain and Channel coasts.

Autolytus prolifer (O. F. Müller) Fig. 6.17
As *A. aurantiacus* but 5–15 mm long; about 60 segments. Median antenna scarcely longer than lateral ones (A). Fused palps hardly extend beyond head.

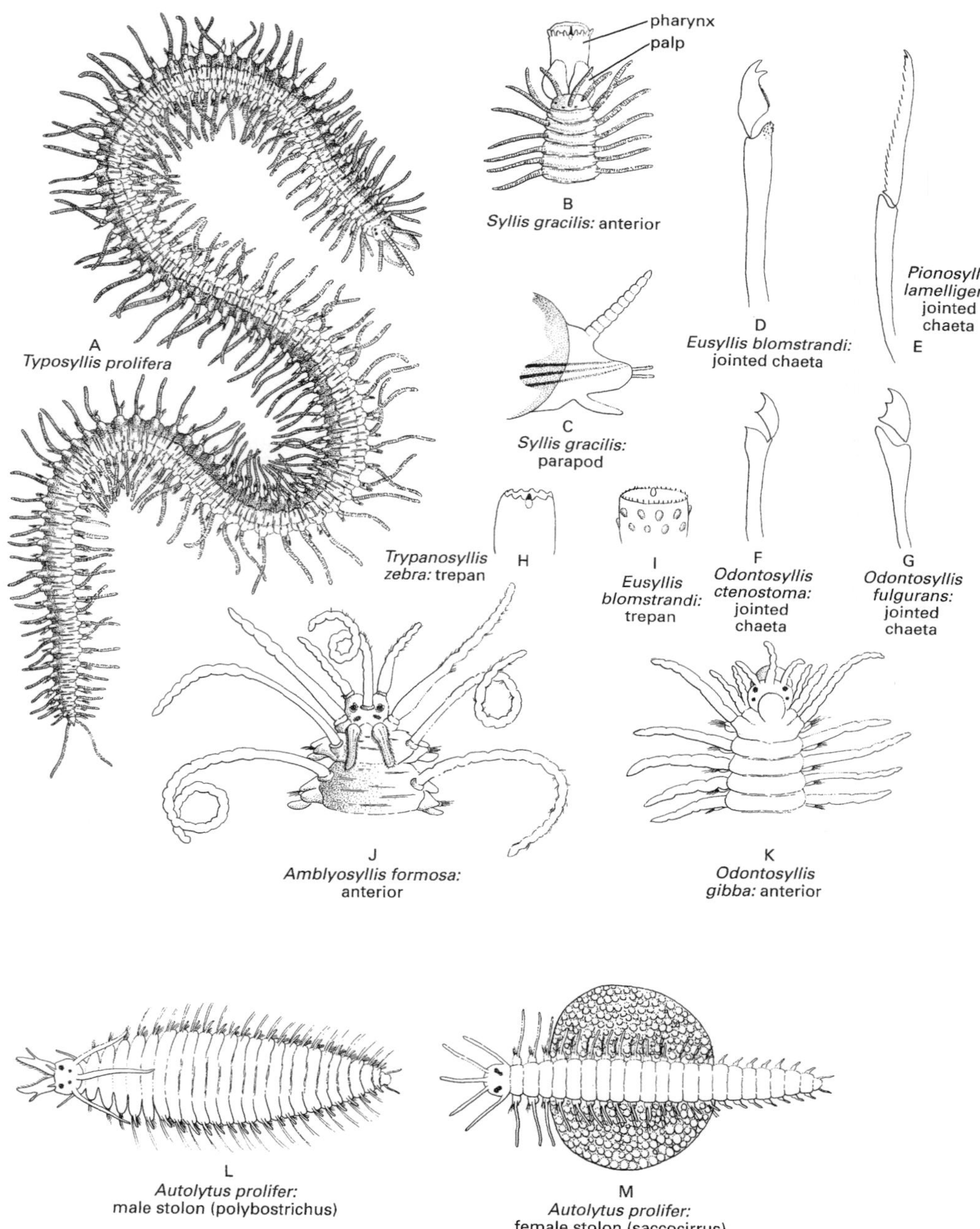

Fig. 6.16

Retracted pharynx curled in an 'S' shape, crowned with about 10 large, unequal teeth (F). Dorsal cirri of first segment much longer than antennae, subsequent ones (A,B) shorter (all about equal to body width). Whitish or yellowish, with numerous rounded flecks, colourless or orange.

In kelp holdfasts and amongst stones or old shells in the shallow sublittoral. Around most of

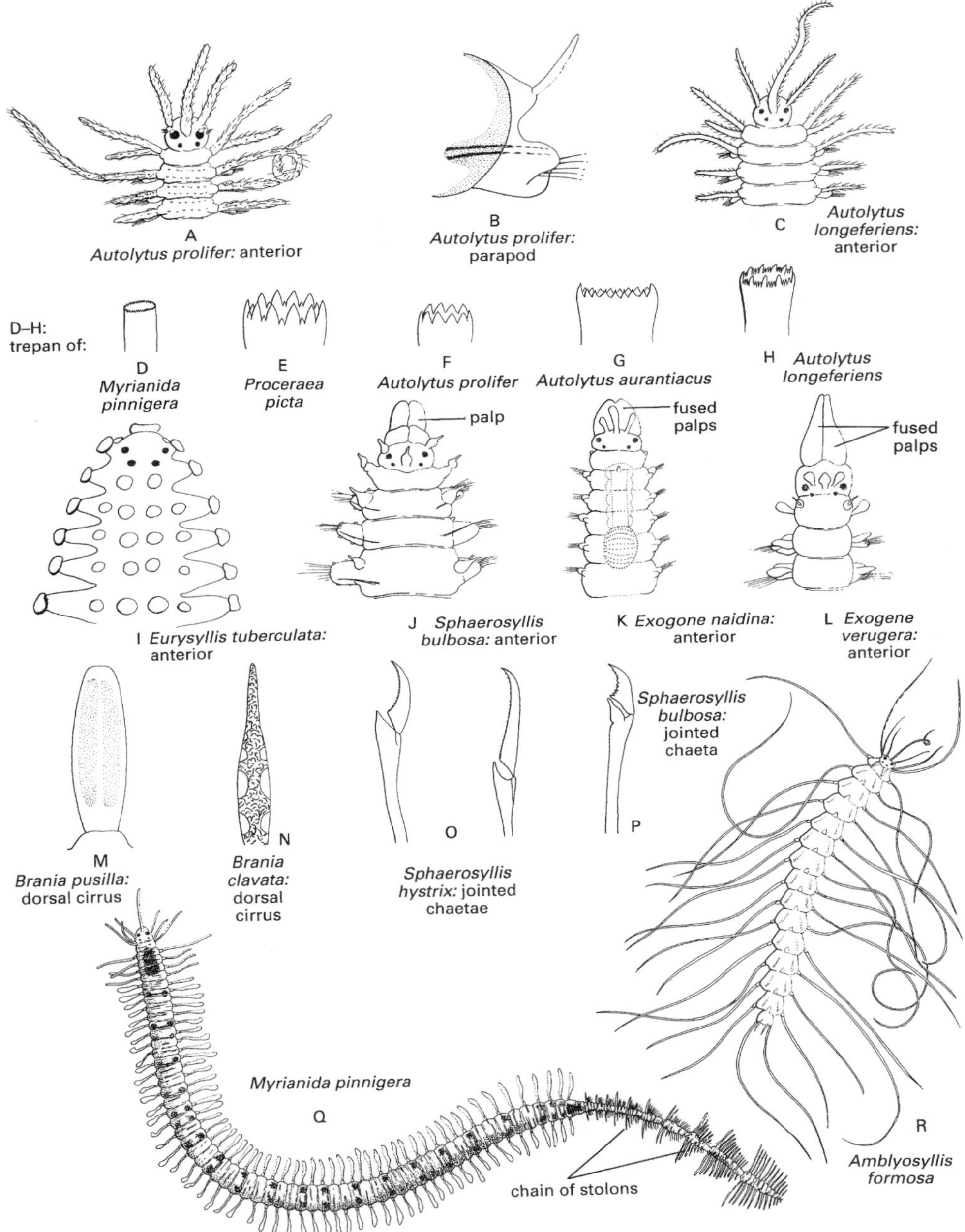

Fig. 6.17

the British coast, especially in the south and west; North Sea, north-east Atlantic, Mediterranean

Brania pusilla (Dujardin) Fig. 6.17
Body 1–2.5 mm long; 28–35 segments. Antennae small, swollen at base Palps fused, rather hidden. Pharynx with a single tooth anteriorly. Dorsal cirri spindle-shaped, with a square-cut end, containing two elongated yellowish bodies (M). Jointed chaetae with blades terminating in a single tooth, some long

and some short. At maturity, the female develops long swimming chaetae and carries eggs or embryos ventrally, two per segment. Colourless, with a brown pharynx; eggs pink.

In kelp holdfasts, on smaller algae and amongst shells or serpulid tubes on the lower shore and in the shallow sublittoral. Around most of the British coast, Europe to the Mediterranean.

Brania clavata (Claparède) Fig. 6.17
As *B. pusilla* but 2–3 mm long. Antennae slightly swollen at base, extending well beyond palps, which project beyond head. Pharynx with a single tooth one-third of the distance back. Dorsal cirri elongated, fusiform (N). Jointed chaetae all with a long blade terminating in two teeth. At maturity, the female carries eggs in a transparent membrane dorsally, 3–5 per segment. Colourless, with a transverse brown band across several of the segments; mature males pale orange.

Amongst small algae and in kelp holdfasts on the lower shore and in the shallow sublittoral. South-west Britain, Europe to the Mediterranean.

Langerhansia cornuta (Rathke)
Body 10–15 mm long. Antennae annulated, subequal, extending well beyond the well-developed and widely separated palps. Pharynx very long when extended, with a single tooth inserted anteriorly. Dorsal cirri annulated, alternately long and of medium length. Blade of jointed chaetae very long, or short and stoutly double-hooked—both sorts occur together in each parapod. Yellowish white or colourless.

At low water and in the shallow sublittoral, amongst serpulid tubes, bryozoans and algae; very often in empty shells, commensal with hermit crabs or sipunculids. Around most of the British coast, Arctic, north and south Atlantic, Mediterranean; Indian Ocean.

Eurysyllis tuberculata Ehlers Fig. 6.17
Body 3–5 mm long: about 65 segments. Antennae globular, around anterior edge of head. Palps fused, visible only from ventrally. Tentacles globular; tentacular segment also bears two dorsal tubercles (I). Pharynx surrounded by six large papillae anteriorly; at its opening, a circle of 10 small teeth of equal size, together with a single large tooth. Four spherical tubercles distributed across dorsal surface of each segment; dorsal cirri are also spherical (I).

At low water and in the shallow sublittoral, in kelp holdfasts, often encrusted with mud. South and west coasts of Britain, Europe to the Mediterranean.

Eusyllis blomstrandi Malmgren Fig. 6.16
Body 6–120 mm long; about 50 segments. Median antenna longer than lateral ones. Palps long, oval, and well separated. Pharynx surrounded by soft papillae anteriorly and crowned by a dark, finely toothed chitinous ring; also a single, large, conical tooth (I). Nuchal fold across posterior edge of prostomium. Upper pair of tentacles much longer than lower. First dorsal cirri longer than width of body, posterior ones shorter; antennae and anterior cirri with fairly distinct annulations, posterior cirri smooth. Orange or yellowish; tips of cirri brownish.

Amongst bryozoans, hydroids, and algae, often in a tube of mucus. Northern and western coasts of Britain, Arctic, north-east Atlantic, Mediterranean.

Exogone naidina (Oersted) Fig. 6.17
Body 2–4 mm long; 24–33 segments. Antennae, tentacles, and dorsal cirri very short, club-shaped. Palps fused in a triangle, slightly indented in front (K). Pharynx narrow, with papillae anteriorly and a single conical tooth. Dorsal cirri missing from chaetiger 2. Colourless or yellowish, eggs orange-red.

At low water and in the shallow sublittoral amongst algae, bryozoans, and ascidians. South and west coasts of Britain, Arctic, north and tropical Atlantic, Mediterranean, north Pacific.

Exogone hebes (Webster and Benedict)
As *E. naidina* but 8–10 mm long; 30–44 segments. Palps well developed, fused into a rigid, pointed snout. Creamy white.

At low water and in the shallow sublittoral, on sand or shell gravel and in sea-grass beds. On most British coasts, southern North Sea, north-east Atlantic.

Exogone verugera (Claparède) Fig. 6.17
As *E. naidina* but 3–8 mm long; 35–45 segments. Palps fused into a cone (L). Colourless.

At low water and in the shallow sublittoral. Along most of the British coast, southern North Sea, north-east Atlantic, Mediterranean; Pacific.

Haplosyllis spongicola (Grube)
Body 20–50 mm long. Antennae annulated, fairly short. Palps large, close together at base but not fused. Pharynx long when extended, with 10 soft papillae and a single terminal tooth. Dorsal cirri annulated, alternately long and of medium length. No jointed chaetae. Orange or yellowish; on each side of last few segments, a large purplish blotch ventrally. Eggs violet, mature males purplish-red.

At low water and in the shallow sublittoral, especially amongst sponges. Around most of the British coast; cosmopolitan.

Myrianida pinnigera (Montagu) Fig. 6.17
Body 15–30 mm long; 66 segments (Q). Median antenna long and flattened, lateral ones about half this length. Palps fused, folded back ventrally and extending a little beyond the head. Pharynx long and sinuous, making one or two complete turns when retracted, crowned by 50–60 little teeth of equal size (D). Dorsal tentacles flattened, longer than the wedge-shaped ventral ones. Dorsal cirri also flattened, longer than width of body. Jointed chaetae with a small blade having two teeth. Stolons may form a long chain of 15–30 individuals (Q). White with large yellow, orange, or red blotches.

South-western coasts of Britain, Europe to the Mediterranean.

Odontosyllis ctenostoma Claparède Fig. 6.16
Body thick, 10–20 mm long; 40–100 segments. Median antenna slightly longer than lateral ones. Palps broad but short, folded back ventrally, separate to the base. Pharynx long, crowned with six large teeth and with two lateral folds resembling ears. Tentacular segment has an occipital flap extending forwards over prostomium dorsally. Dorsal cirri thick, blunt-ended, alternately long and short, not annulated but often lightly grooved transversely. Jointed chaetae with a short, single-hooked blade (F). Green or greenish-yellow, with a speckled grey band across each segment; bioluminescent.

At low water and in the shallow sublittoral, amongst algae. Around most of the British coast. North-east and tropical Atlantic, Mediterranean.

Odontosyllis fulgurans (Audouin and Milne Edwards) Fig. 6.16
As *O. ctenostoma* but body slim, 10–40 mm long; 35–100 segments. Median antenna longer than lateral ones. Pharynx very short. Occipital flap not well developed. Dorsal cirri smooth. Blade of jointed chaetae short, with two small hooks (G). Pale yellow, orange or reddish; eggs violet. Very bright green luminescence.

At low water and in the shallow sublittoral, amongst algae. South and west coasts of Britain, Europe and the Mediterranean.

Odontosyllis gibba Claparède Fig. 6.16
As *O. ctenostoma* but 5–25 mm long; 40 segments. Antennae almost equal in length. Palps fairly prominent. Pharynx short. Occipital flap prominent (K). Dorsal cirri fusiform, often weakly annulated. Jointed chaetae with a long, finely toothed blade. Opaque white, blotched with brown and violet.

At low water and in the shallow sublittoral, under stones and in kelp holdfasts. West coast of Britain, Europe to the Mediterranean.

Pionosyllis lamelligera de Saint-Joseph Fig. 6.16
Body 6–7 mm long; 50–60 segments. Median antenna longer than lateral ones. Palps long, divergent although fused at base. Pharynx with 10 soft papillae and a single tooth anteriorly. Dorsal cirri of chaetiger 1 much longer than following ones, which are shorter than width of body. Antennae and cirri smooth. Some jointed chaetae have a long toothed blade with two small terminal teeth (E), others a very short bidentate hook. Colourless, sometimes with three purple-black bands across back of each segment (these persist in alcohol).

At low water, in tide-pools and also in the shallow sublittoral, in kelp holdfasts and amongst ascidians. Around most of the British coast, north-east Atlantic.

Proceraea picta Ehlers Fig. 6.17
Body 10–25 mm long; 60–100 segments. Median antenna a little stouter and longer than lateral ones. Palps not visible from above. Pharynx 'S'-shaped when retracted, crowned with 10 large pointed teeth alternating with 10 small ones (E). Dorsal cirri of chaetiger 1 larger than median antenna, rolling up spirally; those of chaetiger 2 much shorter, but still two or three times longer than following ones, which are of equal length (less than width of body). Jointed chaetae with a short, broad, bidentate blade. Ventral surface pale pink, dorsal surface with violet/brown markings divided by white lines

longitudinally and transversely. Antennae and tips of tentacles orange-brown.

At low water and in shallow sublittoral amongst algae and sponges or under stones. Around most of the British coast, north and tropical Atlantic, Mediterranean; Australasia.

Sphaerosyllis bulbosa Southern Fig. 6.17

Body 5–6 mm long; 48 segments. Antennae very short, subequal, with a very swollen base. Palps fused together, long and broad (J). Pharynx with a single tooth anteriorly. Dorsal cirri short, with a spherical base, then cylindrical, missing from chaetiger 2. Small papillae on parapods and anal segment. Jointed chaetae few, with a short, toothed blade (P). Fawn or creamy white.

In the intertidal and shallow sublittoral. South and west coasts of Britain, north-east Atlantic.

Sphaerosyllis hystrix Claparède Fig. 6.17

As *S. bulbosa* but 3–5 mm long, 30–40 segments; covered with little papillae. In each parapod from chaetiger 4 rearwards, a gland in the form of a rounded capsule filled with small rod-like particles. Jointed chaetae with a narrow finely toothed blade ending in a single large tooth, some long and others short (O). Colourless or greyish.

On algae or amongst shells on muddy shores, sometimes encrusted with sand or mud. Around most of the British coast, Europe to the Mediterranean.

Syllis gracilis Grube Fig. 6.16

Body slim, 20–50 mm long. Antennae short, annulated; palps separate and stout. Pharynx long and slender with an irregular rim, bearing a single tooth anteriorly (B). Dorsal cirri annulated, alternately short and of medium length. Ventral cirri conical (C). In anterior region, jointed chaetae have a broad blade with a finely toothed edge and long terminal tooth—they become shorter more posteriorly. In mid-region, thick simple chaetae with a bifid end resembling a bird's open beak also occur. Pale yellow-brown, occasionally with lines of brown spots dorsally across anterior segments.

On the lower shore and in the shallow sublittoral, amongst sedentary organisms or old shells and in crevices. Around most of the British coast; cosmopolitan.

Trypanosyllis coeliaca Claparède

Body flattened, 8–12 mm long, with 60–90 segments. Antennae annulated, short, subequal. Palps rounded, divergent. Pharynx surrounded by 10 soft papillae and crowned by a trepan of 10 inward-pointing teeth, also with a single large conical tooth (as Fig. 6.16H). Dorsal cirri short, with a few annulations, all of similar length. Jointed chaetae with a bifid blade. Yellow; dorsal cirri full of yellow/green corpuscles.

On the lower shore, amongst old shells or in kelp holdfasts. South and west coasts of Britain, Europe to the Mediterranean.

Trypanosyllis zebra (Grube) Fig. 6.16

As *T. coeliaca* but 30–60 mm long. Antennae annulated, median one longer than laterals. Palps elongated, well separated. Dorsal cirri annulated, alternately long and of medium length. Whitish anteriorly, with violet stripes across each segment; posteriorly, more yellowish. Cirri white or violet.

In the shallow sublittoral amongst old shells or in kelp holdfasts. South-western coasts of Britain, north-west Atlantic, Mediterranean; Indian Ocean, Japan.

Typosyllis armillaris (Müller)

Body 25–50 mm long, with many segments. Antennae annulated, median one longer than laterals, extending well beyond palps, which are oval, separate, but very close together. Pharynx with 10 soft papillae and a single large tooth inserted well forwards. Dorsal cirri annulated, short (no more than body width). Jointed chaetae with a stout blade; those in mid-region have only a single terminal tooth but, towards front and rear ends of the worm, it is more or less clearly bifid. May bear poorly developed stolons. Yellowish, occasionally with pink bands, or pinkish overall. Eggs violet, mature males bright pink.

On the lower shore and sublittorally, under stones or in kelp holdfasts. Around most of the British coast; cosmopolitan.

Typosyllis krohni (Ehlers)

As *T. armillaris* but 15–30 mm long. Dorsal cirri alternately short and long (long ones also thicker, with clubbed ends, tending to turn inwards across the back). Anterior segments marked with brown/purple bands; cirri with scattered white spots. Eggs and sperm pink.

On the lower shore and sublittorally, amongst algae and under stones. South and west coasts of Britain, northern seas, Europe to the Mediterranean.

Typosyllis prolifera (Krohn) Fig. 6.16

As *T. armillaris* but 10–25 mm long (A). Palps well separated. Pharynx with a single large tooth about one-third of the way back. Dorsal cirri alternating short and long. Colour very variable: greyish, brownish, or reddish, sometimes with pink or orange markings anteriorly. Antennae and cirri with dark spots. Eggs and sperm violet/brown.

On the lower shore amongst algae, also in the shallow sublittoral and on the bottom offshore. South and west coasts of Britain, Europe to the Mediterranean; ? Indo-Pacific.

25. TEREBELLIDAE

Body with many segments. Divided into an inflated anterior region in which most segments have both notopods (with bundles of simple bristles) and neuropods (with uncini set in one or more rows down swollen ridges), and usually a series of glandular cushions along its ventral surface, and a more slender posterior region which more often has only uncini in neuropods. Head-end truncated, surrounded by various lobes, with many long, filiform, mobile tentacles which can be shortened but not withdrawn. Posterior to these are normally one to three pairs of gills whose usual form resembles a much-branched bush. Uncini resemble birds' heads; the teeth are small and hard to see. These worms are common on the lower shore and in the shallow sublittoral, occupying a mucus-lined gallery or fragile tube loosely encrusted with detritus under a stone or in a large algal holdfast (occasionally a more permanent tube of cemented sand grains and shell fragments); usually only their tentacles can be seen.

A large family with 48 species recorded from the British sea area, representing no fewer than 28 genera. Some authorities recognize a family Trichobranchidae, for *Octobranchus*, *Terebellides*, and *Trichobranchus*.

---

1. Fifteen to 25 (in most, 17) anterior segments with bundles of bristles dorsally—the remaining segments (more than half body length) with only uncini inserted on ridges or lappets ventrally. Gills variable **2**

   At least 30 anterior segments have bristles dorsally as well as uncini ventrally, extending for at least half body length. Gills, if present, form two or three rows of numerous unbranched filaments **11**

---

2. Gills bush-like, with a main stem repeatedly branching into finer filaments (Fig. 6.18B), inserted in pairs on two or three segments **3**

   Gills either cirriform and unbranched or bear a series of lamellae (Fig. 6.18E, F) **13**

---

3. Two pairs of gills **4**

   Three pairs of gills **7**

---

4. Stem of gills long, filaments forming a rounded tuft at end (Fig. 6.18D); segments bearing gills also have a small flap on each side where, subsequently, the neuropod is located ***Pista cristata***

   Gills branch from most of length of stem; no lateral flaps on the segments bearing gills **5**

---

5. Neuropods are low swollen ridges throughout body. Anus surrounded by small papillae. Nephridial papillae (on some anterior segments between notopod and neuropod) not very prominent. Bristles have a slightly sickle-shaped end, finely toothed along inside of curve. Posterior half of body rolls into a tight spiral ***Amphitritides gracilis***

   Neuropods become progressively elongated into projecting lappets posteriorly. Rim of anus smooth. Nephridial papillae on each side of segments 6 and 7 may be prominent long tubes (Fig. 6.18C). Bristles have a simple tip like a narrow spear-head **6**

---

6. Seventeen anterior segments with notopodial bristles. Gills large, with a long main stem ***Nicolea venustula***

Fifteen anterior segments with notopodial bristles. Gills small, main trunk very short ***Nicolea zostericola***

---

7. Seventeen anterior segments with notopodial bristles. Neuropods become progressively elongated into projecting lappets posteriorly **8**

Twenty-four to 25 anterior segments with notopodial bristles. Neuropods are short swollen ridges throughout body ***Neoamphitrite figulus***

---

8. Two large triangular lobes (like a wing collar) partly enclose head end. In last 10 segments of anterior region, uncini are inserted in a double row back-to-back (with teeth pointing away from each other) (Fig. 6.19G). Tube strongly constructed from sand grains and shell fragments, with a fringe at its mouth (Fig. 6.19E) ***Lanice conchilega***

No large buccal lobes. Uncini of last 10 anterior neuropods form either a double row face-to-face or a single row in which the teeth face in alternate directions. Tube, if present, is fragile and lacks a fringe **9**

---

9. Large (up to 300 mm long, 20 mm wide) with 11–13 glandular cushions along ventral surface and nine small nephridial papillae along each side (between notopods and neuropods). The bristles have a slightly sickle-shaped end, finely toothed along inside of curve (*sim.* Fig. 6.19B) ***Amphitrite edwardsi***

Medium-sized (not exceeding 150 mm), with 14–15 glandular cushions and six pairs of nephridial papillae. Bristles with a simple tip like a narrow spear-head **10**

---

10. Up to 150 mm long, 8 mm wide, body soft and fragile. Pink, orange or brown coloration, spotted all over with white. Beak of anterior uncini surmounted by two large teeth, then 1–5 smaller ones (Fig. 6.19J) ***Eupolymnia nebulosa***

Up to 60 mm long, 3 mm wide, body more rigid and quite strong. Coloration uniform, without white spots. Beak of anterior uncini surmounted by a single large tooth, then three smaller ones (Fig. 6.19I) ***Eupolymnia nesidensis***

---

11. Gills absent. Some tentacles thread-like, others thicker and club-shaped; head partly surrounded by a prominent pleated collar. Integument smooth and translucent ***Polycirrus caliendrum***

Numerous gill filaments present. Tentacles of more or less uniform size and shape; cephalic lobe may be pleated but is ill-developed. Integument opaque and warty, especially after preservation **12**

---

12. Two rows of gill filaments across dorsal surface, inserted on each side of segments 2 and 3 but not meeting in the middle ***Thelepus cincinnatus***

Three rows of gill filaments across dorsal surface, inserted on each side of segments 2–4, the first meeting in the middle ***Thelepus setosus***

---

13. A single gill, made up of a short, stout main stem with four branches, each bearing a series of kidney-shaped lamellae (Fig. 6.18E) ***Terebellides stroemi***

Six long cirriform gills arranged in pairs (Fig. 6.18F) ***Trichobranchus glacialis***

Amphitrite edwardsi (Quatrefages) Fig. 6.19
Body up to 250–300 mm long, with 100–150 segments; 17 chaetigers in anterior region. Three pairs of many-branched gills. Body segments 2–4 bear

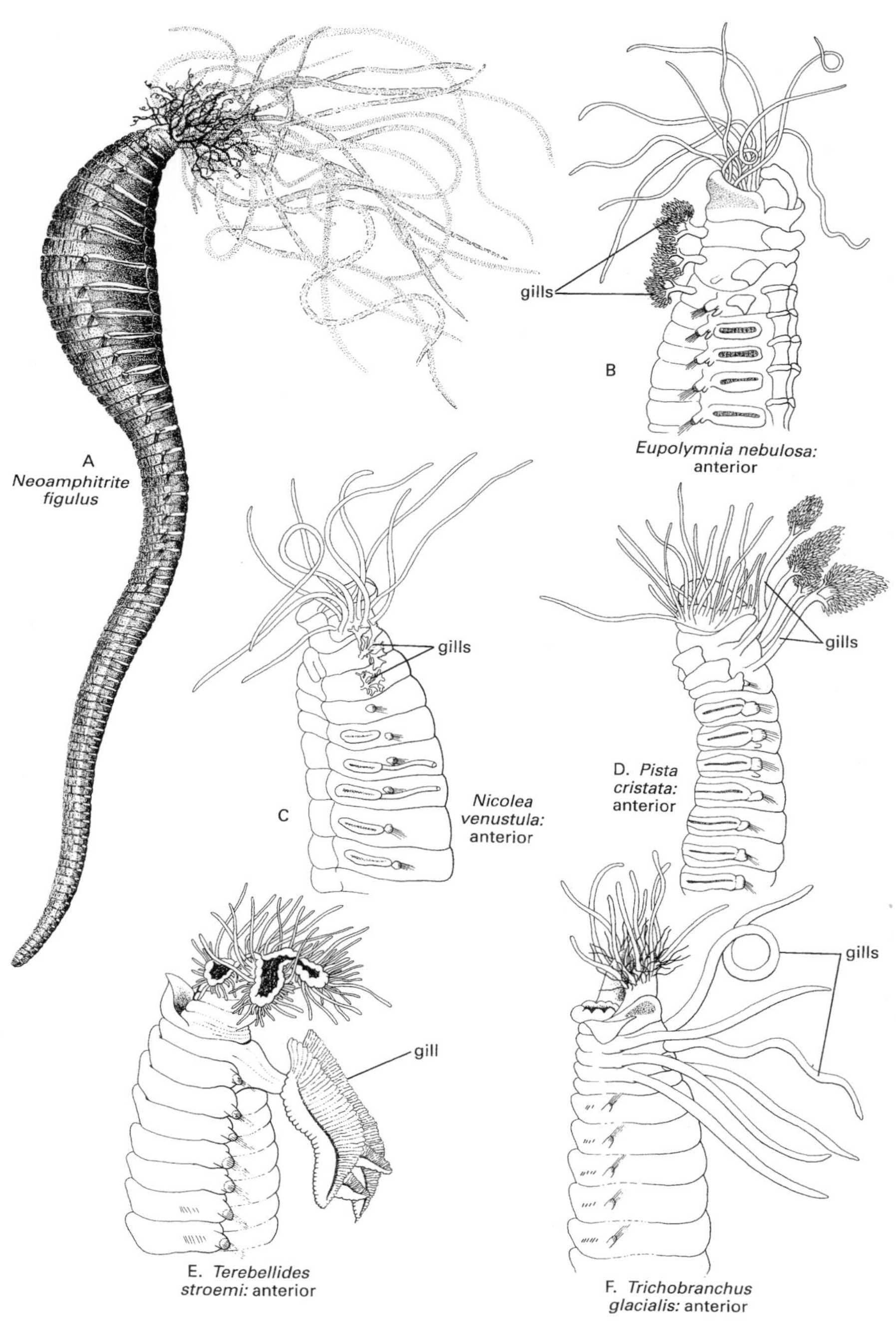
A
Neoamphitrite figulus
gills
B
Eupolymnia nebulosa: anterior
gills
C
Nicolea venustula: anterior
gills
D. *Pista cristata:* anterior
gill
E. *Terebellides stroemi:* anterior
gills
F. *Trichobranchus glacialis:* anterior

Fig. 6.18

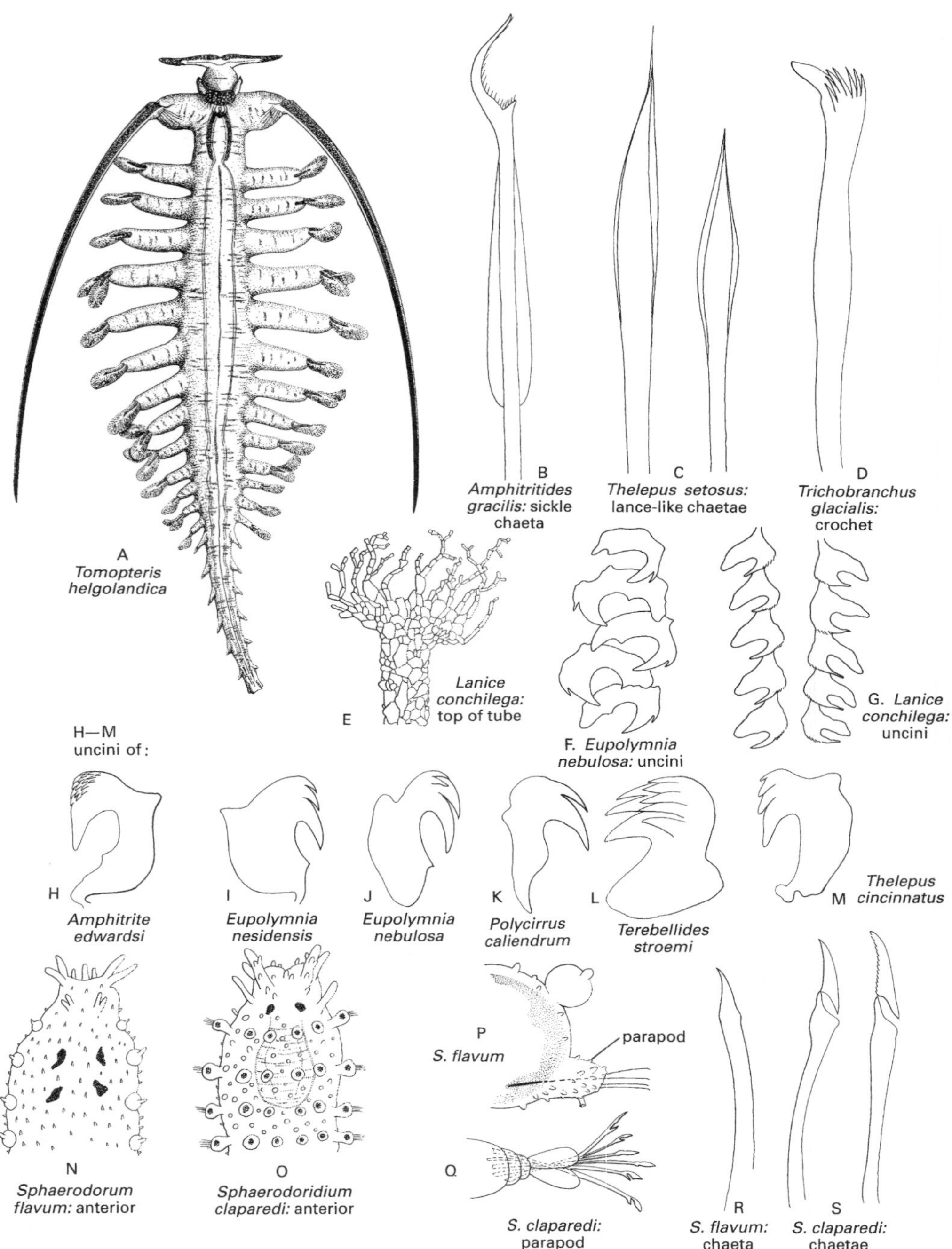
A
Tomopteris helgolandica
B
Amphitritides gracilis: sickle chaeta
C
Thelepus setosus: lance-like chaetae
D
Trichobranchus glacialis: crochet
E
Lanice conchilega: top of tube
F. Eupolymnia nebulosa: uncini
G. Lanice conchilega: uncini
H—M uncini of:
H
Amphitrite edwardsi
I
Eupolymnia nesidensis
J
Eupolymnia nebulosa
K
Polycirrus caliendrum
L
Terebellides stroemi
M
Thelepus cincinnatus
N
Sphaerodorum flavum: anterior
O
Sphaerodoridium claparedi: anterior
P
S. flavum
parapod
Q
S. claparedi: parapod
R
S. flavum: chaeta
S
S. claparedi: chaetae

Fig. 6.19

lateral flaps; 11–13 glandular cushions ventrally, nine pairs of nephridial papillae laterally on anterior segments. Posteriorly, neuropods become projecting lappets. Anus with a toothed rim. Bristles with a sickle-shaped, toothed end. Beak of uncini surmounted by six rows of smaller teeth (H). Salmon-pink or fawn overall, with dark red gills and orange-yellow tentacles.

Tube a cylindrical gallery, dug down into muddy sand (often amongst sea grasses) at very low water. South and west coasts of Britain.

Neoamphitrite figulus (Dalyell) Fig. 6.18
As *A. edwardsi* but up to 150–250 mm long, 90–100 segments; 24 (rarely 25) chaetigers in anterior region (A). Lateral flaps only rudimentary on body segment 4; 13–14 glandular cushions, 17–18 pairs of nephridial papillae. Neuropods not becoming lappets posteriorly. Beak of uncini surmounted by five large and then three rows of smaller teeth. Yellowish fawn or brown, with red gills.

Tube a simple gallery in muddy sand or the mud filling rock-crevices at low water; also amongst kelp holdfasts. Around much of Britain, north-west Europe, Mediterranean; north-west Atlantic, north-west Pacific.

Amphitritides gracilis (Grube) Fig. 6.19
Body 60–120 mm long, with 100–200 segments, very soft; posterior part tends to roll up tightly like a corkscrew; 17–19 chaetigers in anterior region. Two pairs of many-branched gills. No lateral flaps in gill region; 11–13 glandular cushions ventrally, nine pairs of nephridial papillae laterally on anterior segments. Neuropods not developed into lappets posteriorly. Small papillae surround the anus. Bristles with a sickle-shaped, toothed end (B). Uncini very small. Pale red or yellowish-grey, with bright red gills and paler tentacles.

Makes a sinuous gallery in mud beneath stones or filling rock crevices at low water. South and west coasts of Britain, southern North Sea, north-east Atlantic, Mediterranean, Black Sea.

Eupolymnia nebulosa
(Montagu) Figs 6.18, 6.19
Body very rounded, soft and fragile, 50–150 mm long. About 100 segments, 17 chaetigers in anterior region. Cephalic lobe forms a collar bearing very many little dark eyespots (6.18B). Three pairs of many-branched gills. Lateral flaps well developed on body segments 2 and 3, smaller on 4. Four to 15 glandular cushions ventrally, six pairs of nephridial papillae laterally on anterior segments (6.18B). Posteriorly, neuropods become projecting lappets. Beak of uncini surmounted by two large teeth and then 1–5 small ones (6.19F,J). Orange-grey, pink or pale brown, spotted all over with white. Gills bright red, often also with white spots; tentacles pink or whitish with chalky rings.

Sandy tube without a fringe, attached to the underside of stones and old shells at low water and sublittorally. Around most of Britain, north-west Europe, Mediterranean; Red Sea, Arabian Gulf, Indian Ocean, Pacific.

Eupolymnia nesidensis (Delle Chiaje) Fig. 6.19
As *E. nebulosa* but 30–60 mm long, firm and fairly strong; 50–90 segments. Fifteen glandular cushions ventrally. Beak of uncini surmounted by a single large tooth, then three smaller ones (I). Uniformly dark brown, reddish, dark orange, yellowish-green, or olive, with red gills and yellow tentacles; not spotted with white.

Tube encrusted with sand but fragile, entangled amongst algae or hydroids at low water or sublittorally. Around most of Britain, northern seas, north-west Europe, Mediterranean; north Pacific.

Lanice conchilega (Pallas)
The sand mason Fig. 6.19
Body with 150–300 segments, up to 250–300 mm long, posterior region slender, fragile and soft; 17 chaetigers in anterior region. Two erect triangular lobes, joined ventrally by a collar, surround base of tentacles. Three many-branched gills; a pair of otocysts and two foliaceous lobes occur in gill region; 14–20 glandular cushions, prominent but not very distinct from each other, ventrally; five pairs of nephridial papillae laterally on anterior segments. Posteriorly, neuropods become projecting lappets. In the more rearward anterior segments, uncini are inserted in two rows with teeth facing away from each other (G). Pink, yellowish, or brownish; when breeding, whitish or greenish. Glandular cushions dark red posteriorly. Gills blood red, tentacles white, prominent chalky band appears down each side.

Tube very characteristic, made from cemented sand grains and shell fragments with a ragged fringe at the mouth end (E). At low water and sublittorally to well offshore, may be very numerous on sandy beaches; also occurs in mud or mud-filled rock-crevices but appears favoured by moderate exposure to wave action. European shores from the Arctic to the Mediterranean; Arabian Gulf, Pacific.

Nicolea venustula (Montagu) Fig. 6.18
Body 30–60 mm long, with 50–70 segments; 17 chaetigers in anterior region. Tentacles of varied size (C). Two pairs of gills with a long main stem. No lateral flaps in gill region. Thirteen to 17 glandular cushions ventrally, nephridial papillae laterally very small on segment 3 but, in adult males, forming two long tubes extending from segments 6 and 7 (C). Posteriorly, neuropods become projecting fins. Brick red, dorsal surface marked with little spots; gills bright red, tentacles violet.

Tube thin and transparent, encrusted with sand grains and algae; amongst algae, hydroids and old shells at low water and below tide marks. European shores from the northern North Sea to the Mediterranean, circum-Arctic; Red Sea, southern Africa.

Nicolea zostericola (Oersted)
As *N. venustula* but 15–20 mm long (although reaching 65 mm in Arctic waters) with 40–50 segments; 15 chaetigers in anterior region. Gills with a very short main stem. Uniform pink or fawn colour, with no white spots; gut bright red, visible by transparency.

At low water amongst red algae, hydroids or sea grasses to which the slender sandy tube is attached. Cosmopolitan in northern seas.

Pista cristata (O. F. Müller) Fig. 6.18
Body 39–90 mm long; 70–100 segments, 17 chaetigers in anterior region. Two pairs of gills with a long main stem and small branches forming a rounded tuft at the end (D). One gill is always noticeably larger than the other three. Lateral flaps present on segments 2 to 4; 17–20 glandular cushions ventrally, two pairs of nephridial papillae laterally on anterior segments (D). Posteriorly, neuropods become projecting fins. More or less uniform dark red, with brownish gills.

Tube membranous, covered with mud, shell fragments, and algae, in fine sandy or muddy bottoms at low water and sublittorally. Cosmopolitan in northern seas, also South Atlantic and Antarctic.

Polycirrus caliendrum Claparède Fig. 6.19
Body 30–100 mm long; with 70–90 segments, of which 30–60 (sometimes up to 75) bear dorsal bristles; readily autotomizes, particularly around body segments 8 to 10, sealing the break so well that it does not appear to have been mutilated. Tentacles long and numerous, some thread-like, others thicker and club-shaped, surrounded at their base by a wide pleated collar. No gills. Eight to 11 paired glandular cushions ventrally. Uncini have many small sharp teeth above the beak in the anterior region, only one tooth in addition to the beak posteriorly (K). Yellowish-orange, eggs darker; luminesces blue or violet.

Temporary mucous tube, occurring amongst algae, hydroids, serpulid tubes and old shells sublittorally, also in kelp holdfasts at low water. Western coasts of Britain and Europe, Mediterranean; Black Sea, north-east Pacific.

Terebellides stroemi M. Sars Figs 6.18, 6.19
Body 30–60 mm long, firm; 50–60 segments, 18 chaetigers in anterior region. Dorsally a single gill, consisting of four branches bearing a series of kidney-shaped lamellae fused to a fat cylindrical main stem (6.18E). No glandular cushions; neuropods become projecting fins posteriorly. The most anterior uncini are more like crochets but, in the following 12 segments, acquire a large beak surmounted by numerous smaller teeth and posteriorly are of more usual design, with four to five teeth in a semicircle surmounting the beak (6.19L). Flesh-coloured; adult males pale, females more pink. Gills blood-red.

Tube membranous, covered with mud or sand. At low water and sublittoral to well offshore, on bottoms of sand or mud, often amongst sea grasses. Circum-Arctic, British and European shores to the Mediterranean; Australasia, Antarctic.

Thelepus cincinnatus (Fabricius) Fig. 6.19
Body 100–200 mm long; about 100 segments, the most anterior 30 or more bear dorsal bristles. Two rows of simple filamentous gills on each side of dorsal surface of body segments 2 and 3. Four pairs of nephridial papillae laterally on segments 4 to 7. Ventrally, glandular cushions very indistinct. Neuropods become projecting fins posteriorly.

Uncini with a beak surmounted by a pair of teeth, then one in the middle flanked by two smaller ones (M). Brownish, pink, or orange-yellow, paler on the underside. Gills bright red, darker at the tips; tentacles orange or pale pink, sometimes with red spots.

Tube membranous and sinuous, encrusted with loose particles, amongst old shells and algal holdfasts at low water and in the sublittoral to well offshore. Circum-Arctic, British and European shores to the Mediterranean; west Atlantic, Caribbean, Antarctic.

Thelepus setosus (Quatrefages) Fig. 6.19
As *T. cincinnatus* but with 80–120 segments, of which 30–60 anterior ones have dorsal bristles and the 20–50 most posterior bear only uncini. Scattered white warts, more or less prominent, occur posteriorly and become much more marked after preservation. First row of gills meet in the mid-line. Glandular cushions more or less distinct. Colour variable—fawn, yellowish-brown, or yellow, often with fine white spots posteriorly. Gills bright red, tentacles orange-brown.

At low water under stones or around sea grasses; sublittorally amongst serpulid tubes, old shells, and stones. South and west coasts of Britain, Channel and Atlantic coasts of Europe; Red Sea, Indian Ocean, Pacific.

Trichobranchus glacialis Malmgren Figs. 6.18, 6.19
Body 20–30 mm long, 60–70 segments; 15 chaetigers in anterior region. Six long, thick cirriform gills arranged in pairs on body segments 2 to 4; small lateral flaps on segments 2 and 3. No glandular cushions. Neuropods become projecting triangular fins posteriorly. They contain crochets with many teeth (6.19D), becoming very small posteriorly. Orange-red; gills bright red, tentacles violet, eggs yellow.

Tube membranous, covered with mud or fine sand, attached to algae sublittorally. Circum-Arctic, scattered records along European coasts to the Mediterranean; north-west Africa, sub-Antarctic.

26. TOMOPTERIDAE

Small (2–60 mm long), transparent pelagic worms. Body divided into three regions—head, trunk, and tail. In addition to normal anterior antennae (divergent, as 'frontal horns'), the head carries a pair of tentacles (each enclosing a very long aciculum) which curve backwards down each side of the body, reaching at least half its length (a smaller, more anterior pair is present only in juveniles). Eyes usually well developed, head also bears ciliated nuchal organs. Pharynx eversible, may be quite large but lacks teeth. Trunk usually slender or flattened, without obvious segmentation other than elongated parapods, which have paddle-like distal extensions but lack bristles. The tail may be short, bearing rudimentary parapods, or long, without prominent appendages. About a dozen species of the genus *Tomopteris* (*Enapteris* is probably a juvenile stage of *Tomopteris*) may be found in plankton hauls off the coasts of north-west Europe; their identity is ultimately established by the presence and arrangement of various groups of glands at the tip of the parapods.

They swim vigorously by rapid beating of the parapods, and may undergo diurnal vertical migrations, rising to the surface at night. They are voracious predators of other planktonic organisms, with a wide geographical distribution.

Tomopteris helgolandica Greeff Fig. 6.19
Body 12–17 mm long, with 18–21 pairs of parapods; a quarter of these are rudimentary, occurring in the tail region (A). Tentacles extend back for about two-thirds body length. Parapods as elongated cones with terminal fins in various shapes, usually rounded.

North Sea, Irish Sea, Western Approaches, north and south Atlantic, Mediterranean.

27. SABELLARIIDAE

Thick-walled tubes of cemented sand or shell gravel, sometimes aggregated and reef forming. Operculum of modified chaetae. Numerous tentacular filaments near mouth on ventral side. A pair of ciliated branchiae dorsally on most segments of thorax and anterior abdomen. Abdomen distinguished by lateral fleshy flaps bearing short saw-edged chaetae (uncini) and, ventral to these, tufts of fine chaetae. Post-abdomen lies folded forward in a ventral abdominal groove, with the anus facing anteriorly.

---

1. Opercular chaetae in middle row angular, with distal parts lying transverse to axis of body
Sabellaria alveolata

Middle row chaetae sickle-shaped, each with leaf-like distal part projecting anteriorly ***Sabellaria spinulosa***

Sabellaria alveolata (Linnaeus) Fig. 6.20
Length 30–40 mm; thorax with three pairs of flattened chaetal sheaths; inner and middle rows of opercular chaetae with asymmetrically angular spines pointing distally and transversely, respectively, outer row each with a distal part flattened, flexible, and with about five serrations across blunt distal margin. Tube made of cemented sand grains, often densely aggregated, their apertures forming a honeycomb pattern.

On lower shore and shallow sublittoral rocks adjacent to a sand table; southern species; both sides of Britain as far as Firth of Clyde and Berwick; western Ireland; locally abundant.

Sabellaria spinulosa Leuckhart Fig. 6.20
Length 20 to 30 mm; as *S. alveolata* except opercular chaetae in middle row point distally, and outer ones taper with several serrations each side. Others with less taper, with or without a median barb have been recorded; tubes similar to *S. alveolata*.

On rocks; sublittoral (occasionally LWS Menai Straits); boreal and warm temperate species; all coasts; locally abundant.

28. SABELLIDAE

Tubes thin-walled of mucoprotein with or without silt, sand, or fine shell, or thick-walled and gelatinous; no operculum; distal crown of pinnate radioles, some species with a few apinnate tentacular filaments ventrally; most species with a membranous collar on first segment; thorax with dorsal bundles of long and short chaetae (shorter ones sometimes characteristic), and glistening ventral rows (tori) of small, hooked chaetae (uncini), the latter absent from 1st segment; minute companion chaetae (each associated with a thoracic uncinus) in many genera. Positions of chaetae and uncini reversed in abdomen; faecal groove passes from ventral to dorsal surface via the right side just behind thorax, otherwise bilaterally symmetrical; glandular surfaces often restricted to ventral surface forming subquadrangular 'shields'; some genera incubate their embryos. Found with all types of substrata, e.g. rock, mud, sand, algal holdfasts, and encrusting fauna.

---

1. Abdominal tori short (e.g. less than a third the width of the abdomen); radioles with or without webbing; with or without collar; tube—thin mucoid membrane often covered by muco-silt layer **2**

   Abdominal tori forming nearly complete girdles in a groove around each abdominal segment; radioles webbed for most of their length; segment 1 without collar; tube gelatinous **25**

---

2. Abdomen longer than thorax, more than 12 segments **3**

   Abdomen shorter than thorax, with only 3–6 segments **20**

---

3. Most abdominal chaetae (all except far posterior ones) arranged in tight pencil-like tufts **4**

   Abdominal chaetae arranged in short transverse rows **8**

---

4. Abdominal chaetae, arranged in a spiral (best seen when tips are removed) e.g. Fig. 6.22, *Sabella* **5**

   Such chaetae arranged in a C-shape with additional chaetae partially filling the arc, e.g. Fig.6.20, *Bispira* **7**

---

5. Abdominal chaetae numerous, forming spiral of many whorls; tips of radioles short and blunt **6**

   Such chaetae fewer (spiral not more than two whorls); tips of radioles with a bulbous subterminal swelling ***Sabella discifera***

---

6. Segment 1 twice as long as following thoracic segments; lateral collar margins not reaching base of crown;

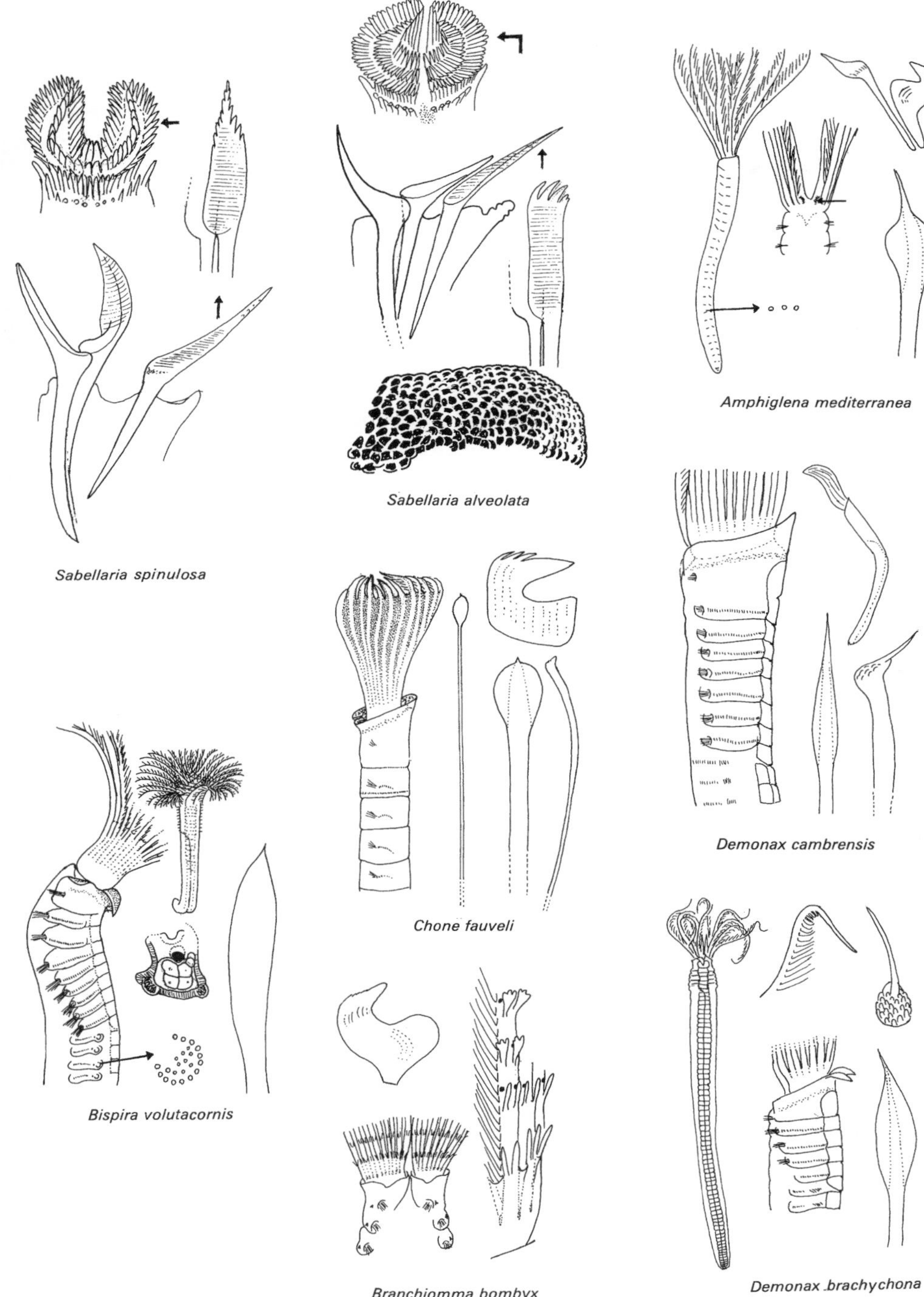
Sabellaria spinulosa
Sabellaria alveolata
Amphiglena mediterranea
Demonax cambrensis
Chone fauveli
Bispira volutacornis
Branchiomma bombyx
Demonax brachychona

Fig. 6.20

one side of crown sometimes larger than other side ***Sabella pavonina***

Segment 1 about as long as following thoracic segments; lateral collar margins extending forward to groove which marks base of crown; one side of crown usually larger, involuted to form 1–5 whorls ***Sabella spallanzanii***

---

7. Outer surface of each radiole bearing paired epithelial flaps (stylodes), interspersed with paired eyes ***Branchiomma bombyx***

   Outer surface smooth except for slightly raised paired eyes ***Bispira volutacornis***

---

8. Collar absent ***Amphiglena mediterranea***

   Collar distinct **9**

---

9. Dorsal collar margins free and widely separated **10**

   Collar fused to mid-dorsal groove **13**

---

10. Thoracic tori fairly short with a gap between their ventral ends and ventral shields; shorter thoracic chaetae spoon-shaped with small distal point; companion chaetae with leaf-shaped tips (indistinct) ***Perkinsiana rubra***

    Thoracic tori long, their ventral ends indenting sides of shields; shorter thoracic chaetae lance-shaped (narrow or wide); companion chaetae with a fairly distinct bulbous head with distal filament **11**

---

11. Lance-shaped heads of thoracic chaetae narrow, about twice width of shaft; tips of radioles (apinnate part) somewhat flanged (triangular in cross-section) ***Demonax cambrensis***

    Such chaetae broader, three times width of shaft; radiolar tips thick or tapered **12**

---

12. Tips of radioles with parallel sides ***Demonax langerhansi***

    Tips of radioles tapered ***Demonax brachychona***

---

13. Radioles with compound eyes in lower half or subterminally **14**

    Radioles without eyes **15**

---

14. Eyes domed and rust coloured, on dorsal sides of proximal half of most radioles ***Pseudopotamilla reniformis***

    Eyes bulbous and blackish-brown, one near tip of each radiole, those on two most dorsal radioles particularly conspicuous ***Megalomma vesiculosum***

---

15. Radioles without web **16**

    Radioles webbed **17**

---

16. Collar cleft at mid-dorsal groove; pinnate radioles frequently dehisce just above an enlarged basal pinnule, but tentacular filaments remain ***Jasmineira elegans***

    Collar entire dorsally (no mid-line groove); whole crown may be shed ***Laonome kroyeri***

---

17. Posterior abdomen forming a postero-ventral funnel **18**

    Abdomen without subterminal funnel **19**

---

18. Funnel with about 12 segments; dorsal abdomen with distinct transverse glandular bands ***Euchone rubrocincta***

Funnel with about four segments; dorsal abdomen without such bands **_Euchone southerni_**

19. Posterior abdomen with a fine short filament; collar straight with a mid-ventral notch at apex of peristome; crown with a few long ventral tentacular (apinnate) filaments **_Chone filicaudata_**

Abdomen obtuse; crown without long filaments; collar oblique, extending beyond peristome as a blunt mid-ventral lobe **_Chone fauveli_**

20. Five abdominal segments; abdominal uncini almost square with anterior margin rasp-like **21**

Three abdominal segments; uncini elongate with short toothed margin **22**

21. Collar margin level with anterior edge of segment 1 **_Oriopsis armandi_**

Collar margin well below anterior edge of segment 1 **_Oriopsis hynensis_**

22. Crown with radioles, each bearing a few pairs of pinnules, with or without two ventral non-vascularized filaments **23**

Crown without usual radioles, bearing four pairs of enlarged ciliated 'pinnules' and two vascularized filaments arising from a common base **_Manayunkia aestuarina_**

23. Crown without paired ventral apinnate filaments; segment 1 without collar but with distinct fleshy ventral triangular lobe **_Fabricia stellaris_**

Crown with long ventral filaments; with or without collar or lobe **24**

24. Anterior of segment 1 with prominent fleshy triangular ventral lobe **_Fabriciola baltica_**

Segment 1 with a deep collar **_Fabriciola berkeleyi_**

25. Thorax indistinct (1–4 segments); distinct eyespots (1–4) each side of each segment; worm a few mm long **_Myxicola aesthetica_**

Thorax distinct (7–9 segments) eyespots absent; more than 2 cm long **26**

26. Thoracic chaetae 100 plus in small circular pads, small and scarcely visible under dissecting microscope; tips of radioles dull purple **_Myxicola infundibulum_**

Thoracic chaetae few but longer (clearly visible with dissecting microscope); tips of radioles not differentially pigmented **_Myxicola sarsi_**

Amphiglena mediterranea (Leydig) Fig. 6.20
Up to 15 mm long; collar absent; 8–10 thoracic and up to 30 abdominal segments; six radioles each side; a crescent-shaped pigment patch (arrowed) at base of each dorsal lip but fading in formalin; shorter thoracic chaetae spoon-shaped with distal filament; short rows of compact S-shaped uncini with companion chaetae; as few as three abdominal chaetae in each group arranged transversely (arrowed); tube of mucus, readily renewed.

In crevices of rock, holdfasts, sponges, etc.; intertidal and shallow water; mainly southern, reaching Isle of Man; fairly common.

Bispira volutacornis (Montagu) Fig. 6.20
Up to 100 mm long, 8–9 thoracic and up to 100 abdominal segments; crown of larger specimens bispiral with up to 200 radioles; radioles with paired longitudinal ridges (whitish), most with one or more paired composite eyes (see cross-section) scattered

on outer surface; collar with two ventral lappets (deeply pigmented except for a white border) and a wide gap between dorsal margins; shorter chaetae from more posterior thoracic segments of two types, one with a slender knee (see *Megalomma*) but the outer ones flattened and obtuse distally (knife-shaped); thoracic uncini and companion chaetae as in *Amphiglena* but forming long rows, especially in posterior half of thorax; abdominal chaetae arranged in a C-shape partially infilled with chaetae (arrowed); tube with a silt-mucoid outer layer.

Under rocky overhangs, sublittorally and in deep pools; mainly southern, reaching Isle of Man; locally abundant.

Branchiomma bombyx (Dalyell) Fig. 6.20
Up to 50 mm long; eight thoracic and about 70 abdominal segments; up to 25 radioles each side, the outer surface bearing several paired composite 'eyes' and paired epithelial flaps (stylodes), usually shaped like shoe-horns but distal ones may have scalloped ends; collar short but covering base of crown, with a shallow cleft ventrally, fused to mid-line groove dorsally and with or without a dorso-lateral notch; thoracic uncini truncated; companion chaetae absent; thoracic chaetae like *Megalomma*: arrangement of abdominal chaetae as in *Bispira*.

Under stones and shells; LWS and sublittoral; all coasts; seldom abundant.

Chone fauveli McIntosh Fig. 6.20
Up to 120 mm long; eight thoracic and about 80 abdominal segments; crown webbed for about three-quarters of length with up to 36 radioles each side (each with a flanged elongate triangular tip) and without ventral tentacular filaments; collar oblique in side-view with margin covering base of crown, fused to mid-dorsal groove and without a ventral cleft; shorter thoracic chaetae almost spoon-shaped; thoracic uncini with very long shafts; abdominal uncini truncated with an angular boss below peg.

Dredged, soft bottom; North Sea, Clyde Sea, Celtic Sea; locally common.

Chone filicaudata Southern
Up to 11 mm long; eight thoracic and 13 abdominal segments; crown webbed for two-thirds of length, with eight radioles and about five ventral tentacular filaments on each side; anal region with short, fine filament; other details like *C. fauveli*.

Dredged, soft substratum; western Ireland, Cumberland, Northumberland, Scotland; fairly common.

Demonax cambrensis
Knight-Jones and Walker Fig. 6.20
Up to 86 mm long; eight thoracic and about 160 abdominal segments; about 20 radioles each side with tapered tips; lance-like tips of shorter thoracic chaetae 7 times as long as wide; companion chaetae (side-view figured) and uncini as *D. saxicola*.

Grab samples in mud, sand, and gravel 26–27 m; Liverpool Bay and lochs of west Scotland; locally common. *Demonax langerhansi* Knight-Jones, up to 10 mm long, with only 5–7 thoracic segments and seven radioles on each side, occurs among epilithic faunas on south-western shores.

Demonax brachychona (Claparède) Fig. 6.20
Up to 25 mm long; 4–8 thoracic and up to 130 abdominal segments and 7–9 radioles on each side, with tapered tips; collar with two pointed ventral lappets and a fairly wide gap between dorsal margins; tips of shorter thoracic chaetae lance-like, about three times as long as wide; thoracic and abdominal uncini S-shaped (as *Amphiglena*) but thoracic companion chaetae with a bulbous toothed head and a fine filament at right angles to a fairly short shaft. One of several warm, temperate species misidentified as *Potamilla torelli* in temperate waters. *Demonax langerhansi* Knight-Jones is very similar but with thicker tips to radioles.

In crevices of rock, lithothamnion, and other organisms; LW and shallow sublittoral; southern species extending to English Channel, South Wales, and southern Ireland.

Euchone rubrocincta (Sars) Fig. 6.21
Up to 17 mm long; eight thoracic and about 30 abdominal segments; crown partially webbed, with tapered tips; tips of shorter thoracic chaetae lance-like, four or five times as long as wide; thoracic uncini with long shafts; 11 or 12 posterior segments form funnel on ventral side, with a membranous anterior margin; glandular area mainly ventral but characteristically extends to the dorsal surface of abdomen as narrow, pale, transverse bands crossing the reddish surface of body.

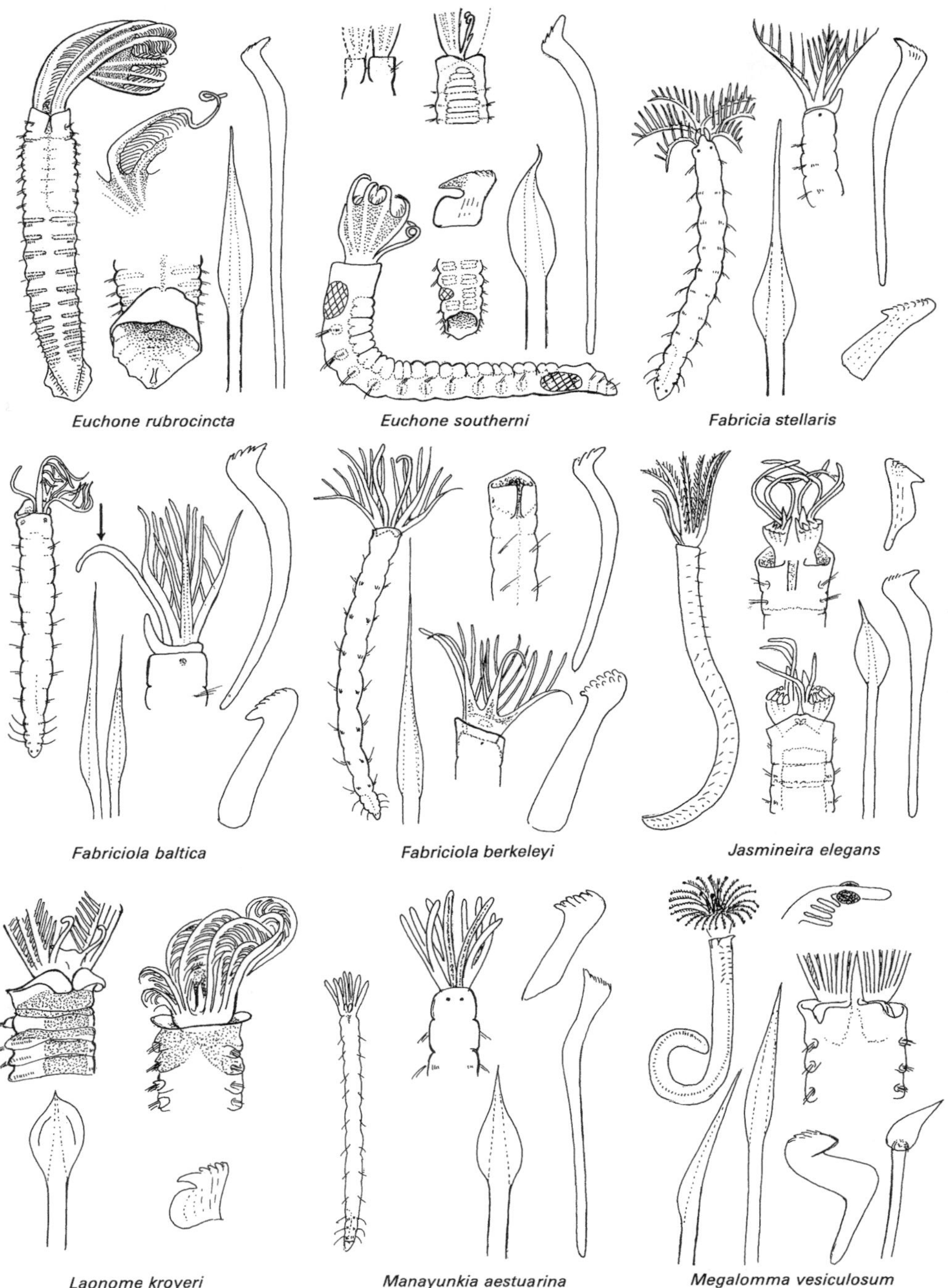
Euchone rubrocincta
Euchone southerni
Fabricia stellaris
Fabriciola baltica
Fabriciola berkeleyi
Jasmineira elegans
Laonome kroyeri
Manayunkia aestuarina
Megalomma vesiculosum

Fig. 6.21

Sublittoral soft substratum; all coasts; locally common.

Euchone southerni Banse Fig. 6.21
Only 4.4 mm long; eight thoracic and 12 abdominal segments, crown webbed for about two-thirds of length, with five radioles each side (long tapered tips) and three pairs of ventral tentacular filaments; collar fused to dorsal groove with ventral margin slightly indented; tips of shorter thoracic chaetae lance-like, about three times as long as wide; thoracic uncini with long shafts; last four segments (perhaps more) form a simple ventral funnel.

Soft substratum; depth 2–6 m; western Ireland.

Fabricia stellaris (Blainville) Fig. 6.21
(=*F. sabella* (Ehrenberg))
Up to 3 mm long; seven thoracic and three abdominal segments; crown with three short radioles each side (each with about six pairs of pinnules and a tapered tip) and without ventral tentacular filaments; without membranous collar, but with a ventral 'mobile' triangular lobe projecting anteriorly; one pair of eyes flanking anal region and another pair on the first segment; shorter thoracic chaetae geniculate, the knee bulbous in both 'face' and side views; thoracic uncini with long shafts; abdominal uncini with shorter, thicker, but flattened shafts (figured bottom right); tube of mucus quickly secreted as worm crawls backwards.

In muddy holdfasts of turf algae; intertidal; all coasts; abundant.

Fabriciola baltica Friedrich Fig. 6.21
Like *Fabricia stellaris* (in having a ventral triangular lobe) but with long, paired tentacular filaments (one arrowed) and shorter thoracic chaetae with a more slender knee (side- and face-views figured).

Surface mud, brackish water; sublittoral; head of Loch Etive, western Scotland; very abundant.

Fabriciola berkeleyi Banse Fig. 6.21
As *F. baltica* but with a distinct collar, distal margin covering base of crown, and long dorsal margins fused to mid-line groove. Often intermixed with *Fabricia stellaris*.

In muddy holdfasts of turf algae; intertidal; South Wales; locally common.

Jasmineira elegans Saint-Joseph Fig. 6.21
Up to 20 mm; 8–9 thoracic and about 30 abdominal segments, six radioles and 2–5 ventral tentacular filaments each side, frequently sheds radioles but not the filaments nor the sturdy base of crown; collar with mid-dorsal groove and without ventral cleft; tips of shorter thoracic chaetae lance-like, three times as long as wide; hooked thoracic uncini with long shafts; abdominal uncini almost S-shaped and generically characteristic (top right of figure).

In kelp holdfasts and crevices (lower shore) and shell/gravel (sublittoral); all coasts; fairly common.

Laonome kroyeri Malmgren Fig. 6.21
Up to 6 mm long; 8–12 thoracic and about 100 abdominal segments; 6–8 radioles and a pair of ventral tentacular filaments joined by a basal membrane; collar cleft ventrally and low but entire dorsally; shorter thoracic chaetae almost spoon-shaped with thoracic and abdominal uncini similar in having a rounded breast and a truncated 'shaft'; companion chaetae absent; glandular areas ventral but extending dorsally in first two segments.

Soft substratum; sublittoral; northern species; west and east coasts of Scotland, western Ireland; not common.

Manayunkia aestuarina (Bourne) Fig. 6.21
Up to 6 mm long; eight thoracic and three abdominal segments, no pinnate radioles but instead eight sturdy ciliated 'pinnules' arising in pairs and two ventral filaments with green blood sinuses; collar absent; tips of shorter thoracic chaetae lance-like, three times as long as wide; other chaetae as *Fabricia stellaris*.

In surface mud of estuaries; LW: Thames, Tyne, Tamar, Burry Inlet, Liffey; locally abundant.

Megalomma vesiculosum (Montagu) Fig. 6.21
Up to 120 mm long; 8–9 thoracic and up to 200 abdominal segments; up to 30 radioles each side, each with a bulbous, subterminal eye, those on two dorsal radioles larger; collar with ventral cleft and dorso-lateral notches, and fused to sides of mid-dorsal groove, thoracic chaetae slender (side- and face-views figured); thoracic (and abdominal) uncini S-shaped, and companion chaetae with leaf-shaped blades.

In mud with stones and sand; lower shore and sublittoral in sheltered bays; southern species; south-west Wales, south-west England and Cornwall, western Ireland; locally abundant.

Myxicola aesthetica (Claparède) Fig. 6.22
About 26 mm long; thorax indistinct with 1–4 segments, abdomen with about 30, each with lateral eyespots; seven radioles each side, webbed for most of length, with flat triangular unpigmented tips; chaetae in minute hair-like tufts which are difficult to distinguish; thoracic and abdominal uncini as figured for other *Myxicola* species but the abdominal ones with more teeth on crest.

In rock crevices and kelp holdfasts; lower shore and sublittoral; South Wales, south Devon and Isle of Man; locally abundant.

Myxicola infundibulum (Montagu) Fig. 6.22
Up to 200 mm (contracts by half when disturbed), round in cross-section; up to 25 radioles each side of crown, webbed for most of length, leaving free purple/brown triangular tips; eight thoracic and over 100 abdominal segments; thoracic chaetae numerous, fine, and small (difficult to see), fascicles indicated by circular pads; collar absent; anterior margin of first segment with a shallow cleft on each side and a triangular ventral lobe; hooked thoracic uncini (these may be absent in mature specimens) with long shafts and a smooth crest as in *M. sarsi*; abdominal uncini truncated; tube (to left of figure), thick, gelatinous, transparent, carrot-shaped and renewable.

In estuarine mud; lower shore and sublittoral; south-west Wales, south-west England, south-west Ireland; locally common.

Myxicola sarsi Krøyer Fig. 6.22
Up to 60 mm long; eight thoracic and about 70 abdominal segments; other characters as in *M. infundibulum* except thorax relatively wider laterally; fewer radioles with tips not differentially coloured; anterior margin of first segment lacks lateral notches and the small ventral lobe is emphasized by a transverse groove; thoracic tufts of chaetae fewer and easily seen.

In fine muddy sand; sublittoral; northern species; north and east coasts of Scotland, north-east England, Isle of Man; fairly common.

Oriopsis armandi (Claparède) Fig. 6.22
Up to 6 mm long; eight thoracic and five abdominal segments; crown with three slightly flanged radioles each side (each with about six pairs of pinnules and a long tapered tip) and a pair of ventral tentacular filaments; collar shallow and with a dorsal gap; all thoracic chaetae slender; thoracic uncini long, shafted, and similar to those of *Fabricia* but abdominal uncini very different, almost square in side-view.

In muddy holdfasts of turf algae and muddy rock-crevices; intertidal; southern species; Anglesey, South Wales, south-west England, western Ireland.

Oriopsis hynensis Knight-Jones Fig. 6.22
As *O. armandi*, except that radioles have more distinct flanges; collar with distal margin well below the oblique anterior margin of first segment; abdominal uncini with a gouge-like boss beneath anterior tooth.

In algal holdfasts, e.g. *Laurencia* and *Codium*; just below LWM: Lough Hyne, southern Ireland; locally common.

Perkinsiana rubra (Langerhans) Fig. 6.22
Up to 80 mm long and slender; 6–11 thoracic and about 150 abdominal segments; 5–10 radioles each side (blood not green as in many sabellids but reddish) with fine tapered tips; collar with ventral cleft and a wide gap between dorsal margins; shorter thoracic chaetae almost spoon-shaped; thoracic uncini somewhat S-shaped but with a shaft twice as long as distance between breast and crest; adjacent companion chaetae with leaf-shaped blades; abdominal uncini with shorter shafts; each abdominal chaeta with bulbous knees; tubes with covering of sand and shell fragments. Often intermixed with *Pseudopotamilla reniformis* and *Demonax langerhansi* and has been misidentified as *Potamilla torelli* in warm and temperate waters.

Bores into limestone rock or lithothamnion; pools, lower shore, and sublittoral; southern species; South Wales; locally abundant.

Pseudopotamilla reniformis (Bruguière) Fig. 6.22
Up to 120 mm long, slender; 9–12 thoracic and up to 200 abdominal segments; up to 13 radioles each side of crown, with tapered tips and 1–3 rust-

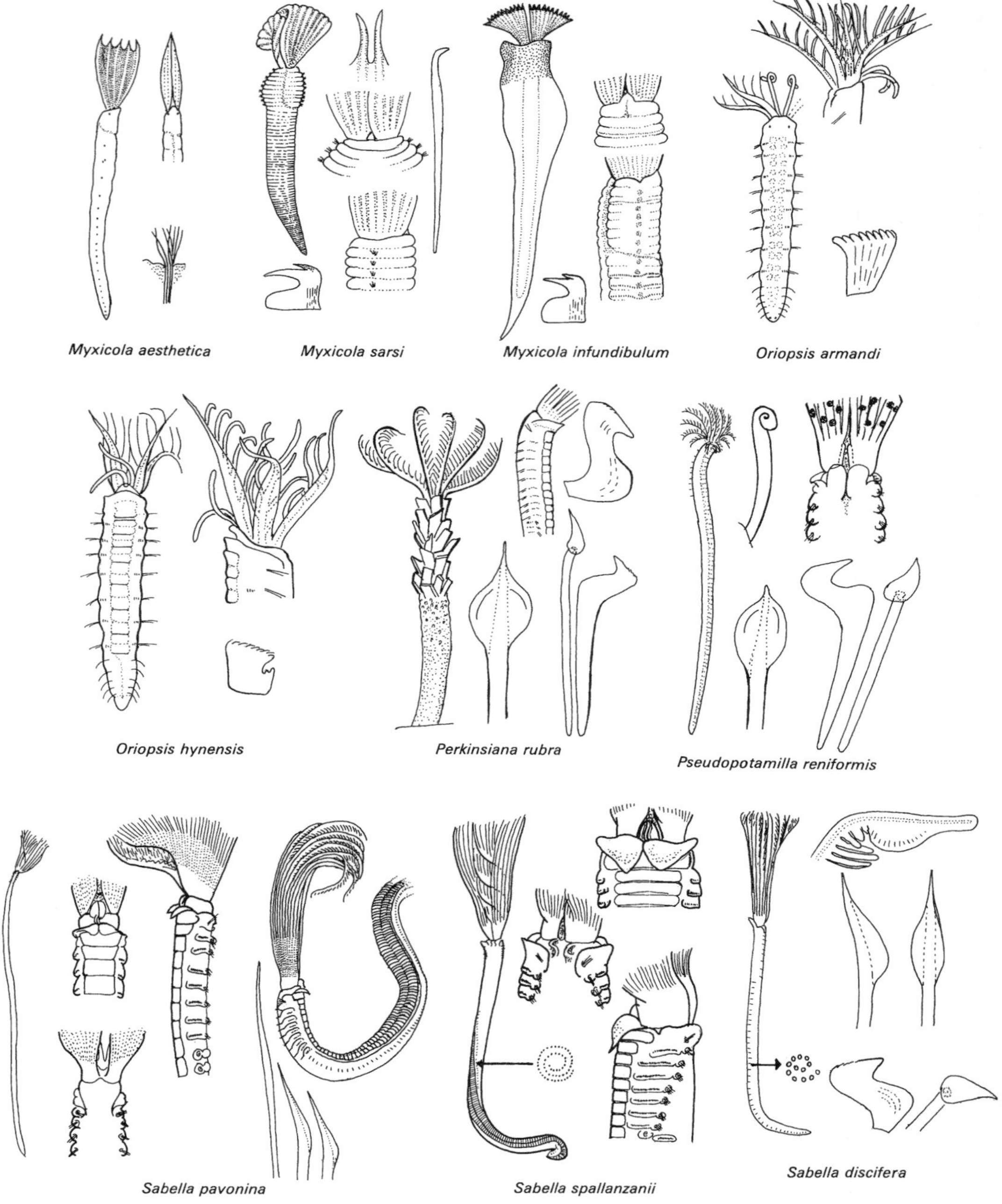
Myxicola aesthetica
Myxicola sarsi
Myxicola infundibulum
Oriopsis armandi
Oriopsis hynensis
Perkinsiana rubra
Pseudopotamilla reniformis
Sabella pavonina
Sabella spallanzanii
Sabella discifera

Fig. 6.22

coloured compound eyes on dorsal side of most radioles; ventral radioles shorter than others (top left) with few or no eyes; collar with ventral lappets and dorso-lateral notches above pockets formed by fusion of collar to sides of mid-dorsal groove; shorter thoracic chaetae almost spoon-shaped; thoracic uncini and companion chaetae with moderately long shafts; tube of mucoid-silt, unusual in that distal part curls ventro-proximally when crown is withdrawn.

In narrow rock-crevices or boring in limestone, shells or lithothamnion; pools, lower shore, and sublittoral; widespread; all coasts; abundant locally.

Sabella discifera Grube Fig. 6.22

Up to 52 mm long, 7–8 thoracic and 50 abdominal segments; each side of crown with about 10 radioles webbed basally, each with a very small, whitish subterminal swelling of thickened epithelium; collar with long ventral lappets and a wide gap between dorsal margins; shorter thoracic chaetae very bulbous at the knee (both side and face views); uncini S-shaped with short 'shafts'; companion chaetae with leaf-like blades; abdominal chaetae arranged in a spiral (arrowed); tube of mucoid silt; multiplies by fission.

On rocks, stones, and gorgonians; sublittoral in strong currents; South Wales, south Devon, Scilly Isles, and Roscoff; locally abundant.

Sabella pavonina Savigny Fig. 6.22

Long slender body up to 300 mm long, 4 mm wide, with up to 300 segments of which 8–16 are thoracic; crown webbed basally, one side bigger and spiralling in from ventral edge in larger specimens (e.g. 40 radioles on one side and 50 the other); lateral collar margins not reaching base of crown (top left and right), dorsal margins separated by a wide gap (centre bottom) and ventral margins forming two fleshy lappets below well-developed ventral sacs (centre top); uncini as figure for *S. discifera* but spiral arrangement of abdominal chaetae with more whorls; all thoracic chaetae slender.

Found on all coasts, with big populations in Menai Strait, Swansea Bay, Poole Harbour, and estuaries of Essex and Plymouth rivers. On stones in sand and mud; sublittoral, locally abundant.

Sabella spallanzanii (Gmelin) Fig. 6.22

Up to 200 mm long and plump (figured specimen 80 × 6 mm); crown very asymmetrical in mature specimens, left or right side forming up to three whorls; first thoracic segment short, lateral margins of collar reaching base of crown; dorsal collar margins curve towards ridges flanking mid-line groove; all but anterior thoracic tori with just a small gap between their ventral ends and ventral shields; uncini as in *S. discifera*; thoracic chaetae as *S. pavonina*; abdominal chaetae arranged in 2–3 whorls (arrowed).

Sublittoral; Channel Isles, Roscoff, and Cherbourg.

29. SERPULIDAE

As Sabellidae, with distal crown and abdominal inversion of the thoracic arrangement and faecal groove passing around right side, but tubes are calcareous and worms lie with dorsal side against substratum (usually hard). Most species with characteristic opercula which arise on the left side but alternate between left and right in successive regenerations in some genera; collar extends symmetrically as 'thoracic membranes' down the dorsal side close to the left and right fascicles of chaetae, sometimes reaching only as far as segment 2, usually extending throughout thorax and often posteriorly to thorax as a ventral apron; collar chaetae often generically characteristic, distal surface with or without teeth and with or without a prominent boss or 'fin'. Some species incubate embryos but most have planktonic larvae; tubes usually attached to rocks, ships' hulls or kelp holdfasts, seldom to algal fronds.

---

1. Tube not adhering to substratum, tusk-shaped and circular in cross section **_Ditrupa arietina_**

   Tube adhering to substratum **2**

---

2. Lateral margins of thoracic membrane extending only as far as second chaetigerous segment **_Filogranula calyculata_**

   Thoracic membrane extending to last thoracic segment **3**

---

3. Operculum present **4**

   Operculum absent **9**

---

4. Opercular peduncle with paired, pointed lateral wings **5**

Peduncle without lateral wings **6**

5. Opercular ampulla somewhat cup-shaped and deeper than distal calcareous cap, which may be flat or domed (seldom conical), with or without distal projections ***Pomatoceros lamarcki***

Ampulla a shallow dish-shape, rarely as deep as distal cap, which is usually domed, often conical, and with or without distal projections ***Pomatoceros triqueter***

6. Peduncle with pinnules **7**

Peduncle without pinnules **10**

7. Tube about 1 mm diameter, generally solitary; operculum spherical, often paired, one smaller than the other ***Apomatus similis***

Tubes generally aggregated forming a thin tracery over surface; operculum cup- or bowl-shaped **8**

8. With two opercula, each an asymmetrical cup; tube diameter 0.5 mm ***Filograna implexa***

Single symmetrical operculum, with a small but bulbous ampulla and a distal cup; tube diameter 0.2 mm ***Josephella marenzelleri***

9. Tube about 2 mm diameter, generally solitary; radioles tapered distally, fairly smooth with bright red blotches (in life) and single lenticular units along their length ***Protula tubularia***

Tube about 0.5 mm diameter, often numerous, forming thick rope-like twisted assemblages in shallow water; radioles with swollen tips and scalloped margins, each protuberance containing groups of lenticular units ***Salmacina dysteri***

10. Opercular peduncle D-shaped or flattened in cross-section **11**

Peduncle round in cross-section **12**

11. Peduncle smooth, stiff, with the flat surface showing dorsally; operculum with many dark-brown teeth; tube smooth ***Ficopomatus enigmaticus***

Peduncle ribbon-like and flexible, with oval annuli when preserved (contraction); operculum smooth, symmetrically domed or subconical; tube with rough or toothed ridges ***Metavermilia multicristata***

12. Operculum asymmetrically domed or subconical; prominent glandular pad on dorsal side of last abdominal segments ***Vermiliopsis striaticeps***

Operculum funnel-like, with serrated circumference, with or without a central crown of spines **13**

13. Operculum a single straight-sided funnel about 4 mm across with 20–40 marginal serrations; crown red (young *Hydroides* opercula are similar, but crown is usually pale and banded with brown) ***Serpula vermicularis***

Operculum a double funnel, the basal one *Serpula*-like, the verticil (the distal one) made of spines(*Hydroides*) **14**

14. Verticil radially symmetrical **15**

Verticil not radially symmetrical ***Hydroides dianthus***

---

15. Spines of verticil without lateral spinules **Hydroides ezoensis**

    Spines with spinules **16**

---

16. Centre of verticil with a small projection; collar chaetae with a finely toothed boss **Hydroides elegans**

    Centre of verticil smooth; collar chaetae bearing one or usually two large teeth **Hydroides norvegica**

Apomatus similis Marion and Bobretzsky Fig. 6.23
Up to 30 mm long (13 mm figured); operculum a globular transparent vesicle on one or both second dorsal radioles, which have short pinnules like the others in the crown; radioles seven to 20 each side, each rachis with about seven paired eyes consisting of irregular groups of eight to ten lenticular units (bottom of figure); collar chaetae scarcely geniculate with a slender knee, finely toothed and tapered distally; tube just over 1 mm across, tapers, meanders and bears fine transverse growth lines and indistinct longitudinal ridges.

On rock and shells; sublittoral; southern species: south-west Wales, Plymouth and Channel Isles.

Ditrupa arietina Müller Fig. 6.23
Tube about 15 mm long and characteristic in being tusk-shaped, made of two layers (see transverse section), the outer thick and vitreous, the inner thin and opaque; operculum with a flattish lens-shaped plate; collar chaetae absent.

Lives unattached to substrata (sand or mud); sublittoral; Shetland, Celtic Sea, south-west Ireland.

Ficopomatus enigmaticus (Fauvel) Fig. 6.23
Operculum without a distal rim but with dark chitinous inwardly curved spines (figured specimen 1.2 mm diameter); collar chaetae slightly geniculate with large teeth at knee and smaller but distinct ones along the distal taper; tube with fine transverse growth rings and occasional shelf-like annuli distally.

On hard substrata in the warmer and sometimes brackish water of estuaries, docks, lagoons, and power station outfalls; sublittoral; southern species, west and south coasts of England and Wales.

Filograna implexa Berkeley Fig. 6.23
Two soft, cup-shaped, asymmetrically developed opercula (figured specimen 0.2 mm diameter) each on a pinnate peduncle; collar chaetae with toothed boss and toothed distal part separated by a gap, as in spirorbid *Spirorbis tridentatus*; tube fine with occasional growth rings, often forming shallow incrustations.

Crevices in rock, *Pentapora*, etc.; sublittoral; south-west Wales and Plymouth (further records uncertain as *Salmacina dysteri* has often been synonymized with this species); locally abundant.

Filogranula calyculata (Costa) Fig. 6.23
Operculum with asymmetrical cup-shaped 'plate' (figured specimen 0.2 mm diameter), with or without a central projection which bears irregular protuberances; collar chaetae as *Filograna implexa*; tube with a median and two lateral rows of pointed teeth positioned so as to form transverse rows of three teeth, anterior portion may become erect and growth then proceeds in spurts, resulting in formation of fluted nesting funnels.

Under edges of piled (silted) shale slabs; sublittoral; Abereiddy marine quarry, south-west Wales; southern species; locally common.

Hydroides dianthus (Verrill) Fig. 6.23
Opercular basal funnel (figured specimen 1.5 mm diameter) edged with 30–40 angular serrations and surmounted by a verticil of large irregular spines curved towards radioles; collar chaetae scarcely geniculate, with two large teeth at knee and smooth taper distally; tube round in cross-section; with meandering growth rings irregularly coiled in one plane, or in aggregations tightly parallel to each other with apertures towards open water.

On hard substrata in or near harbours; sublittoral; on Japanese oysters in France; Swanwich Marina near Hayling Island, southern England; locally abundant.

Hydroides elegans (Haswell) Fig. 6.23
Operculum and tube as *H. dianthus* but verticil with a small central projection and spines regular with

median teeth and lateral spinules; collar chaetae with a finely toothed boss.

On hard substrata in many harbours world-wide, e.g. Swansea, Shoreham; sublittoral.

Hydroides ezoensis Okuda Fig. 6.23
Operculum as *H. elegans* but verticil without central projection and spines without lateral spinules; collar chaetae and tube as *H. dianthus*.

On hard substrata, Japanese immigrant abundant in the Solent; sublittoral.

Hydroides norvegica (Gunnerus) Fig. 6.23
Operculum and tube as *H. elegans*, but verticil without central projection, and spines fused to each other for less than one-third of their length, collar chaetae as in *H. dianthus*.

On hard substrata; usually sublittoral; all coasts; common.

Josephella marenzelleri
Caullery and Mesnil Fig. 6.23
Just 2 mm long; operculum a distal cuticular cup (about 0.1 mm diameter), with longitudinal ridges, scalloped distal margin, heart-shaped ampulla, and pinnate peduncle; collar chaetae slender (as figured for *Apomatus*); tube about 0.2 mm in diameter and slightly flexible, usually part of a delicate tracery of tubes that are often branched (reproduction by scissiparity).

In rock clefts; sublittoral; southern species; Isle of Man, Dale and Abereiddy (south-west Wales), and Plymouth.

Metavermilia multicristata (Phillipi) Fig. 6.23
Distal part of opercular ampulla domed (figured specimen 0.5 mm diameter) or conical and brownish; differs from *Vermiliopsis striaticeps* in that the long peduncle (arising between radioles 1 and 2) is flattened and oval in cross-section, and the glandular surface of the dorsal side of the last few segments does not form a distinct pad; collar chaetae as in *Apomatus* but more slender; tube with five rough ridges or five longitudinal rows of sharp teeth.

On rock (shale); sublittoral; southern species; Abereiddy marine quarry, South Wales; locally common.

Pomatoceros lamarcki (Quatrefages) Fig. 6.24
Operculum with a cup-shaped ampulla, the distal plate concave or convex (figured specimens 1.2 mm diameter), with or without an eccentric projection, which can be with or without distal protuberances; peduncle thick and smooth with a tapered wing on each side; collar chaetae, when present, like those figured for *Apomatus*; tube with a well-defined median keel and, in adults, two more vestigial ridges, one on each side.

On rocks; lower shore and shallow sublittoral; warm temperate species; widespread on rocky shores throughout Britain; very abundant.

Pomatoceros triqueter (Linnaeus) Fig. 6.24
As *P. lamarcki* except that the ampulla is shallow (dish-shaped) and the distal part often conical (with or without projections); tube with only a median ridge.

On rocks; mainly sublittoral; widespread species; all coasts; abundant, particularly adjoining deep water.

Protula tubularia (Montagu) Fig. 6.24
Operculum absent; collar chaetae as figured for *Apomatus*; radioles have 15–25 paired red blotches concealing single lenticular units (side-view detail figured on right); thoracic membranes extensive; tube 2–8 mm diameter, smooth except for indistinct growth rings.

On hard substrata; sublittoral, Shetlands, Scotland, west and south coasts of England, Wales, Ireland, and Channel Isles; widespread but often solitary.

Salmacina dysteri (Huxley) Fig. 6.24
Operculum absent; collar chaetae as in *Filograna*, radiole tips swollen (centre top), each rachis with paired scalloped 'ridges' with about 13 pairs of tightly packed lenticulate units (figured in plan-view, centre bottom, and side-view, to the right); tubes, up to 0.5 mm diameter, similar to those of *Filograna* except that they commonly make dense aggregations in the form of a rope-like mesh with tunnels.

On rock, lower shore, but mainly sublittoral in less sheltered areas; widespread on rocky coasts from Shetlands to Channel Islands; abundant.

Serpula vermicularis Linnaeus Fig. 6.24
Operculum funnel-shaped (figured specimen 2.2 mm diameter) with radial grooves and 20–40 marginal serrations; collar chaetae with a boss bearing two large teeth and a smooth taper distally; tube with longitudinal ridges, but fairly smooth when erect and aggregated.

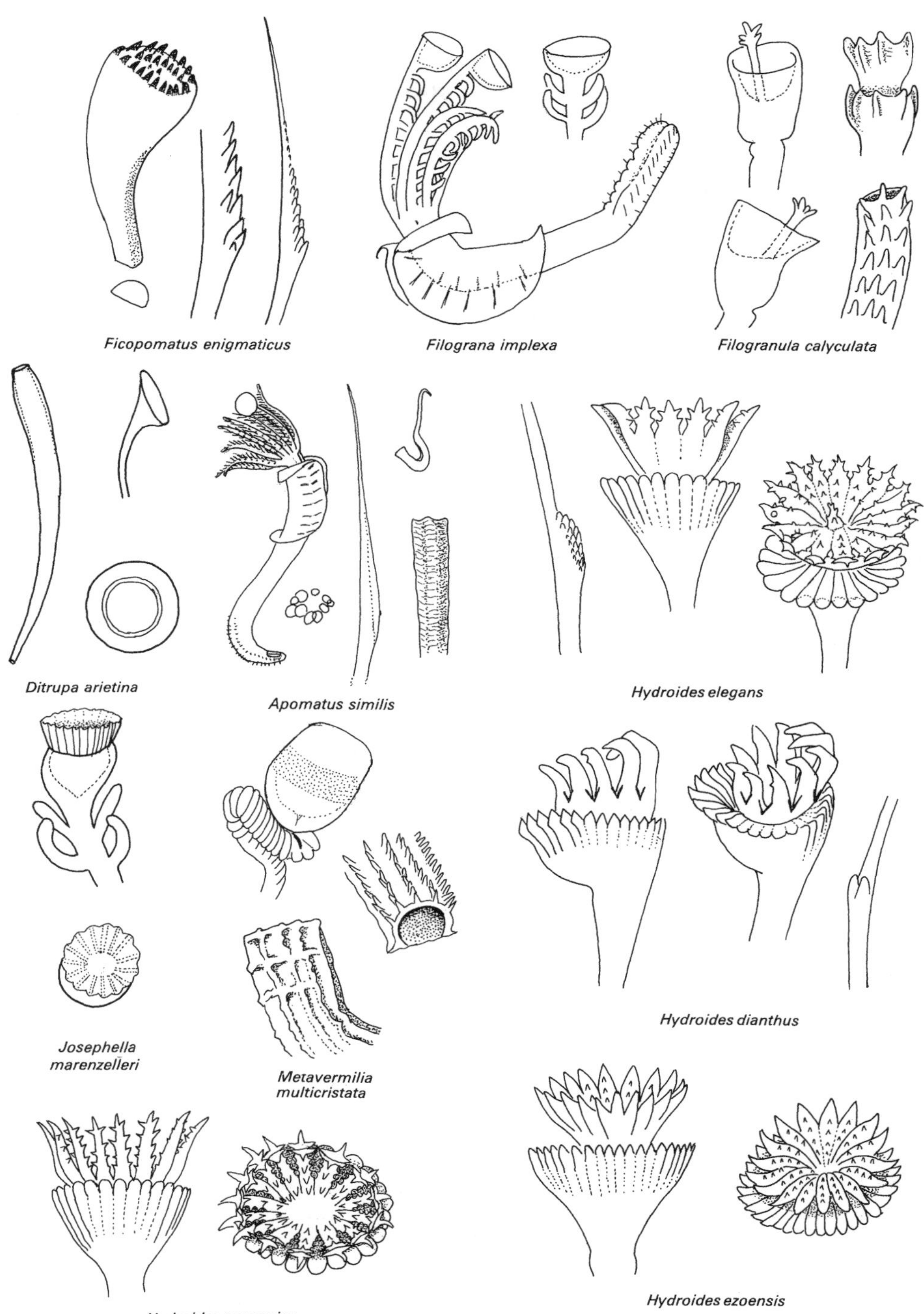

Ficopomatus enigmaticus
Filograna implexa
Filogranula calyculata
Ditrupa arietina
Apomatus similis
Hydroides elegans
Josephella marenzelleri
Metavermilia multicristata
Hydroides dianthus
Hydroides norvegica
Hydroides ezoensis

Fig. 6.23

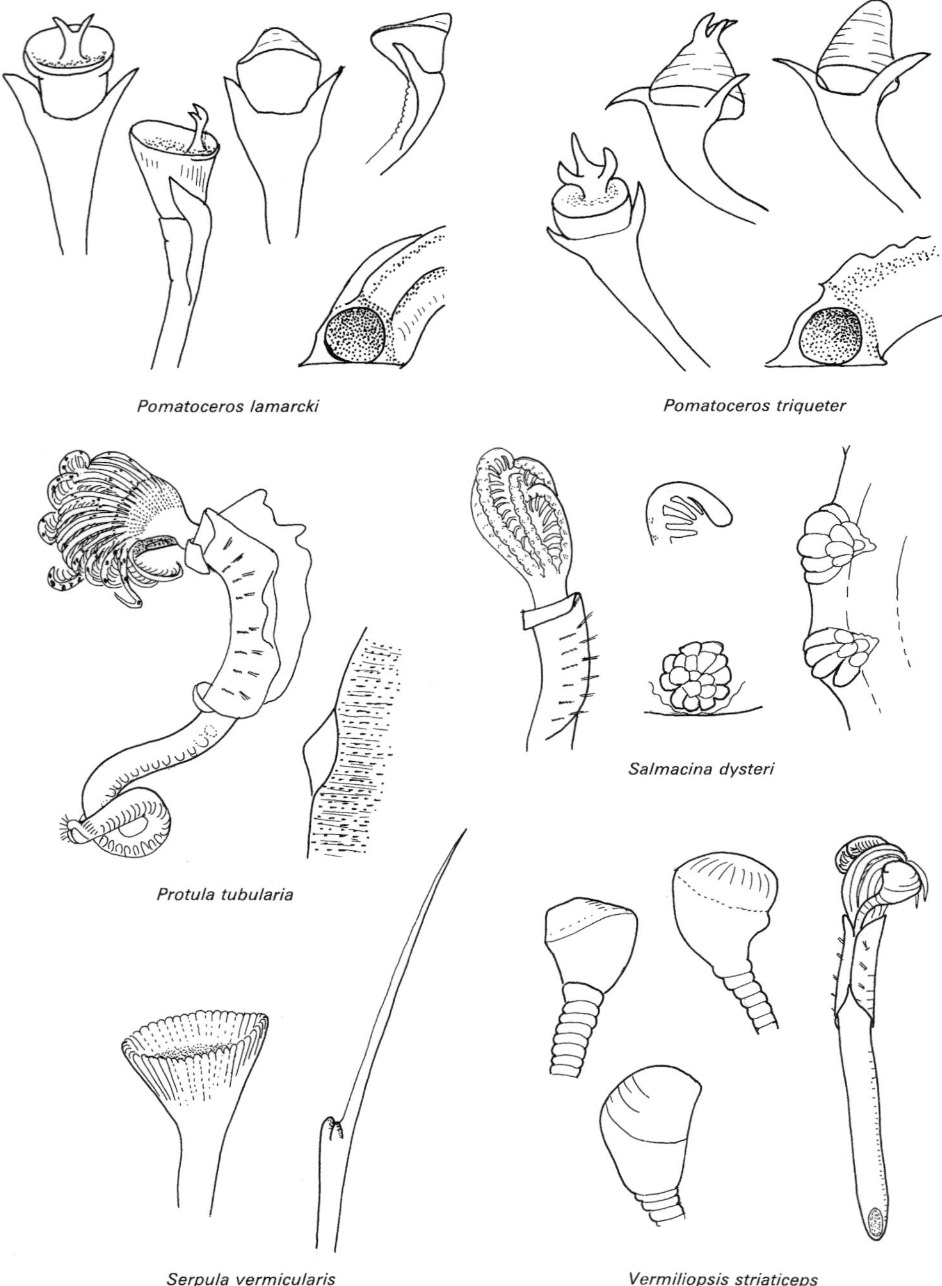

Fig. 6.24

On rocks and shell; sublittoral in lagoons, inlets, fjords, and natural harbours; Shetlands, west and south coast of Britain, western Ireland, and Channel Isles; often abundant.

Vermiliopsis striaticeps (Grube) Fig. 6.24
Operculum asymmetrically domed or subconical (figured opercula less than 1 mm diameter), often with longitudinal striations; peduncle slender, stiff

but annulated, arising near the dorsal mid-line; collar chaetae like those figured for *Apomatus*; tube about 2 mm diameter with about four longitudinal ridges and indistinct growth rings; dorsal side of the last 12 abdominal segments bears a prominent glandular pad. This character, together with the shape and position of the peduncle, easily separates this species from *Metavermilia multicristata*.

On hard substrata (shell and shale); sublittoral; Falmouth harbour and Roscoff.

30. SPIRORBIDAE

As Serpulidae, but tubes coiled or helical, rarely evolute, sinistrally (clockwise) coiling species with operculum on left and change of direction of faecal groove on right, vice versa in dextrally coiling species. Mostly with three thoracic chaetigers, but can be more. Thorax and abdomen separated by an achaetous region; sickle chaetae (figured for *Janua*) common, but absent in *Circeis* and *Neodexiospira*. Dorsal thoracic membranes usually with free margins, but these are fused to form a tunnel in *Neodexiospira* (easily seen in live material, and when preserved can be demonstrated by a fine pin, Fig. 6.25). Clove oil (after dehydration) helps to distinguish finer sculptural details of large opercular talons but should not be used for opercular brood chambers.

Spirorbids have a wide selection of substrata, including algal fronds, but are intolerant of sand scour. Larvae settle and metamorphose very quickly, with no planktotrophic stage.

Opercula drawn to same scale (e.g. that of *Circeis armoricana*, 0.4 mm diameter) except that of *Jugaria quadrangularis*, which is figured at half the scale of the others.

---

1. Tube coiled dextrally (anticlockwise when viewed from above) 2

   Tube coiled sinistrally 9

---

2. Tube glassy or porcellaneous, eggs attached to inside of tube wall 3

   Tube opaque, somewhat chalky; incubation in an opercular brood chamber, but non-brooding operculum may be present; two thoracic tori on concave side 7

---

3. Tube glassy or translucent, with longitudinal ridges, operculum bowl-shaped; three thoracic tori on concave side of adults; on stones **Paradexiospira vitrea**

   Tube opaque or translucent, without ridges; operculum oblique in side-view with blunt talon towards (beneath) lower edge; two thoracic tori on concave side (*Circeis*) 4

---

4. Tube translucent, often helical; talon longer than radius of plate; on hydroids and bryozoans **Circeis spirillum**

   Tube opaque; talon shorter than radius of plate 5

---

5. On macro-algae **Circeis armoricana fragilis**

   On crustaceans and shell 6

---

6. On lobsters and crawfish **Circeis armoricana**

   With hermit crab *Pagurus bernhardus*; on abdomen or on surrounding shell **Circeis paguri**

---

7. Thoracic membrane not fused dorsally; non-brooding operculum with eccentric pin-shaped talon; brooding operculum with shiny transparent wall **Janua pagenstecheri**

   Thoracic membrane fused dorsally to form a tunnel; non-brooding operculum with a flat peripheral talon; brooding operculum with partially opaque wall (*Neodexiospira*) 8

---

8. Non-brooding operculum with axe-shaped talon; brood chamber not much longer than diameter of plate; collar chaetae without cross-striations **Neodexiospira brasiliensis**

Non-brooding operculum with narrow talon; brood chamber longer than broad; collar chaetae with cross-striations **_Neodexiospira pseudocorrugata_**

---

9. Tube glassy or translucent and transversely ridged **_Protolaeospira striata_**

 Tube opaque, porcellaneous or chalky, without prominent transverse ridges **10**

---

10. Operculum with talon **11**

 Operculum forming brood chamber **16**

---

11. Talon central, cone- or pin-shaped **12**

 Talon eccentric or peripheral, and vestigial, spatulate, or massive **13**

---

12. Three thoracic tori on concave side (adult); eggs incubated in tube; tube with one obtuse median ridge; talon pin- or cone-shaped **_Paralaeospira malardi_**

 Two thoracic tori on concave side; tube with three indistinct longitudinal ridges; no eggs in tube. Talon subconical. Could be the non-brooding operculum of **_Pileolaria berkeleyana_**

---

13. Talon peripheral and flat; may or may not be attached to brood chamber **14**

 Talon not peripheral, but eccentric and either vestigial or massive; embryos, when present, in a string attached to the tube wall **17**

---

14. Talon subquadrangular with a ventral spur (not attached to brood chamber); brownish patch on dorsal body surface (black in preservative). Non-brooding operculum of **_Pileolaria militaris_**

 Talon subtriangular or elongate and tapered proximally (usually attached to brood chamber); whitish, indistinct patch on dorsal body surface **15**

---

15. Opercular plate flat or concave, talon short (half diameter of plate) and subtriangular; after brood chamber formation, this operculum remains attached only by the proximal margin of talon **_Jugaria granulata_**

 Opercular plate convex, talon longer than diameter of plate, winged, and narrow, and bifid proximally; this operculum becomes completely fused to brood chamber formed below **_Jugaria quadrangularis_**

---

16. Brood chamber helmet-shaped, with distal spines **_Pileolaria militaris_**

 Brood chamber asymmetrically bilobed, without distal spines **_Pileolaria berkeleyana_**

---

17. Opercular plate with a slight thickening towards lower edge **18**

 Plate with bilobed or massive talon **19**

---

18. Tube smooth with small flange; body greenish-brown; generally on *Fucus* **_Spirorbis spirorbis_**

 Tube without flange; body red; on stones or lithothamnion **_Spirorbis rupestris_**

---

19. Talon bilobed or like a horse's hoof; tube without longitudinal ridges **20**

 Talon rounded or pointed terminally; tube usually with longitudinal ridges **21**

---

20. Usually on *Corallina*; dorsal collar flaps only slightly asymmetrical **_Spirorbis corallinae_**

Usually on other algae; dorsal collar flap on convex side much longer than other **_Spirorbis inornatus_**

21. Tube usually with three distinct smooth ridges forming three projections at mouth; talon with three indistinct lobes; on stones, often on shore **_Spirorbis tridentatus_**

Tube with three indistinct rough ridges, the inner one most prominent; talon with three pointed processes; on sublittoral stones, never on shore **_Spirorbis cuneatus_**

Circeis armoricana St. Joseph Fig. 6.25
Opercular plate oblique, slightly concave (figured specimen 0.4 mm diameter); talon broad but short, with an indentation proximally; collar chaetae geniculate with a sharp knee and teeth confined to the distal margin; larvae with two pairs of eyes, the lateral ones larger and bilobed; tube thick-walled and opaque, coil diameter up to 2 mm.

On *Homarus* and *Palinurus*, particularly in crevices on undersides; sublittoral; Swansea, Lundy, Anglesey, north-west Ireland.

Circeis armoricana fragilis
Knight-Jones and Knight-Jones Fig. 6.25
Like *C. armoricana* but talon longer and occasionally serrated; tube up to 1.5 mm coil diameter, slightly translucent, showing faintly the orange pigment of the worm within.

On *Laminaria* and *Saccorhiza* sublittorally in sheltered cold water; Isles of Scilly and Brittany (southern limit), Plymouth Sound, Fishguard and Holyhead harbours, Loch Sween (Scotland), Lough Hyne (southern Ireland).

Circeis paguri
Knight-Jones and Knight-Jones Fig. 6.25
Opercular plate markedly concave, the talon relatively long; collar chaetae as *C. armoricana*, but larvae characteristic in having three separate pairs of small, rounded eyes; tube 0.6–1.3 mm coil diameter, translucent but colour of worm (pale yellow) does not readily show through.

With hermit crab *Pagurus bernhardus*, attached to telson or to inside of shell that it inhabits; sublittoral; Swansea and Plymouth.

Circeis spirillum (Linnaeus) Fig. 6.25
Similar to *C. armoricana fragilis* but talon more slender in side-view; collar chaetae less geniculate, with teeth not confined to margin and arranged in serried ranks across side (cross-striations); tube usually helically coiled as it grows away from narrow substrata, with translucent walls showing the orange body within.

On hydroids and bryozoans in strong currents; sublittoral; a widespread northern species; Guernsey, Plymouth, Swansea, Anglesey, and Scotland.

Janua pagenstecheri (Quatrefages) Fig. 6.25
Non-brooding operculum with eccentric pin-shaped talon; first brood chamber of a series includes this talon (most without), wall around embryos transparent and distal plate either flat or asymmetrically convex; collar chaetae geniculate with a rounded, finely toothed knee; sickle chaetae (on right of figure) in third thoracic fascicle, with a similar knee but with a flat, ribbed, and marginally toothed blade distally; dorsal margins of collar not fused to form a tunnel; tube less than 2 mm coil diameter, dextral, with indistinct longitudinal ridges and a sloping flange.

On hard substrata and algae, intertidal and sublittoral. All coasts, common.

Jugaria granulata (Linnaeus) Fig. 6.25
Non-brooding operculum with a flat or concave distal plate and peripheral subtriangular talon, proximal margin of which remains attached to outer edge of brood chamber. This is flat-topped with a proximal helmet-shaped calcification; collar chaetae finely toothed, without cross-striations and with finely toothed boss ('fin') at knee; largest abdominal tori three-quarters of the way along abdomen; tube usually without longitudinal ridges, coil diameter 2 mm.

Under rocks and shells; sublittoral, in shallow water only in sheltered conditions; north temperate species; all coasts; common.

Jugaria quadrangularis (Stimpson) Fig. 6.25
Like *J. granulata* but talon wider and longer, with lateral wings and indentations, and with a domed distal plate. This becomes fully incorporated into the wall of the brood chamber; largest abdominal tori lie anteriorly.

On *Laminaria* sublittorally; near Inveraray, south-west Scotland; locally common.

Neodexiospira brasiliensis (Grube) Fig. 6.25
Non-brooding operculum with a peripheral axe-shaped talon, first brood chamber of a series includes this talon in its wall but most are without such talons; chamber wall cylindrical and calcified except for a proximal zone just above basal disc; collar chaetae geniculate and with barely visible cross-striations associated with the toothed margin; sickle chaetae (see *Janua*) absent; thoracic membranes fused dorsally to form a tunnel (demonstrated by a fine pin in figure); tube 2 mm across with three indistinct longitudinal ridges.

On *Sargassum* and moored pontoons; sublittoral; widespread in warm temperate waters; Portsmouth Harbour; locally common.

Neodexiospira pseudocorrugata (Bush) Fig. 6.25
Like *N. brasiliensis* except: talon narrow and flat; brood chamber wall longer; collar chaetae with fairly distinct cross-striations and coarse teeth along the outer margin; tube smaller, 1.5 mm across.

Mainly on *Laminaria saccharina*, *Cystoseira*, and sublittoral Rhodophyceae; locally common in Channel Isles, Scilly Isles, Plymouth, Lundy, and south-east Ireland. Not found off Wales and Isle of Man.

Paradexiospira vitrea (Fabricius) Fig. 6.25
Orange pigment can be seen through the glassy, usually ridged tube, 2.5 mm in coil diameter; operculum bowl-shaped and transparent except at slightly convex centre; three pairs of thoracic tori on concave side (adults); collar chaetae like those of *Jugaria granulata* but with distal cross-striations.

In crevices under rocks in clear water; mostly sublittoral; Arctic and north temperate species; Channel Isles, Scilly Isles, Plymouth (outside break-water), Lizard, Isle of Man, west Scotland; not known from Wales; locally common.

Paralaeospira malardi Caullery and Mesnil Fig. 6.25
Opercular talon variable from a pin- to cone-shape; collar chaetae like those of *Jugaria granulata*; three thoracic tori on concave side (adults); tube 2.3 mm across with a single obtuse longitudinal ridge.

Under rocks, usually sublittoral, particularly on molluscs in *Laminaria* zone; possibly an immigrant from South Africa; Scilly Isles, Channel Isles, Cornwall, Plymouth, south-west Wales, Isle of Man, northern Scotland, southern Ireland; locally common.

Pileolaria berkeleyana (Rioja) Fig. 6.26
Non-brooding operculum with central talon forming a cone; brood chamber calcification helmet-shaped and bilobed distally, the side away from the substratum being slightly more protuberant; live material readily distinguished from *P. heteropoma* by the large diffuse patch of milky orange-pink granules across the dorsal surface (fades in preservative); collar chaetae with coarse serrations, distinct oblique cross-striations and a toothed boss; tube with longitudinal ridges, coil diameter just over 1 mm.

On *Sargassum* in Portsmouth Harbour, and rocks at 3–7 m depth off Plymouth Hoe and in Falmouth Harbour. Widespread in warm and temperate waters.

Pileolaria militaris Claparède Fig. 6.25
Non-brooding opercular talon (bottom left) with a ventral spur which may be with or without distal teeth (top left); brood chamber with a helmet-shaped calcification and a ventro-distal semicircle of spines; dorsal surface of body bears a single rust-coloured patch, which becomes and remains nearly black in preservative; collar chaetae like those of *P. berkeleyana*; tube up to 3 mm, fairly smooth with the last whorl rapidly widening.

Mainly on sublittoral Rhodophyceae; widespread in warm waters; Channel Isles, Plymouth, Gulland Rock, and Lundy Island (Bristol Channel); not common.

Protolaeospira striata (Quievreux) Fig. 6.26
Operculum with an oblique, slightly concave distal plate and a massive talon with lateral knobs and a

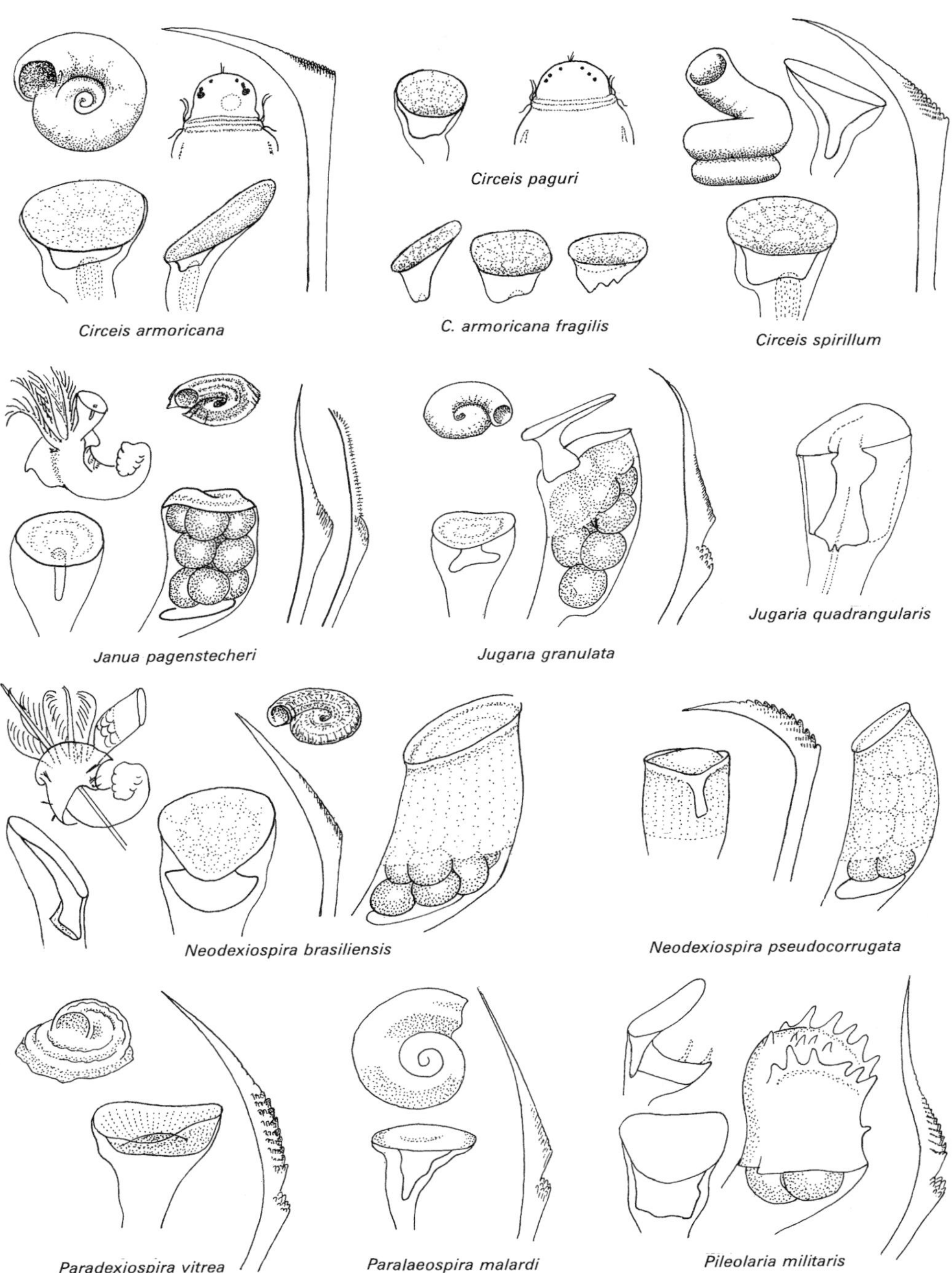
Circeis paguri
Circeis armoricana
C. armoricana fragilis
Circeis spirillum
Janua pagenstecheri
Jugaria granulata
Jugaria quadrangularis
Neodexiospira brasiliensis
Neodexiospira pseudocorrugata
Paradexiospira vitrea
Paralaeospira malardi
Pileolaria militaris

Fig. 6.25

small ventral spur; eggs incubated in a brood sac attached to an epithelial stalk arising from the thorax; collar chaetae similar to those of *Pileolaria*; tube over 2 mm diameter, transparent and easily overlooked.

Under stable rocks; sublittoral; widespread in warm European waters; Channel Isles and Scilly Isles, not common; northern record Gulland Rock, Padstow.

Spirorbis cuneatus Gee Fig. 6.26

Opercular plate slightly oblique, talon with three distinct points in face view, subconical in side-view (bottom left) but slimmer in younger specimens (centre top); collar chaetae with indistinct cross-striations and basal boss; tube opaque, coil diameter just over 1 mm, with a dull surface, three longitudinal ridges and a flange sloping from the outer ridge.

On rocks and shells; sublittoral; Mediterranean and east Atlantic species; south-west England, south-west Wales and Scotland; locally common.

Spirorbis corallinae
de Silva and Knight-Jones Fig. 6.26

Opercular plate flat (concave in young specimens) or domed, talon bilobed but conical in side-view (left); collar chaetae like the largest chaeta figured for *S. tridentatus*; tips of radioles similar in length to adjacent pinnules; larvae with two pairs of eyes, the lateral ones bilobed and anterior ones small and round; tube about 1 mm coil diameter, smooth and with the last whorl partially above the previous ones (owing to narrow substratum), walls porcellaneous.

Almost exclusively on *Corallina officinalis*; rock pools and shallow water; Scilly Isles, and Channel Isles to northern Scotland and Ireland.

Spirorbis inornatus
L'Hardy and Quievreux Fig. 6.26

As *S. corallinae* except that opercular plate remains concave in mature specimens; bare tips of radioles up to twice as long as adjacent pinnules; larvae lack the round pair of anterior eyes; porcellaneous tube nearly 2 mm across, usually coiled in one plane with a small flange close to substratum.

On *Laminaria* at St. Michael's Isle (Cornwall), Milford Haven, Isle of Man and Oban; on *Chondrus, Gigartina, Himanthalia* buttons, *Laminaria, Fucus serratus*, and other algae where there is plenty of water movement; Channel Isles, south-west England, Roscoff, Wales, Ireland, Scotland, and Northumberland.

Spirorbis rupestris
Gee and Knight-Jones Fig. 6.26

Opercular plate very oblique, concave with shallow talon towards lower edge; body orange-red; tips of radioles two or three times as long as adjacent pinnules; collar chaetae similar to larger chaeta figured for *S. tridentatus*; tube up to 4.5 mm across when coiled in one plane, sometimes evolute and frequently with the last whorl ascending, especially when amongst lithothamnion.

On rock particularly associated with *Phymatolithon polymorphum* L.; intertidal, common behind 'hanging' *Fucus*; a north temperate species; Scilly Isles, Channel Isles, Cornwall to Swanage (Dorset), south-west and north-west Wales, Isle of Man, Scotland, and Ireland.

Spirorbis spirorbis (Linnaeus) Fig. 6.26

As *S. rupestris* except: body greenish with brown stomach; tips of radioles not much longer than adjacent pinnules; collar asymmetrical forming a large 'hinged' flap at dorso-convex margin, collar chaetae similar to larger chaetae figured for *S. tridentatus*; tube about 3 mm across, evenly coiled, shiny, with a small peripheral flange.

On *Fucus*, especially *F. serratus*, occasionally on other algae, e.g. *Laminaria* and *Himanthalia* buttons, and stone; intertidal and shallow water; north temperate species; most rocky coasts; abundant.

Spirorbis tridentatus Levinsen Fig. 6.26

Tube shiny usually with rounded longitudinal ridges, coil diameter 3 mm, the last whorl widening abruptly towards mouth; talon conical in side-view (narrow in young specimens) and with indistinct lateral protuberances in face-view; collar chaetae on convex side with a toothed boss separated by a gap from the finely toothed distal portion, which lacks cross-striations, those on concave side smaller with a reduced gap or none at all.

Under stones and in rock-crevices; lower shore and (more sparsely) to a depth of 20 m; north temperate species; abundant on most coasts (tolerates some sand abrasion).

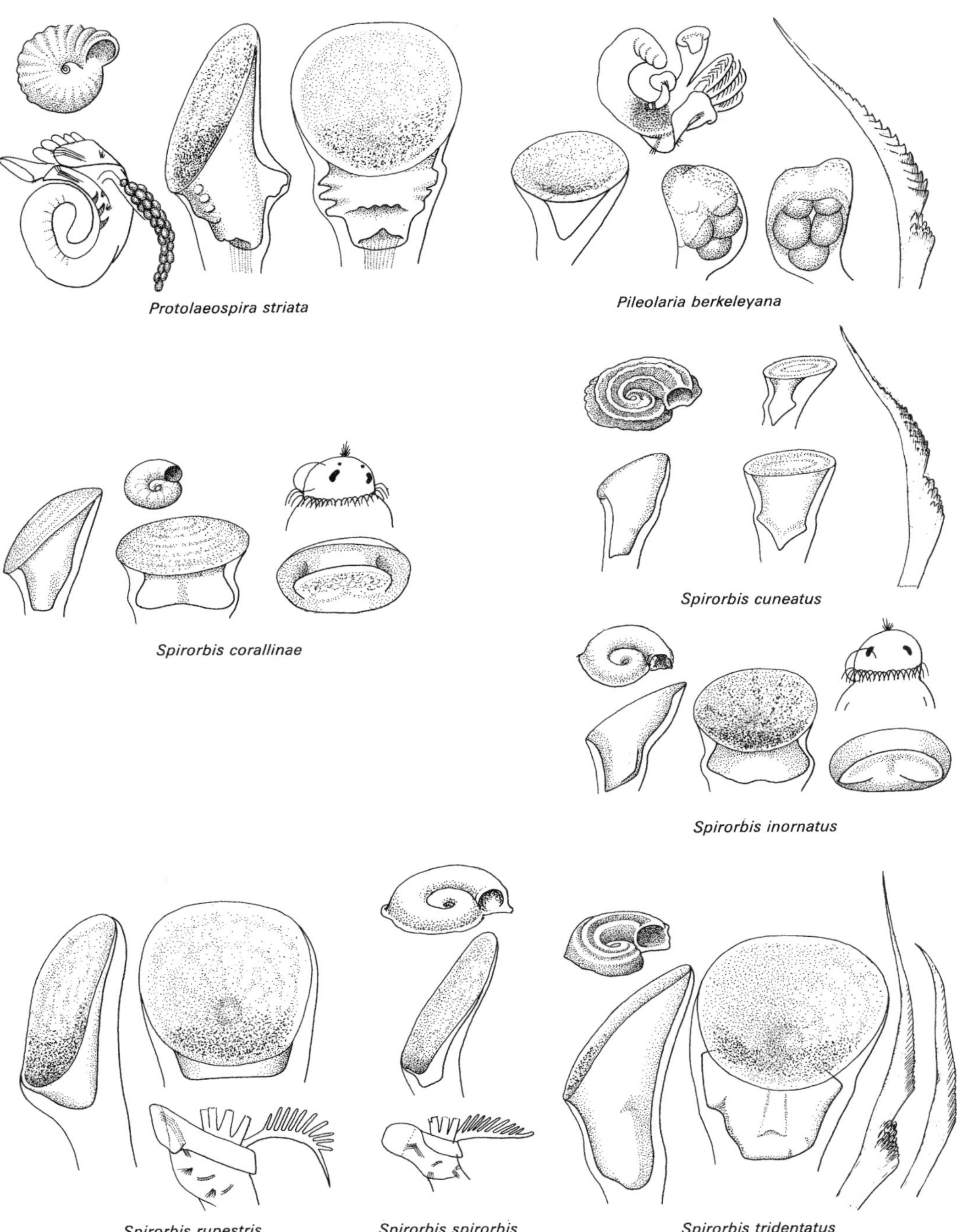
Protolaeospira striata
Pileolaria berkeleyana
Spirorbis corallinae
Spirorbis cuneatus
Spirorbis inornatus
Spirorbis rupestris
Spirorbis spirorbis
Spirorbis tridentatus

Fig. 6.26

Class Oligochaeta

This group includes the familiar earthworms, but most members are aquatic. The majority have been recorded largely from fresh waters and thus occur in the marine environment around streams which flow across sea-shores, in estuaries, or from the inner parts of the Baltic Sea. Some predominantly terrestrial forms may also extend on to upper levels of the intertidal. Few truly marine species feature in British fauna lists, but many of the greater number recorded from continental Europe probably also occur around Britain.

Aquatic oligochaetes are generally similar to earthworms, but much smaller and proportionally more slender (Fig. 6.27). The anterior end usually lacks appendages or obvious sense organs, although the prostomium may be elongated in a proboscis (separate from the mouth) and some have simple eyespots. Typically, each segment behind the prostomium carries four bundles of chaetae, one dorso-lateral and one ventro-lateral on each side; sometimes they are missing from a certain region. In many species, the body-wall is translucent, so that internal organs can be seen. The location of external and internal structures is indicated by giving the segments roman numbers, starting with the peristomium as I. Sexually mature worms develop a *clitellum* similar to the 'saddle' of an earthworm, but less conspicuous, covering several segments in the region V–XIII. Oligochaetes are hermaphrodite; the nature and position of their complex reproductive structures are important in identification.

Marine and brackish-water oligochaetes, mostly associated with soft sediments, are small enough to be lost by normal sieving methods: an 0.5 mm or smaller mesh size is recommended. The seawater–ice method for extracting meiofauna is unreliable for small oligochaetes. Unfortunately, literature for their identification is scattered and incomplete; it is impossible to separate some quite large groups into species unless they are sexually mature, or without dissection and microscopic examination. Details given here are thus restricted to the general characteristics of the relevant three families.

1. ENCHYTRAEIDAE Fig. 6.27

Predominantly terrestrial ('pot worms'). Many marine species are small and whitish. Bristles in all bundles are the same—simple and pointed or, rarely, with a bifid end; occasionally they are few or absent. Testes (Ts) in XI, ovaries (Ov) and male pores (*M) in XII, spermathecae (Sp) in V. There are no penes or genital bristles.

Common genera: *Enchytraeus, Fridericia, Grania, Lumbricillus, Marionina.*

2. NAIDIDAE Fig. 6.27

Some freshwater species have a proboscis or gills and several bear eyespots. Small, delicate, almost transparent worms with various types of bristle: dorsally, usually long hair-like chaetae and short, stouter ones; ventrally, mostly bifid crochets with a swelling where they emerge from the body. Most reproduce asexually by pygidial budding and thus may trail a short chain of smaller individuals; where, rarely, gonads are present, they occupy segments V–VII. There are no penes but genital bristles may be present. The few truly marine species are mainly epibenthic, swim actively, and are not found far offshore.

Common genera: *Amphichaeta, Chaetogaster, Paranais.*

3. TUBIFICIDAE Fig. 6.27

Medium-sized reddish 'sludge-worms', living in muddy sea-shore or estuarine sediments enriched by decaying organic matter, which tend to coil in a spiral when irritated, or very small white worms living in the meiobenthos and recorded well offshore. Dorsal and ventral bristles are of different forms (hair-like, crochets) unless all are bifid. Testes and spermathecae are normally in X, ovaries and penes in XI. Chitinous penis sheaths usually present, often also specialized genital bristles. Asexual reproduction may be by fragmentation (giving rise to atypical worms regenerating missing sections) but not by budding.

Common genera: *Clitellio, Limnodrilus, Tubifex, Tubificoides (= Peloscolex).*

Class Hirudinea

Leeches are highly evolved annelids, with some cephalization, a constant number of segments, and a sucker at each end, used in 'geometrid' locomotion. They may be grouped with oligochaetes in the Clitellata, being hermaphrodite, with direct development within a cocoon secreted by a glandular girdle, the clitellum. All temperate marine leeches are Piscicolidae and they comprise more than half of

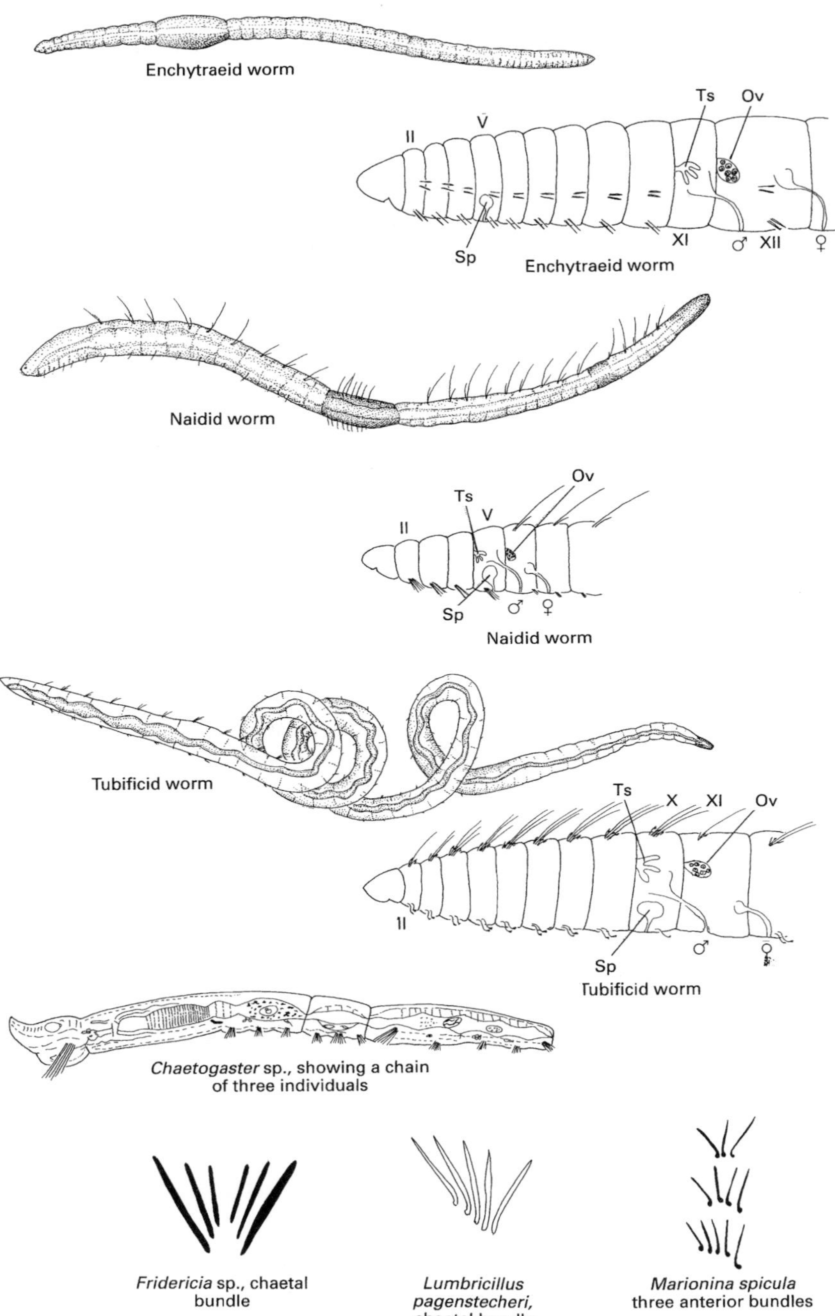
Enchytraeid worm
Ts
Ov
II
V
Sp
XI
♂
XII
♀
Enchytraeid worm
Naidid worm
Ov
Ts
II
V
Sp
♂
♀
Naidid worm
Tubificid worm
Ts
X
XI
Ov
II
♂
♀
Sp
Tubificid worm
Chaetogaster sp., showing a chain of three individuals
Fridericia sp., chaetal bundle
Lumbricillus pagenstecheri, chaetal bundle
Marionina spicula three anterior bundles

Fig. 6.27

that family. They lack jaws but have a protrusible pharynx and are thus Rhynchobdellida. They differ from other rhynchobdellids (the Glossiphoniidae, confined to fresh water) chiefly in having more conspicuous oral suckers, being blood-sucking rather than predatory. All known British species feed on fishes, occasionally leaving them to attach cocoons to rocks, shells, etc. Some occupy characteristic positions on their hosts (e.g. *Hemibdella* on *Solea* and *Sanguinothus* on *Taurulus*), but others are less specific. *Myoxocephalus scorpii* is attacked by at least four species of small leeches, which look rather similar but are placed in different genera.

The anterior part of a piscicolid (the *trachelosome*) consists of the *head* (= six fused segments), plus three *praeclitellar* and three *clitellar segments*. Its hind margin may be marked by a *clitellar constriction*. The rest of the body (the abdomen) has six *testicular*, six *caecal*, and three *anal segments*. The drawing of *Hemibdella* shows the gonads (lying intersegmentally) which characterize the testicular region. There are often only five pairs of testes, leaving more room for the ovary in the anterior part of that region. Between adjacent pairs of testes the gut may give off a succession of *segmental caeca*, which are easily seen during life in freshly collected specimens of transparent species (see drawings of *Oceanobdella*), because they are filled with red blood. Behind the testicular region there is a pair of particularly large caeca ('*postcaeca*'), which extends posteriorly almost to the anus. These may remain separate, or may fuse below the intestine to form a single ventral *caecum*. Fusion may be incomplete, leaving segmental fenestrae, through which muscles run dorso-ventrally.

Leeches should be examined while alive, for then the tissues may be transparent, revealing eyes (e.g. on fore sucker), spots (black spots, '*ocelli*', may occur on the hind sucker), and other markings which may not be visible after death. Some leeches have *pulsatile vesicles* segmentally arranged along each side of the abdomen. *Brumptiana* has these on all 12 segments of the testicular and caecal regions. *Branchellion, Calliobdella, Piscicola*, and *Trachelobdella* generally have them on the first 11 of those segments. From such external clues to segmentation there are obviously three annuli per mid-body somite in *Branchellion*, four in *Pontobdella*, six in most *Calliobdella*, and 13 or 14 in *Brumptiana*. Without segmental markings annulation is difficult to determine, but the testes can be seen after clearing (e.g. in cedar wood oil) and their centres mark the boundaries of somites, as shown in the drawing of *Heptacyclus*. Some leeches, however, would be difficult to identify or describe adequately without a complete series of transverse sections. Good sections are obtained after narcotizing with menthol and fixing in seawater Bouin. For fixation the specimen should be laid out flat and straight, but not compressed unnaturally.

Taxonomically useful internal characters include the coelomic and genital systems. The former may include dorsal, ventral, lateral, and transverse lacunae linked with pulsatile vesicles. Genera that have a large male bursa may extrude it during coitus or fixation (figured here in *Platybdella*).

1. PISCICOLIDAE

Body usually more or less divided into trachelosome and abdomen. Both suckers usually cup-shaped.

---

1. Body with many conspicuous branchiae or tubercles — **2**

   Body smooth or with a single fin or row of small vesicles on each side — **5**

---

2. Body with leaf-like branchiae in a row along each side (*Branchellion*) — **3**

   Body bears numerous tubercles (*Pontobdella*) — **4**

---

3. Branchiae 31 pairs. Fore sucker two-thirds diameter of hind sucker — ***Branchellion borealis***

   Branchiae 33 pairs. Fore sucker half diameter of hind sucker — ***Branchellion torpedinis***

---

4. Fore sucker much wider than trachelosome. No tubercles in ventral mid-line — ***Pontobdella muricata***

   Fore sucker not much wider than trachelosome. Annulus with largest tubercles in each segment bears an unpaired tubercle mid-ventrally — ***Pontobdella vosmaeri***

---

5. Body with flattened fin along each side ***Pterobdellina jenseni***

Body without lateral fins **6**

6. Hind sucker about same width as hind end of body **7**

Hind sucker much wider than hind end of body **8**

7. Fore sucker very large, nearly as wide as short, flattened abdomen ***Ganymedebdella cratere***

Both suckers very small. Attached to spine on scale of *Solea vulgaris* ***Hemibdella soleae***

8. With a row of 11–13 pulsatile vesicles along each side of body **9**

Pulsatile vesicles absent or indistinguishable **14**

9. Abdomen broad, somewhat flattened and very distinct from narrow trachelosome. Suckers narrower than abdomen. Trachelosome bears four pairs of dorso-lateral vesicles. In abdomen the four anterior pairs of vesicles are lateral or ventro-lateral, the remainder are dorso-lateral ***Trachelobdella lubrica***

Abdomen cylindrical or, if flattened, bearing a large hind sucker. Vesicles confined to abdomen and all lateral in position **10**

10. Fore sucker often with cross of pigment. Four eyes, often crescentic. Abdomen with transverse pigment bands. Hind sucker with 14 radiating pigment bands and a paramarginal ring of 14 black spots (ocelli). Fresh or brackish water ***Piscicola geometra***

Fore sucker without pigment cross. Eyes round, four or absent. Abdomen without transverse bands. Hind sucker without black spots. Marine **11**

11. Eyes four. Annuli 13 or 14 between successive vesicles. Hind sucker no wider than body. Abdomen with longitudinal streaks of pigment dorsally and a pair of small red spots associated with each pulsatile vesicle ***Brumptiana lineata***

Eyes four or absent. Annuli three or six between successive vesicles. Hind sucker wider than body **12**

12. Eyes four. Length about 20 mm. On shore fishes, e.g. *Taurulus* at Roscoff ***Calliobdella punctata***

Eyes absent. Length 25 mm. Sublittoral **13**

13. Length 25–50 mm. Hind sucker three or four times breadth of fore sucker. Abdomen swollen and trachelosome well defined. On *Lophius piscatorius* ***Calliobdella lophii***

Length *c.* 30 mm. Hind sucker twice breadth of fore sucker. Abdomen rounded and trachelosome ill-defined ***Calliobdella nodulifera***

14. Annuli seven (14) per mid-body somite ***Heptacyclus myoxocephali***

Annuli two or three (six) per mid-body somite **15**

15. Annuli two per somite. Eyes four, each made up of several black spots ***Janusion scorpii***

Annuli three (six) per somite. If eyes are present, each is a simple, rounded, or crescentic spot of pigment **16**

16. Hind sucker compressed laterally and capable of folding round a finray. Body uniformly reddish, but with a narrow white band adjoining hind sucker. On *Taurulus bubalis* fins
Sanguinothus pinnarum

Hind sucker circular in outline **17**

17. Eyes absent. No ocelli on hind sucker. Male bursa large and thick-walled, resembling a penis when everted. Postcaeca fused for most of their length, with four fenestrae. On gills of *Anarhichas lupus*
Platybdella anarrhichae

Eyes six or four. Paramarginal ocelli usually present on posterior sucker. Male bursa short and thin-walled, forming only a slight prominence even when everted. Postcaeca fused in not more than one or two places, often completely separate **18**

18. Fore sucker inconspicuous, less than half width of hind sucker. Skin mostly transparent. Clitellar constriction separates trachelosome from abdomen (*Oceanobdella*) **19**

Fore sucker discoid, at least half width of hind sucker. Skin opaque and pigmented. Clitellum not constricted, so trachelosome merges with abdomen (*Malmiana*) **21**

19. Usually attached below head of Cottidae. Body unpigmented
Oceanobdella microstoma

Usually attached behind pectoral fin of Blenniidae. Body with transverse stripes **20**

20. On *Lipophrys pholis*
Oceanobdella blennii

On *Zoarces* (and *Pholis*)
Oceanobdella sexoculata

21. Six eyes. Hind sucker twice fore sucker diameter, without paramarginal ocelli
Malmiana bubali

Four eyes. Hind sucker one and a half times fore sucker, with a paramarginal ring of 14 ocelli ***Malmiana yorki***

Branchellion borealis Leigh-Sharpe Fig. 6.28
Genus has each pair of branchiae arising from an annulus, and from the pulsatile vesicles wherever those occur, but this species has the first pair of vesicles lacking branchiae, which total only 31 pairs. Fore sucker bears light spots, in two semicircles of five and four, and is two-thirds width of hind sucker. Body length about 30 mm. Central annulus in each of the four anterior segments of trachelosome bears a row of four small warts dorsally.

On Rajidae. South-west Britain.

Branchellion torpedinis Savigny Fig. 6.28
Branchiae 33 pairs, the first arising from the anterior pulsatile vesicles. Fore sucker bears six eyes in a curved line, and is about half width of hind sucker. Body about 40 mm long. Central annulus in each segment of abdomen bears six light spots dorsally and four ventrally.

On Rajidae and occasionally teleosts. South-west Britain, Mediterranean, and West Africa.

Brumptiana lineata
Llewellyn and Knight-Jones Fig. 6.28
Fore sucker bears two pairs rounded eyes, four white patches with posterior streaks, and two white spots. Hind sucker large (fore : hind ratio 2 : 3), but no wider than abdomen, bearing 14 radiating markings with no ocelli. Body nearly 10 mm long, pale brown. Abdomen with dark pigment forming seven longitudinal lines dorsally. Pulsatile vesicles (12 pairs) each flanked by two red specks. Pairs 4, 6, 8,

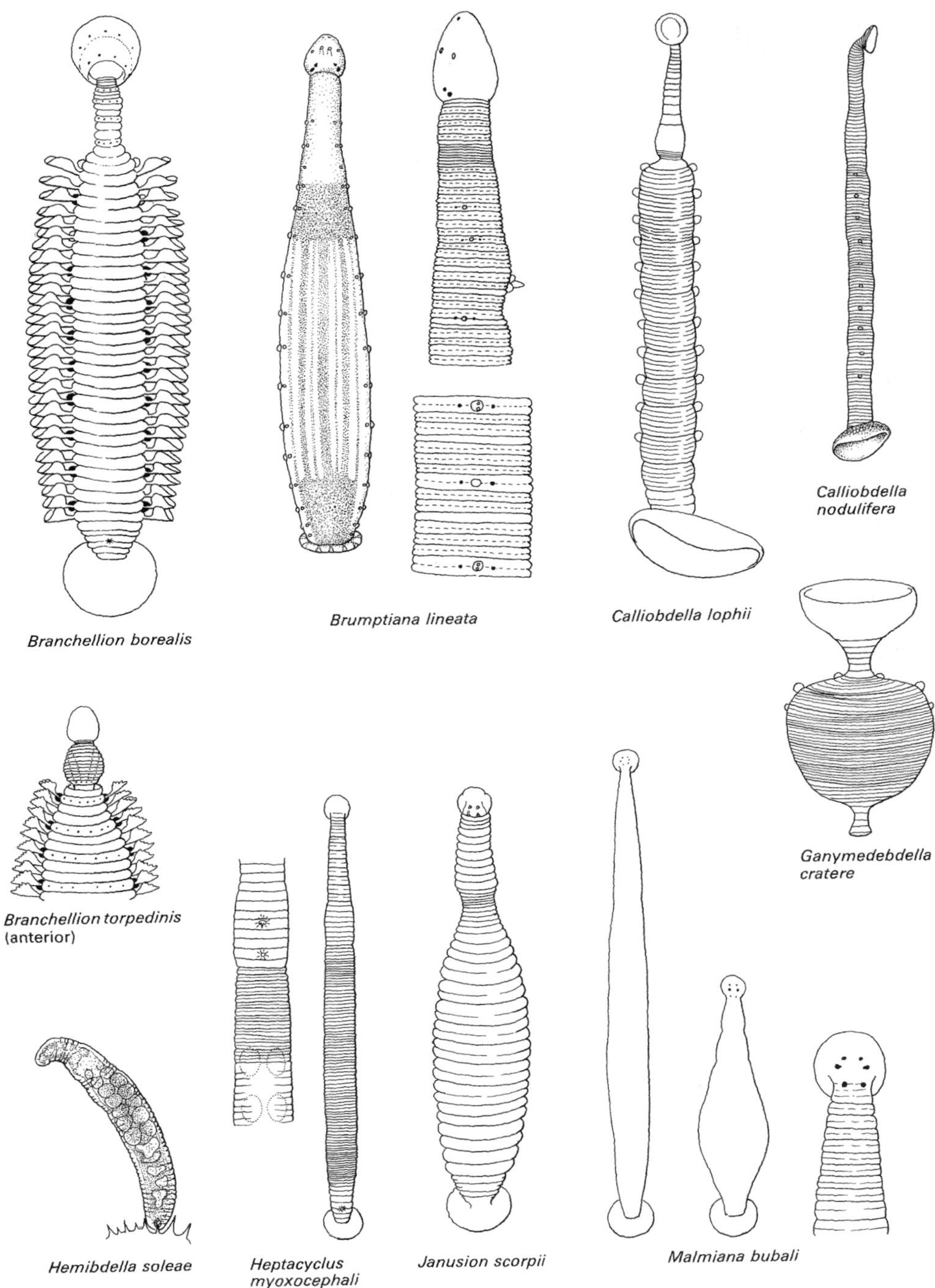
Branchellion borealis
Brumptiana lineata
Calliobdella lophii
Calliobdella nodulifera
Ganymedebdella cratere
Branchellion torpedinis (anterior)
Hemibdella soleae
Heptacyclus myoxocephali
Janusion scorpii
Malmiana bubali

Fig. 6.28

and 10 are white, and there are similar white spots on the trachelosome (two pairs), flanked by red specks. Enlarged side-views show annulation and a spermatophore protruding from male aperture.

In mouth and nose of *Raja*. South-west Britain.

Calliobdella lophii Beneden and Hesse — Fig. 6.28

Length 50 mm. Both suckers large (fore : hind ratio 1 : 4). Without eyes or ocelli. Orange ring girdles body near genital apertures. Bursa eversible, to form penis. Abdomen somewhat flattened and wide, but less so than discoid hind sucker. Pulsatile vesicles 11 pairs. Annuli six per mid-body somite.

Ventro-laterally on *Lophius*. Norway to Mediterranean.

Calliobdella nodulifera (Malm) — Fig. 6.28

Like *C. lophii*, but smaller, fore : hind sucker ratio 1 : 2 and abdomen thinner, more cylindrical and merging into trachelosome imperceptibly.

Externally on a variety of fishes, especially gadoids. Iceland, Norway, Scotland.

Calliobdella punctata Beneden and Hesse

Length 20 mm. Fore sucker, with white spots and two pairs of eyes, about two-thirds width of hind sucker, both suckers oval and relatively small. Body rust-coloured dorsally, lightly speckled with black, paler and pink ventrally; cylindrical and thin (breadth : length ratio 1 : 30), with a slight constriction near the genital apertures. Description incomplete. Pulsatile vesicles 14 or 15 pairs?

On *Taurulus*. Brittany. Probably absent from mainland Britain, but *Trypanosoma cotti*, of which it is the vector, infects *Taurulus* on Isles of Scilly.

Ganymedebdella cratere (Leigh-Sharpe) — Fig. 6.28

Length 8 mm. Fore sucker bowl-shaped, as wide as body (3 or 4 mm) and five times width of the small terminal hind sucker. Abdomen flattened and broadened, with three pairs of pulsatile vesicles. Postcaeca separate.

On a fish 'with a pronounced anal papilla'. Only one found? South Ronaldsway, Orkney.

Hemibdella soleae (Beneden and Hesse) — Fig. 6.28

Length 10 mm. Both suckers small and terminal. No eyes. Body cylindrical. Annuli 12 per mid-body somite. Bursa large and protrusible. Coelom typical, but without vesicles. Postcaeca fused.

On *Solea vulgaris*, with hind sucker clamped around spine on ctenoid scale, as figured. West Europe and Mediterranean. Abundant off south-west Britain.

Heptacyclus myoxocephali Srivastava — Fig. 6.28

Length 13 mm. Both suckers discoid, attached obliquely. Fore : hind sucker ratio 2 : 3. Body cylindrical. Annuli seven (14) per mid-body somite. Bursa small. Testes five pairs, separated by minor enlargements of crop. Dorsal, ventral, and lateral lacunae, but no transverse communications or vesicles. Postcaecal cavities unfused.

On *Myoxocephalus scorpii*. Millport, Firth of Clyde.

Janusion scorpii (Malm) — Fig. 6.28

Length 15 mm. Fore : hind sucker ratio 1 : 2. Eyes four, each composite. Distinct clitellar constriction. Annuli two per mid-body somite. Description incomplete.

On *Myoxocephalus*. Only three found? Plymouth.

Malmiana bubali Srivastava — Fig. 6.28

Length 12 mm. Like *Heptacyclus* (see above), but with six eyes, abdomen more flattened (unusually so for *Malmiana* spp.), annuli three (six) per mid-body somite and four pairs of testes, which are separated by crop caeca and by transverse communications linking the dorsal and lateral lacunae.

On *Taurulus*. Only one found? Robin Hood's Bay, Yorkshire.

Malmiana yorki Srivastava — Fig. 6.29

Like *M. bubali*, but with four eyes anteriorly, a ring of 14 paramarginal ocelli on hind sucker, five pairs of testes, sphincter muscles around the crop, and dermal cells more rounded.

On *Taurulus*. Only one found? Scarborough Bay, Yorkshire.

Oceanobdella blennii (Knight-Jones) — Fig. 6.29

Length 12 mm. Fore sucker 0.7 mm across when expanded. Eyes six. Hind sucker 1.7 mm, with some paramarginal spots, including one on each side of a mid-dorsal brown stripe. Body flattened when resting, often in sinuous but immobile curves (see small-scale figures), more cylindrical and thread-like when extended, marked in adults by transverse brown segmental stripes. White dermal cells show through transparent epidermis, especially dorso-

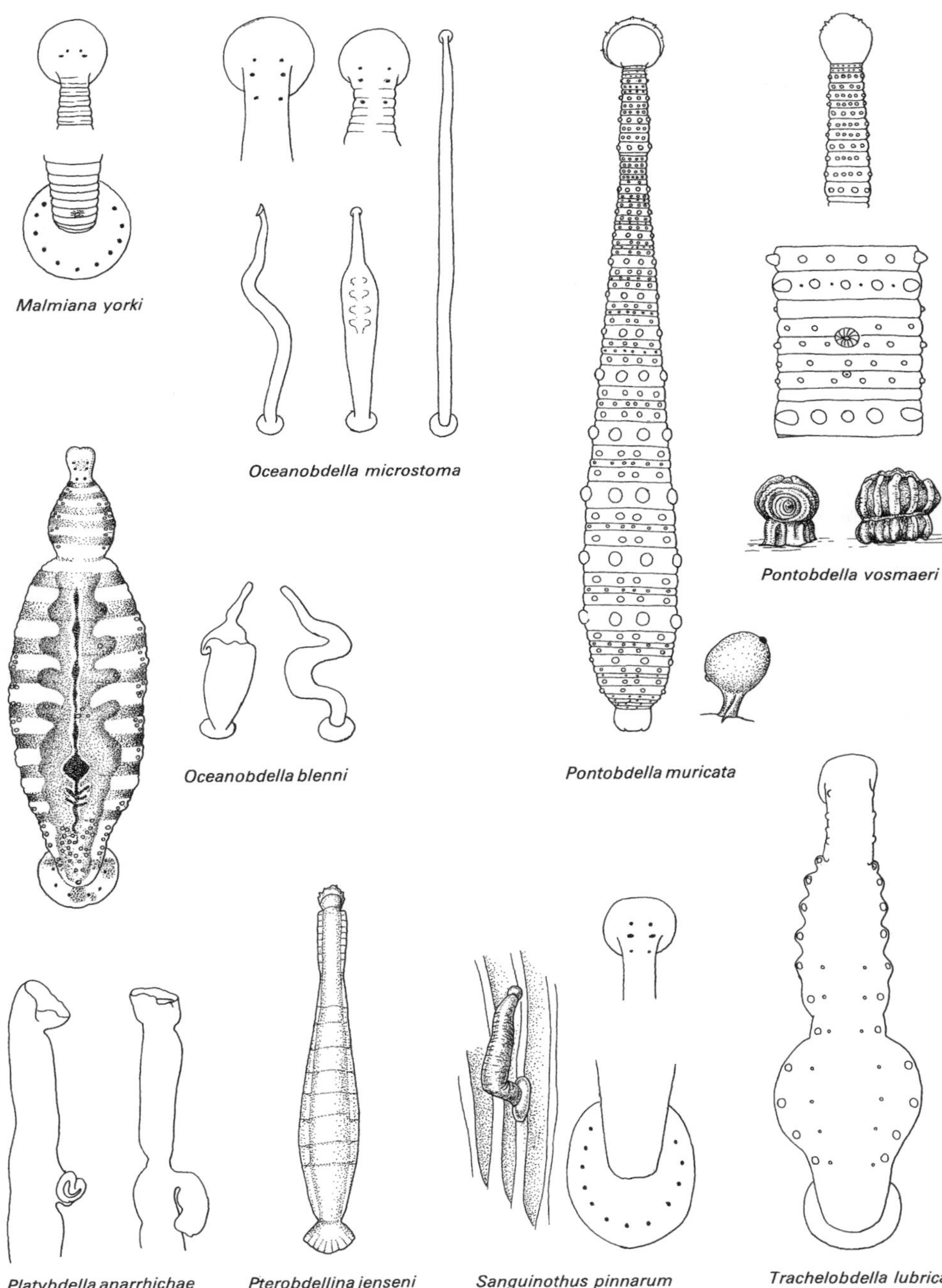

Fig. 6.29

laterally; so does the crop, with its caeca. Annuli three (six) per mid-body somite. Bursa small. Testes usually five pairs. Transverse communications as in *Malmiana*. Postcaeca usually separate.

On *Lipophrys pholis*, usually behind pectoral fins or (as unpigmented juveniles) in gill chamber. Locally common from Firth of Forth to Scarborough, and Mull to Crackington Haven (north Cornwall).

Oceanobdella microstoma (Johansson) Fig. 6.29
Like *O. blennii*, but while resting it does not become so flattened, broadened, or sinuous, and it has a complete circle of about 12 paramarginal ocelli on hind sucker; otherwise lacks pigment, except for numerous creamy bodies seen through the skin (especially dorsally) and red blood in the crop caeca.

Ventrally under head of *Taurulus, Myoxocephalus*, and other Cottidae. Occasionally on shore in spring at Plymouth, Anglesey, and St. Andrews. Also west Sweden, Spitzbergen, Iceland, Greenland, Newfoundland, and Maine.

Oceanobdella sexoculata (Malm)
Leeches very like *O. blennii*, but smaller and not so flattened, occur behind pectoral fins of *Pholis* at St. Andrews and occasionally in Northumberland. If dislodged they reattach to *Pholis* and not to *Lipophrys*. Malm's species was from west Sweden, on *Gadus morhua* and *Zoarces*. It had a ring of ocelli on the hind sucker, with one on each side of a green mid-dorsal stripe. In Scottish material the ring is usually incomplete. Newfoundland material has a complete ring, but each ocellus is in the middle of a radiating dark stripe.

Piscicola geometra (Linnaeus)
Length 20–50 mm. Eyes four. Fore : hind sucker ratio 1 : 2. Hind sucker with ring of 14 ocelli, alternating with 14 radiating dark stripes. Annuli 14 per mid-body somite. Pulsatile vesicles 11 pairs. Bursa small, and spermatophores are deposited externally on a ventral copulation area developed posterior to clitellum when breeding.

Holarctic in fresh water, and on *Pleuronectes flesus* and *Myoxocephalus* in Baltic.

Platybdella anarrhichae (Diesing) Fig. 6.29
Length 20–30 mm. Suckers discoid, without eyes or ocelli, fore:hind ratio 1:2 or 2:3. Abdomen cylindrical, merging with trachelosome. Annulation three (six) or indeterminate. No pulsatile vesicles, or lateral or segmental lacunae. Bursa large and may be extruded like a penis, as in the figures. Testes five pairs. Postcaeca fused, with five fenestrae. Epidermis lacks the small, flask-shaped gland cells present in most piscicolidae.

On gills and body of *Anarhichas lupus*. North Sea, Iceland, and Greenland.

Pontobdella muricata (Linnaeus) Fig. 6.29
Length 100–200 mm. Fore sucker equal to or much larger than hind, much wider than trachelosome, and with a distinct rim marked off by a groove. Annuli four per mid-body somite, with the main annulus larger than the others and bearing larger tubercles (eight), none of them mid-ventral. Cocoons smooth and subspherical, on stalks (as figured).

On *Raja* and occasionally *Pleuronectes*. North Sea, Mediterranean, Iceland, Greenland.

Pontobdella vosmaeri Apathy Fig. 6.29
Length 80 mm. Like *P. muricata* but fore sucker not much wider than trachelosome, lacking paramarginal groove. Fore : hind ratio 1 : 2. Main annuli bear four tubercles dorsally and five ventrally, including a mid-ventral tubercle (enlarged drawing shows those near genital pores). Cocoons smaller than those of *P. muricata*, on broad pedestals and banded with corrugations (figured).

On *Raja*. Plymouth, Roscoff, Capri.

Pterobdellina jenseni Bennike and Bruun Fig. 6.29
Length 27–39 mm. Fore sucker with thickened rim, which bears 4–6 papillae on each side. Body subcylindrical, with a fin-like membrane along each side. Annulation seven (14). Testes five pairs.

On *Raja*. Off Faeroes, depths >400 m.

Sanguinothus pinnarum de Silva and Burdon-Jones Fig. 6.29
Length 7–12 mm. Hind sucker oval, usually clamped round a fin-ray (as figured). Fore : hind ratio about 1 : 2, with six eyes and 14 ocelli, respectively. Body uniformly reddish-brown, with pale, almost white margins to the suckers and around the body next to the hind sucker. Annulation three (six) but indistinct. Bursa small. Testes five pairs. No pulsatile vesicles and no lateral or segmental lacunae. Postcaeca separate.

On fins of *Taurulus*, especially February to May. Firth of Clyde, Berwick-on-Tweed, West Anglesey.

Trachelobdella lubrica (Grube) Fig. 6.29

Genus has a small bursa, thick-walled vagina, flattened and broadened abdomen, 11 pairs of pulsatile vesicles, and a small terminal hind sucker. In this species the latter is radially marked, there are 3–4 pairs of 'non-pulsatile' vesicles dorso-laterally on the trachelosome, and the first four pairs of vesicles on the abdomen are located less dorsally than subsequent pairs. Length 4–30 mm. Specimen drawn here was from a goby near Naples.

On various teleosts. Mediterranean, West Africa, south Australia, and (doubtfully) North Sea.

7. PRIAPULIDS, SIPUNCULANS, ECHIURANS, AND ENTOPROCTS

PHYLUM PRIAPULIDA

Unsegmented worms with a chitinous *cuticle*, small surface spines and *papillae*, a large body cavity, and near-radial symmetry, except for paired gonads and nephridia, and a nerve cord regarded as ventral. The anterior body forms an *introvert*, in which the mouth is central. This is alternately pushed out and withdrawn to burrow, and prey is drawn in and swallowed during withdrawal, having been caught by curved teeth on the everted mouth and pharynx lining. The gut is simple and the anus terminal, except that a few genera have long tails arising ventrally to the anus. These bear many vesicular branches, which contain extensions of the coelom and are thought to be respiratory. Sexes are separate and cleavage radial, producing loricate larvae which stay in the adult environment.

Priapulids live in soft sediments. Only 15 living species are known and only one family occurs in cool seas. The others (Tubiluchidae and Maccabeidae) are meiofaunal, and tropical or Mediterranean so are not treated here.

1. PRIAPULIDAE

Macrobenthic, with sperm of the type associated with external fertilization. Three of the four genera have tails. Two west European species, only one of which occurs around the British Isles.

Priapulus caudatus Lamarck — Fig. 7.1

Length usually 20–75 mm, sometimes more; mouth with seven rings each of five teeth; introvert one-fifth to one-third of total length, marked by 25 longitudinal ridges; rest of body with many rings, sometimes swelling posteriorly so that base of tail is partially enclosed.

In grey mud, sublittoral and up to mid-shore. All British coasts, locally common, e.g. in Holyhead harbour, Menai Strait, and the Solent.

PHYLUM SIPUNCULA

Unsegmented worms, with an anterior *introvert* which can be withdrawn into the *trunk*. When this is everted, by pressure of coelomic fluid, the *mouth* (m in Fig. 7.1) appears at the end. Dorsal to the mouth is a sensory prominence, the *nuchal organ* (no), which lies over the brain. A *ciliated fringe*, which is often emarginated to form *tentacles*, forms a ring round the mouth and/or a horseshoe round the nuchal organ. Extension of the tentacles involves an independent fluid pressure system powered by one or two *contractile vessels* (cv) which accompany the oesophagus, lying attached to its wall. The gut is long and U-shaped, with the two halves of the U twisted together, often round an axial strand, the *spindle muscle*. The *anus* (a) is dorsal, near the anterior end of the trunk. Two *nephridiopores* (np) open at about the same level, but ventro-laterally. The sac-like *nephridia* (n), the dorsal and ventral pairs of *retractor muscles* of the introvert (drm, vrm), and the mid-ventral nerve cord can be seen internally, on opening up the coelom and moving aside the gut-loops; so can the *longitudinal muscle bands* formed in two British species on the inside of the body wall, but usually those body wall muscles form a continuous layer.

Cleavage is radial and some of the larvae resemble trochophores, but the group is isolated. It is remarkably homogeneous, but two classes are recognized.

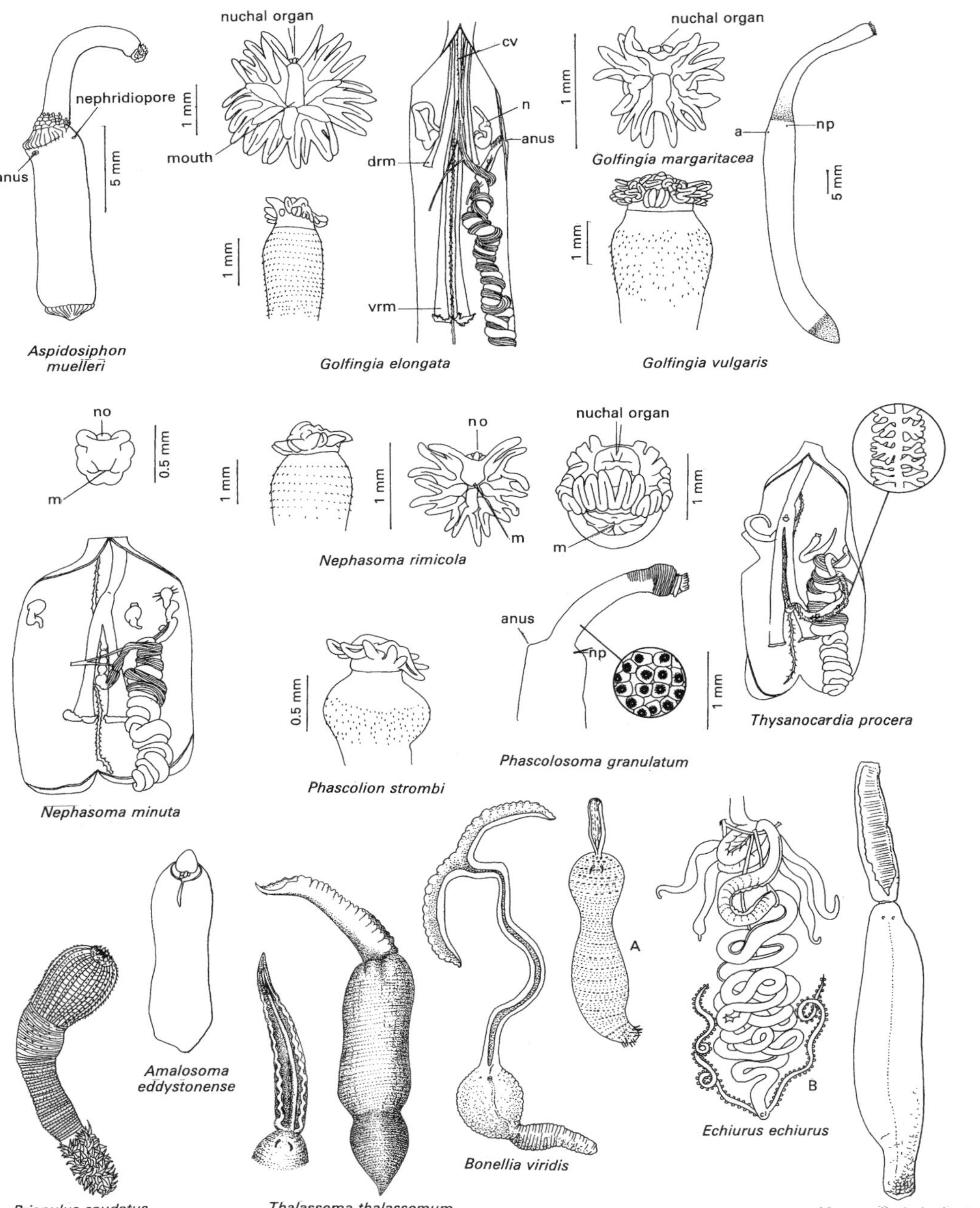

Fig. 7.1

Sipunculans burrow in mud, sand, or gravel on the lower shore or sublittorally. *Phascolion strombi* lives in old gastropod shells, and *Nephasoma minuta* in rock-crevices, with some other Golfingiidae. To narcotize, place the worm in a small quantity of cool seawater and add powdered crystals of menthol. After 1 or 2 h add 80% alcohol a few drops at a time, until the worm no longer responds to touch. Fix in 5% formaldehyde and store in 70% alcohol. To study the anatomy, open by a dorsal cut along the length of the trunk, passing to one side of the anus.

KEY TO BRITISH FAMILIES

1. Trunk with hard shield at each end — **1. Aspidosiphonidae**

   Trunk without shields — **2**

2. Tentacles all dorsal to mouth, almost enclosing nuchal organ — **2. Phascolosomatidae**

   Tentacles surround mouth or may be absent — **3**

3. Either inhabiting old gastropod shells; or with trunk rounded, its length less than twice its breadth and half length of slender extended introvert. Nephridium single — **4. Phascolionidae**

   Not in shells. Trunk length at least three times breadth, usually equalling length of introvert. Two nephridia — **3. Golfingiidae**

Class Phascolosomatidea

Tentacles all lie dorsal to mouth, forming a horseshoe round the nuchal organ. Single muscle attached to body wall posteriorly. Two orders, each with one family.

Order Aspidosiphonida

With horny epidermal shields at one or both ends of trunk.

1. ASPIDOSIPHONIDAE

Ends of trunk chitinized to form anal shield anteriorly and caudal shield posteriorly, but otherwise like Phascolosomatidae. Two genera. One British species.

Aspidosiphon muelleri Diesing — Fig. 7.1

Length up to 80 mm; tentacles 10–12, short and united for much of their length, arranged round nuchal organ in a crescent; anal shield acts like operculum when introvert is withdrawn, so this species does not cement mouths of shells as does *Phascolion*.

In gastropod shells, serpulid tubes, and crevices, e.g. amongst coralline algae; from LW to 1000 m depths. Rare in Britain, but recorded from Shetlands and Norway to Mediterranean.

Order Phascolosomatida

Without epidermal shields.

2. PHASCOLOSOMATIDAE

Differs from later families in having spindle muscle attached posteriorly and all tentacles enclosing nuchal organ. Three genera. One British species.

Phascolosoma granulatum Leuckart — Fig. 7.1

Length up to 100 mm; tentacles 12–60, increasing with size, all forming dorsally open horseshoe; introvert with many hooks anteriorly, arranged in 10–60 rings; the more posterior of these rings are often incomplete because of wear; body covered with domed papillae (shown enlarged) each capped by a dark ring round a clear spot; two nephridia; four retractor muscles; longitudinal muscles of body wall tend to form 20–30 anastomosing bands, but these are not readily seen in small specimens.

In muddy sand and gravel from shore crevices to 90 m depth. Common along Irish west coast, scarce in Hebrides, Orkneys, Shetland, and Norway. Ranges to Mediterranean, Cape Verde Islands, and Indo-West Pacific.

Class Sipunculidea

Tentacular fringe primarily surrounds mouth, although some dorsal tentacles may form a horseshoe round nuchal organ.

Order Golfingiida

Longitudinal muscles form continuous layer inside body wall, not split up into bands. Three families.

3. GOLFINGIIDAE

Tentacles surround mouth or are mere rudimentary lobes. Spindle muscle attached anteriorly, near anus, but not posteriorly. Two nephridia. Three genera.

---

1. Everted introvert at least twice as long as trunk. Contractile vessel bears many branching villi **Thysanocardia procera**

   Introvert not much longer than trunk. Contractile vessel without villi **2**

---

2. Introvert bears many spine-like hooks near anterior end **3**

   Introvert without hooks **Golfingia margaritacea**

---

3. Usually with two pairs of retractor muscles, a ventral pair attached in posterior trunk on each side of nerve cord, and a shorter dorsal pair attached more anteriorly (*Golfingia*) **4**

   Lacking dorsal pair of retractors (*Nephasoma*) **5**

---

4. Trunk pigmented at each end. Introvert hooks arranged irregularly **Golfingia vulgaris**

   Trunk not pigmented at ends. Introvert hooks arranged in rings **Golfingia elongata**

---

5. Tentacles rudimentary. Nephridiopores posterior to anus **Nephasoma minuta**

   Tentacles well formed. Nephridiopores anterior to anus **Nephasoma rimicola**

Golfingia elongata (Keferstein) Fig. 7.1

Length up to 150 mm; tentacles increase from eight to 36 during growth; introvert hooks arranged in up to 20 rings in juveniles less than 30 mm long (as figured), but adults may have only three or four as result of wear; retractor muscles usually four, but one or both of dorsal pair may be missing in aberrant individuals; specimens with both missing are distinguishable from *Nephasoma rimicola* by having more than 20 tentacles at lengths greater than 30 mm.

In muddy sand or gravel from LW to 170 m depth. Juveniles up to 30 mm occasional in shore crevices with *Nephasoma* species. Common off Plymouth. Recorded from Skagerrak to Mediterranean.

Golfingia margaritacea (Sars) Fig. 7.1

Length in British populations up to 20 mm, with 8–16 tentacles; introvert hooks absent; retractor muscles two pairs.

In muddy sand or gravel from LW crevices to 4600 m depth. Uncommon on shores of south-west Britain and west Scotland. Holarctic, perhaps bipolar. Off Norway reaches length of up to 300 mm, with over 100 tentacles, since additional tentacles develop in radial or longitudinal rows behind margin of oral disc.

Golfingia vulgaris (de Blainville) Fig. 7.1

Length up to 200 mm; tentacles increasing from 20 in a single circle at body length 10–15 mm, to 60–150 in three or more circles at lengths over 30 mm; introvert hooks arranged irregularly as figured; retractor muscles two pairs.

In muddy sand or gravel from LW to 2000 m depth, widespread round Britain and from northern Norway to west Africa and Mediterranean.

Nephasoma minuta (Keferstein) Fig. 7.1
Length up to 15 mm; oral disc with two lobes flanking nuchal organ, and 2–6 round mouth; introvert hooks arranged irregularly, but may be lost through wear; retractor muscles one pair, ventral, and these are fused for much of their length.

Very common in rock-crevices from mid-shore to 50 m depth; Shetland, Sweden, Britain, Brittany.

Nephasoma rimicola (Gibbs) Fig. 7.1
Length up to 60 mm; tentacles increasing from eight in juveniles, and 16 in most adults over 20 mm long (as figured), to 20 in large specimens; introvert hooks arranged in 6–10 rings (figured); only the ventral pair of retractor muscles is developed.

Fairly common in LW rock-crevices in south-west Britain.

Thysanocardia procera (Möbius) Fig. 7.1
Length up to 60 mm, but trunk only 15 mm; skin of trunk with irregular pattern of zigzag, mainly longitudinal folds; tentacles in radial rows and fairly numerous, up to 70; nuchal organ separated into two lobes by a longitudinal groove; introvert hooks absent; only the ventral pair of retractors is developed; distal half of contractile vessel bears many branching villi (see enlarged insert).

In muddy sand at depths 2–200 m; Skagerrak; Kattegat; northern North Sea; western Scotland; western Ireland; Milford Haven.

4. PHASCOLIONIDAE

Like Golfingiidae, but lacking left nephridium, spindle muscle, and some retractors. Two genera. A single species of *Phascolion* occurs commonly around Britain. Two species of *Onchnesoma* occur rarely off western coasts in sand and mud, below 25 m.

Phascolion strombi (Montagu) Fig. 7.1
Up to 50 mm long; tentacles 8–16 when half-grown, 40–50 in large specimens which develop multiple circlets; introvert hooks irregularly arranged as figured; skin covered with papillae which are enlarged, conical, and pigmented in a band round middle of trunk; contractile vessel often as wide as oesophagus; main retractor muscle formed by complete fusion of dorsal pair: ventral pair narrower and similarly fused, except for a posterior division which allows attachment on each side of nerve cord.

In gastropod shells, cementing particles round shell mouth to leave narrow hole through which introvert may be protruded. Very common round all British coasts; panatlantic; Mediterranean; bipolar.

PHYLUM ECHIURA

Coelomate worms which may be distantly related to annelids, because most possess at least a pair of *chaetae* and some have trochophore larvae. Adults, however, are unsegmented. The body is plump, with a posterior *anus*, and a *proboscis* extending forward from the anterior *mouth*. Like an annelid prostomium, the proboscis includes the brain, but it is long, soft, muscular, easily torn away, and often coloured differently from the rest of the worm. Its ventral surface forms a shallow ciliated channel along which mucus and detritus are brought to the mouth. The mouth is at the proboscis base and enclosed by the edges of the channel. On the ventral surface, close behind the mouth, there are usually two hooked chaetae and the pore(s) of one or more *nephridia*, which lie in the coelom, attached to the ventral wall, and act as genital ducts. The gut is much longer than the body, so follows a convoluted zigzag course. Two or more *anal vesicles*, bearing ciliated funnels or tubules, also lie in the coelom and connect with the cloaca.

Echiurans burrow into sand or mud, mostly at great depths, but some are found in shallow water, especially in rock-crevices or under stones. They can be narcotized with menthol, fixed by 5% formaldehyde in seawater, and preserved in 70% methanol. For study of internal anatomy the body is opened throughout its length by a mid-dorsal incision.

KEY TO BRITISH FAMILIES

---

1. Without sexual dimorphism. Anal vesicles elongated, with many small ciliated funnels attached directly to their surfaces. With at least a pair of antero-ventral chaetae. One genus has posterior chaetae as well
1. Echiuridae

Most, perhaps all, with dwarf males living on or in females. Anal vesicles sack-like, with ciliated funnels borne on tubules, which usually branch. Some lack chaetae. None has posterior chaetae
2. Bonelliidae

Order Echiurida

Longitudinal muscle fibres of body wall lie between layers of outer circular and inner oblique muscle fibres. Blood vessels present. Both sexes of similar size. One family.

1. ECHIURIDAE

Includes nine genera and nearly 80 species, but few around Britain.

---

1. Two rings of chaetae surround anal region. Proboscis tip not bilobed. Occurs in shallow water **_Echiurus echiurus_**

Posterior chaetae absent. With two pairs of nephridia **_Thalassema thalassemum_**

Echiurus echiurus (Pallas) Fig. 7.1
Proboscis 30–40 mm long, orange with brown streaks; body 70–110 mm long, greyish-yellow, bearing 21–23 rings of papillae, each alternating with four or five rings of smaller papillae. With two anterior chaetae and two posterior circles, each of between five and nine chaetae. Figure B shows a dissection of the convoluted gut, with the pair of long anal vesicles. Holarctic, extending to Kattegat and North Sea. One recorded from shore debris at St. Andrews.

Thalassema thalassemum (Pallas) Fig. 7.1
Body length 20–70 mm; extensible proboscis 10–200 mm long, tapering to a point. Figures show whole animal from right side, with a ventral view of the anterior end to show the proboscis channel and the pair of ventral chaetae found in all species of the family. Chaetal sacs connected internally by a transverse interbasal muscle. Two pairs of nephridia. Anal vesicles also like those in *Echiurus*.

Mediterranean, extending to southern Britain. Very common in Devon and Cornwall, in rock-crevices at LWST. Rarer in South Wales. One dredged off Isle of Man.

Order Bonelliida

Like echiurids, but with dwarf males parasitic on the females. One family. Three species recorded for Britain.

2. BONELLIIDAE

With 23 genera, from 11 of which dwarf males have been described, living on or inside the females. Males not yet known from remaining genera.

---

1. Females lacking chaetae, but with a deep genital groove on mid-ventral surface, between mouth and nephridio-pores **_Amalosoma eddystonense_**

Females without distinct genital groove, but with a pair of antero-ventral chaetae connected internally by an interbasal muscle **2**

---

2. Proboscis bifid. Single nephridium **_Bonellia viridis_**

Proboscis truncate but not bifid. A pair of nephridia **Maxmuelleria lankesteri**

Amalosoma eddystonense Stephen Fig. 7.1
Proboscis ovoid. Body of female up to 144 × 25 mm, mostly greenish, covered with darker papillae. One pair of nephridia which may be packed with ova. Male nearly 2 mm long with two curved chaetae, by which it is attached within genital groove.

Described from several incomplete specimens dredged near the Eddystone Lighthouse and Rame Head, Plymouth, and from two dredged at over 4000 m depth in Gulf of Gascony.

Bonellia viridis Rolando Fig. 7.1
Body up to 150 mm long, green and covered with papillae. Proboscis longer, extensible to over 1 m, bifid. One nephridium usually on right side. Anal vesicles sac-like, with branches that end in ciliated funnels. Males nearly 2 mm long, lacking chaetae, ciliated all over, attached initially to female proboscis, later within pharynx, and eventually inside nephridium.

Very rare off Ireland and Norway, but common in shallow water of Mediterranean, e.g. off Chios.

Maxmuelleria lankesteri (Herdman) Fig. 7.1
Body up to 120 × 300 mm, green and papillate. Proboscis nearly as long, greenish. Two ventral chaetae with a strong interbasal muscle. Two nephridia. Anal vesicles sac-like, covered with densely aggregated tubules. Male unknown.

Dredged sparsely in Irish Sea, west Scotland, Kattegat, Skagerrak. Not clearly distinguished from the Mediterranean form *Maxmuelleria gigas* (M. Müller), which was described earlier.

PHYLUM ENTOPROCTA

A clearly defined group of small, sessile, filter-feeding animals, many of which are colonial. The simplest entoprocts, however, are non-colonial although often gregarious, and found associated with sponges, bryozoans, polychaetes, and sipunculids, in all of which the host animal generates a feeding or respiratory water current of benefit to the commensal.

Each entoproct consists typically of a visceral region, or *calyx*, and a supporting *stalk*; in colonial forms the individuals, or *zooids*, are united by a *stolon*, which grows from its apices, branching and budding off successive zooids. In non-colonial forms daughter individuals are budded directly from the calyx, in time breaking free and settling nearby. The calyx is roughly globular and bears a ring of ciliated tentacles around the margin, the number increasing during life by additions made each side of a posterior gap until the maximum is reached. The tentacles are flexible and can be expanded for feeding or folded down to lie inside the tentacular membrane which joins their bases. Contraction of a muscular sphincter within this membrane can then close the calyx tightly. The calyx contains the looped alimentary canal. It lies in the median plane, with both mouth and anus opening inside the tentacles. The anus is raised on a short papilla, leaving a space, the *atrium*, between it and the mouth. Embryos develop in the atrium. The stalk is muscular and flexible, and the longitudinal muscles continue, with or without a break, into the calyx. The body cavity of stalk, calyx, and tentacles is a pseudocoel filled with gelatinous mesenchymatous tissue.

Commensal entoprocts can be found by collecting and examining suitable host animals (see below). There are many British species, and some are common. Colonial entoprocts can be searched for on hydroids, bryozoans (such as *Scrupocellaria*), and delicate algae low on the shore, where they may be seen as a whitish fuzz over the surface. More often, however, they will be seen first when sorting material under a microscope. Particularly striking are the jerky, sweeping movements made by zooids from the base of their stalk. The structure of the stalk is important and separates two families. The number of tentacles is frequently employed in the discrimination of all entoproct species. For study purposes, therefore, some individuals (or zooids) must be preserved with the tentacles expanded. Isotonic magnesium chloride (7.5% $MgCl_2{\cdot}6H_2O$ in distilled water) added to an equal volume of seawater is a suitable narcotic. Muscles are best seen in unfixed animals, but can be shown up clearly only by the use of phase contrast, interference, or polarizing microscopes.

KEY TO FAMILIES

1. Non-colonial. Small entoprocts (often about 0.5 mm or less) generally found in aggregations commensally on worms, sponges, or bryozoans. Buds produced from the calyx **1. *Loxosomatidae***

   Colonial and stolonate. Freeliving. Buds not produced from the calyx **2**

2. Zooid stalk of uniform structure, muscular and flexible **2. Pedicellinidae**

   Zooid stalk at least partly rigid, supported by a muscular base and, in many cases, divided into two or more slender segments (internodes) by swollen muscular joints (nodes) **3. Barentsiidae**

Order 1 Solitaria

Small, non-colonial forms often less than 0.5 mm high. The longitudinal muscles are continuous throughout both stalk and calyx. Daughter individuals are budded, and separate off, from the oral wall of the calyx. Commensals; although sometimes found free living in aquarium systems or on panels used for fouling studies.

1. LOXOSOMATIDAE

Non-colonial and normally commensal; with the characters of the order. Two principal genera, separated as follows:

1. Stalk terminating in a sucking disk; adult animal capable of movement from place to place. No foot gland in bud ***Loxosoma***

   Adult cemented to host or substratum by base of the stalk; incapable of movement from place to place. A foot (cement) gland present in the bud (Fig. 7.2), persisting or disappearing in the adult ***Loxosomella***

Loxosomella is the larger genus. No species are described here, but some of the commoner ones are listed below with their host organisms:

Loxosoma pectinaricola Franzén (Fig. 7.2), on *Pectinaria belgica* (Polychaeta)

Loxosoma rhodinicola Franzén, on *Rhodine loveni* (Polychaeta)

Loxosomella atkinsae Bobin and Prenant, on *Phascolion strombi* (Sipuncula)

Loxosomella claviformis (Hincks), on *Aphrodita aculeata* and *Hermione hystrix* (Polychaeta)

Loxosomella compressa Nielsen and Ryland, on notopodial chaetae of *Gattynana cirrosa* and *Lagisca extenuata* (Polychaeta)

Loxosomella discopoda Nielsen and Ryland, on *Amphilepis norvegica* (Ophiuriodea)

Loxosomella fauveli Bobin and Prenant, on *Aphrodita aculeata* and *Hermione hystrix* (Polychaeta)

Loxosomella murmanica (Nilus), on *Phascolion strombi* (Sipuncula)

Loxosomella nitschei (Vigelius)(Fig. 7.2), on bryozoans

Loxosomella nordgaardi Ryland, on bryozoans

Loxosomella obesa (Atkins), on *Aphrodita aculeata* (Polychaeta)

Loxosomella phascolosomata (Vogt), on *Golfingia* spp. (Sipuncula)

Loxosomella varians Nielsen, on *Nephtys* spp. (Polychaeta)

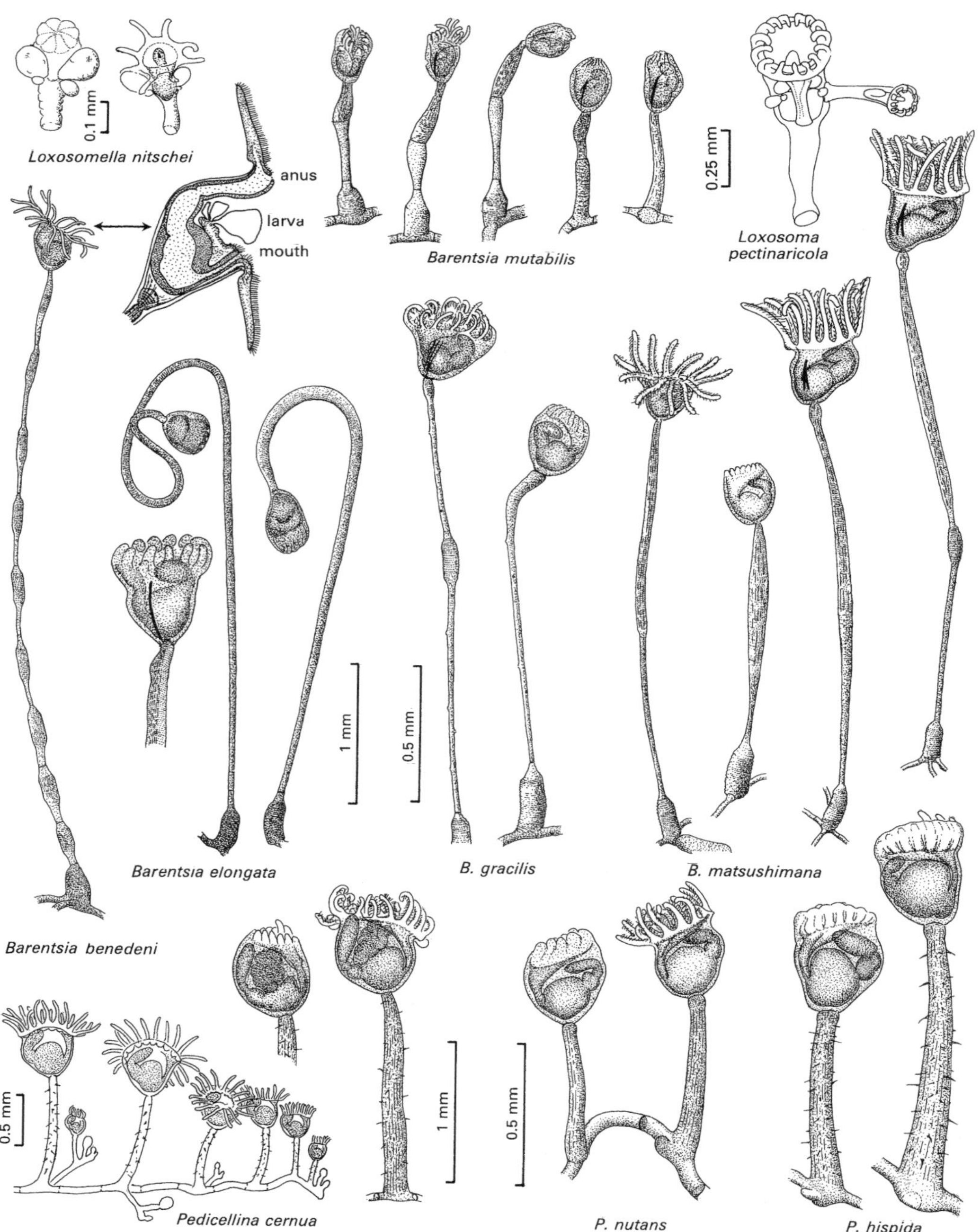

Fig. 7.2

Order 2 Coloniales

Colonial forms with zooids, often about 2 mm high, arising from a common basal plate (suborder Astolonata) or stolon (suborder Stolonata). The genera considered in this book belong to the latter. The calyx is separated from the stalk by a constriction, within which lies the *pump-organ*, consisting of a stack of flat, stellate, contractile cells. Longitudinal muscles of the calyx (*atrial retractors*) and stalk are separate.

2. PEDICELLINIDAE

Colonial, with zooids arising from a creeping, branching stolon. Stolons usually septate, of alternating zooid-bearing and barren segments. Zooid stalks unjointed, flexible, entirely muscular. Muscles in stalk and calyx not continuous; atrial retractor muscles disposed all round the oral end of the calyx. British species are referable to the genus *Pedicellina*.

---

1. Zooids reaching or exceeding 2 mm in height; calyces strongly asymmetric; up to 24 tentacles **Pedicellina cernua**

   Zooids usually less than 1.5 mm in height, occasionally reaching 2 mm; calyces only slightly asymmetric; maximum number of tentacles between 15 and 21 **2**

---

2. Stalk smooth; adult tentacle number 14–16 (–17) **Pedicellina nutans**

   Stalk bristly; adult tentacle number 16–21 **Pedicellina hispida**

Pedicellina cernua (Pallas) Fig. 7.2

Colonies often dense. Stolons creeping, about 60 μm in diameter, stalk parallel-sided or tapering very slightly towards calyx; its height rarely more than twice that of calyx; glabrous or hispid. Calyx about 0.5–0.6 mm high, large in relation to stalk, somewhat compressed, strongly asymmetric in side-view; up to 24 tentacles; anal papilla upright or reflected; glabrous or hispid.

The commonest and most conspicuous entoproct in British waters; on algae, hydroids, and bryozoans; low on the shore and from shallow water; all coasts.

Pedicellina hispida Ryland Fig. 7.2

Stolons creeping, about 60 μm in diameter, often branching in cruciform manner at the zooid bases. Zooid stalks 1.0–1.5 mm in height, tapering from a basal width of about 200 μm to 80–100 μm at the calyx; two and a half to three times calyx height. Calyx 250–500 μm high, somewhat asymmetric; about 20 tentacles.

On hydroids and bryozoans; from low on the shore on English Channel coasts. Perhaps present elsewhere but confused with the other species.

Pedicellina nutans Dalyell Fig. 7.2

Colonies sometimes very dense. Stolons creeping, about 70 μm in diameter. Zooids generally about 1–2 mm in height. Stalks tapering, their diameter at calyx about half that at base: base diameter about 160 μm in a stalk 1.5 mm high and 100 μm in one 0.6 mm high. Calyx usually 200–300 μm high (400–500 μm in some colonies), one-third to one-fifth of total height; bell-shaped, slightly asymmetric in side view; 12–15 (-17) tentacles; anal papilla reflected.

Found in great abundance on small algae near ELWS on certain rocky shores; probably widespread but confused with *P. cernua*.

3. BARENTSIIDAE

Colonial, with zooids arising from a creeping or erect, branching stolon. Stolons septate, of alternating zooid-bearing and barren segments. Zooid stalks divided into alternating rigid sections (*internode*) and flexible muscular joins (*nodes*). Atrial retractor muscles in two orolateral series, converging V-like towards base of calyx. British species are referable to the genus *Barentsia*.

---

1. Stalk relatively short, two to three times as long as calyx; internode short, widening distally and becoming flexible, narrowing again towards calyx base; some zooids often with an undifferentiated (*Pedicellina*-like) stalk **Barentsia mutabilis**

   Stalk relatively long, usually many times longer than calyx (if short, then internode thin, not or scarcely widening distally, flexible just below calyx base); all zooids with a muscular base and variously differentiated stalk **2**

---

2. Adult zooids usually with rigidly cuticularized internodes separated by joint-like nodes; if nodes absent, then stalk above base rigid and terminating below calyx in a sharply demarcated, flexible joint, never constituting more than one-fifth of stalk length **3**

All zooids without intercalated nodes above base; lower portion (one-third to one-half) rigid; changing gradually to a flexible upper portion **4**

---

3. Nodes and internodes numerous (5–20); internodes lacking pores; atrial retractor muscles in a broad band, splaying-out basally; usually in brackish water ***Barentsia benedeni***

   Usually one or two nodes (occasionally none or more than two); cuticle of internodes sparsely porous; 2–5 slender atrial retractor muscles on each side; marine only ***Barentsia gracilis***

---

4. Zooids very tall (up to 15 mm); upper part of stalk flexible, with musculature confined to oral side; very motile, with stalk typically curling distally ***Barentsia elongata***

   Zooids quite short, usually about 1–2 mm high (up to 4 mm); upper part of stalk flexible, but with symmetrical musculature; very motile, but the stalk not typically curling distally ***Barentsia matsushimana***

Barentsia benedeni (Foettinger) Fig. 7.2
Colonies dense; stolons creeping, although sometimes arising from the zooid stalks. Adult zooids up to 10 mm in height. Stalk with 5–10 (–25) nodes; internodes 250–650 μm long, shorter, as long as, or a little longer than nodes. Calyx trigonal, oral side strongly convex, tapering towards its base; 14–20 long tentacles. Atrial retractor muscles in a broad band on each side, fanning out basally.

On various substrata; euryhaline and tolerant of polluted water; most likely to be encountered on pier pilings, etc. in estuaries and harbours.

Barentsia elongata Jullien and Calvet Fig. 7.2
Colonies dense, stolons creeping. Zooids very tall, 5–15 mm. Stalk with a muscular base but without nodes; long, thin, but slightly increasing in thickness distally; rigid proximally, flexible towards calyx, the flexible part with longitudinal muscles confined to oral side, enabling upper region to curl, sometimes in several convolutions. Calyx bell-shaped, slightly asymmetric in lateral view; (10–) 12–16 (–20) tentacles. Atrial retractor muscles in a thin band on each side.

On firm substrata in shallow water; probably widely distributed. Has been found in aquaria.

Barentsia gracilis (M. Sars) Fig. 7.2
Stolon creeping. Zooids 2–3 (–5) mm high. Stalk slender, without or with one or two nodes; internodes much longer than nodes, about 30 (–40) μm in diameter. Calyx up to about 350 μm high, more or less ovoid, quite strongly asymmetric in lateral view; 12–18 tentacles. Atrial retractor muscles in 1–4 slender bands on each side of calyx.

On shells (e.g. between ribs of *Pecten* and *Cardium* valves), crabs, hydroids, bryozoans, and algal stems; stenohaline, on the shore and in shallow water; all coasts; widespread, the commonest *Barentsia* on British coasts.

Barentsia matsushimana Toriumi Fig 7.2
Colonies often extensive. Adult zooids small, seldom longer than 1–1.5 (–4)mm. The stalk slender, with a muscular base but no nodes; rigid proximally, flexible and transversely wrinkled distally; usually poreless; musculature never restricted to the oral side only. Calyx asymmetrical in lateral view; 12–14 tentacles. Atrial retractor muscles small, short, not reaching base of calyx, and widest proximally.

On firm substrata and algae from low on the shore into shallow water; apparently widely distributed though rarely reported. (North Sea coasts and Isle of Man).

Barentsia mutabilis (Toriumi) Fig 7.2
Stolon creeping, 25–45 μm in diameter. Zooids to about 0.9 (–1.6) mm in height. The stalk 500–600-μm high, less than three times calyx height, usually with a muscular base, a biconcave rigid internode and a flexible subcalycal portion; internode usually with pores; stalk sometimes wholly non-rigid and muscular (as in *Pedicellina*). The two kinds of stalk may occur within the same colony. Calyx about 200 μm high, bell-shaped, very slightly asymmetrical in side view; 10–11 (–14) tentacles. Atrial retractor muscles one or two on each side of calyx, converging towards the base of the calyx.

On firm substrata; low in the intertidal zone or in shallow water; apparently uncommon.

8 CRUSTACEANS

The crustaceans (Phylum Crustacea) exhibit a great diversity of structure, adaptation, and development. Freshwater and terrestrial species are common, but the majority of species are marine. Crustaceans occur in all marine habitats, from the supralittoral zone to the abyss; additionally, most holoplanktonic animals belong to the Crustacea, and the meroplankton largely consists of the larval stages of benthic Crustacea.

The crustacean body (Fig. 8.1) is segmented and organized into distinct regions. The segments, or *somites*, are typically compressed or depressed to a varying degree, but the dorsal *tergum*, ventral *sternum*, and lateral *pleuron* are usually recognizable. The body wall is basically chitinous and is usually reinforced by calcium carbonate to form a rigid exoskeleton. Growth involves periodic moulting of the exoskeleton, necessitating hormone-controlled resorption of calcium salts, and their redeposition in the new outer chitinous skeleton. The body regions comprise a *head*, *thorax*, and *abdomen*, with an additional tail piece, the *telson*. In many crustaceans a number of anterior thoracic segments is fused with the head. The abdomen is variously developed throughout the phylum, and is often much reduced. In many groups a fold derived from the head extends posteriorly to form a characteristic *carapace* which covers most or all of the thorax.

Primitively, each body segment bears a pair of biramous appendages. In its simplest form the crustacean head has three pairs of appendages: the *antennule* (or *antenna 1*), the *antenna* (*antenna 2*), and the *mandibles*. Typically, the head is fused with two anterior thoracic segments, the appendages of which are modified as accessory feeding structures, the *maxillule* (*maxilla 1*) and the *maxilla* (*maxilla 2*). Thoracic and abdominal appendages show a wide range of modification throughout the phylum as a whole.

The crustacean limb (Fig. 8.1) consists of two segmented rami, the *exopodite* and the *endopodite*, rising from a peduncle of two articles, the *coxa* and the *basis*. Simple biramous limb structure persists in some primitive crustaceans, but in most groups the exopodite comprises the major functional unit of the limb, while the endopodite is reduced, or lost, or adapted to serve a separate function, such as feeding, cleaning, or respiration. In some higher crustaceans, additional respiratory structures (*podobranchs* and *pleurobranchs*) are associated with the thoracic limbs.

Sexes are separate, with the exception of the sessile Cirripedia in which hermaphroditism is generally the rule. The characteristic larval type is the *nauplius* (Fig. 8.1); it occurs as the first free-swimming larval stage in many crustaceans, while in others it occurs only in the earliest, brooded stages of development. A succession of often very specialized larval instars occurs in the later development of most crustaceans.

Crustaceans range in size from less than 0.5 mm to more than 1 m. Their morphological diversity, and wealth of adaptive features make them a difficult phylum to define satisfactorily and crustacean classification has been the subject of much dispute. Morphological terms, consequently, are often confusing when used in comparative accounts of different classes or orders. A single, unified terminology for the Crustacea as a whole is desirable, but there is no consensus on this problem. Morphological terminology is simplified here with the sole intention of aiding the task of identification.

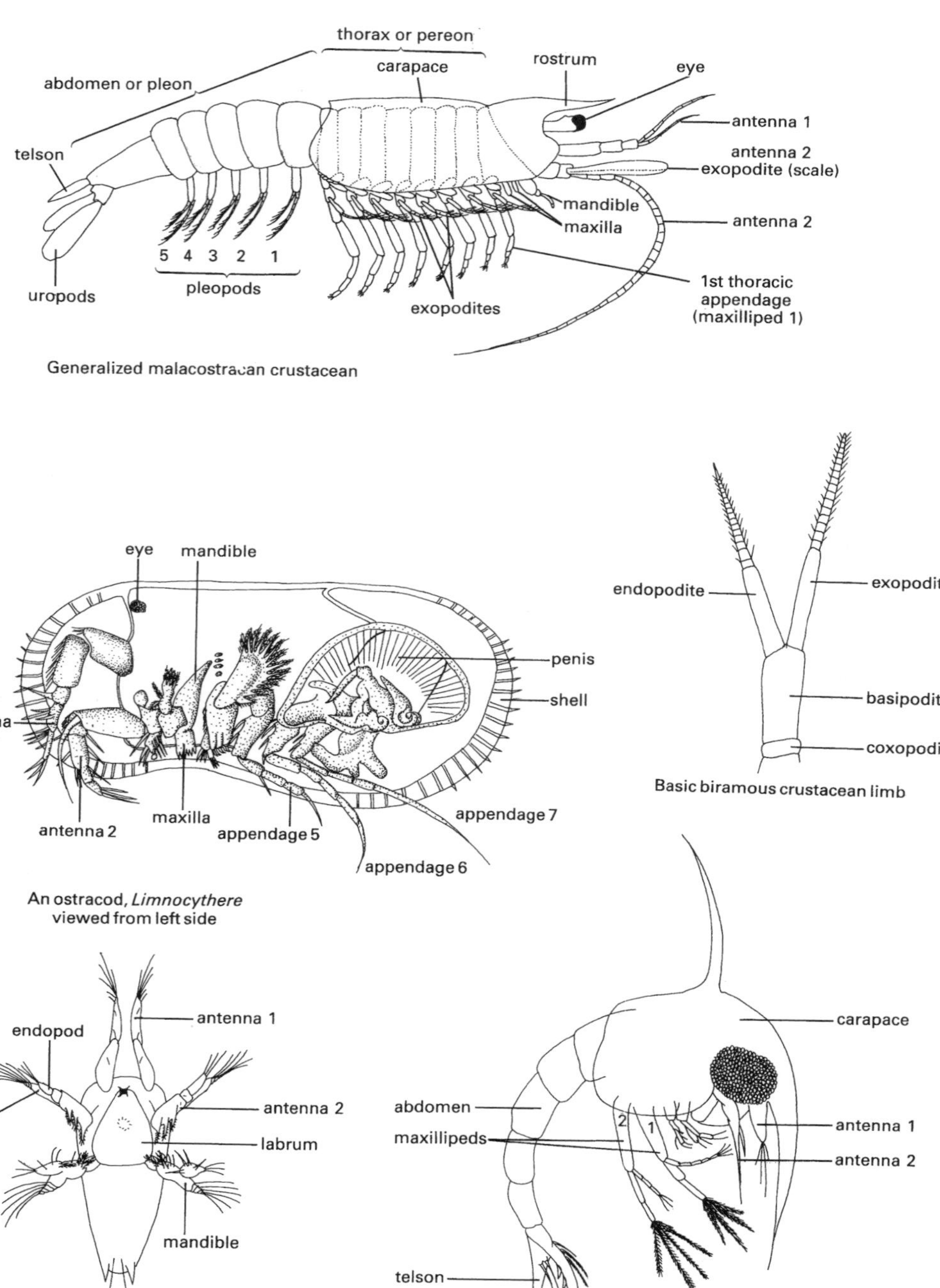

Generalized malacostracan crustacean

Basic biramous crustacean limb

An ostracod, *Limnocythere* viewed from left side

Third nauplius larva of a copedod, *Calanus*

First zoea larva of *Carcinus maenas*

Fig. 8.1

Class Ostracoda

In the Ostracoda the crustacean carapace is present as a bivalved shell into which the animal can withdraw completely (Fig. 8.1). The two valves of the shell are closed by an adductor muscle which attaches to the centre of each valve. The shell is part of the exoskeleton and is thus shed and reformed at each moult. The ostracod body is highly modified, reduced in size, and largely coalesced, with no externally visible segmentation. It bears 5–7 pairs of appendages, the terminology and homology of which vary between orders. Typically, there are two pairs of *antennae* and a pair of *mandibles*, with palps, arising from the indistinct head region. The first thoracic appendage is the *maxilla*, or *maxilla 1*, while the second (appendage 5) may be developed as *maxilla 2* or as the first walking leg. The remaining two pairs of appendages are developed as thoracic walking legs. The abdomen is devoid of appendages, but an unpaired, terminal *caudal furca* is situated anterior or posterior to the anus.

Ostracods are widespread in marine, brackish, and freshwater habitats. Most marine species are benthic, burrowing in, or crawling on, the substratum, but some are free swimming and a few are planktonic. The British marine fauna includes more than 140 species. Identification is not easy and generally requires specialist literature. However, the following key will permit identification of common, intertidal species.

KEY TO COMMONER SHORE SPECIES

1. Oral cone terminating with sucking disk (Fig. 8.2) **2**

   Oral cone without terminal disk **3**

2. Powerful well-developed mandibles; sucking-disk incomplete ***Cytherois fischeri***

   Mandibles less well-developed; sucking-disk complete ***Paradoxostoma***

3. Shell rhomboidal in shape, with small pits on surface; ***Loxoconcha impressa***

   Shell of different shape **4**

4. Shell with a rounded or obtuse posterior protuberance **5**

   Shell without this protuberance **6**

5. Shell with white tips and a white band across the back; posterior protuberance relatively short ***Cytherura gibba***

   Shell with white tips only; posterior protuberance longer ***Semicytherura nigrescens***

6. Shell plump, flattened at posterior end. Female with brood chamber within the shell. Antennae and legs with relatively moderate setae ***Xestoleberis aurantia***

   Shell less plump, with rounded posterior end. Female without brood chamber. Antennae and legs with stout setae **7**

7. Eyes fused; shell relatively elongate ***Leptocythere***

   Eyes separate; shell less elongate **8**

8. Shell dark reddish brown, with relatively sparse hairs ***Cythere lutea***

   Shell lighter or patchy in colour, hairs more dense **9**

9. Shell yellowish grey with irregular dark patches. Limbs dark yellow ***Heterocythereis albomaculata***

   Shell light brown. Limbs colourless ***Hirschmannia viridis***

1. CYTHERIDAE

Shell ovate, subreniform, or sub quadrangular; 0.4–0.9 mm long; right valve often overlapping left dorsally. Sculpture variable. A single V-shaped frontal scar, and a vertical row of four adductor scars. Three pairs of similar walking legs; antenna 1 with five or six joints, bearing bristles and claws; antenna 2 with well-developed spinneret bristle. Caudal furca small, sometimes indistinct, but always present, with two or three setae.

Cythere lutea Müller Fig. 8.2
Shell subreniform, slightly tapered posteriorly, and somewhat angular, anterior margin smoothly rounded; up to 0.74 mm long, male smaller than female; smoothly calcified, with few large punctae.

Intertidal and shallow sublittoral, often in rock pools, ranging into estuaries. All British coasts.

2. CYTHERURIDAE

Shell variable, usually dorsoventrally flattened, with caudal process; 0.3–0.7 mm long. Sculpture variably developed, often including longitudinal or vertical ribs. A single, V-shaped frontal scar, and a vertical row of four adductor scars. Three pairs of similar walking legs; antenna 1 with six joints (rarely five); antenna 2 with four or five joints, a spinneret bristle and two terminal claws. Caudal furca with three bristles or more.

Cytherura gibba (Müller) Fig. 8.2
Shell elongate oval, greatest height close to anterior end, dorsal and ventral borders almost parallel; anterior border smoothly rounded, posterior border with short caudal process; 0.56 mm long; surface with bold reticulate ridges.

In brackish water habitats. All British coasts.

Semicytherura nigrescens (Baird) Fig. 8.2
Shell elongate, greatest height close to mid-line, posterior border with well-marked caudal process, ventral border straight, dorsal border gently convex; 0.35–0.45 mm long; smooth surfaced, with fine punctae or striations.

Intertidal and shallow sublittoral; among algae, often in rock pools, and extending into estuaries and brackish habitats. All British coasts.

2. HEMICYTHERIDAE

Shell elongate, subrectangular or oblong, obliquely rounded anteriorly; 0.5–1.2 mm long. Well calcified, smooth, or with variable sculpture of pits, reticulations, or longitudinal ribs. With two or three anterior frontal scars, and a vertical row of four adductor scars. Three pairs of similar, geniculate, walking legs; antenna 1 with five joints; antenna 2 with well-developed exopod.

Heterocythereis albomaculata (Baird) Fig. 8.2
Shell elongate oval, broadly fusiform in dorsal view; smoothly rounded anteriorly, slightly tapered posteriorly, 0.85–0.98 mm long; surface finely punctate, with larger, conspicuous pores.

Intertidal and shallow sublittoral, among algae. All British coasts.

4. LEPTOCYTHERIDAE

Shell slender, small, oval to subquadrangular, with distinct posterior caudal angle; well calcified, smooth or highly sculptured. A single V-shaped frontal scar, and a vertical row of four adductor scars. Three pairs of walking legs, increasing in length posteriorly. Antenna 1 with five articles; antenna 2 with two terminal claws and a well-developed spinneret bristle; caudal furca with two setae.

Leptocythere Fig. 8.2
Shell elongate, up to 0.7 mm long; surface with coarse punctulation.

Intertidal and shallow sublittoral, sometimes occurring in brackish and estuarine waters.

5. LOXOCONCHIDAE

Shell short, rhomboidal, posterior margin obliquely rounded or with a dorsal caudal process; 0.4–0.8 mm long. Sculpture weakly developed. Three pairs of slender walking legs. Antenna 1 with five or six articles, slender, with weak claws; antenna 2 with two terminal claws. Caudal furca with one or two setae.

Loxoconcha impressa (Baird) Fig. 8.2
Shell obliquely oval, greatest height close to midline; 0.6–0.7 mm long; light yellow.

Intertidal and shallow sublittoral, amongst seaweeds, common and often abundant; also in freshwater and brackish habitats. All British coasts.

Hirschmannia viridis Müller Fig. 8.2
Shell reniform, parallel-sided in dorsal view, tapered anteriorly, rounded posteriorly; 0.45–0.55 mm long, surface finely pitted. Four adductor scars, the dorsal-most U-shaped; frontal scar U-shaped.

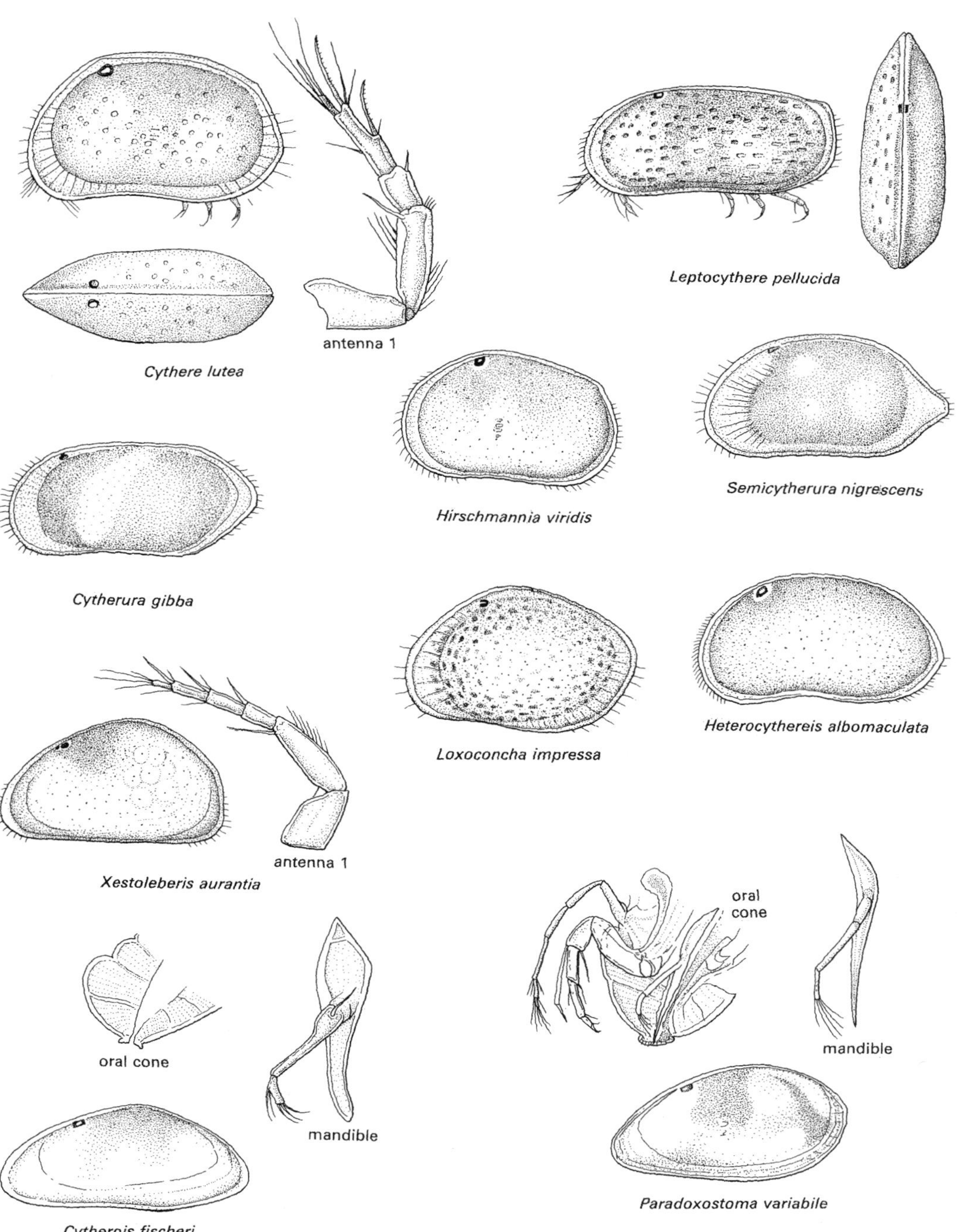
antenna 1
Cythere lutea
Leptocythere pellucida
Hirschmannia viridis
Semicytherura nigrescens
Cytherura gibba
Loxoconcha impressa
Heterocythereis albomaculata
antenna 1
Xestoleberis aurantia
oral cone
mandible
oral cone
mandible
Cytherois fischeri
Paradoxostoma variabile

Fig. 8.2

Intertidal and shallow sublittoral, among algae; often in brackish water. All British coasts; distributed from northern Norway to Biscay.

6. PARADOXOSTOMATIDAE

Shell elongate, laterally compressed and very slender, lightly calcified, usually smooth; 0.3–0.9 mm long. No frontal scar, a vertical or oblique row of four or five adductor scars. Walking legs slender, of differing lengths. Antenna 1 with five or six joints; antenna 2 with well-developed spinneret bristle. Mouthparts modified, often forming a sucking disc.

Cytherois fischeri (Sars) Fig. 8.2
Shell slender, elongate, markedly tapered anteriorly, 0.5–0.65 mm long.

Intertidal and shallow sublittoral, amongst algae; also in estuarine and brackish-water habitats; common and often abundant. All British coasts.

Paradoxostoma variabile (Baird) Fig. 8.2
Shell subovate, greatest height posterior to mid-line, 0.65–0.8 mm long. Colour variable, black to deep blue, brown, or red to grey.

Very common intertidally and sublittorally among algae. Also occurs in estuaries. All British coasts.

7. XESTOLEBERIDAE

Shell subovate, dorsoventrally flattened, with curved dorsal margin; broadest posteriorly or with caudal process; left valve larger than right; 0.4–0.7 mm long. Lightly calcified, smooth or with weak sculpture. A single V-shaped frontal scar, and a vertical row of four adductor scars. Three pairs of walking legs, each with short, hooked claws. Antenna 1 with six articles, the terminal article stout; antenna 2 with strong spinneret bristle. Caudal furca with two setae.

Xestoleberis aurantia (Baird) Fig. 8.2
Shell thinly calcified, smooth, subreniform; strongly dimorphic, female larger and higher than male; 0.38–0.5 mm. Brownish-orange when living.

Among seaweeds, often in rock pools. Common on all British coasts.

Class Copepoda

The Copepoda is the second largest class of the Crustacea (after the Malacostraca) and includes free-living, commensal, semi-parasitic, and parasitic forms. Most species are very small, often less than 0.5 mm. The body is more or less distinctly segmented, except in some parasitic genera, with a *head*, *thorax*, and *abdomen*. Two pairs of *antennae* are usually well developed, and often used for locomotion or prehension. The mouthparts consist of a pair of *mandibles*, *maxillae 1* and *2*, and *maxillipeds*. The mandibles usually have a palp, which is often biramous.

There are six thoracic segments, the first usually fused into the head. The last thoracic segment bears the genital apertures and is therefore often called the *genital segment*. The abdomen has five segments, some of which are nearly always fused. Very often the female apparently has one abdominal segment less than the male of the same species, due to a fusing of the first and second segments. The last abdominal segment bears two *furcal rami*, which are usually setose.

There is usually only one major articulation in the body, either between the fourth and fifth thoracic segments or fifth and sixth thoracic segments. It is therefore easier to regard the part of the body anterior to this division as the *metasome*, and the part posterior as the *urosome*. Each of the thoracic segments may bear a pair of appendages (legs), although those on the fifth and sixth segments are often modified or absent—indeed the pair on the sixth segment is nearly always lacking in free-living forms.

More than 1200 species of Copepoda occur in the British sea area. Accurate identification requires access to a specialist library, and even a representative selection of species is beyond the scope of a laboratory handbook or field guide, such as this. However, the following keys will aid identification of common planktonic or benthic genera or species of intertidal pools, together with the most frequently occurring parasitic species.

KEY TO COPEPODA IN SHORE POOLS

1. Antenna 1 reaching at least as far as last segment of metasome **_Eurytemora_**

   Antenna 1 not reaching any further than segment 1 of metasome **2**

2. Antenna 1 consisting of at least 10 segments ***Cyclopina***

Antenna 1 consisting of less than 10 segments **3**

3. Antenna 2 with one ramus ***Halicyclops***

Antenna 2 with two rami (Harpacticoida) **4**

4. Segments of body sharply demarcated, giving a scalariform appearance **5**

Body not as described **6**

5. Caudal rami short and lamellar ***Asellopsis***

Caudal rami longer and cylindrical ***Laophonte***

6. Cephalic segment very large, almost half the body length; other segments very short. Bright red in colour ***Metis ignea***

Body not as described **7**

7. Endopod of swimming leg 4 absent ***Huntemannia jadensis***

Endopod of swimming leg 4 with two segments ***Mesochra***

Endopod of swimming leg 4 with three segments **8**

8. Endopod of leg 1 with two segments (A in Fig. 8.3) **9**

Endopod of leg 1 with three segments (B in Fig. 8.4) **11**

9. Exopod of leg 1 with one or two segments (A in Fig. 8.3) **10**

Exopod of leg 1 with three segments ***Microthalestris***

10. Exopod of leg 1 shorter than endopod ***Westwoodia***

Exopod of leg 1 longer than endopod ***Zaus***

11. Exopod of leg 1 with one or two segments ***Harpacticus***

Exopod of leg 1 with three segments (C in Fig. 8.4) **12**

12. Exopod of leg 1 longer than endopod. Bright orange in colour ***Tigriopus fulvus***

Exopod of leg 1 shorter than endopod (D in Fig. 8.4) **13**

Exopod and endopod of leg 1 more or less equal in length (E in Fig 8.4) **19**

13. Division between metasome and urosome well marked, with abrupt narrowing at thoracic segment 5 **14**

Division between metasome and urosome unclear; thoracic segments 4 and 5 of similar width **18**

14. Basal segment of endopod of leg 1 longer than entire exopod (D in Fig. 8.4) **15**

This segment equal to or shorter than exopod (F in Fig. 8.4) **16**

15. Exopod of antenna 2 with three segments ***Dactylopusia***

Exopod of antenna 2 with one segment ***Diosaccus***

16. Middle segment of endopods of legs 2–4 each with two setae ***Tisbe***

These segments with one seta each (G in Fig. 8.4) **17**

These segments without uniform setation **18**

---

17. Segments of urosome coarsely spinulose at hind edge; last segment with dorsal series of spinules ***Nitocra***

    Urosome segments less coarsely spinulose; last segment without dorsal series of spinules ***Ameira***

---

18. Middle segment of endopod of leg 2 with one seta, that of legs 3 and 4 with two setae. Exopod of antenna 2 with three segments ***Dactylopusia***

    This segment of legs 2 and 3 with two setae, that of leg 4 with one seta. Exopod of antenna 2 with one segment ***Diosaccus***

    This segment of leg 2 with two setae, those of legs 3 and 4 with 1 seta each. Exopod of antenna 2 with two segments ***Parathalestris***

---

19. Exopod of antenna 2 with one segment; rostrum very small and immobile ***Nitocra***

    Exopod of antenna 2 with two segments; rostrum larger and may be mobile ***Parathalestris***

Order Calanoida

Eurytemora Fig. 8.3
Elongate; some species with a pair of spines or points on posterior tips of the last (sixth) metasome segment. Urosome with three segments in female, five in the male. Furcal rami elongate. A brackish-water genus, often found in pools on salt marshes. Length 1.15–2.2 mm.

Order Cyclopoida

Halicyclops Fig. 8.3
Length about 0.75 mm; body typical cyclopoid shape with short first antennae consisting of only six segments. Metasome of four segments, the urosome of five. In brackish water; often in rock pools.

Cyclopina Fig. 8.3
More elongate than the preceding genus. The metasome consists of four segments, the urosome of five. The furcal rami are elongate in some species. *C. gracilis* Claus is sometimes found in rock pools. Length 0.43–0.90 mm.

Order Harpacticoida

Asellopsis Fig. 8.3
The body tapers very gradually, with the urosome almost as wide as the metasome. There is a pronounced rostrum in front of a well-developed eye. Length about 0.58 mm.

Laophonte Fig. 8.3
No obvious difference in width between metasome and urosome. Constrictions between the segments very marked. Inner ramus of leg 1 with a large claw-like, terminal segment. Single ovisac. Some species may be found in brackish water. Length 0.42–1.30 mm.

Metis ignea Philippi Fig. 8.3
Body more or less pear-shaped, with no obvious demarcation between the metasome and urosome. First (cephalic) segment nearly half the body length. First pair of legs very robust with claw-like setae Bright red in colour. Single ovisac. Length 0.55 mm.

Huntemannia jadensis Poppe Fig. 8.3
Body tapers gradually; a well-defined rostrum, with a setose, rounded tip. Furcal ramus tipped with an extremely short spine which is about equal in length to the ramus. Length 0.96 mm.

Mesochra Fig. 8.3
Body tapering, urosome with five segments. Eye obvious, but not particularly well developed. Rostrum small and rounded from dorsal view. Some species live in brackish water. Length 0.38–0.67 mm.

Westwoodia Fig. 8.3
Body somewhat pear-shaped, with large first (cephalic) segment. All metasome segments project ventrally at the sides, giving rise to plate-like structures. Single ovisac. Length 0.35–0.87 mm. Often in pools and brackish ponds.

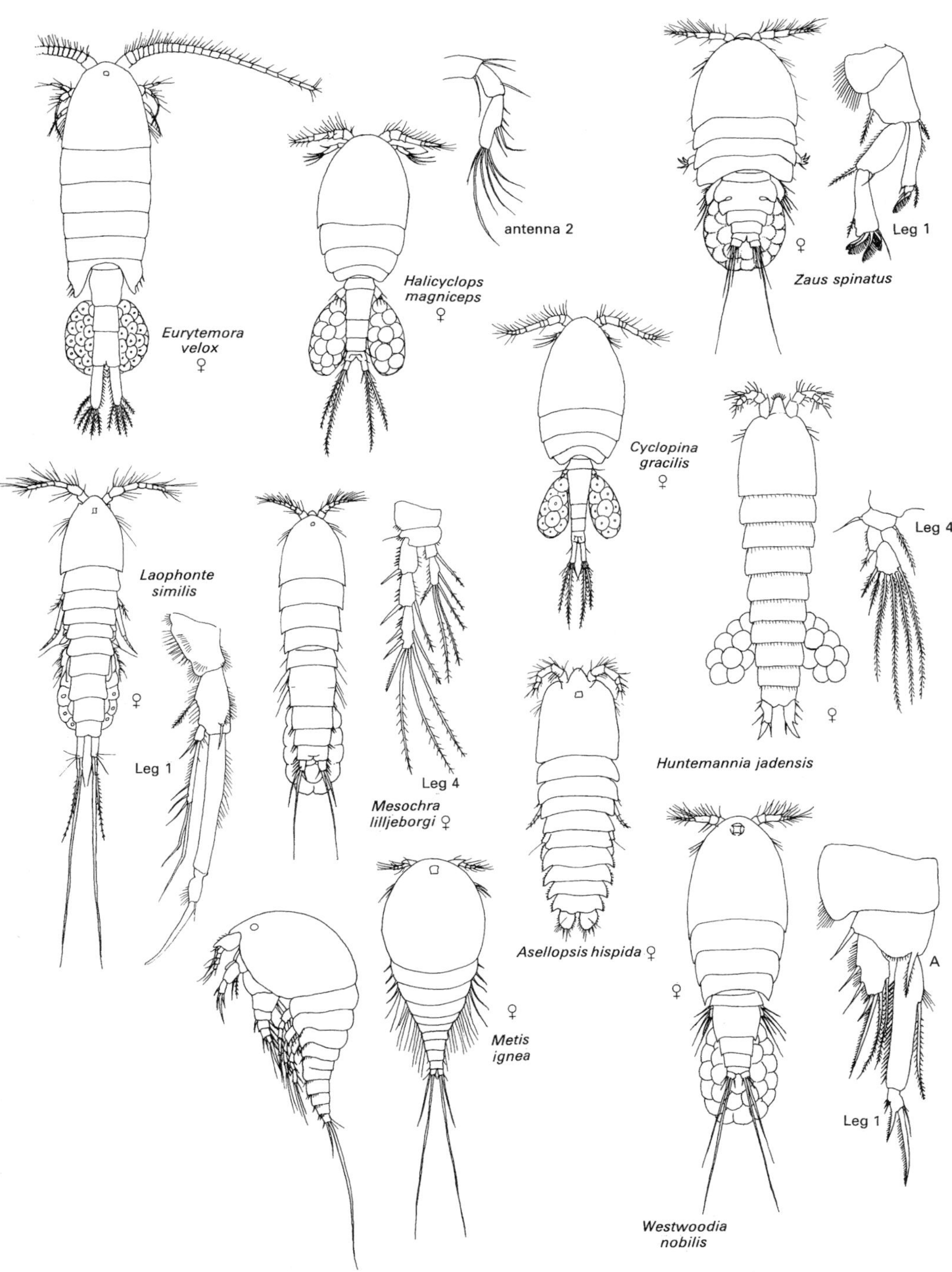

Eurytemora velox ♀
antenna 2
Halicyclops magniceps ♀
Zaus spinatus ♀
Leg 1
Cyclopina gracilis ♀
Huntemannia jadensis ♀
Leg 4
Laophonte similis ♀
Leg 1
Mesochra lilljeborgi ♀
Leg 4
Asellopsis hispida ♀
Metis ignea ♀
Westwoodia nobilis ♀
A
Leg 1

Fig. 8.3

Zaus Fig. 8.3
A broad body with the metasome/urosome division very obvious. Metasome segments produced laterally, and body dorso-ventrally flattened. Eye very obvious. Single ovisac. Length 0.44–1.40 mm.

Microthalestris Fig. 8.4
Body cylindrical, urosome similar in width to metasome, but the division between is evident. Rostrum very obvious in dorsal view. Last (sixth) urosome segment with a dorsal prominence between furcal rami, most obvious in lateral view. Single ovisac. In rock pools, often above high water. Length 0.58 mm.

Harpacticus Fig. 8.4
Body tapering, with an obvious metasome/urosome division. Rostrum broad in dorsal view. Outer ramus of leg 3 of male very robust. Single ovisac. Often in pools on the shore. Length 0.66–1.33 mm.

Tigriopus fulvus (Fischer) Fig. 8.4
Metasome/urosome division obvious. Eye quite prominent; rostrum wide and rounded. Single ovisac. A bright orange copepod found in rock pools, often above high water. Length 1.2 mm.

Dactylopusia Fig. 8.4
Body tapers more or less uniformly. Fourth metasome segment with somewhat pointed postero-lateral edges. Rostrum broad and rounded in dorsal view. Ovisac single. Often in tidal pools, or amongst seaweeds. Length 0.63–0.85 mm.

Diosaccus Fig. 8.4
First metasome segment longer than next three combined; all metasome segments with lateral margins produced ventrally and posteriorly. Rostrum extremely large. Metasome/urosome division pronounced. Two ovisacs. *D. tenuicornis* (Claus), length 0.8 mm, often in tidal pools or amongst seaweeds.

Tisbe Fig. 8.4
Metasome much wider than urosome, somewhat pointed anteriorly, with a small rostrum, not obvious in dorsal view. Metasome segments extended laterally. Eye obvious, but not prominent. Ovisac single. Often in tidal pools or amongst seaweeds. Length 0.57–1.50 mm.

Nitocra Fig. 8.4
Body more or less cylindrical, with metasome/urosome division not obvious. Eye small. Urosome segments with postero-lateral spines; last segment with spines across the postero-dorsal margin. Single ovisac. In brackish water or tidal pools. Length 0.64–0.76 mm.

Ameira Fig. 8.4
Similar to preceding genus, but last abdominal segment without spines across the postero–dorsal margin. Single ovisac. In pools or amongst seaweeds. Length 0.44–0.64 mm.

Parathalestris Fig. 8.4
Body tapers more or less evenly, with the metasome/urosome division not obvious. Eye large. Ovisac single and very large, reaching as far as, or beyond the end of the furcal rami. Amongst seaweeds and sometimes in pools. Length 0.73–2.20 mm.

COMMENSAL AND PARASITIC COPEPODA

Commensalism and parasitism occur in most taxonomic groups within the Copepoda and certain orders are entirely parasitic. The degree of consequent morphological adaptation varies greatly from family to family. In some this is expressed as a reduction of the abdomen, and an enlargement and modification of the thoracic appendages for grasping. In many other groups, however, the adult form is not readily recognisable as a copepod. In particular, among permanently attached species appendages are frequently reduced or lost, body segmentation is not apparent, and the organism appears as little more than a simple, sac-like structure with grossly enlarged reproductive organs. The taxonomic affinity of such species is only revealed by study of their larval and juvenile stages.

There are very many species of parasitic copepod. Those associated with fishes are relatively well studied but those parasitizing marine invertebrates are still poorly known. A comprehensive treatment of commensal and parasitic copepods is beyond the scope of this book. The following accounts are intended to illustrate the taxonomic and morphological diversity of this group of animals. Representative genera are figured, and the species most frequently occurring on the shore are noted,

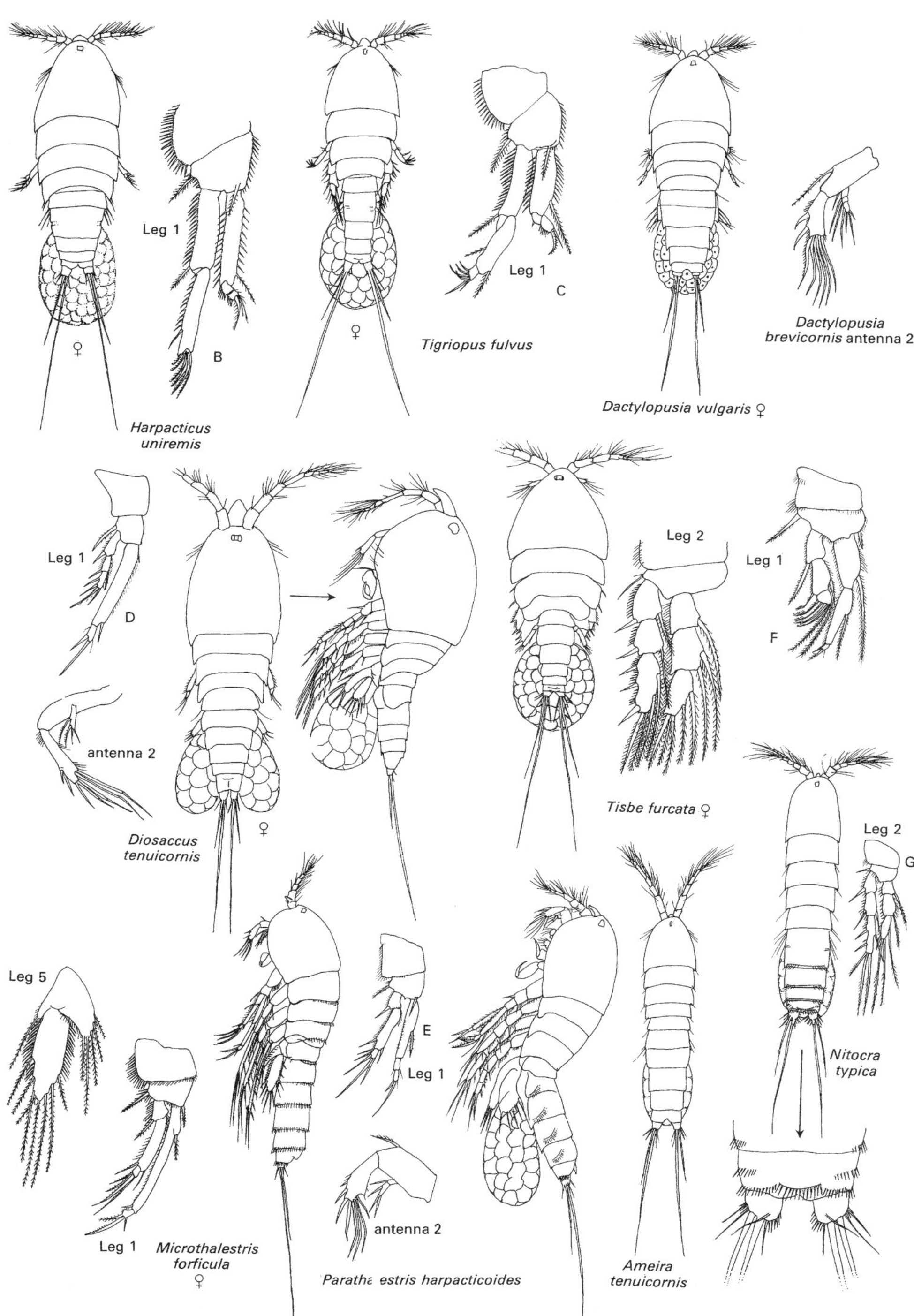

Leg 1
B
Harpacticus uniremis
Leg 1
C
Tigriopus fulvus
Dactylopusia brevicornis antenna 2
Dactylopusia vulgaris ♀
Leg 1
D
antenna 2
Diosaccus tenuicornis
Leg 2
Tisbe furcata ♀
Leg 1
F
Leg 2
G
Nitocra typica
Leg 5
Leg 1
Microthalestris forficula
E
Leg 1
antenna 2
Parathɛ estris harpacticoides
Ameira tenuicornis

Fig. 8.4

together with their usual hosts. For accurate identification of most species it will be necessary to consult specialist texts given in the reference list.

Order Poecilostomatoida

Acanthochondria Fig. 8.5

Particularly associated with fish. The type species, *A. cornuta* (Müller) is primarily a parasite of flatfish in the North Atlantic. In British waters it is commonly reported on the plaice, *Pleuronectes platessa* (L.); flounder, *P. flesus* (L.) and dab, *Limanda limanda* (L.).

Lichomolgus Fig. 8.5

This large genus is particularly associated with ascidians. In the British sea area, *L. poucheti* Canu occurs in *Morchellium argus* (Milne Edwards); *L. forficula* Thorell in *Phallusia mamillata* (Cuvier) and *Ascidia mentula* Müller; *L. tenuifurcatus* Sars in *Diplosoma listerianum* (Milne Edwards) and the echinoderm *Labidoplax digitata* (Montagu), and *L. furcillatus* Thorell in *Ciona intestinalis* (L.). Both *C. intestinalis* and *Clavelina lepadiformis* (Müller) are host to *L. canui* Sars. *Lichomolgus agilis* (Leydig) parasitizes a large number of opisthobranch molluscs, including *Doto coronata* (Gmelin), *Janolus cristatus* (Delle Chiaje), *Archidoris pseudoargus* (Rapp), and *Aeolidia papillosa* (L.), while *L. leptodermata* Gooding is found in the Norway Cockle, *Laevicardium crassum* (Gmelin).

Modiolicola Fig. 8.5

Three species occur in British waters. *Modiolicola insignis* Aurivillus occurs in the mantle cavity of both the common mussel, *Mytilus edulis* (Linnaeus), and the horse mussel, *Modiolus modiolus* (L.). *Modiolicola inermis* Canu and *M. maxima* Thompson are both associated with the Queen Scallop, *Chlamys opercularis* (L.).

Mytilicola Fig. 8.5

The single British species, *M. intestinalis* Steuer, is particularly associated with the common mussel, *Mytilus edulis* (L.), some populations of which may have very high levels of infestation. It has also been reported from *Dosinia exoleta* (L.).

Splanchnotrophus Fig. 8.5

Species of *Splanchnotrophus* have been reported from various opisthobranch sea slugs, including *Ancula gibbosa* (Risso), *Coryphella verrucosa* (Sars), *Aeolidia papillosa* (L.), and *Aeolidiella glauca* (Alder and Hancock). *Splanchnotrophus gracilis* Hancock and Norman has been most frequently recorded, attached to *Acanthodoris pilosa* (Abildgaard, in Müller).

Order Cyclopoida

Ascidicola Fig. 8.5

The only British species, *A. rosea* Thorell, is a parasite of the ascidians *Pyura squamulosa* (Alder), *Corella parallelogramma* (Müller), *Ascidia virginea* (Müller), and *Ascidiella aspersa* (Müller).

Botrillophilus Fig. 8.5

Species of *Botrillophilus* are found in association with compound ascidians. *Botrillophilus ruber* Hesse has been reported from many species, including *Aplidium proliferum* (Milne Edwards), *A. punctum* (Giard), *Botryllus schlosseri* (Pallas), *Botrylloides leachi* (Savigny), *Clavelina lepadiformis* (Müller), and *Morchellium argus* (Milne Edwards). *Botrillophilus brevipes* (Sars) has been recorded only from *B. schlosseri*.

Doropygus Fig. 8.5

Two species are most frequently reported, *D. normani* (Brady) in *Pyura squamulosa* (Alder), and *D. pulex* Thorell in *Ascidiella scabra* (Müller).

Notodelphys Fig. 8.5

Species of this genus are among the commonest parasites of the larger solitary ascidians. In British waters *N. allmani* Thorell has been reported from *Ascidia mentula* Müller and *Ascidiella aspersa* (Müller), *N. agilis* Thorell from *Corella paralellogramma* (Müller), *N. prasina* (Thorell) from *Phallusia mamillata* and *A. mentula*, and *N. elegans* Thorell from *Ciona intestinalis* (L.).

Notopterophorus Fig. 8.5

The female of *Notopterophorus papilio* Hesse is immediately recognised by its striking morphology, but the smaller and rather nondescript male may be overlooked. *Notopterophorus papilio* occurs, sometimes in large numbers, in the branchial cavity of *Ascidia mentula* (Müller).

Order Siphonostomatoida

Artotrogus Fig. 8.5

Body very broad. Female of *A. orbicularis* Boeck parasitizes dorid nudibranchs, but is also found free living among algae.

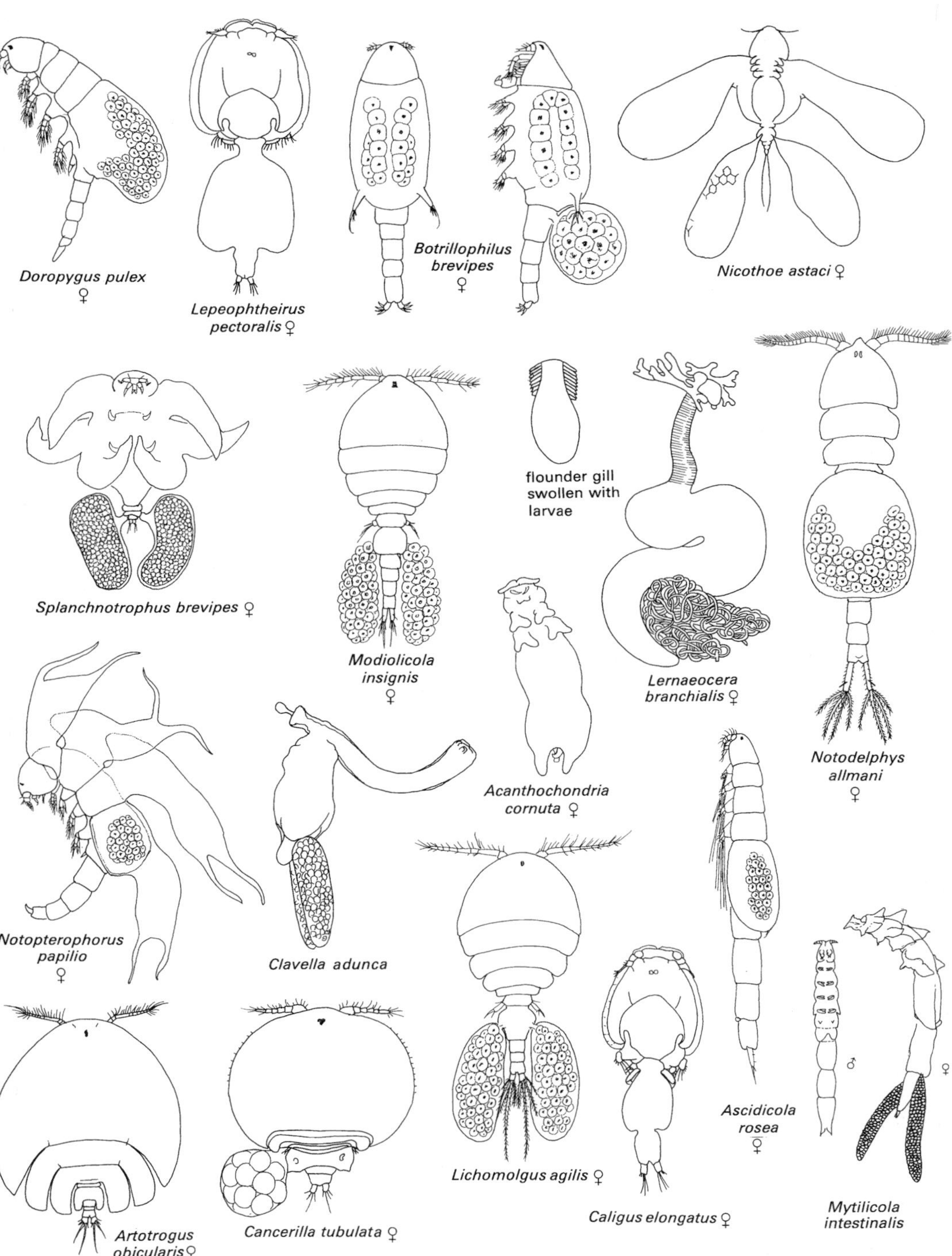
Doropygus pulex
♀
Lepeophtheirus
pectoralis ♀
Botrillophilus
brevipes
♀
Nicothoe astaci ♀
Splanchnotrophus brevipes ♀
Modiolicola
insignis
♀
flounder gill
swollen with
larvae
Lernaeocera
branchialis ♀
Notodelphys
allmani
♀
Acanthochondria
cornuta ♀
Notopterophorus
papilio
♀
Clavella adunca
Lichomolgus agilis ♀
Ascidicola
rosea
♀
♂
♀
Artotrogus
obicularis ♀
Cancerilla tubulata ♀
Caligus elongatus ♀
Mytilicola
intestinalis

Fig. 8.5

Caligus Fig. 8.5

A large genus of successful fish parasites, *Caligus* species are recognised by the broadly rounded cephalothorax. *Caligus curtus* Müller is common on gadoid fish, as well as percomorphs, flat fish, and even elasmobranchs. *Caligus elongatus* Nordmann has been recorded from more than 80 species of fish. *Caligus diaphanus* Nordmann is equally catholic in its choice of host, while *C. labracis* Scott is known only from the ballan wrasse, *Labrus bergylta* Ascanius, and the cuckoo wrasse, *L. mixtus* L., and only from British waters.

Cancerilla Fig. 8.5

Cancerilla tubulata (Dalyell) occurs on the brittlestar *Amphipholis squamata* (delle Chiaje) and *Ophiocomina nigra* (Abildgaard).

Clavella Fig. 8.5

Three species of this genus of teleost parasites occur in British waters. *Clavella adunca* (Strøm) is a common parasite of gadoid fish, including cod, *Gadus morhua* L., whiting, *Merlangius merlangus* (L.), haddock, *Melanogrammus aeglefinus* (L.), and bib, *Trisopterus luscus* (L.).

Lepeophtheirus Fig. 8.5

Very similar to *Caligus*. Species of *Lepeophtheirus* parasitize marine teleosts. Seven species occur in British waters. *Lepeoptheirus pectoralis* (Müller) is a parasite of pleuronectid flat fish, in the north-east Atlantic; records from other species of fish probably result from transference of the parasite in trawled catches. *Lepeophtheirus hippoglossi* (Krøyer) is principally associated with the atlantic halibut, *Hippoglossus hippoglossus* (L.), but may occur rarely on turbot, *Psetta maxima* (L.), and brill *Scophthalmus rhombus* (L.). The usual parasite of the latter two fish is *L. thompsoni* Baird.

Lernaeocera Fig. 8.5

The life cycle of *Lernaeocera* involves two hosts. The copepodite stage of the parasite, little modified from the typical copepod plan, is usually associated with a flatfish, on which it undergoes a series of metamorphoses before moving as an impregnated female to its final host, a gadoid fish. This final, highly modified form of the parasite attaches to the ventral gill arches of the fish and penetrates as far as the ventral aorta; multiple infestation is usually lethal to its host. *Lernaeocera branchialis* (Linnaeus) chooses a pleuronectid flatfish as its intermediate host, and the mature females are most frequent on cod, *Gadus morhua* L., pollack, *Pollachius pollachius* (L.), coalfish, *P. virens* (L.), haddock, *Melanogrammus aeglefinus* (L.); and whiting, *Merlangius merlangus* (L.). *Lernaeocera lusci* (Bassett-Smith) is similar to *L. branchialis*, although smaller; its only known intermediate host is the sole, *Solea solea* (L.), but the final stages occur on numerous gadoid and other fishes. *Lernaeocera minuta* (Scott) occurs on the sand goby, *Pomatoschistus minutus* (Pallas).

Nicothoe Fig. 8.5

The common lobster, *Homarus gammarus* (L.), is host to the only British species, *N. astaci* Audouin and Milne Edwards.

Class Cirripedia

Cirripedia (barnacles and their allies) are highly modified crustaceans. Their free-swimming cyprid larva (Fig. 8.6) has six pairs of thoracic swimming appendages, and a pair of first antennae used for selecting and attaching to a substrate. What happens next varies in the different orders.

In the Thoracica or barnacles proper, the cyprid undergoes metamorphosis, the body performing a forward somersault, bringing the thoracic limbs into the adult position as feathery *cirri*, used for plankton feeding. The bivalved *carapace* of the cyprid becomes the multi-plated *shell* of the barnacle, enclosing a mantle cavity in which the *prosoma* (body) is suspended. The exoskeleton of barnacles is unique in not being completely shed at each moult, its calcareous plates, like the shells of molluscs, thus exhibit concentric growth lines. The prosoma and appendages have a flexible integument which is regularly moulted in typical fashion.

Appendages of the adult comprise the *cirri*, *mandibles*, and two pairs of *maxillae*. The first and second cirri are usually short and function as maxillipeds; the third to sixth are elongate and setate. Cirri can be completely retracted within the mantle cavity—coiling action is mediated by contraction of their offset longitudinal muscles. Extension is a hydraulic process, muscular contraction within the prosoma forcing coelomic fluid into the cirri. The abbreviated abdomen bears paired terminal *caudal appendages* flanking the median penis.

The most generalised suborder of Thoracica is the Lepadomorpha (Fig. 8.6) in which the mantle

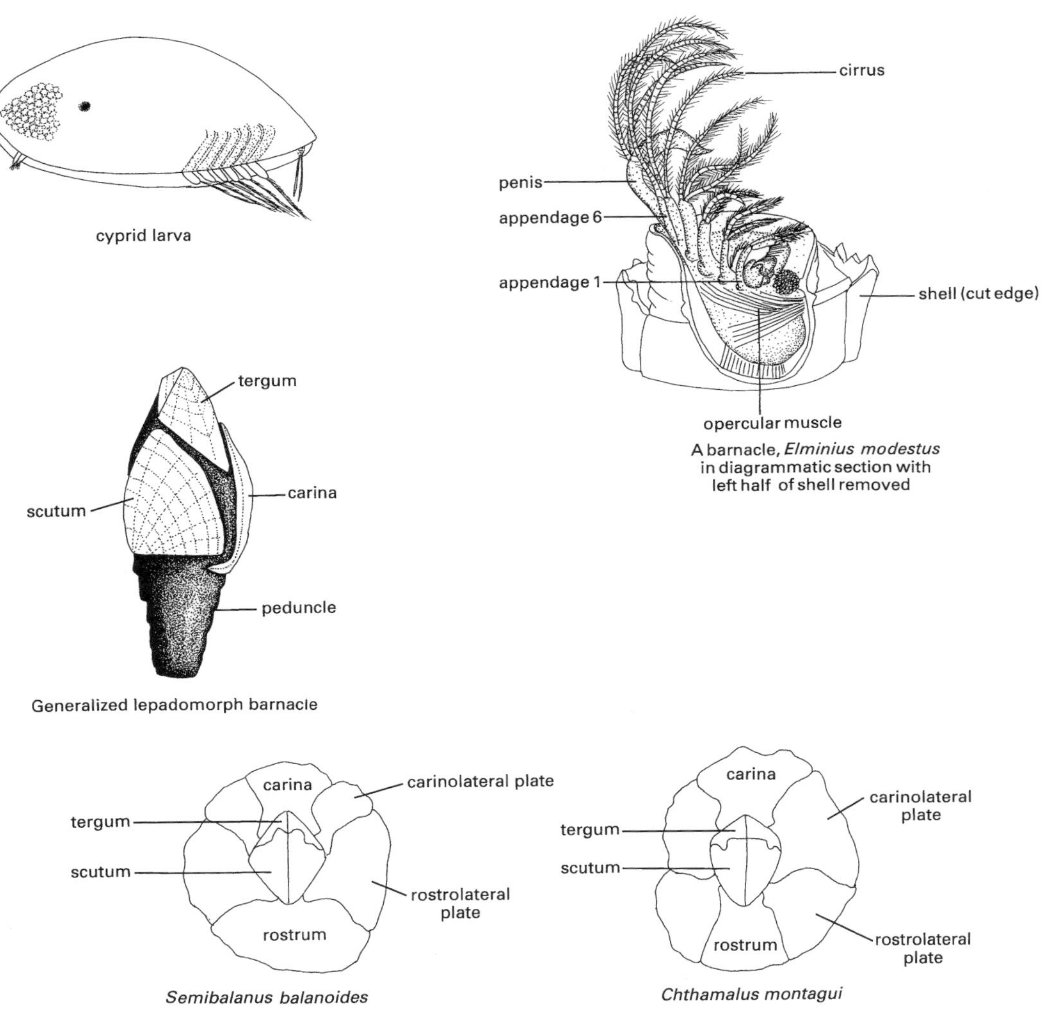

Fig. 8.6

cavity with its shell plates forms the head, or *capitulum*, demarcated from the stalk, or *peduncle*, anchored to the substratum. The peduncle may or may not be armoured with plates. It has no cavity and contains the ovary. The Lepadomorpha comprises two families.

The other two suborders of the Thoracica are the acorn, or stalkless barnacles (sometimes referred to as sessile barnacles). They are characterised by a low conical shape on a roughly circular base. The rigid wall plates surround a central orifice protected by two pairs of movable plates—the operculum. The first of these suborders is the Verrucomorpha, a small, distinctive group with asymmetrical shape, incorporating an operculum of only two plates, allowing the cirri to emerge at one side—suggested to be an adaptation for crevice dwelling. Balanomorpha, the third suborder of Thoracica, comprise the familiar barnacles, including both the Chthamalidae and Balanidae. Normally there are six overlapping wall plates (Fig. 8.6) but these may be reduced in number or supplemented by accessory plates. The *carina* and *rostrum* are median; the other four are paired *carinolaterals* and *rostrolaterals*. The central sectors of plates are referred to as *parieties*, the overlapping parts as *alae*, and the overlapped sectors as *radii*. The base, or *basis*, of a barnacle, where it is attached to the substratum, may be cuticular or, like the plates, calcified. The operculum of the balanomorph is symmetrical and consists of left and right 'doors' which open by moving apart, their outer edges tipping downwards within the wall plates. Each door consists of two plates—the *tergum* at the carinal end, and the anterior *scutum*—hinged together (Fig. 8.6). The margins of the two doors, where they meet in the mid-line, are edged with a flexible *tergoscutal flap* which ensures a close fit. In life, these flaps are often distinctively coloured and aid in identification.

Finally, there are other orders of Cirripedia which are not immediately recognizable as barnacles. The Acrothoracica have plankton-catching cirri but lack shell plates. They are small, reduced cirripedes found in burrows which they excavate in mollusc shells, corals, etc. The order Rhizocephala have the distinctive cirripede nauplius and cyprid larvae but as adults are highly modified animals parasitic on crabs, hermit crabs, or even other barnacles. Another group of parasitic Crustacea, the Ascothoracida, once thought to be cirripedes, are now regarded as separate.

KEY TO FAMILIES

1. Cirripede which is a recognizable barnacle, with a carapace incorporating calcareous plates varying from massive to extremely reduced **2**

   Cirripede totally lacking calcareous plates. Either a minute form burrowing in gastropod shell; OR highly modified endoparasite on malacostracan host; lacking cirri but dispersing by cyprid larva **6**

2. Barnacle with a distinct flexible peduncle demarcated from the capitulum (stalked barnacle, Fig. 8.6) **3**

   Barnacle without a flexible stalk (but may have rigid cup-shaped base); and having a distinct operculum (acorn barnacles, Fig. 8.6) **4**

3. Stalked barnacle with usually five plates to the capitulum, these sometimes very reduced, perhaps only two visible. Goose barnacles **2. Lepadidae**

   Stalked barnacle with 13 or more capitular plates **1. Scalpellidae**

4. Operculum of only two plates opening to the side of the barnacle **3. Verrucidae**

   Operculum of four plates placed more or less centrally on top of barnacle **5**

5. Acorn barnacle in which the anterior wall plate or rostrum is little wider than the operculum and is overlapped by the flanking plate (Fig. 8.6) **4. Chthamalidae**

   Acorn barnacle in which the rostrum, if distinct, is usually much wider than the operculum and always overlaps the flanking plates **5. Balanidae**

6. Cirripede occupying a burrow it has excavated in the columella of the empty shell of a whelk (*Buccinum*) used by hermit crab **6. Alcipidae**

Cirripede parasitizing a crab, squat lobster, hermit crab, or a barnacle, recognizable as an external swelling usually attached to abdomen of host **Rhizocephala** (p. 311)

Order Thoracica

Suborder Lepadomorpha

1. SCALPELLIDAE

Capitulum with more than five plates. Peduncle also armoured with many fine plates. Caudal appendages reduced or absent. Hermaphroditic or gonochoristic with dwarf males. Larvae planktotrophic or lecithotrophic.

Many other species of *Scalpellum* and *Arcoscalpellum* occur at greater depths.

1. Capitulum with 13 distinct plates; carina with distinct angle near mid point ***Scalpellum scalpellum***

Capitulum with more than 13 plates, including many small ones near junction with peduncle ***Pollicipes pollicipes***

Scalpellum scalpellum (Linnaeus) Fig. 8.7
Capitulum distinctively shaped, larger than peduncle, strongly compressed laterally and bearing 13 plates with variously aligned umbos. Peduncle with transverse rows of small imbricating plates. Overall length up to 40 mm.

Attached to hydroids, bryozoa, and rocks, etc.; sublittoral 10–500 m. East Atlantic, Iceland, Norway to west Africa, Azores, Mediterranean. British Isles, south and west coasts, not uncommon.

Pollicipes pollicipes (Gmelin) Fig. 8.7
Peduncle generally much shorter than capitulum; overall length up to 50 mm. Capitulum laterally compressed with five prominent plates and a variable number of medium-sized and smaller, spine-like plates, all with terminal umbones. Peduncle tapers downwards to an expanded attachment point; surface of peduncle armoured with uniformly small plates.

On rocks or often anchored within empty *Balanus perforatus* shells, MTL to MLWN, on moderately exposed oceanic shores. France and Iberia, common at remote locations; formerly recorded from south-west Cornwall.

2. LEPADIDAE

Five plates to capitulum (in a few species they are greatly reduced). Peduncle, which may be very long and contractile, never plated. Tergum with apical umbo; scutum and carina with basal umbones. Larvae planktotrophic. British fauna includes seven species in the genera *Lepas*, *Conchoderma*, and *Dosima*.

1. Capitulum with five large calcareous plates **2**

Capitulum with 2–5 widely separated small plates or none at all ***Conchoderma auritum***

2. Carina (single median plate) angled at umbo, and expanded into disc-shaped base ***Dosima fascicularis***

Carina not angled, and forked at base **3**

3. Capitulum plates smooth or finely marked ***Lepas anatifera***

Capitulum plates, especially terga, strongly striated ***Lepas pectinata***

Lepas anatifera Linnaeus Fig. 8.7
Capitulum length up to 50 mm. Plates translucent white and usually smooth, integument between plates dark, even black. Peduncle 40–850 mm, purplish brown, darkest adjacent to scutum.

Attached to large or small floating objects, cosmopolitan in warmer seas; the commonest species of *Lepas* to be stranded. On south and west coasts of

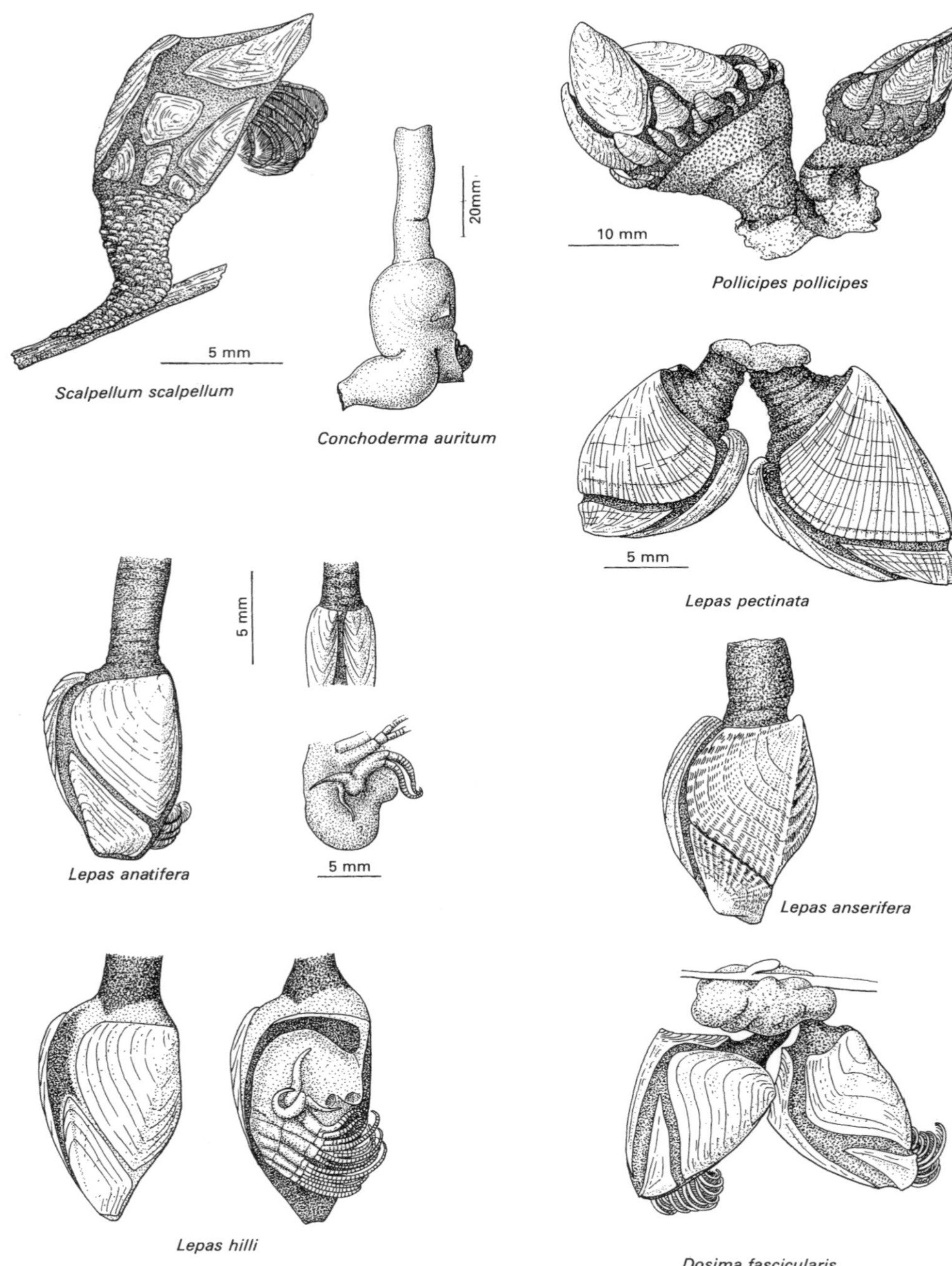
5 mm
Scalpellum scalpellum
20mm
Conchoderma auritum
10 mm
Pollicipes pollicipes
5 mm
Lepas pectinata
5 mm
5 mm
Lepas anatifera
Lepas anserifera
Lepas hilli
Dosima fascicularis

Fig. 8.7

British Isles, rarely on east coast. Also west coasts of France, Spain, and Portugal.

Lepas hilli (Leach) (Fig 8.7), a similar but rarer species, is distinguished by a pale collar between stalk and capitulum.

Lepas pectinata Spengler Fig. 8.7
Small, capitulum only 15 mm. Plates with radiating and transverse ribbing. Carina wide near top, narrow near basal fork. Tergum triangular, also sometimes barbed. Scutum radially ridged, sometimes barbed.

On large, or more often small, floating objects such as seaweed. Cosmopolitan in warmer seas. Often stranded on coast of British Isles (south and west) and Atlantic coasts of France, Spain, and Portugal.

Lepas anserifera L., a rare species occasionally stranded on west coasts of Britain, is similar but larger, with capitulum to 40 mm long (Fig. 8.7).

Dosima fascicularis (Ellis and Solander).
Buoy barnacle Fig. 8.7
Capitulum usually 15–20 mm (up to 30 mm). Plates thin, smooth, and distinctively curved to give whole capitulum streamlined shape. Peduncle shorter than capitulum; pale yellowish to purplish-brown; often has secreted float with texture of expanded polystyrene.

Attached to small or very small objects such as feathers or straws, supplemented in larger specimens by own float. Not often on ships. Cosmopolitan in warmer seas. Often stranded on south and west coasts of British Isles, and west coasts of France, Spain, and Portugal, sometimes in large numbers.

Conchoderma auritum (Linnaeus) Fig. 8.7
Capitulum about 30 mm long, with two distinctive tubular outgrowths ('ears'). Shell plates extremely reduced. Peduncle longer than capitulum.

Characteristically found attached to the whale barnacle *Coronula* (in turn on humpback and other whales); also on ship hulls and buoys off British Isles and west European coasts.

Suborder Verrucomorpha

3. VERRUCIDAE

Shell plates that compose the rigid wall include one tergum and one scutum; the other two wall plates are generally assumed to be the carina and rostrum. The operculum consists of a single, rigid, flat plate comprising the other tergum and scutum. Both right- and left-handed specimens are found. One British species.

Verruca stroemia (O. F. Müller) Fig. 8.8
Up to 5 mm diameter. Shell has depressed shape, white to brownish. Hinge of operculum located on a diameter. All plates have radiating crests which interlock at interplate sutures. Tergoscutal flaps pink or red.

On undersides of boulders, in crevices, on shells, algal holdfasts, etc., LWST and sublittoral to over 500 m. Abundant on all coasts in British Isles; elsewhere Norway to Mediterranean, not Baltic.

Suborder Balanomorpha

4. CHTHAMALIDAE

Worldwide family. The British species have six solid wall plates.

---

1. Orifice clearly kite-shaped with distinct obtuse angle at carinal end; length of tergum less than half length of scutum **_Chthamalus montagui_**

   Orifice more rounded at carinal end; tergum more than half length of scutum **_Chthamalus stellatus_**

Chthamalus montagui Southward Fig. 8.8
Basal diameter up to 10 mm. Brownish or greyish. Usually low conical. Surface nearly always corroded, with sutures often obliterated. Opercular opening nearly always kite-shaped, junction between tergum and scutum close to carinal edge. Scutum much longer than broad.

Upper shore (above the next species) but with scattered individuals down to LWNT. South and west coasts of British Isles.

Chthamalus stellatus (Poli) Fig. 8.8
Basal diameter up to 15 mm. Light grey, often discoloured. Conical or flattened, usually corroded and sutures obliterated. Opercular opening broadly kite-shaped in juveniles, becoming oval or circular with age. Scutum apex usually a right angle; length usually less than 1.5 × width.

Middle shore (below *C. montagui* but wide overlap); scattered individuals down to LWST.

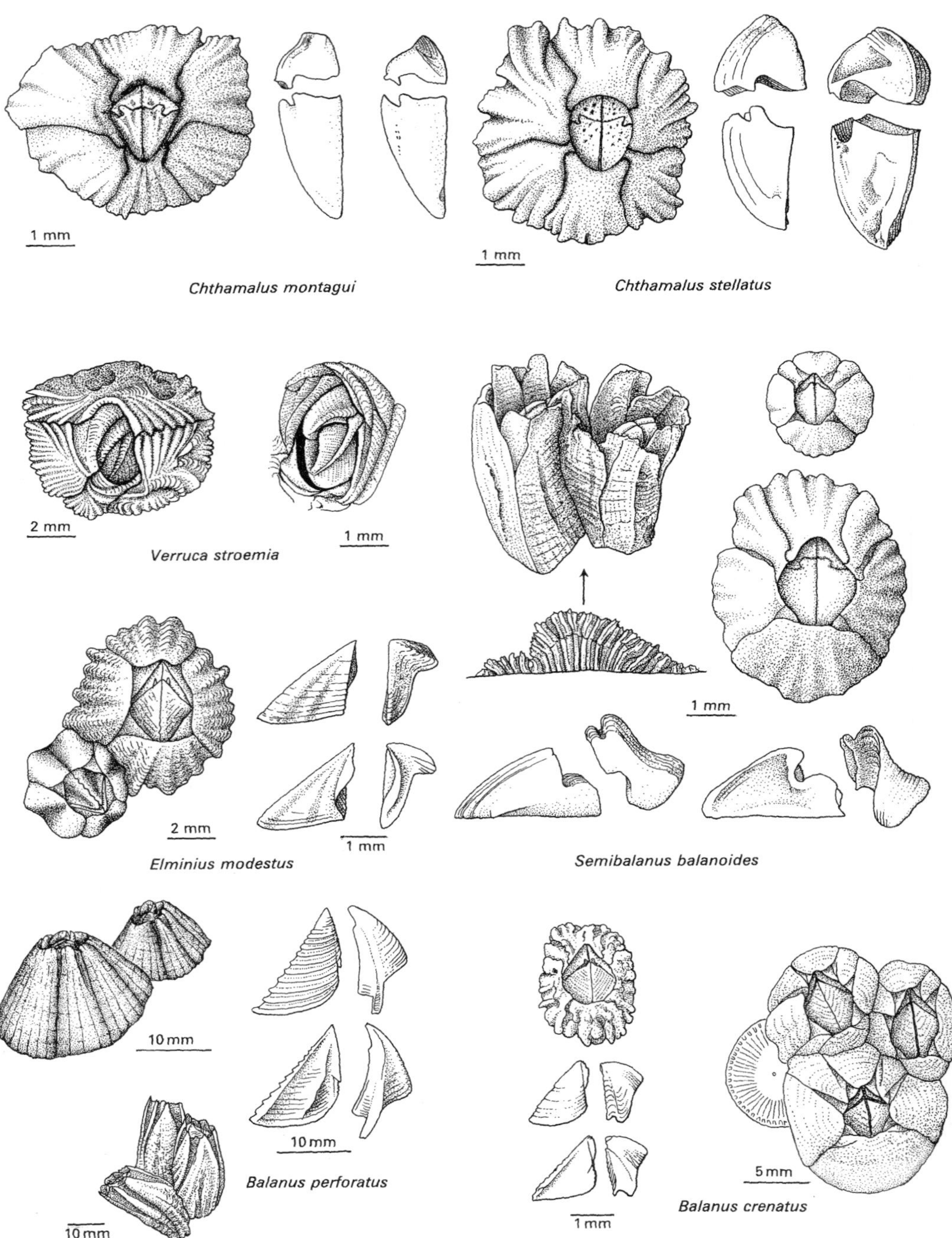
1 mm
1 mm
Chthamalus montagui
Chthamalus stellatus
2 mm
1 mm
Verruca stroemia
1 mm
2 mm
1 mm
Elminius modestus
Semibalanus balanoides
10 mm
10 mm
Balanus perforatus
10 mm
1 mm
5 mm
Balanus crenatus

Fig. 8.8

Abundant in Devon, Cornwall, and Channel Isles and exposed headlands including west coasts of Scotland and Ireland. Progressively less common elsewhere.

5. BALANIDAE

Wall of four or six plates, with rostrum overlapping the laterals. Carinolaterals narrow, sometimes absent. British fauna includes 11 species.

---

1. Barnacle not attached to solid substrata but enclosed in live sponge **_Acasta spongites_**

   Barnacle attached to a solid surface **2**

---

2. Barnacle attached to live coral **_Megatrema anglicum_**

   Barnacle not on coral **3**

---

3. Barnacle wall consisting of only four plates **_Elminius modestus_**

   Wall of six plates, although the dividing sutures are not always obvious **4**

---

4. Base of barnacle forms a calcareous layer on the substratum (*Balanus*) **5**

   Base non-calcified, membranous **_Semibalanus balanoides_**

---

5. Wall shell plates striped or coloured with pink, purple, or brown (but may be corroded grey) **6**

   Wall plates white (but may be corroded grey when old) **7**

---

6. Whole shell purple or pinkish when young (sometimes slightly striped), obscured by grey corrosion in old specimens. Large conical barnacle with very small orifice. Sutures often obscure **_Balanus perforatus_**

   Half of barnacle (carinal half), uniformly dark pink; other half nearly white; clear sutures **_Balanus spongicola_**

---

7. Tergum blunt **8**

   Tergum sharply pointed **9**

---

8. Overlapping interapical areas on tergum and scutum.Tergoscutal flaps with yellow edges, followed by longitudinal purple- brown stripe **_Balanus crenatus_**

   No interapical areas; tergoscutal flaps speckled **_Balanus improvisus_**

---

9. Shell walls strongly ribbed, orifice relatively small. Tergoscutal flaps with yellowish-white edges followed by brown stripe **_Balanus balanus_**

   Shell walls smooth, orifice very large. Tergoscutal flaps white **_Balanus hameri_**

Balanus spongicola Brown Fig. 8.9

Up to 15 mm diameter. Spectacular coloration: carinal half of shell strong reddish pink or purple, other half paler pink or white. Orifice rather small and sharply edged. Operculum somewhat sunk.

Sublittoral, sometimes embedded in sponges but more often attached to shell of *Chlamys opercularis*. British Isles, south-west; quite common in English Channel but South Wales is its northern limit. Extends south to South Africa.

Balanus perforatus Bruguière Fig. 8.8

Large, 15–50 mm diameter; height similar. Various shades of purple to pink (partly as stripes) to white; dull grey colour when eroded. Wall plates slightly ribbed, very strongly built with narrow radii. Orifice usually very small but quite large in sublittoral specimens. Operculum often deeply sunk, tergum with well-marked beak and narrow spur. Tergoscutal flaps purple or brown with patches of blue and white.

MTL to sublittoral on rocks and on artificial structures including ships. Often abundant in

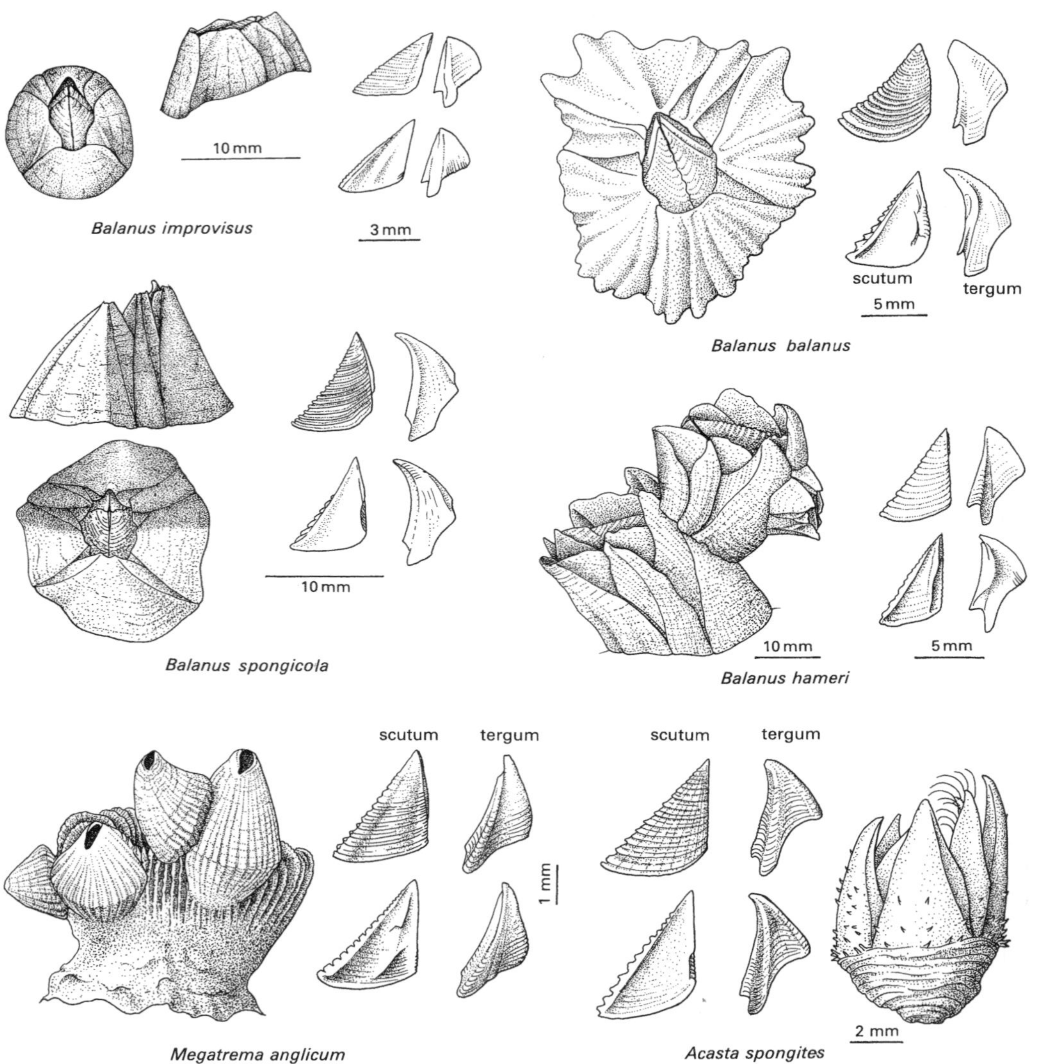

Fig. 8.9

south-west England. Northern limit is South Wales; not found east of Isle of Wight; not in Ireland. Southwards to Mediterranean and west Africa.

Balanus balanus (Linnaeus) Fig. 8.9
Up to 25 mm diameter. White or tinged brown. Conical shape, circular with irregular edge. Orifice moderately sized, acutely angled at tergal end, rounded anteriorly. Operculum raised.

On rocks, stones, and shells; LWST to several hundred metres deep; occasionally on ships. British Isles, all coasts except Cornwall; abundant. Elsewhere, Arctic circle to North and Celtic Seas.

Balanus hameri (Ascanius) Fig. 8.9
Up to 30 mm diameter and similar height. White or cream. Almost cylindrical. Clumps composed of young specimens settled on pre-existing ones. Tergoscutal flaps large, usually white.

On stones or shells, especially in currents; also on buoys and ships. 20–200 m; British Isles, all coasts. Elsewhere: Arctic to North and Celtic Seas.

Balanus improvisus Darwin Fig. 8.9
Up to 15 mm diameter; height less. White or cream, often superficially dirty. Low cone-shaped; slightly ovate; simple outline. Plates narrow, obliquely edged radii. Orifice narrow and diamond-shaped. Tergum with narrow spur. Tergoscutal flaps speckled white and pale purple.

On stones, algae, and artificial substrates including ships; brackish water only, including estuaries and ports; MTL to sublittoral. British Isles, all suitable coasts, common, although retreating where in competition with *Elminius*. Elsewhere: Baltic (only barnacle there), Norway to Spain, Mediterranean to Black Sea.

Balanus crenatus Bruguière Fig. 8.8
Up to 20 mm diameter. White, not eroded. Conical, or tubular when crowded. Wall plates usually smooth when young, irregularly ribbed when adult. Carina in profile slightly concave towards tip. Radii and alae with oblique rough edges. Orifice rather large, operculum flush; tergum blunt, with broad spur.

Lower shore to sublittoral on rocks, small stones, shells, artificial structures, and ships. British Isles, all coasts; abundant. Arctic to west coast of France, as far south as Bordeaux.

Semibalanus balanoides (Linnaeus) Fig. 8.8
Up to 5–10 mm diameter, rarely 15 mm, height varies with form. Shell white or cream to grey-brown. Low cone varying to tall tubular shape when crowded. Round to ovate outline with irregular margin. Wall plates smooth or ribbed; alae and radii oblique-edged. Orifice moderately sized, diamond shaped; operculum nearly flush. Tergoscutal suture characteristically stepped, especially in adults. Tergum with very broad spur. Tergoscutal flaps white with brown spot at central groove.

The commonest middle-shore barnacle of rocky shores in the British Isles. On a wide variety of natural and artificial substrates, including ships; sometimes found sublittorally.

Elminius modestus Darwin Fig. 8.8
Diameter 5–10 mm. Opalescent greyish-white when young, but may be drab greyish-brown when old and eroded. Low conical shape, with an irregular margin and roughly circular when adult. Young specimens show clearly four symmetrically placed wall plates, each with a median notch. Plates smooth when young. Orifice rather large, diamond-shaped; operculum flush; tergum with distinctively shaped spur; scutum with darker area radiating from umbo. Tergoscutal sutures straight when young; more sinuous, approaching shape in *Semibalanus* when old. Tergoscutal flaps white with orange-brown spot at centre, grey towards rostrum.

Intertidal and shallow sublittoral, especially estuarine but also open coast on a wide range of substrata, including rocks, stones, shells, other crustacea, algae, and artificial structures, including ships. Native in New Zealand. Introduced in Britain about 1940. Now in most north European ports.

Acasta spongites Darwin Fig. 8.9
Diameter up to 8 mm. Shell white. Tall cone rising from calcified cup-shaped basis. Wall plates fragile, spiny. Tergoscutal flaps white.

Always associated with the sponge *Dysidea fragilis* (Montagu) in which it is embedded. Sublittoral. South-west British Isles to west Africa; common.

Megatrema anglicum (Leach) Fig. 8.9
Diameter up to 6 mm. Shell pink. Conical shell with circular outline arises from cup-shaped basis. Operculum depressed. Sublittoral, 5–50 m depth. Associated with the coral *Caryophyllia smithi* (Stokes and Broderip). West and south-west coasts of British Isles to west Africa.

Order Acrothoracica

Tiny, burrowing, shell-less cirripedes with 3–5 pairs of cirri. Sexes separate.

6. ALCIPIDAE

Mantle enclosed in chitinous sac, inside burrow, opens by a slit. Four pairs of cirri. First cirri biramous, remaining three pairs are uniramous. Gonochoristic, dwarf males settle in or near female.

Trypetesa lampas (Hancock)
Occupies burrows in whelk shells (*Buccinum* and *Neptunea*) inhabited by the hermit crab *Pagurus bernhardus* (Linn.) Burrows open on surface of columella of the body whorl of the shell. Sublittoral. Probably all British coasts: also southern Scandinavia to Mediterranean.

Order Rhizocephala

Rhizocephala are always parasites of other crustaceans, mostly decapods. The *soma* (body) of the parasite usually ramifies through the host and produces a large external reproductive sac, or *externa*, with a characteristic shape.The soma and externa represent the female adult, having developed from a female cyprid. The virgin externa is variously impregnated by a large male cyprid. Eggs of two sizes hatch to produce male and female nauplius larvae, lacking gut and being gonochoristic and lecithotrophic rather than hermaphroditic and planktotrophic as in barnacles proper.

A hyperparasite—the cryptoniscid isopod *Liriopsis*—is frequently present as an attendant sac attached to the rhizocephalan, especially in *Clistosaccus paguri*, *Peltogaster paguri*, *Tortugaster boschmai*, and *Lernaeodiscus ingolfi*.

KEY TO FAMILIES

1.	On prosoma (body) of barnacle	**1. Chthamalophilidae**
	On decapod crustacean	**2**
2.	On hermit crab	**3**
	Not on hermit crab	**7**
3.	Body 1–3 mm with circular outline, no mantle opening (immature)	**3. Clistosaccidae**
	Body cylindrical with subterminal stalk	**4**
4.	No cuticular shield around stalk	**5**
	Cuticular shield prominent on dorsal surface	**4. Peltogastridae**
5.	Stalk broad, almost one-fifth of externa length	**3. Clistosaccidae**
	Stalk narrower, less than one-tenth of externa length	**6**
6.	Width of body, less than one-third of length	**4. Peltogastridae**
	Body short and fat (immature)	**4. Peltogastridae**
7.	On shrimps or prawns	**2. Sylonidae**
	On burrowing prawns, crabs, or squat lobsters	**8**
8.	On burrowing prawns (Thalassinidea)	**4. Peltogastridae**
	Not on burrowing prawns	**9**
9.	On true crabs (Brachyura)	**6. Sacculinidae**
	On squat lobsters	**10**

10. Stalk on one side of externa **4. Peltogastridae**
Stalk at opposite end to mantle opening **5. Lernaeodiscidae**

1. CHTHAMALOPHILIDAE

Externa more or less rounded and featureless. Parasites of balanomorph barnacles. Two species.

1. On *Chthamalus*, internal body lobed ***Chthamalophilus***
On *Balanus improvisus*, branching roots within host ***Boschmaella***

Chthamalophilus delarge Bocquet-Vedrine Fig. 8.10
Length 1–1.5 mm, white; nearly spherical, mantle opening almost opposite stalk. Normally solitary, but up to five at different stages on one host, suggesting gregariousness. One to 5% of hosts infected.

On *Chthamalus* spp., attached to side of body near cirri. Only recorded so far from north-west France.

Boschmaella balani (Bocquet-Vedrine)
Very similar to *Chthamalophilus* but slightly larger and slightly elongate. Infection rate up to 90%, usually gregarious, up to 40 per host.

On *Balanus improvisus*, attached in mantle cavity. So far only recorded from western Europe.

2. SYLONIDAE

Externa slightly elongated parallel to axis of host. Two very small (left and right) mantle openings. Larvae hatch as cyprids.

One species.

Sylon hippolytes Kröyer Fig. 8.10
Adults 4 mm on small host species, up to 15.2 mm long on large host. Whitish to pink. Initially spherical, broader and longer when adult. Stalk short and broad.

Occurs attached to sternite of abdomen of a wide range of shrimps and prawns. Detailed distribution unknown.

3. CLISTOSACCIDAE

Body cylindrical. Small anterior mantle opening. Parasitic on hermit crabs. One species.

Clistosaccus paguri Lilljeborg Fig. 8.10
Size varies greatly with that of host: 2.5–26.0 mm in length. White. Originally circular, eventually elongate cylindrical with terminal mantle opening. Wide stalk two-thirds of length from mantle opening.

On abdomen of host, usually on left side. Up to 3% hosts infected. Half of these with only one, otherwise up to 10 on a single host.

On *Pagurus bernhardus* (Linn.), *P. pubescens* Krøyer. Northern distribution.

4. PELTOGASTRIDAE

Diverse family. Body spherical to twisted elongate. Stalk always arises between posterior end and half way along dorsal side. Mantle opening at anterior extremity. Parasites of hermit crabs, squat lobsters, and burrowing prawns. Six species are keyed here, but details are given of only the three most common species.

1. On hermit crabs **2**
Not on hermit crabs **5**

2. Shield of thickened cuticle on parasite around stalk (*Peltogaster*) **3**
No shield **4**

3. Body covered in small spines ***Peltogaster curvatus***
Body smooth ***Peltogaster paguri***

4. Body length about three times diameter, usually gregarious ***Peltogastrella sulcata***

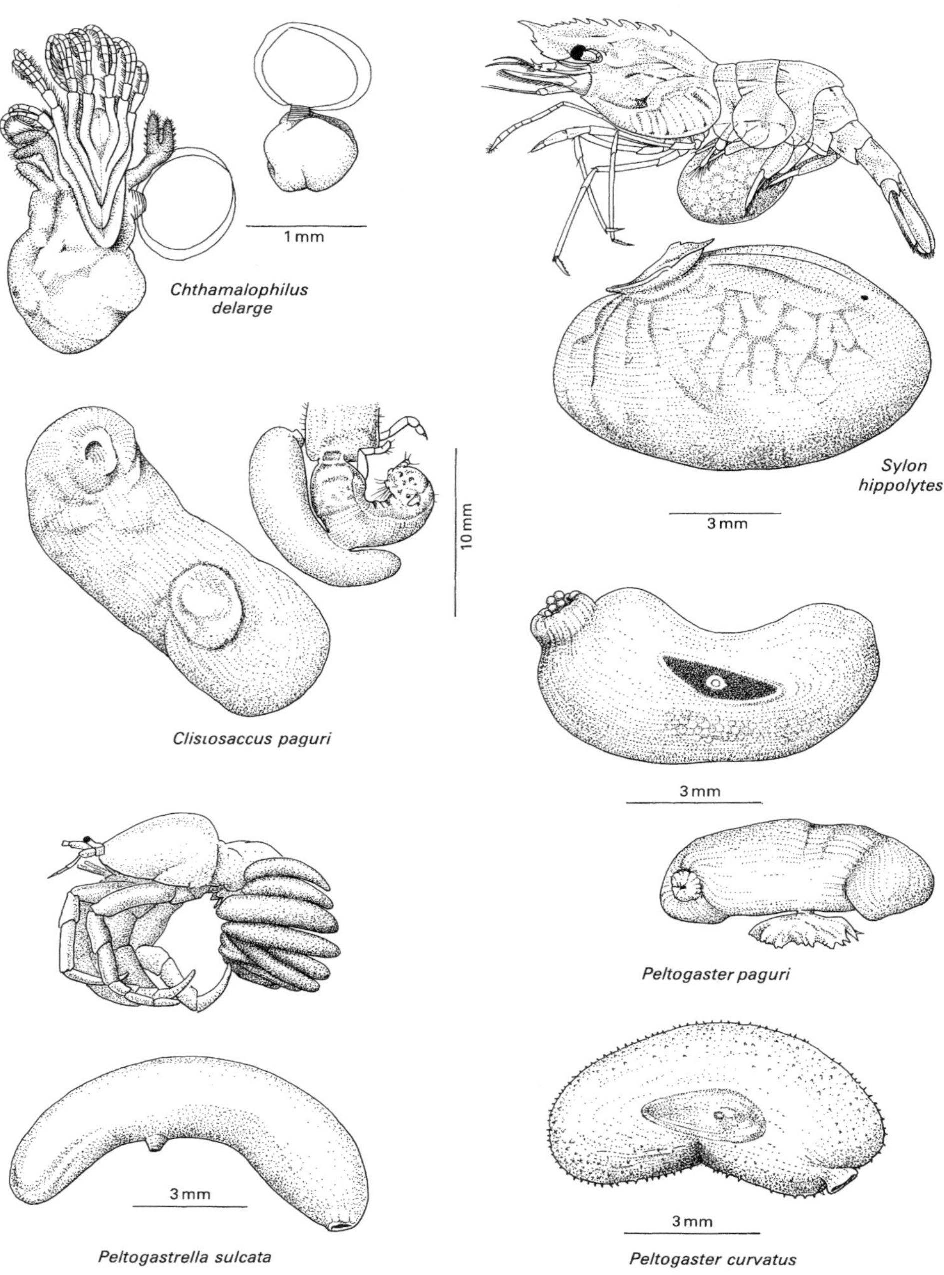
1 mm
Chthamalophilus delarge
Sylon hippolytes
3 mm
10 mm
Clistosaccus paguri
3 mm
Peltogaster paguri
3 mm
Peltogastrella sulcata
3 mm
Peltogaster curvatus

Fig. 8.10

Body not greatly elongated <4 mm juvenile ***Peltogaster***

---

5. On burrowing prawns (Thalassinidea) (Fig. 8.11) ***Parthenopea subterranea* Kossmann**

Not on Thalassinidea **6**

---

6. Body symmetrical or nearly so. On *Galathea* (Fig. 8.11) ***Galatheascus striatus* Boschma**

Strongly asymmetrical body; stalk on left side. On *Munida* (Fig. 8.11) ***Tortugaster boschmai* Brinkman**

Peltogaster paguri Rathke Fig. 8.10
Length 2–15 (exceptionally 26) mm, brown-red colour, greenish when old. Elongate cylindrical, length 3 × width. Mantle aperture an elevated terminal and often lobulate structure. Narrow stalk, about half way from mantle opening. Prominent fusiform dark area (shield) around stalk on parasite. Up to 5% (usually less) of hosts infected. Usually single, multiples by chance up to four.

Attached to abdomen of hermit crabs, usually left side; recorded on *Anapagurus chiroacanthus* (Lilljeborg), *P. cuanensis* Thompson.

Peltogaster curvatus Kossman Fig. 8.10
Length 4–16 mm, apricot-red. Like *P. paguri* but body armed with small spines. Up to 5.5% infection of *Pagurus cuanensis*. Gregarious, up to one-third of hosts carrying 2–4 parasites.

On *Pagurus cuanensis* Thompson and *P. prideauxi* Leach (not *P. bernhardus* (Linn.)); scattered records, England and Iceland.

Peltogastrella sulcata (Lilljeborg) Fig. 8.10
Up to 11 mm; pinkish-white, later apricot-red. Very long and thin, cylindrical, tapering posteriorly; curved; narrow stalk on concave side—smooth. Mantle opening terminal. Up to 14% of *Pagurus cuanensis* Thompson infected (Norway). Gregarious, occasionally single, but usually several, up to 30.

Not on *Pagurus bernhardus* (Linn.). On *Anapagurus chiroacanthus* (Lilljeborg), *A. hyndmanni* Thompson, *A. laevis* (Bell), *Pagurus cuanensis* Thompson, and *P. prideauxi* Leach.

5. LERNAEODISCIDAE

Body compact. Mantle opening opposite stalk or slightly displaced to one side. Parasites of squat lobsters.

---

1. Externally symmetrical. On *Munida*, whitish ***Lernaeodiscus ingolfi***

Mantle opening off centre, to left. On *Munida* ***Triangulus munidae***

Lernaeodiscus ingolfi Boschma Fig. 8.11
Maximum width 17 mm. Whitish colour, sometimes with red patch. Mantle opening very large and directed anteriorly on host. Short, posteriorly situated stalk, attached usually to lst–4th abdominal segment. Recorded up to 0.5% of *Munida sarsi* Brinkman infected. Occurs singly.

On *Munida sarsi* Brinkman and less often, on *M. rugosa* (Fabricius). *Lernaeodiscus squamiferae* Perez is smaller, up to 5 mm, reddish, and occurs on *Galathea squamifera* Leach.

Triangulus munidae Smith Fig. 8.11
Up to 15 mm wide, colour orange. Asymmetrical, mantle opening displaced towards left of parasite. Usually single, occasionally two or three; on 2nd–4th abdominal sterna.

On *Munida rugosa* (Fabricius). *Triangulus galatheae* (Norman and Scott) is up to 5.5 mm wide, yellow to orange, and occurs on *Galathea* spp.

6. SACCULINIDAE

Body compact with mantle opening nearly opposite stalk. Parasites of true crabs. Rather featureless; the following species may be keyed by reference to gross shape and host. Further details are given for two common species.

---

1. On *Macropodia* or *Inachus* ***Drepanorchis neglecta***

On other true crabs (*Sacculina*) **2**

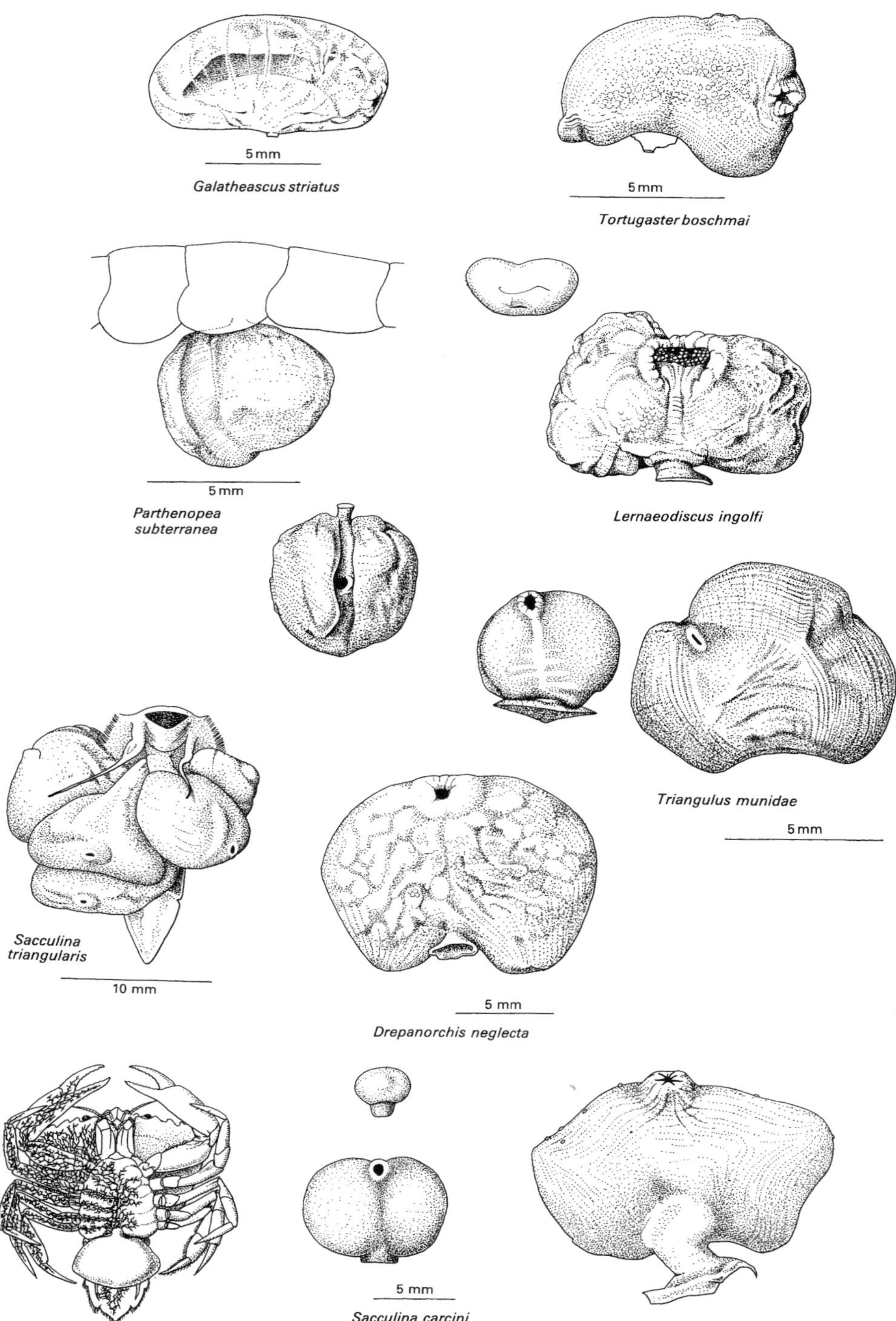
5 mm
Galatheascus striatus
5 mm
Tortugaster boschmai
5 mm
Parthenopea subterranea
Lernaeodiscus ingolfi
Triangulus munidae
5 mm
Sacculina triangularis
10 mm
5 mm
Drepanorchis neglecta
5 mm
Sacculina carcini

Fig. 8.11

2. On *Cancer pagurus*, stalk grades into body (Fig. 8.11) ***Sacculina triangularis*** **Anderson**

Not on *C. pagurus*, stalk usually inconspicuous **3**

3. On *Ebalia tuberosa*, body peripherally lobed ***Sacculina lobata*** **Boschma**

Not on *Ebalia* **4**

4. Body evenly rounded **5**

Body with shoulders **6**

5. On the crab *Pisa*, mantle translucent ***Sacculina gibbsi*** **(Hesse)**

Not on *Pisa* 7

6. On *Xantho pilipes* ***Sacculina bourdoni*** **Boschma**

On *Carcinus, Necora, Pilumnus, Bathynectes, Liocarcinus,* and *Pirimella* ***Sacculina carcini***

7. On spider crab (Majidae) **8**

On *Xantho incisus, Pilumnus hirtellus,* or *Atelecyclus rotundatus* ***Sacculina gerbei*** **Bonnier**

8. Thick lips to mantle opening, on *Hyas* ***Sacculina inflata*** **Leuckart**

No such lips, on *Dorynchus thomsoni* ***Sacculina atlantica*** **Boschma**

Sacculina carcini Thompson Fig. 8.11
Up to 26 mm. Cream when young, brown later. The best known British rhizocephalan. Symmetrical. Mantle aperture raised and slightly to left. Infection rate varies with host species, up to 50% or more; multiple infections (up to five) random. Infected *Carcinus* migrate to deeper water.

On *Carcinus maenas* (Linn.) and *Liocarcinus holsatus* (Fabricius), and a wide range of other Portunidae and Pirimelidae. Rarely found on *Necora puber* (Linn.) or *L. depurator* (Linn.).

British Isles, abundant all coasts.

Drepanorchis neglecta (Fraise) Fig. 8.11
Up to 14 mm, usually less yellowish, becoming browner. Infection rates up to 4.6% recorded in Irish Sea; 70% in Scandinavia—where multiple (up to four externae) observed.

On *Macropodia rostrata* (Linn.), *M. tenuirostris* (Leach), *Inachus dorsettensis* (Pennant), *I. phalangium* (Fabricius), *I. leptochirus* (Leach).

Class Malacostraca

The Malacostraca is the largest class of the Crustacea, and is also the most familiar crustacean group, including the edible and economically important crabs, lobsters, shrimps, and prawns, as well as a number of significant wood-destroying pests.

The malacostracan body plan (Fig. 8.12) comprises a primitive, five-segmented *head*, often fused with the *thorax*, or *pereon*, of eight segments; the *abdomen*, or *pleon*, consists of a further six segments, or rarely seven. The head bears two pairs of antennae, the first pair typically biramous; there are two pairs of *maxillae* and up to three pairs of *maxillipeds*; compound eyes are usually present, often stalked, but may be absent. Each thoracic segment bears a pair of appendages, basically biramous, with a gill at the base of each appendage; the first three pairs are employed in feeding, while the rest form the locomotor *pereopods*. Each abdominal segment is similarly equipped with a pair of biramous appendages, the *pleopods* and *uropods*, modified for swimming or reproduction, or both functions, and occasionally for respiratory purposes. The last pair of abdominal appendages is often broadened, and with the last abdominal segment, the *telson*, constitutes a tail fan.

Malacostracans are mostly dioecious, although a few hermaphrodite species are known. Following copulation a considerable delay may lapse before fertilization and spawning. Eggs are usually retained by the female, within a thoracic brood chamber or attached to the pleopods. Life histories may be relatively simple, with direct development resulting in the hatching of juveniles, or very complex, involving often lengthy series of larval stages.

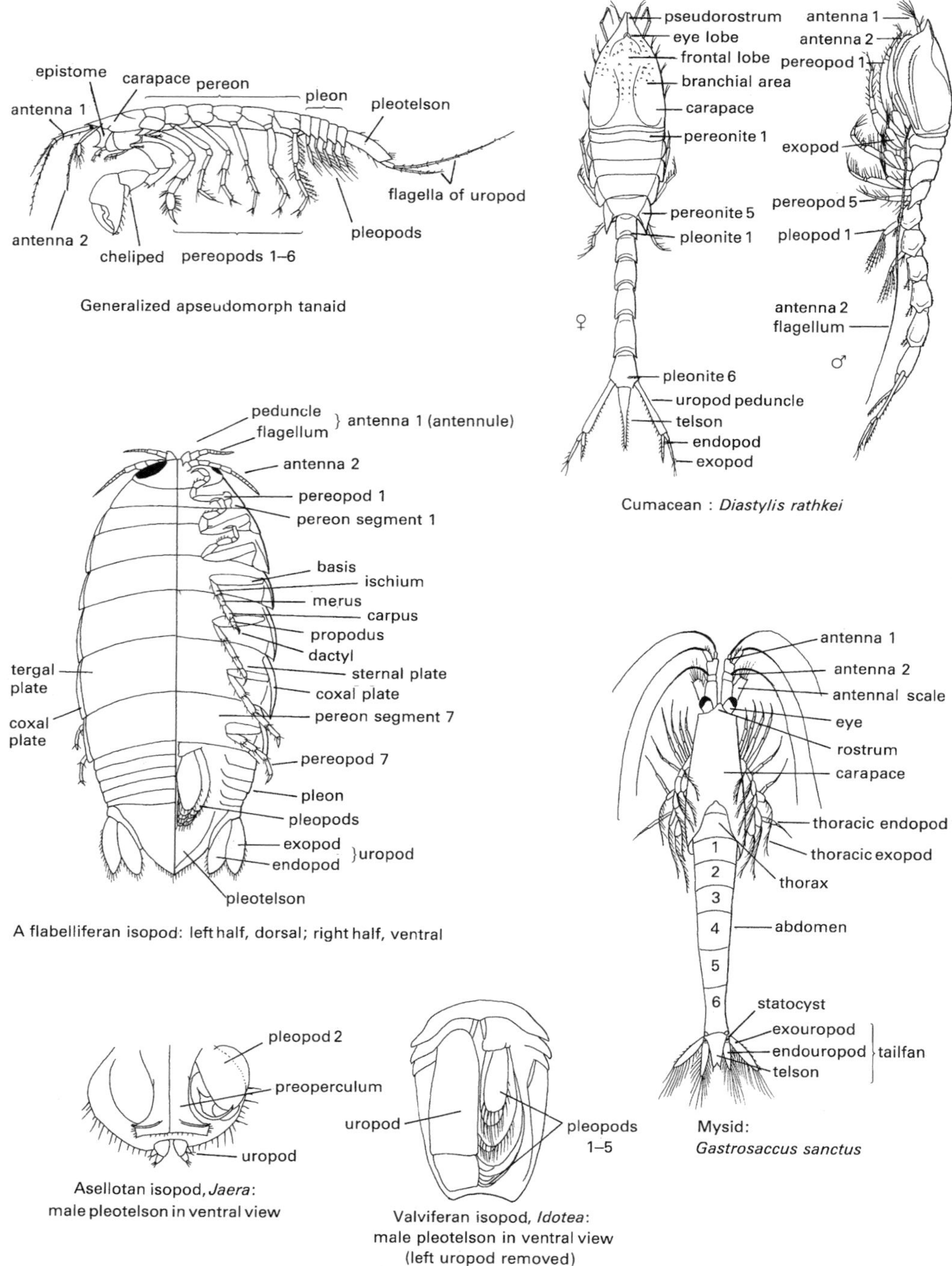
epistome
carapace
pereon
pleon
pleotelson
antenna 1
flagella of uropod
antenna 2
cheliped
pereopods 1–6
pleopods
Generalized apseudomorph tanaid
pseudorostrum
eye lobe
frontal lobe
branchial area
carapace
pereonite 1
pereonite 5
pleonite 1
pleonite 6
uropod peduncle
telson
endopod
exopod
antenna 1
antenna 2
pereopod 1
exopod
pereopod 5
pleopod 1
antenna 2
flagellum
♀
♂
Cumacean : Diastylis rathkei
peduncle
flagellum
} antenna 1 (antennule)
antenna 2
pereopod 1
pereon segment 1
basis
ischium
merus
carpus
propodus
dactyl
sternal plate
coxal plate
pereon segment 7
pereopod 7
pleon
pleopods
exopod
endopod
}uropod
pleotelson
tergal plate
coxal plate
A flabelliferan isopod: left half, dorsal; right half, ventral
antenna 1
antenna 2
antennal scale
eye
rostrum
carapace
thoracic endopod
thoracic exopod
thorax
abdomen
1
2
3
4
5
6
statocyst
exouropod
endouropod
telson
tailfan
Mysid:
Gastrosaccus sanctus
pleopod 2
preoperculum
uropod
Asellotan isopod, Jaera:
male pleotelson in ventral view
uropod
pleopods
1–5
Valviferan isopod, Idotea:
male pleotelson in ventral view
(left uropod removed)

Fig. 8.12

Malacostracan crustaceans occur abundantly in all marine habitats and display wide ecological diversity. Although a majority are particle-feeding scavengers, many feeding specialisations occur, from planktonic filter feeders to macrocarnivores.

Six superorders are recognised within the Malacostraca, only three of which are covered here. The Phyllocarida is a primitive group of small, predominantly filter-feeding malacostracans, of which only the order Leptostraca contains living representatives. In the Peracarida, exemplified by isopods and amphipods, the thoracic segments are typically distinct and a carapace is absent or only scarcely developed. Eggs are brooded in the *marsupium*, a brood chamber formed of a series of plates developed from the *coxae* of the thoracic limbs, and give rise to juveniles through direct development, with no intervening larval stages. The Eucarida comprises the crabs and their allies. In these the thorax is fused dorsally to a rigid carapace which displays no sign of segmentation. Eggs are typically brooded attached to the pleopods of the female and hatch as post-naupliar larvae which pass through a series of larval stages before metamorphosis to adult form.

Superorder Phyllocarida

I Order Leptostraca

The leptostracans are the most primitive members of the Malacostraca. Only 10 living species are known, although many fossil species have been described. The body is small, and the thorax and part of the abdomen is enclosed in a large bivalved *carapace* hinged along the mid-dorsal line. The carapace is attached to the head region of the animal by an adductor muscle. Anteriorly, a hinged and movable *rostral plate* covers the head. The long antennae protrude anteriorly from the carapace, and in males the antennal flagellum is as long as the body. There are eight thoracic segments, each with a pair of biramous leaf-like appendages used for swimming and feeding, and in the females they may be modified for brooding eggs. The pleon comprises seven segments and terminates with a pair of long *furcal rami*; the first four pleon segments bear well-developed biramous appendages, segments five and six have reduced uniramous appendages, and segment seven is without appendages.

A single leptostracan occurs commonly in British coastal waters.

1. NEBALIIDAE

Nebalia bipes (Fabricius) Fig. 8.13

Body up to 12 mm long, light brown, greenish, or dark brown, with red eyes; distinctly shrimp-like in appearance. Carapace smooth, proportionately shorter in female than male. Female antenna 2 flagellum about twice length of antenna 1 flagellum. Male antenna 2 flagellum very long, extending almost to tip of furcal rami.

Intertidal and shallow sublittoral; under stones, in decaying seaweed, in shaded rock pools where they are easily trapped in the surface film on disturbance, frequently in crab pots or bait; often abundant. All British coasts; distributed from North Sea to Mediterranean.

Superorder Peracarida

II Order Cumacea

The Cumaceans are small peracaridan crustaceans, easily recognized by their tadpole-like shape (Fig. 8.12). The broad anterior part of the body comprises the *carapace*, formed from the dorsal fusion of the head and the first three thoracic segments. The carapace is produced into lateral folds which enclose the anterior appendages of the animal, and bears at the front two *pseudorostral lobes*, constituting the *pseudorostrum*. Below the pseudorostrum, on each side, the edge of the carapace has a distinct *antennal notch*. The single eye is situated posterior to the pseudorostrum, just anterior to the dorsal *frontal lobe* of the carapace. The remaining five thoracic segments comprise the *pereon*; the slender abdomen, or *pleon*, consists of six segments and a terminal *telson*. The anterior appendages consist of one or two pairs of *maxillae*, three pairs of *maxillipeds*, and five pairs of *pereopods*. The pleon may have up to five pairs of *pleopods* in males, but female cumaceans lack pleopods entirely. The first antenna may have a brush of modified, jointed setae at the tip of the flagellum; referred to as *aesthetascs*, these appear to be sensory in function.

Sexual dimorphism is very marked in the Cumacea, and both carapace shape and sculpture may differ between the sexes. In the female, antenna 2 is rudimentary and there are no pleopods. The adult male has a well-developed antenna 2 with

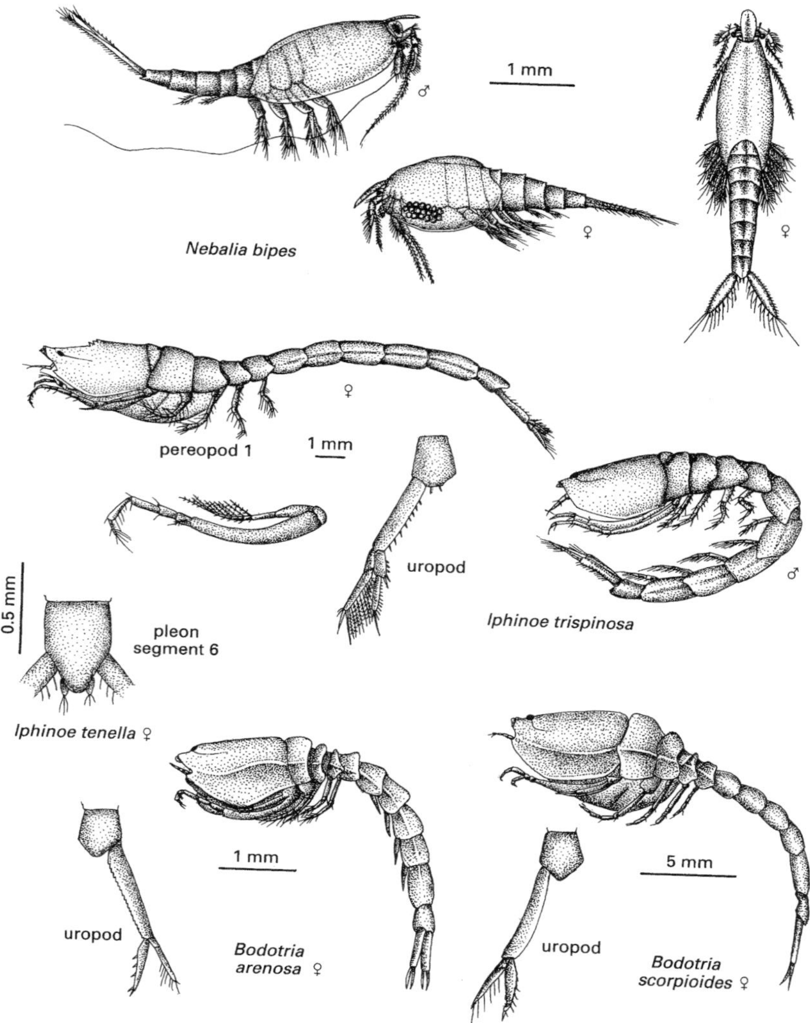

Fig. 8.13

a long flagellum, and, with the exception of the Nannastacidae, has a number of pleopods. Immature males resemble the female, and the flagellum of antenna 2 increases in length with age.

Intertidal species of cumaceans are most easily found on the shore in areas of wet sand, particularly between ripple marks where water lies on the sand surface. Most species are retained by a 1 mm sieve. Alternatively, specimens may be obtained by puddling wet sand with an open hand, when the animals will emerge and swim in the surface water before reburying themselves. Some species are caught in inshore or surf plankton hauls, particularly at night when they may be attracted by a light. Plankton light traps, employing a towed net with a light source close to the mouth, will also catch cumaceans, particularly the males of certain species. Offshore, fine-meshed bottom dredges may be used with good results.

The British fauna includes more than 40 species, in six families. Many of these are found only well offshore, often in deep water, and only common intertidal species are keyed out here.

KEY TO FAMILIES

1. Telson absent, or so well fused with the last abdominal segment as to be indistinguishable separately **2**

   Telson present and obvious, if only as a small plate **3**

2. Male with five pairs of pleopods. Female with exopods on either pereopod I, or on pereopods 1, 2, and 3 **1. Bodotriidae**

   Male without pleopods. Female with exopods on pereopods 1 and 2 only **2. Nannastacidae**

3. Inner ramus of uropod with only one article **3. Pseudocumatidae**

   Inner ramus of uropod with two or three articles **4. Diastylidae**

1. BODOTRIIDAE

No free telson on abdomen. Pleopods with a process on outer edge of inner ramus. Male with five pairs of pleopods, or occasionally only two or three pairs. Exopodites present on maxilliped 3 in both sexes. Male with exopodites on pereopod 1, or on pereopods 1–4, or only on pereopods 1, 2, and 3. Female with exopodites on pereopods 1–3, or pereopod 1 only, or on first two or first four pairs. Thoracic segments often reduced.

1. Exopodites on pereopod 1 only **2**

   Exopodites on pereopods 1–3 or 1–4, sometimes rudimentary **5**

2. First thoracic segment visible from above **3**

   First thoracic segment fused with the carapace and not visible **4**

3. Basis of pereopod 1 longer than the remaining articles together; last abdominal segment with two terminal setae ***Iphinoe trispinosa***

   Basis of pereopod 1 shorter than the remaining joints together; last abdominal segment with six terminal setae ***Iphinoe tenella***

4. Two joints on inner ramus of uropod ***Bodotria scorpioides***

   One joint on inner ramus of uropod ***Bodotria arenosa***

5. Exopods of pereopods 2 and 3 rudimentary **6**

   These exopods well developed ***Vaunthompsonia cristata***

6. Carapace with two lateral folds each side ***Cumopsis goodsiri***

   Carapace without lateral folds ***Cumopsis fagei***

Iphinoe trispinosa (Goodsir) Fig. 8.13

Body slender, up to 10 mm long, whitish or straw coloured. Adult female with 2–6 small teeth in middle of dorsal ridge of carapace; pseudorostrum acutely pointed. Male with dorsal ridge of carapace unarmed, and pseudorostrum blunt. Adult male with median spinous tubercle, and paired lateral tubercles, on ventral surface of pereon segment 2. Basis of pereopod 2 with bifid spur on inner edge.

On fine sand; intertidal and coastal, but also offshore to 150 m. All British coasts; distributed from Norway to the Mediterranean, and the Canary Isles.

Iphinoe tenella Sars — Fig. 8.13

Body up to 8 mm long. Carapace in both sexes with prominent teeth along whole of dorsal crest, increasing in size anteriorly. Adult male without ventral tubercles on pereon segment 2. No spur on basis of pereopod 2.

On muddy sand; intertidal and shallow sublittoral, to 30 m. Reported from Channel coasts, and southwards to the Mediterranean and west Africa.

Bodotria arenosa Goodsir — Fig. 8.13

Body up to 7 mm, brownish-yellow with a transverse white band on carapace. Well-marked longitudinal dorsal ridge, and paired longitudinal lateral ridges, extending along length of carapace, and posteriorly to pereon segment 5; particularly well marked in male.

In coarse sand; lower shore and shallow sublittoral, to 120 m. All British coasts; distributed from south-west Norway to Brittany.

Bodotria scorpioides (Montagu) — Fig. 8.13

Body up to 7 mm long, light yellow with brown speckles. Carapace with dorsal longitudinal ridge and paired lateral ridges, extending posteriorly on to first five pereon segments.

In fine sand, lower shore and shallow subtidal to 100 m. All British coasts; distributed from Norway to western France.

Vaunthompsonia cristata Bate — Fig. 8.14

Body slender, up to 5 mm in male, 6 mm in female. Male with smooth carapace and shallow antennal notch; female carapace with dorsal median crest of two rows of small teeth, and conspicuous serrated antennal notch. Pereopods 1–4 of male, 1–3 of female, each with well-developed exopodites. Antenna 2 flagellum of male extending posteriorly no further than pereon segment 5.

Lower shore and shallow sublittoral, to 40 m, on coarse sand and gravel. Southern and western shores only; distributed south to the Mediterranean and west Africa.

Cumopsis fagei Bacescu — Fig. 8.14

Body up to 6 mm. Carapace without lateral folds, pereon segments smooth dorsally. Female uropod as long as combined length of three posterior pleon segments, proximal segment of exopod longer than distal segment, which terminates with a strong spine but lacks spines on inner edge.

On exposed sandy beaches, predominantly intertidal but also in shallow sublittoral. Distributed from Channel Isles to Morocco, and perhaps occurring on extreme south-west shores of Britain.

Cumopsis goodsiri (Van Beneden) — Fig. 8.14

Body up to 6 mm in female, 5 mm in male, with purplish-brown patches on carapace and pleon segment 5. Both sexes with a pair of longitudinal ridges on each side of carapace, occasionally faint; female with two semicircular folds on dorsal side of pereon segment 2. Antenna 2 flagellum long, extending beyond end of abdomen. Female uropod as long as posterior three segments of abdomen, peduncle and inner ramus bearing variable number of spines.

In fine sand, from mid-tidal level just into shallow subtidal on sheltered beaches. All British coasts; distributed from North Sea to Mediterranean and north-west Africa.

2. NANNASTACIDAE

Pleon lacking both pleopods and telson. Male with exopodites on maxilliped 3 and pereopods 1–4, rarely on pereopods 1, 2, and 3 only. Female with exopodite on maxilliped 3 (rarely, absent) and on pereopods 1, 2, rarely on pereopods 1–3, or lacking completely. Uropod with inner ramus of one article only.

Nannastacus unguiculatus (Bate) — Fig. 8.14

Body short and broad, up to 2 mm long; carapace with pronounced antero-lateral processes, with numerous blunt or acute tubercles, and short setae; female with prominent series of blunt tubercles on postero-lateral borders of carapace. Dorsal keel of abdomen with distinct spines or coarse serrations. Female uropod longer than combined length of posterior two pleon segments; inner ramus more than twice length of peduncle, and three times longer than outer ramus.

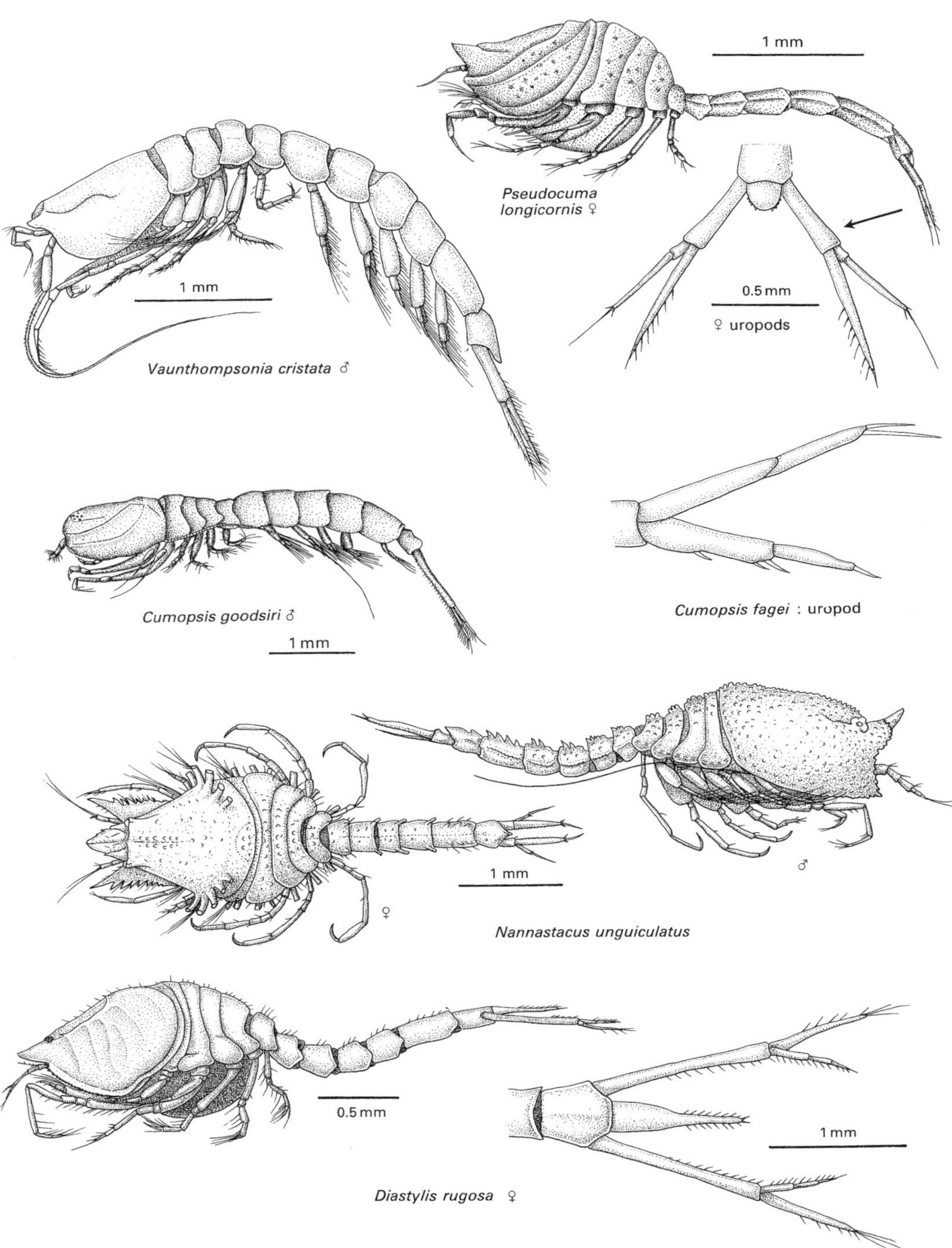
1 mm
Pseudocuma
longicornis ♀
0.5 mm
♀ uropods
1 mm
Vaunthompsonia cristata ♂
Cumopsis goodsiri ♂
1 mm
Cumopsis fagei : uropod
1 mm
♂
♀
Nannastacus unguiculatus
0.5 mm
1 mm
Diastylis rugosa ♀

Fig. 8.14

In coarse deposits, lower shore and shallow sublittoral. All British coasts; distributed from Shetland to the Mediterranean.

3. PSEUDOCUMATIDAE

Small telson present. Male with exopodites on maxilliped 3 and pereopods 1-4; female with exopodites on maxilliped 3 and pereopods 1,2, poorly developed on pereopods 3,4. Male with two rudimentary pairs of pleopods, or sometimes only a single pair. Pleopods without external process on inner ramus. Uropod inner ramus with one article only.

Pseudocuma longicornis (Bate) Fig. 8.14
Body up to 4 mm long, marked with dark brown patches. Carapace with three obliquely longitudinal folds on each side; dorsal edge convex in male, slightly concave in female; pseudorostrum acute, antero-lateral corners without teeth. Pereon segments lack dorsal keel or projections. Telson semicircular to semioval. Antenna 2 flagellum of male extending posteriorly at least as far as pleon segment 5. Uropod peduncle as long as outer ramus in female, shorter in male; inner ramus longer than outer, with about 12 spines on inner edge.

Intertidal and shallow sublittoral, occurring in brackish water and tolerant of very low salinities. All British coasts; distributed from northern Norway to the Mediterranean.

4. DIASTYLIDAE

Telson present, often large, with two apical spines, or none. Pleopods sometimes small, without external process on inner ramus. Male with two pairs of pleopods, or none. Exopodites on maxilliped 3 and pereopods 1–4, or 1–2, in male; in female present or absent on maxilliped 3, present on pereopods 1–2, and occasionally as rudiments on pereopods 3–4. Uropod with inner ramus of three articles, or occasionally only two (or one) articles. Sexual dimorphism often marked.

Diastylis rugosa Sars Fig. 8.14
Body up to 9 mm, greyish white. Female carapace with 3–5 transverse folds on each side, and two pairs of conspicuous acute teeth fronto-dorsally. Male carapace with two transverse folds on each side, associated with a smooth, longitudinal ridge. Postero-lateral corners of pereon segment 5 rounded in female, acutely pointed in male. Telson slightly shorter than uropod peduncle,with up to nine pairs of lateral spines.

Lower shore and shallow sublittoral, to 90 m; in muddy sand, and often in *Laminaria* beds. All British coasts; distributed from Norway to the Mediterranean.

III Order Tanaidacea

Tanaids are small peracaridan crustaceans superficially resembling isopods, with which they were once classified. The tanaid body is dorso-ventrally flattened, or cylindical, and tends to be rather elongate. Anteriorly, the head and first two segments of the thorax are fused to form a *cephalothorax*, covered by a carapace which is produced into lateral folds enclosing a *branchial chamber*. Sessile eyes may be present, and the anterior edge of the carapace may be produced as a *rostrum*. The cephalothorax is followed by a six-segmented *pereon*, and a *pleon* of five segments, which in some species are fused. The posterior *pleotelson* is formed from the fusion of a sixth pleon segment with the *telson*. The second fused thoracic segment bears a pair of chelate appendages, the *chelipeds*, and the following six pereon segments each bears a pair of ambulatory *pereopods*, usually all alike. Each pleon segment has a pair of *pleopods*, used in swimming, although these may be missing in the females of some species. Finally, a single pair of uniramous or biramous *uropods* is borne by the pleotelson. Sexes are separate and sexual dimorphism is often marked. Eggs are brooded in a *marsupium* on the ventral surface of the female and young are released as advanced juveniles, or *mancas*. Tanaids are small and often live in burrows or tubes in soft sandy or muddy substrate, or under stones, in crevices, and amongst red algae such as *Corallina* and *Laurencia*. They are generally inconspicuous and frequently overlooked. There are 30 shallow-water species around the British Isles, eight of which may be found intertidally. Below 200 m species diversity is much higher.

KEY TO FAMILIES

1. Antenna 1 with two flagella. Pereopod 1 stout, with flattened propodus. Male with single genital cone on pereon segment 6 **1. Apseudidae**

Antenna 1 with only one flagellum. Pereopod 1 slender, similar to rest of pereopods. Male with two genital cones on pereon segment 6 **2**

---

2. Three pairs of pleopods present in both sexes. Pereopods without ischium. Uropods uniramous **2. Tanaidae**

   Five pairs of pleopods present, may be lacking in females. Pereopods with ischium. Uropods usually biramous **3**

---

3. Maxilliped bases distinct, not fused medially. Endopod of pleopod with conspicuous proximal seta on inner margin **3. Paratanaidae**

   Maxilliped bases fused medially to greater or lesser extent. Endopod of pleopod not as above **4**

---

4. Maxilliped bases fused proximally, separate distally. Female antenna 1 with three or four articles, male antenna 1 with up to seven articles **4. Leptognathiidae**

   Maxilliped bases fused to form a single plate. Female antenna 1 with four articles, male antenna 1 with six articles **5. Nototanaidae**

---

1. APSEUDIDAE

Body dorso-ventrally flattened, posteriorly tapered. Cephalothorax with prominent rostrum; ocular lobes distinct, eyes present or absent. Pereon segment 1 closely joined to carapace, all other pereon and pleon segments distinct. Antenna 1 with two flagella, antenna 2 with one flagellum. Chelipeds often sexually dimorphic. Pereopod 1 flattened distally, with numerous propodial spines.

---

1. Rostrum parallel sided with rounded apex, and apical spine ***Apseudes latreillii***

   Rostrum triangular, with fine marginal denticulation ***Apseudes talpa***

Apseudes talpa (Montagu) Fig. 8.15

Up to 8 mm long, opaque white. Cephalothorax as broad as long. Eyes present. Antero-lateral margins of pereon segments produced as short, rounded processes. Pleon segments with numerous setae; produced sharply laterally. Pleotelson as long as pleon. Chelipeds stouter in male than female, though not markedly dimorphic.

Middle and lower shore, into shallow sublittoral; in crevices, beneath stones, in *Laminaria* holdfasts and coralline algal turfs. South and west coasts of Britain, and western Ireland, southwards to north Africa.

Apseudes latreillii (Milne Edwards) Fig. 8.15

Body up to 7 mm. Cephalothorax slightly longer than broad. Antero-lateral margins of pereon segments rounded. Pleotelson shorter than pleon. Male cheliped stout, with prominent teeth on inner margins of propodus and dactylus; female cheliped slender, without teeth.

Middle and lower shore, and shallow subtidal; in crevices, in muddy gravel, beneath stones, in *Laminaria* holdfasts, coralline algal turfs, and the roots of *Zostera*. Reported from the Isles of Scilly and the Channel Isles, from Channel coasts, and from north-east England and eastern Scotland.

2. TANAIDAE

Body fusiform, anteriorly truncate, posteriorly rounded. Eyes well developed. Three to five pleon segments. Antenna 1 with peduncle of three articles, with or without a small flagellum. Antenna 2 with 6–8 articles. Pereopods without ischium; pleopods 4–6 with distinct claw. Three pairs of pleopods present. Uropods uniramous.

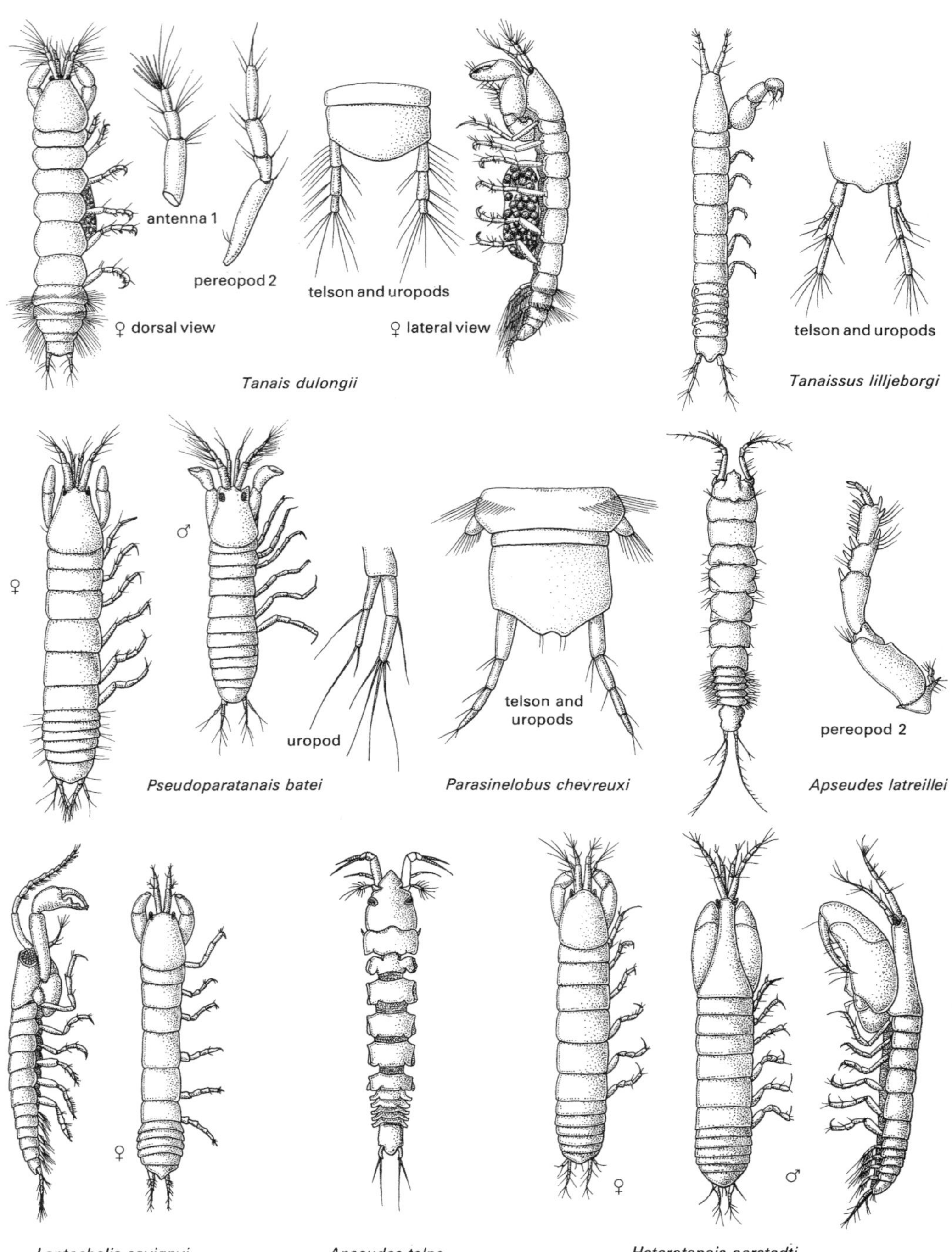
antenna 1
pereopod 2
telson and uropods
♀ dorsal view
♀ lateral view
Tanais dulongii
telson and uropods
Tanaissus lilljeborgi
♀
♂
uropod
telson and uropods
pereopod 2
Pseudoparatanais batei
Parasinelobus chevreuxi
Apseudes latreillei
♂
♀
♀
♂
Leptochelia savignyi
Apseudes talpa
Heterotanais oerstedti

Fig. 8.15

1. Uropod with three articles
   Tanais dulongii

   Uropod with four articles
   Parasinelobus chevreuxi

Tanais dulongii (Audouin) Fig. 8.15
Up to 5 mm long, often mottled with grey. Cephalothorax slightly narrowed posteriorly. Pereon segment 1 much shorter than succeeding segments. Antenna 2 with seven articles, the distal two very short. Cheliped typically with a tooth on inner margin of both propodus and dactylus, sometimes one or both missing; male often with more robust cheliped than female. Pereopods 4–6 with claw bearing a comb of stiff setae. Uropod with three articles.

Builds tube of sand, mud, and silt; intertidal and shallow subtidal, among barnacles, in crevices, under stones, among algal turfs, particularly *Corallina* and *Laurencia*; often on pier piles. Widespread and common, all British coasts; from northern Norway to the Mediterranean, the east coasts of North and South America, and Australia, a distribution suggestive of association with fouling communities.

Parasinelobus chevreuxi (Dollfus) Fig. 8.15
Up to 6 mm long. Cephalothorax broadest posteriorly. Pereon segments 1–3 of similar size. Antenna 2 with six articles, articles 2 and 4 bearing distal fringe of stiff setae. Cheliped typically larger in male than female. Pereopods 4–6 without setal comb on claw. Uropod with four articles.

Builds long, parchment-like tubes in intertidal crevices; often abundant from middle to upper shore. South Devon, Wales, and western Ireland; southwards to north-west Africa.

3. PARATANAIDAE

British species display marked sexual dimorphism. Antenna 1 with three or four articles in female; with peduncle of two or three articles, and multi-articled flagellum, in male. Maxillipeds not fused medially. Cheliped with coxa situated behind proximal projection of basis. Pereopods with ischium; pereopods 4–6 with distal claw. Five pleon segments, each with biramous appendages. Uropods biramous.

1. Antenna 1 with ventral spine on article 2. Uropod outer ramus with one article only; inner ramus with four articles
   Leptochelia savignyi

   Antenna 1 without ventral spine on article 2. Uropod outer ramus with two articles; inner ramus of 4–7 articles
   Heterotanais oerstedti

Heterotanais oerstedti (Krøyer) Fig. 8.15
Up to 2 mm long; rounded posteriorly, carapace anteriorly truncate in female, abruptly narrowed and elongate in male. Eyes prominent in both sexes. Pereon and pleon segments laterally rounded. Antenna 1 with three articles, lacking flagellum, in female; with three-articled peduncle and two-articled flagellum in male. Male with massively enlarged chelipeds, with prominent projection on carpus; female with smaller chelipeds.

Builds tubes attached to algae, stones, and other hard substrata, or on mud surfaces, in sheltered conditions; lower shore and shallow sublittoral, to 10 m. Tolerant of low and varying salinity, from almost freshwater to fully marine. Coastal, estuarine, and fluviatile habitats on south and east coasts of England, and South Wales; distributed from Norway to northern France.

Leptochelia savignyi (Krøyer) Fig. 8.15
Male up to 1.5 mm, female 3 mm; rounded posteriorly, cephalothorax fusiform in female, more nearly quadrate in male. Pereon segments broadest posteriorly in male, more or less parallel-sided in female. Antenna 1 in female three- or four-articled, with single-articled flagellum, in male with two-articled peduncle and multi-articled flagellum. Male cheliped large, with two large teeth on inner margin of propodus and numerous spines on inner margin of dactylus; female cheliped relatively small.

Builds tubes, on seaweed-covered surfaces, or among encrusting seaweeds and sea grasses; intertidal and shallow subtidal. Recorded from south-west England, the Channel Isles, and south and west coasts of Ireland.

4. LEPTOGNATHIIDAE

Antenna 1 of female with three or four articles, in male with 4–7 articles. Antenna 2 with 5–7 articles in both sexes. Maxilliped bases fused to a variable extent. Chelipeds similar in males and females. Pleopods present or absent, highly variable.

Numerous species of this family have been described sporadically from British coasts. One common species is included here; Holdich and Jones (1983) describe and illustrate all species recorded from British waters.

Pseudoparatanais batei (Sars) Fig. 8.15
Female up to 2 mm, male 1 mm; female more or less parallel-sided, male distinctly waisted between pereon segments 3 and 4. Cephalothorax broadest posteriorly in female, broadened anteriorly in male; eyes well developed, particularly large in male. Antenna 1 with three articles in female, six in male; antenna 2 with five articles in both sexes. Uropod biramous, each ramus with two articles.

Lower shore to 100 m, among algae, in kelp holdfasts, hydroid clumps, crevices, muddy gravel, and shell deposits. Most British coasts, except south-east England; Norway and Iceland south to the Mediterranean.

5. NOTOTANAIDAE

Antenna 1 in female with three or four articles, in male with five or six articles; antenna 2 with six articles. Maxilliped bases completely fused. Pleopods reduced or well developed. Uropods biramous.

Tanaissus lilljeborgi (Stebbing) Fig. 8.15
Body up to 2.5 mm long, elongate and slender, rounded posteriorly; cephalothorax abruptly tapered anteriorly. No eyes. Antenna 1 with four articles in female, six articles in male. Chelipeds large; propodus elongate in female, with dactyl forming chelate closure, shorter and broader in male with dactyl forming subchelate closure. Uropod biramous, each ramus of two articles, the inner ramus twice length of outer.

In sand, lower shore and shallow subtidal, often abundant. Probably present on most British coasts; distributed from southern North Sea to northern France.

IV. Order Mysidacea

Small, shrimp-like crustaceans, 5–25 mm long, free swimming or resting on the bottom. Usually translucent but may be black or coloured. All the British coastal species belong to the family Mysidae. They have a pair of statocysts present in the tail fan; telson wider at the base than at the apex; legs without chelae (except one small pair in the uncommon sedentary species of *Heteromysis*, Fig. 8.18F).

Injury followed by healing or regeneration can cause atypical morphology. In doubtful cases this should be checked by reference to appropriate characteristics; for example, the endouropod often provides a good diagnostic feature.

The body size given here is the maximum length, measured from the anterior margin of the carapace (excluding the rostrum) near the insertion of the eyestalks, to the tip of the telson.

Forty species of Mysid are recorded for Britain; 29 occur in coastal waters and may be identified with the following key.

KEY TO COASTAL SPECIES OF BRITISH MYSIDAE

1. Blade of telson (excluding teeth and spines) distinctly forked, bifid or notched at apex, forming an apical cleft (Figs 8.16J, 8.17S, but some species have a smaller cleft than these). It is important to distinguish the last lateral spines from blade of telson when comparing this category with the next **2**

   Telson without apical cleft, more or less rounded, pointed or truncate but not notched at apex (Figs 8.16M,N, 8.17T–X). Note: under low magnification apical spines can give a false impression of a cleft **17**

2. Exouropod with spines, not setae, on outer margin (Figs 8.16D. 8.17L, M) **3**

   Exouropod with setae, not spines, on outer margin (Fig. 8.16L) **6**

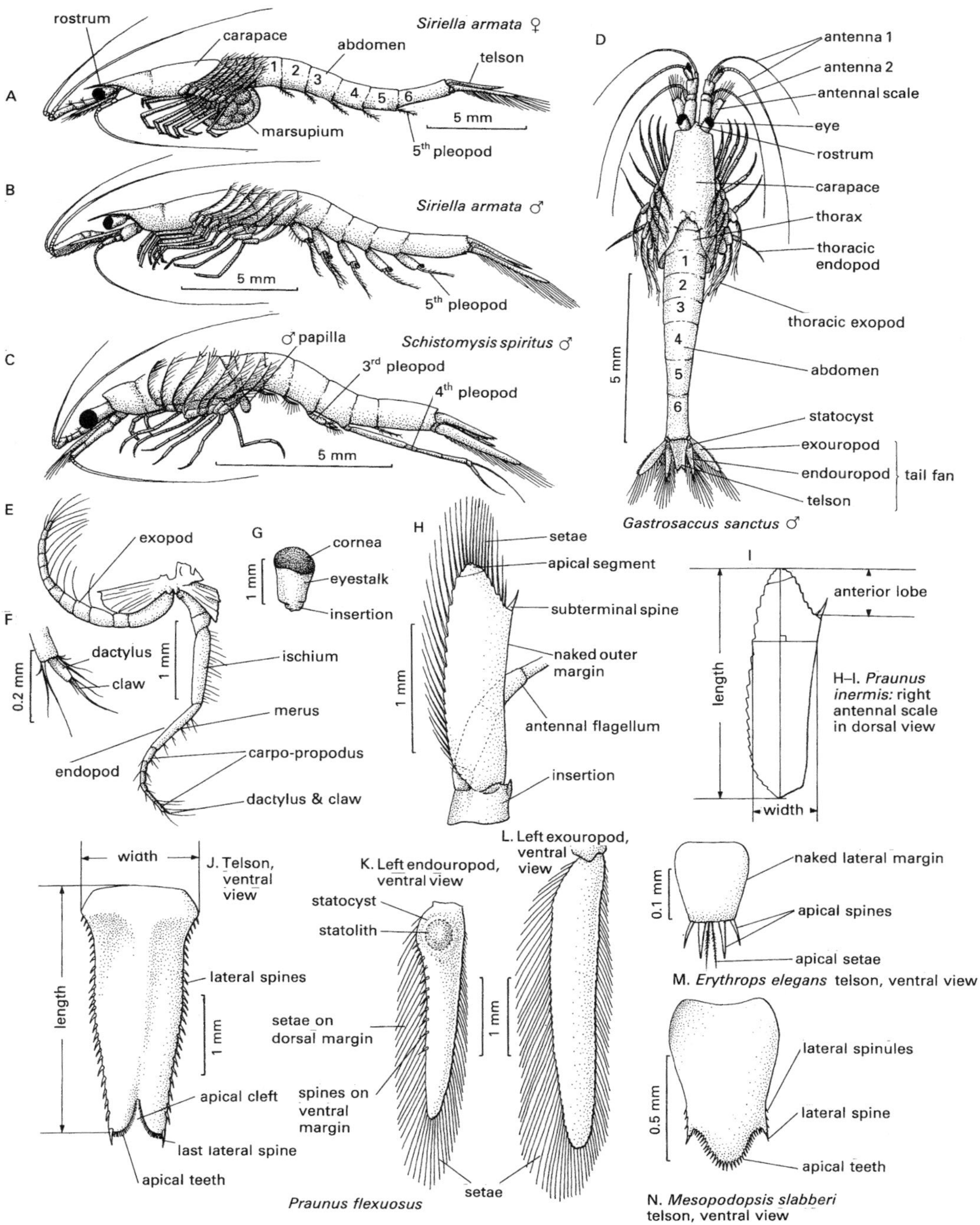
A
rostrum
carapace
abdomen
Siriella armata ♀
telson
1 2 3 4 5 6
5 mm
marsupium
5th pleopod
B
Siriella armata ♂
5 mm
5th pleopod
C
♂ papilla
Schistomysis spiritus ♂
3rd pleopod
4th pleopod
5 mm
D
antenna 1
antenna 2
antennal scale
eye
rostrum
carapace
thorax
thoracic endopod
thoracic exopod
abdomen
statocyst
exouropod
endouropod
telson
tail fan
5 mm
Gastrosaccus sanctus ♂
E
exopod
F
dactylus
claw
0.2 mm
1 mm
ischium
merus
endopod
carpo-propodus
dactylus & claw
G
cornea
eyestalk
insertion
1 mm
H
setae
apical segment
subterminal spine
naked outer margin
antennal flagellum
insertion
1 mm
I
anterior lobe
length
width
H–I. Praunus inermis: right antennal scale in dorsal view
width
J. Telson, ventral view
length
lateral spines
1 mm
apical cleft
last lateral spine
apical teeth
K. Left endouropod, ventral view
statocyst
statolith
setae on dorsal margin
spines on ventral margin
1 mm
L. Left exouropod, ventral view
setae
Praunus flexuosus
naked lateral margin
0.1 mm
apical spines
apical setae
M. Erythrops elegans telson, ventral view
lateral spinules
lateral spine
0.5 mm
apical teeth
N. Mesopodopsis slabberi telson, ventral view

Fig. 8.16

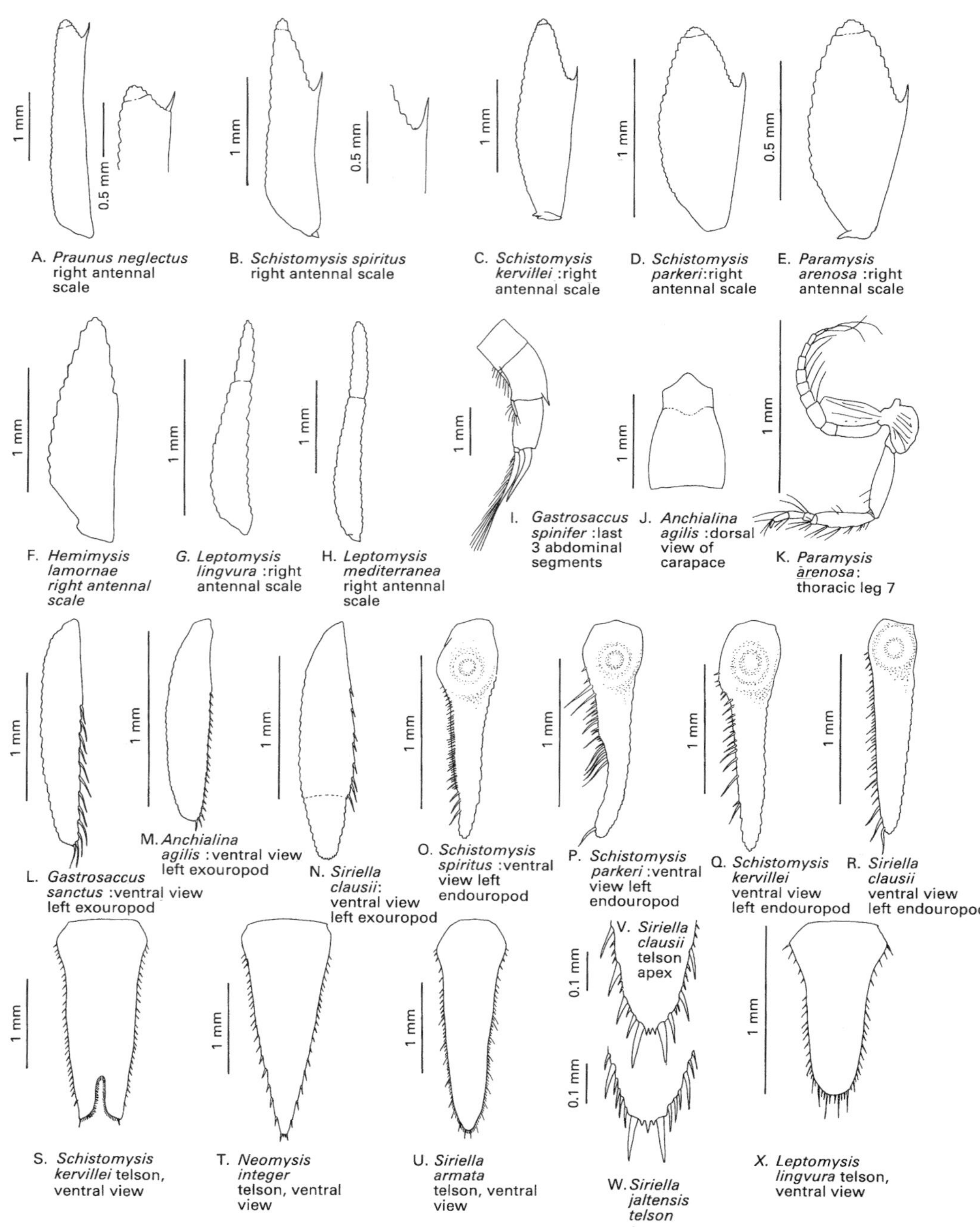

1 mm
0.5 mm
0.1 mm
A. *Praunus neglectus* right antennal scale
B. *Schistomysis spiritus* right antennal scale
C. *Schistomysis kervillei* :right antennal scale
D. *Schistomysis parkeri*:right antennal scale
E. *Paramysis arenosa* :right antennal scale
F. *Hemimysis lamornae right antennal scale*
G. *Leptomysis lingvura* :right antennal scale
H. *Leptomysis mediterranea* right antennal scale
I. *Gastrosaccus spinifer* :last 3 abdominal segments
J. *Anchialina agilis* :dorsal view of carapace
K. *Paramysis arenosa*: thoracic leg 7
L. *Gastrosaccus sanctus* :ventral view left exouropod
M. *Anchialina agilis* :ventral view left exouropod
N. *Siriella clausii*: ventral view left exouropod
O. *Schistomysis spiritus* :ventral view left endouropod
P. *Schistomysis parkeri* :ventral view left endouropod
Q. *Schistomysis kervillei* ventral view left endouropod
R. *Siriella clausii* ventral view left endouropod
S. *Schistomysis kervillei* telson, ventral view
T. *Neomysis integer* telson, ventral view
U. *Siriella armata* telson, ventral view
V. *Siriella clausii* telson apex
W. *Siriella jaltensis telson apex*
X. *Leptomysis lingvura* telson, ventral view

Fig. 8.17

3. Abdominal segment 5 rather narrow viewed from above. In profile showing a dorsal ridge and a spine-like median dorsal process at posterior margin (Fig. 8.17I) ***Gastrosaccus spinifer* (Goës)** (Fig. 8. 18B)

   Easily recognized by finger-like dorsal abdominal 'spine'. Fairly large, up to 21 mm, common but sub-tidal. Lives on bottom from shallow to deep water.

   Abdominal segment 5 normal, or narrowed but without a dorsal ridge or median dorsal process (Figs 8.16D, 8.18A,C) **4**

4. Telson with 20 or more lateral spines on each side. Outer margin of exouropod with 20 or more relatively small spines (Fig. 8.17M). Dorsal posterior margin of carapace straight (Figs 8.17J, 8.18A) ***Anchialina agilis* (G. O. Sars)** (Fig. 8.18A)

   Appears short and stout due to ample carapace with abruptly transverse, straight, hind margin. Usually transparent, small (9 mm) and delicate, with red or brown eyes. Subtidal, uncommon, from shallow water to 100 m. Normally near the bottom, swims actively, moving upwards and shorewards at night. Occasionally found in nocturnal net collections.

   Telson with up to 14 lateral spines on each side. Outer margin of exouropod with 15 or fewer, relatively massive spines (Fig. 8.17L). Dorsal posterior margin of carapace deeply concave (Figs 8.16D, 8.18B, C) **5**

5. Abdominal segment 5 narrow; nearly or quite as narrow as segment 6 (Fig. 8.16D). Telson about twice as long as broad, with five or six large lateral spines on each side ***Gastrosaccus sanctus* (van Beneden)** (Fig. 8.16D)

   Stout-bodied, with relatively thick cephalothoracic region, and abruptly narrowed posterior abdominal segments. Burrows into soft substrata, but emerges and swims freely upwards at night and may be netted in large numbers. Adults up to 15 mm. Mainly south and west coasts, rare west Scotland; south to Mediterranean.

   Abdominal segment 5 not as narrow as 6. Telson about three times as long as broad, with eight or more lateral spines on each side ***Gastrosaccus normani* G. O. Sars** and ***G. lobatus* Nouvel** (Fig. 8.18C)

   These two species are not easily distinguished and were not separated until recently. Both are smaller than *G. sanctus*, reaching about ll mm, with a more normal shape of abdomen. They occur on the bottom offshore, sometimes together. Distribution of both uncertain, but mainly south-west coasts, southwards to Mediterranean.

6. Antennal scale with setae all round (similar to Figs 8.17G, H) **7**

   Antennal scale with a rather long, smooth, and naked outer margin (Figs 8.16H, I, 8.17A–F). Remainder of edge, medial to this, bears setae **8**

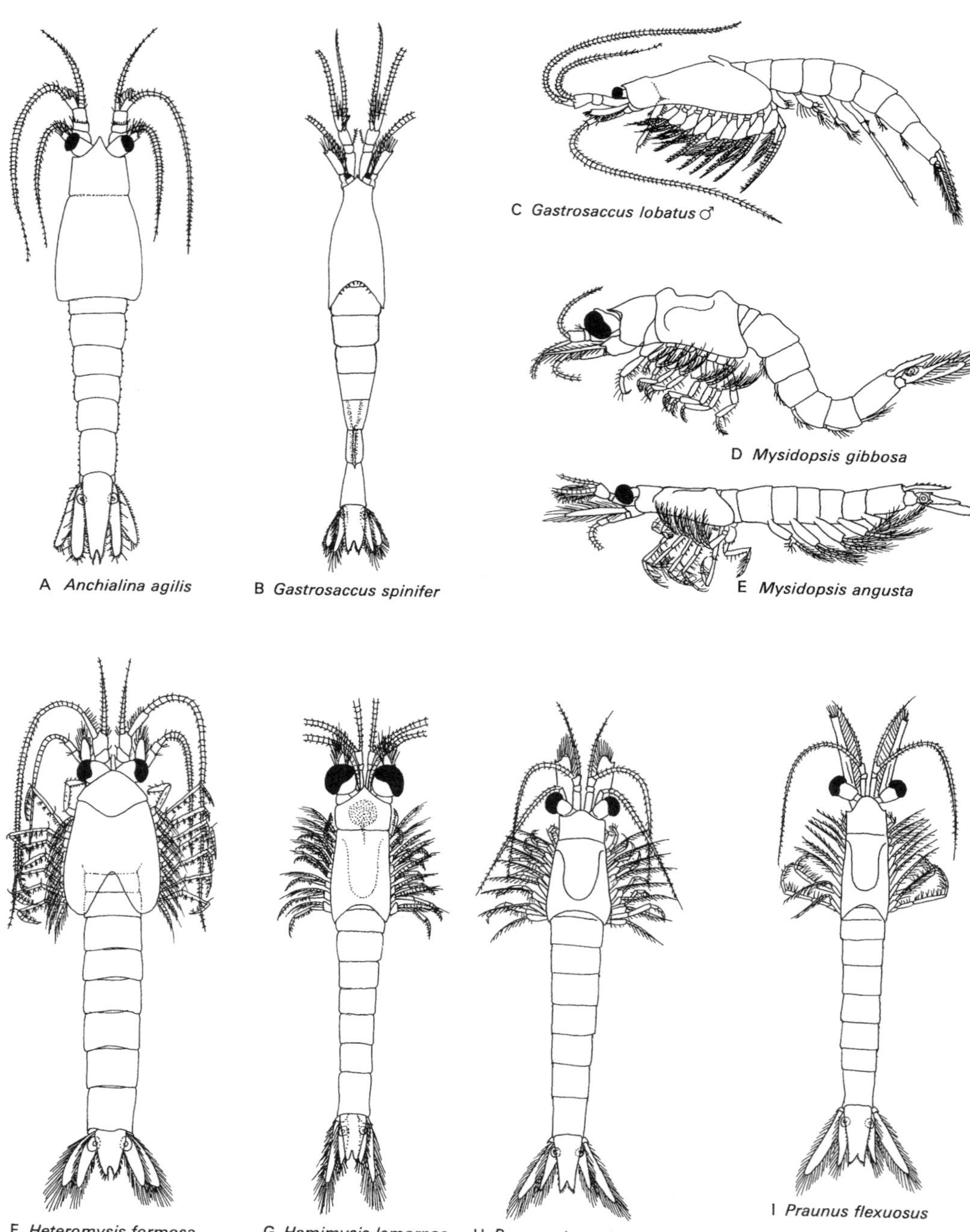
A *Anchialina agilis*
B *Gastrosaccus spinifer*
C *Gastrosaccus lobatus* ♂
D *Mysidopsis gibbosa*
E *Mysidopsis angusta*
F *Heteromysis formosa*
G *Hemimysis lamornae*
H *Praunus inermis*
I *Praunus flexuosus*

Fig. 8.18

7. Apical cleft of telson with teeth inside. Endopod of thoracic limb 3 short and stout, with a prehensile terminal claw
***Heteromysis formosa* Smith** (Fig. 8.18F)

Up to 8 mm long. Dwells more or less permanently in small cavities under stones, shells, and so on, usually on mud. Endopod of thoracic limb 3 different from the others (and unlike other mysids) in being shorter and stouter with a prehensile terminal claw. *Heteromysis armoricana* Nouvel, is very similar and has recently been added to the British list. Both occur in shallow water on south and west coasts and may be more common than previously thought.

Apical notch in telson without teeth or setae, smooth-edged inside. Endopod of third thoracic limb normal, like the others
***Mysidopsis angusta* G. O. Sars** (Fig. 8.18E)

Telson spoon-shaped with a small terminal notch. The statocyst is large in comparison with the small endouropod which contains it; the endouropod bears a strong spine on its inner margin. An uncommon species in shallow water; southern half of Britain, as far north as Moray Firth on east, Clyde Sea on west. *Mysidopsis gibbosa* G. O. Sars, which also occurs in shallow water, has no notch in the telson (see couplet 22). *Mysidopsis didelphys* (Norman) is a northern species found only in deep water.

---

8. Smooth, naked outer margin of antennal scale (about half total length) does not end in a spine but runs straight into the setose part (Fig. 8.17F). Telson has lateral spines restricted to apical half. Eyestalk very short, shorter and smaller than larger black corneal part of eye. Fresh specimens usually red
***Hemimysis lamornae* (Couch)**. Midge shrimp (Fig. 8.18G)

Length 13 mm. Swims actively but keeps under cover among rocks and weed. Breeds in some marine aquaria. Shallow sublittoral, and offshore to at least 100 m. All British coasts; ranges Norway south to Mediterranean.

Smooth naked outer edge of antennal scale ending in a small subterminal spine (Figs 8.16H, I, 8.17A–E). Telson has lateral spines from base to apex (as in Figs 8.16J, 8.17S). Eyestalk may be small but in most cases as large as corneal part of eye (Fig. 8.16G). Fresh specimens not normally red **9**

---

9. Antennal scale very elongate and parallel-sided, from five to eight times as long as broad, with very short anterior lobe which projects only slightly forwards beyond base of subterminal spine, (by less than twice the length of the spine) (Fig. 8.17A) **11**

Antennal scale only moderately elongated, or broad and rounded, with a substantial anterior lobe projecting forward well beyond base of subterminal spine (by more than twice the length of the spine) (Figs 8.16H.I. 8.17B–E) **10**

10. Antennal scale rather oblong, about four times as long as broad, anterior lobe projecting fowards beyond base of subterminal spine by about twice length of spine. Base of spine weakly constricted or articulated when seen under high magnification (Figs 8.16H, I)

***Praunus inermis* (Rathke)** (Fig. 8.18H)

Beware confusion with *Schistomysis* sp. because of the substantial anterior lobe of antennal scale. Length 15 mm. Among weeds in shallow water, but seems to keep well clear of tidal influence and is less inclined to swarm than the other two species of *Praunus*. All British coasts, common; Brittany northwards to White Sea.

Antennal scale with rounded or curving margins and a substantial anterior lobe projecting forwards beyond the subterminal spine by more than twice the length of the spine (Figs. 8.17B–E). Base of spine shows no trace of constriction or articulation **12**

11. Antennal scale only five or six times as long as broad, anterior lobe slightly longer than subterminal spine (Fig. 8.17A). Telson with 18–24 lateral spines. Only three pairs of chromatophores on ventral surface of thorax, between legs. Usually four chromatophores on telson and each exouropod, two on each endouropod. Living specimens almost colourless or pale to dark green

***Praunus neglectus* (G. O. Sars)**

Length 21 mm. In deeper water than *P. flexuosus* but occasionally obtained in large numbers by netting. If the chromatophores are contracted they provide the easiest means of identification. All British coasts, common.

Antennal scale seven or eight times as long as broad, anterior lobe at most as long as subterminal spine. Telson with 22–28 lateral spines (Fig. 8.16J). A separate pair of chromatophores on ventral surface of each thoracic segment, between legs. Usually more than four chromatophores on telson and four or more on each exouropod, three or four on each endouropod. Living specimens blackish, grey-brown or yellowish to almost colourless

***Praunus flexuosus* (Müller)**. Chameleon Shrimp (Fig. 8.18I)

Common. One of the larger British species, adults up to 25 mm long. Often found near artificial constructions such as docks, tolerant of salinity variations. All British coasts; northwards to Norway.

12. Eyestalk longer than wide, like a slightly tapering cylinder, so that eyes are carried appreciably beyond the margin of the carapace (Fig. 8.16C). Distal third of endouropod very slightly incurved (Fig. 8.170)

***Schistomysis spiritus* (Norman)**. Ghost Shrimp (Fig. 8.19A)

Rostrum short or virtually wanting. Antennal scale long and narrow, with naked outer margin, strong subterminal spine and well-developed anterior lobe. Telson with 24–30 lateral spines on each side. Adult body length 18 mm. The most easily recognised member of the genus. Glassily transparent except for eyes, though with chromatophores and may be partly coloured red-brown. Near the bottom in shallow water

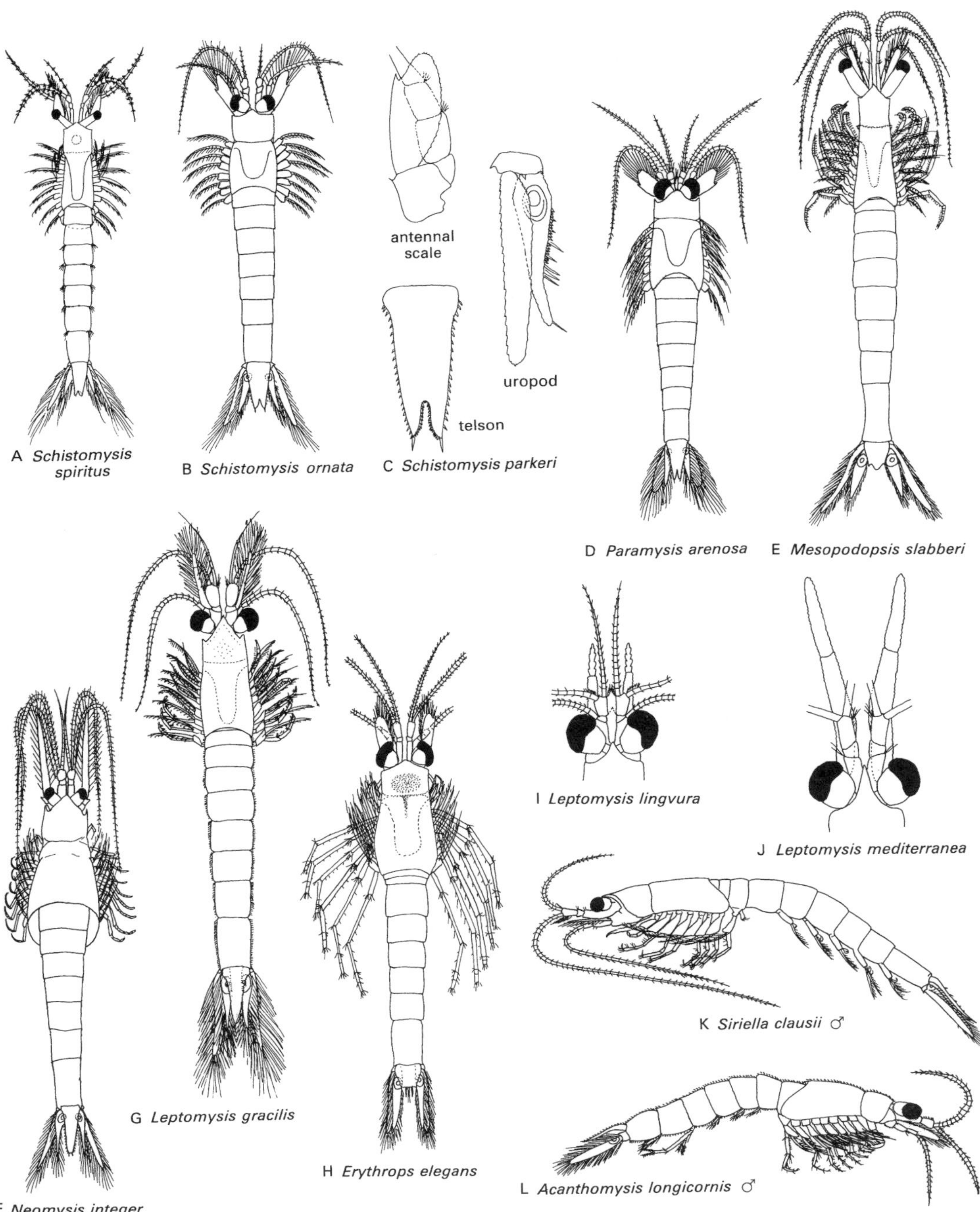
antennal scale
uropod
telson
A Schistomysis spiritus
B Schistomysis ornata
C Schistomysis parkeri
D Paramysis arenosa
E Mesopodopsis slabberi
I Leptomysis lingvura
J Leptomysis mediterranea
K Siriella clausii ♂
G Leptomysis gracilis
H Erythrops elegans
L Acanthomysis longicornis ♂
F Neomysis integer

Fig. 8.19

and estuaries. Common on all British coasts, often abundant; northwards to Norway.

Eyestalk no longer than its width, eyes therefore not reaching appreciably beyond lateral margins of carapace. Endouropod either straight or with distal third incurving markedly (Figs 8.17P, Q) **13**

---

13. Distal third of endouropod incurving markedly (Fig. 8.17P). Spines absent from inner margin of this part, which bears only setae, but strong spines proximal to it and one on the apex

***Schistomysis parkeri* Norman** (Fig. 8.19C)

Adult body length 10 mm. The only species of *Schistomysis* with a broad, ovate antennal scale and relatively short anterior lobe. Less slender than *S. spiritus*. Shallow water, particularly on south-west coasts including South Wales; rare, but reported from several places recently. South to West Africa.

Endouropod straight. Spines along most of its inner margin from near base to near apex (similar to Fig. 8.17Q) **14**

---

14. Antennal scale more than three times as long as broad (Fig. 8.17C). Telson of adult with more than 24 lateral spines on each side (Fig. 8.17S). Eyestalk, when seen from above, almost or quite as large as corneal part of eye. Combined carpo-propodus of thoracic endopods divided into five or six subsegments (Fig. 8.16E). The most basal one of these normal, longer than the following subsegment (Fig. 8.16E) **15**

Antennal scale oval, less than three times as long as broad (Fig. 8.17E). Telson of adult with less than 24 lateral spines on each side. Eyestalk seen from above only about half as big as corneal part of eye. Combined carpo-propodus of thoracic endopods divided into four or five subsegments. The most basal one of these shorter than the following subsegment and very slightly bulbous (Fig. 8.17K) **16**

---

15. Subterminal spine of antennal scale little more than half way along total length of scale. Spines on inner margin of endouropod extending, though sparsely, almost to apex and increasing steadily in size from base to apex. Telson with about 26 lateral spines, evenly spaced, the last one not particularly separated from the others

***Schistomysis ornata* (G. O. Sars)** (Fig. 8.19B)

Length 19 mm, more robust than *S. spiritus*. Anterior lobe of antennal scale relatively longer. Endouropod virtually straight, inner margin with rather sparse (about 16) spines extending to very near apex. Near the bottom from deep water to close inshore, and estuaries, but not extending up to LW. All British coasts; Iceland to West Africa. Common.

Subterminal spine of antennal scale at about two-thirds total length of scale from base (Fig. 8.17C). Spines on inner margin of endouropod irregular in size near base, both long and short, and stopping well short of apex

(Fig. 8.17Q). Telson with about 30 lateral spines, a disproportionate gap between last lateral spine and the preceding series (Fig. 8.17S)

Schistomysis kervillei **(G.O. Sars)**

Length 16 mm. Difficult to separate from *S. ornata*. The gap preceding the last lateral spine of the telson is not always as distinct as in Fig. 8.17S. The position of the subterminal spine on the antennal scale is usually distinctive, but not entirely reliable. The same is true of the other characters, including the number of sub-segments (6) in the carpo-propodus, although collectively they are likely to provide a determination. In shallower water than *S. ornata*, extending to LW. Few records, but known from Kent through Wales to north-west Scotland.

---

16. Cleft of telson about 1.5 times as deep as width between the last lateral spines. Inner margin of endouropod bears about 28 spines, a few of which are much larger than the others, in a row extending almost or quite to apex. Telson with 17–23 lateral spines

Paramysis arenosa **(G.O. Sars)** (Fig. 8.19D)

Rostrum very short or practically wanting. Adult 10 mm long. Body transparent with variable superimposed colours. Half buries itself in sand. From LW to 20 m or so. Fairly common, south and west coasts to Scotland.

Cleft of telson relatively wide and shallow; depth 1 to 1.5 times width between the last pair of lateral spines. Inner margin of endouropod bears 13–18 spines in an irregular row, not nearly reaching apex. Telson with 13–18 lateral spines on each side ***Paramysis nouveli*** **Labat**

Adults may reach 11 mm but are very similar to *P. arenosa*. Eyestalks relatively slightly larger. Coloration similar to *P. arenosa*. Littoral and shallow water. Distribution similar to *P. arenosa* but rare.

---

17. Telson short, somewhat truncated, uropods more than twice as long. Apex of telson flat (Fig. 8.16M) or with prominent pointed 'shoulders' (Fig. 8.16N) **18**

Telson of normal length, much longer than wide and more or less pointed or rounded at apex (Figs 8.17T–X). Nearly as long as uropods **19**

---

18. Telson characteristic: short, with a rounded, toothed apex flanked by angular shoulders (Fig. 8.16N). A strong lateral spine marks the point of each shoulder. Antennal scale setose all round. Eyestalks conspicuously long, cylindrical ***Mesopodopsis slabberi*** **(van Beneden)** (Fig. 8.19E)

Transparent, slender, up to 15 mm. Common in estuaries, sometimes swarming in brackish water, may appear transiently according to tidal influence. All British coasts; distributed Scandinavia to the Mediterranean and West Africa.

Telson short, strongly truncated, without prominent shoulders, almost trapezoidal (Fig. 8.16M). Antennal scale with naked outer margin (similar to Fig. 8.16H). Eyestalks short, eyes red in fresh specimens

Erythrops elegans **(G.O. Sars)** (Fig. 8.19H)

Very small (5–6 mm), body transparent with pigment spots. Corneal part of eyes prominent, wider than eyestalk and dorso-ventrally compressed, thus appearing bean-shaped. Benthic, on mud, sand or gravel; sublittoral, offshore on most British coasts, not common. Range Norway to Mediterranean. Four more species of *Erythrops* occur offshore. One of them, *E. serrata* (G. O. Sars) might be found rarely in shallow water. It is larger (11 mm) and has the naked outer margin of the antennal scale strongly saw-toothed.

---

19. Telson as a whole long, sides becoming more or less straight distally and tapering gradually to a narrow, almost pointed, apex (Fig. 8.17T). The uropods less than twice as long **_Neomysis integer_ (Leach)** (Fig. 8.19F)

    Medium size (17 mm), translucent or pale grey. A short, pointed rostrum. Antennal scale very long and narrow, tapering to a point, with setae all round. This species seems to be particularly subject to injury, causing atypical morphology which may lead to misidentification. In estuaries, brackish pools, freshwater or hypersaline pools; rarely on the open coast. All British coasts, common, abundant at some localities. Range Arctic Norway to Spain.

    Telson long and rounded, the sides curving to form a broad base and rounded apex (Figs 8.17U–X) **20**

---

20. Exouropod divided (by a weak suture) into a shorter distal portion and a longer proximal portion (as in Fig. 8.17N). Outer margin of proximal portion bearing spines, distal segment with setae instead of spines. (Suture may be obscure, but division into distal and proximal portions is evident from shape and armature of exouropod) **21**

    Exouropod not divided, setose all round (similar to Fig. 8.16L) **22**

---

21. Rostrum long, tapering; normally projecting beyond anterior margin of eyes; reaching second basal segment of antenna 1 (Figs 8.16A, B). Eyestalk elongated, carrying eyes beyond lateral margins of carapace. Telson with four or five (rarely three) tiny equal apical spines between two large last lateral ones (Fig. 8.17V). Exouropod rather long and parallel-sided, endouropod slightly incurving. Antennal scale long and almost parallel-sided **_Siriella armata_ (Milne-Edwards)** (Figs. 8.16A, B)

    Up to 22 mm or more. Easily identified by elongate rostrum. Antennal scale with naked outer margin and strong subterminal spine. Pleopods of female reduced to small, simple, setose digits. Male pleopods fairly large, biramous. Endouropod narrow and tapering, inner margin armed with a regular series of spines. Usually pale and transparent. Swims near the bottom; littoral to 20 m. Common on most British coasts; south to Mediterranean.

    Rostrum short; not, or barely, reaching anterior margin of eyes; reaching almost to middle, or less, of first basal segment of antenna 1. Eyestalks short and broad, not longer than wide. Telson with three tiny, apical spines between the two large last lateral ones (Figs 8.17V,W). Exouropod with slightly convex margins (Fig. 8.17N). Endouropod straight (Fig. 8.17R). Antennal scale subovate

***Siriella clausii* G. O. Sars** (Fig. 8.19K)
***S. jaltensis* Czerniavsky** and ***S. norvegica* G. O. Sars**

A group of three closely similar species which are separated from *S. armata* by their very short rostrum. Of these three, *S. clausii* is the smallest (11 mm) and the one most likely to be found in shallow water. It is generally common, extending from the littoral to 20m. *Siriella jaltensis* is slightly larger (15 mm) and favours deeper water, although they are sometimes found together. The largest is *S. norvegica* which attains a length of 21 mm but is not likely to be found near the shore.

---

22. A dorsal finger-like process on eyestalk, projecting outwards over edge of corneal part of eye. Two large median dorsal humps on carapace (Fig. 8.18D). Telson short, not much longer than its maximum width; with up to 18 very short lateral spines on each side

***Mysidopsis gibbosa* G.O. Sars** (Fig. 8.18D)

Abdomen strongly curved, giving the body (length 7 mm) a more marked sigmoid shape in lateral view than other mysids. Telson deeply concave and scoop-shaped dorsally, smoothly rounded at the tip with a prominent pair of apical spines and small lateral ones. Pigmentation variable, usually dark. Generally stays near the bottom in shallow water. Mainly southern Britain, north to the Firth of Forth in the east, and to the Clyde Sea on the west. Range southern Norway to Mediterranean.

No finger-like process on eyestalk, no humps on carapace. Telson more than twice as long as wide, with 20 or more lateral spines, which may be of varying size, on each side (similar to Fig. 8.17X) **23**

---

23. Telson without lateral spines on basal third, except for one or two near base. Distal two-thirds strongly spinose. Endouropod with few spines near base on inner margin; apical half without spines

***Acanthomysis longicornis* (Milne-Edwards)** (Fig. 8.19L)

Length 9 mm. Whole integument typically with minute scales, looking furry, but this character is variable (see also *Leptomysis gracilis*). Rostrum very short, pointed. Eyes prominent. Antennal scale narrow and tapering, but not quite pointed. Telson long, tongue-shaped, with rounded apex, lateral spines numerous on distal two-thirds. Base of endouropod, including statocyst, large in relation to distal portion. Pigmentation sparse, dark brownish. Uncommon, living near the bottom in moderate depths. A southerly distribution but recorded from Scotland; south to Mediterranean.

Telson with lateral spines from base to apex (Fig. 8.17X). Inner margin of endouropod spinose from near base to apex **24**

---

24. Integument microscopically scaly, looks bristly. A notch in carapace on each side of base of rostrum. Rostrum triangular with a slight convexity on each side behind tip, appearing to bulge near the middle

***Leptomysis gracilis* (G.O. Sars)** (Fig. 8.19G)

Leptomysis species are separated on small characters of the integument, rostrum, and antennal scale. All have a smoothly rounded telson, with numerous, close, small and large spines from base to apex, lateral and apical spines not clearly separable, forming a continuous series. Body pale or translucent with some pigmentation, or may be coloured reddish. In *L. gracilis* the integument is minutely scaled all over, appearing hairy (see also *Acanthomysis longicornis*). Length 15 mm. Apical segment of antennal scale with six setae or fewer on each side. Telson usually with a slight constriction at the base of a pair of large spines near the apex. Offshore, in moderate depths, but also in deep estuaries where it lives pelagically near the bottom. All British coasts; Norway to Mediterranean.

Integument smooth. No notches in carapace near base of rostrum. Edges of rostrum normal, more or less concave all along (Figs 8.19I, J) **25**

---

25. Apical segment of antennal scale with 10 setae or more on each side (Fig. 8.17H). Rostrum relatively long, extending almost whole length of the first basal segment of antennal

Leptomysis mediterranea G. O. Sars (Fig. 8.19J)

Length 18 mm. Similar to *L. gracilis* but with a smooth integument and a simple rostrum without notches in the carapace. Antennal scale narrow, setose all round. On or near the bottom, in shallow water, from tidal to deeper levels. Southern and western coasts, rarely Scotland; south to Mediterranean.

Apical segment of antennal scale with six setae or fewer on each side (Fig. 8.17G). Rostrum short, extending only half way along first basal segment of antennule **Leptomysis lingvura (G. O. Sars)** (Fig. 8. 19I)

Length 17 mm. Similar to *L. mediterranea*, but distinguished by apical segment of antennal scale. Rostrum distinctly shorter than in the other two species. Often coloured reddish. May be found in rock pools, netted at tidal levels, or at the bottom in shallow water. Common on all British coasts. Range Norway to Mediterranean.

V. ORDER ISOPODA

The isopod body is usually dorso-ventrally flattened, and lacks a carapace. It comprises a head, *(cephalon)*, a thorax *(pereon)* of seven segments, and an abdomen *(pleon)* of six (Fig. 8.12). The head is fused with one (rarely two) pereon segments bearing *maxillipedes*. The remaining pereon segments (except in the Arcturidae) bear more or less similar, uniramous *pereopods*, or walking legs. The *coxal plates* of the pereopods are visible from above. The pleon and telson are wholly or partly fused, forming a *pleotelson* with five pairs of biramous *pleopods* and a pair of uniramous or biramous *uropods*.

Sexes are separate, and sometimes dimorphic. The male pleopod 2 bears an *appendix masculina*, or is sometimes combined with pleopod 1 to form a copulatory structure. In the female pleopod 1 is missing and pleopod 2 is modified as a flat *operculum*. Embryos are brooded, either in a chamber of flattened plates *(oostegites)* developed basally on pereopods 1–5, or in internal body pouches. The young are released as miniature adults, and a succession of growth stages, or *instars*, is often recognisable.

About 10 000 species occur world-wide, classified in nine suborders, in terrestrial, freshwater, and marine habitats, and as free-living, commensal, and parasitic forms. Marine intertidal species occur among barnacles, in crevices, beneath stones, on or under large macro algae, within turfs of coralline algae and rhodophytes, among sessile animals such as sponges, hydroids, and bryozoans, and in sandy beaches. Very similar species may occupy adjacent, but distinct microhabitats. Specimens may be fixed in 4% seawater formalin and stored in 70% alcohol.

KEY TO FAMILIES

I.	Free living, or parasitic on fish, not parasitic on Crustacea	**2**
	Entirely parasitic on Crustacea (Suborder Epicaridea)	**15**
2.	Adults with five pairs of pereopods (Suborder Gnathiidea)	**I. Gnathiidae**
	Adults with seven pairs of pereopods	**3**
3.	Uropods lateral or ventral (Fig. 8.12)	**4**
	Uropods terminal (Fig. 8.12)	**9**
4.	Uropods ventral, hinged ventro-laterally to pleotelson to form opercular plates covering pleopods (Fig. 8.12) (Suborder Valvifera)	**5**
	Uropods lateral, flattened and with pleotelson forming a tail fan (Fig. 8.20) (Suborder Flabellifera)	**6**
5.	Pereopods all more or less alike	**6. Idoteidae**
	Pereopods 1–4 not ambulatory; resembling mouthparts and quite unlike 5–7	**7. Arcturidae**
6.	Body markedly attenuated; uropod bases extending dorsally above telson with caudal fan somewhat cup-shaped	**2. Anthuridae**
	Body robust; uropod bases not extending above telson	**7**
7.	Pleon with fewer than five distinct segments, more than one fused with telson	**5. Sphaeromatidae**
	Pleon with five distinct segments plus one fused with telson	**8**
8.	Uropod rami tubular, outer ramus claw-like	**3. Limnoriidae**
	Uropod rami flattened, fan-like	**4. Cirolanidae**

9. Terrestrial, pleon usually with six distinct segments (Suborder Oniscidea) **10**

Aquatic, pleon consisting of fewer than six segments (Suborder Asellota) **12**

---

10. Antennal flagellum of two or three articles **11**

Antennal flagellum of 10 or more articles **12. Ligiidae**

---

11. Antennal flagellum of two articles **13. Armadillidiidae**

Antennal flagellum of three articles **14. Halophilosciidae**

---

12. Uropods lacking peduncle; pereon with last three segments usually much smaller than first four; eyes when present on lateral extensions of head **13**

Uropods with peduncle; pereon segments all subequal, no exaggerated posterior narrowing; eyes not on lateral extension of head **14**

---

13. Molar process of mandible normal, strong and truncated. Eyes present, on lateral extensions of head **9. Munnidae**

Molar process of mandible weak and pointed (British genus, *Pleurogonium*, lacks eyes and head has no lateral extensions) **10. Pleurogonidae**

---

14. Antennal flagellum longer than peduncle **8. Janiridae**

Antennal flagellum shorter than peduncle **11. Jaeropsidae**

---

15. Parasitic in decapod crustaceans (crabs, prawns, etc.) **16**

Parasitic in other crustaceans (notably barnacles, isopods, ostracods, mysids, euphausiids) **17**

---

16. In gill chamber or attached to pleon of decapods. Female body with distinct segments, more or less asymmetrical and with seven pairs of pereopods. Oostegites present **15. Bopyridae**

In visceral cavity of decapods. Female body without distinct segments, symmetrical and lobed; pereopods rudimentary or absent. Oostegites present **16. Entoniscidae**

---

17. On body or in brood chamber of mysids and euphausiids. Pereopods well developed, numbering five pairs crowded near mouth. Oostegites present **17. Dajidae**

Notably from barnacles, isopods and ostracods. Without pereopods and, uniquely in the suborder, without oostegites **18. Cryptoniscidae**

---

1. GNATHIIDAE

Male, female, and young (*praniza*) all of different form; each with only five pairs of ambulatory pereopods. Cephalon fused with two pereon segments, limbs of second fused segment modified as flattened *pylopods* (gnathopods). Last pereon segment reduced and without limbs. Male with large forcep-like mandibles which project in front of cephalon. Females and late pranizas with pereon segments 3–5 fused and inflated, particularly in females which incubate the eggs internally. Adults benthic, pranizas often ectoparasitic on fish.

---

1. Male pylopod of five articles; adults usually in estuarine mudbanks **Paragnathia formica**

   Male pylopod of two or three articles; adults usually in marine crevice-like situations **2**

---

2. Male with ridge over each eye **Gnathia oxyuraea**

   Male without ridge over the eyes **3**

---

3. Lateral mandibular spine diverging from mandible **Gnathia dentata**

   Lateral spine not diverging from mandible **4**

---

4. Front of cephalon with a shallow central concavity and a slight median forward projection **Gnathia maxillaris**

   Front of cephalon with a shallow central concavity and a slight median forward projection **Gnathia vorax**

Gnathia dentata Sars Fig. 8.20

Male cephalon centrally tridentate at front, without a concavity; lateral corners acutely pronounced. Male body length 2.8–3.8 mm, females up to 4.3 mm.

In *Laminaria* holdfasts LWST and below, south and west coasts.

Gnathia maxillaris (Montagu) Fig. 8.20

Male cephalon with shallow central concavity at front, with only a slight rounded median forward projection; lateral corners square. Male body length usually 4.5–5.0 mm or less.

Adults in rock-crevices, dead barnacles and *Laminaria* holdfasts; commonest intertidal gnathiid on south coasts.

Gnathia oxyuraea (Lilljeborg) Fig. 8.20

Male cephalon with median tooth at front bordered by a shallow concavity on each side; lateral corners square. Male body length 2.4–5.4 mm; females 3.9 mm.

Mainly sublittoral; common; all British coasts. Barents Sea to Mediterranean.

Gnathia vorax (Lucas) Fig. 8.20

Male cephalon with deep central concavity at front, and a fairly acute median forward projection. Body length 5–7 mm.

Mainly sublittoral; sporadic, all British coasts. Greenland to Mediterranean.

Paragnathia formica (Hesse) Fig. 8.20

Males, females and late pranizas 2.5–5 mm body length; in salt marsh mud cliffs around MHWN. Early pranizas parasitic on inshore or estuarine flatfish. All British coasts. Widely distributed between Morocco and Scotland.

2. ANTHURIDAE

Body long and narrow, subcylindrical. Pereon somites longer than wide. Pleon relatively short with uropod exopods arching dorsally and medially over the telson.

---

1. Telson subquadrate; eyes large; male antennule large **Anthura gracilis**

   Telson narrowing to a rounded apex; eyes small; male antennule small as in female **Cyathura carinata**

Anthura gracilis (Montagu) Fig. 8.20

Females up to 11 mm body length, males 4 mm. Marine, in crevices, among kelp holdfasts and in empty tubes of serpulid worms; LWST and shallow water; south and west coasts only. Clyde Sea to Mediterranean.

Cyathura carinata (Krøyer) Fig. 8.20

Up to 14 mm long, off-white with red-brown mottling. Brackish water habitats, usually in mud. Southern British coasts, southwards to the Mediterranean.

3. LIMNORIIDAE

British forms wood-boring. Uropod exopod much shorter than endopod and with an apical claw; endopod elongate, apex blunt, lacking claw.

---

1. Dorsal surface of rim of pleotelson tuberculate; central area of pleotelson with three tubercles Pleon somite 5 with tubercles, not ridged **_Limnoria tripunctata_**

   Dorsal surface of posterior border of pleotelson smooth, not tuberculate; central area of pleotelson with four tubercles or with an inverted Y-shaped keel. Pleon somite 5 ridged **2**

---

2. Central dorsal surface of pleotelson with an inverted Y-shaped keel. Pleon somite 5 with simple dorsal ridge **_Limnoria lignorum_**

   Central dorsal surface of pleotelson with four tubercles. Pleon somite 5 with Y-shaped dorsal ridge **_Limnoria quadripunctata_**

Limnoria lignorum (Rathke) Fig. 8.20

Antennal flagellum of four articles. Lateral crests of pleotelson slightly tuberculate, dorsal surface of posterior margin not so. Body length up to 3.5 mm.

Prefers bases of exposed pilings where wood is wet and cool; all coasts. Norway to south of British Isles, and on the east and west coasts of North America to 40°S.

Limnoria quadripunctata Holthuis Fig. 8.20

Antennal flagellum of five articles. Lateral crests of pleotelson and the postero-dorsal margin lack tubercles. Body length up to 3.5 mm.

Tolerates exposure to air and occurs higher on pilings than *L. lignorum*; south and west coasts only, from Kent to the Isle of Man. Recorded also from New Zealand, South Africa, and California.

Limnoria tripunctata Menzies Fig. 8.20

Antennal flagellum of five articles. Lateral crests and postero-dorsal margin of pleotelson tuberculate. Body length up to 4 mm.

Locally common on south and west coasts, in timbers near power-station discharges. Widely distributed in temperate and tropical waters around the world.

4. CIROLANIDAE

Body with distinct coxal plates on pereon segments 2–7. Pleon with five distinct segments, plus one (the uropodal segment) fused with the telson. Uropods lateral, not arching dorsally, and with the pleotelson forming a terminal fan. Subfamily Cirolaninae well represented intertidally and inshore.

---

1. Antennal peduncle of five distinct articles; uropod peduncle produced backwards medial to endopod **2**

   Antennal peduncle of four distinct articles; uropod peduncle not produced backwards medial to endopod **3**

---

2. Pleopod 1 with endopod normal, not heavily chitinized; peduncle broader than long; body not markedly elongate **_Cirolana cranchii_**

   Pleopod 1 with both rami heavily chitinized and forming an operculum; peduncle longer than broad; body markedly elongate **_Conilera cylindracea_**

---

3. Pleotelson posterior border narrow and concave, with two large spines at each corner **_Eurydice spinigera_**

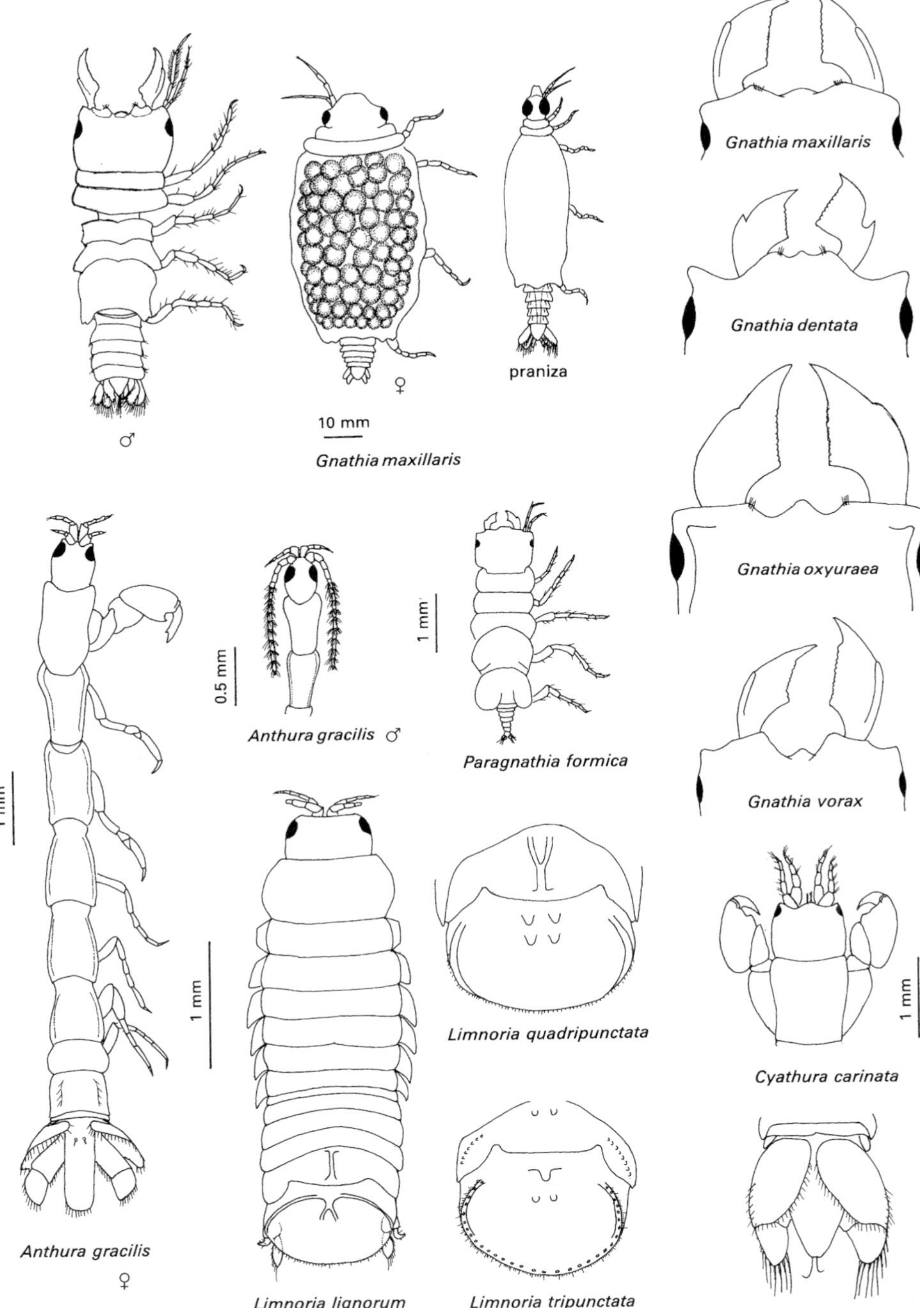

Gnathia maxillaris
Gnathia dentata
Gnathia oxyuraea
Gnathia vorax
♂
♀
praniza
10 mm
Gnathia maxillaris
1 mm
0.5 mm
Anthura gracilis ♂
1 mm
Paragnathia formica
1 mm
Limnoria quadripunctata
1 mm
Cyathura carinata
Anthura gracilis
♀
Limnoria lignorum
Limnoria tripunctata

Fig. 8.20

Pleotelson posterior border broad and slightly convex, with a pair of small spines towards each corner among plumose setae **4**

---

4. Pereon segment 6 with posterior angle of coxal plate produced sharply backwards; black chromatophores on dorsal, lateral, and ventral surfaces ***Eurydice pulchra***

   Pereon segment 6 with posterior angle of coxal plate acute but not produced sharply backwards; black chromatophores on dorsal surface only ***Eurydice affinis***

Cirolana cranchii Leach Fig. 8.21
Few plumose swimming setae on last three pereopods. Frontal lamina, between bases of antennae and antennules, small, pentagonal, less than twice as long as broad, and not visible from above.

Predominantly offshore but also on sand at LWST; south and west coasts only. Clyde Sea southwards to the Mediterranean.

Conilera cylindracea Fig. 8.21
Characteristic cylindrical shape. Predominantly sublittoral in shell gravel; locally common with *Cirolana cranchii* on fish bait in lobster pots.

Eurydice affinis Hansen Fig. 8.21
Pleotelson resembling *E. pulchra*, but body somewhat smaller and paler, with black chromatophores restricted to dorsal surface.

Among *E. pulchra* in Bristol Channel and North Wales; locally common. Reported from Holland and the Atlantic coast of France, and widely from the Mediterranean.

Eurydice pulchra Leach Fig. 8.21
Males up to 8 mm; females up to 6.5 mm. Colour darker than *E. affinis* with many black chromatophores extending around the pleon.

Up to HWNT in intertidal sand, swimming freely with the rising tide; all British coasts; common. Norway to Morocco, absent from the Mediterranean.

Eurydice spinigera Hansen Fig. 8.21
Body length up to 9 mm. Predominantly sublittoral, but also at LWST in intertidal sand and among surf plankton. British south and west coasts only, southwards to the Mediterranean.

5. SPHAEROMATIDAE

Body oval in outline, readily rolling into a ball. Five anterior pleon segments all fused, usually with three sutures, the last two of which, or all three, are incomplete; pleon segment 6 fused with telson as a pleotelson which articulates freely with pleon. Coxal plates of pereon segments 2–7 fused with segment but suture lines may be evident. Uropods lateral; endopod rigidly fused with peduncle; exopod movable. Sexual dimorphism apparent in some species, males of some of these bearing one or two backwardly directed projections on the posterior border of pereon segment 6.

---

1. Uropods with one ramus **2**

   Uropods with two rami **3**

---

2. Pereon segment 6 without projections female ***Campecopea hirsuta***

   Pereon segment 6 with posterior border bearing a single, large, backwardly directed median projection male ***Campecopea hirsuta***

---

3. Pereon segment 6 with posterior border bearing two backwardly directed projections (male *Dynamene*) **5**

   Pereon segment 6 without projections **4**

---

4. Pleotelson with posterior border notched, appearing tridentate or with a semicircular foramen in the mid-line **6**

   Pleotelson posterior border not notched; smoothly rounded or slightly acute **9**

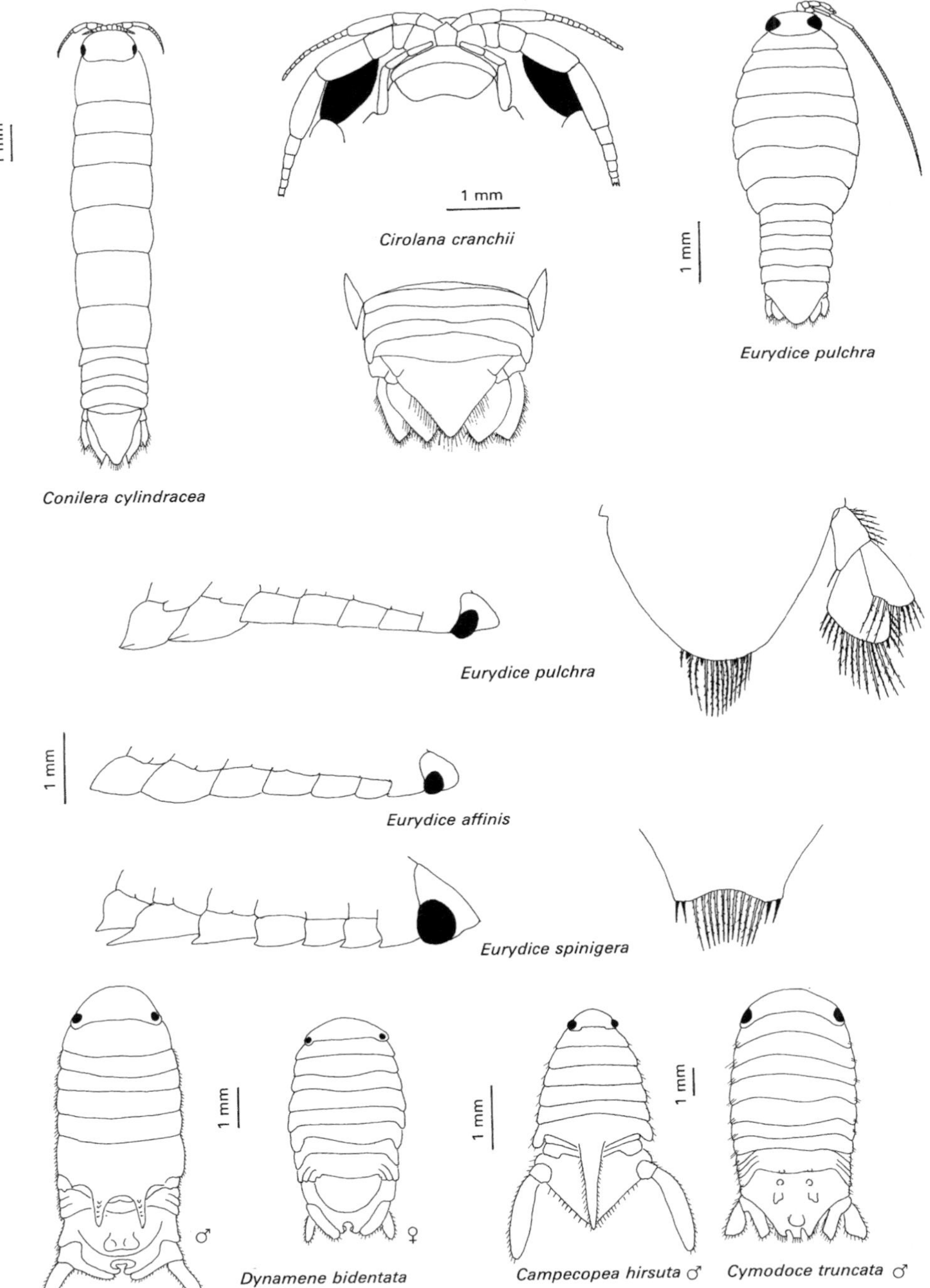
1 mm
1 mm
Cirolana cranchii
1 mm
Eurydice pulchra
Conilera cylindracea
Eurydice pulchra
1 mm
Eurydice affinis
Eurydice spinigera
♂
1 mm
♀
Dynamene bidentata
1 mm
1 mm
Campecopea hirsuta ♂
Cymodoce truncata ♂

Fig. 8.21

5. Pleotelson with two prominent hemispherical bosses which join at bases male ***Dynamene bidentata***

Pleotelson with two bosses joined by a short stem; each boss with a prominent, apical, subsidiary projection male ***Dynamene magnitorata***

6. Pleotelson posterior border with a semicircular foramen in mid-line at the posterior end of a channel in the mid-ventral line (female *Dynamene*) **7**

Pleotelson posterior border slightly or markedly tridentate **8**

7. Pleotelson smoothly rounded in side-view female ***Dynamene bidentata***

Pleotelson with a dorsal projection in side-view female ***Dynamene magnitorata***

8. Pleotelson posterior border with a marked tridentate process at the posterior end of a channel in the mid-ventral line male ***Cymodoce truncata***

Pleotelson posterior border slightly tridentate (viewed from above median tooth partly obscures lateral teeth) at the posterior end of a channel in the mid-ventral line female ***Cymodoce truncata***

9. Exopod of adult uropod serrated **10**

Exopod of adult uropod at most slightly crenulate **11**

10. Pereopods 1–3 and maxilliped palp with plumose setae. Maxilliped palp without lobes on articles 2–4 ***Sphaeroma serratum***

Pereopods 1–3 and maxilliped palp with sparsely plumose setae. Maxilliped palp with prominent lobes on articles 2–4 ***Lekanesphaera levii***

11. Pleotelson dorsal surface with two rows of elongate tubercles. Pereopod 1 propus lacking distal setae ***Lekanesphaera hookeri***

Pleotelson dorsal surface with small tubercles. Pereopod 1 propus with two distal setae ***Lekanesphaera rugicauda***

Campecopea hirsuta (Montagu) Fig. 8.21
Sexually dimorphic. Pleotelson rounded, without prominent tubercles, posterior border not notched, and without a posterior mid-ventral respiratory channel. Pleopods 1–5 lamellar; uropod of both sexes with a large exopod and lacking an endopod. Length up to 3.5 mm in female and 4.0 mm in male.

Exposed shores; MTL–HWNT among barnacles and *Lichina*, or in rock-crevices. South and west coasts only; locally common. Southwards to West Africa.

Cymodoce truncata Leach Fig. 8.21
Sexually dimorphic pleotelson. Pleopods 4 and 5 with exopod lamellar, endopod thick, fleshy and with transverse corrugations. Adult male body length 11 mm or more; female smaller.

Sometimes among algae, but adults characteristically in crevices and old mollusc borings; LWS. Sporadically distributed on south and west coasts from Isle of Wight, as far north as Shetland.

Dynamene bidentata (Adams) Fig. 8.21
Markedly sexually dimorphic. Length up to 7 mm. Male with segment 6 overlapping 7 and produced into a prominent, backwardly directed, bifid process. Pleotelson smooth in female, rugose in male, with a dorsal bilobed boss; respiratory channel sometimes closing ventrally.

MTL–LWS. Juveniles among intertidal algae upon which they feed, and which they resemble in colour; adults in rock-crevices or empty *Balanus perforatus* tests. Each male usually with several females, which may be white or colourless when ovigerous. South and west coasts only; locally common.

Dynamene magnitorata Holdich
Similar to *D. bidentata* but male with characteristic pleotelson bosses joined by a short stalk and each with a subsidiary apical projection. Female with a mid-dorsal projection at posterior margin of pleotelson. Length up to 6 mm in male, 4.8 mm in female.

Juveniles amongst lower shore and upper sublittoral algae, particularly *Chondrus crispus*, and corallines, the colour of which the isopod often matches. Adults in crevices, in tests of *Balanus crenatus*, amongst clumps of ascidians, and particularly within channels of encrusting sponges such as *Halichondria*; also in sponges associated with *Zostera* meadows. Mediterranean and Atlantic coasts of Africa, Spain, and France as far north as Guernsey, where it is common on rocky coasts.

Lekanesphaera hookeri (Leach) Fig. 8.22
Pereopod 1 with up to about 20 smooth setae on ischium and on merus; setae absent from distal region of propus. Uropods with outer border of exopod almost smooth; pleotelson with two longitudinal rows of elongate tubercles, one on each side of mid-line, posterior border not extending beyond uropods. Male up to 10.5 mm, female 8.5 mm.

In brackish ditches near limit of EHWS; most coasts except north-east; patchily distributed from the Mediterranean to northern Scotland.

Lekanesphaera levii
(Argano and Ponticelli) Fig. 8.22
Pereopod 1 with up to 50 smooth setae on ischium, about 60 on merus, and 3–8 in a distal row on the propus. Uropods usually with 6–7 slight striations on outer margin of exopod; pleotelson with smooth dorsal surface, posterior border extending beyond uropod in large males. Male up to 12 mm, female 8 mm.

Rock-crevices and under stones, in brackish localities, between MTL and LWNT. Southern coasts only, patchily distributed.

Lekanesphaera rugicauda (Leach) Fig. 8.22
Pereopod 1 ischium with up to 30 smooth setae, merus with up to about 20, and propus distal row with two (occasionally one). Uropod with outer edge of exopod almost smooth; pleotelson dorsal surface covered with very small tubercles, extending beyond uropods in adult males. Males up to 10 and females 7.5 mm.

In sheltered estuarine situations up to EHWS, usually in salt marsh pools. All British coasts, very common.

Sphaeroma serratum (Fabricius) Fig. 8.22
Pereopod 1 ischium with up to about 60 plumose setae in two rows, propus with a distal row of 15–20. Uropods with 4–7 well-defined serrations on outer edge of exopod; pleotelson with smooth dorsal surface and not extending beyond uropods. Females up to 10 mm and male to 11.5 mm.

In rock-crevices or under stones around MTL, usually on open marine coasts, sometimes near freshwater seepages. Sometimes in crevices and under stones around HWS in sheltered locations. South and west British coasts only, locally common, southwards to the Mediterranean and the Azores.

6. IDOTEIDAE

Body oval, oblong, or elongate; pereon segments and pereopods all more or less alike; pleon somites variously coalesced depending upon genus; uropods usually uniramous.

---

1. Pleon with no distinct segments; partial sutures (not visible from above) indicate almost complete fusion of pleon with telson **2**

   Pleon of two or three distinct segments; remainder fused or partially fused with telson **3**

---

2. Pereon lateral borders appearing serrated; pleotelson expanded laterally in the middle ***Synisoma lancifer***

   Pereon lateral borders straight and parallel; pleotelson narrowing evenly to an acute point ***Synisoma acuminatum***

---

3. Pleotelson with three complete segments and one partial suture. Isopod usually occupying a 'case' ***Zenobiana prismatica***

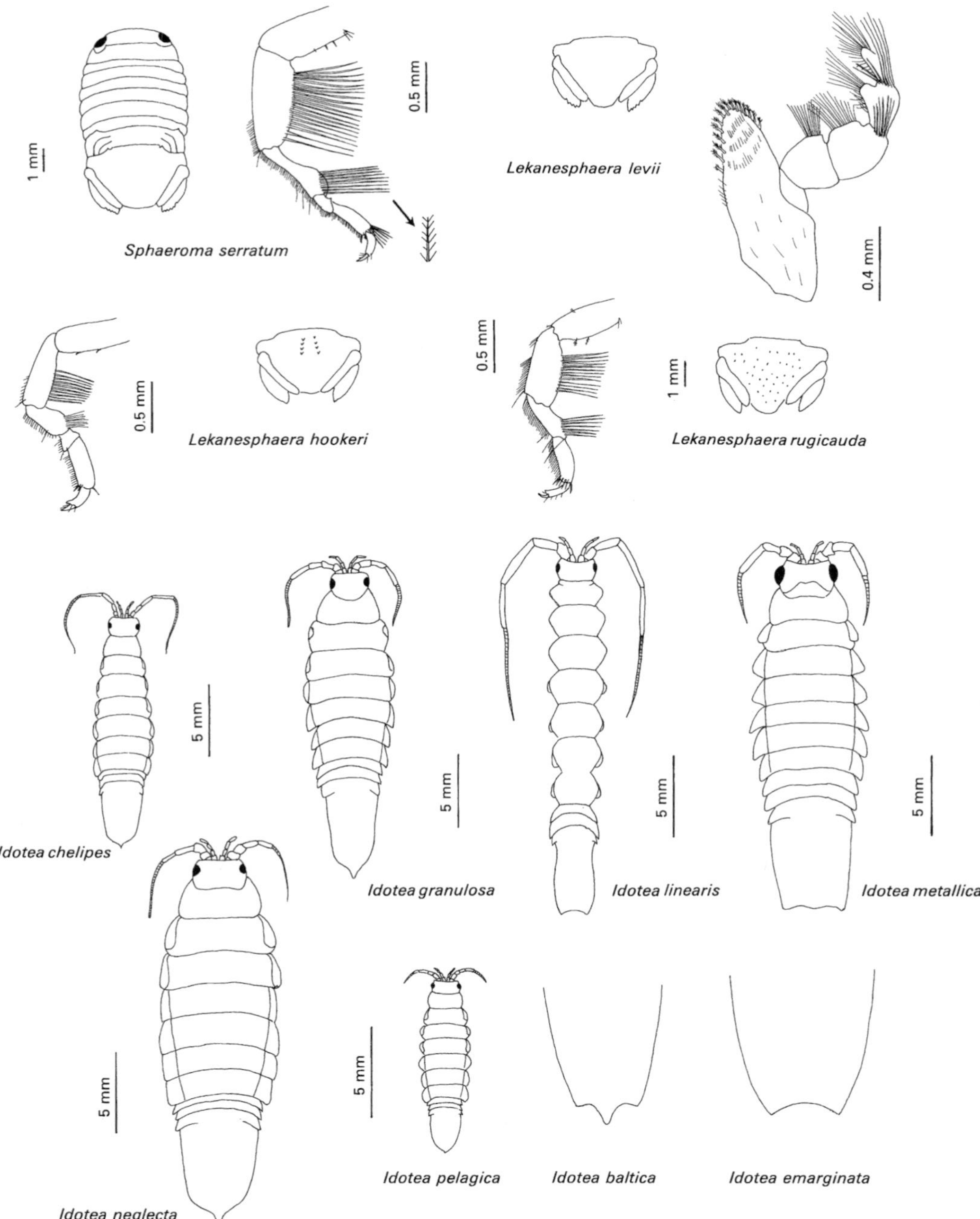
1 mm
0.5 mm
Sphaeroma serratum
Lekanesphaera levii
0.4 mm
0.5 mm
Lekanesphaera hookeri
0.5 mm
1 mm
Lekanesphaera rugicauda
5 mm
Idotea chelipes
5 mm
Idotea granulosa
5 mm
Idotea linearis
5 mm
Idotea metallica
5 mm
Idotea neglecta
5 mm
Idotea pelagica
Idotea baltica
Idotea emarginata

Fig. 8.22

Pleotelson with two complete segments and one partial suture. Free living **4**

4. Pleotelson apical border straight or concave **5**

Pleotelson apical border produced and more or less angulate in centre **7**

5. Body markedly elongate; coxal plates very small, not reaching posterior border on any pereon segment ***Idotea linearis***

Body not markedly elongate, coxal plates 3–7 reaching posterior border of pereon segment **6**

6. Adult pleotelson with apical border concave, sides somewhat rounded. Cephalon lacking complete suture behind eyes ***Idotea emarginata***

Adult pleotelson apical border and sides straight. Cephalon with distinct sinuous suture behind pronounced eyes ***Idotea metallica***

7. Adult pleotelson apical border tridentate or nearly so ***Idotea baltica***

Adult pleotelson apical border not tridentate, almost invariably rounded laterally **8**

8. Adult pleotelson sides slightly concave, apical margin with a pronounced, acute median process ***Idotea granulosa***

Adult pleotelson sides straight or slightly convex, apical margin with at most an indistinct, blunt median process **9**

9. Antenna with flagellum much shorter than peduncle; flagellum densely covered with fine setae in males ***Idotea pelagica***

Antenna with flagellum longer than peduncle; flagellum not densely covered with fine setae in males **10**

10. Body slender; length four or five times width in all but ovigerous females, which are wider. Antennule aesthetasc series with a single distal aesthetasc. In brackish water ***Idotea chelipes***

Body more robust; length little more than three times width. Antennule aesthetasc series begins with a distal pair. Sublittoral but often stranded among drift weed on the shore ***Idotea neglecta***

Idotea baltica (Pallas) Fig. 8.22

Adult pleotelson dorsally keeled, with more or less straight sides. Male 10–30 mm long; female 10–18 mm. Colour sometimes uniformly green or brown but often with white spots or longitudinal lines; female often darker than male.

Generally offshore, but occasionally among attached algae on the shore and often cast up in large numbers among drift weed. All British coasts. Widely distributed in north-east Atlantic region.

Idotea chelipes (Pallas) Fig. 8.22

Adult pleotelson with sides subparallel, slightly keeled posteriorly in mid-dorsal line. Males 5–15 mm long; females 6–10 mm. Colour mostly uniformly green or brown, sometimes with white markings; females often darker.

A brackish-water species; occurs among intertidal algae in sheltered estuaries or on green algae in sheltered brackish pools at or above high-water mark. All British coasts; patchy and not common. Norway southwards to the Mediterranean.

Idotea emarginata (Fabricius) Fig. 8.22
Pleotelson with sides slightly convex. Males 7–30 mm; females 9–19 mm. Males often uniformly brown in colour, though sometimes with white markings; females generally darker in background colour, often with longitudinal lateral white bands, or alternating white and darker transverse bands.

Most common sublittorally on accumulations of detached algae; occasionally between tide marks on attached algae or drift weed. All British coasts; locally abundant. Western European coasts from Norway to Spain.

Idotea granulosa Rathke Fig. 8.22
Adult pleotelson narrowing sharply at first, with concave lateral sides. Males 5–20 mm; females 6–13 mm. Mostly uniformly brown, red or green, depending on the alga inhabited; occasionally with longitudinal white markings.

The common resident idoteid between tide marks on all open but not excessively exposed coasts, large specimens on *Ascophyllum* and *Fucus*, and small specimens preferring smaller algae such as *Cladophora* and *Polysiphonia*. On all British coasts. Arctic Norway to northern coast of France.

Idotea linearis (Linnaeus) Fig. 8.22
Body and appendages markedly slender. Adult pleotelson with sides slightly concave anteriorly. Males 15–40 mm, females smaller. Colour green or brown, with darker or lighter longitudinal stripes; adult female often darker than male, frequently with paler markings around edges.

A sublittoral species occasionally cast up on the shore and often found swimming near the water's edge on sandy shores at low tide. South and west coasts of Britain, southwards to the Mediterranean, Morocco, and the Canary Isles.

Idotea metallica Bosc Fig. 8.22
Pleotelson with sides and posterior border straight. Male 8–30 mm, female 9–18 mm. Colour uniformly greyish or brown.

Typically in Gulf Stream debris which may be stranded on open, west-facing coasts; rare.

Idotea neglecta Sars Fig. 8.22
Adult pleotelson with fairly straight sides converging posteriorly. Male 8–30 mm, female 10–16 mm. Colour often uniformly brownish, sometimes with white longitudinal lateral markings, and occasionally with white marbling over the whole dorsal surface; adult females mostly darker than males.

Generally sublittoral on accumulations of detached algae or fish waste, but also between tide marks on attached algae or among cast-up drift weed. All British coasts, locally common. West European coasts from Norway to France.

Idotea pelagica Leach Fig. 8.22
Body short and stout. Pereopods all very robust, terminal claw relatively larger than in any other species. Pleotelson broadly rounded, not keeled. Males 4–11 mm, females 7–10 mm. Colour merges well with typical background of barnacles, mostly dark purple-brown, with white diamond-shaped patches or elongated stripes down the mid-dorsal line, and with white markings along dorsal edges; females often darker.

Patchily distributed on exposed shores among mussels, barnacles, and stunted fucoid algae; HWNT–LWST. All British coasts. West European coasts from Norway to France.

Synisoma acuminatum Leach Fig. 8.23
Body long, narrow, and subcylindrical. Coxal plates of pereon very small, barely visible from above. Up to 25 mm in length and rather less than 5 mm in breadth.

Typically among the alga *Halidrys siliquosa* in pools on rocky shores around MTL. South and west coasts only, locally frequent, southwards to the Mediterranean.

Synisoma lancifer (Dollfus) Fig. 8.23
Pereon with coxal plates triangular, lateral borders appearing more or less symmetrically serrated. Males up to 22.5 mm long, females smaller. Among algae and under stones particularly around LWST. Southwestern coasts only, rare; southwards to the Mediterranean.

Zenobiana prismatica (Risso) Fig. 8.23
Body slender, parallel-sided. Up to 13.5 mm long. Typically in a case of hollowed-out plant material such as *Zostera* stems, or in old worm tubes such as those of *Pomatoceros*; south and west coasts only, rare. Clyde Sea southwards to the Mediterranean.

7. ARCTURIDAE

Body subcylindrical and elongate, middle pereon segment enlarged. Pereopods 1–4 elongate, with

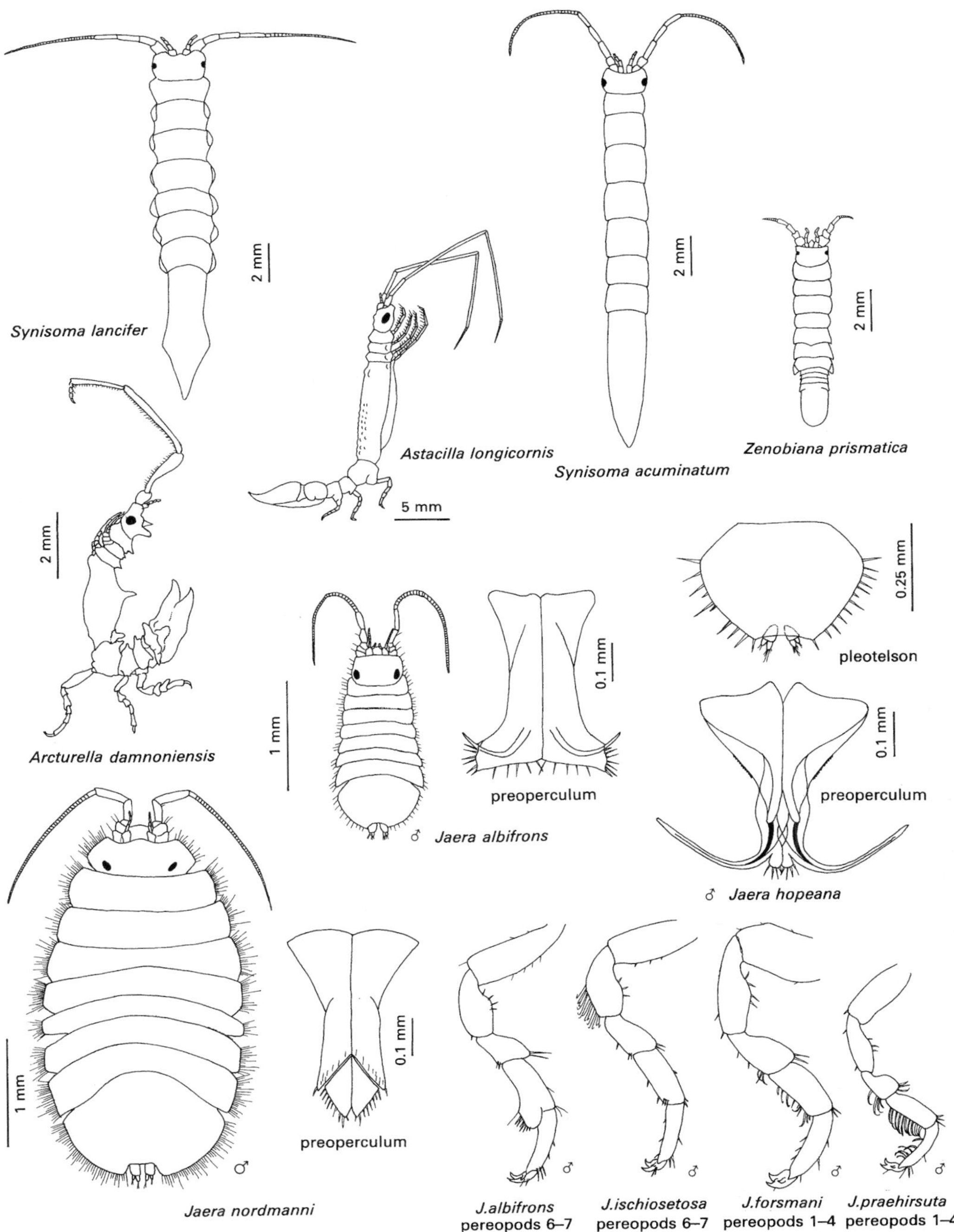

Synisoma lancifer
2 mm
Astacilla longicornis
5 mm
2 mm
Synisoma acuminatum
2 mm
Zenobiana prismatica
2 mm
Arcturella damnoniensis
0.25 mm
pleotelson
1 mm
0.1 mm
preoperculum
♂ Jaera albifrons
0.1 mm
preoperculum
♂ Jaera hopeana
1 mm
0.1 mm
preoperculum
♂
Jaera nordmanni
♂
J.albifrons
pereopods 6–7
♂
J.ischiosetosa
pereopods 6–7
♂
J.forsmani
pereopods 1–4
♂
J.praehirsuta
pereopods 1–4

Fig. 8.23

long plumose setae, and directed toward the mouth; pereopods 5–7 short, stout and adapted for clinging.

---

1. Middle pereon somite long and cylindrical. Dorsal surface smooth or with only small tubercles **_Astacilla longicornis_**

   Middle pereon somite long, not cylindrical but with angled corners. Dorsal surface with large tubercles **_Arcturella damnoniensis_**

Arcturella damnoniensis (Stebbing) Fig. 8.23
Head with three conical tubercles; each pereon segment with at least one median tubercle. Pereopods 1–4 sparsely setose. Pleotelson with lateral borders deeply serrated. Female more strongly tuberculate than male and with large triangular lobes on each side of segment 4. Up to about 10 mm long.

Sublittoral, occasionally intertidal; south-west coasts only.

Astacilla longicornis (Sowerby) Fig. 8.23
Female covered with small tubercles, two of which are conspicuous anteriorly on dorsal surface; male smooth and cylindrical. Pereopods 1–4 very densely setose. Females up to 30 mm body length; males up to 10 mm.

Occasional off most coasts with sporadic intertidal records. Norway southwards to Portugal.

8. JANIRIDAE

Body oval or elongate. Eyes, when present, dorsal. Pleon of one small, inconspicuous segment, others fused with telson to form a large, shield-shaped pleotelson. Uropods terminal or subterminal, with a peduncle, generally biramous; uropods styliform, articles cylindrical. Female pleopods covered by a plate-like operculum; males with a copulatory pre-operculum.

---

1. Without eyes; interstitial, body narrow **_Microjaera anisopoda_**

   With eyes; not interstitial. Body wider, oval **2**

2. Uropods well developed, slightly shorter or slightly longer than pleotelson, postero-lateral borders of which are serrated **3**

   Uropods small, barely, if at all, projecting beyond the pleotelson notch in which they are inserted. Pleotelson postero-lateral borders not serrated **4**

---

3. Antennae longer than body. Uropods longer than pleotelson. Pereon segments 2 and 3 with sinuous lateral borders **_Janira maculosa_**

   Antennae shorter than body. Uropods shorter than pleotelson. Pereon segments 2 and 3 with smooth lateral borders **_Janiropsis breviremis_**

---

4. Posterior border of pleotelson barely invaginated where uropods insert; ectocommensal on *Sphaeroma* **_Jaera hopeana_**

   Posterior border of pleotelson with deep invagination where uropods insert; free living **5**

---

5. Body widely oval, densely fringed with spines. Male preoperculum narrow and pointed **_Jaera nordmanni_**

   Body narrower, sparsely fringed with spines. Male preoperculum T-shaped **6**

---

6. Male pereopods 6 and 7 with prominent lobes (fringed with spines) on carpus **_Jaera albifrons_**

   Male pereopods 6 and 7 without carpal lobes **7**

---

7. Male pereopods 6 and 7 with a cluster of curved setae on distal portion of ischium **_Jaera ischiosetosa_**

   Male pereopods 6 and 7 without curved setae **8**

8. Male pereopods 1–4 sparsely covered with curved setae on propus, carpus, and merus. Pereopod 6 with carpal spine poorly developed, carpal spine of pereopod 7 well developed **_Jaera praehirsuta_**

   Pereopods 1–4 sparsely covered with curved setae on propus, carpus, and merus. Pereopods 6 and 7 both with carpal spine well developed **_Jaera forsmani_**

Jaera albifrons Leach Fig. 8.23
Body narrowly oval in females, males smaller than females and broadest across posterior pereon segments; lateral margins of both sexes fairly sparsely fringed with spines. Eyes relatively large, situated laterally, particularly in males. Males up to 2.4 mm long, females up to 5.0 mm.

Typically estuarine; most common at HWNT beneath stones in areas which retain surface water at low tide. All British coasts, very common. French coasts northwards to Arctic waters.

Jaera forsmani Bocquet Fig. 8.23
Like *J. albifrons* except for pereopods of adult males: 1–4 have a few (usually six or less) curved setae on propus, carpus, and merus; 6 and 7 have no carpal lobe but have well-developed spines in its place. Males up to 3.4 mm, females up to 6.0 mm.

Least tolerant of reduced salinity; thinly distributed beneath stones in well-drained areas from HWNT to LWST. South and west coasts only; not common.

Jaera ischiosetosa Forsman Fig. 8.23
Like *J. albifrons* except for pereopods of adult male: 1–4 normal; 6 and 7 each with a dense cluster of curved setae on distal portion of the ischium. Males up to 2.7 mm, females up to 5.0 mm.

Very tolerant of reduced salinity; beneath stones in strong streams flowing over sheltered shores; LWNT to above HWNT. Most British coasts, locally common. Russia southwards to Brittany.

Jaera praehirsuta Forsman Fig. 8.23
Like *J. albifrons* except for adult male pereopods: 1–4 with many curved setae on propus, carpus, and merus; 7 with a large spine on carpus; unlike *J. forsmani* the carpal spine of pereopod 6 is small. Males up to 3.0 mm body length, females up to 4.5 mm.

Among algae such as *Fucus serratus*; sheltered marine shores and estuaries; HWNT–LWST. Most British coasts; occasional. France to Norway.

Jaera hopeana Costa Fig. 8.23
Body fairly narrow, with very prominent spines fringing lateral margins. Males about 1.5 mm body length; ovigerous females up to 2.0 mm.

Ectocommensal on *Sphaeroma serratum*; HWNT–MTL; rare, south-west England.

Jaera nordmanni (Rathke) Fig. 8.23
Body broadly oval, flat, with sides densely setose. Eyes small and situated fairly medially on the head. Males up to about 4.5 mm; females up to 3.5 mm.

Very tolerant of reduced salinity; MTL–EHWS, under stones in freshwater streams flowing down the shore. South and west coasts only, locally very common. Western Scotland south to the Mediterranean.

Janira maculosa Leach Fig. 8.24
Body oblong and flattened; anterior pereon segments laterally excavate, with bilobed coxal plates. Adult male with pereopod 1 normal, subequal. Male body length up to 10 mm; females up to 7 mm.

Among sponges, ascidians, hydroids, bryozoans, *Laminaria* holdfasts; LWST and below; most coasts, not uncommon. Norway to Atlantic coasts of France.

Janiropsis breviremis Sars Fig. 8.24
Body oblong and flattened; outline of anterior pereon segments smooth. Adult male pereopod 1 and antennal peduncle larger than in female. Male body length 6 mm; females 4 mm.

Among sponges, hydroids, bryozoans, tunicates, *Laminaria* holdfasts, usually below tide marks. Rare, reported sporadically from north-east, south, and west coasts.

Microjaera anisopoda Bocquet and Levi — Fig. 8.24

Body elongate, head subrectangular, without eyes. Pereon segment 1 large and partly embracing head. Pereopod 1 larger than remainder. Pleon oval and shield-shaped, with a sharp terminal projection. Females 1.4 mm long, males a little smaller; length six times width.

Interstitial in coarse sand and gravel; from MTL into sublittoral; extreme south-west coasts only; rare.

9. MUNNIDAE

Body short, narrowing sharply posteriorly; generally with pereon segments 5–7 much smaller than 1–4 and sharply marked off from them. Eyes on lateral extensions of head. Uropods small, lacking peduncle. Female pleopods covered by a plate-like operculum; males with a copulatory preoperculum.

---

1. Appendages short; antennae less than half body length — ***Paramunna bilobata***

   Appendages long; antennae at least as long as body — **2**

---

2. Antennae as long as body — ***Munna kroyeri***

   Antennae twice body length — ***Munna minuta***

Munna kroyeri Goodsir — Fig. 8.24

Antenna flagellum shorter than peduncle. Pleotelson with 2–6 large spines visible from above on each lateral border, posterior border not serrated. Females up to 3 mm long, males 1.7 mm.

LWST and below; among sessile hydroids, bryozoans, *Laminaria* holdfasts. Reported sporadically from south-west, west, and north-west coasts; rare.

Munna minuta Hansen

Like *M. kroyeri*, but with relatively more slender appendages. Antennae almost twice length of body. Pleotelson with no more than one large spine on each lateral border; postero-lateral borders serrated. Females up to 3 mm body length; males smaller.

LWST and below; among hydroids and bryozoans. Recorded on most British coasts, but rare.

Paramunna bilobata Sars

Like *M. kroyeri* but appendages short; antennae less than half body length; head with pronounced bilobed frontal border. Up to 1 mm long.

Predominantly sublittoral, among gravel. Rare; west coasts.

10. PLEUROGONIDAE

Small isopods resembling Munnidae but with molar process of mandible reduced to a narrow point, and not as a normal grinding process.

Pleurogonium rubicundum Sars

Resembles *Munna*, but with diagnostic lack of eyes. Sublittoral, off north-east and north-west coasts; rare.

11. JAEROPSIDAE

Resembling Janiridae but with molar process of mandible reduced, elongate, and with no grinding surface. Antennae very short, with the antennal flagellum shorter than peduncle; coxal plates not visible from above.

Jaeropsis brevicornis Koehler — Fig. 8.24

Cephalon with bilobed mid-frontal region and median, jointed rostral projection. Antennule basal article large, inner distal border with a transparent serrated carina. Uropods very small, hardly visible from above; peduncle with a serrated carina on the outer border. Females 2–2.5 mm body length; males 1.5 mm.

LWST and below, among ascidians, encrusting algae, bryozoans, sponges. South and west coasts only; not common.

12. LIGIIDAE

Antennal flagellum with 10 or more articles. Endopod or uropod joined to basis well beyond telson tip; exopod and endopod subequal and elongate.

Ligia oceanica (Linnaeus) — Fig. 8.24

Largest British oniscid isopod, up to 25 mm body length. In rock-crevices, caves, groynes, and quays

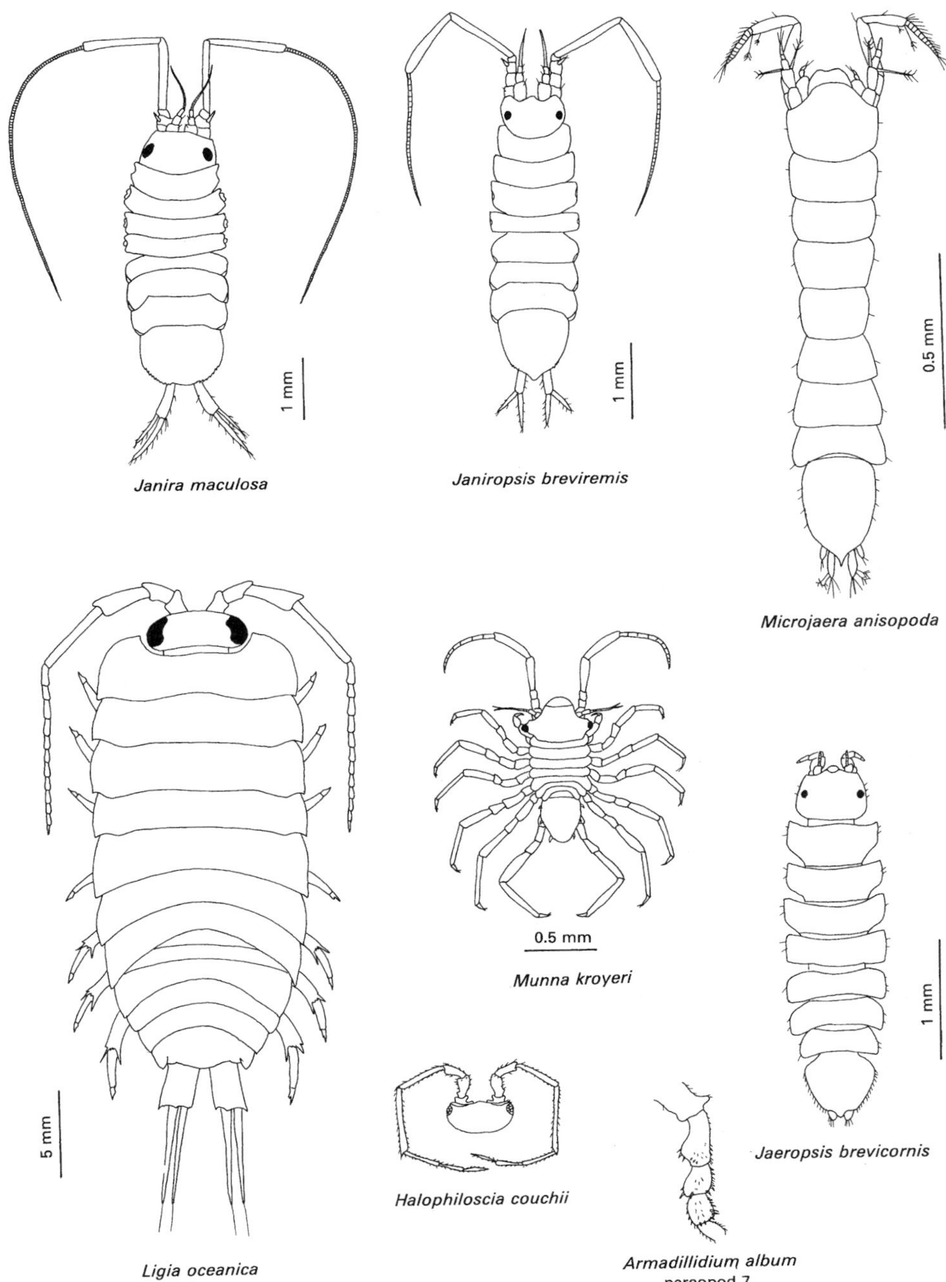
1 mm
Janira maculosa
1 mm
Janiropsis breviremis
0.5 mm
Microjaera anisopoda
0.5 mm
Munna kroyeri
1 mm
Jaeropsis brevicornis
5 mm
Ligia oceanica
Halophiloscia couchii
Armadillidium album
pereopod 7

Fig. 8.24

at or above HWM; all British coasts, widespread and abundant.

13. ARMADILLIDIIDAE

Antennal flagellum of two articles. Uropod endopod fused to basis beneath telson, exopod rectangular, projecting only slightly beyond telson. Rolls into a ball.

Armadillidium album Dollfus Fig. 8.24
Resembling common inland species *A. vulgare* (Latreille), but with diagnostic projection of distal border of pereopod 7 basis. Body length 6 mm.

Among debris and in burrows on sandy shores around strand line; Cumberland, southwards and eastwards to Norfolk and Yorkshire, and south-east Ireland.

14. HALOPHILOSCIIDAE

Antennal flagellum of three articles. Uropod endopod fused to basis beneath telson; exopod and endopod fairly elongate, unequal.

Halophiloscia couchii (Kinahan) Fig. 8.24
Antennae very long. Body brown in colour. Under stones around HWST on rocky shores; south-west coasts; rare.

15. BOPYRIDAE

Parasitic in the branchial cavity or on the abdomen of decapods, mainly Macrura and Anomura. Body of female sometimes asymmetrical; pereon with distinct segments, seven pairs of pereopods and five pairs of oostegites, the first pair bilobed; pleon segments more or less distinct. Males small, with all seven pereon segments distinct and pleon segments distinct or fused; often found on female pleon.

Parasitic on	Parasite
Spirontocaris, Hippolyte	*Bopyroides hippolytes*
	Bopyrina giardi
Palaemon	*Bopyrus fougerouxi*
Callianassa	*Ione thoracica*
	Pseudione callianassae
Upogebia	*Gyge branchialis*
Galathea	*Pleurocrypta longibranchiata*
	Pleurocrypta marginata
Pisidia	*Pleurocrypta porcellanae*
Pagurus, Anapagurus	*Athelges paguri*
	Pseudione hyndmanni
	Pseudione proxima
Diogenes	*Pseudione diogeni*

Athelges paguri (Rathke) Fig. 8.25
Pleotelson with four pairs of biramous, paddle-like pleopods. Females up to 11 mm body length; males 4 mm. Usually attached to abdomen, but occasionally in gill chambers, of the hermit crabs *Pagurus bernhardus* (Linn.) and *Anapagurus laevis* (Bell). All British coasts: local.

Bopyrina giardi Bonnier Fig. 8.25
Female resembling *Bopyrus*, but with only three or four (not five) pairs of uniramous pleopods. Females 1.7 mm; males 0.7 mm. Typically in the branchial chamber of *Hippolyte varians* Leach. South and west coasts only; local.

Bopyroides hippolytes Krøyer Fig. 8.25
Resembling *Bopyrus* but pleopods present only as five pairs of raised ridges. Females up to 11 mm, males 2.5 mm. Typically offshore on *Spirontocaris*, but has been recorded intertidally from *Hippolyte varians* Leach. South and west coasts only; local.

Bopyrus fougerouxi Giard and Bonnier Fig. 8.25
Female with five pairs of uniramous lamellar pleopods. Females up to 11 mm body length; males 2 mm. In the branchial chamber of *Palaemon serratus* (Pennant), where it gives rise to the so-called 'face ache' condition of the prawn. Most British coasts; locally not uncommon.

Gyge branchialis Cornalia and Panceri Fig. 8.25
Female with very small, vesicle-like uniramous pleopods. Females up to 12 mm; males 5 mm. Recorded from the gill chamber of the mud-burrowing decapods *Upogebia deltaura* (Leach) and *U. stellata* (Montagu); LW and below. South-west coasts only; rare.

Ione thoracica Montagu Fig. 8.25
Female pleon with six pairs of characteristic branched lateral processes, and five pairs of pleopods each with a lamellar endopod and cylindrical exopod. Female body length 6–7 mm, males

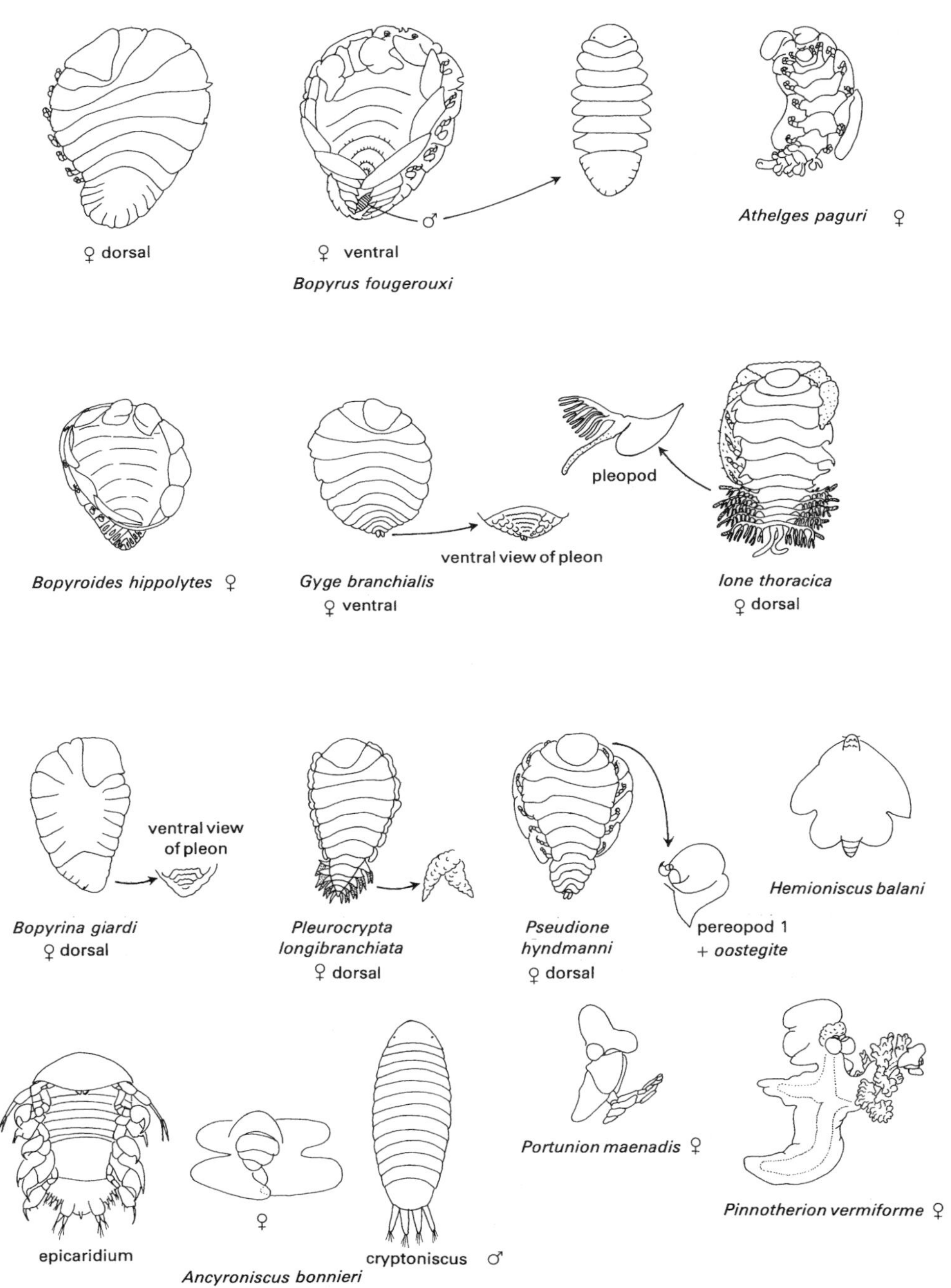

♀ dorsal
♀ ventral
♂
Bopyrus fougerouxi
Athelges paguri ♀
pleopod
ventral view of pleon
Bopyroides hippolytes ♀
Gyge branchialis
♀ ventral
Ione thoracica
♀ dorsal
ventral view
of pleon
Hemioniscus balani
Bopyrina giardi
♀ dorsal
Pleurocrypta
longibranchiata
♀ dorsal
Pseudione
hyndmanni
♀ dorsal
pereopod 1
+ oostegite
Portunion maenadis ♀
Pinnotherion vermiforme ♀
epicaridium
♀
Ancyroniscus bonnieri
cryptoniscus ♂

Fig. 8.25

smaller. Recorded from the gill chamber of *Callianassa subterranea* (Montagu), a predominantly sublittoral decapod burrowing in muddy sand. South and west coasts only; rare.

Pleurocrypta longibranchiata (Bate and Westwood) Fig. 8.25

Female pleon with prominent lateral processes and five well-developed, biramous, tuberculate pleopods. Females up to 8 mm, males 2 mm. Probably the most regular parasite of intertidal *Galathea squamifera* Leach. South-west, west, and north coasts. local.

Pleurocrypta marginata Sars

Resembles *P. longibranchiata* but female cephalon has a broad, flat frontal border, and the pereon has very wide, flattened, contiguous lateral plates. Pleopods short and smooth. Recorded from *Galathea squamifera* Leach and *G. intermedia* Lilljeborg. South and west coasts; rare.

Pleurocrypta porcellanae Hesse

Resembles *P. longibranchiata* but recorded from the gill chamber of *Pisidia longicornis* (Linn.) taken subtidally. All British coasts; locally distributed and rare.

Pseudione callianassae Kossman

From the burrowing prawn *Callianassa subterranea* (Montagu). Predominantly sublittoral; south-west coasts; rare.

Pseudione diogeni Popov

Recorded from the hermit crab *Diogenes pugilator* (Roux). Channel Islands; rare.

Pseudione hyndmanni (Bate and Westwood)

Mature female with five pairs of short, smooth pleopods, biramous and lanceolate. In the branchial cavity of the hermit crabs *Pagurus bernhardus* (Linn.) and *P. pubescens* Krøyer. South, west, and north coasts; local.

Pseudione proxima Bonnier

Differs from *P. hyndmanni* in having long, tuberculate pleopods projecting beyond sides of pleon; the postero-lateral border of the maxilliped has two short processes; oostegite 1 with posterior process rounded and incurved. One British record, from intertidal *Pagurus bernhardus* (Linn.), at Plymouth.

16. ENTONISCIDAE

Almost always endoparasitic in the visceral cavity of Anomura and Brachyura, surrounded by a membrane of host origin and communicating with the branchial cavity of the host by a pore. Female body with only indistinct traces of segmentation, pereopods rudimentary or entirely absent, and with five pairs of oostegites arising somewhat dorsally owing to the considerable reduction of the dorsal body surface. Twelve species recorded for Britain.

Parasitic on	Parasite
Eurynome	*Entionella monensis*
Pinnotheres	*Pinnotherion vermiforme*
Carcinus	*Portunion maenadis*

Entionella monensis Hartnoll

Recorded from the spider crab *Eurynome aspera* (Pennant). Predominantly sublittoral; Irish Sea; rare.

Pinnotherion vermiforme Giard and Bonnier Fig. 8.25

Length 1–3.5 mm. Recorded from *Pinnotheres pisum* (Linn.) Infecting the mussel *Mytilus*. Cornwall; local.

Portunion maenadis Giard Fig. 8.25

Size 3.2 mm. Single British record from *Carcinus maenas* (Linn.); Plymouth.

17. DAJIDAE

In incubatory chamber, or on dorsal surface, of mysids and euphausiids; rarely on Brachyura. Female body symmetrical; segments, when apparent, visible only in mid-dorsal region; pereopods well developed but numbering only five pairs and crowded near mouth. Oostegites present but small, with brood pouch often appearing bilateral. Pleopods small or absent.

A possible record of *Prodajus ostendensis* Gilson has been noted from the mysid *Gastrosaccus spinifer* (Goës) on the east coast.

18. CRYPTONISCIDAE

Recorded particularly from barnacles, isopods, and ostracods. Female stage sac-like, with few distinct segments and lacking pereopods. Male (cryptoniscus) internal with basal article lamellar and often serrated; pereopods 3–7 usually slender with setiform dactyl; pleopods and uropods biramous. Unlike the Bopyridae, Entoniscidae, and Dajidae, which comprise the superfamily Bopyrina, the Cryptoniscidae lack oostegites and incubate the brood internally. They are protandrous hermaphrodites, the cryptoniscus stage being male and the final, highly modified stage being female.

Parasitic on	Parasite
Dynamene	*Ancyroniscus bonnieri*
Balanus, Eliminius	*Hemioniscus balani*
Peltogaster	*Liriopsis pygmaea*

Ancyroniscus bonnieri
Caullery and Mesnil Fig. 8.25

Female a bilobed sac attached to the gut, but protruding into the brood pouch, of the host isopod *Dynamene bidentata*, where it feeds on the brood. On hatching, a microniscus stage and then an epicaridean stage attach to planktonic hosts before moulting to the cryptoniscus which locates the isopod host. South and west coasts; local.

Hemioniscus balani (Spence Bate) Fig. 8.25

Female with middle part of body sac-like and lobed. Body up to 8 mm. In the mantle cavity of shore barnacles; south and west coasts; locally abundant in *Semibalanus balanoides* (Linn.).

Liriopsis pygmaea Rathke

Female a constricted sac with an anterior lobe buried in the host and a projecting posterior lobe. Parasitizes *Peltogaster paguri* Rathke, a parasitic barnacle itself attached to the abdomen of the hermit crabs *Anapagurus laevis* (Bell) and *Pagurus cuanensis* Thompson. South and west coasts; rare.

VI. Order Amphipoda

The amphipod body is typically laterally compressed and divided into three regions: *head*, *pereon* (thorax), and *pleon* (abdomen), with a terminal *telson* attached dorsally to the last pleon segment (Fig. 8.26). The head bears six pairs of appendages: *antennae 1*, *antennae 2*, *mandibles*, *maxillae 1*, *maxillae 2*, and *maxillipeds*. The pereon has seven visible segments, each with a pair of uniramous appendages, the *pereopods*; pereopods 1 and 2 are modified as *gnathopods*. The pleon consists of six segments; the first three constitute the *pleosome*, and each bears a pair of biramous *pleopods*, the last three comprise the *urosome*, which typically bears three pairs of uniramous or biramous *uropods*. Females are recognised by a ventral brood pouch formed from a variable number of paired plates arising from the bases of pereopods 2–6.

Body form varies widely throughout the Amphipoda, but the most marked divergence from the basic pattern occurs in the suborder Caprellidea (Fig. 8.40). These are recognized by their very long, cylindrical bodies, and a reduction in number and type of appendages. The head is immovably fused to the first two segments of the thorax; pereon segment 1 remains visible, however, and as usual bears the first pair of gnathopods. The pereon consists of seven segments but the pleon is rudimentary or absent; there are two pairs of gnathopods and three posteriorly situated pairs of pereopods (pereopods 5–7); pereopods 3 and 4 are variously developed or absent. The Hyperiidea (Fig. 8.39) are plump, rounded amphipods with enormously developed eyes. They are pelagic, and in British waters are frequently encountered in the gastric pouches of some jellyfishes (e.g. *Rhizostoma*).

KEY TO SUBORDERS

1. Body long and slender, pleon usually vestigial. Head fused with pereon segment 1 (bearing gnathopod 1). Three pairs of terminal pereopods; pleopods and uropods lacking. Two or three pairs of gills; two pairs of brood lamellae
   Suborder Caprellidea (p. 403)

   Body with pleon and pereon distinct. Head not fused with pereon segment I. Pleopods and uropods present. More than three pairs of gills; four or five pairs of brood lamellae **2**

2. Body rounded. Eyes often very large, occupying most of head. Maxillipeds without a palp. Coxae small or absent. Pelagic **Suborder Hyperiidea** (p. 401)

Eyes not occupying most of head, of variable size, sometimes apparently absent. Coxae present and typically relatively large. Maxillipeds with a palp. Benthic **Suborder Gammaridea** (this page)

Suborder Gammaridea

Gammaridean amphipods most usually have laterally compressed body shapes, although some may be more nearly cylindrical, and a few are distinctly dorso-ventrally compressed, or flattened. The body consists of a head, seven thoracic segments forming the pereon, and six abdominal segments, comprising the pleon (Fig. 8.26).

In shore and shallow-water species the head usually bears a pair of round or reniform laterally situated eyes. These may be indistinct in some species, fused dorsally, or supplemented by one or two further pairs. Eyes are lacking in some species. Between the eyes the head may have a median, dorsal projection called the *rostrum*. *Antenna 1* has a basal *peduncle* of three articles, and a slender *flagellum* of a few or many articles. In many species antenna 1 also bears an *accessory flagellam* at the distal tip of the third peduncle article; it may consist of one to many articles, and may be small or well developed. *Antenna 2* has five peduncle articles and a flagellum, but no accessory flagellum. The two pairs of antennae often differ in size between males and females of the same species.

Amphipod mouthparts consist of an upper lip, lower lip, and one pair each of *mandibles*, *first maxillae*, *second maxillae*, and *maxillipeds* (Fig. 8.26). The mandibles are attached laterally to the mouth, adjacent to the lips, and in front of the remaining mouthparts. They usually have a series of teeth, or *incisors*, at the antero-distal end, and a *molar process* on the medio-ventral surface. In some families the molar processes are reduced and smooth, and in others they are absent. Most gammarideans have single tri-articulate *mandibular palps* attached dorso-laterally to each mandible.

The seven pereon segments carry the *pereopods*, the *gills*, and, in females, the *brood lamellae*. Each pereopod consists of seven articles (Fig. 8.26), the first, closest to the body is the *coxa*, and is followed by the *basis* (article 2), *ischium* (article 3), *merus* (article 4), *carpus* (article 5), *propodus* (article 6), and *dactylus* (article 7). The coxae resemble ventral extensions of the pereon segments and contribute to the laterally compressed body shape. Gills are generally attached to the inner surfaces of coxae 2–7, and in females the brood lamellae are situated on coxae 2–5. In young females the lamellae are small buds but they become longer and more heavily setose as the animal grows, and the setae interlock to form a cradle enclosing the eggs.

The first two pairs of pereopods constitute the first and second *gnathopods* and usually one or both may be distally *chelate*, *subchelate*, or *carpochelate* (Fig. 8.26). A chelate gnathopod has the form of a pincer similar to that of a crab. In the subchelate condition the *dactylus* (article 7) is folded back onto the posterior distal border (*palm*) of the *propodus* (article 6); enlargement of the propodus and the presence of processes from the *carpus* (article 5) may give a *carpochelate* structure. The gnathopods are usually larger than the other pereopods and may show considerable sexual dimorphism, being larger and/or more complex in the male.

The three pleon segments may bear dorsal teeth, and carry paired, biramous *pleopods* with strongly setose, multi-segmented rami. The *epimeral plates* project ventrally, overhanging the pleopods, from each side of the pleon segments.

The three *urosome* segments are of more or less equal length; they may be fused to a greater or lesser degree, and one segment may be especially elongate. The dorsal surface of the urosome may bear teeth or groups of spines and/or setae which are important in identification. Each urosome segment has a pair of biramous *uropods*. The third uropods may be *aramous*, *uniramous*, or *biramous*, with mono- or bi-articulate, equal or unequal rami. The rami may have hooks, spines, and/or setae; rarely, one pair of uropods may be absent.

The *telson* is attached above the anus on the postero-dorsal surface of the third urosome segment; it may be entire, notched, cleft, or emarginate (Fig. 8.26), of variable size, and carry hooks, spines, or setae.

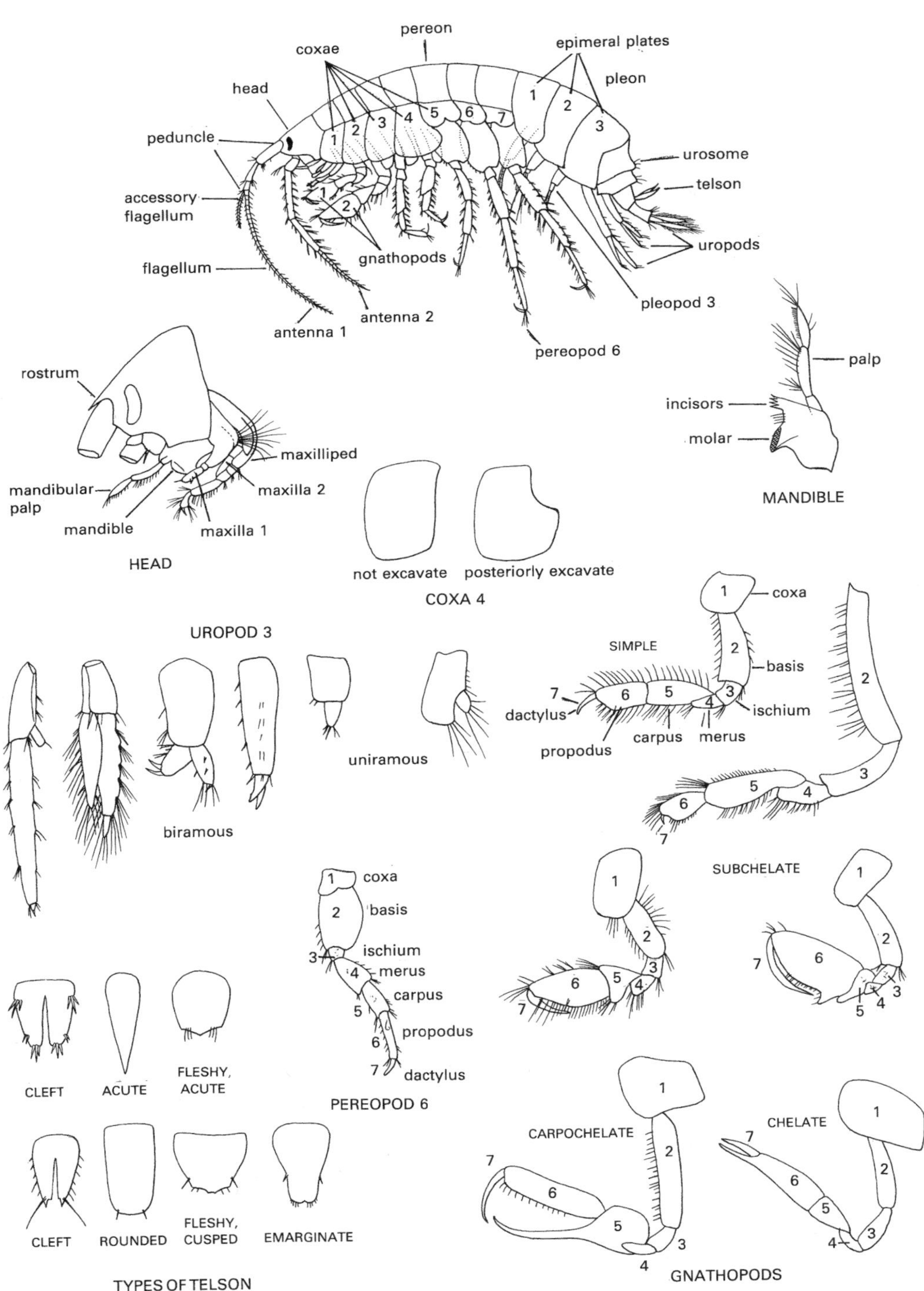
pereon
coxae
epimeral plates
head
pleon
peduncle
urosome
telson
accessory flagellum
uropods
gnathopods
flagellum
pleopod 3
antenna 2
antenna 1
pereopod 6
rostrum
palp
incisors
molar
maxilliped
mandibular palp
maxilla 2
mandible
maxilla 1
MANDIBLE
HEAD
not excavate
posteriorly excavate
COXA 4
coxa
UROPOD 3
SIMPLE
basis
ischium
dactylus
carpus
merus
propodus
uniramous
biramous
SUBCHELATE
coxa
basis
ischium
merus
carpus
propodus
dactylus
CLEFT
ACUTE
FLESHY, ACUTE
PEREOPOD 6
CARPOCHELATE
CHELATE
CLEFT
ROUNDED
FLESHY, CUSPED
EMARGINATE
GNATHOPODS
TYPES OF TELSON

Fig. 8.26

KEY TO FAMILIES

Approximately 300 species of gammaridean amphipods, in 36 families, have been recorded from brackish waters, intertidal, and subtidal (to 200 m) habitats around the British Isles. This key will allow identification of 84 species, representing 22 families, which can be found in intertidal and shallow subtidal habitats. If a specimen does not conform exactly to the characters used in the main key and the family keys, and does not match both the illustrations and diagnostic notes given for any of the selection of species included here, then it should be identified using one of the more comprehensive works listed in the Reference section (p. 758).

---

1. Body elongate, flattened, or sub cylindrical; urosome depressed, the segments often coalesced. Telson short, entire. Coxae short, spaced apart, or just touching. Mandible with a small palp. Uropod 3 very small, uniramous or without rami; sometimes absent (Fig. 8.27L) **21. Corophiidae**

   Characters not combining as above **2**

---

2. Body cylindrical and fat. Rostrum small or medium sized, never hood-like. Accessory flagellum well developed. Pereopod 7 as long or longer than 6. Pereopods broad, flattened, spinose and setose. Telson cleft (Fig. 8.27G) **11. Haustoriidae**

   Characters not combining as above **3**

---

3. Body cylindrical, slender. Head elongate, dorso-ventrally flattened with a large hood-like rostrum covering base of antenna I. Pereopods spiny; pereopod 7 distinctly shorter than 6. Telson cleft (Fig. 8.27J) **14. Phoxocephalidae**

   Characters not combining as above **4**

---

4. Head with conspicuous downturned rostrum, eyes more or less coalesced dorsally. Accessory flagellum absent or rudimentary. Pereopod 7 very much longer than 6. Ramus of uropod 3 scarcely longer than peduncle. Telson small and entire (Fig. 8.27I) **13. Oedicerotidae**

   Characters not combining as above **5**

---

5. Head with conspicuous downturned rostrum; eyes large, not coalesced. Mouthparts forming a conical bundle pointing ventrally. Last pereon and all pleon segments with large paired teeth on dorsal side. Epimeral plate 3 with two large teeth on hind border. Antennae subequal, accessory flagellum rudimentary. Gnathopods small. Telson entire and subquadrate (Fig. 8.27C) **3. Acanthonotozomatidae**

   Characters not combining as above **6**

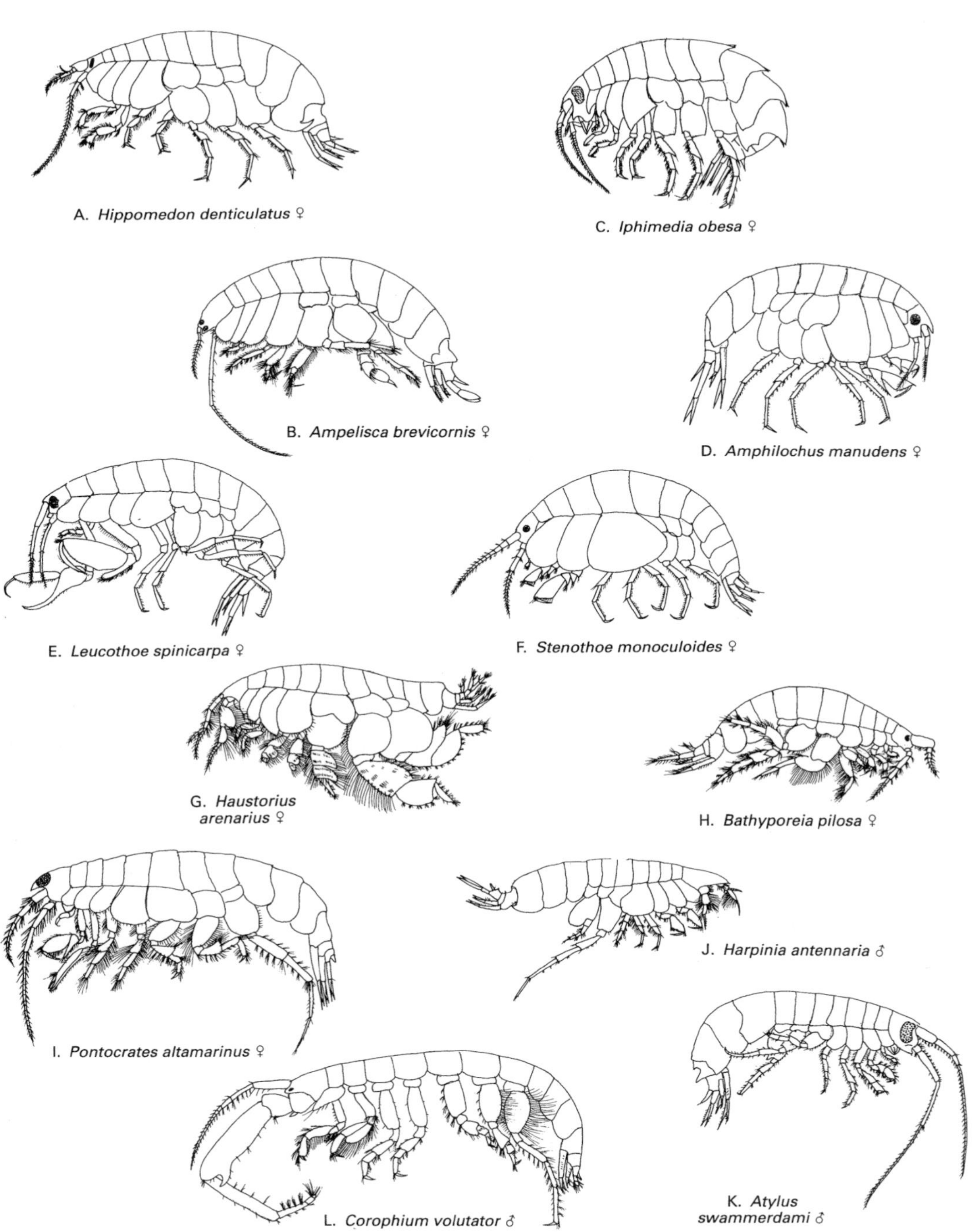
A. *Hippomedon denticulatus* ♀
C. *Iphimedia obesa* ♀
B. *Ampelisca brevicornis* ♀
D. *Amphilochus manudens* ♀
E. *Leucothoe spinicarpa* ♀
F. *Stenothoe monoculoides* ♀
G. *Haustorius arenarius* ♀
H. *Bathyporeia pilosa* ♀
I. *Pontocrates altamarinus* ♀
J. *Harpinia antennaria* ♂
K. *Atylus swammerdami* ♂
L. *Corophium volutator* ♂

Fig. 8.27

6. Pereopods 5–7 broad, flattened, spiny, and setose; pereopod 7 not much longer than 6. Antenna 1 short, with stout first peduncle article, as large as head, with remaining articles and flagellum projecting at right angle from lower surface; with small, two-articled accessory flagellum. Telson cleft (Fig. 8.27H) **12. Pontoporeiidae**

   Characters not combining as above **7**

7. Body usually short and fat. Antenna 1 short, or very short; peduncle stout with article 1 much larger and fatter than subsequent articles; flagellum scarcely longer than peduncle. Accessory flagellum present. Gnathopods usually small; gnathopod 2 with swollen propodus, rudimentary dactylus, and elongate ischium. Mandible with three-articled palp (Fig. 8.27A) **I. Lysianassidae**

   Characters not combining as above **8**

8. Body usually short, fat, or compressed. Antenna 1 and 2 medium or short, almost equal length. Accessory flagellum absent. Pereopod 1 coxa very small or rudimentary, partially hidden; pereopod 2–4 coxae grossly enlarged. Gnathopods variable but never chelate. Urosome segment 1 without a dorsal tooth. Uropod 3 uniramous or biramous. Telson entire **9**

   Characters not combining as above **10**

9. Uropod 3 longer than uropod 2; biramous, each ramus consisting of just one article (Fig. 8.27D) **4. Amphilochidae**

   Uropod 3 slightly shorter than uropod 2; uniramous, the ramus with two articles (Fig. 8.27F) **6. Stenothoidae**

10. Gnathopod 1 large, carpochelate; gnathopod 2 larger than 1, subchelate. Antenna 1 longer than 2. Accessory flagellum small, with just one article. Body slender. Coxae large. Uropod 3 biramous. Telson entire, triangular (Fig. 8.27E) **5. Leucothoidae**

    Characters not combining as above **11**

11. With two, four, or six eyes. Urosome segments 2 and 3 coalesced; urosome usually with at least one sharp or blunt dorsal tooth. Uropod 3 biramous. Telson medium or long, deeply cleft. Accessory flagellum absent, or if present, vestigial. Gnathopods small, subchelate or simple **12**

    Characters not combining as above **14**

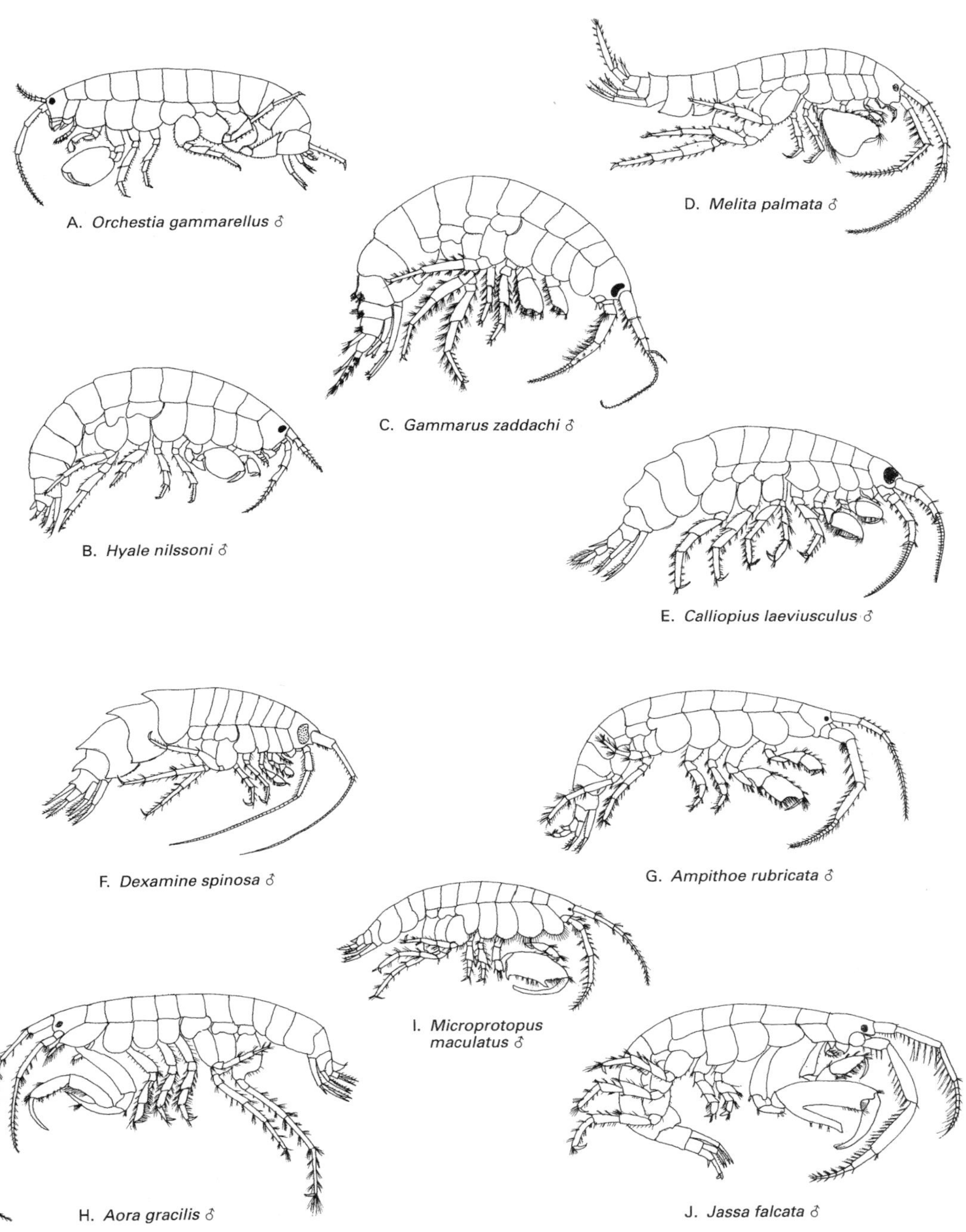
A. Orchestia gammarellus ♂
B. Hyale nilssoni ♂
C. Gammarus zaddachi ♂
D. Melita palmata ♂
E. Calliopius laeviusculus ♂
F. Dexamine spinosa ♂
G. Ampithoe rubricata ♂
H. Aora gracilis ♂
I. Microprotopus maculatus ♂
J. Jassa falcata ♂

Fig. 8.28

12. Head distinctly flattened, elongate, rostrum absent. Usually four, sometimes two or six eyes. Body strongly compressed, generally smooth dorsally. Pereopod 7 shorter and of different structure to pereopod 6 (Fig. 8.27B) **2. Ampeliscidae**

Head not as described; rostrum present, large or small. Two eyes. Pereopods 6 and 7 broadly similar in structure **13**

13. Antenna 1 shorter than 2. Rostrum large. Mandible with palp. Often with dorsal teeth on pleon and last pereon segment. Urosome segment 1 with two teeth dorsally: small anterior tooth separated from much larger posterior tooth by distinct cleft (Fig. 8.27K) **16. Atylidae**

Antennae variable in length. Rostrum small. Mandible without palp. Teeth sometimes present on pleon. Urosome segment 1 with single tooth, or nearly smooth (Fig. 8.28F) **17. Dexaminidae**

14. Antenna 1 shorter than 2; accessory flagellum absent or rudimentary. Epimeral plate 3 with sharp or blunt point at postero-ventral corner, often with denticulate or angular posterior border. Coxa 4 with concave posterior edge. Uropod 3 biramous; rami usually at least twice length of peduncle, inner rami generally longer than outer. Telson entire but not fleshy, often triangular, usually about twice as long as wide, sometimes notched apically (Fig. 8.28E) **15. Calliopiidae**

Characters not combining as above **15**

15. Accessory flagellum absent. Uropod 3 uniramous, the ramus with just one articler shorter than, or subequal to, peduncle. Mandible without a palp **16**

Characters not combining as above **17**

16. Telson short, entire. Antenna 1 shorter than peduncle of antenna 2. Large body size (10 mm). On strandline (Fig. 8.28A) **7. Talitridae**

Telson short, cleft into two lobes. Antenna 1 longer than peduncle of antenna 2 (Fig. 8.28B) **8. Hyalidae**

17. Accessory flagellum present or absent. Uropod 3 biramous, rami small or very small, always shorter than peduncle; outer ramus frequently with one or two large hooked spines distally, or slightly curved and with only small teeth on disto-lateral border. Coxa 4 not concave posteriorly. Telson entire, fleshy, and short; length never much more than width **18**

Characters not combining as above **19**

18. Antenna 1 with peduncle article 3 very short, less than half length of article 2. Uropod 3 with fat, blunt rami armed with large hooks (Fig. 8.28G) **18. Ampithoidae**

Antenna 2 with peduncle article 3 more than half length of article 2. Uropod 3 with triangular rami armed with either one hooked spine or with small disto-lateral teeth (Fig. 8.28J) **22. Ischyroceridae**

---

19. Telson short and fleshy, quadrate or nearly circular, entire, or with a weak notch. Coxa 4 not concave posteriorly. Accessory flagellum present or absent **20**

Telson elongate, flat, entire or cleft. Coxa 4 with concave posterior edge, except when small **21**

---

20. Gnathopod 1 usually larger than 2, complex and markedly subchelate, or carpochelate in male. Antenna 1 with peduncle article 3 shorter than article 1 (Fig. 8.28H) **19. Aoridae**

Gnathopod 1 equal to or smaller than 2. Antenna 1 with peduncle article 3 equal to or longer than article 1, or if article 3 shorter than I, then uropod 3 uniramous (Fig. 8.28I) **20. Isaeidae**

---

21. Telson with shallow apical notch, but not obviously cleft. Coxa 4 with deeply concave posterior edge. Accessory flagellum well developed, with three or more articles. Uropod 3 biramous, short, not exceeding length of uropod 1 **15. Calliopiidae** (part)

Telson cleft for at least one-third of its length **22**

---

22. Urosome with small groups of dorsal spines. Head with small rostrum, eyes usually large and reniform. Gnathopods 1 and 2 of similar size. Coxae large, overlapping (Fig. 8.28C) **9. Gammaridae**

Urosome spines sparse or lacking. Head without rostrum, eyes small and rounded. Gnathopod 2 much larger than 1. Coxae short, not overlapping (Fig. 8.28D) **10. Melitidae**

---

1. LYSIANASSIDAE

Body compact and usually smooth dorsally. Antenna 1 short or very short, peduncle article 1 much larger than subsequent articles. Accessory flagellum usually present, consisting of three or more articles, occasionally vestigial or with fewer articles. Mandible with three-articled palp. Gnathopods normally small; gnathopod 2 mitten-shaped terminally with article 3 (ischium) elongate, article 5 (carpus) bulbous and scaled posteriorly, article 6 (propodus) swollen and scaly, article 7 (dactylus) small or rudimentary. Uropod 3 with lanceolate rami; outer ramus often bi-articulate, sometimes longer than inner. Telson entire or cleft.

A very large family with more than 30 British species. The majority occur in relatively deep water and only the commoner inshore species are keyed out below.

---

1. Telson short and broad, with straight distal edge, not notched. Gnathopod 1 smaller than 2, with simple dactyl. Antenna 1 with stout tooth on inner distal edge of article 1 ***Lysianassa ceratina***

   Telson deeply or shallowly cleft **2**

---

2. Epimeral plate 3 produced postero-ventrally as a long, sharp point. Head small, partly hidden by coxal plate 2. Gnathopods slender, propodus much shorter than carpus. Uropod 2 inner ramus without median constriction ***Hippomedon denticulatus***

   Epimeral plate 3 may be acutely angled postero-ventrally, but not produced into a point **3**

---

3. Telson shallowly cleft, for less than one-third of its length. Gnathopod 2 propodus long and slender ***Orchomene humilis***

   Telson cleft for at least half of its length **4**

---

4. Gnathopod 1 slender; ischium triangular, produced anteriorly into a point. Antenna 1 and 2 short, subequal in both sexes ***Socarnes erythrophthalmus***

   Gnathopod 1 stout; ischium not produced anteriorly. Antenna 2 longer than antenna 1 in male ***Orchomene nana***

Hippomedon denticulatus (Bates) Figs 8.27A, 8.29

Length 14 mm. Antenna 1 short, peduncle robust, articles 1 and 2 with short point at front edge, accessory flagellum very small, indistinct; antenna 2 much longer than 1. Gnathopod 1 slender, simple; gnathopod 2 subchelate. Uropod 2 shorter than 1 and 3; uropod 3 with subequal rami, outer ramus bi-articulate. Telson deeply cleft.

Sublittoral, in shallow water, burrowing in sand. All coasts.

Lysianassa ceratina (Walker) Fig. 8.29

Body up to 10 mm; yellow with white flecks. Head with prominent lateral lobes; eyes large, oval to reniform. Antenna 1 peduncle stout, flagellum of about nine articles, slightly shorter than peduncle. Antenna 2 in female as long as antenna 1, with flagellum of about 10 articles; in male longer than body, with stout peduncle and very slender, filiform flagellum. Gnathopod 1 simple, propodus distally tapered with several short spines on inner posterior margin. Gnathopod 2 very slender, minutely sub-chelate. Uropod 1 elongate, with subequal rami; uropod 2 shorter than 1, with inner ramus distally constricted; uropod 3 with broad peduncle, its outer ramus with only one article.

Common among algae in shallow water. All coasts, locally abundant.

Genus *Orchomene*

First coxae not tapered distally. Antennae short and subequal in females, antenna 2 much longer than 1 in males. Gnathopod 1 subchelate, usually short and strong; ischium short but not triangular, carpus shorter than propodus, dactylus sometimes toothed on interior margin. Gnathopod 2 minutely chelate or subchelate. Uropod 3 with bi-articulate outer ramus. Telson variable, entire or deeply cleft.

Orchomene humilis (Costa) Fig. 8.29

Length 8 mm, whitish yellow speckled with red. Epimeral plate 3 quadrate, with crenulate posterior margin. Gnathopod 1 dactylus without a tooth on interior margin; gnathopod 2 propodus long and slender. Uropod 3 with few spines in female, with plumose setae in male. Telson cleft for less than half its length, shallowly and widely in female, more narrowly and deeply in male.

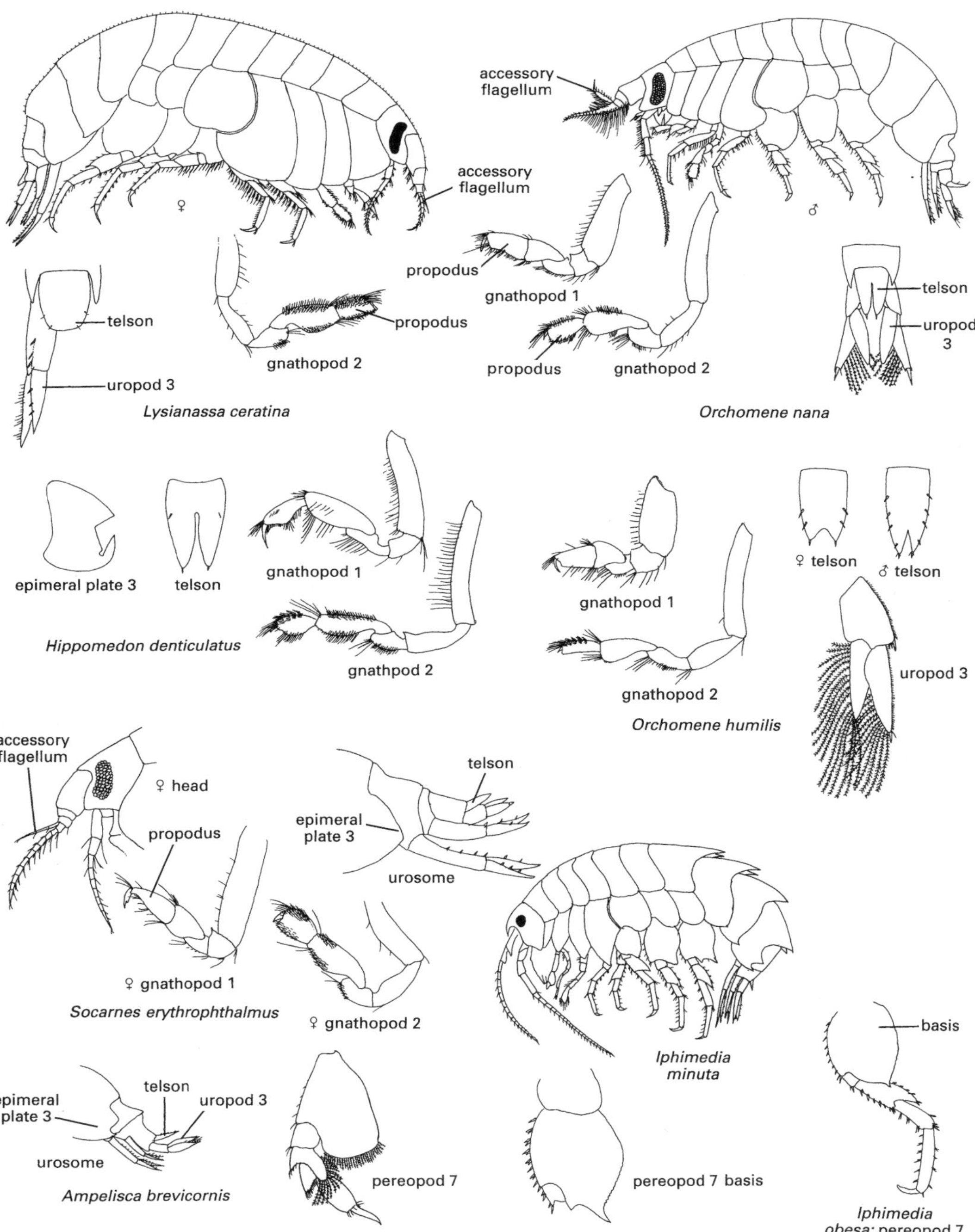
accessory flagellum
♀
telson
uropod 3
propodus
gnathopod 2
Lysianassa ceratina
accessory flagellum
♂
propodus
gnathopod 1
propodus
gnathopod 2
telson
uropod 3
Orchomene nana
epimeral plate 3
telson
gnathopod 1
Hippomedon denticulatus
gnathpod 2
gnathopod 1
gnathopod 2
Orchomene humilis
♀ telson
♂ telson
uropod 3
accessory flagellum
♀ head
propodus
♀ gnathopod 1
Socarnes erythrophthalmus
telson
epimeral plate 3
urosome
♀ gnathopod 2
Iphimedia minuta
basis
epimeral plate 3
telson
uropod 3
urosome
Ampelisca brevicornis
pereopod 7
pereopod 7 basis
Iphimedia obesa: pereopod 7

Fig. 8.29

Amongst sublittoral epibenthos and shells, sometimes in *Laminaria* holdfasts. All coasts.

Orchomene nana (Krøyer) Fig. 8.29
Length 5 mm, greyish-white. Epimeral plate 3 with smooth posterior margin and rounded postero-ventral corner. Gnathopod 1 dactylus toothed on inner margin; gnathopod 2 propodus short, oval, and half length of carpus. Telson narrowly cleft nearly to its base.

Lower shore and sublittoral, on sandy and coarse grounds, in *Laminaria* holdfasts, or on dead crabs. All coasts.

Socarnes erythrophthalmus Robertson Fig. 8.29
Length 4 mm, translucent green, pink, or yellow-orange. Head with prominent, triangular, lateral lobes; eyes large, oval, bright red. Coxae large. Gnathopod 1 simple, carpus and propodus of equal length, ischium triangular with short anterior point; gnathopod 2 minutely subchelate, propodus half length of carpus. Pereopod 7 not longer than 6. Epimeral plate 3 rounded posteriorly. Uropod 3 outer ramus bi-articulate. Telson deeply cleft.

Intertidal and shallow sublittoral, under stones or in sand. South and west coasts.

2. AMPELISCIDAE

Body laterally compressed. Accessory flagellum absent. Head elongate, without a rostrum. Two, four, or six eyes. Anterior coxae long. Gnathopods simple, 2 longer and thinner than 1. Pereopod 7 shorter than, and structurally different from, 6. Urosome segments 2 and 3 coalesced. Uropod 3 biramous. Telson short or medium, deeply split. Builders of limp tubes on sandy and muddy sediments. Perhaps 19 British species, in three genera, but the taxonomy is unclear. Only the following species is common intertidally.

Ampelisca brevicornis (Costa) Figs 8.27B, 8.29
Length 12 mm, translucent white with scattered brown and/or yellow spots. Head with a concave ventral margin. Antenna 1 very short in female, not reaching peduncle article 5 of antenna 2; in male just reaching peduncle article 5 of antenna 2. Pereopod 7 merus equal length to ischium, with long postero-ventral lobe; basis with broad posterior lobe. Urosome segment 1 with small, rather rounded, dorsal keel. Epimeral plate 3 with long, curved acute point at postero-ventral corner; posterior margin sinuous with median lobe as long as postero-ventral point. Telson without spines on dorsal surface, but with marginal setae.

Intertidal and sublittoral, on sandy mud. Common on all coasts.

3. ACANTHONOTOZOMATIDAE

Body short, stout, last pereon and all pleon segments each with a pair of pointed teeth dorsally. Antennae short to medium length, subequal. Accessory flagellum rudimentary or missing. Head with well-developed rostrum. Mouth parts grouped into a ventrally pointing conical bundle; mandible with palps. Coxae 1–4 medium to large, convex anteriorly with single postero-ventral point. Gnathopods small. Basis of pereopods 5–7 expanded, with one or two points on posterior border. Epimeral plate 3 with two large, pointed teeth on posterior border. Uropod 3 rami longer than peduncle. Telson entire, slightly notched. Eight British species, several of which have only recently been recognized.

---

1. Epimeral plate 3 with two smooth, straight, equal-sized teeth on posterior border ***Iphimedia obesa***

   Epimeral plate 3 with two crenulated, unequal teeth on posterior border, the medio-distal distinctly curved. Pleon segment 3 with large, triangular dorso-lateral teeth ***Iphimedia minuta***

Iphimedia minuta Sars Fig. 8.29
Body up to 6 mm; colour variable, generally white with yellow, orange or brown markings; eyes red. Pereon segment 7 and pleon segments 1–3 each with a pair of strong dorsal processes. Antenna 1 about one-third body length, flagellum with about 17 articles; antenna 2 slightly longer than 1. Pereopods 5–7 with basis rounded proximally; pereopod 5 basis rounded distally; pereopod 6 basis with acute postero-distal angle; pereopod 7 basis with produced, acute distal angle, and serrated posterior margin.

Lower shore and shallow sublittoral, amongst algae and hydroids. Common on all British coasts.

Iphimedia obesa Rathke Figs 8.27C, 8.29
Body stout, up to 13 mm; colour variable, white or yellow with red or pink bands. Pereon segment 7 and pleon segments 1–3 with paired dorso-lateral

teeth. Pereopods 5 and 6 basis with posterior margin evenly rounded, without teeth; pereopod 7 basis with triangular process or small tooth on postero-distal margin. Epimeral plate 3 with two smooth, almost equal-sized teeth on posterior border, separated by wide gutter. Uropod 3 with inner ramus little longer than outer.

Shallow sublittoral, on hydroids and amongst algae. Common on all British coasts.

4. AMPHILOCHIDAE

Body usually short and fat. Antennae medium to short, subequal. Accessory flagellum absent, rarely rudimentary. Rostrum conspicuous. Coxa I very small, partially hidden; coxae 3 and 4 large or immense, 4 concave posteriorly. Mandible usually with palp of three articles. Gnathopods small or medium sized, subchelate or nearly simple. Pereopods slender. Uropod 3 biramous, with elongate peduncle, longer than uropod 2. Telson entire, unarmed; usually long, triangular, sometimes immense. Fifteen species, in six genera, recorded from Britain.

I. Coxae 3 and 4 very large, with contiguous borders, completely hiding coxae 1 and 2. Telson greatly enlarged, fin-like **_Peltocoxa damnoniensis_**

Coxae 3 and 4 moderately overlapping, not hiding coxae 1 and 2 **2**

---

2. Gnathopods 1 and 2 small, simple, or feebly subchelate **_Gitana sarsi_**

Gnathopod 2 large, subchelate; gnathopods 1 and 2 with propodus bearing a sharp tooth anteriorly **_Amphilochus manudens_**

Amphilochus malnudens Bate Figs 8.27D, 8.30

Length 5 mm, reddish-brown with dark markings. Antennae short. Coxal plates 2–4 large, overlapping, with denticulate margins. Gnathopods subchelate; gnathopod 2 larger than 1. Epimeral plate 3 produced postero-ventrally as a small, sharp tooth. Uropod 3 peduncle longer than rami. Telson three times as long as wide, terminating in a sharp point.

Sublittoral on silty grounds and amongst hydroids. All coasts.

Gitana sarsi Boeck Fig. 8.30

Length 3 mm, white with broad brownish bands. Coxa 2 with distal crenulation. Gnathopods small, subequal, simple. Pereopods and uropods slender. Telson lanceolate, minutely tridentate apically, about two-thirds length of uropod 3 peduncle.

Intertidal and sublittoral, amongst algae, in *Laminaria* hold fasts and amongst epibenthos. All coasts.

Peltocoxa damnoniensis (Stebbing) Fig. 8.30

Up to 3 mm, red to purple. Antennae very short, robust; antenna 1 peduncle stout; flagellum scarcely longer than peduncle article 1, with ventral fringe of long setae; accessory flagellum rudimentary, minute, inarticulate. Gnathopods 1 and 2 dactylus with serrated margin.

Lower shore and sublittoral, among algae on rocky shores. Southern and western coasts.

5. LEUCOTHOIDAE

Body slender, slightly cylindrical. Antenna 1 as long as or longer than 2; both with long peduncle and relatively short flagellum. Accessory flagellum uniarticulate, very small. Mandible with palp, without molar process. Coxae large; coxa 1 not concealed. Gnathopods large. Gnathopod 1 distinctively carpochelate; gnathopod 2 powerfully subchelate, slightly smaller than 1. Gnathopod 2 propodus larger and more distinctly toothed in male than female. Uropod 2 shorter than 1 and 3. Uropod 3 biramous, with styliform rami shorter than peduncle. Telson entire, triangular. Four British species in the genus *Leucothoe*, two of which occur commonly intertidally.

I. Adult gnathopod 1 dactylus more than one-third length of propodus. Epimeral plate 3 with quadrate postero-ventral corner, lacking a distinct tooth. Male gnathopod 2 propodus with crenulate margin to palm **_Leucothoe spinicarpa_**

Adult gnathopod 1 dactylus less than one-third length of propodus, carpus with large, regularly spaced teeth. Epimeral plate 3 with well-marked tooth on postero-ventral corner **_Leucothoe incisa_**

Leucothoe spinicarpa
(Abildgaard) Figs 8.27E, 8.30
Length 12 mm; pinkish, greenish, or white, translucent or opaque, eyes red. Coxa 4 without antero-ventral tooth. One-third to one-half length of propodus. Gnathopod 2 with small, blunt teeth on palm of propodus and a projection above articulation with dactylus, no bifid distal tip on carpus. Epimeral plate 3 almost rectangular, without distinct tooth postero-ventrally.

Littoral and sublittoral, amongst algae and frequently in ascidians and sponges. All coasts.

Leucothoe incisa Robertson Fig. 8.30
Length 7 mm, greenish- or yellowish-white, eyes intense brick-red. Coxa 4 rounded antero-ventrally. Dactylus of adult gnathopod 1 one-quarter to one-fifth length of propodus. Gnathopod 2 propodus with projection above articulation with dactylus; carpus with bifid distal tip, bearing a small spine. Epimeral plate 3 with deeply concave posterior margin and large, curved tooth at postero-ventral corner.

Lower shore and sublittoral, amongst rocks and sand. All coasts.

6. STENOTHOIDAE

Body short, fat, or compressed, usually smooth dorsally. Head with inconspicuous rostrum. Antennae subequal; accessory flagellum absent or vestigial. Coxa 1 very small, largely hidden; coxae 2–4 very large, plate-like; coxa 4 not concave posteriorly. Gnathopods usually powerful, subchelate; gnathopod 1 rarely simple. Pereopod 5 basis slender. Urosome segment 1 not toothed dorsally. Uropod 3 uniramous, rami bi-articulate. Telson entire, medium length, separate from last urosome segment. A large family with more than 20 British species, in seven genera.

---

1. Mandible without a palp **2**

   Mandible with a palp, either minute, uni-articulate, or moderately well developed with two or three articles (Fig. 8.26) **3**

---

2. Antenna 1 peduncle with articles 1 and 2 produced antero-dorsally. Coxa 2 triangular. Pereopod 7 merus markedly expanded posteriorly and postero-distally ***Parametopa kervillei***

   Antenna 1 peduncle articles 1 and 2 not produced antero-dorsally. Coxa 2 quadrangular. Pereopod 7 merus expanded only postero-distally. Uropod 3 peduncle much shorter than rami ***Stenothoe monoculoides***

---

3. Mandible with tiny uni-articulate palp. Coxa 4 exceptionally large, extending posteriorly as far as coxa 7. Telson elongate oval, unarmed ***Stenula rubrovittata***

   Mandible with small but distinct two- or three-articled palp. Gnathopod 2 propodus with small palmar process or spine, palm minutely crenulate ***Metopa pusilla***

Metopa pusilla Sars Fig. 8.30
Length 3 mm, greenish-white with brown patches. Antennae more or less equal, about one-half body length; no accessory flagellum. Gnathopod 1 simple, slender, gnathopod 2 subchelate, moderately powerful in male, smaller in female; propodus broad distally with oblique, almost straight, slightly crenulate palm delimited by small tooth. Telson oval with two or three pairs of spines dorso-laterally.

Among algae and hydroids in the shallow sublittoral. Probably all coasts.

Parametopa kervillei Chevreux Fig. 8.30
Length 5 mm. Antennae short and subequal; antenna 1 peduncle articles 1 and 2 slightly produced antero-dorsally. Accessory flagellum absent. Coxa 2 large, triangular. Gnathopod 2 larger than 1; male gnathopod 1 simple, 2 subchelate; both gnathopods subchelate in female. Pereopods 6 and 7 with basis and merus expanded posteriorly and postero-distally. Telson oval, with two pairs of spines.

Shallow sublittoral, frequently in *Saccorhiza* holdfasts. South-west coasts.

Stenothoe monoculoides
(Montagu) Figs 8.27F, 8.30
Length 3 mm, whitish marked with red. Antennae short, subequal; no accessory flagellum. Gnathopod

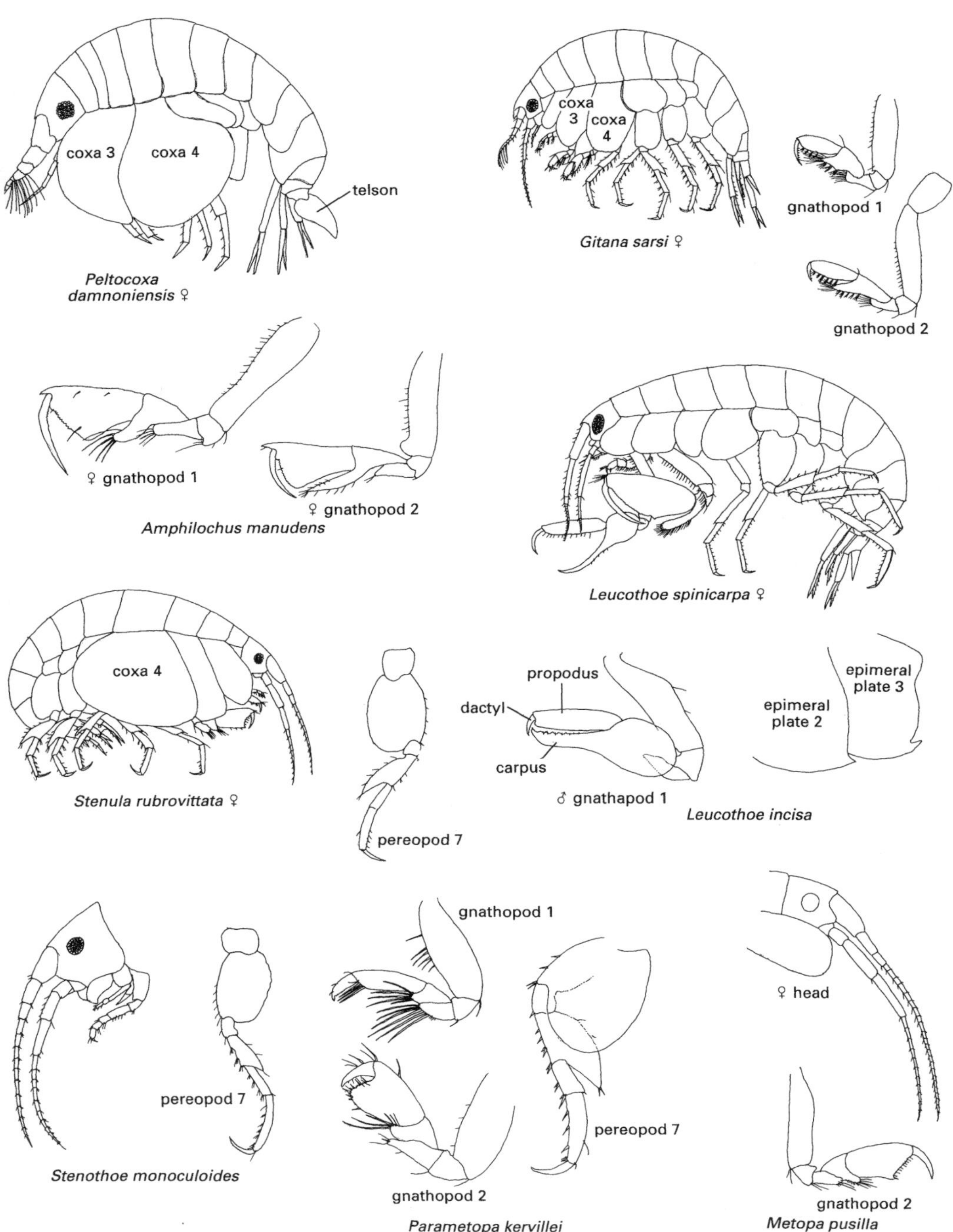
coxa 3
coxa 4
telson
Peltocoxa damnoniensis ♀
coxa 3
coxa 4
Gitana sarsi ♀
gnathopod 1
gnathopod 2
♀ gnathopod 1
♀ gnathopod 2
Amphilochus manudens
Leucothoe spinicarpa ♀
coxa 4
Stenula rubrovittata ♀
pereopod 7
propodus
dactyl
carpus
♂ gnathapod 1
epimeral plate 3
epimeral plate 2
Leucothoe incisa
gnathopod 1
pereopod 7
pereopod 7
Stenothoe monoculoides
gnathopod 2
Parametopa kervillei
♀ head
gnathopod 2
Metopa pusilla

Fig. 8.30

1 and 2 subchelate, 2 larger than 1; both with rectangular propodus, palm slightly oblique, convex, smooth, delimited by small group of spines. Pereopod 7 merus only slightly expanded. Uropod 3 peduncle only half length of rami. Telson oval and unarmed.

Intertidal and sublittoral, amongst algae and hydroids. Locally common. All coasts.

Stenula rubrovittata (Sars) Fig. 8.30

Length 5 mm, whitish banded with red. Antennae short and stout. Accessory flagellum absent. Coxa 4 exceptionally large. Gnathopod 1 weak, feebly subchelate. Gnathopod 2 much larger, subchelate. Pereopods 6 and 7 with basis expanded. Telson oval, unarmed.

Typically associated with hermit crabs (*Pagurus bernhardus* (Linn.)), also amongst sublittoral algae and hydroids. All coasts.

7. TALITRIDAE

Body laterally compressed, smooth dorsally. Antenna 1 shorter than antenna 2 peduncle. Accessory flagellum absent. Mandible without a palp. Coxae medium sized. Gnathopods variable; gnathopod 2 often large in male. Urosome segments not coalesced. Uropods short, relatively powerful; uropod 3 very short, uniramous, ramus shorter than, or subequal to, peduncle. Telson short, entire. Nine British species, in four genera, but only the following occur commonly.

---

1. Antenna 2 with some flagellum articles produced into small teeth at the joints **3**

   Antenna 2 with flagellum articles smoothly jointed **2**

---

2. Ramus of pleopods about half length of peduncle. Male gnathopod 2 propodus with palm convexly curved, not longer than ventral margin, delimited by spine. Telson with very small posterior notch ***Orchestia gammarellus***

   Ramus of pleopods as long as, or longer than, peduncle. Male gnathopod 2 propodus with sinuous palm margin, oblique to and longer than ventral margin. Telson without posterior notch ***Orchestia mediterranea***

---

3. Uropod 3 rami each with terminal spine, as long as ramus. Male with gnathopod 1 simple, gnathopod 2 small, mitten-shaped ***Talitrus saltator***

   Uropod 3 rami each termlnated by a few short spines. Male gnathopod large, subchelate, propodus produced antero-ventrally into a large hook. Female gnathopod 2 merus without a tooth. Uropod 3 rami shorter than peduncle. Telson longer than wide, with slight notch posteriorly ***Talorchestia deshayesii***

Genus *Orchestia*

Antenna 1 shorter than antenna 2 peduncle. Antenna 2 without teeth on flagellum articles, peduncle article 5 a little longer than 4, with 5–7 groups of spines dorsally. Coxa 1 broadly triangular. Gnathopod 1 small, subchelate; gnathopod 2 small, mitten-shaped in female, large and subchelate in male. Female gnathopod 2 basis with few or no spines on posterior border. Telson entire.

There are two common British marine species, but they show few morphological differences and females are especially difficult to distinguish.

Orchestia gammarellus (Pallas) Figs 8.28A, 8.31

Length 18 mm; colour variable, red-brown to brown-green, marked with red stripes. Eyes round. Antenna 1 almost reaching distal end of peduncle article 4 of antenna 2. Pereopod 7 merus and carpus expanded in older adult females. Uropod 3 rami with two dorsal spines.

At and above strandline on open coasts and in estuaries, on shingle and rocky shores, in crevices, beneath algae and strand line debris; also common in salt marshes. All coasts.

Orchestia mediterranea da Costa Fig. 8.31

Length 17 mm; colour variable but usually brownish-green. Rami of pleopods as long as or longer than peduncles, total length of each pleopod considerably greater than in *O. gammarellus*.

At and below the strandline, mainly on shingly and rocky shores, often in crevices. Often with *O. gammarellus*, but less common in estuaries and

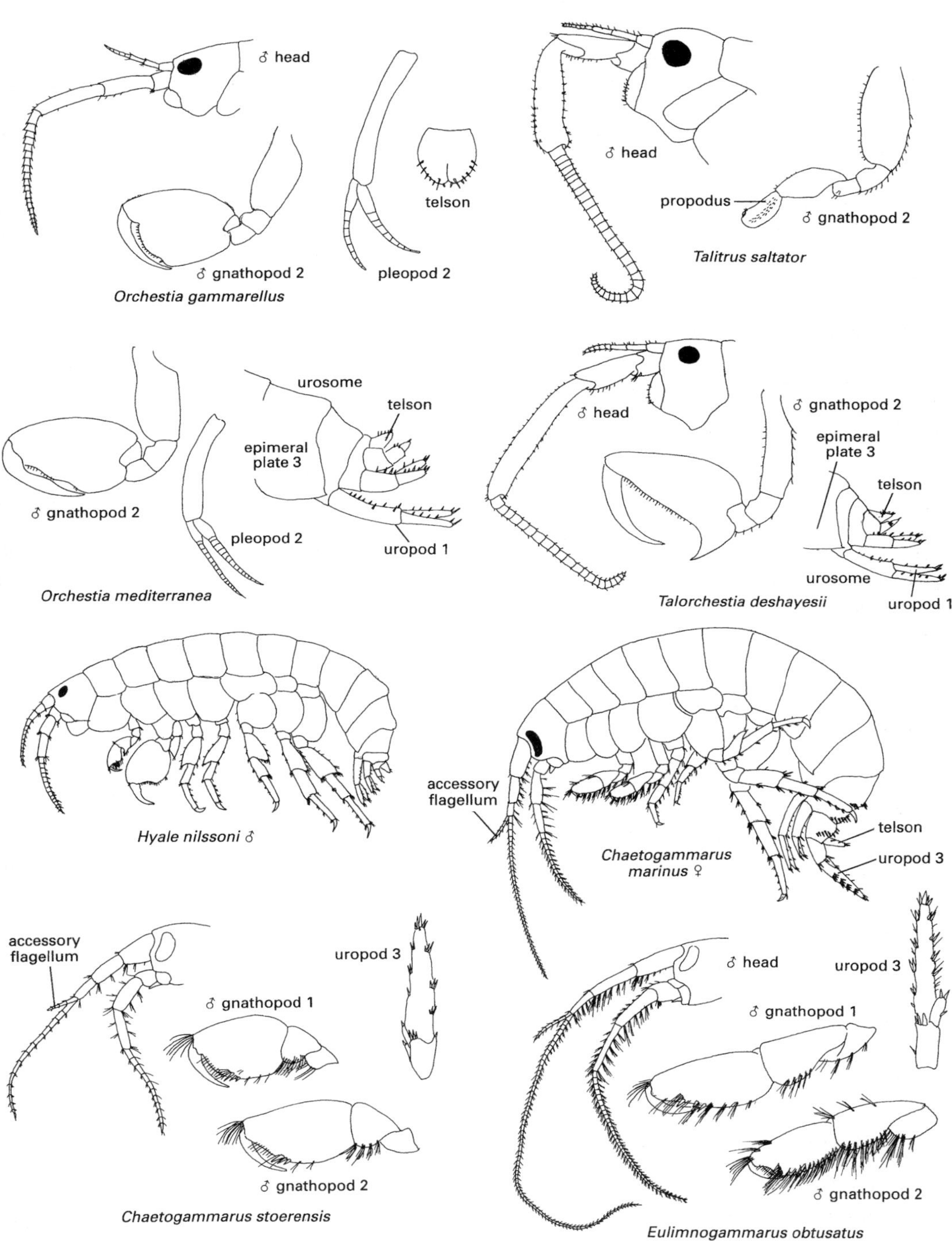

♂ head
♂ gnathopod 2
pleopod 2
telson
Orchestia gammarellus
♂ head
propodus
♂ gnathopod 2
Talitrus saltator
urosome
telson
epimeral plate 3
♂ gnathopod 2
pleopod 2
uropod 1
Orchestia mediterranea
♂ head
♂ gnathopod 2
epimeral plate 3
telson
urosome
uropod 1
Talorchestia deshayesii
Hyale nilssoni ♂
accessory flagellum
telson
uropod 3
Chaetogammarus marinus ♀
accessory flagellum
uropod 3
♂ gnathopod 1
♂ gnathopod 2
Chaetogammarus stoerensis
♂ head
uropod 3
♂ gnathopod 1
♂ gnathopod 2
Eulimnogammarus obtusatus

Fig. 8.31

salt-marshes. Probably on all coasts except perhaps the far north.

Talitrus saltator (Montagu) Fig. 8.31
Length 20 mm, brown-grey. Antenna 1 just reaching distal end of peduncle article 4 of antenna 2. Flagellum of antenna 2 with teeth on some articles. Coxa 1 broadly rectangular. Uropod 3 with several stout spines on peduncle; rami with four dorsal spines, and one terminal spine equal to length of ramus.

On sandy shores, at strandline, beneath algae or other debris. All coasts.

Talorchestia deshayesii (Audouin) Fig. 8.31
Length 10 mm, pale brown with darker stripes, eyes black. Antenna 1 shorter than antenna 2 peduncle. Flagella articles of antenna 2 produced as teeth, peduncle article 5 longer than 4 and very spiny. Epimeral plate 3 with a slight point at postero-ventral corner. Uropod 3 rami shorter than peduncle. Telson longer than wide, slightly notched posteriorly.

Beneath strandline algae and other debris on sandy and shingly shores, often extending above HWS. All coasts.

8. HYALIDAE

Body stout. Head without rostrum. Antenna 1 shorter than 2, but longer than antenna 2 peduncle. Gnathopods 1 and 2 subchelate; male with gnathopod 2 enlarged. Uropods short, stout; uropods 1 and 2 biramous, uropod 3 uniramous. Telson short, cleft into two lobes, unarmed or with few small spines. Six species recorded for Britain.

Hyale nilssoni (Rathke) Figs 8.28B, 8.31
Length 7 mm, brown to vivid green. Gnathopod 2 male propodus broad, rather rectangular, palm only a little longer than ventral margin. Gnathopod 2 female propodus with approximately parallel dorsal and ventral margins, carpus with a fat lobe which is as wide as long. Pereopods 5–7 basis with only slight crenulations on posterior border.

Among intertidal algae on rocky and muddy shores from the *Pelvetia* level downwards. All coasts, including lower reaches of estuaries. Often common.

9. GAMMARIDAE

Body compressed, smooth dorsally except for the urosome which may bear small, often conspicuous, spines and/or setae. Head with small rostrum, large post-antennal sinus; eyes usually large, reniform. Antenna 1 and 2 well developed; accessory flagellum elongate, multi-articulate. Mandible with elongate, three-articled palp. Coxae moderately large, overlapping; coxa 4 concave posteriorly; coxae 5–7 with anterior lobe longer than posterior. Gnathopods 1 and 2 approximately equal-sized, subchelate, relatively well developed. Pereopods with broad basis. Uropods 1–3 biramous; uropod 3 with bi-articulate outer ramus, equal to, rarely much larger than, inner ramus. Telson cleft. Sexual dimorphism marked, reflected in antennae, gnathopods, pereopods, and uropods. Fifteen species recorded for British Isles.

---

1. Uropod 3 inner ramus at least one-third length of outer ramus **2**

   Uropod 3 inner ramus small, less than one-third length of outer ramus **7**

---

2. Eyes small, more or less round. Antenna 2 male with a few short setae on peduncule, each less than twice width of peduncle segment 4. Freshwater species, sometimes washed into brackish areas
[***Gammarus pulex*** (**L.**) and ***G. lacustris*** **Sars**]

   Eyes elongate, oval, or reniform, twice as long as wide. Antenna 2 male with many long setae on peduncle, often more than twice width of peduncle segment 4. Brackish water and marine species **3**

---

3. Pereopod 7 basis with postero-distal corner distinct from ischium, bearing only setae **4**

   Pereopod 7 basis with postero-distal corner indistinct, bearing one or two spines as well as setae **5**

---

4. Uropod 3 inner ramus only one-third length of outer ramus. Epimeral plates 2 and 3 bearing few small setae, not set in notches. Mainly marine
Gammarus finmarchicus

Uropod 3 inner ramus more than half length of outer ramus. Epimeral plates 2 and 3 with few, relatively long setae posteriorly, set in small notches. Freshwater and brackish
Gammarus duebeni

---

5. Mandibular palp article 3 with irregularly alternating short and long setae on ventral margin **6**

Mandibular palp article 3 with regular, comb-like row of short to medium setae on ventral margin. Pereopod 7 basis about 1.5 times as long as wide. Urosome with prominent, angular dorsal humps. Epimeral plate 3 with several small setae on posterior margin
Gammarus locusta

---

6. Male pereopods 6 and 7 with posterior marginal setae and spines on merus and carpus, about equal length. Adult male antenna 1 with five or six groups of setae on ventral surface of peduncle article 2, and two or three groups ventrally on article 3 (excluding terminal groups). Marine or brackish water
Gammarus salinus

Male pereopods 6 and 7 with posterior marginal setae, on merus and carpus, much longer than spines. Adult male antenna 1 with 6–10 groups of setae on ventral surface of peduncle article 2, and 3–6 groups ventrally on article 3 (excluding terminal groups)
Gammarus zaddachi

---

7. Gnathopod 1 propodus larger than gnathopod 2 propodus. Gnathopod 2 carpus elongate
Eulimnogammarus obtusatus

Gnathopod 1 propodus not larger than gnathopod 2 propodus. Gnathopod 2 carpus short, more or less triangular **8**

---

8. Antenna 2 and uropod 3 outer ramus with dense setae. Urosome with dorsal spines in transverse rows
Chaetogammarus marinus

Antenna 2 with few setae, uropod 3 outer ramus lacking large setae
Chaetogammarus stoerensis

Genus *Chaetogammarus*

Antenna 1 longer than 2. Eyes elongate, usually reniform. Gnathopods subchelate; gnathopod 2 larger than 1. Coxae, pereopods 5–7, and dorsal surface of urosome more or less without setae. Pereopods spiny. Urosome with groups of spines dorsally. Uropod 3 outer ramus long, spiny, often with many setae; inner ramus small, scale-like.

Chaetogammarus marinus Leach Fig. 8.31

Length up to 25 mm, dark green, sometimes marked with red or yellow. Antenna 2 densely setose, especially flagellum. Pereopods 3 and 4 with few setae in male, rather more in female. Pereopod 7 basis with postero-distal corner not produced, merus about three times as long as wide. Uropod 3 inner ramus small, with spines and setae; male outer ramus with spines and many setae on both margins, but lacking plumose setae; female outer ramus with fewer setae, some of which may be plumose.

Intertidal, on sheltered shores beneath stones and amongst algae; sometimes in estuaries. All coasts.

Chaetogammarus stoerensis (Reid) Fig. 8.31

Length 8 mm, pale green to greenish-blue. Antennae 1 and 2 short, with few setae. Pereopod 7 basis with postero-distal corner slightly produced; propodus distinctly longer than carpus. Uropod 3 inner ramus tiny, outer ramus broad, long, spiny, but lacking setae.

Intertidal, between HWN and LWN, on clean gravel in areas of freshwater runoff. All coasts.

Eulimnogammarus obtusatus (Dahl) Fig. 8.31

Length 20 mm; brown to olive green, sometimes tinged purple. Antenna 1 longer than antenna 2. Urosome with small humps and groups of short spines dorsally. Uropod 3 inner ramus less than one-fifth length of outer.

Intertidal, lower shore, on clean gravel and shingle, and amongst algae. All coasts.

Genus *Gammarus*

Head with lateral lobes truncated or rounded; eyes small and round, or large and reniform. Gnathopod 1 about equal size to, or smaller than, gnathopod 2. Epimeral plates 2 and 3 round, quadrate, or slightly acute. Urosome with groups of spines and setae dorsally. Uropod 3 rami foliaceous, setose and/or spiny; inner ramus more than half length of biarticulate outer ramus. Telson cleft, with lateral and apical spines, with or without setae.

A large, mainly freshwater genus. Ten marine and two freshwater species occur in Britain but identification is difficult as many diagnostic features are seen only in adult males.

Gammarus duebeni Liljeborg — Fig. 8.32

Up to 16 mm, yellowish-green. Eyes reniform. Antenna 1 slightly longer than 2. Epimeral plate 3 with seta (not spines) set in notches along nearly straight posterior margin. Urosome with only small dorsal humps. Uropod 3 inner ramus more than half length of outer.

In brackish water, and in small streams crossing rocky shores; freshwater in Ireland, Isle of Man, Western Isles. All coasts.

Gammarus finmarchicus Dahl — Fig. 8.32

Adult male up to 20 mm, brown to yellow. Head with rather rounded lateral lobes, eyes narrowly reniform. Antenna 1 much longer than 2, with relatively few groups of peduncle setae; accessory flagellum about as long as peduncle article 2. Epimeral plate 3 with postero-ventral margin nearly straight, bearing only a few small setae. Uropod 3 with inner ramus about one-third to half length of outer.

Intertidal, in pools. All coasts.

Gammarus locusta (Linnaeus) — Fig. 8.32

Length 33 mm. Head with lateral lobes sloping forward, upper angles produced; eyes moderately large, reniform. Antenna 1 peduncle article 1 with, or without, one ventral group of setae; article 2 with one or two groups, article 3 with or without one ventral group of setae (excluding apical tufts). Antenna 2 with numerous groups of peduncular setae. Epimeral plates 2 and 3 acutely produced postero-ventrally, posterior margin concave with numerous short setae. Urosome with prominent, angular dorsal humps. Uropod 3 rami of nearly equal length, with numerous groups of spines and setae.

Marine. Middle shore to sublittoral, often abundant under stones and amongst algae. All coasts.

Gammarus salinus Spooner — Fig. 8.32

Length 22 mm, brown or greenish-brown with light banding. Head with lateral lobes broadly truncated; eyes large, reniform. Antenna 1 peduncle with numerous groups of setae vent rally. Urosome with only small dorsal humps. Uropod 3 inner ramus just shorter than article 1 of outer ramus.

Brackish waters, usually in more saline conditions than *G. zaddachi*. All coasts.

Gammarus zaddachi Sexton — Figs 8.28C, 8.32

Length 22 mm, light grey, green, or yellow with darker bands. Head with truncated lateral lobes; eyes moderately large, reniform. Antenna 1 peduncle, in male, with at least 15 groups of setae ventrally (excluding apical tufts). Antenna 2 peduncle, in male, with numerous groups of long setae. Epimeral plates 2 and 3 with acute postero-ventral corners, posterior margins with several short setae. Urosome with small dorsal humps. Uropod 3 inner ramus about three-quarters length of article 1 of outer ramus.

Marine and brackish waters, lower shore and shallow sublittoral. All coasts.

10. MELITIDAE

Body elongate, compressed; pleon toothed or smooth dorsally, pereon and urosome usually smooth. Head without rostrum; eyes typically small and round. Antennae well developed; antenna 1 usually much longer than 2; accessory flagellum present, often conspicuous. Mandible with large molar process, and long, three-articled palp. Coxae contiguous, often short. Gnathopods subchelate; gnathopod 2 much larger than 1, especially in male. Pereopods variable. Epimeral plate 3 frequently toothed posteriorly. Uropods biramous; uropod 3 ramus variable. Telson cleft, often spiny. Sexual dimorphism apparent in antennae, gnathopod 2, and pereopods. A diverse family, with 20 British species in 11 genera. Most are sublittoral in distribution and few are common.

---

1. Uropod 3 inner ramus short, less than half length of outer — **2**

   Uropod 3 with rami more or less equal in length, less than twice length of

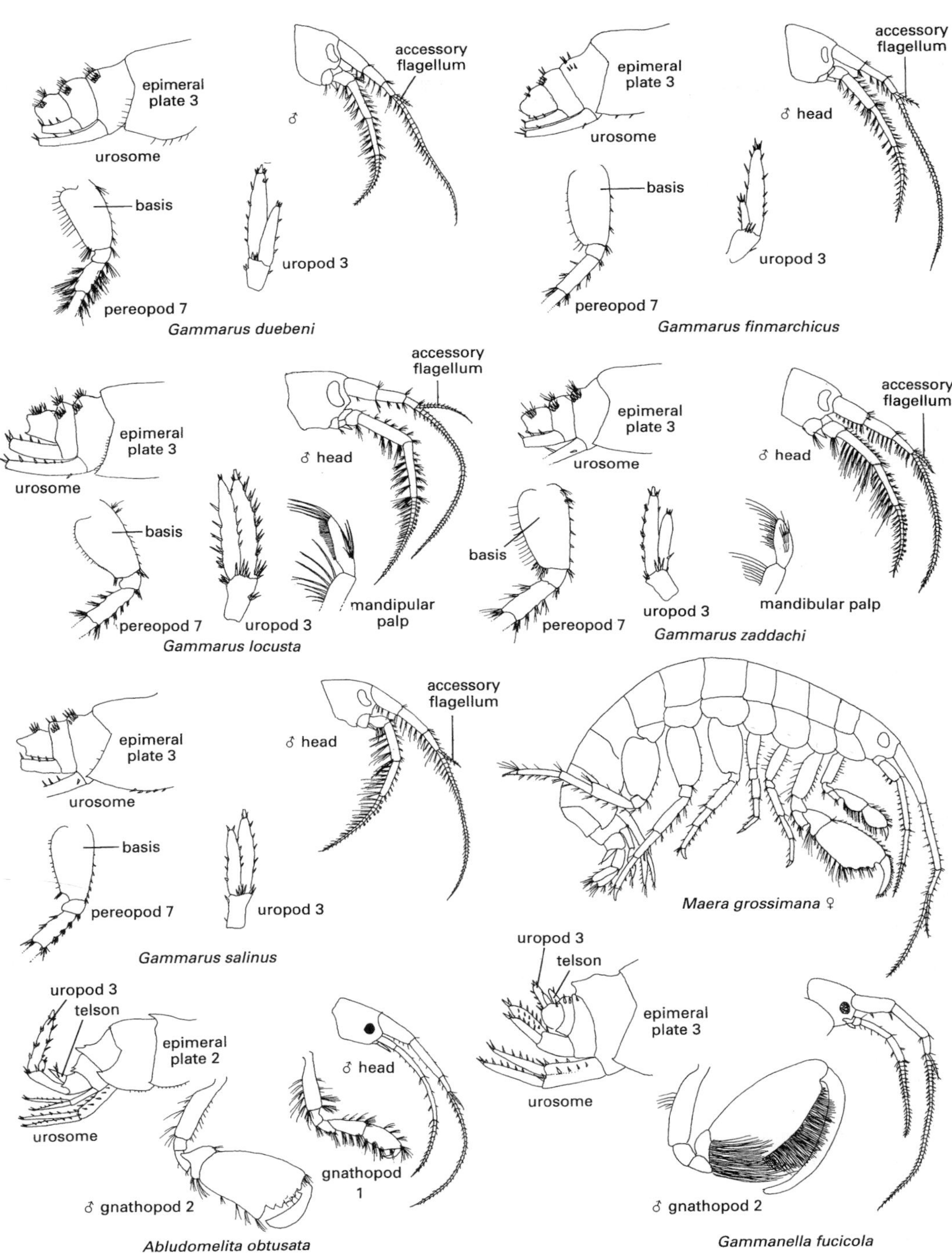

epimeral plate 3
accessory flagellum
♂
urosome
basis
uropod 3
pereopod 7
Gammarus duebeni
epimeral plate 3
accessory flagellum
♂ head
urosome
basis
uropod 3
pereopod 7
Gammarus finmarchicus
accessory flagellum
epimeral plate 3
♂ head
urosome
basis
mandipular palp
pereopod 7
uropod 3
Gammarus locusta
epimeral plate 3
accessory flagellum
urosome
♂ head
basis
uropod 3
mandibular palp
pereopod 7
Gammarus zaddachi
accessory flagellum
epimeral plate 3
♂ head
urosome
basis
pereopod 7
uropod 3
Gammarus salinus
Maera grossimana ♀
uropod 3
telson
epimeral plate 2
♂ head
urosome
gnathopod 1
♂ gnathopod 2
Abludomelita obtusata
uropod 3
telson
epimeral plate 3
urosome
♂ gnathopod 2
Gammanella fucicola

Fig. 8.32

peduncle. Epimeral plate 3 with smooth margins, except for a distinct tooth at postero-ventral corner. Gnathopod 2 propodus approximately quadrate **_Maera grossimana_**

---

2. Uropod 3 small, not projecting much beyond uropod 1; outer ramus about equal length to peduncle **_Gammarella fucicola_**

   Uropod 3 large, projecting well beyond uropod 1; outer ramus much longer than peduncle **3**

---

3. Pleon without teeth dorsally **4**

   Pleon with dorsal teeth on one or more segments. Epimeral plate 3 with posterior and postero-ventral borders smooth, or only slightly toothed **_Abludomelita obtusata_**

---

4. Head with both lateral lobes and post-antennal angles produced, with a cleft between. Telson with apical spines **_Melita palmata_**

   Post-antennal lobes of head not produced, cleft absent. Telson without apical spines **_Melita hergensis_**

Abludomelita obtusata (Montagu) Fig. 8.32
Length 9 mm, brownish. Pleon and urosome segments variably toothed dorsally; pleon segment 2 usually with one large median and two smaller lateral teeth, segments 4 and 5 with small median teeth and two larger lateral teeth. Epimeral plate 3 with postero-ventral corner acutely produced but weakly toothed. Uropod 3 outer ramus spiny, less than twice length of peduncle; inner ramus minute.

Sublittoral. All coasts.

Gammarella fucicola (Leach) Fig. 8.32
Length 10 mm, yellow to brown. Body robust; pereon and pleon smooth dorsally, urosome with a keel. Head with truncated lateral lobes. Antenna 1 much longer than 2; accessory flagellum with four or five articles. Gnathopod 2 male very large. Epimeral plate 3 posterior margin more or less straight, with small tooth at postero-ventral corner. Urosome segments 2 and 3 with paired dorsal spines.

Intertidal and sublittoral, amongst algae. All coasts.

Maera grossimana (Montagu) Fig. 8.32
Length 10 mm; yellowish, tinged with red. Eyes moderately large, oval. Antenna 1 peduncle article 1 longer than head; accessory flagellum with up to eight articles. Coxae short, with smooth margins. Gnathopod 2 propodus expanded, more or less quadrate, with a distinct palm. Uropod 1 peduncle lacking median and proximal submarginal spine. Uropod 3 rami broad, truncated, with numerous long spines.

Intertidal and sublittoral, under stones. South and south-west coasts.

Genus *Melita*
Body compressed and slender. Pleon and urosome often with dorsal teeth and spines. Antenna 1 longer than 2; accessory flagellum variable, with two or more articles. Gnathopods subchelate; gnathopod 2 male larger than 1. Uropod 3 extending well beyond uropod 1; outer ramus uniarticulate, or with spine-like second article, inner ramus very small. Telson deeply cleft.

Melita hergensis Reid Fig. 8.33
Length 10 mm; blue-black, often with pale bands on limbs. Antenna 2 peduncle much longer than antenna 1 peduncle; accessory flagellum with 2–3 articles. Gnathopod 2 male similar to *M. palmata*, but propodus less broad. Female coxa 6 with elongate anterior lobe. Epimeral plate 3 with postero-ventral corner acutely produced, postero-ventral margin weakly toothed. Telson cleft, with dorsolateral spines; apical spines absent.

Intertidal, lower shore; on clear, fairly exposed, fully marine rocky shores. Southern coasts.

Melita palmata (Montagu) Figs 8.28D, 8.33
Length 16 mm; pale yellow to blue-grey, limbs banded. Antenna 2 peduncle just longer than antenna 1 peduncle; accessory flagellum with 2–4 articles. Male gnathopod 2 with broad propodus,

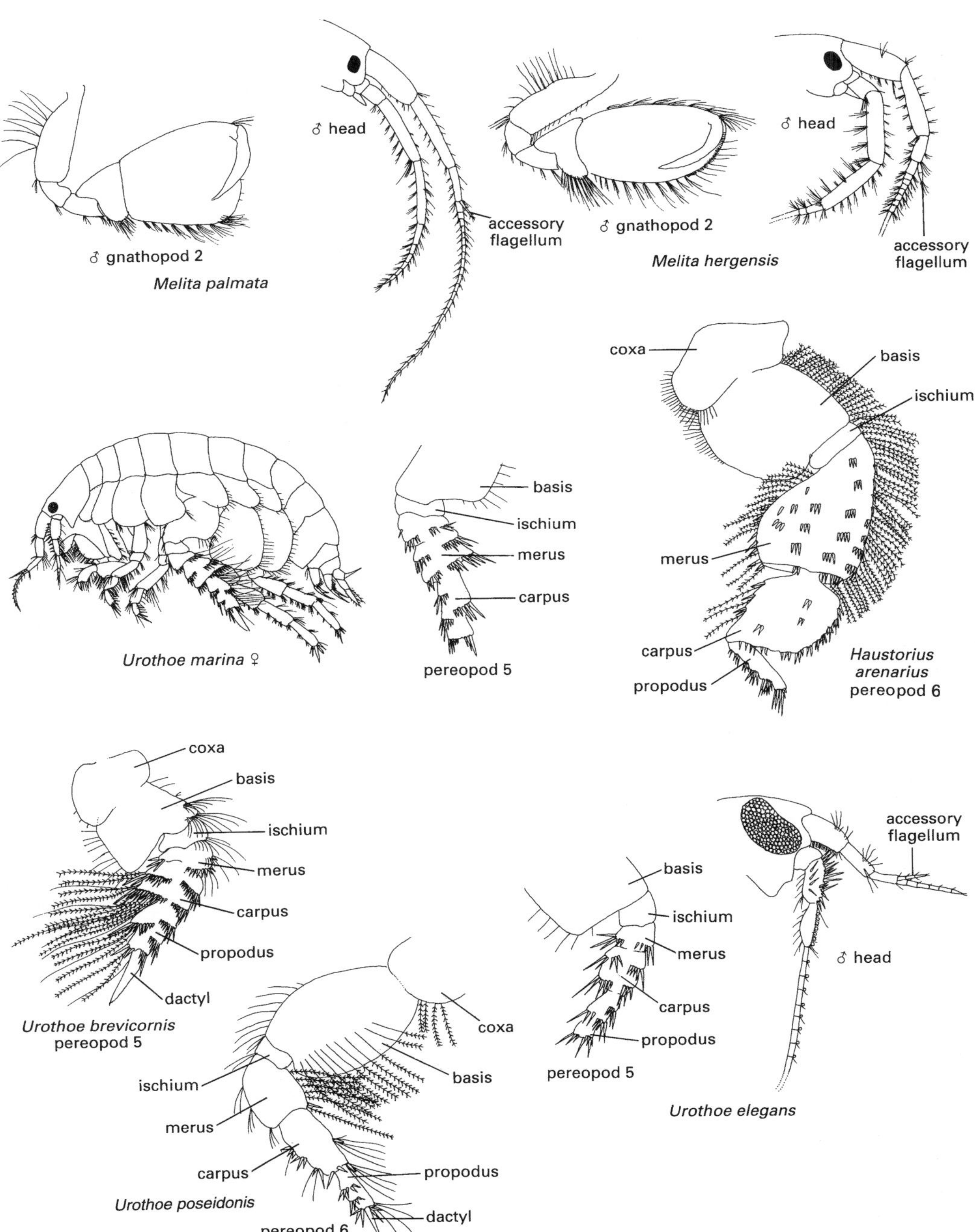
♂ gnathopod 2
Melita palmata
♂ head
accessory flagellum
♂ gnathopod 2
Melita hergensis
♂ head
accessory flagellum
coxa
basis
ischium
merus
carpus
propodus
Haustorius arenarius pereopod 6
Urothoe marina ♀
basis
ischium
merus
carpus
pereopod 5
coxa
basis
ischium
merus
carpus
propodus
dactyl
Urothoe brevicornis pereopod 5
coxa
basis
ischium
merus
carpus
propodus
dactyl
Urothoe poseidonis
pereopod 6
basis
ischium
merus
carpus
propodus
pereopod 5
accessory flagellum
♂ head
Urothoe elegans

Fig. 8.33

palm crossed by dactylus. Female coxa 6 with anterior lobe forming a hook. Epimeral plate 3 with postero-ventral corner produced as a small tooth, postero-ventral margin slightly toothed. Telson cleft, with apical and internal marginal spines.

Marine and brackish water; intertidal and shallow sublittoral, among silty, stony habitats. All coasts.

11. HAUSTORIIDAE

Body rounded, plump. Antenna 1 shorter than antenna 2. Accessory flagellum present. Head with a small rostrum. Some pereopods adapted for digging, with articles flattened, strongly spinose and setose. Gnathopods feeble. Telson wholly or partly split. Burrowers in sand. Six British species, in two genera. Some authorities place *Urothoe* in a separate family Urothoidae.

---

1. Pereopods 3–7 with only five articles, dactylus absent. Pereopod 6 propodus making forward facing acute angle with distal border of carpus. Eyes indistinct **Haustorius arenarius**

   Pereopods 3–7 with six articles, dactylus more or less lanceolate. Eyes distinct **2**

---

2. Uropod 1 rami strongly curved, smooth **Urothoe marina**

   Uropod 1 rami straight, or only slightly curved, often bearing spines or setae **3**

---

3. Pereopod 5 carpus more than twice as wide as long, much broader than merus **Urothoe poseidonis**

   Pereopod 5 carpus only slightly wider than long, if at all **4**

---

4. Antenna 1 accessory flagellum less than half length of primary flagellum **Urothoe elegans**

   Antenna 1 accessory flagellum more than half length of primary flagellum **Urothoe brevicornis**

Haustorius arenarius (Slabber) Figs 8.27G, 8.33

Length 13 mm. Body rounded, fat, with very small urosome. Eyes indistinct. Pereopods without dactylus but with other articles enlarged. Uropod 3 short, rami equal. Telson short, partially cleft.

In clean, medium to coarse sands. Intertidal, generally close to the highest point of emergence of water table. All coasts, including sandy estuaries. Common.

Genus *Urothoe*

Body rounded, urosome not especially reduced. Antenna 2 flagellum short in female, in male more than three-quarters body length. Eyes distinct. Pereopods with more or less lanceolate dactylus. Uropod 3 rami of equal length. Telson split.

Urothoe brevicornis Bate Fig. 8.33

Length 8 mm. Antenna 1 peduncle article 3 shorter than article 1; articles 1 and 2 subequal; flagellum with eight articles. Accessory flagellum with seven articles. Telson a little longer than wide, with two pairs of median, marginal setae, two pairs of terminal setae, and one pair of fine terminal spines.

On wet, clean, medium to fine sand from about MTL into the sublittoral. Probably stenohaline. Frequent but rarely numerous, more common in the south.

Urothoe elegans Bate Fig. 8.33

Length 4 mm. Antenna 1 peduncle article 3 shorter than article 1; articles 1 and 2 subequal; flagellum with six articles. Accessory flagellum with three articles. Telson with two pairs of median marginal setae, two pairs of terminal setae and one pair of thick terminal spines.

On muddy sand and fine sediments, sublittoral. All coasts.

Urothoe marina (Bate) Fig. 8.33

Length 8 mm. Antenna 1 flagellum with nine articles. Accessory flagellum with 5–7 articles. Antenna 2 peduncle robust, spiny; flagellum of female two-articled, as long as peduncle article 5; in male as long as body. Pereopods 5 carpus only slightly broader than merus, slightly broader than long. Uropod 1 peduncle short, rami smooth, strongly curved. Telson as wide as long, each lobe with one apical spine and three or four setae.

Mostly sublittoral, sometimes at extreme low water, in coarse sand or gravel. All coasts.

Urothoe poseidonis Reibisch Fig. 8.33
Length 6 mm. Antenna 1 flagellum small, shorter than peduncle article 3, with five articles. Antenna 2 flagellum of female small, two-articled; in male as long as body. Uropod 1 outer ramus slightly curved, with one or two groups of spines, inner ramus straight with up to three long setae. Telson broader than long, each lobe with two short apical setae.

Lower shore and shallow sublittoral, sometimes common on sheltered sandy beaches. Reported from the Wash, the English Channel, Channel Isles, and Bristol Channel.

12. PONTOPOREIIDAE

Body laterally compressed. Head without rostrum, eyes large and distinct. Antenna 1 shorter than 2, with grossly enlarged article 1; remaining peduncle articles and flagellum attached at disto-ventral tip of article 1, forming a right angle (geniculate); accessory flagellum small, with two articles. Gnathopods simple, setose. Pereopods 3, 4 short; pereopods 5–7 elongate, with basis broad and spiny. On British coasts this family is represented by eight species in the single genus *Bathyporeia*.

---

1. Dorsal surface of urosome segment 1 with spines directed posteriorly and bristles directed anteriorly **2**

   Dorsal surface of urosome segment 1 with anteriorly directed bristles only **6**

---

2. Adults small (<3.5 mm). Epimeral plate 3 with only a single group of spines just above ventral margin ***Bathyporeia nana***

   Adults larger. Epimeral plate 3 with more than one group of spines just above ventral margin **3**

---

3. Epimeral plate 3, in adult female and juvenile male, with a well-developed tooth at postero-ventral corner. Adult male with tooth reduced, may be indicated only by uneven border **4**

   Epimeral plate 3 evenly rounded at postero-ventral corner **5**

---

4. Epimeral plate 3 with well-developed tooth at postero-ventral corner, extending beyond vertical margin of posterior border (reduced in males). Antenna 1 peduncle article 1 with more or less rounded tip; coxae 2 and 3 with tooth at postero-ventral corner ***Bathyporeia guilliamsoniana***

   Epimeral plate 3 with small tooth almost at postero-ventral corner, not extending beyond vertical margin of posterior border (reduced in males). Antenna 1 peduncle article 1 with angular tip and more or less vertical anterior border; coxae 2 and 3 without tooth on postero-ventral corner ***Bathyporeia pelagica***

---

5. Antenna 1 peduncle article 1 with sharply angular tip; coxae 2 and 3 with well-developed tooth at postero-ventral corner ***Bathyporeia tenuipes***

   Antenna 1 peduncle article 1 with rounded tip; coxae 2 and 3 with small tooth on postero-ventral corner ***Bathyporeia elegans***

---

6. Antenna 1 peduncle article 1 with round, narrow tip. Epimeral plate 3 with not more than three groups of spines just above ventral margin ***Bathyporeia pilosa***

   Antenna 1 peduncle article 1 with semi-rounded, broad tip. Adult epimeral plate 3 with 4–6 groups of spines just above ventral margin ***Bathyporeia sarsi***

Genus *Bathyporeia*
Body laterally compressed. Eyes distinct. Antenna 1 geniculate between peduncle articles 1 and 2. Female with antenna 2 flagellum short; male with antenna 2 flagellum more than half body length.

Uropod 3 inner ramus very short, scale-like, outer ramus much longer. Telson completely split.

Bathyporeia elegans Watkin Fig. 8.34
Length 5 mm. Dorsal surface of urosome segment 1 with one pair of posteriorly directed spines and anteriorly directed bristles. Epimeral plate 3 evenly rounded at postero-ventral corner, with about four groups of spines just above ventral border. Segment 2 of accessory flagellum about one-third length of segment 1.

In wet, fine, and often muddy sand from MLWN into the sublittoral. All coasts, common or abundant.

*Bathyporeia guilliamsoniana (*Bate) Fig. 8.34
Length 8 mm. Dorsal surface of urosome segment 1 with a pair of posteriorly directed spines and anteriorly directed bristles. Epimeral plate 3 with a well-developed tooth at postero-ventral corner extending beyond vertical margin of posterior border in female and juvenile male, shorter in adult male. First segment of antenna 1 peduncle with more or less rounded tip.

In wet, fine, and medium sand at and below MLWN but most common in the shallow sublittoral. All coasts, often abundant.

Bathyporeia nana Toulmond Fig. 8.34
Length 3 mm. Dorsal surface of urosome segment 1 with one pair of posteriorly directed spines and anteriorly directed bristles. Epimeral plate 3 rounded at postero-ventral corner in adults (somewhat angular in juvenile male, with a distinct tooth in juvenile female), with a single group of two or three spines just above ventral border.

In wet, fine, and often muddy sand from MLWS into the sublittoral. Sometimes abundant. First found on the Atlantic coast of France, now recorded from the Irish Sea, Bristol Channel, English Channel and the North Sea.

Bathyporeia pelagica (Bate) Figs. 8.34
Length 6 mm. Dorsal surface of urosome segment 1 with a pair of posteriorly directed spines and anteriorly directed bristles. Epimeral plate 3 with a small tooth almost at postero-ventral corner not extending beyond vertical margin of posterior border in female and juvenile male; in the adult male the tooth is reduced to an uneven border around postero-ventral corner.

In wet, clean, fine to medium sand from above MTL into the shallow sublittoral. Stenohaline. All coasts, often abundant.

Bathyporeia pilosa
Lindström Figs 8.27H, 8.34
Length 6 mm. Dorsal surface of urosome segment 1 with anteriorly directed bristles only. Epimeral plate 3 smoothly rounded at postero-ventral corner, not more than three groups of spines just above ventral border.

In fine and medium sand from MHWN downwards but rarely sublittoral. Favours damp rather than wet sand and is normally commonest above the highest point of emergence of the beach water table. Euryhaline. All coasts, common to very abundant (>10 000/m^2).

Bathyporeia sarsi Watkin Fig. 8.34
Length 8 mm. Dorsal surface of urosome segment 1 with anteriorly directed bristles only. Epimeral plate 3 smoothly rounded at postero-ventral corner, with four or more groups of spines just above ventral border in adult; juvenile with three or fewer groups of spines.

In fine and medium sand from MHWN downwards, sometimes into the sublittoral. Usually at and below highest point of emergence of the beach water table, usually commonest below *B. pilosa* but above *B. pelagica*. Common to abundant from west Wales southwards and around into the North Sea. Not common in Irish Sea and apparently absent from north-west Scotland.

Bathyporeia tenuipes Meinert Fig. 8.34
Length 6 mm. Dorsal surface of urosome segment 1 with paired posteriorly directed spines and anteriorly directed bristles. Epimeral plate 3 rounded at postero-ventral corner, with numerous groups of spines just above ventral border.

In clean, fine sand, almost always sublittoral. From West Ireland southwards and around into the North Sea. Infrequent.

13. OEDICEROTIDAE

Body slightly or moderately cylindrical. Head with downcurved rostrum, usually conspicuous. Eyes distinct, partly or wholly coalesced. Accessory flagellum rudimentary or absent. Pereopods setose, 7 much longer than 6. Uropod 3 rami scarcely

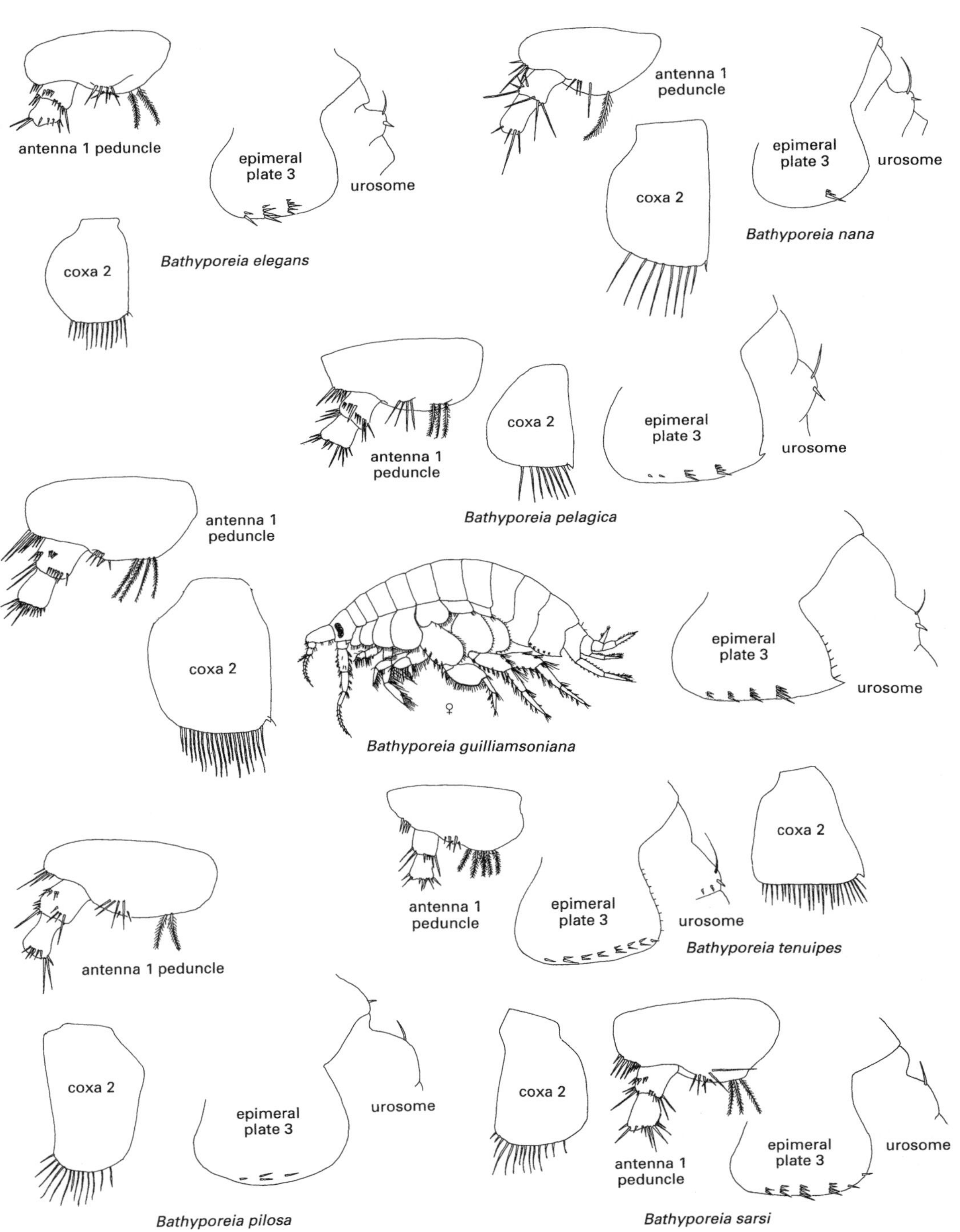
antenna 1 peduncle
epimeral plate 3
urosome
coxa 2
Bathyporeia elegans
antenna 1 peduncle
coxa 2
epimeral plate 3
urosome
Bathyporeia nana
antenna 1 peduncle
coxa 2
epimeral plate 3
urosome
Bathyporeia pelagica
antenna 1 peduncle
coxa 2
♀
epimeral plate 3
urosome
Bathyporeia guilliamsoniana
antenna 1 peduncle
epimeral plate 3
urosome
coxa 2
Bathyporeia tenuipes
antenna 1 peduncle
coxa 2
epimeral plate 3
urosome
Bathyporeia pilosa
coxa 2
antenna 1 peduncle
epimeral plate 3
urosome
Bathyporeia sarsi

Fig. 8.34

longer than peduncle. Telson short, entire or with shallow cleft or crenulations. Fifteen species recorded from Britain, in seven genera.

---

1. Gnathopod 1 subchelate, stout; propodus with smooth distal border. Gnathopod 2 chelate, slender, carpus produced distally as long, slender process beneath propodus **2**

   Gnathopods 1 and 2 subchelate, slender; similar, both with carpus produced distally as long, slender process underlying elongate propodus ***Perioculodes longimanus***

---

2. Gnathopod 1 propodus narrowly oval, only a slight angle between palm and hind margin. Pereopod 7 basis with distinctly convex posterior border ***Pontocrates altamarinus***

   Gnathopod 1 propodus flatter, somewhat quadrangular, with distinct angle between palm and hind margin. Pereopod 7 basis with straight or sinuous posterior border ***Pontocrates arenarius***

Perioculodes longimanus
(Bate and Westwood) Fig. 8.35
Length 5 mm, translucent white. Eyes completely coalesced, reddish. Rostrum short, downcurved. Antenna 2 as long as body in male, less than half body length in female. Telson entire.

In clean or muddy fine sand, from MLWS into sublittoral. Common. All coasts.

Genus *Pontocrates*

Eyes coalesced but with median line of separation sometimes visible, situated normally. Antenna 2 equal to body length in male, less than half body length in female. Gnathopod 1 subchelate, stout; distal border of propodus smooth, underlain by distal carpal process. Gnathopod 2 chelate, long and slender; with long distal carpal process underlying propodus. Telson entire or with very slight posterior cleft.

Pontocrates altamarinus
(Bate and Westwood) Figs 8.27I, 8.35
Length 7 mm; translucent white with orange/brown patches which usually persist even in preserved specimens. Antenna 1 female relatively long, equal to combined length of head and pereon segments 1–3; peduncle article 2 slender, as long as article 1; flagellum with eight articles, shorter than peduncle. Telson oval, distal border smooth or slightly cleft, with about five setae.

In fine and medium sand, lower shore and sublittoral. All coasts.

Pontocrates arenarius (Bate) Fig. 8.35
Length 6.5 mm, translucent white. Antenna 1 female relatively short, about equal to twice length of head; peduncle article 2 shorter and much thinner than article 1; flagellum with nine articles, equal to length of peduncle. Telson smoothly oval, with about two distal setae.

In sand from middle shore to sublittoral. All coasts.

14. PHOXOCEPHALIDAE

Body cylindrical, tapering at either end. Antennae short; accessory flagellum well developed, with several articles. Head elongate with large hood-like rostrum enveloping bases of first antennae. Eyes separate or absent. Mandible with a palp. Gnathopods similar, small to medium sized, subchelate. Pereopods spiny, adapted for burrowing. Pereopod 7 dissimilar to and shorter than 6; basis very expanded posteriorly. Uropod 3 biramous, external ramus bi-articulate, longer than internal. Telson short or medium sized, always deeply split. Ten British species, in four genera.

---

1. Pereopod 5 basis about as wide as ischium. Epimeral plate 3 with large sharp tooth at postero-ventral corner. Posterior border of pereopod 7 basis with eight or nine indistinct teeth. Coxae 1 and 2 without a tooth at postero-ventral corner ***Harpinia antennaria***

   Pereopod 5 basis more than twice width of ischium. Rostrum pointed, very long, extending to end of antenna 1 peduncle. Eyes indistinct ***Phoxocephalus holbolli***

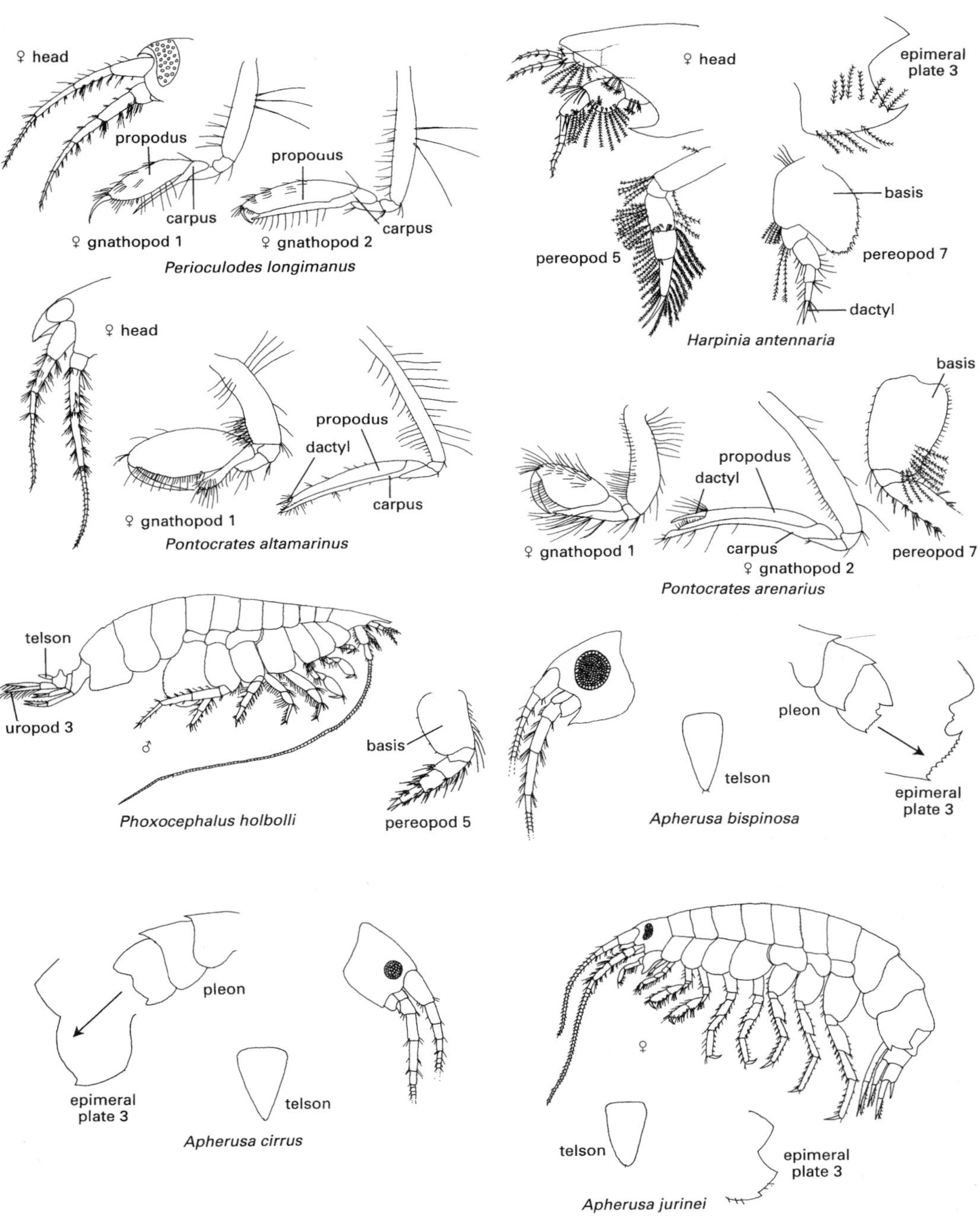

♀ head
propodus
carpus
♀ gnathopod 1
propodus
carpus
♀ gnathopod 2
Perioculodes longimanus
♀ head
epimeral plate 3
basis
pereopod 5
pereopod 7
dactyl
Harpinia antennaria
♀ head
propodus
dactyl
carpus
♀ gnathopod 1
Pontocrates altamarinus
basis
propodus
dactyl
♀ gnathopod 1
carpus
♀ gnathopod 2
pereopod 7
Pontocrates arenarius
telson
uropod 3
♂
Phoxocephalus holbolli
basis
pereopod 5
pleon
telson
epimeral plate 3
Apherusa bispinosa
pleon
epimeral plate 3
telson
Apherusa cirrus
♀
telson
epimeral plate 3
Apherusa jurinei

Fig. 8.35

Harpinia antennaria
Meinert Figs 8.27J, 8.35
Length 5 mm, translucent white. Eyes absent. Antennae very short in both sexes. Antenna 2 peduncle article 4 with ventral lobe. Pleon with many very short setae dorsally. Uropod 3 short, rami unequal in female, subequal in male. Telson short with rounded lobes.

On muddy and silty sand, and mixed ground; sublittoral, rarely at ELWS. All coasts.

Phoxocephalus holbolli (Krøyer) Fig. 8.35
Length 7 mm; pale brown to orange, with white flecks. Head with narrow, elongate rostral hood, extending to end of antenna 1 peduncle; eyes indistinct. Antenna 1 peduncle article 1 longer than articles 2 and 3 combined, flagellum of six articles, accessory flagellum elongate, with four articles. Antenna 2 slightly longer than 1 in female, with short, six-articled flagellum; in male with filiform flagellum as long as body. Uropod 3 female inner ramus three-quarters length of outer ramus article 1, sparsely setose; in male inner ramus about equal length to outer, with plumose setae.

Lower shore and sublittoral, in fine sand and mud. All coasts.

15. CALLIOPIIDAE

Body laterally compressed, often with a few teeth dorsally. Accessory flagellum present or absent. Antenna 1 slender, medium to short, usually shorter than antenna 2. Rostrum small. Mandible with a palp. Coxae small to medium sized, coxa 4 almost always posteriorly concave. Gnathopods feeble or occasionally moderately powerful, and more or less subchelate. Uropod 3 with elongate lanceolate rami, inner usually longer than outer. Telson entire, or slightly notched apically. Nine species recorded from Britain, in three genera.

---

1. Accessory flagellum well developed. Eyes large, occupying more than half side of head
Gammarellus angulosus

Accessory flagellum vestigial or absent **2**

---

2. Gnathopods relatively feeble; carpus almost as long as propodus. With or without dorsal teeth **3**

Gnathopods more robust, carpus distinctly shorter than propodus. Body smooth dorsally
Calliopius laeviusculus

---

3. Body smooth dorsally. Epimeral plate 3 with angular posterior border
Apherusa jurinei

Body with some dorsal teeth. Epimeral plate 3 with denticulate or convex posterior border **4**

---

4. Epimeral plate 3 smoothly convex posteriorly, with a tiny postero-ventral tooth ***Apherusa cirrus***

Epimeral plate 3 with complexly denticulate border. Pereopod 7 basiswith denticulate posterior border. Gnathopods slender; gnathopod 1 female with carpus longer than propodus ***Apherusa bispinosa***

Genus *Apherusa*

Antenna 1 usually shorter than 2, less than half body length. Accessory flagellum absent. Gnathopods with carpus almost always longer than or equal to propodus. Gnathopod 1 with carpus lacking a posterior lobe. Body often with teeth dorsally on pereon and/or pleon. Epimeral plate 3 often with denticulate posterior border. Telson pointed, notched, or rounded.

Apherusa bispinosa (Bate) Fig. 8.35
Length 6 mm. Colour variable, often deep red or violet with red eyes, or translucent white marked with black and/or green. Eyes large and oval. Coxa 1 angled antero-distally. Epimeral plate 3 with a sharp, sometimes bifid, proximal tooth followed by a gap and a number of distinct denticles ending in an acute postero-ventral tooth. Telson

triangular, twice as long as broad, with two distal setae.

In *Laminaria* hold fasts and among coralline algae, both littorally and sublittorally. All coasts.

Apherusa cirrus (Bate) Fig. 8.35

Length 6 mm. Reddish with small, round black eyes. Coxae 1–3 rounded anteriorly. Epimeral plate 3 with smoothly convex posterior border and a tiny tooth at postero-ventral corner. Telson an elongated triangle, rounded at tip, with two distal setae.

Amongst algae, especially *Halidrys* and *Laminaria*. All coasts.

Apherusa jurinei (Milne Edwards) Fig. 8.35

Length 6 mm. Colour variable, whitish or yellow with bands of brown, crimson or green, sometimes distinctly striped. Body smooth dorsally. Eyes red, large, and reniform. Coxae 1–3 quadrangular. Epimeral plate 3 posterior border angled in mid-region, with a small postero-ventral tooth. Telson rounded–triangular, with 2–4 setae.

Common in *Laminaria* holdfasts and among other littoral and sublittoral algae. All coasts.

Calliopius laeviusculus (Krøyer) Figs 8.28E, 8.36

Length 8 mm. Antennae more or less equal, short and thick; antenna 1 peduncle article 3 produced antero-ventrally. Epimeral plate 3 with smooth, convex, posterior border and a small, sharp postero-ventral tooth. Pereopod 7 basis with distinctly crenelate posterior border. Telson rounded, triangular.

Among algae, lower shore and shallow sublittoral. Common. All coasts.

Gammarellus angulosus (Rathke) Fig. 8.36

Length 15 mm; yellowish, mottled with red or brown. Antenna 1 and 2 of equal length, accessory flagellum well developed. Gnathopods 1 and 2 subchelate, similar. Dorsal keels not pronounced, scarcely produced posteriorly. Uropods 1 and 2 biramous, spiny; uropod 3 rami elongate, lanceolate, setose, uni-articulate. Telson entire, with terminal notch.

Among algae, lower shore and shallow sublittoral. All coasts.

16. ATYLIDAE

Body compressed, pereon segments short, pleon segments long. Some body segments with keel-like, posteriorly pointed dorsal teeth. Urosome segments 2 and 3 coalesced. Rostrum prominent. Eyes distinct and separate. Accessory flagellum vestigial or absent. Mandible with a distinct palp. Gnathopods subchelate, small. Uropod 3 biramous. Telson medium length, deeply split or cleft. In Britain represented by four species in the single genus *Atylus*.

---

1. Distinct dorsal teeth present only on urosome segments. Pereopod 3 dactylus shorter than propodus **_Atylus swammerdami_**

   Distinct dorsal teeth present on whole of pleon and last pereon segments **2**

---

2. Pereopod 5 basis with postero-ventral corner produced into a hook; pereopod 6 basis produced at distal angle **_Atylus vedlomensis_**

   Pereopod 5 basis with postero-ventral corner produced into a triangular point; pereopod 6 basis not produced at distal angle **_Atylus guttatus_**

Atylus guttatus (Costa) Fig. 8.36

Length 10 mm; reddish-white with brown patches dorsally and on coxal plates. Eyes large. Rostrum nearly straight. Pereon and pleon with low dorsal keel. Coxae serrated distally; coxae 3 and 4 broad. Telson with a deep median cleft, the two halves diverging distally.

Intertidal, and sublittoral to 75 m. A southern species, present on the Mediterranean and Atlantic coasts of western Europe, extending northwards to Irish Sea and southern North Sea.

Atylus swammerdami (Milne Edwards) Figs 8.27K. 8.36

Length 8 mm. Translucent white with brown patches. Pereon, and pleon segments 1 and 2, smooth dorsally; pleon segment 3 sometimes with a very small, dorsal, keel-like tooth. Urosome with two distinct teeth, the first small, the second large, separated by a deep notch. Pereopods 3 and 4 short, similar; 5, 6, and 7 with propodus as long as, or longer than, carpus.

Lower shore and sublittoral, generally on sand or amongst algae. Ubiquitous and often common.

Atylus vedlomensis
(Bate and Westwood) Fig. 8.36

Length 8 mm; translucent, tinged yellow, with patches and irregular streaks of orange on the back, sides, and limbs. A keel-like tooth present dorsally on last pereon segment and on each pleon segment. Arrangement of urosome teeth similar to *A. swammerdami*. Rostrum somewhat down-curved. Coxae slightly serrated distally. Telson with deep median split, the two halves not diverging distally.

On fine shelly sand and mud in shallow water. All British coasts, but commonest in the south.

17. DEXAMINIDAE

Body laterally compressed. Antennae usually medium or long, rarely very short. Accessory flagellum absent. One pair of eyes. Rostrum small. Mandible without a palp. Pleon usually with dorsal teeth. Urosome segments 2 and 3 coalesced. Gnathopods feeble, subchelate. Uropod 3 biramous. Telson deeply cleft, medium or long. Four species recorded for Britain, in three genera.

I. Pereopod 7 basis with posterior margin expanded ***Dexamine spinosa***

Pereopod 7 basis with parallel margins, not expanded ***Dexamine thea***

Genus *Dexamine*

Urosome segment 1 terminating in a sharp postero-dorsal tooth. Antenna 1 longer than 2 in female, shorter or subequal in male. Rostrum small. Coxae medium-sized, rather rounded. Gnathopod 1 smaller than 2. Pereopods 3–7 with merus shorter than combined length of carpus and propodus. Epimeral plate 3 with a long postero-ventral point. Telson much longer than wide.

Dexamine spinosa
(Montagu) Figs 8.28F, 8.36

Length 12 mm; whitish, marked with red, brown, violet, and orange. Eyes brown. Pleon segments 1–3 with single postero-dorsal tooth. Uropod 3 rami twice length of peduncle.

Littoral and sublittoral; among *Laminaria* holdfasts and other algae, in *Zostera*, and on sand or mud. All British coasts.

Dexamine thea Boeck Fig. 8.36

Length 5 mm; yellowish-green marked with white. Eyes red. Much smaller than *D. spinosa*, with pleon teeth less obvious. Uropod 3 rami only a little longer than peduncle.

Intertidal and sublittoral, among algae, especially *Laminaria* holdfasts. South and west coasts, from Portland to the Clyde Sea; also coasts of Ireland.

18. AMPITHOIDAE

Body smooth dorsally. Antenna 1 as long as or longer than 2. Accessory flagellum present or absent. Rostrum absent. Mandible with or without a palp. Coxae small to medium, coxa 4 not concave posteriorly. Gnathopods well developed, subchelate, 2 larger than 1, especially in male. Pereopod 5 frequently much shorter than 6 and 7. Uropod 3 biramous, with short quadrate rami always shorter than peduncle; outer ramus armed apically with 1–3 large, hooked spines. Telson short, entire, fleshy; often with small cusps. Six species recorded for Britain, in three genera.

I. Mandible with palp **2**

Mandible without palp. Pereopod 5 basis broadly expanded. Telson short, broad, truncated at apex ***Sunamphitoe pelagica***

---

2. Pereopods 5–7 propodus scarcely expanded distally. Telson with two small apical cusps. Gnathopods 1 and 2 without distal lobe on ischium. Gnathopod 2 basis with a small lobe antero-dorsally, propodus with only a slightly concave palm ***Ampithoe rubricata***

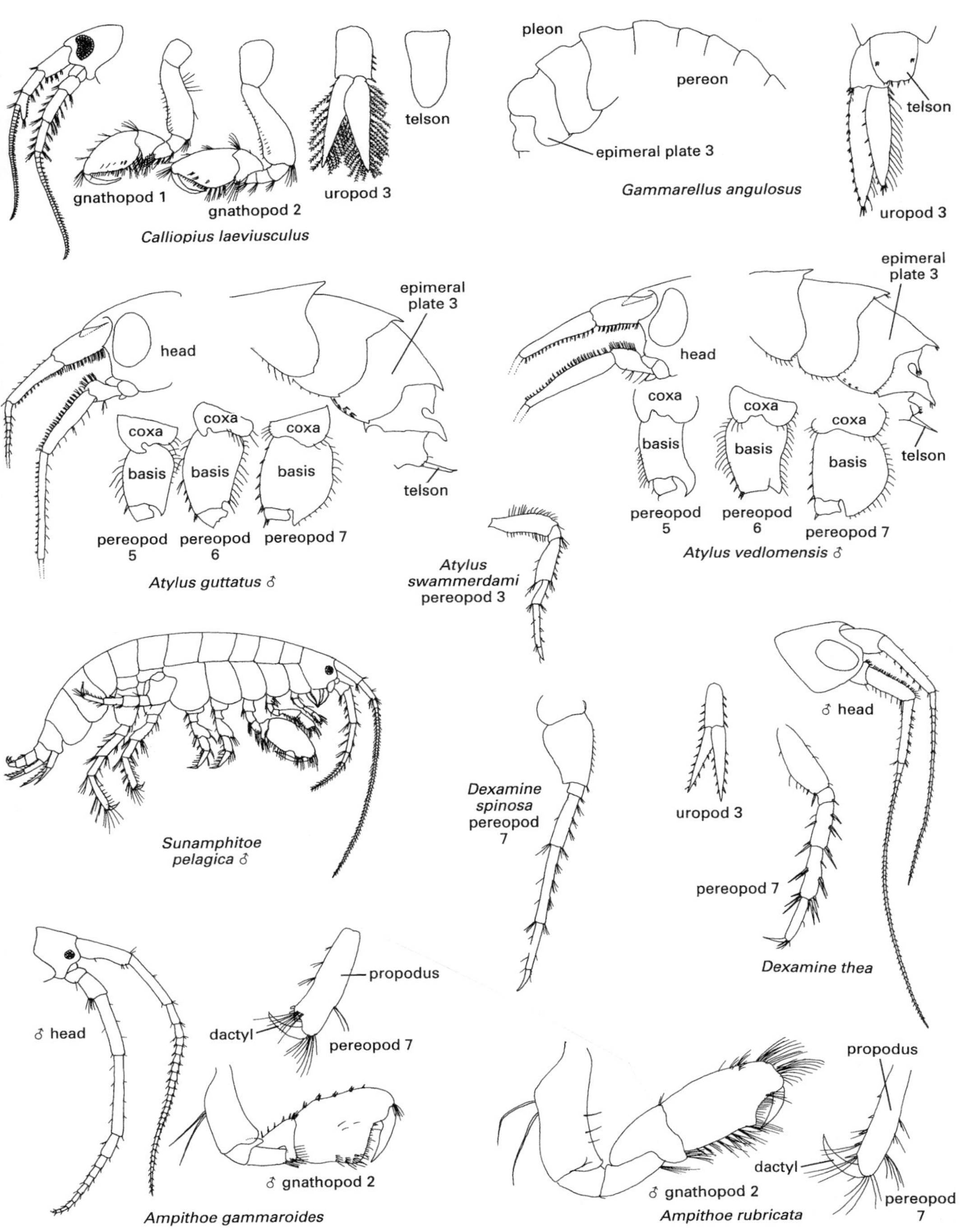
gnathopod 1
gnathopod 2
uropod 3
telson
Calliopius laeviusculus
pleon
pereon
epimeral plate 3
Gammarellus angulosus
telson
uropod 3
head
epimeral plate 3
coxa
basis
coxa
basis
coxa
basis
telson
pereopod 5
pereopod 6
pereopod 7
Atylus guttatus ♂
Atylus swammerdami pereopod 3
epimeral plate 3
head
coxa
basis
coxa
basis
coxa
basis
telson
pereopod 5
pereopod 6
pereopod 7
Atylus vedlomensis ♂
Sunamphitoe pelagica ♂
Dexamine spinosa pereopod 7
uropod 3
♂ head
pereopod 7
Dexamine thea
♂ head
propodus
dactyl
pereopod 7
♂ gnathopod 2
Ampithoe gammaroides
♂ gnathopod 2
Ampithoe rubricata
propodus
dactyl
pereopod 7

Fig. 8.36

Pereopods 5–7 propodus strongly expanded distally. Telson with two prominent curved spines and cusps. Gnathopod 2 basis and ischium with distinct antero-distal lobes **_Ampithoe gammaroides_**

Genus *Ampithoe*
Accessory flagellum absent. Mandible with a slender three-articled palp. Gnathopods subchelate, 2 usually larger than 1. Uropod 3 outer ramus armed with a pair of large, curved spines. Telson short, fleshy, with two apical cusps.

Ampithoe rubricata
(Montagu) Figs 8.28G, 8.36
Length 20 mm, red to green, with light or dark markings. Eyes small. Uropod 3 outer ramus with group of long setae dorsally. Telson convex apically.

Littoral and sublittoral, very common, making tubes amongst algae and under stones. All British coasts.

Ampithoe gammaroides Bate Fig. 8.36
Length 8 mm. Male gnathopod 2 propodus with straight palm. Antennae often exceeding half body length, with few setae. Telson with two large curved hooks and a number of short setae.

Littoral and sublittoral, amongst algae. Probably on all coasts.

Sunamphitoe pelagica
(Milne-Edwards) Fig. 8.36
Length 10 mm; greenish-yellow or yellow with red markings. Accessory flagellum absent. Gnathopod 2 male very large, much larger than 1; female with gnathopods subequal. Pereopods 5–7 propodus only slightly swollen distally; pereopod 5 basis considerably expanded; pereopods 6–7 basis only moderately expanded. Uropod 3 outer ramus with two apical hooks, inner ramus with one or two apical spines and several setae.

Intertidal and sublittoral, on algae. All coasts.

19. AORIDAE.

Body smooth dorsally. Antenna 1 usually longer than 2, usually with an accessory flagellum. Rostrum vestigial. Mandible with a palp. Coxae of variable size and shape, coxa 4 not concave posteriorly. Gnathopods large and subchelate, sometimes complexly so, 1 always larger than 2. Pereopod 7 conspicuously elongate. Uropod 3 biramous, short, rami as long as or longer than peduncle. Telson entire, fleshy, and short, nearly circular or square.

Adult males may be readily identified by the structure of their gnathopods. In a number of species, however, identification of females is exceptionally difficult. The British fauna includes 17 species in five genera.

---

1. Gnathopod 2 basis with fringe of very long, plumose setae. Gnathopod 1 dactylus extending slightly beyond palm margin of propodus. Pereopods 5–7 with basis at most only a little expanded posteriorly **_Leptocheirus pilosus_**

   Gnathopod 2 basis with very few, short setae **2**

---

2. Male gnathopod 1 with merus produced anteriorly as long, pointed process. Pereopod 7 with basis expanded, and concave posterodistally **_Aora gracilis_**

   Male gnathopod 1 with normal merus **3**

---

3. Male gnathopod 1 carpus with two or more stout processes on posterodistal margin **_Microdeutopus gryllotalpa_**

   Male gnathopod 1 carpus without postero-distal processes. Pereopods 3 and 4 with dactylus only three quarters as long as propodus **_Lembos websteri_**

Aora gracilis (Bate) Figs 8.28H, 8.37)
Length 6 mm; translucent yellow, or greenish-white marked with brown. Eyes black. Gnathopod 1 very long and slender. Gnathopod 1 female almost simple, merus not produced, but with propodus longer than carpus. Gnathopod 2 in both sexes subchelate, not especially setose. Epimeral plate 3 with posterior border convex, and a small tooth at postero-ventral corner. Uropod 3 rami subequal, longer than peduncle.

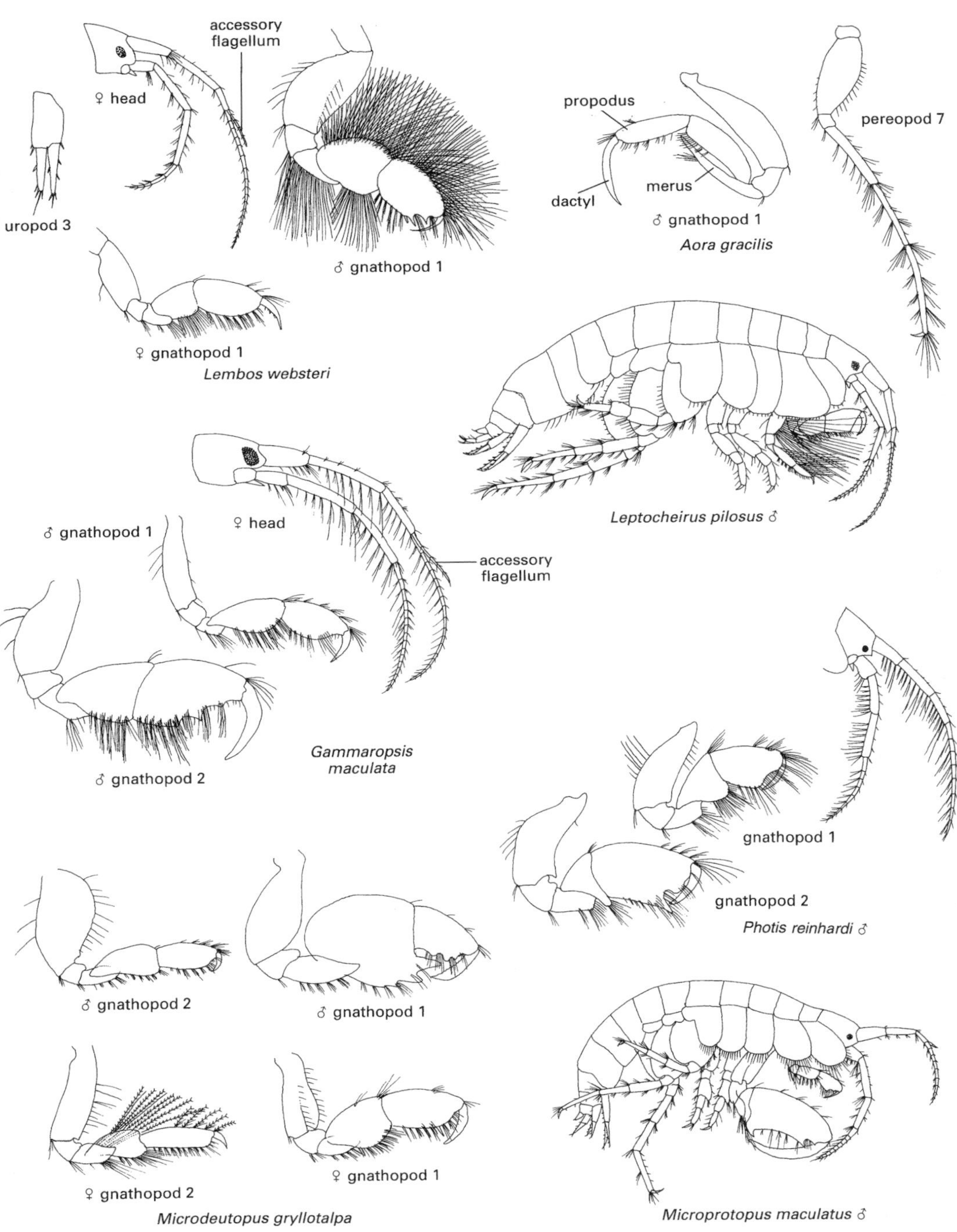
accessory flagellum
♀ head
uropod 3
♂ gnathopod 1
♀ gnathopod 1
Lembos websteri
propodus
dactyl
merus
♂ gnathopod 1
Aora gracilis
pereopod 7
Leptocheirus pilosus ♂
♂ gnathopod 1
♀ head
accessory flagellum
♂ gnathopod 2
Gammaropsis maculata
gnathopod 1
gnathopod 2
Photis reinhardi ♂
♂ gnathopod 2
♂ gnathopod 1
♀ gnathopod 2
♀ gnathopod 1
Microdeutopus gryllotalpa
Microprotopus maculatus ♂

Fig. 8.37

Intertidal and sublittoral, building tubes amongst algae and hydroids. All British coasts.

Lembos websteri Bate — Fig. 8.37

Length 4 mm, yellowish-brown with brown marks. Accessory flagellum with four articles. Gnathopod 1 male subchelate, with basis, carpus and propodus enlarged; palm of propodus with one sharp and one blunt tooth, and delimited by a spine. Gnathopod 1 female subchelate, with propodus larger than carpus, with an oblique palm delimited by a spine. Uropod 3 rami about 1.5 times length of peduncle.

Intertidal and sublittoral, frequently in *Laminaria* holdfasts, and amongst epibenthos. All British coasts.

Leptocheirus pilosus Zaddach — Fig. 8.37

Length 4 mm, brownish-red. Eyes black. Antenna 1 longer than 2. Accessory flagellum very small, uni-articulate. Coxa 1 and 2 large, of similar size. Gnathopod 1 subchelate, 2 simple. Gnathopod 1 female with propodus a little longer and broader than carpus, palm margin convex and just shorter than dactylus. Gnathopod 1 male with basis, carpus, and propodus large. Uropods 1 and 2 with long dorsal tooth and peduncle. Uropod 3 very small, with rami shorter than peduncle.

In brackish water, amongst *Ruppia maritima* or *Cordylophora lacustris*. Reported from the Wash and Norfolk, Plymouth, Channel Isles, Anglesey, and the Isle of Man; also from Fastnet, Cork, and the northern Hebrides

Microdeutopus gryllotalpa Costa — Fig. 8.37

Length 10 mm. Antenna 1 much longer than 2. Accessory flagellum with three articles. Gnathopod 1 male powerful, carpus nearly as broad as long, with two or four teeth postero-distally, the distal longest; propodus short with sinuous posterior margin. Gnathopod 1 female with carpus and propodus subequal; carpus broadest. Gnathopod 2 male with basis expanded, crenulate and convex anteriorly; carpus and propodus subequal, carpus broadest. Gnathopod 2 female with basis a little expanded and crenulate anteriorly; carpus elongate, subequal to propodus; merus, carpus, and propodus with feathery setae anteriorly.

In rock pools amongst algae, amongst *Zostera*, in salt marshes; also sublittorally with a variety of epibenthos, especially in docks. All British coasts but commonest in the south-west.

20. ISAEIDAE

Body smooth dorsally. Antenna 1 usually with peduncle article 3 equal to or longer than 1. Rostrum absent. Mandible with a palp. Coxa 4 not concave posteriorly. Gnathopods usually powerful, subchelate, 1 equal to or smaller than 2. Uropod 3 short, biramous, rarely uniramous; rami as long as, or longer than, peduncle, without hooks or teeth. Telson entire, short, fleshy; nearly circular or square. Fifteen species, in six genera, recorded for Britain.

---

1. Uropod 3 uniramous. Gnathopod 2 male very large, with propodus palm bearing one small proximal and two distal teeth. Gnathopod 2 female subchelate, propodus with distinct palm — ***Microprotopus maculatus***

   Uropod 3 biramous — **2**

---

2. Uropod 3 with inner ramus much shorter than outer. Male gnathopod 2 basis with antero-distal lobe. Coxa 2 evenly rounded; coxa 3 weakly convex anteriorly — ***Photis reinhardi***

   Uropod 3 with subequal rami. Accessory flagellum with six articles. Antenna 1 peduncle with article 3 as long as, or longer than, article 1 — ***Gammaropsis maculata***

Gammaropsis maculata (Johnston) — Fig. 8.37

Length 10 mm, pale yellow with darker bands. Antenna 1 longer than 2. Gnathopod 1 propodus equal to carpus, oval, without a well-defined palm. Gnathopod 2 large, propodus twice length of carpus. Uropod 3 rami a little shorter than peduncle.

Intertidal and sublittoral among hydroids, algae, and *Laminaria* holdfasts. Mostly southern shores, often common in south-west.

Microprotopus maculatus Norman — Fig. 8.28I

Length 2 mm, dark brown with pale bands. Antennae short, accessory flagellum bi-articulate. Pereopods 3 and 4 with basis slightly enlarged.

Pereopods 5–7 with basis very enlarged, oval. Epimeral plate 3 prolonged, rounded, slightly crenulate posteriorly. Telson small, quadrangular.

Among algae and on sand, littoral and shallow sublittoral. All British coasts.

Photis reinhardi Krøyer Fig. 8.37

Length 5 mm. Head lateral lobes produced, angular terminally; eyes not quite at extremity. Antennae short, subequal; no accessory flagellum. Gnathopod 1 propodus somewhat quadrangular, with a well-defined palm. Gnathopod 2 male propodus enlarged, palm sinuous and delimited by a large tooth. Gnathopod 2 female similar to male but with propodus palm lacking tooth. Epimeral plate 3 rounded postero-ventrally. Telson short, triangular.

Lower shore and sublittoral, in *Laminaria* holdfasts and among algae. North Sea.

21. COROPHIIDAE

Body dorso-ventrally flattened, occasionally elongate or cylindrical. Urosome segments always distinctly depressed, some or all coalesced. Antennae variable, accessory flagellum usually absent, or present but small. Mandibular palp present, sometimes reduced to two or one article(s). Coxae short, usually separate from each other or just touching. Gnathopods variable, one pair often complexly subchelate. Uropod 3 small, variably uniramous, sometimes with hooked rami, or with peduncles produced distally to give a pseudobiramous appearance. Telson entire, fleshy; circular, very broad and short, or symmetrically trapezoidal. Many species construct tubes while a few make semi-permanent burrows in the sediment.

Fourteen British species in three genera. Some are marine, a number are exclusively estuarine. In many species it is convenient to key out males and females separately, using the structure of the antennae.

---

1. Gnathopod 2 merus and carpus elongate, about equal size, longitudinally fused. Antenna 1 peduncle article 3 shorter than 1; accessory flagellum absent. Mandibular palp with two articles **2**

   Gnathopod 2 merus and carpus not elongate, or longitudinally fused; carpus smaller than propodus. Accessory flagellum present. Mandibular palp with three articles. Epimeral plates 1–3 acutely pointed ***Unciola crenatipalma***

---

2. Urosome segments separate and freely articulating **3**

   Urosome segments fused **4**

---

3. Uropod 1 with single row of spines along outer edge of peduncle, not replaced proximally by setae. Uropod 3 rami not eccentrically placed on peduncle ***Corophium volutator***

   Uropod 1 with two rows of spines along outer edge of peduncle, outer row replaced proximally by setae. Uropod 3 rami not eccentrically placed on peduncle ***Corophium arenarium***

   Uropod 1 with one or two rows of spines on outer edge of peduncle, never replaced proximally by setae. Uropod 3 rami eccentrically attached to peduncle ***Corophium multisetosum***

---

4. Lateral lobes of head acutely pointed, eyes small. Burrowers in sand ***Corophium crassicorne***

   Lateral lobes of head truncate or rounded, eyes well developed. Tube builders on hard substrata **5**

---

5. Antenna 2 with two terminal teeth on peduncle article 4, the smaller above the larger ***Corophium sextonae*** male

   Antenna 2 peduncle article 4 armed only with spines **6**

---

6. Antenna 2 peduncle article 4 with 4–6 spines set in a single row on a flange ***Corophium sextonae*** female

Antenna 2 peduncle article 4 with two pairs of spines and one terminal spine
Corophium bonellii female

Genus *Corophium*
Antenna 1 with peduncle article 3 shorter than 1 and shorter than the flagellum. Accessory flagellum absent. Antenna 2 equal to or longer than 1, thickened and, especially in male, with flagellum much shorter than peduncle article 5; peduncle article 4 usually with a distal tooth. Mandibular palp with two articles. Coxae short and separate. Gnathopods small; 1 subchelate; 2 simple, very setose, with merus and carpus elongate and fused longitudinally. Uropod 3 flattened, ramus equal in length to peduncle. Telson trapezoidal.

Corophium arenarium Crawford Fig. 8.38
Length 6 mm, whitish with brown markings. Eyes small and black. Rostrum triangular, lateral lobes of head rounded and short. Antenna 1 female peduncle article 1 with two spines proximally on lower edge, followed by three setae and a terminal spine. Antenna 2 female with peduncle article 4 bearing a small spine on inner surface and a spiny process terminally which extends just beyond end of article. Antenna 1 male with lower edge of peduncle article 1 armed with two small spines and three setae; inner edge slightly crenulate. Antenna 2 male with powerful articles: article 4 armed terminally with a long tooth.

A builder of semi-permanent burrows in sand or muddy sand rather than mud. Generally littoral, often estuarine. Southern Britain (Norfolk to Irish Sea).

Corophium bonellii (Sars) Fig. 8.38
The male is unknown and the species may well be parthenogenetic. The following description therefore applies to the female.

Length 5.5 mm. Rostrum short, triangular. Antenna 1 peduncle article 1 shorter than combined length of articles 2 and 3, armed ventrally with three large, straight spines distally, and one or two sharply curved spines proximally; inner margin of article 1 also armed with 1–3 spines, the proximal one sharply curved and short; flagellum with a maximum of eight articles. Antenna 2 peduncle article 5 usually with two spines, proximal spine may be as large as distal, or smaller, or absent.

A tube-builder, common in *Laminaria* holdfasts and on other algae and hydroids, littorally and sublittorally on all coasts. Displaced by *C. sextonae* in the Plymouth area.

Corophium crassicorne Bruzelius Fig. 8.38
Length 5 mm, brown. Rostrum short and pointed. Antenna 2 female with peduncle article 4 nearly crescent-shaped in section, the concavity facing inwards and upwards; with a row of 6–8 spines on ventral margin. Antenna 2 male peduncle article 4 with two terminal teeth, the smaller above the larger; peduncle article 5 with a single small tooth on the ventral margin.

A burrower in muddy sand, LWS and subtidal. All British coasts.

Corophium multisetosum Stock Fig. 8.38
Length 8 mm, whitish with brown markings. Eyes black. Rostrum short and triangular. Female with lateral lobes of head rounded, shorter than rostrum; male lateral lobes more triangular, with rounded tips, longer than rostrum. Antenna 1 female with lower edges of peduncle article 1 armed with three spines, and about 14 setae between second and third spines. Antenna 2 female with peduncle article 4 armed with one spine and a very long terminal tooth, one-third length of article 5. Antenna 1 male with peduncle article 1 armed on lower edge with 3–5 small spines and numerous setae. Antenna 2 male large, peduncle article 4 armed terminally with a long tooth.

A builder of semi-permanent burrows in mud or sand, and a tube-builder on sessile objects. Inhabits the upper regions of estuaries in salinities often <1% and rather lower than those preferred by *C. volutator* and *C. arenarium*. Southern Britain (Norfolk to Wales).

Corophium sextonae Crawford Fig. 8.38
Length 4.5 mm; greyish-white, with two darker bars across each segment, and on the antennae and head. Eyes large and black. Rostrum short, acutely pointed. Antenna 2 female peduncle article 5 with one or two ventral spines; flagellum with three articles. Antenna 1 male with five-articled flagellum, shorter than or equal to peduncle article 2. Antenna 2 male with few setae, flagellum with three articles.

A tube-builder common in shallow dredgings and *Laminaria* holdfasts from south and south-west coasts. Also recorded from The Netherlands.

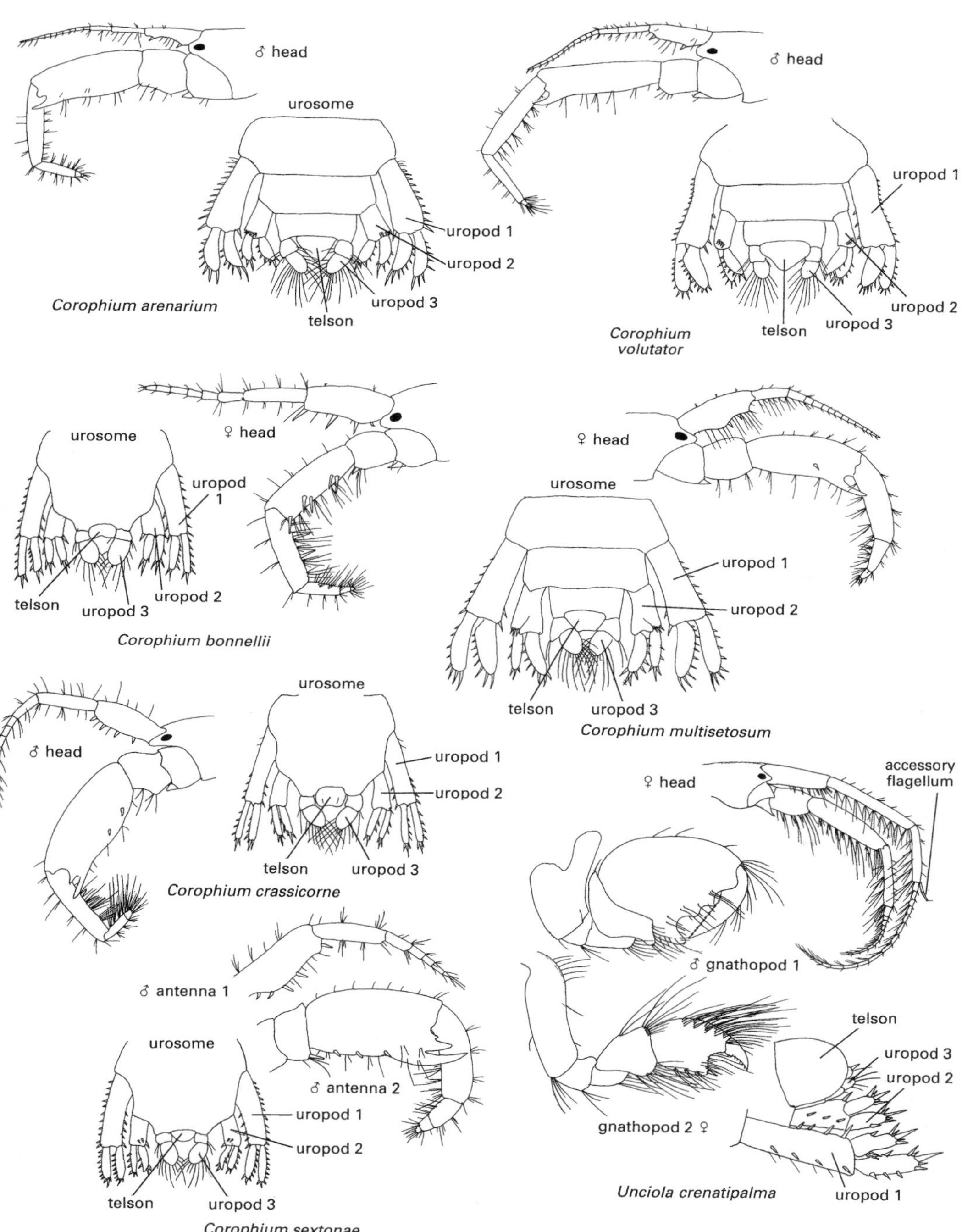
♂ head
urosome
uropod 1
uropod 2
uropod 3
telson
Corophium arenarium
♂ head
uropod 1
uropod 2
uropod 3
telson
Corophium volutator
♀ head
urosome
uropod 1
telson
uropod 3
uropod 2
Corophium bonnellii
♀ head
urosome
uropod 1
uropod 2
telson
uropod 3
Corophium multisetosum
♂ head
urosome
uropod 1
uropod 2
telson
uropod 3
Corophium crassicorne
♀ head
accessory flagellum
♂ gnathopod 1
♂ antenna 1
urosome
♂ antenna 2
uropod 1
uropod 2
telson
uropod 3
Corophium sextonae
telson
uropod 3
uropod 2
gnathopod 2 ♀
Unciola crenatipalma
uropod 1

Fig. 8.38

Corophium volutator (Pallas) Figs. 8.27L, 8.38

Length 8 mm, whitish with brown markings. Eyes small and black. Rostrum and lateral lobes of head short and rounded. Antenna 1 female with two, rarely three well-developed spines, separated by three to four setae on the lower edges of peduncle article l. Antenna 2 female without a spine on inner surface of peduncle article 4, terminal tooth on article 4 strong, extending just beyond end of article. Antenna 1 male with peduncle article 1 bearing two, rarely three, small spines on lower edge and distinct crenulations on inner edge. Antenna 2 male with powerful peduncle articles; article 4 with a terminal tooth.

Intertidal. A builder of semi-permanent burrows in mud, usually in estuaries. All British coasts.

Unciola crenatipalma (Bate) Fig. 8.38

Body slender, elongate; length 7 mm, yellow and orange, fading to translucent white. Antenna 1 flagellum more or less equal to peduncle. Accessory flagellum present but small. Antenna 2 stronger, although slightly shorter, than l; flagellum more or less equal to length of peduncle article 5. Gnathopods subchelate, 1 larger than 2. Uropod 3 with peduncle somewhat produced distally, a little longer than the eccentrically positioned rami. Telson rounded and short.

Sublittoral, usually among muddy stones and shells. Quite common in southern Britain, apparently absent in the north.

22. ISCHYROCERIDAE

Body laterally compressed, smooth dorsally. Antenna 1 peduncle article 3 more than half length of 2. Accessory flagellum present or absent. Rostrum absent. Mandible with a palp. Coxae 4 not concave posteriorly. Gnathopods powerfully subchelate. Uropod 3 usually not projecting beyond 1 and 2, uniramous or biramous with small or very small, rather triangular rami, always shorter than peduncle; outer ramus with one hooked spine distally and/or small distolateral teeth. Telson fleshy, subcircular or nearly square.

Adult males may be identified using the following key. Juvenile males and females are very difficult to identify and may not key out readily. The British fauna includes 12 species, in six genera.

1. Uropod 3 uniramous **2**

   Uropod 3 biramous **3**

2. Gnathopod 1 basis slender; gnathopod 2 carpus with two teeth (in old males the inner tooth may be just a rounded process) ***Ericthonius punctatus***

   Gnathopod 1 basis broad, with anterior flange; gnathopod 2 with one slender tooth, or with one main tooth and a small accessory tooth ***Ericthonius difformis***

3. Accessory flagellum an indistinct tubercle. Coxae 1 and 2 equal in length ***Parajassa pelagica***

   Accessory flagellum distinct, bi-articulate. Coxae 1 and 2 not equal in length **4**

4. Male gnathopod 2 often with distal process on palm of propodus, but lacking proximal tooth. Uropod 3 outer ramus with straight terminal spine and a number of small distolateral teeth ***Ischyrocerus anguipes***

   Male gnathopod 2 propodus with small distal tooth and large proximal process. Uropod 3 outer ramus with hooked spine distally, and 1–3 distolateral teeth ***Jassa falcata***

Genus *Ericthonius*

Antenna 1 with peduncle article 1 shorter than 3, flagellum more or less equal to peduncle; accessory flagellum absent. Antenna 2 slender, as long as antenna 1, with flagellum as long as peduncle. Mandibular palp with three articles. Coxae short, at most just touching serially. Female gnathopods subchelate; male gnathopod 2 much enlarged, carpochelate. Uropod 3 with hooked rami. Telson very short and wide with a spiny patch on each side.

Ericthonius difformis Milne Edwards Fig. 8.39

Length 10 mm; body very slender, greyish with brown spots. Eyes large. Antenna 2 shorter than 1, peduncle article 5 longer than 4, flagellum longer than article 5. Male gnathopod 1 coxa small, rounded triangular, widely separate from coxa 2. Gnathopod 2 coxa much broader than long, with anterior stridulating ridges; basis slender, broadened distally; carpus slightly convex antero-proximally, postero-distal process moderately deflected, with one large tooth, and sometimes a small accessory tooth distally; propodus with irregular posterior margin, sometimes with distinct proximal lobe, lacking cutting edge; dactylus at least three-quarters length of propodus.

Shallow sublittoral, amongst algae and hydroids, and often associated with *Zostera*. Distribution incompletely known; reported from south-western coasts, but probably more widely distributed.

Ericthonius punctatus (Bate) Fig. 8.39

Length 10 mm. Eyes large. Antenna 1 flagellum about as long as peduncle. Antenna 2 slightly longer than 1, peduncle article 5 slightly longer than 4, flagellum about as long as articles 4 and 5 together. Male gnathopod 1 coxa small and rounded, widely separate from coxa 2. Gnathopod 2 coxa rounded, with stridulating ridges on distal margin; basis broadened distally; carpus strongly convex antero-proximally, postero-distal process strongly deflected, with two teeth separated by rounded depression; propodus lacking posterior marginal lobes or serrations; dactylus elongate.

Sublittoral, building tubes amongst algae and hydroids. Common on all ratify coasts.

Ischyrocerus anguipes (Krøyer) Fig. 8.39

Length 8–10 mm; colour variable, usually banded. Similar to, and difficult to distinguish from, species of *Jassa*. Accessory flagellum bi-articulate and relatively elongate, almost as long as article 1 of antenna 1 flagellum. Coxa 1 more than three-quarters length of 2; coxa 5 longer than 6. Male gnathopod 2 propodus lacking a large proximal projection. Uropod 3 outer ramus with or without an unhooked terminal spine, usually with small, blunt, recurved, lateral denticles near tip.

Abundant in *Laminaria* holdfasts (north-east coast), also among intertidal and sublittoral epibenthos. All British coasts.

Jassa falcata (Montagu) Figs 8.28I, 8.39

Length 7 mm; yellow-grey, strongly marked with brown, red, or black, depending on habitat colour. Eyes small, round, and dark. Accessory flagellum bi-articulate, distinct, although clearly shorter than article 1 of antenna 1 flagellum. Coxa 1 angular, elongated anteriorly; coxa 2 longer than deep but with anterior margin much shorter than posterior margin of coxa 1. Epimeral plate 3 with a minute tooth at postero-ventral corner. Uropods 1 and 2 with outer ramus shorter than inner. Uropod 3 with peduncle much longer than rami. Telson small, triangular, with two setae on each side of apex.

The antennae and gnathopods show considerable variation. In the adult male, gnathopod 2 propodus is greatly enlarged, with a proximal projection from the palm which is never bifid. In adult females also the propodus is enlarged, but the palm margin is concave rather than straight and there is no proximal projection.

Intertidal and sublittoral. Builds tubes among algae and hydroids, and on solid structures. An important fouling species. Often abundant on buoys, ships, and in harbours. In *Laminaria* holdfasts and similar habitats. Widespread and common. All British coasts.

Parajassa pelagica (Leach) Fig. 8.39

Length 5 mm; greyish with bands of brown. Eyes small. Antennae short and robust, with dense whorls of long setae. Antenna 1 about one-third body length, accessory flagellum represented by an indistinct tubercle; antenna 2 much longer than 1, very robust. Coxae 1 and 2 equal in length; coxa 5 longer than 6. Epimeral plate 3 rounded posteriorly. Telson approximately triangular.

Common in *Laminaria* holdfasts and similar habitats. All British coasts.

Suborder Hyperiidea

Hyperiid amphipods have short, typically fat or swollen bodies. The head is very deep and rounded with large, often strikingly coloured, eyes occupying the whole of each side of the head. The pereon is often deep, rounded laterally and dorsally, and compressed laterally; the pleon is slender and the urosome well developed, often elongate. The male has elongate, slender antennae, while in the female both pairs of antennae are very short. Gnathopods are poorly developed and differ between sexes;

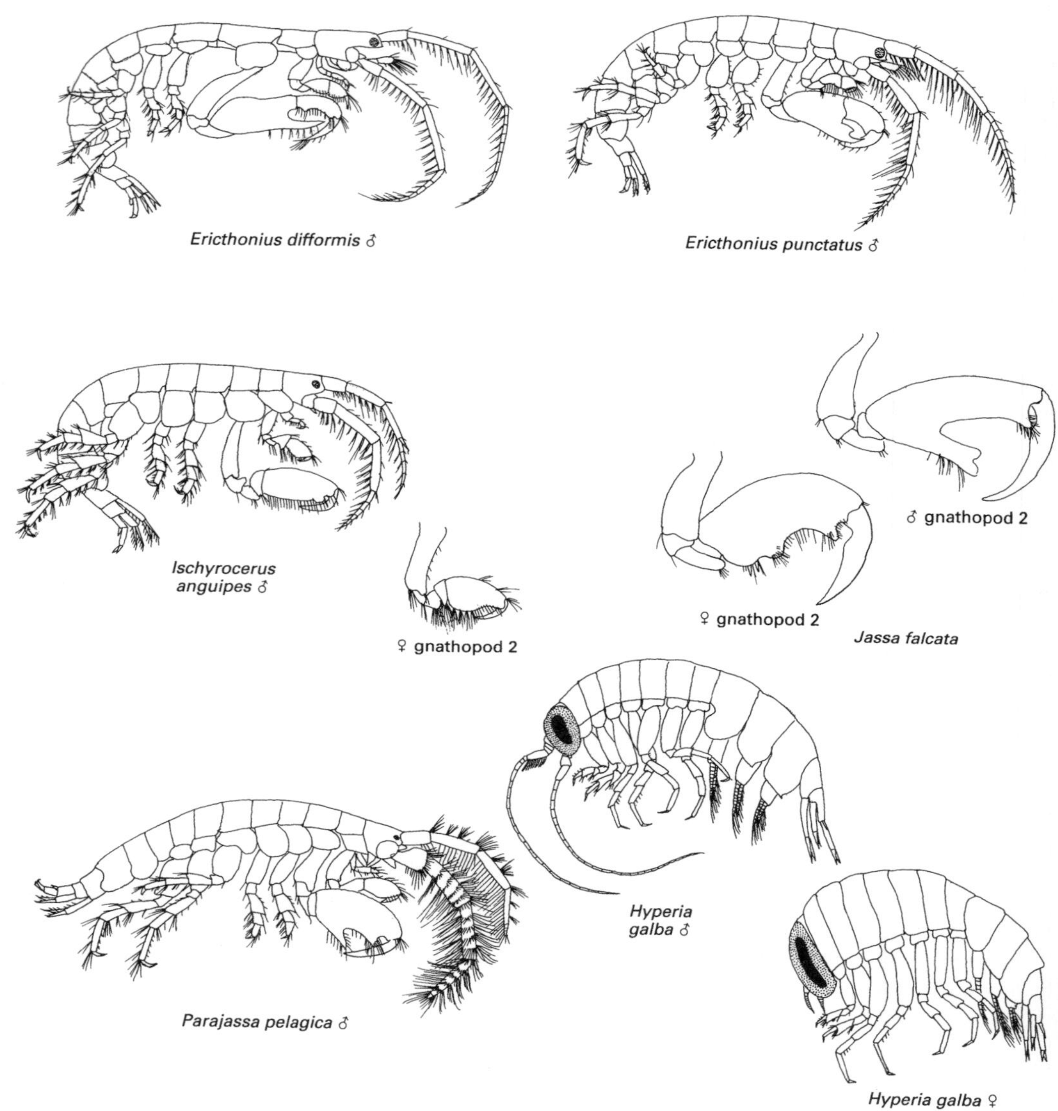

Fig. 8.39

pereopods 3–7 are slender, and in all pereopods the coxa is very small, or absent. Pleopods are well developed, often powerfu]. Uropods are laminar, biramous, and lack spines.

Approximately 50 species of hyperiids have been recorded around the British Isles, but most are rare or restricted to deep water. A single species is commonly encountered associated with large jellyfish.

Hyperia galba (Montagu) Fig. 8.39
Length 12 mm; light translucent brown, with enormous green eyes. Female with very short, subequal antennae: peduncle of three articles attached to

immovable basal segment, flagellum scarcely longer than peduncle. Male with long slender antennae; antenna 2 longer than 1, about two-thirds body length. Gnathopods 1 and 2 small, simple; carpus with acute, projecting, disto-ventral lobe. Uropods 1–3 broad, laminar; rami lanceolate, with finely denticulate margins.

In *Rhizostoma*, *Aurelia*, and other scyphozoa. All British coasts. Widespread and common.

Suborder Caprellidea

Caprellids are elongate and slender, with appendages reduced in number. They tend to be slow-moving, and are often found clinging to algae, hydroids, and bryozoa. Certain species are opportunistic feeders whereas others may actively prey on small Crustacea; some are known to be commensal with other marine invertebrates, such as starfish.

The animal may be divided into a *head*, a thorax (or *pereon*), and an *abdomen*. The head and pereon segment 1 may be completely or partially fused and the length of segment 1 may vary appreciably. *Antenna 1* is longer than *antenna 2*, with a three-jointed peduncle and a multi-articulate flagellum. Antenna 2 has a peduncle of three or four joints and a shorter flagellum, normally of two articles. A *mandibular palp* may be present, projecting upwards between the peduncles of antennae 2. The pereon may be smooth or may bear dorsal and/or lateral *tubercles* or sharp *spines*. A pair of round or club-shaped *gills* is normally found on the ventro-lateral borders of segments 3 and 4, and may be present on segment 2 in some species. The *gnathopods* are normally larger than other pereopods and modified into grasping claws, with a movable finger and an opposing cutting edge. The pereopods may be completely absent from segments 3 and 4, or may be reduced to minute appendages. Pereopods 5, 6, and 7 are normally of approximately equal length and are used for holding on to the substratum while the animal feeds with the gnathopods. The *propodus* of the posterior pereopods may bear one or two *grasping spines* on its inner surface. The small abdomen may be furnished with lobes and/or articulated appendages, or may be without either, differences between the sexes being usual. Female caprellids develop paired *lamellae* on the ventral borders of pereon segments 3 and 4, and these enlarge to form a *brood-pouch* for the developing eggs.

Twenty-two species are recorded for Britain, classified in three families

KEY TO SPECIES

1. Gills on segments 2, 3, and 4. Pereopods 3 and 4 six-segmented, pereopod 5 five-segmented **_Phtisica marina_**

   Gills on segments 3 and 4 only. Pereopods 3 and 4 greatly reduced or absent **2**

2. Pereopods 3 and 4 minute, reduced to two segments. Mandibular palp of three segments. Head and pereon segments 1 and 2 with strong, forwardly directed spines **_Pseudoprotella phasma_**

   Pereopods 3 and 4 absent, or represented only by a single minute seta. Mandibular palp absent **3**

3. Pereopod 5 with six segments, about equal in size to 6 and 7. Pereopods 3 and 4 absent **4**

   Pereopod 5 reduced, minute, with only two segments. Pereopods 3 and 4 represented only by a single seta. Often in association with starfish **_Pariambus typicus_**

4. Antenna 2 with short setae ventrally, not arranged in parallel rows. Head bulbous and skull-shaped **_Caprella acanthifera_**

   Antenna 2 with long setae ventrally, arranged in parallel rows **5**

5. Prominent ventral spine arising between insertions of gnathopod 2 on each side **_Caprella equilibra_**

No ventral spine between insertions of gnathopod 2. Head may bear one small tubercle, or pair of tubercles
Caprella linearis

---

1. PHTISICIDAE

Phtisica marina Slabber Fig. 8.40
Up to 25 mm long; body smooth, slender; head rounded. Antenna 1 from half to equal body length, flagellum with up to 20 articles. Antenna 2 as long as antenna 1 peduncle; flagellum of 3–6 articles. Gnathopod 1 propodus with four or five grasping spines, margin without serrations but with numerous short spines; dactyl not serrate. Gnathopod 2 propodus often with two unequal grasping spines, sometimes only larger spine present; palm with numerous short spines and few setae; propodus often with inflated appearance in older males. Pereopods 6 and 7 six-segmented. Abdomen of male with two pairs of articulate appendages and one pair of pyriform appendages. Female abdomen with two pairs of articulate appendages, one pair of lobes, and an anterior raised projection.

Usually sublittoral, on hydroids and algae, or attached to floating objects and buoys. Known to swim occasionally, especially at night. All British coasts.

Pseudoprotella phasma (Montagu) Fig. 8.40
Length up to 25 mm. Head with large, anteriorly directed spine dorsally. Pereon segment 1 with similar spine; segment 2 with two spines in middle, one spine on posterior margin. Pereon segment 3 with small triangular tubercle dorsal to insertion of pereopod, and two small rounded dorsal tubercles mid-way along segment. Small rounded dorsal tubercle, or occasionally a spine, at junction of pereon segments 3 and 4; tubercles on segment 4 as on 3; segment 5 with small lateral triangular tubercle one-third length along each side. Antenna 1 about two-thirds length of body, flagellum of 18–26 articles. Antenna 2 flagellum with two articles. Pereopods 5, 6, and 7 six-segmented, about equal length. Gills on segments 3 and 4 only. Abdomen with two pairs of non-jointed appendages in male, without appendages in female.

Usually sublittoral, often on algae and large hydroids. All British coasts, probably more common in south.

2. CAPRELLIDAE

Pariambus typicus (Krøyer) Fig. 8.40
Length up to 7 mm. Head and body smooth; female stout, male more slender. Antenna 1 about one-third to half length of body, flagellum of eight articles. Antenna 2 with short ventral setae; about as long as antenna 1 peduncle. Gnathopod 1 with short obtuse spine on distal merus, plus one or two distal setae; carpus triangular. Gnathopod 2 inserted near middle of segment; palm of propodus armed with strong proximal poison spine and large sub-median tooth; dactyl slightly sinuous on inner border. Gnathopod 2 female without sub-median tooth. Pereopods 6 and 7 with six articles. Abdomen in male with plaque and pair of vestigial lobes, female without lobes.

Sublittoral, but may occur on the lower shore. On coarse sandy substrata, but often found on the aboral surface of the starfish *Asterias rubens* and *Crossaster papposus*. All British coasts.

Caprella acanthifera Leach Fig. 8.40
Length up to 13 mm. Head smooth, domed, and skull-shaped. Body stout, with varying numbers of small, paired dorsal tubercles on all segments except 1. Posterior regions of segments 2, 3, and 4 often with additional single tubercle. Antenna 1 about two-thirds body length, flagellum of two articles. Gnathopod 1 propodus with proximal spine. Gnathopod 2 often massive in large males; palm of propodus concave, with proximal spine and median tooth. Pereopods 5–7 with pair of grasping spines on propodus, and 2–4 accessory spines proximal to grasping spines. Abdomen with pair of lobes and pair of bi-articulate appendages in male only.

Mid-shore level to sublittoral. Frequent in rock pools and gullies, on algae, hydroids, and bryozoa. All British coasts, possibly more frequent in north.

Caprella equilibra Say Fig. 8.40
Up to 23 mm long. Head and body smooth dorsally, except for paired tubercles occasionally present on pereon segment 5. Head flattened anteriorly. Strong lateral anteriorly directed spines at base of gnathopod 2. Antenna 1 about half body length, flagellum of 13–15 articles. Antenna 2 one-fifth to one-third body length, flagellum of two articles. Gnathopod 1 propodus with two proximal grasping spines; margin of dactyl and propodus serrate. Gnathopod 2 with short, stout basis; distal anterior

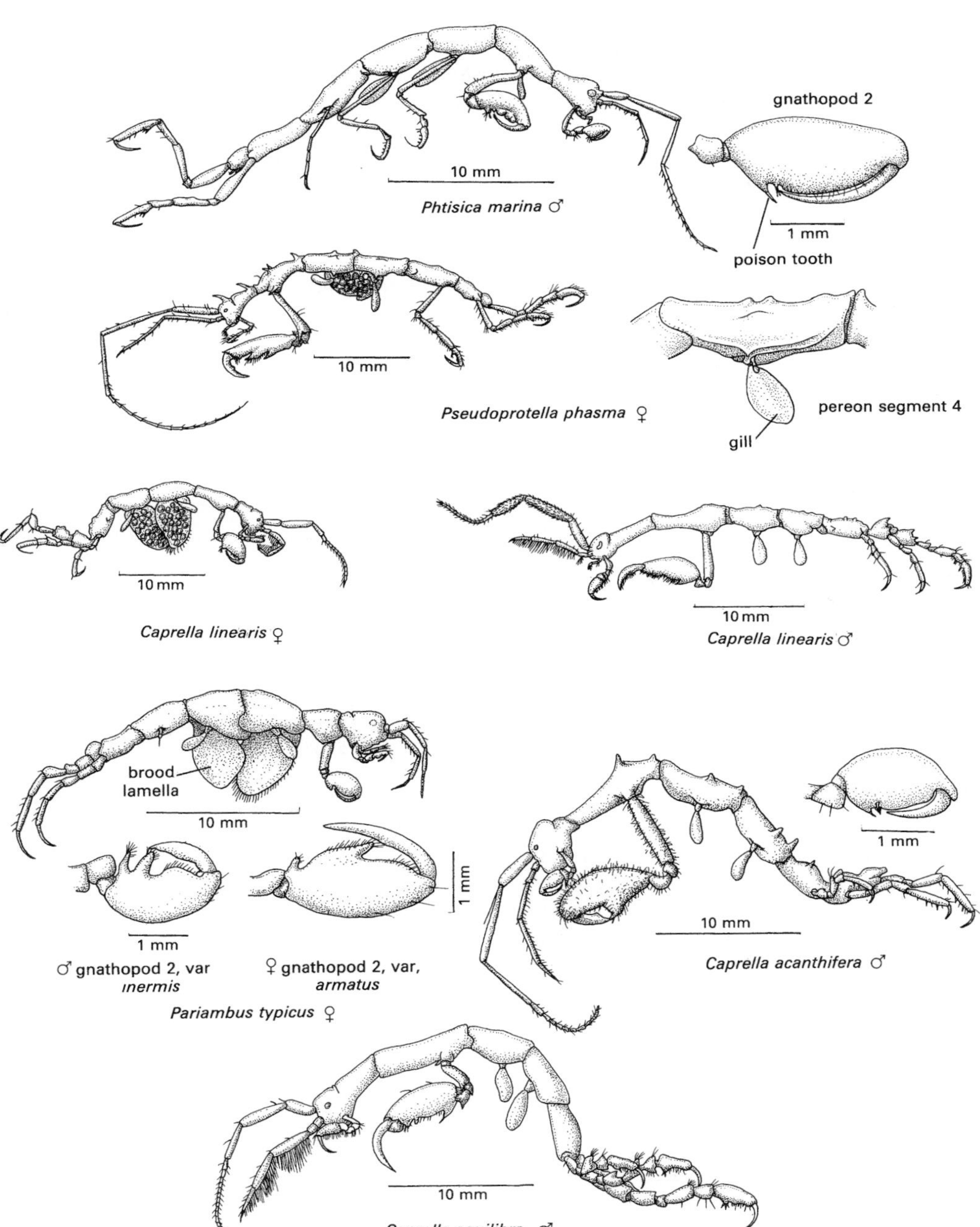
gnathopod 2
10 mm
1 mm
poison tooth
Phtisica marina ♂
10 mm
pereon segment 4
Pseudoprotella phasma ♀
gill
10 mm
Caprella linearis ♀
10 mm
Caprella linearis ♂
brood lamella
10 mm
1 mm
1 mm
1 mm
♂ gnathopod 2, var inermis
♀ gnathopod 2, var, armatus
Pariambus typicus ♀
10 mm
Caprella acanthifera ♂
10 mm
Caprella equilibra ♂

Fig. 8.40

margin produced into triangular projection; palm of propodus with single proximal grasping spine, large rectangular distal tooth, and more proximal tooth. Pereopods 5–7 propodus with two proximal grasping spines. Abdomen of male with one pair of appendages and one pair of lobes; female with lobes only.

Sublittoral, usually associated with floating buoys, pontoons. All British coasts.

Caprella linearis (Linnaeus) Fig. 8.40

Up to 22 mm long. Head and body segments normally smooth; tubercules paired when present. Pereon segment 5 with pair of antero-lateral tubercles, two small antero-dorsal tubercles, and four postero-dorsal tubercles. Pereon segments 1 and 2 elongate in male. Antenna 1 often with setose peduncle in male, and flagellum of up to 15 articles; female flagellum up to 20 articles. Antenna 2 shorter than segments 1 and 2 of antenna 1, flagellum of two articles. Gnathopod 2 propodus with small tooth adjacent to poison tooth, and two distal projections separated by a deep cleft. Distal projections smaller and closer together in female. Pereopods 5–7 propodus with pair of proximal grasping spines.

ELWS to sublittoral. Often on buoys among hydroids and bryozoa. All British coasts.

Superorder Eucarida

VII Order Decapoda

The Decapoda is the largest natural grouping within the Malacostraca. In all decapods the thoracic segments are fused dorsally to a *carapace*, a fold of which extends ventrally on each side of the animal, enclosing the gills and constituting a *branchial chamber*. The maxilla has a large, modified endopodite, the *scaphognathite*, which drives water through the branchial chamber by rhythmic beating.

Most frequently, the order is divided simply into two suborders, the swimming decapods, or Natantia, and the walking decapods, or Reptantia, and this arrangement is maintained by many authorities. However, a more natural classification recognises the suborders Dendrobranchiata and Pleocyemata. In dendrobranchiate decapods the gills are subdivided to form arborescent tufts, the first three pairs of pereopods are chelate, and eggs, which are not carried by the female, hatch as nauplii larvae. The Pleocyemata have unbranched gills, show variable chelation of the second and third pereopods, and their eggs, which develop attached to the female pleopods, hatch as zoeae larvae. The Dendrobranchiata comprises the single infraorder Penaeidea, which together with three infraorders of the Pleocyemata constitute the Natantia. In British waters natant decapods are represented by the infraorder Caridea only. The remaining four infraorders of the Pleocyemata constitute the Reptantia, and representatives of all four groups occur in the British sea area.

Suborder Pleocyemata

Infraorder Caridea

[Natantia]

The natant decapods have an elongate body divided into three regions: head, thorax (or *pereon*) and abdomen (or *pleon*) (Fig. 8.41). Fusion of some head segments with the pereon gives a *cephalothorax*. This and the remaining pereon segments are usually covered by the *carapace*, which also extends ventrolaterally to cover the gills, viscera, and the bases of the thoracic limbs. The anterior extension of the carapace into a 'beak', or *rostrum*, is a common feature amongst natantians. The rostrum is often armed with teeth on its dorsal and ventral borders, and it may be straight, or curving upwards or downwards. Anteriorly the head possesses a pair of stalked *eyes*, a pair of sensory *antennules* (or first antennae) and a pair of *antennae* (or second antennae). The antennules are usually biramous, although in the Paleamonidae they are triramous. The basal three segments of the antennule form the *peduncle* and the first of these is usually produced anterolaterally into a process called the *stylocerite*. The antennae are usually much longer than the antennules, the exopod being produced into a scale-like process, the *scaphocerite*. The mouthparts consist of one pair of *mandibles*, two pairs of *maxillae*, and three pairs of *maxillipeds*. The mandibles grind and crush food, and usually bear a *molar process*, plus an *incisor process* and a *palp*. The three maxillipeds are derived from the first three pairs of thoracic limbs of the primitive form, and maxilliped 3 is usually the least modified. The five pairs of *pereopods* are used for locomotion; the anterior two or three pairs usually terminate in a claw (or *chela*), which assists in feeding, and in offence and defence. In some families, (e.g. Pasiphaeidae) all the pereopods bear *exopods*, whereas in others (e.g. Hippolytidae) exopods are absent, or present only on some limbs

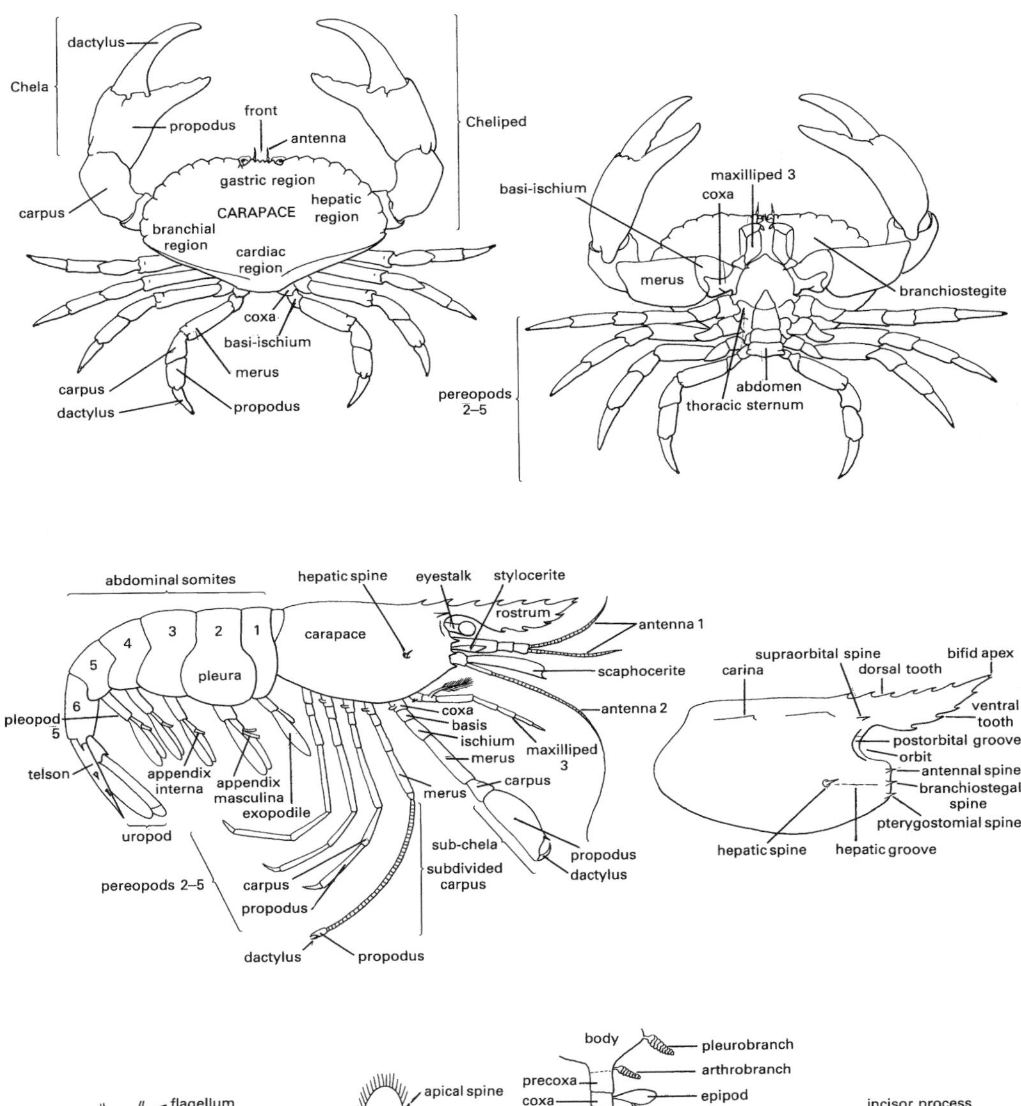
dactylus
Chela
propodus
front
antenna
Cheliped
gastric region
carpus
CARAPACE
hepatic region
branchial region
cardiac region
coxa
basi-ischium
merus
carpus
dactylus
propodus
basi-ischium
maxilliped 3
coxa
merus
branchiostegite
pereopods 2–5
abdomen
thoracic sternum
abdominal somites
hepatic spine
eyestalk
stylocerite
rostrum
antenna 1
1
2
3
4
5
6
carapace
pleura
scaphocerite
antenna 2
pleopod 5
coxa
basis
ischium
merus
maxilliped 3
telson
appendix interna
appendix masculina
exopodile
merus
carpus
uropod
sub-chela
propodus
dactylus
subdivided carpus
pereopods 2–5
carpus
propodus
dactylus
propodus
supraorbital spine
bifid apex
carina
dorsal tooth
ventral tooth
postorbital groove
orbit
antennal spine
branchiostegal spine
pterygostomial spine
hepatic spine
hepatic groove
body
pleurobranch
arthrobranch
precoxa
coxa
epipod
basis
podobranch
ischium
merus
exopodite
carpus
propodus
dactylus
flagellum
stylocerite
RIGHT ANTENNA 1
apical spine
flagellum
scaphocerite
peduncle
RIGHT ANTENNA 2
PEREPOD
incisor process
molar process
palp
MANDIBLE

Fig. 8.41

(e.g. Crangonidae). *Epipods* may also be present (e.g. *Alpheus*) or absent (e.g. *Crangon*). The carpus of pereopod 2 is divided in some families (e.g. Hippolytidae) but not in others (e.g. Palaemonidae). Each pleon segment bears a pair of biramous *pleopods*; the *endopods* of pleopods 2–5 usually have a short *appendix interna*, bearing hooks and presumably assisting in locomotion by coupling together adjacent pleopods during swimming. The sixth pleopods are flattened and expanded to form the *uropods*, which, with the tail (or *telson*), make up the tail-fan of the animal.

The sexes are separate and secondary sexual features are noticeable in adult natantians. The first pleopods differ, and usually the endopod of pleopod 2 of the male bears an *appendix masculina* which aids in copulation. In most cases the female of a species is larger than the male. Eggs are carried externally by the females, attached to long setae which develop on the pleopods for this purpose. Planktonic *zoeae* larvae metamorphose through *post-larval* stages to the adult form, the number of zoeae and post-larvae varying between species.

Natant decapods occur in freshwater, brackish water and estuarine environments, although most are exclusively marine. *Palaemonetes* inhabits brackish pools in northern Europe, whereas in Mediterranean regions it is found in freshwater habitats. Some species are adapted to life in a fluctuating estuarine environment, e.g. *Crangon crangon* and *Palaemon longirostris*. Many species are encountered on the shore and in the shallow sublittoral zone, although natant decapods are also well represented in catches from the bathypelagic regions. Members of many families are benthic in habit, whereas others are pelagic and make extensive diurnal vertical migrations. A few natantians are found in specific associations with other animals (e.g. *Typton*). Most natant decapods are predacious, many being omnivorous and often scavenging for food. They are preyed upon to a great extent by fish, and to a lesser extent by other invertebrates, such as coelenterates and octopods.

KEY TO FAMILIES

1. Pereopod 1 chelate or simple — **2**

   Pereopod 1 subchelate — **7. Crangonidae**

2. Fingers of chelae of pereopods I and 2 slender, with serrated cutting edges — **I. Pasiphaeidae**

   Cutting edges of chelae not serrated — **3**

3. Pereopod 2 carpus subdivided into two or more segments — **4**

   Pereopod 2 carpus not subdivided. Pereopod 1 with well-developed chela — **2. Palaemonidae**

4. Pereopod 1 chela distinct, at least on one side — **5**

   Pereopod 1 chela microscopically small or absent — **6. Pandalidae**

5. Both of first pair of pereopods chelate — **6**

   Only one of first pair of pereopods chelate, the other ending in simple curved dactyl — **5. Processidae**

6. First pair of chelae short, not swollen; tips of fingers usually dark coloured. Eyes free, not covered by carapace. Pereopod 2 carpus with more or less than five segments, but never with five segments **4. Hippolytidae**

Tips of fingers of first pair of chelae not dark coloured. Pereopod 1 stronger than 2, chelipeds swollen and often unequal. Eyes partly or wholly covered by carapace. Pereopod 2 carpus with five segments **3. Alpheidae**

---

1. PASIPHAEIDAE

Rostrum short, sometimes represented by a spine arising behind frontal margin of carapace. Mandibular palp present or absent. Maxilliped 2 with article 7 attached terminally to article 6, exopod rudimentary or absent. First two pairs of pereopods longer and much stouter than the others; chelae elongate, with slender dactyl and propodus; carpus short, unsegmented. Exopods on all pereopods, often very long; present (often very small) on maxilliped 3; often forming chief part of maxilliped 1. Eggs often large, development abbreviated. Three species recorded for Britain. *Pasiphaea tarda* Krøyer occurs rarely in deep water off south-west Ireland and north-west Scotland.

---

1. Pleon with distinct dorsal keel. Rostrum slender, spine-like with lower edge concave. Pereopod 2 basis with 7–12 spines. Telson cleft at apex ***Pasiphaea multidentata***

Pleon without dorsal keel. Rostrum short, triangular, and upturned, apex pointed. Pleon segment 6 with dorsal posteriorly directed spine. Telson slightly convex at apex ***Pasiphaea sivado***

Pasiphaea multidentata Esmark Fig. 8.42

Length up to 110 mm; milky white, smaller specimens transparent. Stylocerite twisted, acutely pointed. Scaphocerite (including apical spine) more than half length of carapace; outer margin evenly convex throughout length with long, stout apical spine exceeding lamellar portion. Maxilliped 3 extends to distal end of scaphocerite.

Off west and south-west coasts, 10–2000 m.

Pasiphaea sivado (Risso) Fig. 8.42

Length up to 100 mm; translucent, with pink or red spots. Stylocerite twisted, acutely pointed, with spine at apex. Outer margin of scaphocerite slightly convex, apical spine stout, exceeding lamellar portion. Maxilliped 3 about 1.25 × length of scaphocerite.

West coasts. Not in Channel or North Sea, 20–500 m.

2. PALAEMONIDAE

Rostrum compressed, usually dentate. Carapace with antennal spine, with or without hepatic and branchiostegal spines. Eyes well developed. Mandible with or without palp; incisor process and molar process separated by cleft. Maxilliped 2 with article 7 attached lateral to article 6. Maxilliped 3 with exopod, with or without arthrobranch or pleurobranch. Pereopod 1 with well-developed chela, but smaller than that of pereopod 2; pereopod 2 carpus unsegmented. Pereopods without epipods; maxillipeds 1 and 2 with epipods. Telson tapering. Six or seven gills plus two epipods. Eight species recorded from Britain.

---

1. Rostrum very short, unarmed. Antennules biramous, pereopod 2 asymmetrical, with swollen chelae. Living in sponges ***Typton spongicola***

Rostrum well developed, with teeth on dorsal and ventral borders. Antennules triramous; pereopod 2 symmetrical, chelae slender **2**

---

2. Mandible with palp **3**

Mandible without palp. Rostrum straight, 4–6 dorsal teeth, two ventral

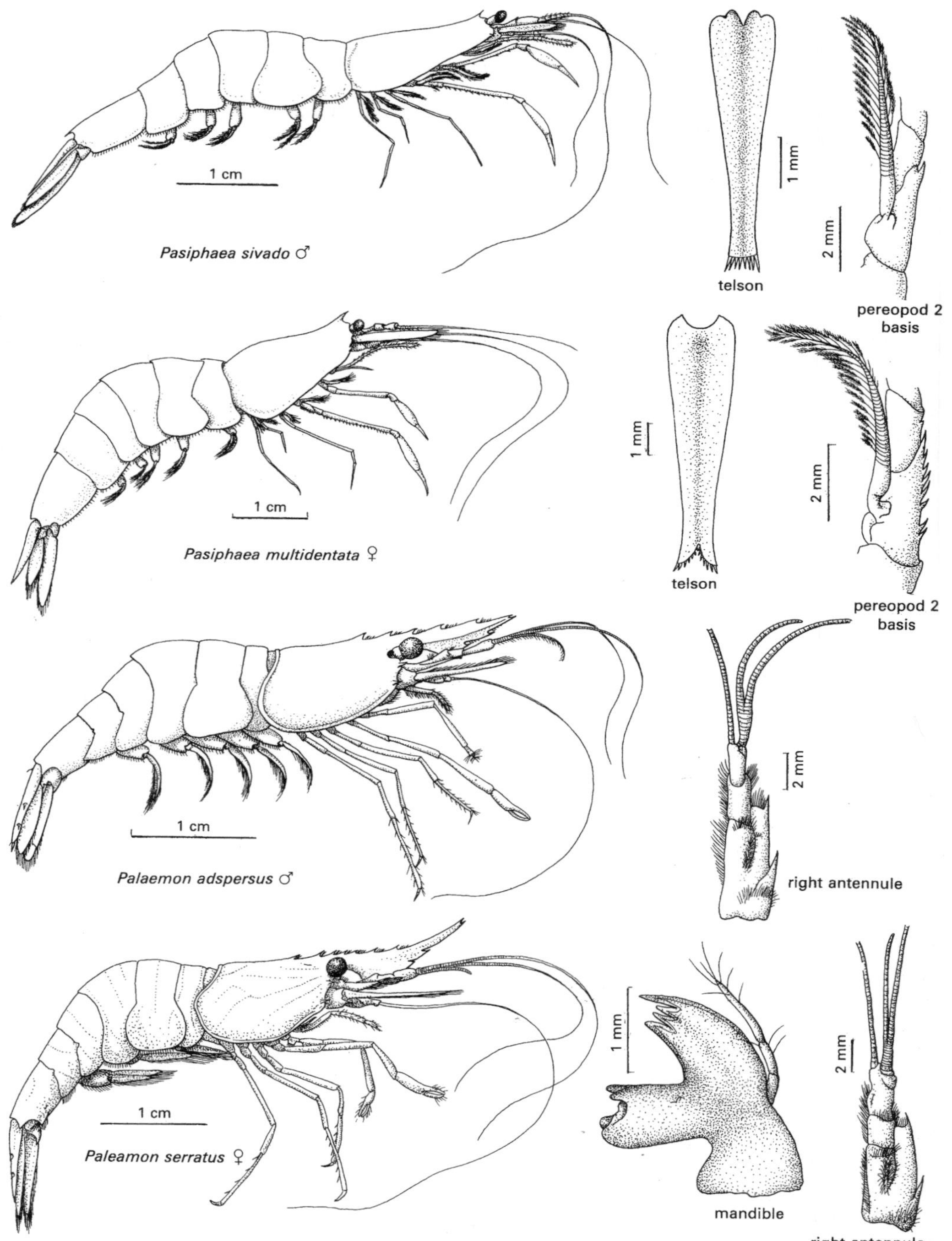
1 cm
Pasiphaea sivado ♂
1 mm
telson
2 mm
pereopod 2
basis
1 cm
Pasiphaea multidentata ♀
1 mm
telson
2 mm
pereopod 2
basis
1 cm
Palaemon adspersus ♂
2 mm
right antennule
1 cm
Paleamon serratus ♀
1 mm
mandible
2 mm
right antennule

Fig. 8.42

teeth. One dorsal tooth behind posterior edge of orbit ***Palaemonetes varians***

---

3. Mandibular palp with three segments **4**

Mandibular palp with two segments. Rostrum straight, or very slightly upcurved; 7–9 dorsal teeth, three (rarely 2–4) ventral teeth. Three (occasionally two) of the dorsal teeth behind posterior edge of orbit. Dactyl of pereopod 2 one-third length of propodus ***Palaemon elegans***

---

4. Rostrum straight, with dorsal teeth extending into distal third. Pereopod 2 carpus equal to or slightly longer than merus **5**

Rostrum with distinct upward curve, dorsal teeth not extending into distal third. Six or seven dorsal teeth, four or five ventral teeth; two of the dorsal teeth behind posterior edge of orbit. Pereopod 2 merus 1.25 × length of carpus ***Palaemon serratus***

---

5. Rostrum straight or very slightly upcurved. Seven to eight dorsal teeth, three or four (rarely five) ventral teeth; two of the dorsal teeth behind posterior edge of orbit, second tooth about 1.5 × more distant from first than from next. Pereopod 2 carpus equal to or slightly longer than merus ***Palaemon longirostris***

Rostrum straight, five or six dorsal teeth, three (rarely two or four) ventral teeth; one dorsal tooth behind posterior edge of orbit, second tooth often directly above edge. Pereopod 2 carpus about 1.2 × length of merus ***Palaemon adspersus***

Palaemon adspersus Rathke Fig. 8.42

Length up to 70 mm; body uniform yellowish-grey, pigment spots on lower half of rostrum, distinctive. Carapace with antennal and branchiostegal spine. Antennules triramous; shorter ramus of outer antennule exceeds length of peduncle, fused for one-third its length to longer flagellum. Outer edge of stylocerite very slightly convex, anterior edge convex. Pereopod 2 dactyl little over half length of propodus.

Shallow sublittoral; south and south-east coasts only.

Palaemon elegans Rathke Fig. 8.43

Length up to 63 mm; thorax and pleon usually bearing dark yellow-brown bands. Carapace with antennal and branchiostegal spine. Antennules triramous; shorter ramus of outer antennule about equal in length to peduncle, fused for about two-fifths its length to longer flagellum. Outer edge of stylocerite straight or very slightly concave; anterior border convex, becoming markedly concave adjacent to apical spine. Scaphocerite extends to proximal half of pereopod 2 propodus; apical spine not exceeding lamellar portion. Pereopod 2 dactyl one-third length of propodus; carpus equal to or very slightly longer than merus.

Intertidal, usually under rocks and stones; all coasts, possibly scarcer in the north.

Palaemon longirostris Milne Edwards Fig. 8.43

Length up to 77 mm; usually almost colourless, but speckled with small red chromatophores when viewed closely. Antennules triramous; shorter ramus of outer antennule about two-thirds length of peduncle, fused for about one-third its length to longer flagellum. Stylocerite border slightly concave, anterior border convex. Scaphocerite extends to distal half of pereopod 2 carpus; apical spine not exceeding lamellar portion. Pereopod 2 dactyl about two-fifths length of propodus, but variable.

Estuarine and brackish-water species, may occur in dense shoals; south and south-east coasts only.

Palaemon serratus (Pennant). Common prawn Fig. 8.42

Length up to 110 mm; colour variable, pereon and pleon often banded with brownish-red, plus horizontal lines on pereon. Carapace with antennal and branchiostegal spine. Antennules triramous; shorter ramus of outer antennule about six-sevenths length of peduncle, fused for one-fifth to one-quarter its length to longer flagellum. Outer edge of stylocerite very slightly concave, anterior border virtually flat, spine long and stout. Scaphocerite extends to half length of pereopod 2 dactyl, or to tip of dactyl in juveniles; spine does not exceed lamellar portion. Pereopod 2 dactyl about half length of propodus.

Intertidal to 40 m, mainly in rock pools, under ledges and in weed; frequently on west and south coasts. Scarce on north-east coast north of Thames.

Palaemonetes varians (Leach) Fig. 8.43

Length up to 50 mm; almost colourless. Carapace with antennal and branchiostegal spine. Antennules triramous; shorter ramus of outer antennule about four-fifths length of peduncle, fused for about three-quarters of its length to longer flagellum. 0uter edge of stylocerite slightly concave, anterior border straight, spine short and stout. Scaphocerite extends into proximal half of pereopod 2 propodus (or to half length of dactyl in juveniles); spine not exceeding lamellar portion. Pereopod 2 dactyl little over one-third length of propodus; carpus about 1.2 × length of merus, but occasionally up to 1.7 × length of merus.

Brackish or almost freshwater pools and ditches, saltmarshes, may occur in dense shoals; all coasts, but scarce in northern Scotland.

Typton spongicola Costa Fig. 8.43

Length up to 25 mm; yellow-red or orange. Carapace with prominent supraorbital spine, about two-thirds length of rostrum in adults. Antennular peduncle 1.6–2.3 × length of rostrum; stylocerite very small. Scaphocerite rudimentary, about one-half to two-thirds length of eye. Maxilliped 3 stout, about two-thirds length of antennular peduncle. Pleon segment 6 with pointed lateral process and posteriorly directed process on each side, plus small median posteriorly directed tooth. Telson with two, rarely three pairs of lateral spines; tip slightly convex.

Sublittoral, 8–90 m, living in sponges; south-west coast only, very scarce.

3. ALPHEIDAE

Rostrum small, reduced, or absent, never spinose. Carapace sometimes with supraorbital and pterygostomial spines, but no antennal spine. Anterior projections of carapace usually forming hood over the short stalked eyes, and partially or wholly concealing them in dorsal view. Mandible with two-articled palp and incisor process. Maxilliped 2 with article 7 attached laterally to article 6. Maxilliped 3 with exopod, epipod present or absent. Pereopod 1 usually strong, robustly chelate, often asymmetrical, especially in the male. Pereopod 2 usually with five-segmented carpus, minutely chelate. Pereopod 5 with series of spines on outer, posterior surface of article 6. Pereopods 3–5 with simple dactyl, or with not more than two accessory denticles. Telson linguiform, usually rather short and broad; anal tubercles may be present. An articulated process at base of uropod in some genera. Gills: five pleurobranchs, one arthobranch, rudimentary or absent on maxilliped 3, plus 2–8 epipods. Three British species.

---

1. Eyes completely covered in dorsal view by projecting anterior margins of carapace, forming orbital hoods. Rostrum short, not reaching distal end of basal article of antennular peduncle. No articulate process on segment 6 at base of uropods **2**

   Eyes only partially covered by projecting anterior margins of carapace. Rostrum well developed, reaching beyond distal end of basal article of antennular peduncle. An articulated process on segment 6 at base of uropods ***Athanas nitescens***

---

2. Orbital hoods produced into a short spine over each eye. Chela of pereopod 1 without longitudinal ridge, dactyl articulating with propodus by lateral and oblique movement ***Alpheus macrocheles***

   Orbital hoods with rounded markings, without spines. Chelae of first pereopods markedly dissimilar in size and shape, articulation of dactyl normal ***Alpheus glaber***

Athanas nitescens (Leach) Fig. 8.44

Length up to 20 mm; green, blue, or red-brown, often with white dorsal stripe. Stylocerite acutely pointed, two-thirds to three-quarters length of scaphocerite. Maxilliped 3 about equal in length to scaphocerite.

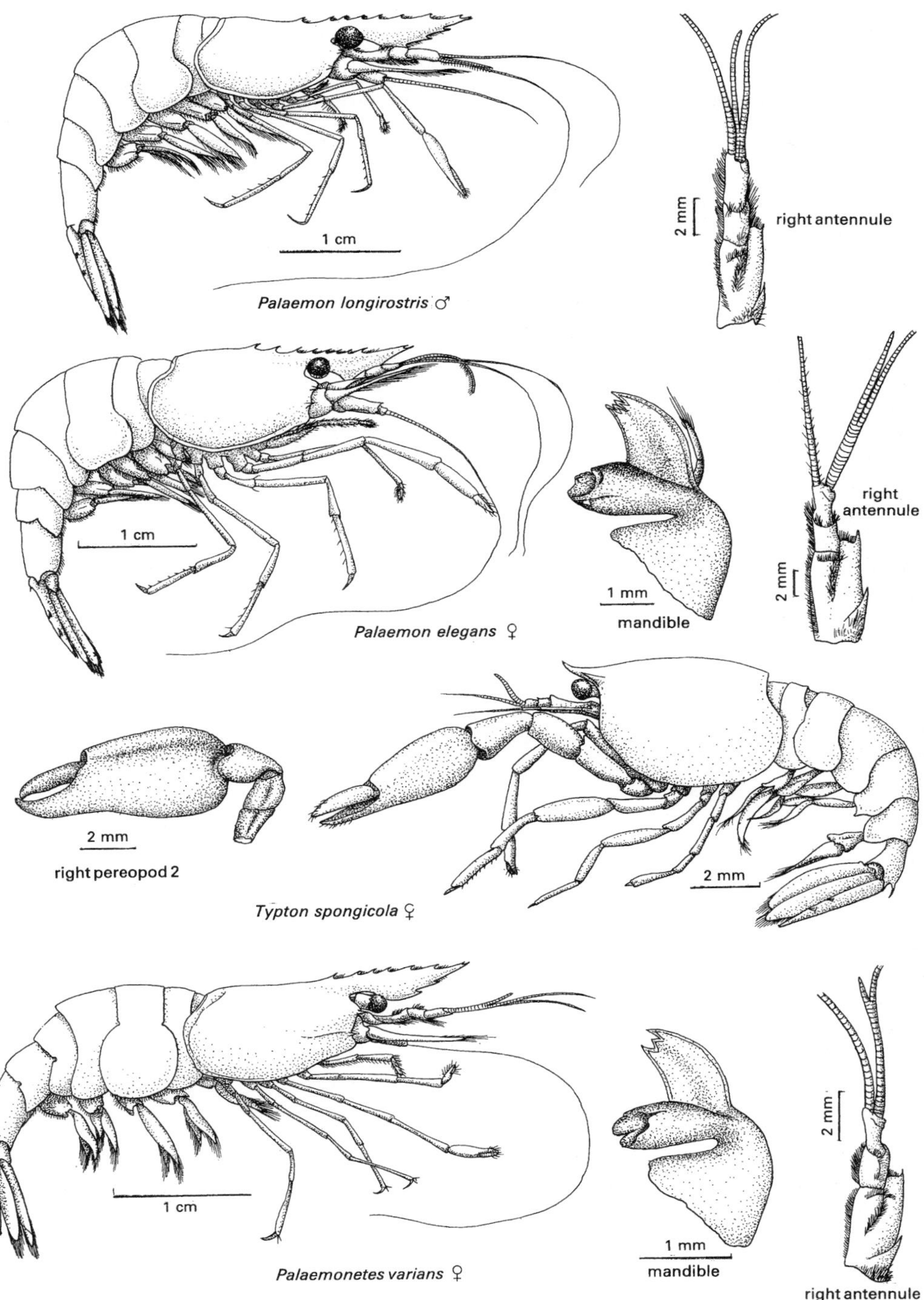
1 cm
Palaemon longirostris ♂
2 mm
right antennule
1 cm
Palaemon elegans ♀
1 mm
mandible
right
antennule
2 mm
2 mm
right pereopod 2
2 mm
Typton spongicola ♀
1 cm
Palaemonetes varians ♀
1 mm
mandible
2 mm
right antennule

Fig. 8.43

Lower shore to about 60 m, often under stones in gravel-floored pools; south and west coasts, scarce in north-east.

Alpheus glaber (Olivi) Fig. 8.44
Length up to 43 mm; usually red dorsally, with lateral borders of carapace and pleon white. Rostrum narrow, straight, unarmed, apex acutely pointed. Stylocerite acutely pointed, about one-third length of scaphocerite. Maxilliped 3 slightly longer than scaphocerite.

Sublittoral, 33–550 m; south and west coasts only, rare.

Alpheus macrocheles
(Hailstone). Snapping prawn Fig. 8.44
Length up to 35 mm; orange-red. Rostrum short, not reaching to distal end of basal article of antennular peduncle. Stylocerite acutely pointed, extending to distal end of basal article of antennular peduncle. Maxilliped 3 about as long as antennal peduncle.

Lower shore to 185 m; south coast only, very scarce.

4. HIPPOLYTIDAE

Rostrum long or short. Supraorbital spine present or absent. Mandible with or without palp, with or without incisor process. Maxilliped 2 with article 7 attached laterally to article 6. Exopod of maxilliped 1 with flagellum. Maxilliped 3 with or without exopod. Exopods absent from all pereopods. One to seven pairs of epipods. Pereopod 1 usually not much longer than others, but sometimes sexually dimorphic; chelate. Pereopod 2 carpus with two or more segments. Telson tapering more or less acutely. Gills 5–6, with varying number of epipods. A large family with 17 species recorded in British waters.

---

1. Pereopod 2 carpus consisting of two segments **5**

   Pereopod 2 carpus consisting of more than two segments **2**

---

2. Pereopod 2 carpus with three segments **6**

   Pereopod 2 carpus with more than three segments **3**

---

3. Pereopod 2 carpus with six segments. Mandibular palp absent. Epipods on pereopods 1 and 2 ***Thoralus cranchii***

   Pereopod 2 carpus with seven segments **4**

---

4. Carapace without supraorbital spines **7**

   Carapace with one or two supraorbital spines **8**

---

5. Rostrum nearly straight, about half length of carapace and always slightly longer than antennular peduncle; 5–10 dorsal teeth (usually seven or eight), one ventral tooth (rarely two or three). Maxilliped 3 always longer than scaphocerite by as much as the last two articles. Sublittoral ***Caridion gordoni***

   Rostrum curving downwards slightly, deeper than *C. gordoni*; slightly shorter than antennular peduncle; 5–7 dorsal teeth, one ventral tooth. Maxilliped 3 only slightly longer than scaphocerite. Shore and shallow sublittoral ***Caridion steveni***

---

6. Rostrum longer than carapace, usually without dorsal tooth at base in adults; often present in small specimens, and in adults as a small tubercle. Rostrum 2.25 × length of antennular peduncle. Maxilliped 3 is 0.25 × length of scaphocerite, little more in large specimens. Scaphocerite 4.5 × longer than broad ***Hippolyte inermis***

   Rostrum almost as long as carapace, always with dorsal tooth at base; 1.5 × length of antennular peduncle, little more in large specimens. Maxilliped 3 half length of scaphocerite, which is 3 × longer than broad ***Hippolyte varians***

   Rostrum almost as long as carapace in adults, less in juveniles. Posterior dorsal border of rostrum with 2–4 teeth; rostrum half to two-thirds length of

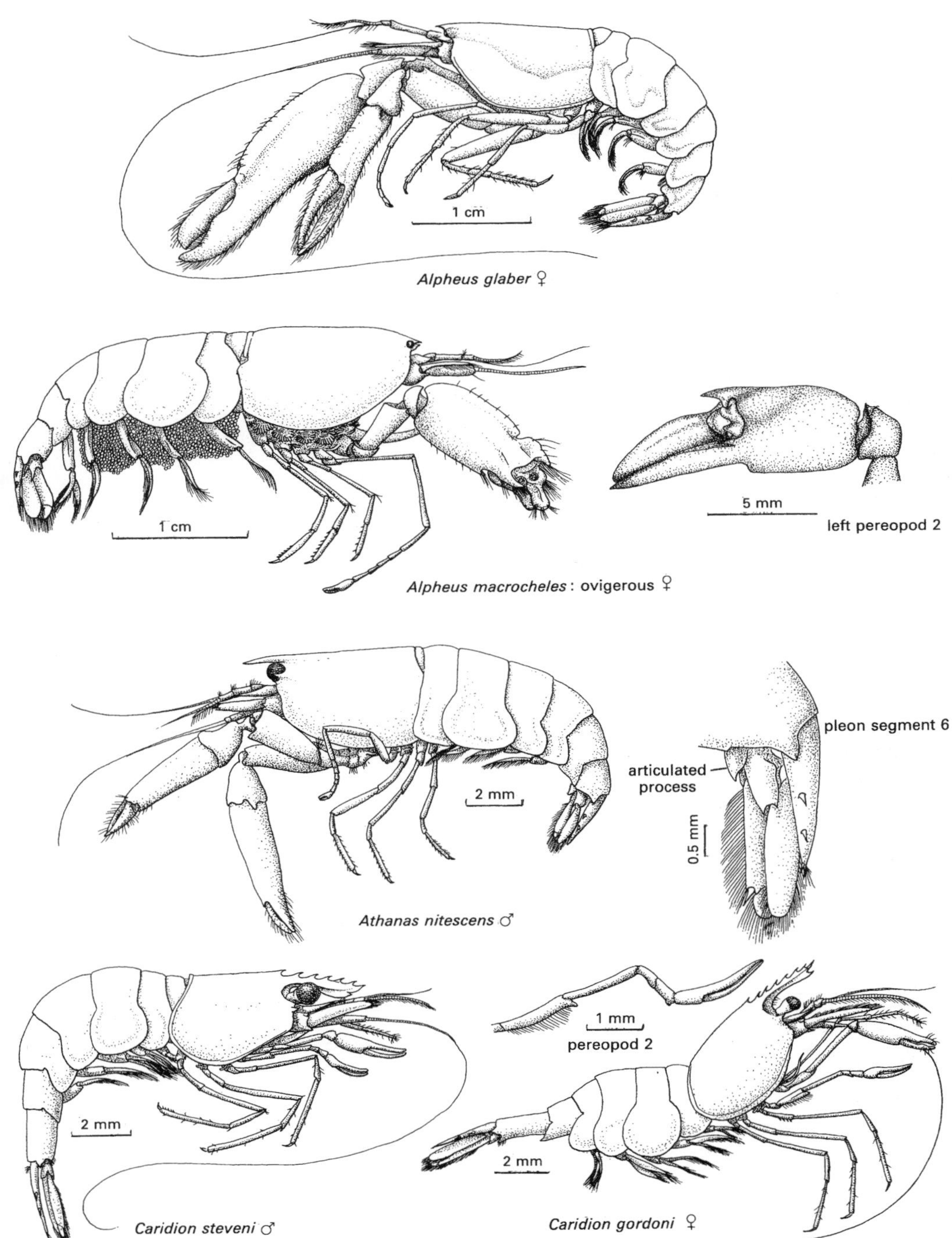
1 cm
Alpheus glaber ♀
1 cm
5 mm
left pereopod 2
Alpheus macrocheles: ovigerous ♀
pleon segment 6
articulated process
0.5 mm
2 mm
Athanas nitescens ♂
1 mm
pereopod 2
2 mm
2 mm
Caridion steveni ♂
Caridion gordoni ♀

Fig. 8.44

antennular peduncle. Maxilliped 3 about half length of scaphocerite, which is about 2.8 × longer than broad. Carapace and pleon segments with tufts of setae dorsally. Very scarce
Hippolyte longirostris

---

7. Rostrum twice length of antennular peduncle, little more in large specimens; 5–7 dorsal teeth, 3–5 ventral teeth. Epipods on pereopods 1 and 2
Eualus gaimardii

Rostrum equal to or little over half length of antennular peduncle; four dorsal teeth; tip bi- or tridentate (rarely a single point). Epipods on pereopods 1 and 2 ***Eualus occultus***

Rostrum half length of antennular peduncle; 2–5 dorsal teeth; tip a single point (rarely bidentate). Epipods on pereopods 1–3 ***Eualus pusiolus***

---

8. Carapace with one supraorbital spine. Maxilliped 3 without exopod. Inner basal part of eyestalk without tubercle. Rostrum with 5–7 dorsal teeth, 2–4 ventral teeth. ***Lebbeus polaris***

Carapace with two supraorbital spines. Maxilliped 3 with exopod. Tubercle on inner basal part of eyestalk. **9**

---

9. Dorsal rostral teeth extend almost to posterior edge of carapace; teeth may have serrated margins in large specimens. Apex of rostrum with two points, with arcuate space or smaller teeth between. Pleon segment 3 produced posteriorly into a very distinct hook over segment 4
Spirontocaris spinus

Dorsal rostral teeth extend only to two-thirds length of carapace. Apex of rostrum with one point, with successive teeth on ventral border posterior to the apical point. Pleon segment 3 only slightly produced over segment 4
Spirontocaris lilljeborgi

Caridion gordoni (Bate) Fig. 8.44

Length up to 27 mm; transparent, with red chromatophores on carapace. Carapace with antennal spine. Stylocerite narrow, acutely pointed, two thirds length of antennular peduncle or slightly less. Scaphocerite outer margin straight or slightly concave, apex acutely rounded, spine not exceeding lamellar portion. Antennal flagellum about three-quarters length of body. Pleon segment 4 usually without ventro-posterior tooth in female. Telson with two pairs of lateral spines.

Sublittoral 10–500 m; all coasts.

Caridion steveni Lebour Fig. 8.44

Length up to 27 mm; red, with diffuse yellow. Carapace with antennal spine. Stylocerite narrow, acutely pointed, half to two-thirds length of antennular peduncle. Scaphocerite outer margin straight, apex less acute than *C. gordoni*, spine not exceeding lamellar portion. Antennal flagellum usually as long as body, sometimes a little shorter. Pleon segment 4 with ventro-posterior tooth in both sexes. Telson with two pairs of lateral spines.

Lower shore to 30 m; all coasts.

Eualus gaimardii (Milne Edwards) Fig. 8.45

Length up to 100 mm; translucent, with reddish-brown spots and markings. Carapace with antennal and pterygostomial spine. Scaphocerite outer margin straight, becoming slightly concave towards apex; twice length of antennular peduncle, apical spine not exceeding lamellar portion. Maxilliped 3 equal to or slightly shorter than scaphocerite. Pleon segment 4 with acute ventro-posterior spine. Telson with 3–5 pairs of lateral spines.

From ELWS to 300 m; north, north-west, and north-east coasts only.

Eualus occultus (Lebour) Fig. 8.45

Length up to 22 mm. Colour variable, may be dark brownish-green, with stripes of red-brown on body and extremities. Carapace with antennal spine. Scaphocerite outer margin slightly concave, apical spine equal to or slightly shorter than lamellar portion; Scaphocerite extends to about half length

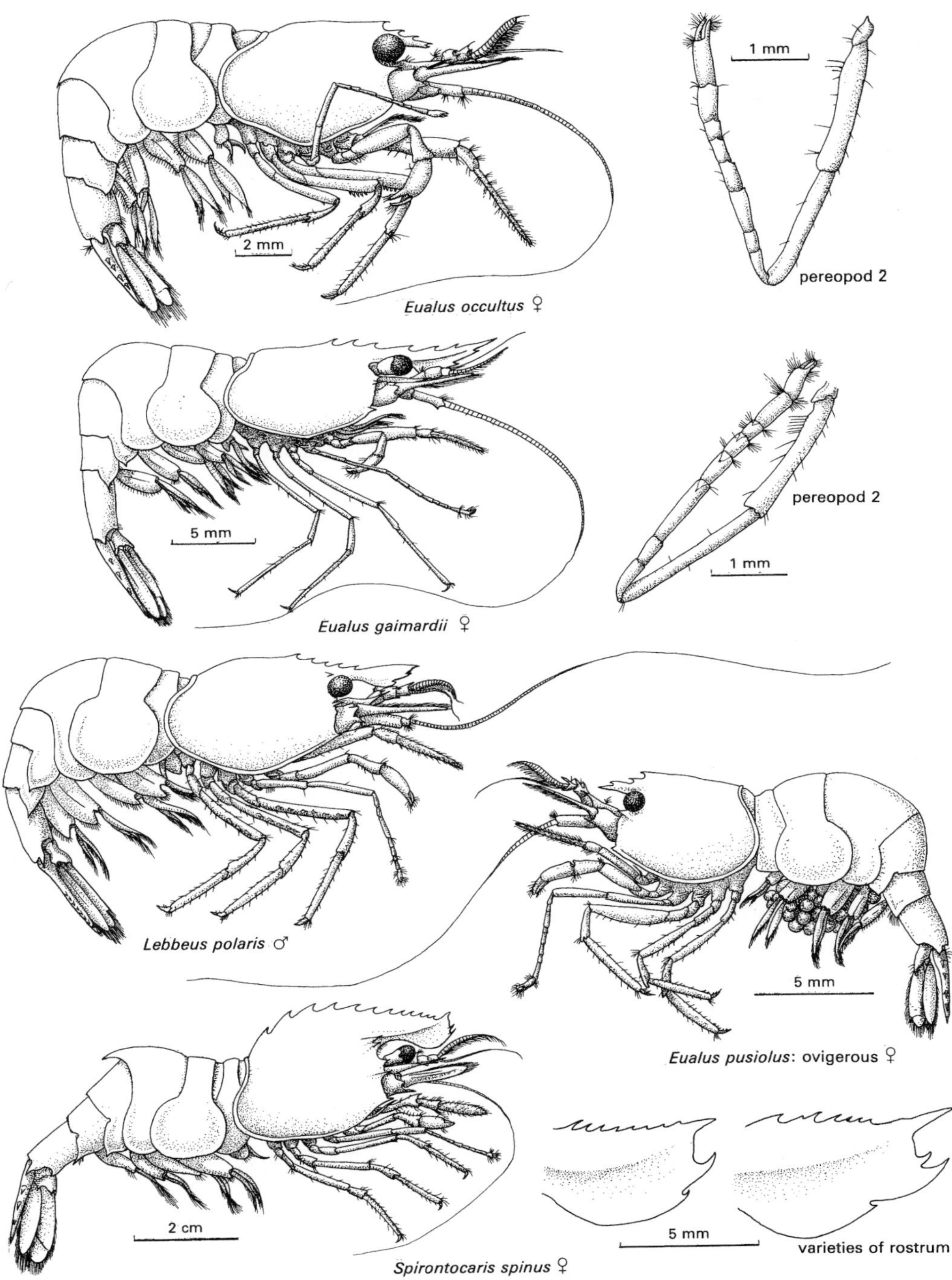
1 mm
pereopod 2
2 mm
Eualus occultus ♀
pereopod 2
5 mm
1 mm
Eualus gaimardii ♀
Lebbeus polaris ♂
5 mm
Eualus pusiolus: ovigerous ♀
2 cm
5 mm
varieties of rostrum
Spirontocaris spinus ♀

Fig. 8.45

of last article of maxilliped 3, sometimes slightly less. Maxilliped 3, 1.5–1.6 × length of scaphocerite. Telson usually with four or five pairs of lateral spines, occasionally two or three pairs.

From LWS to about 80 m; under stones and in pools. South and west coasts only.

Eualus pusiolus (Krøyer) Fig. 8.45
Length up to 28 mm; colour variable, green, red-brown, pink, speckled. Carapace with antennal spine. Scaphocerite outer margin slightly concave; 1.3 × length of antennular peduncle, spine exceeding lamellar portion or equal in length. Maxilliped 3 up to 1.25 × length of scaphocerite. Telson usually with four or five pairs of lateral spines.

From ELWS to about 500 m, under stones and in pools. All coasts, possibly scarcer in south.

Lebbeus polaris (Sabine) Fig. 8.45
Length up to 90 mm, usually 60-70 mm; pale, with red and yellowish markings on carapace and pleon, tips of chelae brownish black. Rostrum almost straight, or very slightly downcurved. Antennular peduncle with strong, anteriorly directed spine on lateral border. Scaphocerite broad, outer border straight or very slightly concave, apical spine not exceeding lamellar portion. Pleon segment 4 with small ventro-posterior spine, segment 6 with strong, acute, postero-ventral projection, and additional projection on each side of telson insertion. Telson with 8–11 pairs of lateral spines; not evenly spaced, and variable, often with different numbers on each side, from four or five to 11 or 12, sometimes in double rows on each lateral border of telson.

Sublittoral, from 0 to 930 m, most usually 30–300 m. Reaches southern limit of its distribution off Shetland; formerly reported off the Hebrides.

Spirontocaris lilljeborgi (Danielssen) Fig. 8.46
Length up to 74 mm; usually bright red, sometimes mottled brownish-red. Rostrum about 1.5 × length of antennular peduncle; dorsal teeth extend along about two-thirds length of carapace, making an angle of about 30° with posterior border of carapace; apex of rostrum normally a single acute point, with series of smaller teeth posterior and ventral to point; ventral border less convex than in *S. spinus*. Stylocerite acutely pointed, three-quarters to five-sixths length of antennular peduncle. Maxilliped 3, 1–1.25 × length of scaphocerite. Telson with three or four pairs of lateral spines.

Sublittoral, 20–1000 m; north-east and north-west coasts, scarce in south.

Spirontocaris spinus (Sowerby) Fig. 8.45
Length up to 60 mm; usually bright red. Rostrum about 1.5 × length of antennular peduncle; dorsal rostral teeth extend almost to posterior border of carapace, making an angle of about 45° with posterior border; dorsal teeth occasionally serrated in large specimens. Apex of rostrum with two points, with arcuate space or series of small teeth between; ventral border deeply convex. Stylocerite acutely pointed, equal to or slightly longer than antennular peduncle. Maxilliped 3 about equal in length to scaphocerite. Telson with 3–5 pairs of lateral spines.

Sublittoral, 20–400 m; north-east and north-west coasts, apparently absent from south and south-west coasts.

Hippolyte inermis Leach Fig. 8.46
Length up to 42 mm; usually green, occasionally crimson or brown specimens. Carapace with one supraorbital spine, plus pterygostomial and hepatic spine. Stylocerite acutely pointed, three-fifths length of antennular peduncle. Telson with two pairs of lateral spines.

LWS, about 50 m; south and west coasts only.

Hippolyte longirostris (Czerniavsky) Fig. 8.46
Length up to 20 mm; colour variable, greenish-brown, almost transparent with flecks of red-brown. Carapace with one supraorbital spine, plus antennal and hepatic spines. Stylocerite acutely pointed, about half length of antennular peduncle, less in juveniles. Telson with two pairs of lateral spines.

Lower shore, in weed, pools, and along shore edge. Channel coast only, probably more frequent in south-west.

Hippolyte varians Leach.
Chameleon prawn Fig. 8.46
Length up to 32 mm; colour variable; red, brown, green, flecked reddish-brown, almost transparent. Carapace with one supraorbital spine, plus antennal and hepatic spine; fascigerous tufts occasionally present on dorsal carapace and pleon segments. Stylocerite acutely pointed, little over half length of antennular peduncle. Telson with two pairs of lateral spines.

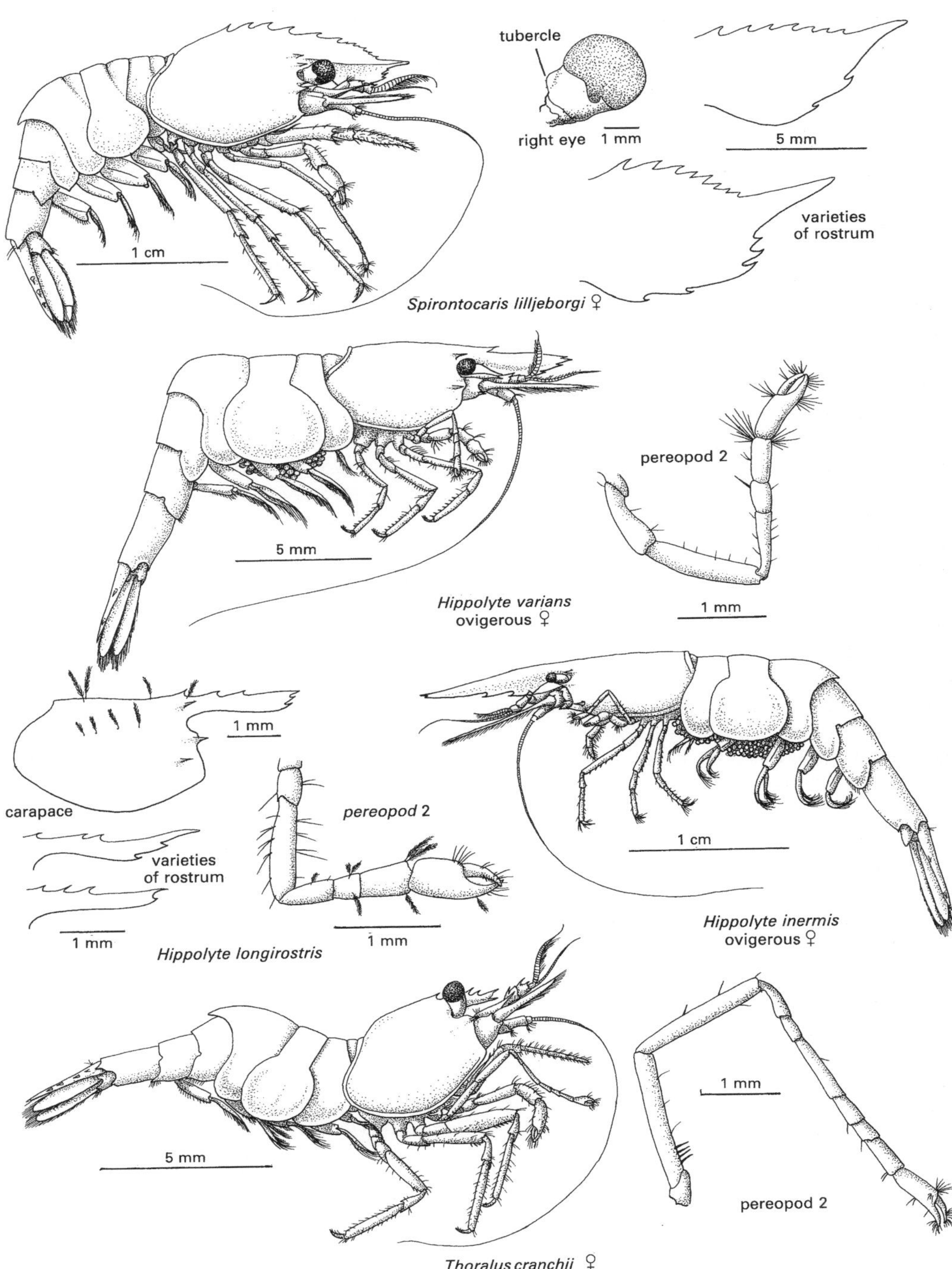
tubercle
right eye
1 mm
5 mm
1 cm
varieties of rostrum
Spirontocaris lilljeborgi ♀
pereopod 2
5 mm
1 mm
Hippolyte varians ovigerous ♀
1 mm
carapace
pereopod 2
varieties of rostrum
1 mm
1 mm
Hippolyte longirostris
1 cm
Hippolyte inermis ovigerous ♀
1 mm
5 mm
pereopod 2
Thoralus cranchii ♀

Fig. 8.46

Lower shore to 150 m, but most frequent intertidally, in weed, pools, and along shore edge; all coasts.

Thoralus cranchii (Leach) Fig. 8.46

Length up to 22 mm; usually semi-transparent with purplish-brown blotches; occasionally dark brownish-green. Rostrum little over half length of antennular peduncle; three or four (rarely five or six) dorsal teeth, tip bi- or tridentate (rarely a single point). Carapace without supraorbital spines, but with antennal spine. Scaphocerite outer margin straight or very slightly concave; apical spine only very slightly shorter than lamellar portion. Maxilliped 3 extends to just beyond tip of scaphocerite. Telson with 2–7 pairs of lateral spines, usually with four pairs.

From lower shore to about 70 m, under stones and in pools; all coasts.

5. PROCESSIDAE

Rostrum short, unarmed. Mandible without incisor process or palp. Maxilliped 2 with article 7 attached laterally to article 6. Maxilliped 3 large, pediform, with exopod. Exopods absent from all pereopods, or present only on pereopod 1. Pereopods without epipods. First pair of pereopods asymmetrical, one simple and one chelate. Second pair of pereopods usually unequal, one much longer than the other, both chelate; carpus and merus multi-articulate. Telson channelled. Gills: five plus two epipods. Six species recorded in British waters.

---

1. Stylocerite with tooth on anterior external corner. Ventro-posterior corner of pleon segment 5 rounded or angular, but without distinct tooth **2**

   Stylocerite without tooth on anterior external corner. Ventro-posterior corner of pleon segment 5 with at least one tooth directed posteriorly. Pereopod 2 *left*: merus 5–7 articles, carpus 17–20 (to 22) articles. Pereopod 2 *right*: merus 12–18 (to 21) articles, carpus 30–34 (to 39) articles **_Processa edulis_ subsp. _crassipes_**

---

2. Second pereopods equal in length, merocarpal articulation reaching half-way along eye. Pereopod 2 merus of four or five articles, carpus of 11 (to 15) articles. Tooth of stylocerite below lamellar portion **_Processa modica_**

   Second pereopods clearly unequal in length **3**

---

3. Rostrum, in profile, deepest in middle. Pleon segment 5 ventrally convex; segment 6 with prominent ventro-posterior spine. Scaphocerite reaching to distal end of antennular peduncle or little further; sides nearly parallel **_Processa nouveli_ subsp. _holthuisi_**

   Rostrum deepest, in profile, in posterior half. Pleon segment 5 ventrally straight; segment 6 with short ventro-posterior tooth. Scaphocerite reaching beyond antennular peduncle by half maximum width; inner margin sinuous, outer margin slightly convex **_Processa canaliculata_**

Processa canaliculata Leach Fig. 8.47

Length up to 74 mm; dull whitish, with greenish tinge at anterior end of carapace. Carapace with antennal spine. Maxilliped 3 about 1.3 × length of scaphocerite. Pereopod 2 *right*: carpus 41–43 segments, merus 16–19 (rarely up to 24) segments; pereopod 2 *left*: carpus 18–22 (rarely up to 28) segments; merus 4–8 (rarely up to 11 segments).

Shallow sublittoral, often on sandy shores, to about 200 m; west, south-west, and north-east coasts.

Processa nouveli subsp. *holthuisi* Al-Adhub and Williamson Fig. 8.47

Length up to 51 mm; whitish, spotted variably with purple and red. Carapace with antennal spine. Maxilliped 3 about 1.5 × length of scaphocerite. Length of eye 1–1.25 × length of rostrum; width of eye in dorsal view 1.5–1.8 × width of scaphocerite. Pereopod 2 *right*: carpus 29–42 segments, merus 10–20 segments; pereopod 2 *left*: carpus 15–19 segments, merus 5 segments.

Sublittoral, 30–230 m; west coast, central and northern North Sea.

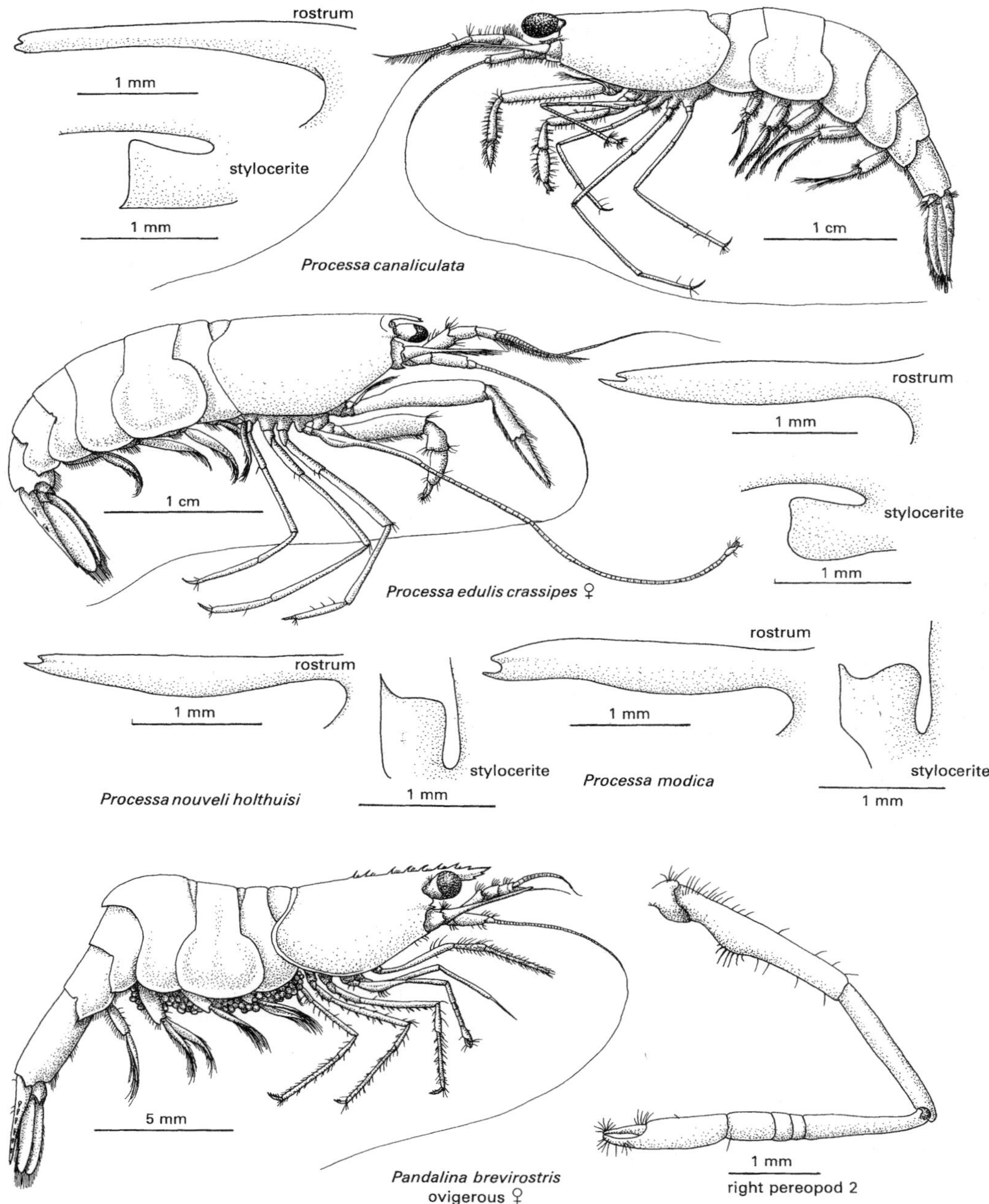

Fig. 8.47

Processa edulis subsp. *crassipes*
Nouvel and Holthuis Fig. 8.47

Length up to 44 mm; whitish, with pale green hue. Rostrum slightly downturned, little shorter than eye; ventral tooth at tip longer than dorsal. Carapace with antennal spine. Scaphocerite with outer margin virtually straight, lamellar portion exceeding apical spine. Maxilliped 3 about 1.75 × length of scaphocerite. Eye globular, as wide as long. Merocarpal articulation of right pereopod 2 not reaching beyond distal extremity of penultimate segment of maxilliped 3.

Lower shore to 20 m, often on sandy shores; south and west coasts only.

Processa modica Williamson Fig. 8.47

Length up to 33 mm. Rostrum slightly down curved at tip, dorsal tooth slightly shorter than ventral. Carapace with antennal spine. Pleon segment 6 with prominent postero-ventral spine.

Shallow sublittoral to 100 m; North Sea, not known in Channel or Irish Sea.

6. PANDALIDAE

Rostrum well developed. Mandible deeply cleft into molar and incisor portions, palp usually three-articled. Maxilliped 2 with article 7 attached laterally to article 6. Exopod of maxilliped 1 with flagellum. Maxilliped 3 with or without exopod; exopods absent from all pereopods. Epipods on maxillipeds 1–3, present or absent on pereopods 1–4. Pereopod 1 simple, or microscopically and imperfectly chelate; pereopod 2 minutely chelate, with bi-, tri-, or multiarticulate carpus; pereopods 3–5 long and slender. Telson acute. Five British species.

---

1. Rostrum as long as, or longer than carapace, curving upwards **2**

   Rostrum not more than half carapace length, virtually straight, with seven or eight dorsal teeth and two or three ventral teeth. Right pereopod 2 carpus with four segments ***Pandalina brevirostris***

---

2. Maxilliped 3 without an exopod **3**

   Maxilliped 3 with exopod. Rostrum with 8–10 dorsal teeth and 6–8 ventral teeth. Right pereopod 2 carpus with five segments (rarely six or seven) ***Dichelopandalus bonnieri***

---

3. Right pereopod 2 carpus with 20 or more segments **4**

   Right pereopod 2 carpus with five segments. Rostrum with eight or nine dorsal teeth and six or seven ventral teeth. Scaphocerite narrowing towards tip, about one-third to one-half length of rostrum, with outer edge slightly concave ***Pandalus propinquus***

---

4. Rostrum with 12–16 dorsal teeth and 6–8 ventral teeth. Dorsal teeth extend into anterior third of rostrum. Lamellar portion of scaphocerite extends beyond apical spine. Right pereopod 2 carpus with 23–36 segments ***Pandalus borealis***

   Rostrum with 10–12 dorsal teeth and 5–7 seven ventral teeth. Dorsal teeth do not extend beyond middle of rostrum. Apical spine of scaphocerite exceeds lamellar portion. Right pereopod 2 carpus with 20–22 segments ***Pandalus montagui***

Pandalina brevirostris (Rathke) Fig. 8.47

Length up to 33 mm; whitish, with many red and yellow chromatophores. Carapace with antennal and pterygostomian spine. Stylocerite broadly rounded, shorter than eye. Scaphocerite outer margin slightly convex, apical spine exceeding lamellar portion. Maxilliped 3, 1–1.2 × length of scaphocerite. Carpus of left pereopod 2 with 14–20 segments. Telson with 6–9 pairs of lateral spines.

Sublittoral, 10–100 m; all coasts.

Pandalus borealis Krøyer Fig. 8.48

Length up to 120 mm or larger; pale red, pleon often deeper red. Carapace with strong antennal spine, small pterygostomian spine. Stylocerite broadly rounded, shorter than eye. Outer margin of scaphocerite very slightly convex. Carpus of left pereopod 2 with 50–60 segments. Telson with 8–11 pairs of lateral spines.

Sublittoral, 20–600 m; north-east coast only.

Pandalus montagui Leach.
Aesop prawn Fig. 8.48

Length up to 160 mm, but usually less than 100 mm; semi-translucent with patches of red on carapace and pleon. Carapace with strong antennal spine and small pterygostomian spine. Stylocerite broadly rounded, shorter than eye. Carpus of left pereopod 2 with 50–65 segments. Telson with 5–7 pairs of lateral spines.

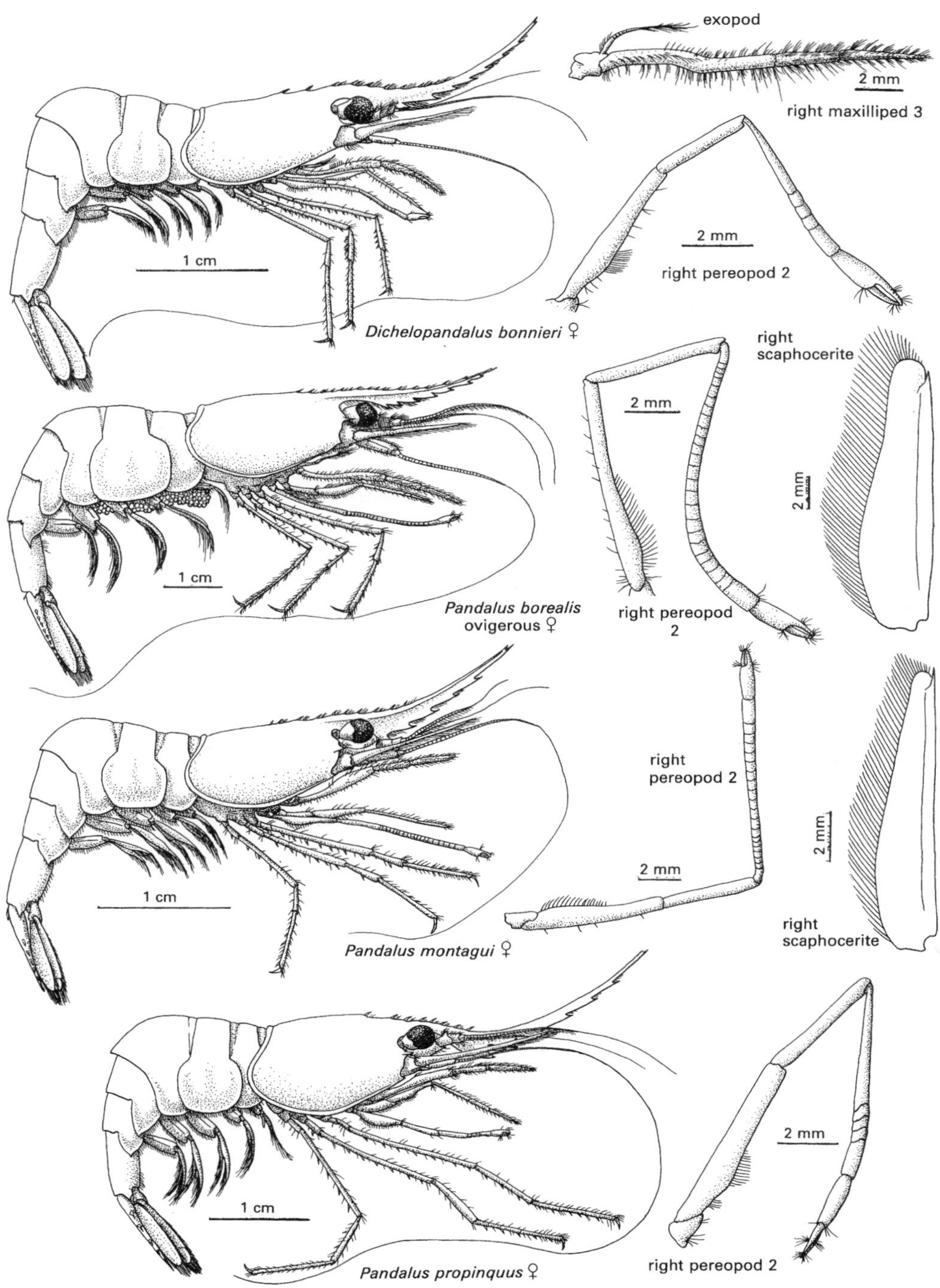
exopod
2 mm
right maxilliped 3
2 mm
right pereopod 2
1 cm
Dichelopandalus bonnieri ♀
right scaphocerite
2 mm
2 mm
1 cm
Pandalus borealis ovigerous ♀
right pereopod 2
right pereopod 2
2 mm
2 mm
1 cm
Pandalus montagui ♀
right scaphocerite
2 mm
1 cm
Pandalus propinquus ♀
right pereopod 2

Fig. 8.48

Sublittoral, 5–230 m, usually 20–150 m, juveniles sometimes intertidal; all coasts.

Pandalus propinquus G. O. Sars Fig. 8.48
Length up to 150 mm, but usually less than 100 mm; pale red. Carapace with strong antennal spine, small pterygostomial spine. Stylocerite broadly rounded, shorter than eye. Carpus of left pereopod 2 with 25–30 segments. Telson with five or six pairs of lateral spines.

Offshore, 40–2000 m; all coasts, possibly scarcer in north-east.

Dichelopandalus bonnieri Caullery Fig. 8.48
Length up to 160 mm, usually less than 100 mm; pale reddish, tip of rostrum bright red. Carapace with strong antennal spine, small pterygostomian spine. Stylocerite broadly rounded, shorter than eye. Scaphocerite outer margin straight, but often slightly concave in juveniles; apical spine slightly exceeds lamellar portion. Carpus of left pereopod 2 with 38–45 segments, although normally about 30 in juveniles. Rarely these pereopods are reversed, i.e. right with 38–45 segments. Telson with 6–8 pairs of lateral spines.

Sublittoral 33–400 m; south and west coasts, less common in North Sea, not in Channel.

7. CRANGONIDAE

Rostrum short or spiniform. Carapace sometimes more or less sculptured. Eyes well developed. Mandible simple, without palp. Maxilliped 2 with article 7 small, attached obliquely at apex of article 6. Maxilliped 3 with exopod, epipod present or absent. Pereopod 1 strong, subchelate; pereopod 2 slender, sometimes reduced, in one genus absent; pereopod 3 slender, pereopods 4 and 5 more robust, sometimes with dilated dactyls. Telson tapering. No epipods on pereopods. Exopods on pereopods, if present, on pereopod 1 only. Gills: 5–8 plus two or three epipods. Eleven species are recorded for Britain.

---

1. Pereopod 2 extends to about three-quarters length of pereopod 1 propodus. Pereopod 2 dactyl less than half length of propodus **8**

   Pereopod 2 extends at most to proximal quarter of pereopod 1 propodus, often less. Pereopod 2 dactyl half length of propodus or more **2**

---

2. Pereopod 1 with small exopod. Stylocerite acutely pointed **3**

   No exopod on pereopod I. Stylocerite rounded or truncate **4**

---

3. Two spines on first lateral ridge of carapace, one spine on second ridge ***Pontophilus norvegicus***

   Three spines on first lateral ridge of carapace, two spines on second ridge ***Pontophilus spinosus***

---

4. Rostrum with rounded or triangular apex **5**

   Rostrum with truncate or emarginate apex **7**

---

5. Carapace with one or three spines on median line **6**

   Carapace with two spines on median line, posterior spine may be reduced to a tubercle ***Philocheras bispinosus***

---

6. Carapace with one spine on median line, slightly in advance of lateral spine on each side. Rostrum broadly triangular at apex ***Philocheras trispinosus***

   Carapace with three spines on median line, small tubercle usually before anterior spine. Six (rarely five) spines on first lateral ridge, two (rarely three) spines on second lateral ridge ***Philocheras echinulatus***

---

7. Rostrum broadly truncate at apex. One spine on median line of carapace. Stylocerite broadly rounded. Endopods of pleopods 2–5 without appendix interna in both sexes ***Philocheras fasciatus***

Rostrum truncate and emarginate at apex. One spine, plus posterior depressed spine, on median line of carapace. Usually five peaks on first lateral ridge; spine and posterior peak on second lateral ridge. Scaphocerite with spine half-way along outer edge. Endopods of pleopods 2–5 with appendix interna in male, pleopods 2–4 with appendix interna in female
Philocheras sculptus

---

8. Pleon segment 6 smooth on dorsal side
Crangon crangon

Pleon segment 6 with deep dorsal longitudinal groove and parallel ridges
Crangon allmanni

Crangon allmanni Kinahan Fig. 8.49
Length up to 75 mm, usually less; brownish-grey. Rostrum about half length of eye, narrow with rounded apex. Maxilliped 3, 1–1.2 × length of scaphocerite. Endopod of pleopod 2 one-fifth to one-quarter length of exopod in male, and one-quarter to one-third length of exopod in female.

Sublittoral, 10–250 m; all coasts.

Crangon crangon (Linnaeus)
Common Shrimp Fig. 8.49
Length up to 90 mm; mottled grey or brownish. Rostrum about half length of eye, or slightly more, narrow with rounded apex. Maxilliped 3 about equal in length to scaphocerite. Endopod of pleopod 2 one-fifth length of exopod in male or one-quarter length of exopod in female.

From MTL to about 50 m on sandy shores or in sandy-bottomed pools; all coasts.

Pontophilus norvegicus (M. Sars) Fig. 8.49
Length up to 75 mm; mottled reddish-brown. Rostrum narrow, triangular, with small, anteriorly directed tooth on each lateral border; extends to anterior margin of eye. Carapace with three anteriorly directed spines on median line. Apical spine of scaphocerite slightly shorter than lamellar portion. Pereopod 2 extends to half length of pereopod 1 merus; pereopod 2 dactyl about half length of propodus.

Sublittoral, 50–500 m; northern and western coasts.

Pontophilus spinosus (Leach) Fig. 8.49
Length up to 52 mm; mottled reddish-brown with greenish markings on carapace. Rostrum triangular, with small anteriorly directed tooth on each lateral border; extends to anterior margin of eye. Carapace with three anteriorly directed spines on median line. Apical spine of scaphocerite slightly longer than lamellar portion. Pereopod 2 extends to half length of pereopod 1 merus; pereopod 2 dactyl about half length of propodus.

Sublittoral, 12–150 m; all coasts.

Philocheras bispinosus Hailstone Fig. 8.49
Length up to 26 mm; pale mottled grey or reddish. Rostrum narrow, triangular, with rounded apex, extending to about two-thirds length of eye. Antennal and pterygostomian spine present. Stylocerite broad diamond shape, rounded. Apical spine of scaphocerite equal to or slightly longer than lamellar portion. Pereopod 2 extends to proximal quarter of propodus of pereopod l; pereopod 2 dactyl two-thirds length of propodus, often little more.

Sublittoral, 5–400 m; all coasts.

Philocheras echinulatus (M. Sars) Fig. 8.49
Length up to 45 mm; greenish, with red-brown gastric region. Rostrum triangular, narrow, with rounded apex; extends almost to anterior margin of eye. Stylocerite broad, rounded, diamond-shaped, slightly longer than eye. Scaphocerite with outer margin slightly concave, apical spine stout, exceeding lamellar portion. Pereopod 2 extends to proximal quarter of propodus of pereopod l; pereopod 2 dactyl about five-sixths length of propodus.

Offshore, 60–900 m; west and north-east coasts.

Philocheras fasciatus (Risso) Fig. 8.49
Length up to 19 mm; whitish, with usually dark-brown band on pleon segments 4 and 6. Rostrum extending almost two-thirds length of eye. Scaphocerite apical spine not exceeding lamellar portion. Pereopod 2 extends to proximal quarter of propodus of pereopod l; pereopod 2 dactyl three-quarters length of propodus.

Lower shore and shallow sublittoral to about 50 m on sandy shores and in sandy-bottomed pools; all coasts.

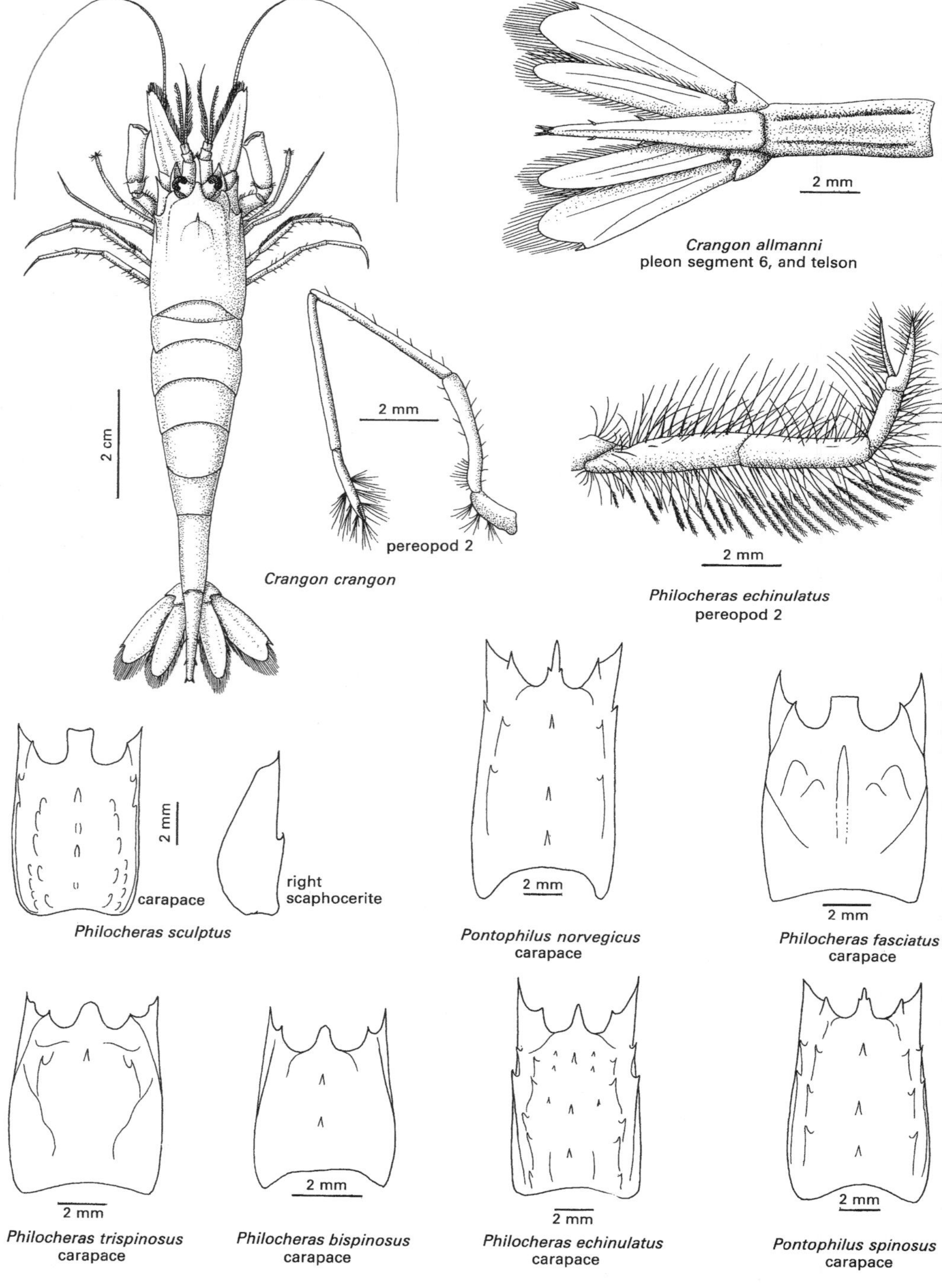
2 mm
Crangon allmanni
pleon segment 6, and telson
2 cm
2 mm
pereopod 2
Crangon crangon
2 mm
Philocheras echinulatus
pereopod 2
2 mm
carapace
right
scaphocerite
Philocheras sculptus
2 mm
Pontophilus norvegicus
carapace
2 mm
Philocheras fasciatus
carapace
2 mm
Philocheras trispinosus
carapace
2 mm
Philocheras bispinosus
carapace
2 mm
Philocheras echinulatus
carapace
2 mm
Pontophilus spinosus
carapace

Fig. 8.49

Philocheras sculptus (Bell) Fig. 8.49

Length up to 24 mm; colour variable, brick-red, brown, pleon sometimes whitish. Rostrum extends three-quarters length of eye. Stylocerite rounded, broadly diamond-shaped, slightly longer than eye. Scaphocerite with apical spine exceeding lamellar portion. Subchelar spine of pereopod 1 mobile. Pereopod 2 extends to proximal quarter of propodus of pereopod 1; pereopod 2 dactyl three-quarters length of propodus.

Sublittoral, 10 to about 230 m; all coasts, possibly more frequent in south.

Philocheras trispinosus Hailstone Fig. 8.49

Length up to 27 mm; yellowish-brown, mottled. Rostrum extends half to three-quarters length of eye. Stylocerite broadly rounded, slightly longer than eye. Scaphocerite apical spine not exceeding lamellar portion. Pereopod 2 extends to distal end of carpus of pereopod 1; pereopod 2 dactyl about three-quarters length of propodus.

From lower shore to about 40 m, on sandy shores and sandy-bottomed pools; all coasts.

[Reptantia]

The reptant decapods comprise four infraorders of widely different form. They are predominantly marine, although estuarine, freshwater, and coastal–terrestrial species occur. Marine Reptantia are most abundant in the shallow waters of the continental shelf seas, where they are primarily scavengers, but some are predators of molluscs, fish, and other crustaceans, while others are microphagous detritus feeders, or even filter feeders.

The infraorders Astacidea and Palinura (Fig. 8.50), comprising the lobsters and crawfish (or crayfish), are characterized by a cylindrical carapace, and an elongate, cylindrical, or compressed abdomen enclosed within a heavily calcified exoskeleton. Typically, they have a well-developed tail fan. In the crawfish (Palinura) the first four pairs of pereopods usually bear simple, terminal claws. In the lobsters (Astacidea) the first three pairs of pereopods are chelate, and the chelae of the first pereopods are very large. The infraorder Anomura includes the porcelain crabs, squat lobsters, and hermit crabs. As in the Astacidea, the first pair of pereopods often bear large chelae, and many anomurans are superficially similar either to lobsters, or to the true crabs. However. in all anomurans the fifth pair of pereopods is reduced in size and generally concealed by the carapace. The Brachyura, or true crabs, are characterised by a flattened, heavily calcified carapace (Fig. 8.41). The abdomen is small, folded beneath the carapace. The first pereopods are developed as massive chelae, and the remainder end in simple claws.

KEY TO FAMILIES

1.	Abdomen long and symmetrical with well-developed tail fan	3
	Abdomen short, or if long, curved asymmetrically	2
2.	Fifth pereopods (and sometimes fourth) much smaller than anterior ones	8
	Fifth pereopods similar to anterior ones; abdomen comparatively small, flattened, lacking pronounced tail fan	11
3.	Chelae present on pereopods 1–3; eyes large and movable	**1. Nephropidae**
	No chelae on third pereopods	4

4. Antenna ending in long flagellum; carapace dorsally convex, spiny; shelf over eyes **2. Palinuridae**

Carapace smooth apart from spiny rostrum, no shelf over eyes **5**

5. Pleura of abdominal segments large, no thalassinian line **3. Axiidae**

Pleura generally small; thalassinian line on carapace (Fig. 8.51) **6**

6. Expodite of uropod with articulated outer article **4. Laomediidae**

Uropod exopodites (and endopodites) simple **7**

7. First pereopods (chelae) equal in size; large triangular rostrum **6. Upogebiidae**

First pereopods unequal; very small rostrum **5. Callianassidae**

8. Abdomen asymmetrical and usually soft; uropods, if present, asymmetrical **9**

Abdomen reduced and curved ventrally, but provided with distinct pleura and terga and symmetrical uropods **10**

9. Abdomen modified for occupying gastropod shell **7. Paguridae**

Abdomen asymmetrical and tightly reflexed under cephalothorax as in Brachyura; no tail fan **8. Lithodidae**

10. Carapace longer than broad; abdomen only loosely reflexed; rostrum large, triangular, and spiny **9. Galatheidae**

Crab-like. Carapace almost circular; abdomen tightly reflexed under sternum; rostrum small **10. Porcellanidae**

11. Third maxillipeds tapering, to fit within the triangular mouth area **12. Leucosiidae**

Third maxillipeds distally wide, fitting within a quadrangular mouth area **12**

12. Last pereopods small and held dorsally; first article of antenna mobile, with conspicuous excretory pore; female opening coxal; pleopods on first female abdominal segment **11. Dromiidae**

Last pereopods held normally; first article of antenna fused to carapace, its excretory pore hidden; female openings sternal; no pleopods on first female abdominal segment **13**

---

13. Carapace drawn forward into a pronounced, often bifid rostrum. Antenna has enlarged second and third articles, fused to underside of carapace edge **13. Majidae**

Front of carapace usually wide, without prominent rostrum **14**

---

14. Carapace slightly longer than wide; antenna longer than carapace width **14. Corystidae**

Carapace usually wider than long; antenna shorter than carapace width **15**

---

15. Third maxilliped carpus arises at inner margin of merus **16**

Third maxilliped carpus arises near centre of merus margin **21**

---

16. Last pereopod, especially the dactylus, flattened for swimming **17. Portunidae**

Dactylus of last pereopod not flattened for swimming **17**

---

17. Carapace almost circular, the margin with a dense fringe of hair **15. Atelecyclidae**

Carapace hexagonal, square or broadly oval, not particularly hairy **18**

---

18. Single median lobe to front edge of carapace **19**

Pair of median lobes at front edge of carapace **20**

---

19. Carapace just wider than long; five sharp teeth on each side **17. Portunidae**

Carapace much wider than long; about nine or ten broad lobes or teeth on each side **16. Cancridae**

---

20. Eyestalk length about one-third width of carapace **19. Goneplacidae**

Eyestalk length much less than one-third width of carapace. Front of carapace with central pair of broad lobes **18. Xanthidae**

21. Carapace nearly circular; eyes very small. Commensal crab **20. Pinnotheridae**

Carapace nearly square; eyes conspicuous. Free-living crabs **21. Grapsidae**

Infraorder Astacidea

1. NEPHROPIDAE

Carapace with postcervical (transverse) and branchiocardiac (oblique) grooves, which are sometimes contiguous. Last thoracic segment fused to carapace. Chelae on first three pereopods. Eyes movable. Two familiar species in British waters.

1. Eyes very large, kidney-shaped, broader than eyestalks; antennal scale leaf-like ***Nephrops norvegicus***

Large eyes, the same width as eyestalks; spine-like antennal scale ***Homarus gammarus***

Nephrops norvegicus (Linnaeus). Norway lobster, Dublin Bay prawn, scampi, langoustine Fig. 8.50
Length up to 240 mm; pale orange. Carapace with distinct groove and longitudinal spinose keels. Rostrum long and spinose. Abdomen with transverse grooves. First legs with very long slender keeled chelae.

In shallow burrows, in soft sediments, at 20–800 m; British Isles, all coasts, common; elsewhere Norway and Iceland to Morocco and Mediterranean.

Homarus gammarus (Linnaeus).
Common lobster Fig. 8.50
Length variable, up to 500 mm; blue-coloured above with coalescing spots; yellowish below. Carapace and abdomen generally lacking strong spines or ridges and only slightly granular. Rostrum rather short and spiny. Strong gastro-orbital groove with, below it, the cervical groove. Chelae large.

Rocky substrata, LWST to 60 m; British Isles, all coasts, common; elsewhere Lofoten Isles to Morocco, Mediterranean, Black Sea.

Infraorder Palinura

2. PALINURIDAE

Carapace slightly compressed, without lateral ridges. Antenna base without scaphocerite; fused to epistome; antennal flagellum very long and strong. Pereopods with chelae. Two British species.

Palinurus elephas (Fabricius). Common crawfish, spiny lobster, langouste Fig. 8.50
Length up to 500 mm, usually 250–350 mm; adults reddish-brown with yellow spots. Carapace covered with forward-directed spines; supra-orbital spines particularly prominent. Anternal stalks very heavy and spiny; flagellum stout, tapering, and longer than body. Pereopod 1 subchelate; row of spines on upper crest of merus characteristic. Reddish eggs, September to October; hatching in 6 months as phyllosoma larvae.

Mainly rocky bottoms from 20 to 70 m. South and west coasts of the British Isles, common; elsewhere south to Azores and Mediterranean.

Palinurus mauretanicus Gruvel (Fig. 8.50) is distinguished from *P. elephas* by pereopod 1, which is no stouter than its other pereopods. It occurs off south-west Ireland, and southwards to the Mediterranean and Senegal, but deeper than 200 m.

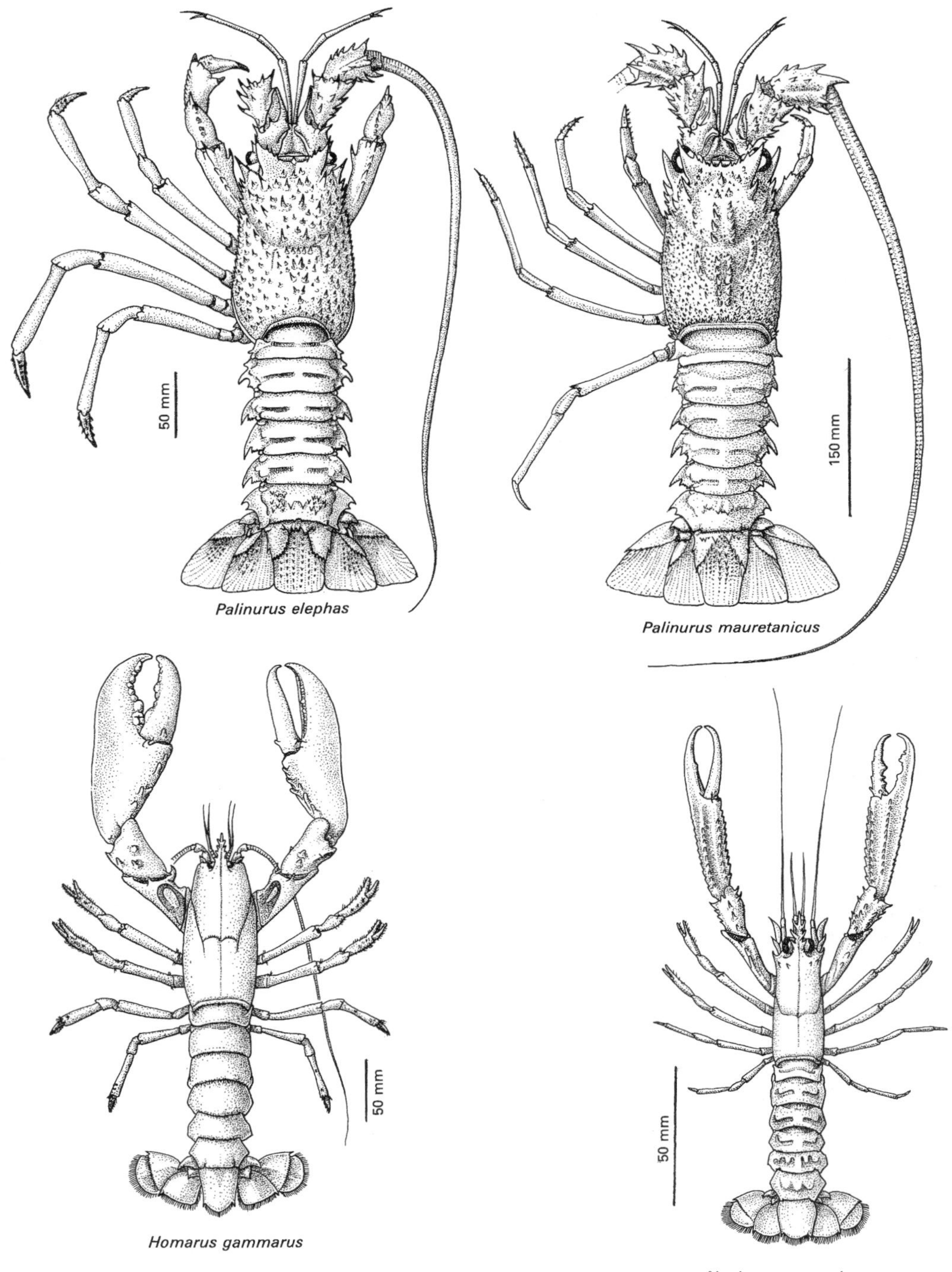

Palinurus elephas

Palinurus mauretanicus

Homarus gammarus

Nephrops norvegicus

Fig. 8.50

Infraorder Anomura

3. AXIIDAE

The first of four families of the superfamily Thalassinoidea. Members of this superfamily are all small, mostly burrowing Anomura with smooth carapaces; there may be a few spines associated with the rostrum but never the thorn-like extensions of the carapace over the eyes seen in the Palinuridae. The Axiidae lack the *thalassinian line* (characteristic of the other three families). The abdominal pleura are large. The first pereopods bear large, often asymmetrical, chelae. The third pereopods are not chelate.

Two genera are represented in British waters, each by a single species.

---

1. Body laterally compressed; no dorsal ridge on carapace; no suture on uropod exopod; eyes pigmented **_Axius stirhynchus_**

   Body cylindrical; suture on uropod exopod; eyes unpigmented; median dorsal ridge extends full length of carapace **_Calocaris macandreae_**

Axius stirhynchus Leach Fig. 8.51
Length about 72 mm. Rostrum triangular, margins slightly ridged. Chelae unequal; tips of fingers cross when closed; dactyl bears a well-marked ridge and is more hairy than finger of propodus. Pereopod 4 longest.

Burrows in mud or sand, at LWST to shallow sublittoral. Confined to south-west British Isles; uncommon; elsewhere to Spain and Mediterranean.

Calocaris macandreae Bell Fig. 8.51
Up to 40 mm long. Rostrum slightly upturned, its margins continuing on to the carapace; eyes large. Chelipeds long and unequal; fingers up to twice length of manus, compressed and covered with tufts of setae.

Burrows in mud at 35–1400 m; western coasts of British Isles; elsewhere ranges from Iceland and Norway to Mediterranean; also North America. Arabian Gulf, Indian Ocean, and Pacific.

4. LAOMEDIIDAE

The second family of the Thalassinoidea, the Laomediidae are characterised by possessing a *thalassinian line* dorso-laterally on the carapace, parallel to the mid-line. Unlike the Callianassidae and Upogebiidae, the Laomediidae bear *transverse sutures* to the *exopods* and *endopods* of the *uropods*.

Represented in Britain by a single species, *Jaxea nocturna*.

Jaxea nocturna (Chiereghin) Fig. 8.51
Length 40–60 mm; pinkish-white with yellow or chestnut hairs. Typical thalassinoid with enormous equal-sized chelae. Eyes minute and hidden by the rostrum.

Burrows in mud, sublittoral, 10–50 m. Off west coasts of British Isles, scarce; most often seen as fragments in fish stomachs or as planktonic larvae. Elsewhere south to Mediterranean, particularly in the Adriatic.

5. CALLIANASSIDAE

The third family of the Thalassinoidea, characterised by a *thalassinian line* which extends to the front of the carapace. Chelipeds are unequal, and pleopods 3, 4, and 5 have a plate-like subterminal article (propodus) fringed with hairs. Two British species.

---

1. Telson much shorter than uropod; maxilliped 3 as wide as long **_Callianassa tyrrhena_**

   Telson as long as uropod; maxilliped 3 slender **_Callianassa subterranea_**

Callianassa tyrrhena (Petagna) Fig. 8.52
The larger of the British species; up to 67 mm long; whitish with pink or blue spots, sometimes greenish-grey. Rostrum practically absent, eyes small and close together. The third maxillipeds form an operculum, their lower articles (merus and ischium) being broad and flat. Chelipeds unequal, left sometimes the larger; fingers cross at tips.

Burrows in muddy sand, 5–20 m or deeper. Perhaps all coasts of British Isles, not uncommon; elsewhere south Norway to Mediterranean.

Callianassa subterranea (Montagu) Fig. 8.52
The smaller species, up to 40 mm long; pale puce, sometimes quite dark. Rostrum minute. Third maxillipeds leg-like, not forming an operculum. Chelipeds massive, the finger-tips not crossing markedly.

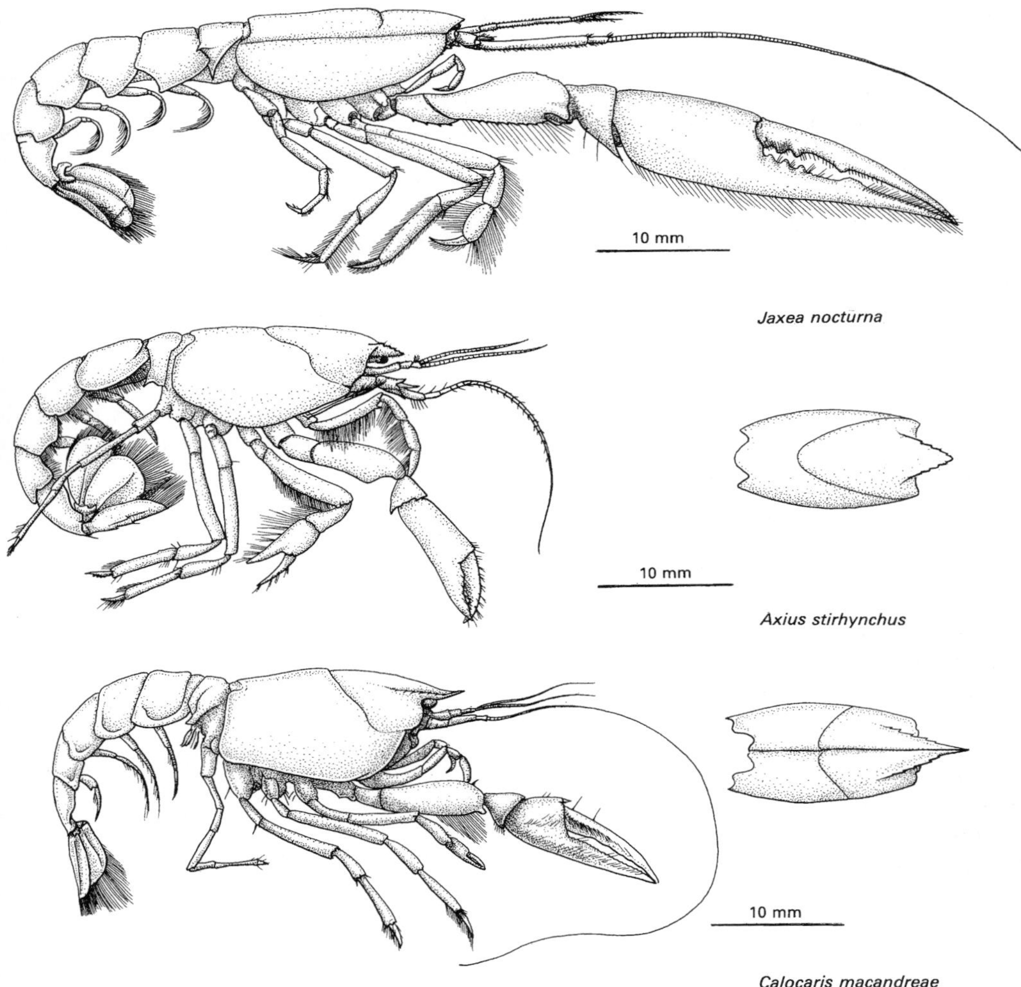

Fig. 8.51

Burrows in sandy mud; LWST down to 20 m; southern coasts of British Isles. commom; elsewhere southwards, probably to Mediterranean.

6. UPOGEBIIDAE

These have resemblances to the callianassids but the cervical groove (rather than the thalassinian line) extends to the front edge of the carapace and chelipeds are similar in size. Three British species.

---

1. No ocular spine; thumb of chela as long as dactyl **_Upogebia deltaura_**

   Ocular spine present, thumb much shorter than dactylus of chela **2**

---

2. Propodus of chela wide; thumb subterminal **_Upogebia pusilla_**

   Propodus of chela narrow; thumb almost terminal **_Upogebia stellata_**

Upogebia deltaura (Leach) Fig. 8.52

Largest of the three species; length 80 mm or exceptionally 150 mm. Colour dirty yellow, tinged white, green, or red. Similar to *U. stellata*, but lacks ocular spines; front of carapace including rostrum hairy above. Chelipeds equal and quite robust; the movable dactyl only slightly longer than 'thumb', no pronounced fringe of hairs along edge of manus, and only a short spine on outer extremity of carpus; abdomen broader and more softly membranous than *U. stellata*.

Uses burrows of other animals; LWST to 40 m; perhaps all coasts of British Isles, common, often with *U. stellata*. Elsewhere, Norway to Spain and Mediterranean, and Black Sea.

Upogebia pusilla (Petagna) Fig. 8.52
Length up to about 45 mm. Carapace more slender than the other species; there is a well-marked ocular spine and the lateral margin is strongly notched where the cervical groove meets it. Chelipeds distinctively long, with relatively heavy chelae; dactyl more than twice length of thumb, which is set back subterminally on the manus; a short dense fringe of hairs behind thumb.

Burrows in sand and muddy sand, shallow sublittoral; British Isles: uncommon, probably only in south; elsewhere Norway to Spain, Mediterranean and Black Sea.

Upogebia stellata (Montagu) Fig. 8.52
Adult body length up to 50 mm; yellowish-white, often orange spotted. Carapace with strong cervical groove; anterior lateral border with a small ocular spine; well-developed hairy rostrum. Chelipeds equal in size, long and slender; ventral margin of manus with dense rows of hairs; outer extremity of carpus with a long spine. Abdomen narrower and more rigidly chitinized than in *U. deltaura*.

In burrows, LWST to shallow sublittoral; British Isles, all coasts, commonest of three species. Elsewhere south-west Norway to Spain and Mediterranean.

7. PAGURIDAE

The Paguridae, or hermit crabs, are anomurans adapted to living in gastropod shells; growing crabs move to progressively larger shells. The abdomen is asymmetrical and twisted to fit the shell's dextral coil. The last pleopod on the left side is a hook for anchoring the body to the spiral shell. The right pleopods are usually absent in both sexes, those of the left (apart from the last) are lost in males but present for egg-carrying in females. The abdomen of pagurids is soft. The first pereopods are chelate, the right being larger than the left in most species. The fourth and fifth pereopods are generally poorly developed.

In some genera the chelae may be equal or subequal in size (exceptionally the left is larger, as in *Diogenes*). The action of the chelae fingers is parallel to the longitudinal plane of the body rather than across it. Eighteen species are recorded for Britain, in seven genera. Some authorities recognize two additional families, Diogenidae (for *Diogenes* and *Clibanarius*) and Parapaguridae (for *Parapagurus*).

The hermit crab and its shell are frequently the habitat for specific associated animals; coelenterates or sponges on the outside, polychaetes sharing the shell cavity, and spirorbids or acrothoracican cirripedes occupying the columella.

---

1. Third maxillipeds contiguous at base **2**

   Third maxillipeds widely separate **3**

---

2. Chelipeds subequal
   Clibanarius erythropus

   Left cheliped much larger than the right
   Diogenes pugilator

---

3. Chelae almost equal in length; both vas deferens protrude
   Catapaguroides timidus

   Left chela smaller than right; neither, or only left, vas deferens projecting **4**

---

4. Vas deferens not protruding; more than four spines (except in *P. pubescens*, with four) on each side of rear margin of telson **7**

   Four spines on each side of telson; rostrum wide and rounded **5**

---

5. Eyestalk short, about half length of hard carapace ***Anapagurus laevis***

   Eyestalk long, more than half length of hard carapace **6**

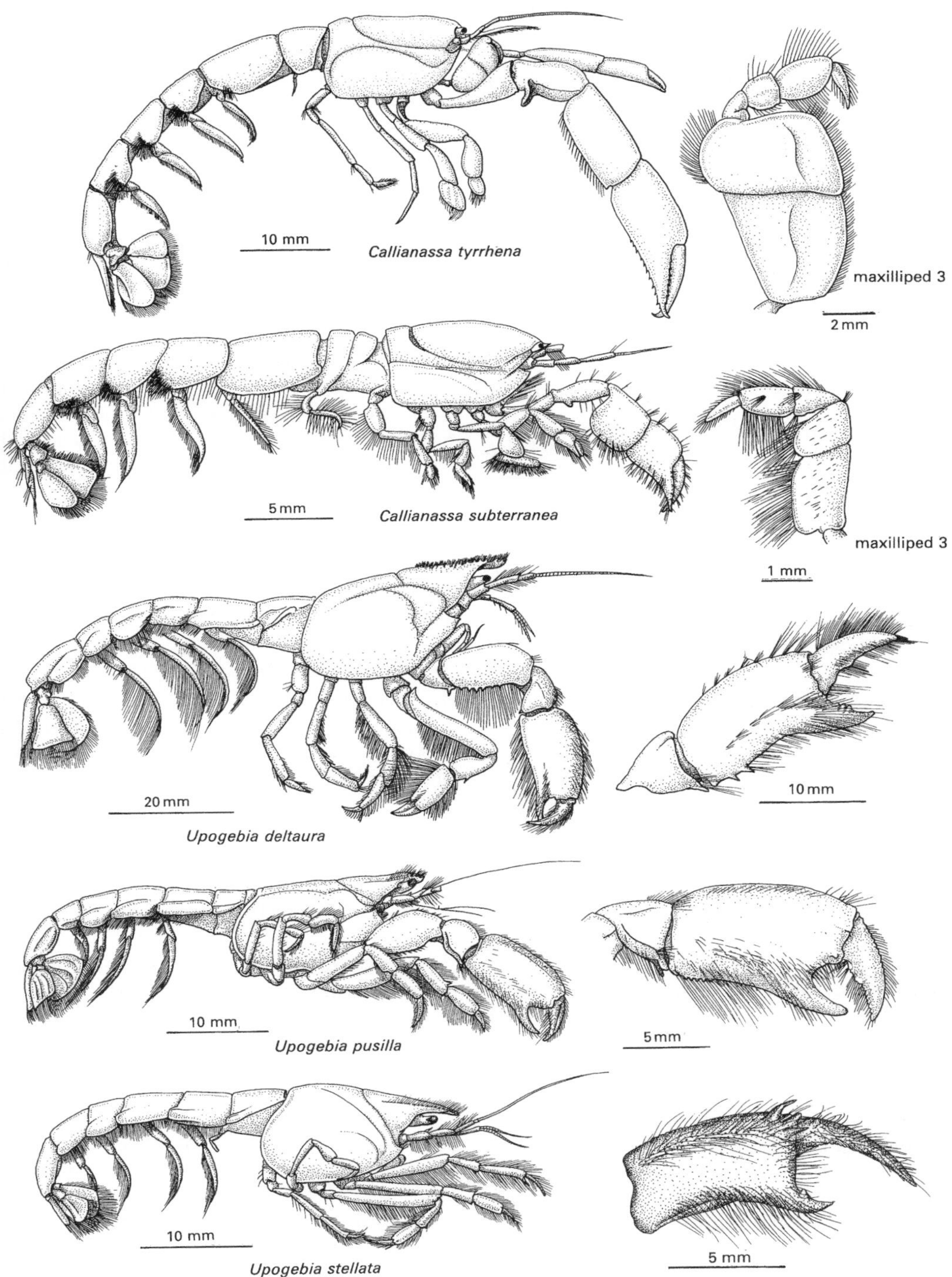
10 mm
Callianassa tyrrhena
maxilliped 3
2 mm
5 mm
Callianassa subterranea
maxilliped 3
1 mm
20 mm
Upogebia deltaura
10 mm
10 mm
Upogebia pusilla
5 mm
10 mm
Upogebia stellata
5 mm

Fig. 8.52

6. Stalk of antennule three times length of eyestalk; right chela almost hairless **_Anapagurus hyndmanni_**

Antennule stalk one and a half times length of eyestalk; right chela hairy **_Anapagurus chiroacanthus_**

7. Right chela setose **8**

Right chela bald **10**

8. Manus of right chela with two depressions **_Pagurus forbesii_**

Manus of right chela without depressions **9**

9. Rostrum rounded; right chela enclosed in dense plumose setae, with carpus as long as manus **_Pagurus cuanensis_**

Rostrum pointed; short setae of right chela arranged in minute groups arising from tubercles; carpus as long as whole chela **_Pagurus pubescens_**

10. Sharp pointed rostral region **_Pagurus bernhardus_**

Rostral region rounded **11**

11. Manus of left chela with toothed keel **_Pagurus variabilis_**

Manus of left chela without a toothed keel **_Pagurus prideauxi_**

Diogenes pugilator (Roux) Fig. 8.53
Carapace length up to 11 mm; greenish tinted. Left chela considerably larger than right. Propodus and carpus of chela short, covered with sharp teeth.

Inhabits fairly sheltered sandy bottoms; LWST to 35 m: south and west coasts of British Isles, common. Elsewhere from Holland to Angola, Mediterranean, Black Sea, and Red Sea.

Catapaguroides timidus (Roux) Fig. 8.53
A small hermit crab, carapace length about 5 mm. Rostrum triangular, similar sized to antero-lateral teeth; ocular peduncles subcylindrical, about as long as antennular peduncles. First pleopods lacking in females. The males have sexual tubes arising from the usual openings; the left one is short, conical, and curved; the right is strong, wide, and curved from right to left over the base of the abdomen.

Amongst *Zostera* spp., *Posidonia* sp., and algae; intertidal to 80 m. Extreme south-west British Isles, Channel Isles (?), uncommon; elsewhere Roscoff to Canaries; abundant in Mediterranean.

Clibanarius erythropus (Latreille) Fig. 8.53
Carapace length about 15 mm. Rostrum small, acute, and slightly protruding. Chelae with thick fingers bearing horny, black tips and covered by wide, blunt tubercles and hair. Eyes extend slightly beyond antennal peduncles.

In intertidal pools and on sand, gravel, and algae in shallow sublittoral, to 40 m; Channel Isles only, uncommon; formerly sporadic South Devon and Cornwall; elsewhere Brittany to Azores, Mediterranean and Black Sea.

Anapagurus chiroacanthus (Lilljeborg) Fig. 8.53
Carapace length about 6 mm. Rostrum rounded, flanked by rounded, antero-lateral teeth to carapace. Abundant hair covering carapace, chelae, and legs. Sexual tube of male on left side only; slightly curved.

Shallow, sublittoral species; British Isles: localized in northern North Sea, and apparently Guernsey; elsewhere from Norway to Spain and Mediterranean.

Anapagurus hyndmanni Thompson Fig. 8.53
Carapace length about 10 mm. Gently rounded rostrum flanked by sharp antero-lateral teeth. Right cheliped has carpus almost as long as propodus. Last article of antennular peduncle longer than in other species.

Associated with sand, mud, or gravel in shallow sublittoral, LWST to 35 m. British Isles, all coasts, uncommon. Elsewhere north Biscay to Portugal.

Anapagurus laevis (Bell) Fig. 8.53
Carapace length about 7 mm. Antero-lateral edge of carapace has very conspicuous tooth. Eye reaches base of article 3 of antennular peduncle, which is

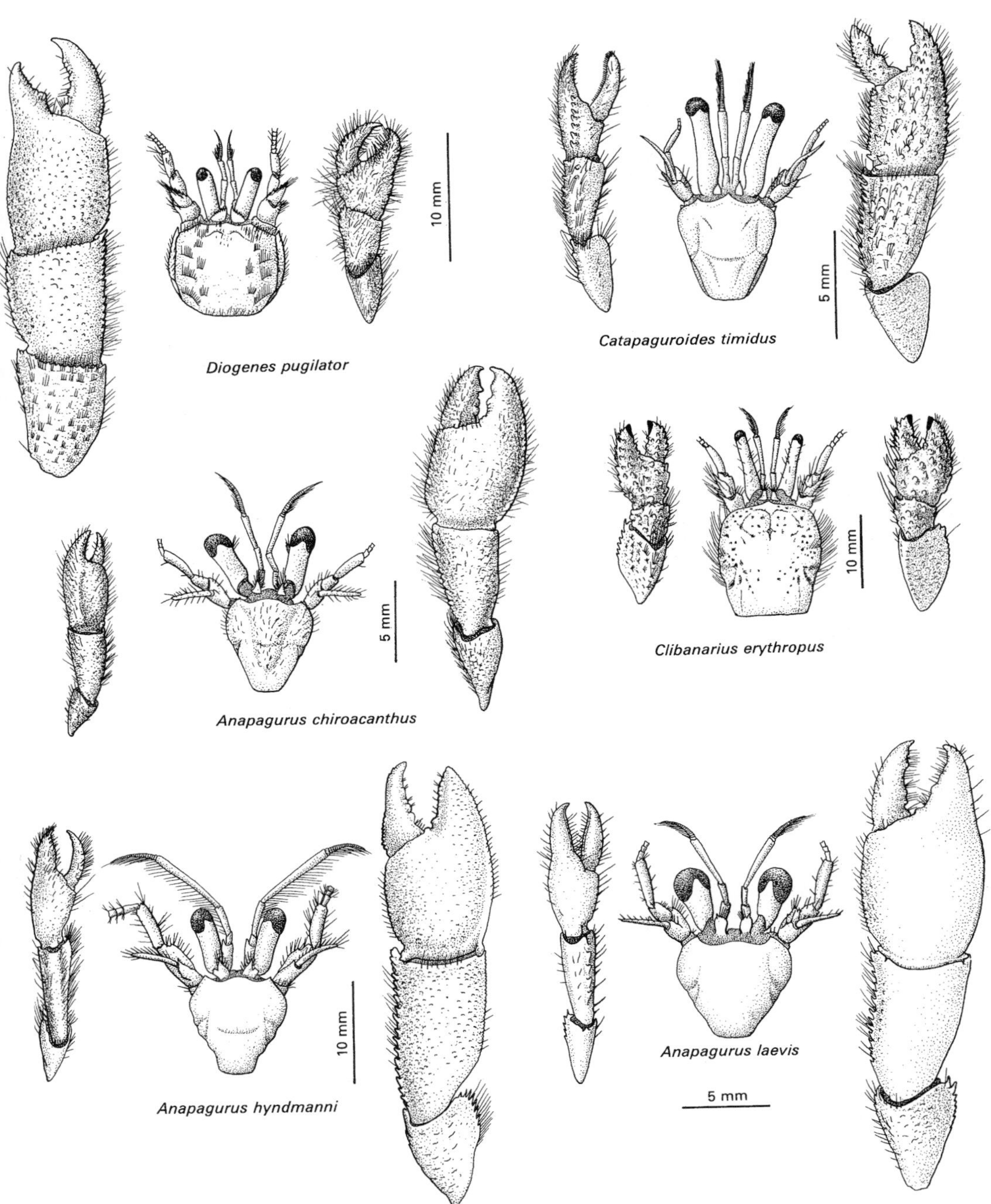
10 mm
Diogenes pugilator
5 mm
Catapaguroides timidus
5 mm
Anapagurus chiroacanthus
10 mm
Clibanarius erythropus
10 mm
Anapagurus hyndmanni
Anapagurus laevis
5 mm

Fig. 8.53

three times longer than article 2. Right cheliped of male reaches four times length of thorax.

On substrata of muddy sand or gravel, 20–200 (400) m. All British coasts, very common. Elsewhere Norway to Senegal, Azores, Mediterranean.

Pagurus bernhardus (Linnaeus) Fig. 8.54
Carapace length about 35 mm, intertidal specimens much smaller; reddish. Chelipeds covered with uniformly distributed granules or small teeth. Propodus of right chela with two rows of larger granulations, starting at the base and converging towards the middle.

On rocky and sandy substrata; MTL to 140 m (occasionally 500 m). Very common off all British coasts. Elsewhere from Iceland and Norway to Portugal; also on American Atlantic coasts.

Any small gastropod shell may be occupied; large specimens are usually found in shells of *Buccinum undatum* Linn. Commensals include *Calliactis parasitica* (Couch), *Hydractinia echinata* (Fleming), *Nereis fucata* (Savigny), *Trypetesa lampas* (Hancock); parasites: *Clistosaccus paguri* Lilljeborg and *Peltogaster curvatus* Kossman.

Pagurus cuanensis Thompson Fig. 8.54
Carapace length about 15.6 mm; pale reddish-brown with darker spots and some white, legs predominantly red. Upper surface of right chela strongly hairy, covered with conical teeth, more developed in the mid-line.

Shallow sublittoral, LWST to 175 m. All British coasts, common; elsewhere Norway to South Africa, Mediterranean.

Pagurus forbesii Bell Fig. 8.54
Carapace length about 8.5 mm; reddish-yellow with darker spots in the gastric region. Upper surface of right chela granulated and slightly hairy; propodus with two characteristic depressions separated by a large elongated ridge extending on to finger.

A shallow sublittoral species: 20–70 m; off south and south-west coasts of British Isles, not uncommon. Elsewhere south to Senegal, Mediterranean.

Pagurus prideauxi Leach Fig. 8.54
Carapace length about 14 mm; muddy red colour with paler spots, chelae salmon pink. Upper surface of right chela regularly convex with a slight protruding part, and a slightly raised longitudinal region; blunt tubercles and small granules distributed evenly over both dorsal surface and sides.

On sand, mud, or gravel, LWST to 40 m (exceptionally 400 m); all British coasts, locally very common, elsewhere Norway to Cape Verde; Mediterranean. Adults often accompanied by commensal cloak anemone, *Adamsia carciniopados* (Otto).

Pagurus pubescens Krøyer Eig. 8.54
Carapace length about 20.5 mm; chela (propodus) 20 (31) mm. Front of carapace with a central, acutely pointed rostrum flanked laterally by two slightly shorter and less acute antero-lateral teeth. Right cheliped larger than left, three times length of carapace; propodus and carpus about equal length.

Shallow sublittoral, 8–50 (100) m; sand, mud, or rock, not uncommon. A northern species, north and west coasts of Britain, including Irish Sea; elsewhere Norway, Iceland.

Pagurus variabilis
(Milne Edwards and Bouvier) Fig. 8.54
Carapace length 15–19 mm. Antero-lateral teeth of carapace in advance of rounded rostrum. Article 2 of the antennal peduncle drawn out as a spine which reaches beyond base of its last article. Right chela bears a median keel.

Offshore, 40–50 m; all British coasts, common; elsewhere Norway to White Cape, Mediterranean.

8. LITHODIDAE (STONE CRABS)

Large and crab-like with a well-calcified carapace. Right chela larger than left; fifth pereopods very small and hidden. Abdomen, particularly of female, strongly asymmetrical, reflecting the pagurid ancestry. Mostly in deep water. One British species.

Lithodes maja Linnaeus Fig. 8.56
Carapace width up to 120 mm. Carapace almost circular with a well-defined rim of irregular spines; scattered spines on dorsal surface and also on limbs.

Offshore; from 10 to 600 m; locally common; northern species extending not much further south than Isle of Man.

9. GALATHEIDAE

Carapace longer than wide; rostrum triangular or elongated. Sternum of last thoracic segment free from carapace. Abdomen curved under cephalothorax.

Eight species recorded for Britain.

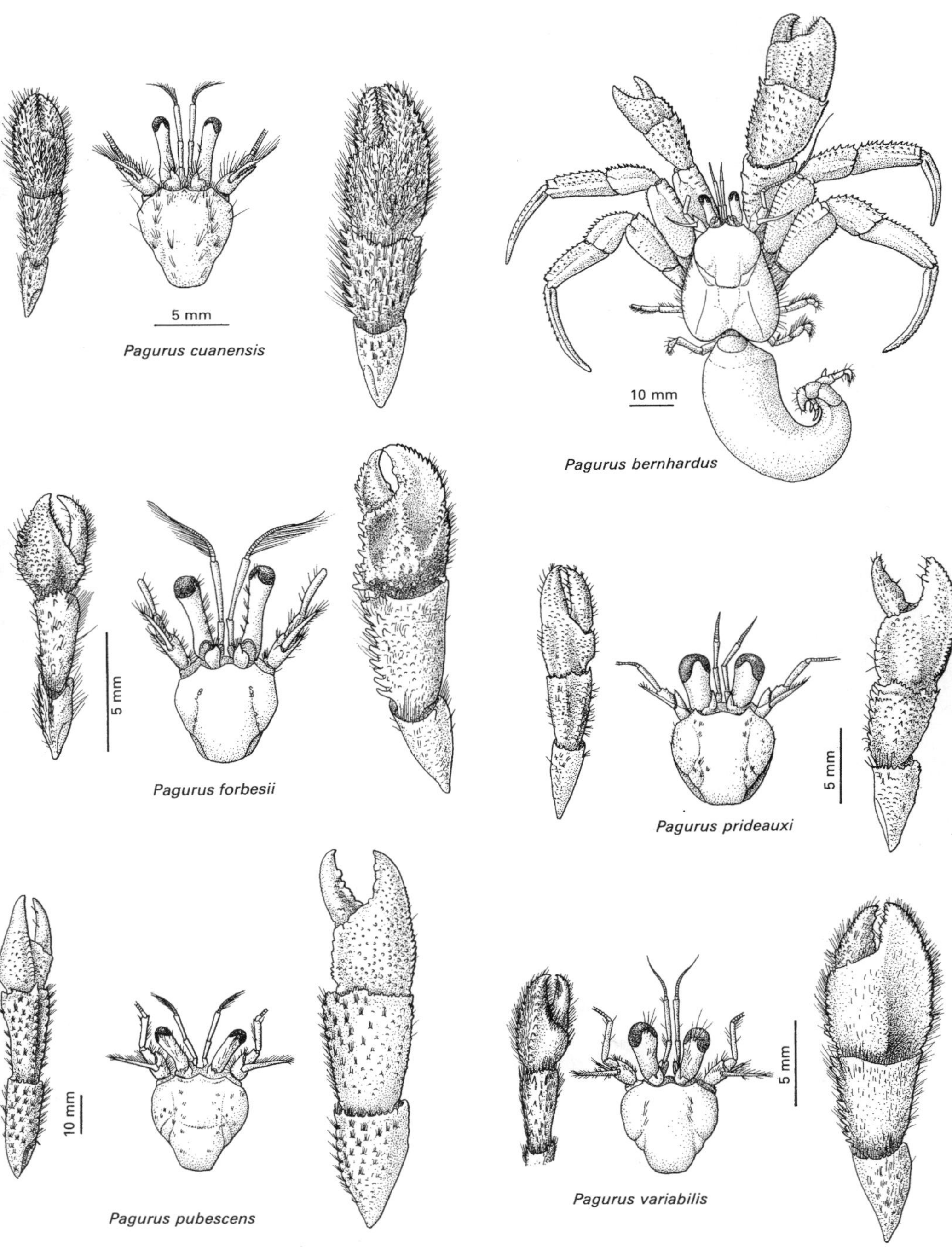
5 mm
Pagurus cuanensis
10 mm
Pagurus bernhardus
5 mm
Pagurus forbesii
5 mm
Pagurus prideauxi
10 mm
Pagurus pubescens
5 mm
Pagurus variabilis

Fig. 8.54

1. Rostrum broad, flattened, triangular, and toothed on each side **2**

   Rostrum a simple spine accompanied by long supraorbital spines; chelipeds very hairy; intense reddish colour **Munida rugosa**

2. Basal joint of antennule with three massive spines **3**

   Basal joint of antennule with two such spines **Galathea intermedia**

3. Manus of chela with few spines; pereopods 2–3 with epipodite **4**

   Manus covered with spines; pereopods 2–5 without epipodite **Galathea strigosa**

4. Merus of maxilliped 3 longer than ischium, chela with scaly tubercles **Galathea squamifera**

   Merus of maxilliped 3 about same length as ischium; chelae covered with hairs or scales fringed with short setae **5**

5. Abdominal segments with single transverse furrow; rostrum concave above, few setae **Galathea nexa**

   Abdominal segments with three transverse furrows; rostrum thickly clothed in setae or scales **Galathea dispersa**

Munida rugosa (Fabricius) Fig. 8.55
Overall length, with abdomen extended, about 60 mm; carapace length, including rostrum, about 30 mm; pinkish-yellow, with edges and transverse grooves reddish, some red spines. Carapace with few spines on posterior margin. Chelipeds very long.

Stony bottoms, LWST to 150 m; all British coasts, fairly common; elsewhere Norway to Madeira, Mediterranean.

Galathea dispersa Bate Fig. 8.55
Length 35–45 mm; red to dull yellow, sometimes with white spotting, never any blue. Rostrum narrow and pilose; tubercles at base of apical tooth hidden by hairs. Antennal peduncle without a spine on article 2. Merus of maxilliped 3 with central spine smaller than in *G. nexa*; chelipeds closely covered with fringed scales. Eggs carried in spring.

Sublittoral, depths of 10–500 m; all British coasts, common; elsewhere, Norway and south Iceland to Madeira and Canaries, Mediterranean.

Galathea intermedia Lilljeborg Fig. 8.55
Small, length up to 18 mm; carapace 8.5 mm; mainly salmon-red. Rostrum narrow, especially in males; paired lateral teeth protruding very little from edge. Transverse post-rostral groove arched forward centrally, with a pair of distinctive setae in a separate smaller groove just behind it.

Sublittoral, 15–20 m (exceptionally 25 m); all British coasts, very common; elsewhere Norway to Dakar and Mediterranean.

Galathea nexa Embleton Fig. 8.55
Length, male up to 40 mm, female 30 mm; reddish-green with darker spots, bluish spots on labrum. Rostrum relatively shorter than in *G. dispersa*, each lateral tooth bears a single seta; a few tubercles are just visible on concave upper surface posterior to apical tooth. Article 2 of antenna bears a spine (lacking in *G. dispersa*). Spines of chelae and pereopods stronger than in *G. dispersa*.

Sublittoral, 25–270 m; south and west coasts of British Isles, uncommon; elsewhere southwards to Tenerife and Mediterranean.

Galathea squamifera Leach Fig. 8.55
Length up to 65 mm; carapace 32 mm; dark chestnut-brown with greenish tinge; spine tips red; juveniles reddish. Apical spine of rostrum stands out from, but is larger than, the lateral spines; carapace shiny between grooves and scattered short hairs, chelae closely covered in scale-like tubercles; merus of maxilliped 3 bears a distal row of spines, the end ones longest.

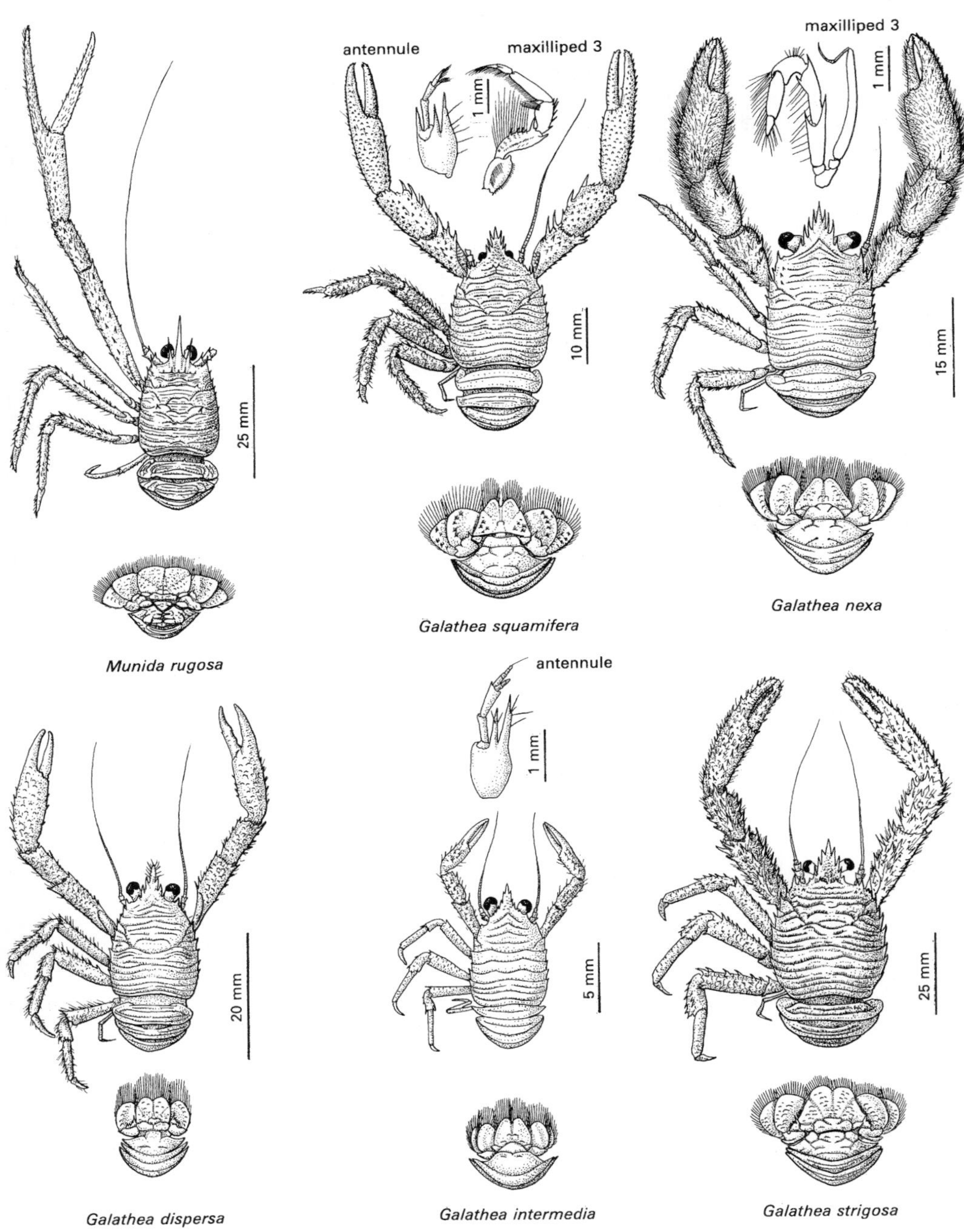

Munida rugosa

Galathea squamifera

Galathea nexa

Galathea dispersa

Galathea intermedia

Galathea strigosa

Fig. 8.55

LWST to shallow sublittoral, juveniles deeper (30–70 m); all coasts of British Isles, common; elsewhere Norway to Azores and Mediterranean.

Galathea strigosa (Linnaeus) Fig. 8.55
Large, overall body length about 90 mm, carapace up to 53 mm; vivid red with patches and bands of bright blue. Rostrum has a long apical spine and is covered with fringed scales; three epigastric spines on each side, middle one longest; carapace grooves densely pilose. Merus of maxilliped 3 with two strong spines.

From shallow sublittoral to 600 m; gravelly and rocky bottoms; all British coasts, very common; elsewhere from North Cape, Scandinavia to Spain, Canaries, Mediterranean, and Red Sea.

10. PORCELLANIDAE

Carapace nearly circular, crab-like, flat, smooth; front quite wide. Antennae very long. Maxilliped 3 with flattened ischium and merus. Abdomen very reduced and thin; flexed under cephalothorax. Pereopod 5 extremely thin and tucked under carapace. Two British species.

---

1. Chelipeds with dense fringe of long setae on outer edge; chelae very large **_Porcellana platycheles_**

   Chelipeds devoid of setae; chelae narrow **_Pisidia longicornis_**

Porcellana platycheles (Pennant).
Porcelain crab Fig. 8.56
Carapace length up to 15 mm; greyish-brown above, underside dirty yellowish/white. Carapace slightly longer than broad, with setose posterior margins; front with slightly advanced, acute median tooth, flanked by smaller acute submarginals. Chelipeds unequal, massive, compressed; propodus densely fringed with long setae. Pereopods heavily setose. Distal edges of abdominal segments setose.

Undersides of boulders; intertidal; all British coasts, north as far as Shetland, abundant; elsewhere North Sea, Holland to White Cape, Canaries, Mediterranean.

Pisidia longicornis (Linnaeus) Fig. 8.56
Carapace length 8 (10) mm. Dark maroon or olive. Carapace slightly convex, smooth, and bald; front three-lobed; median lobe with longitudinal furrow, serrated terminally. Antennae very long and slender. Eyes deeply sunk. Chelipeds unequal, large, compressed, not setose; fingers slightly twisted, meet only at tips.

Rock or gravel substrata, also in the brozoan *Pentapora fascialis* (Pallas), and amongst other colonial forms; intertidal to 100 m; abundant on all British coasts; elsewhere, Norway to Angola, Canaries, and Mediterranean.

Infraorder Brachyura

11. DROMIIDAE

Carapace with shallow orbits, convex, more or less circular, front with three teeth, the middle one small and deflexed. Antennal stalk with four articles. Pereopod 1 with strong chelae, pereopods 4 and 5 reduced. Uropods vestigial. Abdomen wide, usually with seven segments.

Only one genus and species in British Isles.

Dromia personata (Linnaeus).
Sponge crab Fig. 8.56
Carapace of male up to 53 mm long, breadth 67 mm; dark brown with pink chelae. Body often covered with a sponge. Carapace length/breadth ratio 4/5; whole body and legs covered with velvet-textured pile. Chelae stout and equal, smaller in female. Abdomen of both sexes with all segments free.

Rocky or stony substrata, 10–30 m (to 100 m); uncommon around British Isles: south coasts, northern North Sea, Anglesey; elsewhere south to West Africa, Azores, Mediterranean.

12. LEUCOSIIDAE

Eyes and orbits very small. Third maxillipeds completely filling buccal frame. Five British species in the genus *Ebalia*.

---

1. Abdominal segments 3–6 fused in both sexes. Surface of carapace raised in the form of a 'plus'-sign. Pleopod 1 of male with terminal tufts of hair **_Ebalia tuberosa_**

   Abdominal segments 3–5 fused in male, 4–6 in females. No 'plus'-shaped elevation. Male pleopod 1 without hairs **2**

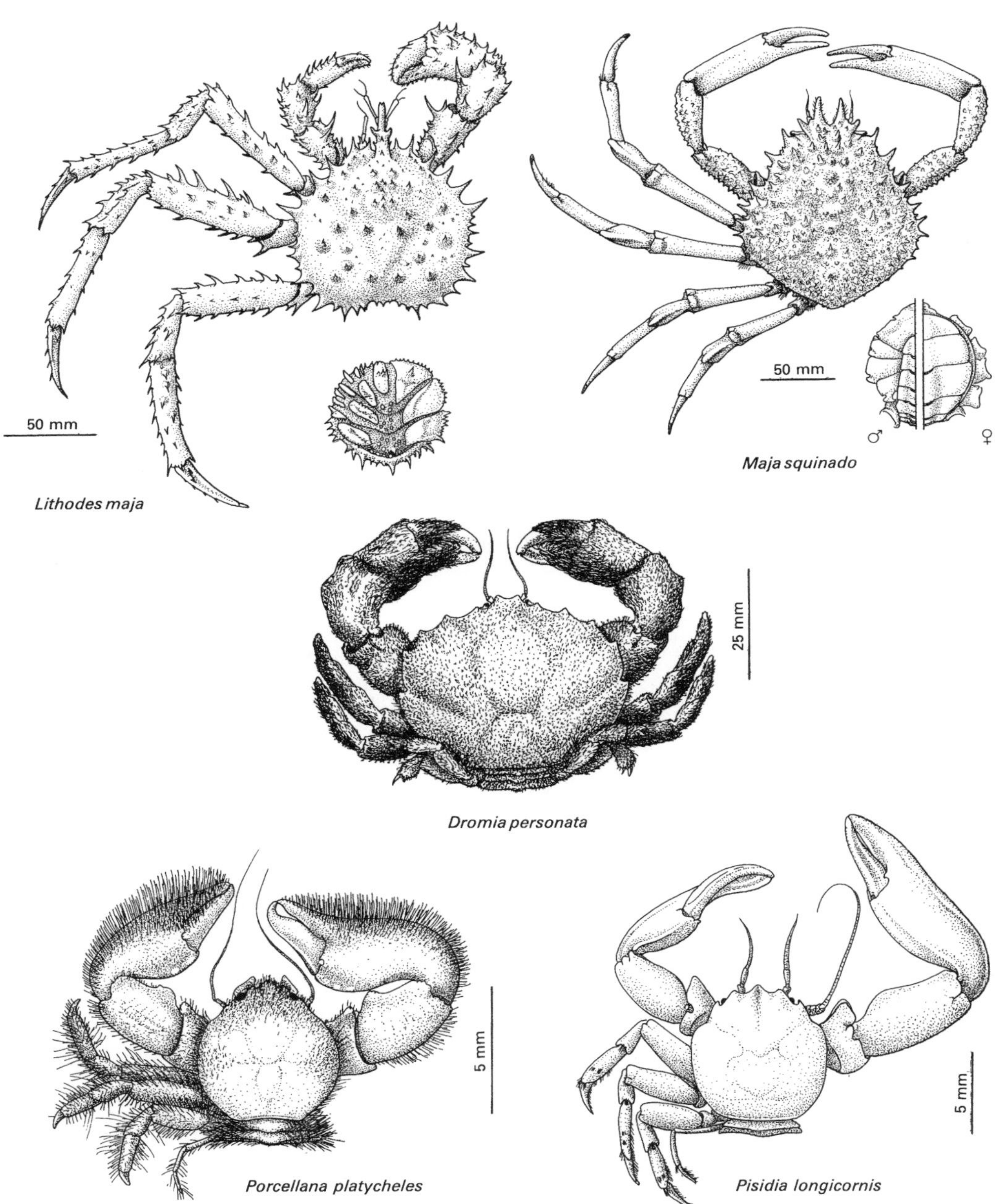
50 mm
Lithodes maja
50 mm
♂
♀
Maja squinado
25 mm
Dromia personata
5 mm
Porcellana platycheles
5 mm
Pisidia longicornis

Fig. 8.56

2. Pereopods furnished with teeth or long tubercles **Ebalia granulosa**

   Pereopods finely granular, not toothed **3**

3. Dactylus of chela very much shorter than manus. Lateral margins of carapace edged with large tubercles **Ebalia nux**

   Dactylus of chela not much shorter than manus of propodus. Carapace not edged with large tubercles **4**

4. Carapace slightly broader than long; branchial region strongly inflated **Ebalia tumefacta**

   Carapace about as long as broad; branchial region elevated but not inflated; granular **Ebalia cranchii**

Ebalia tuberosa (Pennant).
Pennant's nut crab Fig. 8.57
Carapace length 13 (17) mm. Colour variable, light orange to reddish brown, sometimes spotted white. Carapace rhomboid, slightly broader than long; surface with many prominent tubercles, smaller at front; widest part anterior of halfway. Chelipeds more robust and longer in male.

On gravel and stony bottoms; intertidal (rare) to 190 m; all British coasts, common; elsewhere, west Norway to north-west Africa, Azores, and Mediterranean.

Ebalia nux Milne Edwards Fig. 8.57
Carapace length 6 (12) mm. Brick-red with darker red spots. Carapace as broad as long, widest halfway back, with numerous bead-like tubercles; front straight, with small central notch; orbits inset from margin in female. Chelipeds equal, very long and slender, more robust in male. Pereopods 2–5 slender, strongly tuberculate.

Deep water on mud and sand, 80–3000 m; British Isles, Celtic Sea, and off Hebrides; elsewhere south to Cape Verde, Azores, and Mediterranean.

Ebalia tumefacta (Montagu).
Bryer's nut crab Fig. 8.57
Carapace length 10 (12) mm. Colour variable, reddish- to yellowish-grey, often spotted red; otherwise variegated brown (black), or banded pink or orange. Carapace broader than long, minutely tuberculate; antero-lateral margin forming an almost straight line. Chelipeds equal, propodus swollen, longer in male.

On muddy sand, gravel, or stones, 2–15 m; probably all British coasts, very common; elsewhere south Norway to north-west Africa; not Mediterranean.

Ebalia granulosa Milne Edwards Fig. 8.57
Carapace length 9.5 (11) mm. Yellowish with two spots on carapace. Carapace as broad as long, with low tubercles; front notched, biconvex; antero-lateral margins sinuous and cristate. Prominences similar to *E. tumefacta* but lower. Chelipeds equal, stout; margins of merus and propodus sharply cristate, merus with poorly defined tubercles. Chelipeds longer, and propodus more inflated in males. Pereopods 2–5 thin; merus often with spine-like tubercles.

Shallow sublittoral, 20–3000 m; off west Scottish coast, west Ireland, rare; elsewhere south to Spain and Mediterranean.

Ebalia cranchii Leach.
Cranch's nut crab Fig. 8.57
Carapace length 8 (11) mm. Colour variable, reddish-yellow to reddish-white with dark-red spots; legs yellowish. Carapace may be slightly longer than broad; front variable, often only faintly notched. Surface tubercles flat to bead-like. Distinctive prominences often capped by tubercles. Chelipeds equal, stout, moderately tuberculate; merus margin sharp but not cristate; longer, with propodus more inflated in male.

On muddy sand and gravel, 5–100 m; all British coasts, not uncommon; elsewhere west Norway to West Africa, Mediterranean.

13. MAJIDAE

Antennal peduncle well developed, article 2 elongated and fused to underside of carapace. Chelipeds not longer than other pereopods but very mobile and used for (amongst other things) attaching pieces of weed, sponge, and other sedentary organisms to

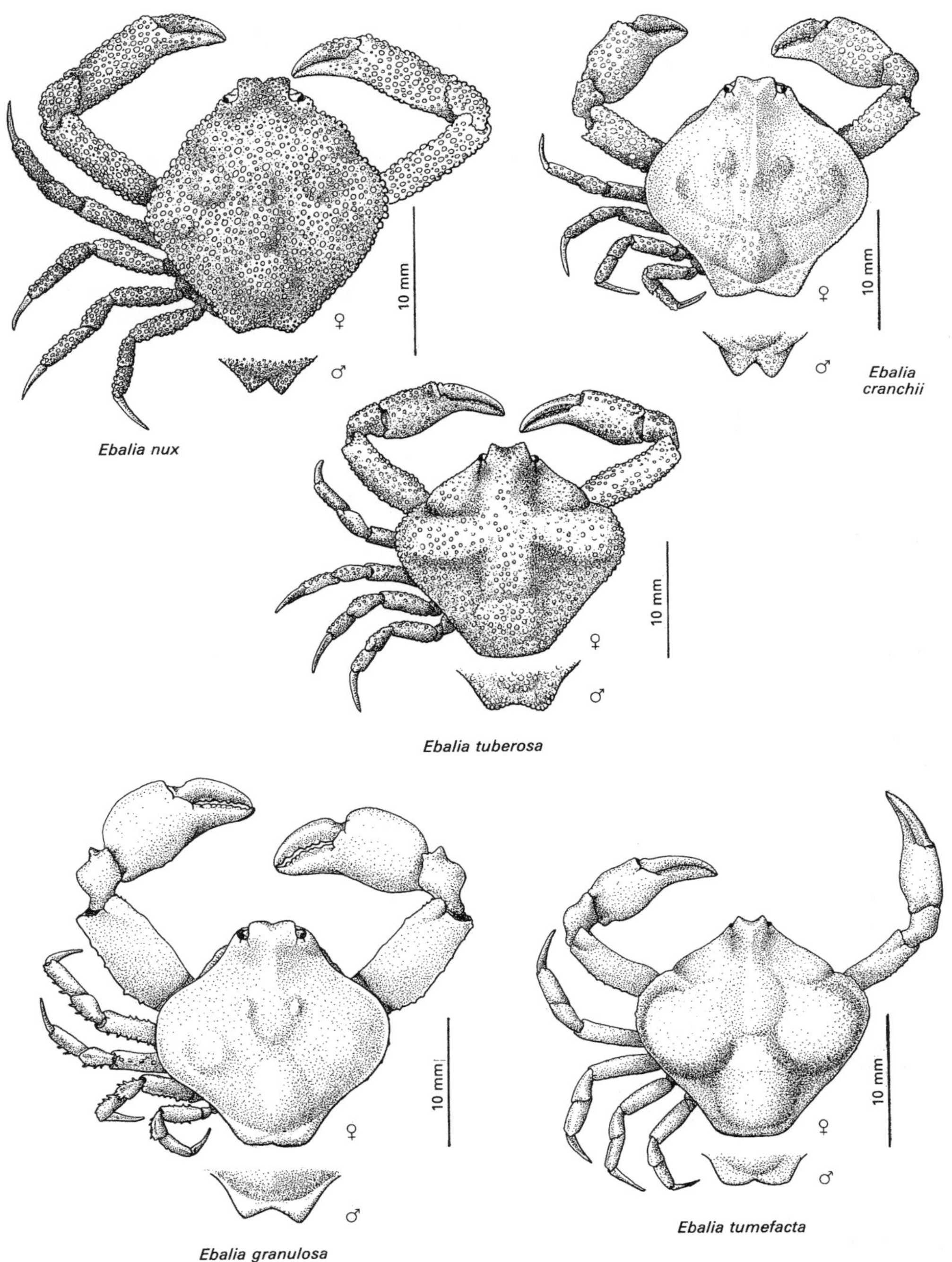
10 mm
♀
♂
Ebalia nux
10 mm
♀
♂
Ebalia cranchii
10 mm
♀
♂
Ebalia tuberosa
10 mm
♀
♂
Ebalia granulosa
10 mm
♀
♂
Ebalia tumefacta

Fig. 8.57

the hooked setae. A large family, with 17 species recorded around Britain.

1. Basal article of antennal peduncle (segments 2 and 3) more than twice as long as broad **5**

   Length of basal article of antenna less than twice width **2**

2. Eye retractile into orbit, concealing cornea when viewed from above. Very broad basal article to antenna ***Maja squinado***

   Cornea of retracted eye still at least partly visible from above **3**

3. Post-orbital spine flattened and expanded laterally; telson of male very wide, fitting into a slot in abdominal segment 6 **4**

   Post-orbital spine not flattened or expanded laterally, telson of male rounded–triangular **12**

4. Hepatic region not dilated laterally, only a slight swelling ***Hyas araneus***

   Hepatic region dilated laterally, forming with post-orbital region a lyriform shelf; carapace margin contracted between it and branchial region ***Hyas coarctatus***

5. Orbital spines present; eyestalks retractile **7**

   No orbital spines; eyestalks not retractile **6**

6. Rostrum short, extending to end of first free segment of antenna ***Achaeus cranchii***

   Rostrum extending well beyond end of first free segment of antenna **9**

7. Gastric region of carapace with four small tubercles in a transverse row, behind which is a strong median spine ***Inachus dorsettensis***

   Gastric region with only two small tubercles in transverse row, behind which is a median spine **8**

8. Distinct U-shaped cleft separating rostral horns, male thoracic sternum with callosity ***Inachus leptochirus***

   Very narrow slit between rostral horns. Male thoracic sternum lacks callosity ***Inachus phalangium***

9. Anterior edge of antennule socket visible from above, between rostrum and antennal peduncle. Dactyl of pereopods 4 and 5 very strongly arched, nearly forming a semicircle. Rostrum very short, upswept ***Macropodia linaresi***

   Anterior edge of antennule socket not visible from above between rostrum and antennal peduncles. Dactyl of pereopods 4 and 5 moderately arched **10**

10. Rostrum upswept, extending beyond extremity of antennal peduncle. Always some strong spine-like teeth on basal article of antenna ***Macropodia tenuirostris***

    Rostrum at most extending to extremity of antennal peduncle. Basal articles of antenna smooth or with some tubercles or variously developed teeth **11**

11. Rostrum straight or slightly upswept; basal article of antenna always smooth ***Macropodia rostrata***

Rostrum more or less arched downwards; basal article of antenna nearly always bearing teeth **_Macropodia deflexa_**

---

12. Supra-orbital eave extends forward as a pre-orbital spine **14**

Supra-orbital eave not extended so **13**

---

13. Tubercles at posterior margin of carapace usually fused into a crescent-shaped group **_Eurynome aspera_**

Tubercles at posterior margin not fused **_Eurynome spinosa_**

---

14. One branchial tooth **_Pisa armata_**

Two or more branchial teeth **_Pisa tetraodon_**

Hyas araneus (Linnaeus).
Great spider crab Fig. 8.58
Carapace, of large male, length 105 mm, breadth 83 mm; reddish-brown; dirty white underside. Carapace rounded posteriorly, uneven and tuberculated. Rostrum horns close-set. Chelipeds shorter than pereopods 2–5. Scattered hairs, some hooked, on carapace and limbs.

On hard and sandy bottoms, LWST to 50 m (occasionally to 350 m); all British coasts, common. A northern species; Spitzbergen, Iceland, Norway to English Channel; also Greenland and North America.

Hyas coarctatus Leach Fig. 8.58
Carapace, of a large male, length 61 mm, breadth 44 mm; similar colour to *H. araneus*. Rostral horns slightly longer and less close than in *H. araneus*, converging at tips. Chelipeds longer than in *H. araneus*.

On hard and sandy bottoms, LWST to 50 m (occasionally to 500 m), often deeper than *H. araneus*. All British coasts, common. Another northern species, Spitzbergen and Norway to Brittany; also Greenland and North America.

Inachus dorsettensis (Pennant).
Scorpion spider crab Fig. 8.58
Carapace, of a large male, length 30 mm, breadth 27 mm; greyish or light brown with reddish spots. Carapace surface with very consistent pattern of spines, including a distinctive transverse row of four. Chelipeds equal, more swollen in males.

Stony, sand, or mud substrata; shallow sublittoral, 6–100 m; all British coasts, common; elsewhere, north Norway to South Africa, Mediterranean.

Inachus phalangium (Fabricius).
Leach's spider crab Fig. 8.58
Carapace, of large male, length 20.5 mm, breadth 17.5 mm; brownish yellow. Similar to *I. dorsettensis* but carapace spines less prominent.

On stony, sandy, and shell substrata, shallow sublittoral, 11–55 m; all British coasts, frequent; elsewhere, Norway to West Africa, Cape Verde, Mediterranean.

Inachus leptochirus Leach.
Slender-legged spider crab Fig. 8.58
Carapace, of large male, length 28 mm, breadth 24 mm; colour variable, yellowish or greyish-brown. Very similar to *I. phalangium*. Distinct but narrow U-shaped gap between rostral horns.

On mud and muddy sand, 32–230 m; all British coasts, scarce; elsewhere, Faeroes, to West Africa, Azores, Mediterranean.

Achaeus cranchii Leach.
Cranch's spider crab Fig. 8.58
Carapace, of typical male, length 11 mm, breadth 9 mm; pale reddish-brown. Carapace with distinctive outline, swollen in hepatic and branchial regions; no orbital or hepatic spines. Eyes on long stalks. Chelipeds spiny, equal, fatter in males. Pereopods shorter than those of *Inachus*, last three pairs ending in a sickle-like dactyl. Many hooked setae on body.

Offshore 20–70 m; west coasts only, rare; elsewhere, south to West Africa, Azores, and Mediterranean.

Macropodia rostrata (Linnaeus).
Long-legged spider crab Fig. 8.59
Carapace, of large male, length 22 mm, breadth 15 mm; brown tinged with grey, yellow, or red.

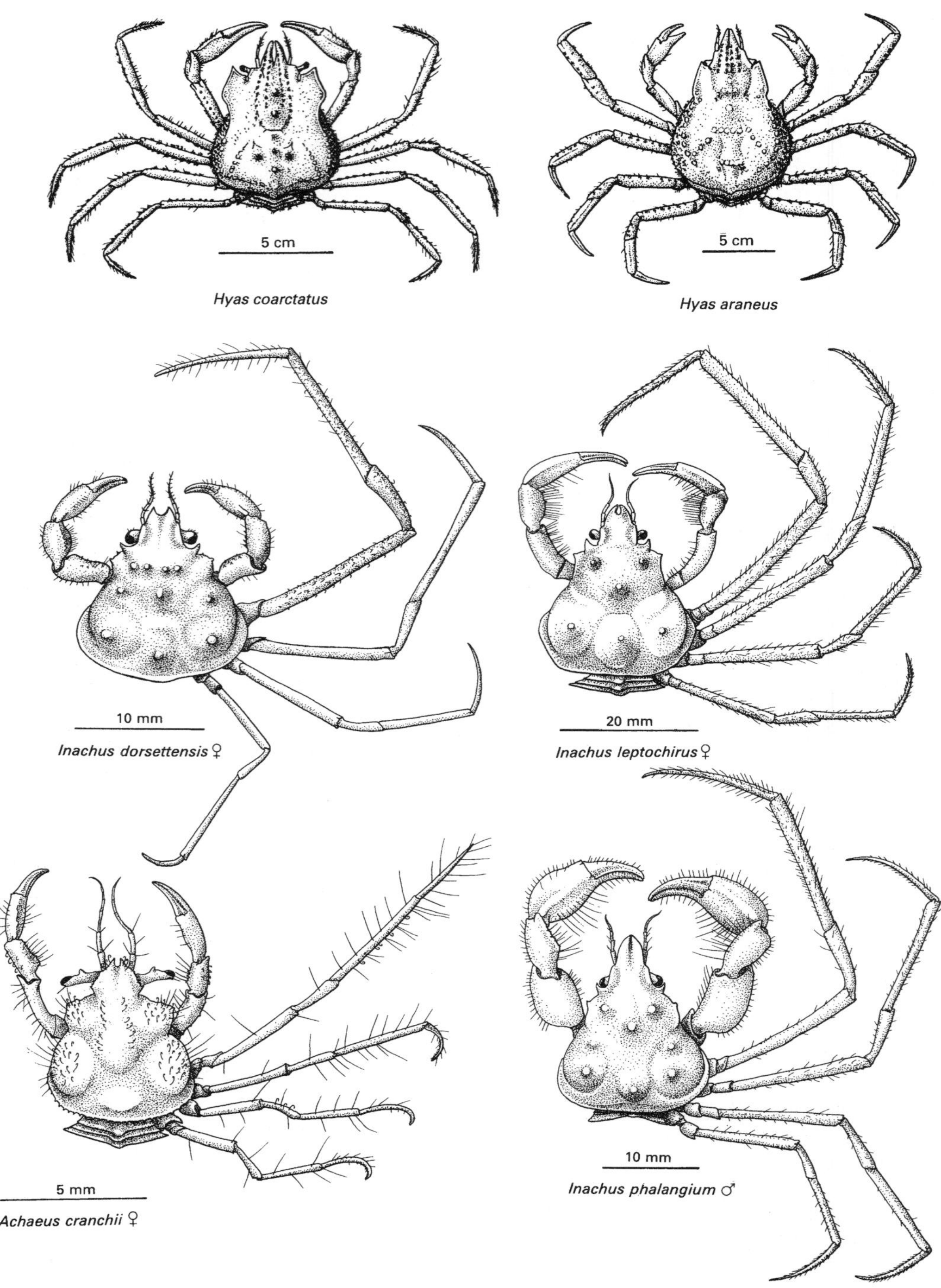
5 cm
Hyas coarctatus
5 cm
Hyas araneus
10 mm
Inachus dorsettensis ♀
20 mm
Inachus leptochirus ♀
5 mm
Achaeus cranchii ♀
10 mm
Inachus phalangium ♂

Fig. 8.58

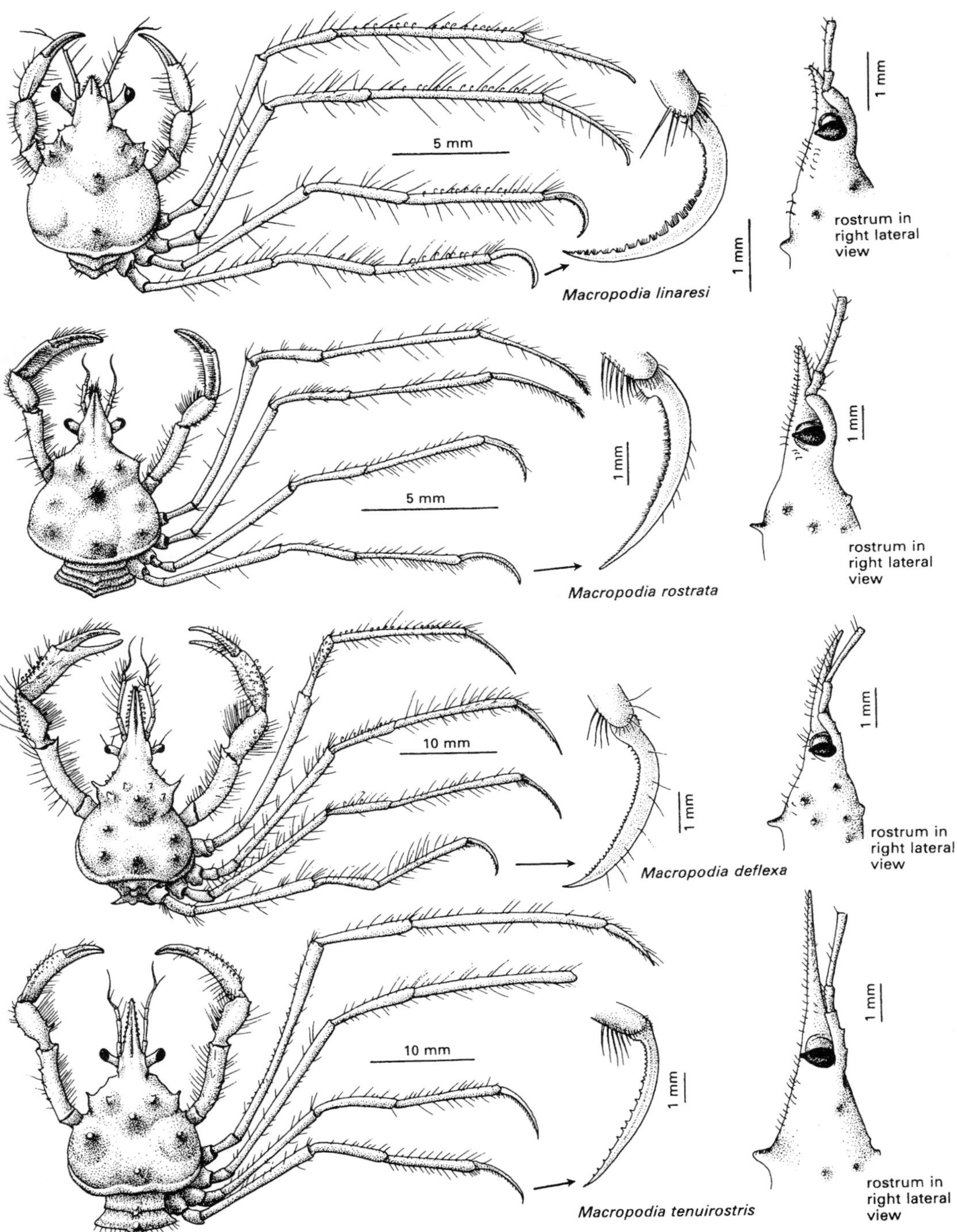
5 mm
1 mm
1 mm
rostrum in right lateral view
Macropodia linaresi
5 mm
1 mm
1 mm
rostrum in right lateral view
Macropodia rostrata
10 mm
1 mm
1 mm
rostrum in right lateral view
Macropodia deflexa
10 mm
1 mm
1 mm
rostrum in right lateral view
Macropodia tenuirostris

Fig. 8.59

On hard or mixed substrata, shallow sublittoral, 4–90 m; all British coasts, common; elsewhere, Norway to West Africa, Azores, Mediterranean.

Sometimes infected with the rhizocephalan cirripede *Drepanorchis neglecta* (Fraise), as are other species of the genus.

Macropodia linaresi
Forest and Zariquiey Fig 8.59
Carapace length, average 7.7 mm. Shallow sublittoral, south coasts of British Isles, not uncommon; elsewhere south to Spain, Mediterranean.

Macropodia deflexa Forest Fig. 8.59
Carapace, average male length 29 mm; average female length 25 mm. Colour greyish-brown; young striped pink. Shallow sublittoral 15–90 m; on sandy and weedy grounds. South and all west coasts of British Isles, elsewhere probably southwards.

Macropodia tenuirostris (Leach)
Slender spider crab Fig. 8.59
Carapace, large female, length 19 mm, breadth 11 mm; reddish brown. Shallow water, LWST to 168 m; all British coasts, common; elsewhere, Faeroes Bank southwards at least to Portugal.

Eurynome aspera (Pennant).
Strawberry crab Fig. 8.60
Carapace, of large male, length 17 mm, breadth 10 mm; light rose intermixed with a tint of bluish-grey. Carapace longer than broad, covered with large, irregular warty tubercles; distinctive amongst these is a fused group in the gastric region and a partly fused series bordering the posterior margin; an oval group in the cardiac region has two large tubercles at its anterior end. Rostral horns taper from a widened basal region to form a V-shaped cleft. Long chelipeds, and merus of rather short pereopods 2–5 also tuberculate.

On stony sand and shell substrata, sublittoral, mostly 12–40 m (occasionally 2–120 m); all British coasts, common; elsewhere, west Norway to West Africa; Mediterranean.

Eurynome spinosa Hailstone (Fig. 8.60) is similar but lacks the posterior group of fused warty tubercles.

Pisa tetraodon (Pennant).
Four-horned spider crab Fig. 8.60
Carapace length (excluding rostrum) of typical male, 34 mm; brownish-red. Carapace with three large spines each side between post-orbitals and posterior-laterals.

Shallow sublittoral, down to 50 m; south and south-west coasts of British Isles, not uncommon; elsewhere to Gibraltar and Mediterranean.

Pisa armata (Latreille) (Fig. 8.60) lacks the three large spines on lateral border of carapace.

Maja squinado (Herbst).
Common spider crab Fig. 8.56
Carapace, of large specimen, length 200 mm, breadth 150 mm; cheliped 450 mm; red, brownish-red, or yellowish. Very large crab often covered with attached algae. Circular, convex carapace bordered by strong tapering spines. Chelipeds not much thicker than pereopods 2–5 except in larger males

On various substrata, down to 50 m; abundant on west and south-west British coasts, less common in North Sea; elsewhere south to Cape Verde, Mediterranean.

14. CORYSTIDAE

Burrowing crabs with the carapace longer than broad. Antennae long, stiff, and hairy; interlocking to form an inhalant respiratory water tube. One species in British waters.

Corystes cassivelaunus (Pennant).
Masked crab Fig. 8.60
Carapace, of large female, length 39 mm, breadth 29 mm; pale red, grading into yellowish-white. Dorsal surface minutely granulated; markings sometimes give impression of a face; lateral margin with four teeth, of which the second is largest. Chelipeds of male more than twice length of carapace; those of female about as long as carapace.

Sandy, soft bottoms, LWST to 90 m; all British coasts, very common; elsewhere, Sweden to Portugal, Mediterranean.

15. ATELECYCLIDAE

Carapace almost circular. Antero-lateral margins with well-developed teeth. Three British species.

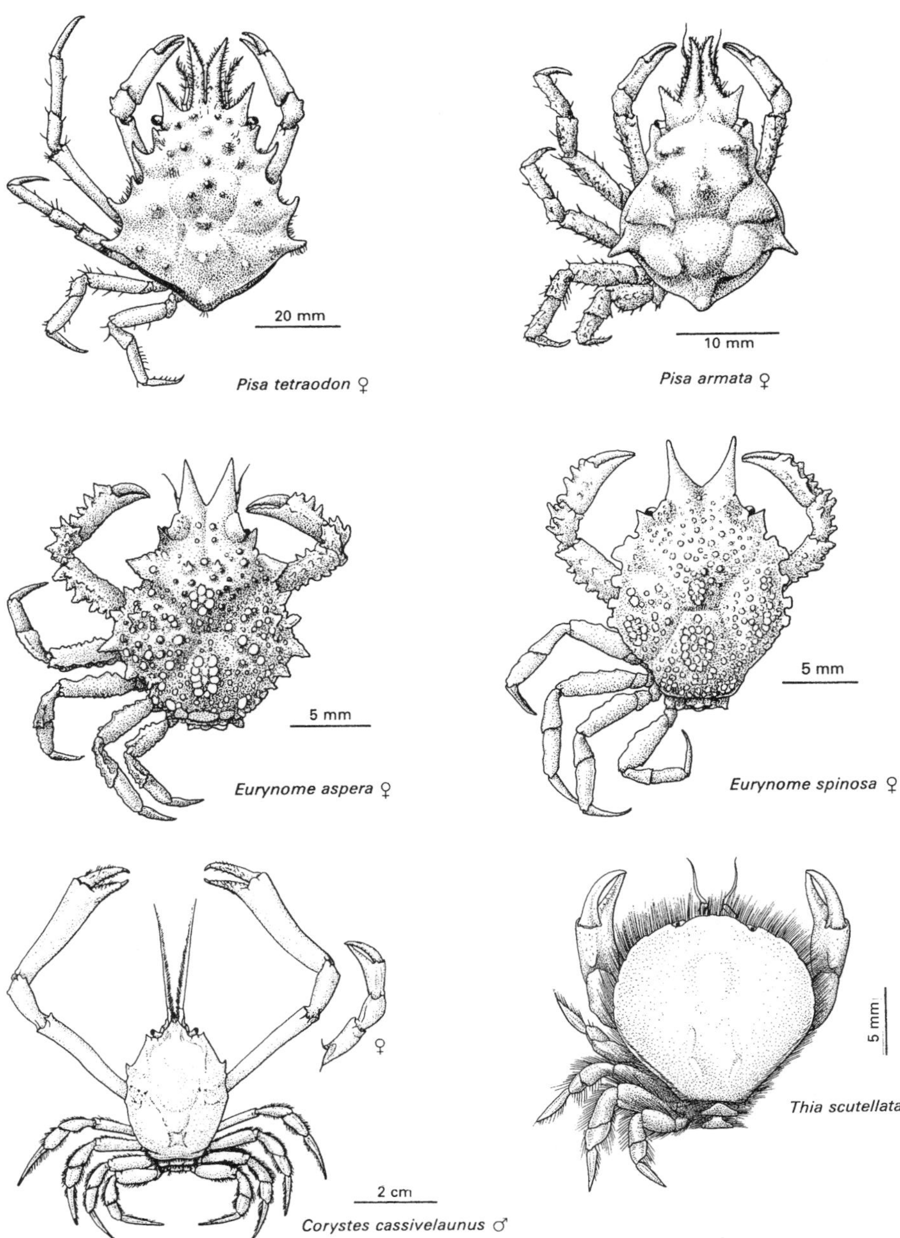
20 mm
Pisa tetraodon ♀
10 mm
Pisa armata ♀
5 mm
Eurynome aspera ♀
5 mm
Eurynome spinosa ♀
♀
2 cm
Corystes cassivelaunus ♂
5 mm
Thia scutellata

Fig. 8.60

Some authorities recognise a separate family, Thiidae, for *T. scutellata*.

---

1. Frontal region of carapace toothed **2**

   Frontal region of carapace without teeth ***Thia scutellata***

---

2. Carapace covered with coarse granules bearing hairs; merus of maxilliped 3 as long as ischium; carapace wider than long ***Atelecyclus undecimdentatus***

   Carapace with very small granulations, confined to gastric region, and very few hairs; merus of maxilliped 3 much shorter than ischium, carapace at most slightly wider than long ***Atelecyclus rotundatus***

Atelecyclus rotundatus (Olivi).
Circular crab Fig. 8.61

Carapace length 28 (39) mm. Reddish-brown, legs light brown; chelae fingers black. Carapace almost circular, about as broad as long, with numerous, short, transverse, granular striae; front with three acute teeth, central the longest; antero-lateral margin with 9–11 acute teeth with tuberculate edges. Chelipeds equal, male with two spines on upper edge of manus. Pereopods 2–5 moderately developed, slightly flattened.

Sandy and gravel bottoms, sublittoral, 12–300 m; probably all British coasts, not uncommon; elsewhere Norway to South Africa.

Atelecyclus undecimdentatus (Herbst) (Fig. 8.60) is a rare species reported from the English Channel south to West Africa.

Thia scutellata (Fabricius).
Thumbnail crab Fig. 8.60

Carapace length 20 (22) mm; pale pink, with brown markings. Carapace heart-shaped, slightly broader than long, convex, smooth, with densely setose margins. Chelipeds equal, slightly compressed.

LWST and shallow sublittoral to 45 m; burrowing in sand and mud; all British coasts (and North Sea?), not uncommon; elsewhere, Sweden, Holland to West Africa, Mediterranean.

16. CANCRIDAE

Carapace wide and oval; front to lateral border lobed. Two British species. *Cancer bellianus* Johnson occurs in deep water off western coasts of Britain.

Cancer pagurus Linnaeus.
Edible crab Fig. 8.61

Carapace, of typical male, up to length 92 mm, breadth 150 mm, large specimens up to breadth of 250 mm; reddish-brown; chela fingers black. The wide, oblong-shaped carapace is distinctively marked along its fronto-lateral margins with 10 rounded lobes. Chelae slightly unequal in shape and toothing; somewhat smaller in females. Tufts of stiff hairs in rows on pereopods, dactyls ending in spine-like tips.

Mid-tide to shallow sublittoral (100 m); rocky bottoms; all British coasts, abundant; elsewhere, north Norway to West Africa, Mediterranean.

17. PORTUNIDAE

Carapace usually broader than long, broadest at last pair of anterolateral teeth; front margin usually with an uneven number of teeth, often three; usually five antero-lateral teeth, but up to nine. Eyes and orbits prominent. Carpus of chelipeds usually with strong acute carpal process. Dactyl of last pereopod usually flattened, often oval-shaped, forming a paddle used for swimming. Fifteen species recorded around Britain.

---

1. Dactyl of pereopods 2–4 slightly flattened **4**

   Dactyl of pereopods 2–4 long and conical **2**

---

2. Dactyl of pereopod 5 narrowly conical, as other dactyls ***Pirimela denticulata***

   Dactyl of pereopod 5 flattened **3**

---

3. Dactyl of pereopod 5 slightly broader than that of pereopods 2–4 ***Carcinus maenas***

   Dactyl of pereopod 5 broad and flat **6**

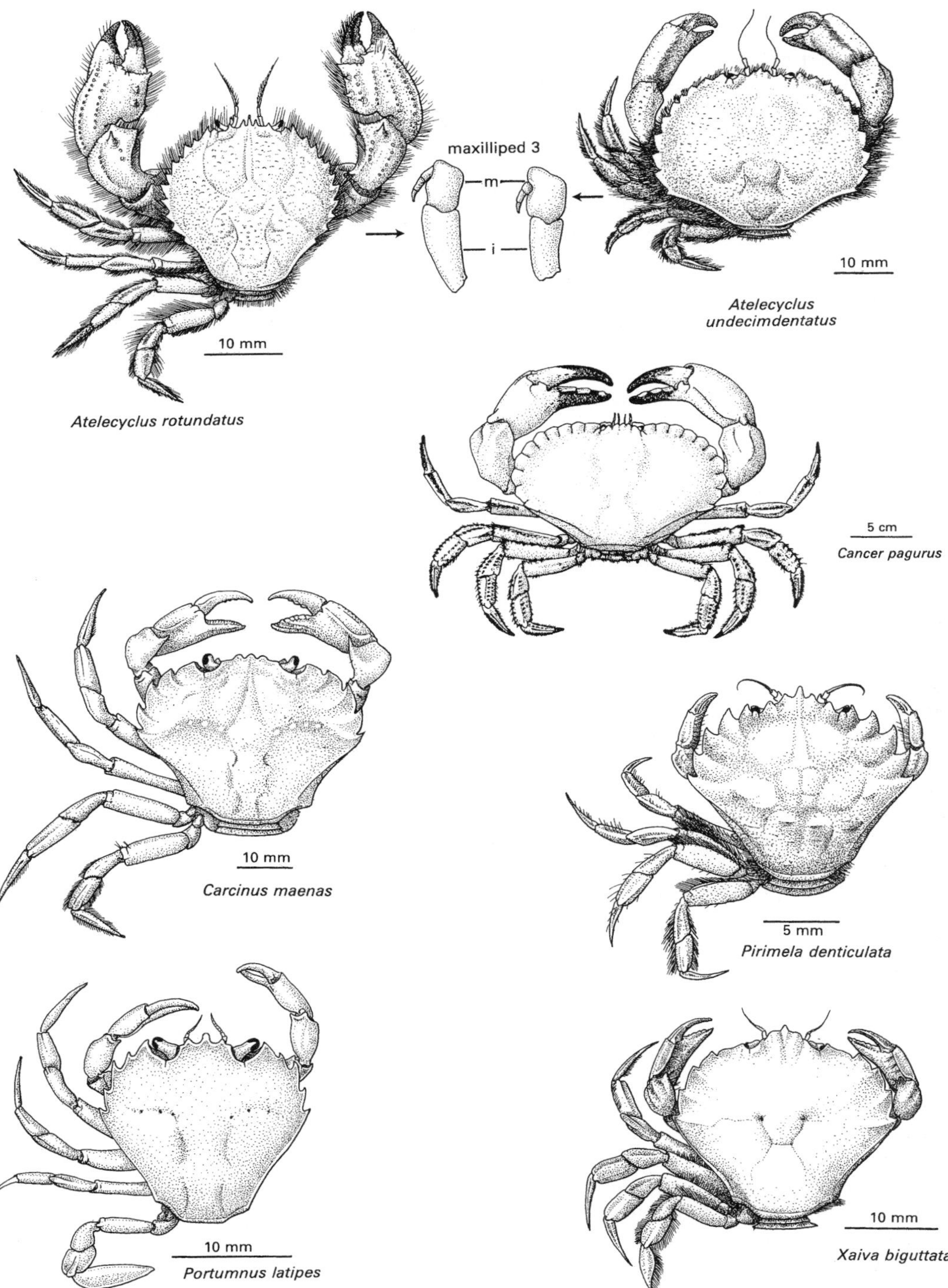
maxilliped 3
m
i
10 mm
Atelecyclus rotundatus
10 mm
Atelecyclus undecimdentatus
5 cm
Cancer pagurus
10 mm
Carcinus maenas
5 mm
Pirimela denticulata
10 mm
Portumnus latipes
10 mm
Xaiva biguttata

Fig. 8.61

4. Carapace wider than long, front with single strong tooth, not trilobed **Xaiva biguttata**

Front of carapace trilobed, or with five small teeth between eyes **5**

5. Carapace slightly wider than long, with five small teeth between eyes **Polybius henslowi**

Carapace slightly longer than wide, front trilobed **Portumnus latipes**

6. Front of carapace with 7–10 unequal teeth **Necora puber**

Front with three teeth or lobes, or toothless **7**

7. Front entire (toothless) **Liocarcinus arcuatus**

Front with three lobes or teeth **8**

8. Front with three blunt lobes, the middle one longest **9**

Front with three teeth which may be blunt or sharp, all similar **10**

9. Front projecting; dactyl of pereopod 5 without distinct median rib; carapace usually smooth and almost hairless **Liocarcinus pusillus**

Front lobed, not projecting; dactyl of pereopod 5 with strong median rib; carapace and parts of limbs covered by transverse rows of hairs arising from crenulated streaks **Liocarcinus corrugatus**

10. Carapace with minute rows of hairs arising from transverse granular lines **Liocarcinus depurator**

Carapace smooth **11**

11. Front with three similar teeth, the middle often the shortest; cheliped carpus has no tooth on outer edge **Liocarcinus marmoreus**

Middle of three front teeth just longest; cheliped carpus with tooth at outer angle **Liocarcinus holsatus**

Pirimela denticulata (Montagu) Fig. 8.61
Carapace length 15 mm, breadth 18 mm; blackish-violet, margins red, whitish tubercles; chelipeds and legs banded brown and yellow. Chelipeds equal; rather small, especially in females; carpal processes small.

LWST to 180 m, gravel and sand; all British coasts, common; elsewhere, Norway to West Africa, Mediterranean.

Polybius henslowi Leach.
Henslow's swimming crab Fig. 8.62
Carapace length 39 (44) mm. Reddish-brown. Carapace flat and circular, five low, broad teeth on antero-lateral margin; five nearly similar teeth on front between orbits. Chelipeds strong, equal. Propodus and dactyl of last pereopod paddle-shaped, length/breadth ratio 1.7, obtusely rounded terminally.

Truly a swimming crab; shallow inshore waters. British Isles, especially south coasts, common; elsewhere to Morocco, Mediterranean.

Necora puber (Linnaeus).
Velvet fiddler or devil crab Fig. 8.62
Carapace length 50 (65) mm, breadth about 66 mm. Blue, obscured by brown pubescence, red prominences. Carapace dorsal surface flattened and pubescent; frontal margin with up to 10 narrow unequal teeth, middle two often larger; antero-lateral margin with five sharp, forward-pointing teeth. Chelipeds equal, strong, pubescent.

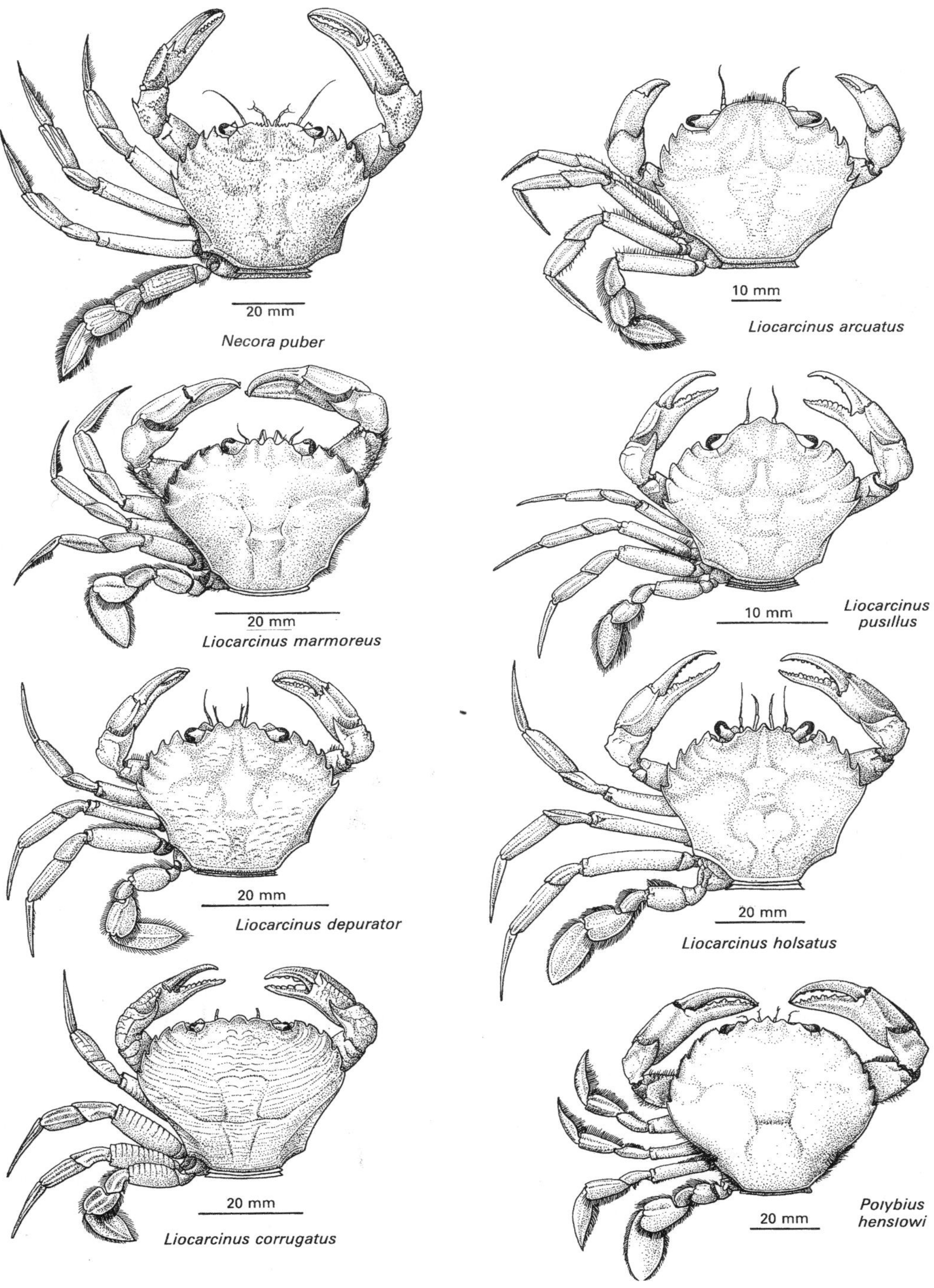
20 mm
Necora puber
10 mm
Liocarcinus arcuatus
20 mm
Liocarcinus marmoreus
10 mm
Liocarcinus pusillus
20 mm
Liocarcinus depurator
20 mm
Liocarcinus holsatus
20 mm
Liocarcinus corrugatus
20 mm
Polybius henslowi

Fig. 8.62

On stony and rock substrata; intertidal and shallow water, occasionally to 70 m. All British coasts, widespread and very common; elsewhere western Norway to West Africa, Mediterranean, Black Sea.

Liocarcinus arcuatus (Leach) Fig. 8.62
Carapace length 22 (29) mm, breadth about 30 mm; brown, limbs lighter. Carapace gently convex, with minute transverse ridges; frontal margin between eyes entirely without teeth, hairy.

On sandy and stony substrata, 2–108 m; south and west coasts of British Isles, common.

Liocarcinus corrugatus (Pennant).
Wrinkled swimming crab Fig. 8.62
Carapace length 33 (43) mm, breadth about 41 mm. Brownish-red with patches of red or yellow. Carapace with numerous, strong, transverse, hairy ridges; frontal margin with prominent, rounded, median lobe, flanked by broad sub-median lobes reaching the orbits.

On stones and gravel, 1–100 m; south and west coasts of British Isles, uncommon; elsewhere south to North Africa, Mediterranean.

Liocarcinus pusillus (Leach) Fig. 8.62
Carapace length 20 (23) mm. Variable colour; yellow, variegated with red-brown to uniform brown. Carapace gently convex, frontal margin protruding, with a low median lobe and broader submedian lobes.

On stones, gravel, and rough ground, also sand; intertidal to 100 m; all British coasts, common; elsewhere, north Norway to West Africa, Canaries, Mediterranean.

Liocarcinus depurator (Linnaeus).
Harbour crab Fig. 8.62
Carapace length about 40 mm, breadth about 51 mm. Pale reddish-brown, tip of last pereopod violet. Carapace relatively flat, with numerous transverse, hairy, crenulations; front of carapace with a median lobe slightly more prominent than two similar flanking lobes; orbits wide; posterior margin of carapace broad.

Soft, sandy, and mixed bottoms, LWST to 450 m; all British coasts, very common; elsewhere, Norway to West Africa, Mediterranean.

Liocarcinus holsatus (Fabricius) Fig. 8.62
Carapace length 30 (37) mm, breadth about 38 mm. Brownish-grey, tinged with green. Very similar to *L. depurator* but lacking transverse crenulations on carapace; three similar-sized lobes on frontal margin, central one sometimes shortest; orbits smaller than in *L. depurator*; posterior edge of carapace narrower than in *L. depurator*.

On hard and mixed bottoms; 6–350 m; all British coasts, widespread and often very common; elsewhere north Norway to Spain and Canaries; not Mediterranean. Host to the parasitic cirripede *Sacculina carcini* Thompson in some areas.

Liocarcinus marmoreus (Leach).
Marbled swimming crab Fig. 8.62
Carapace length 30 (35) mm, breadth about 35 mm. Marbling patterns of reddish-yellow, light brown. Carapace surface smooth; frontal margin with three similar-sized lobes.

On fine sand and gravel, LWST to 200 m; east and west coasts of British Isles, common; elsewhere Denmark to Spain and Azores. Apparently replaced by closely related species in Mediterranean.

Carcinus maenas (Linnaeus).
Shore crab, green crab Fig. 8.61
Carapace length 55 (60) mm, breadth about 73 mm; males slightly larger than females, maximum size (after terminal anecdysis) varies with locality. Colour variable, adults dark green, mottled to grey, with green or puce (brick red) legs and underside; juveniles variable, often speckled or with central white or black triangle or other strong pattern. Carapace much broader than long, minutely granular; frontal margin with three gently rounded lobes, central one slightly advanced, flanked on each side by a lower smaller prominence which forms the mesial angle of the orbit; orbital margin incised. Chelipeds subequal. Moderately strong dactyl of last pereopod scarcely wider than first three but compressed with setose margin.

Ubiquitous intertidally, including splash-zone pools, saltmarshes, and estuaries; also in shallow sublittoral down to 200 m; all British coasts, abundant; elsewhere north Norway and Iceland to West Africa, north-east Americas and Indo-west Pacific.

Sacculina carcini Thompson, a parasitic cirriped, is common, especially on sublittoral specimens.

Xaiva biguttata (Risso) Fig. 8.61

Carapace length about 23 mm, breadth about 23 mm. Colour variable, dull greenish-yellow to white with large brown patches; dactyl sometimes violet. Carapace frontal margin protrudes as three obtusely rounded lobes, the median slightly in advance; orbits and eyes small; the first of five antero-lateral teeth has a concave outer face. Last pereopod lanceolate, dactyl bearing a median keel.

On gravel and sand, intertidal or shallow sublittoral; south and west coasts of British Isles, sparse; elsewhere south to Cape Verde, Mediterranean.

Portumnus latipes (Pennant).
Pennant's swimming crab Fig. 8.61

Carapace length about 20 mm; reddish with white shading. Carapace as broad as long, heart-shaped, smooth; frontal margin with three small subacute teeth, the median slightly in advance; orbits large. Chelipeds slightly unequal. Dactyl of last pereopod broadly lanceolate.

Sandy bottoms, LWST to shallow sublittoral, 150 m; all British coasts, except for north, common; elsewhere, North Sea to North Africa, Mediterranean. Black Sea.

18. XANTHIDAE

Carapace broader than long; oval or subrectangular. Frontal region broad. Seven British species.

---

1. Outer surface of cheliped carpus spiny or tuberculated **2**

   Chelipeds without spines or tubercles on carpus **3**

---

2. Carapace largely covered by hairs ***Pilumnus hirtellus***

   Carapace not hairy ***Monodaeus couchi***

---

3. Antero-lateral margin of carapace with five teeth, the first forming the outer angle of the orbit **4**

   Antero-lateral margin of carapace with four teeth similarly counted **5**

---

4. Pereopods 2–5 have thick fringe of hairs on carpus (and merus) ***Xantho pilipes***

   Pereopods 2–5 have very few hairs on carpus ***Xantho incisus***

---

5. Basal segment of antenna (articles 2 and 3) not in contact with margin of front ***Rhithropanopeus harrisii***

   Basal segment of antenna in contact with front ***Neopanope sayi***

Pilumnus hirtellus (Linnaeus).
Hairy crab Fig. 8.63

Carapace length 9 (15) mm. Brownish-red (or purplish); fingers light brown; legs banded purple and yellowish white. Carapace smooth with conspicuous club-shaped setae; front straight, with median notch and serrated edges. Chelipeds large and very unequal, carpus spiny; manus of larger chela smooth or slightly tuberculate; manus of smaller chela always tuberculate or spiny. Chelipeds and legs with long club-shaped setae.

Rock and stony bottoms, LWNT to 70 m; all British coasts, very common intertidally; elsewhere, west Norway to North Africa, Cape Verde, Mediterranean, Black Sea.

Rhithropanopeus harrisii (Gould).
Dwarf crab Fig 8.63

Carapace, of typical male, length 12 mm, breadth 15 mm. Yellowish-green, with black spots. Fingers of chelipeds whitish. Carapace front edge bearing two rows of minute tubercles. Chelipeds unequal, rather smooth. Margins of pereopods 2–5 hairy.

North-east American, brackish water, species introduced into south Baltic, Dutch, and other European estuaries, and particularly docks warmed by power-station effluents, e.g. Swansea and Southampton.

Neopanope sayi (Smith). Mud crab Fig. 8.63

Carapace length 13 (23) m. Generally reddish-brown; fingers of chelae dark brown to black. Carapace oval, convex, minutely granular; front with small median notch. Chelipeds very unequal in male only.

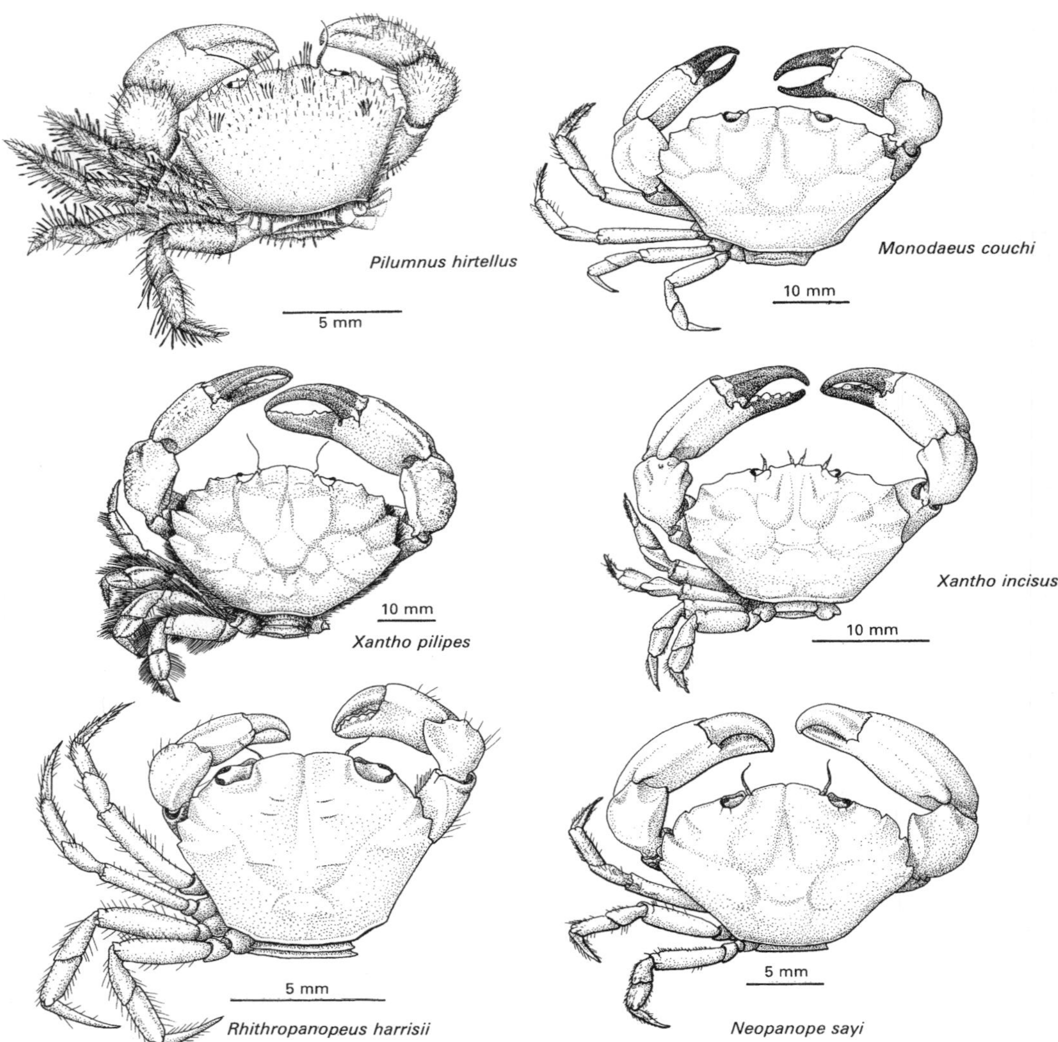

Fig. 8.63

North American species introduced into various European coastal situations frequented by shipping.

Monodaeus couchi (Couch).

Couch's crab Fig. 8.63

Carapace length 13 (18) mm. Reddish-brown, yellowish-grey below, Carapace oval, relatively flat. Cheliped fingers black or brown. Chelipeds quite stout, unequal in male; carpus and propodus with tuberculate keels.

On sand, sandy mud, and gravel, 10–100 m, south and south-west British coasts, Irish Sea, rare; elsewhere, south to Cape Verde and Azores, Mediterranean.

Xantho pilipes Milne Edwards.

Risso's crab Fig. 8.63

Carapace length 15 (21) mm. Yellowish with red markings, fingers of chelipeds brown. Carapace antero-lateral margin with four, more or less obtuse teeth, all hairy on underside, last two with fringed hairs.

Stone, shell, and sand bottoms, LWST to 110 m; Shetland and all west coasts, south coast as far as Suffolk, very common; elsewhere, west Norway, Sweden to West Africa, Mediterranean.

Xantho incisus Leach.
Furrowed crab Fig. 8.63
Carapace length 16 (22) mm. Yellowish-brown with darker patches, fingers of chelipeds black. Very similar to *X. pilipes*. Antero-lateral teeth more obtuse, none fringed with hairs.

On rocky shores, LWST to 40 m; more southern than last species, south and west British coasts, widespread and often abundant; elsewhere south to Cape Verde, Azores, and Canaries, Mediterranean.

19. GONEPLACIDAE

Crabs with a wide rectangular carapace and very long chelipeds. One species in British waters.

Goneplax rhomboides (Linnaeus).
Angular crab Fig. 8.64
Carapace, of typical male, length 20 mm, breadth 34 mm, cheliped 100 mm; female, length 11 mm, breadth 18 mm. Reddish-yellow. Eyes on long retractable peduncles. Chelipeds of males four or five times length of carapace; shorter in female.

Burrows in muddy sand; 8–80 m; all British coasts, common; elsewhere, south to South Africa, Mediterranean.

20. PINNOTHERIDAE

Carapace often thin and translucent. Eyes small. Ischium of maxilliped 3 sometimes fused with merus. Commensal with other invertebrates.

---

1. Dactyl of pereopods 2–5 strongly curved, half length of propodus. Front of male carapace extended forward **Pinnotheres pisum**

   Dactyl of pereopods 2–5 slightly curved, more than half length of propodus. Front of male carapace not extended **Pinnotheres pinnotheres**

Pinnotheres pisum (Linnaeus).
Pea crab Fig. 8.64
Carapace, of typical male, length 6 mm, female length 13 mm. Male pale yellowish-grey with symmetrical darker markings. Female translucent, revealing yellow internal organs and red gonads. Male carapace hard, opaque with projecting front margin; chela with swollen manus; pereopods 2–5 with long setae when young. Female carapace translucent, front deflected downwards; chela manus not swollen.

Parasitic; in mantle cavity of live lamellibranchs: *Modiolus modiolus* (Linn.) abundant; *Mytilus edulis* Linn. rare. Also occasional in *Spisula*, *Glycymeris*, *Venus* and *Venerupis*. All British coasts; elsewhere, Norway to West Africa and Mediterranean.

Pinnotheres pinnotheres (Linnaeus).
Pinna pea crab
Carapace average length, male 7 mm, female 12 mm. Brown. Male carapace hard and opaque; front hardly projecting or deflected. Female carapace poorly calcified, soft, translucent; front deflected downwards.

Parasitic in mantle cavity of lamellibranchs: commonest in *Pinna*, also in *Modiolus* and oysters. Also in tunicates: *Ascidia mentula* Müller, *Ascidiella aspersa* (Müller). South-west British Isles, scarce; elsewhere, south to Gabon, Mediterranean.

21. GRAPSIDAE

Carapace usually more or less rectangular; lateral margins not clearly separable into anterior and posterior regions. There are no native species of Grapsidae; two exotic species have been established in Britain, and two others occur sporadically, viz. *Pachygrapsus marmoratus* (Fabricius) and *Planes minutus* (Linn.).

---

1. Mat of hairs on manus of chela; carapace narrowing anteriorly **Eriocheir sinensis**

   Manus of chela without mat of hairs; carapace narrowing posteriorly **Brachynotus sexdentatus**

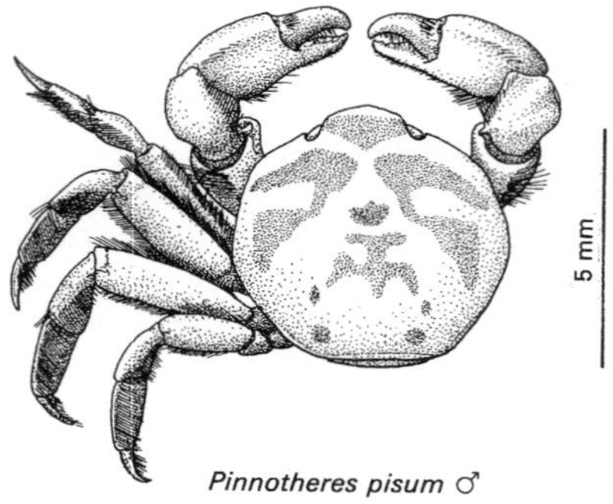

Pinnotheres pisum ♂

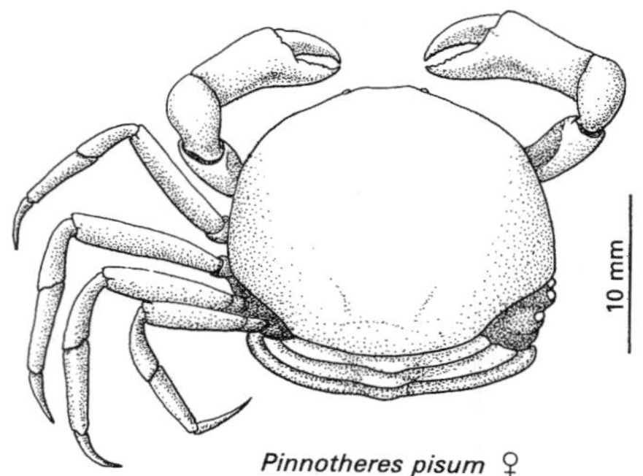

Pinnotheres pisum ♀

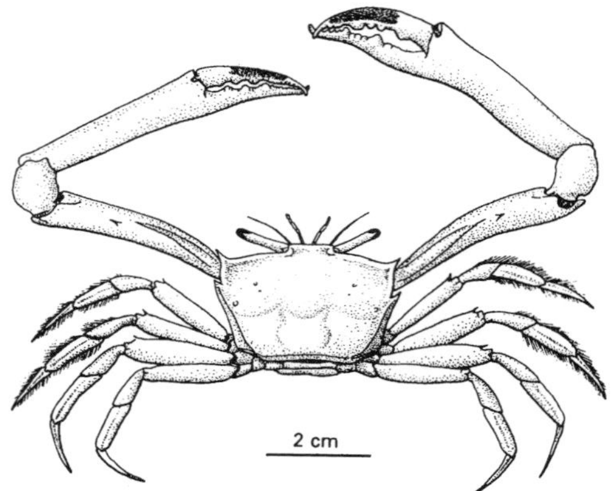

Goneplax rhomboides

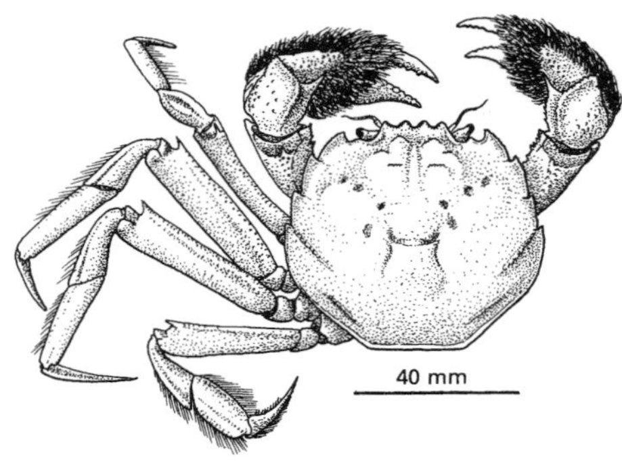
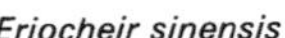

Eriocheir sinensis

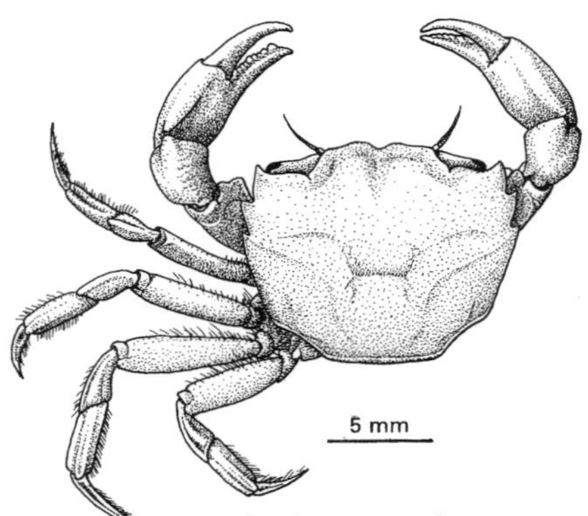

Brachynotus sexdentatus

Fig. 8.64

Eriocheir sinensis Milne Edwards.
Mitten crab Fig. 8.64

Carapace length 56 (62) mm. Greyish-green to dark brown, legs lighter. Carapace almost square, convex; front with four conspicuous teeth separated by three curved notches. Chelipeds robust, equal, palm with dense tuft of long setae.

Freshwater, river crab; returns to saline water in estuaries for breeding; British Isles: various east-coast estuaries, including River Thames, River Ouse, increasingly common; elsewhere Baltic Sea to France; introduced to Europe in 1912 by shipping trade with China. A pest species, causing damage to river banks and fishing nets.

Brachynotus sexdentatus (Risso).
Mediterranean crab Fig. 8.64

Carapace length 10 (18) mm. 0live green, speckled With black, chelipeds and legs light olive or grey, mottled black. Orbits wide; three lateral teeth. Chelipeds equal; much larger in male.

Mediterranean species, littoral to 90 m. British Isles: survives in docks warmed by power-station effluents; presumably introduced by shipping.

9 ARACHNIDS, INSECTS, MILLIPEDES, AND SEA SPIDERS

TERRESTRIAL ARTHROPODS IN THE MARINE ENVIRONMENT

Animals of terrestrial origin on the seashore include insects (Phylum Hexapoda), arachnids (Phylum Chelicerata), and myriapods (Phylum Myriapoda). These usually occupy very different habitats from animals of marine origin, because most require periodic exposure to air. In contrast to their terrestrial relatives, their main problems are respiration, maintenance of position, and osmoregulation. They are restricted to habitats which allow them to survive tidal immersion, spray, wave action, desiccation, and temperature extremes, although one family of mites, the Halacaridae, cannot tolerate exposure to air and remain permanently submerged, or at least wetted, and occur even in the sublittoral. Terrestrial arthropods occur in the splash or supralittoral zone, which is wetted by seaspray, and in the intertidal or littoral zone which is covered and uncovered by the diurnal tides. The number of terrestrial species declines dramatically below these zones, which are barriers for both terrestrial and marine animals. The terrestrial fauna is associated with particular habitats within each zone.

SUPRALITTORAL ZONE

Mites and small insects live under or within lichens, where particles of soil may also become trapped. Decomposing marine algae, terrestrial vegetation, and other debris deposited by the tide at high water mark may also contain very diverse animal communities. These include predatory spiders, beetles, flies, and mites living on drier surface debris; and fly larvae, nematodes, and mites deep within the decomposing material.

LITTORAL ZONE

Mites, flies, and beetles may forage on open substrata (rock, algae, or sand) during periods of tidal emersion, retreating before the advancing tide to refuges in the supralittoral, or to air pockets in the littoral zone, in rock-crevices, between and within barnacle shells, or below larger stones. Halacarid mites occur in fully submerged littoral habitats protected from wave action, for example in larger crevices, on algae and lichens.

Shore-dwelling arachnids and insects require special collecting techniques. Larger, active spiders and beetles may be caught along the strandline with pit-fall traps. Smaller mites and insects running on open surfaces may be captured with a paintbrush dipped in alcohol, or with an entomologists aspirator ('pooter'). Crevices containing mites and false scorpions are best opened with a hammer and cold chisel, the visible animals collected with an aspirator, and the sediment containing microscopic specimens scraped up with a scalpel or knife and bottled for extraction in the laboratory. The study of crevice faunas should, however, be done with restraint, as it permanently destroys the habitat. Mites and spiders on or under dry stones may be dislodged by brushing the stones on to a clean sheet of paper or into a bag, while mites under wet stones or in barnacle-covered stones can be collected by scrubbing the material in a bucket of seawater, and then filtering the debris through a bronze 125 μm screen. Specimens may be extracted from sediment samples suspended in seawater to which is added sufficient chloroform to make a concentration of 3–5%. The mixture is shaken and allowed to settle. The surface liquid may then be poured off into a suitable dish and searched for mites with a microscope. Mites are best extracted from littoral algae and lichens using hypersaline made up from cooking salt dissolved in water to a specific gravity of 1.2 at 20 °C. The sample is placed in at least 20 times its volume of hypersaline and stirred vigorously for 30 seconds. The mites float to the surface, and can be collected by filtration through a very fine nylon mesh (60 or 120 μm).

Arachnids may be maintained alive in vials containing a little moist tissue paper, but beware that many are cannibalistic. Spiders and false scorpions are best preserved in a mixture of 50% ethyl alcohol, 10% formalin, 5% glycerol and 35% distilled water but viewed in 70% ethyl alcohol. Mites should be fixed in a mixture of 10% glacial acetic acid, 50% glycerol, and 40% distilled water, to prevent them becoming brittle. They can be rendered transparent (cleared) for identification in 50–80% lactic acid at 50 °C for 4–10 days and then mounted in lactic acid on a cavity slide for viewing under a microscope. Mites should be stored in a solution of 70% aqueous alcohol and 5% glycerol.

PHYLUM CHELICERATA

Class Arachnida

Arachnids are air-breathing chelicerate arthropods, possessing six pairs of appendages, including lobster claw-like two-piece mouthparts (*chelicerae*), used to capture and/or cut up food prior to ingestion, a pair of *palps* and, in adults, four pairs of legs. The three shore-frequenting orders may be differentiated as follows:

---

1. Palps chelate (Fig. 9.1A, B) **False Scorpions: Pseudoscorpiones**
   Palps not chelate **2**

---

2. Body with a distinct pedicel or waist (Fig. 9.1C, D) **Spiders: Araneae**
   Pedicel absent (Figs 9.2, 9.3) **Mites: Acari**

Order Pseudoscorpiones (false scorpions)

False scorpions (Fig. 9.A, B) resemble true scorpions by having large chelate *palps* but are much smaller (under 5 mm long), and lack the characteristic scorpion tail. All are predators. The body is divided into two regions, an anterior *prosoma* bearing two pairs of simple eyes (*ocelli*) and six pairs of appendages, and a posterior segmented *opisthosoma*. The *chelicerae* bear structures used in feeding, grooming, and silk production. Each palp *chela* comprises a hand/fixed digit and a movable 'finger' (Fig. 9.1A), long sensory hairs or *trichobothria*, and a subterminal opening to a venom gland. Functions of the palp include sensory (vibration) reception, courtship, defence, fighting, nest building, and prey capture and immobilization.

Only one species commonly occurs on British shores, *Neobisium maritimum* (Leach), in upper littoral crevices on the south and west coasts. It is frequently found in silk cocoons where it hibernates, moults, and broods eggs. Two others, *Chthonius halberti* Kew and *C. tetrachelatus* (Preyssler), from the supralittoral and extreme upper littoral zone among tidal debris and patches of lichens, also occur extensively in terrestrial habitats.

Order Araneae (spiders)

The body of a spider is divided into an anterior *prosoma* and a larger, posterior unsegmented *opisthosoma*, by a narrow *pedicel* or waist. The prosoma bears six pairs of appendages and is covered with an armoured *carapace* bearing four pairs of *ocelli* (simple eyes) arranged on an anterior 'face'. The *chelicerae* are suspended below the face, and each consists of a large basal segment with a number of distal spines associated with the pointed and hinged *fang* (Fig. 9.1E). Prey is seized by the chelicerae and paralysed by venom injected by the fang. The palps are simple and leg-like in females and distally dilated to resemble boxing gloves in the male, where they are used in courtship and sperm transfer. The terminal leg segments have two or three claws, while that of leg IV may also have a dorsal comb of stout curved

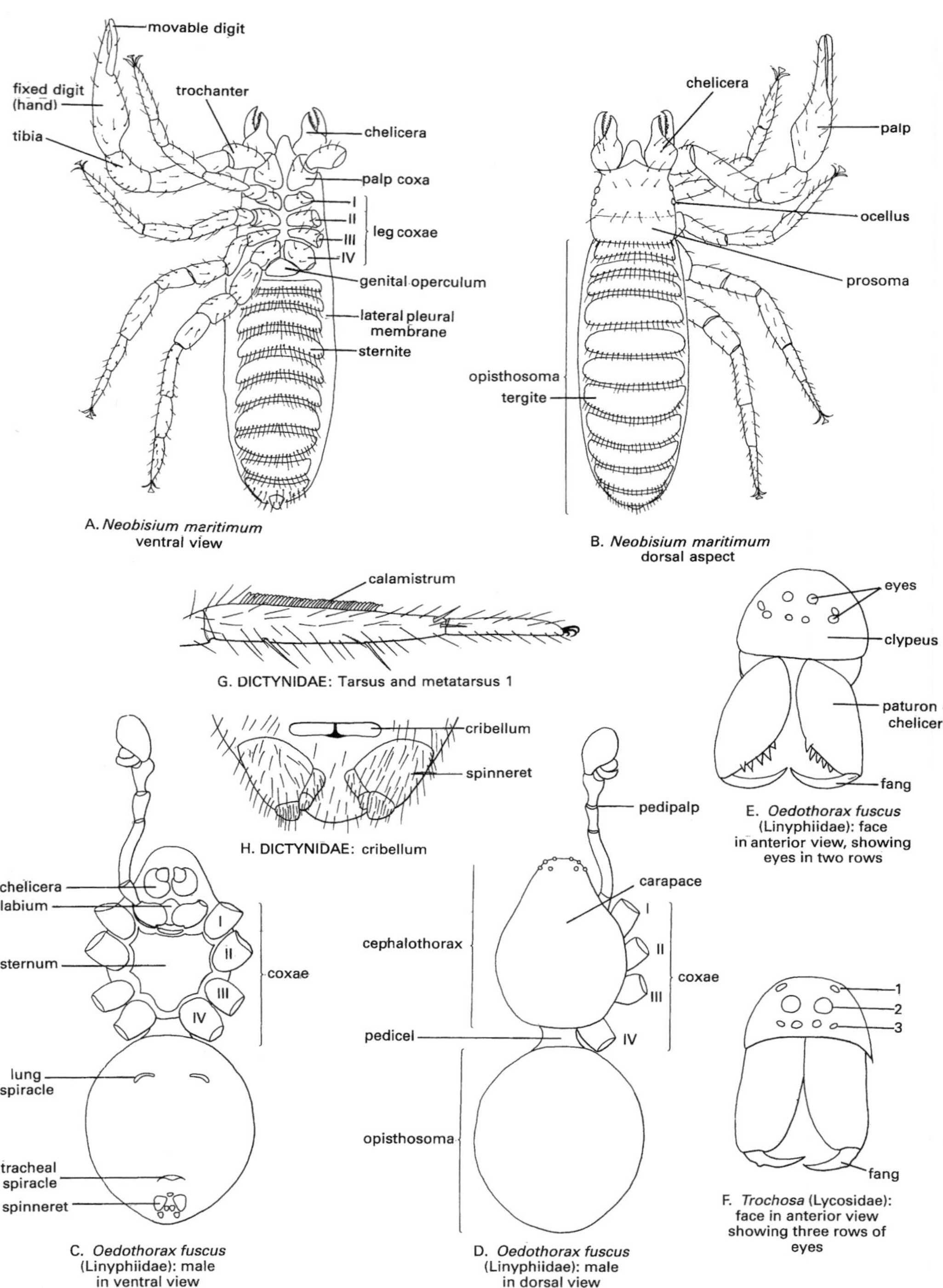
movable digit
fixed digit (hand)
trochanter
tibia
chelicera
palp coxa
I
II
III
IV
leg coxae
genital operculum
lateral pleural membrane
sternite
A. Neobisium maritimum ventral view
chelicera
palp
ocellus
prosoma
opisthosoma
tergite
B. Neobisium maritimum dorsal aspect
calamistrum
G. DICTYNIDAE: Tarsus and metatarsus 1
eyes
clypeus
paturon of chelicera
fang
E. Oedothorax fuscus (Linyphiidae): face in anterior view, showing eyes in two rows
cribellum
spinneret
H. DICTYNIDAE: cribellum
pedipalp
carapace
cephalothorax
coxae
pedicel
opisthosoma
D. Oedothorax fuscus (Linyphiidae): male in dorsal view
chelicera
labium
sternum
coxae
lung spiracle
tracheal spiracle
spinneret
C. Oedothorax fuscus (Linyphiidae): male in ventral view
1
2
3
fang
F. Trochosa (Lycosidae): face in anterior view showing three rows of eyes

Fig. 9.1

hairs termed a *calamistrum* (Fig. 9.1G). The spherical to elongate opisthosoma is equipped with two or three pairs of silk-producing *spinnerets*, which may form a 'dragline', web, or swathing threads to entangle and immobilize prey. A plate or *cribellum* may occur anterior to the spinnerets, but only if leg IV is equipped with a calamistrum (Fig. 9.1H).

British shore-dwelling spiders are restricted to tidal debris, sand dunes, and gravel in the supralittoral and extreme upper littoral zones. A few species forage in saltmarshes, where they climb up vegetation to avoid wetting by flood tides but the majority of shore-dwelling species are also characteristic of sand dunes and heathland habitats.

KEY TO FAMILIES OF ARANEAE

1. Calamistrum present on leg IV (Fig. 9.1G); cribellum present immediately anterior of the spinnerets (Fig. 9.1H). Length 2.5–3.5mm — **1. Dictynidae**

   Calamistrum and cribellum absent — **2**

2. Legs each with three terminal claws, middle claw present — **3**

   Legs each with two terminal claws, middle claw absent — **4**

3. Small spiders, length 1–3mm; two rows of eyes (Fig. 9.1E) — **2. Linyphiidae**

   Large spiders, length 4–14mm; three rows of eyes (Fig. 9.1F) — **3. Lycosidae**

4. Three rows of eyes; carapace massive and square-fronted; body generally black and shiny. Length 2.5–3.0 mm — **4. Salticidae**

   Two rows of eyes; carapace flat and broad; body covered in fine grey hairs, giving it a mousy appearance. Length 3–4 mm — **5. Gnaphosidae**

1 DICTYNIDAE — Fig. 9.1G,H

One species, *Dictyna major* Menge has been recorded from the supralittoral among tidal debris and sand dunes. North of England and Scotland only.

2 LINYPHIIDAE
Sheet-web or money spiders — Fig. 9.1C–E

Small spiders with complex and variable morphology. Usually occur among tidal debris and vegetation; supralittoral, under stones and boulders. Ten species occur on British shores.

3 LYCOSIDAE
Wolf spiders — Fig. 9.1F

Large, robust spiders which are usually active at night, especially in summer. Occur in tidal debris, among sand dunes and shingle. The genera *Trochosa* and *Lycosa* have been recorded on saltmarshes and mud flats. Four species have been reported around Britain.

4 SALTICIDAE. Jumping spiders

One species, *Euophrys browningi* Milledge and Lockett, has been recorded from the supralittoral among tidal debris and gravel, on sheltered shores on the south-east coast of England.

5 GNAPHOSIDAE

One species, *Haplodrassus minor* Cambridge, has been recorded from the supralittoral among tidal debris and shingle, on the south coast of England.

Order Acarina (mites)

The Acari, or mites, are small, highly specialized arachnids, in which the body is fused into one unit,

Table 9.1 Guide to identification of suborders of Acarina (unique key characters are in bold)

Group	Mesostigmata	Astigmata	Cryptostigmata	Prostigmata
Size	Up to 1.5 mm	Up to 0.8 mm	Up to 1 mm	Up to 2 mm
Armour	**Discrete shields**	None or very little	**Dull and leathery or sclerotized and shiny**	Usually none, but some with shields
Colour	White to brown	White or pale brown but blue when cleared	**Black or dark brown**	Very variable but may be **bright orange or pink**
Gnathosoma	**Conspicuous**	Inconspicuous	Often hidden	Variable
Palps	**Long with basal claw**	Very short	Short	Variable, may be small or **very large and modified**
Other distinctive characters	**Large aggregations**	**Very long hairs; blue when cleared**	Very slow moving	Some move very rapidly

with no trace of division or segmentation. The body, or *idiosoma*, has an anterior *gnathosoma*, bearing the *chelicerae* and *palps*, and the legs are directly attached to the idiosoma (Fig. 9.2A, B). The form of all appendages varies considerably throughout the group and the idiosoma may be variously armoured or completely soft. The life cycle includes a six-legged larva, up to three nymphs, and adult. Many species occur in the supralittoral, fewer species in the littoral zone, and only one family in the sublittoral zone.

Mite identification is very difficult and requires some expertise. However, specimens may be taken to one of the four major groups (suborders) by reference to at least one of the characters in Table 9.1

Suborder Mesostigmata (gamasid mites)

Large numbers of supralittoral and littoral species, which mostly feed on other mites, insects, nematodes, and eggs of such animals. The gnathosoma is always conspicuous, while the idiosoma is armoured with discrete *shields*, the arrangement of which is diagnostic. Many species and families on the shore, but only three are conspicuous.

6 CYRTOLAELAPIDAE — Fig. 9.2A,B

Small brown mites with a single dorsal shield. The only Mesostigmata common in littoral crevices, often in large aggregations, also among barnacles, under stones and foraging on the open shore down to almost LWS. For example, *Hydrogamasus*, 0.4–0.8 mm.

7 HALOLAELAPIDAE — Fig. 9.2C

Small mites with two shields on the dorsal surface, the posterior one has a notch in the front end. Usually in the lower littoral. For example, *Halolaelaps*, 0.6–0.8 mm.

8 MACROCHELIDAE — Fig. 9.2D

These mites are easily recognized because they are large, do not have any claws on the tip of the first leg, and the single dorsal shield does not extend to the posterior tip of the idiosoma, leaving a white rim around the rear of the body. Occur only in

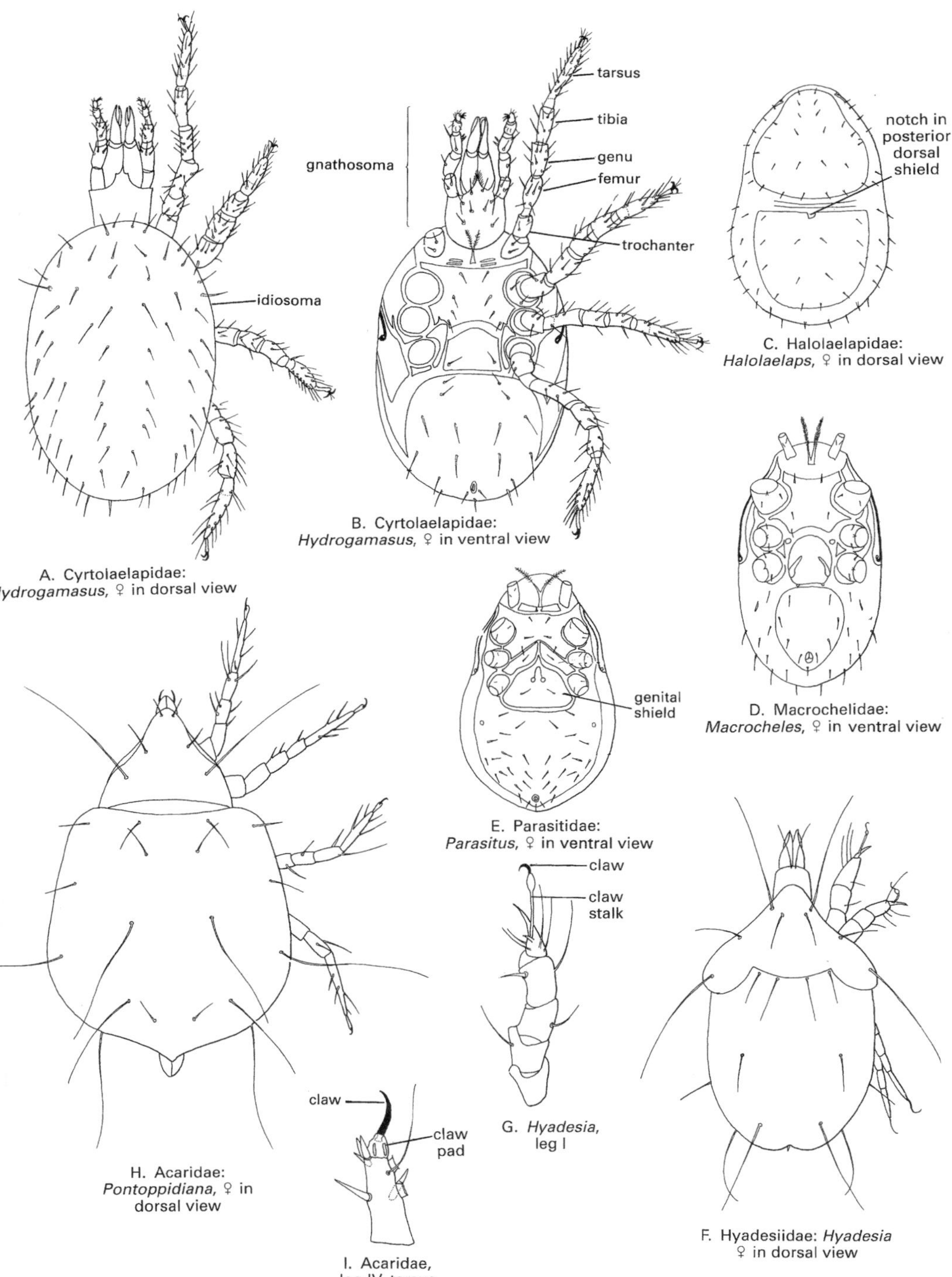

A. Cyrtolaelapidae: *Hydrogamasus*, ♀ in dorsal view

B. Cyrtolaelapidae: *Hydrogamasus*, ♀ in ventral view

C. Halolaelapidae: *Halolaelaps*, ♀ in dorsal view

D. Macrochelidae: *Macrocheles*, ♀ in ventral view

E. Parasitidae: *Parasitus*, ♀ in ventral view

F. Hyadesiidae: *Hyadesia* ♀ in dorsal view

G. *Hyadesia*, leg I

H. Acaridae: *Pontoppidiana*, ♀ in dorsal view

I. Acaridae, leg IV, tarsus

Fig. 9.2

tidal debris, sometimes in large numbers. For example, *Macrocheles*, 0.8–1.5 mm.

9 PARASITIDAE Fig. 9.2E

One shield on the dorsal surface. The female has in addition a unique, almost triangular (*genital*) shield in the centre of the ventral surface. A very large family with a number of species in the supralittoral and upper littoral zones among tidal debris and crevices. For example, *Parasitus*, *Vulgarogamasus*, 0.6–0.8 mm.

Suborder Astigmata

Small, almost unarmoured mites with very small pedipalps. Only two families occur on the shore, one in the littoral and one in the supralittoral.

10 HYADESIIDAE Fig. 9.2F,G

Legs I and II are very thick with pointed distal segments and a very small claw on a thin stalk. Restricted to the littoral zone among algae and barnacles. Probably feed upon bacterial, algal, or fungal films. These mites turn a characteristic blue or mauve colour when cleared in lactic acid. For example, *Hyadesia*, 0.3–0.5 mm.

11 ACARIDAE Fig. 9.2H,I

Legs not heavy; all claws are small and set in a cushion-like pad; body may have very long hairs (setae). In the supralittoral among tidal debris and patches of lichens, but very common in the terrestrial environment. Detritivores or herbivores. For example, *Pontoppidiana*, *Tyrophagus*, *Rhizoglyphus*, 0.3–0.8 mm.

Suborder Cryptostigmata (beetle or oribatid mites)

A very large group of free-living mites well represented in the terrestrial environment. They are common in supralittoral lichen patches and sheltered, well-decomposed tidal debris. Only one family occurs in the littoral zone. Adults are usually completely armoured with a leathery or sclerotized cuticle and the gnathosoma is small and well hidden. Most are fungivorous, a few are herbivorous. The most common families are listed below.

12 AMERONOTHRIDAE Fig. 9.3A,B

The cuticle is characteristically dull and leathery. The only Cryptostigmata which occur in the littoral zone, usually among barnacles, lichen, and *Mytilus* clumps, with a few species in the supralittoral. Length 0.5–0.7 mm.

13 CERATOZETIDAE Fig. 9.3C

Small, shiny, globular mites with downward-pointing wing-like pteromorphae on the sides of the idiosoma. Length 0.3–0.7 mm.

14 ORIBATULIDAE Fig. 9.3D

Small, shiny, globular mites without pteromorphae. Length 0.4–0.6 mm.

Suborder Prostigmata

An enormous and highly diverse group of mites, both in terms of habitat and morphology. Generally soft-bodied and occasionally brightly coloured. Feeding is variable, but most are predators.

15 BDELLIDAE. Snout mites Fig. 9.3E,F

Big, soft-bodied, bright-pink or red mites, with a characteristic tapering gnathosoma and long 'elbowed' palps. Forage at all levels down to MTL. Predatory. Length up to 3.5 mm.

16 ERYTHRAEIDAE Fig. 9.3G,H

Large, bright-red, soft-bodied mites with straight pedipalps equipped with two small, terminal digits (*thumb-claw*). The body is covered in a dense carpet of barbed setae. Forage on dry surfaces in the supralittoral and upper littoral. Predatory. Length 1–2 mm.

17 NANORCHESTIDAE Fig. 9.3I

Very small, soft-bodied mites which look like pink dust in the field. Under the compound microscope high magnification reveals characteristic shrub-like body setae. Usually in upper littoral crevices. Herbivorous or fungivorous. Length 0.1–0.4 mm.

18 RHAGIDIIDAE Fig. 9.3J

Usually bright orange, pink or red, soft-bodied mites, which can move very fast on open rocks. They have long legs, prominent palps, and massive jaw-like chelicerae. Predatory. On open rock surfaces, among debris and lichens in the supralittoral but may go down to MTL. Length up to 1.5 mm.

19 HALACARIDAE. Sea Mites Fig. 9.3K–N

These have armour plates which are visible under the compound microscope. The gnathosoma and palps are large, distinct, and occasionally modified

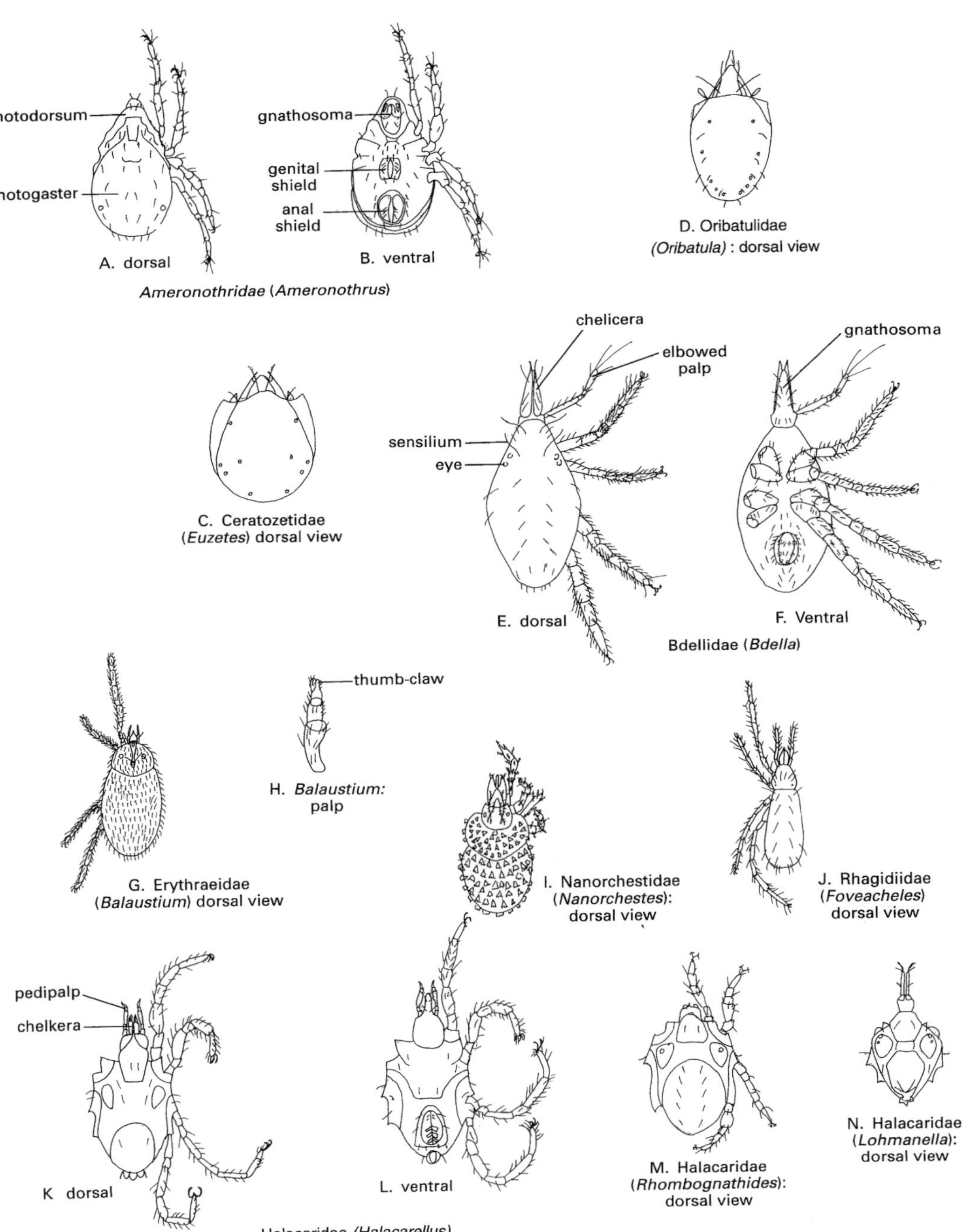
notodorsum
notogaster
A. dorsal
gnathosoma
genital shield
anal shield
B. ventral
Ameronothridae (Ameronothrus)
D. Oribatulidae
(Oribatula) : dorsal view
C. Ceratozetidae
(Euzetes) dorsal view
chelicera
elbowed palp
sensilium
eye
E. dorsal
gnathosoma
F. Ventral
Bdellidae (Bdella)
thumb-claw
H. Balaustium: palp
G. Erythraeidae
(Balaustium) dorsal view
I. Nanorchestidae
(Nanorchestes): dorsal view
J. Rhagidiidae
(Foveacheles)
dorsal view
pedipalp
chelkera
K dorsal
L. ventral
Halacaridae (Halacarellus)
M. Halacaridae
(Rhombognathides):
dorsal view
N. Halacaridae
(Lohmanella):
dorsal view

Fig. 9.3

to form grasping structures, in predatory species (Fig. 9.3K,N), but are small and indistinct in herbivorous ones (Fig. 9.3M). Usually occur on littoral/sublittoral algae, among barnacles, under stones and in pools, occasionally in sediments. Length 0.1–1.5 mm.

TERRESTRIAL MOIETY OF LITTORAL ARTHROPODS: MYRIAPODA AND HEXAPODA

The myriapods (millipedes and centipedes) and hexapods (insects and related forms) are all mandibulate arthropods, with unbranched (uniramous) legs and the head with one pair of antennae.

---

1. Adult body with two tagmata: head and trunk; most trunk segments with a pair of ambulatory appendages (legs) (**MYRIAPODA**) **2**

   Adult body with three tagmata: head, thorax, and abdomen, thorax with three pairs of ambulatory appendages (**HEXAPODA**) **3**

---

2. All but two trunk segments with one pair of walking legs, first pair modified to form maxilliped. Gonopore near posterior end of body **Class Chilopoda**

   Each trunk segment except anterior four have two pairs of walking legs. Gonopore near anterior end of body **Class Diplopoda**

---

3. Mouthparts in pouch (Entognathous). Adult, abdomen of six segments **Class Collembola**

   Mouthparts not in pouch (Ectognathous). Adult, abdomen with more than six segments (**Class Insecta**) **4**

---

4. Wings absent. Thorax with three almost equal independent segments, abdomen with rudimentary appendages ventrally **Subclass Apterygota**

   Wings present in adults, or occasionally secondarily lost. Thoracic segments closely united, in particular the second and third segments forming the pterothorax. Abdomen of adults without rudimentary appendages. Mandibles with two condyles unless modified for piercing. **Subclass Pterygota**

PHYLUM MYRIAPODA

Class Chilopoda

Littoral and supralittoral representatives in Order Geophilomorpha. Worm-like, burrowing. Predaceous, first pair of walking legs are poison claws. On upper littoral under rejectamenta or in crevices. Each segment has dorsal tergite, ventral sternite. May be divided into presternite and metasternite.

---

1. Coxal pores distributed over ventral surface of coxa
   Strigamia maritima (Leach) (Fig. 9.4C)

Coxal pores largely concentrated along edge of adjacent metasternite or opening into pits **2**

---

2. Two coxal pores on each side ***Hydroschendyla submarina*** (Grube) (Fig. 9.4A)

More than two coxal pores on each side ***Geophilus fucorum*** Brolemann (Fig. 9. 4B)

Class Diplopoda

Worm-like, tubular shape, herbivorous. Head has paired antennae, groups of *ocelli*, a pair of mandibles and lower lip or *gnathochilarium* formed from the first maxillae (Fig. 9.4D). First trunk segment is *collum*.

Thalassisobates littoralis (Silvestri). pale greyish-white; under stones between HWN and HWS in fine gravel; associated with the chilopods *Geophilus fucorum* and *Strigamia maritima*. *Ommatoiulus sabulosus* (L.): rust brown or dark red-brown, two straw-coloured bands each side of median dorsal band.

PHYLUM HEXAPODA

Class Collembola

Two suborders: the Symphypleona—characterized by short, globular, indistinctly segmented bodies—and the Arthropleona—with elongate, cylindrical, clearly segmented bodies. Most littoral species belong to the second suborder. Body consists of three tagmata: head, thorax of three segments, and abdomen of six. Head has pair of antennae, usually with four segments, often 1–8 *ocelli* and a *postantennal organ* (Fig. 9.4E). Mouthparts in pouch. '*Ventral tube*' on first abdominal segment. May have '*springing organ*' on fourth abdominal segment.

KEY TO FAMILIES

---

1. Pronotum with some setae, body generally stout, antennae short **2**

   Pronotum small, without setae, usually hidden under mesonotum; body slender; antennae rather long **4**

---

2. Pseudocelli present (Fig. 9.4E), antennal segment 3 with complex sense-organ; eyes absent. Usually white **1. Onychiuridae**

   Pseudocelli absent **3**

---

3. Chewing mouthparts, with well-developed molar plate (Fig 9.4F) not projected in a cone **2. Hypogastruridae**

   Mouthparts usually without molar plate, often projected in a cone **3. Neanuridae**

---

4. Body with smooth or ciliated setae. Abdominal segments 3 and 4 usually subequal, 4–6 sometimes fused. Furca often reduced **4. Isotomidae**

   Body often with scales or densely ciliated. Abdominal segment 4 distinctly longer than 3. Furca well developed **5. Entomobryidae**

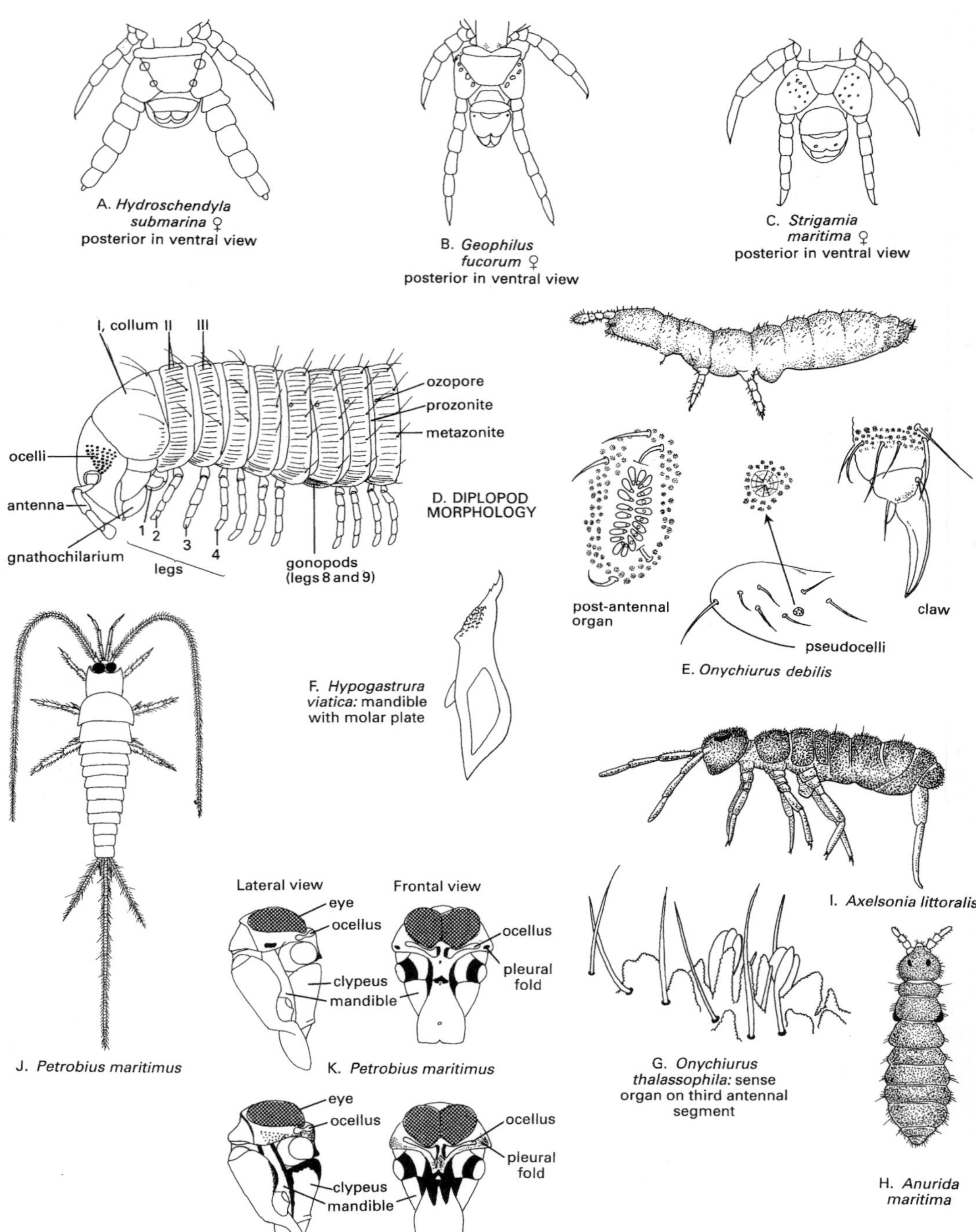
A. *Hydroschendyla submarina* ♀ posterior in ventral view
B. *Geophilus fucorum* ♀ posterior in ventral view
C. *Strigamia maritima* ♀ posterior in ventral view
I, collum
II
III
ozopore
prozonite
metazonite
ocelli
antenna
gnathochilarium
1
2
3
4
legs
gonopods (legs 8 and 9)
D. DIPLOPOD MORPHOLOGY
post-antennal organ
pseudocelli
claw
E. *Onychiurus debilis*
F. *Hypogastrura viatica:* mandible with molar plate
J. *Petrobius maritimus*
Lateral view
Frontal view
eye
ocellus
clypeus
mandible
ocellus
pleural fold
K. *Petrobius maritimus*
eye
ocellus
clypeus
mandible
ocellus
pleural fold
L. *Petrobius brevistylis:* head
I. *Axelsonia littoralis*
G. *Onychiurus thalassophila:* sense organ on third antennal segment
H. *Anurida maritima*

Fig. 9.4

1 ONYCHIURIDAE

Pronotum with some setae, body generally stout. Antennae short. Third antennal segment with complex sense-organ (Fig. 9.4G). Pseudocelli present, eyes absent. Usually white. Four species recorded around Britain, mainly on northern coasts. For example, *Onychiurus debilis* (Moniez) (Fig. 9.4E), *O. thalassophila* (Bagnall) (Fig. 9.4G).

2 HYPOGASTRURIDAE

Body generally stout, with short antennae. No pseudocelli. Chewing mouthparts with well-developed molar plate (Fig. 9.4F). Pronotum with some setae. *Hypogastrura viatica* (Tulberg) is common on all coasts, among seaweed and in salt-marshes.

3 NEANURIDAE

Body stout, with short antennae. Setae on pronotum. No pseudocelli. Mouthparts usually without molar plate. Widespread on Atlantic coasts. Six British species, the most common being *Anurida maritima* (Guerin) (Fig. 9.4H), blue-black in colour, and *Anuridella marina* (Willem), which is yellowish-white.

4 ISOTOMIDAE

Small pronotum without setae, usually hidden under mesonotum. Body slender, antennae rather long. Abdominal segments 3 and 4 usually subequal; 4–6 sometimes fused; furca often reduced. Usually blue-grey colour. All coasts. Six British species For example, *Axelsonia littoralis* (Moniez) (Fig. 9. 4I).

5 ENTOMOBRYIDAE

Pronotum small, without setae, usually hidden under mesonotum. Antennae rather long, body slender. Body often has scales or is densely ciliated. Abdominal segment 4 longer than 3. Furca well developed. A single species, *Pseudosinella halophila* (Bagnall), occurs on southern coasts.

Class Insecta

Subclass Apterygota

Order Archaeognatha

Mandibles with a single condyle. Large compound eyes, ocelli present, coxae of legs have short articulated appendages, *styli*. Spiracles present on 2–8 abdominal segments. Cerci shorter than terminal filament. Body cylindrical or laterally compressed. Often referred to as machilids or jumping bristletails. The other apterygote order, Thysanaura is not represented in the littoral zone.

Two species occur on British coasts above mean HWN.

Petrobius maritimus (Leach) Fig. 9.4 J,K

Length 10–15 mm. Annuli or rings on antennae paler than other segments. Dark spot on pleural fold of head. Sub-coxae of male abdominal segment 8 not prolonged into rounded lobes.

Usually under stones and on screes in situations where there is a reasonable amount of air movement.

Petrobius brevistylis Carpenter Fig. 9.4.L

Length 10–15 mm. Annuli of antennae longer but of same colour as other segments. Often darker coloured, no dark spot on pleural fold. Frontal scales present, area of pigmentation along lacinia of galea. Sub-coxae of male abdominal segment 8 prolonged into rounded lobes.

Prefers areas of unbroken rock surfaces with at least 0.5 m between crevices.

Subclass Pterygota

1. Adults with one pair of wings, hind ones modified to halteres (Fig. 9.5A). **Order Diptera**

   Adults with two pairs of wings, anterior pair often thickened forming elytra or hemelytra **2**

2. Biting mouthparts with well-developed mandibles (Fig. 9.6A) **Order Coleoptera**

   Mouthparts piercing, sucking (Fig. 9.5M) **Order Hemiptera: Suborder Heteroptera**

Order Diptera

Large order. Most flies feed on liquids, though mouthparts vary. Large compound eyes, antennae, and venation of single pair of wings important in classification (Fig. 9.5A–C). In some a facial depression bounded by a *ptilinal suture*. May also be a crescent-shaped mark above bases of antennae, the *lunule*. The bulk of the thorax consists of the wing-

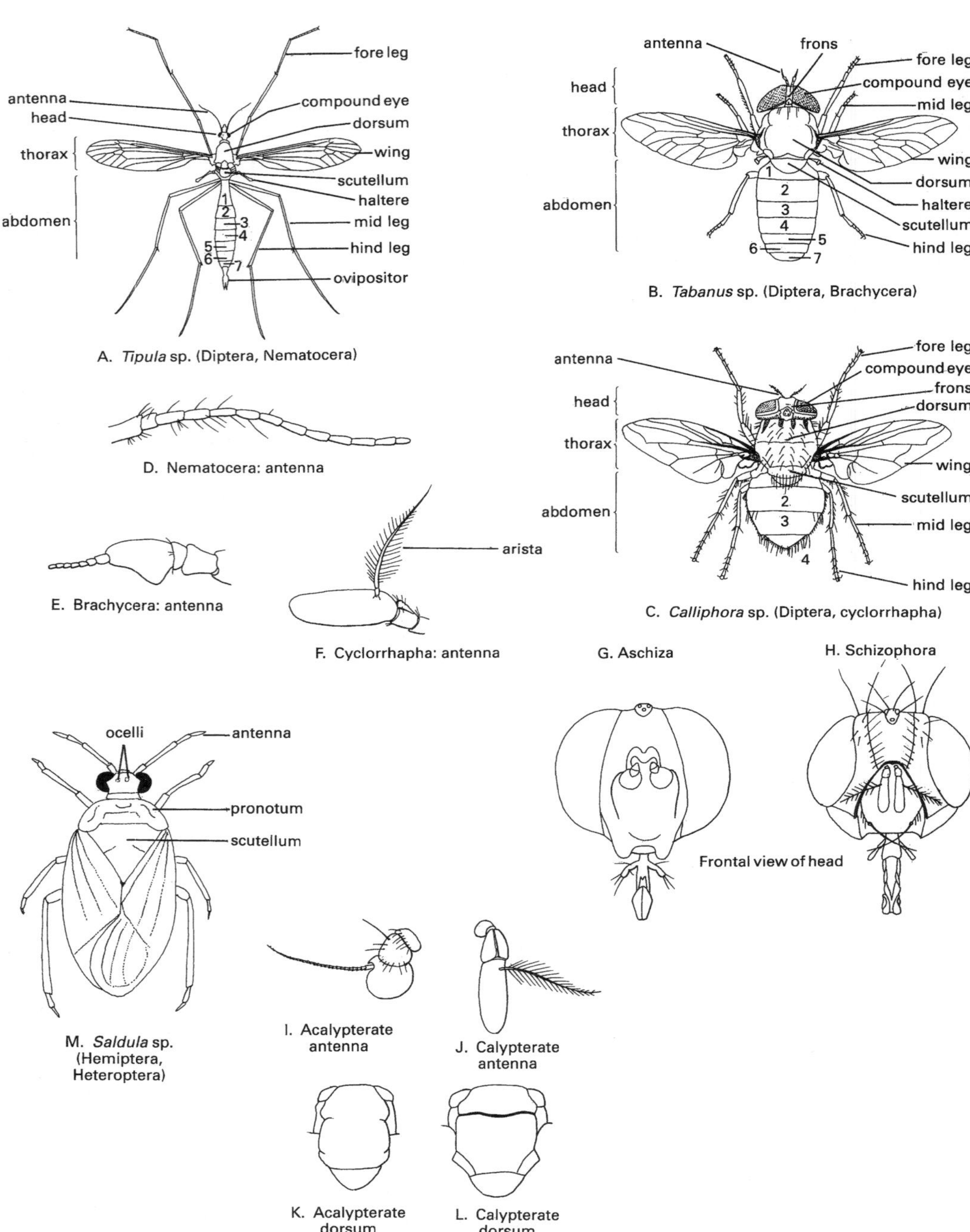
fore leg
antenna
head
compound eye
dorsum
thorax
wing
scutellum
haltere
abdomen
1
2
3
4
5
6
7
mid leg
hind leg
ovipositor
A. *Tipula* sp. (Diptera, Nematocera)
antenna
frons
fore leg
head
compound eye
mid leg
thorax
wing
dorsum
abdomen
haltere
scutellum
hind leg
B. *Tabanus* sp. (Diptera, Brachycera)
antenna
fore leg
compound eye
frons
head
dorsum
thorax
wing
scutellum
abdomen
mid leg
hind leg
C. *Calliphora* sp. (Diptera, cyclorrhapha)
D. Nematocera: antenna
arista
E. Brachycera: antenna
F. Cyclorrhapha: antenna
G. Aschiza
H. Schizophora
Frontal view of head
ocelli
antenna
pronotum
scutellum
M. *Saldula* sp. (Hemiptera, Heteroptera)
I. Acalypterate antenna
J. Calypterate antenna
K. Acalypterate dorsum
L. Calypterate dorsum

Fig. 9.5

bearing mesothorax. The lateral aspect of the thorax has a number of clearly defined areas, the *pleural plates*, or *pleurites*. The wings arise from the *pteropleuron*. Dorsal surface divided into *prescutum*, *scutum*, and *scutellum*. The two former may be separated by a transverse *suture*. Base of wing sometimes has a number of small flaps, the *squamae*.

Antennae many segmented (Fig. 9.5D), usually longer than head and thorax, no frontal lunule or ptilium **Suborder Nematocera**

Representatives of five families occur on the shore: *Clunio marinus* Haliday (Chironimidae), Tipulidae, Dixidae, Culicidae, and Ceratopogonidae.

Antennae of imago shorter than thorax (Fig. 9.5E), no frontal lunule or ptilium **Suborder Brachycera**

Representatives of four families can be found: Stratiomyidae, Asilidae, Empididae, Dolichopodidae.

Antenna of imago three-segmented with dorsally positioned seta (Fig. 9.5F), *arista*. Head with frontal lunule and usually a ptilium **Suborder Cyclorrhapha**

Frontal suture absent, lunule often indistinct or absent, ptilium absent (Fig. 9.5G)

Section A. Aschiza, e.g. Phoridae, Syrphidae

Frontal suture and lunule distinct, ptilium present (Fig.9.5H)

Section B. Schizophora

Antennal segment 2 without distinct external groove (Fig. 9.5I); haltere not covered by squama

Acalyptratae: Sepsidae, Coleopidae, Chamaemyiidae, Ephydridae, Tethinidae, Heleomyzidae, Sphaeroceridae, Canacidae.

Antennal segment 2 with distinct external groove (Fig. 9.5J); halteres covered by squamae

Calyptratae: Scathophagidae, Anthomyiidae, Muscidae

Order Coleoptera

Head with biting mouthparts, *mandibles*, *maxillae*, and *labium*, sometimes on a projection of the head, the *rostrum*. Usually two pairs of wings, front pair forming elytra, which meet in the mid-line (Fig. 9.6A). Hind ones sometimes reduced or absent. Elytra cover abdomen in most, but in some genera are short, leaving several abdominal segments exposed. A number of families have one or two species in the littoral zone but most belong to either the Carabidae or Staphylinidae. In the latter the abdomen is flexible, with seven or eight abdominal segments visible. Some species occur well down the intertidal zone. *Micralymma marinum* (Strøm) (Fig. 9.6G), in rock-crevices; *Diglotta marina* (Haliday), on rocky and sandy shores; *Bledius spectabilis* (Kraatz) (Fig. 9.6F), in saltmarshes. Other species occur on strandlines. Carabids, *Aepopsis robinii* (Laboulbene) (Fig. 9.6E) and *Aepus marinus* (Strøm) (Fig. 9.6D), live in crevices or beneath stones, both are 2 mm long, yellow-brown, with small eyes and slightly abbreviated elytra. *Cillenus laterale* (Samouelle) is abundant in saltmarshes, *Eurynebria complanata* (Linn.) (Fig. 9.6B) and *Broscus cephalotes* (Linn.) (Fig. 9.6C) occur beneath strandline debris on sandy shores.

Order Hemiptera

Suborder Heteroptera

Representatives of the Heteroptera occur on rocky shores and saltmarshes. Many belong to the Saldidae (Fig. 9.5M). A few are wingless but most have two pairs of wings, the hind pair membranous, the front pair partially hardened, termed the *hemelytra*. The anterior segment of the thorax, the *prothorax* is large and prominent; the *pronotum* is the only part of the thorax visible when viewed from above. A triangular region, the *scutellum*, part of the mesothorax, is apparent between the wings when folded. Antennae never more than five segments. Mouthparts characteristic; the mandibles and maxillae are needle-like stylets, held in a groove of the labrum; the composite structure is called a *rostrum*. The Saldidae are oval-shaped, with large coxae. Tarsi three segmented, claws apical and symmetrical without arolia. Short, wide head with ocelli and large compound eyes, long four-segmented antenna and four-segmented rostrum. Dull black or brown, *Saldula palustris* (Douglas) occurs on estuarine mudflats. *Aepophilus bonnairei* (Signoret) (Fig. 9.6H) is unique in the Saldidae in that the hemelytra do not cover the abdomen, it lacks ocelli and has relatively small eyes. It occurs in crevices near low water. Representatives of several other families occur in the littoral zone.

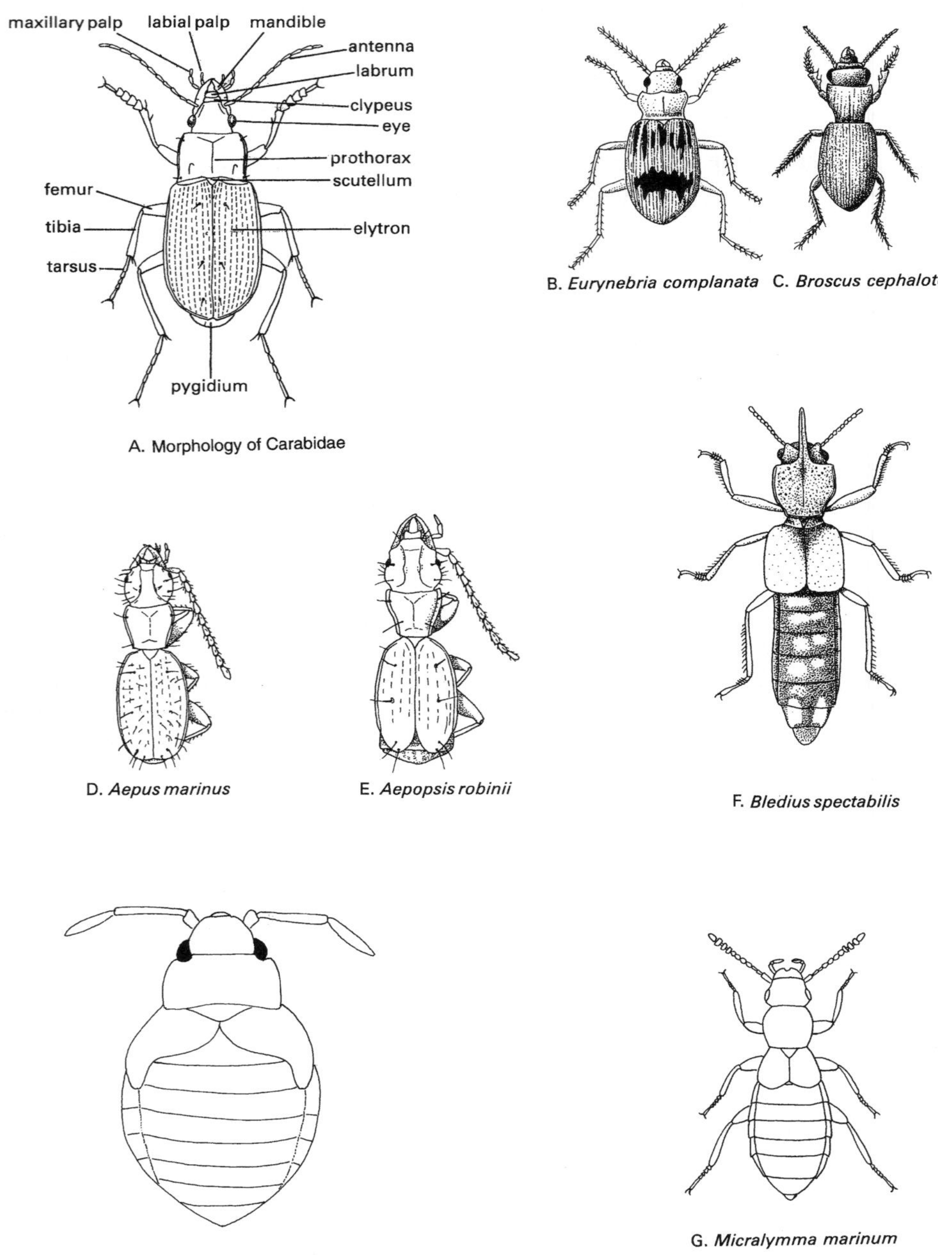

A. Morphology of Carabidae

B. *Eurynebria complanata* C. *Broscus cephalotes*

D. *Aepus marinus*

E. *Aepopsis robinii*

F. *Bledius spectabilis*

G. *Micralymma marinum*

H. *Aepophilus bonnairei*

Fig. 9.6

PHYLUM PYCNOGONIDA

Pycnogonids, often referred to as sea spiders, occur commonly on rocky shores, which provide a stable substratum for attachment and the pycnogonid's principal foods—hydroids and bryozoans. Most have a narrow elongate body divided into a number of segments (Fig. 9.7). At the anterior end there is a *cephalon* with a *proboscis*, bearing a terminal mouth. On the dorsal side of the cephalon there is a *tubercle* bearing the eyes, one pair directed forwards and the other pair posteriorly. The first *trunk* segment is fused with the cephalon and the last segment has an *abdomen* or *anal process*, with a terminal anus. In most adults the trunk segmentation is visible externally.

The appendages consist of a pair of sensory *palps*, *chelifores* which may have terminal *chelae*, used for food gathering or perhaps gripping the substratum, *ovigerous legs*, and four pairs of *ambulatory* (walking) or *natatory* (swimming) *legs*. Some juvenile stages have fewer than four pairs of legs and cannot be identified with this key. Some species are without palps or chelifores, or both. In some species the ovigerous legs are present only in the males, whereas in others, although present in females, they are reduced. The terminal segments of the ovigerous legs may bear a number of serrated spines. The body segment has a prominent *lateral process* with which the legs articulate. These consist of three *coxae*, a *femur*, two *tibiae*, a *tarsus*, a *propodus*, and a *terminal claw*, with some *auxiliary claws* associated with it or the propodus.

During the breeding season, females can be identified readily by the presence of eggs in their femurs, but at other times of the year some species are difficult to sex, and males and females can only be distinguished by the relative length of their ovigers.

Littoral pycnogonids can be collected among small seaweeds, hydroids, bryozoans, and algal holdfasts. These should be allowed to stand in bowls of seawater before sorting, as many of the pycnogonids will walk out as the oxygen content of the water falls. Specimens should be narcotized with a few drops of ethyl acetate added to the water and then, after 10–15 min, placed in hot Bouin's fixative. They can thus be preserved in an extended state and stored in 70% alcohol, or mounted on slides with Canada balsam after dehydration with alcohol.

KEY TO FAMILIES

Chelifores (see Fig. 9.7) can be observed from the dorsal surface, and palps from the ventral surface.

---

1. Chelifores and palps present **2**

   Palps, or both chelifores and palps, absent **3**

---

2. Palps five-segmented. Chelae of chelifores conspicuous, over-reaching proboscis **1. Nymphonidae**

   Palps eight- or nine-segmented. Chelae small, chelifores shorter than proboscis **2. Acheliidae**

---

3. Chelifores present, palps lacking **4**

   Both chelifores and palps lacking **5**

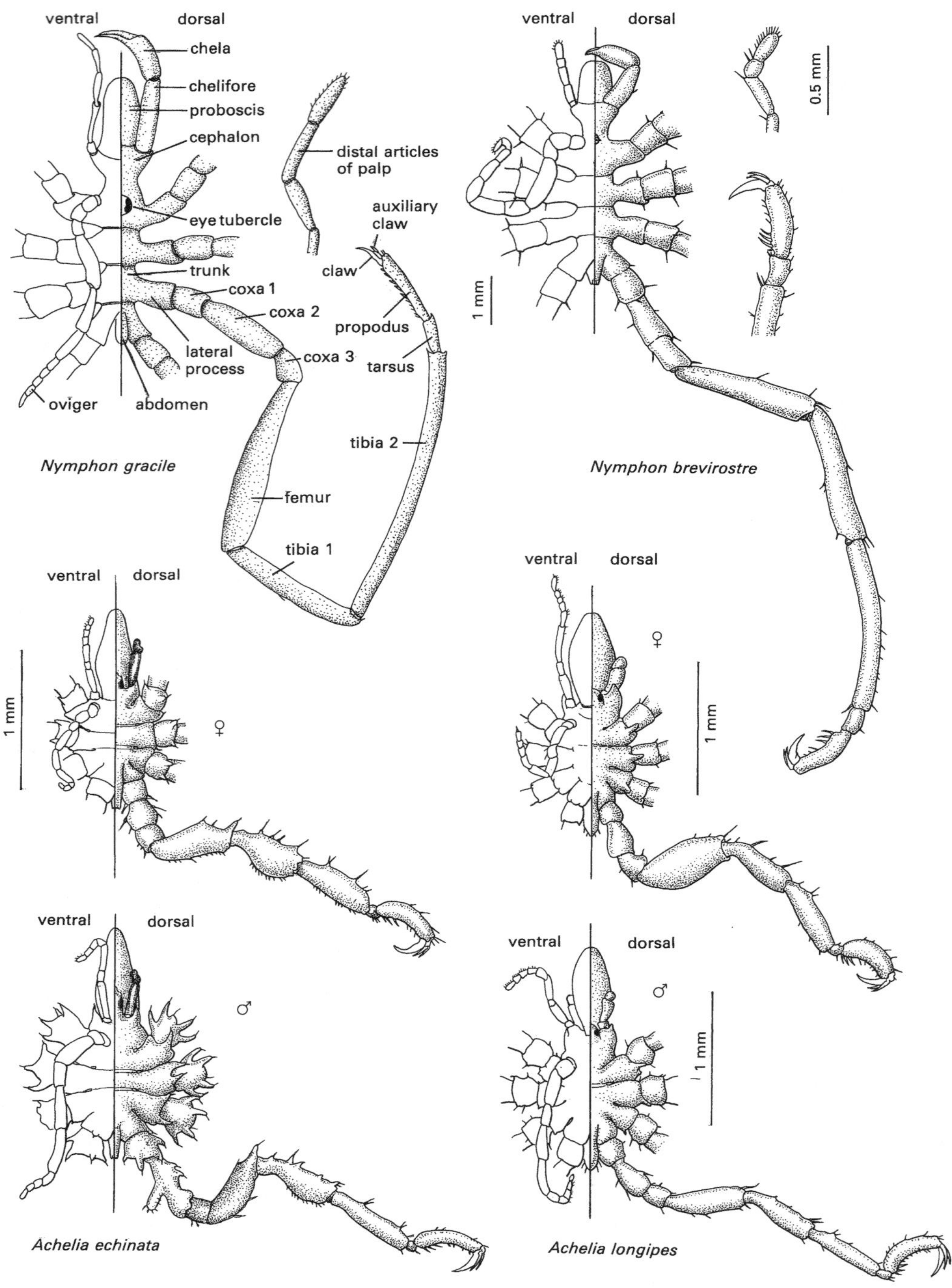
ventral
dorsal
chela
chelifore
proboscis
cephalon
eye tubercle
trunk
coxa 1
coxa 2
lateral process
coxa 3
oviger
abdomen
distal articles of palp
auxiliary claw
claw
propodus
tarsus
tibia 2
femur
tibia 1
Nymphon gracile
ventral
dorsal
0.5 mm
1 mm
Nymphon brevirostre
ventral
dorsal
1 mm
♀
Achelia echinata
ventral
dorsal
♂
ventral
dorsal
♀
1 mm
ventral
dorsal
♂
1 mm
Achelia longipes

Fig. 9.7

4. Ovigerous legs ten-segmented in both sexes **6. Callipallenidae**

Ovigerous legs five- or six-segmented in males; absent in females **6**

5. Body slender, legs about twice as long as body. Auxiliary claws present. Ovigerous legs seven-segmented (present in male only) **4. Endeidae**

Body stout, legs stout, little longer than the body. Auxiliary claws absent. Ovigerous legs nine-segmented (present in males only) **5. Pycnogonidae**

6. Auxiliary claws laterally placed, small compared with size of main claw, or absent **7. Anoplodactylidae**

Auxiliary claws dorsally placed, large compared with size of main claw **3. Ammotheidae**

1 NYMPHONIDAE

Body slender with a relatively short and wide proboscis. Chelifores are two-jointed, having functional chelae with fingers bearing prominent teeth. Palps and ovigerous legs are present in both sexes. Several species have been observed swimming and are amongst the most active pycnogonids. Eight British species, only two of which are frequently recorded in coastal waters.

1. Terminal and penultimate segments of palps equal in length ***Nymphon gracile***

Terminal segments of palps approximately twice as long as subterminal segments ***Nymphon brevirostre***

Nymphon gracile Leach Fig. 9.7

Slender, smooth body, 4 mm long with elongate limbs three or four times as long as body. Proboscis elongate, twice as long as wide. Propodus and tarsus subequal in length; propodus has a number of spines with three slightly longer ones half-way along length.

Nymphon gracile is a shallow-water species around most European Atlantic coasts from Norway to Morocco. Occurs in the littoral zone during the summer months.

Nymphon brevirostre Hodge Fig. 9.7

Legs three and a half times as long as body. Proboscis short and stout. Thoracic segments broader than long. Propodus longer than tarsus, slightly bent, with a series of spines consisting of a few short spines, and three longer ones, near its proximal end.

All British coasts; distributed from the Arctic to Brittany.

2 ACHELIIDAE

Palps with eight or nine segments. Chelifores shorter than proboscis, with small, non-functional chela in adults. Six British species.

1. Palps with nine segments. Male coxal projections, bearing genital openings, small. Only one complete suture on trunk segments. Chelifores at most half as long as proboscis ***Achelia longipes***

Palps with eight segments. Male coxal projections prominent on second coxae of legs 3 and 4. No suture between segments 3 and 4. Lateral processes and legs with spine-bearing projections ***Achelia echinata***

Achelia longipes Hodge Fig. 9.7

Body approximately 2 mm long, legs 8 mm long. Body smooth, not spiny, joint between segments 2 and 3 frequently indistinct. A prominent pointed dorsal tubercle on each lateral projection. Proboscis with widest point near middle of length, the three lips around mouth, although beak-like, somewhat blunter than in *A. echinata*.

This species is thought to be more southern in its distribution than *A. echinata*. Reported from west coasts of Ireland, South Wales, North Wales, east coast of Britain, Plymouth.

Achelia echinata Hodge Fig. 9.7

Body 2 mm long. Two distinct sutures visible externally. Proboscis as long as body, broadest one-third of distance from proximal end, tapering towards mouth. Legs with a number of spines, characteristically one pair on each of lateral processes and two pairs on each first coxa; larger in male which, in addition, has a prominent projection on second coxae of third and fourth pairs of legs. Ovigerous legs with nine segments, present in both sexes, though slightly smaller in female. Ocular tubercle subconical with a sharp terminal spine.

Of wide occurrence throughout the British Isles.

3 AMMOTHEIDAE

No palps. Chelifores extend beyond end of proboscis. Ovigerous legs with 10 segments, present in both sexes. Two British species.

Phoxichilidium femoratum (Rathke) Fig. 9.8

Body length 2–3 mm, with legs almost three times as long. Proboscis cylindrical, almost constant width throughout its length. Chelifores with chelae reaching beyond tip of proboscis. Ovigerous legs with five segments. Abdomen short, hardly as long as either of the hind pair of lateral processes. Heel of propodus armed with four large, single teeth, and a pair which is usually smaller. There are well-developed, dorsally situated auxiliary claws.

Widespread from the north Atlantic, Greenland, and Norway to France.

4 ENDEIDAE

Body and legs elongate. Adults with neither chelifores nor palps, though juvenile six-legged and early eight-legged forms have long, slender chelifores with a small chela. Ovigerous legs seven-segmented, in males only. Two British species.

---

1. Auxiliary claws at least half as long as main claws. Mouth surrounded by numerous spines ***Endeis charybdaea***

   Auxiliary claws less than half as long as main claws. Mouth with few spines around it ***Endeis spinosa***

Endeis spinosa (Montagu) Fig. 9.8

Body 3 mm long. Smaller than *E. charybdaea*. Tip of proboscis slightly tapered, circular in cross-section, with fewer spines around mouth. Auxiliary claws relatively shorter compared with length of main claw. Male femurs with 19–20 cement-gland ducts.

Occurs in the littoral zone, and to depths of 12 m in the sublittoral zone. Widespread on British coasts.

Endeis charybdaea (Dohrn) Fig. 9.8

Body 5–6 mm long. Proboscis cylindrical proximally, widening distally, with a swollen region almost half-way along it. Apex oval in cross-section, then tapered slightly before being further enlarged to form a bulbous tip. Mouth surrounded by concentric rows of irregularly arranged spines.

Usually occurs in sublittoral zone, below 13 m, mainly around south-west coasts.

5 PYCNOGONIDAE

Body squat. With neither palps nor chelifores. Ovigerous legs with nine segments, in males only. One British species.

Pycnogonum littorale (Strøm) Fig. 9.9

Body 5 mm long, sturdy. No chelifores or palps in either of juveniles; nine-segmented ovigerous legs present only in males, appearing at an early stage of development as buds, number of segments increasing as growth proceeds. Proboscis conical, never longer than trunk. Abdomen truncate at posterior end. Legs slightly shorter than body, terminating with a claw; no auxiliary claws. Genital apertures situated on ventral surface of second coxae of hind legs of male, but on dorsal surface of same coxae in females.

Widespread in the British Isles.

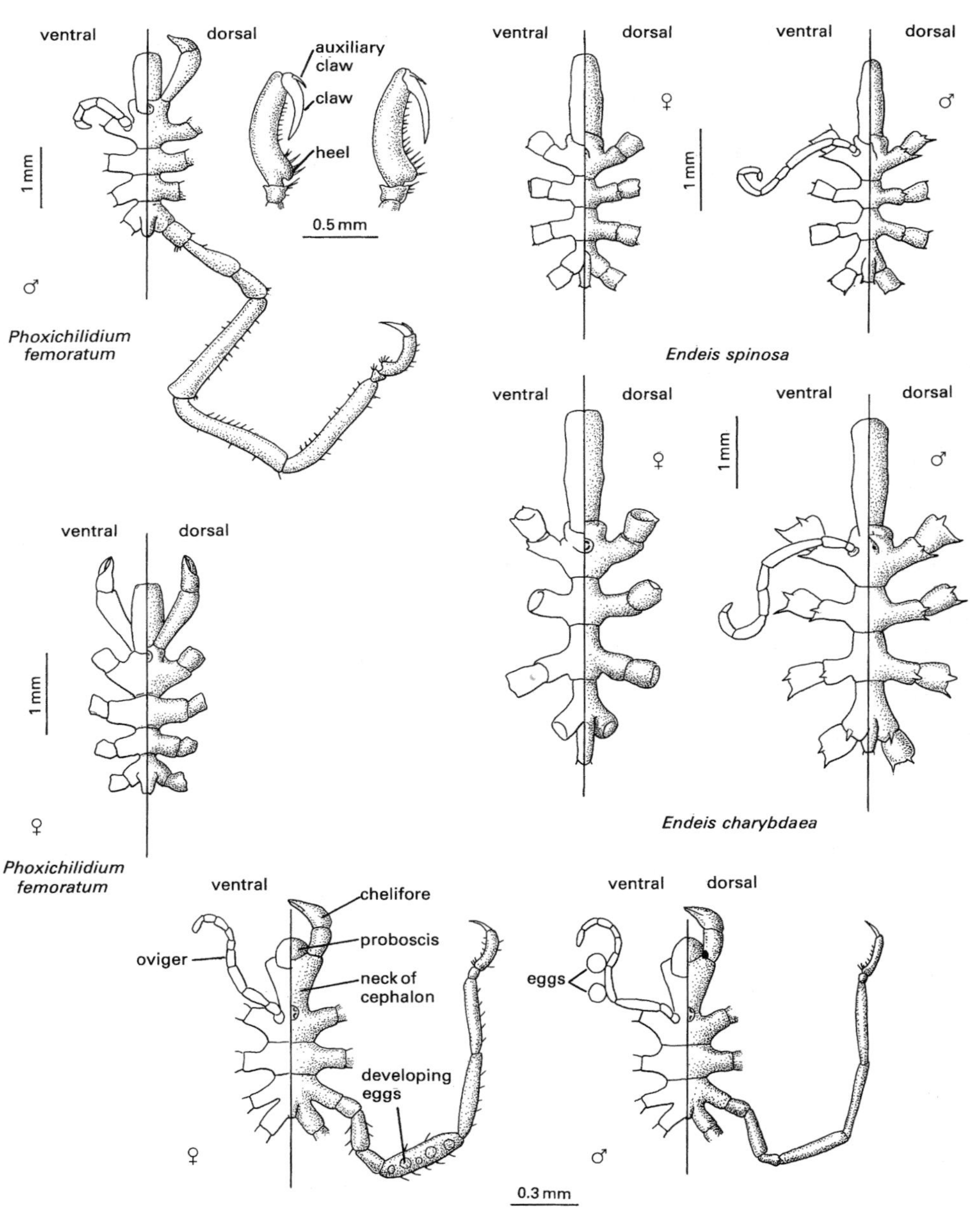
ventral
dorsal
auxiliary claw
claw
heel
1 mm
0.5 mm
♂
Phoxichilidium femoratum
ventral
dorsal
♀
1 mm
ventral
dorsal
♂
Endeis spinosa
ventral
dorsal
♀
1 mm
ventral
dorsal
♂
Endeis charybdaea
ventral
dorsal
1 mm
♀
Phoxichilidium femoratum
ventral
chelifore
proboscis
oviger
neck of cephalon
developing eggs
♀
ventral
dorsal
eggs
♂
0.3 mm
Callipallene brevirostris

Fig. 9.8

6 CALLIPALLENIDAE

Body small, elongate. Chelifores present, palps lacking. Four British species.

---

1. Sole of propodus straight, auxiliary claws at most half as long as principal claw, usually less. Neck long and slender with distal end distinctly set off from rest ***Callipallene phantoma***

   Sole of propodus curved. Auxiliary claws more than two-thirds as long as principal claw ***Callipallene brevirostris***

Callipallene brevirostris (Johnston) Fig. 9.8
Body 1–5 mm long; legs four times as long as body. Proboscis short, rounded at distal end. Cephalon long, wider anteriorly. Trunk segment 2 short and wide, segments 3 and 4 fused, abdomen relatively small. Ovigerous legs with 10 segments, all short except segments 3 and 4; this appendage longer in male than in female. Femur of female proportionately wider than in male.

Essentially sublittoral, but occurs occasionally in the littoral zone. Widespread around Britain.

Callipallene phantoma (Dohrn)
Body 2 mm long, a considerable proportion being taken up by the neck, which is long and narrow but widens considerably at anterior end. Legs four times length of body. Proboscis cylindrical, with a bluntly rounded distal end. Trunk segment 2 considerably longer than wide, segments 3 and 4 fused; with a short abdomen. Femur shorter than tibia on leg 2 but longer than tibia on leg 1. Ovigerous legs with 10 segments, all short except segments 3, 4, and 5; segment 3 shorter than either segments 4 and 5.

Essentially sublittoral but occurs occasionally in the littoral zone. Widespread around Britain.

7 ANOPLODACTYLIDAE

Usually slim bodied and of small size. Proboscis tubular, with rounded tip. Legs relatively long. Chelifores present; no palps. Five or six segmented ovigenous legs present in males. Auxiliary claws laterally placed, small in comparison to main claw; absent in some species.

Four British species, mostly with southern distribution. *Anoplodactylus virescens* Hodge and *A. angulatus* (Dohrn), both of which lack protuberances on the lateral processes, are least common, with south-west distributions.

---

1. Anterior part of cephalon long and narrow, overhanging posterior part of proboscis. Auxiliary claws small and laterally positioned. Protuberances on lateral processes without a spine at apex. Cutting lamellae of propodus short, preceded by 4–6 small teeth ***Anoplodactylus petiolatus***

   Cephalon short and wide; does not overhang proboscis. Auxiliary claws absent. Protuberances on lateral processes with spine at apex. Cutting lamellae of propodus long, preceded by one or two small teeth ***Anoplodactylus pygmaeus***

Anoplodactylus pygmaeus (Hodge) Fig. 9.9
Body length 0.7–1 mm; legs two and a half times as long as body. Proboscis rounded at end, with a subterminal constriction. Cephalon broad, with a long ocular process extending anteriorly. Ovigerous legs with six segments. General shape more compact than *A. petiolatus*. Abdomen approximately twice as long as last pair of lateral processes. Small dorsal protuberance on each lateral process, each with a terminal spine. Propodus with two spines on heel, and one or two spines, and a long cutting lamella, on sole. No auxiliary claws.

A common shallow-water species, distributed from Denmark to the Azores.

Anoplodactylus petiolatus (Krøyer) Fig. 9.9
Slender body, 1–1.5 mm long; legs three times as long as body. Proboscis with a rounded end. Ovigerous legs with six segments. Long neck, with a long ocular tubercle. Body segments broad, with wide lateral processes; abdomen long and narrow, reaching well beyond last lateral processes, which have a dorsal protuberance which lacks a terminal spine. Propodus with two large spines and a smaller pair, with 4–6 spines and a short cutting lamella on sole. There are small, laterally placed, auxiliary claws.

Widely distributed around Britain.

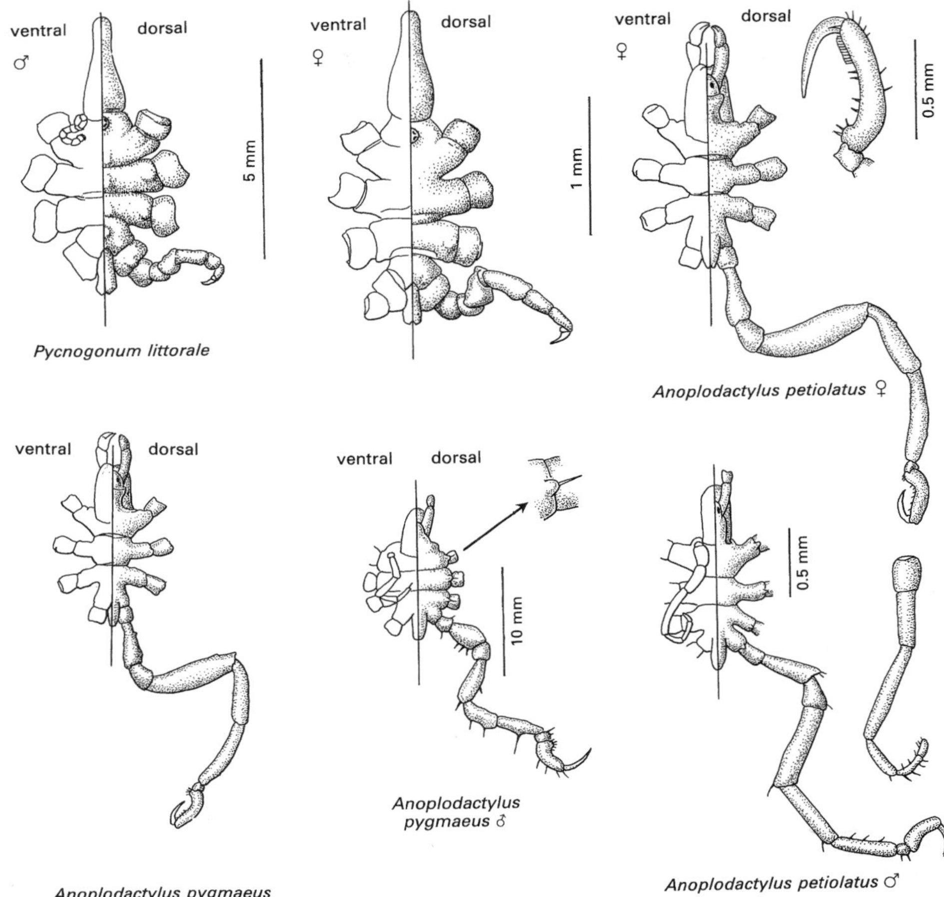

ventral
dorsal
♂
5 mm
ventral
dorsal
♀
1 mm
ventral
dorsal
♀
0.5 mm
Pycnogonum littorale
Anoplodactylus petiolatus ♀
ventral
dorsal
ventral
dorsal
10 mm
0.5 mm
Anoplodactylus pygmaeus ♂
Anoplodactylus pygmaeus
Anoplodactylus petiolatus ♂

Fig. 9.9

10 MOLLUSCS

The molluscs (Phylum Mollusca) are one of the largest groups of marine organisms and, with the exception of the Monoplacophora, representatives of all classes can be found in British waters. Living molluscs occur in terrestrial, freshwater, and marine ecosystems, and in the latter case have become adapted to every kind of habitat from abyssal oozes to oceanic surface currents.

Molluscs are unsegmented, bilaterally symmetrical animals. The body consists of a muscular *foot*, a variously developed *head*, and a soft, non-muscular *visceral mass*. The foot may be adapted for grasping the substratum, for locomotion, burrowing or feeding, and is often closely associated with the head. In the gastropods the head is usually well developed, with paired sensory organs and often specialized feeding apparatus. The buccal cavity contains a complex, eversible feeding structure, the *odontophore*, associated with a chitinous, toothed ribbon, the *radula*. In bivalves the head has regressed, and feeding and sensory functions are carried out by other parts of the body. In cephalopods the head and foot are integrated in the anterior, tentacle-bearing end of the animal. The main body organs of the mollusc are concentrated in the visceral mass, primitively situated dorsally towards the posterior end of the animal. The visceral mass is enclosed by a wide fold of body wall, the *mantle* or *pallium*. The mantle edge secretes the calcareous external shell, characteristic of molluscs, although secondarily absent in many taxonomic groups. The shell is three-layered, with an organic outer layer, the *periostracum*, a middle layer of columnar calcite, and an inner layer of laminated calcite, often nacreous. The mantle skirt covers the whole of the animal and typically incorporates a *mantle* or *pallial cavity*, primitively posterior in position. In primitive gastropods it contains the anus, the paired openings of the gonads and kidneys, and paired gills, or *ctenidia*, together with sensory organs, *osphradia*, and a mucus-secreting *hypobranchial* gland. Circulation and sanitation within the pallial cavity are aided by cilia and mucus-secreting cells. In the bivalves, in which the whole animal lies within an enlarged mantle cavity, ciliary/mucus gill cleansing has been adapted to microphagous feeding; the head has largely disappeared, and the mouth is equipped with pronounced *labial palps* for the collection of detritus-laden mucus.

In all molluscs the coelom is represented by a small cavity surrounding the heart and gonads. The blood circulatory system is open, and both blood and coelomic fluid circulate through extensive haemocoelic spaces which serve as an efficient hydrostatic skeleton. In primitive molluscs sexes are separate, fertilization is external, and embryos develop as planktonic *trochophore* larvae. Although most marine molluscs are still dioecious, hermaphroditism is the rule in terrestrial groups; simultaneous or consecutive hermaphrodites are known among certain families of marine molluscs, while the subclass Opisthobranchia is entirely monoecious.

Class Polyplacophora

Chitons have an oval, limpet-like body and a dorsal shell of eight, arched transverse plates, interlocked along their anterior–posterior edges and attached laterally to the mantle. The mantle skirt is developed as a toughened *girdle* around the whole periphery of the animal, often fringed with spines and with calcified granules, spines, or small plates on its upper surface. The small head is hidden from dorsal view by the girdle. The larger part of the ventral surface is formed by the sole of the foot. The mantle cavity is essentially posterior but in all species is continuous anteriorly with deep channels on each side of the animal between the girdle and the foot. In most families the gills occur in paired series along part or all of the lateral pallial groove.

The inner layer of the chiton shell is usually produced anteriorly in each of the six intermediate plates as an articulating flange, the *articulamentum*. It may also be produced laterally as notched *insertion plates*, which attach the valve plates to the body of the animal. The head and tail valves may have a similar series of projecting insertion plates around the convex anterior (or posterior) border. In some cases it is necessary to remove and clean shell plates to be certain of the species identity. Other features of taxonomic importance are valve sculpture and the granulation or spinulation of the girdle.

All polyplacophorans are herbivorous grazers living on hard substrata. Like limpets they have considerable powers of adhesion and are able to withstand wave shock. Some genera, such as *Tonicella*, may be particularly associated with certain crustose algae. Sexes are separate, embryos develop as short-swimming lecithotrophic larvae.

KEY TO FAMILIES

1. Girdle with erect tufts of bristles; up to 10 pairs — **4. Acanthochitonidae**

   Girdle smooth, granular, or scaly; with or without spines, but not with tufts of bristles — **2**

2. Girdle covered with small, flat scales, sometimes interspersed with fine, recumbent spines, and fringed with spines — **3**

   Girdle with fine or coarse granulation, or appearing smooth; with or without recumbent spines, usually fringed with spines — **3. Ischnochitonidae**

3. Intermediate valves rounded, without a keel; or, if keeled then without spines on the girdle. No insertion plates — **1. Lepidopleuridae**

   Intermediate valves keeled, girdle with randomly distributed long spines on its surface, and a fringe of equally long spines. Head valve with insertion plates — **2. Hanleyidae**

1 LEPIDOPLEURIDAE

Small chitons with relatively narrow girdle bearing spines or scales; shell valves enclosing most of animal, typically with granular ornamentation. Articulamentum only slightly produced; insertion plates poorly developed or absent. Gills posterior.

Two common species are included here. *Leptochiton scabridus* (Jeffreys) (Fig. 10.1) may be found around the Channel Isles (and rarely S. Cornwall); it is recognized by the red colour of its foot.

1. Shell valves with strong keels and prominent posterior beaks, sculpture fine. Girdle with rectangular, longitudinally striated scales — ***Leptochiton asellus***

   Shell valves smoothly and strongly arched, without keels; regular, fine granulation giving chequered effect. Girdle with fine flat scales with curled tips — ***Leptochiton cancellatus***

Leptochiton asellus (Gmelin) Fig. 10.1

Up to 18 mm, broadly oval, about 1.5 × as long as wide. Valves off-white or yellow, often with black or brownish encrustations; with distinct keel and small posterior beaks. Surface finely granular,

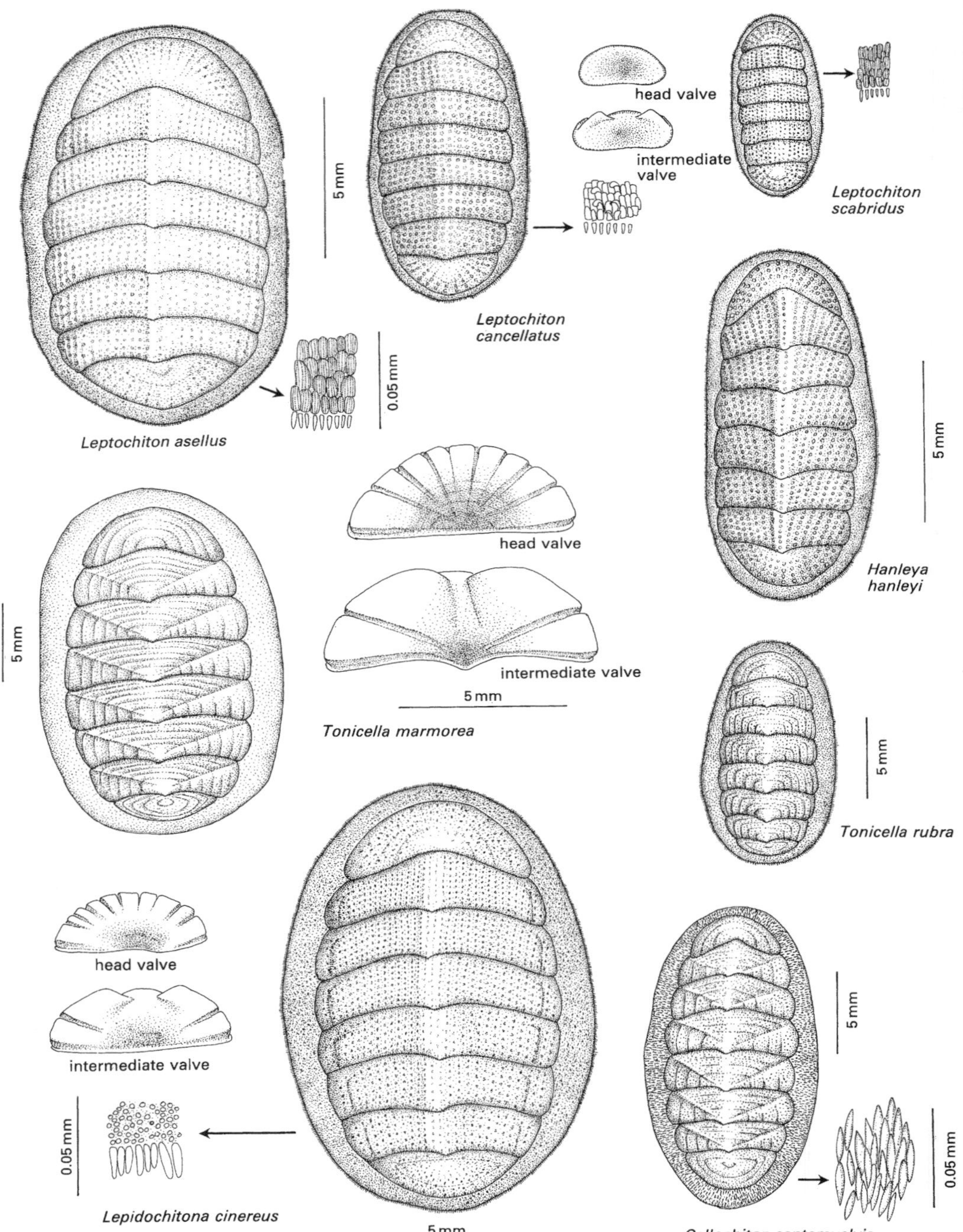
head valve
intermediate valve
Leptochiton scabridus
5 mm
Leptochiton cancellatus
0.05 mm
Leptochiton asellus
5 mm
head valve
intermediate valve
5 mm
Tonicella marmorea
Hanleya hanleyi
5 mm
Tonicella rubra
head valve
intermediate valve
0.05 mm
Lepidochitona cinereus
5 mm
5 mm
0.05 mm
Callochiton septemvalvis

Fig. 10.1

coarser towards margins. Girdle narrow, covered with overlapping, elongate rectangular scales, with longitudinal striations; fringed with delicate marginal spines.

Lower shore and sublittoral. All coasts, widespread and common.

Leptochiton cancellatus (Sowerby) Fig. 10.1

Up to 9 mm long, elongate oval, about twice as long as broad. Valves yellowish-white to fawn, smoothly and strongly arched, rounded dorsally, without keels. Surface finely granular, appearing chequered, especially towards margins. Girdle narrow, covered with overlapping thin, flat, scales, their free edges often curled; fringed with delicate marginal spines.

Lower shore and shallow sublittoral. Predominantly southern and western, extending as far north as Shetland, perhaps absent from north-east coasts.

2 HANLEYIDAE

Small to medium-sized chitons; the relatively narrow girdle bearing spicules and granules, often interspersed with spines, and a dense fringe of slender spines around the edge. Valves highly arched, often keeled; with variable, often coarse, granulation. Head valve with an insertion plate, lacking notches, but intermediate valves without insertion plates. Articulamentum well developed.

Two species recorded for Britain; only the following is at all frequent.

Hanleya hanleyi (Bean in Thorpe) Fig. 10.1

Up to 22 mm long, elongate oval, about 2.5 × as long as broad; off-white to yellowish. Sculpture of coarse granules, in longitudinal series, conspicuous laterally, and developed as protruding tubercles on head and anal valves. Girdle narrow, covered with small, flat scales interspersed with fine, recumbent spines; fringed with similar spines.

Lower shore and sublittoral. Northern coasts only, south to Northumberland in the east, and to Isle of Man and western Ireland in the west.

3 ISCHNOCHITONIDAE

Small to medium-sized chitons; girdle relatively broad, bearing granules or small, recumbent spines. Shell valves smooth or finely granular, typically with diagonal ridges on intermediate valves; head valve usually longer than tail valve. Articulamentum well developed, with up to 12 marginal slits on head and tail valves; lateral insertion plates with 1–3 notches.

British fauna includes four common species described below, and the rare *Ischnochiton albus* (Linn.) (Fig. 10.2), which is found infrequently on northern coasts.

---

1. Girdle covered with densely packed, recumbent spines. Head valve with about 15 notches on anterior ventral rim. Intermediate plates with three pairs of notches in edge of insertion plates **_Callochiton septemvalvis_**

   Girdle with dense or sparse covering of granules, spines limited to fine marginal fringe only. Head valve with up to 10 notches on anterior ventral rim. Intermediate plates with no more than two notches on each side **2**

---

2. Up to 15 pairs of gills, restricted to posterior parts of pallial grooves. Girdle with densely packed, ovoid granules. Valves almost smooth **_Tonicella rubra_**

   More than 15 pairs of gills, extending anteriorly to occupy greater part of pallial groove on each side **3**

---

3. Shell valves smooth and glossy. Girdle broad, smooth and leathery, with minute, widely scattered granules **_Tonicella marmorea_**

   Shell valves finely granular. Girdle relatively narrow, with conspicuous, rounded granules, of variable size, fringed with blunt, fusiform spines **_Lepidochitona cinereus_**

Lepidochitona cinereus (Linnaeus) Fig. 10.1

Up to 24 mm long, broadly oval, about 1.5 × as long as wide. Colour variable: red, brown, yellow to green, in patches and bands. Valves smoothly arched, rather low, with distinct, low keel and pronounced beaks. Surface with regular, fine granulation, slightly coarser towards margins where growth lines may be enhanced. Girdle relatively narrow, densely covered with fine rounded granules; fringed with fusiform spines.

Middle to lower shore, and sublittoral. All coasts, widespread and common.

Tonicella marmorea (Fabricius) Fig. 10.1

Up to 40 mm long, broadly oval, about 1.5 × as long as wide. Valves smooth and glossy, chestnut brown with variable white or fawn marbling; not highly arched, keel scarcely developed but beaks pronounced. With distinct growth lines and some very fine granulation. Girdle rather broad, thin, leathery, reddish-brown, with minute, sparsely distributed granules, fringed with tiny, somewhat flattened marginal spines.

Lower shore and sublittoral. Predominantly northern, ranging south to Yorkshire coast in the east, to west Scotland and Irish Sea on west coast.

Tonicella rubra (Linnaeus) Fig. 10.1

Up to 20 mm long, broadly oval, about 1.5 × as long as wide. Valves smooth and glossy, pale buff to deep chestnut brown, marbled with white; gently arched, keel scarcely developed but beaks well marked. Sculpture more distinct than in *T. marmorea*: very fine granulation dorsally, becoming rather pronounced marginally; growth lines conspicuous. Girdle marbled brown and white, densely covered with spherical granules, fringed with slightly flattened spines.

Lower shore and sublittoral. All coasts, except south-east.

Callochiton septemvalvis (Brown) Fig. 10.1

Up to 30 mm long, oval, about twice as long as broad. Shell smooth, greenish- or reddish-brown with white or greenish-white marbling; gently arched, keel distinct, beaks well marked. With distinct growth lines, and fine dorsal granulation, becoming coarser marginally. Girdle relatively broad, covered with densely packed, overlapping spines, and with a fringe of similar spines.

Lower shore and sublittoral. Widespread, all coasts.

4 ACANTHOCHITONIDAE

Small to large chitons, with broad girdle thickly covered with coarse spines, and typically bearing conspicuous tufts of large bristles. Shell valves partially overlapped by girdle; coarsely granular, with well-marked keel. Articulamentum well developed; about five slits in margin of head valve, lateral insertion plates with a single notch.

---

1. Valve sculpture of large pyriform tubercles, unevenly distributed
   Acanthochitona crinitus

   Valve sculpture of fine, rounded granules, densely and regularly distributed
   Acanthochitona fascicularis

Acanthochitona crinitus (Pennant) Fig. 10.2

Up to 34 mm long, elongate oval, about two-and-a-half times long as broad. Colour variable: off-white, grey, yellowish, chestnut, green, olive, brown and shades of blue; in patches and streaks, or marbled, rarely unicolourous. Valves strongly arched, keel moderate, rounded, beaks pronounced; longitudinal ridges on keel, laterally with large, flat topped, pyriform granules, varying in size and unevenly spaced. Girdle with 18 dense tufts of bristles (up to 1 mm long), one on each side, close to junction between each pair of plates, and four in an arc anterior to head plate; surface with fine, irregular granules interspersed with recumbent spines, fringed with dense series of long acicular spines (up to 0.5 mm).

Lower shore and sublittoral. All coasts, common.

Acanthochitona fascicularis (Linnaeus) Fig. 10.2

Up to 60 mm long, elongate oval, twice as long as broad. Colour variable, marbled with off-white, grey, yellowish, or brown. Valves strongly arched, with rounded keel and prominent beaks; longitudinal ridges on keel, laterally with fine, densely packed, oval or rounded granules. Girdle with 18 dense tufts of bristles (up to 1.5 mm long), one on each side at posterior plate boundaries, and four in an arc anterior to head plate; surface covered with densely packed, recumbent spines (feeling velvety to touch) and fringed with dense series of longer spines (up to 1 mm).

Lower shore and sublittoral. Recorded distribution from north Kent, south and west to Channel Isles, northwards to Anglesey, and on west coast of Ireland.

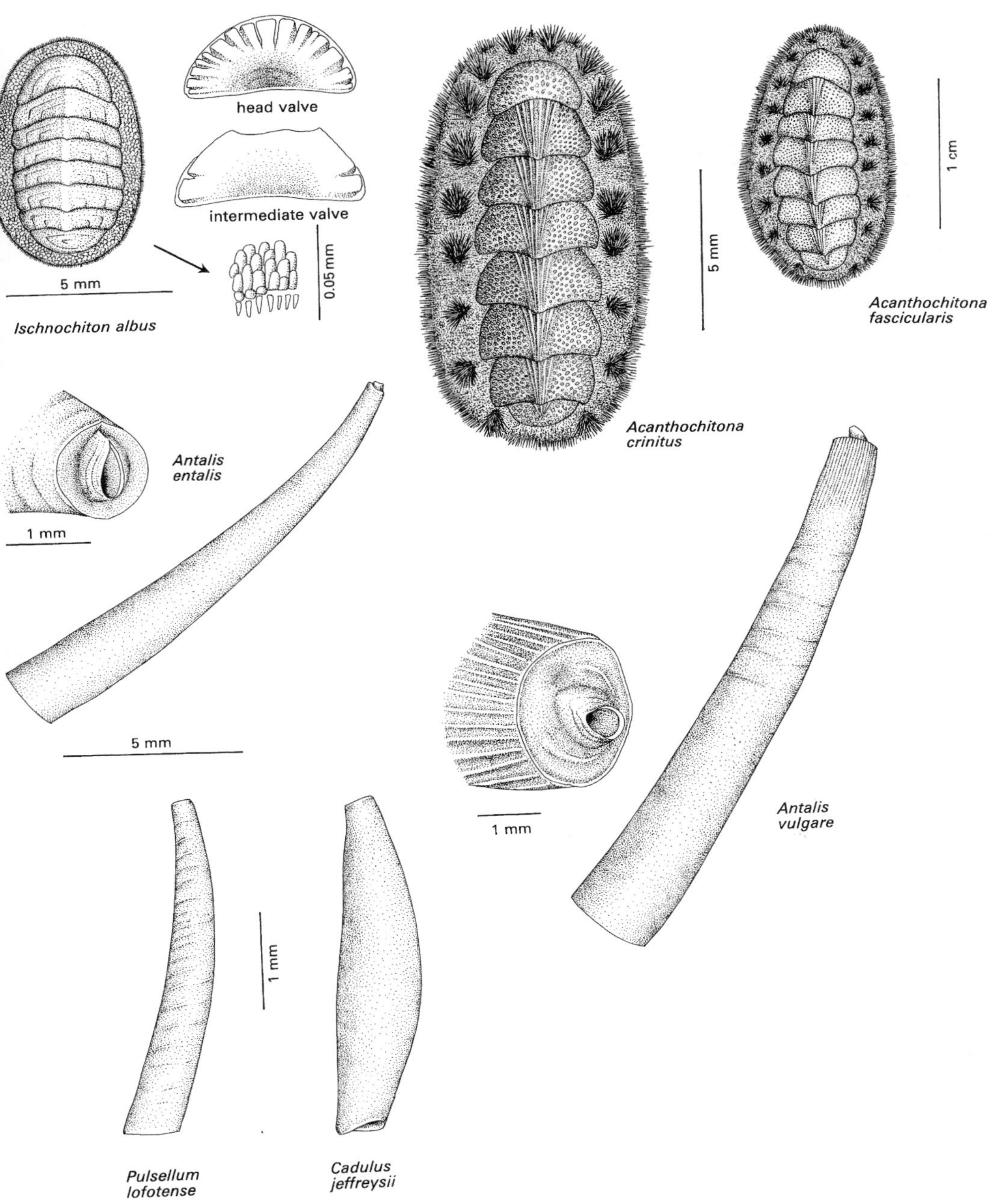
head valve
intermediate valve
0.05 mm
5 mm
Ischnochiton albus
5 mm
Acanthochitona crinitus
1 cm
Acanthochitona fascicularis
Antalis entalis
1 mm
5 mm
1 mm
Antalis vulgare
1 mm
Pulsellum lofotense
Cadulus jeffreysii

Fig. 10.2

Class Scaphopoda

The scaphopod body is antero-posteriorly elongate; the mantle is fused along almost the whole of the ventral side and secretes a tubular shell, open at each end. The head and short, cylindrical foot protrude from the broad, anterior end of the shell. The mantle cavity extends the whole length of the shell, and both inhalant and exhalant respiratory currents pass through the narrower, posterior end. The mouth is surrounded by short, ciliated tentacles or *captacula*. Scaphopods are microphagous deposit feeders, and live sedentary lives in offshore, soft-bottom deposits. The broad anterior end of the shell is buried vertically in the substratum, with the posterior end projecting above it.

---

1. Shell solid, thickly calcified, slightly tapered and gently curved. Narrow end with distinct longitudinal striations **Dentalium vulgare**

   Not as described **2**

---

2. Shell fairly thin, smooth, large (up to 35 mm); markedly tapered and distinctly curved. Narrow end occluded by a septum with a central spout **Dentalium entalis**

   Shell scarcely tapered; not, or scarcely, curved. Less than 10 mm long **3**

---

3. Straight, gently tapered at each end, more or less bottle-shaped **Cadulus jeffreysii**

   Straight, gently tapered to posterior end. Rather thick, with marked concentric growth lines **Pulsellum lofotense**

---

1 DENTALIIDAE

Shell smooth or sculptured, curved, regularly tapering. Foot conical, with laterally expanded sheath.

Antalis entalis Linnaeus Fig. 10.2

Up to 40 mm long, maximum width 5 mm at anterior end; glossy white, translucent, smooth, with no growth rings or ridges; sharply tapered to posterior end and with a pronounced curve. Anterior aperture circular; posterior end obliquely truncate, aperture oval with a notch on convex side, occluded by a calcified septum with a central tube (*pipe*), having a pyriform orifice.

Offshore in sand. All coasts, but uncommon in south; not present around Channel Isles.

Antalis vulgare (da Costa) Fig. 10.2

Up to 60 mm long, maximum width of 6 mm at anterior end; dull, opaque white, with a few marked, oblique concentric growth lines or ridges; gently tapered towards posterior end, less curved than *D. entalis*. Posterior portion of shell with fine, closely spaced, longitudinal striations. Anterior aperture circular; posterior end obliquely truncate, aperture circular with smooth rim, occluded by septum, with central pipe bearing a circular orifice.

Offshore, in sand. South-western coasts only, in the Channel and off western Ireland. Not common.

2 SIPHONODENTALIIDAE

Shell small, smooth, typically tapered at both ends. Foot simple, vermiform, or distally expanded to form a disk.

Three species are recorded for Britain but all are rare. The following species is the most likely to be found, but then only in offshore, deepwater dredging.

Pulsellum lofotense (M. Sars) Fig. 10.2

Up to 6 mm long, with maximum width of 0.5 mm at anterior end; transparent, white, with well-marked oblique, concentric growth lines; tapered gently to posterior, but scarcely curved. Anterior and posterior apertures circular; no septum or pipe at posterior end, rim of aperture with four indistinct notches.

Offshore. Rarely reported; known from western coasts of Ireland and Scotland, and from Shetland.

3 CADULIDAE

Shell small, shaped like a banana, being broadest at the middle. Foot vermiform, with a broad, terminal pedal disc.

Cadulus jeffreysii (Monterosata) Fig. 10.2

Up to 4 mm long; transparent, white, smooth, and glossy. Fusiform in shape: ventral side straight, dorsal arched; tapering towards both ends, anterior aperture scarcely wider than posterior.

Offshore. Rarely reported; known from Shetland, Hebrides, and south-west Ireland. *Cadulus subfusiformis* (M. Sars) is as long as *C. jeffreysii*, but much more slender, and very inconspicuous.

Class Gastropoda

Subclass Prosobranchia

The prosobranch body typically is enclosed in a coiled shell (Fig. 10.3). A *dextrally* coiled shell when viewed from above the tip or *apex*, spirals in a clockwise direction; an anticlockwise spiral is referred to as *sinistral*. The coils, or *whorls*, meet at *sutures* and vary in profile from flat-sided to strongly convex, or tumid. The apical whorls represent the larval shell, or *protoconch*; it is often eroded in some shells, but in others it remains as a prominent and often diagnostic feature. The lower whorls comprise the adult shell, or *teleoconch*; the last whorl terminates in the *aperture*. In dextral shells the aperture is on the right side of the axis of coiling, or *columella*; to its right is an *outer lip* and to the left an *inner lip*, these may form a continuous apertural rim, or *peristome*. Either lip may be internally thickened and bear ridges, or *teeth*. In some shells the peristome may be interrupted *adapically* (towards the apex) by an *anal sinus*, or *abapically* (away from the apex) by a *siphonal canal*, or both. The columella may be hollow, and open basally as an *umbilicus*.

The shell surface may bear successive growth lines, generated at the margin of the outer lip as the shell grows, or a series of regular, raised, longitudinal ridges, *ribs* or *costae*. Reflected or flared ribs are termed *varices*, and the outer lip may be bounded by a *labial varix*. Variations in calcareous secretion along the outer lip produces spiral ridges, or *striae*, and grooves. Ribs and striae, when present together, interact to form a *reticulate* or *decussate* pattern of squared or rectangulate blocks.

The prosobranch body can be withdrawn head first into the shell, and the aperture is then sealed by a horny *operculum*, carried on the postero-dorsal surface of the foot. The operculum is calcified in the family Tricoliidae. The foot may also bear *epipodial tentacles* (Fig. 10.3), arising from a lateral *epipodial fold*; in some families one or two *metapodial tentacles* (Fig. 10.3) originate from the opercular fold or posterior tip of the foot. The head has a pair of *cephalic tentacles*, and eyes set on lateral eyestalks which project for a variable distance alongside, and may fuse with the tentacles (Fig. 10.3). Behind and dorsal to the head is the *mantle cavity* in which the gills, or *ctenidia*, and their associated sense organ (the *osphradium*) and mucus gland (the *hypobranchial gland*) are housed, together with the genital/renal and anal papillae. The penis is often conspicuous on the right of the mantle cavity in some species. The mantle cavity of limpets consists of a shallow *pallial groove* around the base of the foot, and a deeper *nuchal cavity* behind the head. In some species the free edge of the mantle skirt bears *pallial tentacles*; most rissoaceans have one or two, turritellids many. In some mesogastropods and in neogastropods an extension of the mantle, the *siphon*, forms an inhalant channel within a *siphonal canal* at the base of the shell (Fig. 10.3).

KEY TO FAMILIES

1. Shell external; in active animals possibly covered by mantle folds which withdraw when touched — **2**

   Shell internal, permanently hidden beneath papillate mantle; animal resembles dorid nudibranch (p. 556) but has smooth tentacles and lacks a dorsal gill circlet — **26. Lamellariidae**

2. Shell with marginal slit, or with one or several holes in addition to aperture — **3**

   Shell imperforate apart from aperture — **4**

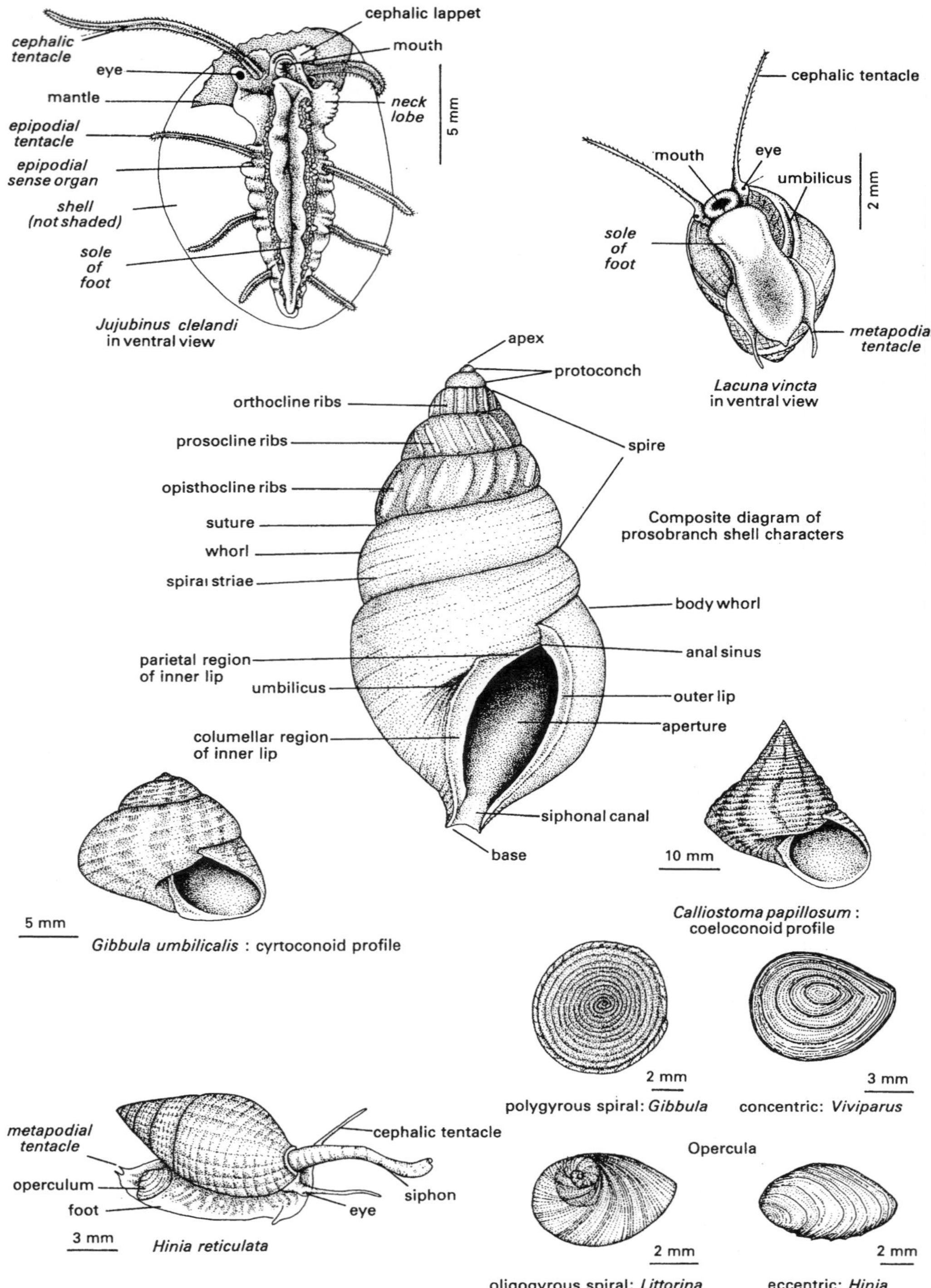
cephalic tentacle
eye
mantle
epipodial tentacle
epipodial sense organ
shell (not shaded)
sole of foot
cephalic lappet
mouth
neck lobe
5 mm
Jujubinus clelandi in ventral view
cephalic tentacle
mouth
eye
umbilicus
2 mm
sole of foot
metapodial tentacle
Lacuna vincta in ventral view
apex
protoconch
orthocline ribs
prosocline ribs
spire
opisthocline ribs
suture
whorl
spiral striae
Composite diagram of prosobranch shell characters
body whorl
parietal region of inner lip
anal sinus
umbilicus
outer lip
columellar region of inner lip
aperture
siphonal canal
base
10 mm
5 mm
Gibbula umbilicalis : cyrtoconoid profile
Calliostoma papillosum : coeloconoid profile
2 mm
3 mm
polygyrous spiral: Gibbula
concentric: Viviparus
Opercula
metapodial tentacle
operculum
foot
cephalic tentacle
siphon
eye
3 mm
Hinia reticulata
2 mm
2 mm
oligogyrous spiral: Littorina
eccentric: Hinia

Fig. 10.3

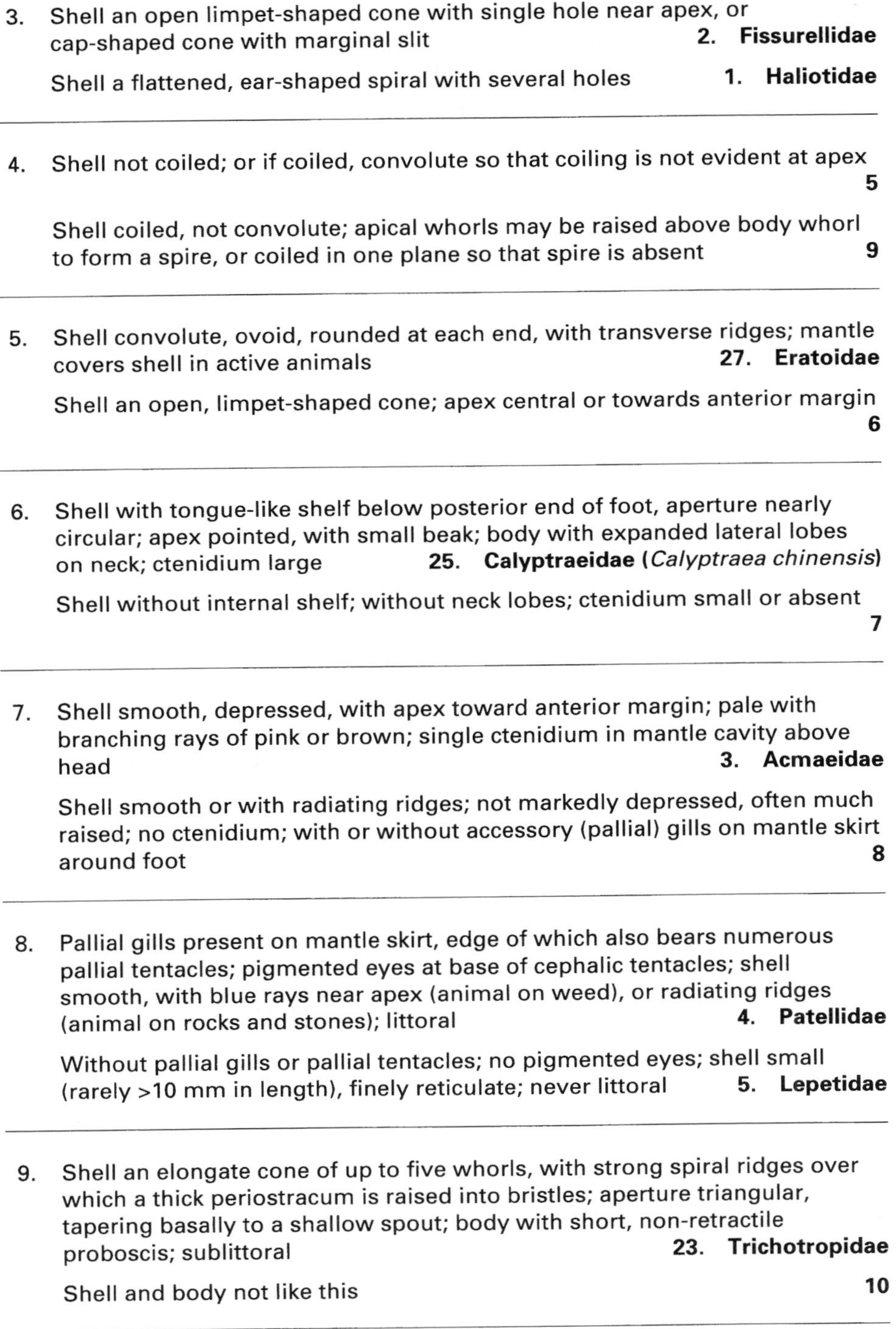

3. Shell an open limpet-shaped cone with single hole near apex, or cap-shaped cone with marginal slit — **2. Fissurellidae**

   Shell a flattened, ear-shaped spiral with several holes — **1. Haliotidae**

---

4. Shell not coiled; or if coiled, convolute so that coiling is not evident at apex — **5**

   Shell coiled, not convolute; apical whorls may be raised above body whorl to form a spire, or coiled in one plane so that spire is absent — **9**

---

5. Shell convolute, ovoid, rounded at each end, with transverse ridges; mantle covers shell in active animals — **27. Eratoidae**

   Shell an open, limpet-shaped cone; apex central or towards anterior margin — **6**

---

6. Shell with tongue-like shelf below posterior end of foot, aperture nearly circular; apex pointed, with small beak; body with expanded lateral lobes on neck; ctenidium large — **25. Calyptraeidae (*Calyptraea chinensis*)**

   Shell without internal shelf; without neck lobes; ctenidium small or absent — **7**

---

7. Shell smooth, depressed, with apex toward anterior margin; pale with branching rays of pink or brown; single ctenidium in mantle cavity above head — **3. Acmaeidae**

   Shell smooth or with radiating ridges; not markedly depressed, often much raised; no ctenidium; with or without accessory (pallial) gills on mantle skirt around foot — **8**

---

8. Pallial gills present on mantle skirt, edge of which also bears numerous pallial tentacles; pigmented eyes at base of cephalic tentacles; shell smooth, with blue rays near apex (animal on weed), or radiating ridges (animal on rocks and stones); littoral — **4. Patellidae**

   Without pallial gills or pallial tentacles; no pigmented eyes; shell small (rarely >10 mm in length), finely reticulate; never littoral — **5. Lepetidae**

---

9. Shell an elongate cone of up to five whorls, with strong spiral ridges over which a thick periostracum is raised into bristles; aperture triangular, tapering basally to a shallow spout; body with short, non-retractile proboscis; sublittoral — **23. Trichotropidae**

   Shell and body not like this — **10**

10. Shell glossy, porcellaneous, with short spire and large, rather inflated last whorl; aperture long, narrow, with or without teeth on inner and outer lips; colourful mantle covers shell in active animals; no operculum **27. Eratoidae**

Shell and body not like this **11**

11. Shell without siphonal canal **12**

Shell with siphonal canal (Fig. 10.3) **44**

12. Shell with elevated spire **24**

Spire very depressed (not rising much above body whorl), or absent **13**

13. Shell a minute (<1 mm diameter), biconcave disc, spire absent **16. Omalogyridae**

Shell not like this **14**

14. Shell with umbilicus **15**

Shell without umbilicus **20**

15. Umbilicus wide, funnel-shaped, exposing all older whorls; shell fragile, minute (<2.5 mm diameter), almost disc-like, with three or four whorls and a low spire **16**

Umbilicus deep, with narrow round or oval mouth, or chink-like with ∧-shaped umbilical groove **17**

16. Shell smooth with irregular growth lines; deep chestnut brown **15. Skeneopsidae**

Shell with spiral ridges often confined to base of body whorl; colourless or white; foot with three pairs of epipodial tentacles **7. Skeneidae**

17. Umbilicus chink-like, with ∧-shaped umbilical groove; aperture height almost equal to shell height; foot with two pairs of metapodial tentacles **9. Lacunidae**

Umbilicus deep, with narrow round or oval mouth; foot without metapodial tentacles **18**

18. Shell smooth, globular, polished; umbilicus deep, partly occluded by callus-like extension of inner lip; foot with greatly enlarged propodium

forming plough-like shield over head and front of shell; creeps through and over sand **28. Naticidae**

Shell and body not like this **19**

---

19. Shell minute (<2 mm with three or four whorls), globose (height = breadth); aperture with complete peristome; foot without epipodial tentacles; operculum with nucleus on columella margin, with peg-like process and three radiating ridges on underside **17. Rissoellidae**

Shell small (>4 mm with four or five whorls), depressed globose (higher than broad); aperture without complete peristome; foot with five or six pairs of epipodial tentacles **6. Trochidae**

---

20. Shell cap-shaped, almost hemi-ellipsoid, with slightly coiled sub-terminal apex; internally a large shelf-like partition lies at posterior end below apex; body with expanded neck lobes and large ctenidium; often found in stacks of ascending size **25. Calyptraeidae** (*Crepidula fornicata*)

Shell and body not like this **21**

---

21. Shell minute (2 mm or less), ear-shaped with up to two and a half rapidly expanding whorls; without periostracum; aperture very large, though animal cannot fully withdraw into shell; no operculum; in crevices and empty barnacle tests on upper shore PULMONATA (*Otina*) (p. 572)

Shell and body not like this **22**

---

22. Shell thinly calcified, with thick periostracum; no operculum; usually sublittoral **23**

Shell solid, without periostracum; with operculum; common intertidally **10. Littorinidae**

---

23. Shell cap-shaped, with large, almost circular aperture applied to substratum so that apex coils posteriorly; periostracum around aperture forms fringe, with earlier fringes evident on expanded last whorl; shell not covered by mantle **24. Capulidae**

Shell ear-shaped, periostracum not drawn out into frills; mantle edge swollen, partly covering shell in active animals **26. Lamellariidae**

---

24. Spire tall, shell awl-shaped with at least 10 whorls in adult **25**

Spire short, fewer than 10 whorls; shell more broadly conical **29**

---

25. Shell smooth, highly polished, whorls flat-sided **31. Eulimidae**

Shell with obvious ornament of ribs or spiral ridges, or both **26**

---

26. Sculpture of spiral ridges only; shell large (up to 20 whorls, 55 mm high); mantle edge with pinnate tentacles, operculum with fringed margin; no umbilicus **18. Turritellidae**

Sculpture of ribs, with or without spiral ridges **27**

27. Sculpture of ribs only **28**

Sculpture of ribs and spiral ridges producing reticulate or tuberculate surface; base of last whorl with spiral ridges only **19. Cerithiidae**

28. Shell large (up to 40 mm high), whorls very tumid, linked at sutures by thin, widely spaced ribs; aperture circular, with complete peristome **29. Epitoniidae**

Shell small (up to 8 mm high), whorls almost flat-sided, ribs numerous and flexuous; aperture ear-shaped; apex of shell sinistral OPISTHOBRANCHIA: **Pyramidellidae** (*Turbonilla*) (p. 549)

29. Shell delicate, purple or pale violet, broadly conical or globose; pelagic, usually washed ashore on south-western beaches **30. Janthinidae**

Shell not like this **30**

30. Shell smooth, globular, and polished; umbilicus deep, partly occluded by a callus-like extension of inner lip; foot with greatly enlarged propodium forming plough-like shield over head and front of shell; creeps through and over sand **28. Naticidae**

Shell and body not like this **31**

31. Shell oval–conical, smooth, with 4–6 tumid whorls patterned with zigzag streaks of variable colour; operculum calcareous, white, convex; foot with three pairs of epipodial tentacles **8. Tricoliidae**

Shell not like this; operculum horny, foot with or without epipodial tentacles **32**

32. With at least three pairs of epipodial tentacles; operculum circular with central nucleus; shell depressed-globular or pyramidal (breadth usually greater than height); aperture about 45° or more to vertical, without complete peristome, usually with nacreous lining; protoconch with broad point on tip of initial whorl **6. Trochidae**

Without epipodial tentacles; operculum not circular, with few turns; shell globose–conical or a more oval or elongate cone (higher than broad) **33**

33. Shell smooth, polished, oval or barrel-shaped with incised sutures; aperture long, narrow with tooth on columella; pinkish or brown with white bands; shallow burrower in sand OPISTHOBRANCHIA: **Acteonidae** (p. 544)

Shell not like this **34**

---

34. Shell with chink-like umbilicus and ∧-shaped groove on columella; foot with two prominent metapodial tentacles (Fig. 10.3) **9. Lacunidae**

Shell with or without umbilicus, but if present, without ∧-shaped groove leading to it; foot with or without metapodial tentacles **35**

---

35. Shell turreted, with fine spiral ridges and ribs below thick periostracum; foot with two prominent metapodial tentacles **9. Lacunidae**

Shell and body not like this **36**

---

36. (*In this dichotomy it is important that shell size be related to the number of whorls*) Shell rarely up to 9 mm, commonly <6 mm in height at five or six whorl stage, 2 mm or less at 3–4 whorls; oval–conical or globular; smooth, or with ribs, or spiral ridges, or both; aperture with complete peristome; foot with or without metapodial tentacle, mantle with one or two pallial tentacles or none **37**

Shell minimum 9 mm, frequently larger at 5–6 whorl stage, 4 mm or more at 3–4 whorls; drop-shaped with short spire of flat-sided whorls, or more globular with tumid whorls; smooth or with spiral ridges, never ribbed; aperture without distinct peristome, the inner lip forming a thin glaze on the body whorl; without metapodial or pallial tentacles **10. Littorinidae**

---

37. Apex of shell sinistral, spire otherwise dextral; smooth or with fine decussation, with or without tooth on columella; cephalic tentacles grooved on outer sides, with eyes set between them, a shelf-like projection (*mentum*) lies between the underside of head and propodium (Fig. 10.17) OPISTHOBRANCHIA: **Pyramidellidae** (p. 546)

Shell and body not like this **38**

---

38. Shell smooth, oval, or spindle-shaped with flat-sided whorls and shallow sutures; last whorl large (>60% of shell height), aperture narrow, scimitar-shaped with two or three prominent teeth on columella; no operculum, cephalic tentacles short, retractile, with eyes at base; in crevices, usually on upper shore PULMONATA: **Ellobiidae** (p. 572)

Shell and body not like this **39**

---

39. Shell smooth, with five or six slightly tumid whorls; usually cream with single spiral band of red (may be uniform red or white); operculum crimson

with internal peg-like process; foot with red opercular lobes and slightly bifid posterior tip; no metapodial or pallial tentacles **13. Barleeidae**

Shell and body not like this **40**

---

40. Mantle edge with pallial tentacle on right (protrudes from adapical angle of aperture), with or without similar tentacle on left **41**

    Mantle edge without pallial tentacle on left or right **42**

---

41. Metapodial tentacle present, though it may not extend to posterior tip of foot; shell smooth, or with ribs or spiral ridges, or both; marine **12. Rissoidae**

    Metapodial tentacle absent, shell smooth; in estuarine and brackish waters **11. Hydrobiidae**

---

42. Adult shell 5–6 mm with six whorls; thin periostracum, occasionally drawn out into peripheral keel of bristles, often with black encrustations on spire; umbilicus at most a minute chink; in fresh and brackish waters only **11. Hydrobiidae** (*Potamopyrgus*)

    Adult shell minute (<2 mm with 3–4 whorls), without periostracum; umbilicus usually obvious, chink-like or deep and round; foot with prominent opening of mucus gland medially on sole; marine **43**

---

43. Shell whitish, glossy and slightly iridescent, its transparency revealing distinctive dark spots on mantle; umbilicus a narrow chink or deep and round; snout deeply bifid, each half tentacle-like; cephalic tentacles single or bifid; operculum concentric with nucleus at middle or columella edge, a peg-like process and three radiating ridges on its inner surface **17. Rissoellidae**

    Shell cream or light horn colour with three red-brown bands on body whorl (young shells may be uniformly brown); with or without umbilical chink; snout bifid but not tentacle-like, cephalic tentacles single, foot with short triangular metapodial tentacle, operculum spiral, without ridges **14. Cingulopsidae**

---

44. Shell dextral **45**

    Shell sinistral **21. Triphoridae**

---

45. Shell with tall spire of tumid whorls bearing crescentic ribs and fine spiral ridges; mature shells with palmate extension to outer lip, juveniles (up to eight whorls) with sharply pointed siphonal canal but without palmate outer lip **22. Aporrhaiidae**

    Shell not like this **46**

---

46. Shell elongate, needle-like, of up to 14 whorls, each with three rows of tubercles; aperture small with short siphonal notch at base **20. Cerithiopsidae**

Shell not like this **47**

---

47. Shell smooth, or with growth lines only **33. Buccinidae**

Shell with ribs or spiral ridges or both **48**

---

48. Shell with spiral ridges only **49**

Shell ribbed, with or without spiral ridges **50**

---

49. Spire short, last whorl >80% of shell height, outer lip thick, with internal teeth in mature shells **32. Muricidae** (*Nucella*)

Spire long, last whorl <80% of shell height **33. Buccinidae**

---

50. Spiral ridges absent **32. Muricidae** (*Boreotrophon*)

Spiral ridges present, though may be fine **51**

---

51. Shell with 12 or more ribs on penultimate whorl **52**

Penultimate whorl with fewer than 12 ribs **56**

---

52. Shell breadth equals 50% or more of total height **53**

Shell breadth less than 50% of height **54**

---

53. Shell breadth >60% of height, whorls tumid, ribs strongly crescentic, fading below periphery of last whorl; aperture broadly oval, without teeth on outer lip; shell large (up to 110 mm high) **33. Buccinidae**

Shell less than 30 mm in height, ribs not strongly crescentic, outer lip thickened and ridged internally; foot with two metapodial tentacles **34. Nassariidae**

---

54. Siphonal canal long (equal to aperture length, or nearly so), straight, and narrow **32. Muricidae** (*Trophonopsis*)

Siphonal canal not as described **55**

---

55. Shell up to 30 mm, spire tall, almost flat-sided, with reticulate pattern; aperture may have teeth on inner and outer lips, siphonal canal separated

from base of shell by deep spiral groove; foot with two metapodial tentacles, operculum present **34. Nassariidae**

Shell not as above, <30 mm in height; foot without metapodial tentacles, with or without operculum **35. Turridae**

---

56. Shell breadth equal to or greater than half the height **57**

Shell breadth less than half the height; adult commonly up to 10 mm, aperture with anal sinus; no operculum **35. Turridae**

---

57. Shell <13 mm with 7–9 whorls; no operculum **35. Turridae**

Shell >15 mm with seven or eight whorls; with operculum **32. Muricidae**

Order Archaeogastropoda

1 HALIOTIDAE

Only one British representative of this family.

Haliotis tuberculata Linnaeus
Ormer Fig. 10.4
Shell flattened, ear-shaped, with a series of funnel-shaped holes, the most recent 5–7 forming respiratory openings through which the mantle extends as short exhalant siphons. Outer surface mottled; inner iridescent mother-of-pearl. Up to 90 × 60 mm. Foot large, with numerous epipodial tentacles. No operculum.

Mediterranean north to Channel Islands, on rocky shores from ELWST to 13 m. Absent from British and Irish mainlands.

2 FISSURELLIDAE

Shell conical or cap-shaped with reticulate sculpture; perforated apically or subapically by a single hole or marginally by a narrow slit. No operculum. Epipodial tentacles well developed. Five British species, in three genera, only two of which are common.

---

1. Shell cap-shaped with marginal slit ***Emarginula fissura***

Shell conical with single apical hole ***Diodora graeca***

Emarginula fissura (Linnaeus)
Slit limpet Fig. 10.4
Cap-shaped with posteriorly directed, recurved apex and marginal exhalant slit. Up to 10 × 8 × 6 mm. Foot shield-shaped, with 10 epipodial tentacles on each side; an additional tentacle on back of head on right side.

Lower shore to 265 m, on rocks and boulders bearing sponges, upon which it feeds. All suitable west European coasts from Norway and Faroe Islands to Canary Islands.

Diodora graeca (Linnaeus)
Keyhole limpet Fig. 10.4
Conical, oval in plan view, apex perforated by a single oval opening. Up to 25 × 16 × 10 mm. Foot oval, with 30–35 epipodial tentacles; thick papillose mantle skirt hangs below margin of shell. Body cream-coloured, deep orange, or red.

Under boulders and overhangs on rocky shores from LWST to 250 m. Distributed from Faroes to Canary Islands. Absent from west coast of Norway, North Sea coasts of England, and eastern parts of Channel.

3 ACMAEIDAE

Shell a shallow, smooth cone with apex toward anterior margin. Cavity behind head with single prominent gill, protruding beyond shell margin in active animals; pallial gills absent. Animals firmly attached to hard surfaces; foot lacks epipodial tentacles and operculum.

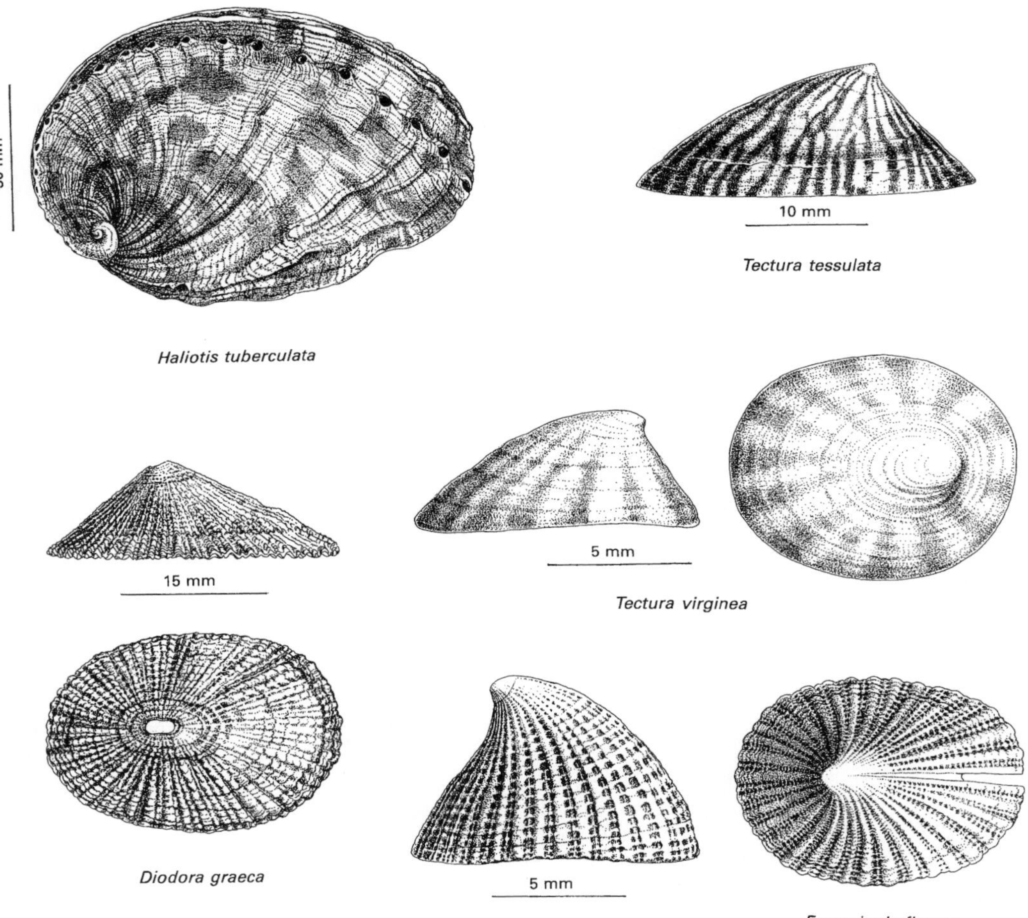

Fig. 10.4

1. White or pinkish with pink or brownish rays, head scar white; lips around mouth smooth; edge of mantle without tentacles, white with pink bands **_Tectura virginea_**

   White or greenish with chestnut brown rays, head scar chocolate brown; lips around mouth fringed; edge of mantle cuprous green, with short tentacles **_Tectura tessulata_**

Tectura virginea (Müller)
White tortoiseshell limpet Fig. 10.4

Pale pink, yellowish or white, with broad, often branching rays of pink or brown; occasionally chequered or unicolorous; pure white shells may occur. Up to 10 × 9 × 4 mm. Lips around mouth form a broadly lobed frill, drawn out posterolaterally into extensile lobes. Mantle edge minutely papillate, studded with chalk-white repugnatorial glands.

On boulders and small, smooth stones, particularly those encrusted with red algae such as *Lithothamnion*. LWST to 100 m on all British coasts. Distributed from Iceland to the Azores but absent from the west coast of Jutland and the Baltic.

Tectura tessulata (Müller)
Tortoiseshell limpet Fig. 10.4

White or pale green with irregularly branched brown bands radiating from apex to produce a tortoiseshell pattern. Up to 30 × 24 × 10 mm, smaller in littoral habitats. Lips around mouth lobed, with narrow, somewhat stiff, outer fringe. Mantle edge cuprous green with two staggered rows of short tentacles.

On boulders and small, smooth stones, particularly those encrusted with red algae such as *Lithothamnion*. LWNT to 50 m. A circumpolar species extending south to Dublin, Liverpool, and the Humber; absent from west coast of Ireland.

4 PATELLIDAE

Shell conical or cap-shaped with no spiral coiling; operculum absent. Foot large and oval, with horse-shoe-shaped muscle attachment to shell. Mantle cavity a shallow groove around base of foot, deepening to nuchal cavity behind head. Mantle edge bears pallial gills; no gill in nuchal cavity. Animals firmly attached to rock (*Patella* spp.) or weed (*Helcion*).

---

1. Shell with radial ribs; pallial gills along entire length of mantle skirt; on rock **2**

   Shell usually smooth or with marked concentric ridges; pallial gills do not extend around anterior margin of nuchal cavity; on weed **4**

---

2. Pallial tentacles white **3**

   Pallial tentacles translucent or grey; foot olive, grey, or khaki; inside of shell grey/green, with bluish iridescence ***Patella vulgata***

---

3. Pallial tentacles translucent white; foot apricot-yellow or cream; inner surface of shell porcellaneous, often with cream or orange head scar ***Patella ulyssiponensis***

   Pallial tentacles chalky white; foot dark olive-black; inner surface of shell cream-orange with chocolate brown marginal rays ***Patella depressa***

---

4. Shell small, translucent, depressed, with bright blue rays radiating from apex; on fronds of *Laminaria* ***Helcion pellucidum pellucidum***

   Shell solid, conical, with ledged profile; blue rays on apex of shell only; from *Laminaria* holdfasts ***Helcion pellucidum laevis***

Patella vulgata Linnaeus
Common limpet Fig. 10.5

With irregular radiating ribs; apex central or slightly anterior; inner surfaces grey-green. Up to 60 × 50 × 30 mm. Mantle skirt fringed with translucent pallial tentacles arranged in three series of different lengths.

From between MHWST and MHWNT downward to ELWST on all suitable rocky shores; upper limit of distribution increased by shade and exposure. Tolerant of low salinity but generally confined to habitats with salinity >25‰. Distributed from Mediterranean to Lofoten Islands, Norway.

Patella ulyssiponensis Gmelin
China limpet Fig. 10.5

Apex noticeably anterior to mid-line; radiating ridges finer than in *P. vulgata*, tending to show characteristic pattern of alternating single and triple ridges. Shell interior porcellaneous, often tinted orange apically. Up to 50 × 40 × 20 mm. Foot cream-orange, pallial tentacles translucent white, arranged in two series of different sizes.

On rocky shores at MLWNT and below, favouring wet areas such as gullies and rock pools; most common on exposed shores, avoiding extreme shelter and low salinities. A southern species distributed from the Mediterranean to the British Isles, where it achieves its northern limit. Absent from British shores between the Isle of Wight and the Humber, and from continental shores east of Barfleur (near Cherbourg).

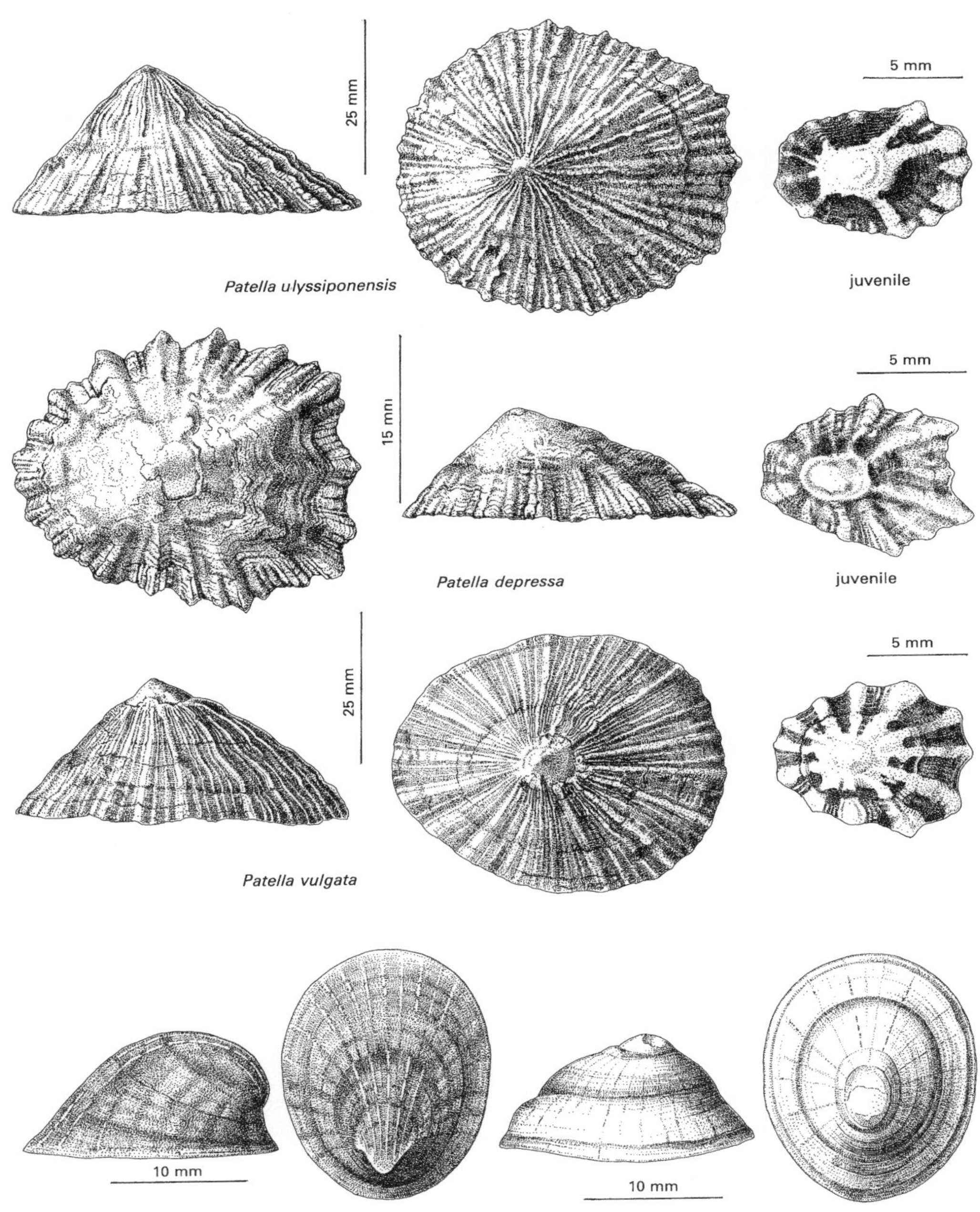
25 mm
5 mm
Patella ulyssiponensis
juvenile
15 mm
5 mm
Patella depressa
juvenile
25 mm
5 mm
Patella vulgata
10 mm
10 mm
Helcion pellucidum pellucidum
Helcion pellucidum laevis

Fig. 10.5

Patella depressa Pennant
Black-footed limpet Fig. 10.5
Shell a flattened cone with fine radiating ribs; distinctive orange-brown marginal rays on inner surface. Up to 30 × 25 × 12 mm. Foot dark olive or black, pallial tentacles chalky white; body otherwise similar to *P. vulgata*.

On exposed rocky shores only, from MHWN to MLWN, prefers vertical surfaces. A Mediterranean species, reaching its northern limit on shores of south-west England and Wales; rare on the west coast of Ireland.

Helcion pellucidum (Linnaeus)
Blue-rayed limpet Fig. 10.5
Largely confined to *Laminaria* plants, where it occurs in two forms, *H. p. pellucidum* on the fronds (less frequently on fronds of *Fucus serratus*) and *H. p. laevis* within the holdfast.

Helcion p. pellucidum is thin-shelled, translucent, smooth; with bright blue rays from apex to posterior margin. Up to 20 × 15 × 7 mm.

Helcion. p. laevis is taller and more robust with a ledged profile, apical region exhibiting features of the *pellucidum* shell. Elsewhere, dull and devoid of blue rays. Up to 20 × 8 × 10 mm.

Intertidally on *Laminaria*, and occasionally on low level (LWST) *Fucus serratus*; sublittorally to 27 m. All British and Irish coasts. Distributed from Iceland to Portugal; absent from Belgium, Holland, east coast of Denmark, and Baltic Sea.

5 LEPETIDAE

Shell small, cap-shaped or conical, with or without spirally whorled protoconch; sculpture finely reticulate. No operculum. Ctenidia and pallial gills absent. Animals sublittoral, attached to stones.

---

1. Shell with recurved apex bearing spiral protoconch; internally a small septum below apex **_Propilidium exiguum_**

   Shell apex towards anterior margin, without spiral protoconch; no internal septum **_Iothia fulva_**

Iothia fulva (Müller)
Tawny limpet Fig. 10.6
Conical, orange-yellow (occasionally white), opaque; apex near anterior margin, with blister-like protoconch; no internal apical septum. Up to 7 × 5 × 2.5 mm.

Sublittoral, on stony ground from 5 to 200 m. Distributed from Iceland southwards to the Azores. Recorded off west coast of Scotland, Irish Sea, west coast of Ireland, and British (not continental) North Sea coasts.

Propilidium exiguum (Thompson)
Curled limpet Fig. 10.6
Cap-shaped, white, semi-transparent; apex central, recurved posteriorly, with spiral protoconch. Internally, a small septum projects from posterior face of shell below apex. Up to 4 × 3 × 2.5 mm.

Sublittoral only, on hard substrata, 7 to 170 m. Off Shetlands, west coast of Scotland and Ireland. Distributed from Norway south to Azores and Mediterranean; absent from Baltic, North Sea, and English Channel.

6 TROCHIDAE

Shell spiral, conical (often pyramidal), frequently with umbilicus; 'mother of pearl' inside often revealed at eroded apex. Plane of aperture markedly inclined. Foot with extensile epipodial tentacles; operculum multi-spiralled and circular. Detrital sweepers, with the exception of *Calliostoma* which is probably a microphagous carnivore.

---

1. Shell a sharply pointed cone; whorls flat, with strong spiral ridges **2**

   Shell bluntly pointed, rather convex in profile; whorls tumid **3**

---

2. Shell height approximately equal to breadth; four or five pairs of epipodial tentacles **_Calliostoma zizyphinum_**

   Shell height approximately twice breadth; three pairs of epipodial tentacles **_Jujubinus striatus_**

---

3. Columella with tooth or bulge; outside of shell not nacreous, or only so at eroded apex; animal with three pairs of epipodial tentacles **4**

   Columella a smooth simple curve, without bulge; shell externally nacreous;

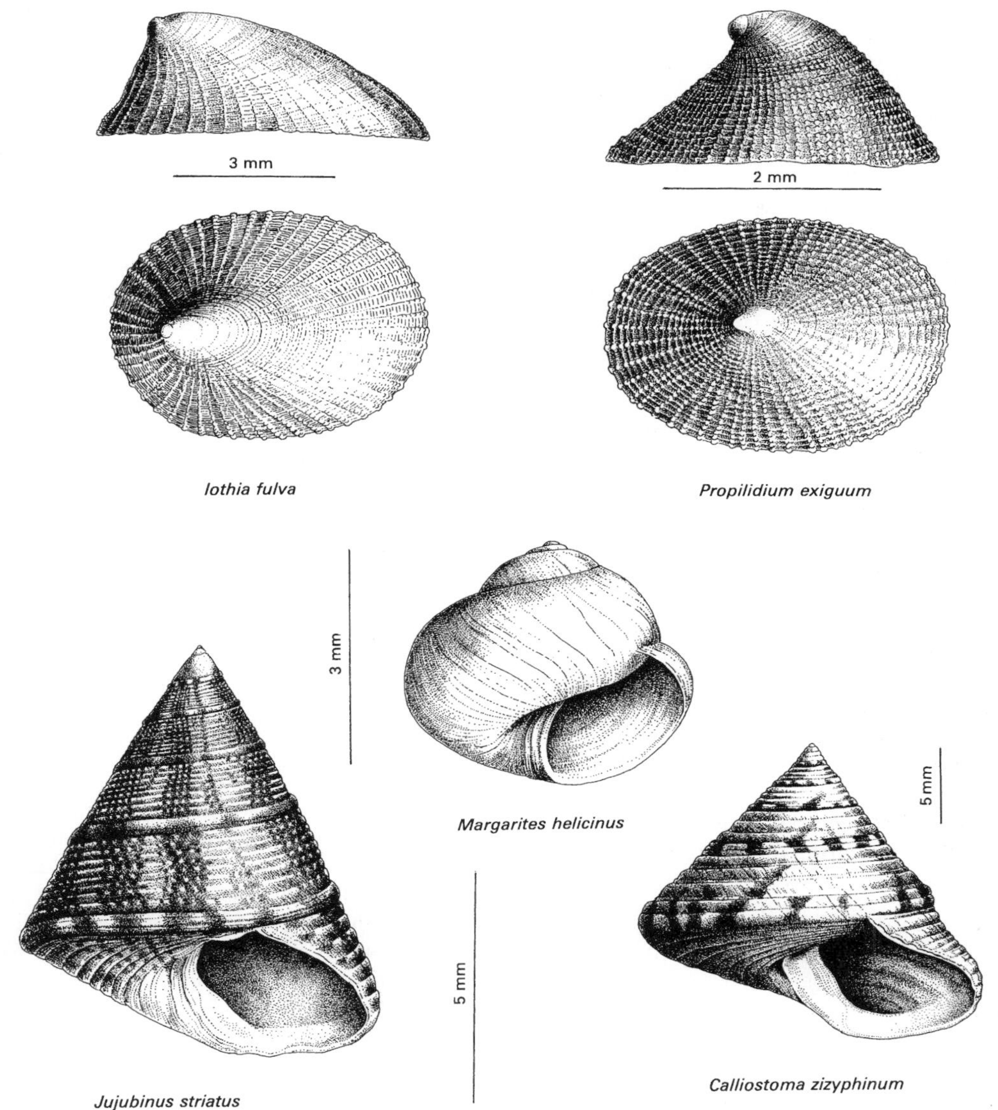

Fig. 10.6

animal with six pairs of epipodial tentacles ***Margarites helicinus***

---

4. Umbilicus a small depression; finely patterned dark shell often eroded to underlying nacre at apex; height > breadth; columella tooth prominent ***Monodonta lineata***

Umbilicus open or closed; shell height ⩽ breadth; bulge on columella does not form prominent tooth **5**

---

5. Upper part of each whorl with nodular ribs; shell profile stepped; umbilicus wide and deep, its margins limited by a

spiral ridge arising from base of columella **Gibbula magus**

Upper part of each whorl without ribs **6**

---

6. Shell profile slightly stepped; body whorl with 16–18 spiral ridges above and 14–18 below periphery; umbilicus deep but occluded to varying degree by inner lip; mainly sublittoral **Gibbula tumida**

   Shell not stepped, spiral ridges less numerous; umbilicus variable or absent; mainly littoral **7**

---

7. Shell light grey/yellowish in colour with fine (0.5 mm or less) striped pattern of darker colour; umbilicus narrow (occasionally absent in older shells) **Gibbula cineraria**

   Shell light greenish/horn-coloured with broad stripes (1 mm or more) of red or purple; more depressed, convex profile **8**

---

8. Umbilicus round and deep (occluded in some older specimens); reddish purple stripes usually entire on base of shell **Gibbula umbilicalis**

   Umbilicus in young specimens only; coloured stripes deeper purple and often give rise to a chequered pattern of rectangular, pigmented spots on base of shell. Channel Isles only **Gibbula pennanti**

Margarites helicinus (Fabricius) Fig. 10.6
Translucent, globular, with 4–5 tumid whorls dipping sharply into sutures. Umbilicus oval, conspicuous, partly occluded by inner lip. Orange-red or horn-coloured with green or purple refringency; often with brown band on last whorl. Up to 9 × 11 mm. Six (sometimes five) pairs of epipodial tentacles.

Under stones, in pools and on weed, low on rocky shores, and sublittoral. Circumboreal, extending south to Yorkshire on east coast and the Isle of Man and North Wales on the west. Locally abundant. Four other species of *Margarites* occur rarely in offshore, northern waters.

Gibbula cineraria (Linnaeus)
Grey top shell Fig. 10.7
Profile slightly convex, last whorl angulate, base flat. With fine spiral ridges and grooves, 10–17 (commonly 13–14) on base of last whorl. Umbilicus small, oval or egg-shaped, deep. Greyish or straw-coloured with many thin stripes of reddish-brown or purple. Up to 16 × 15 mm.

On rocky shores, on weed and under stones, below LWNT (may occur higher in rock pools), sublittoral to about 130 m. Distributed from Gibraltar to north Norway and Iceland. Abundant on all British coasts.

In *G. cineraria* from the Channel coasts of Devon and Cornwall the umbilicus may be wholly or partially occluded by the inner lip. This tendency seems to increase with age.

Gibbula umbilicalis (da Costa)
Flat top shell Fig. 10.7
Similar to *G. cineraria* but distinguished by a markedly convex profile (especially in older shells), a more angular last whorl (especially in younger shells), fewer (8–11) spiral ridges on base of last whorl, a larger and more circular umbilicus and a cream or greenish background with broad stripes or bands of red or reddish-purple. Up to 16 × 22 mm. Non-umbilicate forms have been recorded from south-east Cornwall and south-west Devon.

Habitat similar to that of *G. cineraria*, but more tolerant of emersion and thus occurs higher up shore (MHWS–MLWS). Distributed along Atlantic coast of France and west coast of Britain, Ireland, and northwards to Orkney. Absent from North Sea, and British coasts in eastern basin of Channel.

Gibbula pennanti (Philippi)
Pennant's top shell Fig. 10.7
Similar to *G. umbilicalis* but with upper whorls tending to be more tumid and sutures more prominent. Umbilicus open in young shells, closing later to form a depression against inner lip; occasionally a small umbilical chink may persist. Colour similar but with reticulate rather than striped pattern on base of last whorl; some shells homogeneous coppery red, green or black, with or without red or white spiral bands at periphery. Up to 16 × 15 mm.

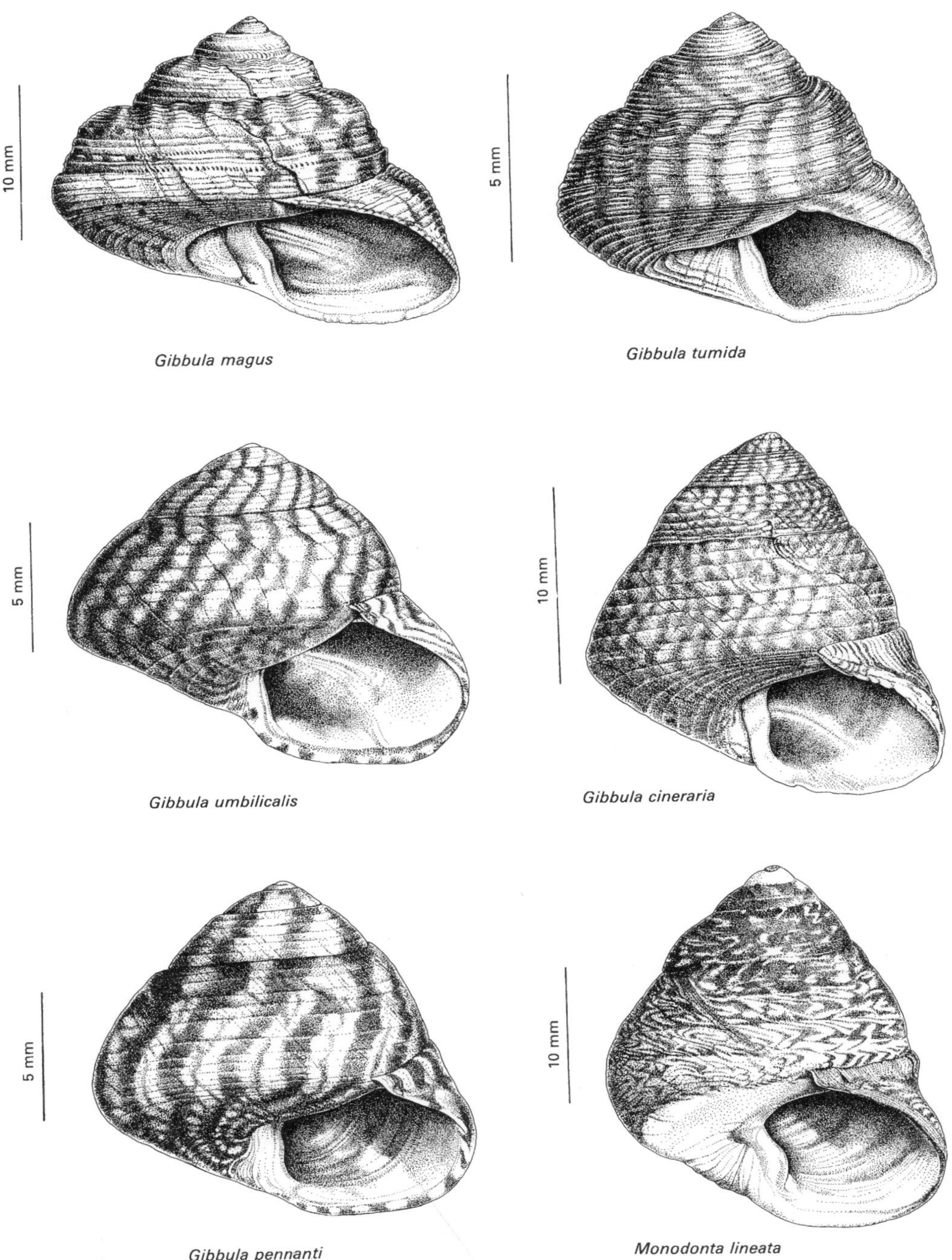

Gibbula magus

Gibbula tumida

Gibbula umbilicalis

Gibbula cineraria

Gibbula pennanti

Monodonta lineata

Fig. 10.7

Habitat similar to *G. cineraria* and *G. umbilicalis*; particularly common in low shore pools. A southern species occurring on Atlantic coasts of Spain and France and in the Channel Islands. Absent from mainland Britain and French Channel coasts east of Barfleur (near Cherbourg).

Gibbula magus (Linnaeus)
Turban top shell Fig. 10.7
Shell with markedly stepped profile and prominent peripheral keel on last whorl; with short, nodular ribs on upper part of each whorl, above a variable number of spiral ridges and grooves. Umbilicus large and deep, edged by thickened spiral ridge. White, grey, or yellowish, with irregular brown, red, or purple streaks and blotches. Up to 30 × 35 mm.

On weed, under stones, and on shingle beaches at ELWST. Sublittoral, on muddy gravel, to 70 m. Distributed from Mediterranean and Azores to western basin of English Channel, west coast of Britain, and north to Shetland.

Gibbula tumida (Montagu) Fig. 10.7
With stepped profile of 6–7 slightly tumid, flat-shouldered whorls. Fine growth lines traverse spiral grooves and ridges. Umbilicus open and deep in young specimens, occluded by inner lip with age, and in larger shells appearing as an oval opening or small chink. Light fawn with broken brown bands across whorls; usually with greenish-blue iridescence. Up to 9 × 10 mm.

Sublittoral only; on *Laminaria*, and down to 1200 m on stones on sandy-muddy bottoms. Iceland and north Norway south to British, Irish, French, and Iberian coasts.

Monodonta lineata (da Costa)
Thick top shell Fig. 10.7
Profile slightly convex, sculpture of fine growth lines and occasional distinct growth disturbances; young specimens with thin spiral ridges and grooves on first three whorls, older shells often abraded to reveal underlying 'mother-of-pearl'. Columella thick, with prominent bulge or 'tooth' near base; umbilicus a slight depression or chink in adult shells, often open in juveniles. Cream or horn-coloured, heavily overlain with reddish-brown, purple, or green zigzag streaks; base of shell less heavily pigmented. Up to 30 × 25 mm.

On rocky or boulder shores from MHWS to MLWS, avoiding excessive exposure and deposits of shingle or sand. Patchily distributed in the British Isles: western basin of the English Channel, Welsh coast north to Anglesey, south and west coasts of Ireland.

Calliostoma zizyphinum (Linnaeus)
Painted top shell Fig. 10.6
Regularly conical, straight-sided, up to 12–13 whorls; with regular spiral grooves and ridges; last whorl with prominent peripheral keel bearing two broad ridges; base of shell rather flat. Yellowish, pale pink or violet with streaks and blotches of brown, red, or purple. Occasionally pure white or violet. Up to 30 × 30 mm.

Below LWNT on weed-covered rocky shores, sublittoral on hard and soft bottoms to 300 m; tolerant of salinities down to 21‰. Distributed from Mediterranean, Canaries, and Azores to Norway. All British and Irish shores. The northern *C. formosa* (McAndrew and Forbes) and the southern *C. papillosum* (da Costa) are also recorded from Britain.

Jujubinus striatus (Linnaeus)
Grooved top shell Fig. 10.6
Conical with 6–8 flat-sided whorls; with spiral ridges, often slightly nodular in apical and basal areas of shell. Umbilicus closed in adult shells, open in young. White, creamy, or greenish, with brown blotches and stripes, often pink apically. Up to 10 × 8 mm. Three pairs of epipodial tentacles.

On weeds, especially *Zostera*, at LWST and down to 200 m. Mediterranean northward to south-west British Isles, Isle of Man, and Channel Isles where it is common. Three other species of *Jujubinus* are recorded for Britain, but none is common.

7 SKENEIDAE

Minute white or colourless shells with 3–4 tumid and rapidly expanding whorls; often with sculpture of fine spiral ridges, which may be confined to basal areas of last whorl. Aperture with distinct peristome; umbilicus open. Foot with three pairs of epipodial tentacles, operculum multi-spiralled. Often overlooked because of small size. Eight species occur in British waters.

Skenea serpuloides (Montagu) Fig. 10.8
Thin and glossy; last whorl constitutes most of shell, smooth above, sharp spiral ridges below. Aperture almost circular, with complete peristome; umbilicus wide, revealing underside of spire. Up to

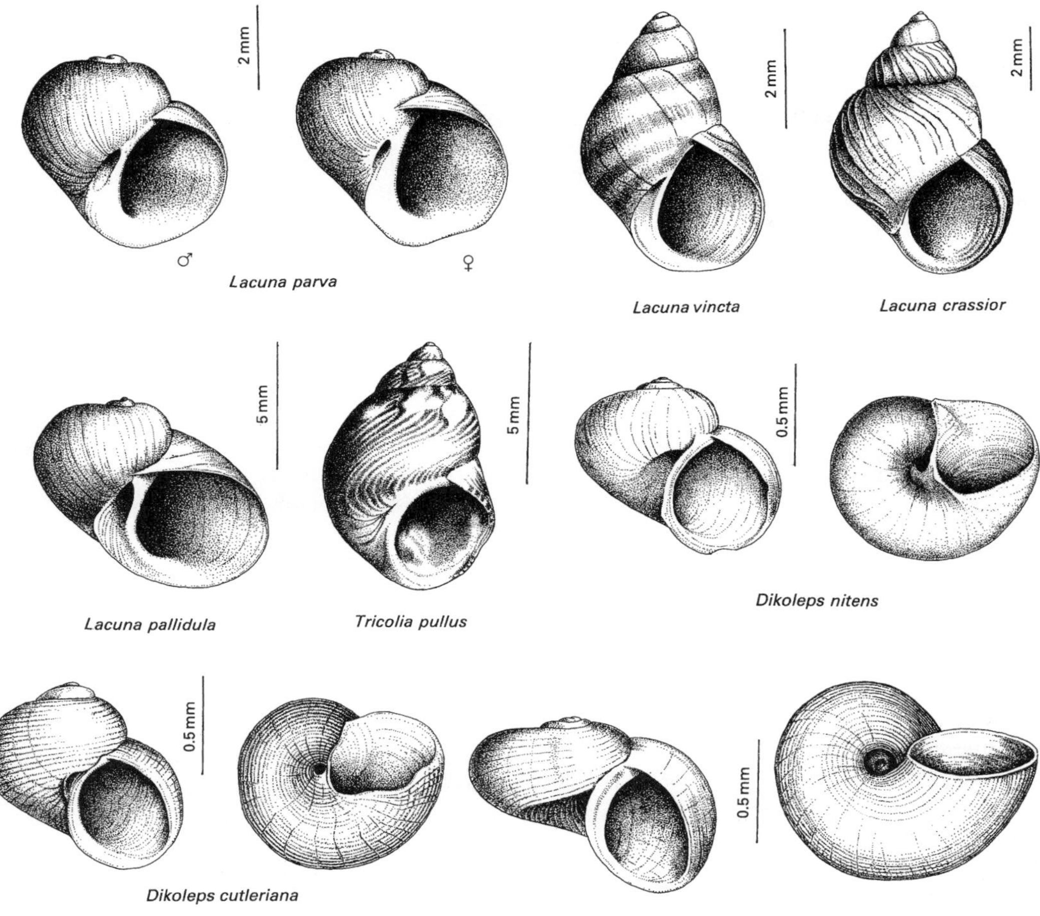

Fig. 10.8

1.5 × 1.2 mm. Head with additional post-optic tentacle behind right cephalic tentacle.

On rocky shores among weeds and detrital sediments at LWST; sublittoral on gravelly and shelly sand to 50 m. A southern species extending north to British Isles; recorded off west coast of Ireland, south-west England, and North Sea coast as far south as Scarborough.

Dikoleps nitens (Philippi) and *D. cutleriana* (Clark) occur in similar habitats to *S. serpuloides* but are less common. Both are shown in Fig. 10.8.

8 TRICOLIIDAE

Only one British representative of this family.

Tricolia pullus (Linnaeus)
Pheasant shell Fig. 10.8
Conical or turban-shaped, glossy; last whorl approximately two-thirds shell length. Cream or white with irregular red or purple streaks and blotches, often producing a zigzag pattern; rarely plain. Operculum milk-white, calcareous, markedly convex. *Tricolia pullus* is the only British prosobranch with such an operculum. Up to 9 × 5 mm.

On rocky shores near LWST and sublittoral to 35 m; most abundant on tufted red algae such as *Lomentaria*, *Laurencia*, and *Chondrus*. Mediterranean northward to west and south-west coasts of Britain and Ireland; rare in the north.

Order Mesogastropoda

9 LACUNIDAE

Shell a tumid or turreted spire, or globular with last whorl prominent and spire much reduced; usually small, thin, and smooth, often with an umbilical chink or groove on the columella. Foot with two prominent metapodial tentacles; operculum a rapidly expanding spiral of two or three turns, thin and horny.

---

1. Shell with 5–7 tumid or turreted whorls; aperture drawn out, angulate below columella **2**

   Shell globular, spire much reduced; aperture not drawn out at base, or only slightly so **3**

---

2. Shell solid; 6–7 turreted whorls with fine spiral ridges and ribs below thick periostracum; no colour banding, with or without umbilicus ***Lacuna crassior***

   Shell thin; 5–6 smooth, tumid whorls, usually with brown banding; umbilicus obvious ***Lacuna vincta***

---

3. Aperture nearly equal to shell height; without colour bands on last whorl ***Lacuna pallidula***

   Aperture much less than this, usually with red-brown banding on last whorl ***Lacuna parva***

Lacuna vincta (Montagu)
Banded chink shell — Fig. 10.8

Umbilicus ∧-shaped with prominent groove or chink, aperture below drawn out and slightly angular. Pale horn-colour with two or four red-brown bands on last whorl; upper whorls with two colour bands, becoming faint towards apex which is generally purplish. Banding occasionally faint or absent. Up to 10 × 5 mm. Foot with two short, flat metapodial tentacles projecting from below operculum.

Common on seaweeds, occasionally on *Zostera*, at LWST, sublittoral to 40 m. Tolerant of salinities down to 20‰. Circumboreal, extending south as far as the Channel and Atlantic coasts of France. On suitable weed-covered shores on all British and Irish coasts.

Lacuna crassior (Montagu)
Thick chink shell — Fig. 10.8

Shell strong, turreted, yellow, with thick periostracum forming a series of irregular folds on last whorl. Umbilicus occasionally closed and umbilical groove reduced; aperture drawn out and angulate at base of columella. Up to 14 × 10 mm. Metapodial tentacles white, less flattened than those of *L. vincta*.

Uncommon. On mud/gravel at LWST and sublittoral to 90 m. Circumboreal; recorded from Dorset, Milford Haven, and the Celtic Sea, and north-east coasts north of Scarborough.

Lacuna parva (da Costa)
Small chink shell — Fig. 10.8

Globular, with three rapidly expanding whorls; sutures deep. Apparently smooth but with fine spiral ridges and growth lines. Last whorl occupies most of spire; aperture rounded in male; more ovate, with basal flare or spout, in female. Umbilicus as in *L. vincta*; umbilical groove with longitudinal ridges. Red-brown or horn-coloured, often with rufous bands on last whorl and lilac tinge to apex. Up to 4 × 4 mm.

On seaweeds, lower shore (occasionally higher, where it may occur on *Fucus vesiculosus*) and sublittoral to 40 m. Distribution similar to that of *L. vincta* but extending south to Spanish coasts.

Lacuna pallidula (da Costa)
Pallid chink shell — Fig. 10.8

Globular, smooth, with three or four rapidly expanding whorls; sutures deep. Last whorl occupies most of spire; aperture very large, almost equal to shell height. Umbilicus deep, open, revealing much of interior surface of spire; umbilical groove wide (wider in males than in females), with longitudinally ridged surface. Yellowish-green, occasionally white; unbanded. Up to 12 × 6 mm in females, 6 × 3 mm in males.

On seaweeds at LWST, sublittoral to 70 m. Distribution similar to that of *L. vincta*.

10 LITTORINIDAE

Shell solid, with or without a moderately high spire, often pigmented. Sculpture of growth lines, with or without spiral ridges; never ribbed. No umbilicus or umbilical groove. Foot shield-shaped, without epi- or metapodial tentacles. Operculum a spiral of 2–3 rapidly expanding turns.

Because of the degree of overlap in characters between some species it may be necessary to examine both shell and body features of live specimens in order to identify them.

The taxonomy of the *L. saxatilis* complex has long been disputed. The separation of *L. nigrolineata*, *L. neglecta*, and *L. arcana* from *L. saxatilis* has been generally accepted, but there is still some uncertainty over the taxonomic status of *L. saxatilis*, *L. rudis*, and *L. tenebrosa*.

---

1. Spire of shell low or absent; apex blunt or completely flat **2**

   Shell with well-developed spire; apex pointed **3**

---

2. Shell with small but distinct spire; outer lip arises below level of apex; aperture wide ***Littorina obtusata***

   Spire very low or absent; outer lip arises level with apex; aperture narrow, with constricted throat ***Littorina mariae***

---

3. Spire of flattened or slightly tumid whorls and shallow sutures; outer lip meets last whorl tangentially **4**

   Spire of tumid whorls and deep sutures; outer lip meets last whorl almost at right angle **5**

---

4. Adult shell large (up to 38 mm high); usually with slight spiral ridges (these may be more obvious in young specimens); columella white; cephalic tentacles with transverse black stripes ***Littorina littorea***

   Adult shell small (up to 9 mm high), smooth; columella pigmented; periostracum extends beyond outer lip; cephalic tentacles with longitudinal black stripes ***Littorina neritoides***

---

5. Shell small (usually <5 mm), globular, with short spire; thin and superficially smooth, with brown spiral band entering aperture; floor of mantle cavity dark; penis with long tip; ovoviviparous ***Littorina neglecta***

   Shell with well-developed spire; often with spiral ridges, which may be eroded or otherwise indistinct; no brown band entering aperture; floor of mantle with diagonal white band; oviparous or ovoviviparous **6**

---

6. Animal found on sheltered brackish shores, commonly on weeds, often sublittoral; shell superficially smooth, dark, and thin ***Littorina tenebrosa***

   Animals on upper levels of fully saline shores, in crevices and under stones; shell usually with spiral ridges, solid **7**

---

7. Shell with strap-like spiral ridges often with a shallow central furrow; grooves between ridges often darkly pigmented; penis with very short tip and up to 12 glands; oviparous ***Littorina nigrolineata***

   Shell usually with sharper spiral ridges, though strap-like ridges may occur; penis with moderately long tip; oviparous or ovoviviparous **8**

---

8. Oviparous; prostate does not reach level of anus and has a small ciliated field alongside; penis with tapering tip; found with *L. rudis* from which it is indistinguishable in most cases in respect of shell form ***Littorina arcana***

   Ovoviviparous; prostate reaches beyond anus and has a large ciliated field alongside; penis with blunt tip **9**

---

9. Spiral ridges usually with crest at their adapical edge; aperture with prominent

everted area at base of columella; in *Pelvetia* zone, among weeds and on stones **_Littorina rudis_**

Spiral ridges usually with central crest; only small everted area at base of aperture; in high shore crevices (*Pelvetia* zone and above) **_Littorina saxatilis_**

Littorina littorea (Linnaeus)
Common periwinkle Fig. 10.9
Spire prominent, pointed, with slightly tumid whorls; last whorl up to 85% of shell height; upper whorls may be flat-sided or even concave. Appears smooth, especially in older specimens, but has numerous, slight spiral ridges which may be strap-like on last whorl. Aperture ear-shaped, outer lip arising tangentially to last whorl. Black or dark grey-brown, often lighter towards apex, with heavier pigmentation between spiral ridges; occasionally red, orange, or white. Columella region white. Up to 32 × 25 mm. Cephalic tentacles with many transverse black stripes and, ventrally, a single longitudinal line.

Common on all but the most exposed rocky shores; mainly intertidal, the upper range increasing with exposure, but sublittoral to 60 m in the north. In sheltered areas it makes feeding excursions over muddy shores; tolerant of brackish water. Distributed from northern Spain to White Sea. All British coasts, though rare or absent in Isles of Scilly and Channel Isles.

Littorina obtusata (Linnaeus)
Flat periwinkle Fig. 10.9
With blunt apex and very much reduced spire, last whorl tumid, comprising >90% of shell height; appears smooth but is very finely reticulate. Aperture a tear-drop shape; outer lip thickens rapidly toward throat in adult shells. Colour extremely variable and a number of colour morphs have been recognized: *olivacea* (uniform light to dark green), *citrina* (uniform yellow to orange, with white throat) and *reticulata* (green, brown, or black, with reticulate pattern) are most common. Up to 15 × 17 mm. Body colour tends to follow that of shell; cephalic tentacles with two longitudinal lines and no transverse stripes.

On weeds, especially *Fucus vesiculosus* and *Ascophyllum nodosum*, from MHWN to MLWS; occasionally in shallow sublittoral. Tolerant of low salinity and a wide range of exposure. Distributed from western Mediterranean to northern Norway; on all British coasts.

Littorina mariae Sacchi and Rastelli
Flat periwinkle Fig. 10.9
Shell similar to *L. obtusata*. Spire reduced or absent, last whorl constituting most or all of shell. Aperture with a more restricted throat; outer lip impinges high on last whorl, almost at right angle to it. Up to 11 × 12 mm.

Tends to live lower on the shore than *L. obtusata*, otherwise habitat and distribution similar.

Littorina neritoides (Linnaeus)
Small periwinkle Fig. 10.9
With prominent, flat-sided spire and pointed apex; periostracum imparting a surface bloom, rather like that on black grapes, to all but youngest shells. Aperture tapers sharply towards apex; outer lip meets body whorl tangentially, with thin periostracal extension along margin. Black or dark brown; last whorl occasionally with spiral banding or vertical stripes. Uniformly pale shells may occur. Operculum and interior of aperture dark. Up to 9 × 7 mm. Cephalic tentacles lack transverse stripes and have two longitudinal dark lines on dorsal surface.

In crevices and empty barnacle cases on upper shore and in splash zones; often several metres above EHWST on exposed shores. Distributed from Black Sea to western Norway. Absent from southern North Sea; on British Channel coast it does not penetrate eastwards beyond Isle of Wight. Local distribution may be patchy.

Littorina nigrolineata Gray Fig. 10.9
Shell with well-developed spire of moderately tumid whorls; with strap-like spiral ridges, those on last whorl with a shallow central groove. Outer lip meets last whorl almost at right angle, margin slightly crenulate, internally thickening rapidly towards throat. Whitish, yellow, or deep orange-red, often with dark-brown pigmentation between spiral ridges; peristome white, throat brown. Up to 30 ×

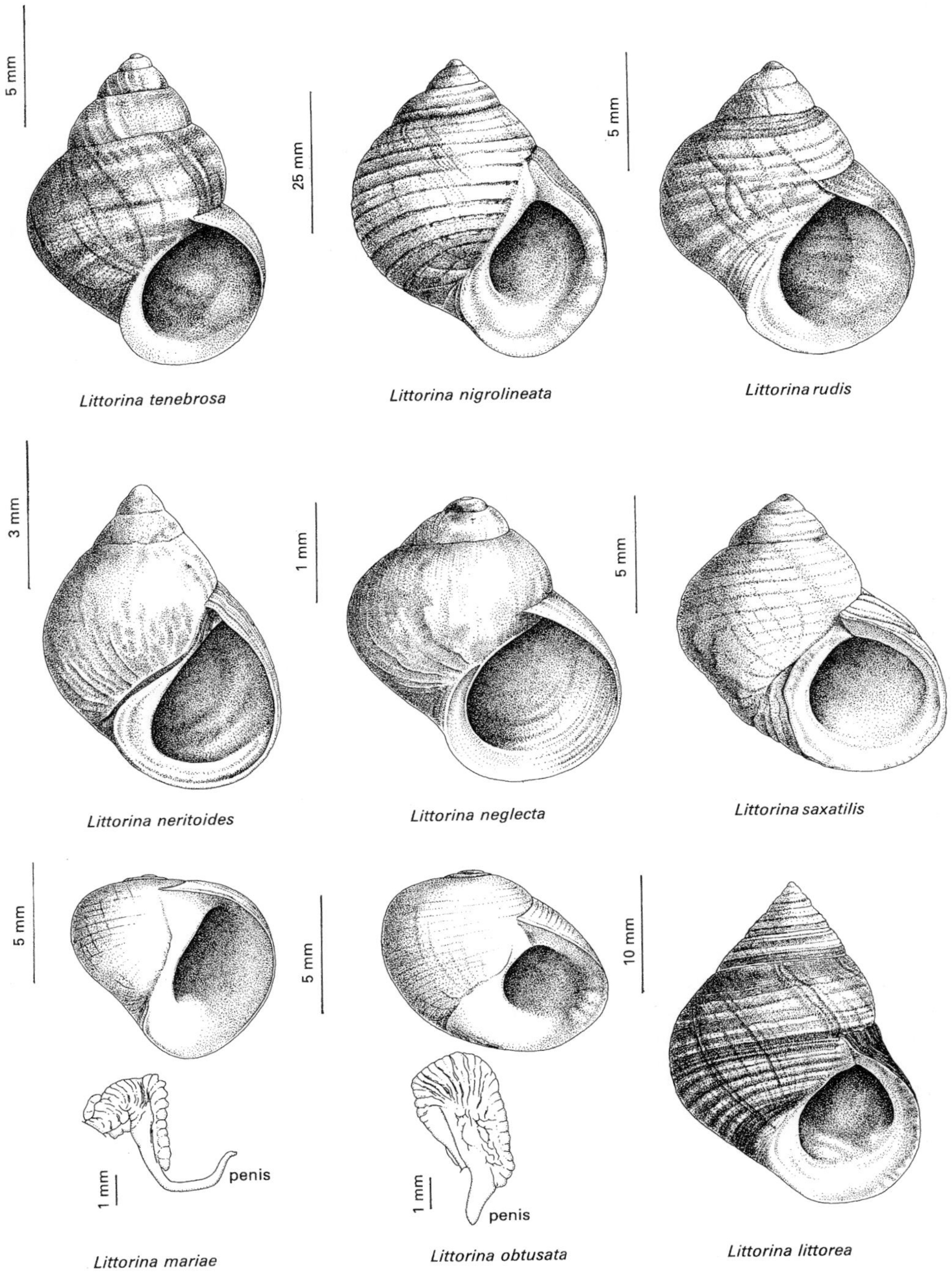
5 mm
Littorina tenebrosa
25 mm
Littorina nigrolineata
5 mm
Littorina rudis
3 mm
Littorina neritoides
1 mm
Littorina neglecta
5 mm
Littorina saxatilis
5 mm
penis
1 mm
Littorina mariae
5 mm
penis
1 mm
Littorina obtusata
10 mm
Littorina littorea

Fig. 10.9

27 mm, often less. Penis with very short tip; female oviparous, ovipositor conspicuous below right eye.

Among weeds and stones; higher on exposed shores (barnacle/*Fucus spiralis* zone), extending down to *Ascophyllum* belt in sheltered areas. Distribution imperfectly known; probably occurs on all British coasts other than south-east.

Littorina neglecta Bean Fig. 10.9

Globular, rather thin, superficially smooth, with short, blunt spire. Aperture very rounded, outer lip thin, meeting last whorl almost at right angle. Pale with dark, often tessellated markings, or brown with light areas; frequently a dark spiral band, bordered by cream, follows periphery of last whorl and enters aperture. Columella region and throat brown, base of aperture white. Up to 5 × 4.5 mm. Cephalic tentacles without transverse stripes, penis with elongate tip, female ovoviviparus, with brood pouch.

Most common on exposed or moderately exposed shores, in empty barnacle shells and crevices towards lower limits of barnacles, and below (recorded from *Laminaria* holdfasts and *Corallina*). Less common on sheltered shores but may be abundant in saltmarshes. Distribution imperfectly known; on western British coasts but may be more widespread.

Littorina saxatilis (Olivi)
Rough periwinkle Fig. 10.9

With three or four tumid whorls. Sculpture variable, either superficially smooth or with strong ∧-shaped spiral ridges, often symmetrical in profile. Aperture large, rounded, outer lip everted slightly where it meets columella. Colour variable, usually with some pattern (tessellation or spiral lines) and with dark throat. Up to 18 × 14 mm. Ovoviviparous. Extensive ciliary field (red in live animals) between genital tract and edge of columella muscle; prostate extends beyond anus, penis with elongate tip and variable number (4–26) of small glands arranged in one or two rows along its length but rarely extending to its curved base.

In *Pelvetia* zone and above, in crevices and empty barnacle shells. Widely distributed, from Mediterranean to Arctic.

Littorina rudis (Maton) Fig. 10.9

With 4–5 tumid whorls. Usually with strong spiral ridges, strap-like or ∧-shaped in profile; ∧-shaped ridges frequently asymmetrical, with crest of ridge lying towards shell apex. Aperture with prominent spout where outer lip meets columella. Shell usually monochrome, without patterning or dark throat. Up to 18 × 14 mm.

In *Pelvetia* zone, lower than *L. saxatilis*, preferring damp places among stones rather than dry crevices. Widely distributed, from Mediterranean to Arctic.

Littorina tenebrosa (Montagu) Fig. 10.9

Shell thin, sometimes transparent, with 5–6 tumid whorls, deep sutures and somewhat turreted profile. Superficially smooth but with growth lines and very fine spiral ridges. Aperture slightly squared, outer lip everted below columella but not forming prominent spout as in *L. rudis*. Usually dark, often with a tessellated pattern; columella and throat brown. Body similar to *L. rudis*, of which it may be a brackish water ecomorph.

Characteristically on weed in sheltered areas with reduced salinities; often permanently submerged. Distribution uncertain, probably widespread on British and European coasts.

Littorina arcana Hannaford Ellis

Shell essentially like *L. rudis*. Female oviparous; presence of glandular oviduct and absence of brood pouch can be used to separate females of this species from those of *L. saxatilis* and *L. rudis*, with which it is found. In males the prostate does not extend to level of anus, penial glands frequently extend to, or around, curved base of penis; ciliated field adjacent to genital tract is neither extensive nor red, as in *L. saxatilis* or *L. rudis*.

In upper barnacle and *Verrucaria* zone, particularly on exposed shores. Probably widely distributed.

11 HYDROBIIDAE

Shell small, conical, smooth; aperture ear-shaped or oval, with complete peristome; operculum spiral and horny. Animal with long, bifid snout; cephalic tentacles long, slender, setose; mantle edge with or without pallial tentacle; foot folding transversely into anterior and posterior halves, latter with a median mucus gland.

---

1. Umbilicus absent or very much reduced, whorls tumid, sometimes with spiral keel and/or periostracal bristles; snout

dark with light transverse band anteriorly, tentacles pale, dark at base. Mantle edge without pallial tentacle. In fresh or brackish water **Potamopyrgus jenkinsi**

Shell with umbilical chink; mantle edge with pallial tentacle on right side. In brackish water **2**

---

2. Whorls flat-sided or only slightly tumid, last whorl angulate; tentacles with black ring near tip, snout with black transverse bar. Often on open shores **Hydrobia ulvae**

   Whorls more tumid, slightly transparent; pigmentation not as above. In more confined brackish situations **3**

---

3. Whorls distinctly tumid; tentacles with longitudinal black line near tip; snout with triangular black area between eyes; penis with flagellum **Hydrobia ventrosa**

   Whorls moderately tumid; spire moderately slender; tentacles with conical black marks near tip; penis without flagellum **Hydrobia neglecta**

Hydrobia ulvae (Pennant)
Laver spire shell Fig. 10.10
Elongate, conical, with 6–7 flat-sided or slightly tumid whorls, apex rounded; outer lip arises tangentially to last whorl, which is slightly angulate at periphery, especially in younger shells; inner lip reflected over columella, leaving a small umbilical chink. Periostracum horn-coloured, shell translucent white. Up to 6 × 3 mm. Body with long bifid snout, often with transverse band of dark pigment anteriorly and a shield-shaped pigmented area between cephalic tentacles; a single, small pallial tentacle projects from mantle edge near junction of last whorl and outer lip. Cephalic tentacles with dark pigmented band near tip; left tentacle thicker than right.

A brackish-water species favouring firm mud and muddy sand; most abundant above MTL but sublittoral to 20 m. Widely distributed from northern Norway to Senegal.

Hydrobia ventrosa (Montagu) Fig. 10.10
With five or six markedly tumid whorls; outer lip abuts almost at right angle to last whorl, not angulate at periphery; inner lip reflected over umbilicus, leaving a shallow groove. Colour as in *H. ulvae*. Up to 5 × 2 mm. Snout lacks dark transverse band; no dark ring on tentacles, a dark dorsal line often present near tips.

Not as common as *H. ulvae*, occurs in more brackish waters, preferring isolated lagoons, creeks, and ditches to open coasts. Patchily distributed from Norway and the Baltic to the Mediterranean and Black Sea.

Hydrobia neglecta Muus Fig. 10.10
More slender than *H. ventrosa*, whorls less tumid, apex blunt; spire slightly convex in profile. Up to 4 × 1.5 mm. Tentacles with cone-shaped areas of black pigmentation set less than their own length away from tentacle tips.

Habitat similar to that of *H. ventrosa*, though appears to be less tolerant of salinities below 10‰ and seems to favour the presence of *Potamogeton* or *Zostera*. First described in 1963 and subsequently recorded from North Sea coasts, Guernsey, and the western coasts of Scotland and Ireland.

Potamopyrgus jenkinsi (Smith)
Jenkins' spire shell Fig. 10.10
With six tumid whorls, often encrusted with black deposit. Last whorl large; aperture ear-shaped or oval; inner lip occludes umbilicus. Some shells, notably from brackish-water populations, have a spiral keel, with or without periostracal bristles; some have a spiral row of bristles but lack keel. Beneath black encrustations shell is pale horn colour. Up to 6 × 3 mm. Snout evenly and darkly pigmented, with pale transverse band anteriorly. Cephalic tentacles pale with darker base; mantle edge without pallial tentacle.

Common in freshwater and brackish (<16‰) habitats, often with *H. ventrosa* in the latter. Most populations comprise females only, which reproduce parthenogenetically, and males are rare. Widely distributed throughout Europe. In the British Isles it is scarce or absent in mid-Wales and large parts of the Scottish Highlands.

12 RISSOIDAE

Shell small (typically 4–5 mm or less), with conical spire; aperture ear-shaped or oval, usually with

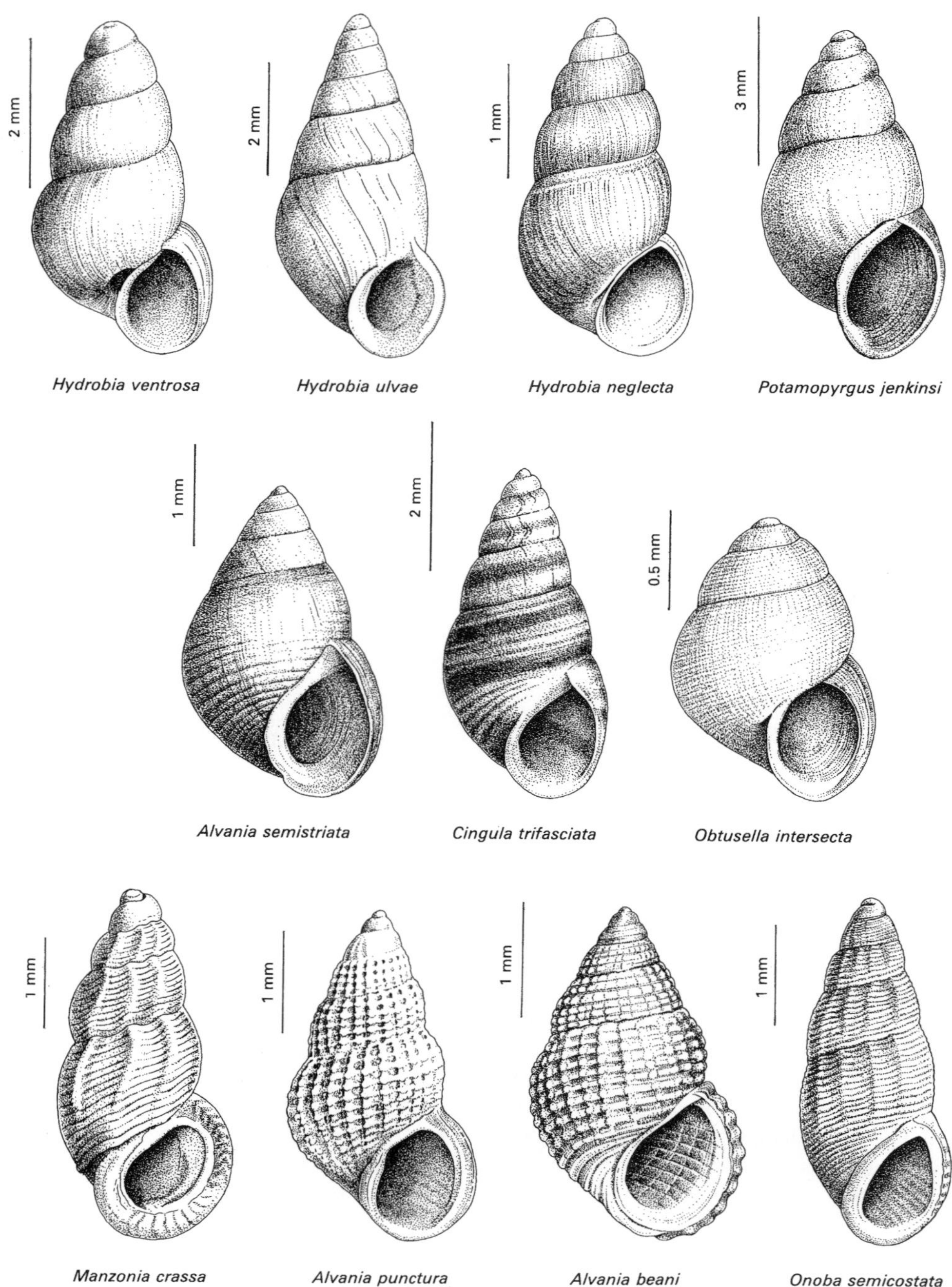
2 mm
Hydrobia ventrosa
2 mm
Hydrobia ulvae
1 mm
Hydrobia neglecta
3 mm
Potamopyrgus jenkinsi
1 mm
Alvania semistriata
2 mm
Cingula trifasciata
0.5 mm
Obtusella intersecta
1 mm
Manzonia crassa
1 mm
Alvania punctura
1 mm
Alvania beani
1 mm
Onoba semicostata

Fig. 10.10

complete peristome; operculum spiral and horny. Shell sculpture variable: smooth, or with spiral ridges, ribs, or reticulations; may vary within species according to environmental conditions. Body with long bifid snout; mantle edge smooth, with one or two pallial tentacles. Foot folds transversely, the posterior sole having a median mucus gland; a single metapodial tentacle projects from below rear of operculum. A large family with more than 30 species recorded from Britain. All are small, generally less than 5 mm long, and often inconspicuous.

---

1. Shell with reticulate pattern of pits on last whorl; with or without ribs **_Rissoa violacea_**

   Shell ornament not so **2**

---

2. Shell without ornament, or with spiral ridges, sometimes with ribs on subsutural parts of whorls only **3**

   Shell with ribs and spiral ridges, the latter conspicuous or microscopic; ribs extend from suture to suture on whorls of spire and reach at least to periphery of last whorl; uppermost whorls of spire may lack ribs **10**

---

3. Shell slender, breadth <50% of height **4**

   Shell oval–conical, breadth ⩾ 50% of height **5**

---

4. Shell with conspicuous, strap-like spiral ridges and short, subsutural ribs; last whorl >60% of shell height **_Onoba semicostata_**

   Shell thin, smooth (spiral ridges microscopic), without ribs; last whorl <50% of shell height; brackish habitats **_Rissoa membranacea_**

---

5. Shell uniform yellowish or white, with numerous, microscopic spiral ridges; globose, with blunt apex **_Obtusella intersecta_**

   Shell with spiral or transverse colour bands (or both) on last whorl; apex blunt or sharp **6**

---

6. Shell yellowish-white, usually with three spiral bands of reddish-brown on last whorl, no transverse pigmentation; spiral ridges on last whorl, conspicuous at base; outer lip arises tangentially to last whorl **_Cingula trifasciata_**

   Shell without this combination of characters **7**

---

7. Shell with two rows of transverse, reddish-brown marks on last whorl; whorls almost flat-sided, with fine strap-like spiral ridges below sutures and at base of last whorl; outer lip arises tangentially to last whorl **_Alvania semistriata_**

   Shell without this combination of characters **8**

---

8. Shell often with prominent, white varix behind outer lip, traversed in most mature shells by a brown comma-shaped mark (juveniles lack mark and varix, aperture with angulate periphery); colour variable, spiral bands with or without transverse marks, apex diffuse lilac; up to 5 mm long and eight whorls **_Rissoa parva_ var _interrupta_**

   Shell without this combination of characters; transverse brown marks on penultimate and last whorls; up to 3 mm long, five or six whorls **9**

---

9. *(In their non-ribbed form the next two species are difficult to separate on shell alone)*

   Apex bronze; shell slightly globular, whorls tumid, sutures deep; often with labial varix; metapodial tentacle flattened **_Rissoa sarsi_**

Apex dark purple; shell not globular, whorls slightly tumid; often without labial varix; metapodial tentacle cylindrical ***Rissoa inconspicua***

---

10. Uppermost whorls without transverse ornament, other whorls ribbed to a variable degree; metapodial tentacle single and filiform; pallial tentacle on right **11**

All whorls with transverse ornament (may be eroded in worn shells); metapodial tentacle single or triple; pallial tentacle on left and right **15**

---

11. Shell markedly convex, whorls flat-sided, last whorl about 60% or more of height; peristome everted, throat thickened, columella with small tooth opposite umbilical region; labial varix large, white; marine habitats to 35–20‰ ***Rissoa membranacea***

Shell not like this **12**

---

12. Last and penultimate whorls with 18–30 thin ribs and numerous fine spiral ridges; periphery of whorls may be reticulate; protoconch dark purple ***Rissoa inconspicua***

Last whorl with 8–25 broad ribs, smooth or with microscopic spiral ridges in between **13**

---

13. Mature shells usually with comma-shaped mark crossing labial varix (juveniles lack mark and varix, periphery of aperture angulate); apex lilac; ribs broad-based, with narrow crests, those on last whorl extending to spiral ridge at periphery ***Rissoa parva***

Shell never with comma-shaped mark; aperture never angulate **14**

---

14. Shell solid; upper whorls flat-sided, spire concave; ribs broad, with broad summit, 10 on penultimate whorl; lilac tinge to apex and peristome; south-west shores only ***Rissoa guerini***

Shell semi-transparent; whorls tumid, spire slightly convex; ribs do not link with spiral ridge at periphery of last whorl; apex bronze, penultimate and last whorl often with brown transverse bands ***Rissoa sarsi***

---

15. Ribs strong and flexuous, crossed by narrow, strap-like spiral ridges (may be eroded at crests of ribs); base of last whorl with two strong spiral keels; peristome thick and ridged ***Manzonia crassa***

Ribs and spiral ridges of equal prominence; shell with reticulate pattern **16**

---

16. Mature shell with ridges on inside of outer lip; rather coarsely reticulate, with strong labial varix; no umbilicus ***Alvania beani***

Inside of outer lip not ridged; shell finely reticulate, labial costa slight; umbilical chink ***Alvania punctura***

Cingula trifasciata (Montagu) Fig. 10.10
Regularly conical, with six or seven slightly tumid whorls, apex rather blunt; last whorl with spiral ridges, often restricted to, and more obvious at, base. Aperture ear-shaped, with shallow anal sinus where outer lip abuts tangentially to last whorl; umbilicus occluded by inner lip, though umbilical groove remains. Yellow-brown, with three spiral brown bands on last whorl, one or two on whorls of spire; occasionally uniform black or dark purple; var. *rupestris* uniform cream or white. Up to 4 × 2 mm.

Common in silty crevices or under stones from MTL downwards. The var. *rupestris* is confined to crevices and appears to favour exposed shores; it does not seem to mix with banded shells. Bay of Biscay to west coast of Norway. On Channel and British North Sea coasts; absent from continental shores of North Sea.

Obtusella intersecta (Wood) Fig. 10.10
Globose-conical, with four or five tumid, semi-transparent whorls; spire profile convex, with blunt tip. Fine reticulate sculpture visible with × 10 hand lens. Distinct umbilical groove leads to small umbilical chink. Uniform buff or pale horn. Up to 2 × 1.4 mm.

At LWST and below; locally common amongst algae and on sandy and gravelly bottoms from 10 to 60 m. Distributed from Iberia to Norway; absent from Baltic and eastern shores of North Sea.

Onoba semicostata (Montagu) Fig. 10.10
Cylindrical with blunt apex; five or six slightly tumid whorls, with spiral ridges and, on apical parts of whorls, truncated ribs; ribs absent in var. *ecostata*. Aperture small, oval, with thickened peristome. Inner lip reflected over umbilical groove; umbilicus may be present in young shells. Off-white, often with two red-brown bands on last whorl. Periostracum may be darker, even black. Up to 4 × 1.8 mm, most commonly 2 × 1 mm.

Gregarious, under stones, among corallines and in other silty areas on rocky shores, from LWM and sublittoral to 100 m. From Mediterranean to Norway; common, especially in summer, on all suitable British shores.

Manzonia crassa
(Kanmacher in J. Adams) Fig. 10.10
Conical, white, glossy, with six tumid, semi-transparent whorls; apical whorls smooth or finely reticulate, contrasting with strong flexuous ribs and spiral ridges on lower whorls. Spiral ridges accentuated on base of shell, forming two prominent spiral keels; aperture small, oval, with peristome and thickened labial varix; a deep umbilical groove but no umbilicus. Up to 3.5 × 2 mm. Pallial tentacle on left and right.

Fairly common among weeds and under stones at LWM; more common sublittorally, to 50 m, on sandy bottoms. Distributed from Mediterranean to Norway; absent from continental coasts of North Sea.

Alvania beani (Thorpe) Fig. 10.10
Oval–conical, with up to seven rather flat-sided whorls; sutures deep, profile markedly convex. With strap-like ribs and spiral ridges producing conspicuous reticulations. Aperture pear-shaped, outer lip with well-developed varix crossed by spiral ridges, internally thickened and ridged in mature shells; inner lip reflected over umbilical region leaving only slight groove; no umbilicus. Pale orange or cream, often with two darker spiral bands on last whorl. Up to 3.5 × 2 mm. Pallial tentacle on left and right.

On weeds and stones in sediment-rich areas; sublittoral to 50 m but occasionally at LWST. Not uncommon. Distributed from Canary Islands and Mediterranean to northern Norway.

Alvania punctura (Montagu) Fig. 10.10
Oval–conical, with six tumid whorls, apex pointed; profile slightly convex. Apical whorls lightly tuberculate, contrasting sharply with decussate pattern of lower whorls; ribbed element of decussation reduced below periphery of last whorl. Aperture oval or egg-shaped, outer lip with slight varix, not thickened or ridged internally; inner lip reflected over umbilical groove leading to narrow, elongate umbilicus. White, cream, or tinged reddish-brown, often darkened by conspicuous periostracum; red-brown blotches in three series on last whorl, umbilical region similarly stained. Up to 3 × 1.8 mm. Pallial tentacle on left and right.

Locally common among *Laminaria* at LWST, and sublittoral on sandy bottoms to 100 m. Distributed from Mediterranean to northern Norway; absent from Baltic and eastern basin of English Channel.

Alvania semistriata (Montagu) Fig. 10.10
Oval–conical, with five or six slightly tumid whorls. Fine strap-like spiral ridges visible mainly below sutures on spire and toward base of last whorl. Aperture oval, with thickened peristome; outer lip arises tangentially to last whorl; inner lip reflected over umbilical groove, umbilicus absent. Whitish or buff with two rows of red-brown, comma-shaped marks on last whorl. Up to 3 × 1.6 mm. Pallial tentacle on left and right.

Gregarious. Below stones, in fine weeds and in silted areas of rock pools below MTL; sublittoral to 100 m. Distributed from Mediterranean to Norway; on all British coasts, more common in south.

Rissoa parva (da Costa) Fig. 10.11
Oval–conical with seven or eight glossy, slightly tumid whorls. Two basic forms recognized: ribbed and smooth, the latter often referred to as var. *interrupta*. Uppermost whorls of ribbed shells rarely ornamented, lower whorls with strong, steep-sided ribs; alternation of smooth and ribbed whorls may

be erratic, and on some ribbed shells the last whorl may be smooth. Both ribbed and smooth shells have fine spiral ridges on most whorls. Aperture often angulate peripherally in young shells and with varix in mature shells. No umbilicus. Cream or buff to horn-coloured or chocolate-brown; apical whorls often tinged violet. Background colour of lower whorls may be enhanced by darker spiral streaks and blotches, in most mature specimens a prominent comma-shaped mark present on last whorl, commencing behind and passing through varix. Up to 5 × 3 mm. Pallial tentacle on right of mantle only.

Abundant below MTL and sublittoral to 15 m, under stones and on weeds on rocky shores; most commonly associated with corallines and other finely branched algae in which diatoms and detritus collect. The proportion of ribbed : smooth shells is higher on more sheltered shores, particularly in summer populations. Distributed from Mediterranean to northern Norway; absent from Baltic.

Rissoa inconspicua Alder Fig. 10.11

Oval-conical, with five or six glossy, slightly tumid whorls. Profile flat-sided to appreciably convex. With fine spiral ridges, most prominent at periphery of last whorl; with or without numerous (19–30), fine blunt ribs on all but upper three whorls. Ribs and ridges produce a fine reticulate pattern, most obvious at periphery of last whorl. A small umbilical groove but no umbilicus. Aperture oval, with distinct anal sinus. White or horn-coloured, with deep-purple apex. Younger whorls often with a series of red-brown streaks below suture; peristome red-brown. Up to 3 × 2 mm.

On weeds in low-shore rock pools, though more common sublittorally on algae and shell gravel to 100 m. Shows limited tolerance of brackish waters. Distributed from the Azores and Mediterranean to Arctic waters; in the Baltic as far as Lubeck Bay.

Rissoa sarsi Lovén Fig. 10.11

Oval–conical, with six tumid whorls; often with globular or inflated body whorl and short spire. With very fine spiral ridges and irregular growth lines, sometimes with ribs on younger whorls. Ribs usually weak, or even absent on youngest parts of last whorl, though a varix may be present. Ribs less numerous on penultimate whorl (9–16) than in *R. inconspicua*. Aperture oval, angulate, slightly drawn out towards apex; small umbilical chink. White, cream, or horn-coloured, with apical whorls often darker and sometimes with faint violet tinge (never dark purple/violet as in *R. inconspicua*). Penultimate and last whorl often with red-brown bands (lying between ribs in ribbed shells), which may or may not connect basally to a spiral brown band on last whorl; usually a dark-brown streak on columella.

On rocky shores among weed, especially *Codium* and *Zostera*. Occasionally intertidal but more usually sublittoral to 15 m. Shows some tolerance to brackish waters. Distributed from Mediterranean to northern Norway.

Rissoa guerini Récluz Fig. 10.11

Oval–conical, with eight whorls and slightly concave profile; upper four or five whorls flat-sided, or nearly so, younger whorls more tumid. Strong ribs confined to youngest three or four whorls; number variable on last whorl and may fade out toward aperture; 10 ribs on each of two preceding whorls. Varix broad, smooth, set a short distance back from outer lip. Fine spiral ridges between ribs, absent from upper, non-ribbed whorls. Slight umbilical groove. Cream, light horn-coloured, or brown with lighter ribs. Apical whorls often slate-coloured or lilac, may be streaked with reddish-brown. Lilac coloration also present in aperture. Up to 6 × 3 mm.

Locally common below stones, and on algae and *Zostera* at LWST and below. A southern species distributed along eastern Atlantic coasts between the Canary islands and south-western coasts of England and Ireland.

Rissoa violacea Desmarest Fig. 10.11

Rissoa violacea is an aggregate species confined to the Mediterranean. The segregates *R. lilacina* Recluz, *R. rufilabrum* Alder, and *R. porifera* Lovén occur on British shores. All are distinguished from other *Rissoa* species by the fine reticulation of the shell surface.

R. lilacina Récluz: oval–conical, with six or seven slightly tumid whorls, the youngest two or three ribbed. Ribbing incomplete on third whorl: apertural view will show ribs on last and penultimate whorls only; apart from a wide varix, ribs are slight or absent on lower part of last whorl. Whitish with longitudinal orange-brown stripes between ribs; apex and aperture with lilac tinge. Up to 5 × 2.75 mm.

Occasional or locally common on weeds and amongst sandy gravel from LWST to 50 m. A southern species extending to the western Channel

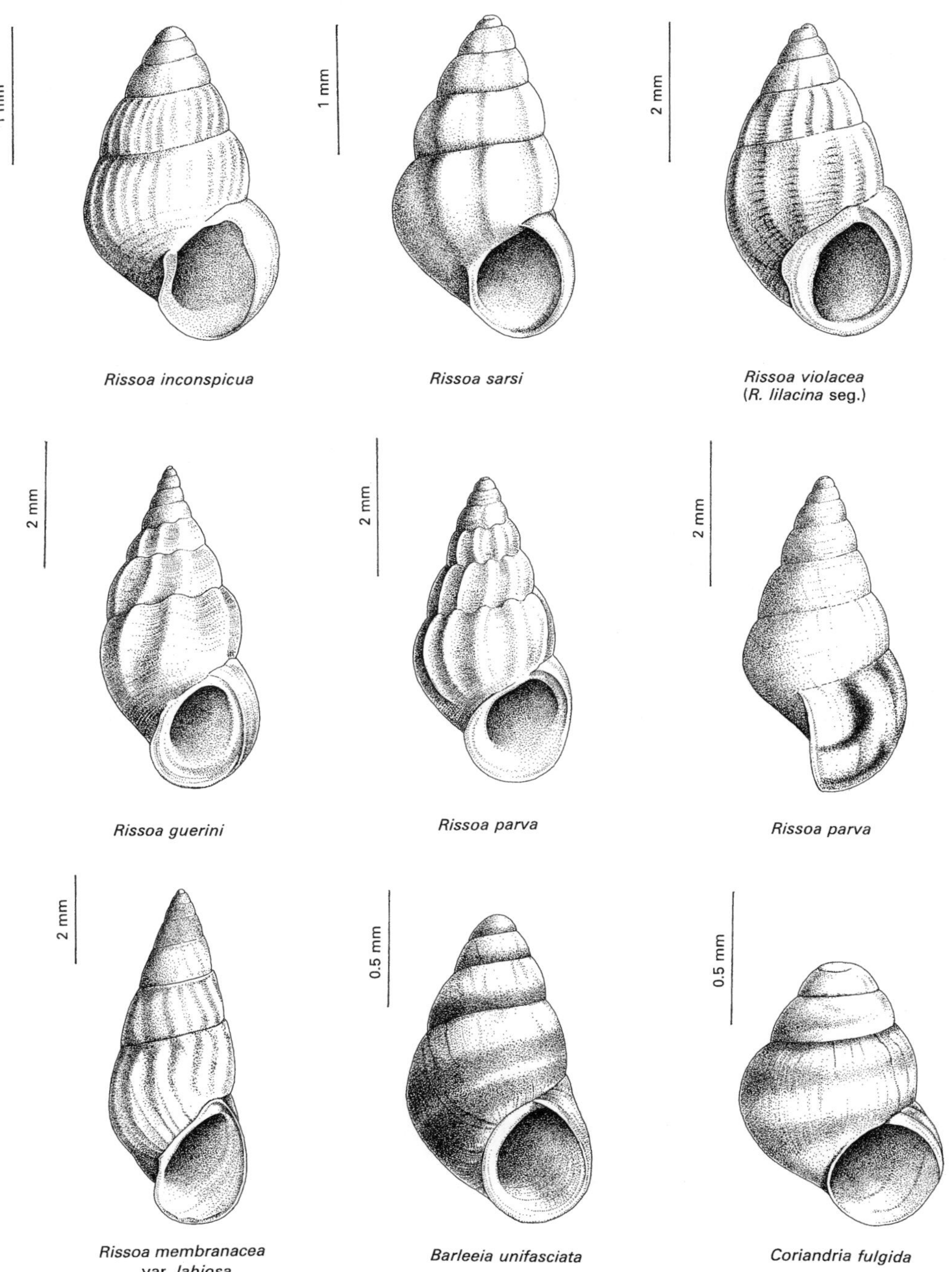

Rissoa inconspicua *Rissoa sarsi* *Rissoa violacea* (*R. lilacina* seg.)

Rissoa guerini *Rissoa parva* *Rissoa parva*

Rissoa membranacea var. *labiosa* *Barleeia unifasciata* *Coriandria fulgida*

Fig. 10.11

basin, the Welsh coast, western shores of Scotland, and the south and west coasts of Ireland.

R. rufilabrum Alder: ribs fewer (11–24 compared to 22–36 in *R. lilacina*) and, in apertural view, appearing to be confined to last whorl; ribs occasionally reduced or absent. Up to 5 × 2.75 mm. Habitat as for *R. lilacina*; distributed from Channel Islands to Norway.

R. porifera Lovén: without ribs, reticulate pattern prominent, visible on most whorls; varix slight. Horn-coloured, varix white, apex and peristome tinged brown or reddish-lilac. Up to 5 × 2.75 mm. Habitat as for *R. lilacina*; from western Ireland to Shetlands, Norway, and in the Kattegat.

Rissoa membranacea (J. Adams) Fig. 10.11
Shell form is extremely variable and several varieties have been described. Some variability relates in part to environmental conditions, though it is likely that *R. membranacea* is an aggregate species of which some of the major varieties are segregates.

Shell thin, with 8–9 flat-sided, or very slightly tumid whorls; apex pointed. With microscopic spiral ridges; lower whorls often with sinuous ribs, up to 10 on last whorl. Aperture rather triangular, with thin, everted peristome and broad, white varix; columella often with slight tooth-like prominence in umbilical region; inner lip reflected over umbilical groove. Light buff, greenish, or pale-horn coloured, with longitudinal yellow-brown markings; darker markings between ribs; apical region often tinged with violet. Up to 9 × 3 mm. Northern Britain, in brackish waters with salinities of 15–20‰.

var. *labiosa* Montagu: more solid with 5–7 flattened whorls. Last whorl with 15–18 ribs; tooth on columella pronounced, throat thickened. Southern Britain, in more marine situations (salinity 20–35‰).

var. *octona* Nilsson: slender, with spire of 10–12 tumid whorls; last whorl with 10 ribs, or none, comprising 40% of shell height. Ribbing, when present, may involve all but most apical whorls. Aperture oval without thickening of throat; columella tooth poorly developed. Northern Britain, in brackish waters (salinity 12–16‰).

var. *cornea* Lovén: with five or six slightly tumid whorls; last whorl approx. 60% of shell height, lightly ribbed. Aperture oval; columella tooth clear; throat thickened. Northern Britain in brackish waters (salinity 6–7‰).

All varieties are associated with *Zostera*, or weeds of similar form, at LWST and below, and are tolerant of a wide range of salinities. Widely distributed from Norway to Canary Islands, extending into the Baltic as far as Rugen Island.

13 BARLEEIDAE

Only one British representative of this family.

Barleeia unifasciata (Montagu)
Red spire shell Fig. 10.11
Oval–conical, smooth, with blunt apex and five slightly tumid whorls. Aperture oval; small umbilical groove, no umbilicus. Banded brownish-red and white, or uniform red, white, or crimson. Up to 2 × 1.5 mm. Mantle edge without pallial tentacles. Foot with dark-purple opercular lobes; operculum dark red, ear-shaped, and swollen.

Not uncommon among algae and in rock pools near LWM. Mainly confined to south-west British Isles and Ireland, but recorded from Scotland.

14 CINGULOPSIDAE

Only one British representative of this family.

Coriandria fulgida (J. Adams) Fig. 10.11
Globular, with three or four tumid whorls, last whorl two-thirds of shell height; sutures deep, apex blunt. Aperture large, almost circular; small umbilical groove, often with umbilicus. White, cream, or pale horn, with red-brown banding; usually three bands on last whorl, two on penultimate. Young shells mostly uniform dark-horn colour. Up to 1.5 × 0.9 mm. Foot with anterior and posterior mucus glands, animals often found suspended from weed or water surface by mucous strings; operculum spiral, yellow.

In rock pools and on fine weeds, MTL and below; more common in summer. Distributed from west coast of Scotland to Mediterranean. Absent from eastern Channel and North Sea; common on south and west coasts.

15 SKENEOPSIDAE

Only one British representative of this family.

Skeneopsis planorbis (Fabricius) Fig. 10.12
Very small (less than 2 mm diameter), semi-transparent, glossy; a depressed spire of four tumid, rapidly expanding whorls; umbilicus deep, wide. With irregular, sinuous growth lines, sometimes with fine spiral lines on upper surface of last whorl. Aperture almost circular, with thin-edged peristome. Red-brown or pale-horn coloured, with pale operculum. Up to 0.75 × 1.5 mm.

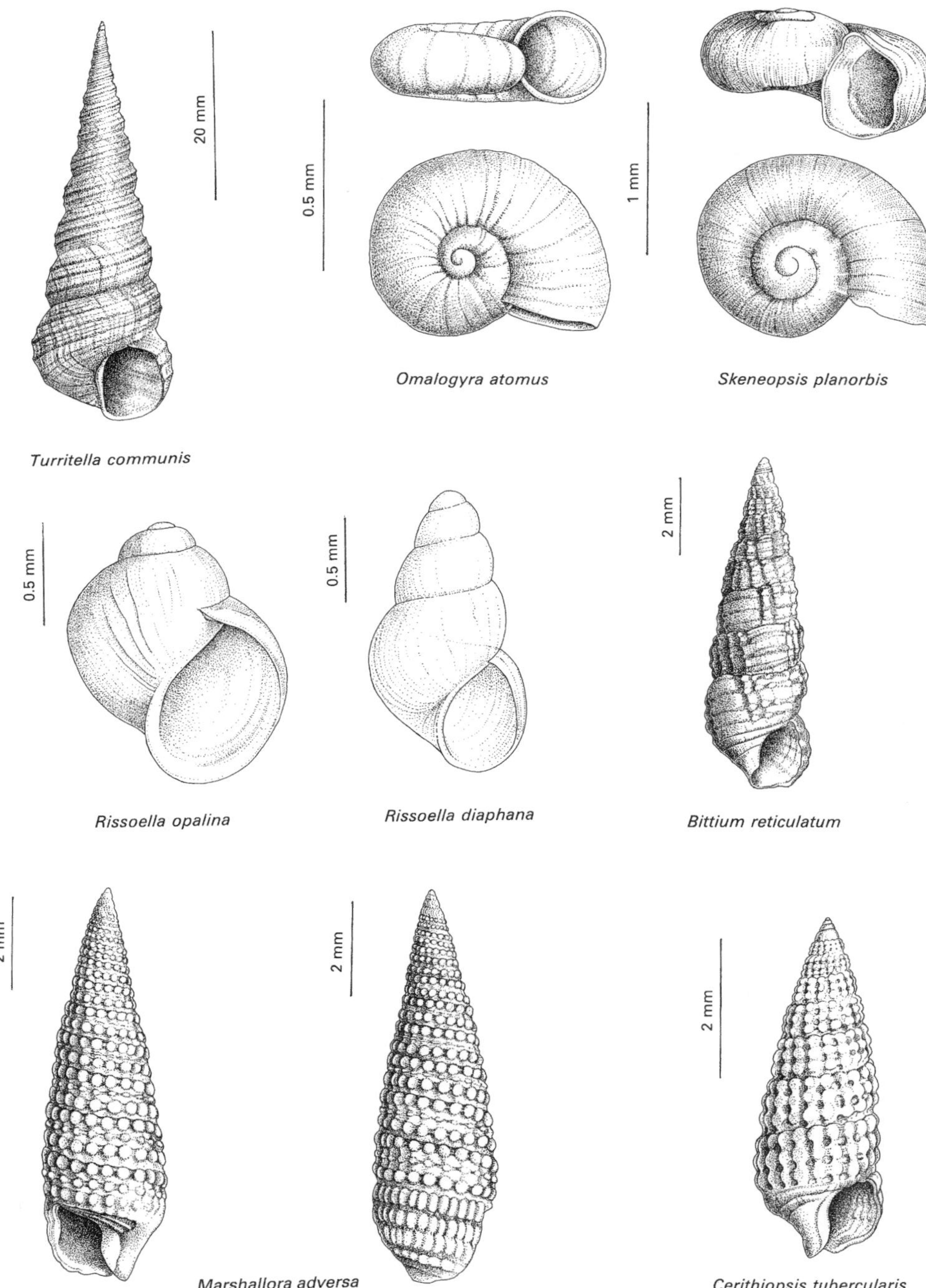
20 mm
0.5 mm
1 mm
Omalogyra atomus
Skeneopsis planorbis
Turritella communis
0.5 mm
0.5 mm
2 mm
Rissoella opalina
Rissoella diaphana
Bittium reticulatum
2 mm
2 mm
2 mm
Marshallora adversa
Cerithiopsis tubercularis

Fig. 10.12

On fine algae and in coralline pools from MTL to 70 m. Abundant in summer, rare in winter. Often found suspended from weeds or water surface by mucus. Widely distributed, from Azores to Arctic. On all suitable British shores; absent from eastern shores of North Sea and Baltic.

16 OMALOGYRIDAE

Shell very small (1 mm diameter), a biconcave planospiral of up to four whorls. Aperture circular. Snout broad, bilobed, and ciliate, with eyes at base of each lobe; cephalic tentacles absent. Foot bilobed anteriorly, with median pedal gland opening on sole. No ctenidium.

Omalogyra atomus (Philippi) Fig. 10.12

Shell with 2.5–3 rapidly expanding, tumid whorls of almost circular cross-section; with irregular growth lines and perhaps low spiral ridges. Aperture with peristome; nearly circular in outline, slightly flattened or concave where it abuts penultimate whorl. Reddish-brown, cream, or yellow. Up to 0.4 × 1.0 mm.

On algae, particularly *Enteromorpha* and *Ulva*; common in pools below MTL, often suspended by mucous threads from surface; extends to 20 m. Distributed from Azores to Norway; absent from Baltic and continental shores of North Sea.

17 RISSOELLIDAE

Shell very small (less than 2 mm), transparent and smooth; short, tumid, and conical, or globular. Umbilicus a shallow chink, or deep and round; operculum with peg-like process articulating with columella. Snout bifid, cephalic tentacles single or bifid. Foot with posterior mucus gland connected by groove to posterior margin of foot. Transparency reveals distinctive dark spots on mantle. Three species recorded for Britain; *R. globularis* (Jeffreys) is known only from Shetland and north-west Scotland.

---

1. Shell short, tumid, conical; aperture <50% shell height; body whorl ≤70% shell height. Tentacles single. Snout deeply bifid, almost tentaculate **_Rissoella diaphana_**

   Shell globular; aperture ≥50% shell height; body whorl ≥80% shell height; umbilicus a narrow chink; tentacles bifid **_Rissoella opalina_**

Rissoella diaphana (Alder) Fig. 10.12

Shell with 4.5 tumid whorls, slightly stepped at sutures; apex blunt. Aperture egg-shaped; inner lip slightly reflected over small umbilical chink. White; dark spots on mantle skirt particularly distinctive. Up to 1.5 × 0.85 mm. Snout deeply bifid, appearing tentaculate; cephalic tentacles cylindrical, finely setose. Mantle edge without pallial tentacles.

Common in summer among filamentous red algae and in coralline rock pools between MTL and LWST. Distributed from Mediterranean to Norway; absent from Baltic and most of North Sea.

Rissoella opalina (Jeffreys) Fig. 10.12

Globular, with three or four tumid, rapidly expanding whorls. Aperture almost D-shaped, columella straight, with small umbilical chink. Pale yellow or horn-coloured, with opalescent lustre; three dark spots on mantle show through in live shells. Up to 2 × 1.4 mm. Snout bifid, slightly lobed; cephalic tentacles bifid, setose.

Locally common in summer on filamentous red algae and in rock pools near LWST; often with *R. diaphana*. Distributed along western shores of British Isles from Channel Isles to Shetland; absent from North Sea coasts, east Channel and Baltic.

18 TURRITELLIDAE

Only one British representative of this family.

Turritella communis Risso.
Tower shell Fig. 10.12

Tall, sharply pointed, conical; 16–20 tumid whorls. With numerous spiral ridges, three (or up to six) usually more prominent than others, may have beaded appearance. Aperture small, angulate; outer lip crenulate. No umbilicus. Buff-coloured with lighter spiral ridges, may have lilac tinge on base. Up to 55 × 16 mm. Mantle edge with numerous pallial tentacles, associated with filter-feeding habit. Operculum small, circular, and concave, with numerous marginal bristles.

Locally abundant sublittorally, to 200 m, on soft bottoms; sometimes at ELWST. Distributed from North Africa and Mediterranean to Lofoten Isles; rare or absent from east Channel and southern North Sea.

19 CERITHIIDAE

Only one British representative of this family.

Bittium reticulatum (da Costa)
Needle whelk Fig. 10.12

Elongate, conical, up to 16 (usually 10–12) moderately tumid whorls; apex pointed. With strong ribs and spiral ridges, interacting to give oval tubercles: four rows on lower whorls, three on upper whorls; no tubercles on lowest spiral ridge of each whorl, or on basal ridges of last whorl. Aperture oval, drawn out to a short sinus basally; outer lip crenulate; no umbilicus. Buff to light chestnut-brown; tubercles, varices, and columella lighter. Up to 15 × 6 mm.

On soft substrata, associated with *Zostera* and other weeds; from LWST to 250 m. Locally abundant. Widely distributed from Canaries to Lofoten Isles; absent from continental shores of North Sea and east Channel.

20 CERITHIOPSIDAE

Only one species readily encountered in British waters.

Cerithiopsis tubercularis (Montagu) Fig. 10.12

Tall, narrow, conical, with up to 14 flat-sided whorls. Protoconch smooth or nearly so, lower whorls with three rows of close set, rounded tubercles; base of last whorl devoid of tubercles but with two spiral ridges. Aperture small, narrow, with short siphonal canal. Outer lip crenulate; no umbilicus. Chestnut-brown, tubercles and columella pale. Up to 6.5 × 2.25 mm.

On sponges (*Halichondria panicea* and *Hymeniacidon perleve*), particularly where associated with *Lomentaria*, *Corallina*, and other fine red algae; from LWST to 100 m. Most abundant in March and August–September. Distributed from Azores and Mediterranean to Norway; absent from southern North Sea.

21 TRIPHORIDAE

Shell sinistral (coiled anticlockwise); tower-shaped, with many flat sided whorls, sculptured by spiral rows of tubercles; sutures slight. Aperture squarish, with short siphonal canal; operculum multi-spiral.

Marshallora adversa (Montagu) Fig. 10.12

Last whorl with three rows of tubercles above two or three nodular spiral ridges; tubercles more rectangular as they approach outer lip. Penultimate and two or three whorls above with three rows of tubercles, central row being smallest; upper whorls with two rows. Chestnut-brown; tubercles, apex, and edge of outer lip often lighter. Up to 7 × 2 mm.

On, or embedded in, sponges (*Halichondria* and *Hymeniacidon*), under stones or on weeds from LWST to 100 m. Distributed from Biscay to Norway; absent from southern North Sea and eastern Channel basin.

22 APORRHAIIDAE

Shell tall, with ribs and fine spiral ridges. Last whorl in mature shells with prominent spiral rows of tubercles and a palmate extension to outer lip. Last whorl in juveniles (less than eight whorls) with a strong spiral ridge below periphery, and elongate-oval aperture, drawn out basally into a sharp point. Operculum narrow, set across foot. On sublittoral soft substrata.

---

1. Uppermost process of palmate extension does not reach shell apex; basal process hooked
   Aporrhais pespelecani

   Uppermost process of palmate extension projects beyond shell apex; basal process not hooked
   Aporrhais serresianus

Aporrhais pespelecani (Linnaeus)
Pelican's foot Fig. 10.14

Spire tall, whorls tumid, with crescentic ribs and fine, flat spiral ridges; apical whorls finely decussate or with spiral ridges only. Aperture in mature shells dominated by palmate outer lip; uppermost process fuses with lower part of spire, does not reach apex; basal process curved upward towards aperture (in juveniles this is the only process and resembles a sharply pointed siphonal canal). Cream or sandy-coloured, sometimes with purplish stain on back of last whorl; apertural surfaces pearly white. Up to 42 × 28 mm.

Sublittoral; locally common on mud and muddy sand to 180 m. Distributed from Mediterranean to northern Norway and Iceland; not common around Channel Isles and Scilly Isles.

Aporrhais serresianus (Michaud)
Serres' pelican's foot Fig. 10.14
Similar to *A. pespelecani*, though generally smaller and more delicate; adult shells distinguished by shape of palmate outer lip, which projects beyond shell apically and does not curve upward basally. White. Up to 38 × 24 mm.

Sublittoral, on fine mud down to 1000 m. Distributed from Mediterranean to northern Norway and Iceland; in British Isles it occurs off west coasts of Ireland, Scotland, Orkney, and Shetland.

23 TRICHOTROPIDAE

Only one common British representative of this family.

Trichotropis borealis
Broderip and Sowerby Fig. 10.13
Conical, with up to five tumid whorls; covered by a conspicuous periostracum. With prominent, chord-like spiral ridges from which the periostracum is drawn out into stiff bristles or triangular plates. Aperture rather triangular, tapering basally to a V-shaped spout; outer lip thin, crenulate, with periostracal margin; inner lip reflected over columella, forming a narrow umbilical groove. Whitish, periostracum light-horn coloured. Up to 11 × 6 mm.

Sublittoral, on stones and shells from hard bottoms; 10–270 m. Locally common. Circumboreal; occurs in northern North Sea, west coast of Scotland, north and west coasts of Ireland.

24 CAPULIDAE

Only one British representative of this family.

Capulus ungaricus (Linnaeus)
Bonnet shell Fig. 10.13
Cap-shaped, a rapidly expanding planospiral of up to three whorls; covered by periostracum. With numerous spiral ridges and fine growth lines; periostracum projects as fringe to aperture, successive positions of aperture marked by series of similar fringes on last whorl. Aperture large, oval, with finely serrated margin. White, with horn-coloured periostracum. Up to 40 × 25 mm across the aperture and 15 mm high. No operculum.

Sublittoral (occasionally at LWST), on stones and shells (especially those of living scallops and *Turritella communis*) down to 850 mm. Widely distributed from West Africa to northern Norway; less common in North Sea.

25 CALYPTRAEIDAE

Shell limpet-like, with sharp apex, or a depressed hemisphere of variable height; internal septum present. Ciliary feeders, with enlarged mantle cavity and ctenidium; large propodial flap on either side of neck. Foot large, oval, without operculum.

Calyptraea chinensis (Linnaeus)
Chinaman's hat Fig. 10.13
Limpet-like, rather depressed; apex with small nipple-shaped beak, often slightly coiled. With indistinct growth lines, sometimes with small tubercles. Aperture almost circular, with thin, bevelled edge. White or yellowish, with pearly white interior. Up to 15 × 5 mm.

Sublittoral, on stones and shell gravel associated with soft substrata, to 20 m; under stones at LWST on sheltered shores in south-west England. Distributed from north-west Africa and Mediterranean to west coast of Scotland (Loch Ryan); absent from North Sea and east Channel basin.

Crepidula fornicata (Linnaeus)
Slipper limpet Fig. 10.13
Oval, cap-shape; spire set posteriorly and to right; smooth, with irregular growth lines. Aperture elongate-oval, edge thin. Cream or pinkish, with streaks and blotches of reddish-brown; throat chestnut-brown, with white partition. Up to 50 × 25 mm.

On shells and stones on soft substrata; LWST to 10 m, often cast ashore by storms. The animals form curled stacks of up to 12 individuals; large shells at the bottom, becoming progressively smaller towards the apex of the chain. They are pests of oyster beds. An accidental introduction from America, now distributed from Essex along south coast to Bristol Channel, Belfast Lough, and Co. Kerry.

26 LAMELLARIIDAE

Shell wholly or partly covered by papillated mantle lobes; fragile, smooth, with up to three rapidly expanding whorls. No snout; eversible proboscis used to feed on tunicates and coelenterates. Foot with double-edged anterior margin; no operculum.

---

1. Shell with thick periostracum; mantle only slightly reflected over shell. With fine reticulation, aperture almost circular **_Velutina velutina_**

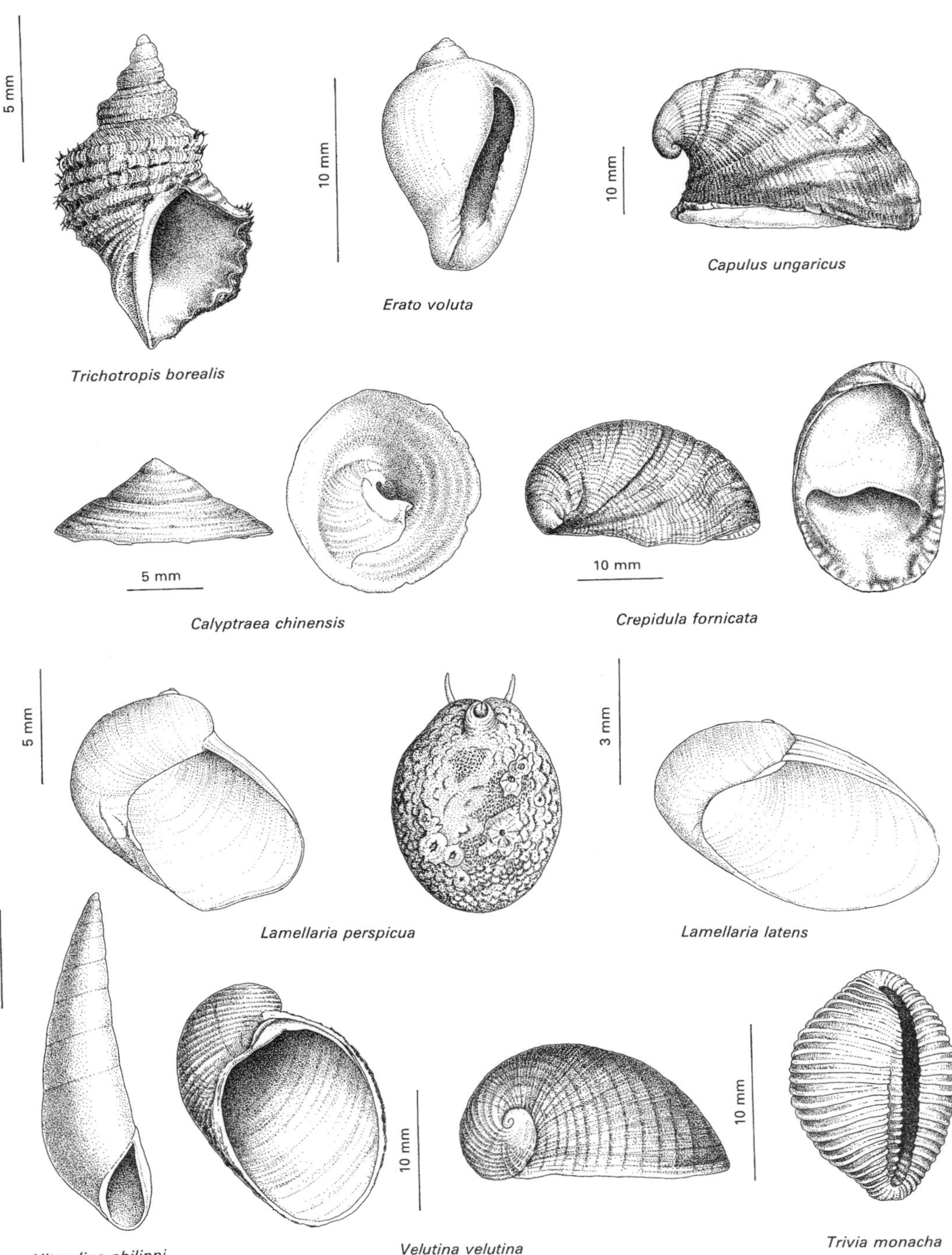
5 mm
10 mm
10 mm
Trichotropis borealis
Erato voluta
Capulus ungaricus
5 mm
10 mm
Calyptraea chinensis
Crepidula fornicata
5 mm
3 mm
1 mm
Lamellaria perspicua
Lamellaria latens
10 mm
10 mm
Vitreolina philippi
Velutina velutina
Trivia monacha

Fig. 10.13

Shell internal, covered by papillated mantle; resembles dorid sea slug but with anterior siphonal notch in mantle, smooth tentacles, and no dorsal circlet of gills **2**

---

2. Up to 20 mm long; dorsal mantle domed, tuberculate, purplish-grey with white flecks, occasionally yellowish; shell with high spire ***Lamellaria perspicua***

   Up to 10 mm long; dorsal mantle rather flat, smoother, yellowish with dark flecks; shell with low spire ***Lamellaria latens***

Velutina velutina (Müller)
Velvet shell Fig. 10.13
Shell fragile, semi-transparent, with up to three rapidly expanding, tumid whorls, thickly covered by periostracum; last whorl comprises most of shell, spire low. With fine spiral ridges and growth lines, producing reticulate pattern. Aperture almost circular, periostracum projects beyond margin of outer lip. White, periostracum dark brown. Up to 20 × 20 mm.

Associated with tunicates, particularly solitary forms such as *Styela*; normally sublittoral, to 1000 m, but may occur at ELWST. Circumboreal, extending down west coasts of British Isles to Mediterranean. *Velutina plicatilis* (Müller) is smaller than *V. velutina*, more slender, with similar distribution but rarer.

Lamellaria perspicua (Linnaeus) Fig. 10.13
Shell permanently covered by mantle; fragile, with two or three rapidly expanding whorls and short spire. Smooth, apart from growth lines, or with fine spiral lines. Aperture rather spoon-shaped, drawn out basally to right; peristome in marked inclined plane. Mantle grey or grey-lilac flecked with white, but buff or orange forms occur, which when flecked with black may resemble *L. latens*. Shell up to 9 × 10 mm; body up to 20 × 12 mm. Mantle edge thickened, with siphonal groove anteriorly; dorsal surface tuberculate.

Fairly common under stones, in pools and on surfaces covered by compound ascidians; LWST to 1200 m. From Mediterranean to northern Norway and Iceland; widely distributed around British Isles.

Lamellaria latens (Müller) Fig. 10.13
Shell permanently covered by mantle; fragile, with two rapidly expanding whorls; spire not obvious, profile rather depressed. Aperture like that of *L. perspicua* but relatively larger. Shell up to 5 × 9 mm. Body as in *L. perspicua*, mantle smoother and less domed in profile; up to 10 × 6 mm. Buff with brown or black flecks.

Habitat and distribution as for *L. perspicua*.

27 ERATOIDAE

Shell convolute (spire overgrown by last whorl) or with very short spire; robust, polished, enveloped by mantle lobes in active animals. Aperture long and narrow. No snout; eversible proboscis used to feed on compound ascidians. No operculum.

---

1. Shell convolute, no spire visible in mature shells; ridged **2**

   Shell with short spire; smooth or with slight spiral ridges **3**

---

2. Shell with three dark spots on abapertural side; body highly pigmented, back end of foot striped ***Trivia monacha***

   Shell without spots; body pale, back end of foot yellow ***Trivia arctica***

---

3. Shell smooth, thick, narrowing basally to give harp-shaped outline; mantle lobes heavily pigmented ***Erato voluta***

   Shell slightly ribbed, thin, broad basally, with oblong outline, mantle lobes rather pale ***Trivia* juvenile**

Erato voluta (Montagu) Fig. 10.13
Smooth, glossy, with three or four slightly tumid whorls and short spire; last whorl about 80% or more of shell height, harp-shaped in profile.

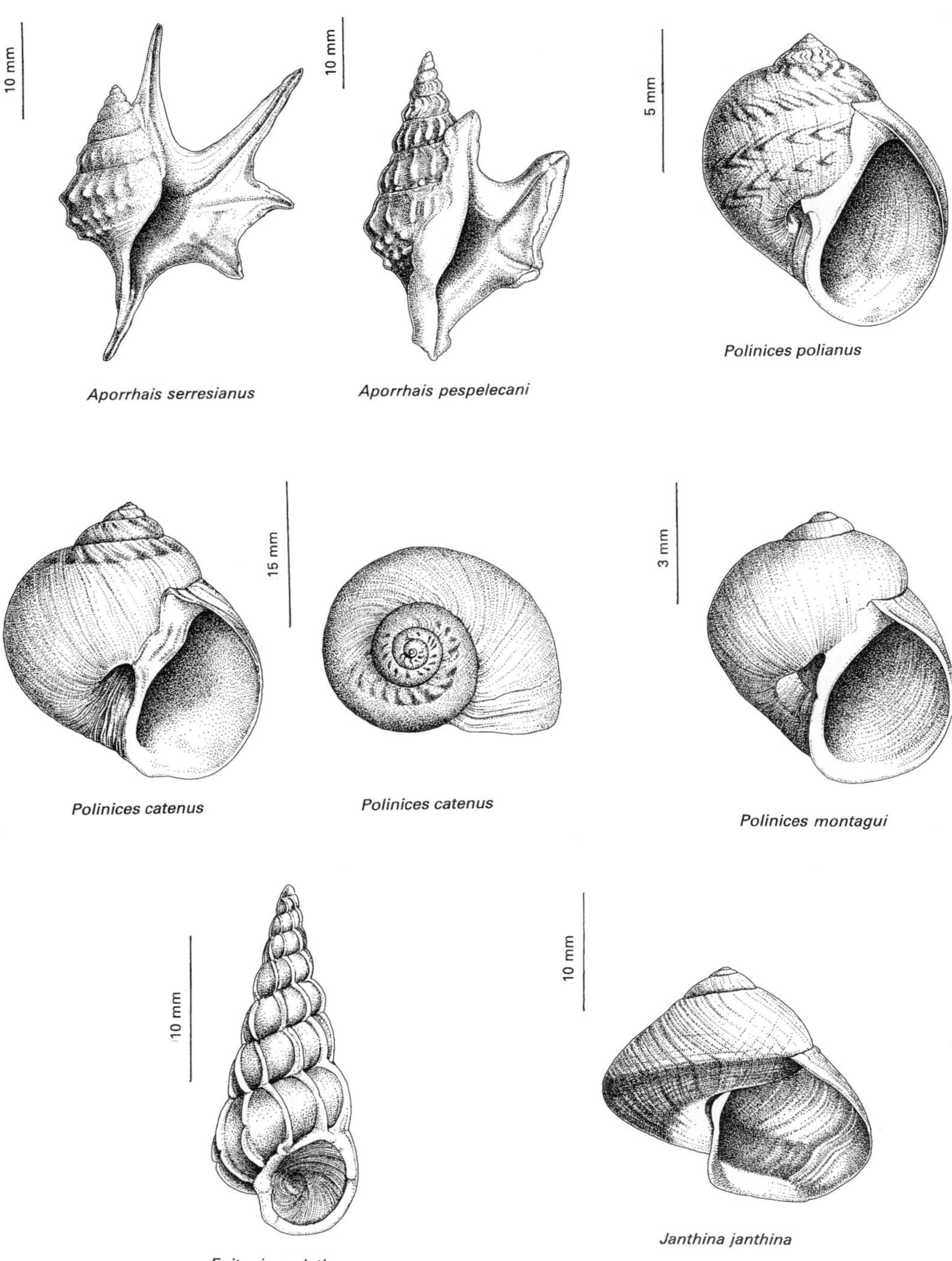
10 mm
10 mm
5 mm
Aporrhais serresianus
Aporrhais pespelecani
Polinices polianus
15 mm
3 mm
Polinices catenus
Polinices catenus
Polinices montagui
10 mm
10 mm
Epitonium clathrus
Janthina janthina

Fig. 10.14

Aperture long, narrow, with parallel sides and short siphonal canal; outer lip thick, with internal row of teeth; base of columella grooved. White. Up to 12 × 7.5 mm. Mantle drawn out into long siphon anteriorly, laterally into large tuberculate lobes which envelop much of shell in active animals.

Sublittoral, associated with ascidians on hard substrata from 20 to 150 m. Distributed from Mediterranean to Norway; widely distributed around British Isles, more common in southern parts of range.

Trivia monacha (da Costa)
Spotted cowrie Fig. 10.13
Adult convolute, egg-shaped, with flattened apertural side. With strong, often branching ridges and grooves, those at ends of shell almost U-shaped. Juveniles (up to 5 mm) with low spire of two or three tumid whorls; smooth at first, developing spiral ridges in older shells. Aperture narrow, running whole length of shell; turned to left (i.e. towards swollen body whorl) at ends, and ridged on both sides. Juvenile aperture widely open, narrowing towards apex, not ridged. Juveniles white, adults pinkish or reddish-brown, with three dark spots; apertural side white. Up to 12 × 8 mm. Mantle drawn out into long siphon anteriorly, extending laterally over all or most of shell in life.

Fairly common on rocky shores, under stones and ledges at LWST; associated with compound ascidians on which it feeds. Distributed from Mediterranean to British Isles, more common in south.

Trivia arctica (Pulteney) Northern cowrie
Similar to *T. monacha* but lacks pigment spots and is usually smaller. Up to 10 × 8 mm.

Mainly sublittoral, to 100 m in north of range, 1000 m in south; occasionally at LWST, associated with compound ascidians. Distributed from Mediterranean to Norway; all British coasts, more common in north.

28 NATICIDAE

Shell smooth, globular or oval, with low spire and enlarged last whorl; aperture semicircular, umbilicus prominent. Foot enlarged, partially covering shell in creeping animals; operculum large, horny in British species. Animal ploughs through soft substrata in search of bivalve prey, whose shells it bores. Spawn a characteristic collar or necklace-shaped ribbon, stiffened with sand; commonly known as necklace shells. Six species, in two genera, recorded from Britain, but only three *Polinices* species common.

---

1. Outer lip arises more or less at right angle to last whorl; sutures well defined; umbilicus rounded; last whorl with subsutural row of brown marks (young shells may have additional, but very faint, rows of similar marks) ***Polinices catenus***

   Shell not like this **2**

---

2. Outer lip arises tangentially to last whorl; sutures slight; umbilicus elongate; last whorl with five spiral rows of brown marks ***Polinices polianus***

   Outer lip separated from last whorl by deep notch; sutures deep, channelled; columella crossed by deep groove from umbilicus to aperture; shell pale, without pigmented spots ***Polinices montagui***

Polinices polianus (Della Chije)
Alder's necklace shell Fig. 10.14
Globose, with six or seven slightly tumid whorls; last whorl accounts for most of shell height; sutures shallow, spire almost flat-sided. With numerous, fine, growth lines; smooth to naked eye. Aperture large, almost semicircular, with peristome; outer lip arises tangentially to last whorl; inner lip partially occludes rather elongate umbilicus, and swells to fill angle between outer lip and last whorl. Buff or light-horn coloured, columella and base pale; five spiral rows of brown marks on body whorl, one row on whorls of spire. Up to 16 × 14 mm. Foot large, with expanded propodial and opercular lobes; in active animals propodium forms a ploughshare, covering anterior of shell, opercular lobes being drawn up around posterior.

On sand and gravel; usually sublittoral from 10 to 50 m (recorded from 2000 m) but occasionally at LWST. Distributed from Mediterranean to Norway; common on all British coasts.

Polinices catenus (da Costa)
Spotted necklace shell Fig. 10.14
Similar to *P. polianus* Spire more elevated, with stepped profile and well-defined sutures. Outer lip arises almost at right angle to last whorl; umbilicus deep, round, occlusion by inner lip very slight. Buff, lighter basally and on inner lip, with subsutural row of reddish-brown marks. Younger shells may have additional, but very faint, rows of marks on last whorl. Up to 30 × 30 mm. On sand; LWST to 125 m. Distributed from Mediterranean to Skagerrak; common on all British coasts; absent from coasts of Norway and Sweden.

Polinices montagui (Forbes)
Montagu's necklace shell Fig. 10.14
Similar to *P. polianus.* Spire rather low, with stepped profile and deeply channelled sutures, producing notch between outer lip and last whorl; inner lip crossed by groove passing from umbilicus to aperture. Light fawn, with darker throat; no spiral rows of brown marks. Up to 8 × 8.5 mm. Sublittoral, on sandy and muddy bottoms; 15–200 m. A northern species extending down western coasts of British Isles to Mediterranean. More common in north of range.

29 EPITONIIDAE

Shell tall, narrow, conical, with many tumid whorls. Sutures deep, or whorls completely separate, connected only by thin ribs. Aperture small, round. Carnivorous, most often feeding on anthozoans. Five British species.

Epitonium clathrus (Linnaeus)
Wentletrap Fig. 10.14
Up to 15 whorls, circular in section, not meeting at sutures but held together by fusion of thin, crescent-shaped ribs; shell smooth between ribs, which are widely spaced and prominent. Aperture with thickened margin; inner lip reflected over bases of seven ribs. Off-white, often with brown spiral bands. Up to 40 × 12 mm.

Sublittoral, on sandy or muddy substrata to 70 m; may be found at LWST when spawning (long threads of pyramidal capsules). Distributed from Mediterranean to Norway and Kattegat; widely distributed, not uncommon around the British Isles. Three other species are known from Britain but none is common.

30 JANTHINIDAE

Shell violet, globose, thin, and fragile. The animals are pelagic, drifting at the surface of the sea by means of a mucous float secreted by the foot; gregarious, feeding primarily on the pelagic coelenterates *Porpita* and *Velella.*

Janthina janthina (Linnaeus)
Violet snail Fig. 10.14
Globose, last whorl with bluntly angulate periphery. Appears smooth, but with numerous regular growth lines and irregular spiral grooves. Aperture angulate at periphery and at base; outer lip with shallow peripheral sinus. Violet or bluish-purple, darker on base. Up to 40 × 30 mm. Foot with broad depression anteriorly from which the float is secreted; no operculum in adults.

Pelagic, with circumequatorial distribution; cast up on western shores of Britain and Ireland after prolonged south-westerly gales. Three other species of *Janthina* occur rarely on western coasts of Britain.

31 EULIMIDAE

Shell long, awl-shaped, smooth, and highly polished; whorls numerous, flat-sided. Animals ectoparasitic on echinoderms, feeding by means of a long, eversible proboscis. Numerous species of eulimid have been recorded from Britain, almost all of them are associated with echinoderms. They are inconspicuous animals, most are uncommon and only one is considered here.

Vitreolina philippii (Rayneval and Ponzi)
Philippi's glass snail Fig. 10.13
Shell spire curved, often markedly; very glossy, whitish, and transparent when fresh. Up to 8 × 2.5 mm.

Common on soft substrata from LWST to 200 m, apparently associated with various echinoderms. Distributed from Mediterranean to Norway; absent from eastern Channel and perhaps also from southern North Sea.

Order Neogastropoda

32 MURICIDAE

Shell with ribs or spiral ridges; or both, giving rise to conspicuous tubercles. Aperture small, oval, with siphonal canal: often long and, in some species, forming a tube in mature shells; outer lip thickened, with internal teeth in mature shells. Proboscis eversible; no snout. Foot with accessory boring organ (ABO) anteriorly on sole. On hard substrata, boring the shell of their prey, often other molluscs, barnacles, and tube worms, by a combined action of radula and ABO. Most lay vase-shaped egg-capsules attached to hard surfaces.

---

1. Shell without spiral ridges **Boreotrophon truncatus**

   Shell with spiral ridges **2**

---

2. Shell without transverse ribs, often with prominent growth lines; outer lip curves without sharp inflection to tip of short siphonal canal **Nucella lapillus**

   Shell with transverse ribs; outer lip with marked inflection at base of siphonal canal **3**

---

3. Siphonal canal long (about 25% shell height), narrow, almost straight; penultimate whorl with 12 or more ribs **Trophonopsis muricatus**

   Siphonal canal long or short, not straight, open or closed in mature shells; penultimate whorl with fewer than 12 ribs **4**

---

4. Spire markedly turreted, ribs with distinct shoulder, which may be raised above level of suture in mature shells; siphonal canal long (about 25% shell height), broad and open in juveniles, closed and somewhat shorter in mature shells; two or three major spiral ridges on penultimate whorl, far apart, with minors in between **Ocenebra erinacea**

   Shell not like this **5**

---

5. Spiral ridges numerous (9–10 on penultimate whorl), approximately equal in width to intervening grooves; siphonal canal open; shell light-horn coloured, body cream; south-east coasts **Urosalpinx cinerea**

   Spiral ridges less numerous (6–8 on penultimate whorl), broader than intervening grooves; siphonal canal closed in mature shells; shell reddish-buff, body red; south-west coasts **Ocinebrina aciculata**

Boreotrophon truncatus (Ström) Fig. 10.15
Sharply pointed, slightly turreted, with sinuous, laminar ribs, bearing sharp, wave-like crests. Siphonal canal short (about 50% of aperture height), bending away from plane of aperture distally; outer lip thin, smooth internally. Yellowish with pink tinge, throat darker. Up to 15 × 7.5 mm.

On stony and gravelly bottoms, from LWST to 200 m. Not uncommon in dredged samples. Circumboreal, extending south into northern North Sea and down western coasts of British Isles to Biscay. Subfossil specimens of the Arctic species *B. clathratus* (Linn.) are also found around Britain.

Trophonopsis muricatus (Montagu) Fig. 10.15
Narrow conical, whorls tumid. With ribs, spiral ridges and fine growth lines; ribs numerous (15–20 on penultimate whorl), narrow, interacting with cord-like spiral ridges to give a deep reticulation set with raised tubercles. Siphonal canal long, straight, about equal to aperture height; outer lip thin, crenulate and grooved internally. Yellowish, often with spiral brown bands; sometimes colourless. Up to 19 × 8 mm.

Sublittoral, on muddy, sandy, and gravelly bottoms from 20–300 m. Not uncommon. Distributed from Mediterranean to south and west coasts of British Isles. The coarsely ribbed *T. barvicensis* (Johnston) occurs around northern Britain.

Nucella lapillus (Linnaeus)
Dog whelk Fig. 10.15
Oval–conical, with short spire; whorls slightly tumid, last whorl large (80% or more of shell height). With broad, low spiral ridges, becoming prominent basally as a keel over siphonal canal; in juveniles and in animals from sheltered habitats growth lines may be raised to give frilled surface to shell (var. *imbricata*). Siphonal canal short, partially closed by columellar fold; outer lip curves in a more

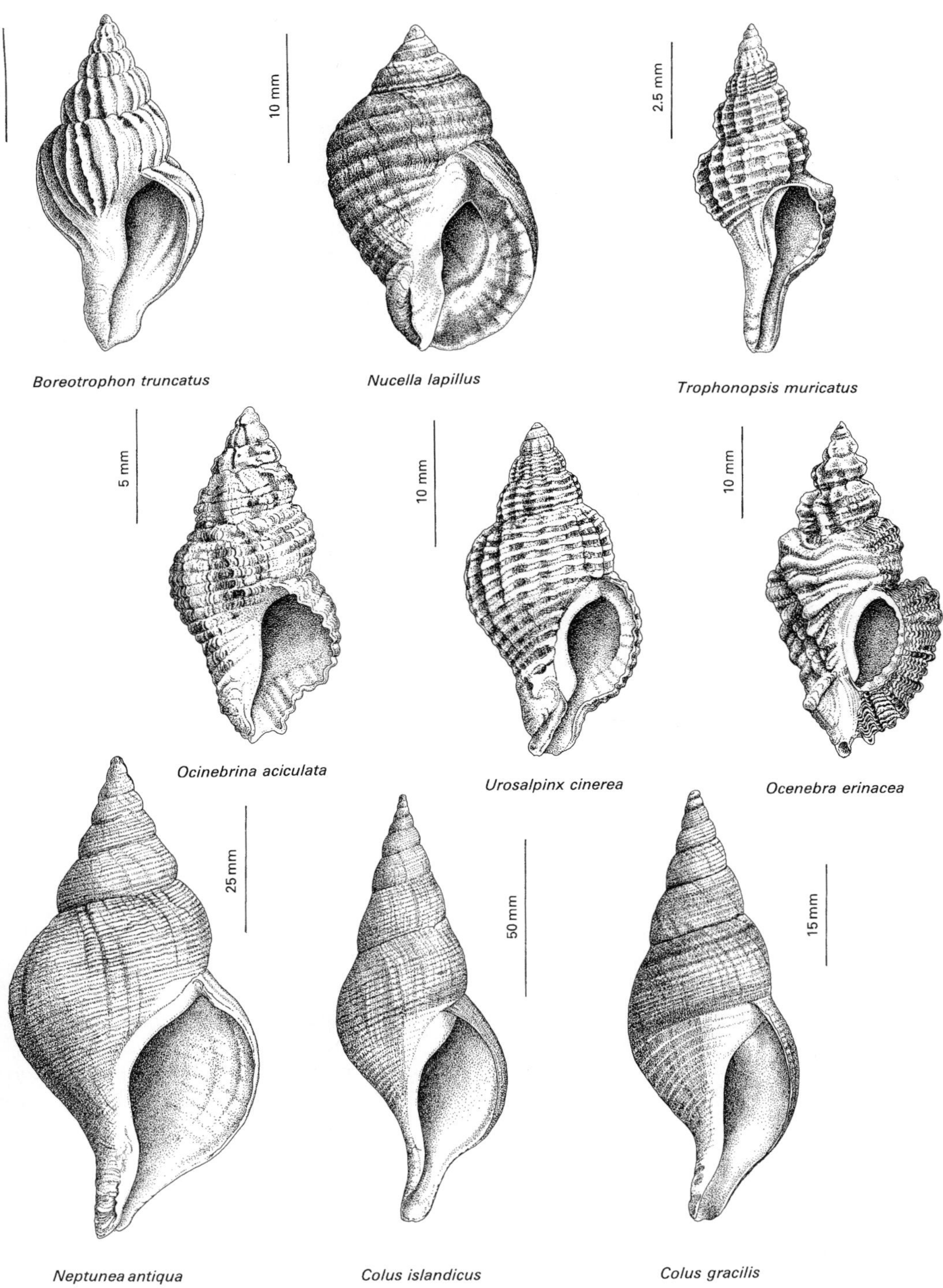
5 mm
10 mm
2.5 mm
Boreotrophon truncatus
Nucella lapillus
Trophonopsis muricatus
5 mm
10 mm
10 mm
Ocinebrina aciculata
Urosalpinx cinerea
Ocenebra erinacea
25 mm
50 mm
15 mm
Neptunea antiqua
Colus islandicus
Colus gracilis

Fig. 10.15

or less constant arc to tip of siphonal canal, thin and crenulate in juveniles, thickening later and developing internal teeth. Whitish, with brown or purplish throat but a range of monochrome and striped shell forms also occur. Up to 42 × 22 mm.

Abundant on all rocky shores, from MHWNT to MLWST, less common sublittorally to 40 m. Avoids low salinities and excessive weed cover, but tolerant of a wide range of exposure. Widely distributed from Straits of Gibraltar to Arctic.

Urosalpinx cinerea (Say)
American oyster drill Fig. 10.15
Apex sharply pointed, whorls tumid. With prominent broad ribs and numerous low, narrow spiral ridges which may be eroded at crests of ribs and do not form a prominent keel over siphonal canal. Siphonal canal short, open; outer lip thin (thickened internally in mature shells), with sharp inflection at base of siphonal canal. Yellowish, off-white, or buff, with darker throat. Up to 40 × 20 mm.

On oysters, from lower intertidal to 12 m, hibernating in mud at 7 °C or less. An accidental introduction from the eastern USA, presently confined to Essex and Kent coasts.

Ocenebra erinacea (Linnaeus)
Oyster drill, sting winkle Fig. 10.15
Turreted, apex sharply pointed, whorls tumid. With strong, buttress ribs, cord-like spiral ridges, and growth lines. Ribs with marked shoulder, which in mature shells may lie above the suture. Siphonal canal long, equal to aperture height; open and slightly curved in juveniles, closed over in mature shells. Outer lip thin, crenulate in young shells, often with hollow varix behind and internally grooved; thickened internally in mature shells. Cream, often with rusty brown patches. Up to 50 × 25 mm.

On rocky shores, in silty crevices and beneath stones from LWST to 150 m; intertidal range increased in summer. Distributed from the Azores and Mediterranean to south and west coasts of British Isles; becoming rare in north.

Ocinebrina aciculata (Linnaeus) Fig. 10.15
Apex pointed, whorls tumid. With broad ribs, broad spiral ridges and growth lines. Spiral ridges stronger on base of last whorl, forming a spiral keel on siphonal canal. Growth lines numerous, raised to produce a finely laminate surface, particularly on last whorl. Siphonal canal short, closed in mature shells. Outer lip thin, crenulate, without varix; thickened and internally fluted in mature shells. Up to 10 × 15 mm.

On rocky shores, usually under stones with *Balanus perforatus*; LWST down to 15 m. Distributed from Mediterranean to Scillies and western Channel.

33 BUCCINIDAE

Shell with tall spire, last whorl 60–70% of shell height; adult shell of most species large (commonly 50–100 mm or more). Sculpture of spiral ridges, which may or may not be prominent, with or without ribs; when present, ribs very prominent. Aperture commonly large, broadly oval (more elongate in *Colus*); siphonal canal frequently short, never closed. Periostracum on juvenile shells, often retained, at least in part, on adults. Proboscis eversible; no snout. Siphon long, commonly projecting well beyond tip of canal in active animals. Foot like that of muricids, but lacks ABO. Common on soft substrata. Sixteen species recorded from Britain.

---

1. Shell with ribs and spiral ridges; ribs markedly crescentic; spiral ridges numerous, of variable size; aperture very broad, without labial varix **_Buccinum undatum_**

   Shell with spiral ridges only, though these may be faint **2**

---

2. Whorls tumid, with slight subsutural ridge; columella with pronounced spiral ridge (siphonal fasciole) extending from base to umbilical region **_Neptunea antiqua_**

   Whorls slightly tumid or almost flat-sided, sutures shallow, slightly excavated, but without subsutural ridge; no siphonal fasciole **3**

---

3. Protoconch often bulbous, wider than next whorl; whorls tumid; spiral ridges strong (except where eroded), fewer than 40 on last whorl; siphonal canal rather long, straight, and narrow **_Colus islandicus_**

Protoconch not bulbous, diameter equal to or less than that of next whorl; whorls almost flat-sided; spiral ridges less strong, up to 60 on last whorl; siphonal canal rather short, broad, and bent to left ***Colus gracilis***

Colus gracilis (da Costa) Fig. 10.15

Fusiform, with 8–10 almost flat-sided whorls. Numerous, narrow, strap-like spiral ridges and crescentic growth lines. Periostracum thin, often slightly hispid, usually worn away from a triangular area to left of apex of aperture. Protoconch swollen, up to 3 mm diameter, set obliquely on spire. Siphonal canal broad, rather short, inclined sharply to left; outer lip arises between spiral ridges 20–30 on last whorl. White, with light horn-coloured periostracum. Up to 70 × 28 mm.

On sandy and muddy substrata from 30 to 80 m; occasionally intertidal in north of range, where it is also more common. Distributed from Norway to Portugal; recorded all round British Isles, rare in southern North Sea and Channel.

Colus islandicus (Mohr) Fig. 10.15

Similar to *C. gracilis*. Protoconch bulbous, often of greater diameter than next whorl and set more obliquely on spire; whorls more tumid. Spiral ridges often more pronounced, fewer than 40 on last whorl. Aperture with long narrow siphonal canal, not markedly inclined to left; outer lip arises between spiral ridges 10–12 on last whorl. White, with light horn-coloured periostracum. Up to 150 × 50 mm.

On soft substrata, from 10 to 3000 m. A northern species extending south to Shetland and the north coast of Scotland; subfossil shells may be encountered in more southern waters.

Neptunea antiqua (Linnaeus)
Red whelk Fig. 10.15

Slightly concave spire of seven tumid whorls; protoconch rather bulbous, last whorl about 70–80% shell height; sutures slightly accentuated by narrow subsutural ridge, more conspicuous on last whorl. Numerous, fine spiral ridges and flexuous growth lines. A thick spiral keel (siphonal fasciole) runs from shell base to umbilical region. Aperture broadly oval or egg-shaped, pointed adapically; basally a short, broad siphonal canal. Yellowish, sometimes with reddish tinge, throat usually more heavily pigmented. Up to 100 × 50 mm, occasionally larger.

Sublittoral, from 15 to 1200 m; mainly on soft substrata. Widely distributed from Biscay to Arctic; locally common around much of British Isles, less so in south and off west coast of Ireland.

Buccinum undatum Linnaeus
Common whelk Fig. 10.16

Spire of seven or eight tumid whorls, last whorl large (about 70% of shell height). With strong crescentic ribs, spiral ridges, and growth lines. Ribs do not extend to base of last whorl. Spiral ridges numerous, major ridges separated by variable number of minor ridges to form a regular pattern; interaction with growth lines produces fine reticulation. A thick spiral keel (siphonal fasciole) runs from base of shell to umbilical region. Aperture broadly oval, tapering adapically to a point; outer lip describes a high arc to tip of short, siphonal canal; inner lip reflected over columella and large area of parietal region of last whorl. Buff, with darker bands above and below periphery; throat and peristome white. Up to 110 × 68 mm.

On hard and soft substrata, occasionally at LWST but usually sublittoral down to 1200 m. Tolerates brackish water to 15‰. Distributed from Iceland and northern Norway to the Bay of Biscay; common and often abundant around British Isles, except Isles of Scilly.

34 NASSARIIDAE

Shell oval–conical, whorls rather flat-sided, or tumid, apex sharp. Sculpture of flexuous ribs, growth lines, and strap-like spiral ridges; often markedly reticulate. With or without distinct periostracum. Aperture with short, deep siphonal canal, never closed; a deep spiral channel runs from tip of canal to umbilical region of columella. Outer lip with labial varix, thickened internally, with row of teeth, in mature shells.

Carrion feeders, with long eversible proboscis and no snout; siphon extends long way from canal and carried aloft in active animals. Foot with well-developed antero-lateral horns and two metapodial tentacles; female with opening of ventral pedal gland on sole; no accessory boring organ (ABO).

---

1. Adult shell large (up to 30 mm); whorls almost flat-sided; ribs and spiral ridges of upper whorls of equal height,

producing squared, block-like reticulation; usually with complete periostracum **Hinia reticulata**

Adult shell small (less than 15 mm); whorls tumid; ribs higher than spiral ridges on all whorls, reticulation not block-like or conspicuous; periostracum not obvious **2**

---

2. Penultimate whorl with 8–10 spiral ridges; up to 12 narrow ridges on boss below basal spiral channel; dark brown mark on roof of siphonal canal **Hinia incrassata**

Penultimate whorl with four or five spiral ridges; five narrow ridges on boss below basal spiral channel; no dark brown mark on roof of siphonal canal; sublittoral only **Hinia pygmaea**

Hinia reticulata (Linnaeus)
Netted dog whelk Fig. 10.16

Whorls almost flat-sided, spire a little convex, slightly stepped at sutures; block-like reticulation particularly conspicuous on spire. Outer lip arises tangentially to last whorl. Cream or buff, often with darker subsutural bands; periostracum tan-coloured, though often discoloured by sulphides. Up to 30 × 14 mm.

Common near LWST, in sedimentary areas of rocky shores and sublittorally to 15 m on soft substrata; often covered in sediment, with only siphon exposed. Distributed from Canaries and Azores to Norway; common on all British coasts.

Hinia incrassata (Ström)
Thick-lipped dog whelk Fig. 10.16

With up to eight tumid whorls; periostracum confined to grooves between ridges. Ribs more prominent than ridges on all whorls, interacting to form oblong tubercles, without marked reticulation; last whorl with strong labial varix. Outer lip arises at about 45° to last whorl, inner lip toothed or ridged in mature shells. Buff or light tan, often with brown bands below sutures and at base of last whorl; labial varix and columella white, siphonal canal with dark brown mark on roof. Up to 12 × 6 mm.

Common near LWST in silted areas of rocky shores, below stones and in crevices; more abundant in shallow sublittoral, extending to 200 m. Distributed from Mediterranean to northern Norway and Iceland; on all British coasts.

Hinia pygmaea (Lamarck)
Small dog whelk Fig. 10.16

Similar to *H. incrassata*, but ribs less flexuous, with more pronounced reticulation; whorls of spire may have varices; aperture with nine or ten teeth internally on outer lip (seven in *H. incrassata*); inner lip not extending so far over last whorl; with brown siphonal canal, but lacking dark mark on roof. Up to 14 × 8 mm.

Sublittoral only, on sandy bottoms from 1 to 1200 m. Not common. Distributed from Mediterranean, north along western coasts of British Isles to Shetland; absent from Irish and North Seas.

35 TURRIDAE

Shell an elongate, often spindle-shaped cone; generally small (up to 15 mm), though some species may reach 25 mm in length. Sculpture variable; strong ribs and spiral ridges, ribs and fine spiral ridges, or spiral ridges only. Aperture long and narrow, with siphonal canal and anal sinus. Operculum present or absent. Twenty species described from Britain, but most are uncommon.

---

1. Shell with narrow, minutely granular subsutural band; penultimate whorl with 12 or more ribs. Ribs extend beyond periphery of last whorl; spiral ridges cord-like, slightly swollen where they cross ribs; outer lip thickened internally **Raphitoma leufroyi**

Penultimate whorl with fewer than 12 ribs **2**

---

2. Penultimate whorl with 10–11 ribs crossed by 4–7 narrow, cord-like spiral ridges; outer lip thickened internally, with up to nine teeth in mature shells; protoconch with diamond-shaped reticulation; pale, with chestnut-brown spiral ridges **Raphitoma linearis**

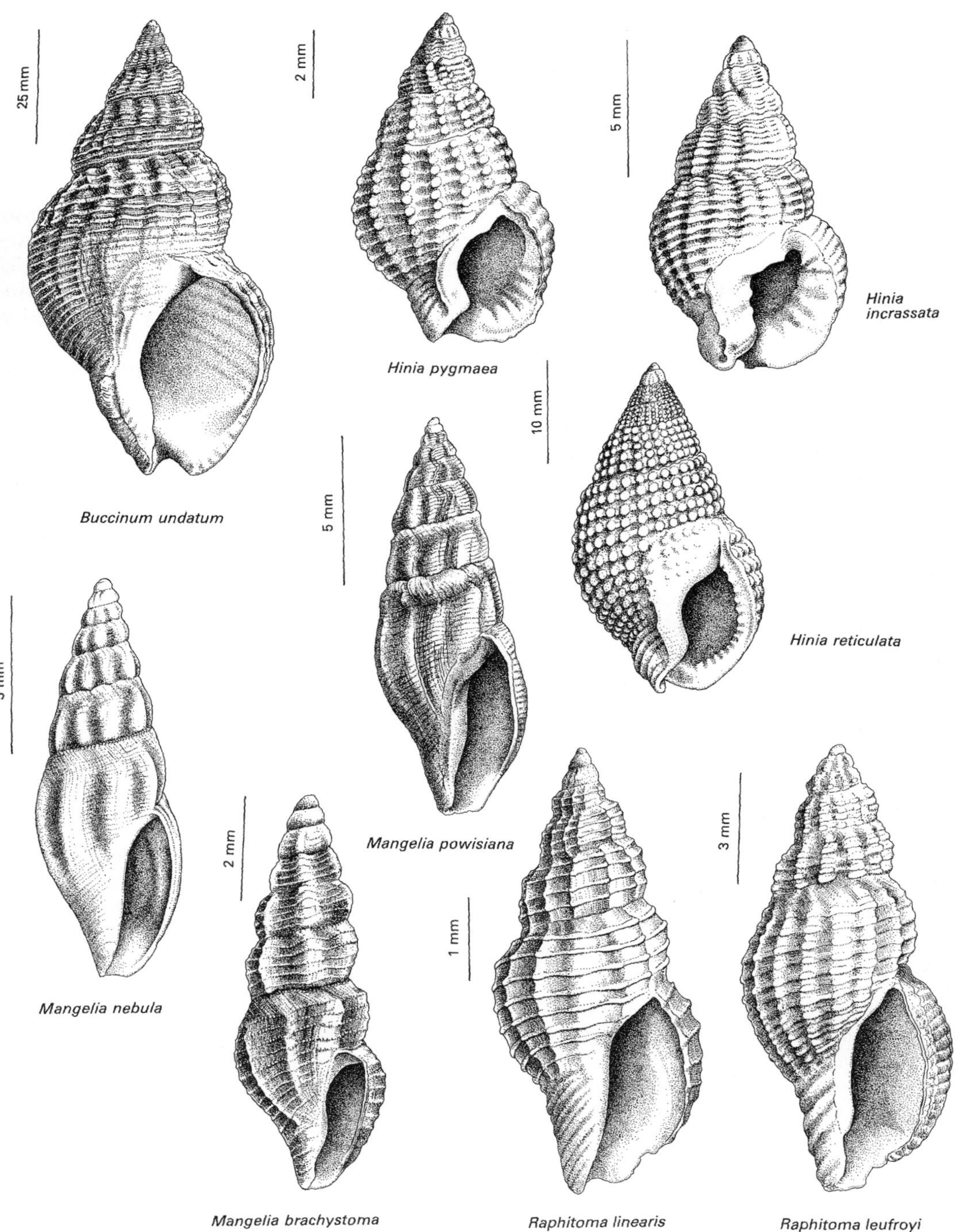
25 mm
2 mm
5 mm
Hinia incrassata
Hinia pygmaea
10 mm
5 mm
Buccinum undatum
Hinia reticulata
5 mm
2 mm
3 mm
Mangelia powisiana
1 mm
Mangelia nebula
Mangelia brachystoma
Raphitoma linearis
Raphitoma leufroyi

Fig. 10.16

Shell not like this **3**

3. Whorls only slightly tumid, profile of shell rather flat-sided; subsutural band prominent, broad, swollen, and beaded; spiral ridges numerous, regular (often of two alternating sizes), with beaded tubercles which are not often eroded or scale-like ***Mangelia powisiana***

   Whorls tumid, profile turreted; subsutural band less well developed; spiral ridges narrow, less regular, tubercles often eroded or scale-like **4**

4. Penultimate whorl with 8–10 ribs and 5–6 major (many more minor) spiral ridges; subsutural band narrow with strongly beaded spiral ridge immediately below suture; tubercles on spiral ridges very prominent, scale-like; rarely exceeds 7 × 3 mm ***Mangelia brachystoma***

   Penultimate whorl with 7–9 ribs and numerous close set, narrow spiral ridges; subsutural band slightly swollen, often beaded but without a prominent, strongly beaded ridge next to suture ***Mangelia nebula***

Mangelia nebula (Montagu) Fig. 10.16
Narrow, with eight or nine tumid whorls; sutures deep, with a slightly swollen, often beaded subsutural band. Strong, flexuous ribs, spiral ridges, and growth lines present; ridges, finely beaded, numerous, with major ridges separating small groups of minors. Protoconch of three or four whorls, initially smooth, becoming tuberculate. Aperture elongate–oval, with anal sinus and short, wide siphonal canal. Light-horn coloured or reddish-brown. Up to 14 × 15 mm.

On sand and muddy gravel bottoms from LWST (rarely) to 50 m. Distributed from Mediterranean to Norway; locally common, more frequent in south of British Isles; absent from eastern Channel and southern North Sea.

Mangelia brachystoma (Philippi) Fig. 10.16
Similar to *M. nebula*; profile more turreted, subsutural band more prominent. Major spiral ridges more prominent than in *M. nebula*; interaction with growth lines produces conspicuous, erect, scale-like tubercles. Pale cream to chestnut-brown, perhaps with lilac tint. Up to 7 × 3 mm.

Not intertidal; on sand and sandy mud from 4 to 60 m. Locally common. Distributed from Mediterranean to Norway; off most British coasts; absent from southern North Sea.

Mangelia powisiana (Dautzenberg) Fig. 10.16
Slender, with flat-sided or slightly tumid whorls; sutures deep, bordered by broad, swollen subsutural band. Spire profile more straight-sided and last whorl curving more gently towards siphonal canal than in *M. nebula*. With flexuous ribs, spiral ridges, and growth lines. Ribs much reduced where they cross subsutural band; spiral ridges numerous, often alternately broad and narrow; interaction with growth lines gives them a finely beaded appearance and a fine reticulate pattern to shell. Pale chestnut-brown, with lighter ribs and spiral ridges; apex and throat dark chestnut. Up to 15 × 6 mm.

On sand from LWST (rarely) to 100 m. Locally common. Distributed from Biscay to Norway; perhaps confined to west coasts of Britain and Ireland.

Raphitoma linearis (Montagu) Fig. 10.16
Elongate, turreted, with 7–9 tumid whorls; sutures deep, sinuous, with narrow, minutely granular subsutural band. Ribs strong, rather narrow, about 11–12 on last whorl. Spiral ridges narrow, slightly nodular where they cross ribs, much reduced on subsutural band; 16–20 on last whorl, 4–7 on penultimate. Protoconch finely reticulate, with distinctive diamond pattern. Aperture elongate–oval, with short siphonal canal and shallow anal sinus. Outer lip thin, crenulate, with a sharp inflexion at junction with siphonal canal; older shells with a row of nine teeth internally. Cream with lilac tinge to apex or light-horn colour; spiral ridges often darker-horn coloured and visible through outer lip. Up to 12 × 7 mm.

Occasionally under stones or in pools at LWST; more commonly sublittoral on sand, shell, and gravel bottoms at 10–200 m. Distributed from Mediterranean to northern Norway; off most British coasts but intertidal only on southern shores.

Raphitoma leufroyi (Michaud) Fig. 10.16
Similar to *R. linearis*; profile more convex, subsutural band shelf-like, minutely granular. Ribs and spiral ridges on last whorl more numerous (14–17

ribs, up to 25 spirals) than in *R. linearis*; spiral ridges slightly expanded where they cross ribs. Outer lip without sharp inflexion at junction with siphonal canal; thickened internally in older shells but without row of teeth. Light-horn coloured, often with chestnut-brown spiral ridges; reddish-brown patches may be evident, particularly on last whorl; first whorl of protoconch white. Up to 15 × 6 mm.

Quite common on sandy, shelly, and stony bottoms; LWST to 150 m. Distributed from Mediterranean to northern Norway; widely distributed in British Isles; absent from southern North Sea and east Channel.

Subclass Opisthobranchia

The opisthobranchs are divided into nine orders. The Cephalaspidea (Fig. 10.17A) usually have an external shell, though this may be internal or absent, and may have an *operculum*. The head forms a flattened *cephalic shield*, and the foot may bear expanded *parapodial lobes*, or *parapodia*. The Pyramidellidae (Fig. 10.17B) are ectoparasites with solid tall-spired shells with the *aperture* in the *body whorl*. The Thecosomata and Gymnosomata are pelagic and are not included here. The Anaspidea (Fig. 10.17C), or sea hares, have enlarged parapodial lobes and an internal shell, sometimes absent, located between them; the elongated head bears rolled *oral tentacles* and *rhinophores*. In the Notaspidea (Fig. 10.17D) the body is flattened, with a *gill* visible on the right side; the shell may be internal, external, or absent; the head bears an *oral veil* with rolled oral tentacles and rhinophores. The interstitial Acochlidiacea are small animals, less than 10 mm long, lacking both shell and operculum. In the Sacoglossa (Fig. 10.17E–G) the shell is external, internal, or absent; they usually possess rolled rhinophores, and oral tentacles are very small or absent. Finally, the Nudibranchia (Fig. 10.17H–L) lack a shell and have extremely varied body forms. Flattened oval animals usually have *papillae* or *tubercles* on the *mantle*, often referred to as the *dorsum* or *notum*. The *hyponotum* is the underside of the notum, not the foot. Other species may be smooth and limaciform (slug-like), with gills located on the dorsum, often together in a ring or arc around the *anal papilla*. In many nudibranchs finger-like extensions of the body wall, *cerata*, are present on the *pallial margin*. These contain lobes of the digestive gland and may have terminal *cnidosacs*. In the Dotoidae (Fig. 10H) the cerata function as gills and some have flap-like basal extensions called *pseudobranchs*. The nudibranch head often bears both oral tentacles and rhinophores; the latter may be smooth, tuberculate, wrinkled, swollen, branched, or lamellated, and are sometimes retractable into simple or elaborate sheaths. The foot is arbitrarily divided into the *metapodium* (posterior) and *propodium* (anterior), the latter often bearing *propodial tentacles*.

All opisthobranchs are hermaphroditic: the reproductive organs are present in all adults on the right-hand side of the body. Copulation is usually reciprocal, two individuals fertilizing one another. The spawn normally consists of a large number of eggs contained within a gelatinous substance and is laid in ribbons or capsules of many colours, shapes, and sizes. Most species hatch as free-swimming planktonic veligers after an embryonic period that may be as short as 5 days or as long as 50.

Most opisthobranchs are carnivorous, although sacoglossans and aplysiids are herbivorous. The pyramidellids are equipped with a long *proboscis*, armed with a *stylet* for their ectoparasitic mode of feeding on tubiculous polychaetes and molluscs. Most British Opisthobranchia are epifaunal. Some, such as *Acteon tornatilis*, are found on sandy beaches, and the majority of the other shelled bullomorphs occur on sandy or muddy bottoms. The larger nudibranchs are easier to find and usually occur low on rocky shores. Small nudibranchs pose greater difficulties because they often form part of the epifaunal community of specialized microhabitats centred on hydroids, sponges, and seaweeds, for example, and are extremely well camouflaged. Sacoglossans and aplysiids are also well camouflaged amongst the algae upon which they feed; together with notaspideans, they may be found in clear, shallow, rocky habitats, and in rock pools.

KEY TO FAMILIES

1.	Shell external	2
	Shell internal or absent	6

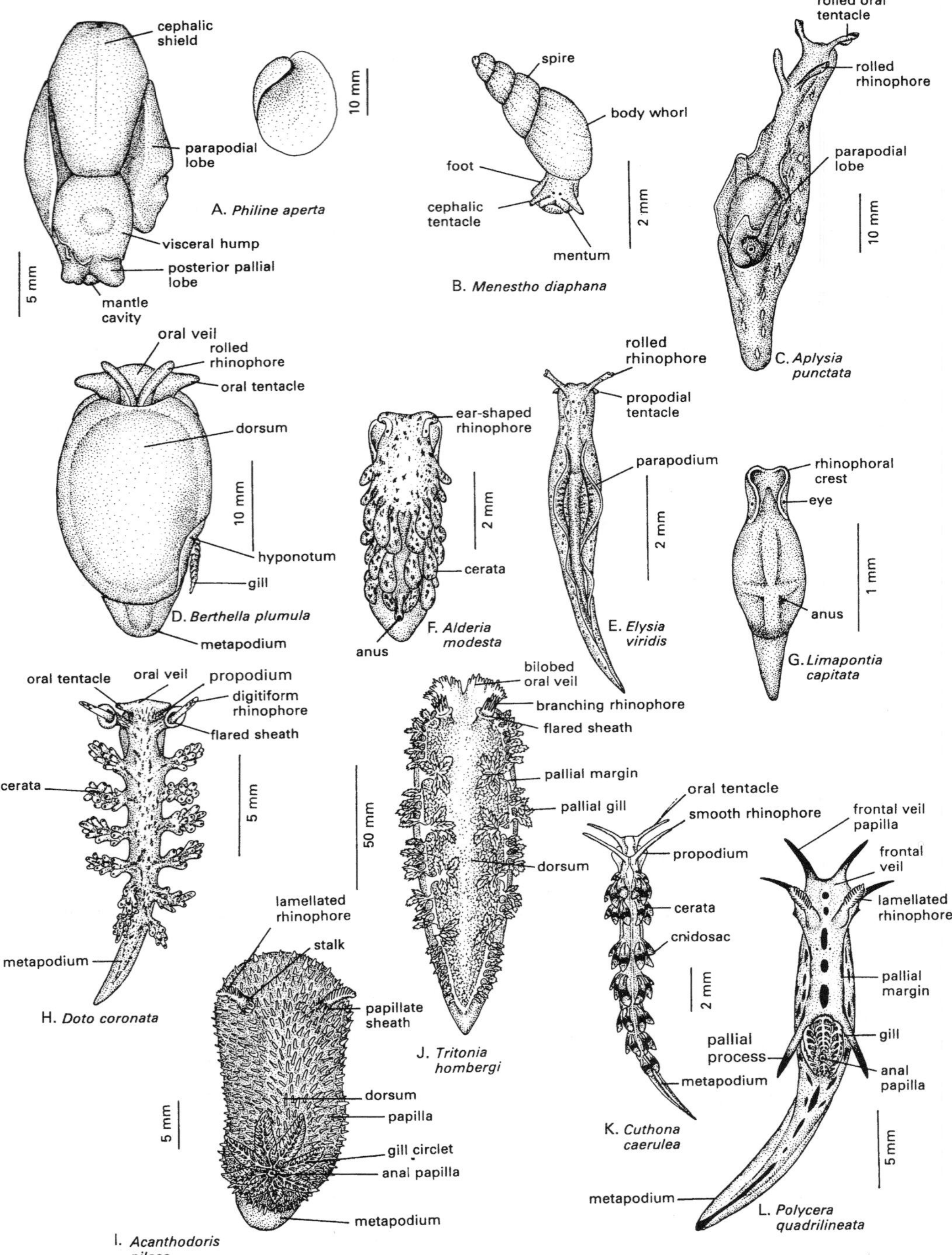
cephalic
shield
10 mm
parapodial
lobe
A. Philine aperta
visceral hump
posterior pallial
lobe
5 mm
mantle
cavity
spire
body whorl
foot
cephalic
tentacle
2 mm
mentum
B. Menestho diaphana
rolled oral
tentacle
rolled
rhinophore
parapodial
lobe
10 mm
C. Aplysia
punctata
oral veil
rolled
rhinophore
oral tentacle
dorsum
10 mm
hyponotum
gill
D. Berthella plumula
metapodium
ear-shaped
rhinophore
2 mm
cerata
F. Alderia
modesta
anus
rolled
rhinophore
propodial
tentacle
parapodium
2 mm
E. Elysia
viridis
rhinophoral
crest
eye
1 mm
anus
G. Limapontia
capitata
oral tentacle
oral veil
propodium
digitiform
rhinophore
flared sheath
cerata
5 mm
metapodium
H. Doto coronata
50 mm
bilobed
oral veil
branching rhinophore
flared sheath
pallial margin
pallial gill
dorsum
J. Tritonia
hombergi
lamellated
rhinophore
stalk
papillate
sheath
5 mm
dorsum
papilla
gill circlet
anal papilla
metapodium
I. Acanthodoris
pilosa
oral tentacle
smooth rhinophore
propodium
cerata
cnidosac
2 mm
metapodium
K. Cuthona
caerulea
frontal veil
papilla
frontal
veil
lamellated
rhinophore
pallial
margin
gill
pallial
process
anal
papilla
5 mm
metapodium
L. Polycera
quadrilineata

Fig. 10.17

---

2. Operculum present **3**

   Operculum absent **4**

---

3. Large, up to 20 mm. Spire less than one-sixth shell height. Aperture elongate, obtusely joined to last whorl, length about six-sevenths of last whorl, two-thirds of total shell height **1. Acteonidae**

   Very small, largest species up to 8 mm, usually 3–4 mm. Spire more than half total shell height. Aperture rounded, acutely joined to last whorl, length about half last whorl **7. Pyramidellidae**

---

4. Cephalic shield with ear-like tentacles **3. Retusidae**

   Cephalic shield without conspicuous tentacles **5**

---

5. Fragile, bubble-like shell **8. Akeridae**

   Strong solid shell **4. Scaphandridae**

---

6. Shell internal **7**

   Shell absent **11**

---

7. Body consisting of four parts: cephalic shield, visceral hump, paired parapodial lobes **5. Philinidae**

   Body not as described **8**

---

8. Shell obvious, comprising larger part of animal **2. Diaphanidae**

   Shell inconspicuous **9**

---

9. Animal oval in shape, hard, with an obvious siphon anteriorly. Variously coloured and patterned PROSOBRANCHIA: **Lamellariidae** (p. 526)

   Animal soft, with two sets of anterior tentacles, paired rhinophores and either an oral veil or oral tentacles **10**

---

10. 'Sea hares', with enlarged parapodial lobes **9. Aplysiidae**

    Slugs with soft, flat bodies, lacking distinct parapodial lobes **10. Pleurobranchidae**

---

11. Gills absent. Body tuberculate, tough, black. Anterior tentacles tubular with terminal knobs. South-west rocky shores only
PULMONATA: **Onchidiidae** (p. 572)

Gills present or absent, body not as described **12**

---

12. Gills situated posteriorly, on right side under mantle **6. Runcinidae**

Gills dorsal, not postero-lateral; or absent **13**

---

13. Large obvious eyes, usually set in unpigmented region **14**

Eyes absent or, if present, small, set far back or at posterior base of rhinophores **16**

---

14. Body smooth, elongate, with no projections **12. Limapontiidae**

Body with lateral parapodial lobes or dorsal cerata **15**

---

15. Conspicuous parapodial lobes **11. Elysiidae**

Cerata arranged in transverse rows **13. Stiligeridae**

---

16. Circlet of gills surrounding anal papilla **17**

Gills, when present, not as described **23**

---

17. Gills retractile, completely withdrawing into a deep cavity **18**

Gills separately contractile, not completely withdrawn **21**

---

18. Mantle smooth or faintly papillate **21. Cadlinidae**

Mantle with conspicuous papillae **19**

---

19. Papillae very small and uniform **20**

Papillae large and obvious, of different sizes. Head with very short oral tentacles. Mantle with blotchy markings of various colours
23. Archidorididae

---

20. Mantle velvety; papillae spiculose, with retractile central projection
24. Kentrodorididae

Papillae blunt. Coloured patch present between rhinophores
22. Rostangidae

21.	Mantle fully developed	**18. Onchidorididae**
	Mantle very reduced, foot obvious in dorsal view	**22**
22.	Body smooth, with elongate dorsal processes	**19. Goniodorididae** (part). **20. Polyceridae**
	Body keeled, with reduced mantle skirt	**19. Goniodorididae**
23.	Pleated sensory organ (caruncle) present between rhinophores; fusiform cerata continuous anteriorly	**25. Janolidae**
	No caruncle; cerata, if present anteriorly, not fusiform	**24**
24.	Rhinophores with conspicuous sheaths	**25**
	Rhinophores not as described	**28**
25.	Rhinophores simple, finger-like, with flared sheaths	**17. Dotoidae**
	Rhinophores divided or lamellated	**26**
26.	Rhinophores branched	**14. Tritoniidae**
	Rhinophores lamellated	**27**
27.	Rhinophore sheaths and dorsal processes arborescent	**16. Dendronotidae**
	Rhinophore sheaths not as described	**15. Lomanotidae**
28.	Rhinophores lamellated, sometimes indistinctly so	**29. Facelinidae**
	Rhinophores wrinkled, swollen, or smooth	**29**
29.	Rhinophores with up to three bulbous swellings	**30. Favorinidae**
	Rhinophores without swellings	**30**
30.	Propodium extended, forming conspicuous propodial tentacles	**31**
	Propodium not as described	**32**
31.	Body thin, elongated; cerata in clusters	**31. Flabellinidae**
	Body broader; cerata in closely packed rows	**26. Aeolidiidae**

32. Oral tentacles absent **27. Embletoniidae**

Oral tentacles present **28. Eubranchidae**
32. Tergipedidae

1 ACTEONIDAE

External shell with up to eight whorls; able to accommodate whole of animal. Operculum present. Cephalic shield with two pairs of lobes. One British species.

Acteon tornatilis (Linnaeus) Fig. 10.18
Shell solid, glossy, opaque, pink with white bands, each with a narrow darker pink edge on either side; two white bands on body whorl, one on each of other whorls. Up to 20 mm long; aperture two-thirds of shell height. Body creamy white; foot and head divided anteriorly; head with large lobes, foot with small, blunt propodial tentacles.

Burrows in clean, fine sand, below MTL and in shallow sublittoral. Known to feed on the polychaetes *Lanice conchilega* and *Owenia fusiformis*.

2 DIAPHANIDAE

Shell external or internal, colourless, fragile, several whorls. No operculum. Cephalic shield and propodium with short tentacular processes. Three species recorded from Britain, but two are very rare.

Colpodaspis pusilla M. Sars Fig. 10.18
Shell internal but obvious; large, bubble-like, completely covered by mantle, up to 3.2 mm long. Body white with opaque white speckling, extending beyond mantle edge and shell at both ends; posterior lobe is an extension of the mantle; foot completely hidden, Y-shaped, with bilobed propodium. Total length up to 5 mm.

Offshore among hydroids and erect bryozoans. Widely distributed on west coasts of England and Ireland.

3 RETUSIDAE

External shell rather fragile, able to accommodate whole animal, with or without short spire. No operculum. Cephalic shield with rounded or pointed, postero-lateral, tentacular processes. Four species recorded from British waters.

1. Shell up to 10 mm long, aperture shorter than last whorl. Cephalic tentacles rounded ***Retusa obtusa***

Shell cylindrical, up to 5 mm long, aperture larger than last whorl Cephalic tentacles pointed ***Retusa truncatula***

Retusa obtusa (Montagu) Fig. 10.18
Shell translucent white, glossy, and sculptured, usually elongated. Body whitish, extending beyond shell only anteriorly. Cephalic shield square, slightly indented in mid-line anteriorly. Posteriorly, two rounded tentacles point backwards and outwards. Foot very small, squat, shield-like shape in ventral view, approximately same size as head, indented anteriorly. Total length up to 15 mm.

In mud or muddy sand, below surface in first few centimetres, from lower shore to 300 m; feeds on *Hydrobia ulvae*. All British coasts, on sheltered muddy shores, locally abundant.

Retusa truncatula (Bruguière) Fig. 10.18
Shell translucent white or yellowish, glossy; cylindrical, with longitudinal striations. Body white. Cephalic shield indented anteriorly at mid-line, cephalic tentacles pointed. Foot elongate, with rounded corners, indented anteriorly. Total length up to 7 mm.

In sand, muddy sand, and mud; lower intertidal to about 50 m. Feeds on *Hydrobia ulvae* and perhaps foraminiferans and small lamellibranchs. All British coasts, except extreme south-east.

4 SCAPHANDRIDAE

External shell stout and rounded, usually unable to accommodate entire animal. No operculum. Cephalic shield with or without tentacular processes; parapodial lobes present. Five British species. *Roxania utriculus* (Brocchi) occurs offshore on muddy sand; *Scaphander* and *Cylichna* each include a second, rare species.

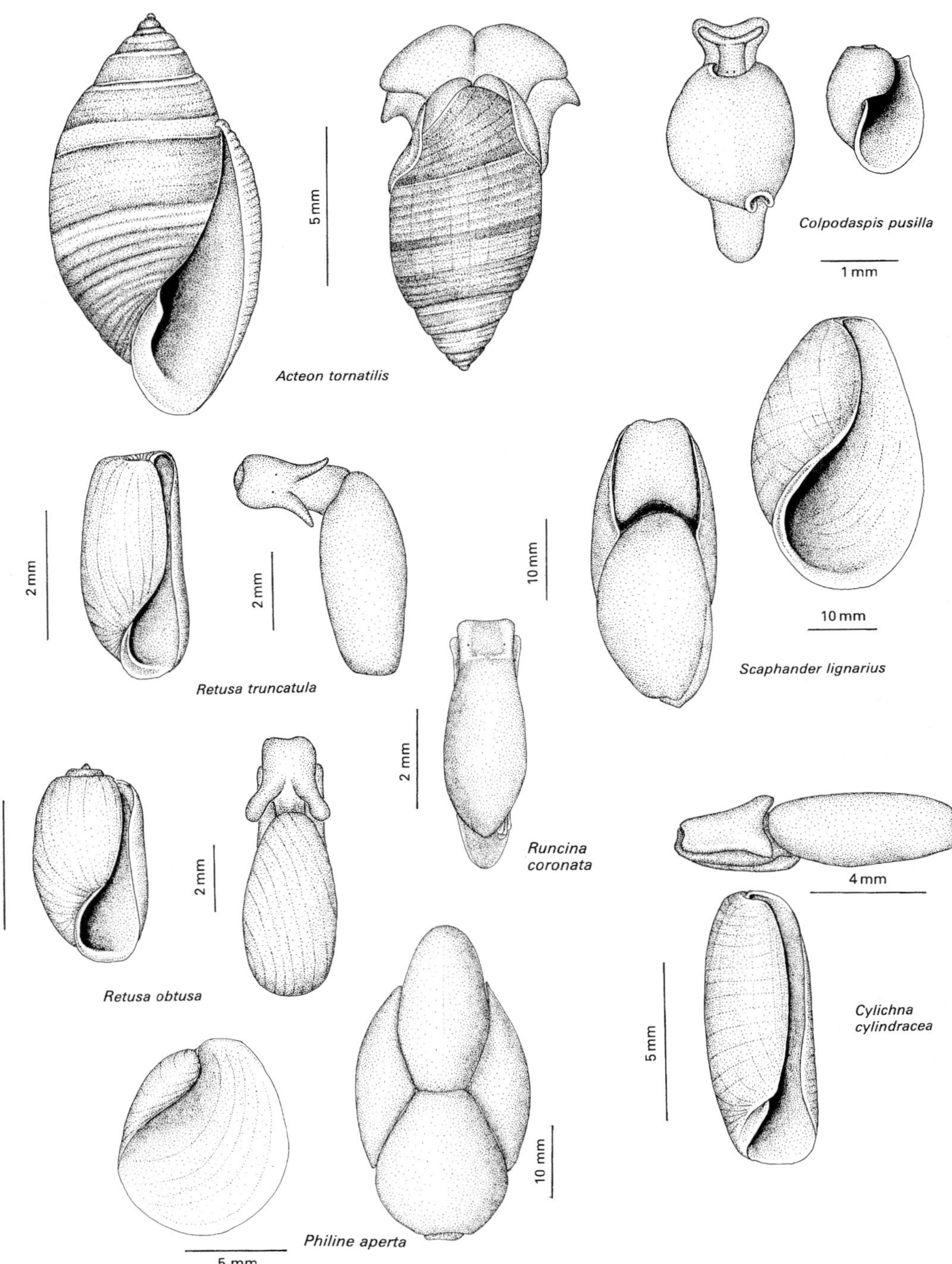
5 mm
Acteon tornatilis
Colpodaspis pusilla
1 mm
2 mm
2 mm
Retusa truncatula
10 mm
10 mm
Scaphander lignarius
2 mm
Runcina coronata
2 mm
2 mm
Retusa obtusa
4 mm
5 mm
Cylichna cylindracea
10 mm
Philine aperta
5 mm

Fig. 10.18

1. Shell large (45 mm), inflated, pyriform; aperture three to four times wider at base than apex
Scaphander lignarius

Small shell (less than 15 mm) elongated, cylindrical; aperture as long as last whorl, of almost constant width
Cylichna cylindracea

Cylichna cylindracea (Pennant) Fig. 10.18
Shell cylindrical, solid, glossy, opaque, whitish with areas of brown. Body white to pale buff. Cephalic shield indented anteriorly; smooth, short, rounded tentacles posteriorly. Foot a little larger than head, with very small parapodial lobes between head and shell.

Burrows in fine sand, in the top 30 mm. Rarely on lower shore; sublittoral to 1500 m. All British coasts.

Scaphander lignarius (Linnaeus) Fig. 10.18
Shell solid, glossy, opaque green, yellow or brown; sculpture of spiral lines, crossed by longitudinal lines. Body white or orange. Head very short, squat, indented anteriorly; tentacular processes smooth, short, conical. Parapodial lobes extending from anterior end of head shield to about middle of shell; pallial lobe also visible. Total length up to 60 mm.

Burrows in muddy sand; sublittoral to 700 m. Feeds on infaunal worms and lamellibranchs. Common, but absent from east coast of Britain.

5 PHILINIDAE

Shell internal, with large aperture. Parapodia large and fleshy but not meeting mid-dorsally. Posterior pallial lobe conspicuous, extending beyond the metapodium. Eight species of *Philine* are recorded from muddy sand habitats around Britain, but all are rather uncommon, with the following exception.

Philine aperta (Linnaeus) Fig. 10.18
Shell large, up to half body length, open, inflated, covering visceral hump, enclosed by folds of mantle. Body up to 70 mm, about twice as long as wide, translucent white to pale yellow, with white dots, quadripartite. Cephalic shield longer than wide, indented posteriorly, may or may not be indented anteriorly; parapodial lobes fleshy, rounded on outer sides, moulding to form of head and visceral hump on inner sides. Mantle cavity widely open posteriorly, indented dorso-medially.

In sand and muddy sand, feeding on *Pectinaria*, *Echinocyamus*, foraminiferans, and small infaunal lamellibranchs and gastropods. Lower shore and shallow sublittoral, common. All British coasts.

6 RUNCINIDAE

Shell reduced or absent. No operculum. Cephalic shield united with mantle, no tentacular processes. No parapodial lobes. Body smooth and elongated. Two British species. *Runcina ferruginea* Kress is presently known only from Plymouth, south-west Ireland, and north-west Spain.

Runcina coronata (Quatrefages) Fig. 10.18
Body smooth, elongate, up to 6 mm; brown with white speckles; head separated by a light collar lacking brown pigment, a similar oval patch at posterior end of mantle. Mantle edge translucent with orange tinge. Foot translucent cream, visible on each side of head and neck; eyes barely visible on edge of brown pigment. Metapodium extending beyond mantle, bordered with white spots. Three gills protrude beneath posterior edge of mantle on right side of body.

Intertidal and shallow sublittoral, in clear coralline rock pools. Feeds on *Codium*. English Channel, west coasts, northwards to Orkney.

7 PYRAMIDELLIDAE

Small animals, with spiral conical shell, into which entire body may be withdrawn, closed by an operculum. Protoconch sinistral, rest of shell dextrally coiled. A portion of the foot, the *mentum*, projects anteriorly between head and propodium, slightly indented in mid-line. Tentacles with a concave surface. Eyes located between tentacles (cf. prosobranchs with stalked eyes lateral to tentacles). Pyramidellids are most often found in association with molluscs and tubiculous polychaetes. As many as 40 species occur around the British Isles but most are rare, or only infrequently recorded.

1. Shell a slender spire of 10 or more whorls **2**

Shell with fewer than 10 whorls, rather short and fat **3**

2. Shell with strong, straight ribs, with concentric grooves between **_Turbonilla crenata_**

Shell with thick, sinuous oblique ribs, no concentric sculpture **_Turbonilla lactea_**

3. Shell with well-marked ribs; spiral ridges on abapical portion of body whorl **_Partulida spiralis_**

Shell without ribs, smooth, or with some fine spiral striae **4**

4. Shell small, comprising protoconch and three whorls **_Ondina diaphana_**

Shell of more than three whorls **5**

5. Last whorl smoothly rounded, without a keel **6**

Up to five whorls; last whorl angular at broadest point, with a faint keel; aperture quadrate **_Odostomia unidentata_**

6. Shell slender, up to six whorls; last whorl comprising less than half total shell height. Columellar tooth well marked, extending as ridge into aperture **_Odostomia plicata_**

Shell rather fat, of five whorls or less. Last whorl comprising about half total shell height **7**

7. Columellar tooth weakly developed; umbilicus distinct **_Brachystomia_ scalaris**

Columellar tooth strongly developed, extending as ridge into aperture; umbilicus indistinct **_Brachystomia eulimoides_**

Partulida spiralis (Montagu) Fig. 10.19
Shell usually long and slender, comprising four whorls and protoconch, up to 2.8 mm; white, solid, and deeply ridged. Abapical part of last whorl with spiral grooves between flat ridges. Aperture thickened, rather square, with slightly flared lip; tooth on columella variably developed, most prominent on shells with thickened apertural rim.

All British coasts except extreme south-east.

Ondina diaphana (Jeffreys) Fig. 10.19
Shell fusiform, slender, up to 3.4 mm; comprising three whorls and protoconch; smooth, white, and rather shiny, with some very fine striations on abapical part of last whorl. Aperture elongate, thickened, columellar tooth variably developed, umbilicus large.

An obligate parasite of the sipunculid *Phascolion strombi*, reported from scattered localities around British Isles.

Brachystomia eulimoides Hanley Fig. 10.19
Shell with long conical spire, up to 5.2 mm, with five whorls and protoconch; body whorl comprising approximately half total shell height. White to yellowish-white, no sculpture; weak growth lines visible on last whorl; whorls flat-topped at sutures. Aperture elongate oval, with thickened outer lip; columellar tooth well developed, extending as ridge into aperture; umbilicus shallow, obscured by columella but visible on broken shells.

Parasitic on lamellibranchs, and also the prosobranch *Turritella communis*. Scattered localities around British Isles.

Brachystomia scalaris MacGillivray Fig. 10.19
Shell with short fat spire, up to 4.4 mm, with five whorls and protoconch; white, shiny, and smooth. Whorls rounded, curving into sutures basally, with a few faint growth lines. Aperture oval, almost as long as last whorl, with flattened outer lip. Columellar tooth weakly developed; umbilicus small but distinct.

All British coasts. Hosts are many prosobranchs and several lamellibranchs, especially *Mytilus edulis*.

Odostomia plicata (Montagu) Fig. 10.19
Shell with tall slender spire, up to 3.8 mm, with six whorls and protoconch; white, smooth, and glossy, with faint growth lines. Sutures well marked, whorls flat-topped. Aperture drop-shaped, with gently

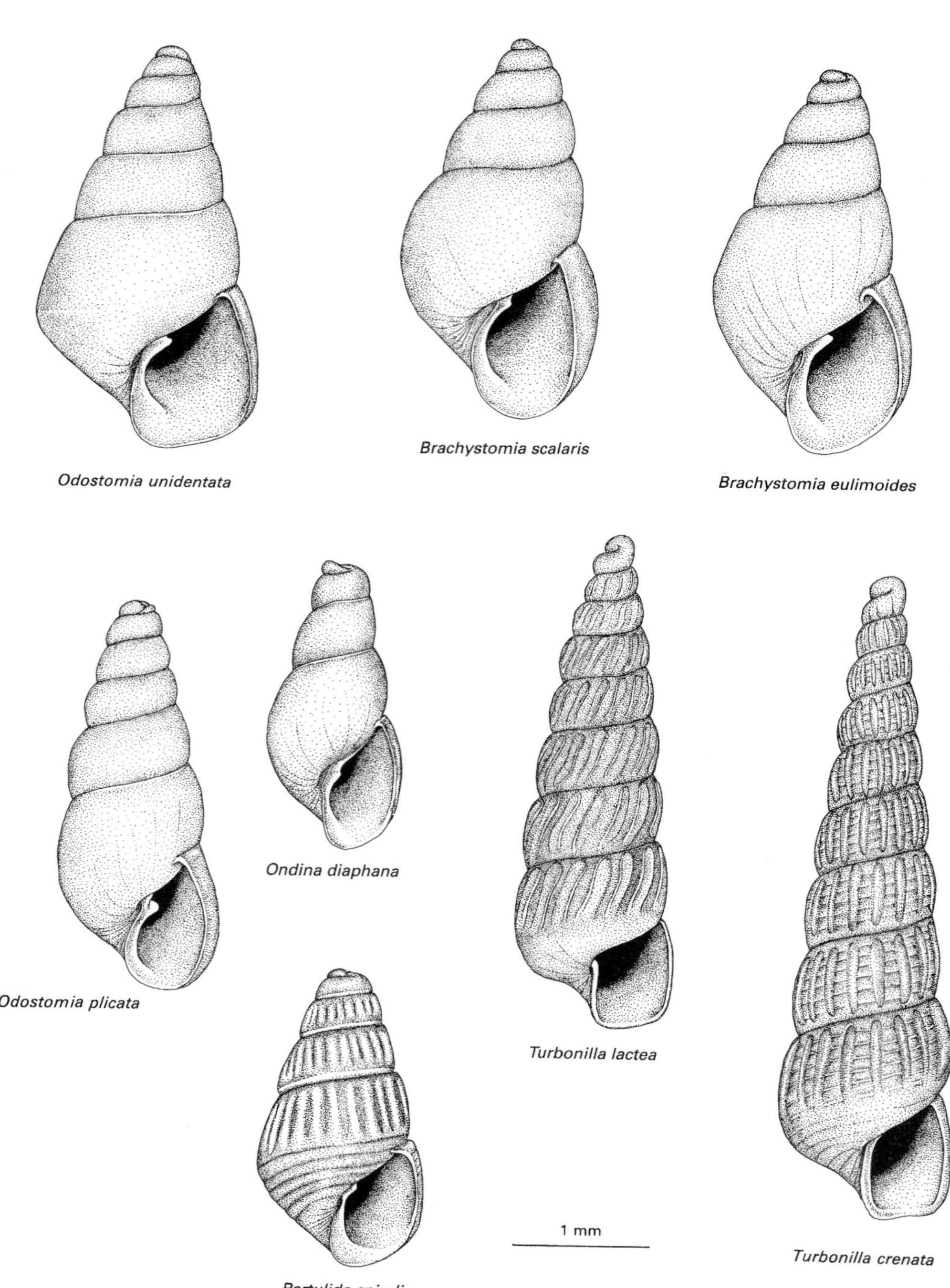
Odostomia unidentata
Brachystomia scalaris
Brachystomia eulimoides
Odostomia plicata
Ondina diaphana
Turbonilla lactea
Turbonilla crenata
Partulida spiralis
1 mm

Fig. 10.19

curved, thickened outer lip. Columellar tooth well marked, extending as ridge into aperture; umbilicus small, sometimes obscured by columellar lip.

All British coasts except extreme south-east.

Odostomia unidentata (Montagu) Fig. 10.19

Shell with tall spire, up to 3.8 mm, with five whorls and protoconch; white to off-white or greyish, smooth except for few growth lines and faint keel on last whorl. Whorls dipping very slightly into sutures. Aperture elongate, rather angular; columellar tooth well developed and conspicuous.

All British coasts.

Turbonilla crenata (Brown) Fig. 10.19

Shell with elongate, slender spire; up to 8 mm, with 11 whorls and protoconch; creamy white to yellow, banded with yellowish-brown. With strong, closely spaced ribs, and concentric grooves; sutures deep and angular; a series of concentric grooves on abapical part of last whorl. Protoconch obvious, smooth, slightly bulbous. Aperture rather squared, lip thickened, may be turned out in old individuals. Umbilicus small, usually obscured by columellar lip.

Recorded from western and northern coasts.

Turbonilla lactea (Montagu) Fig. 10.19

Shell with slender, elongate spire, up to 9 mm with 10 whorls and protoconch. Whorls almost straight-sided with deep, flat sutures; with thick, flat, slightly sinuous oblique ribs, indistinct on first whorl, prominent thereafter; abapical portion of body whorl smooth with few very faint growth lines; no concentric grooves between ribs. Protoconch smooth and bulbous. Aperture rather square, thickened in older shells; no umbilicus. Deep orange operculum.

Lower shore and subtidally, feeding on sedentary worms, such as *Cirriformia* and *Amphitrite*. Chiefly on west coast of Britain.

8 AKERIDAE

With oval external shell of several whorls, too small to accommodate whole animal. Operculum absent. Large parapodia, which meet mid-dorsally. One British species.

Akera bullata (Müller) Fig. 10.20

Shell large, bubble-like, fragile, up to 40 mm; glossy, white to pale brown, up to six whorls forming a rather flat spire. Body up to 60 mm, pale grey to orange, often streaked anteriorly with blotchy lines of purplish-brown, covered with white and dark spots. Head indented anteriorly, expanded laterally to form flattened tentacular lobes. Parapodial lobes enlarged, used for evasive swimming action; covering lateral and dorsal areas of shell when at rest.

In and on soft, fine mud, in sheltered bays formerly favoured by *Zostera*. Herbivorous. Lower shore to 370 m, all British coasts.

9 APLYSIIDAE

Shell internal or absent. Head elongated, with rolled oral tentacles and rolled rhinophores. Parapodial lobes present. A single British species. Two additional species of *Aplysia* have been recorded on a very few occasions from south-west England.

Aplysia punctata (Cuvier)
Sea hare Fig. 10.20

Shell fragile, transparent, pale amber; situated between parapodia and visible through mantle foramen. Body up to 300 mm, olive-green, brown, or purplish-black in adults, pinkish-red in juveniles, often marbled. Head with paired oral tentacles, and paired, rolled rhinophores, eyes just anterior to rhinophores. Parapodia widely spaced anteriorly, joined posteriorly; margins dark brown or blackish in adults, with white line on very edge, in juveniles violet with white edge. Metapodium smooth, quite high but without a keel, evenly pointed.

Herbivorous, feeding especially on *Plocamium*, *Enteromorpha*, *Ulva*, *Delesseria*, and *Laminaria*. Colour may be related to diet. Lower shore and shallow subtidal, juveniles may occur in weedy rock pools. Common on all British coasts.

10 PLEUROBRANCHIDAE

Internal shell. Oral tentacles smooth; oral veil obvious. Rhinophores rolled. Gill present as rachis with double row of tubercles, situated on right side between foot and mantle. Three British species. *Berthellina citrina* (Rüppell and Leuckhart) has occurred rarely on south-west coasts.

---

1. Mantle pale brown, with darker patches; papillate
 Pleurobranchus membranaceus

 Mantle pale yellow to orange; smooth
 Berthella plumula

Pleurobranchus membranaceus (Cuvier) Fig. 10.20

Shell up to 50 mm long; thin, fragile, colourless to pale amber. Body up to 120 mm, pale brown with patches of dark brown between soft, conical, retractile mantle papillae. Mantle much smaller than foot, may lift and curl to form an exhalant siphon. Foot smooth, indented at head region. Head flattened, with smooth, elongate oral tentacles; rhinophores smooth, rolled, protruding from beneath anterior end of mantle. A posterior metapodial gland on dorsal side of pre-sexual animal at body length of 40 mm. Mantle epidermis contains gland cells which secrete sulphuric acid on disturbance.

Lower shore and shallow sublittoral, to 70 m, in clean rocky areas where it preys on ascidians. May be common at certain, unpredictable, times of year. Western coasts, from the Channel to Orkney.

Berthella plumula (Montagu) Fig. 10.20

Shell up to 30 mm long, completely covered by mantle, which extends just a little beyond it, all around. Body up to 60 mm, smooth, cream to orange, often with reticulate markings in middle of mantle. Head flat, large oral veil lying between propodium and mantle; oral tentacles slightly rolled, rhinophores enrolled. Mantle edge may be raised on right side, above gill, to facilitate respiration. A metapodial gland present dorsally, appearing when body is about 30 mm.

A specialized predator of compound ascidians, commonly observed on *Botryllus*; under stones and in rock pools, lower shore and sublittoral to 10 m. Common off all British coasts.

11 ELYSIIDAE

Shell absent. Body smooth, elongate, with parapodial lobes. Oral tentacles absent. Rolled rhinophores. One British species.

Elysia viridis (Montagu) Fig. 10.20

Up to 45 mm long, vivid green to brown or bright red, with brilliant spots of blue, red, and green. Parapodia extending along almost whole body length, with green chloroplasts visible. Propodial tentacles conspicuous; rhinophores rolled, opening ventrally, distal ends usually white. Black markings sometimes present, white patches may surround eyes. Anus dorso-lateral, below right rhinophore.

On algae, especially *Codium fragile*, in shallow water and rock pools. The chloroplasts derived from *Codium* are retained undamaged by *Elysia* and continue to photosynthesize, to the benefit of the slug. All British coasts.

12 LIMAPONTIIDAE

Shell absent. Small, less than 10 mm in length, no dorsal papillae or parapodia. Smooth, simple body form. Cephalic tentacles present in varying forms or absent. Three British species.

---

1. Head without ridges or tentacles. In saltmarshes, on *Vaucheria* ***Limapontia depressa***

   Head with crests or tentacles. In coralline rock pools **2**

---

2. Head with rhinophoral crests, bearing eyes on their sides ***Limapontia capitata***

   Head with smooth, tapering rhinophores; eyes in pale areas behind rhinophores ***Limapontia senestra***

Limapontia capitata (Müller) Fig. 10.20

Commonly about 4 mm long, rarely up to 8 mm; dark brown or black, velvety. Rhinophoral crests forming dorso-lateral flaps, pale externally and around eyes. Pale areas also on head, metapodium, and sometimes middle of dorsum. Anal and renal openings situated to right of mid-line, close together.

Common in clean coralline rock pools from MTL to ELWS, feeding on *Bryopsis*, *Cladophora*, *Enteromorpha*, and *Chaetomorpha*, algae whose frond diameter approximates to width of foot of the slug. All British coasts.

Limapontia depressa (Alder and Hancock) Fig. 10.20

Up to 6 mm; dark, often with tiny pale spots on dorsum, and white patches around eyes. The var. *pellucida* Kevan is bright yellow with pale areas on sides of head, occasional bluish-white and black specks on dorsum, with greenish lobes of digestive gland visible through skin. No tentacles; body shape

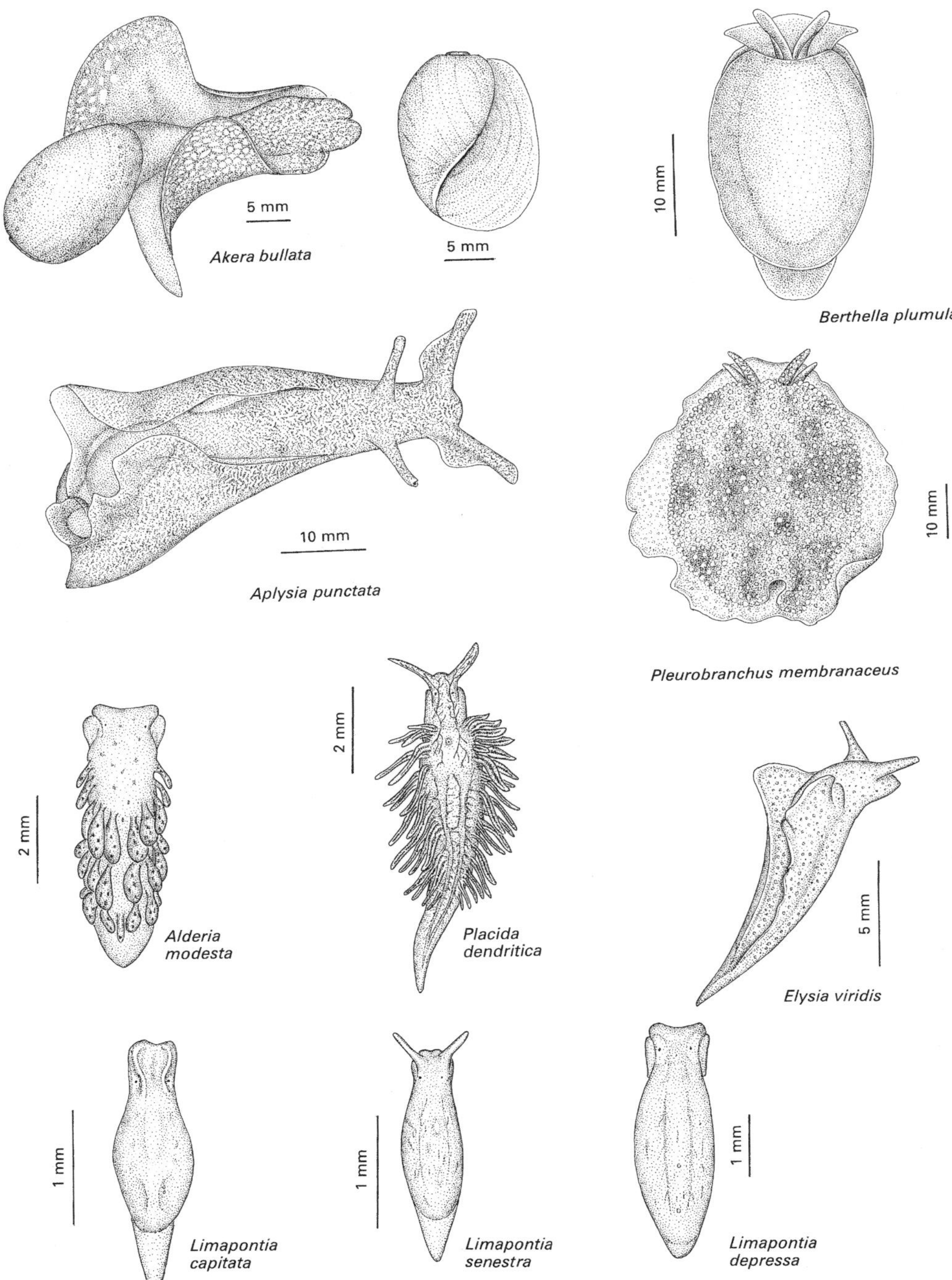
5 mm
Akera bullata
5 mm
10 mm
Berthella plumula
10 mm
Aplysia punctata
10 mm
Pleurobranchus membranaceus
2 mm
2 mm
Alderia modesta
Placida dendritica
5 mm
Elysia viridis
1 mm
Limapontia capitata
1 mm
Limapontia senestra
1 mm
Limapontia depressa

Fig. 10.20

more flattened than in other *Limapontia* species; anal and renal openings widely spaced, dorsal, close to mid-line. Metapodium short.

On saltmarshes, feeding on *Vaucheria*. Common on all British coasts.

Limapontia senestra (Quatrefages) Fig. 10.20
Up to 6 mm; marbled olive-brown to black, with paler metapodium and rhinophores, and pale patches around eyes. Digitiform tentacles present, lacking in juveniles shorter than 2 mm. Anal and renal openings located near mid-line, close together. Ventrally, propodium indented on mid-line and head bilobed. Distinctive bursal swelling on right side. The var. *corrugata* Alder and Hancock has a wrinkled dorsum.

Intertidal, in coralline rock pools, feeding on *Cladophora rupestris* and *Enteromorpha*. All British coasts.

13 STILIGERIDAE

Shell absent. Body elongated, with cerata lacking cnidosacs. No oral tentacles. Rhinophores flattened, ear-like, rolled, or digitiform structures. Six species recorded from Britain, in four genera.

---

1. Rhinophores absent, or forming short ear-shaped projections. Ratio of foot width to body length 1 : 3. Short swollen cerata; anus located posteriorly **Alderia modesta**

   Rhinophores rolled, with open slit on ventral side. Ratio of foot width to body length 1 : 8. Body-greenish white. Cerata cylindrical, smooth, rounded at tip **Placida dendritica**

Alderia modesta (Lovén) Fig. 10.20
Body flattened, up to 10 mm, pale fawn with patches of white, green, and brown. Up to seven transverse paired rows of broad, round-topped cerata. Anal papilla postero-dorsal, on mid-line.

In saltmarshes, on damp mud clear of brackish water, associated with species of *Vaucheria*. All British coasts.

Placida dendritica
(Alder and Hancock) Fig. 10.20
Body up to 11 mm, greenish-white, with green or brown digestive glands branching into cerata, across head and into rhinophores. Cerata elongate, smoothly pointed, with white tips; eight rows on each side, up to four per row. Rhinophores enrolled, with ventral opening. Propodium simple, lobed. Metapodium flattened and smoothly pointed.

Intertidal and shallow sublittoral, on green algae, particularly *Codium tomentosum* and *Bryopsis plumosa*. All British coasts, but more frequent on west.

14 TRITONIIDAE

Body elongated, quadrangular in cross-section. Pallial gills present. Oral veil often bilobed and produced into digitiform processes. Rhinophores issuing from dilated sheaths and with branched subterminal processes. Five species of *Tritonia* are recorded for Britain, but two are known only from a few specimens collected off south-west coasts.

---

1. Oral veil bilobed, with numerous digitiform processes. Very large, up to 200 mm **Tritonia hombergi**

   Oral veil not obviously bilobed; with fewer, longer processes. Smaller, up to 34 mm **2**

---

2. Body translucent white, with opaque white dorsal markings. Oral veil with four long tapering processes **Tritonia lineata**

   Body white to brown, with green and brown mottling. Oral veil rounded, with six processes, more or less equal length, but sometimes broken off **Tritonia plebeia**

Tritonia hombergi (Cuvier) Fig. 10.21
Juveniles translucent white with opaque white pigment on dorsum, darkening with age to dark purplish-brown mottling; ventral surface lighter. Oral veil large, bilobed; translucent white, or peach with denser white or coloured spots. Rhinophores with conspicuous trumpet-shaped sheaths, thick, branching into feathery processes, tips tinted greenish- or yellowish-brown. Mantle with soft tubercles and translucent white or peachy feathery gills; larger primary gills paired, with up to six branches, flexed towards mid-line, secondary gills with fewer branches, projecting laterally. Gill number increases with age. Foot translucent, with white spots.

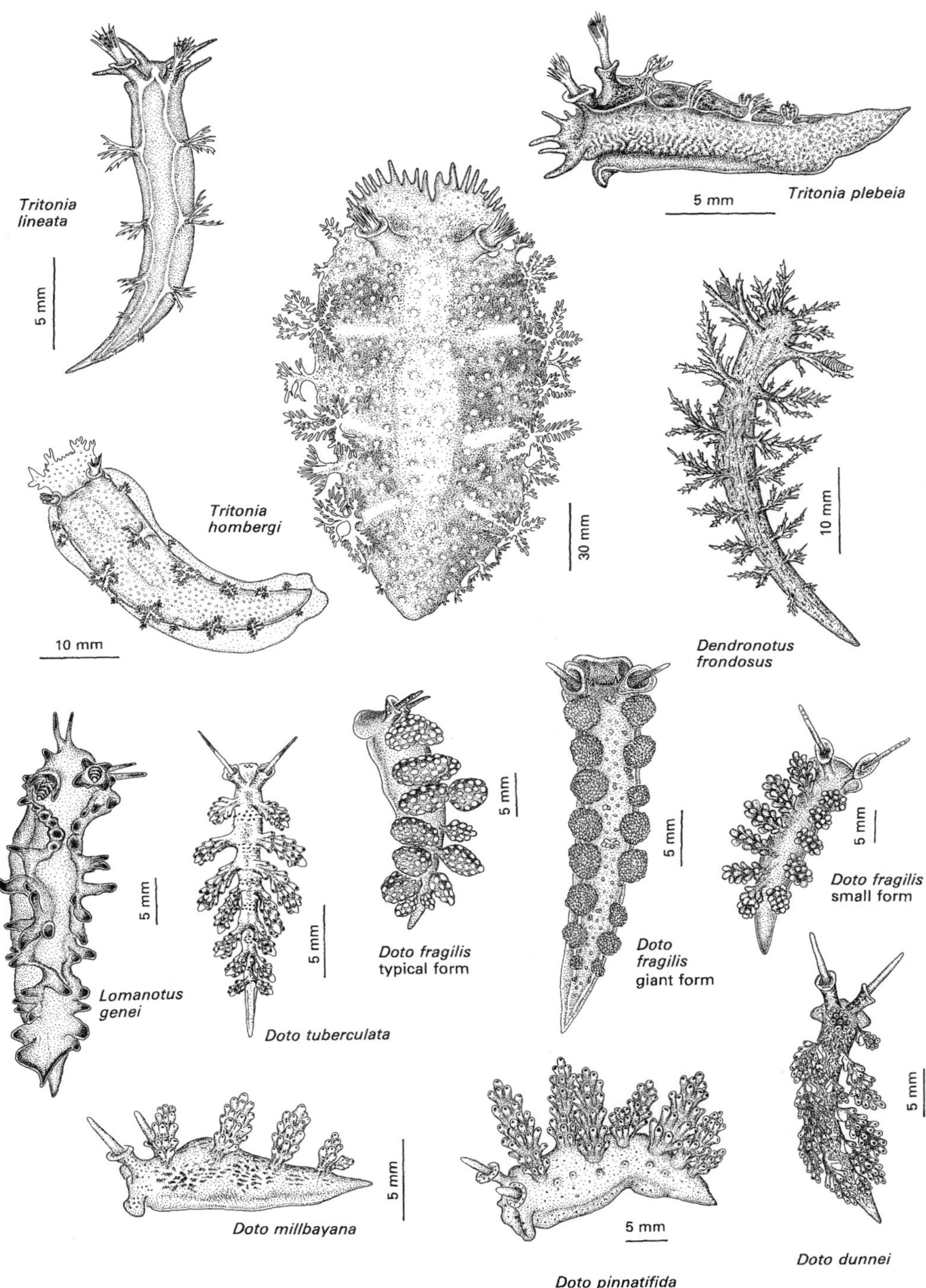
Tritonia lineata
5 mm
Tritonia plebeia
5 mm
Tritonia hombergi
10 mm
30 mm
10 mm
Dendronotus frondosus
Lomanotus genei
5 mm
Doto tuberculata
5 mm
Doto fragilis typical form
5 mm
Doto fragilis giant form
5 mm
Doto fragilis small form
5 mm
Doto dunnei
5 mm
Doto millbayana
5 mm
Doto pinnatifida
5 mm

Fig. 10.21

This is the largest British nudibranch and can attain 200 mm length. It is always found in association with the soft coral *Alcyonium digitatum*, on which it feeds. Shallow sublittoral, and offshore to 80 m; all British coasts but more common in the north.

Tritonia lineata
Alder and Hancock Fig. 10.21
Up to 34 mm; white, tinted rose from internal organs. A brilliant line of opaque white on each side of dorsum, meeting in a long point posteriorly. Oral veil processes white distally. Rhinophore tips sometimes tinged brown. Gills paired, 4–6 on each side, consisting of two main branching stems; increasing in size anteriorly, white. Foot translucent, lobed anteriorly, pointed posteriorly.

Diet uncertain, but found on *Kirchenpaueria pinnata*, on gorgonians and small octocorals such as *Sarcodictyon roseum*. Shallow sublittoral, to 40 m; less common than *T. hombergi*, localized, but occurring off all British coasts.

Tritonia plebeia Johnston Fig. 10.21
Up to 30 mm; white to pale yellow or brownish, with green and brown mottling. Oral veil translucent white with brown speckling. Mantle flatter than in other *Tritonia* species; pallial margin conspicuous, usually pale, with dark brown mottling on either side. Gills, up to 10 pairs, translucent whitish or yellowish.

Feeds on the soft corals *Alcyonium digitatum* and *Paramuricacea placonis*, and the gorgonian *Eunicella verrucosa*. Shallow sublittoral to 129 m, all British coasts, although there are no records for the extreme south-east.

15 LOMANOTIDAE

Elongated; conspicuous pallial rim produced laterally, with wavy margin. Cerata present. Bilobed oral veil with two pairs of digitiform tentacles. Rhinophores swollen and lamellated, issuing from dilated sheaths. Two British species, only one of which is common. *Lomanotus marmoratus* (Alder and Hancock) is known from a few localities around south-west England.

Lomanotus genei Verany Fig. 10.21
Up to 94 mm; colourless, or brightly coloured, pale orange-red, deep crimson or purple; unicolorous, without spots, cerata may be tipped with yellow or orange. Cerata comparatively short and flat, often curved at the ends and bearing a series of longitudinal wrinkles on the under surface.

Shallow sublittoral, feeding on *Nemertesia ramosa*. All British coasts.

16 DENDRONOTIDAE

Body limaciform with 4–8 pairs of pallial cerata. Oral veil with 2–5 pairs of digitiform tentacles, usually subdivided. Long, swollen, lamellated rhinophores, sheaths with branched rims.

One British species.

Dendronotus frondosus (Ascanius) Fig. 10.21
Up to 100 mm. Colour varying with age: juveniles to 4 mm pale; from 4 to 30 mm, uniform or mottled white to orange or reddish-brown, often in contrasting bands and stripes; specimens larger than 30 mm becoming rather drab. Dorso-pallial rim not as conspicuous as in tritoniids but produced into richly arborescent gills: up to nine pairs of large gills, with smaller ones in between. Rhinophore sheaths and anterior margin of head similarly arborescent.

Juveniles feed on calyptoblastic hydroids, such as *Hydrallmania*, *Obelia*, and *Sertularia*, adults on larger gymnoblastic species, especially *Tubularia indivisa* and *T. larynx*. In shallow inshore waters off all British coasts.

17 DOTOIDAE

Body with tuberculate, swollen pallial cerata; usually a single row down each side; *pseudobranchs* (small gill-like excrescences) on the mesial face of largest cerata in some species. Rhinophores long and tapering. Sheaths tall and flared. Twelve species recorded for Britain, in the single genus *Doto*, but it is probable that others may yet be recognized.

1. Tubercles on cerata with coloured distal spots **2**
   Not as described **5**

2. Body very densely pigmented with blackish-red spots of varying shapes. Rhinophore sheaths trumpet-shaped, speckled inside. Pseudobranchs present as well as crests on head ***Doto dunnei***
   Body pale with streaks or mottling **3**

3. Distal tubercular spots large, dark red to purple-red; body with red streaks or blotches. Five to eight pairs of cerata **_Doto coronata_**

Distal tubercular spots brown or black **4**

4. Black or brown spots. Lateral papillae present between cerata, each tipped with a black spot. Prominent crests running anteriorly from rhinophores **_Doto pinnatifida_**

Black spots, except on terminal tubercle. Pseudobranchs present. Transverse rows of black-tipped tubercles on dorsum between pairs of cerata; lateral longitudinal row of tubercles outside cerata **_Doto tuberculata_**

5. Small head. Large pseudobranchs present on the cerata; scattering of brown or black spots on cerata. Rhinophore sheath rims white **_Doto millbayana_**

Pale fawn to orange body; rhinophores ochre. Form, colour, and size of cerata variable **_Doto fragilis_ complex**

Doto coronata (Gmelin) Fig. 10.17

Up to 12 mm; translucent white or pale yellow, with crimson, red, maroon, or purple pigmented spots coalescing to form streaks or blotches, covering much of dorsum, sometimes concentrated at bases and sides of cerata, with a dark maroon streak in centre of each. Rhinophores translucent white with spots of opaque white. Head relatively large with only slight crests; rhinophores long, smooth, digitiform; sheaths either trumpet-shaped with smooth rims slanting upwards from median side, or with a tongue-like protrusion of anterior margin. Five to eight pairs of cerata, each with three to four rings of tubercles; a round, dark red or maroon spot on each tubercle. Oral veil dilated into distinct lateral lobes, with smooth contours. Propodium lobed, metapodium elongate, mantle fusing into it.

Intertidal and sublittoral, to 180 m; common off all British coasts. Feeds on gymnoblastic and calyptoblastic hydroids, including *Dynamena pumila*, *Clava multicornis*, *Obelia* spp., *Hydrallmania falcata*, *Sertularia* spp., *Tubularia larynx*.

Doto dunnei Lemche Fig. 10.21

Up to 25 mm; fawn-yellow to brownish, with very dense pigmentation of maroon to black spots. Seven or eight pairs of light brown cerata, each with seven rings of 5–8 tubercles; pyriform, decreasing in size basally. Tubercles lighter than cerata, with distinct spot at tip. Wing-like pseudobranchs on each ceras, with 3–5 leaves on each side. Rhinophores very long, rounded distally; sheaths wide and trumpet-shaped, speckled within, more so on outer anterior side. Oral veil with distinct rounded lateral flaps, notched basally; head with crests. Propodium simply rounded, not lobed, notched on mid-line.

In shallow water, on *Kirchenpaueria pinnata*. Western shores of Britain and Ireland.

Doto fragilis (Forbes) Fig. 10.21

Three forms are recognized, perhaps representing three separate species. Typical form up to 26 mm, brownish-yellow, sometimes with orange tinges. Up to nine pairs of fusiform cerata, unspotted, stalk darker than body, with up to seven concentric rows of pale tubercles. Rhinophores elongate, digitiform, three times length of sheaths, ochre with cream to yellow tips; sheaths flared, with white, slightly wavy rims. Oral veil with paired lateral lobes. Found on *Nemertesia* spp.

Giant form up to 34 mm, orange-brown with distinctive translucent white foot and metapodium. Cerata rounded and compressed; tubercles numerous, tightly packed, cream-coloured; additional clumps of tubercles scattered between cerata and towards mid-line. Rhinophores digitiform, only twice length of sheaths; flared, crenulate rhinophore sheaths, orange-brown with paler rims. On *Halecium muricatum*.

Small form uniform pale creamy-orange. Cerata less rounded than in giant form, but less elongate than typical form, slightly darker than body; tubercles in about five concentric rows, each with pigmented end. Rhinophores ochre with white tips, sheaths half length of rhinophore. On *Halecium halecinum*.

All three forms occur commonly around the whole British coast, from the lower shore to about 200 m.

Doto millbayana Lemche Fig. 10.21
Up to 14 mm, whitish, digestive gland showing yellowish through body wall, brown in cerata; scattered blackish-brown spots on bases of rhinophore sheaths and on cerata. Distal ends of rhinophores and rims of sheaths white. Six pairs of cerata, each with four or five rings of 5–8 low, rounded tubercles; well-developed pseudobranchs, each with five processes in feather-like arrangement, spotted at tips. No papillae on body. Oral veil with very small, almost semicircular flaps, directed antero-laterally. Prominent head crests. Foot with a slight transverse furrow anteriorly, but no distinct corners or propodial tentacles. Propodium broader than metapodium, which tapers sharply to acute end.

Feeds on *Plumularia setacea*, a small hydroid sometimes associated with *Nemertesia ramosa*, which is also occasionally eaten. Deposits flat, broad, winding spawn ribbon. Not common. Recorded around the Cornish peninsula, and from the west coast of Ireland, and Orkney.

Doto pinnatifida (Montagu) Fig. 10.21
Up to 29 mm; cream or pale fawn to pale orange, with brown to black mottling down centre. Up to 10 pairs of cerata, unicolorous with body; tubercles in five or six concentric rows, long, conical, spread apart, nearly all tipped with a dark brown or black spot. Lateral papillae between larger cerata, also tipped black. Rhinophores translucent, with white flecks along most of length; sheaths flared, rims with black spots, sometimes undulate. Oral veil smoothly rounded. Head region appearing concave due to white speckled lateral ridges extending from rhinophores.

Shallow sublittoral, feeding on *Nemertesia antennina*. On western coasts of Britain and Ireland, may be locally common.

Doto tuberculata Lemche Fig. 10.21
Up to 19 mm; pale fawn, orange or greenish-yellow, with finely dispersed darker brownish speckling on dorsum. Foot, ceratal tubercles, and rhinophores without speckling. Usually eight pairs of cerata, slightly compressed, swollen in middle, each with three or four rings of simple or complex, short, conical tubercles; all, except terminal one, tipped with a black dot. Pseudobranch, in larger specimens, forming a ridge with five small lobes, the largest with a distal spot. Two to four black-tipped, raised tubercles in rows across body between each two pairs of cerata. Pallial border just visible as a yellowish crest running longitudinally between cerata, fading towards last pairs. Foot rounded anteriorly, slightly notched medially, expanding and then narrowing to merge with lateral margins. Oral veil with a small rounded lobe on each side. Rhinophores pigmented white distally, about three times length of sheaths which are flared, with undulating edge.

Shallow sublittoral, on *Sertularella gayi* and *Abietinaria abietina*. South-west England and north-west Ireland.

18 ONCHIDORIDIDAE

Phanerobranchiate dorids (gills separately contractile), flattened and oval with mantle skirt usually larger than foot. Rhinophores lamellated, no sheaths. Feed on encrusting bryozoans. Eleven species recorded for Britain.

---

1. Brightly coloured dorsum: white with central red region and yellow submarginal band
   Onchidoris luteocincta

   Body pigment not as described **2**

---

2. Body uniformly pigmented **3**

   Body large, up to 40 mm, with brown blotches; up to 29 gills
   Onchidoris bilamellata

---

3. Body usually >20 mm, varying in colour, with long, soft, pointed papillae; rhinophore sheaths low, papillate
   Acanthodoris pilosa

   Body <17 mm, yellow or white with tubercles **4**

---

4. Dorsum usually white, with stout, flattened tubercles; rhinophores and gills colourless; gills in horseshoe around anal papilla
   Onchidoris muricata

   Body usually yellow, with fusiform tubercles which elongate at periphery; gills and lamellate portion of

rhinophores usually darker; gills form circlet around anal papilla
Adalaria proxima

Acanthodoris pilosa (Abildgaard in Müller) Fig. 10.22
Commonly about 30 mm, but up to 70 mm; white, pale grey, or brown, to dark purplish-black or charcoal grey. Dorsum covered with long, soft, pointed papillae, of uniform length. Rhinophores with long conical stalks, the distal, lamellate ends posteriorly directed. Up to nine large gills encircling anal papilla. Metapodium visible posterior to mantle.

Intertidal and shallow sublittoral; usually found on its bryozoan prey. *Alcyonidium* spp., *Flustrellidra hispida*, *Callopora* spp. Common on all British coasts.

Adalaria proxima (Alder and Hancock) Fig. 10.22
Up to 17 mm, white or yellow. Dorsum covered with club-like papillae, becoming elongate towards periphery. Rhinophores digitiform, diverging and directed posteriorly; stalk translucent, lamellae often darker yellow than dorsum. Up to 12 gills encircling anal papilla, often darker than dorsum; frequently with tubercles within gill circlet. Gills small in comparison with *Acanthodoris pilosa*. Often confused with *Onchidoris muricata*, which has rather uniform, mushroom-shaped, mantle tubercles, which do not change form towards periphery. Rhinophores are blunter and the digestive gland extends further to the anterior in *A. proxima*.

Feeds on a variety of encrusting bryozoans, especially *Electra pilosa*; intertidal and shallow sublittoral to 60 m. South Cornwall, Bristol Channel, Irish Sea, western Ireland, and north coasts of Britain.

Onchidoris bilamellata (Linnaeus) Fig. 10.22
Up to 40 mm, dull white with brown blotches of varying intensity, decreasing towards periphery. Dorsum with rough, club-like tubercles, becoming more elongate towards margins. Rhinophores more densely pigmented basally than distally; lamellae rough, uneven, with irregular edges. Gill circlet relatively large; gills speckled brown around edges; tubercles inside gill circlet.

Feeds on barnacles; intertidal and shallow sublittoral to 20 m. All British coasts.

Onchidoris luteocincta (M. Sars) Fig. 10.22
Up to 11 mm, dorsum white, with red central region and yellow sub-marginal band. Tubercles sharply conical. Metapodium extending beyond dorsum; with obvious keel and crenulated edge marked in opaque white, tip often yellow.

Intertidal and shallow sublittoral, feeds on the bryozoans *Smittoidea reticulata*, *Cellepora pumicosa*, and *Crisia*.

Onchidoris muricata (Müller) Fig. 10.22
Up to 14 mm, white, or rarely pale yellow. Mantle noticeably spiculose, with abundant, stout, flattened, spiculose tubercles. Rhinophores colourless. Up to 11 colourless gills in a horseshoe around anal papilla. Frequently confused with *Adalaria proxima* but generally paler in colour, with more pointed rhinophores; tubercles more flattened than in *A. proxima*, and digestive gland, in ventral view, situated farther to posterior.

Lower shore and shallow sublittoral, to 15 m, feeding on encrusting bryozoans. All British coasts, but more common in north.

19 GONIODORIDIDAE

Phanerobranchiate, soft and limaciform (may be broad—*Okenia*) with reduced mantle which often has long papillae. Rhinophores unsheathed, lamellate, sometimes with long papillae. Feed on encrusting bryozoans and ascidians. Nine British species, only three of which are common.

---

1. Dorsum keeled, without conspicuous projections **2**

   Dorsum not as described **3**

---

2. Translucent milky-white ground colour; dorsum with luminescent white spots, yellow tinges around margins and on rhinophores and oral tentacles
Goniodoris nodosa

   White, with crimson dorsal blotches and yellow submarginal band
Onchidoris luteocincta (this page)

---

3. Broad, elongate mantle projections issuing from a raised rim around

rhinophores and gills, longest anteriorly. Gills feathery, numerous. Colourful: white, red, and yellow ***Okenia elegans***

Elongate, slender and smooth dorsal projections: posteriorly, a ring or a pair beside three tripinnate gills, anterior projections issuing from rhinophore bases. Ring of pallial processes white with brilliant orange or yellow tips ***Ancula gibbosa***

Goniodoris nodosa (Montagu) Fig. 10.22
Up to 27 mm, translucent white with almost luminescent opaque white spots. Mantle reduced, showing foot; edges curled, translucent, often tinged yellow at margin. Areas around rhinophores, gill circlet, anterior foot, most of oral tentacles and proximal region of gills without spots. Notum with median keel. Metapodium shortened, yellow distally, with median keel. Rhinophores evenly lamellated, with faint yellow tinge and white spots. Oral tentacles smooth, yellowish, with dense white pigment at tips. Up to 13 fat, fleshy gills.

All British coasts, shallow inshore waters to 120 m, common. Juveniles on encrusting bryozoans such as *Alcyonidium gelatinosum* and *Callopora dumerilii*; adults on compound ascidians such as *Botryllus schlosseri*, *Dendrodoa grossularia*, and *Diplosoma listerianum*.

Okenia elegans (Leuckart) Fig. 10.22
Up to 80 mm. Mantle much reduced, but with tentacles around whole of margin; typically dark red, marginal processes red or orange with yellow or white tips. A mid-dorsal row of processes posterior to rhinophores, terminating with a pair anterior to gills. Two long, slender tentacles anteriorly. Foot broad, translucent white or pink, with yellow edge. Rhinophores same colour as mantle at base, pale just proximal to lamellae, dark red on distal third, with yellow terminal knob. Up to 21 gills, white basally, grading from red to yellow distally.

Usually in association with its prey, solitary ascidians such as *Ciona*, *Molgula* spp, and *Polycarpa* spp.

Ancula gibbosa (Risso) Fig. 10.22
Up to 33 mm, translucent cream or white. Rhinophores and gills with coloured tips; tips of pallial processes and rhinophoral processes bright yellow or yellow-orange, rarely white. Rhinopores with crinkly stalks, cylindrical in region of lamellae, swollen medially, each with 10 lamellae, flattening distally, ending in a long blob; elongate processes issuing from rhinophore stalk, not from head. Three pinnate gills, encircled by 14 pallial processes. Oral tentacles elongate, finger-like, directed antero-laterally. Metapodium sometimes with a yellow tip, sometimes with a median row of yellow pigment. Foot margins may be coloured yellow.

This species is known to eat the compound ascidians *Diplosoma listerianum*, *Botryllus schlosseri*, and *Botrylloides leachi*. It has been found among erect bryozoans and may feed on these as well. Under rocks in shallow water, offshore to 110 m; all British coasts.

20 POLYCERIDAE

Soft, limaciform phanerobranchiates. Pallial margin often with elongated papillae laterally and anteriorly. Rhinophores lamellated, sometimes sheathed. Feeding on encrusting and erect bryozoans. Seven British species.

---

1. Frontal veil with digitiform or tubercular papillae **2**

   Frontal veil smooth, dorsum with black and yellow or orange spots ***Thecacera pennigera***

---

2. Frontal veil papillae long and digitiform **3**

   Frontal veil papillae short. Body bright orange with blue spots ***Greilada elegans***

---

3. Frontal veil papillae smooth, tapering. With paired, smooth, digitiform processes flanking gills. Dorsal oval blotches of orange on slightly raised mounds ***Polycera quadrilineata***

   Frontal veil papillae feathery, mantle with paired digitiform processes ***Limacia clavigera***

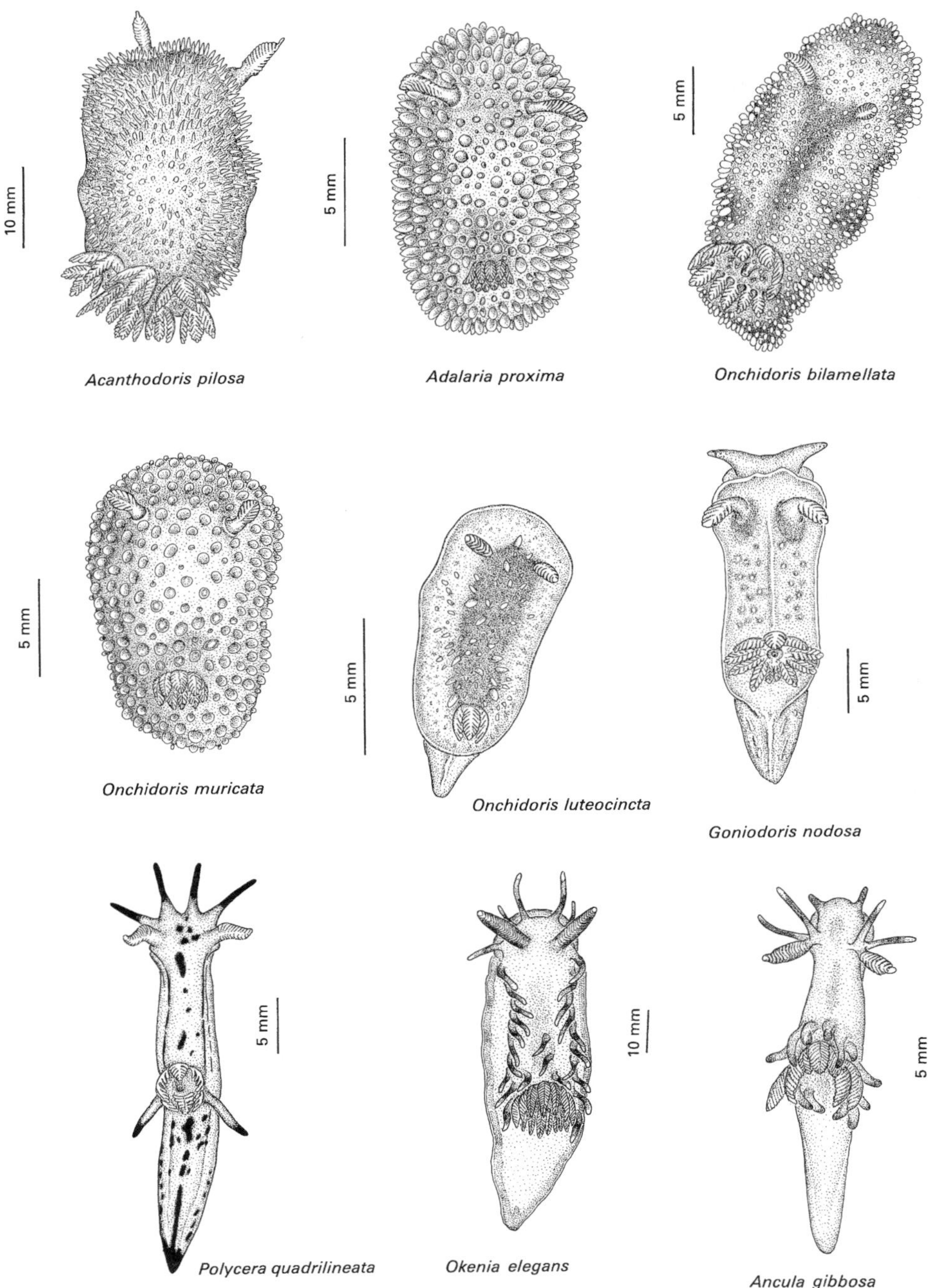

10 mm
5 mm
5 mm
Acanthodoris pilosa
Adalaria proxima
Onchidoris bilamellata
5 mm
5 mm
5 mm
Onchidoris muricata
Onchidoris luteocincta
Goniodoris nodosa
5 mm
10 mm
5 mm
Polycera quadrilineata
Okenia elegans
Ancula gibbosa

Fig. 10.22

Polycera quadrilineata (Müller) Fig. 10.22

Up to 39 mm; translucent white with patches of yellow and orange, typically with a row of oval blotches along mid-line, merging posteriorly with a line along metapodium; occasionally with black streaks, blotches, or fine spots on dorsum. Pallial margin edged with yellow-orange, usually with spots on flank. Four frontal veil papillae, rarely more, coloured yellow. Pallial process on each side of gills elongate, pointed, long and thin; up to 11 pinnate gills. Distal tips of pallial processes, and gill rachides, yellow. Rhinophores with thick stalks, scarcely narrower than lamellate portions, which are held angled to stalks, terminated with cylindrical knob; distal portion of rhinophore yellow.

Intertidal and sublittoral to about 60 m, on encrusting Bryozoa. All British coasts. Some specimens may be almost entirely black, or grey with black speckles.

Greilada elegans Bergh Fig. 10.23

Up to 43 mm; bright orange with iridescent blue spots (rarely with white spots). Frontal veil produced anteriorly into small digitiform papillae. Pallial margin tuberculate, most usually white, fusing posterior to gills to form a keel; metapodium also keeled, with a few blue tubercles, blue mounds on flanks also continuing on to metapodium. Rhinophores large, lamellate portion angled to stalk, bright orange. Up to seven large, feathery gills, anterior to anal papilla, each white-edged, with blue dots on inner surfaces.

Feeds on the bryozoan *Bugula*, especially *B. turbinata*. Reported from south-west England and western Ireland, from shallow sublittoral habitats.

Thecacera pennigera (Montagu) Fig. 10.23

Up to 30 mm, white with small black and yellow spots, and larger, irregular orange blotches. Three to five pinnate gills, and posteriorly two long, white-tipped digitiform processes. Rhinophores lamellate, with distinctive smooth rimmed sheaths produced posteriorly, and with lateral, median openings. Head rounded, lacking oral tentacles but with small, paired propodial tentacles.

Feeds on the bryozoan *Bugula plumosa*, on which it also lays its spawn. Shallow sublittoral. Rare; recent records are from Dover, South Cornwall, and the Bristol Channel.

Limacia clavigera (Müller) Fig. 10.23

Up to 18 mm, white with yellow or orange dorsal papillae; rhinophores, gills, processes, and cerata also tipped with yellow or orange. Pallial margin with a paired series of projecting cerata; head fringed with similar, smaller processes with feathery ends. Rhinophores long, lamellate, with short stalks issuing from low, thick-rimmed sheaths; oral tentacles long, conspicuous, with dorsal grooves. Three to five pinnate gills anterior to anal papilla. Foot white, except at tip where marginal and median pigment bands merge.

Shallow sublittoral, to 80 m, on encrusting bryozoans, including *Membranipora*, *Electra*, *Cryptosula*, *Callopora*, and *Schizoporella*. All British coasts, except extreme south-east.

21 CADLINIDAE

Cryptobranchiate (gills retractile simultaneously into common branchial pit); broad, with ample mantle skirt. Rhinophores lamellate. Propodium bilaminate. Feeding upon encrusting sponges. One British species.

Cadlina laevis (Linnaeus) Fig. 10.23

Up to 32 mm; translucent white, with opaque white or yellow edge to mantle. Dorsum with opaque, conical tubercles of varying size; mantle skirt with distinctive, glistening, white or yellow subepidermal glands. Rhinophores short, tapered, white, often with yellow tips; sheaths low with crenulate rims. Five, rarely six or seven, tripinnate gills.

Intertidal and shallow sublittoral, feeding on sponges. Predominantly northern but recorded from all British coasts.

22 ROSTANGIDAE

Cryptobranchiate; broad, with ample mantle skirt. Dorsum finely tuberculate. Rhinophores lamellate. Propodium bilaminate, anterior lobe notched medially. Feeding on encrusting sponges. One British species.

Rostanga rubra (Risso) Fig. 10.23

Up to 15 mm; bright red, with scattered black spots and a rough, yellowish patch between rhinophores. Mantle coarsely tuberculate. Head with short, lateral oral tentacles; rhinophores short and swollen, cream-coloured, often with red spots on rachides. Up to 10 short, pinnate gills.

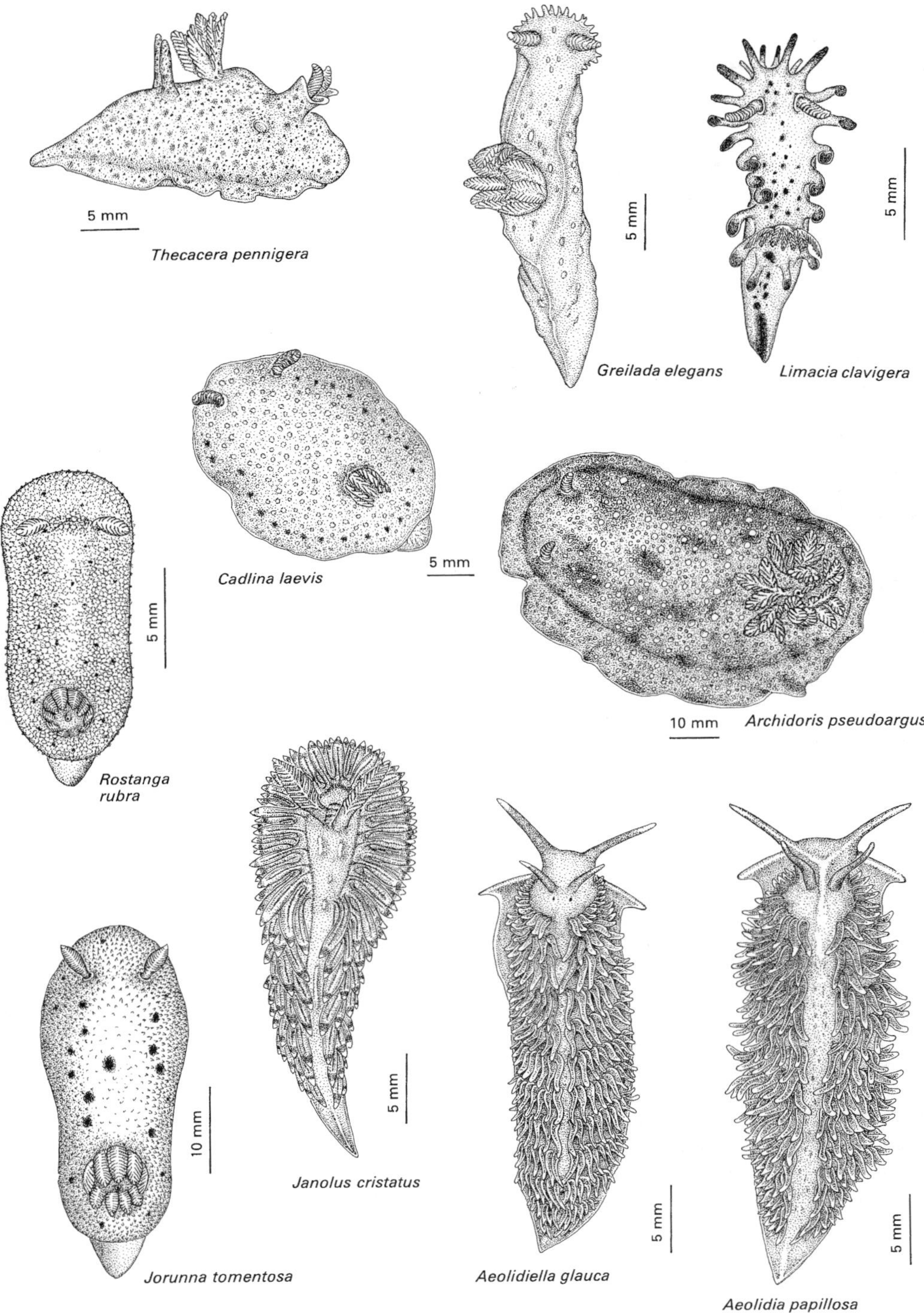
5 mm
Thecacera pennigera
5 mm
Greilada elegans
5 mm
Limacia clavigera
Cadlina laevis
5 mm
5 mm
10 mm
Archidoris pseudoargus
Rostanga rubra
5 mm
Janolus cristatus
10 mm
Jorunna tomentosa
5 mm
Aeolidiella glauca
5 mm
Aeolidia papillosa

Fig. 10.23

Intertidal and shallow sublittoral, feeding on red sponges such as *Microciona atrasanguinea* and *Ophlitaspongia seriata*. West coasts only.

23 ARCHIDORIDIDAE

Cryptobranchiate; broad, with ample mantle. Dorsum tuberculate, some tubercles coalescing. Rhinophores lamellated. Sheaths sometimes present. Propodium bilaminate. Feeding on encrusting sponges. Two species in Britain, but *Atagema gibba* Pruvot-Fol is known from just a single Cornish locality.

Archidoris pseudoargus (Rapp)
Sea lemon Fig. 10.23

Up to 120 mm long, oval, firm; variously mottled and blotched with yellow, green, brown, and red. Mantle coarsely tuberculate, tubercles densely spaced, rounded and of varying sizes. Head with vestigial oral tentacles; rhinophores short, conical, pale yellow. Up to nine tripinnate gills, translucent, with brown pinnules.

Juveniles tend to be nondescript creamy yellow, but always with purple-brown patch of pigment between tubercles, anterior to gills. Dark specimens may be confused with *Discodoris planata* (Alder and Hancock), which is distinguished by white, stellate dorsal markings, and small but distinct oral tentacles.

The most common, and perhaps familiar, of the British nudibranchs. Lower shore, and offshore to 300 m; feeds mainly on the sponge *Halichondria panicea* or, at greater depths, on other siliceous sponges. All British coasts.

24 KENTRODORIDIDAE

Cryptobranchiate; broad, ample mantle skirt. Dorsum finely tuberculate. Rhinophores lamellated. Head with short oral tentacles; propodium bilaminate, anterior lobe notched. Feeding on encrusting sponges. One British species.

Jorunna tomentosa (Cuvier) Fig. 10.23

Up to 55 mm, elongate oval, posterior portion of foot visible in dorsal view; cream, fawn, or sandy brown, almost always with a loosely paired series of dark-brown blotches, as well as speckles of light brown. Mantle with small, uniform, spiculose tubercles, with a soft, velvety texture; rhinophore and gill pockets slightly raised. Head with slender oral tentacles; rhinophores darker or lighter than mantle, with brown flecks, but always white tipped. Up to 17 tripinnate gills, pale coloured.

Lower shore and shallow sublittoral, feeding on sponges, including *Halichondria panicea* and species of *Haliclona*. All British coasts, widespread and common.

25 JANOLIDAE

Elongated with tapering tail. Cerata present; digitiform, continuous around anterior. Rhinophores lamellate, or wrinkled and papillate. Three British species.

Janolus cristatus (Delle Chiaje) Fig. 10.23

Up to 80 mm, tapered posteriorly, with thickly distributed lateral cerata, continuous around whole of head. Transparent, with white patches on dorsum; cerata with brown central streak of digestive gland, tipped bluish-white; often with brilliant metallic or iridescent sheen. Rhinophores large, light brown or orange, with white tips; linked basally by a wrinkled light brown or orange sensory caruncle.

Offshore, to 40 m; feeding on the bryozoans *Bugula*, *Callopora*, and *Cellaria*. Locally common on all British coasts, except south-east.

26 AEOLIDIIDAE

Nudibranchs with cerata in rows, each containing lobes of the digestive system, and terminal cnidosacs storing unfired nematocysts obtained from the prey. Propodial tentacles and long oral tentacles present. Rhinophores smooth (British species). Feeding on actinians (British species). Four species are recorded from Britain, but only two are common.

---

1. Body large (up to 120 mm); colour variable, with a white band spreading over head to oral tentacles; rhinophore tips pale. A median dorsal zone lacking cerata **_Aeolidia papillosa_**

   Body up to 40 mm; no white pattern on head; rhinophore tips pigmented white. More substantially covered with cerata **_Aeolidiella glauca_**

Aeolidia papillosa (Linnaeus) Fig. 10.23

Up to 120 mm, commonly smaller; wide, flattened, tapered posteriorly, with prominent oral tentacles and rhinophores, and abundant, closely set, cerata; mid-dorsal region lacking cerata. Cream, fawn, brown, orange-pink, or dark purple-brown; usually a distinctive, pale V-shaped patch on head, between rhinophores and oral tentacles. Rhinophores and oral tentacles speckled with brown, tipped with white; rhinophores usually darkest towards base. Cerata elongate, straight or with recurved ends, always tipped with white.

The largest North Atlantic aeolid, *A. papillosa* is widespread and common on all British coasts, on the shore and in the shallow sublittoral. It feeds on sea anemones, including species of *Actinia*, *Anemonia*, *Metridium*, *Sagartia*, and *Tealia*.

Aeolidiella glauca
(Alder and Hancock) Fig. 10.23

Up to 40 mm, broad, with prominent oral tentacles, and numerous cerata in distinct rows; translucent white, pale yellow, grey-brown, or orange-pink, depending on diet. Dorsum, cerata, and tentacles covered with white spots or blotches. Cerata with creamy white tips, a pinkish subterminal tinge, and pale brown to greenish-brown digestive gland.

Lower shore, and shallow sublittoral, feeding on various species of actinians, including *Sagartia elegans*. All British coasts.

27 EMBLETONIIDAE

Flattened aeolidaceans with swollen cerata in single series down each side of body. Ceratal tips rounded (British species); cnidosacs absent and replaced by terminal pads of nematocysts. Smooth rhinophores; dilated, bilobed oral veil; propodial tentacles absent. One British species.

Embletonia pulchra
(Alder and Hancock) Fig. 10.24

Up to 7 mm; translucent white or pink with white epidermal speckling or blotches. Cerata in a single row of up to seven on each side; digestive gland pale yellow to orange, red, or brown. Rhinophores smoothly tapering, widely spaced.

Shallow sublittoral; inconspicuous and frequently overlooked. Perhaps feeds on common shallow-water hydroids, such as *Hydrallmania*, *Nemertesia*, *Dynamena*, and *Tubularia*. All British coasts.

28 EUBRANCHIDAE

Aeolidaceans with swollen cerata of varying shapes. Arranged in simple or branched rows. Smooth rhinophores and oral tentacles; propodial tentacles absent. Feeding on hydroids. Seven British species, in the single genus *Eubranchus*, but two are known from just two or three records.

---

1. Cerata inflated **2**

Cerata smooth and digitiform, or knobbly, up to 10 rows, olive-green or brown, sometimes banded ***Eubranchus cingulatus***

---

2. Cerata flattened, not in rows; each with white tip and yellow subterminal ring ***Eubranchus tricolor***

Cerata not as described **3**

---

3. Cerata not obviously in rows, white with orange or yellow subterminal ring; rhinophores and oral tentacles orange-tipped ***Eubranchus farrani***

Not as described **4**

---

4. Maximum of 10 cerata per side, sparse, urn-shaped, with brown pigment usually forming interrupted bands ***Eubranchus exiguus***

Many cerata, in closely packed rows, white mottled with brown and orange, with yellow subterminal ring ***Eubranchus pallidus***

Eubranchus cingulatus
(Alder and Hancock) Fig. 10.24

Up to 29 mm; greyish-brown with white spots, and variably developed spots, blotches, or large patches of olive-green or brown, particularly between bases of cerata. Up to 10 rows of up to 13 long, digitiform cerata; slightly inflated, and appearing knobbly when contracted, but not distinctly tuberculate. Tips of cerata transparent, with white subterminal band, and two or three incomplete bands of olive-green or brown; white spots usually present, digestive gland lobes white, buff, pale yellow, or pale brown. Rhinophores about twice length of oral tentacles; both tipped with white, with subterminal band of olive green or brown, and sparse or dense speckles of white and olive or brown.

Feeds on the hydroid *Kirchenpaueria pinnata*. Records at present unreliable through confusion with *E. doriae* (Trinchese). This species is smaller than *E. cingulatus* (up to 8 mm long), similar in coloration, but with inflated, tuberculate cerata, in up to seven rows, each with at most eight cerata.

Eubranchus exiguus
(Alder and Hancock) Fig. 10.24
Up to 10 mm, translucent greyish- or yellowish-white with speckles or patches of olive-brown on oral tentacles, head, rhinophores, dorsum, and cerata. Opaque white dots scattered all over body. Metapodium long and flattened with definite edge between foot and mantle; translucent white with olive-brown marginal pigmentation. Rhinophores almost smooth, white tipped, with olive-brown pigment, which may form a distinct band. Oral tentacles smooth, about half length of rhinophores, with similar pigmentation. Cerata few, usually 10 or fewer, not in rows; variable shape, usually urn-like, but never knobbly; each with clear tip above olive-brown bands; up to three colour bands present on each ceras. Hepatic lobes pale, often colourless.

On *Obelia* spp., *Laomedea flexuosa*, and other small, shallow-water calyptoblastic hydroids, often together with *Tergipes tergipes*. All British coasts.

Eubranchus farrani
(Alder and Hancock) Fig. 10.24
Up to 20 mm, slender; up to 10 rows of cerata, with a maximum of five per row, inflated and not flattened. Occurs in five colour forms; intermediate forms perhaps occur and juveniles may be completely colourless except for hepatic contents:

1. White or greyish-white, with ends of cerata, rhinophores, and oral tentacles suffused with orange or orange-yellow; tips clear, with white glands below. Scattered patches of orange and orange-yellow on dorsum and head, and sometimes on cerata.
2. Chocolate-brown, with or without orange bands on cerata, oral tentacles, and rhinophores. Tips of cerata translucent, usually patches of orange on dorsum, white glands not usually visible.
3. Golden, yellow, or orange; cerata with translucent tips and white glandular rings. Body usually pale; cerata rather slender.
4. Translucent white; cerata with white cnidosacs and tips; hepatic lobes pale brown, buff, or orange, clearly visible. With white speckling; rhinophores and oral tentacles often with orange pigment, cerata and body may have a few orange patches.
5. Translucent, with no coloured pigmentation; rhinophores and oral tentacles translucent; cerata with white tips. Brownish hepatic lobes visible in cerata.

Feeds on calyptoblastic hydroids, such as *Obelia*, attached to kelps; occurs in shallow, sublittoral water off all British coasts.

Eubranchus pallidus
(Alder and Hancock) Fig. 10.24
Up to 23 mm; translucent greyish-white, with markings varying from yellow-orange to orange, crimson, orange-brown, and dark brown; small white spots always present, sometimes forming larger patches. Cerata inflated, up to ten rows of seven; with pale or translucent tips, and subterminal white bands, often interspersed with, or partially obscured by, a yellow or gold ring; covered with superficial white specks and orange, red, or brown patches or spots. Propodium wider than in other *Eubranchus* species. Oral tentacles translucent with white or clear tips, speckled with white; rhinophores about twice as long as oral tentacles, band of orange to red present subterminally.

Offshore, on shell gravel and stones, feeding on various calyptoblastic and gymnoblastic hydroids; adults are found most often on *Tubularia*, juveniles on smaller hydroids attached to *Tubularia*. All British coasts.

Eubranchus tricolor Forbes Fig. 10.24
Up to 45 mm, pale yellow or greyish-white. Cerata flattened, inflated, not arranged in rows; unicolorous with body, with subterminal ring of yellow or orange-yellow, white cnidosac extending from tip to just beyond yellow ring. Digestive gland extensions smooth, brown or red-brown, usually darker brown or violet at tip, just below cnidosac. Rhinophores smooth, tapering, suffused with pale yellowish-brown, except at tips.

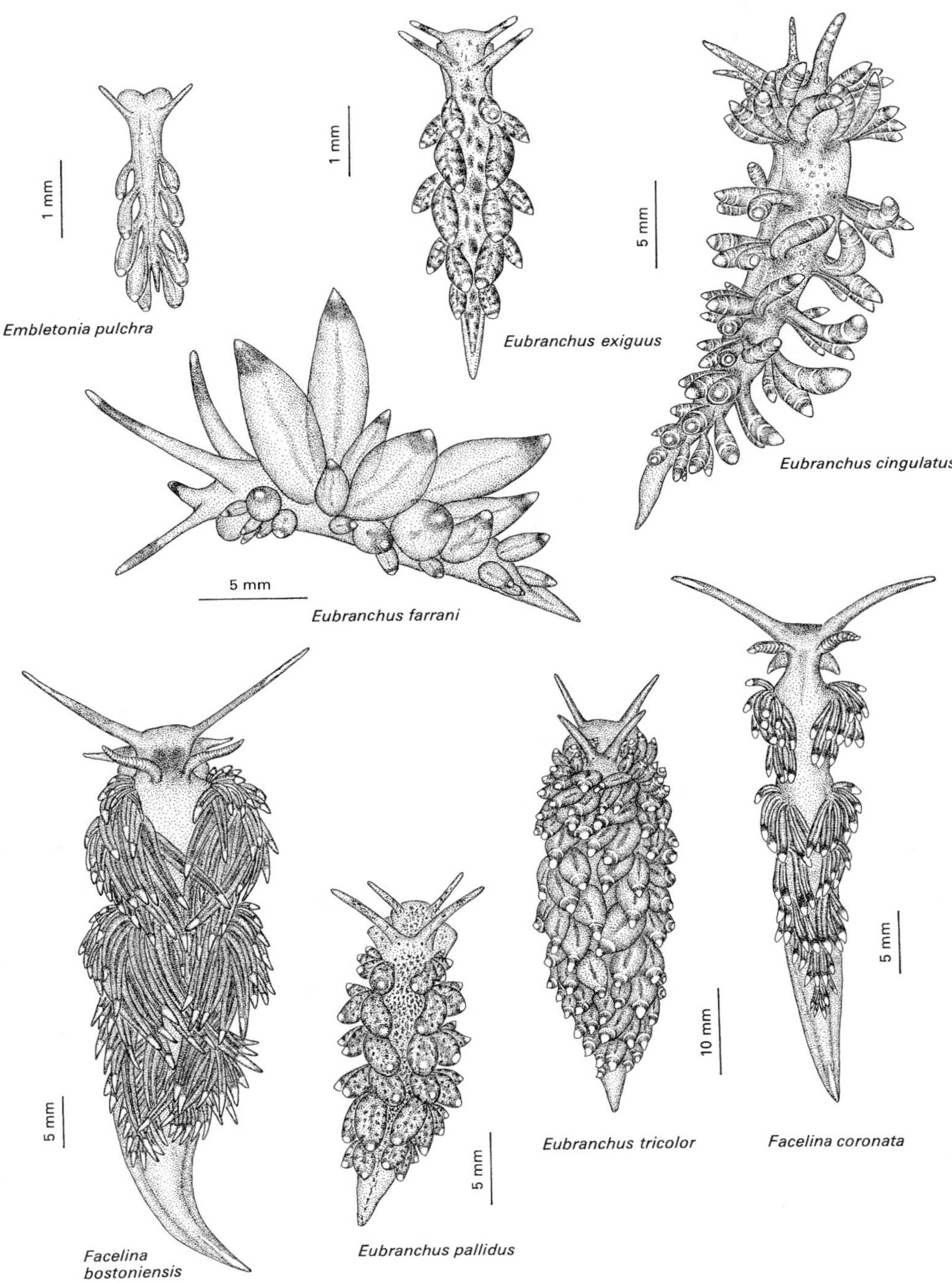
1 mm
Embletonia pulchra
1 mm
Eubranchus exiguus
5 mm
Eubranchus cingulatus
5 mm
Eubranchus farrani
5 mm
Facelina bostoniensis
5 mm
Eubranchus pallidus
10 mm
Eubranchus tricolor
5 mm
Facelina coronata

Fig. 10.24

Offshore, to about 80 m, on various hydroids, especially *Obelia geniculata*. All British coasts.

29 FACELINIDAE

Nudibranchs with regular rows of tapering cerata. Long lamellated rhinophores, lamellae as rings, or meeting obliquely. Propodial tentacles and long mobile oral tentacles both present. Propodium bilaminate, anterior lamina notched. Five species recorded from Britain.

---

1. Typically with blue iridescence over head and cerata; subterminal white on cerata extending basally on anterior side. Digestive lobules darker, or more reddish, at distal end
   Facelina coronata

   Translucent white body, with rose tinge. Cerata long, overlapping, not in clusters. White patch between rhinophores
   Facelina bostoniensis

Facelina bostoniensis (Couthouy) Fig. 10.24
Up to 55 mm, translucent white, tinged rose pink; a white patch between rhinophores, others may be present on head and pericardium; oesophagus usually clearly visible, red. Oral tentacles up to half body length, distal third sometimes white; rhinophores lamellated, tips with white pigment. Cerata in loosely defined clusters, anterior ones overlapping posterior; translucent, with distinct subterminal white areas; hepatic lobe light to dark brown, or greenish-brown, clearly visible. Foot wider than body.

Usually sublittoral, on *Tubularia*; intertidal on *Clava multicornis*, but will feed on a variety of hydroids. Common on all British coasts except extreme south-east.

Facelina coronata
(Forbes and Goodsir) Fig. 10.24
Up to 38 mm; translucent white with pink tint, usually with blue iridescence on head and cerata. Rhinophores half length of oral tentacles, light to dark brown, with ring-like lamellae; anterior surfaces and distal parts of oral tentacles with superficial white pigmentation. Red oesophagus always visible posterior to rhinophores. Cerata in distinct rows, with gap between rows one and two; iridescent blue with white subterminal bands extending basally on anterior side. Digestive lobules clearly visible, light brown to dark brownish-red, darkening or reddening distally. Foot little wider than body.

Intertidal and shallow sublittoral, on various hydroids, especially *Tubularia indivisa*. All British coasts, except south-east.

30 FAVORINIDAE

Similar to Facelinidae but with cerata in arcs or horseshoe shapes, not rows. Rhinophores smooth or with 1–3 bulbous swellings. Recurved propodial tentacles; large flexible oral tentacles present. Three British species, but only two are commonly encountered.

---

1. Rhinophores with three bulbous swellings near base; white pigment on oral tentacles extending on to head and dorsum as a streaky line
   Favorinus blianus

   Rhinophores with subterminal swelling and brown pigment below swelling; white pigment on propodial tentacles and head, forming a rhombus shape posterior to rhinophores
   Favorinus branchialis

Favorinus blianus
Lemche and Thompson Fig. 10.25
Up to 30 mm, pale straw yellow. Propodial tentacles with superficial white pigment; oral tentacles with white pigment extending on to head, and posteriorly as a streaky line on dorsum and metapodium. Rhinophores yellowish, with three bulbous swellings near base, a dark brown stripe on posterior side broadens over each swelling. Cerata digitiform, curved towards centre of body, arranged in horseshoe pattern, the first placed very close to rhinophores; white tipped, with white streaks distally on anterior side, hepatic lobes pale yellow-brown.

On lower shore, and subtidally to about 35 m; feeds on *Tubularia larynx* and the eggs of other nudibranchs. Western coasts of Britain and Ireland.

Favorinus branchialis (Rathke) Fig. 10.25
Up to 15 mm, delicate, slender; white, digestive gland varying in colour according to prey. Up to six clusters each of six or seven cerata; clusters

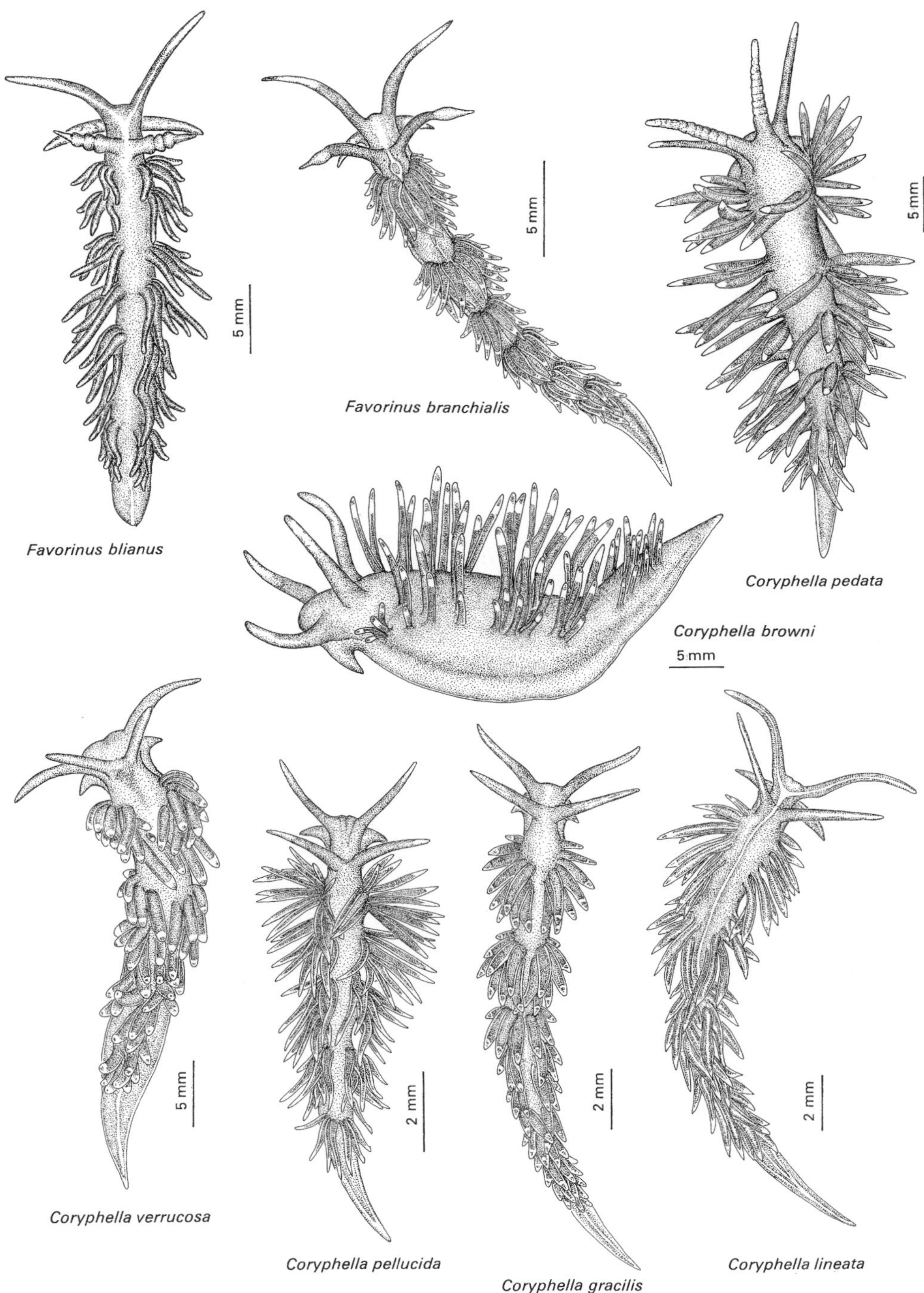
5 mm
Favorinus blianus
5 mm
Favorinus branchialis
5 mm
Coryphella pedata
Coryphella browni
5 mm
5 mm
Coryphella verrucosa
2 mm
Coryphella pellucida
2 mm
Coryphella gracilis
2 mm
Coryphella lineata

Fig. 10.25

V-shaped anteriorly, linear posteriorly, cerata long and tapering, white with olive-green or brown subterminal band, covered with white spots. Oral tentacles with subterminal half white. Rhinophores slender, smooth, with translucent white, elongate tip, a cream or pale yellow subterminal swelling (rarely two), and brown basal pigmentation. White pigment between oral tentacles, and dorsal part of propodial tentacles, extends posteriorly between rhinophores, widens to rhombus shape posterior to rhinophores, and narrows again continuing as a line down dorsum to metapodium. Juveniles may lack both pigment and rhinophoral swellings.

Lower shore and shallow subtidal, to about 20 m; usually feeding on eggs of other opisthobranchs, but also on calyptoblastic hydroids such as *Obelia geniculata* and *Sertularia argentea*. On western coasts, extending farther to north and south than *F. blianus*.

31 FLABELLINIDAE

Aeolidaceans with cerata in rows, either on a notal ridge or on slight expansions of the pallial ridge (British species), which separates the sides from the dorsum. Rhinophores wrinkled or smooth (British species), distinct propodial tentacles; long, mobile oral tentacles. Feed on hydroids. Six British species.

---

1. Body (excluding cerata) purple — ***Coryphella pedata***

   Body (excluding cerata) translucent white — **2**

---

2. Body with opaque white lines — **3**

   Body not as described — **4**

---

3. Opaque line continuous from tips of oral tentacles to metapodium; also on sides of body — ***Coryphella lineata***

   With a single opaque line from behind rhinophores to tip of tail — ***Coryphella verrucosa***

---

4. Body length up to 15 mm, narrow subterminal band on cerata; well-developed pallial ridge — ***Coryphella gracilis***

   Body length up to 50 mm, broad subterminal band on cerata; pallial ridge not obvious — ***Coryphella browni***

Coryphella browni Picton — Fig. 10.25

Up to 50 mm, white, with a broken streak of white pigment on dorsum from posterior cerata to tip of metapodium. Oral tentacles long, rhinophores wrinkled, both streaked with white on distal third; propodial tentacles pointed and grooved. Up to eight clusters of cerata in transverse rows: seven precardial and 15 postcardial, each row with up to five cerata. Only first two groups of cerata are clearly separated by a pallial ridge, which disappears between other clusters. A broad white subterminal band encircles cnidosac on each ceras (cf. *C. gracilis*); digestive gland varies from bright red to chocolate-brown.

Frequently in large numbers on the hydroid *Tubularia indivisa*, often with *C. lineata*. All British coasts. *C. pellucida* (Alder and Hancock) (Fig. 10.25) is similar, but has long slender cerata with solid white tips instead of subterminal bands, and smooth rhinophores. It also occurs on *T. indivisa*, but is most frequent on northern British shores.

Coryphella gracilis (Alder and Hancock) — Fig. 10.25

Maximum length 15 mm; translucent white, more opaque when developing ova are present. Up to six clusters of cerata, all originating from a distinct pallial ridge; each ceras with a thin band of white pigment subterminally (cf. *C. browni*), digestive gland bright red to bright green. Propodial tentacles rather thick; oral tentacles smooth, rhinophores slightly wrinkled, both streaked distally with white. Metapodium also with white streak.

Feeds on *Eudendrium ramosum*, and possibly also on *Clytia johnstoni* and *Halecium* spp. In shallow waters, all British coasts.

Coryphella lineata (Lovén) — Fig. 10.25

Up to 50 mm, translucent white. Oral tentacles long, smooth, rhinophores slightly longer, wrinkled; both colourless except for a thin white stripe running from white tips. Stripes from oral tentacles meet on head and continue posteriorly along dorsum to tip of colourless metapodium. Lateral white stripes originate anterior to rhinophores and unite with median stripe on metapodium. Propodium notched, transversely grooved; propodial tentacles moderately long. Cerata elongated, arranged in 5–8 clusters on a distinct pallial ridge. Each ceras with a subterminal white ring extending basally as longitudinal stripe, anterior one thickest; white pigment sometimes poorly developed. Digestive gland lobes reddish-orange or reddish-brown, more rarely dark brown, crimson, or green.

Shallow sublittoral, between 20 and 40 m; preferred diet is *Tubularia indivisa*. Common on all British coasts.

Coryphella pedata (Montagu) Fig. 10.25
Up to 48 mm; distinguished by its violet or magenta coloration. A maximum of seven clusters of cerata, all arising from expansions of pallial ridge. Each ceras with white subterminal band, digestive gland lobe pinkish-orange to red. Oral tentacles and rugose rhinophores darker violet than body, with white tips. Propodial tentacles rather long, tapering, recurved; paler than oral tentacles and rhinophores, sometimes tipped with white.

Shallow sublittoral, to 40 m, usually on the hydroid *Eudendrium ramosum*. Common on all British coasts.

Coryphella verrucosa (M. Sars) Fig. 10.25
Up to 35 mm, translucent white. Cerata stout, in 5–7 clusters on an expanded pallial ridge, which is absent between clusters. Each ceras with white subterminal band, digestive gland lobes light brown, crimson, or maroon. A white dorsal stripe extends from pericardium to tip of metapodium. Rhinophores rugose, shorter than oral tentacles; both with white distal streaks. Propodium notched medially and transversely grooved, the groove extending on to short, stout propodial tentacles.

Sublittoral, feeding on *Tubularia indivisa* and other hydroids, and on the scyphistomae of *Aurelia aurita*. A boreo-arctic species, occurring in the North Sea and on the coasts of Scotland.

32 TEGIPEDIDAE

Small aeolidaceans with cerata in even rows (reduced to single row in *Tergipes*); cerata clavate, digitiform, or fusiform. Propodial tentacles reduced or absent; oral tentacles smooth (reduced in *Tenellia*). Twelve species recorded from Britain.

---

1. Head rounded, with short oral tentacles **_Tenellia adspersa_**

   Head with long oral tentacles **2**

---

2. Cerata sparse: arising singly down each side **_Tergipes tergipes_**

   Cerata in obliquely transverse rows **3**

---

3. Rhinophores banded with orange, crimson, olive, or brown **4**

   Rhinophores otherwise pigmented **5**

---

4. Tentacles with olive or brown bands and white tips; no pigment on body other than white speckling **_Cuthona amoena_**

   Tentacles with orange or red bands; crescent-shaped patches of orange, red, or brown between rhinophores to first cerata (sometimes one between oral tentacles and rhinophores) **_Cuthona foliata_**

---

5. Rhinophores with white pigmented ends **6**

   Rhinophores orange or brown, darker than body, darkening distally **_Cuthona gymnota_**

---

6. Digestive gland, within cerata, dark green with black specks **_Cuthona viridis_**

   Digestive gland pink to brown. Cerata numerous, extending in front of rhinophores; club-shaped with white tips. Body suffused with pink posteriorly **_Cuthona nana_**

Cuthona amoena (Alder and Hancock) Fig. 10.26
Up to 10 mm; with scattered, small, creamy white spots, often on raised tubercles, especially on head between rhinophores and on prominent pericardial swelling; with faint brown or olive-brown mottling. Rhinophores smooth, with opaque white tips and an olive or brown band (rarely two) subterminally, its width equal to its distance from tip; oral tentacles

short, pigmentation similar to rhinophores. Three rows of cerata anterior to pericardial swelling, five rows posteriorly; each ceras with scattered reddish-brown and white spots, darkening basally; small cnidosac visible at tip, usually a subterminal ring of red-brown spots. Digestive gland lobes sandy-brown to olive-green, often darkest basally.

Sublittoral, to 40 m, on *Halecium halecinum* and *H. beani*. All British coasts.

Cuthona foliata (Forbes and Goodsir) Fig. 10.26

Up to 11 mm; yellowish, with distinctive crescentic patches of orange, red, or brown posterior to rhinophores, often with a red or orange patch on each side of head between oral tentacles and rhinophores, and on dorsum posterior to prominent pericardial swelling. Rhinophores short and stubby, banded with orange or red; oral tentacles with less distinct banding, speckled with white dots extending on to head. Cerata fusiform, in 7–8 rows, each of up to five cerata; subterminal yellow speckling, fading towards base, digestive gland lobe green, brown, or yellow-brown.

Intertidal and shallow subtidal, feeding on calyptoblastic hydroids, especially *Abietinaria abietina*, *Sertularella polyzonias*, and *S. gayi*. All British coasts.

Cuthona gymnota (Couthouy) Fig. 10.26

Up to 22 mm; translucent white or colourless, or tinted orange or rose. Rhinophores deeper orange or brown, fading basally; oral tentacles may be pale or dark. Cerata digitiform or fusiform, rather stout; up to 12 rows, with up to seven cerata in the largest rows close to pericardium. Each ceras with large white cap, and subterminal band of dull white, yellow, or orange; digestive gland lobe pinkish-orange to red-brown. Metapodium short and rounded.

Lower shore and shallow subtidal, to 30 m; preferred prey is *Tubularia indivisa* but on more exposed shores feeds on *T. larynx*. All British coasts, common.

Cuthona nana (Alder and Hancock) Fig. 10.26

Up to 28 mm; translucent white, sometimes suffused with pink posterior to pericardial swelling. Head broad; oral tentacles short, digitiform, often white distally; rhinophores smooth, bluntly pointed, also tending to whiten distally. Cerata abundant, more than 20 rows arranged in up to 16 clusters; club-shaped, white-tipped, digestive gland lobes brown in starved specimens, tending to pink.

Intertidal and sublittoral, to 30 m, in sheltered waters; feeds predominantly on *Hydractinia echinata*. Recorded from scattered localities on all British coasts.

Cuthona viridis (Forbes) Fig. 10.26

Up to 30 mm; white or pale yellow. Oral tentacles and rhinophores white distally. Cerata elongated, cylindrical, up to nine rows of seven; cnidosacs white or pale yellow, almost filling tip of each ceras, a subterminal ring of white or pale yellow, streaking anteriorly in larger specimens. Digestive gland lobes dark green with black specks.

Intertidal, and offshore to 100 m. Feeds on hydroids, including *Abietinaria abietina*, *Sertularella* spp., *Nemertesia* spp., and, intertidally, *Clytia johnstoni*. All British coasts, except extreme south-east.

Tenellia adspersa (Nordmann) Fig. 10.26

Up to 9 mm, pale yellow to dull brown with black speckling of varying density. Head domed and spatulate, oral tentacles, if present, very short; rhinophores smooth, tapered to blunt points, white distally. Cerata fusiform, up to six rows, each of up to three cerata; pale yellow to orange, cnidosacs distinct close to tips, sometimes with subterminal band of white dots.

Intertidal and shallow sublittoral; euryhaline, often in harbours, estuaries, and canals. Feeds on hydroids, including *Laomedea longissima*, *L. loveni*, *Cordylophora lacustris*, *Protohydra leuckarti*. Rare, the few British records are from the Bristol Channel and from Fleet and Portishead, on the south coast.

Tergipes tergipes (Forskal) Fig. 10.26

Up to 8 mm; translucent white, with distinct greenish digestive gland; streaks of red-brown originating on sides of head, at bases of oral tentacles, run posteriorly on each side; similarly coloured band extends from base of rhinophores on each side to meet lateral line. Up to eight, rarely 10, fusiform cerata in two single, longitudinal series: anterior two paired, successive ones alternating; each with relatively large white tip, reddish-brown subterminal ring overlying cnidosac.

Shallow sublittoral, to 290 m, on small calyptoblastic hydroids growing on kelps or moored structures. All British coasts, common.

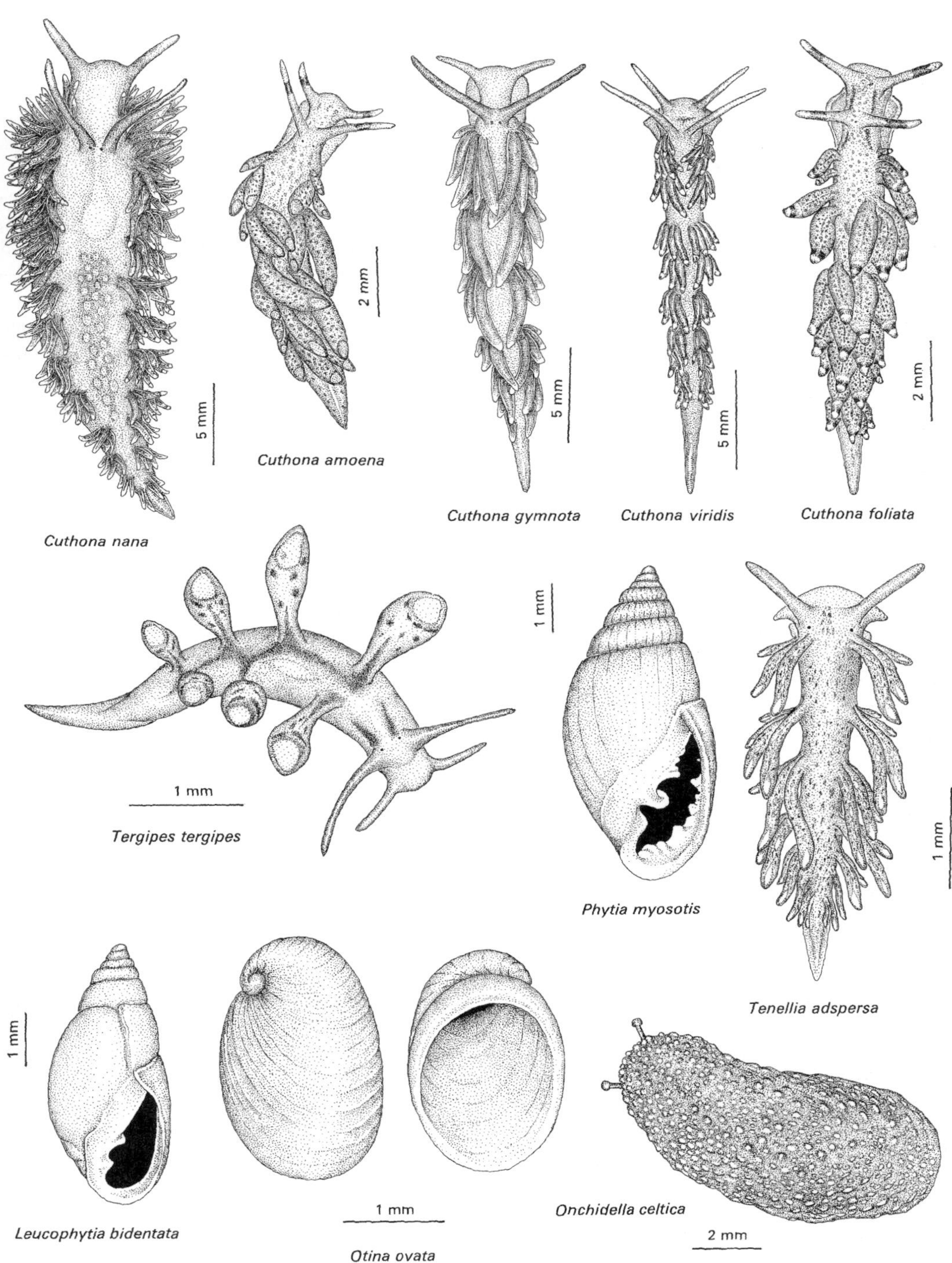
5 mm
Cuthona nana
Cuthona amoena
2 mm
Cuthona gymnota
5 mm
Cuthona viridis
5 mm
Cuthona foliata
2 mm
1 mm
Tergipes tergipes
1 mm
Phytia myosotis
1 mm
Tenellia adspersa
1 mm
Leucophytia bidentata
1 mm
Otina ovata
Onchidella celtica
2 mm

Fig. 10.26

Subclass Pulmonata

The pulmonate gastropods are primarily adapted to terrestrial and freshwater habitats, perhaps having evolved from a number of estuarine or supralittoral stocks. Pulmonates exhibit detorsion, lack ctenidia, and display a major modification of the mantle cavity which functions as a *pulmonary sac* or *lung*. The opening of the mantle cavity is reduced as a contractile *pneumostome*. A shell may be present, spirally coiled and large enough to accommodate the whole of the animal, or small and cap-like, and covering only the posterior portion of the body. Certain genera possess a large limpet-like external shell, while others have a small internal shell. Shelled pulmonates do not have an operculum. In a majority of land slugs a shell is lacking altogether.

Some groups of pulmonates have returned to marine habitats, and a few species may be found on rocky coastlines in all parts of the world. Snail-like, limpet-like, and slug-like genera occur, which may be confused with similar prosobranch or opisthobranch genera. A few representatives of marine pulmonate snails and slugs occur commonly on British shores.

1 ELLOBIIDAE

Shell fusiform, spirally coiled, with large last whorl and thickened, toothed aperture.

---

1. Shell slender; aperture with two teeth on columella, outer lip untoothed **_Leucophytia bidentata_**

   Shell broadly oval; aperture with three or more teeth on columella, outer lip with three or more teeth **_Phytia myosotis_**

Leucophytia bidentata (Montagu) Fig. 10.26
Shell rather solid, glossy white, slender fusiform; up to 6 mm, with six or seven whorls, the last whorl comprising about two-thirds of total shell height. Aperture elongate oval; outer lip thin, slightly flared abapically; inner lip thickened, with two strong folds on columella, projecting into aperture as well-marked teeth.

Upper shore, in crevices and among seaweeds, particularly on sheltered estuarine shores. Reported from scattered localities around southern half of Britain, and on all Irish coasts. May be locally common.

Phytia myosotis (Draparnaud) Fig. 10.26
Shell thin, rather short and broad, greyish-brown; up to 8 mm, with seven whorls, the last whorl comprising about three-quarters of total shell height, smooth. Aperture oval, with distinctly thickened inner and outer lips, each bearing at least three distinct teeth.

Upper shore, particularly in estuaries, among seaweed, saltmarsh plants, and rotting debris. Southern and western coasts only, locally common.

2 OTINIDAE

Shell cap-shaped, with minute spire of two whorls and a broad, shallow last whorl.

Otina ovata (Brown) Fig. 10.26
Shell minute, oval, up to 2 mm long, consisting largely of a shallow last whorl, and completely covering the body of the animal; reddish-brown to purple, smooth, with faint concentric growth lines.

Upper shore, in barnacle tests and among short, dense weeds such as *Lichina pygmaea*. Predominantly southern in distribution; perhaps locally common, but inconspicuous.

3 ONCHIDIIDAE

Pulmonate sea slugs, lacking a shell, the head and foot hidden by a tuberculate mantle. Head with a pair of retractile tentacles bearing eyes at tips. Pneumostome posteriorly situated close to anus.

Onchidella celtica (Forbes and Hanley) Fig. 10.26
Body solid and fleshy, oval, domed, up to 12 mm long, deep olive to black. Mantle covered with large, coarse, well-spaced tubercles, obscuring head and foot completely when the animal is motionless. Two short, thick tentacles, bearing eyes at tips, visible anteriorly when the animal is crawling.

Between tidemarks on exposed rocky shores, active during periods of total emersion, crawling on small algae. Retreats to crevices upshore as tide rises. Coasts of Cornwall and south Devon, locally abundant.

Class Bivalvia

The bivalve molluscs are sedentary animals, living firmly attached to fixed substrata, or hidden in cracks and crevices, or boring into a variety of rocks, wood, or organic carbonates. Many burrow into bottom deposits, some simply lying in the very top layers, others inhabiting more permanent burrows. With a few exceptions, such as scallops and file shells, mobility is limited to movement within the burrow or the immediate environment.

Externally the bivalve body appears very simple. There is no definite head; the gut and visceral mass are enclosed laterally by two flaps of body wall, the *mantle*, which also encloses a space, the *mantle cavity*, containing a pair of large, and often elaborately folded *ctenidia* or gills. The mantle edge secretes the two valves of the shell, producing both the calcium carbonate component and its organic matrix, *conchiolin*. Dorsally, an uncalcified portion of conchiolin forms a *ligament*. The elasticity of the ligament tends to force the valves apart and closure is achieved by paired, internal *adductor muscles*. The outer surface of the shell is covered by a horny *periostracum*; this may be thin and smooth, or coarse, and is often produced as periostracal spines. The bivalve *foot* is typically wedge shaped, and in active burrowers is used for probing and digging into the substratum. The edges of the mantle flaps are often fused to a greater or lesser extent, and posteriorly may be developed as tubular *siphons*. The presence, type, and development of the siphons reflect the mode of life of the animal, being short in shallow-burrowing cockles, longer in the Veneracea. The dorsal, *exhalant*, and ventral, *inhalant*, tubes may be separate, elongate, and highly flexible in deposit-feeding species, or partly or wholly fused, and in static deep burrowers may be relatively rigid.

The shell may be bilaterally symmetrical (*equivalve*), or one valve may be smaller than the other (*inequivalve*). The oldest part of the valve is dorsal in position, where it forms a convex *umbo*. It may be symmetrical about a dorso-ventral mid-line through the umbo (*equilateral*), or the umbo may be in front of or behind the mid-line (*inequilateral*). The two valves articulate along the dorsal *hinge line*, which frequently bears a number of interlocking teeth. These are situated below the umbo (*cardinal teeth*) or to either side of it (*lateral teeth*), and both types may be present in the same species. The *ligament* is always posterior to the umbones; it may be external to the hinge line, or partly or wholly internal, with the internal element in a pit, or spatulate depression (*chondrophore*). The outer surface of the valve may bear concentric rings or ridges marking successive growth increments, and a concentric sculpture of ridges or grooves is also present in many species. Radiating striations, or *ribs*, often bearing sharp spines, may also occur. The inner surface of the shell bears prominent anterior and posterior *adductor muscle scars*, varying in shape and relative sizes between species. The attachment of the mantle to the inner surface of the valve is marked by the *pallial line*, extending between the two adductor scars. Posteriorly, this may show an indentation (*pallial sinus*) indicating the position of the retracted siphons in the closed shell.

It is important to orientate a shell correctly to distinguish left and right valves: the pallial sinus (when present) and the ligament are always posterior in position, umbones tend to be inclined anteriorly and, when muscle scars are unequal, the anterior is the smaller. The most highly modified shells will still be difficult to orientate and notes on these are introduced where appropriate.

KEY TO FAMILIES

1. Hinge line of shell with numerous small, identical teeth, alternating with sockets (Taxodont: Fig.10.28) **2**

   Hinge line with or without cardinal and/or lateral teeth, but not taxodont **4**

2. Shell almost circular in outline. Light or dark reddish-brown, typically with zigzag patterns **3. Glycymeridae**

Shell triangular, oval, or irregular in outline **3**

3. Ligament external, extending across a broad, grooved cardinal area between widely spaced umbones. Hinge teeth in a continuous series **2. Arcidae**

Ligament internal, in a small indentation on the hinge line, separating the teeth in each valve into posterior and anterior series **1. Nuculidae**

4. Hinge line of shell drawn out into projecting ears on one or both sides of umbones. Typically fan-shaped or elongate oval, with radiating ridges (Fig. 10.30) **5**

Hinge line not forming paired ears **6**

5. Shell length half, or less, depth; outline elongate-oval, ears small **8. Limidae**

Shell length about equal to depth; curved edge of margin forming at least a semicircle, ears prominent **7. Pectinidae**

6. Each valve with one adductor muscle scar only, left valve with or without associated byssal scars. Shell inequivalve, often irregular in outline with a corrugated surface **7**

Each valve with at least two adductor scars **8**

7. Shell thin and brittle. Right valve flat, with a large hole close to umbo. Left valve convex, the single round adductor scar adjacent to, or partly fused with, one or two smaller byssal scars **4. Anomiidae**

Shell solid, often very thick and rugose. Right valve flat, without a hole, left valve convex; each with a single, elliptical adductor scar, often deeply recessed **6. Ostreidae**

8. Hinge teeth absent. Ligament reduced, internal. Shell equivalve, inequilateral and irregular in outline, frequently gaping. Anterior dorsal margin reflected in front of umbones (Fig. 10.41); accessory plates present dorsally or at tips of siphons. Borers in wood, rock, peat, etc. **9**

Not as described above **10**

9. Shell elongate, oval or quadrate, gaping to a varying extent along anterior, ventral, or posterior margins. Each valve with a slender fragile process (*apophysis*) curving out from beneath the umbo (look for stump if broken) **33. Pholadidae**

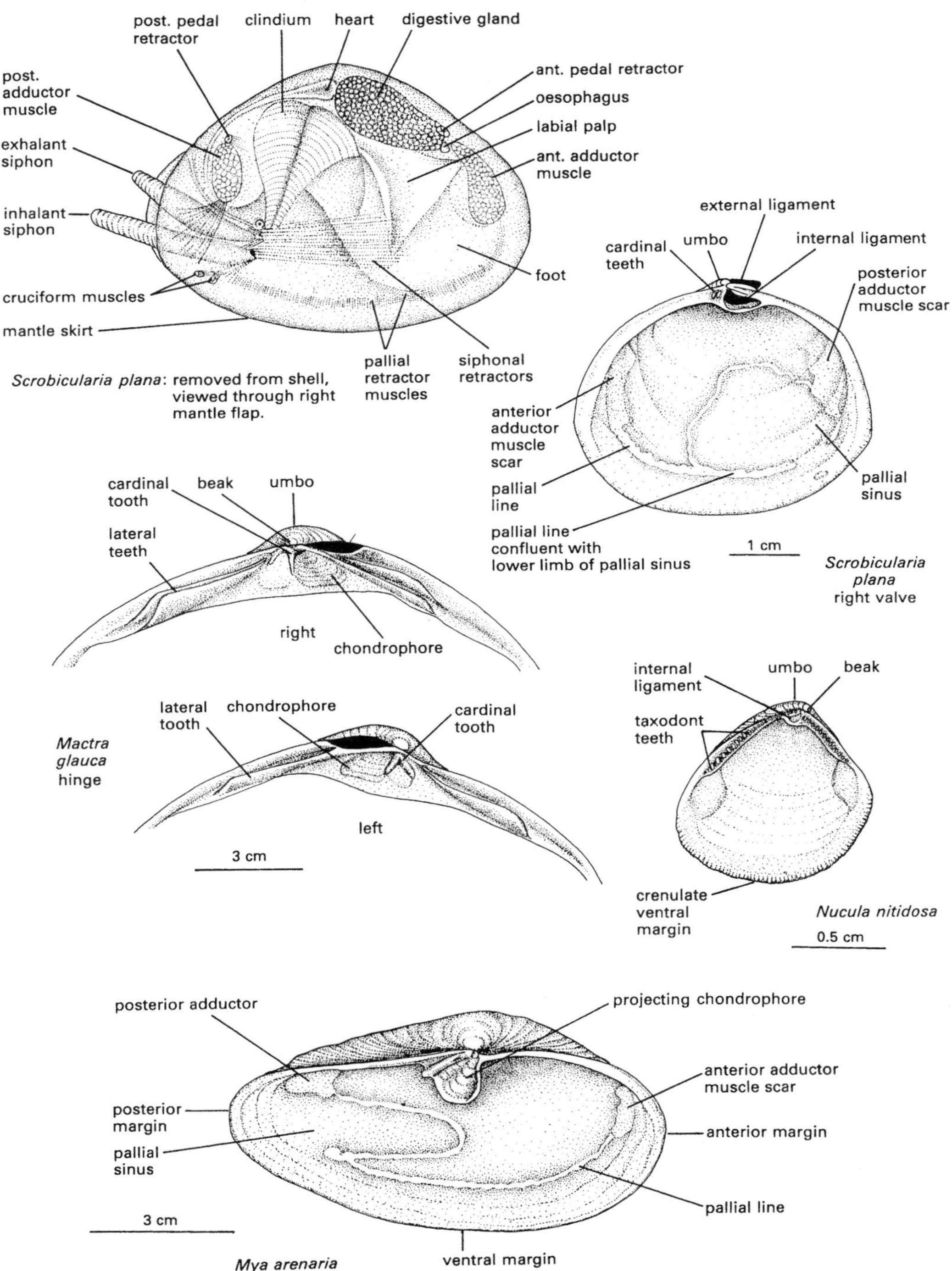
post. pedal retractor
clindium
heart
digestive gland
post. adductor muscle
exhalant siphon
inhalant siphon
cruciform muscles
mantle skirt
ant. pedal retractor
oesophagus
labial palp
ant. adductor muscle
foot
pallial retractor muscles
siphonal retractors
Scrobicularia plana: removed from shell, viewed through right mantle flap.
external ligament
cardinal teeth
umbo
internal ligament
posterior adductor muscle scar
anterior adductor muscle scar
pallial line
pallial line confluent with lower limb of pallial sinus
pallial sinus
1 cm
Scrobicularia plana right valve
cardinal tooth
beak
umbo
lateral teeth
right
chondrophore
lateral tooth
chondrophore
cardinal tooth
Mactra glauca hinge
left
3 cm
internal ligament
umbo
beak
taxodont teeth
crenulate ventral margin
Nucula nitidosa
0.5 cm
posterior adductor
projecting chondrophore
anterior adductor muscle scar
posterior margin
anterior margin
pallial sinus
pallial line
3 cm
Mya arenaria
ventral margin

Fig. 10.27

Shell highly modified; short and globular, formed from several distinct lobes (Fig.10.41) **34. Teredinidae**

---

10. Hinge line without teeth or chondrophore **11**

    Hinge line with either or both teeth and chondrophore **13**

---

11. Shell markedly inequilateral, subtriangular, oval or bean-shaped; umbones anterior, terminal or subterminal (Fig.10.29). Valves not gaping **5. Mytilidae**

    Shell only slightly inequilateral, or almost equilateral; umbones close to mid-line, or subterminal above a large, oval ventral gape **12**

---

12. Shell inequilateral, umbones subterminal; a broad oval gape extending whole length of shell. Boring in limestone, sandstone, or organic concretions **32. Gastrochaenidae**

    Umbones close to mid-line of shell. If ventral gape present, then valves almost equilateral. Shell with distinct radiating ribs **5. Mytilidae**

---

13. Shell very elongate, straight or gently curved, with dorsal and ventral margins almost or quite parallel (Fig.10.39). Umbones and ligament anterior, subterminal in position **28. Solenidae**

    Not as described above **14**

---

14. Pallial line with a posterior sinus; sometimes small and indistinct, but always present **15**

    Pallial line without a sinus **29**

---

15. Shell subtriangular with truncate posterior margin. Right valve more convex than left, enclosing it, leaving a broad border of right valve covered only by a fringe of left periostracum (Fig.10.40). Left valve with a deep pit adjacent to the chondrophore, right valve with a corresponding tooth **30. Corbulidae**

    Not as described above **16**

---

16. Ligament external only, no chondrophore **17**

    Ligament internal and external, chondrophore present **24**

---

17. Right valve with one cardinal tooth, left valve with two, or none. All teeth liable to be worn. Shell fragile, often irregular in outline **31. Hiatellidae**

    Not as described above **18**

---

18. Shell elongate, cylindrical, with anterior and posterior gape. Cardinal teeth peg-like: two right and one left, or two left and one right **27. Solecurtidae**

Not as described above **19**

---

19. Both valves with three cardinal teeth. Shell typically thick and solid, subtriangular. No cruciform muscle scars; pallial line usually distinct from border of pallial sinus **19. Veneridae**

One or both valves with only two cardinal teeth. Shell typically thin, fragile. Cruciform muscle scars present or absent **20**

---

20. Two cardinal teeth in each valve. Shell often thin and fragile. Cruciform muscle scars present (Fig.10.37); pallial line often confluent with lower border of pallial sinus **21**

Three cardinal teeth in left valve, two in right. Shell thin, circular, or oval. No cruciform muscle scars; pallial line distinct from border of pallial sinus **20. Petricolidae**

---

21. Hinge line with cardinal and lateral teeth, the latter conspicuous in right valve, but in the left appearing simply as slender ridges close to the margin **22**

Hinge line with cardinal teeth only **23**

---

22. Inner margin of shell crenulate or finely serrate **23. Donacidae**

Inner margin of shell smooth **24. Tellinidae**

---

23. Shell almost circular, colour variable but often in concentric bands. Intertidal only, often abundant in soft estuarine deposits **24. Tellinidae** (*Macoma balthica*)

Shell elongate-oval, anterior end not regularly semicircular, often keeled or ridged posteriorly. Sublittoral, occasionally on lower shore. Not estuarine **26. Psammobiidae**

---

24. One or more cardinal teeth present **25**

No cardinal teeth **27**

---

25. Left valve with two cardinal teeth joined to form a single ∧-shaped structure. Two anterior and two posterior lateral teeth may be present in right valve. No cruciform muscle scar **26**

Left valve with one or two cardinal teeth, but not joined. Right valve without lateral teeth, or with single, small anterior and posterior teeth. Cruciform muscle scars present **25. Scrobiculariidae**

---

26. Shell subtriangular or elliptical, without a gape. Umbones on or close to mid-line **21. Mactridae**

    Shell elongate, gaping at both ends. Umbones in anterior half **22. Lutrariidae**

---

27. Internal ligament supported by a projecting, spatulate chondrophore in left valve and a smaller, concealed chondrophore in the right. Shell with a posterior gape **29. Myacidae**

    Internal ligament supported by a concave chondrophore inset into the hinge line of each valve. Shell without a posterior gape **28**

---

28. Each valve with a short crack in the calcification extending posteriorly from the umbo **35. Periplomatidae**

    Valves without a crack **36. Thraciidae**

---

29. Shell globular, equivalve. Umbones prominent, enrolled, curving away from the hinge line. Cardinal and lateral teeth parallel to hinge line **17. Glossidae**

    Not as described above **30**

---

30. Shell equivalve; valves strongly convex, with radiating ridges often well developed. Ligament external **18. Cardiidae**

    Radiating ribs rarely developed, sculpture of concentric lines or ridges. Ligament external or internal **31**

---

31. Shell oval or circular, often large; heavy and solid, with prominent umbones. Periostracum thick, brown or black. Ligament external. Three cardinal teeth in each valve **16. Arcticidae**

    Ligament external or internal. Three cardinal teeth in one valve only, or both valves with fewer than three **32**

---

32. Anterior adductor muscle scar larger than posterior. Ligament external **33**

    Both muscle scars of similar size. Ligament external or internal **34**

---

33. Anterior adductor scar comma-shaped, completely fused with pallial line along its anterior edge. Shell smooth, often silky, sculptured with very fine concentric lines between distinct growth stages **11. Diplodontidae**

Anterior adductor scar elongate-oval, separated from pallial line along a varying proportion of its anterior edge. Shell dull, with concentric lines or ridges **10. Lucinidae**

---

34. Ligament external. Lateral teeth absent. At least two cardinal teeth in each valve, three may be present in left valve. Shell typically subtriangular, solid **9. Astartidae**

Ligament internal. Lateral teeth may be present. Cardinal teeth variable, never more than two in each valve. Shell typically fragile and small (<12 mm) **35**

---

35. Shell thin and flat, fragile, approximately square in outline, outer surface densely punctate **14. Leptonidae** (*Lepton squamosum*)

Shell often thin and fragile, but distinctly convex and broadly oval in outline. Without punctulations **36**

---

36. Right valve with one cardinal tooth, left valve with two; one posterior lateral tooth in each valve, but no anterior lateral **12. Kelliidae**

Cardinal teeth present or absent. Anterior lateral teeth present in one or both valves **37**

---

37. Anterior lateral teeth present in one or both valves, but no posterior laterals **38**

One anterior and one posterior lateral tooth present in each valve **39**

---

38. Shell with fine radiating ridges, most distinct about the umbones. White. Single anterior lateral tooth in each valve, other teeth absent **15. Montacutidae** (*Montacuta substriata*)

Shell with concentric lines only. White, usually stained with red deposit. Left valve with one anterior lateral tooth, right valve with one cardinal tooth **15. Montacutidae** (*Tellimya ferruginosa*)

---

39. Shell with fine radiating striations. Each valve with single cardinal tooth, and single anterior and posterior laterals **14. Leptonidae** (*Epilepton clarkiae*)

Shell with fine concentric lines only, or with few striations irregularly developed about the umbones **40**

---

40. Shell elongate oval, with low umbones, cardinal teeth absent from both valves. Distributed from the lower shore to the outer continental shelf **15. Montacutidae** (*Mysella bidentata*)

Shell plump, broadly oval, with prominent umbones. Left valve with single cardinal tooth. Between tidemarks only, attached by byssus **13. Lasaeidae**

---

1 NUCULIDAE. NUT SHELLS

Shell subtriangular or oval, smooth, equivalve, inequilateral, with umbones posterior to mid-line. Hinge line with taxodont teeth in two distinct series, anterior more numerous than posterior. Ligament internal, in an elliptical chondrophore, below and just in front of beaks. Adductor muscle scars about equal; pallial line faint, without a sinus. The British fauna includes six species in two genera. The families Nuculanidae and Yoldiidae, also taxodont, comprise another six small, rare species.

---

1. Shell surface chequered: distinct at hinge line and forming coarse, transverse corrugations anteriorly across lunule ***Nucula sulcata***

Shell not chequered; lunule smooth, without transverse ridges **2**

---

2. Periostracum very glossy, often with bold concentric bands of grey or greenish-yellow. Anterior and posterior hinge lines at right angle to each other ***Nucula nitidosa***

Periostracum dull or matt. Anterior and posterior hinge lines at greater than 90° to each other ***Nucula nucleus***

Nucula nitidosa Winckworth Fig. 10.28

Glossy olive or yellow-olive, often with concentric bands of light yellow; dead shell white or grey with bluish growth lines; with fine radiating striations and fainter concentric lines. Length 10–13 mm. Anterior hinge line right-angled to posterior; 20–30 hinge teeth anteriorly, 10–14 posteriorly; lunule and escutcheon poorly defined. Margin of shell finely crenulate.

On silt and fine sand. Common, and often abundant, on all British coasts, down to 100 m. Distributed from Norway south to the Mediterranean and west Africa.

Nucula nucleus (Linnaeus) Fig. 10.28

Dull light yellow, light brown or greenish-yellow, with darker or lighter concentric bands, dead shell white or grey; with fine radiating striations and coarser concentric lines. Up to 10 mm long, rarely to 12 mm. Anterior and posterior hinge lines angled at greater than 90°; 16–25 hinge teeth anteriorly, 10–14 posteriorly; lunule poorly developed, escutcheon distinct with strongly pouting hinge line. Margin of shell crenulate.

On coarse sand and fine gravel, often common. All British coasts, to about 150 m. Distributed from Norway south to the Mediterranean, and South Africa.

Nucula sulcata Bronn Fig. 10.28

Dull light yellow to brown, often with darker patches, and brown or black encrustations about hinge line, dead shell white or grey; fine radiating striations and coarser concentric lines give a chequered effect, distinct at hinge line. To 15 mm long, but may exceed 20 mm. Anterior hinge line right-angled to posterior; 20–30 hinge teeth anteriorly, 10–14 posteriorly; lunule with strong transverse corrugations, escutcheon distinct, with depressed border and pouting hinge line. Margin of shell crenulate.

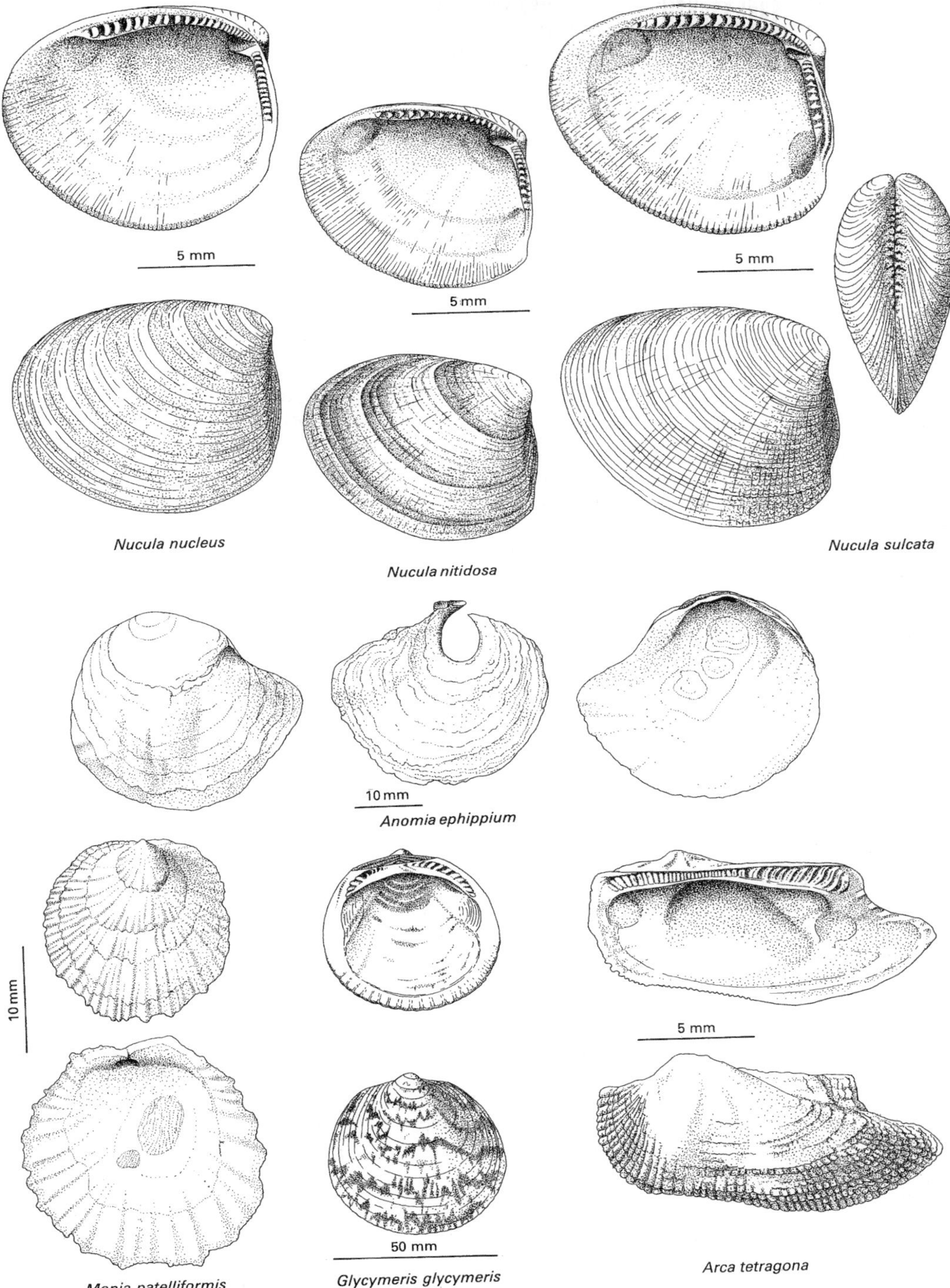
5 mm
5 mm
5 mm
Nucula nucleus
Nucula nitidosa
Nucula sulcata
10 mm
Anomia ephippium
10 mm
5 mm
50 mm
Monia patelliformis
Glycymeris glycymeris
Arca tetragona

Fig. 10.28

On mud and sand, from 10 to 200 m. Mostly off west coasts. Distributed from northern Norway to Mediterranean and west Africa.

2 ARCIDAE

Shell thick and strong, oval, quadrate or irregular in outline; equivalve, inequilateral, umbones anterior to mid-line, widely separated. Ligament external, extending between umbones across a broad, grooved, diamond-shaped cardinal area. Hinge line straight, with a continuous series of identical teeth, alternating with sockets. Adductor muscle scars about equal. No pallial sinus. Four species recorded from Britain but only the following is at all frequent.

Arca tetragona Poli. Ark shell Fig. 10.28

Elongate, to 50 mm long; off-white or yellow, with brown periostracum. Anterior end deeper than posterior, umbones prominent and widely spaced, each with a sharp, raised ridge from beak to posterior margin. Cardinal area with distinct chevron ridges, partly obscured in older specimens. With fine radiating ridges; older shells worn to a rough, chalky white surface, coarse and rugose between umbones. Hinge line with 40–50 teeth. Ventral margin irregular and variable, finely crenulate, with a distinct byssal gape ventral to umbones.

Sessile, attached by a thick green byssus; in crevices, among rocks and shell debris, often encrusted with other organisms, or abraded by the substratum. Lower shore to about 100 m, all British coasts. Distributed from Norway south to Mediterranean and west Africa.

3 GLYCYMERIDAE

Shell equivalve, equilateral, umbones situated on mid-line. Ligament external, in a narrow, elongate cardinal area. Hinge teeth in a continuous series in young shells, separated by a smooth area below umbones in older individuals. Adductor muscle scars about equal, no pallial sinus. One British species.

Glycymeris glycymeris (Linnaeus)
Dog cockle Fig. 10.28

Shell thick, robust; almost circular, anterior hinge line curving more abruptly downwards than posterior. Up to 65 mm long; dull brown, yellow, or light purplish-red, uniformly coloured or in irregular concentric zigzags on a cream background; periostracum remaining as a broad band around margin, dark brown, coarse, velvety to touch, usually with fine concentric lines and radiating striae, growth stages clear.

A shallow burrower in fine shell gravels, or sandy/muddy gravels, offshore to about 100 m; common on all British coasts. Distributed from Norway south to the Mediterranean and west Africa.

4 ANOMIIDAE

Shell thin and brittle, approximately circular in outline, often irregular; attached by a byssus passing through a hole in right valve, with left valve uppermost. Inequilateral, umbones rounded, indistinct. Inequivalve; left valve convex, right valve flat or concave. Right valve with a single adductor muscle scar; left valve with a single adductor scar, adjacent to, or fused with, one or two byssal attachment scars. No hinge teeth. Ligament internal, in a deep pit below umbone of right valve. Inner surface of shell nacreous, glossy, no pallial line. Three British species. The smallest, *Heteranomia squamula* (Linn.), is a northern species which ranges south to Biscay.

---

1. Left valve with three closely spaced scars in the concavity of the shell **_Anomia ephippium_**

   Left valve with two discrete scars **_Monia patelliformis_**

Anomia ephippium Linnaeus
Common saddle oyster Fig. 10.28

White or grey, occasionally with a pink or bluish tinge. Up to 60 mm. Left valve convex with three distinct scars in an oblique line in concavity of shell. Right valve thin, flat or irregularly concave, with elliptical byssal aperture dorsally; single adductor scar postero-ventral to byssal aperture. Sculpture of wavy concentric lines, often raised as irregular scales.

From the lower shore to at least 150 m. Attaches to various hard substrata, including other molluscs, and to algal holdfasts. Widespread and common on all British coasts. Occurs in Mediterranean; distributed from Iceland to west Africa, and to the South Atlantic Islands.

Monia patelliformis (Linnaeus)
Ribbed saddle oyster Fig. 10.28

Thin, circular or oval, sometimes irregular; white or grey, occasionally with patches or streaks of pale brown. Usually less than 40 mm long. Umbones marginally anterior to mid-line, set back from hinge. Left valve with a larger, oval byssal scar immediately below umbo, and a circular, postero-ventral muscle scar. Right valve byssal aperture large, oval, or drop-shaped.

Attached to hard substrata, offshore to at least the edge of the Continental Shelf. Widespread off all British coasts; distributed from Norway to the Mediterranean.

5 MYTILIDAE

Shell elongate, equivalve; inequilateral, beaks at anterior end, terminal or subterminal; rarely almost equilateral. Periostracum usually conspicuous, typically darker than shell, with or without spines. Hinge line without teeth, rarely with small crenulations, continuous with shell margin. Ligament internal, or external in a narrow groove and inconspicuous. Adductor muscle scars unequal, anterior scar reduced. No pallial sinus. Attached by a diffuse, fibrous byssus. The British fauna includes 13 species.

---

1. Shell smooth, with or without concentric or radiating lines, but without raised ribs **2**

   Shell with two series of radiating ribs, separated by a smooth area extending from umbones to ventral margin **6**

---

2. Umbones terminal, at extreme anterior end of shell **3**

   Umbones subterminal, with anterior margin extending beyond beaks **4**

---

3. Umbones rounded, not downturned. Ventral margin convex or pouting. Mantle edge typically light yellow-brown ***Mytilus edulis***

   Umbones beaked or pointed, downturned. Ventral margin slightly concave posterior to umbones. Mantle edge typically purple or violet ***Mytilus galloprovincialis***

---

4. Hinge line with an oval, ridged area close to umbones (visible with hand lens). Periostracum hairy, covered with smooth spines. Usually less than 20 mm long ***Modiolula phaseolina***

   Hinge line of shell smooth. Periostracum with or without spines. Often longer than 20 mm **5**

---

5. Periostracum with fringes of broad serrated spines, particularly well developed posteriorly. Light yellow-red, with darker yellow-brown periostracum ***Modiolus barbatus***

   Periostracum without spines, or with smooth spines in small individuals. White, or blue/purple ***Modiolus modiolus***

---

6. Umbones very prominent, shell very fat. Dorsal margin downcurved towards posterior. Up to 18 ribs anteriorly, 20–35 posteriorly ***Modiolarca tumida***

   Umbones distinct, but not prominent, shell more slender. Dorsal margin upcurved, posterior end as deep as mid-line **7**

---

7. Elongate oval. Periostracum light or dark greenish-brown, with streaks and blotches of green and reddish-brown, often with red brown chevrons along hinge line. 20–30 ribs posteriorly ***Musculus costulatus***

   Oval to rhomboidal. Periostracum pale green or brown, often unicolorous, or with a few darker patches. 30–40 ribs posteriorly ***Musculus discors***

Mytilus edulis Linnaeus
Common mussel Fig. 10.29

Oval, pyriform or subtriangular, with umbones at apex. Length commonly 50–100 mm, but many rarely reach 30 mm, and largest exceed 150 mm. Light horn to blue or purple, dark purplish-blue when older; with pale blue and brown rays on translucent ground when small; periostracum deep blue to black, or dirty brown. Umbones prominent, raised umbonal ridges curving posteriorly and ventrally. Mantle edge of live specimens typically yellow-brown.

Mytilus edulis occurs in dense beds from upper shore and into the shallow sublittoral. Both size and shape are greatly influenced by environmental factors.

Widespread and common on all British coasts; ranges from Arctic waters south to the Mediterranean.

Mytilus galloprovincialis (Lamarck)
Mediterranean mussel Fig. 10.29

Oval, subtriangular, or pear-shaped, umbones terminal, pointed, slightly curved ventrally. Posterior to umbones the ventral margin is slightly concave, the shell edges forming a flattened area. Blue to deep purplish-black, periostracum light brown to blue-black; without colour rays typical of *M. edulis*. Mantle edge of live specimens typically purple.

This species is readily confused with *M. edulis*. It occurs intertidally around south-west England, south Wales, and southern and western Ireland. On the coast of France it extends westward from Cherbourg and south to the Mediterranean. Distinguishing features are the beaked, downturned umbones, the flattening of the ventral margin, and the colour of the mantle edge. Both species of *Mytilus* display great variation in shell morphology and old specimens may be impossible to identify with certainty.

Modiolus modiolus (Linnaeus)
Horse mussel Fig. 10.29

Irregularly oval or rhomboidal; umbones anterior but subterminal, with anterior margin projecting beyond. Dorsal margin convex, ventral margin concave. Umbonal ridges well developed. Up to 100 mm long, or longer. Bluish-white to slate blue, darkening in larger specimens. Periostracum very glossy, light horn to mahogany or dark brown in old shells, usually with a lighter yellow-brown strip along umbonal ridges. Young shells with numerous long, smooth periostracal spines.

Lower shore to about 150 m, on coarse grounds; it may form immense aggregations offshore. All British shores; south to the Bay of Biscay.

Modiolus barbatus (Linnaeus)
Bearded horse mussel Fig. 10.29

Elongate oval; umbones subterminal, above a curved anterior margin. Hinge line acute to umbones; ventral margin slightly concave; umbonal ridges very prominent, rounded, almost parallel with ventral margin. Up to 60 mm long. Yellowish-white, light yellow or reddish-brown with glossy periostracum, yellow, reddish-brown or light mahogany, lighter along umbonal ridges, bearing dense fringes of flattened, serrated spines. Inner surface pale bluish-white, tinged red or light purple.

Lower shore to about 100 m, on coarse grounds. Southern and western, north to Yorkshire on the east coast and the Clyde in the west; south to the Mediterranean and north-west Africa.

Modiolula phaseolina (Philippi)
Bean horse mussel Fig. 10.29

Broadly oval, with concave ventral margin, appearing bean-shaped, up to 20 mm long. Umbones subterminal. Light yellow to light purple; periostracum horn, light brown, or yellow-brown, paler along umbonal ridges, with broad fringes of smooth, slender spines. Inside of shell bluish-white to light purple; small oval knob on hinge line, close to umbones, with ridged surface—visible with hand lens in largest specimens.

Lower shore and subtidal. All British coasts. Ranges from northern Norway to Mediterranean and north-west Africa.

Musculus costulatus (Risso) Fig. 10.29

Elongate oval, dorsal margin curving gently above umbones, posterior end almost truncate, umbones subterminal, distinct. Up to 13 mm long. Off-white; periostracum light or dark greenish-brown, with patches of deep brown, reddish-brown, or purple, umbones frequently darker; often with broad streaks or zigzags, particularly anteriorly, with reddish-brown chevrons along hinge line. Ventral and posterior margins often deep green. About 10 broad, radiating ribs anterior to umbones, 20–30 posteriorly. Umbonal ridges without ribs.

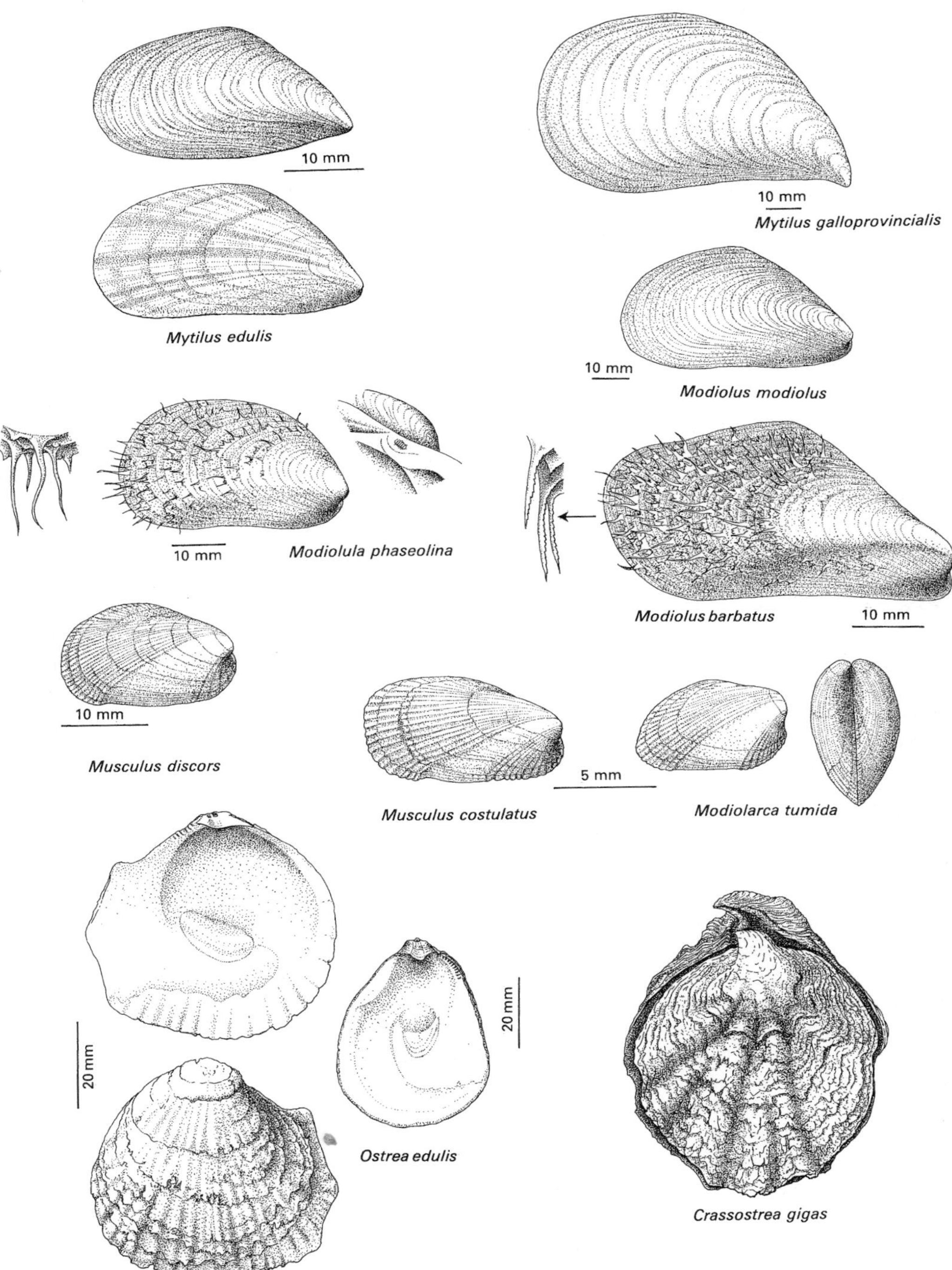

10 mm
10 mm
Mytilus galloprovincialis
Mytilus edulis
10 mm
Modiolus modiolus
10 mm
Modiolula phaseolina
Modiolus barbatus
10 mm
10 mm
Musculus discors
5 mm
Musculus costulatus
Modiolarca tumida
20 mm
20 mm
Ostrea edulis
Crassostrea gigas

Fig. 10.29

On rocky shores, lower shore, and shallow sublittoral. Mostly south and west coasts, south to Mediterranean and north-west Africa.

Musculus discors (Linnaeus)
Green crenella Fig. 10.29
Oval to rhomboidal; dorsal margin curving well above umbones, posterior broadly rounded. Umbones subterminal, distinct. Up to 13 mm long. Light yellow-brown; periostracum light brown, with pale green tinge, often darker along ventral margin. Eight to 12 broad ribs anteriorly, 30–50 more slender ribs posteriorly.

On rocky shores from mid-tidal zone into shallow sublittoral. Widespread and common off all British coasts. Distribution circumboreal, in north-east Atlantic ranging south to Mediterranean and Madeira.

Modiolarca tumida (Hanley)
Marbled crenella Fig. 10.29
Oval, dorsal margin downcurved, tapering posteriorly. Umbones and umbonal ridges very prominent, giving the two valves a very swollen appearance. Anterior margin almost straight. Up to 20 mm long. Off-white to light yellow brown, with horn-coloured or light green periostracum, occasionally with darker brown or red-brown markings. Fifteen to 18 broad ribs anteriorly, 20–35 posteriorly.

Lower shore and shallow sublittoral, among algal holdfasts, attached to undersides of shells and stones, or embedded in the tests of large tunicates, particularly *Ascidia mentula* Müller and *Ascidiella aspersa* (Müller). Widespread and common on all British coasts. South to Mediterranean.

6 OSTREIDAE

Shell thick, rugose; inequivalve, left (lower) valve convex, often forming a deep bowl, frequently overlapping right valve, usually cemented to substratum when small, right valve flat or slightly concave. Inequilateral, umbones anterior to mid-line. Juveniles with small taxodont teeth on each side of umbones, absent in adults. Ligament internal, thick; inner surface nacreous, with single adductor scar, elliptical, distinct, and often recessed. Four species present in Britain, two of which are introduced.

---

1. Left valve sculptured with coarse concentric grooves and ridges, often raised to form scales, and 10 or more fine ribs, worn and indistinct on older parts of shell; right with concentric sculpture but no ribs. Adductor scar white, off-white, or cream ***Ostrea edulis***

   Both valves with coarse concentric grooves and few (six or seven) very prominent ribs. Adductor scar deep purple or brownish ***Crassostrea gigas***

Ostrea edulis Linnaeus. Flat oyster Fig. 10.29
Shell oval or pear-shaped, or circular when young. Off-white, yellowish, or cream, with light brown or bluish concentric bands on right valve. Up to 100 mm long, rarely larger. Left valve with coarse concentric sculpture, often developed as flat scales, and numerous irregular, radiating ribs. Right valve with concentric sculpture but lacking ribs. Inner surfaces pearly, white or bluish-grey, often with darker blue areas marking enclosures of detritus; adductor scar white or yellowish-white.

Widely distributed around British Isles, from lower shore to about 80 m, as indigenous populations or introduced stocks. Occurs naturally from Norway south to Mediterranean.

Crassostrea gigas (Thunberg)
Portugese oyster Fig. 10.29
Shell elongate oval. Up to 180 mm long. Umbones prominent and often enrolled. Left valve with coarse concentric sculpture, raised and frilled, forming flat scales; about six bold raised ribs give grossly crenulate, saw-toothed appearance to margin. Right valve similarly sculptured, ribs corresponding to channels in opposite valve. Off-white, yellowish, or bluish-grey, often with darker purplish patches. Inner surface pearly, white or bluish-white, adductor scar deep purplish-blue or brown.

Introduced to south-east and south-west coasts, breeding only sporadically. Dead shells may occur elsewhere. The American oyster, *C. virginica* (Gmelin), is deep purple or reddish-brown in colour but, unlike *C. gigas*, the shell lacks bold ribs and a crenulate margin.

7 PECTINIDAE

Shell thick and strong, oval or circular, occasionally irregular and distorted, with bold, radiating ribs. Equivalve or inequivalve, one valve often flat or less concave than the other. Hinge line straight without teeth, with short, projecting ears on each side of umbones. Equilateral, or inequilateral with anterior ears longer than posterior. Ligament internal, in a well-marked chondrophore. One adductor scar only. Fifteen species are recorded from Britain, but most are found only in deep water or offshore.

---

1. Both shell valves irregular or distorted; ribs often worn, or obscured by concretions. Right valve strongly convex, usually cemented to substratum in adults **_Chlamys distorta_**

   Both valves regular, without distortion in shape or outline **2**

---

2. Right valve strongly convex, left valve flat; each with up to 17 broad, rounded ribs. Large, up to 150 mm long **_Pecten maximus_**

   Both valves shallowly convex **3**

---

3. Anterior ear of each valve much longer than posterior **4**

   Anterior ear of right valve with a byssal notch; otherwise ears of both valves essentially similar **6**

---

4. Shell sculpture of very fine radiating lines, occasionally prominent at edges of shell as ridges: 30–40 in number, up to 10 particularly broad and prominent, variably developed, but not regular over whole surface. Margin of shell semicircular. Often with coloured bands, rays, and zigzags **_Palliolum tigerinum_**

   Sculpture of raised ribs extending from umbones to margin, always prominent. Margin of shell forming a semi-ellipse **5**

---

5. Up to 35 radiating ribs, of regular width, each with numerous spines, particularly erect near margin, where they may overlap edge. Byssal notch in right anterior ear deep and conspicuous, with distinct teeth on the lower side **_Chlamys varia_**

   Each valve with 35–50 radiating ribs, of irregular width; spines less frequent, often small. Byssal notch in right anterior ear shallow, not particularly conspicuous **_Chlamys distorta_** juv.

---

6. Each valve with about 20 radiating ribs, without spines and smooth on the crest, but separated by transverse corrugations **_Aequipecten opercularis_**

   Sculpture of fine radiating lines and concentric striations, with 3–10 (typically seven) smooth, flattened ribs **_Pseudamussium septemradiatum_**

Pecten maximus (Linnaeus)
Great Scallop Fig. 10.30
Left valve flat, right valve strongly convex, slightly overlapping left along its margin. Ears equal, with a small byssal notch in right anterior ear. Right valve off-white, yellowish, or light brown, often with bands or spots of darker pigment, left valve light pink to reddish-brown. Up to 150 mm long, slightly longer than deep, both valves with about 16 bold, radiating ribs.

Attached by a byssus when small, later free and capable of swimming. Lives in depression in sand or fine gravel with left valve uppermost, offshore to about 100 m. All British coasts; from Norway to the Atlantic coast of Spain.

Aequipecten opercularis (Linnaeus)
Queen scallop Fig. 10.30
Left valve more convex than right. Anterior ear of right valve slightly larger than posterior, overlapping that of left valve, with distinct fold along hinge line; below it a small, rounded byssal notch, finely dentate along lower edge. Up to 90 mm long. Light pink to brown, orange, or yellow, usually with bands, zigzags, rays and spots of darker or lighter shades, right valve often paler than left, frequently with coloured markings on a creamy ground. With about 20 bold, radiating ribs.

Attached by byssus when young, later unattached, lives with left valve uppermost and swims freely and swiftly. Occasionally between tidemarks, offshore to 100 m, on sand and fine gravel. Common off all British coasts. Norway south to Mediterranean and Canary Isles.

Chlamys distorta (da Costa)
Hunchback scallop Fig. 10.30

Irregularly oval. Right valve with small byssal notch in anterior ear, more convex than left. Up to 50 mm long. Dull white or grey, to yellow, red, or brown. Juveniles with 35–50 regular, radiating ribs on each valve; cemented adult shells with up to 70 coarse spiny ribs.

Offshore, to at least 100 m. On hard substrata, at first attached by a byssus, but right valve eventually firmly cemented and becoming very irregular. On all British coasts, common; ranges from northern Norway and Iceland, south to the Mediterranean and west Africa.

Chlamys varia (Linnaeus)
Variegated scallop Fig. 10.30

Oval. Both valves with anterior ear more pronounced, twice as long as posterior. Right anterior ear with distinct, rounded byssal notch and fine teeth on lower border. Up to 60 mm long. Off-white, yellow, or orange, to brick-red, purple, or brown, often with bands or patches of darker or lighter colour. Very young stages usually unicolorous. With 25–35 bold ribs, and a few corrugated concentric lines, raised into prominent spatulate spines along ribs, most pronounced close to margin.

Attached by byssus when young, but may become free later; often grows enclosed by algal holdfasts. Common off all British coasts, lower shore to about 100 m. Present in Mediterranean, south to west Africa.

Palliolum tigerinum (Müller)
Tiger Scallop Fig. 10.30

Broadly oval. Right valve with narrow byssal notch in anterior ear, few poorly developed teeth on ventral border. Both valves with prominent, ribbed anterior ear and negligible posterior ear. Rarely longer than 25 mm. White or cream, light reddish-brown or purple, in radiating or concentric bands or zigzags, or irregular blotches. A common form has a broad median band extending from umbone to margin. Left valve frequently more highly coloured and patterned than right, which is often unicolorous. Up to 50 fine ribs may develop in later growth stages, often of unequal width, giving finely crenulate margin.

Lower shore to about 100 m, most frequent on coarser ground. All British coasts. Norway to Atlantic coasts of Spain, Portugal, and Morocco.

Pseudamussium septemradiatum (Müller) Fig. 10.30

Almost circular. Right anterior ear with very small byssal notch. Up to 50 mm long. Left valve dull brick-red, brown, or purple, with white or cream spots and blotches; right valve pale brick-red with spots in small specimens (<15 mm), becoming paler as animal grows. In large specimens right valve is largely creamy-white with reddish umbones, colour extending, and fading, down the ribs. Each valve with 3–10, most often 5–7, evenly spaced, broad, rounded ribs, increasing rapidly in width from umbone to margin. Right valve may have a group of four or five slender ribs close to anterior margin.

Lives with the left valve uppermost. Offshore to about 300 m, reported from west coast of Scotland and coasts of Northumberland and Durham. Norway to Mediterranean and north-west Africa.

8 LIMIDAE

Shell thin, delicate, typically elongate-oval along a dorso-ventral axis. Equivalve. Equilateral; or inequilateral with anterior ear larger than posterior and curvature of anterior margin more pronounced. Valves often gaping. Umbones spaced apart, with distinct cardinal area between. Sculpture of fine radiating ribs. Usually dull coloured. Ligament internal, in well-marked chondrophore. No hinge teeth. One indistinct adductor muscle scar only.

Limaria hians (Gmelin)
Gaping file shell Fig. 10.30

Obliquely oval. Anterior ears more prominent than posterior, sharply pointed; elliptical dorsal gape on anterior margin and narrower gape along whole of posterior margin. Up to 25 mm long. Off-white, often stained dirty brown. Thick fringe of orange or red pallial tentacles protrudes all around margin of shell.

The animal is able to swim; when settled it uses its byssus threads to construct a nest of gravel and shell debris. On coarse bottoms, from lower shore to about 100 m. Southern and western coasts only, south to Mediterranean and Canary Isles.

9 ASTARTIDAE

Shell thick, strong, oval, or subtriangular; equivalve; inequilateral, umbones just anterior to mid-line. Sculpture of prominent, concentric lines or ridges; periostracum thick. Two cardinal teeth in right valve, two or three in left valve. True lateral teeth

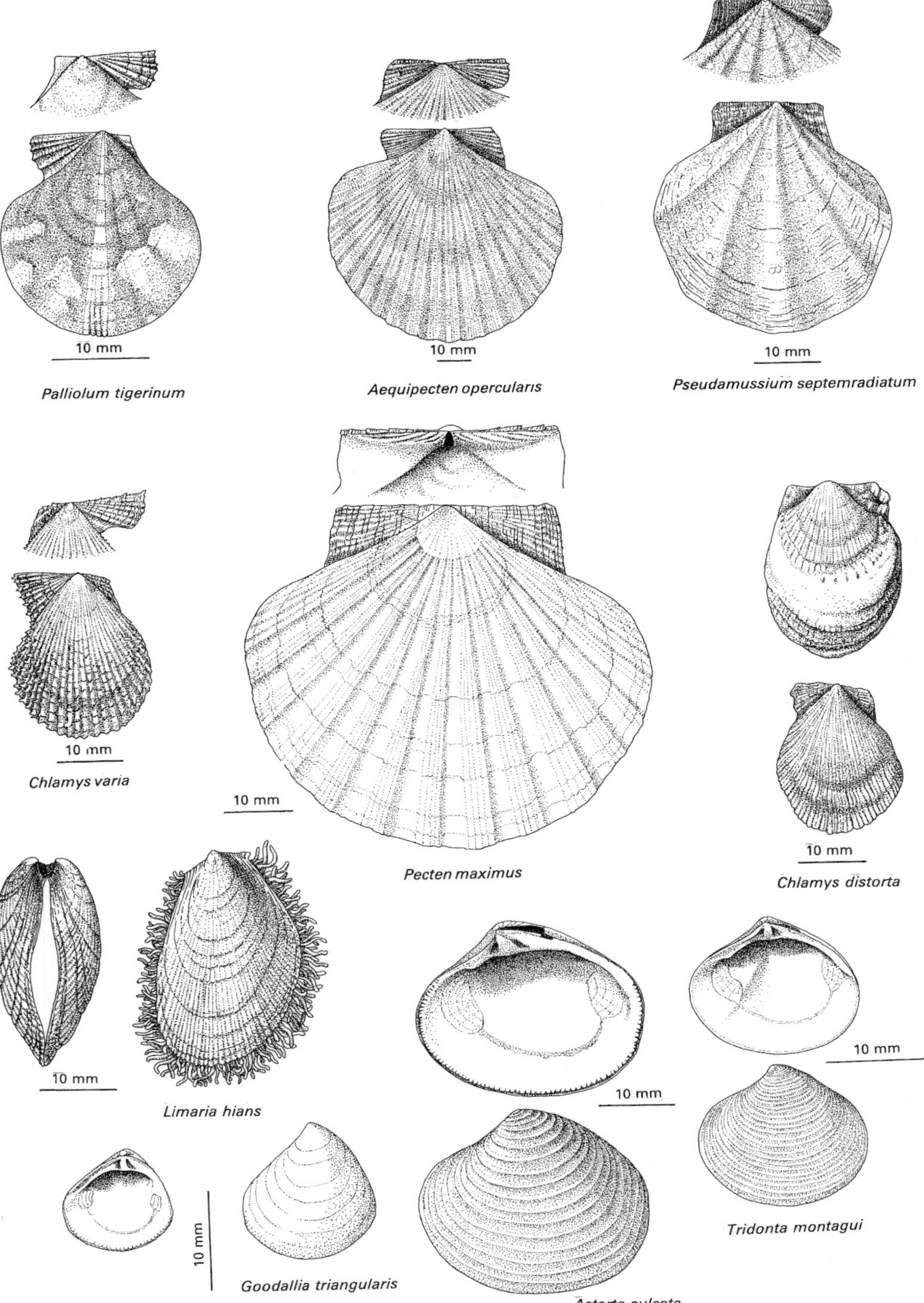
10 mm
Palliolum tigerinum
10 mm
Aequipecten opercularis
10 mm
Pseudamussium septemradiatum
10 mm
Chlamys varia
10 mm
Pecten maximus
10 mm
Chlamys distorta
10 mm
Limaria hians
10 mm
10 mm
Tridonta montagui
10 mm
Goodallia triangularis
Astarte sulcata

Fig. 10.30

absent. Ligament external. Adductor muscle scars about equal; pallial line distinct, without sinus. Six species recorded from Britain, only three of which occur frequently.

---

1. Shell dorso-ventrally elongate, triangular, rarely exceeding 3 mm length **_Goodallia triangularis_**

   Shell more broadly oval, up to 30 mm long **2**

---

2. Inner margin of shell crenulate; with about 20 bold concentric ridges. Up to 30 mm long **_Astarte sulcata_**

   Inner margin of shell smooth; with 30–40 finer concentric ridges. Up to 15 mm long **_Tridonta montagui_**

Astarte sulcata (da Costa) Fig. 10.30
Broadly oval, prominent umbones imparting subtriangular outline. White or pink with thick, mid- to dark-brown periostracum, often encrusted and stained black about umbones. Up to 30 mm long. Lunule and escutcheon distinct, deep. With bold concentric ridges. Left valve with two prominent cardinal teeth below umbones and a third, slender, below ligament. Right valve with very thick anterior cardinal tooth and slender one posteriorly; additionally, slender ridge below escutcheon may be taken for posterior lateral tooth. Adductor scars and pallial line clear, inner shell margin crenulate.

In mud, or fine gravels with sand, shell, or mud, offshore. All British coasts; from southern Arctic waters to Mediterranean and north-west Africa.

Tridonta montagui (Dillwyn) Fig. 10.30
Broadly oval, umbones centred on mid-line, prominent, giving subtriangular outline. Up to 15 mm long. White, with a thick, light- or dark-brown periostracum. Lunule and escutcheon distinct, elongate. With 30–40 raised concentric ridges. Right valve with two cardinal teeth: anterior broad, ridged or bifid, posterior indistinct. Left valve with two equal-sized teeth anteriorly; posteriorly, the third is poorly developed and may appear to be absent. Adductor scars and pallial line indistinct, inner shell margin smooth.

Offshore to about 100 m, in sand and fine gravel. Circumboreal, most frequent off north and west coasts, extending no farther south than the Bay of Biscay.

Goodallia triangularis (Montagu) Fig. 10.30
Triangular; umbones prominent, situated on mid-line. About 3 mm long. White, with light yellowish to brown periostracum. Lunule and escutcheon indistinct, shallow. With numerous fine concentric lines, closely spaced. Right valve with two cardinal teeth, anterior broad and triangular, posterior slender. Left valve with two approximately equal-sized cardinals, posteriorly a third one appears as a ridged projection, like a lateral tooth. Inner shell margin with distinct bead-like crenulations.

Offshore, to at least 100 m, in sand and fine gravel. All British coasts, south to Mediterranean and north-west Africa.

10 LUCINIDAE. HATCHET SHELLS

Shell equivalve; inequilateral, umbones just anterior to mid-line. Hinge line with cardinal and lateral teeth; ligament external, often deeply inset, lunule and/or escutcheon typically well developed. Pallial line without sinus. Adductor scars unequal: anterior larger, often partly or largely separated from pallial line. Shell usually white and dull. Five species recorded from Britain.

---

1. Anterior adductor scar very elongate and narrow, touching pallial line only at its anterior end. Shell with numerous, fine, closely spaced, concentric ridges **_Lucinoma borealis_**

   Anterior adductor scar fused with pallial line along about half of its anterior edge. Shell with concentric lines or ridges **2**

---

2. Shell margin strongly convex anterior to umbones. Escutcheon absent. Sculpture of fine concentric lines **_Loripes lucinalis_**

Shell margin shallowly concave anterior to umbones. Escutcheon distinct, bordered by raised, crenellated margins **_Myrtea spinifera_**

Lucinoma borealis (Linnaeus) Fig. 10.31
Roughly circular; dull white, with thin light-brown periostracum. Up to 40 mm long. Lunule pyriform, escutcheon fusiform, deep. Ligament clearly visible. Left valve with bifid anterior and single posterior cardinal teeth, right valve with single anterior and bifid posterior. Each valve with single anterior and posterior lateral teeth, weakly developed. Pallial line and adductor muscle scars distinct; anterior scar up to five times as long as broad, only its narrow anterior end attached to pallial line.

In sand and gravel from lower shore to about 100 m. Norway to Mediterranean; present on most British coasts.

Loripes lucinalis (Lamarck) Fig. 10.31
Circular, almost equivalve; white or cream, periostracum with very pale beige tints. Up to 20 mm long. Lunule short and deep, no escutcheon; ligament deeply inset, hidden. Left valve with bifid anterior and single posterior cardinal teeth, right valve with single anterior and bifid posterior. Each valve with single anterior and posterior lateral teeth, often distinct. Anterior adductor scar elongate oval, fused with pallial line along dorsal one-third of its anterior border.

In sand and fine gravel, from lower shore to at least 150 m. Reported from the Channel, off Northumberland coast, and from Isle of Man; south to Mediterranean and west Africa.

Myrtea spinifera (Montagu) Fig. 10.31
Oval. White or cream with indistinct, pale beige periostracum. Up to 25 mm long. Dorsal margin anterior to umbones concave, with distinct oval lunule; ligament visible in deep fusiform escutcheon bordered laterally by erect, flared, and crenellated rims. Right valve with single cardinal tooth, and single anterior and posterior lateral teeth. Left valve with two cardinal and single anterior and posterior lateral teeth; ridges may develop along hinge line and appear as extra laterals. Anterior adductor scar elongate oval, roughly shaped like a dumb-bell, touching pallial line along its anterior ventral border.

In mud and muddy sands, offshore to about 100 m. Western coasts from south Devon to Shetland Isles, rarely elsewhere. Norway to Mediterranean and Azores.

11 DIPLODONTIDAE

Shell equivalve; inequilateral, umbones anterior to mid-line. Hinge line with cardinal and lateral teeth, ligament external. Pallial line without sinus. Adductor muscle scars subequal, anterior being the larger. One British species.

Diplodonta rotundata (Montagu) Fig. 10.31
Irregularly oval, posterior half deeper than anterior. White, periostracum light yellowish-brown; surface smooth or silky in unworn specimens. Up to 30 mm long. Lunule indistinct; escutcheon well marked, deep, ligament distinct. Left valve with bifid anterior cardinal tooth and a single posterior cardinal tooth; right valve with single anterior and bifid posterior. Both valves with single anterior and posterior lateral teeth. Anterior adductor scar dorso-ventrally elongate, comma-shaped, closely applied to pallial line.

In sand and gravel, offshore to about 70 m. English Channel and south and west Irish coasts, south to Mediterranean and west Africa.

12 KELLIIDAE

Shell equivalve, inequilateral, small. Ligament internal and external; hinge with cardinal teeth, typically small and indistinct, and poorly developed lateral teeth. Pallial line broad, ill-defined, without sinus. Three British species, two of which are known only from a few scattered records.

Kellia suborbicularis (Montagu) Fig. 10.31
Broadly oval, almost circular; umbones just in front of mid-line. White, with pale horn or yellowish periostracum. Up to 10 mm long. Right valve with single cardinal tooth below beak of umbo, and single posterior lateral; left valve with two cardinal teeth below beak, and single posterior lateral. Small external ligament posterior to beaks; in each valve internal ligament marked by embayment in hinge line posterior to cardinal tooth/teeth.

Lower shore to at least 100 m; in holes and crevices, old shells, barnacle tests. Most frequent on southern and western shores. Norway south to Mediterranean and West Africa.

13 LASAEIDAE

Shell equivalve, inequilateral, small. Ligament internal; hinge line with cardinal and lateral teeth, few and small. Adductor scars about equal, pallial line without sinus. Two species recorded from Britain. *Semierycina nitida* (Turton) is a rare inhabitant of offshore sandy gravel.

Lasaea rubra (Montagu) Fig. 10.31

Obliquely oval, both valves strongly convex; umbones posterior to mid-line. Up to 3 mm long, white, with concentric bands of pale pink or red, darkest towards margins; periostracum light yellow-brown. Right valve with single anterior and posterior lateral teeth, left valve with one small cardinal tooth and single anterior and posterior laterals; in both valves internal ligament is marked by deep embayment below hinge line and broad plate below posterior lateral tooth.

The venus shell *Turtonia minuta* is the same size as *Lasaea rubra*, occupies similar habitats and superficially resembles it. However, *Turtonia* may be distinguished by the group of three cardinal teeth present in each valve.

Intertidal, common on rocky shores all around British Isles. Attached by byssus, in crevices, among *Fucus* clumps or in *Lichina* tufts; often in very large numbers. Norway to Mediterranean and Canary Isles.

14 LEPTONIDAE. COIN SHELLS

Small, round shells, thin and generally rather fragile. Equivalve; equilateral or just inequilateral. Ligament internal. Hinge with one cardinal tooth centrally, and with anterior and posterior lateral teeth. Adductor scars about equal. Pallial line indistinct, without sinus. Five British species. Some authorities restrict the family to *Lepton squamosum*, referring the others to the family Neoleptonidae.

Lepton squamosum (Montagu) Fig. 10.31

Roughly square in outline, umbones small, just anterior to mid-line. Up to 12 mm long. White, translucent, periostracum thin. With fine concentric lines, most conspicuous close to distinct growth stages; both valves with dense punctulations, reminiscent of a brachiopod.

In silty sand and fine gravel from ELWS into shallow sublittoral; in or close to burrows of callianassid shrimps (*Upogebia* spp.). Norway to Mediterranean; in British waters, most frequent off southern and western coasts. *Lepton nitidum* Turton is very similar but is almost circular in shape; it occurs offshore, but is not commensal.

Epilepton clarkiae (Clark) Fig. 10.31

Obliquely oval, umbones posterior to mid-line; up to 2 mm long. Yellowish-white, periostracum indistinct, visible as darker bands about growth stages. With fine concentric lines and irregular, radiating striations; growth stages clear. Adductor scars and pallial line indistinct.

Lower shore into sublittoral, in association with various species of sipunculid. South and west coasts of British Isles; present in Mediterranean.

Neolepton sulcatulum (Jeffreys) and *N. sykesi* (Chester) are similar to *E. clarkiae*. Both are almost circular in outline, with bold concentric lines, and in British waters are known only from the Channel Isles.

15 MONTACUTIDAE

Small, fragile shells. Equivalve; inequilateral. Hinge line with thin anterior and/or posterior lateral teeth, cardinal teeth often absent. Ligament internal. Adductor scars about equal. Pallial line diffuse, without sinus. Nine species known from Britain, most of which occur as commensals of other invertebrates.

Tellimya ferruginosa (Montagu) Fig. 10.31

Elongate oval, umbones posterior to mid-line; up to 8 mm long. White; periostracum thin, light yellowish, usually covered with thick, granular, rusty red deposit. Right valve with single cardinal tooth, left valve with one anterior lateral.

Commensal in burrows of the sea urchin *Echinocardium cordatum*. Widespread and common off all British coasts; Norway to Mediterranean and north-west Africa.

Montacuta substriata (Montagu) Fig. 10.31

Obliquely oval, umbones posterior to mid-line; up to 3 mm long. White, with inconspicuous pale-brown periostracum. Ventral margin often slightly concave medially. Each valve with single anterior lateral tooth only.

Lives in association with sea urchins, usually attached to anal spines. Recorded from various localities, associated with *Spatangus purpureus* and *Echinocardium flavescens*. From ELWS, offshore to about 100 m. Ranges from Norway to Mediterranean.

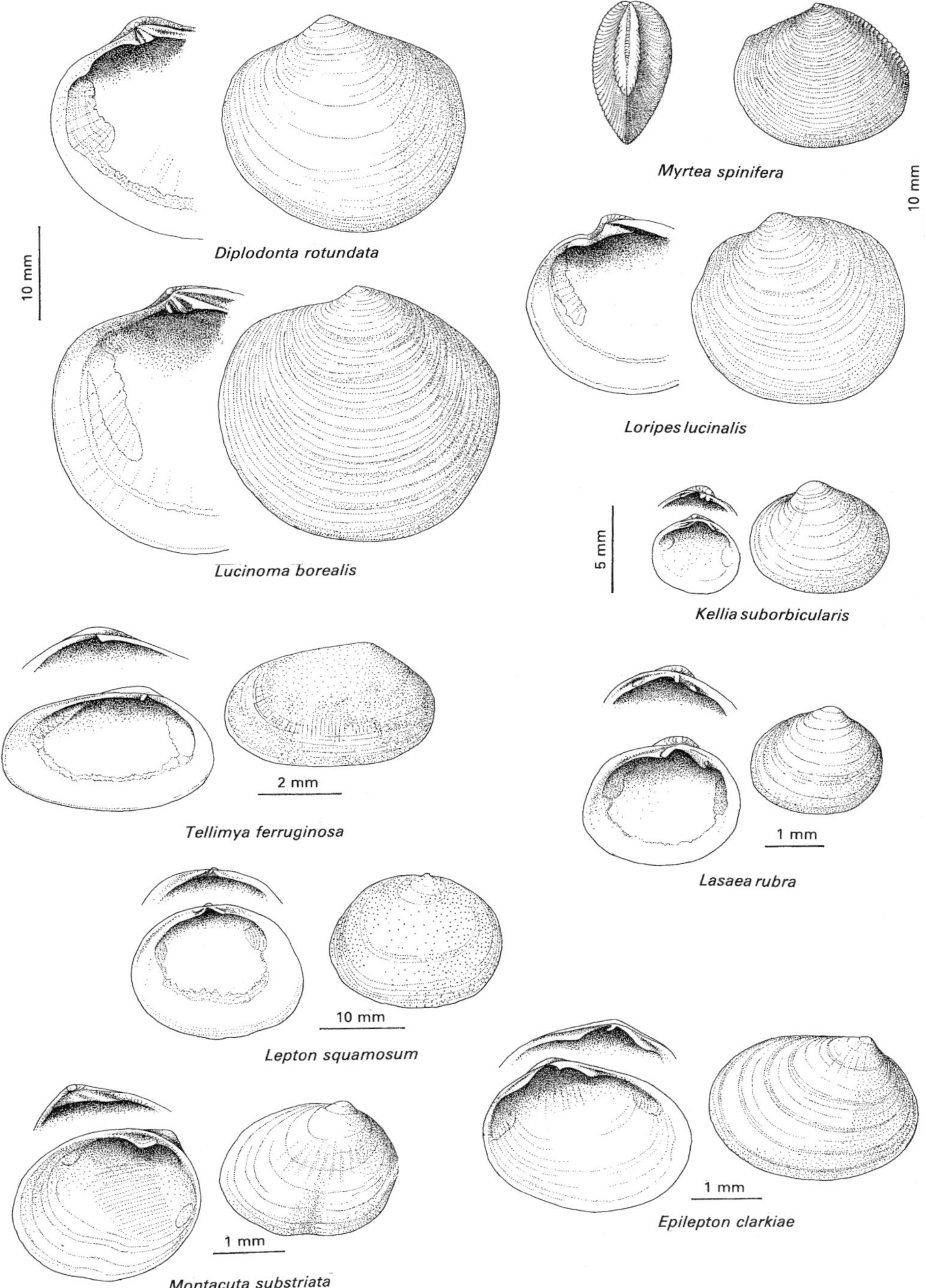
Myrtea spinifera
10 mm
Diplodonta rotundata
10 mm
Loripes lucinalis
Lucinoma borealis
5 mm
Kellia suborbicularis
2 mm
Tellimya ferruginosa
1 mm
Lasaea rubra
10 mm
Lepton squamosum
1 mm
Montacuta substriata
1 mm
Epilepton clarkiae

Fig. 10.31

Mysella bidentata (Montagu) Fig. 10.32
Oval, umbones posterior to mid-line; up to 3 mm long. White or translucent, periostracum light brown or olive. Each valve with single anterior and single posterior lateral tooth.

ELWS to about 100 m, in muddy sand or fine gravel, in crevices of dead oyster valves, in burrows of the sipunculid *Golfingia*, or associated with the ophiuroid *Amphiura brachiata*. Widespread around Britain; ranges from Norway to west Africa and Mediterranean.

16 ARCTICIDAE

Shell equivalve; inequilateral, umbones anterior to mid-line. Periostracum thick, conspicuous; ligament external. Cardinal and lateral teeth present. Pallial line without sinus. One British species.

Arctica islandica (Linnaeus)
Icelandic cyprine Fig. 10.32
Broadly oval, with prominent umbones, up to 120 mm long; dull white with thick periostracum: glossy chestnut-brown in smaller individuals, deep greenish-brown to black in large specimens. Anterior hinge line strongly curved, lunule ill-defined; escutcheon shallow, with thick, brown or black ligament. Right valve with three prominent cardinal teeth and single posterior lateral tooth; triangular pit in front of anterior cardinal, surrounded by small knobs and ridges. Left valve with three cardinals and one posterior lateral, anterior cardinal continuous with series of small ridges and denticulations.

In sand and muddy sand; offshore, perhaps to edge of Continental Shelf. All British coasts, from Arctic to Bay of Biscay.

17 GLOSSIDAE

Shell equivalve, inequilateral, umbones prominent, enrolled and coiled. Ligament external. Hinge line with cardinal and lateral teeth. Pallial line without sinus. One British species.

Glossus humanus (Linnaeus)
Heart Cockle Fig. 10.32
Valves strongly convex, roughly circular, deeper posteriorly than anteriorly, with prominent, rounded, and tightly enrolled umbones, curving away from hinge line; umbones anterior to mid-line. Dull white or beige, periostracum thick, light brown or olive, darkening to deep brown or green in larger specimens. Up to 100 mm long. Three cardinal teeth in each valve: in right both posterior cardinals closely set, almost in continuous line, separated by deep groove from anterior tooth; in left, narrow ridged posterior cardinal is separated by groove from two closely set anterior teeth. Single ridge-like posterior lateral tooth in each valve.

In soft substrata, offshore; mostly southern and western coasts, west Scotland and Ireland to Bristol Channel. Distributed from Norway to Mediterranean.

18 CARDIIDAE

Shell equivalve, both valves strongly convex; inequilateral, prominent umbones just anterior to midline. Sculpture of bold radiating ribs, often with conspicuous spines, tubercles, or scales. Two peg-like cardinal teeth in each valve. Lateral teeth present. Adductor scars about equal, pallial line without sinus. The British fauna includes 11 species. *Plagiocardium papillosum* (Poli) is a rare species known only from the Isles of Scilly and the Channel Isles.

---

1. Radiating ribs inconspicuous, most prominent ventrally, no knobs or spines. Prodissoconch smooth and glossy **_Laevicardium crassum_**

   Radiating ribs conspicuous, visible even on prodissoconch, typically with knobs or spines **2**

---

2. Ribs visible on inner surface of shell as grooves, extending practically to region below umbones **6**

   Ribs visible only on ventral part of inner surface, the grooves fading rapidly beyond pallial line; area beneath umbones completely smooth **3**

---

3. Right valve with two posterior lateral teeth; left valve with single anterior and posterior lateral teeth equally developed **_Cerastoderma edule_**

   Right valve with one posterior lateral tooth; left valve with anterior lateral tooth more prominent than posterior **4**

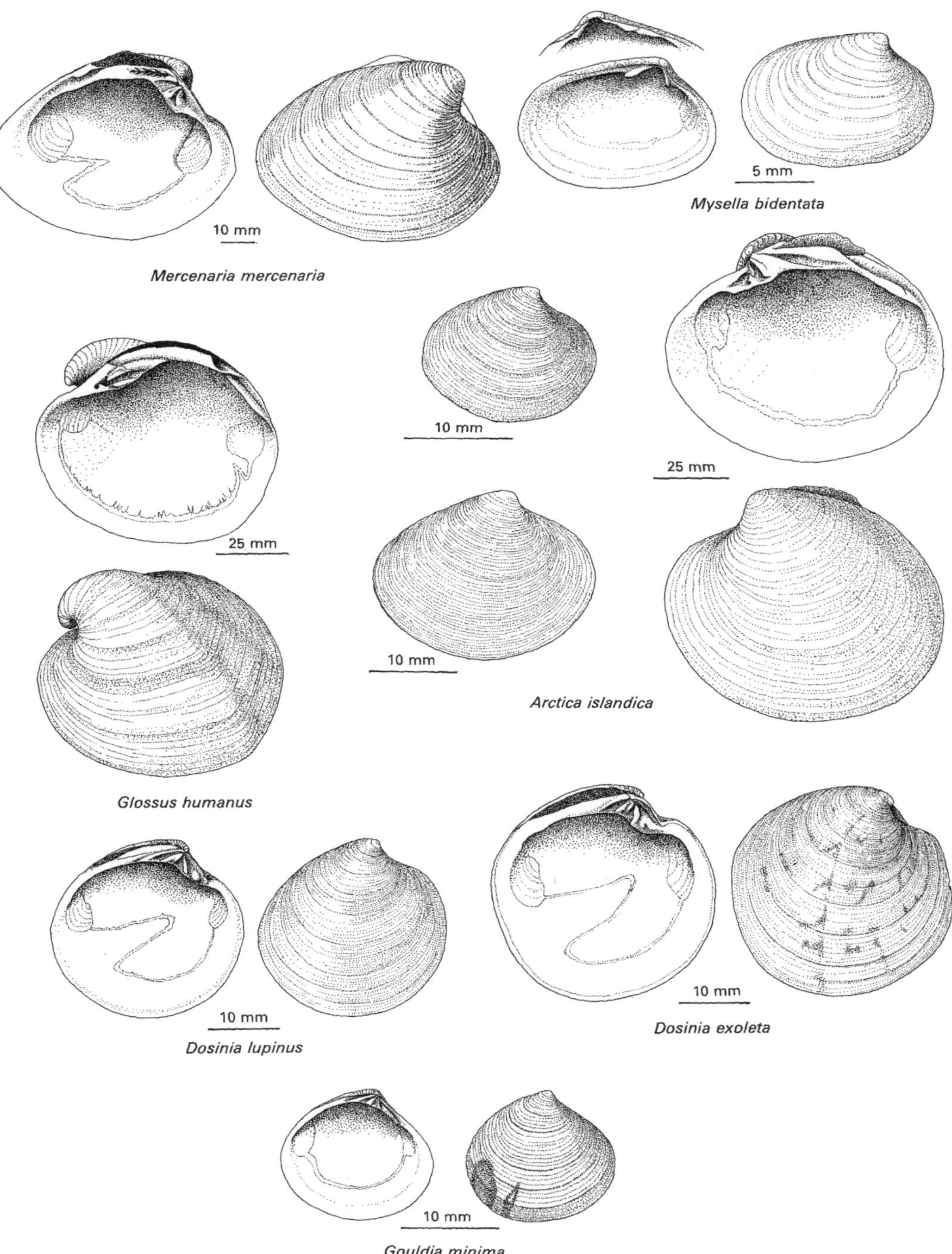
10 mm
Mercenaria mercenaria
5 mm
Mysella bidentata
10 mm
25 mm
25 mm
10 mm
Arctica islandica
Glossus humanus
10 mm
Dosinia lupinus
10 mm
Dosinia exoleta
10 mm
Gouldia minima

Fig. 10.32

4. Up to 20 ribs, each with regularly spaced spines, distinct ventrally but forming sharp spiny ridge dorsally, or on young specimens. Up to 90 mm long **_Acanthocardia tuberculata_**

   20 or more ribs, occasionally with small tubercles close to margins but without spines. Not exceeding 20 mm **5**

5. Up to 28 ribs, each with transverse grooves, developing flat, square tubercles close to margins **_Parvicardium scabrum_**

   Up to 22 ribs, with sparsely developed, conical tubercles close to margins **_Parvicardium exiguum_**

6. Each rib with sharp keel bearing numerous pointed spines (bases visible in worn specimens) **7**

   Ribs with or without blunt tubercles developed close to anterior and posterior margins, but not keeled **8**

7. Posterior margin convex; spines long and prominent anteriorly, usually shorter posteriorly. Cardinal teeth of left valve equal sized **_Acanthocardia echinata_**

   Posterior margin almost straight, spines short and thick anteriorly, longer and sharper posteriorly. In left valve, anterior cardinal tooth marginally longer than posterior **_Acanthocardia aculeata_**

8. Right valve with two posterior lateral teeth. Concentric sculpture extending over surface of ribs, occasionally raised to form flat scales but not tuberculate. Up to 60 mm **_Cerastoderma glaucum_**

   Right valve with one posterior lateral tooth. Concentric sculpture restricted to grooves between ribs. Ribs smooth, with few square tubercles **9**

9. 28 or more ribs, each with short, flattened tubercles; concentric sculpture well developed, forming series of lattices between ribs **_Parvicardium minimum_**

   Up to 26 ribs; tubercles sparsely developed, most prominent close to anterior and posterior margins **_Parvicardium ovale_**

Acanthocardia echinata (Linnaeus)
Prickly cockle Fig. 10.33

Obliquely oval, anterior hinge line sloping gently to convex anterior margin; posterior hinge line more steeply inclined, posterior margin only slightly convex. Up to 75 mm long. Light fawn to deeper brown, in concentric bands, often darker or bluish tinged along posterior margin. With 18–23 ribs, each with sharp central keel, regularly produced into erect, sharp spines, continuous basally with keel; spines most prominent on anterior ribs. Right valve with two anterior and one posterior lateral teeth; left valve with single anterior and posterior laterals, anterior being longer, thicker, and more prominent than posterior. Inner surfaces white, glossy, occasionally bluish beneath umbones. External sculpture visible as grooves which extend whole depth of shell.

In fine sand and gravel, usually with mud, offshore. All British coasts; Norway to Mediterranean and Canary Isles.

Acanthocardia tuberculata (Linnaeus)
Rough cockle Fig. 10.33

Approximately rhombic in shape: hinge line sloping gently from umbones on each side, posterior margin scarcely convex, almost straight, anterior margin more strongly convex. Up to 90 mm long. Off-white, yellow or light brown, often in concentric bands of different shades, frequently darker about umbones. With 18–20 bold ribs, and fine concentric grooves and ridges. Dorsally, each rib has central keel, bearing short, pointed spines; ventrally keel is obscured and spines appear separate. Right valve

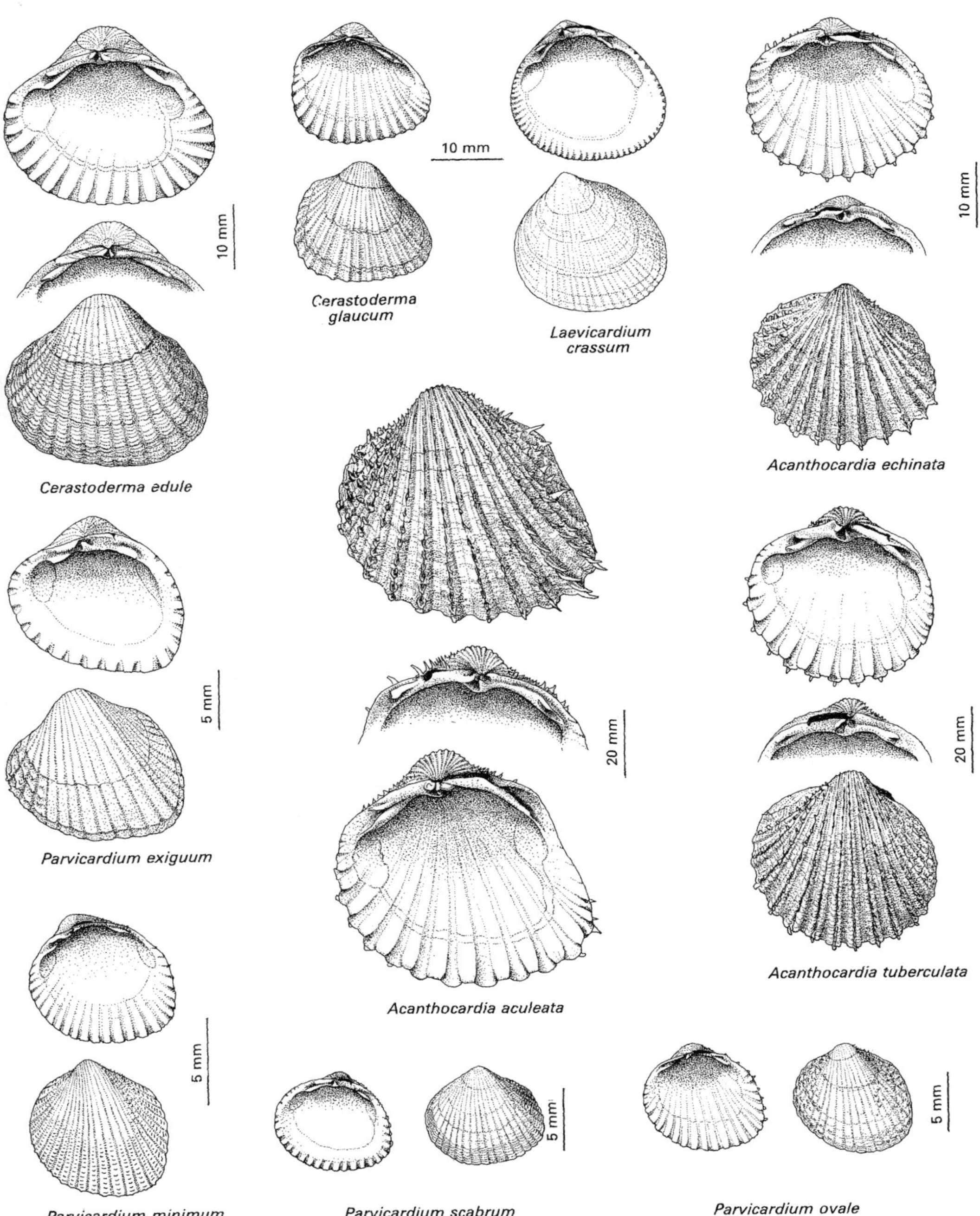
10 mm
10 mm
10 mm
Cerastoderma glaucum
Laevicardium crassum
Acanthocardia echinata
Cerastoderma edule
5 mm
20 mm
20 mm
Parvicardium exiguum
Acanthocardia tuberculata
Acanthocardia aculeata
5 mm
5 mm
5 mm
Parvicardium minimum
Parvicardium scabrum
Parvicardium ovale

Fig. 10.33

with two anterior and one posterior lateral teeth, left valve with single anterior and posterior laterals. Inner surface white, external sculpture visible as grooves extending from ventral margin to pallial line, fading rapidly beyond it and smooth beneath umbones.

On muddy sand and gravel, lower shore and shallow sublittoral. Southern coasts only, south to Mediterranean and north-west Africa.

Acanthocardia aculeata (Linnaeus)
Spiny cockle Fig. 10.33

Shell markedly eccentric: anterior margin regularly convex, continuous with hinge line; posterior margin almost straight, with marked angle between it and posterior hinge line. Up to 100 mm long. Off-white, yellowish, or light chestnut, usually with concentric bands of colour, occasionally with pink tinge. About 20 bold ribs, with distinct concentric grooves and ridges between. Each rib with sharp keel, bearing regularly spaced spines; longest and sharpest spines towards postero-ventral margin, anteriorly they are short, blunt, and recurved towards hinge line. Each valve with two cardinal teeth, in left valve anterior (lower) tooth is larger than posterior; right valve with two anterior and one posterior lateral teeth, left with single anterior and posterior teeth. Inner surfaces white, with pink tints along ribs; sculpture visible as grooves extending through whole of shell.

South and west coasts only, offshore. South to Mediterranean and north-west Africa.

Cerastoderma edule (Linnaeus)
Common cockle Fig. 10.33

Broadly oval, up to 50 mm long; cream, light yellowish, or pale brown. About 24 broad ribs, and closely spaced concentric ridges, ridges on ribs may be developed as flattened, scale-like spines. Right valve with two anterior and two posterior laterals. Inner surfaces dull white, in larger specimens the posterior adductor scar, and areas around it, may be tinted chestnut or light purple. External sculpture visible as grooves extending from ventral margin, fading rapidly beyond pallial line.

In sandy mud, sand, or fine gravel, from mid-tidal level to just below ELWS. Common on all British coasts, tolerant of salinities as low as 10%, and often abundant in sheltered bays and estuaries. Distributed from north-east Norway to west Africa.

Cerastoderma glaucum (Poiret)
Lagoon cockle Fig. 10.33

Shell markedly eccentric: anterior margin shallowly convex, forming sharp junction with hinge line, posterior margin continuous with hinge line, sloping steeply from umbones. Off-white, yellowish, or greenish, with darker brown areas, overlain by greenish-brown periostracum. Up to 50 mm long. About 24 ribs, with closely spaced concentric corrugations, forming small, flattened scales on ribs. Right valve with two anterior and two posterior laterals, left valve with single anterior and posterior laterals. Inner surfaces dull white or brown, ribs visible as distinct grooves throughout whole of shell, extending beneath umbones. Posterior margin not crenulate.

In brackish water, overlapping distribution of *Cerastoderma edule*. Known positively only from the southern parts of the British Isles, from East Anglia to South Wales, but probably present elsewhere. Recorded from many parts of western Europe, the Mediterranean, and Black Sea.

Laevicardium crassum (Gmelin)
Norway cockle Fig. 10.33

Obliquely oval, up to 70 mm long, as deep as its greatest length. Cream, light yellow to fawn, with blotches of pink, chestnut, or brown, particularly about the umbones. With numerous (about 50) faint ribs, most conspicuous ventrally, concentrated on middle regions of shell and absent from anterior and posterior parts. Prodissoconch and first growth stages typically smooth and glossy. Each valve with two cardinal teeth, right posterior considerably larger than right anterior. Right valve with two anterior and one posterior lateral; left valve with single anterior and posterior laterals, anterior much larger than posterior. Inner surfaces smooth, glossy white tinted with pink, darkest close to margin, external ribs visible as grooves extending from margin to pallial line but fading rapidly beyond it.

Offshore, to edge of Continental Shelf, in sand and shelly gravel. All British coasts; Norway to Mediterranean and west Africa.

Parvicardium ovale (Sowerby) Fig. 10.33

Obliquely oval, small, up to 13 mm; yellowish-white. Up to 26 flat ribs; concentric ridges poorly developed. Tubercles sparsely developed on ribs, usually only at anterior and posterior margins; anteriorly forming flat scales, posteriorly more spine-like. Right valve with two anterior and one posterior lateral teeth, left valve with single anterior and posterior laterals, anterior the larger. Inner surfaces glossy white, external sculpture visible as grooves which extend beneath umbones.

In sand and gravel, offshore to about 100 m. All British coasts. From Iceland to Mediterranean and Canary Isles.

Parvicardium minimum (Philippi) Fig. 10.33
Obliquely oval, up to 12 mm; yellowish-white, with 28 or more broad ribs, indistinct concentric ridges between. Each rib with numerous small, closely spaced, flattened spines, present over whole of shell, although sometimes worn on greatest areas of curvature. Right valve with two anterior and one posterior lateral teeth, left valve with single anterior and posterior teeth, anterior the larger. Inner surfaces white, external sculpture visible as grooves extending throughout whole of shell.

In mud, sand, and fine gravel, offshore to about 160 m. Western coasts of Britain and Ireland; from Iceland to Mediterranean and north-west Africa.

Parvicardium exiguum (Gmelin)
Little cockle Fig. 10.33
Obliquely oval, hinge line sloping strongly to anterior, umbones well in front of mid-line. Yellowish-white, with a thick, patchy brown periostracum; to 13 mm long. Up to 22 broad ribs, with well-marked concentric ridges between; small tubercle-like spines present on all ribs in young specimens, later persisting only at anterior and posterior margins, and sparingly ventrally. Right valve with two anterior and one posterior lateral teeth, upper anterior being very small. Left valve with single anterior and posterior laterals, anterior being more prominent. Inner surfaces dull white, or tinted light brown or green; ribs form deep marginal crenulations, not extending into interior of shell.

In sand, gravel, or mud; lower shore and shallow sublittoral, often extending into estuaries. All British coasts. From Norway to Mediterranean and Black Sea.

Parvicardium scabrum (Philippi) Fig. 10.33
Broadly oval, up to 12 mm; yellowish-white with pale brown periostracum. Up to 28 ribs, interstices between almost smooth, with a few concentric lines or grooves. Broad, scale-like spines present on all ribs, but most pronounced along posterior margin. Right valve with two anterior and one posterior lateral teeth, upper anterior very small. Left valve with single anterior and posterior laterals, anterior more prominent than posterior. Inner surfaces white, marginal crenulations extending as grooves as far as pallial line, fading rapidly beyond it.

In sand, gravel, or shell; offshore, perhaps to edge of Continental Shelf. All British coasts; from Norway to Mediterranean and north-west Africa.

19 VENERIDAE

Shell typically thick and strong, equivalve; inequilateral, umbones anterior to mid-line, usually prominent. Sculpture principally of concentric grooves. Characteristically with three cardinal teeth in each valve, occasionally with anterior lateral teeth. Lunule distinct. Ligament external. Adductor scars about equal, pallial line with a sinus. British fauna includes 17 species. Some authorites regard *Venerupis saxatilis* and *V. senegalensis* as a single species, and *Turtonia minuta* has sometimes been referred to a separate family.

---

1. Shell oval, smooth, and plump; minute, less than 3 mm long. Inhabiting crevices between tidemarks, attached by byssus. Cardinal teeth poorly developed, three in each valve but only two in each distinctly visible **_Turtonia minuta_**

   Adult shell considerably larger. Cardinal teeth distinct. Distributed above and below ELWS **2**

---

2. Shell with bold concentric ridges, extending over whole surface of valve, not simply developed at posterior margin **3**

   Shell smooth, or with radiating ridges or striations, or with concentric grooves and lines, occasionally forming ridges posteriorly or around umbones, but without regular concentric ridges **8**

---

3. Shell oblong, elongate oval, or irregular. Concentric sculpture friable, tending to wear. Inhabiting holes and crevices **4**

   Shell oval or triangular, regular, often thick and solid. Concentric sculpture thick, not easily broken **5**

---

4. Concentric sculpture with raised and frilled edges. Shell often irregular. Pallial sinus short, triangular ***Irus irus***

   Concentric sculpture more prominent posteriorly, but not frilled. Shell regularly oblong. Pallial sinus deep, U-shaped ***Venerupis saxatilis***

---

5. Concentric ridges few in number—about 15—broad and flat, varying in width and imparting a stepped appearance to surface of valve ***Clausinella fasciata***

   Concentric ridges numerous, closely spaced, with acute edges **6**

---

6. Concentric ridges of equal height and thickness over whole of valve, generally rather fine. Shell with three broad bands of red-brown or chestnut radiating from umbones ***Chamelea gallina***

   Concentric ridges coarse, prominent, often varying in width and height. Uniform off-white or cream, with or without light-brown patches about umbones **7**

---

7. Concentric sculpture intersecting with radiating grooves or ridges, accentuated posteriorly where the concentric ridges are disrupted to form coarse, raised knobs ***Venus verrucosa***

   No radiating sculpture, concentric ridges entire, most prominent posteriorly ***Circomphalus casina***

---

8. Inner margin of shell with distinct crenulations **9**

   Inner margin of shell smooth **10**

---

9. Sculpture of numerous radiating ribs intersected by concentric grooves, giving a regularly decussate effect. Shell up to 20 mm long, superficially resembling a small cockle ***Timoclea ovata***

   Sculpture of concentric grooves and striations, developed as fine ridges anteriorly and posteriorly, and about umbones. Lunule heart-shaped, very distinct. Up to 120 mm long ***Mercenaria mercenaria***

---

10. Pallial sinus very shallow, represented only by a small indentation in pallial line ***Gouldia minima***

    Pallial sinus deep and distinct, U- or V-shaped **11**

---

11. Shell almost circular, with arched anterior margin **12**

    Shell broadly oval or rhomboidal; anterior margin sloping gently away from umbones **13**

---

12. Lunule as broad as long; in front of it the anterior margin is strongly convex ***Dosinia exoleta***

Lunule longer than broad; in front of it the anterior margin is gently convex **Dosinia lupinus**

---

13. Left valve with a stout anterior lateral tooth, right valve with a corresponding elongate pit. Pallial sinus mamilliform **Callista chione**

    No lateral teeth. Pallial sinus regularly U-shaped **14**

---

14. Shell sculpture of concentric grooves only **Tapes rhomboides**

    Radiating ribs or lines present, as well as concentric grooves or lines **15**

---

15. Pallial sinus extending beyond mid-line of shell, lower edge confluent with pallial line for part of its length **16**

    Pallial sinus not extending beyond mid-line of shell, lower edge distinct from pallial line for whole of its length **17**

---

16. Sculpture of fine radiating striae and concentric grooves and ridges, the latter not prominent. A burrower in sand and gravel **Venerupis senegalensis**

    Sculpture principally of concentric grooves and ridges, often prominent posteriorly. Radiating striae indistinct. A borer in rock **Venerupis saxatilis**

---

17. Sculpture principally of concentric grooves, radiating striae indistinct. Shell elongate-oval **Tapes aureus**

    Sculpture of concentric grooves and light ridges, intersecting with radiating ridges; more pronounced posteriorly, where the shell has a marked decussate appearance **Tapes decussatus**

Venus verrucosa Linnaeus
Warty venus Fig. 10.34

Oval, up to 60 mm long; off-white to light brown, darkest about umbones and antero-dorsal margin, periostracum deeper brown. With 20 or more prominent, concentric ridges; prodissoconch with distinct radiating ridges, early stages appearing decussate, but only posteriorly in later stages where radiating grooves disrupt concentric sculpture, to give erect warty knobs. A small lateral tooth in front of anterior cardinal of left valve, and a corresponding pit in right. Pallial line faint, sinus small and triangular.

Lower shore to about 100 m, in sand and gravel. Predominantly southern and western, from English Channel to western Scotland; south to Mediterranean west and south Africa.

Circomphalus casina (Linnaeus) Fig. 10.34

Broadly oval, up to 50 mm long; dull white, often tinted light pinkish-brown with a few darker blotches; periostracum deep brown. With 25–40 concentric ridges, of variable width; appearing flat but with acute dorsal edges, more prominent posteriorly. With a small accessory knob (lateral tooth) in front of anterior cardinal tooth of left valve, and a corresponding pit in right. Pallial sinus short, triangular.

In coarse sand and gravels, offshore, perhaps to edge of Continental Shelf. Widespread and common off all British coasts. Distributed from North Sea to Mediterranean and west Africa.

Timoclea ovata (Pennant)
Oval venus Fig. 10.34

Subtriangular, up to 20 mm long; white, light yellow, or pale brown, often with streaks and blotches of pink, chestnut, or light purple. With numerous radiating ridges, pronounced near margins, and equally numerous concentric grooves, giving a decussate effect. Pallial sinus small, U-shaped; inner shell margin finely crenulate.

Timoclea ovata superficially resembles a species of *Parvicardium*, but the hinge teeth show it to be a venus shell.

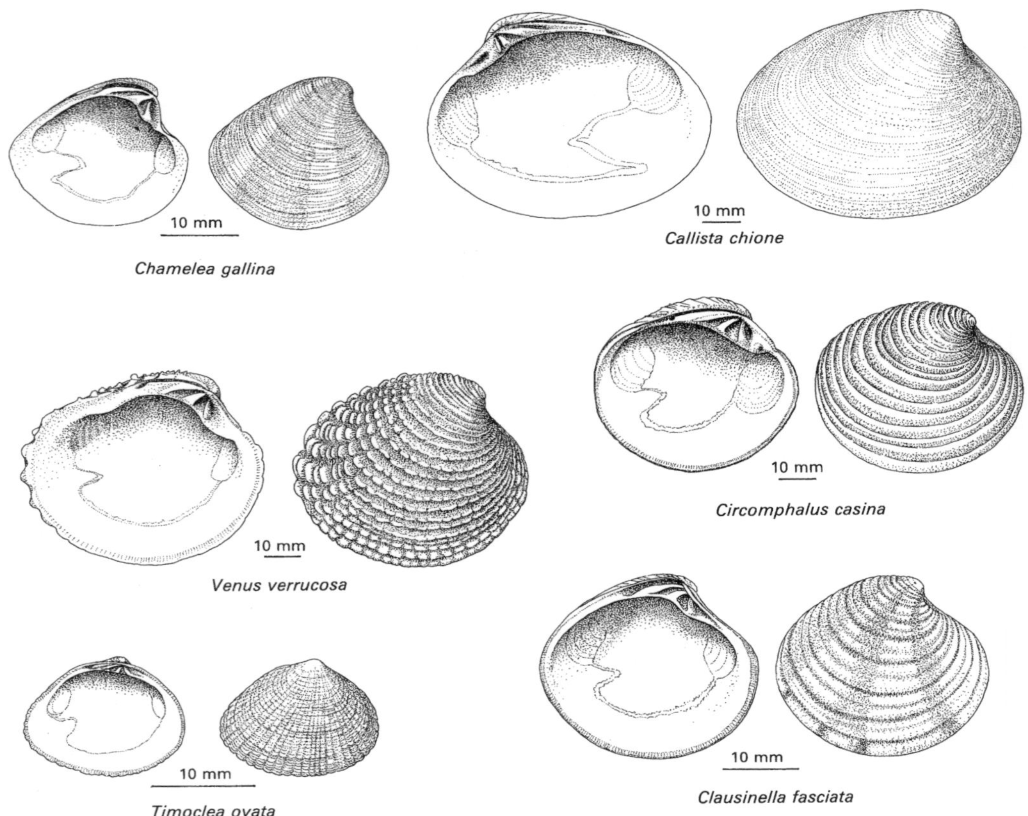

Fig. 10.34

In sand and gravel; offshore, perhaps to edge of Continental Shelf. All British coasts. Distributed from Norway to Mediterranean and Cape Verde Islands.

Clausinella fasciata (da Costa)
Banded venus Fig. 10.34
Subtriangular, rather flat, umbones prominent; up to 25 mm long. Colour variable, white to yellow, brown, pink, or deep purple, typically with bands, rays, or streaks of colour on a paler ground. Up to 15 broad, bold, concentric ridges, round in section with fine concentric striations between. Each growth stage indicated by two closely conjoined ridges. Anterior margin distinctly concave in front of umbones. Pallial sinus small, U-shaped. margin very finely crenulate.

In coarse gravels, usually with sand or shell; offshore to about 100 m. All British coasts. Norway to Mediterranean and Canary Isles.

Chamelea gallina (Linnaeus)
Striped venus Fig. 10.34
Broadly triangular, umbones prominent, dorsal margin sloping steeply posteriorly, concave anteriorly; up to 40 mm long. Off-white or cream, often tinted pale yellowish and often with numerous, very fine chestnut or pinkish streaks; typically with three broad bands of deep chestnut or reddish-brown radiating from umbones, fading in long-dead material. Numerous fine, concentric ridges. Pallial sinus V-shaped. Margin crenulate.

In sand, from lower shore into shallow sublittoral. Common on all British coasts. Distributed from Norway to Mediterranean and Canary Isles.

Mercenaria mercenaria (Linnaeus)
Quahog Fig. 10.32
Subtriangular, umbones prominent, anterior to midline; up to 120 mm long. Light brown to grey, often in varying concentric bands. With numerous

thin concentric ridges, sharp and raised in early growth stages and close to anterior or posterior margins, but usually worn smooth on central areas of larger shells. Three conspicuous cardinal teeth in each valve, with a group of small corrugations behind posterior tooth. Pallial sinus short, triangular; inner margin very finely crenulate.

In mud, lower shore and shallow sublittoral. Introduced into British waters on several occasions since the middle of the nineteenth century. Populations may still be found at several points on the south and south-east coasts, most notably in Southampton Water and the Solent, and at Burnham-on-Crouch.

Dosinia exoleta (Linnaeus)
Rayed artemis Fig. 10.32

Almost circular; umbones small but distinct, just anterior to mid-line. Up to 60 mm long. Dorsal margin gently convex posteriorly, deeply concave immediately anterior to umbones, junction with anterior margin highly arched. Lunule heart-shaped, as broad as long; escutcheon poorly defined. White, yellowish, or light brown, with irregular rays, streaks, or blotches of darker brown or pinkish brown. Pallial sinus deep, narrow, U-shaped, extending well into anterior half of shell.

In muddy gravel or shell gravel; lower shore to about 100 m. Widespread, common off all British coasts. From Norway to Mediterranean and west Africa.

Dosinia lupinus (Linnaeus)
Smooth artemis Fig. 10.32

Approximately circular, umbones small but prominent. Up to 40 mm long. Anterior hinge line shallowly concave below lunule; boundary with anterior margin distinct, but margin not highly arched beyond it. Off-white, fawn, or light brown, umbones often tinted yellow or pink. Lunule deeply impressed, finely striated, heart-shaped, longer than wide. Escutcheon poorly defined. Pallial sinus deep, narrow, and U-shaped, extending into anterior half of shell.

In sandy mud, sand, and shell gravel, from lower shore to at least 120 m. All British coasts; distributed from Iceland to Mediterranean and west Africa.

Gouldia minima (Montagu) Fig. 10.32

Subtriangular, almost equilateral, umbones on mid-line; up to 15 mm long. White to light brown, red, or purple, often with streaks, spots, or rays of darker pigment on a lighter ground. Two anterior lateral teeth in right valve, one in left. Pallial sinus very small, represented by a shallow concavity immediately below posterior adductor scar.

In sandy mud, fine sand, or gravel; offshore to at least edge of Continental Shelf. All British coasts. Norway to Mediterranean and west Africa.

Callista chione (Linnaeus)
Brown venus Fig. 10.34

Oval, with prominent umbones, up to 90 mm long; light reddish-brown to chestnut, in concentric bands, with darker streaks radiating from umbones. With numerous fine, concentric lines; growth stages clear. Left valve with a single, prominent anterior lateral tooth; a corresponding elongate pit in right valve. Pallial sinus broad, with concave borders, mamilliform.

In sand, offshore to at least 100 m. South-west only: from North Wales, south-west England, and Channel Isles, south to Mediterranean, Canary Isles, and the Azores.

Tapes rhomboides (Pennant)
Banded carpet shell Fig. 10.35

Elongate oval, umbones distinctly anterior. Up to 60 mm long. Hinge line sloping anteriorly, almost straight posteriorly, posterior margin smoothly rounded. Cream, fawn, or light reddish-brown, typically with streaks, blotches, or irregular rays of chestnut, light purple, or pinkish-brown. Concentric sculpture of numerous fine grooves, no radiating sculpture. Centre cardinal tooth of left valve, and centre and posterior of right, bifid. Pallial sinus U-shaped, extending to a point below the posterior edge of ligament.

In coarse sands, gravel, and shell gravel, from lower shore, perhaps to edge of Continental Shelf. All British coasts; from Norway to Mediterranean and north-west Africa.

Tapes aureus (Gmelin)
Golden carpet shell Fig. 10.35

Oval to subtriangular, hinge line sloping away from umbones on both sides; regularly convex anteriorly and posteriorly. Up to 45 mm long. Umbones just anterior to mid-line. Off-white, cream, yellow, or light brown, usually with deeper reddish- or purplish-brown markings. With numerous, fine, concentric ridges and grooves, and faint radiating striations. Centre cardinal tooth of left valve, and centre and posterior of right, bifid. Pallial sinus U-shaped, not extending beyond mid-line of shell.

Lower limb of sinus distinct from pallial line for whole of its length.

A shallow burrower in mixed soft bottoms; lower shore into shallow sublittoral. All British coasts, though apparently rarely reported from Scotland; from Norway to Mediterrenean and north-west Africa.

Tapes decussatus (Linnaeus)
Chequered carpet shell Fig. 10.35

Broadly oval or square, umbones distinctly anterior; up to 75 mm long. Posterior hinge line straight, posterior margin truncate; anterior hinge line grading into down-sloping anterior margin. Cream, yellowish, or light brown, often with darker markings. With fine concentric striae and bolder radiating lines; prominent posteriorly, where the shell is conspicuously decussate. Centre cardinal tooth in left valve, and centre and posterior in right are bifid. Pallial sinus U-shaped, not extending beyond mid-line of shell. Lower limb of sinus distinct from pallial line for whole of its length.

Lower shore and shallow sublittoral, in sand, muddy gravel, or clay. Mostly off southern and western coasts. South to Mediterranean and west Africa.

Venerupis senegalensis (Montagu)
Pullet carpet shell Fig. 10.35

Elongate, oval to quadrate, umbones distinctly anterior; hinge line sloping anteriorly, straight posteriorly, forming a sharp angle with posterior margin. Up to 50 mm long. Cream, light fawn to brown, in irregular patches, bands, and rays, usually darkest close to posterior margin. With very fine concentric lines and radiating striations, equally developed, the concentric element predominant posteriorly. Centre cardinal tooth of left valve, and centre and posterior of right, bifid. Pallial sinus deep, U-shaped, extending beyond mid-line of shell; posteriorly, lower limb of pallial sinus may be confluent with pallial line for a short distance.

A shallow burrower in mixed sandy bottoms, usually attached by a byssus. Lower shore into shallow sublittoral. All British coasts; northern Norway to Mediterranean and north-west Africa.

Venerupis saxatilis (Fleuriau)
Carpet shell Fig. 10.35

Roughly oblong, umbones anterior to mid-line and incurved. Posterior hinge line straight, posterior margin truncate; anterior hinge line continuous with regularly convex anterior margin. Up to 40 mm long. Off-white, cream, or light brown. With closely spaced, small corrugated ridges and grooves, most pronounced close to posterior margin, often worn about umbones; fine radiating striations also present. Centre cardinal tooth in left valve, and centre and posterior in right, bifid. Pallial sinus deep, U-shaped, extending well beyond mid-line of shell. Lower limb of pallial sinus posteriorly confluent with pallial line.

Intertidal. Attached by a byssus, in holes and crevices in soft rocks, including those bored by other molluscs. Reported from south coasts of Britain, and west coast as far north as Clyde Sea. Ranges south to north-west Africa.

Irus irus (Linnaeus) Fig. 10.35

Elongate, oblong, or irregular, umbones anterior to mid-line. Up to 25 mm long. Yellowish-white or fawn, usually darker about the umbones. With about 10 thin, concentric ridges, raised and convoluted, conspicuous posteriorly where they form distinct frills. Fine radiating ridges visible between concentric sculpture. Centre cardinal tooth of left valve, centre and posterior of right, bifid. Pallial sinus short, U-shaped.

In holes and crevices in rock, or in *Laminaria* holdfasts; lower shore and shallow sublittoral. Southern and western coasts only; south to Mediterranean and west Africa.

Turtonia minuta (Fabricius) Fig. 10.35

Oval, umbones anterior to mid-line. Very small, not exceeding 3 mm long. Off-white to light brown, sometimes pinkish dorsally. With very fine concentric lines. Three cardinal teeth in each valve, anterior of right and posterior of left very reduced, inconspicuous.

Intertidal only, on rocky shores. Attached by byssus, in holes and crevices, among or in barnacle shells and algae, often very common. All British coasts. South to Mediterranean, and north into boreal/arctic waters, where it is apparently circumpolar in distribution.

20 PETRICOLIDAE

Shell equivalve, inequilateral. Ligament external. Right valve with two cardinal teeth, left with three. No lateral teeth. Adductor scars about equal. Pallial line with sinus. Two British species.

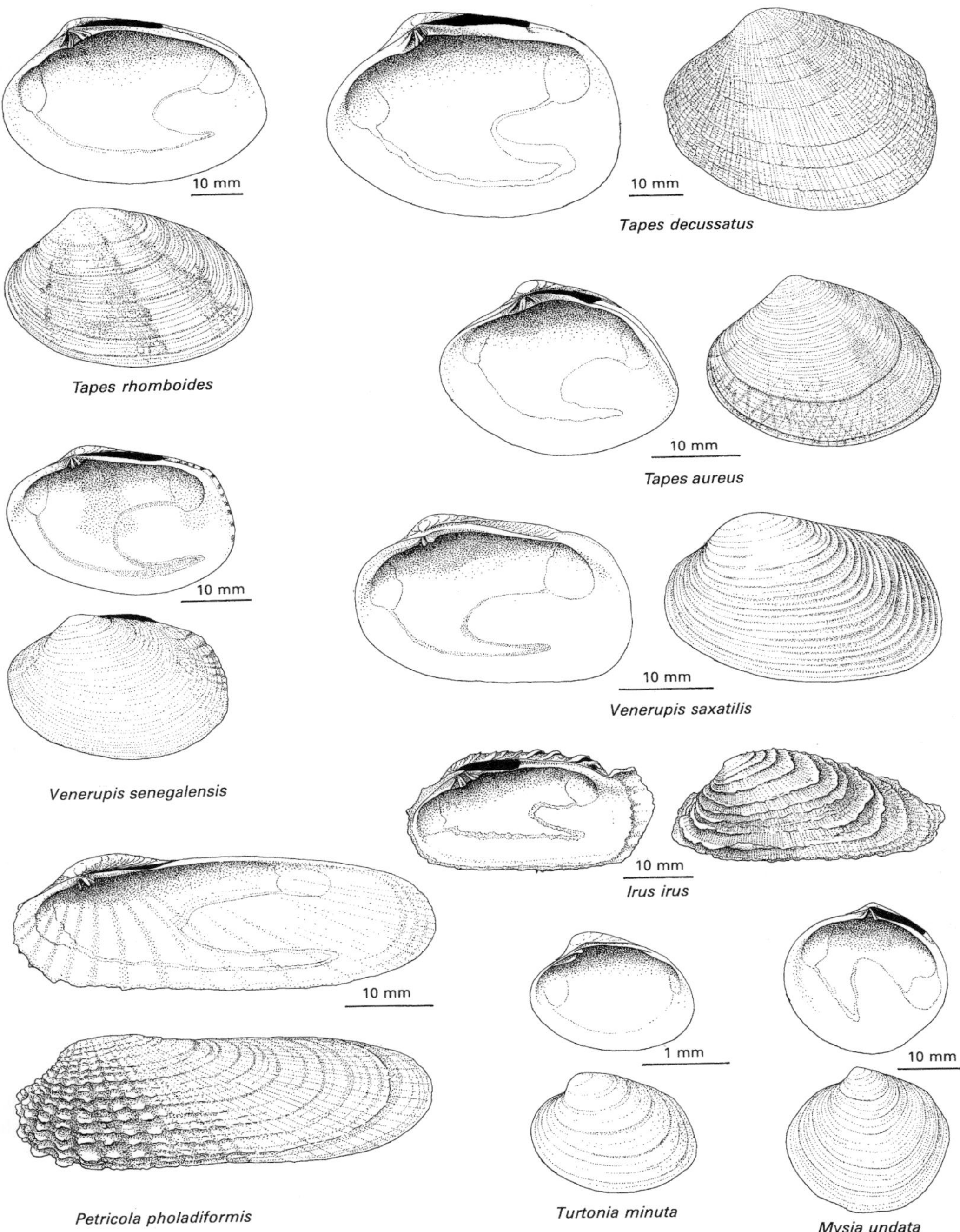
10 mm
10 mm
Tapes decussatus
Tapes rhomboides
10 mm
Tapes aureus
10 mm
10 mm
Venerupis saxatilis
Venerupis senegalensis
10 mm
Irus irus
10 mm
1 mm
10 mm
Petricola pholadiformis
Turtonia minuta
Mysia undata

Fig. 10.35

1. Shell elongate. Sculpture of concentric ridges and radiating lines, developed as pronounced ribs anteriorly **_Petricola pholadiformis_**

   Shell orbicular. Sculpture of fine concentric lines only **_Mysia undata_**

Petricola pholadiformis Lamarck
American piddock Fig. 10.35
Elongate oval, shaped rather like a date; umbones close to anterior margin. Up to 60 mm long. Dull white or fawn, with brown or greyish periostracum. With numerous fine concentric ridges, prominent anteriorly where they may be raised and corrugated, forming flat scales; radiating ridges also present, developed anteriorly into about 10 raised ribs. Pallial sinus deep, U-shaped, extending to mid-line of shell.

Boring into clay, peat, or soft rocks, from middle shore into shallow sublittoral. Southern and south-eastern coasts only, from Lincolnshire to Dorset and Cornwall, but most common off Essex and the Thames estuary. Southern Norway to Mediterranean and west Africa.

Mysia undata (Pennant) Fig. 10.35
Almost circular, umbones just anterior to mid-line. Up to 38 mm long. White or light yellow, periostracum indistinct. Pallial sinus U-shaped, extending into anterior half of shell.

Burrowing in muddy, coarse bottoms. Offshore, but probably extending little deeper than 50 m. All British coasts; distributed from Norway to Mediterranean and Canary Isles.

21 MACTRIDAE

Shell equivalve, equilateral or just inequilateral. Ligament external, thin; and internal, in a small chondrophore recessed within the hinge line. Two or three cardinal teeth in each valve; in the left, two cardinals fused to form a single forked structure. Lateral teeth present. Pallial line with a sinus. British fauna includes six species.

1. Lateral teeth smooth. Inner and outer portions of ligament separated by a distinct calcareous septum on dorsal side of chondrophore. Right anterior cardinal tooth almost parallel to hinge line **_Mactra stultorum_**

   Lateral teeth serrated along interlocking surfaces. Inner and outer portions of ligament not separated **2**

2. Forked cardinal tooth of left valve small, extending only halfway down hinge plate. Shell subtriangular **_Spisula solida_**

   Forked cardinal tooth of left valve larger, extending practically to edge of hinge plate **3**

3. Shell broadly elliptical, anterior and posterior regularly and equally convex **_Spisula elliptica_**

   Shell subtriangular, asymmetrical; posterior hinge line sloping more steeply than anterior, posterior margin somewhat drawn out **_Spisula subtruncata_**

Mactra stultorum (Linnaeus)
Rayed trough shell Fig. 10.36
Oval, umbones just anterior to mid-line. Up to 50 mm long. White, umbones tinted purple, with light brown rays of varying width radiating from umbones. Shell margin prominent at hinge line. Right valve with paired, elongate anterior and posterior lateral teeth. Left valve with single, elongate anterior and posterior lateral teeth. Chondrophore triangular, posterior to cardinal teeth, with a small dorsal septum isolating it from external ligament. Pallial sinus broad and rounded, not extending far into shell.

In clean sand; lower shore and shallow sublittoral. Widespread, and often abundant, on most British coasts. Norway to Mediterranean and west Africa.

Mactra glauca Born is a rare species known from a few offshore localities around south-west England and the Channel Isles. It is larger than *M. stultorum*, and in the right valve the anterior cardinal tooth is at an angle to the hinge line (Fig. 10.36).

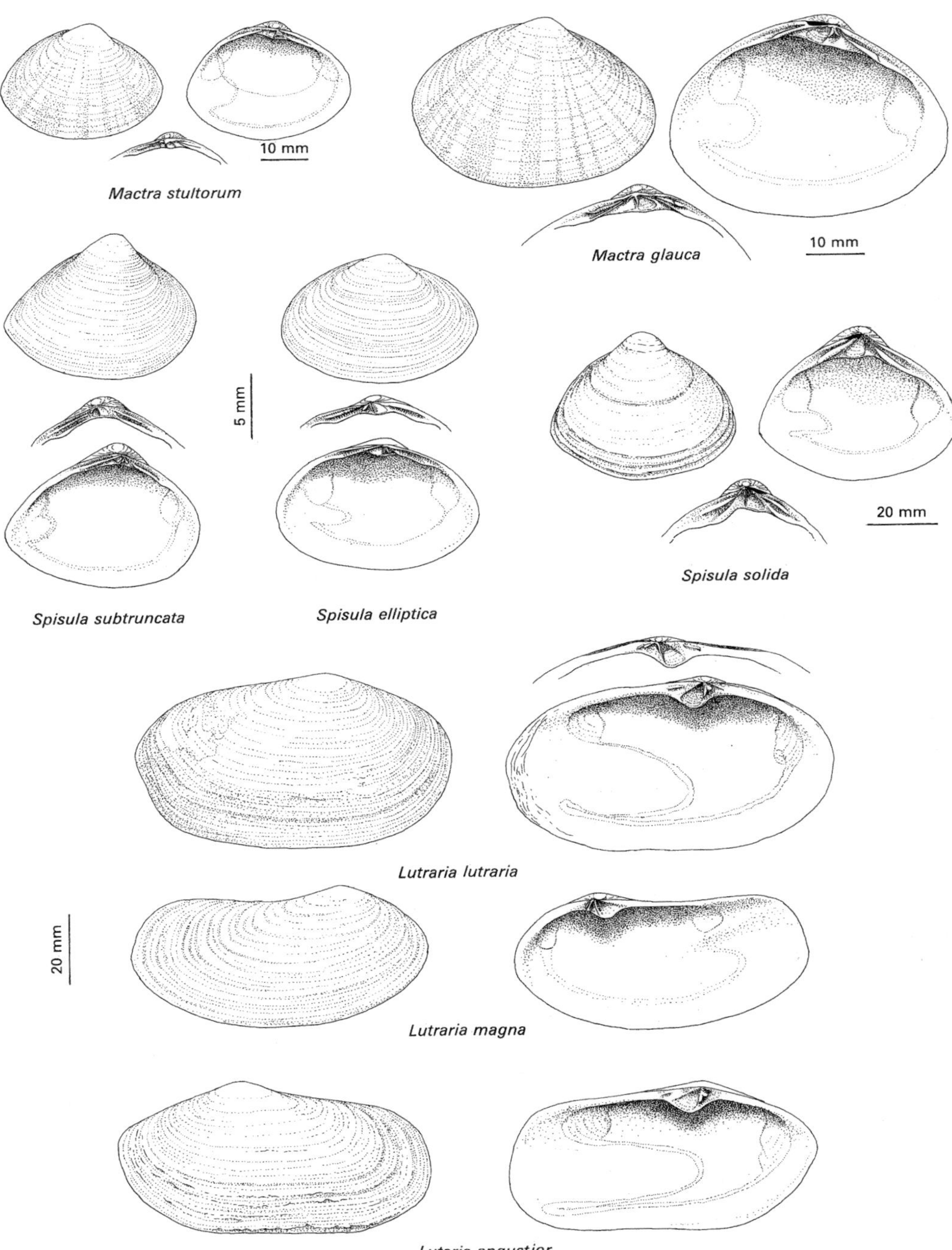
10 mm
Mactra stultorum
10 mm
Mactra glauca
5 mm
20 mm
Spisula solida
Spisula subtruncata
Spisula elliptica
Lutraria lutraria
20 mm
Lutraria magna
Lutaria angustior

Fig. 10.36

Spisula solida (Linnaeus)
Thick trough shell Fig. 10.36
Subtriangular, anterior more regularly convex than posterior but not markedly so; umbones more or less on mid-line, anterior and posterior hinge with approximately equal slope. Up to 50 mm long. Dull white to light fawn; periostracum greyish-brown, usually prominent around margin and along growth lines. Right valve with paired, elongate anterior and posterior lateral teeth. Left valve with single elongate anterior and posterior lateral teeth. Pallial sinus linguiform, extending to a point below and beyond the mid-line of posterior lateral teeth.

Burrows in sand; lower shore and sublittoral. Widespread in British waters; south Iceland and Norway to Spain and Morocco.

Spisula subtruncata (da Costa) Fig. 10.36
Subtriangular but distinctly asymmetrical; umbones close to mid-line, posterior hinge line and margin sloping more steeply than anterior, posterior end appearing slightly drawn out. Up to 30 mm long. Dull white to cream, periostracum greyish-brown. Lunule and escutcheon broad and elongate, latter bounded by low ridges extending from umbones. Right valve with paired, elongate anterior and posterior lateral teeth. Left valve with single elongate, anterior and posterior laterals. Pallial sinus broad and rounded, extending to a point below and behind the mid-line of the posterior lateral teeth.

Burrowing in muddy or silty sand; lower shore and shallow sublittoral. Widespread and common off most British coasts; from Norway to Mediterranean and Canary Isles.

Spisula elliptica (Brown)
Elliptical trough shell Fig. 10.36
Elongate oval, umbones close to mid-line. Up to 30 mm long. Dull white with greenish or greyish-brown periostracum. Right valve with paired, elongate anterior and posterior lateral teeth. Left valve with single anterior and posterior lateral teeth. Pallial sinus oval, extending to a point below and beyond mid-line of posterior lateral teeth.

In mixed soft substrata, offshore to about 100 m. All British coasts. Northwards to the Barents Sea.

22 LUTRARIIDAE

Shell elongate, gaping anteriorly and posteriorly. Equivalve; inequilateral, umbones anterior to mid-line. Ligament external and internal, latter in a chondrophore recessed into hinge line. Cardinal and lateral teeth present, two of left cardinal teeth partly fused, forming a forked structure. Adductor scars about equal. Pallial line with a sinus. Three British species.

---

1. Anterior cardinal tooth of right valve bifid. Dorsal margin distinctly concave posterior to umbones ***Lutraria magna***

   Anterior cardinal tooth of right valve entire. Dorsal margin straight or convex **2**

---

2. Shell elongate oval, dorsal hinge line gently convex. Pallial line distinct from lower edge of pallial sinus ***Lutraria lutraria***

   Shell rather angular, dorsal hinge line more or less straight. Pallial line largely fused with lower edge of pallial sinus ***Lutraria angustior***

Lutraria lutraria (Linnaeus)
Common otter shell Fig. 10.36
Elongate, elliptical. Up to 130 mm long. White or yellowish, often tinted pink or purple, glossy; periostracum olive, thick, brittle, and peeling readily in dead specimens. Right valve with two thin cardinal teeth anterior to chondrophore, and a single, poorly developed posterior lateral tooth. Lower edge of pallial sinus distinct from pallial line.

In mixed soft substrata; lower shore to about 100 m. All British coasts; Norway south to Mediterranean and west Africa.

Lutraria angustior Philippi Fig. 10.36
Elongate, angular, anterior hinge line sloping more steeply than posterior. Up to 200 mm long. Dull white or yellowish, periostracum pale yellowish-brown. With numerous fine grooves, developing as fine ridges close to margin. Right valve with two cardinal teeth and a single, poorly developed, posterior lateral. Lower edge of pallial sinus largely fused with pallial line.

In mixed soft substrata, offshore to about 50 m. Southern coasts only; south to Mediterranean and west Africa.

Lutraria magna (da Costa)
Oblong otter shell Fig. 10.36

Elongate, oblong, dorsal margin distinctly concave posterior to umbones, with almost equal curvature to ventral margin. Posterior margin somewhat truncate. Up to 120 mm long. Light yellow or fawn, with brown periostracum. With fine concentric lines. Right valve with two rather prominent cardinal teeth anterior to chondrophore, the anterior distinctly bifid. Pallial sinus deep, lower edge largely fused with pallial line.

In shell gravels, offshore. South-western coasts only; south to Mediterranean and north Africa.

23 DONACIDAE

Shell equivalve, inequilateral. Ligament external. Hinge line with two cardinal teeth in each valve, anterior and posterior lateral teeth also present. Adductor scars about equal. Pallial line with a sinus. Cruciform muscles present extending between valves, involved in siphon retraction: scars often visible close to lower edge of pallial sinus. Shell margin crenulate.

Donax vittatus (da Costa)
Banded wedge shell Fig. 10.37

Roughly wedge-shaped, umbones posterior to midline. Dorsal margin sloping gently anteriorly from umbones, continuous with rounded anterior margin; posteriorly sloping more steeply, posterior margin bluntly pointed. Up to 35 mm long. White, yellowish, light brown, or purple, frequently lighter about the umbones, with pale radiating rays and often with pigmented bands along growth lines. Periostracum light brown to olive brown, glossy. Right posterior cardinal tooth broad and bifid, left anterior weakly bifid. One anterior and two posterior lateral teeth in right valve; single small, anterior and posterior laterals in left valve. Pallial sinus broadly oval, extending to mid-line of shell.

In sand, lower shore and shallow sublittoral. On most British coasts; distributed from Norway south to Mediterranean and north-west Africa.

Donax variegatus (Gmelin) occurs in similar habitats on the southern and south-west coasts of Britain. More regularly oval than *D. vittatus*, it is distinguished principally by the marginal crenulations which are so fine as to feel smooth to the touch.

24 TELLINIDAE

Shell equivalve, inequilateral, posterior ventral border frequently with a dent or twist. Ligament external. Hinge line with two cardinal teeth in each valve, lateral teeth usually present. Adductor scars about equal. Pallial line with sinus. Cruciform muscles present, linking valves, and leaving scars close to lower edge of pallial sinus. Ten species recorded from British waters.

---

1. Hinge line with cardinal teeth only. Shell almost circular, colour variable, often in concentric bands. Intertidal, often abundant in soft estuarine deposits **_Macoma balthica_**

   Hinge line with cardinal and lateral teeth, the latter usually conspicuous in right valve, but in left appearing simply as slender ridges close to margin **2**

---

2. Shell broadly oval, little longer than deep. Lower edge of pallial sinus distinctly separated from pallial line **_Arcopagia crassa_**

   Shell elongate oval, at least 1.5 times as long as deep. Lower edge of pallial sinus confluent with pallial line **3**

---

3. Umbones close to mid-line of shell. No anterior lateral tooth in left valve **4**

   Umbones clearly situated in posterior half of shell. A small anterior lateral tooth in left valve **6**

---

4. Dorsal and ventral margins concave posteriorly, imparting a characteristic hooked outline to that part of shell **_Angulus squalidus_**

   Posterior part of shell not hooked **5**

---

5. Posterior margins convex, shell subtriangular **_Angulus tenuis_**

   Posterior margin convex ventrally, slightly concave dorsally; shell somewhat attenuated posteriorly **_Fabulina fabula_**

---

6. Posterior dorsal margin of shell sloping gently from umbones, posterior part of shell regularly rounded
 Moerella pygmaea

 Posterior dorsal margin of shell sloping steeply from umbones, posterior part of shell irregularly truncate
 Moerella donacina

Angulus tenuis da Costa
Thin tellin Fig. 10.37
Subtriangular, valves shallowly convex, umbones just posterior to mid-line; anterior dorsal margin sloping gently, forming a continuous curve with anterior margin, posteriorly more steeply inclined, posterior margin slightly truncate. Ligament very conspicuous, a fusiform, brown prominence posterior to umbones. Up to 28 mm long. White, shades of pink, yellow, or orange, often in distinct bands. Periostracum glossy, thin. With fine concentric lines, growth stages clear. Right valve with two cardinal teeth, and single anterior and posterior laterals; left valve with two cardinals, and a single small, posterior lateral. Pallial sinus broadly oval, its lower edge largely fused with pallial line.

In sand, middle shore to shallow sublittoral, often very abundant. Common, on all British coasts. Norway to Mediterranean and north-west coast of Africa.

Angulus squalidus (Pulteney) Fig. 10.37
Broadly oval, posterior end somewhat attenuated. Umbones just posterior to mid-line. Anterior hinge line downward sloping, continuous with anterior margin; posterior margin attenuated, concave dorsally and ventrally, appearing hooked. Posteriorly, ventral margin has a lateral dent, concave on right and convex on left. Up to 45 mm long. Fawn, yellow, orange, or pink, often banded; periostracum light brown, usually conspicuous around margins. With a straight keel extending posteriorly from umbones on each valve. Right valve with two cardinal teeth, and single anterior and posterior laterals; left valve with two cardinals and a single, poorly developed posterior lateral. Pallial sinus deep, lower edge fused with pallial line.

In muddy sand, offshore, though probably not beyond 40 m. Not common, mostly southern and western coasts of Britain, including the Isles of Scilly and Channel Isles. Its distribution elsewhere appears little known.

Fabulina fabula (Gmelin) Fig. 10.37
Elongate oval, umbones close to mid-line; anterior margin broadly rounded, posterior tapering and curved to right. Up to 20 mm long. White, pale yellow, or orange, often in concentric bands. Right valve with wavy striations extending from dorsal anterior to ventral posterior margins. Right valve with two cardinal teeth and single anterior lateral tooth; left valve with two cardinal teeth and small posterior lateral tooth. Pallial sinus very deep, lower edge completely fused with pallial line.

In mixed sandy deposits; lower shore and shallow sublittoral. All British coasts. Norway to Mediterranean and north-west Africa.

Moerella donacina (Linnaeus) Fig. 10.37
Elongate oval, umbones in posterior half of shell. Anterior hinge line sloping gently; posterior hinge line sloping more steeply, forming abrupt junction with posterior margin. Up to 25 mm long. Yellowish-white to cream, with bands of pink or red, and discontinuous rays of cream or red. Right valve with two cardinal teeth and single anterior and posterior laterals; left valve with two cardinals, and single, indistinct anterior and posterior laterals. Pallial sinus very broad and deep, extending almost as far as anterior adductor scar, lower edge fused with pallial line.

In coarse sand and shell gravel, in shallow coastal waters. Southern and western coasts only, south to Mediterranean and west Africa.

Moerella pygmaea (Lovén) Fig. 10.37
Elongate oval, umbones posterior to mid-line; hinge line sloping gently from umbones anteriorly, more steeply posteriorly. Up to 10 mm long. White, light orange, pink, or reddish-brown, often in patches. Right valve with two cardinal teeth and single anterior and posterior laterals; left valve with two cardinal teeth, and single, poorly developed, anterior and posterior laterals. Pallial sinus irregular, extending to just beyond mid-line of shell.

In sand and shell gravel, offshore. South-west and west coasts, and from east and north-east of Scotland. Northern Norway to Mediterranean and west Africa.

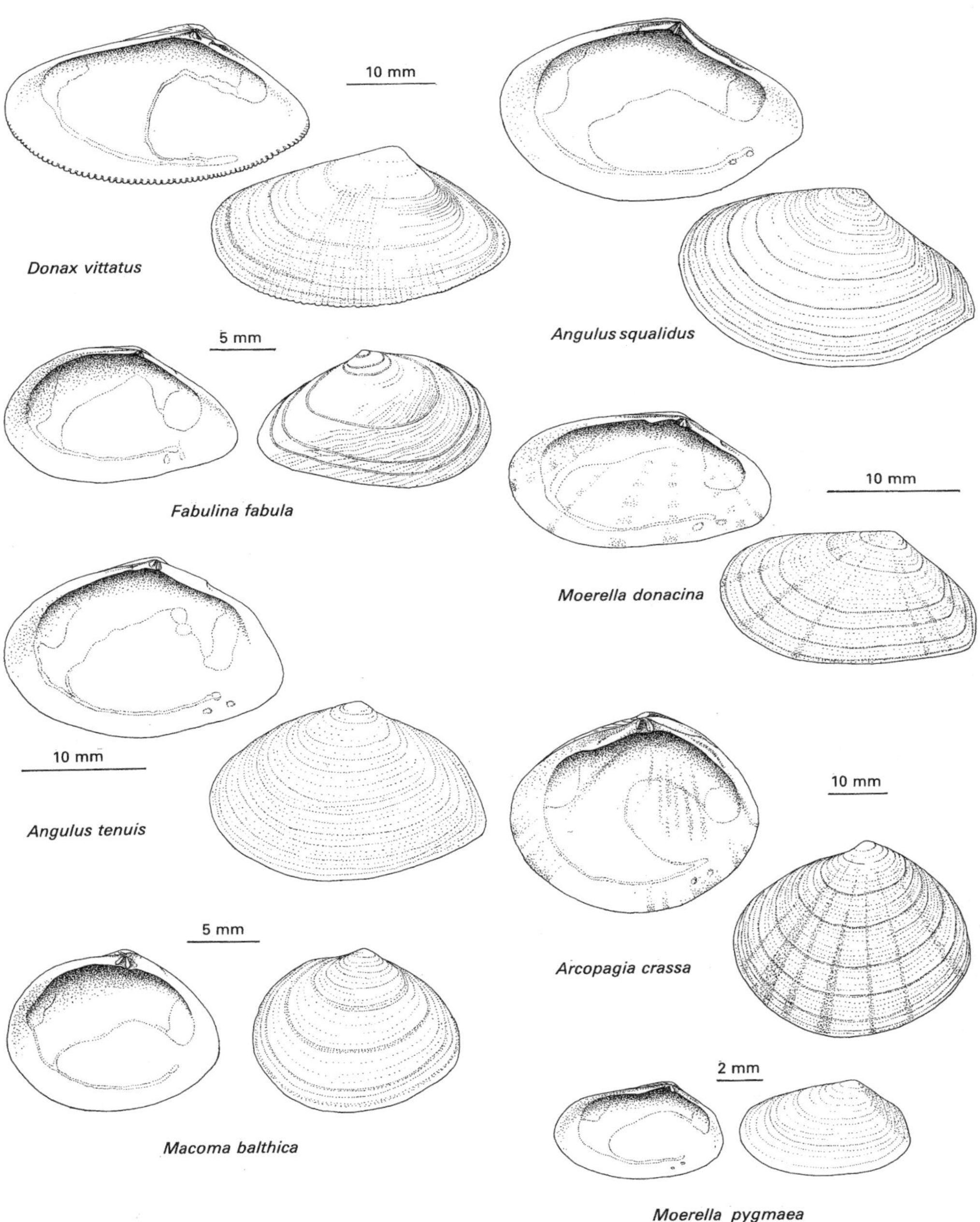

10 mm
Donax vittatus
Angulus squalidus
5 mm
Fabulina fabula
10 mm
Moerella donacina
10 mm
Angulus tenuis
10 mm
Arcopagia crassa
5 mm
Macoma balthica
2 mm
Moerella pygmaea

Fig. 10.37

Arcopagia crassa (Pennant)
Blunt tellin Fig. 10.37

Oval, little longer than deep; umbones posterior to mid-line, conical and prominent. Postero-ventral margin with a shallow dent: concave on left and convex on right. Up to 60 mm long. Off-white or fawn; frequently pale orange or light red about umbones, with a few irregular rays of red. Right valve with two cardinal teeth and single anterior and posterior laterals; left valve two cardinals, and single, small anterior and posterior lateral teeth. Pallial sinus broadly oval, extending deep into interior of shell.

In mixed sandy deposits; lower shore, extending far out into shelf waters. Not common; most frequent on south-western coasts, but apparently occurring around Scotland. Distributed from Norway southwards to Senegal.

Macoma balthica (Linnaeus)
Baltic tellin Fig. 10.37

Broadly oval, umbones more or less on mid-line; anterior hinge line and margin regularly convex, posterior hinge line and margin slightly attenuated. Up to 25 mm long. White, yellow, pink, or purple, in various shades, unicolorous or banded. Periostracum light brown, most conspicuous at margins. Two small cardinal teeth in each valve, no lateral teeth. Pallial sinus irregular, deep, lower edge largely fused with pallial line.

In soft substrata, particularly in estuaries and on tidal flats, where it may be abundant; intertidal only. Present on most British coasts, distributed from the White Sea south to Spain and Portugal.

25 SCROBICULARIIDAE

Shell equivalve, equilateral, or just inequilateral. Ligament with small external element, and larger internal element in a triangular chondrophore recessed within hinge line. Right valve with two cardinal teeth, left with one. Lateral teeth variously developed. Pallial line with a sinus. Cruciform muscles present, scars indistinct. Six British species. Some authorities recognize a separate family, Semelidae, for *Abra*.

---

1. Without lateral teeth. Shell yellow-white to grey, with coarse concentric grooves; up to 60 mm long; intertidal only, often abundant in estuarine muds **_Scrobicularia plana_**

   Lateral teeth present, usually prominent in right valve. Shell white, small (<40 mm); mostly offshore, one intertidal species **2**

---

2. Shell elongate oval or fusiform, about twice as long as deep **3**

   Shell broadly oval or subtriangular, at the most only 1.5 times as long as deep **4**

---

3. Anterior and posterior dorsal margins with approximately equivalent slope from umbones. Beaks almost on the mid-line; proportionately less slender **_Abra nitida_**

   Posterior dorsal margin sloping more steeply from umbones than anterior. Beaks just posterior to mid-line; proportionately more slender **_Abra prismatica_**

---

4. Shell subtriangular, equilateral, umbones prominent. Growth stages indistinct **_Abra tenuis_**

   Shell oval, inequilateral, umbones in posterior half. Growth stages clear **_Abra alba_**

Scrobicularia plana (da Costa)
Peppery furrow shell Fig. 10.38

Oval, umbones just anterior to mid-line. Up to 65 mm long. Dull white, yellowish, or grey, usually darkest about the growth lines. Pallial sinus broad, lower edge largely fused with pallial line; posteriorly, cruciform muscle scars may be apparent.

Intertidal only, burrowing in soft substrata in estuaries and tidal flats, in conditions of fluctuating salinity. Often abundant. Widely distributed around

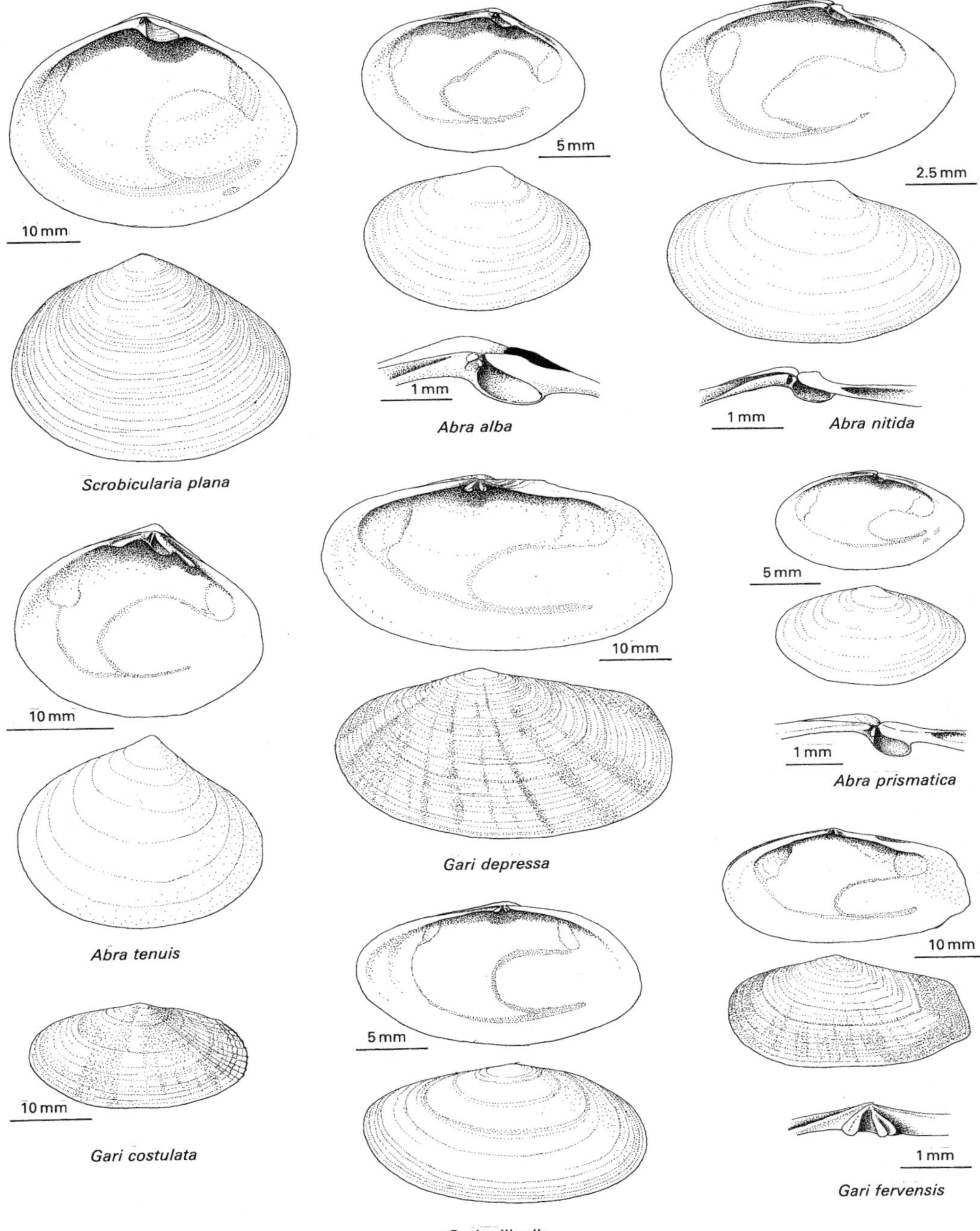
10 mm
Scrobicularia plana
5 mm
1 mm
Abra alba
2.5 mm
1 mm
Abra nitida
10 mm
Abra tenuis
10 mm
Gari costulata
10 mm
Gari depressa
5 mm
Gari tellinella
5 mm
1 mm
Abra prismatica
10 mm
1 mm
Gari fervensis

Fig. 10.38

the British Isles; ranges from Norway to Mediterranean and west Africa.

Abra alba (Wood) Fig. 10.38
Broadly oval, umbones posterior to mid-line. Up to 25 mm long. White and glossy. With very fine concentric lines, growth stages clear. Chondrophore elliptical, posteriorly directed. Pallial sinus deep, rather irregular; cruciform muscle scars visible with hand lens.

In soft substrata; occasionally on lower shore but most abundant in shallow, offshore waters (to about 60 m) where it may be the dominant member of the benthic infauna. Widespread around the British Isles; distributed from Norway to Mediterranean and west Africa.

Abra nitida (Müller) Fig. 10.38
Elongate oval, umbones just posterior to mid-line; anterior and posterior hinge line sloping equally from umbones. Rarely longer than 12 mm. Glossy white. With very fine concentric lines, scarcely visible with hand lens. Pallial sinus deep, lower edge partly confluent with pallial line, cruciform muscle scars indistinct.

Burrowing in mixed soft substrata, offshore to edge of Continental Shelf. Widespread, and often common, around British Isles. Norway south to Mediterranean and north-west Africa.

Abra prismatica (Montagu) Fig. 10.38
Elongate oval to fusiform, approximately twice as long as deep. Up to 13 mm long. Glossy white. With very fine concentric lines, visible with hand lens. Pallial sinus with lower edge partly fused with pallial line.

In mixed sandy bottoms, lower shore to about 60 m. Widespread around British Isles; distributed from Norway to Mediterranean and north-west Africa.

Abra tenuis (Montagu) Fig. 10.38
Subtriangular, umbones prominent, with hinge line sloping away evenly on each side. Up to 13 mm long. Dull white, with pale grey or brown periostracum. Sculpture of numerous fine concentric lines. Pallial sinus deep, its lower edge partly fused with pallial line; cruciform muscle scars distinct.

In soft substrata, in estuaries and tidal flats subject to fluctuating salinity. Intertidal only. East coasts as far north as Northumberland, the south coast, and the west as far north as Scotland. South to Mediterranean and north-west Africa.

26 PSAMMOBIIDAE

Shell equivalve, or slightly inequivalve; almost equilateral, with beaks just in front of mid-line. Ligament external, supported by a distinct plate posterior to beaks. Two cardinal teeth in each valve, usually unequal, one or both typically bifid. No lateral teeth. Pallial line with sinus. Cruciform muscles present, scars indistinct. Four British species.

---

1. Posterior end of shell abruptly truncate, with a raised keel extending from umbone to posterior ventral corner of each valve **_Gari fervensis_**

   Posterior end of shell not abruptly truncate, regularly or irregularly rounded **2**

---

2. Posterior end of shell with a group of fine ribs extending from ligament to posterior margin, which is finely crenulate where they meet it **_Gari costulata_**

   Without a posterior group of ribs **3**

---

3. Posterior dorsal margin almost horizontal, posterior margin sloping steeply from it. Valves with a distinct posterior gape **_Gari depressa_**

   Posterior dorsal margin sloping away from umbones, continuous with gently rounded posterior margin. Without a posterior gape **_Gari tellinella_**

Gari fervensis (Gmelin)
Faroe sunset shell Fig. 10.38
Elongate, more than twice as long as broad. Umbones anterior to mid-line; anterior hinge line forming a smooth continuous curve with anterior margin, posterior hinge line sloping more steeply, forming a sharp angle with truncate posterior margin. Up to 50 mm long. Dull white, yellowish, fawn, or shades of red, pink, or purple; usually in concentric bands of different intensity, with a few white or cream rays radiating from umbones. Right valve with two cardinal teeth, anterior grooved, pos-

terior distinctly bifid; left valve with a large, bifid anterior cardinal, and a slender posterior cardinal. Pallial sinus deep, U-shaped, its lower edge largely fused with pallial line.

In mixed sand or gravelly bottoms; lower shore and sublittoral. Common on all British coasts. Norway to Mediterranean and west Africa.

Gari depressa (Pennant)
Large sunset shell Fig. 10.38
Elongate oval, umbones anterior to mid-line. Anterior hinge line sloping gently, continuous with convex anterior margin; posteriorly almost straight, forming abrupt junction with truncate, oblique, posterior margin. Up to 60 mm long. Glossy white, cream or yellow, often with radiating streaks of reddish-brown or lilac. Right valve with two bifid cardinal teeth, anterior broader than posterior; left valve with a bifid anterior cardinal and thin posterior cardinal. Pallial sinus deep, U-shaped, its lower edge largely fused with pallial line.

In sandy substrata; lower shore and shallow sublittoral. Most frequent on southern and western coasts; south to Mediterranean and west Africa.

Gari costulata (Turton) Fig. 10.38
Elongate oval, umbones just posterior to mid-line. Anterior hinge line sloping very gently from umbones, anterior margin somewhat upcurved; posterior hinge line sloping more steeply, continuous with regularly convex posterior margin. Up to 25 mm long. White, yellowish, or pink, often with patches or rays of red or purple. With about 20 flat ridges extending from umbones to posterior margin. Right valve with two bifid cardinal teeth, anterior larger than posterior; left valve with bifid anterior and thin posterior cardinals. Pallial sinus deep, U-shaped, lower edge largely fused with pallial line.

In muddy sand, offshore though not to great depths. Widespread, though not common, around British Isles. South to Mediterranean and west Africa.

Gari tellinella (Lamarck) Fig. 10.38
Elongate oval, umbones posterior to mid-line. Anterior hinge line sloping very gently, posterior more steeply. Up to 25 mm long. Yellowish-white, orange, pink, chestnut, or dull red, often in contrasting bands, frequently with rays of red or chestnut. Right valve with two, indistinctly bifid, cardinal teeth, anterior larger than posterior; left valve with broad, bifid anterior and thin, indistinct posterior cardinals. Pallial sinus broad and deep, lower border largely fused with pallial line.

In coarse sand and gravel, offshore to about 60 m. Present off most British coasts, but apparently absent from much of the North Sea. Distributed from southern Iceland to Mediterranean.

27 SOLECURTIDAE

Shell elongate, equivalve, inequilateral; gaping anteriorly and posteriorly. Ligament external. Hinge line with peg-like cardinal teeth, no more than two in each valve; lateral teeth present or absent. Pallial line with sinus. Cruciform muscles present, scars indistinct. Three British species.

---

1. Shell 4-5 times as long as deep; dorsal and ventral margins almost parallel, anterior and posterior margins rounded: resembling a pea pod. Ligament elongate, situated medially on dorsal margin ***Pharus legumen***

   Shell elongate oval, about twice as long as deep **2**

---

2. Shell with concentric sculpture only ***Azorinus chamasolen***

   Shell with concentric sculpture, and distinct radiating striations on posterior two-thirds ***Solecurtus scopula***

Solecurtus scopula (Turton) Fig. 10.39
Anterior and posterior margins rounded, shaped rather like a date; umbones close to mid-line. Up to 60 mm long. Dull white, periostracum olive or yellow brown. With two groups of radiating ridges, one in middle region of shell, the other extending posteriorly from umbones. Right valve with two cardinal teeth, posterior larger than anterior, perpendicular to hinge line and curving outwards from beneath umbo; left valve with one identical tooth. Pallial sinus broad, irregular, extending into anterior half of shell.

In sandy or shelly gravel, lower shore and shallow sublittoral. South and west coasts only, as far north as Orkney. Distributed south to the Mediterranean and west Africa.

Azorinus chamasolen (da Costa) Fig. 10.39

Anterior and posterior margins rounded; shaped like a date, umbones just anterior to mid-line. Up to 60 mm long. Dull white or fawn; periostracum dark brown to olive, coarse and corrugated. Right valve with two cardinal teeth, posterior larger than anterior, perpendicular to hinge line, curving outwards from beneath umbo; left valve with one identical tooth. Pallial sinus broad, extending up to but not beyond mid-line of shell.

In mud or mixed muddy bottoms, shallow sublittoral and offshore. South and west coasts of British Isles. Norway to Mediterranean and north Africa.

Pharus legumen (Linnaeus) Fig. 10.39

Anterior and posterior margins rounded, anterior end distinctly tapered. Umbones low and indistinct, sited about one-third of length from anterior margin; ligament immediately posterior to umbones, situated in middle third of shell. Up to 130 mm long. White or light brown, with a glossy, light olive or yellow periostracum, postero-dorsal section of shell fawn. With a group of very fine striae radiating from umbones to middle third of ventral margin. Right valve with a single cardinal tooth projecting perpendicularly from hinge line, a single elongate anterior lateral, and a short, peg-like, posterior lateral tooth. Left valve with two elongate, closely spaced, cardinal teeth; a single, elongate, anterior lateral tooth and a single, projecting, posterior lateral tooth, almost as long as cardinals. Pallial sinus short, quadrate.

Burrowing in sand, lower shore and shallow sublittoral. Off south-west coasts of England, Wales, and Ireland; south to Mediterranean and north-west Africa.

28 SOLENIDAE

Shell elongate, equivalve, inequilateral: umbones at anterior end. Gaping anteriorly and posteriorly. Ligament external. One or two peg-like cardinal teeth in each valve; lateral teeth present or absent. Adductor scars unequal, anterior typically elongate, posterior small. Pallial sinus present. Six British species. Some authorities recognize two families, Solenidae for *S. marginatus* and Cultellidae for *Ensis* and *Phaxus*.

---

1. Ligament in middle of shell, anterior end tapered **_Pharus legumen_** (this page)

   Ligament close to anterior end of shell, which is not tapered **2**

---

2. Each valve with a deep dorsoventral groove externally, close to anterior end **_Solen marginatus_**

   Without such a groove **3**

---

3. Lateral tooth in each valve projecting, similar in size to cardinal teeth and situated close to them. Anterior end rounded, posterior gently truncate **_Phaxas pellucidus_**

   Lateral teeth elongate, ridge-like, clearly differentiated from cardinal teeth. At least one end sharply truncate **4**

---

4. Dorsal and ventral margins of shell practically straight, and parallel **_Ensis siliqua_**

   Either or both dorsal and ventral margins curved, not parallel **5**

---

5. Anterior end of shell rounded, dorsal and ventral margins both equally curved **_Ensis ensis_**

   Anterior end of shell truncate. Ventral margin curved, dorsal margin almost straight **_Ensis arcuatus_**

Solen marginatus Montagu

Grooved razor shell Fig. 10.39

Dorsal and ventral margins of shell straight, parallel, anterior end obliquely truncate, with a distinct dorso-ventral groove close to margin. Up to 120 mm long. White or yellow, periostracum glossy, light olive or brown. A single cardinal tooth in each valve, no laterals.

Burrows in sand, lower shore and shallow sublittoral; south-eastern, southern, and western coasts only. Distributed from Norway to the Mediterranean and north Africa.

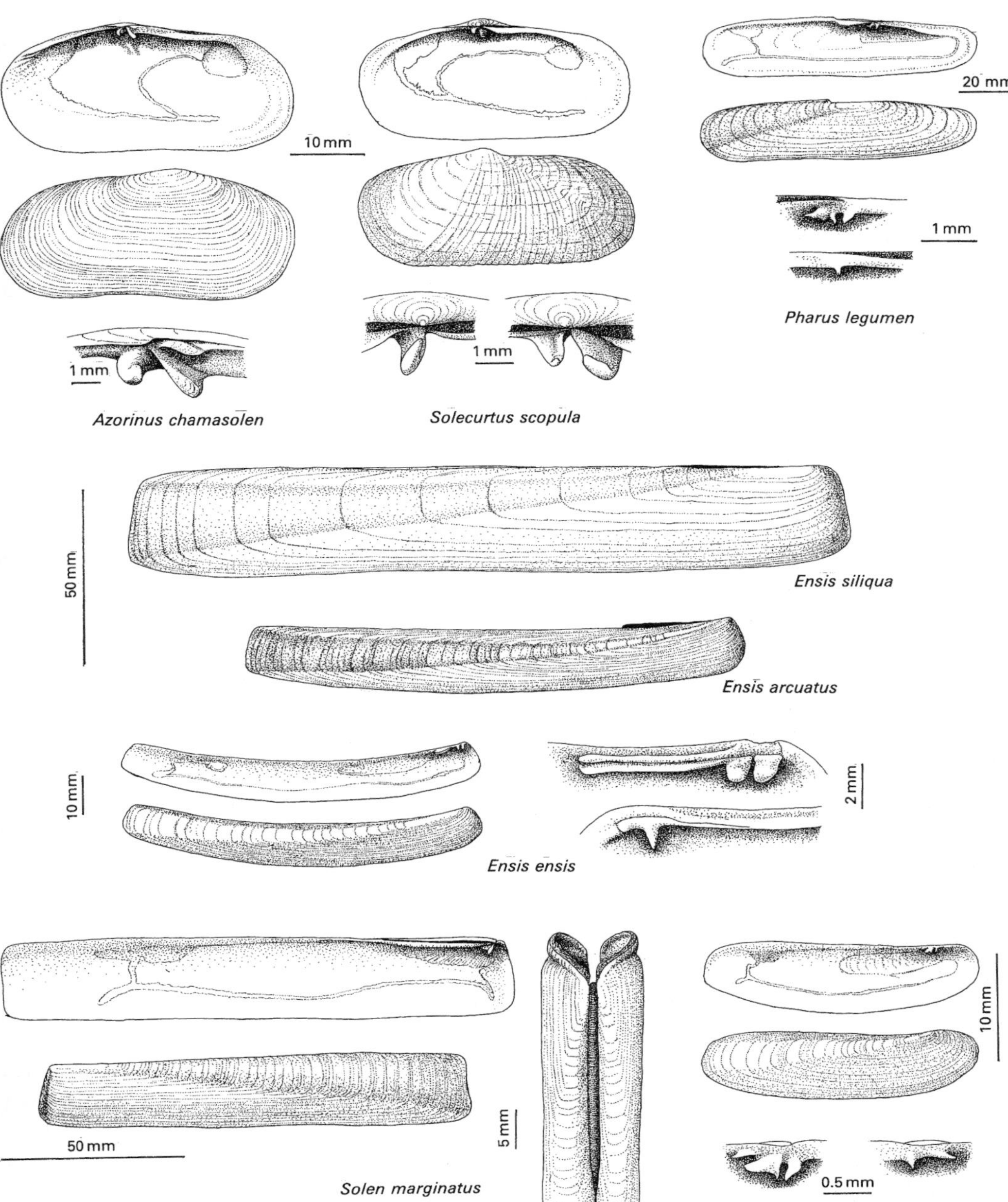
10 mm
1 mm
Azorinus chamasolen
1 mm
Solecurtus scopula
20 mm
1 mm
Pharus legumen
50 mm
Ensis siliqua
Ensis arcuatus
10 mm
2 mm
Ensis ensis
50 mm
5 mm
Solen marginatus
10 mm
0.5 mm
Phaxas pellucidus

Fig. 10.39

Ensis ensis (Linnaeus)
Common razor shell Fig. 10.39
Dorsal and ventral margins of shell symmetrically curved about mid-line; anterior margin rounded. Up to 130 mm long. Dull white or cream, with pale reddish or purplish-brown streaks and spots; periostracum glossy, light to dark olive or green. Left valve with two projecting, peg-like cardinal teeth, and two elongate, posterior laterals; right valve with one short cardinal and a single, elongate posterior lateral.

Burrows in fine sand, lower shore and shallow sublittoral. All British coasts; distributed from Norway south to the Mediterranean and north-west Africa.

Ensis siliqua (Linnaeus)
Pod razor shell Fig. 10.39
Dorsal and ventral margins of shell parallel, almost straight, anterior and posterior margins obliquely truncate, with rounded corners. Up to 200 mm long. White or cream, with pale purple, pink, or reddish streaks; periostracum glossy, light olive or yellow brown, grey-green along posterior dorsal region. Left valve with two projecting, peg-like cardinal teeth and two elongate, posterior laterals; right valve with a single cardinal and a single, elongate posterior lateral.

Burrowing in sand, lower shore and shallow sublittoral. All British coasts, distributed from Norway to the Mediterranean and north-west Africa.

Ensis arcuatus (Jeffreys) Fig. 10.39
Dorsal margin of shell almost straight, ventral margin distinctly curved, shell deepest at mid-line. Anterior and posterior margins obliquely truncate. Up to 150 mm long. White or cream, with reddish or purplish-brown streaks and blotches; periostracum glossy, light olive or yellow brown, darker grey-green along posterior dorsal region. Left valve with two projecting, peg-like cardinal teeth and two elongate, posterior laterals, one above the other; right valve with one cardinal and a single elongate, posterior lateral.

Burrowing in sand and gravel, lower shore and shallow sublittoral. All British coasts, distributed from Norway to Spain.

Phaxas pellucidus (Pennant) Fig. 10.39
Dorsal margin almost straight, ventral margin curved. Anterior end rounded and upturned, posterior slightly truncate. Up to 40 mm long. White or cream, sometimes with dark markings; periostracum glossy, light yellow-brown or olive. Left valve with a group of three projecting teeth: two cardinals, the posterior broad and bifid, and a single lateral. Right valve with two teeth, a single cardinal and a backwardly directed posterior lateral.

In mixed fine substrata, offshore to about 100 m. Widespread, and often abundant, around Britain; distributed from Norway to north-west Africa.

29 MYACIDAE

Shell equivalve, inequilateral, typically with a pronounced posterior gape. Ligament largely internal, supported by a chondrophore—usually projecting and prominent in the left valve. Hinge teeth absent. Pallial sinus present. Three British species.

---

1. Shell thin and brittle, not exceeding 20 mm. Chondrophore small. Umbones anterior to mid-line; posterior hinge line sloping from umbones, gently truncate posteriorly ***Sphenia binghami***

   Shell thick and strong, up to 150 mm long. Chondrophore large, projecting in left valve, concealed beneath umbo in right. Umbones close to or posterior to mid-line **2**

---

2. Posterior half of shell semi-elliptical, margin rounded ***Mya arenaria***

   Posterior half of shell angular, abruptly truncate ***Mya truncata***

Mya arenaria Linnaeus
Sand gaper Fig. 10.40
Oval, umbones just posterior to mid-line, anterior regularly rounded, posterior somewhat tapered. Up to 150 mm long. Off-white, yellowish, or fawn, dark greyish-brown about the umbones; periostracum light brown, often stained by iron deposits. Left valve with a prominent spatulate chondrophore, projecting at a right angle to hinge line, with a distinct tooth-like ridge along its posterior edge. Right valve with a concave, spatulate chondrophore recessed beneath umbo.

In sand, often mixed with mud or gravel, lower shore and offshore to about 20 m. Often very

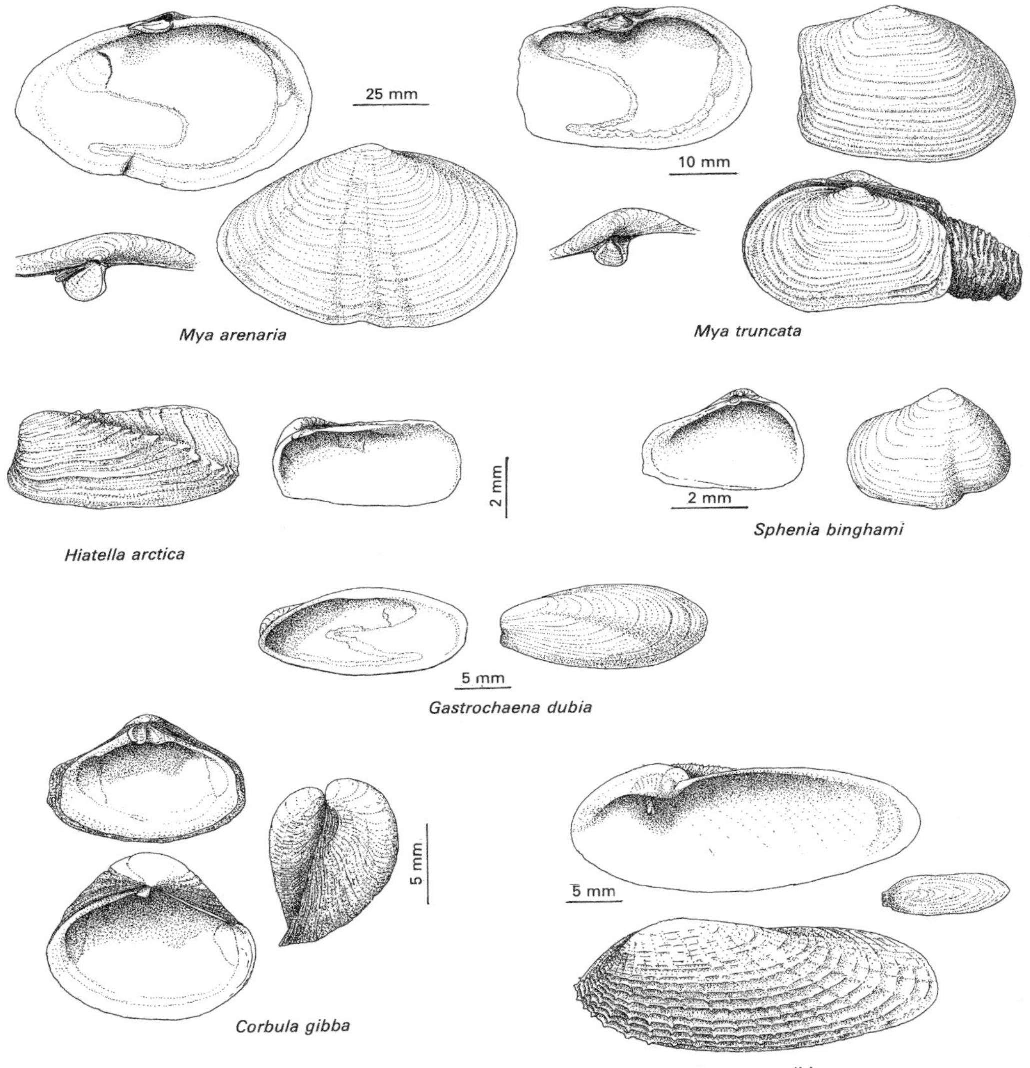

Fig. 10.40

common in estuaries, where it may occur in extensive beds. All British coasts; circumboreal, not reaching the Mediterranean.

Mya truncata Linnaeus

Blunt gaper Fig. 10.40

Anterior end of shell regularly rounded, posterior abruptly truncate, umbones on mid-line. Up to 70 mm long. Dull white to light brown, periostracum light olive to brown. Left valve with a prominent spatulate chondrophore, projecting at a right angle to hinge, bounded anteriorly and posteriorly by a distinct ridge. Right valve with a concave, spatulate chondrophore recessed beneath umbo, a blunt ridge anterior to it.

In mixed sandy substrata, from the lower shore to about 70 m. All British coasts; circumboreal, in the north-east Atlantic extending south to Biscay.

Sphenia binghami Turton Fig. 10.40

Irregularly oval, umbones anterior to mid-line; anterior end rounded, posterior sloping away from umbones, irregularly rounded or gently truncate. Up to 20 mm long. White, with pale green or brown periostracum. Left valve with a short, broad chondrophore projecting horizontally from beneath umbo, continuous posteriorly with the thickened dorsal margin. Left valve with a recessed chondrophore, completely obscured beneath umbo.

Attaches by a byssus, in holes and crevices or algal holdfasts, in shallow sublittoral waters. All British coasts; southwards to Mediterranean and north-west Africa.

30 CORBULIDAE

Shell inequivalve, more or less equilateral. Ligament internal, in a chondrophore recessed within the hinge line. True cardinal teeth absent, lateral teeth present. Pallial line with a very small sinus. A single British species.

Corbula gibba (Olivi)
Basket shell Fig. 10.40

Broadly oval to subtriangular, umbones close to mid-line; right valve convex, enclosing and overlapping left. Up to 15 mm long. Dull white to cream; periostracum coarse, grey-brown. Right valve with a single, massive tooth, anterior to a recessed, triangular chondrophore, and single elongate anterior and posterior lateral teeth. Left valve with a shallow chondrophore posterior to a deep socket, no lateral teeth.

In muddy sand and gravel, occasionally on the lower shore, most abundant offshore. Common and widespread off all British coasts; distributed from Norway south to the Mediterranean and west Africa.

31 HIATELLIDAE

Shell equivalve, inequilateral. Ligament external. Right valve with one cardinal tooth, left with two or none. No lateral teeth. Pallial line a series of disjunct scars, with a small sinus. Three British species, only one of which occurs commonly on the lower shore.

Hiatella arctica (Linnaeus)
Wrinkled rock borer Fig. 10.40

Elongate, rectangular or rather irregular; umbones close to anterior end, anterior margin rounded, posterior truncate. Up to 40 mm long. Dull white; periostracum coarse, yellow-brown. Two distinct ridges extend posteriorly from umbo on each valve: in juveniles these bear a few short, coarse tubercles, worn away in older specimens.

Attaches by a byssus, in holes, crevices, or algal holdfasts; also bores into soft calcareous rocks and shells. Lower shore and well out on to the Continental Shelf; common on all British coasts. Distributed from Arctic waters south to, at least, the Mediterranean and north-west Africa.

32 GASTROCHAENIDAE

Shell equivalve, inequilateral, gaping widely posteroventrally. Ligament external, hinge line without teeth. Pallial sinus present. A single Brittish species.

Gastrochaena dubia (Pennant)
Flask shell Fig. 10.40

Elongate, umbones close to anterior end, both valves strongly convex. Dull white or yellowish, periostracum coarse, light brown; up to 25 mm long. Sculpture of coarse concentric lines.

A borer in sandstone, limestone, and organic carbonates, lower shore and shallow sublittoral. South and south-west coasts only; southwards to the Mediterranean.

33 PHOLADIDAE

Shell equivalve, inequilateral, often elongate or irregular; sculpture often pronounced anteriorly, forming a rasping file which assists the animal in boring. Typically gaping along part or all of the anterior, ventral, and posterior margins. Ligament internal, reduced; hinge teeth absent. Anterodorsal margin reflected, curving back towards umbones. Accessory plates present along the dorsal margin: a *protoplax* anterior to umbones, a *mesoplax* above or immediately posterior to the umbones, and a *metaplax* posteriorly; not always present together. Additional elements may be developed to close an impermanent pedal gape (*callum*), or to protect posteriorly projecting siphons (*siphonoplax*). Foot muscles supported, in each valve, by a long, curved process (*apophysis*) extending ventrally from beneath the umbo. Pallial line with sinus. Eight species recorded from British waters.

---

1. Anterior margin of shell oblique to hinge line, gaping widely **2**

Both ends of shell rounded, gaping posteriorly but not anteriorly **Barnea candida**

---

2. Umbonal reflections supported by distinct vertical septa **Pholas dactylus**

   Umbonal reflections without septa **Zirfaea crispata**

Pholas dactylus Linnaeus
Common piddock Fig. 10.41

Elongate oval, umbones anterior to mid-line; antero-ventral margin deeply concave about a large, elliptical pedal gape, posterior margin regularly rounded, not gaping. Both valves strongly convex anteriorly. Up to 150 mm long. Dull white or grey, periostracum yellowish, often discoloured. Umbonal reflection pronounced, with 10 or more vertical septa linking it with umbone. Dorsal margin with a broad, oval protoplax, a small mesoplax and an asymmetric metaplax overlying right side. Apophysis extending from beneath umbo, halfway towards ventral margin.

Bores into wood, peat, compacted sand, and various soft rocks. Lower shore and shallow sublittoral. South and south-west coasts only, south to the Mediterranean and north-west Africa.

Barnea candida (Linnaeus)
White piddock Fig. 10.40

Elongate oval, shaped rather like a date; anterior and posterior margins rounded, gaping posteriorly. Up to 65 mm long. White, periostracum yellowish or light brown. Umbonal reflections prominent anteriorly, closely applied to umbones posteriorly and without septa. An elongate, oval protoplax dorsally. Apophysis slender, extending one-third of distance to ventral margin.

Bores into wood, peat, and various soft rocks. Lower shore and sublittoral. All British coasts; distributed from Norway to the Mediterranean and west Africa.

Zirfaea crispata (Linnaeus)
Oval piddock Fig. 10.41

Obliquely oval, both valves strongly convex; umbones anterior to mid-line, anterior margins obliquely truncate above a broad pedal gape; posterior gaping widely. Up to 90 mm long. Dull white, periostracum light yellowish-brown. A groove extends from umbones to ventral margin. Umbonal reflections broad, closely applied to, and extending just posterior to, the umbones; a small mesoplax present. Apophysis slender, extending no more than one-third of distance to ventral margin.

Bores into peat, clay, and soft rocks, rarely into waterlogged wood; lower shore and shallow sublittoral. Widespread around British Isles, distributed from Norway to the Bay of Biscay.

34 TEREDINIDAE. SHIPWORMS

Shell thin, equivalve, inequilateral and highly modified, gaping anteriorly and posteriorly, appearing trilobed and composed of several distinct elements: a triangular *anterior lobe*, with fine ridges parallel to the hinge; a deeper *anterior disc*, with fine ridges longitudinally; a narrow groove, the *median disc*, and finally a roughly triangular *auricle*. Umbones prominent, ligament and hinge teeth absent, but a dorsal articulating condyle present below umbo, and a ventral condyle on the ventral shell margin. Animal elongate, enclosed within its burrow, which is lined with calcareous deposits. A pair of accessory plates (*pallets*) present posteriorly at the tips of the siphons. Eleven species have been reported stranded on British shores, but most are rare, and sporadic in occurrence.

---

1. No apophyses present. Shell gaping anteriorly, closed posteriorly. Dorsal mesoplax present **Xylophaga dorsalis**

   Long, slender apophyses present. Shell gaping anteriorly and posteriorly. **2**

---

2. Posterior lobe of shell (auricle) much smaller than anterior. Broad end of pallet truncate, slender handle extending as a ridge along mid-line of blade **Nototeredo norvegica**

   Posterior lobe of shell almost as large as anterior. Broad end of pallet with concave edge, slender handle not extending along mid-line of blade **Teredo navalis**

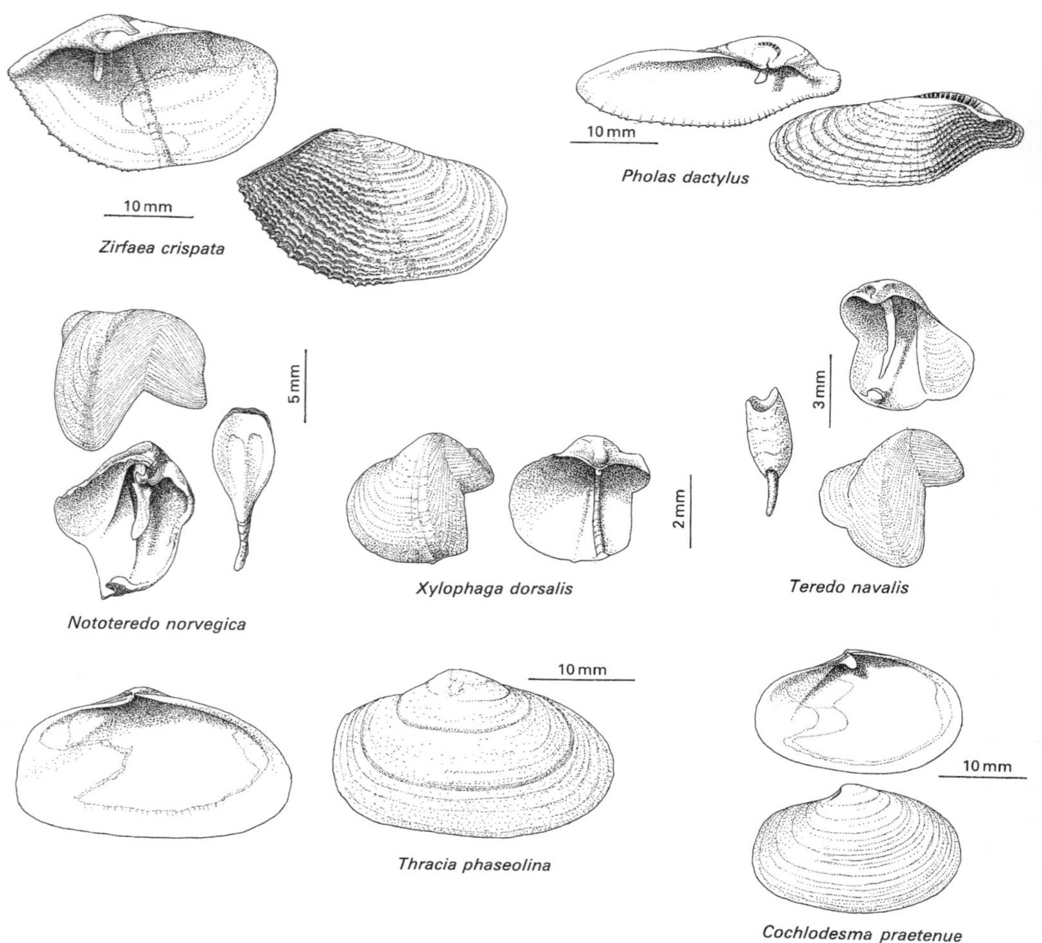

Fig. 10.41

Teredo navalis Linnaeus Fig. 10.41
Up to 10 mm long, white, periostracum light brown. Anterior lobe triangular; anterior disc with fine vertical to oblique ridges; median disc narrow, concave, with concentric sculpture; posterior disc with fine concentric lines; auricle semi-elliptical, almost as large as anterior lobe. Both valves strongly convex. Umbonal reflection closely applied to umbo; posterior hinge line broad, concave, with a bounding ridge. Apophysis curved, slender, extending two-thirds of depth of shell. Pallet blade elliptical with a concave edge, handle short, cylindrical, not extending on to blade.

The borings of *T. navalis* are frequently found in drifted, waterlogged wood, each lined with a thin tubular, calcareous deposit. Widespread around the world, it has been collected on all British coasts.

Nototeredo norvegica (Spengler) Fig. 10.41
Up to 20 mm long. White, periostracum brown, darkening to almost black in larger specimens. Anterior lobe triangular; anterior disc with finer oblique ridges; median disc narrow, concave, with concentric sculpture; posterior disc with concentric lines; auricle smooth, crescent-shaped, much smaller than anterior lobe. Both valves strongly convex. Umbonal reflections small, closely applied to anterior face of umbones; hinge line posterior to umbo broad and concave, delimited by a raised ridge. Apophysis broad, blade-like, extending halfway to opposite ventral border. Pallet concave, blade elliptical with a

straight edge, handle cylindrical, extending along mid-line of blade as a distinct ridge.

Bores into fixed and floating wood; widespread in the north-east Atlantic, from Iceland to the Mediterranean, reported off most British coasts.

Xylophaga dorsalis Turton Fig. 10.41
Shell globular, each valve strongly convex, rounded posteriorly, with a right-angled indentation anteriorly, exposing a pedal gape. Anterior lobe triangular; anterior disc elongate, extending from umbo to ventral margin, forming posterior edge of gape, with regular, oblique, ridges; median disc concave with raised margins and very fine concentric lines; posterior disc comprising more than half shell, with fine concentric lines. Umbonal reflection small, closely applied to umbo posteriorly, obscured by a bipartite mesoplax. No apophyses, although left valve has a prominent chondrophore.

Boring into wood, present off all British coasts. Distributed from Iceland and Norway south to the Mediterranean.

Xylophaga properly belongs to the Pholadidae but so closely resembles species of Teredinidae that it is most usefully keyed out here.

35 PERIPLOMATIDAE

Shell inequivalve, right valve more convex than left; inequilateral. Ligament external and internal, latter component in an oval chondrophore. No hinge teeth. Umbones typically with a crack in the calcification. Pallial line with a sinus. A single British species.

Cochlodesma praetenue (Pulteney) Fig. 10.41
Oval, umbones just posterior to mid-line; anterior margin rounded, posterior somewhat truncate, hinge line sloping posterior to umbones. Up to 40 mm long. Dull white, periostracum light yellowish-brown. Sculpture of fine concentric lines, each valve with a distinct crack for a short distance behind umbones. Postero-dorsal surface of shell finely granular, with coarser periostracum, and often encrusted detritus.

In mixed fine deposits, lower shore to about 100 m. Reported from various localities around the British Isles; distributed from Norway and southern Iceland to the Mediterranean.

36 THRACIIDAE

Shell inequivalve, right valve slightly larger or more convex than left, inequilateral. Ligament external and internal, the latter in a triangular chondrophore; supported by a *lithodesma*, a short calcified peg fitting into a notch anterior to chondrophore. No hinge teeth. Pallial line with a sinus. Six British species, difficult to tell apart, but only the following is at all common on the lower shore.

Thracia phaseolina (Lamarck) Fig. 10.41
Oblong oval, umbones just posterior to mid-line; anterior margin rounded, posterior abruptly truncate. Up to 38 mm long. Dull white, periostracum thin, light yellowish-brown. Chondrophore slender, triangular, posterior to umbo; anterior to chondrophore a small notch receives the short, peg-like lithodesma.

In mixed sand, from the lower shore to about 50 m. Reported from numerous localities around Britain. Geographical range uncertain, but present from Norway to the Mediterranean.

Class Cephalopoda

In the Cephalopoda the molluscan body displays its most highly modified form. The cephalopod body is elongated along a dorso-ventral axis, so that the defined anterior end, bearing the head, is morphologically ventral, while the rounded or pointed posterior end is, in fact, dorsal. In most cephalopods the shell is reduced and internal, or absent. Two significant features of cephalopod morphology derive from the modification of the head and foot, viz. the tentacles or *arms* which surround the mouth, and the *funnel* through which water is expelled from the mantle cavity. The *mantle* is thickened and muscularized, completely enveloping the *visceral mass*. The ventral *mantle cavity* contains the *ctenidia*, anus, and reproductory and excretory orifices. Water is actively pumped into the cavity and expelled through the anteriorly situated *funnel*. This may be flexible; while rapid water expulsion is utilized in jet-propulsion, the funnel may be directed to allow more deliberate, slower movement. In squids and cuttlefishes lateral or terminal fins are employed in normal swimming. A substantial internal shell, the familiar cuttlebone, is present in cuttlefish. A light, laminar calcareous structure, it imparts a firmness and rigidity to the body, and functions as a very finely balanced device for buoyancy control. In the squids, in which neutral buoyancy is achieved through regulation of ionic balances, the shell is represented by a slender, quill-like structure of keratin. Cephalization is marked; enlarged cerebral ganglia are enclosed in a cartilaginous capsule, and the head bears large, complex eyes. The mouth is armed with tough jaws, resembling a parrot's beak, and most cephalopods have modified salivary glands which secrete a venom used to immobilize prey. The tentacles, or *arms*, surround the mouth; they are muscular, prehensile, and bear numerous cup-shaped suckers, which may be armed with chitinous hooks and spines. Octopods have eight equal-sized arms, while cuttlefish and squid have additionally a pair of longer, retractile *tentacles*, typically with the suckers grouped on terminal pads. In most cephalopods the mantle epidermis is richly supplied with chromatophores, and octopods and squids are often strikingly coloured, with a potential for rapid change of both pattern and shade in response to environmental or other cues. A final characteristic of cephalopods is the ink sac; this is discharged when the animal is threatened, the dense ink cloud serving to mask its escape. Sexes are separate, and breeding involves the transfer of packaged sperm by the male to the female, usually following a specific behavioural display. In most species one arm of the male, the *hectocotylous* arm, is specially modified for sperm transfer.

All cephalopods are predatory. The Octopoda are for the most part benthic or bentho-pelagic, living in holes or crevices, beneath large rocks or in caves, and tend to capture passing prey. The bentho-pelagic cuttlefish are active hunters, as are the fast swimming, pelagic squids.

KEY TO FAMILIES

1. Head with eight arms and two tentacles. Suckers stalked **2**

   Head with eight arms only. Suckers sessile **4. Octopodidae**

2. Body broadly oval and rather flattened, or short and rounded. Internal shell substantial, calcified; or smaller, chitinous. Fins always paired, lateral **3**

   Body typically torpedo-shaped. Fins paired, terminal or subterminal. Arms little longer than head. Fins elongate, extending more than half length of mantle **3. Loliginidae**

3. Body flattened. Fins extending along entire length of body, on each side, but not joining posteriorly. Internal shell calcified. Male with left ventral hectocotylized arm **1. Sepiidae**

   Body short, ovoid. Fins forming rounded, projecting lateral lobes, not reaching posterior end of body. Internal shell chitinous, reduced. Male with left, or left and right, dorsal hectocotylized arm **2. Sepiolidae**

1 SEPIIDAE

Inner shell calcified, extending whole length of mantle. Body elongate oval, flattened, with lateral fins extending whole length of body, not joined posteriorly. Male with left ventral arm hectocotylized. Three British species. *Sepia orbignyana* Ferussac is a rare species reported sporadically from south-west England.

1. Body broadly oval. Dorsal anterior mantle edge forming a rounded lobe, extending to just posterior of the eyes. Shell parallel sided, evenly rounded posteriorly ***Sepia officinalis***

Body slender oval. Dorsal anterior mantle edge forming an acute lobe extending between eyes. Shell tapered anteriorly and posteriorly **Sepia elegans**

Sepia officinalis Linnaeus
Common cuttlefish Fig. 10.42
Maximum mantle length 400 mm. Dorsal anterior edge of mantle developed as a blunt, rounded lobe. Arms with four rows of suckers; relatively short, comprising less than half combined length of head and arms; central pair of arms broadly flattened dorso-ventrally, all arms with acute outer edges. Shell elongate oval, almost parallel-sided, smoothly rounded posteriorly.

Shallow sublittoral, and offshore to 250 m; common inshore during summer months. Widely distributed, abundant on all British coasts.

Sepia elegans d'Orbigny Fig. 10.42
Maximum mantle length 80 mm. Dorsal anterior mantle edge produced as an acute lobe extending between eyes. Arms with two rows of suckers; short, comprising less than half combined length of head and arms. Shell with lanceolate outline, distinctly narrowed close to rounded posterior end.

Southern and western coasts only, in similar habitats to *S. officinalis*.

2 SEPIOLIDAE

Inner shell reduced, chitinous. Body rounded; fins paired, lateral, not extending whole length of body, not meeting posteriorly. Male with left, or both left and right, dorsal hectocotylous arm. Ten species occur around Britain, but only two occur commonly in inshore waters.

---

1. Dorsal edge of mantle free, not fused to head **Rossia macrosoma**

   Dorsal edge of mantle fused with dorsal surface of head **Sepiola atlantica**

Rossia macrosoma (Delle Chiaje) Fig. 10.42
Maximum mantle length 60 mm, about one-third total length of animal; dorsal anterior edge of mantle not fused to head, with well-marked lip continuous with rest of mantle opening. Arms about twice length of head, more or less rounded in section; each with two rows of suckers proximally, four or more rows distally. Fins short and rounded, rather fleshy.

Shallow sublittoral and offshore. All British coasts, more common in north and west.

Sepiola atlantica d'Orbigny
Little cuttlefish Fig. 10.42
Maximum mantle length about 20 mm, about half total body length. Dorsal anterior mantle edge fused to dorsal surface of head, between eyes; ventral edge of mantle with straight, thickened edge. Arms more or less rounded in section; with two longitudinal rows of suckers, increasing to 4–8 rows distally on dorsal arms. Tentacle club with eight longitudinal rows of suckers. Fins thin and delicate, each as broad as body at point of attachment, irregularly rounded.

Shallow subtidal and offshore; frequently in intertidal rock pools, often taken on sandy shores in shrimp nets. Widespread and common on all British coasts.

3 LOLIGINIDAE

Internal shell reduced to a slender, horny pen. Body slender, with well-developed fins equivalent to more than half length of mantle. Arms relatively short. Dorsal anterior mantle edge produced, extending between eyes. Male with left ventral arm hectocotylized. Four British species.

---

1. Posterior end of body blunt. Fins angular, forming a diamond shape in dorsal view **2**

   Posterior end of body produced and pointed. Fins with rounded outline, forming a heart shape in dorsal view **Alloteuthis subulata**

---

2. Tentacle club with very large median suckers and much smaller marginal suckers **Loligo vulgaris**

   Tentacle club with median suckers not, or only slightly, larger than marginal suckers **Loligo forbesii**

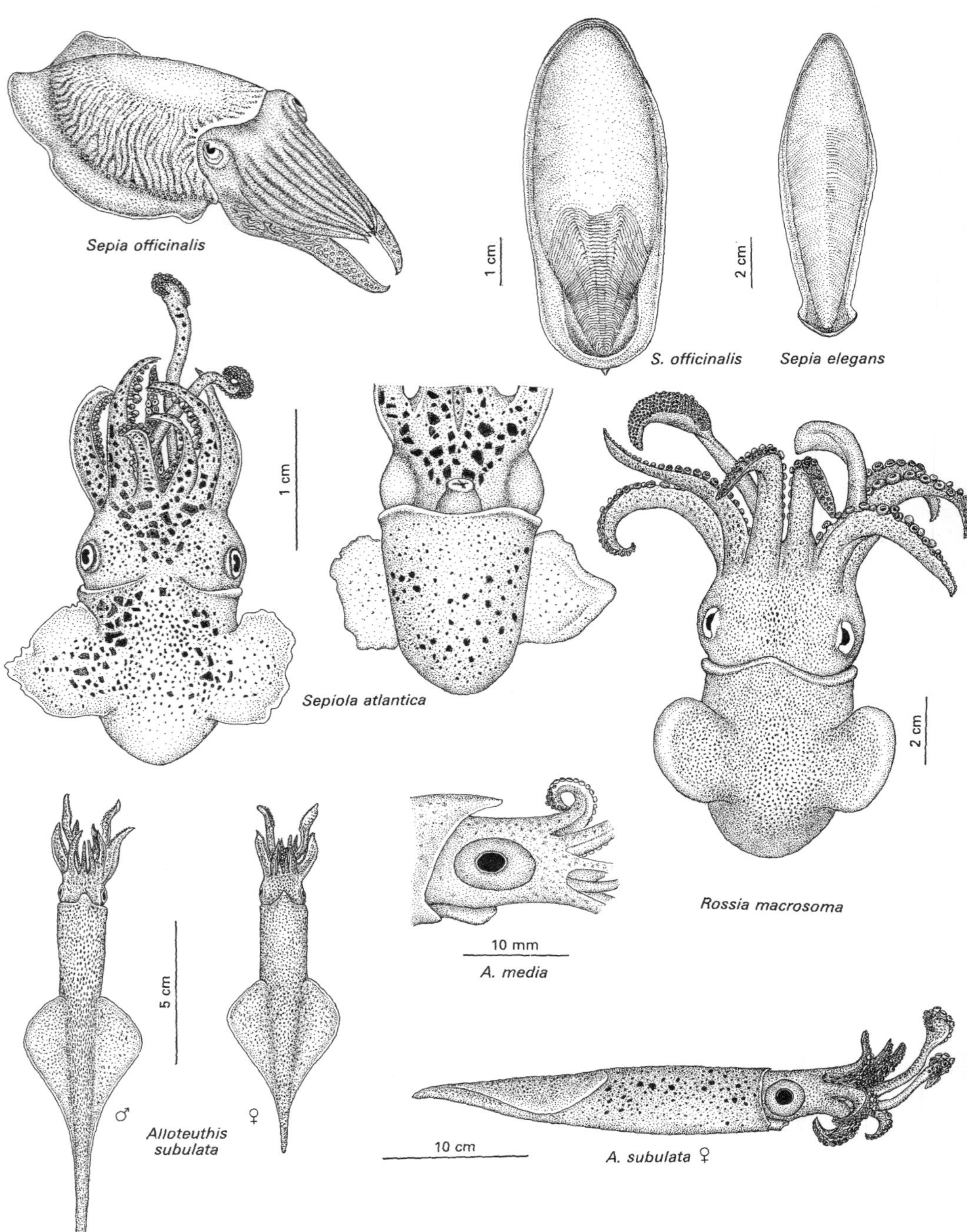
Sepia officinalis
1 cm
S. officinalis
2 cm
Sepia elegans
1 cm
Sepiola atlantica
2 cm
Rossia macrosoma
10 mm
A. media
5 cm
♂
♀
Alloteuthis subulata
10 cm
A. subulata ♀

Fig. 10.42

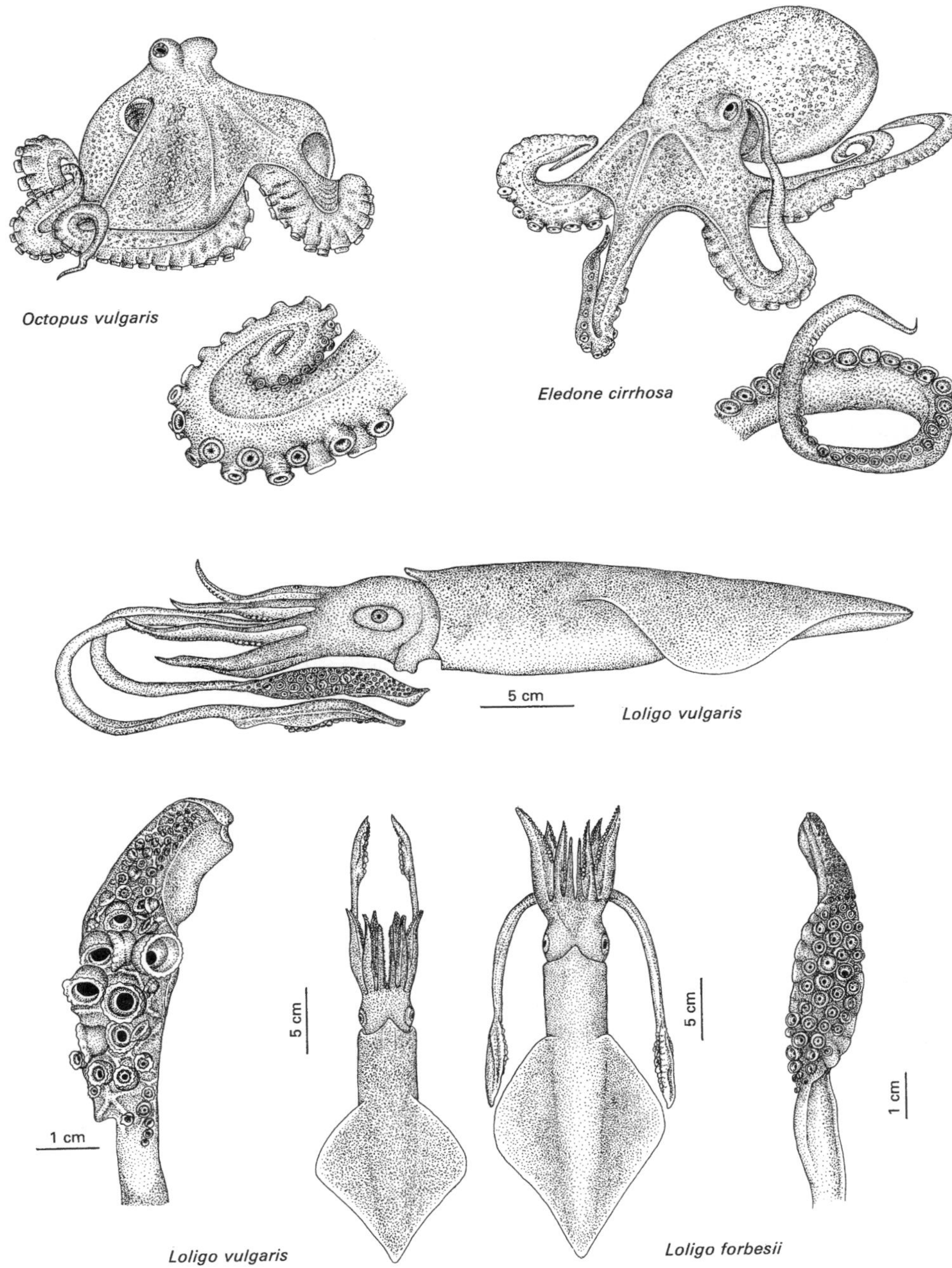

Fig. 10.43

Loligo forbesii Steenstrup

Long finned squid Fig. 10.43

Up to 600 mm long, maximum mantle length about 350 mm. Dorsal surfaces densely pigmented reddish-brown, ventral surfaces lighter. Anterior dorsal edge of mantle produced as a blunt lobe extending to a point just posterior to eyes. Arms about twice length of body, angular in section, ventral pair particularly broad and flat; bearing two longitudinal rows of suckers. Tentacle club with median suckers of the middle region less than twice diameter of adjacent marginal suckers.

Widespread and abundant, in coastal waters and offshore, around the whole of the British Isles.

Loligo vulgaris Lamarck
Common squid Fig. 10.43

Similar to *L. forbesi*, reaching 750 mm total length. Distinguished by the tentacle club, on which median suckers of the middle region are three or four times the diameter of adjacent marginal suckers.

Southern and western coasts only; eastwards as far as the Low Countries, and northwards as far as the Irish Sea.

Alloteuthis subulata (Lamarck) Fig. 10.42

Up to 210 mm, with maximum mantle length 140 mm; finely tapered posteriorly; in males body posterior to fins constitutes one-third to one-half total mantle length, in females less than one-quarter. Dorsal anterior edge of mantle produced as a quadrate projection extending between eyes. Arms up to twice length of head, with flattened, angular section, dorsal pair half length of ventral pair; all with two longitudinal series of suckers. Tentacle club with median suckers of middle region up to four times diameter of adjacent marginal suckers.

Widespread and abundant, in coastal waters and offshore, on all British coasts.

Alloteuthis media (Linnaeus) Fig. 10.42

Similar to *A. subulata*. Arms and tentacles tend to be proportionately more slender, but the chief characteristic by which the two species are separated is the lateral mantle edge, which in *A. media* extends obliquely posteriorly between the dorsal and ventral margins.

Uncommon, but apparently present on all British coasts.

4 OCTOPODIDAE

Body saccular, without an internal shell. Head and mantle fused dorsally, with only a restricted ventral mantle aperture. Eight arms only, bearing sessile suckers without chitinous teeth.

Octopus vulgaris Lamarck
Common octopus Fig. 10.43

Body distinctly warty, up to 1000 mm total length, with maximum arm spread of 3000 mm, but commonly much smaller. Arms thick and stout, bearing two longitudinal rows of suckers.

On rocky coasts, from the shallow sublittoral to 100 m. Widely distributed in the warm temperate Atlantic, *O. vulgaris* reaches its north-eastern limit in south-western Britain and the western English Channel, where it varies in abundance from year to year.

Eledone cirrhosa (Lamarck)
Curled octopus Fig. 10.43

Body smooth or finely tuberculate, up to 500 mm total length, with maximum arm spread of 700 mm. Arms slender, finely tapered distally and curled when at rest, with a single longitudinal row of suckers.

On rocky coasts, often on the lower shore, and subtidally to 120 m. Widely distributed in the north-east Atlantic, present on all British coasts.

11 BRYOZOA

Bryozoans (Phylum Bryozoa) are sessile, colony-forming coelomates. Each colony arises by asexual budding from an *ancestrula* formed by the metamorphosis of a sexually produced larva. The individual units of the colony, or *zooids*, remain in communication with each other through pores. Zooid polymorphism is developed to a varying degree: the normal feeding zooids are termed *autozooids* and specialized morphs are called *heterozooids*. Each autozooid has a circular *lophophore* bearing a series of slender, ciliated tentacles, which can be withdrawn into an anterior introvert of the body, called the *tentacle sheath*. The alimentary canal is deeply looped, so that the anus opens near the mouth, but just outside the lophophore. The skeletal component of the zooid is sometimes termed the *zooecium*, in contrast to the viscera which constitute the *polypide*. Colonies, but not necessarily zooids, are hermaphrodite; embryos are generally brooded, either in the mother zooid or in special brood chambers.

The most primitive zooid shape appears to be cylindrical, as in the common intertidal species *Bowerbankia gracilis* (Fig. 11.1), in which the autozooids are borne along slender stolons. The terminal opening, apparent in the wall of the retracted zooid, is termed the *orifice*.

More commonly, bryozoan zooids are squat and adjacent, forming a continuous encrusting layer. Such zooids have a lower, or *basal*, surface applied to the substratum, and an upper, or *frontal*, surface; in free-rising colonies the zooids retain their distinctive basal and frontal surfaces. In the most fundamental type of zooid the frontal surface, or most of it, remains flexible as a *frontal membrane*, but the lateral and basal walls are calcified (Fig. 11.1). The end of the zooid nearest to the origin of the colony is termed *proximal*, while the end furthest away is *distal*. The orifice is situated near to the distal end of the frontal membrane, and is closed by a muscular sphincter, paired lips, or, most typically, a hinged flap, the *operculum*.

There are two living classes of marine Bryozoa. The class Stenolaemata contains a single order, the Cyclostomatida. In these bryozoans the zooid is typically elongate and tubular, with a circular terminal orifice which is not closed by an operculum. They are usually thickly calcified, and appear minutely speckled through the presence of minute uncalcified spots, or pseudopores. Embryos develop in swollen reproductive chambers or in modified *gonozooids*, but other types of polymorph are absent.

The class Gymnolaemata includes two orders, the calcified, highly polymorphic Cheilostomatida, and the uncalcified Ctenostomatida, in which polymorphism is very limited. Cheilostomates are basically box-shaped, with a frontal orifice, closed by a hinged operculum. Formerly two suborders were recognized, the Anasca, in which the frontal surface of the zooid was partly or wholly membranous, though sometimes shielded by overarching spines, and the Ascophora, in which the frontal surface was completely calcified and the operculum coincided exactly with the zooid orifice. This division is now thought to be artificial and many suborders have been proposed. The Cheilostomatida is especially characterized by specialized heterozooids, such as *avicularia*, and by *ovicells*. Avicularia are generally small, non-feeding zooids, often attached to an autozooid, with a jaw-like *mandible*, homologous with the operculum of an autozooid. Another specialized heterozooid is the *vibraculum* in which the operculum or mandible has evolved into a whip-like seta. Ovicells are more or less globular chambers, which hold a single embryo, produced at the distal end of fertile zooids.

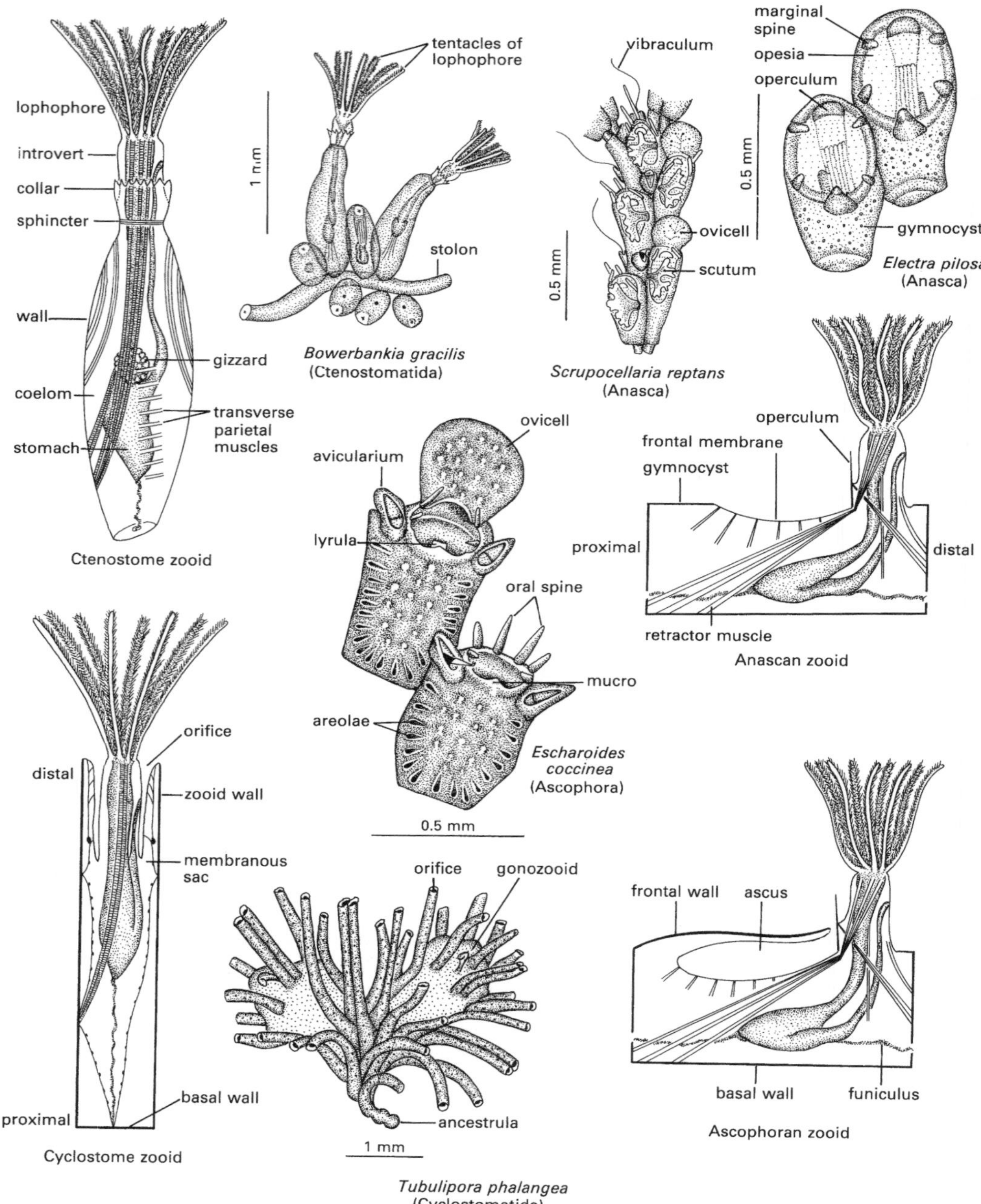

Fig. 11.1

The pores of ascophorines, which may be confined to the margin or extend all over the frontal surface, are merely gaps in the calcification: they are best called *pseudopores*. The frontal wall often rises to a median point (*mucro*) or prominence (*umbo*) proximally to the orifice, this position being referred to as *suboral*. In some families (e.g. Smittinidae) there is commonly a suboral avicularium, but in others (e.g. Schizoporellidae) avicularia are placed beside the orifice. The orifice varies in shape from family to family: in the Hippoporinidae it is oval or cordate with a pair of lateral teeth (*condyles*) on which the operculum pivots. In Schizoporellidae it is suborbicular with a notch (*sinus*) in the proximal margin. The operculum again pivots on condyles, though these are not denticulate, and the sinus represents the exterior opening of the ascus. A development from this is the D-shaped orifice with separate, suboral ascopore found, for example, in the Microporellidae.

In the order Ctenostomatida zooids may be cylindrical or squat, with a frontal or terminal orifice, but are entirely uncalcified. The orifice appears puckered or, rarely, lipped, when closed, and does not have an operculum. Avicularia and ovicells are absent, and polymorphism is limited to stolons.

With few exceptions, such as the very obvious *Membranipora membranacea* colonies found on *Laminaria* fronds, bryozoans are inconspicuous and often hidden from view. A few, such as *Alcyonidium hirsutum*, *A. gelatinosum*, *Bowerbankia imbricata*, *Electra pilosa*, and *Flustrellidra hispida*, are commonly found on intertidal fucoids, where they form tufts, gelatinous crusts, or lacy patches. More species occur on the lower shore, as hard, calcified, white or coloured encrustations in *Laminaria* holdfasts or on stones and below rocks; or as hanging tufts, perhaps among anemones and ascidians, under large boulders or in caves. In general they prefer dark places and protection from desiccation. As with other filter feeders, they are favoured by non-destructive movement of water and may reach spectacular abundance in straits and rapids systems. *Laminaria* fronds may here be smothered with bushy *Scrupocellaria* colonies and innumerable scabby patches of encrusting bryozoans. They will be dredged on rocks and shells.

KEY TO FAMILIES

1. Zooids cylindrical or elongated, arising singly, in clusters or in fused groups from long, creeping or free stolons **2**

   Zooids variously shaped, but not rising upwards from long stolons **4**

2. Zooids calcified, whitish, either rising freely in branching chains or borne singly; long, tubular, and with a lateral membranous area at the free end. Orifice not terminal **3**

   Zooids uncalcified, generally more or less buff coloured, not rising freely in branching chains; if long and cylindrical, then externally symmetrical. Orifice terminal (CTENOSTOMATIDA) **25**

3. Zooids arising singly from stolons, long and cylindrical **5. Aeteidae**

   Zooids arising in chains, roughly diamond-shaped **6. Scrupariidae**

4. Colony firm, gelatinous, encrusting or lobate; or erect and muddy. Non-calcified. Operculate orifices, ovicells, avicularia and vibracula, always absent (CTENOSTOMATIDA) **25**

Colony calcified (though not necessarily hard and brittle). Operculate orifices, ovicells, or other brood chambers, avicularia and vibracula sometimes present **5**

---

5. Zooids cylindrical, perhaps united for greater part of their length but often ending in free distal portion which bears the terminal orifice. Individual zooids often partly obscured by calcification surrounding their basal parts. Chitinous operculum never present. Avicularia and vibracula absent (CYCLOSTOMATIDA) **30**

   Zooids more or less box- or boat-shaped, not long and tubular. Orifice not at end of a free cylindrical portion. A more or less distinct operculum usually present. Avicularia and vibracula may be present (CHEILOSTOMATIDA) **6**

---

6. Upper or frontal surface wholly or largely membranous; polypide visible; operculum semicircular, not greatly differentiated from the membrane. The membrane often surrounded by spines borne on side walls of the zooid, these may overarch and obscure the membrane like a series of ribs. Membrane sometimes underlain by a calcified lamina (and therefore not very apparent), in which case colony bushy and composed of jointed, dichotomously divided branches. Colony often erect and tufted **7**

   Frontal surface, except for operculum, covered by calcified wall, which may be smooth or granular, porous or imperforate; polypide always completely obscured. Orifice a distally situated opening in frontal wall, closed by a distinct operculum, and often partly surrounded by spines. Colony sometimes erect but never bushy or tufted **15**

---

7. Colony encrusting, forming continuous expansions **8**

   Colony not encrusting **11**

---

8. Ovicells and avicularia never present. True spines on rim of membranous area absent (but spinules, small protuberances or pointed tooth-like projections from the rim of membranous area, or a single bristle or thorn-like process proximal to frontal membrane, may be present). Often on algae **8. Membraniporidae**

   Ovicells and avicularia usually present. Rim of membranous area bearing one to many spines, which may arch over frontal membrane and be more or less fused together **9**

---

9. Marginal spines modified to form a cage over frontal membrane, by fusing with one another, the sutures marked by lines of pores **11. Cribrilinidae**

Marginal spines erect, or flattened and overarching frontal membrane, but never fused with adjacent spines on the same side **10**

---

10. Spines erect or sloping slightly over frontal membrane, or with about five pairs of flattened spines arching over the membrane and meeting each other in the mid-line. Ovicell not closed by zooidal operculum **10. Calloporidae**

Generally eight or more pairs of flattened spines arching over frontal membrane and meeting each other in the mid-line. Operculum of zooid also closing ovicell **11. Cribrilinidae**

---

11. Colony in the form of creeping chains of zooids in single series which give rise to similar but erect chains, each like a string of beads **6. Scrupariidae**

Colony not as above **12**

---

12. Branches of colony composed of two single series of zooids joined basally (i.e. back to back) **7. Eucrateidae**

Branches composed of two or more adjacent series of zooids arranged as a single layer, joined back to back with others, or disposed around an axis **13**

---

13. Colony lobate, with wide, unjointed branches. Zooids in two layers, joined back to back **9. Flustridae**

Colony tufted, with narrow branches, generally either jointed or with pedunculate (bird's head) avicularia. Zooids in a single layer **14**

---

14. Branches composed of short series of zooids separated at bifurcations by distinct joints. Pedunculate avicularia absent. Vibracula often present **12. Scrupocellariidae**

Branches not jointed at bifurcations. Pedunculate avicularia usually present. Vibracula absent **13. Bugulidae**

---

15. Zooids heaped and jumbled together, lacking any common orientation. Orifice with a distinct notch or sinus in its proximal margin. Either one short avicularian process suborally or a pair of lateral avicularia raised on processes **21. Celleporidae**

Zooids clearly orientated, arranged in regular series; or separated from each other by short stolons **16**

---

16. A conspicuous median pore (usually the ascopore) present proximal to the orifice **17**

No conspicuous median pore **18**

17. Frontal wall below pore clearly marked with radiating ribs separated by rows of punctures **11. Cribrilinidae**

Pores in frontal wall not arranged in radiating lines. Ascopore situated directly below the orifice **18. Microporellidae**

18. Orifice with a distinct notch or sinus in its proximal margin **19**

Orifice without a proximal sinus **20**

19. Avicularia absent. Zooids distant, not forming a continuous crust; or zooids contiguous but with ovicells borne by differentiated, shorter zooids **16. Hippothoidae**

Avicularia present. Frontal wall generally covered with pores; ovicell porous or imperforate **20. Schizoporellidae**

20. A lyrula (median tooth) present in the orifice **21**

Lyrula absent, colony encrusting or foliaceous **22**

21. Avicularia absent **19. Escharellidae**

Suboral or lateral avicularia present. Frontal wall proximal to orifice, produced into a raised tapering point (mucro) **15. Exochellidae**

22. Orifice flanked each side by an avicularium, and bordered distally (in non-ovicellate zooids at least) by a series of spines. Frontal wall perforated by large pores **11. Cribrilinidae**

Flanking avicularia and spines absent **23**

23. Frontal wall with marginal and central pores. Orifice with condyles on which operculum pivots **22. Hippoporinidae**

Frontal wall imperforate or with marginal pores only. Orifice without condyles **24**

24. Avicularium (if present) suboral. Marginal pores large, separated by buttresses. Ovicells absent. **14. Umbonulidae**

Avicularium distal to the orifice, borne on ovicell of ovicellate zooids. Frontal wall entire. Spaces in the calcification present between adjacent zooids **17. Chorizoporidae**

---

25. Colony a continuous adherent crust, or forming thick, fleshy, free lobes; or erect, branching, and muddy **26**

Colony composed of creeping or free stolons bearing or merging with solitary, clustered, or fused groups of (usually cylindrical) zooids **28**

---

26. Colony erect with spirally disposed branches; zooids cylindrical and covered with muddy particles **25. Nolellidae**

Colony erect with fleshy, free lobes and immersed zooids, or forming a continuous adherent crust **27**

---

27. Zooids bearing spines. Orifice two-lipped. On algae **24. Flustrellidridae**

Zooids without spines or, if apparently spiny, encrusted with particles and growing on hydroids. Orifice not two-lipped **23. Alcyonidiidae**

---

28. Stolons radiating from zooid bases. Zooid walls opaque with fine encrusting particles. New zooids not budded off from sides of existing zooids. Marine **25. Nolellidae**

Colony composed of a network of stolons bearing zooids in clusters, the clusters often coinciding with the sites of stolonal branching **29**

---

29. Zooids very small, less than 0.5 mm in length. Gizzard absent. Expanded tentacles not forming a perfect circle but the dorsalmost pair curving strongly downwards. Branch stolons frequently arising in opposite pairs and diverging from main axis at right angles. Often found on *Corallina* **26. Walkeriidae**

Zooids generally larger, exceeding 0.5 mm in length. Gizzard present. Expanded tentacles forming a perfect circle. Stolonal branches not normally arising in opposite pairs. Various substrata **27. Vesiculariidae**

---

30. Colony erect. Branches composed of a single or double series of zooids, divided at intervals by non-calcified joints. Brood chambers pear-shaped **1. Crisiidae**

Colony adherent; linear, lobed, or circular **31**

---

31. Colony more or less circular or discoidal, sometimes forming secondary colonies by marginal budding **32**

Colony linear or lobed, not circular; adnate, its branches (lobes) as broad as long; zooids generally in transverse rows; gonozooid in conspicuous lobes between the autozooids, its opening on a tube at the base of an autozooid **2. Tubuliporidae**

---

32. Colony flat. Zooids contiguous, not separated by spaces. Calcified walls (especially of brood chambers) punctate. Margin of orifice not spiny **3. Diastoporidae**

Colony convex. Zooids separated by a series of walls and spaces. Calcified walls not punctate. Margin of orifice spiny **4. Lichenoporidae**

Class Stenolaemata

Order Cyclostomatida

1 CRISIIDAE

Colonies of slender branches divided by *joints* into *internodes* of one, two, or several zooids. One or more branches may arise from each internode. In fertile internodes, one or more zooids become transformed into *gonozooids*. Sterile internodes lack gonozooids.

---

1. Sterile internodes consisting of a single zooid **2**

Sterile internodes, apart from the basal ones, consisting of a number of zooids **3**

---

2. Each zooid with a filiform spine; gonozooid free for most of its length ***Crisidia cornuta***

Zooids without spines; gonozooid adherent to an autozooid for most of its length ***Filicrisia geniculata***

---

3. Internodes short, consisting of nine zooids or fewer, distal parts of colony may be a little longer **4**

Internodes long, consisting of 11 zooids or more **5**

---

4. Internodes mostly 5–7 zooids, branches strongly incurved; zooids never bearing spines ***Crisia eburnea***

Internodes often with more than seven zooids, branches not incurved; many zooids bearing spines ***Crisia aculeata***

---

5. Joints black (except near growing points); branch base situated low on zooid bearing it, wedged in between successive zooids on same side; opening of gonozooid transversely elongate, facing forwards ***Crisia denticulata***

Joints pale; branch base not wedged in between successive zooids; opening of gonozooid on an upwardly directed, trumpet-like spout ***Crisia ramosa***

Crisia aculeata Hassall Fig. 11.2

Colonies up to 20 mm; internodes mainly of 5–7 zooids, but longer distally; branches usually arising from first or second zooid, straight, base short; some zooids with long, jointed spines; gonozooid

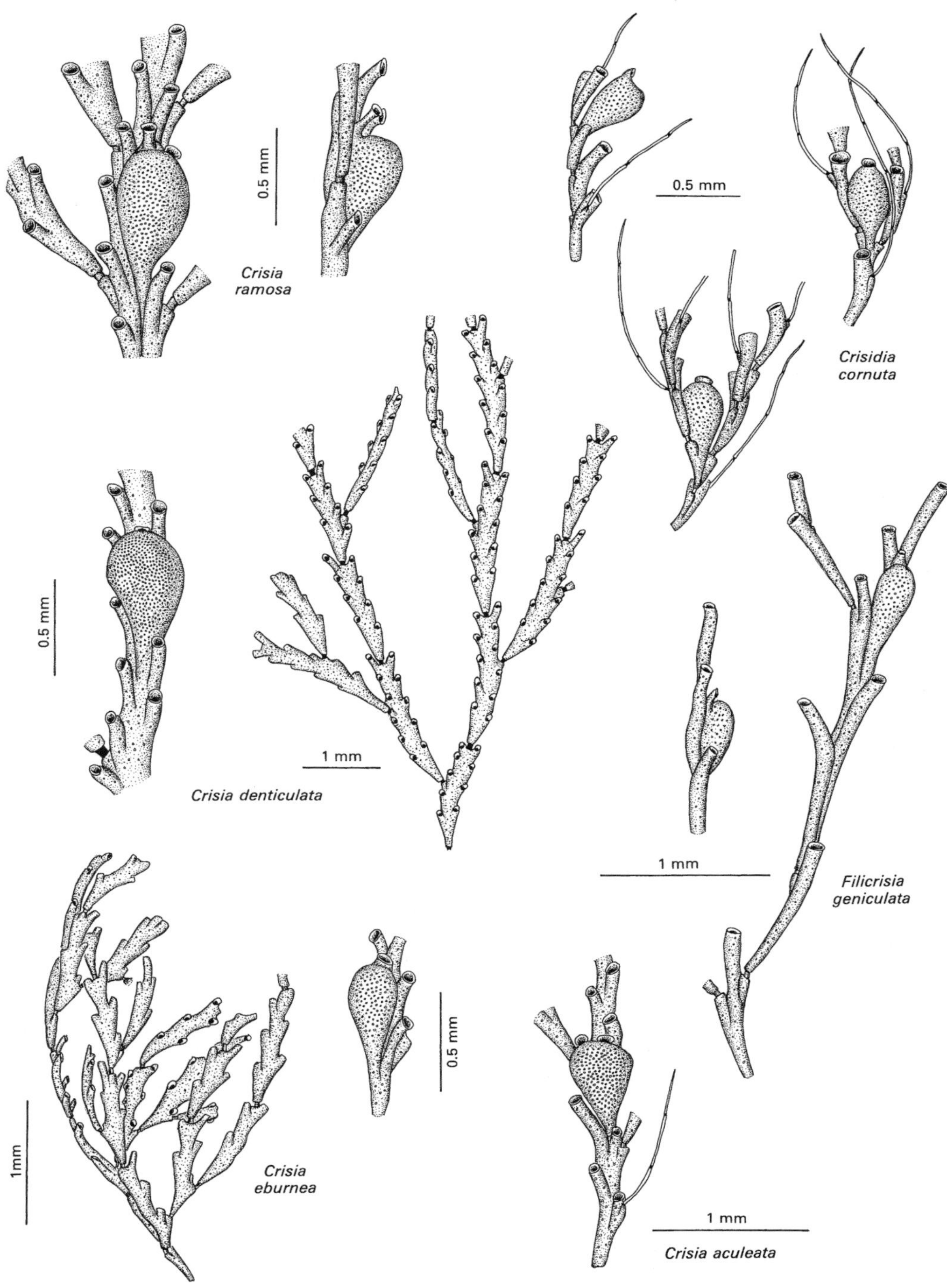
0.5 mm
Crisia ramosa
0.5 mm
Crisidia cornuta
0.5 mm
1 mm
Crisia denticulata
1 mm
Filicrisia geniculata
0.5 mm
1mm
Crisia eburnea
1 mm
Crisia aculeata

Fig. 11.2

situated fairly high in its internode, adnate, inflated distally, then abruptly flattened to inconspicuous, sessile opening.

On algae and rocks; LWST and shallow water; all coasts; quite common.

Crisia denticulata (Lamarck) Fig. 11.2
Colonies up to 25 mm; branches usually alternate along main axes, one from each internode; joints jet black (except near the growing points); internodes long, most often of 11 zooids, base situated low on bearing zooid, wedged in between it and zooid below; gonozooid pyriform, positioned high in its internode, opening facing forwards, not on a long tube.

On hydroids, algae and rocks; LWST and shallow water; not found north of the British Isles; common.

Crisia eburnea (Linnaeus) Fig. 11.2
Colonies up to 20 mm, branches incurved; internodes short, of an odd number of zooids, commonly five or seven; branches usually rise from lowest zooid in each odd-numbered internode, alternating along main axes but characteristically all from same side in distal parts of colony, base short; joints yellowish; spines absent; gonozooid replacing second or third zooid, adnate, its opening subterminal, transversely elongated, facing forwards.

On various substrata, especially red algae; LWST and shallow water; all coasts; common.

Crisia ramosa Harmer Fig. 11.2
Colonies up to 20 mm, much branched; internode length variable, often very long, with numerous zooids, giving rise to up to five branches; joints yellowish; branch base long, reaching but not wedged between next zooid below; gonozooid pyriform, most often in 6th–8th position, its opening circular, on a long funnel-shaped tube.

On algae, bryozoans and rocks; shallow water; mainly southern, but reaching the Skagerrak; frequent.

Crisidia cornuta (Linnaeus) Fig. 11.2
Colonies 10–15 mm high; internodes each a single zooid, separated by yellowish joints; zooids curved, giving rise to next internode plus a long slender spine, or two internodes; gonozooid pyriform, free for most of its length, situated at a bifurcation with a pair of internodes rising from its base; opening of gonozooid terminal, spout-like.

On hydroids, bryozoans, algae, and rocks; LWST and shallow water; all coasts; common.

Filicrisia geniculata (Milne Edwards) Fig. 11.2
Colonies up to 20 mm; sterile internode a single zooid, joints yellowish; zooids long and slender, curved, without spines; fertile internodes of 3–5 zooids, gonozooid a slender club-shape, adherent to an autozooid, its opening on a short, curved, subterminal tube.

On hydroids, bryozoans, rocks, and algae; LWST and shallow water; all coasts, common.

2 TUBULIPORIDAE

Colonies largely adherent, fan-shaped or lobed; zooids free, or in transverse rows resembling miniature organ pipes. Gonozooid an irregular inflation on colony surface, more strongly punctate than autozooids and extending between them; its opening at the end of a short tube. This large family includes numerous offshore species which mostly occur only on hard substrata. Only the three following species occur commonly in inshore waters.

---

1. Opening of gonozooid tubular, flaring, much larger than an orifice, upwards or sideways facing **_Tubulipora plumosa_**

   Opening of gonozooid tubular, but not flaring, smaller or not much larger than an orifice, sideways or downwards facing (the opening itself often invisible from above) **2**

---

2. Opening of gonozooid facing sideways, a little larger than an orifice **_Tubulipora liliacea_**

   Opening of gonozooid facing downwards, completely invisible from above, not larger than an orifice **_Tubulipora phalangea_**

Tubulipora liliacea (Pallas) Fig. 11.3
Colony usually divided into several slender, radiating lobes; strongly flushed with purple when alive. Zooids in each lobe in transverse rows, diverging alternately to each side of mid-line; inner zooids tallest, so that height diminishes towards margin of

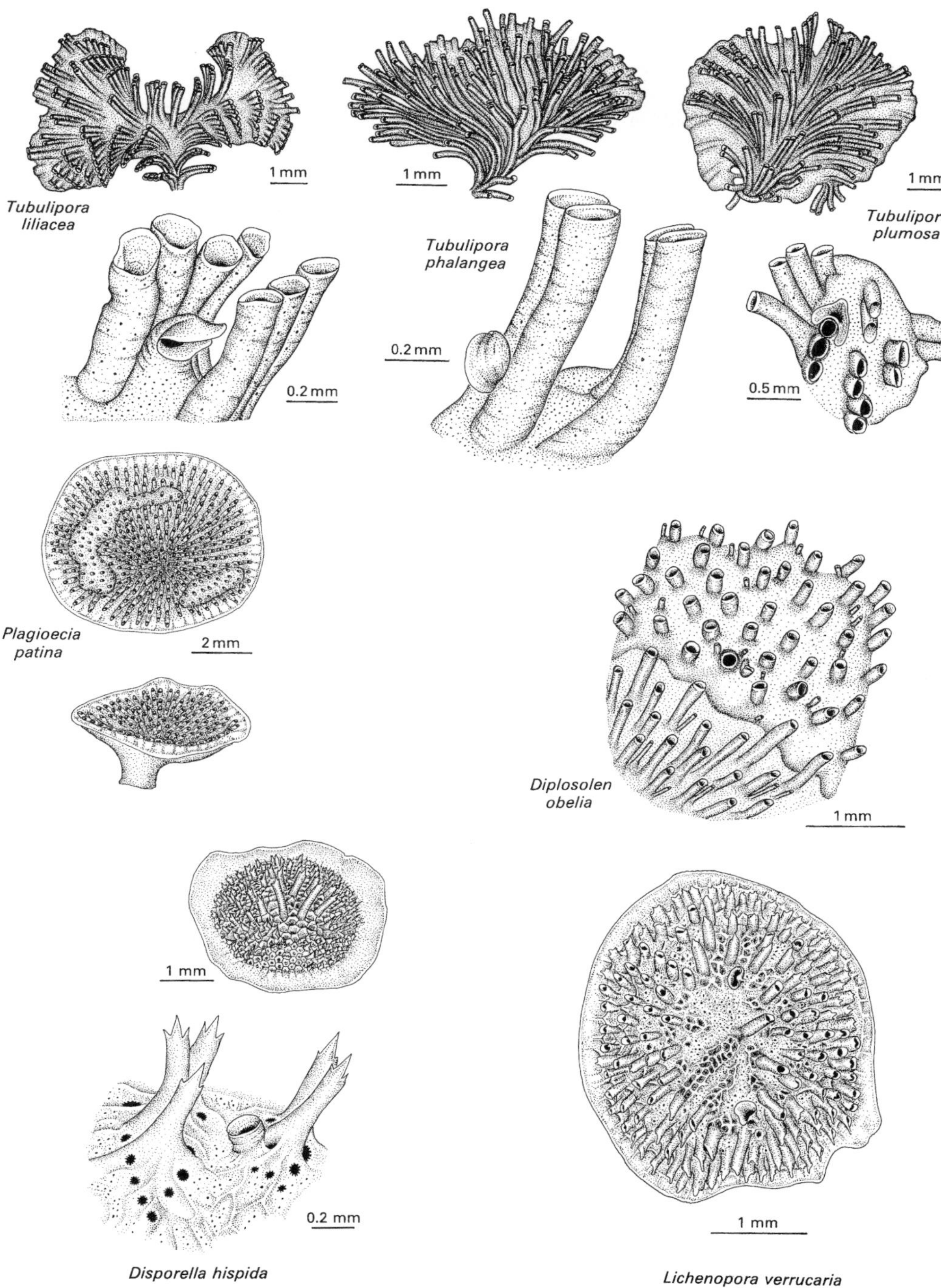
1 mm
1 mm
1 mm
Tubulipora liliacea
Tubulipora phalangea
Tubulipora plumosa
0.2 mm
0.2 mm
0.5 mm
Plagioecia patina
2 mm
Diplosolen obelia
1 mm
1 mm
0.2 mm
1 mm
Disporella hispida
Lichenopora verrucaria

Fig. 11.3

lobe. Orifice diameter about 165 µm. Gonozooid rather zigzag or star-shaped, fitting between rows of zooids; its opening, about 260 µm in diameter, facing horizontally at end of a curved tube, decumbent on an autozooid.

Found particularly on hydroids in shallow water; all coasts; frequent.

Tubulipora phalangea Couch Fig. 11.3

Colonies circular or lobed, up to about 15 mm across, white or lightly flushed with purple. Zooids long and slender, in transverse rows, diverging alternately to each side of mid-line, or radial in the presence of a gonozooid. Orifice diameter about 160 µm. Gonozooid spreading between rows of zooids; its opening about 160 µm in diameter, facing downwards at the end of a curved tube, decumbent on an autozooid.

On stones, shells, and algae; shallow water; all coasts; frequent.

Tubulipora plumosa
Thompson in Harmer Fig. 11.3

Colonies lobed, up to about 20 mm across, white or lightly flushed with purple. Zooids in transverse rows, diverging alternately to each side of mid-line, or radiating from a flat central area. Orifice diameter about 185 µm. Gonozooid spreading between rows of zooids; its opening about 350 µm in its largest diameter, facing directly or obliquely upwards, at the end of a short, flaring tube, decumbent on an autozooid.

On algae, particularly on *Saccorhiza* and *Laminaria* bases in shallow water; all coasts; frequent.

3 DIASTOPORIDAE

Colonies adherent, usually discoid; zooids largely or wholly adnate, with subcircular orifice. Gonozooid an irregular inflation on the surface of the colony, its opening at the end of a short tube.

---

1. Dwarf zooids present among the autozooids ***Diplosolen obelia***

   Dwarf zooids absent ***Plagioecia patina***

Diplosolen obelia (Johnston) Fig. 11.3

Colonies flat, outline variable. Zooids wholly adnate or with distal ends slightly raised, interspersed with dwarf zooids (*nanozooids*). Gonozooid rather small, spreading between several zooids, its opening on a short, free-standing tube.

On stones, shells, and algae; shallow water; all coasts; fairly common.

Plagioecia patina (Lamarck) Fig. 11.3

Colonies discoid, flat or bowl-shaped; surrounded by a white border; about 5 mm across. Zooids in radiating lines, wholly adnate or with distal ends slightly raised; orifice more or less elliptical. Gonozooid roughly crescentic, spreading among several surrounding zooids, its opening on a short, free-standing tube.

On stones, shells, and algae; shallow water; all coasts; common.

4 LICHENOPORIDAE

Colonies discoid, adnate, convex; zooids erect or suberect, radiating from a zooid-free porous (alveolar) central area in which a colonial brood chamber develops by fusion of *alveolae*.

---

1. Spaces (alveoli) between zooids thin-walled, sometimes roofed over; margin of orifice generally produced above as a pointed process; opening of brood chamber a flaring tube ***Lichenopora verrucaria***

   Alveoli thick-walled; margin of orifice often trifid; opening of brood chamber circular, lacking any tube ***Disporella hispida***

Disporella hispida (Fleming) Fig. 11.3

Colony discoid, or composed of several confluent disks. Zooids disposed around central area, orifices raised, frequently with 1–3 slender points. Central alveoli with thick walls, the openings often reduced to small central pores. Opening of brood chamber round, tubeless.

On algae, stones, and shells; shallow water; all coasts; common.

Lichenopora verrucaria (Fabricius) Fig. 11.3

Colony discoid. Zooids disposed around a small central area, the orifices raised, with a single pointed process. Central alveolae with thin walls, sometimes roofed over. Opening of brood chamber on an upright flaring tube.

On algae, hydroids, shells, and stones; LWST to shallow water; northern coasts only.

Class Gymnolaemata

Order Cheilostomatida

5 AETEIDAE

Colonies consisting of a series of cylindrical zooids, proximal part of each recumbent, distal part erect. The adherent portions form a linear series, resembling stolons supporting uprights at intervals. Each upright consists of a stem and a distal region in which a basal surface and frontal membrane may be distinguished; operculum terminal. No avicularia or vibracula. Embryos develop in delicate, evanescent *ovisacs*, not in ovicells.

---

1. Terminal portion of zooid wider than stem, usually downwards facing **_Aetea anguina_**

   Terminal portion not wider than the stem; zooid straight **2**

---

2. Stem region of zooid appearing striated when viewed by transmitted light under a microscope **_Aetea sica_**

   Stem region of zooid not striated when viewed as above **_Aetea truncata_**

Aetea anguina (Linnaeus) Fig. 11.4
Upright part of zooid 0.6–0.8 mm high, arising from a pronounced dilatation. Stem faintly but clearly annulate; distal region one-quarter to one-third of upright length, wider than stem, finely punctate, usually facing downwards.

In *Laminaria* holdfasts and on hydroids, bryozoans, and shells; LWST to shallow water; all coasts; common.

Aetea sica (Couch) Fig. 11.4
Upright part of zooid 0.8–1.8 mm high, arising from pronounced dilatation. Stem faintly but clearly annulate; distal region about one-third of free length; scarcely wider than stem, finely punctate, erect.

Various substrata; LWST to shallow water; all coasts; frequent.

Aetea truncata (Landsborough) Fig. 11.4
Upright part of zooid 0.6–0.8 mm high, rising from weak dilatation. Stem punctate, not striated; distal region one-third to one-half of free length, scarcely wider than stem, finely punctate, erect.

Various substrata; LWST to shallow water; all coasts; frequent.

6 SCRUPARIIDAE

Colonies erect, with zooids arranged uniserially. Zooids elongated, with oval frontal membrane in distal part of zooid. No avicularia or vibracula. Embryos develop in bivalved chambers borne on special reproductive zooids.

---

1. Creeping part of colony a line of zooids. Zooids slender, with frontal membrane almost parallel to zooidal axis **_Scruparia ambigua_**

   Creeping part of colony a stolon. Frontal membrane at marked angle to axis of zooid **_Scruparia chelata_**

Scruparia ambigua (d'Orbigny) Fig. 11.4
Free branching chains of zooids arise from similar but adnate chains. Zooids long and slender, frontal membrane almost parallel to basal wall. Brood chamber subglobular, bivalved, about 200 μm across, scarcely wider than its supporting zooid.

On algae, hydroids, bryozoans, and shells; LWST to shallow water; all coasts; common.

Scruparia chelata (Linnaeus) Fig. 11.4
Free branching chains of zooids arise from weak dilatations in a creeping stolon. Zooids horn-shaped, frontal membrane and basal wall diverging at an acute angle. Brood chamber subglobular, bivalved, about 350 μm across, wider than its supporting zooid.

Various substrata; LWST to shallow water; uncommon in Scotland, frequent elsewhere.

7 EUCRATEIDAE

Colonies erect, tufted. Zooids elongate, paired back to back in branching series. No spines, avicularia or ovicells. Embryos developing in evanescent *ovisacs*.

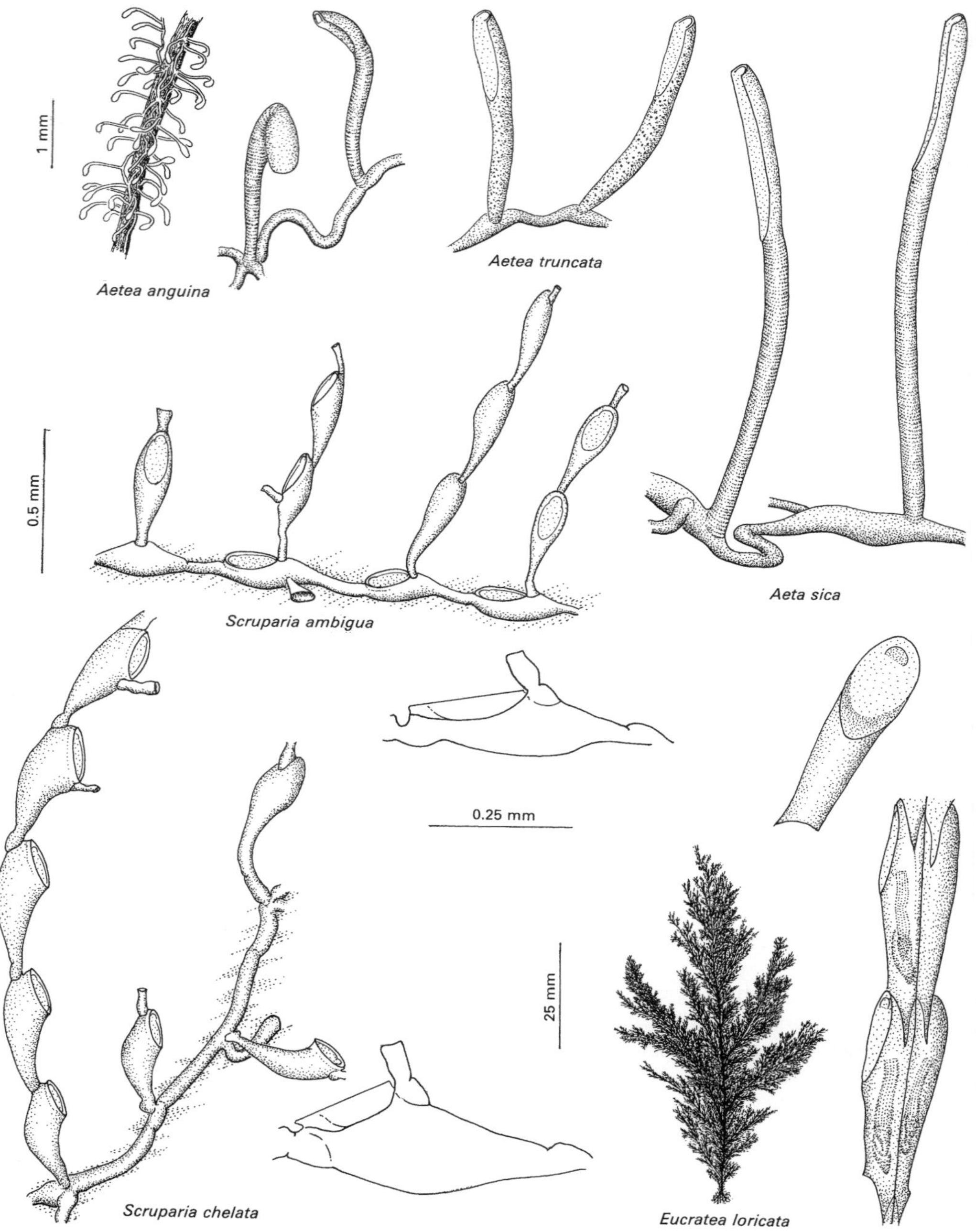

Fig. 11.4

Eucratea loricata (Linnaeus) Fig. 11.4

Colony a dense, slender clump 100 mm or more high; whitish-brown. Zooids elongate, broader distally, with oval frontal membrane; all pairs in a branch having the same orientation.

Sublittoral, but often washed up; all coasts; local.

8 MEMBRANIPORIDAE

Colonies encrusting. Zooids squat, longer than broad, with all or most of frontal surface membranous. No avicularia or vibracula. Embryos not brooded, larva a *cyphonautes*.

---

1. Membranous area oval, occupying about one-half of total zooid length. Proximal part of zooid covered by a punctate wall, and bearing a median tooth or bristle. One or more thorn-like denticles present on each lateral wall **_Electra pilosa_**

   Membranous area occupying all or most of frontal surface **2**

---

2. Membranous area more or less rectangular. Zooid walls bearing a tubercle at each corner. On *Laminaria* (rarely other algae) **_Membranipora membranacea_**

   Membranous area more or less oval. Zooid walls without tubercles. Not on *Laminaria* (often in brackish water) **3**

---

3. Usually a thorn-like spine proximal to membranous area. Operculum calcified and opaque **_Electra crustulenta_**

   No median spine proximal to membranous area. Operculum transparent **4**

---

4. Small triangular spaces usually present in angles between zooids. Sometimes a few delicate spinules on margin of membranous area. Marine or brackish **_Conopeum reticulum_**

   Triangular spaces never present between zooids. Marginal spinules generally absent except for a distal pair. Brackish water only **_Conopeum seurati_**

Conopeum reticulum (Linnaeus) Fig. 11.5
Colonies forming gauze-like encrustations. Zooids oval, about 0.5 × 0.3 mm. Membranous area occupying whole of frontal surface; margin thickened, granular, sometimes with a few delicate spines. Triangular dwarf zooids often present in angles between normal zooids.

On stones, shells, and submerged structures; MLWN to shallow water; open coasts, also estuaries and brackish water (to less than 20‰); widespread and common.

Conopeum seurati (Canu) Fig. 11.5
Colonies forming gauze-like encrustations. Zooids oval, about 0.5 × 0.35 mm. Membranous area occupying whole of frontal surface, margin usually with paired distal spines, and sometimes a few delicate lateral spines. Triangular dwarf zooids absent.

In brackish water (as low as 1‰), on aquatic plants, shells, and submerged structures; locally common, particularly in brackish lagoons, in England southwards to the Mediterranean; northern limit unknown.

Electra crustulenta (Pallas) Fig. 11.5
Colony an encrusting patch, but in enclosed lagoons developing large erect clumps. Zooids ovate hexagonal 0.5 × 0.3 mm. Membranous area occupying most of frontal surface; proximal part non-porous. Operculum calcified, markedly opaque compared to frontal membrane. Margin crenulate, with a single, short, proximal spine.

In brackish water; on available substrata; lower shore and shallow water; estuaries; locally common.

Electra pilosa (Linnaeus) Fig. 11.5
Colonies irregular or stellate greyish patches. Zooids ovate-oblong, 0.45 × 0.3 mm; proximal part porous, distal half or two-thirds occupied by oval frontal area; operculum inconspicuous. Margin with 3–13 short, tooth-like spines, median proximal usually longest, sometimes bristle-like.

Marine, various substrata, particularly *Fucus serratus* and other algae; lower shore and shallow water; all coasts; very common.

Membranipora membranacea (Linnaeus) Fig. 11.5
Colonies extensive lacy sheets on algae. Zooids rectangular, about 0.4 × 0.15 mm, with tubercles or short spines at corners. Membrane covering whole frontal surface. Lateral walls incorporating vertical uncalcified bands to provide flexibility.

On algae, particularly *Laminaria*; LWST to shallow water; all coasts; common and often conspicuous.

9 FLUSTRIDAE

Colonies erect, foliaceous; fronds flexible, made up of two layers of zooids placed back to back; arising from an encrusting base. Zooids ovate–quadrangular, frontal surface membranous. Avicularia usually present, in series with autozooids. Ovicell completely immersed in next distal zooid.

Six British species, but only two occur commonly in inshore habitats.

---

1. Colony generally with broad lobes. Two pairs of spines on most non-ovicellate zooids. Avicularia present **_Flustra foliacea_**

   Colony with rather narrow branches. A single pair of spines present on non-ovicellate zooids. Avicularia absent **_Chartella papyracea_**

Chartella papyracea
(Ellis and Solander) Fig. 11.5

Colonies delicate, whitish, bushy tufts, much divided, with narrow fronds; about 30 mm high. Zooids oblong, with a single, short spine on each distal angle. Ovicell underlying frontal membrane of distal zooid, somewhat pointed from above. Embryos orange.

Below boulders at LWST, and on stones in shallow water; southern Britain only; frequent.

Securiflustra securifrons (Pallas) is similar but has avicularia, large autozooids, distinctive ovicells, and a colony with wedge-shape branches (Fig. 11.5).

Flustra foliacea (Linnaeus)
Hornwrack Fig. 11.5

Colony a coarse greyish tuft, broadly foliaceous or with narrower strap-shaped fronds; commonly about 100 mm high; often smelling of lemon. Zooids narrow proximally, broad and rounded distally, with a pair of spines on each side. Avicularia situated at bifurcation of zooid rows, mandible rounded. Ovicell egg-shaped, its calcified chamber bulging into both the distal and bearing zooids. Embryos pale orange.

On stones and shells in shallow water; all coasts, common. Dead colonies often washed up.

10 CALLOPORIDAE

Colonies encrusting. Zooids with most of frontal surface membranous; sometimes underlain proximally by a calcified shelf (*cryptocyst*). Avicularia generally present. Ovicells globular, prominent, opening above the zooidal operculum. A large family with at least 20 species, in 12 or more genera, presently known from Britain.

---

1. A single large spine beside membranous area, or one to two pairs of distal spines **2**

   Membranous area surrounded by spines (five pairs or more) **4**

---

2. Surface of ovicell uniformly granular. One or two pairs of spines present at distal end of membranous area **_Callopora dumerilii_**

   Surface of ovicell including a distinct and strongly demarcated rounded or triangular area of different texture. Except on marginal zooids, one or two spines present beside membranous area, one often greatly developed, or spines absent **3**

---

3. Differentiated area of ovicell more or less circular or rectangular. Developing embryos orange. Frontal membrane underlain by a conspicuous calcified lamina surrounding a trilobate opening **_Amphiblestrum flemingii_**

   Differentiated area of ovicell triangular. Developing embryos white. No calcified lamina under frontal membrane **_Amphiblestrum auritum_**

---

4. Avicularia stalked, borne on rim of membranous area and replacing marginal spines which they closely resemble **_Cauloramphus spiniferum_**

   Some zooids bearing sessile, triangular avicularia proximal to membranous area **5**

---

5. Spines, apart from distalmost pair(s), flattened, arched over frontal membrane, and meeting in mid-line; like ribs. Developing embryos orange **_Callopora rylandi_**

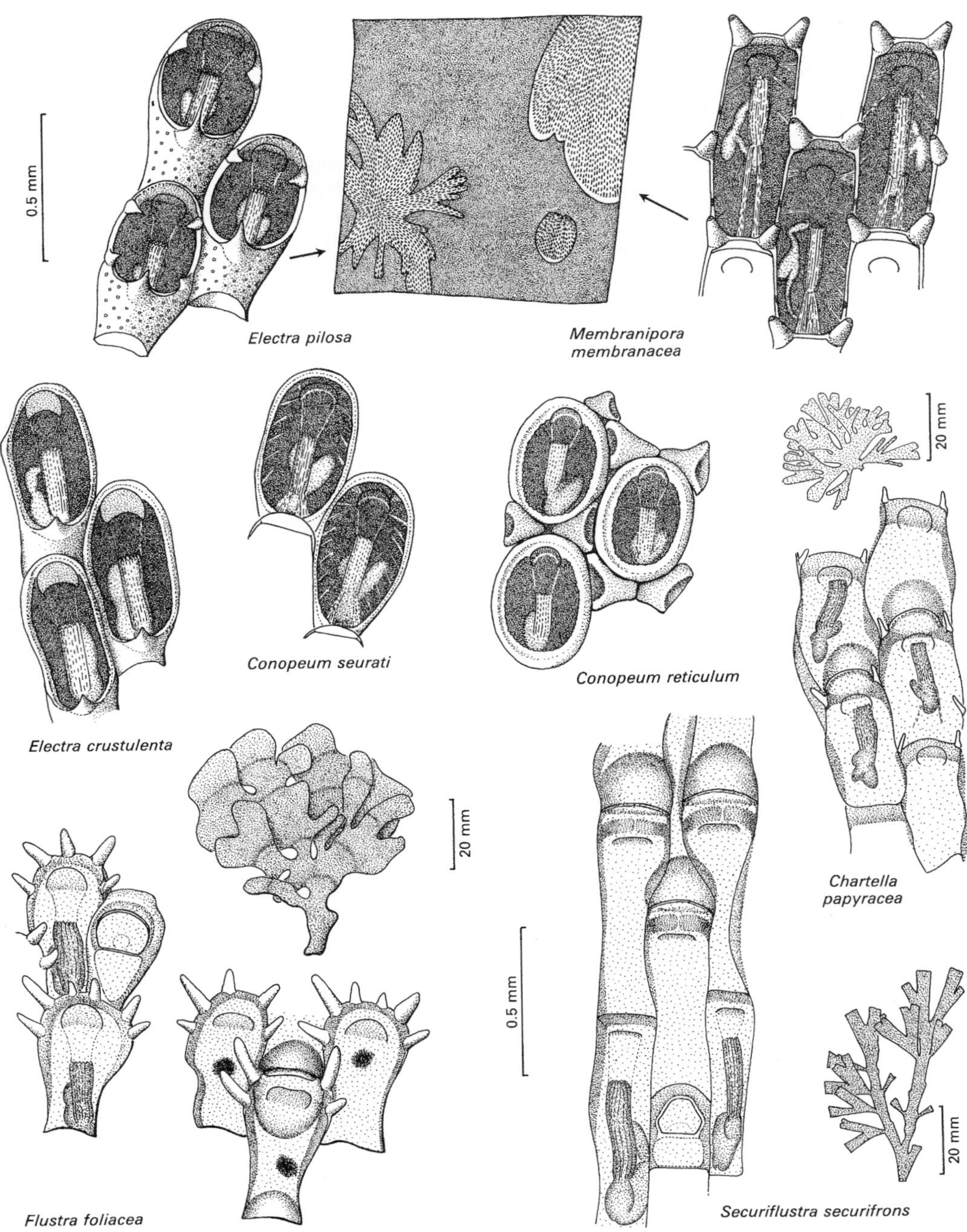
0.5 mm
Electra pilosa
Membranipora membranacea
20 mm
Conopeum seurati
Conopeum reticulum
Electra crustulenta
20 mm
Chartella papyracea
0.5 mm
20 mm
Flustra foliacea
Securiflustra securifrons

Fig. 11.5

Spines not rib-like, more or less erect. Developing embryos pink **6**

---

6. Spines rounded in section, up to 12 in total but usually fewer, almost perfectly erect ***Callopora lineata***

   Spines, except for distalmost pair(s), flattish in section, 12 or more, closely set and sloping inwards across membranous area ***Callopora craticula***

Amphiblestrum flemingii (Busk) Fig. 11.6
Zooids pear-shaped, widest proximally; frontal membrane underlain in its proximal third by a cryptocyst. Often 4–6 spines situated near orifice, sometimes one only: this, or one of the lower spines, greatly enlarged, tall, and curved. Generally two avicularia on proximal part of zooid, variously orientated. Ovicells globular, upper surface often bearing a crescentic ridge enclosing a finely granulated area. Embryos orange.

On stones and shells; LWST and offshore; all coasts; common in shallow water.

Amphiblestrum auritum (Hincks) Fig. 11.6
Zooids ovate, membranous area occupying most of frontal surface; no underlying cryptocyst. Up to four spines situated near orifice, one of them often enlarged. Generally two avicularia on proximal part of zooid, variously orientated. Ovicells globular, upper surface with a rib rising to an apex distally and enclosing a small triangular space. Embryos white.

On stones, shell, and *Laminaria* holdfasts; shore and shallow sublittoral; all coasts, but especially in north.

Callopora craticula (Alder) Fig. 11.6
Zooids hexagonal, 0.3-0.4 × 0.25 mm, oval membranous area occupying most of frontal surface. Twelve to 15 slender spines surrounding area, distalmost pairs erect, remainder slanted inwards towards mid-line of zooid. A single avicularium on proximal part of zooid, variously orientated. Ovicells rather small, smooth, with a transverse rib. Embryos pink.

On shells and seaweed; offshore; Isle of Man, Scotland, north-east England; locally common.

Callopora dumerilii (Audouin) Fig. 11.6
Zooids hexagonal, 0.4–0.5 × 0.25 mm, oval membranous area occupying most of frontal surface. Generally one or two pairs of spines situated near orifice. One or two avicularia on proximal part of zooid. Ovicells with a granular surface; not ribbed. Embryos orange.

Chiefly on stones and shells, below LWST; all coasts; quite common.

Callopora lineata (Linnaeus) Fig. 11.6
Zooids hexagonal, about 0.5 × 0.25 mm, membranous area oval, occupying most of frontal surface; usually 8–10 more or less erect spines around area. A single avicularium on proximal part of zooid, variously orientated. Ovicells fairly large, smooth, with a transverse rib. Embryos pink.

On shells, stones, and algae from the shore and shallow water, sometimes abundant on *Laminaria* fronds in sheltered situations; all coasts; common.

Callopora rylandi
Bobin and Prenant Fig. 11.6
Zooids oval-hexagonal, about 0.5 × 0.35 mm, oval membranous area occupying most of frontal surface. Seven to 11 flattened spines, like ribs, overarch area and meet in mid-line. A single avicularium sometimes present on proximal part of zooid. Ovicell globular, finely granular, with an inconspicuous rim distally. Embryos orange.

On rocks, stones, and around *Himanthalia* holdfasts; lower shore; south and west of the British Isles; frequent.

This species is rather similar to *Membraniporella nitida* (Johnston), which occurs on hard substrata offshore. *Membraniporella nitida* has more ribs (17–21), and in ovicellate zooids the ovicell is closed by the zooid operculum.

Cauloramphus spiniferum (Johnston) Fig. 11.6
Zooids elongate–oval, membranous area occupying most of frontal surface. Fourteen to 16 close-set, slightly incurved spines surrounding area; interspersed among them one or more slender, stalked, club-shaped avicularia; mandible subterminal. Ovicells shallow, with a frontal rib. Embryos orange.

On stones and shells; LWST and offshore; all coasts; quite common, particularly in the south and west.

11 CRIBRILINIDAE

Colonies incrusting. Zooids with frontal membrane spanned by flattened spines, with or without

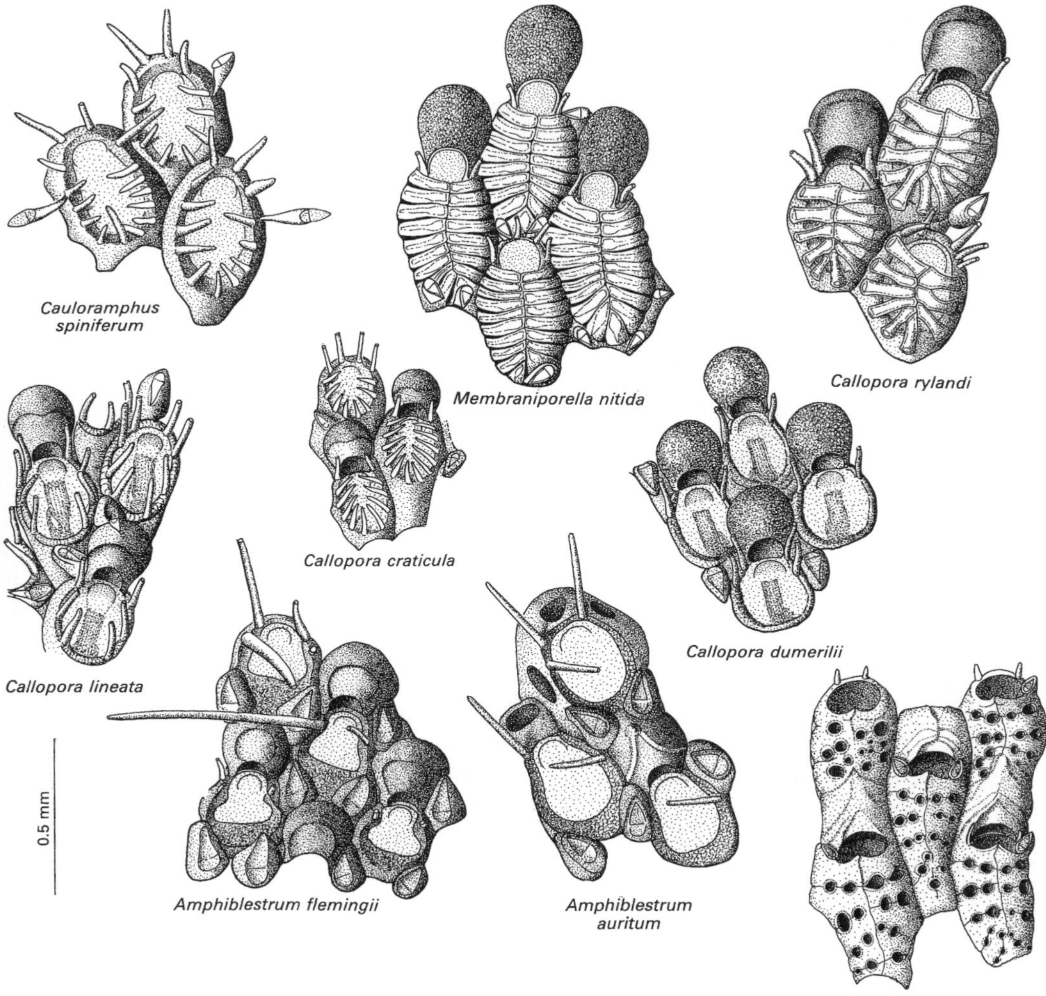

Fig. 11.6

lateral fusions. Avicularia generally present. Ovicells globular, opening below (and closed by) the zooidal operculum.

At least 15 British species, some of which have only been recognized and named recently. Most occur on dead shell or rock offshore, only one occurs inshore and intertidally.

Cribrilina cryptooecium Norman Fig. 11.6
Zooids elongate-oblong, 0.4–0.55 × 0.3 mm. Membranous area covered and obscured by a frontal shield of 7–11 fused ribs; generally two large pores in suture line between adjacent ribs. Distalmost ribs thickened, sometimes produced into a median point (*mucro*) below orifice. Small lanceolate avicularia may be present on one or both sides of orifice. Ovicell not prominent, smooth, imperforate, becoming overgrown by an A-shaped bar. Embryos orange.

On rocks and algae on the shore and in the shallow sublittoral; all coasts; common.

12 SCRUPOCELLARIIDAE

Colonies erect or suberect, with narrow unilamellar, dichotomously dividing, jointed branches; attached to substratum by rhizoids. Zooids in two series, alternately arranged. Basal and lateral walls calcified; frontal wall calcified apart from oval membranous area, which may be partly covered by a modified spine (*scutum*); generally with spines on distal angles. Avicularia and vibracula usually present. Ovicells present, opening above the orifice, recumbent on the distal zooid.

British fauna includes about 15 species, but most are found offshore, or only at the extreme south-west or northern limits of the British Isles. Only three species occur commonly inshore.

---

1. A single vibraculum situated in axil of each bifurcation (view from basal side). Scutum divided like an antler **Scrupocellaria reptans**

   A pair of vibracula in each axil. Scutum more or less undivided, or absent 2

---

2. Scutum absent **Scrupocellaria scruposa**

   Scutum present **Scrupocellaria scrupea**

Scrupocellaria reptans (Linnaeus) Fig. 11.7
Colony a spreading straggling tuft, branches frequently anchored by rhizoids; mostly with 5–7 zooids between bifurcations. Zooids narrow proximally, broader distally, membranous area overarched by a branched scutum; two to three spines on outer angle, one on inner. A rather small lateral avicularium present, and often a larger frontal avicularium situated just proximal to membranous area. Ovicells globular, with small punctures. Embryos red.

On *Flustra*, algae, and submerged structures, and below boulders; LWST and shallow water; all coasts; common. Especially abundant on *Laminaria* fronds in sheltered channels.

Scrupocellaria scrupea Busk Fig. 11.7
Colony an erect tuft about 20 mm high; up to 20 zooids between bifurcations. Zooids narrow proximally, broader distally, membranous area overarched by an oval or reniform scutum; three spines on outer angle, one or two on inner. A large lateral avicularium present, and sometimes a small frontal avicularium situated just proximal to membranous area. Ovicells globular, imperforate.

On algae and stones; offshore; southern Britain; frequent.

Scrupocellaria scruposa (Linnaeus) Fig. 11.7
Colony an erect tuft; 7–11 zooids between bifurcations. Zooids slightly narrower proximally, no scutum over membranous area; two or three spines on outer angle, two on inner. A large lateral avicularium present, small frontal avicularia found on top of ovicells. Ovicells subglobular, imperforate. Embryos orange-red.

On stones, shells, *Flustra*, and submerged structures; LWST and shallow water; all coasts; common, though rarely as prolific as *S. reptans*.

13 BUGULIDAE

Colonies erect and tufted; developing from an upright ancestrula; attached to substratum with rhizoids. Branches unilamellar; dividing dichotomously; without joints. Zooids in two or more series, alternating. Basal and lateral walls lightly calcified, frontal surface mainly or wholly membranous. Orifice closed by a sphincter muscle; operculum absent.

A stalked avicularium, shaped like a bird's head, frequently present. Ovicells present, opening above orifice, not recumbent on distal zooid.

Fifteen or more species recorded from Britain, but only the following are found commonly on the shore or in the shallow sublittoral.

---

1. Branches made up from zooids attached to each other only by their proximal parts. Free distal part occupied mainly by an oval membranous area, and bearing several very long spines. Colony white **Bicellariella ciliata**

   Zooids wholly attached to one another by their inner sides. Colony purplish, buff, brown, or yellow 2

---

2. Avicularia absent. Ovicell attached to inner distal angle of zooid, orientated transversely. Colony a deep purplish-brown **Bugula neritina**

   Avicularia present. Ovicell attached across distal end of zooid, facing

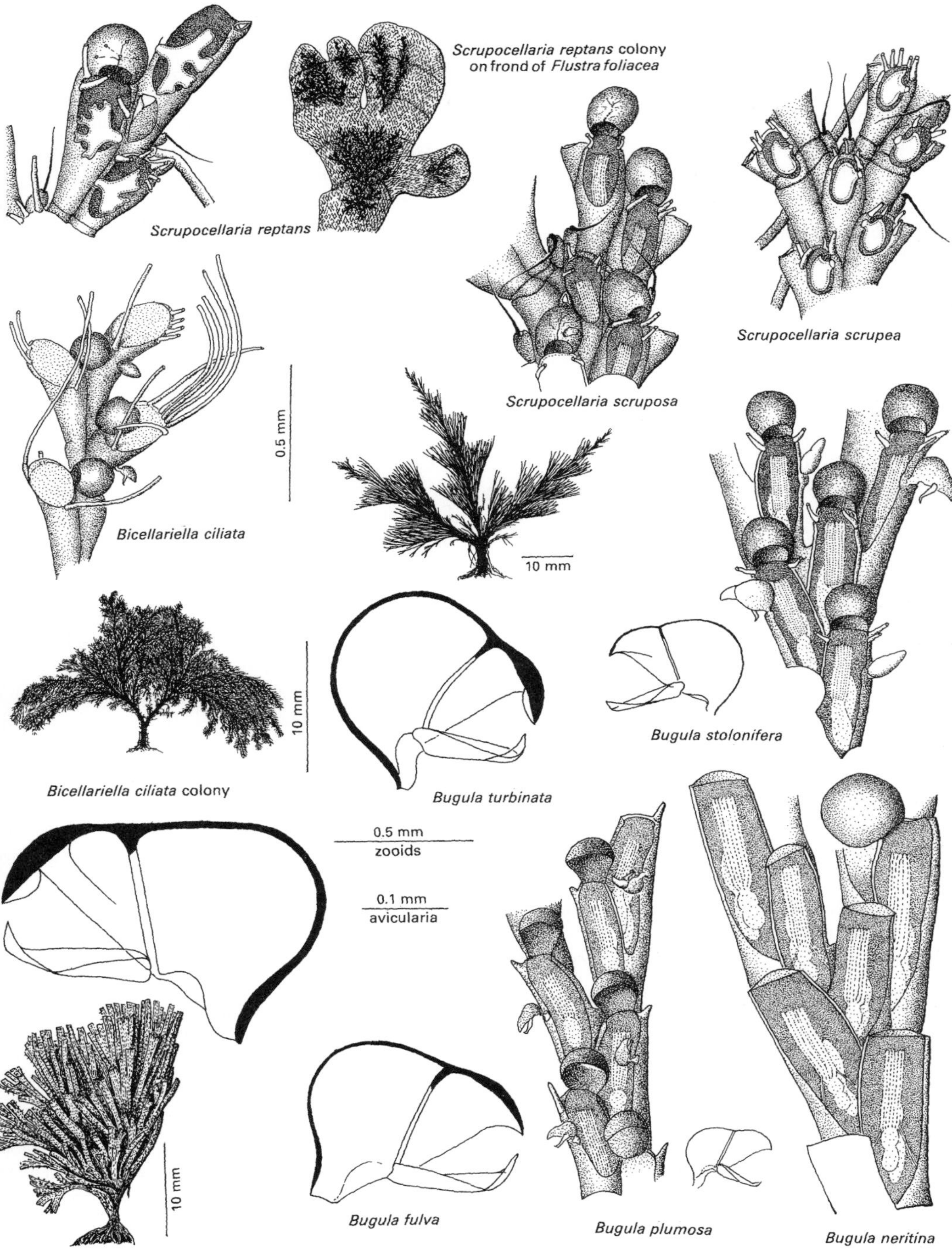
Scrupocellaria reptans colony on frond of Flustra foliacea
Scrupocellaria reptans
Scrupocellaria scrupea
Scrupocellaria scruposa
0.5 mm
Bicellariella ciliata
10 mm
10 mm
Bicellariella ciliata colony
Bugula stolonifera
Bugula turbinata
0.5 mm
zooids
0.1 mm
avicularia
10 mm
Bugula fulva
Bugula plumosa
Bugula neritina
Bugula flabellata

Fig. 11.7

proximally. Colony yellowish to purplish-buff in colour **3**

---

3. Branches comprising two series of zooids (occasionally four series at tips of branches) **4**

   Branches comprising several series of zooids. Ovicells nearly spherical. At least two spines present on each distal angle of all zooids ***Bugula flabellata***

---

4. Inner distal angle of zooid unarmed, outer angle with one projection or spine. Branches of colony disposed in spirals. Avicularia very small, their length less than the width of a zooid ***Bugula plumosa***

   Both distal angles bearing at least one spine. A majority of bifurcations usually involving one or more completely enclosed zooids **5**

---

5. Colony made up of branches clearly disposed in spirals. Outer distal angle bearing one spine only. Beak of avicularium abruptly hooked ***Bugula turbinata***

   Colony tufted, branches not in spirals **6**

---

6. Outer distal angle of zooid bearing two spines, inner angle bearing one ***Bugula stolonifera***

   Both distal angles bearing 2–3 spines ***Bugula fulva***

Bicellariella ciliata (Linnaeus) Fig. 11.7
Colony a white, feathery tuft, 10–20 cm high. Zooids narrow below, broad distally; membranous area occupying about one-third of frontal surface, with 4–7 long, slender, curving spines on outer margin. Avicularia small, attached below membranous area. Ovicells subglobular, helmet-shaped, contracting to a stem-like base. Embryos white.

On algae, hydroids, and bryozoans; LWST and shallow water; all coasts; frequent.

Bugula flabellata Thompson in Gray Fig. 11.7
Colony a dense tuft, 20–50 mm high, buff when living, greyish when dried or preserved. Branches broad, wedge-shaped; zooids about 0.80 × 0.17 mm, arranged in 3–8 series. Membranous area occupying most of frontal surface, three spines on marginal distal angle, two on all inner angles. Large lateral avicularia on marginal zooids, their length exceeding width of bearing zooid; small avicularia present on inner zooids; the beak hooked at a right angle. Ovicell globular; embryos yellow.

In various habitats, but almost invariably attached to other bryozoans; LWST and shallow water; all coasts, but not found north of the British Isles; common.

Bugula fulva Ryland Fig. 11.7
Colony dense, bushy, 20–30 mm high, tawny when living, greyish when dry. Zooids about 0.60 × 0.16 mm, arranged biserially. Membranous area occupying all of frontal surface, three spines on outer distal angle, two on inner, sometimes a sixth spine between the two groups. Avicularia lateral, somewhat longer than width of bearing zooid, with downcurved beak. Ovicells globular; embryos yellow.

In *Laminaria* holdfasts, below boulders and in marine fouling communities; LWST and shallow water; coasts north to Northumberland; locally common.

Bugula neritina (Linnaeus) Fig. 11.7
Colony a luxuriant, purplish-brown tuft, up to 80 mm high. Zooids large, about 0.75 × 0.25 mm, biserially arranged. Frontal area occupying whole of frontal surface; no spines or avicularia. ovicell globular, attached to inner distal angle of zooid, its opening concealed; embryos pale brown.

On pier piles and in marine fouling communities; introduced on shipping to a few ports in southern Britain; locally common.

Bugula plumosa (Pallas) Fig. 11.7
Colony buff, feathery, up to 80 mm high. Zooids about 0.50 × 0.15 mm, biserially arranged. Membranous area occupying three-quarters or more of frontal surface; a single unjointed spine on outer distal angle; inner angle unarmed. Avicularia small, lateral, shorter than width of bearing zooid; beak slightly downcurved. Ovicells appearing globular when containing embryos, deciduous; embryos yellow.

On stones, shells, and submerged structures; LWST and shallow water; widespread and common along coasts of southern Britain.

Bugula stolonifera Ryland Fig. 11.7
Colony dense, 30–40 mm high, greyish-buff. Zooids about 0.72 × 0.17 mm, arranged biserially. Membranous area occupying one-half to three-quarters of frontal surface; two spines on outer distal angle, one on inner. Avicularia lateral, length equal to width of bearing zooid; beak downcurved. Ovicells subglobular; embryos yellow.

Mainly a species of ports and harbours, occurring on pilings and other submerged structures; locally common in southern Britain.

Bugula turbinata Alder Fig. 11.7
Colony buff–orange, bushy, 30-60 mm high. Zooids about 0.55 × 0.18 mm, mainly arranged in two series. Membranous area occupying almost whole of frontal surface; each distal angle with one spine. Large lateral avicularia on marginal zooids, small avicularia on inner zooids; head rounded and broad, beak hooked at a right angle. Ovicells globular; embryos yellow.

Below boulders and in gullies from MLWN into shallow water; southern part of British Isles; common.

14 UMBONULIDAE

Colonies encrusting. Zooids with a calcified frontal wall overarching the frontal membrane; a row of pseudopores around margin. Buttress-like ridges overgrow frontal wall leaving spaces (*areolae*) above the pseudopores. Orifice large, without condyles.

Umbonula littoralis Hastings Fig. 11.8
Colonies orange. Zooids large, about 1.0 × 0.7 mm; ovate. Frontal wall convex, rising to a central prominence from which radiate about 18 ridges. Orifice large, cordate. A small avicularium usually located on distal face of the central prominence. Ovicells absent; red embryos brooded within parent zooids.

Below rocks and on *Laminaria* holdfasts; lower shore; all coasts; usually common.

A second species, *U. ovicellata* Hastings, occurs subtidally off south-western coasts. It is distinguished by the presence of ovicells.

15 EXOCHELLIDAE

Colonies encrusting. Zooids with a calcified frontal wall overarching the frontal membrane; a row of pseudopores around the margin. Orifice without condyles. Spines distal to the orifice and avicularia both present.

Escharoides coccinea (Abildgaard) Fig. 11.8
Colonies reddish-orange. Zooids ovate; frontal wall convex, rough or granular, produced below orifice into a median point (*mucro*) and two smaller, lateral denticles. Orifice subcircular. Six distal spines in non-ovicellate zooids. Large, pointed, outwardly directed avicularia on each side of orifice. Ovicells globular, granular, embryos red.

Below rocks and on *Laminaria* holdfasts; lower shore and in shallow water; all coasts; common.

Escharoides mamillata (Wood) occurs offshore, on hard substrata, around west and south-west Britain.

16 HIPPOTHOIDAE

Colonies encrusting, forming a loose network or a crust. Frontal wall imperforate, calcification generally showing transverse growth lines. Ovicells present. Six species recorded from Britain, but only one is common.

Celleporella hyalina (Linnaeus) Fig. 11.8
Colonies encrusting, forming small circular, glistening patches. Zooids contiguous; frontal wall smooth, silvery or white; sometimes with a prominence just below orifice. Orifice almost circular, with a shallow proximal sinus. No avicularia. Ovicells large, globular, on special dwarf zooids, punctate. Embryos pale yellow.

Most characteristically found on *Laminaria* and other algae, often in great abundance; LWST and in shallow water; all coasts, common.

17 CHORIZOPORIDAE

Colony encrusting. Frontal wall of zooids thin, imperforate; orifice D-shaped, without spines. Ovicells imperforate, closed by the zooidal operculum. Avicularia present. *Pore-chambers* tubular, separating the zooids. One British species.

Chorizopora brongniartii (Audouin) Fig. 11.8
Colony white. Zooids ovate-elongate, distinct, separated from each other by a line of pores; frontal wall smooth and silvery or transversely furrowed, imperforate. Orifice with a small pointed avicularium placed distally to it. Ovicells elongate–globular, imperforate, sometimes keeled and surmounted by an avicularium, opening closed by zooidal operculum. Embryos pinkish-red.

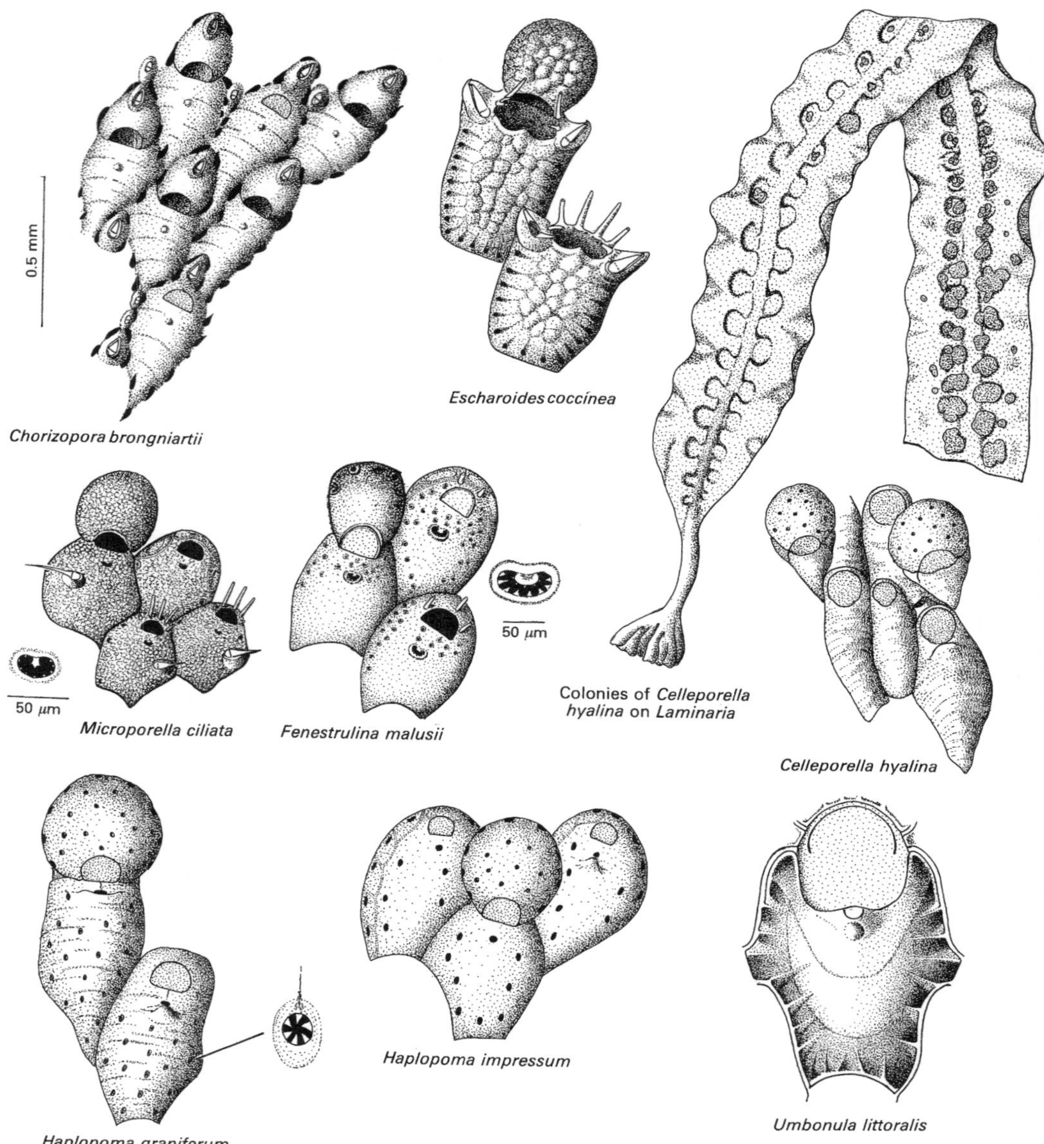

Fig. 11.8

Found particularly on shell; offshore, down to at least 100 m; all coasts; common.

18 MICROPORELLIDAE

Colonies encrusting. Zooid frontal wall punctate. Orifice D-shaped, the *ascopore* proximal to it; with or without distal spines. Avicularia often present. Ovicells present. Nine species recorded from Britain.

---

1. Orifice without associated spines; ascopore oval or round **2**

Orifice with spines on the distal margin (except in presence of ovicell); ascopore crescentic **3**

---

2. Frontal surface bearing very few scattered, simple pores. Usually on red algae ***Haplopoma impressum***

   Frontal surface bearing numerous pores, each one containing small, inwards-projecting teeth. Usually on stones ***Haplopoma graniferum***

---

3. Avicularia (often setiform) present. Developing embryos red ***Microporella ciliata***

   Avicularia absent. Developing embryos yellow ***Fenestrulina malusii***

Fenestrulina malusii (Audouin) Fig. 11.8
Colonies forming white patches. Zooids hexagonal. Frontal wall convex, smooth, punctate. Orifice with 3–4 distal spines; a lunate ascopore near centre of zooid. Ovicells subglobular, with a distal series of alternating ribs and spaces. Embryos deep yellow.

Mainly on stones and shells; offshore; all coasts; frequent.

Haplopoma graniferum (Johnston) Fig. 11.8
Colony white. Zooids elongate–hexagonal; frontal wall convex, rather thick, with or without a prominence (*umbo*) proximal to orifice, punctured by numerous stellate pseudopores. Ovicells more or less globular, with or without keel, punctate. Embryos pink.

On stones and shells; between tidemarks or in shallow water; all coasts; quite common.

Haplopoma impressum (Audouin) Fig. 11.8
Zooids ovate-hexagonal; frontal wall vitreous, translucent in life, with or without a prominence (*umbo*) proximal to orifice, with a few scattered pseudopores. Ovicells subglobular, with or without a keel, sparsely punctate. Embryos pink.

On algae, particularly Rhodophyta; LWST and shallow water; frequent in south and west Britain but unrecorded from the north-east.

Microporella ciliata (Pallas) Fig. 11.8
Colonies forming small white patches. Zooids hexagonal. Frontal wall silvery, frosted, punctate. Orifice with 5–7 spines distally; a lunate ascopore close to its proximal margin. An avicularium, with acute or setiform mandible, frequently located to one side of orifice. Ovicell globular, punctate or granular. Embryos red.

On stones, shells, and algae, sometimes abundant on *Laminaria* fronds; LWST and offshore; all coasts; common.

19 ESCHARELLIDAE

Colonies encrusting. Zooid frontal wall imperforate except for marginal pseudopores; generally raised into a point (*mucro*) at the orifice. Orifice suborbicular, with spines distally and an internal tooth proximally. Avicularia absent. Ovicells present. Ten species recorded from Britain, but only three are common in coastal waters.

---

1. Zooids small (about 0.5 mm in length), sutures between usually obscured by additional calcification. Orifice with six distal spines (four in presence of ovicell). Developing embryos pink ***Escharella immersa***

   Zooids larger (0.6–0.8 mm in length), sutures remaining distinct. Orifice bearing four spines or fewer. Embryos not pink **2**

---

2. Orifice with four distal spines (sometimes two in presence of ovicell). A tall truncate mucro generally present. Embryos white ***Escharella ventricosa***

   Orifice with three spines or fewer. Mucro small, often bifid. Embryos orange ***Escharella variolosa***

Escharella immersa (Fleming) Fig. 11.9
Zooids hexagonal, varying in appearance with age. Frontal wall smooth and shining in young zooids; later covered in heavy calcification except for drop-shaped areolae marking site of marginal pseudopores; raised into a small pointed mucro at the orifice.

Orifice longer than broad, with six spines distally (four in fertile zooids), and a small concave or bifid internal tooth. Ovicell globular; embryos pink.

On stones and shells; LWST and offshore; all coasts; common.

Escharella variolosa (Johnston) Fig. 11.9
Zooids elongate-hexagonal, flat. Frontal wall granular, somewhat covered by additional calcification in older zooids; raised as a pointed or bifid mucro at the orifice. Orifice suborbicular, with 2–5 spines distally (two in fertile zooids), and a small square internal tooth proximally. Ovicells globular, with marginal pseudopores; embryos orange.

On stones and shells; offshore; all coasts; frequent.

Escharella ventricosa (Hassall) Fig. 11.9
Zooids hexagonal. Frontal wall granular, convex, raised into a tall broad mucro. Orifice suborbicular, with four spines distally, and a forked internal tooth proximally. Ovicell globular; embryos white.

On stones and shells; offshore; all coasts; frequent.

20 SCHIZOPORELLIDAE

Colonies encrusting. Frontal wall of zooids usually evenly punctate. Orifice suborbicular, with a clearly defined notch (*sinus*) in its proximal border, delimited from the main part of the orifice by hinge teeth (*condyles*) for the operculum. Avicularia usually present. Ovicells opening above the orifice, often closed by the operculum. More than 20 species of this large family occur around Britain. Most are found offshore, on shell or rock, but a few may be common on the lower shore.

---

1. Many zooids at least bearing a mamilla, below the orifice, to which is articulated a long, pointed spine
Phaeostachys spinifera

   Frontal wall flat or umbonate, sometimes produced as a finger-like projection upon which rests an avicularium, but never bearing a long, articulated spine **2**

---

2. Each zooid bearing one or two avicularia beside orifice, each directed obliquely outwards. Ovicell imperforate
Schizoporella unicornis

   Avicularia suboral or, if present in pairs, facing inwards. Ovicell punctured
Schizomavella linearis

Phaeostachys spinifera (Johnston) Fig. 11.9
Colony silvery or greyish. Zooids ovate to rhomboid, convex; frontal wall lightly punctured. Orifice deeply D-shaped, with a well-developed sinus in the straight proximal margin. Five to seven distal spines, and a single, articulated spine proximal to orifice. Frequently an avicularium, with slender, pointed mandible directed outwards, situated beside the spine. Ovicell prominent, subglobular, imperforate, with fluted margin; embryos pinkish-red.

On *Laminaria* holdfasts, algal stems, and stones; lower shore; all coasts; fairly common.

Schizomavella linearis (Hassall) Fig. 11.9
Colony glistening, reddish. Zooids rhomboid, short; arranged in regular radiating series. Frontal surface flat, with pseudopores all over. Orifice suborbicular, sinus wedge-shaped; two to four spines distal to orifice, except in fertile zooids. Either a single avicularium decumbent on a central prominence, with an upwardly directed, pointed mandible, or a pair of small avicularia, one each side of the orifice, pointed inwards. Ovicell subglobular, resting on distal zooid, punctate, closed by the operculum. Embryos orange-red.

On stones and shells; LWST and offshore; all coasts; common.

Schizoporella unicornis
(Johnston in Wood) Fig. 11.9
Colony whitish or pink. Zooids generally rectangular, arranged in regular series. Frontal wall slightly convex, with pseudopores all over; often a knob (*umbo*) proximal to orifice. Orifice D-shaped, with a well-marked median sinus in the straight proximal edge. A raised avicularium on one or both sides of orifice, tapering mandible directed obliquely outwards. Ovicell globular, prominent, resting on distal zooid; imperforate, with a fluted margin, closed by zooidal operculum. Embryos orange-red.

Usually on stones and under boulders, but on algae locally; lower shore; all coasts; common.

21 CELLEPORIDAE

Colonies encrusting, nodular or erect. Zooids usually erect, superposed in disorderly manner.

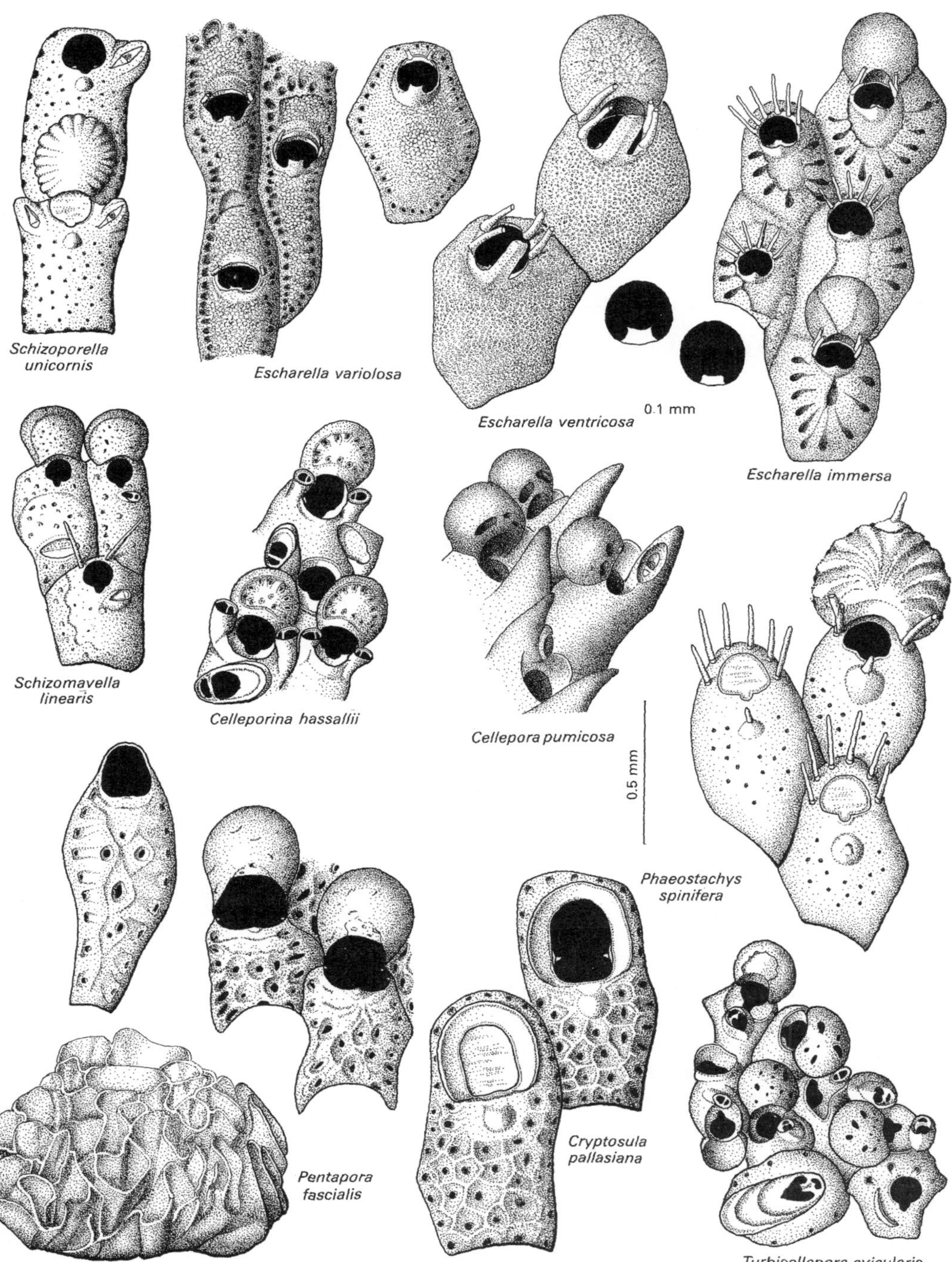
Schizoporella unicornis
Escharella variolosa
Escharella ventricosa
0.1 mm
Escharella immersa
Schizomavella linearis
Celleporina hassallii
Cellepora pumicosa
0.5 mm
Phaeostachys spinifera
Pentapora fascialis
Cryptosula pallasiana
Turbicellepora avicularis

Fig. 11.9

Orifice suborbicular, usually with a clearly defined notch (*sinus*) in its proximal edge; spines absent. Avicularia dimorphic, with large, scattered, interzooidal avicularia and small frontal avicularia mounted on one or two spiky prominences. Ovicells free-standing, punctate, opening above the orifice. Twelve species recorded from Britain.

---

1. Orifice circular, without a proximal sinus **_Cellepora pumicosa_**

   Orifice with distinct sinus **2**

---

2. A single, more or less suboral, avicularium on each zooid. Ovicells with a few large punctures **_Turbicellepora avicularis_**

   Two avicularia, one each side of the orifice. Ovicell with a crescent of fine pores **_Celleporina hassallii_**

Cellepora pumicosa (Pallas) Fig. 11.9
Zooids subcylindrical or ovoid, erect (except near colony margin). Frontal wall convex, smooth, with distinct marginal pores; a slender, acuminate process below the orifice. Orifice circular. Avicularia dimorphic; a small avicularium with pointed mandible on, and directed up, each suboral spike; large avicularia, with rounded spatulate mandibles, scattered among autozooids. Ovicells subglobular, smooth, imperforate, or with a few large pores. Embryos orange-red.

Below rocks and on *Laminaria* holdfasts; lower shore and in shallow water; all coasts; quite common.

Celleporina hassallii (Johnston) Fig. 11.9
Colonies nodular or forming thick multilaminar crusts. Zooids erect, except at colony growing edge. Frontal surface convex, smooth. Orifice suborbicular, with rounded sinus; flanked by a pair of columnar processes, each bearing a small terminal avicularium with a rounded mandible. Spatulate avicularia scattered among the autozooids. Ovicells subglobular, with a porous crescent on upper surface. Embryos red.

On stones, *Laminaria* holdfasts, larger hydroids, etc.; LWST and shallow water; all coasts; common.

Turbicellepora avicularis (Hincks) Fig. 11.9
Colonies nodular or forming thick multilaminar crusts. Zooids erect, except at colony growing edge. Frontal surface convex, smooth. Orifice deeply D-shaped, with a wedge-shaped sinus. Proximal to orifice, and slightly to one side of sinus, a conical prominence bearing an avicularium with a pointed, upwards-directed mandible. Spatulate avicularia scattered among autozooids. Ovicells subglobular, with a few scattered punctures. Embryos deep yellow.

On stones, hydroids, shell, and kelp holdfasts; LWST and shallow water; all coasts.

Turbicellepora magnicostata (Barroso), found intertidally around the Isles of Scilly, has fewer, larger pores in the ovicell, a wider sinus to the orifice, and distinct marginal areolar ribs.

22 HIPPOPORINIDAE

Colonies encrusting or foliaceous. Frontal wall of zooid with pseudopores all over or confined to a marginal series. Orifice usually longer than wide, arched distally, straighter proximally, operculum pivoting on pointed, lateral condyles. Avicularia usually present. Ovicells present or absent. Two British species.

---

1. Colony encrusting. Ovicells never present **_Cryptosula pallasiana_**

   Colony erect and foliaceous. Ovicells present **_Pentapora fascialis_**

Cryptosula pallasiana (Moll) Fig. 11.9
Colonies encrusting, forming white or pinkish patches. Zooids hexagonal, with a thick frontal wall, porous over the whole area, each pseudopore sunk in a small pit. Orifice large, longer than broad, curved distally, narrowed a little below the middle; no spines, but usually a raised rim. Sometimes a small avicularium, with rounded mandible, situated just proximal to orifice. No ovicells; orange embryos develop inside the zooid.

Usually on stones, less often on algae and inert submerged structures; LWST and shallow water; all coasts; common.

Pentapora fascialis (Pallas) Fig. 11.9
Colonies erect, of bilamellar, strongly calcified, foliaceous, anastomosing fronds, forming orange-buff

clumps generally up to about 150 mm across, but occasionally much larger. Young zooids elongate-hexagonal, with scattered pores in frontal wall; later covered with heavy additional calcification, outline becoming rectangular or broadly hexagonal. Orifice cordate. Sometimes a small avicularium, with rounded mandible, just proximal to orifice. Ovicells at first globular, resting on distal zooid, later immersed in additional calcification.

On shells and stones; offshore; south-west Britain; locally quite common.

Order Ctenostomatida

23 ALCYONIDIIDAE

Colonies adnate, forming a network or a crust, or erect and lobate; fleshy or gelatinous. Zooids squat, contiguous or in series. Orifice circular, closed by a sphincter, often raised on a papilla. Embryos developing in clusters in *gonozooids*. A taxonomically difficult family with many species recorded from British waters, most of which are difficult to identify. Three species are readily recognized.

---

1. Colony in the form of free lobes. Surface not tuberculate **_Alcyonidium diaphanum_**

   Colonies encrusting; rarely lobate, the surface then tuberculate **2**

---

2. Surface smooth **_Alcyonidium gelatinosum_**

   Surface tuberculate **_Alcyonidium hirsutum_**

Alcyonidium diaphanum Lamouroux — Fig. 11.10

Colonies lobate, digitiform, usually smooth, occasionally knobbly; firmly gelatinous; pale brown; commonly about 150 mm long, but attaining 500 mm. Orifices slightly raised; 15–17 tentacles.

Attached to stones and boulders, or detached; LWST and shallow water, may be washed up; all coasts; generally common, occasionally present sublittorally in great quantity. Apparently responsible for allergic dermatitis 'Dogger Bank itch' in the North Sea.

Alcyonidium gelatinosum (Linnaeus) — Fig. 11.10

Colonies encrusting, forming glossy gelatinous patches on algae; rarely lobate; pale brown; surface smooth. Interzooidal boundaries not visible. Orifices scarcely raised; 19–20 tentacles. Gonozooids containing small clusters of embryos, appearing yellowish-fawn, during autumn and winter.

On algae, especially *Fucus serratus*; middle and lower shore; all coasts; common unless very exposed.

Alcyonidium hirsutum (Fleming) — Fig. 11.10

Colonies encrusting or lobate, firmly gelatinous; lobes reaching about 100 mm. Surface matt, covered with minute, conical papillae; greyish-brown. Orifices not raised; 16–20 tentacles. Gonozooids with conspicuous rings of white embryos from autumn to spring.

On algae, particularly *Fucus serratus* and *Gigartina stellata*; lower shore; all coasts; common.

24 FLUSTRELLIDRIDAE

Colonies encrusting or erect; gristly. Zooids squat; contiguous. Orifice closed by chitinous lips.

Flustrellidra hispida (Fabricius) — Fig. 11.10

Colony purplish-brown, hispid. Zooids large, with horny spines around orifice, and sometimes around zooid margin. Orifice bilabiate; polypides large, expanded tentacle bells about 1 mm across. Embryos white, developing in summer.

On algae, especially *Fucus serratus* and *Gigartina stellata*; middle and lower shore; all coasts; common.

25 NOLELLIDAE

Colonies erect or creeping. Zooids upright cylinders, not contiguous; branching from their bases, with or without slender stolonic extensions. Zooid walls particle encrusted. This family includes a number of small, cryptic species. Two are more conspicuous and quite common.

---

1. Colony erect, branching, forming a dense, grey-brown muddy tuft up to 100 mm high. Zooids cylindrical, budded directly from each other **_Anguinella palmata_**

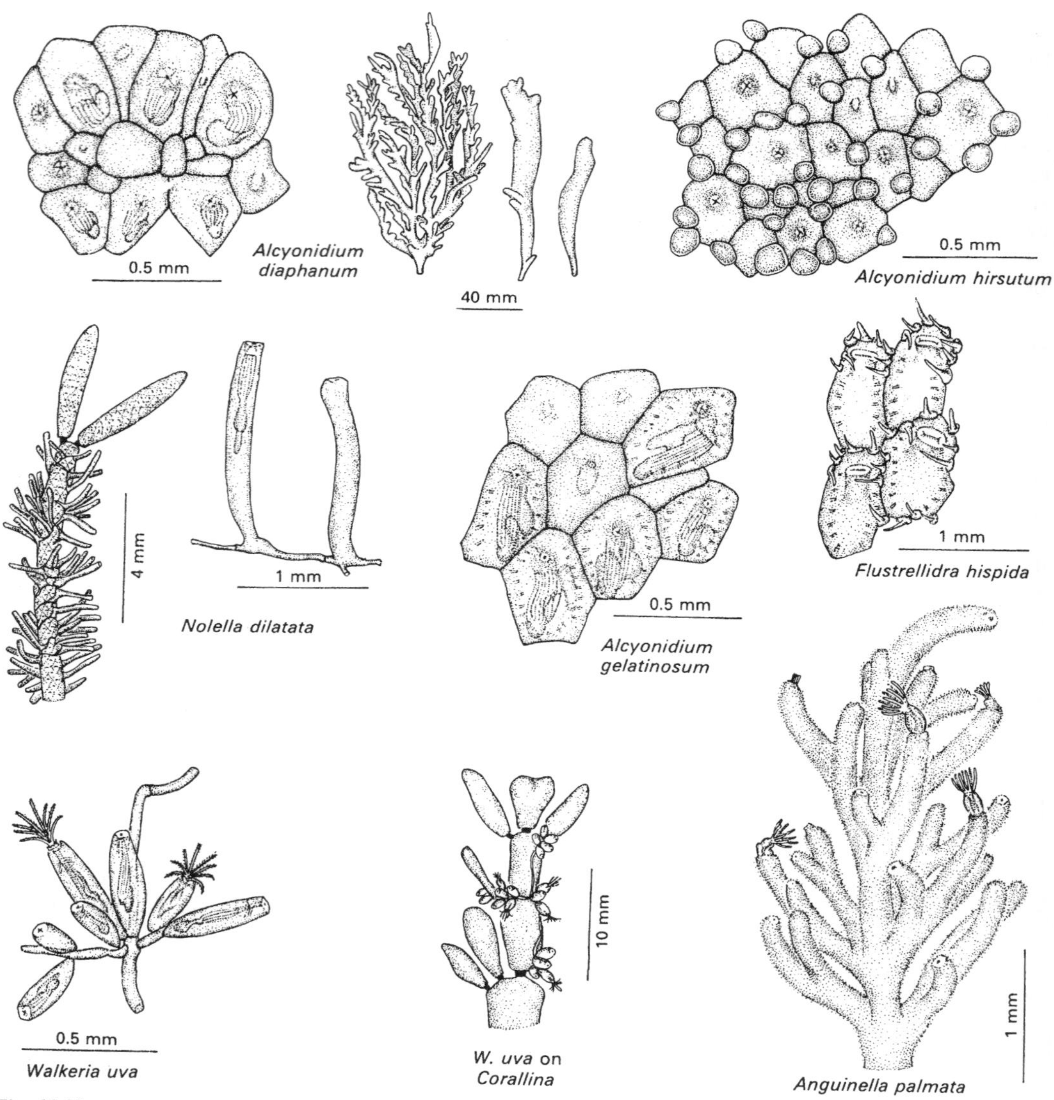

Fig. 11.10

Colony repent, forming a cluster or sward of erect, cylindrical, sandy coloured zooids about 2 mm in height attached to substratum and spreading by stolonic extensions
Nolella dilatata

Anguinella palmata van Beneden Fig. 11.10
Colonies in form of erect stems with spirally arranged branches from which curving, cylindrical zooids are budded. Zooids to 1.5 mm, with 10–11 tentacles; colony surface muddy with accreted particles. Unmistakable but easily overlooked on account of its muddy appearance.

On stones, shells, pier piles, and other solid structures; intertidal in shelter or sublittoral, tolerating fluctuating salinities; southern Britain; probably quite common but biologically almost unknown.

Nolella dilatata (Hincks) Fig. 11.10
Colonies consisting of a network of zooids connected by stolonic extensions from their bases. Zooids cylindrical, up to 1.5 mm in height; walls opaque, appearing sanded with minute particles. 16–20 tentacles.

On other bryozoans, hydroids, ascidians, shells, etc.; LWST and shallow water; all coasts; frequent.

26 WALKERIIDAE

Colonies stolonate, creeping or tufted. Stolons thin; autozooids not contiguous, grouped at distal end of stolonic segments (*kenozooids*). Autozooids cylindrical, the orifice terminal; eight tentacles, when everted not forming a perfect circle, two of them outcurved; no gizzard.

Walkeria uva (Linnaeus) Fig. 11.10
Colonies creeping or tufted. Stolons very slender, branches typically arising in pairs at the nodes, in a small cluster of zooids. Zooids small, less than 0.5 mm in height. Embryos unpigmented.

Mainly on algae (creeping colonies are often found on *Corallina officinalis*; loose tufts may develop on *Halidrys siliquosa*); LWST and shallow water; all coasts; frequent.

27 VESICULARIIDAE

Colonies stolonate, creeping or tufted; stolons bifurcating. Zooids cylindrical, contiguous or not, usually grouped; arising from a budding zone of variable length in each stolonic segment (*kenozooid*). Zooids with terminal orifice; 8–10 tentacles, which form a circle when everted; gizzard present. This large family includes many British species, some of which have only recently been recognized and named. They are taxonomically difficult, but the following six species are all common on the lower shore and easily identified.

---

1. Zooids fused together along their length, forming small groups **_Amathia lendigera_**

   Zooids remote or clustered, but not fused together **2**

---

2. Colony tufted, up to 300 mm in length, attached basally by rhizoids. Zooids in single series, generally more or less remote from each other **_Vesicularia spinosa_**

   Colony tufted or repent. Zooids in double series, generally clustered **3**

---

3. Zooid clusters disposed around stolon in distinct helices. Colony a free-hanging, basally attached tuft **4**

   Zooid clusters not helical. Colony repent or forming dense clumps **5**

---

4. Living polypide pigmented bright lemon yellow **_Bowerbankia citrina_**

   Living polypides buff-coloured **_Bowerbankia pustulosa_**

---

5. Polypide with eight tentacles. Developing embryos pink. Stolon considerably narrower than zooid. Zooids solitary, in pairs, or in small clusters; each often provided with a small proximal 'tail'. On various substrata **_Bowerbankia gracilis_**

   Polypide with 10 tentacles. Developing embryos yellow. Stolon, at least when free from substratum, about as wide as a zooid. Zooids often in dense clusters. Particularly on intertidal fucoids **_Bowerbankia imbricata_**

Amathia lendigera (Linnaeus) Fig. 11.11
Colonies forming wiry tufts; stolons about as wide as zooids. Eight to 16 contiguous zooids in straight, biserial groups below bifurcations; the height in each group decreasing distally. Eight tentacles.

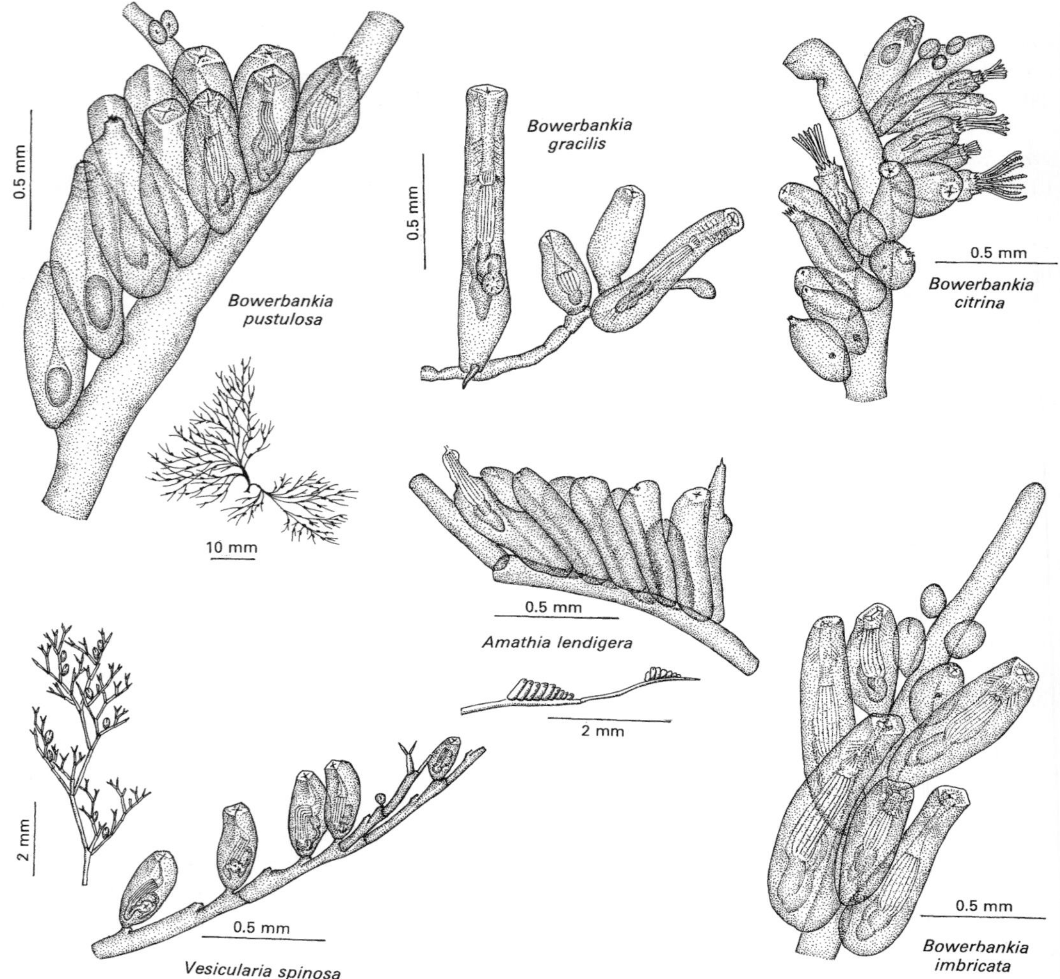

Fig. 11.11

On algae, especially *Halidrys siliquosa*, and other substrata; LWST and shallow water; reaches its northern limit in Scotland; frequent.

Bowerbankia citrina (Hincks) Fig. 11.11
Colonies forming flaccid tufts; stolons about as wide as zooids; zooids disjunct, in biserial, spirally disposed clusters below bifurcations. Polypides (in life) bright lemon-yellow; eight tentacles.

On rocks and boulders; LWST and shallow water; southern coasts; local.

Bowerbankia gracilis Leidy Fig. 11.11
Colonies adnate; stolons less wide than zooids; zooids disjunct, solitary, in pairs or in small clusters, 0.5–1.5 mm in height, often with a caudate process beside the point of origin from the stolon. The colony sometimes very dense, like a pile or sward. Eight tentacles; embryos pink.

On any submerged substratum, in fully saline or brackish water (down to 15‰); LWST and offshore; all coasts; common, but rarely conspicuous.

Bowerbankia imbricata (Adams) Fig. 11.11

Colonies adnate or tufted; stolons slender when adnate, but wider than zooids when free; zooids disjunct, in straight, biserial clusters; 0.8–1.0 mm (occasionally to 1.5 mm) in height; 10 tentacles; embryos yellow.

On intertidal fucoids, especially *Ascophyllum nodosum* and *Fucus vesiculosus*; mainly intertidal; all coasts; common.

Bowerbankia pustulosa (Ellis and Solander) Fig. 11.11

Colonies forming flaccid tufts; stolons about as wide as zooids; zooids disjunct, in biserial, spirally disposed clusters below bifurcations. Polypides unpigmented; eight tentacles; embryos yellow.

On rocks and boulders; LWST and shallow water; all coasts; frequent.

Vesicularia spinosa (Linnaeus) Fig. 11.11

Colonies tufted; stolons much branched, tapering to spine-like points terminally; zooids ovoid to fusiform, disjunct, in single series, more or less remote from each other, irregularly placed on internodes; eight tentacles.

On stones and shells; in shallow waters; all coasts; locally common.

12 SEA URCHINS, STARFISH, AND SEA CUCUMBERS

Echinoderms (Phylum Echinodermata), which include sea urchins, starfish, brittle stars, and sea cucumbers, have a distinctive radial pattern which may take the form of five *arms* radiating from a central disc or a more globular or cylindrical shape, again with structures arranged in five rays (or multiples of five) (Fig. 12.1). A few exceptional types have more than five *radii* and in others a bilateral shape has been superimposed on the basic radial plan. Echinoderms are relatively large invertebrates and, with rare exceptions, they are benthic (bottom living) and very slow moving.

Typically echinoderms have an outer layer of calcareous *skeletal plates*. These may be joined together to form a complete shell or '*test*' as in the sea urchins; more usually they are separated by flexible integument as in the starfishes. The plates, or *ossicles*, have a lattice microstructure with a very efficient strength to weight ratio, and also form the rigid *spines*, and other structures described in appropriate places below.

Along the length of each of the five rays there are usually two rows of mobile structures called *tube feet*, each usually terminating in a circular *sucker-disc*, about 1 mm diameter, used for attaching to a surface. There are often hundreds of tube feet to one ray, and in the starfish they carry out stepping actions to move the animal about. The ray with its complement of tube feet is referred to as an *ambulacrum* or walking surface. Associated ossicles are referred to as *ambulacral* plates; the plates between the rays are referred to as *interambulacral*. In the urchins and sea cucumbers the *ambulacra* and *interambulacra* run all the way from the oral to the anal end of the animal. In the stars they are confined to the lower surface of the animal, the *adoral* surface; the upper surface, lacking ambulacra, is referred to as the *aboral* surface. There is a variety of other surface structures, the first of which are *papulae* or gills. These are most obvious in live starfish as numerous delicate protuberances over the aboral surface. Secondly, there are the unique echinoderm structures called *pedicellariae* found in starfish and urchins. These are minute jawed structures used for removing sediment and rubbish from the surface of the animal. Sometimes they are set on long mobile stalks and are three-jawed, others are sessile and scissor-like. Some species have four or five different types. One type has poison glands for dealing with small intruding crustaceans, etc.

Internally the echinoderms have another unique system, the hydraulic *water vascular system* which controls the action of the tube feet. It consists of a ring-main around the throat and a radiating vessel internal to each ambulacrum. There is also a replenishment vessel (*stone canal*) running to the ring canal from the surface of the animal where a special structure, the sieve plate or *madreporite*, serves to monitor and control the intake of sea water.

Echinoderms occupy a diversity of habitats. They are fully marine and barely penetrate estuaries. Intertidally they avoid desiccation by being confined to rock pools and under boulders. Some small species of brittle stars and asteroids only become evident when algal clumps from rock pools are sorted through. Collecting burrowing forms in intertidal sands requires careful digging. Diving, when possible, is a better method than dredging or trawling for sublittoral specimens since it is much less likely to result in damage.

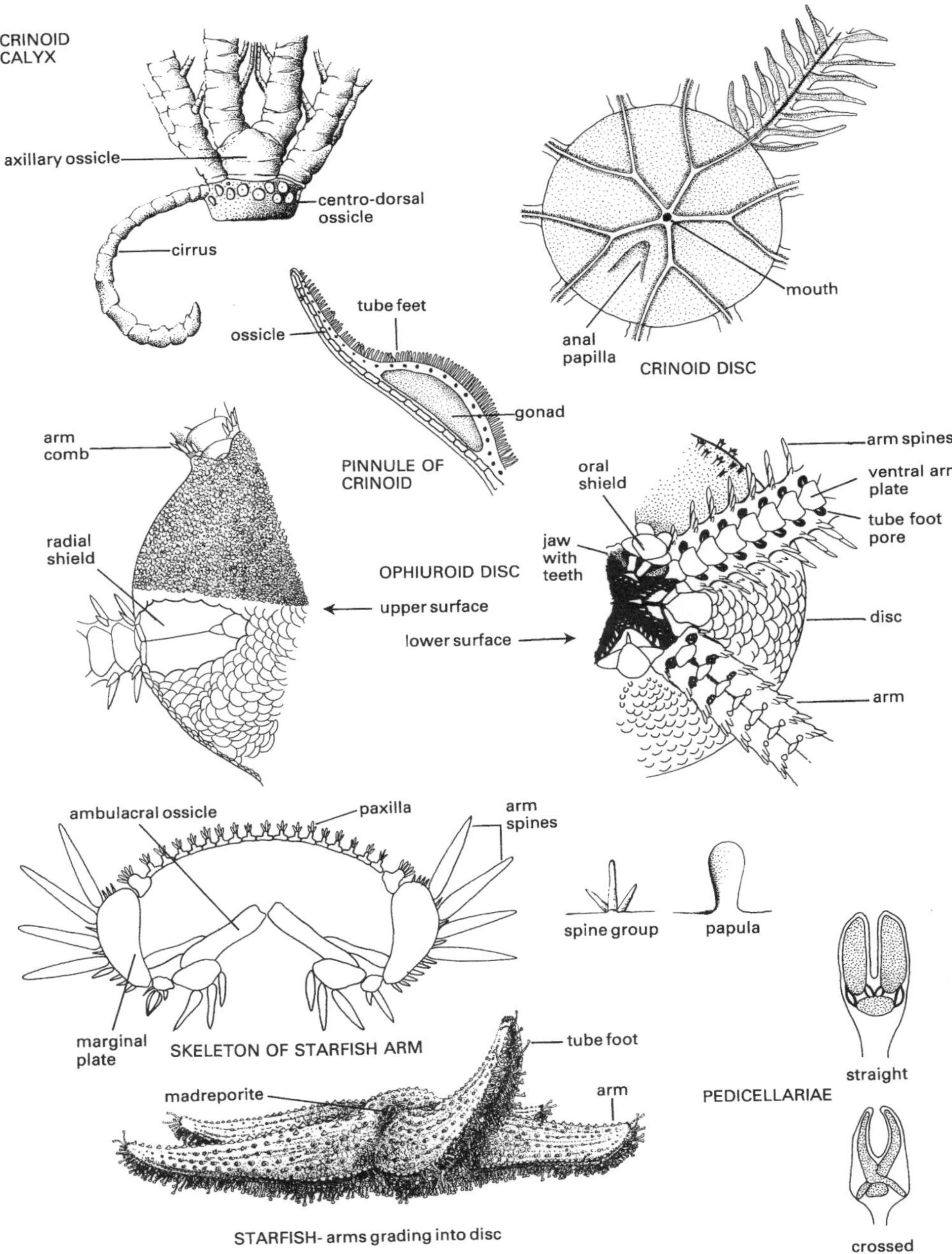
CRINOID
CALYX
axillary ossicle
centro-dorsal
ossicle
cirrus
mouth
anal
papilla
CRINOID DISC
tube feet
ossicle
gonad
PINNULE OF
CRINOID
arm
comb
radial
shield
OPHIUROID DISC
upper surface
lower surface
oral
shield
jaw
with
teeth
arm spines
ventral arm
plate
tube foot
pore
disc
arm
ambulacral ossicle
paxilla
arm
spines
marginal
plate
SKELETON OF STARFISH ARM
spine group
papula
straight
PEDICELLARIAE
crossed
tube foot
madreporite
arm
STARFISH- arms grading into disc

Fig. 12.1

Class Crinoidea

Echinoderms with a cup-shaped body and 10 feathery arms (Figs. 12.1, 12.2) with the mouth and *ambulacral grooves* on the upper surfaces. Some crinoids, the 'sea lilies' (not found on British coasts), live permanently attached by a stalk; feather stars attach temporarily by prehensile cirri. In the feather stars, a large *ossicle*, the *centro-dorsal* (corresponding to the top of the stalk) is surrounded by series of other large ossicles: five *basals* alternating with five *radials*; each radial is succeeded by *first* and *second brachials*. The second brachial is also known as the *axillary*, and is the point at which the ray divides into two, resulting in 10 arms. Each arm is supported throughout its length by a series of articulating *brachial ossicles*. Side branches, or *pinnules*, arise alternately along the arms and are supported by a row of progressively smaller ossicles. The few lowermost, or *oral*, pinnules are usually distinctive. All pinnules are furnished with a ciliated ambulacral food groove with suckerless tube feet. Gonads are present in the proximal pinnules.

The cirri arise from sockets on the centro-dorsal ossicle and are unbranched structures consisting of articulated ossicles. The terminal ossicle is usually claw-shaped.

Order Comatulida

Only one family, the Antedonidae, is represented in British waters. Five species occur, in four genera, but only one is frequent in shallow coastal waters.

1 ANTEDONIDAE

Antedon bifida (Pennant)

Feather star Fig. 12.2

Proximal (oral) pinnules long and modified; about 35 segments. Distal edges of arm ossicles (brachials) prominent. Gonad confined to proximal half of pinnule. Cirri short, up to 20 (rarely 30). Arm length up to 100 mm, usually less. Colour variable, dark reddish to orange and yellow, often mottled or banded.

On hard substrata amongst hydroids, bryozoans, etc. Shallow sublittoral down to 200 m. All coasts except east England, sometimes very abundant; elsewhere Shetland to Portugal.

Frequently parasitized by *Myzostoma*, a hermaphrodite polychaete.

Class Asteroidea

The starfishes. Echinoderms with a star-shaped body plan, the five (or more) arms grading into the central disc. The lower surface bears the central mouth and is referred to as the *oral*, *ambulacral*, or *actinal* surface, the upper being the *aboral* or *abactinal* (Fig. 12.1). The arms are also referred to as *rays* and their axes as *radii*, areas of the disc between the radii are referred to as *interradial*. Running from the mouth to the end of each arm is a deep open furrow, the *ambulacrum* or *ambulacral groove*, in which the two or four rows of *tube feet* are situated. The tube feet may or may not have a terminal disc-shaped attachment sucker; internally each is connected through a pore in the ossicles to a small water-filled *ampulla*, which may be single or double. Paired *ambulacral ossicles* bound the groove internally and are flanked by *adambulacral ossicles* or *plates* which form the sills of the furrow. From these may arise *furrow spines*, *pedicellariae*, and *ventro-lateral spines*. The angle between two ambulacra adjacent to the mouth is sometimes marked by a special ossicle, the *ventral mouth plate*. The edges of the arms may be sharply marked by large or distinctive ossicles known as *marginal plates*; other surfaces usually bear spines of various types which may be prominent and arranged in rows. *Paxillae* are spines tipped with a bunch of smaller elongated ossicles like a little brush. Pedicellariae are always only two-jawed but come in two basic types: *crossed*, in which the jaw ossicles have a scissor-action, and *straight* with a forceps action. *Papulae* are thin-walled outpushings of the body surface that act as gills.

KEY TO FAMILIES

1. Starfish with large marginal plates. Papulae mostly on upper surface; no crossed pedicellariae (Order Phanerozonia) **3**

   Starfish with inconspicuous marginals. Papulae on upper and lower surfaces **2**

2. No crossed pedicellariae, straight ones rare. Spines of upper surface in groups. Normally only two rows of tube feet (Order Spinulosa) **5**

   Crossed pedicellariae present. Spines of upper surface usually arranged singly rather than grouped. Tube feet in four rows (Order Forcipulata) **7. Asteriidae**

---

3. Tube feet without a terminal sucker **4**

   Tube feet with a terminal sucker. Small, indistinct marginal plates. Upper surface naked or with a few short spines **3. Poraniidae**

---

4. Well-developed upper marginal plates; simple papulae; one pair of gonads at base of each arm **1. Astropectinidae**

   Upper marginals replaced by paxillae; papulae bush-shaped; series of paired gonads down length of arms **2. Luidiidae**

---

5. Animal more or less flattened, with sharp edge between upper and lower surfaces **4. Asterinidae**

   No sharp edge between upper and lower surfaces **6**

---

6. Very small disc; five long cylindrical arms, small spines irregularly arranged singly or in small groups like paxillae **6. Echinasteridae**

   Large disc; more than five arms, long, well-formed paxillae **5. Solasteridae**

Order Phanerozonia

1 ASTROPECTINIDAE

Starfish with well-developed upper and lower marginal plates covered with fine spines. Upper surface covered with paxillae. Tube feet pointed and suckerless; provided with double ampullae. Five genera in British seas but only *Astropecten* occurs in shallow coastal waters.

Astropecten irregularis (Pennant) Fig. 12.2
Diameter commonly 100 mm but up to 200 mm, pale violet or yellowish. Five, rather short, stiff, tapering arms meeting at distinct angles. With a continuous fringe of spines, arising four or five to each lower marginal plate, second or third spine of each group being the longest.

Partly buried in sandy substrata; sublittoral, 10–1000 m. Common on all British coasts. Range Norway to Morocco.

2 LUIDIIDAE

Starfish with upper marginal plates represented by marginal paxillae; lower marginals well formed and spiny. Upper surface covered with paxillae. Tube feet lack terminal sucker disc. Only one genus in British sea area.

---

1. Five arms ***Luidia sarsi***

   Seven arms ***Luidia ciliaris***

Luidia sarsi Düben and Koren Fig. 12.2
Diameter up to 340 mm (usually smaller); reddish-brown, with darker bands. Five gently tapering medium length arms. Only lower row of marginal ossicles well defined, each bearing three or four fringing spines. Each two marginal paxillae followed by four lateral paxillae, then 15–20 median paxillae. Bivalved pedicellariae occur on underside.

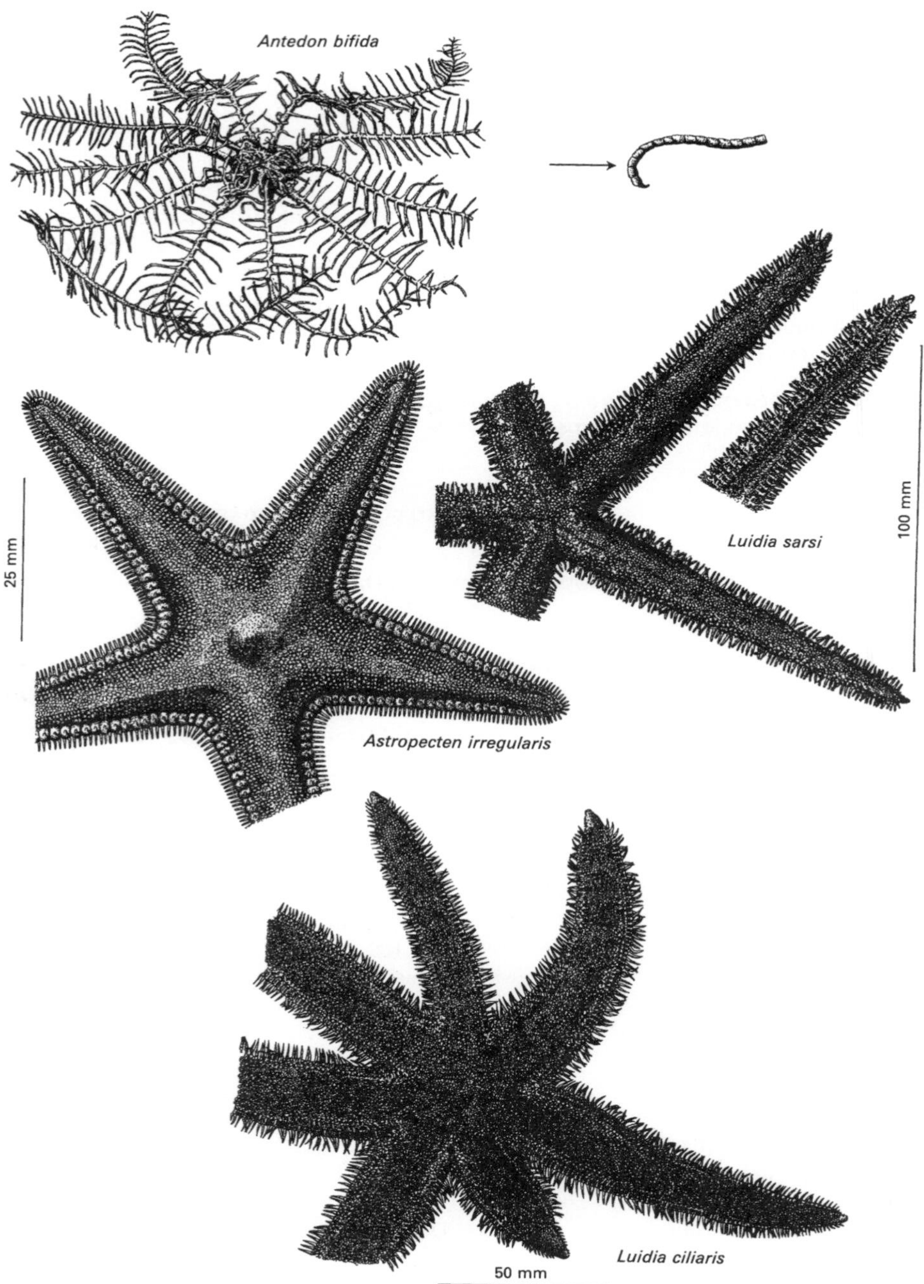
Antedon bifida
25 mm
100 mm
Luidia sarsi
Astropecten irregularis
Luidia ciliaris
50 mm

Fig. 12.2

Feeds on other echinoderms, in addition to polychaetes, molluscs and crustaceans. Sublittoral on muddy substrata 10–600 (1300) m. Common on all British coasts except south. Distributed from Trondheim to Cape Verde and Mediterranean.

Luidia ciliaris (Philippi) Fig. 12.2
Diameter up to 600 mm but usually smaller; brick red. A fine starfish with seven arms, only the distal parts tapering. Lower marginal ossicles each bear four or five spines. Paxillae of upper surface consist of two marginals to every three laterals. Three-valved pedicellariae sometimes occur on ventrolateral plates.

Feeds on other echinoderms, more exclusively than *L. sarsi*, including *Ophiothrix*, *Echinocardium*, and *Asterias*. Sandy substrata sublittorally, shallower than *L. sarsi*, 4–400 m. Common on all British coasts except southern North Sea. Distributed from Faeroes to Cape Verde and Mediterranean.

3 PORANIIDAE

With a distinctively near-naked upper surface; marginal plates somewhat reduced. No pedicellariae.

Porania pulvillus (O.F. Müller) Fig. 12.3
Up to about 110 mm diameter, colour red or yellowish-white. Large disc; five short, wide, tapering arms. Upper surface smooth, with only scattered groups of papulae; oral interradial areas crossed by radiating furrows on otherwise smooth surface. Upper marginal plates smooth and narrow; lower marginals with 3–5 spines forming a fringe.

Muddy substrata, sublittoral, 10–300 (–1000) m. All British coasts except southern North Sea. Distributed from Trondheim to Biscay.

Order Spinulosa

4 ASTERINIDAE

With very short arms giving an almost pentagonal shape to the flattened body.

---

1. Very flat and thin, with radial thickenings indicating ambulacra. 100–120 mm diameter **_Anseropoda placenta_**

   Slightly swollen, no radial thickenings, up to 50 mm **2**

---

2. Two spines on ventral mouth plates, no dark pattern on upper surface. Up to 50 mm **_Asterina gibbosa_**

   Ventral mouth plate spines rare and only in larger specimens, dark star-shaped pattern on upper surface. Up to 15 mm **_Asterina phylactica_**

Asterina gibbosa (Pennant)
Cushion star Fig. 12.3
Up to 50 mm diameter, usually smaller; dark olive green to greeny brown, or pale brown, sometimes mottled; small ones may have dark central pattern but never on green background. Subpentagonal; lower side flat, upper surface swollen. Mouth plates bear two spines.

Intertidal rock pools and sublittoral to about 125 m. Common on all British coasts, except North Sea; southwards to Azores and Mediterranean.

Asterina phylactica
Emson and Crump Fig. 12.3
Up to 15 mm diameter; upper surface grey-green, with central area marked with dark star-shape, resulting from strong red colour of spine groups in radial areas; underside pale. Very similar to *A. gibbosa* but smaller and distinctively coloured. Mouth plates rarely possess spines.

Intertidal rock pools and sublittoral to at least 18 m. Common on south and west coasts of British Isles; also Brittany and Mediterranean.

Anseropoda placenta (Pennant) Fig. 12.3
Up to 200 mm diameter, red above, yellow on underside. Subpentagonal, very thin and flat. No pedicellariae but covered on upper and lower surfaces by fine radiating rows of spine groups.

On sand bottoms, sublittorally, 10–200 (–600) m. Common on all British coasts except southern North Sea. Distributed from Shetland to Mediterranean.

5 SOLASTERIDAE

Usually with more than five arms and the disc rather large. Upper surface strengthened by a reticulum of ossicles. Four species, in three genera, occur around Britain but two are limited to deep water.

---

1. Large marginal paxillae in a single row; 3–5 furrow spines on each adambulacral plate **Crossaster papposus**

   Marginal paxillae form a double row; 2–3 small furrow spines in each group **Solaster endeca**

Crossaster papposus (Linnaeus)
Common sun star Fig. 12.3
Up to 340 mm diameter; disc colour red, arms off-white with broad transverse red band. Disc large; 10–12 tapering arms (rarely 8–16). Marginal paxillae large, arranged in a single row. Three to five furrow spines; adambulacral plates with series of 5–9 larger spines. No pedicellariae.

Infralittoral fringe to 50 (–1200) m. Common on all British coasts. Distributed from Arctic to the English Channel, also on both east and Pacific coasts of North America.

Solaster endeca (Linnaeus)
Purple sun star Fig. 12.3
Up to 400 mm diameter; orange sometimes tinged violet. Disc quite large; nine or ten arms (rarely 7–13), long and tapering. Marginal paxillae small, forming a double row in which upper ones are smaller. Upper surface crowded with small paxillae bearing short spines. Adambulacral plates bear 2–3 furrow spines and rows of 6–8 short outer spines.

Infralittoral fringe to about 450 m. Not uncommon on western and northern coasts; Irish Sea but not southern North Sea or Channel; northwards to Spitzbergen and Greenland, also east and west coasts of North America.

6 ECHINASTERIDAE

Starfish with very small disc and five long cylindrical arms lacking a distinct margin between upper and lower surface. Reticulate skeleton in body wall. Small groups of spines, no paxillae or pedicellariae.

---

1. Spines end in a cluster of long spikes not covered by skin **Henricia sanguinolenta**

   Spines end in a cluster of rather short points covered by thick skin **Henricia oculata**

Henricia sanguinolenta
(O. F. Müller) Fig. 12.4
Maximum diameter 200 mm; red with whitish tips to arms. Very small disc; cylindrical tapering arms. Scattered groups of small spines on upper surface. Spines on adambulacral ossicles form a single graded series, decreasing in size outwards.

Infralittoral fringe and shallow water, down to 1000 (2450) m. Common on north, north-west, and north-east coasts. Distribution circumarctic: Greenland, Iceland, Scandinavia, Baltic, north-west, Atlantic, and north Pacific. Records from Biscay may result from confusion with *Henricia oculata* (Pennant) which is a similar size to *H. sanguinolenta*, red, orange, or yellow, with a very small disc and fine arms. It is found on coarse sand and gravel, infralittoral fringe to about 100 m, on south and west British coasts including Irish Sea.

Order Forcipulata

7 ASTERIIDAE

Body wall strengthened by reticulate pattern of ossicles and armed with longitudinal rows of single spines. Usually both crossed and straight pedicellariae present. Three species, in three separate genera, only two of which are common.

---

1. Many adambulacral spines bear sessile pedicellariae near their tips **Asterias rubens**

   Adambulacral spines do not bear pedicellariae **Marthasterias glacialis**

Asterias rubens Linnaeus
Common starfish Fig. 12.4
Up to 520 mm diameter, but commonly 100–300 mm; basically orange but ranging from pale brown to delicate violet, deep-water specimens pale. Disc small, five arms (rarely 4–8, also often regenerating arms and other abnormalities), broad at base, tapering. Body wall very flexible with numerous groups of papulae in soft areas. Major spines of upper surface often in one or more longitudinal rows, sometimes surrounded by bundles of straight and crossed pedicellariae. Ventro-lateral spines just outside adambulacrals, in oblique rows. Straight pedicellariae on lower surface in ambulacral grooves and attached to furrow spines.

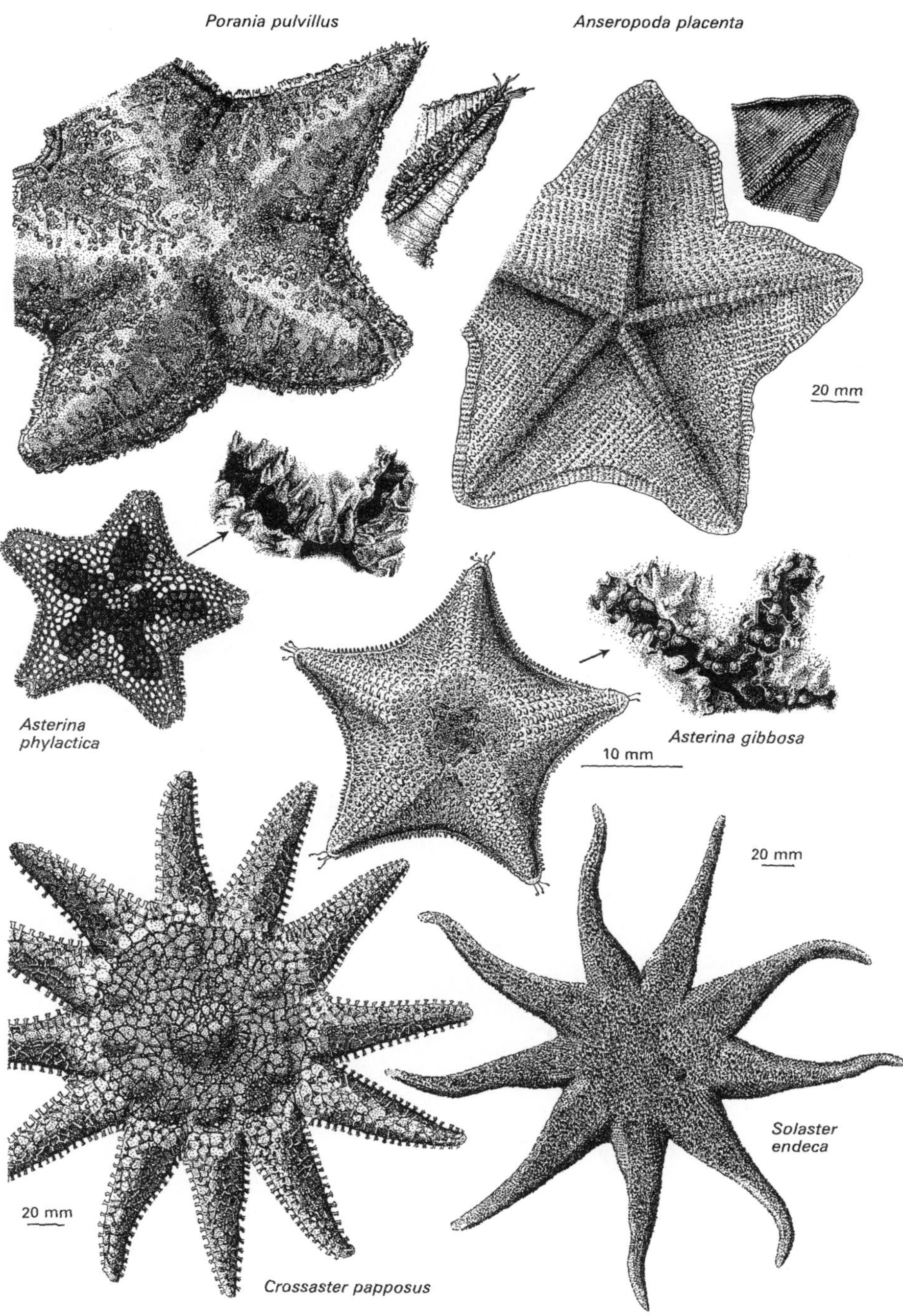
Porania pulvillus
Anseropoda placenta
20 mm
Asterina phylactica
Asterina gibbosa
10 mm
20 mm
Solaster endeca
20 mm
Crossaster papposus

Fig. 12.3

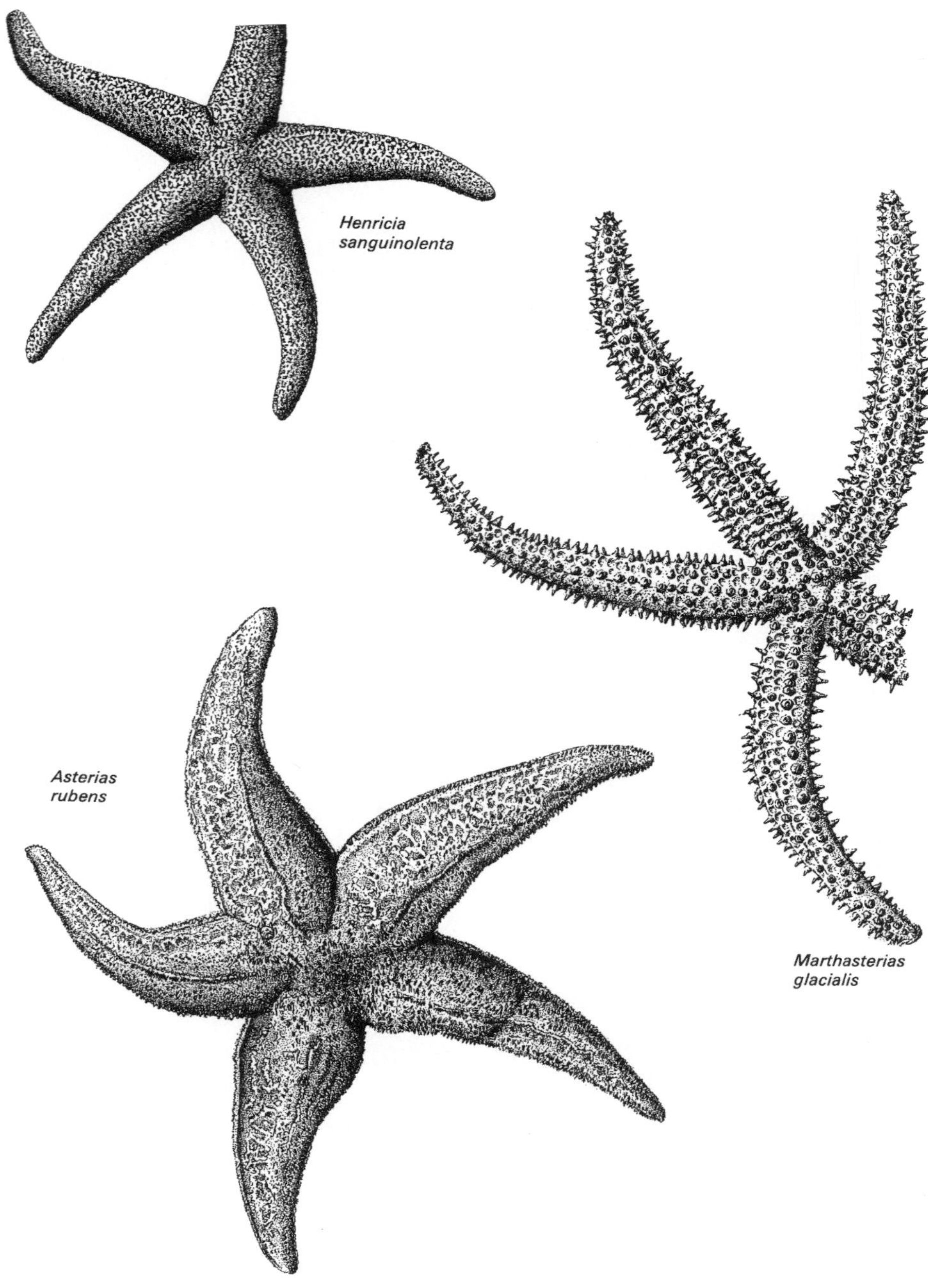
Henricia
sanguinolenta
Marthasterias
glacialis
Asterias
rubens

Fig. 12.4

On a variety of mostly hard substrata; midlittoral, infralittoral fringe, and sublittoral to 650 m. Abundant on all British coasts. Distributed from Iceland and the White Sea southwards to Senegal; not in the Mediterranean.

A caprellid amphipod, *Pariambus typicus* Krøyer is often found attached to spines, also a semiparasitic copepod, *Scottomyzon gibbosum* Scott.

Marthasterias glacialis (Linnaeus)
Spiny starfish Fig. 12.4

Up to 700 mm diameter; pale grey to green, tinged purple, yellowish, or reddish. Disc very small, five evenly tapering long arms, increasing in relative length with age. Well-marked longitudinal rows of spines: mid-dorsal, upper marginal, lower marginal. Thick cushion of crossed pedicellariae at base of each major spine. Single series of adambulacral spines, without pedicellariae. Straight pedicellariae scattered on upper surface and sessile on adambulacral ossicles (but not on their spines).

Infralittoral fringe down to about 180 m. All British coasts except east coast of England and eastern half of Channel. Finmark and Iceland to Cape Verde Isles, Azores, and Mediterranean.

Class Ophiuroidea

This is the largest group of living echinoderms, comprising some 1800 species. They occur sublittorally or on the lower shore, amongst stones and small algae, and also on the surface of, or buried in, muds and sands. The adult consists of a flattened *disc* and five narrow, flexible *arms* which break off easily, hence the common name brittle star. The upper surface of the disc has no apertures but is covered with various scales (Fig. 12.1). The central scale (*primary plate*) is surrounded by concentric rings of *secondary plates*. Adjacent to the base of each arm is a pair of *radial shields*. Some of the dorsal plates of the disc may be covered with scales or small spines. Near the base of each arm there may be a row of papillae (*arm comb*). The lower surface of the disc has a central *mouth* leading to the stomach. Food is ingested through the mouth and undigested remains leave by the same route. The mouth is divided by five interradial *jaws* (Fig. 12.5). At the apex of the jaw are teeth, a single vertical row, or a double row of *teeth papillae*. In some species single or paired *infradental papillae* cover the outer part of the tooth. Along the jaw edge are usually found the *mouth papillae*. Moving from the jaw apex to the interradius, the first pair of plates are the *adoral shields* which partially surround the *oral (mouth) shield*, often the largest plate on the ventral disc. One of the five oral shields houses the *madreporite*. From the jaw angle the first plate to occur is the first *ventral arm plate*. Usually either side of this plate lies the *second foot* (or *tentacle*) *pore*. This may open into the angle of the jaw and is lined with *tentacle scales*. The first tube foot is hidden within the mouth. Lying along either side of the arms on the ventral side of the disc are the *genital slits*. The arms consist of a series of one *dorsal*, one *ventral*, and two *lateral plates* forming an arm segment. The distal edges of the lateral plates may be fringed with spines. On each side of the ventral plate lies the *tube foot pore* which may be covered by a *tentacle scale*.

KEY TO FAMILIES

---

1. Arms simple, unbranched, moving horizontally. Distinct scales covering disk and arms **2**

   Arms simple or branched, moving vertically. Articulation of arm segments by means of hour-glass-shaped surface. No distinct scale covering on disc or arms. (Two deep-water species in families Gorgonocephalidae and Asteryonychidae (Order **Euryale**), not dealt with here)

---

2. Arm spines movable, lying flat against arms when disturbed, held erect at rest; disc scales distinct **1. Ophiolepidae**

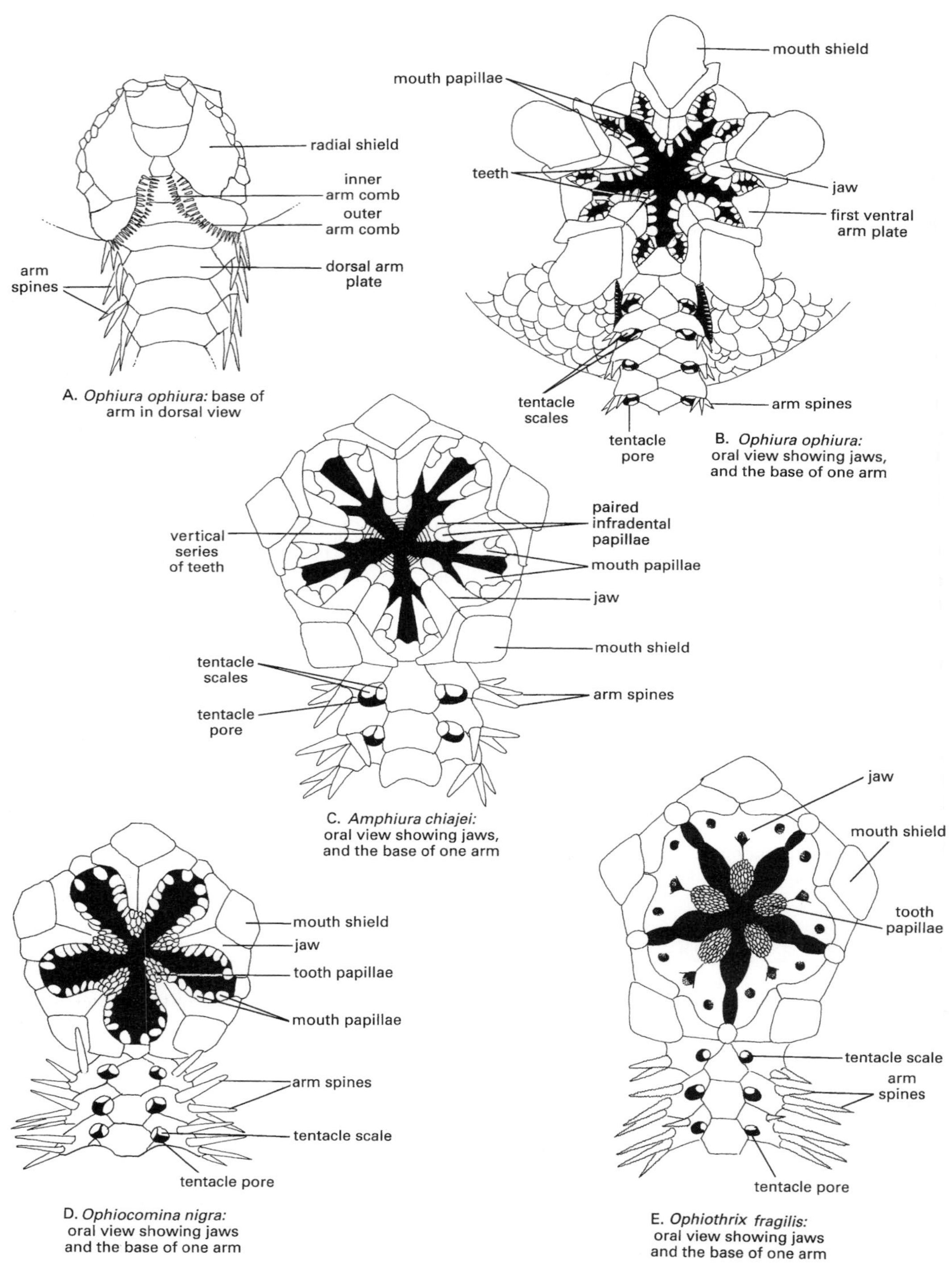
radial shield
inner arm comb
outer arm comb
dorsal arm plate
arm spines
A. *Ophiura ophiura:* base of arm in dorsal view
mouth shield
mouth papillae
teeth
jaw
first ventral arm plate
tentacle scales
arm spines
tentacle pore
B. *Ophiura ophiura:* oral view showing jaws, and the base of one arm
paired infradental papillae
vertical series of teeth
mouth papillae
jaw
mouth shield
tentacle scales
arm spines
tentacle pore
C. *Amphiura chiajei:* oral view showing jaws, and the base of one arm
mouth shield
jaw
tooth papillae
mouth papillae
arm spines
tentacle scale
tentacle pore
D. *Ophiocomina nigra:* oral view showing jaws and the base of one arm
jaw
mouth shield
tooth papillae
tentacle scale
arm spines
tentacle pore
E. *Ophiothrix fragilis:* oral view showing jaws and the base of one arm

Fig. 12.5

Arm spines rigidly erect, not movable **3**

---

3. Tooth papillae present on apex of jaw (Fig. 12.5D,E) **4**

   No tooth papillae, but a single vertical row of teeth (Fig. 12.5C) **5**

---

4. Mouth papillae present around sides of jaw (Fig. 12.5D) **2. Ophiocomidae**

   No mouth papillae around jaw (Fig. 12.5E) **3. Ophiotrichidae**

---

5. Single infradental papilla on apex of jaw **4. Ophiactidae**

   Paired infradental papillae on apex of jaw (Fig. 12.5C) **5. Amphiuridae**

Order Ophiurae

1 OPHIOLEPIDAE

Scales on both sides of disc. Mouth papillae form a continuous series around jaw. Single vertical row of teeth. Arm spines movable, will lie flat against arms when stimulated. Single or double combs at base of arm.

The British fauna includes 11 species in four genera, but only four species of *Ophiura* occur commonly in shallow waters.

---

1. Ventral plates in proximal part of arm separated by pores. About 30 papillae in arm combs ***Ophiura ophiura***

   No pores between proximal ventral plates. Seven to fifteen papillae in outer arm comb **2**

---

2. One or two small tentacle scales; ventral plates widely separated **3**

   Three or more tentacle scales ***Ophiura albida***

---

3. Only three mouth papillae on jaw, outer one plate-like. Inner arm combs meet in centre of arm ***Ophiura affinis***

   Three or four small mouth papillae in jaw ***Ophiura robusta***

Ophiura ophiura (Linnaeus) Fig. 12.6

Disc up to 35 mm diameter, upper side red-brown, lower side pale. Primary plate distinct. Radial shields large, about one-half disc radius; just touching above innermost dorsal arm plate. Two layers of arm combs above each arm; outer comb consists of about 30 fine papillae. Proximal ventral arm plates separated by a pair of pores in mid-line of arm. Mouth shields occupy two-thirds of distance from disc edge to adoral plate. Four to six mouth papillae.

Common all round Britain. Mainly sublittoral on a variety of soft substrata. Distributed from northern Norway to Madeira, and the Mediterranean.

Ophiura albida Forbes Fig. 12.6

Disk up to 12 mm diameter, upper surface red-brown. Primary plates not distinct. Radial shields small. Innermost dorsal arm plates heart-shaped. Outer arm comb of 10–12 short, stout spines; inner comb indistinct. Dorsal and ventral arm plates with convex outer edge. Three arm spines equidistant from each other. Spines equal to one-half lateral plate length. Mouth shield approximately one-third disk radius. Three to five simple mouth papillae.

Common all round Britain. Mainly sublittoral on a variety of soft substrata. From northern Norway to the Azores, and the Mediterranean.

Ophiura affinis Lütken Fig. 12.6

Maximum disc diameter 8 mm. Primary plates conspicuous. Radial shields small, wholly separate, similar in size to central plate. Second tentacle pore with only two tentacle shields. Outermost mouth papillae plate-like. Occasionally rudimentary arm combs may meet across base of arm.

On muddy sand, fine shell, and gravel. All British coasts, but not in the southern North Sea. Distributed from northern Norway to Biscay; also recorded from the Mediterranean.

Ophiura robusta Ayres Fig. 12.6

Maximum disc diameter 10 mm. Scales of aboral surface uniform in size. Radial shields very small, wholly separated. Arm combs consist of stunted papillae between disc and side of arm base. Dorsal arm plates with strongly convex outer edge. Ventral plates have shallow concavity on outer edge, except first ventral plate which has three shallow concavities on outer edge. Three arm spines, lower two short, upper long and stout. Short, broad mouth shields and three or four mouth papillae.

On various substrata; northern Britain only. A predominantly arctic/boreo-arctic species, occurring in both the east and west North Atlantic.

2 OPHIOCOMIDAE

Disc with granules completely covering scales and radial shields. Rigidly erect arm spines. Well-developed dorsal and ventral plates. Mouth papillae present. Tooth papillae present over a series of strong teeth. Second pair of tube feet inside mouth edge. Two species in the genus *Ophiopsila* are recorded from Britain, in addition to the widely distributed *Ophiocomina nigra*.

Ophiocomina nigra (Abildgaard) Fig. 12.6

Disc up to 25 mm diameter, black to grey. Fine scales on dorsal disc surface completely covered by granules, but only part of ventral interradius covered with granules. Mouth shields oval; 10–15 tooth papillae over single vertical row of teeth; four long, thin mouth papillae. Two tentacle scales, outer one larger than inner. Five to seven smooth arm spines.

Mainly on stones or rock, on current-swept bottoms. Common on all south and west coasts and occasionally in the North Sea. Ranges from western Norway to the Azores, and Mediterranean.

3 OPHIOTRICHIDAE

Large, conspicuous radial shields. Spines and thorny towers cover remainder of dorsal surface of disc. Dorsal and ventral arm plates well developed. Well-developed tooth papillae. No mouth papillae. Second pair of tube feet inside mouth edge.

Ophiothrix fragilis Thomson Fig. 12.6

Disc up to 20 mm diameter; colour variable. Large, naked, conspicuous radial shields, equal to two-thirds of disc radius. Keel on naked dorsal arm plates. One small tentacle scale. Seven, often serrated, arm spines. Gonads expand ventral side of disc between arms.

Very common on all British coasts, in lower littoral, and sublittorally on suitable hard bottoms including some sand/shell sediments. Widely distributed in the eastern Atlantic, from northern Norway to the Cape of Good Hope. *Ophiothrix luetkeni* Thomson is a rare species, recorded from deep water off western Ireland.

4 OPHIACTIDAE

Scales visible through spines and granules of dorsal disc surface. Mouth papillae present. Single series of square teeth overlain by infradental papillae. Second pair of tube feet inside mouth edge. Short, erect spines. Four species in three genera recorded from Britain, only one is common.

Ophiopholis aculeata Linnaeus Fig. 12.7

Disc up to 20 mm diameter. Colour variable, often red and purple. Radial shields concealed. Primary plates obvious, surrounded by granules and blunt spines, continuing onto ventral side. Dorsal arm plates oval, surrounded by ring of small plates. Ventral arm plates rectangular. Six to seven arm spines. Three mouth papillae; infradental papillae small; wedge-shaped truncated vertical row of teeth.

On hard substrata, offshore to 1880 m. Common round most of Britain although rarer in the south. A circumpolar cold-water species, reaching its southern limit in the north-east Atlantic in the British Isles.

5 AMPHIURIDAE

One pair of infradental papillae on apex of jaw; a single series of square teeth, no tooth papillae. Mouth papillae on each side of jaw, well separated from or contiguous with infradental papillae. Two pairs of tube feet within mouth. Arms are long, fine, and very flexible. Nine species, in three genera, recorded from Britain.

---

1. One outer mouth papilla on each side of jaw, not contiguous with infradental papillae (Fig. 12.5C) **2**

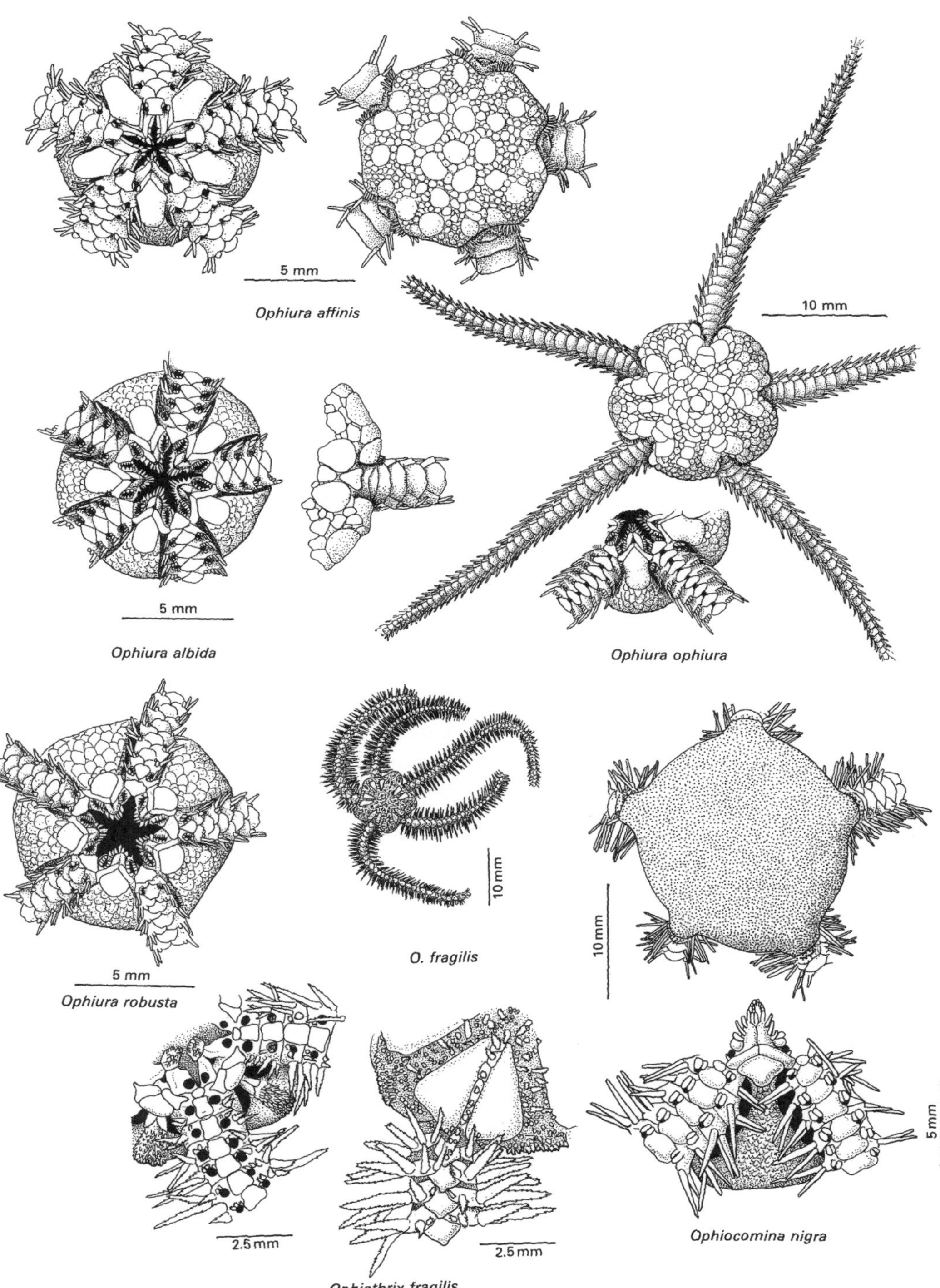
5 mm
Ophiura affinis
10 mm
5 mm
Ophiura albida
Ophiura ophiura
10 mm
10 mm
O. fragilis
5 mm
Ophiura robusta
5 mm
2.5 mm
2.5 mm
Ophiocomina nigra
Ophiothrix fragilis

Fig. 12.6

Two outer mouth papillae on each side of jaw, contiguous with infradental papillae
Amphipholis squamata

---

2. Scales on ventral side, and margin of dorsal side, of disc, each ending in a small tubercle. Radial shields with a transverse furrow. Arms 15 × disc diameter **Amphiura brachiata**

Scales smooth. No transverse furrows on radial shield. Arms 12 × disc diameter, or less **3**

---

3. Two tentacle scales. Underside of disc wholly covered with scales
Amphiura chiajei

No tentacle scales. Underside of disc partly or wholly naked
Amphiura filiformis

Amphipholis squamata Delle Chiaje Fig. 12.7
Disc very small, 5 mm diameter, greyish; dorsal surface covered by small scales. Radial shields equal to one-third disc radius and joined for almost entire length. Dorsal arm plates rounded, triangular, broader than long in proximal part of arm. Ventral plates triangular, truncated at proximal apex. Three to four short, conical arm spines. Mouth shields small rhomboids. Outer mouth papillae elongate and broad.

Common all round Britain from middle shore into sublittoral, mainly under stones or amongst rock-pool weeds, and occasionally on sandy bottoms. Cosmopolitan in temperate and warm temperate seas.

Amphiura brachiata (Montagu) Fig. 12.7
Disc diameter up to 12 mm; with very long, thin flexible arms; brown-grey in colour. Dorsal surface covered by fine scales with small primary plates and separate, small, radial shields. Mouth shields diamond-shaped. Mouth papilla broad and scale-like. Two infradental papillae cover teeth. Dorsal arm plates rhombic; ventral arm plates rectangular, with marked median longitudinal keel in proximal part of arm, and a small lateral keel on each side. Two tentacle scales in proximal part of arm.

Littoral and sublittoral in fine sand, buried in sediment with only distal parts of arms protruding. Often common. Western coasts of Britain and Ireland; also recorded from western edge of Dogger Bank.

Amphiura chiajei Forbes Fig. 12.7
Disc up to 11 mm diameter, covered with fine scales on both dorsal and ventral sides. Primary plates distinct. Radial shields separate. Outer mouth papillae broad and flat. Mouth shields rhombic. Four to six arm spines. Two tentacle scales.

Amphiura chiajei is a deposit feeder living especially in muddy substrata. All British coasts. Distributed from western Norway to the Azores, and the Mediterranean.

Amphiura filiformis O.F. Müller Fig. 12.7
Disc up to 10 mm diameter, red-grey in colour, covered with fine scales, not extending to ventral interradii. Radial shields usually separate, but may join at base. Mouth shields rounded pentagons. Mouth papillae conical, spine-like. Dorsal arm plates oval; ventral arm plates square, with convex outer edge and a small median keel. Five to seven arm spines. No tentacle scales.

Common sublittorally off all British coasts, in suitable muddy sand substrata. This species is a suspension feeder. Ranges from western Norway to the Mediterranean.

Class Echinoidea

Echinoderms with a globular or disc-shaped rigid test consisting of 20 meridional rows of plates running from the apical region above, to the oral region on the underside. Of these rows, five pairs are *ambulacral* plates and five *interambulacral*. The ambulacral plates are each pierced by several pairs of pores, each pore pair corresponding to one tube foot. Otherwise the test is densely furnished with mobile spines between which are three-jawed *pedicellariae* (Fig. 12.1) of several distinctive types: very small *triphyllos* types; larger *ophiocephalous*, with long flexible (snake-like) stalks; long-jawed *tridactyle* pedicellariae; and taxonomically important *globiferous* pedicellariae. The latter have toxin-secreting jaws whose blade-ossicles are distinctive. The top of the test has five radial *ocular* plates and five interradial *genital plates*,

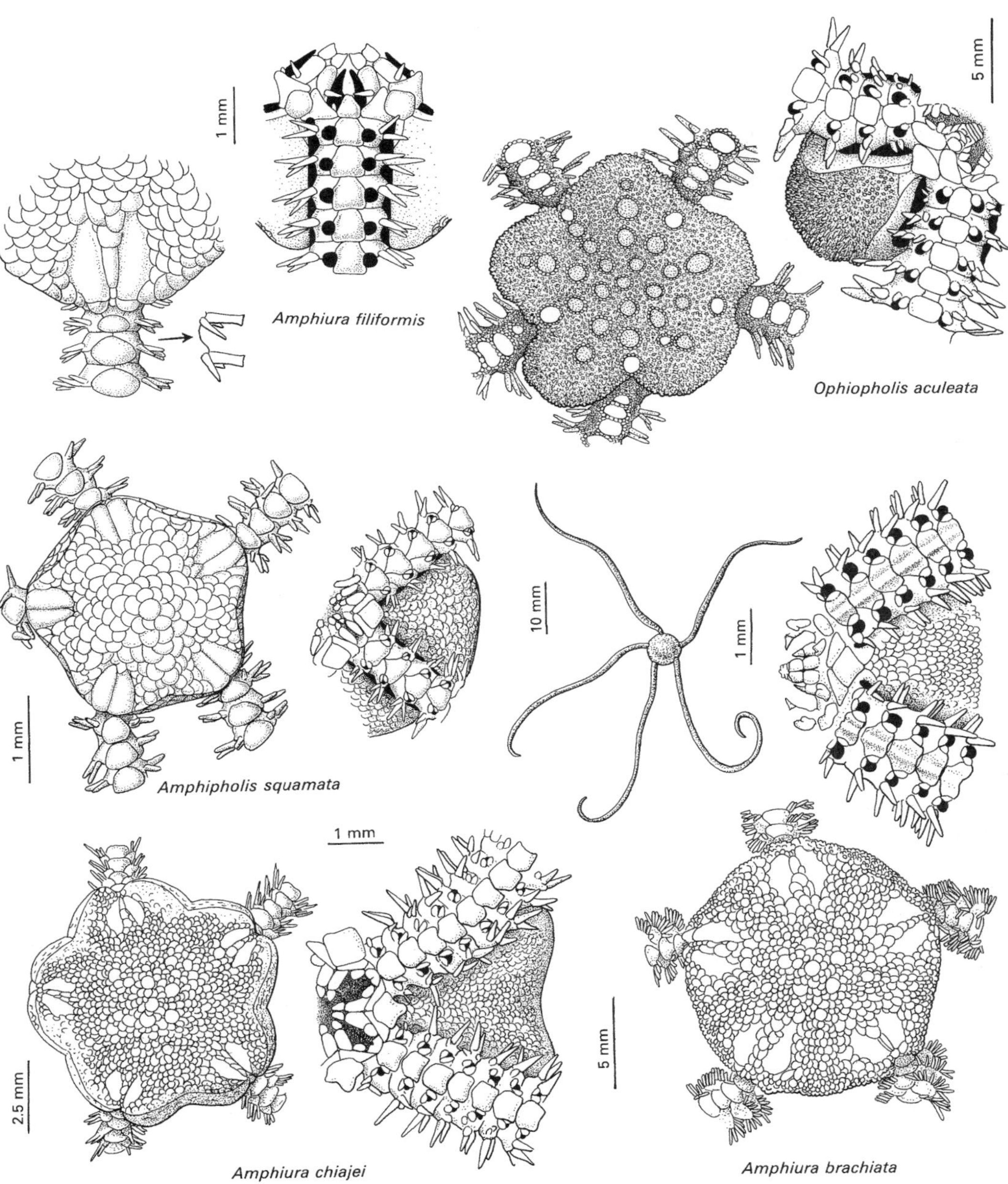
1 mm
Amphiura filiformis
5 mm
Ophiopholis aculeata
10 mm
1 mm
1 mm
Amphipholis squamata
1 mm
2.5 mm
5 mm
Amphiura chiajei
Amphiura brachiata

Fig. 12.7

one of the latter enlarged as the *madreporite*. In some groups there is a complex dental apparatus (*Aristotle's lantern*), which operates five teeth or jaws which project from the mouth. The buccal region surrounding the mouth is then flexible to facilitate the grazing action of the teeth. There are five pairs of buccal tube feet arising from the membrane and five pairs of branched gills.

KEY TO FAMILIES

1. Anus situated within apical system, test radially symmetrical, gills sometimes present on peristome **2**

   Anus outside apical system, test secondarily bilateral, no gills on peristome **3**

2. Globiferous pedicellariae have head attached to stalk without neck; lateral teeth present below terminal tooth of each jaw **1. Echinidae**

   Globiferous pedicellariae have a flexible neck and their jaws lack lateral teeth **2. Strongylocentrotidae**

3. Well-developed teeth and 'lantern' apparatus; mouth central; test usually flattened; some fine tube feet on interambulacra **3. Fibulariidae**

   No teeth or 'lantern'; mouth often slightly forward of centre, tube feet only on ambulacra; unpaired posterior interambulacrum forming a lip to the mouth, followed by two large plates forming a 'plastron'; subanal fasciole present **4. Spatangidae**

Order Diadematoidea

1 ECHINIDAE

Ambulacral plates with three or more pore pairs. Test without depressions. Jaws of the globiferous pedicellariae have one or two pairs of teeth below the single terminal tooth. Three additional species in the genus *Echinus* have been recorded from Britain.

1. Three pore-pairs to each ambulacral plate **2**

   Five or six pore-pairs to each ambulacral plate ***Paracentrotus lividus***

2. Membrane around mouth set with thick plates. Globiferous pedicellariae with grooved jaw blades and a row of strong teeth along each side ***Psammechinus miliaris***

   Membrane around mouth with a few thin plates. Globiferous pedicellariae have tubular jaw blades with only one or two lateral teeth **3**

3. Test globular, reddish, tubercles white. Secondary spines numerous, similar in length to primaries; buccal ossicles bear some small spines ***Echinus esculentus***

   Test subconical; red colour not uniform, often broken by vertical white bands; secondary spines smaller than primaries, no spines on buccal ossicles ***Echinus acutus***

Paracentrotus lividus (Lamarck)

Purple sea urchin Fig. 12.8

Up to 70 mm diameter; spines dark violet to dark olive, rarely brownish; test greenish. Primary spines

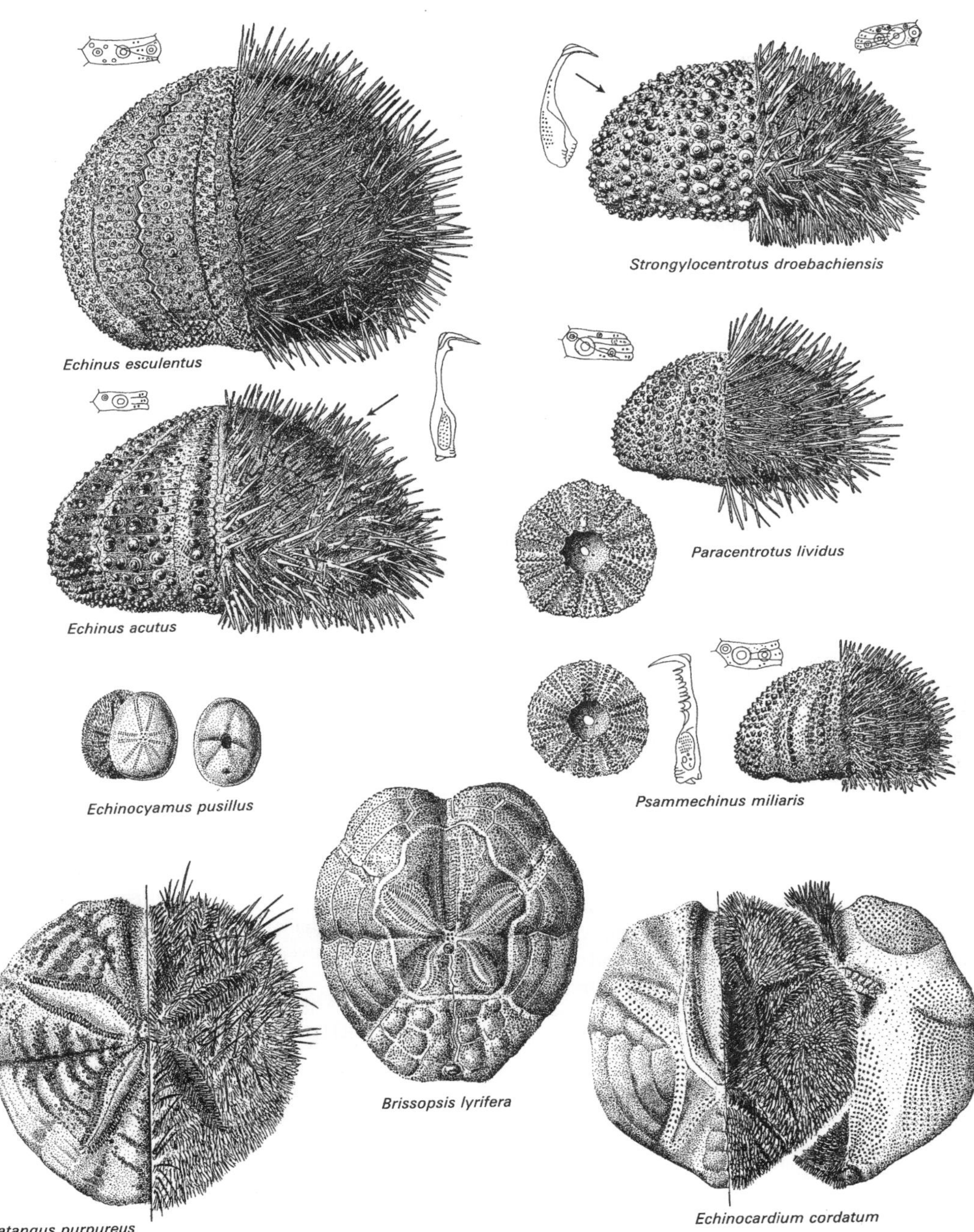
Echinus esculentus
Echinus acutus
Strongylocentrotus droebachiensis
Paracentrotus lividus
Echinocyamus pusillus
Psammechinus miliaris
Brissopsis lyrifera
Spatangus purpureus
Echinocardium cordatum

Fig. 12.8

fairly long if not worn down. Five or six pore pairs per ambulacral plate. Globiferous pedicellariae have grooved jaw blades but these do not have a row of sharp spines each side.

Grazes on algae; breeds in summer. Gonads have commercial value as seafood delicacy in Mediterranean. Intertidal and shallow sublittoral (to 30 m). In rock pools and other holes bored in softer rocks, often in large groups; also amongst *Zostera* and coralline strands. Common south and west Ireland. Rare Devon and Cornwall. Elsewhere southwards to Morocco, Mediterranean, Canaries and Azores.

Psammechinus miliaris
P. L. S. Müller — Fig. 12.8

Up to 50 mm diameter, robust, slightly flattened; greenish with violet tips to spines. Spines mostly of similar shortish size and closely packed. Ambulacral plates have three pore pairs and one primary tubercle. Globiferous pedicellariae abundant, jaw blades grooved, with a row of sharp spines along each side.

Omnivorous, minor pest of oyster beds. Breeds early summer. Frequently harbours a polychaete, *Flabelligera affinis* M. Sars, amongst spines. Under stones and rocks, and amongst *Zostera*; intertidal, occasionally down to about 100 m. Common on all British coasts. Distributed from Iceland, Scandinavia, Baltic, North Sea, southwards to Morocco and the Azores. Not in Mediterranean.

Echinus esculentus Linnaeus
Common, or edible, sea urchin — Fig. 12.8

Up to 160 mm at about 8 years old, maximum 176 mm; large and globular, red or sometimes pale purple verging greenish; spines reddish with violet tips. Primary and secondary spines and tubercles (bosses), of similar size but distinguishable. Ambulacra have a primary tubercle on only every second or third plate. Globiferous pedicellariae large, jaw blades bearing only one lateral tooth each side. Buccal plates with small spines in addition to pedicellariae.

Omnivorous, feeding on *Laminaria*, Bryozoa, barnacles, etc. Gonads regarded as a delicacy in Portugal and elsewhere. On rocky substrata from infralittoral fringe, especially at 10–40 m and down to 1200 m. Common on all British coasts. Elsewhere: Iceland, Finmark, to Portugal. Not Mediterranean.

Echinus acutus Lamarck — Fig. 12.8

Up to 150 mm diameter, less globular than *E. esculentus* and slightly conical; white with brownish stripe down each row of plates; alternating dark and light growth bands to larger plates. Spines usually reddish with white tips. Rather sparsely covered by spines on upper half; secondary spines shorter and more slender than primaries. Only alternate ambulacral plates bear primary spines. Pedicellariae but no spines on buccal ossicles.

On mixed bottoms, 200–1000 m (20–1200 m). Off the south-west and north coasts of British Isles, common. Distributed from Bear Island to North Africa and the Mediterranean.

2 STRONGYLOCENTROTIDAE

Full radial symmetry. More than three pore pairs per ambulacral plate. Globiferous pedicellariae lack lateral teeth to jaw blades which are distinctively tubular. Only one genus known. One species in the British Isles.

Strongylocentrotus droebachiensis
O. F. Müller — Fig. 12.8

Up to 80 mm diameter; greenish-brown; spines variable reddish, greenish, or even violet; tips usually white. Five or six pore pairs to each ambulacral plate. Usually two ocular plates abut on periproct. Primary and secondary spines of similar length. Globiferous pedicellariae large with numerous spicules arranged in two rows in each jaw. Buccal plates lack spines.

Mid- and infralittoral fringe down to 1200 m. Not common around British Isles, confined to east coasts, not on west. A northern, circumpolar species, occurring off Greenland, Spitzbergen, and south to New Jersey; in the North Pacific southwards to Vancouver and Korea.

Order Clypeasteroidea

3 FIBULARIIDAE

Very small sea urchins. Ambulacra are only slightly petaloid. Anus situated on lower surface a short distance behind mouth. One British species.

Echinocyamus pusillus (O. F. Müller)
Green sea urchin — Fig. 12.8

Up to 15 mm diameter, grey or greenish; flattened, slightly elongated, bilaterally symmetrical. Six to nine pore pairs per ambulacral plate. The four genital pores larger than ocular pores.

In coarse sand or fine gravel on sheltered shores, and sublittorally down to 1250 m; common off all British coasts. Elsewhere Finmark, Baltic, Iceland, and south to west Africa, Azores, and Mediterranean.

4 SPATANGIDAE

This family is characterized by ciliated spines (the 'subanal fasciole') situated below the anus and enclosing a group of 'sanitary-drain-building' tube feet. There are six British species; *Spatangus raschi* Loven, *Echinocardium flavescens* and *E. pennatifidum* Norman are relatively rare.

---

1. Only one ring of ciliated spines present in the subanal fasciole **Spatangus purpureus**

   Two rings of ciliated spines **2**

---

2. Ring of ciliated spines enclosing anterior ambulacrum (inner fasciole) **Echinocardium cordatum**

   Ring of ciliated spines enclosing whole petalliferous area (peripetalous fasciole) on upper surface **Brissopsis lyrifa**

Spatangus purpureus O. F. Müller
Purple heart urchin Fig. 12.8
Up to about 120 mm long, deep violet; some of long upper spines white; broad and low, heart-shaped. Labrum rounded. Subanal fasciole about three times as broad as high. Four petaloid ambulacral areas on upper surface not enclosed by a fasciole. Large spine tubercles present on interambulacral, but not on ambulacral, areas.

Occurs shallowly buried in coarse sand or gravel (rarely mud), in sheltered areas in the infralittoral fringe, down to about 900 m. Locally common on British coasts; distributed from North Cape, Norway, to North Africa, the Azores, and the Mediterranean.

The bivalve mollusc *Montacuta substriata* (Montagu) is often found attached to spines.

Echinocardium cordatum (Pennant)
Sea potato Fig. 12.9
Commonly 60 mm long, but up to 90 mm; fawn, spines yellowish. Anterior ambulacrum slightly recessed up to and including part of apical area, both surrounded by ring of ciliated spines ('inner fasciole'). Highest part of test posterior to apical system. Some large spines and their prominent tubercles scattered on anterior interambulacra.

Occurs buried about 80 mm deep (to 150 mm) in sand, mainly intertidal (lower midlittoral and infralittoral fringe) but also sublittoral to 230 m. Common to abundant on all British coasts. Almost cosmopolitan, Norway to South Africa, Mediterranean, Australia, and Japan.

The bivalve mollusc *Tellimya ferruginosa* (Montagu) is often associated with this species, living in its burrow near the spines of the subanal fasciole. The amphipod crustacean *Urothoe marina* (Bate) is another common commensal.

Brissopsis lyrifera (Forbes) Fig. 12.9
Length 40–70 mm; colour reddish-brown. Test rather fragile, slightly longer than wide, notched in position of anterior ambulacrum; all ambulacra slightly recessed above; highest point of test behind apical system. Ambulacra form distinctive petal-shaped areas on upper surface. Petalliferous fasciole of ciliated spines encloses all five ambulacral petals. The posterior petals diverge.

Lives buried in mud, sublittorally 5–500 m. All British coasts except south, sometimes abundant. Elsewhere from Norway and Iceland to South Africa and the Mediterranean, also east coast of North America; not Greenland.

Class Holothurioidea

The holothurians, or sea cucumbers, are soft, elongate, sausage-shaped animals. In cross-section they may be round or slightly dorso-ventrally flattened. The *ventral* surface is often recognized by three longitudinal rows of tube feet, the *trivium*, and the dorsal surface by two longitudinal rows, the *bivium*. The *mouth* is at the anterior end, surrounded by the *tentacles* which may be bush-, shield-, or feather-shaped, or simply digitate. The tentacle shape is different for each order. The skin of holothurians is usually thick, except in the Apoda where it is thin and transparent. There is no rigid skeleton, as found in the other classes of echinoderms, but the skin of holothurians contains numerous delicately structured microscopic *calcareous bodies* (Fig. 12.9), including *tables*, *discs*, *wheels*, *anchors*, and some amorphous types. These are primary taxonomic characters. Under the epidermis is a layer of circular muscle with five *longitudinal muscles* often visible running the full length of the body. Muscles can retract the anterior part of the body (the *introvert*) including the mouth and tentacles. *Tube feet* may occur in the trivium or bivium or all over the body, but never in a furrow as in other echinoderms. In some subfamilies the tube feet are entirely lacking, or occur as a series of papillae round the anus. The *gonads* are found only in the mid-dorsal interradius.

KEY TO FAMILIES

1. Well-developed tube feet; body form various, but usually stout, not worm-like — **2**

   Tube feet absent; body form slender and worm-like (Order Apoda) — **4. Synaptidae**

2. Tentacles shield-shaped. No retractor muscles (Order Aspidochirota) — **1. Holothuriidae**

   Tentacles bush-shaped. Retractor muscles present (Order Dendrochirota) — **3**

3. Body cylindrical or uniform, without well-defined ventral sole — **2. Cucumariidae**

   Body more or less flattened, with obvious ventral sole — **3. Psolidae**

Order Aspidochirota

1 HOLOTHURIIDAE

Only one species occurs in British coastal waters.

Holothuria forskali Delle Chiaje
Cotton spinner Fig. 12.9

Up to 200 mm long, flattened ventral side bearing numerous tube feet in three or four series. Dorsal surface black, ventral surface either brown or yellow. Usually 20 tentacles around a terminal mouth, and well-developed tentacle ampullae. Dorsal surface covered with conical papillae. Calcareous deposits consist of a small disc with four holes in the skin, branched rods in tube feet, and curved rods in tentacles.

In shallow water, west and south-west coasts of Britain; present on all Atlantic coasts of Europe, and in the Mediterranean.

Order Dendrochirota

2 CUCUMARIIDAE

Seven species are keyed out below. The British fauna includes at least 15 species, in 10 genera, but most are rare or limited to deep waters.

1. Tube feet in distinct series along length of body — **2**

   Tube feet scattered over whole body — **6**

2. Body elongate, tapered, curving upwards anteriorly and posteriorly — ***Trachythyone elongata***

   Body not elongate or tapered or curving upwards — **3**

3. Very large, up to 500 mm; thick skin with few deposits — ***Cucumaria frondosa***

   Small, rarely longer than 150 mm; skin with numerous calcareous deposits — **4**

4. Calcareous deposits smooth — **5**

   Calcareous deposits mostly tuberculate — ***Ocnus lactea***

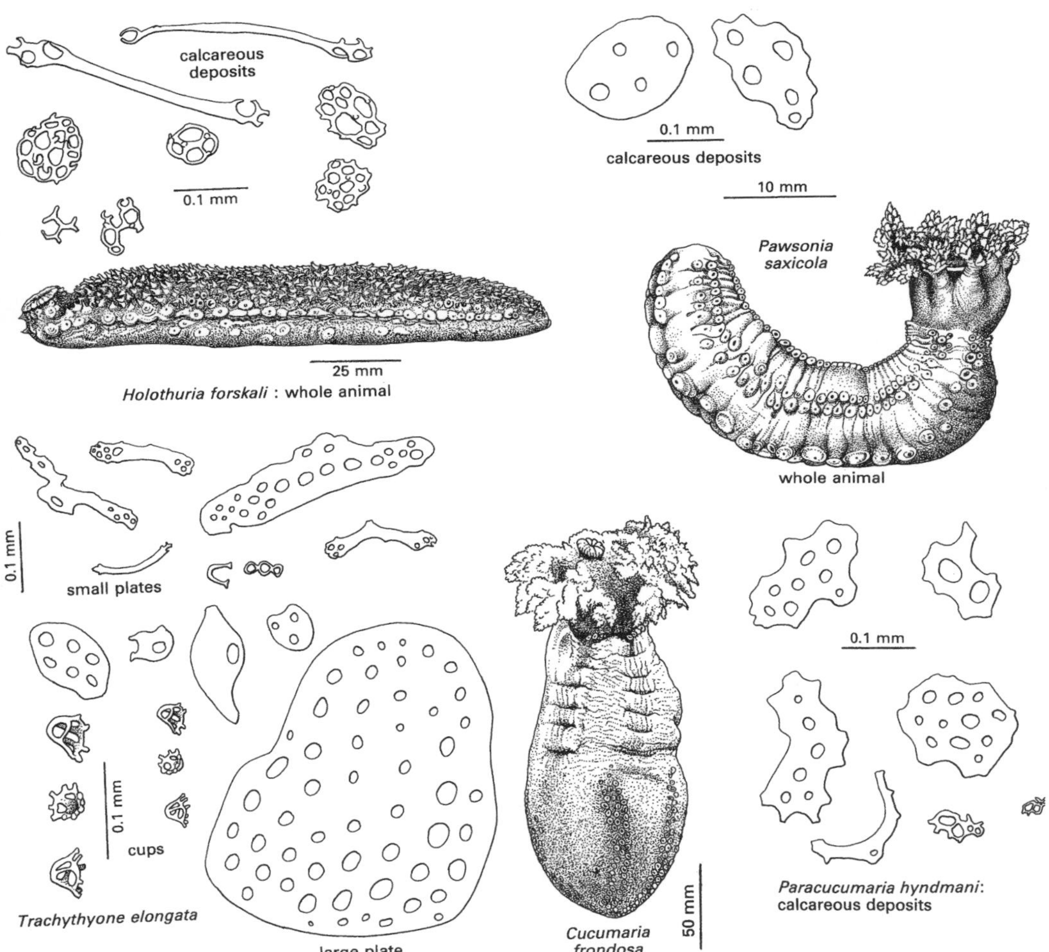

Fig. 12.9

5. Deposits large fenestrated plates, **no** star-shaped plates in outer skin
Paracucumaria hyndmanni

Deposits small, button-like, with four main holes, or holes at either end. Small star-shaped bodies in outer layer of skin
Pawsonia saxicola

6. Deposits with a spire of two rods
Thyone fusus

Deposits with no spire but smooth holes. Body with posterior prolongation
Thyone raphanus

Cucumaria frondosa (Gunnerus) Fig. 12.9

Body cylindrical, dark brown, up to 500 mm long. Tentacles thick, bushy. Skin very thick, with few deposits; fenestrate plates present in young specimens, but older animals almost completely lacking them.

Lower shore and shallow sublittoral, to perhaps 200 m. Often common among kelps. West Scotland, from the Clyde Sea to Shetland; northwards to Arctic waters.

Trachythyone elongata
(Düben and Koren) Fig. 12.9
Body elongate, tapered at both ends, up to 150 mm; dark brown. Tentacles short. Tube feet in five distinct rows; not completely retractile. Three types of calcareous deposits: large, smooth, irregular plates, with numerous holes, in deepest layers of skin; small, rounded plates with few holes, in upper layers of skin; small, cup-shaped deposits close to skin surface.

Shallow sublittoral to 65 m, on muddy bottoms, partly buried in sediment. On east coasts, as far south as Durham, on west from Shetland to the western end of the Channel. Distributed from western Norway to the Mediterranean.

Paracucumaria hyndmanni
(Thompson) Fig. 12.9
Body short and stout, barrel-shaped, tapered posteriorly, up to 50 mm long, greyish-white to pale red. Tube feet in conspicuous double rows, almost completely retractile. Skin thick and smooth, with large fenestrate deposits, closely packed and tending to overlap when the animal is completely contracted.

On hard grounds, offshore, to below 1000 m, usually covered with shell fragments and gravel. On west coasts from Shetland to the west Channel, on the east as far south as St. Andrews. Distributed from Western Norway to the Mediterranean.

Pawsonia saxicola (Brady and Robertson)
Sea gherkin Fig. 12.9
Body cylindrical, white, with dark tentacles, up to 150 mm long. Tube feet in double longitudinal rows ventrally, smaller dorsally and arranged in zigzag series. Skin thin and smooth, with small, button-like deposits, each with four principal holes centrally and one or more small ones at each end; small, star-shaped spicules scattered in outer layers of skin.

Intertidal and shallow sublittoral, in crevices, kelp holdfasts and under stones. Common on south and west coasts of Britain and Ireland. Distributed southwards to the Azores.

Ocnus lactea (Forbes) Fig. 12.10
Body cylindrical, white to brown, up to 40 mm long. Tube feet thinly distributed, in zigzag rows, not completely retractile. Skin thick, smooth, with abundant deposits: typically with four holes and numerous nodules, or large irregular and plate-like; small, cup-shaped spicules in outer layers of skin.

Lower shore and sublittoral, to 100 m; on hard bottoms, among stones, shell, and calcareous algae. West coasts of Britain and Ireland, in the east as far south as Durham. Distributed from western Norway to Brittany.

Thyone fusus (O. F. Müller) Fig. 12.10
Body ovoid, tapered at both ends, up to 200 m long, white to pinkish. Tube feet in longitudinal series, more or less distinct. Skin very delicate, ruptures readily on handling; with few, widely dispersed deposits: typically small, smooth tables, with four holes and a spire of two rods, fused distally and sometimes bearing apical thorns.

On shell gravels, characteristically covering itself with shell fragments. West coasts of Britain and Ireland, on the east as far south as Northumberland. Distributed from west Norway to Madeira and the Mediterranean.

Thyone raphanus
Düben and Koren Fig. 12.10
Body ovoid, tapered posteriorly to form a slender 'tail'; typically curved in a U-shape; yellowish-brown, up to 60 mm long. Tube feet only thinly distributed on dorsal side, and absent completely from posterior part of 'tail'. Skin thick, with numerous, closely spaced, often overlapping, deposits: large, smooth, fenestrate plates, often tuberculate.

Shallow sublittoral, to 1000 m, on sandy or muddy bottoms, typically buried in sediment with only the posterior 'tail' projecting. All British coasts. Distributed from western Norway to the Mediterranean.

3 PSOLIDAE

Only one known British species in this family.

Psolus phantapus (Strussenfelt) Fig. 12.10
Body arched upwards at either end, posterior with a long cortical tail-like prolongation; up to 150 mm long, yellowish-brown to black, with

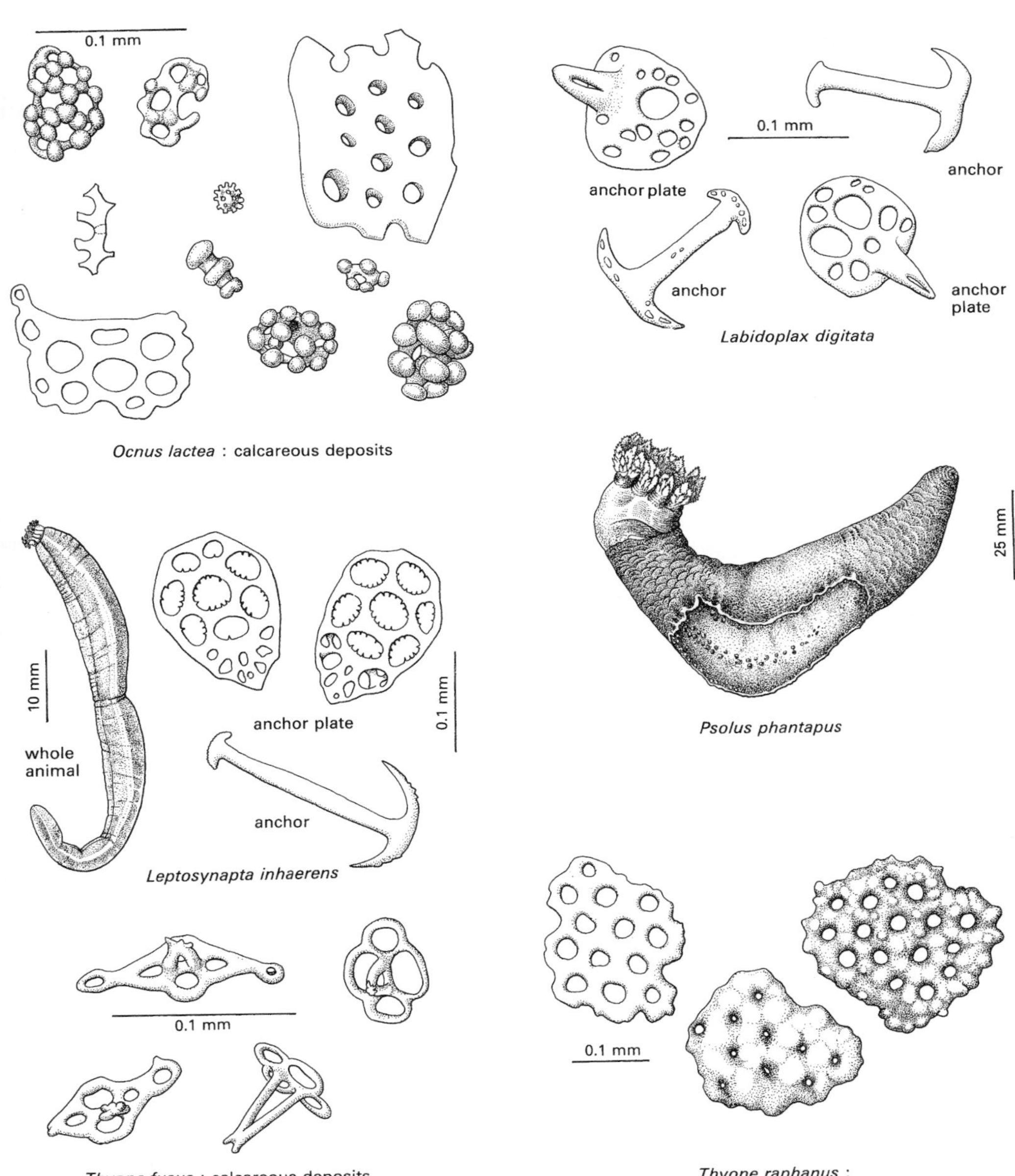
0.1 mm
Ocnus lactea : calcareous deposits
anchor plate
0.1 mm
anchor
anchor
anchor plate
Labidoplax digitata
25 mm
Psolus phantapus
10 mm
whole animal
anchor plate
0.1 mm
anchor
Leptosynapta inhaerens
0.1 mm
Thyone fusus : calcareous deposits
0.1 mm
Thyone raphanus : calcareous deposits

Fig. 12.10

orange tentacles. Ten tentacles around mouth. Well-developed ventral rectangular sole with numerous tube feet but very few calcareous deposits, consisting of either small cups or larger spherical to oval bodies of complicated structure. Apart from the sole, tube feet are found only on the soft-skinned introvert.

Lower shore to 400 m. North-west coasts of Britain and Ireland, on the east coast as far south as Yorkshire; northwards to the Arctic.

Order Apoda

4 SYNAPTIDAE

All species in this family are superficially similar. Four species each in the two genera *Labidoplax* and *Leptosynapta* are known from Britain; only *L. digitata* and *L. inhaerens* are frequently reported, although the three other species of each genus may be locally common.

---

1. Calcareous anchor plates with a distinct narrow handle ***Labidoplax digitata***

   Anchor plate not having a narrow handle ***Leptosynapta inhaerens***

Leptosynapta inhaerens (O. F. Müller) Fig. 12.10

Body slender, soft, pink; up to 300 mm long. Fairly transparent, 12 tentacles each with 5–7 pairs of digits, the large outermost digit being unpaired. Calcareous ring comprises 12 pieces. Calcareous deposits consist of an anchor and its plate, the latter with six (occasionally seven) serrate holes around a central hole.

From lower shore to 50 m, buried in sand or mud. South and west coasts from Plymouth to the Shetlands; on the east as far south as Northumberland. Distributed from northern Norway to Brittany.

Labidoplax digitata (Montagu) Fig. 12.10

Body up to 300 mm long, red or brown, darkest dorsally. Twelve tentacles, each with two pairs of lateral digits and a rudimentary terminal digit. The calcareous ring consists of 12 pieces with perforate radials. Anchor plates are racket-shaped with a handle and four large, smooth central holes. Anchors are up to 0.35 mm long with serrate arms.

In muddy sand from the lower shore down to 70 m. All western coasts, southwards to the Mediterranean.

13 ACORN-WORMS AND SEA SQUIRTS

1. PHYLUM HEMICHORDATA

Hemichordates comprise worm-like Enteropneusta (= Balanoglossida) and tentaculate Pterobranchia. Enteropneusts have gill slits, so are linked with chordates, particularly because they also have a dorsal, sometimes tubular, nerve cord in the *collar*. They lack a notochord, however, so are placed in a phylum separate from the Chordata. The 'tornaria' larvae of one family are like some echinoderm larvae. Pterobranchia are closely related to enteropneusts but the only pterobranchs known from Britain belong to a minute colonial genus, atypical of hemichordates.

In Enteropneusta ('acorn-worms') the preoral *proboscis* may resemble an acorn, while the *collar* looks like the acorn's cup. The long *trunk* that follows consists first of a branchial region, extending for 5–20% of the total body length, marked externally by dorsolateral *gill pores*. Each pore is the opening of a gill pouch, which is associated with a *gill slit* in the pharynx wall. Each gill slit is U-shaped, because of downgrowth of a *tongue bar* from its dorsal margin, but this cannot usually be seen unless the pharynx is opened so as to view the gill slits from the inside (Fig. 13.1). In Ptychoderidae and some Spengeliidae the tongue bars are joined to neighbouring gill bars by many short *synapticulae*. The second part of the trunk is often termed the oesophageal or genital region, though the gonads generally occur throughout the branchial region as well. In *Saccoglossus* and in Spengeliidae there is, on each side, a series of *oesophageal pores*, like reduced gill pores. Behind the oesophagus the gut has *hepatic pouches*, which form external protuberances in Ptychoderidae. Finally, there is a long *intestinal region*, which is laden with muddy sand and usually tears under its own weight if the worm is not supported by water.

All British enteropneusts burrow in sand or mud. When dug up they are remarkably inactive, but very soft-bodied, so entire specimens are rarely obtained. Some species smell characteristically of iodoform.

KEY TO FAMILIES

1. Microscopic colonies of zooids with pinnate tentacles and interconnecting stalks, in branching annulated tubes attached to shells or old coral (PTEROBRANCHIA) **1. Rhabdopleuridae**

   Soft, non-colonial worm, burrowing in sand or mud, with preoral proboscis, short postoral collar, and long trunk bearing (anteriorly) two rows of gill pores (ENTEROPNEUSTA) **2**

2. Proboscis not much longer than collar. Trunk bearing genital folds or wings anteriorly and protuberant hepatic pouches in a middle region **3. Ptychoderidae**

   Proboscis may be longer than collar. Trunk without prominent protuberances **2. Harrimaniidae**

Class Pterobranchia

1 RHABDOPLEURIDAE

Minute colonial zooids, each with a creeping cephalic shield (representing the proboscis of Enteropneusta), an inconspicuous collar bearing a pair of pinnate ciliated tentacles, a sac-like body with U-shaped gut and antero-dorsal anus, and a long stalk connecting with other zooids. The branching tubes (diameter about 0.2 mm) mostly adhere to substrata and bear obliquely transverse ridges, which are interrupted in a criss-cross pattern along the free side. Erect ends of branches, from which undisturbed zooids protrude their tentacles, bear ridges which form complete rings, with no obliquity.

---

1. Tubes in close contact with one another, forming compact encrustations only a few millimetres across
   Rhabdopleura compacta

   Tubes directed away from each other, to form a straggling system of main branches and side branches, often several centimetres long
   Rhabdopleura normani

Rhabdopleura compacta Hincks Fig. 13.1
Zooids lemon-yellow, spotted with reddish-brown and light green.

On concave (lower) surfaces of shells of *Glycymeris*; 20–100 m; near Belfast, Plymouth, and Roscoff.

Rhabdopleura normani Allman
Straggling form of colony is the most practicable distinction from the foregoing species, because the embryonal ring which marks the colony's point of origin is very difficult to find. Zooids described from Norway as spotted with orange and black.

On stones and dead branches of the coral *Lophelia*; 160–900 m; North Atlantic from Barents Sea to Azores. On coralligenous concretions at 20–50 m in Mediterranean (Banyuls).

Class Enteropneusta

2 HARRIMANIIDAE

Up to 300 mm long. Gill bars without synapticulae. Hepatic caeca inconspicuous, not distending body wall. Trunk without circular muscles. Yolky eggs without planktotrophic larvae.

---

1. Length up to 300 mm, with 100–140 pairs of gill pores. Proboscis white, yellowish, or pink, up to 20 mm long. When burrowing only one peristaltic bulge passes down proboscis at a time. Trunk lacks red spots.
   Saccoglossus horsti

   Length up to 200 mm, with 50–95 pairs of gill pores. Proboscis pink, up to 30 mm long. When burrowing each bulge appears at tip before preceding one has reached posterior end of proboscis. Pale brownish trunk bears occasional small red spots
   Saccoglossus ruber

Saccoglossus horsti
Brambell and Goodhart Fig. 13.1
Very like *S. ruber*, but proboscis relatively shorter. Proboscis 13–18 × 4–5 mm, a transverse section through it revealing about nine rings of longitudinal muscle fibres. Length of tongue bars (cut pharynx ventrally and view from within) is reduced suddenly in posterior gill slits, so that only the last two have bar length shorter than its breadth. Egg diameter 0.25 mm.

In mud or muddy sand near low water; reported sporadically on southern coasts of England.

Saccoglossus ruber (Tattersall) Fig. 13.1
Proboscis pink, up to 30 × 3 mm. Collar red or orange, 4 × 3–3.5 mm. Trunk red behind collar and then brown, paler posteriorly, with a few small red spots. Transverse section through proboscis shows 4–6 rings of longitudinal muscle fibres around a less regular central mass. Lengths of tongue bars decrease gradually near hind end of pharynx, length less than breadth in last eight bars.

In sand at LWST on sheltered beaches in Ireland, Scotland, and Wales. Sublittoral in Firth of Clyde.

3 PTYCHODERIDAE

Length may exceed 0.5 m, but only 2–3% is proboscis. Gill bars joined by synapticulae. Oesophageal pores absent. Numerous body-wall

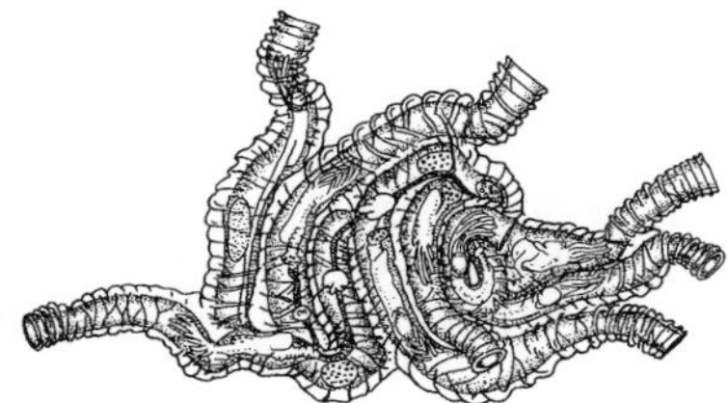
Rhabdopleura compacta

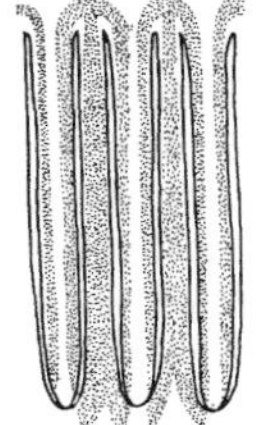
Saccoglossus horsti: tongue bars

S. horsti: posterior tongue bars

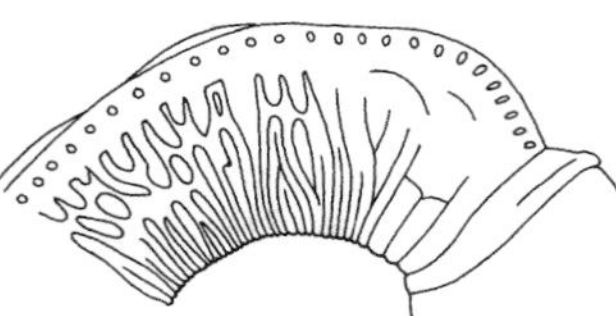
Saccoglossus ruber: anterior trunk

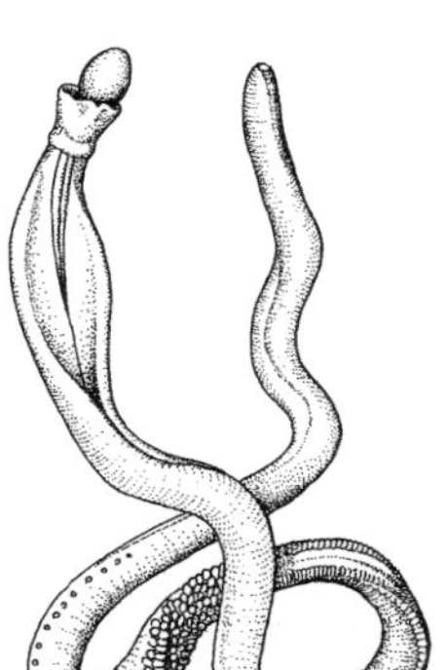
Glossobalanus sarniensis

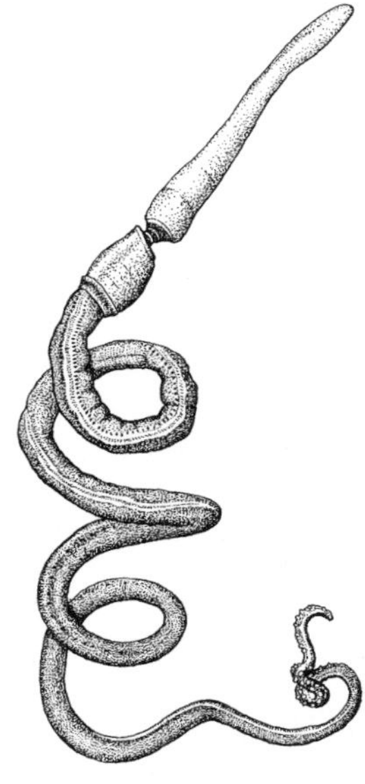
Saccoglossus ruber

Fig. 13.1

protuberances associated with hepatic caeca. A dorsolateral septum, with which the gonads are associated, divides the trunk coelom on each side, within genital folds. A layer of circular muscle usually lies outside the longitudinal muscles of trunk. Eggs small, developing into tornaria larvae.

Glossobalanus sarniensis Koehler Fig. 13.1
May exceed 500 × 10 mm. Proboscis egg-shaped, about 10 mm long. Genital wings separated from hepatic region by a transitional region which lacks prominent protuberances.

In sand at LWST; recorded from Channel Isles, Tresco (Isles of Scilly), Salcombe, Strangford Lough.

PHYLUM CHORDATA: SUBPHYLUM UROCHORDATA (= TUNICATA)

Class Ascidiacea

The tunicate body is contained within a *test*, or *tunic*, secreted by the underlying *mantle*. This test may be thin, thick, clear, opaque, clean, or covered with sand or living organisms. Tunicates are represented on most rocky shores by the ascidians, which are sessile and either colonial or unitary. Some are indefinite in shape, and so hidden within their tunics that they superficially resemble sponges or some of the more fleshy bryozoans or cnidarians (e.g. *Alcyonidium* or *Alcyonium*). When cut open, however, they display a more complicated internal organization (Fig. 13.2), particularly a filter-feeding pharynx, or *branchial sac*, in which mucus is secreted by the ventral *endostyle*. The branchial sac contains perforations, the *stigmata*, in transverse rows. Transverse blood vessels pass between adjacent rows. In some genera the stigmata are arranged in spiral groups.

The feeding current enters the test via an *oral* (or branchial) opening or *siphon*, and passes into the pharynx through a ring of *oral tentacles*. Flow is generated by cilia, passing through the stigmata into the surrounding *atrium*. From there the water passes to the exterior via a dorsal *atrial* (excurrent) *siphon*. Positive pressure builds up inside the atrium, which becomes distended with water. On disturbance, contraction of mantle muscles forces out jets of water through the oral and atrial sphincters, hence the term 'sea squirts' popularly applied to larger ascidians. Unitary ascidians may be 1 cm long or more, and the largest British forms reach about 15 cm.

Colonial ascidians are made up of numerous zooids formed by *budding*. Each zooid is like a small unitary ascidian with few stigmata, but rarely exceeds a few millimeters in length. The colonies may be larger (sometimes as much as 10–20 cm). The zooids may arise from stolons or project from a common test, or be wholly immersed. If the zooids are embedded within the test they are generally arranged in systems; i.e. the atria do not open individually to the exterior but internally with others into a *common cloaca* (Fig. 13.3). The atrial openings may wholly lack siphons, or there may be a *languet* (tongue) projecting from its upper margin. The common cloaca communicates with the exterior through an irregular, circular, or elevated aperture.

Ascidians are hermaphroditic but the location of gonads within the body varies. In the order Enterogona (*enteron*, gut; *gone*, seed) they are essentially associated with the gut loop, though displaced into a *postabdomen* in the long, slender zooids of Polyclinidae. In the order Pleurogona (*pleura*, side) there is no abdomen and both the gut loop and gonads are attached to the mantle wall. Most unitary ascidians shed their eggs into the sea, whereas nearly all colonial species and many small unitary ones incubate their embryos. These will be found in the atria of most colonial species but in the basal test of Didemnidae. Nearly all hatch as tailed *tadpole larvae*.

Ascidians contain no skeletal parts and their identification depends upon features of their internal morphology, e.g. position of the gut and gonads, and structure of the branchial sac. Identification must therefore usually be preceded by dissection. Since contraction caused by death or fixatives distorts the body shape, narcotization is usually essential if diagnostic characters are to be observed. Dissection without fixation is usually impossible so that narcotized specimens have also to be fixed.

Collected specimens should immediately be placed in water, in plastic bags of appropriate size, to which is added a few crystals of menthol. Narcotization will usually occur within a couple of hours. Fix the specimens by adding one-tenth the seawater volume of commercial formalin (40% formaldehyde). Fixed specimens can then be stored in a less noxious preservative such as 1% propylene phenoxytol in distilled water, or transferred to seawater for immediate examination. The colour of specimens is likely to fade in preservative and should be recorded before fixing.

Note (draw if necessary) the external characteristics, particularly the arrangement of systems found in many colonial species. Unitary specimens should be dissected under water in a wax-based dish, observing through a dissecting microscope if necessary. Place the animal on its right side (i.e. its atrial siphon to your right); pin as and when necessary. Remove the test. This may be accomplished either by making a circumferential cut above the base (Fig. 13.2), so allowing the test to be turned inside out over the siphons, or by cutting with scissors in the mid-line, starting at the oral siphon, proceeding down the line of the endostyle, around the base, and up to the atrial siphon; thence back to the oral siphon. Carefully separate the left half of the test from the mantle and remove it. Cut and open the mantle by a dorsal cut from the base to the atrial siphon, carefully freeing the mantle from the branchial sac by cutting the connecting strands (Fig. 13.2).

Compound ascidians should be handled in a Petri dish or watch-glass on the stage of a stereomicroscope. If there are separate zooids, open them with fine scissors, cutting down from the oral siphon. In others (e.g. many polyclinids) zooids are embedded in, but easily dissected out of, the common test. Handling small zooids requires the

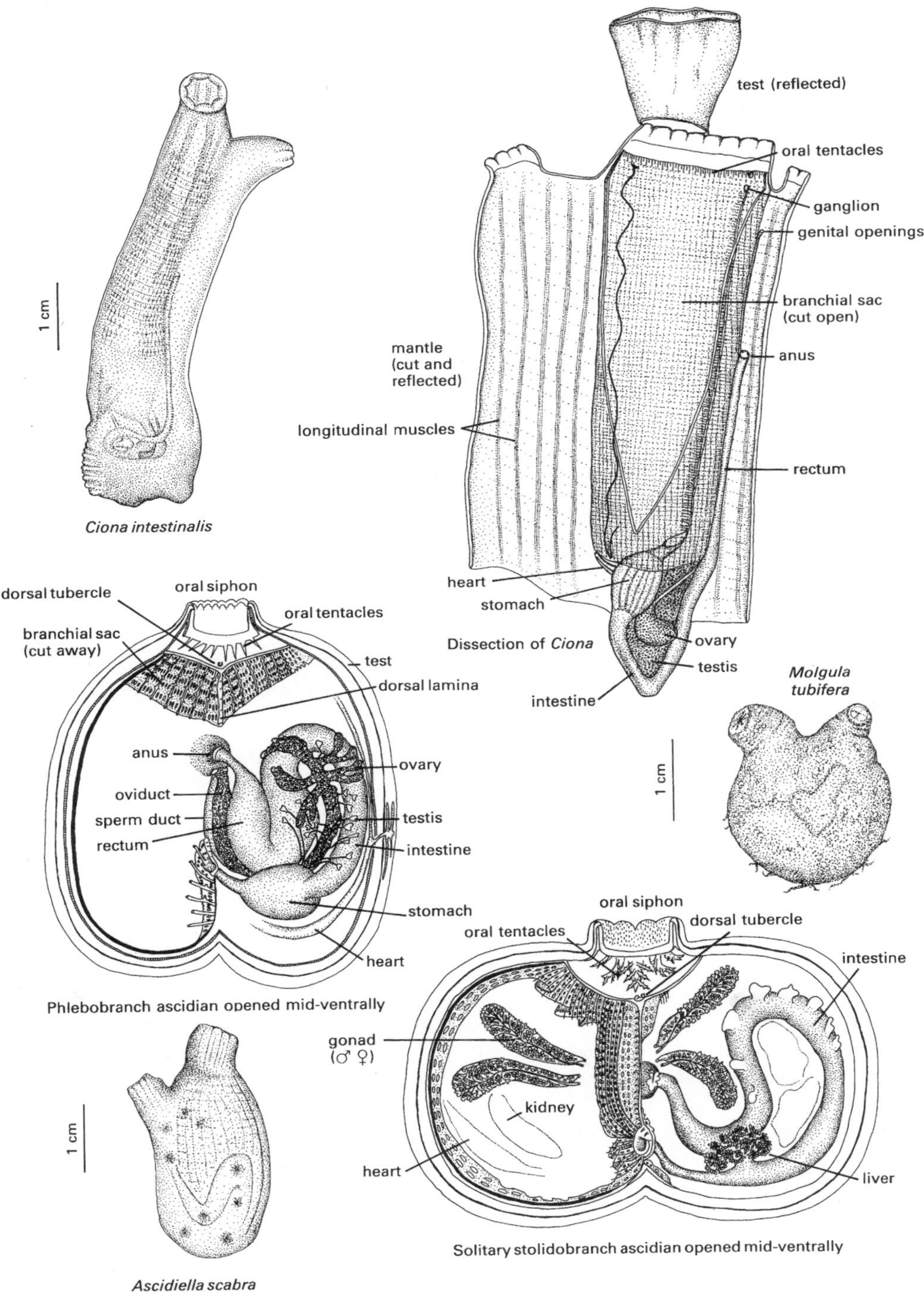
1 cm
Ciona intestinalis
test (reflected)
oral tentacles
ganglion
genital openings
branchial sac
(cut open)
anus
mantle
(cut and
reflected)
longitudinal muscles
rectum
heart
stomach
Dissection of Ciona
ovary
testis
intestine
Molgula
tubifera
1 cm
dorsal tubercle
oral siphon
oral tentacles
branchial sac
(cut away)
test
dorsal lamina
anus
oviduct
sperm duct
rectum
ovary
testis
intestine
stomach
heart
Phlebobranch ascidian opened mid-ventrally
oral siphon
oral tentacles
dorsal tubercle
intestine
gonad
(♂ ♀)
kidney
heart
liver
Solitary stolidobranch ascidian opened mid-ventrally
1 cm
Ascidiella scabra

Fig. 13.2

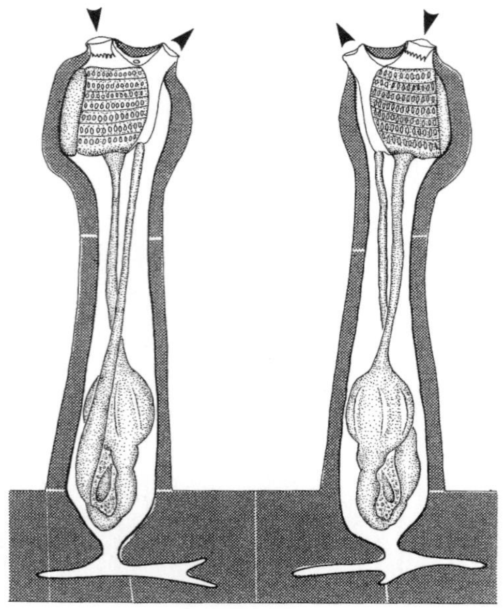

Pycnoclavella

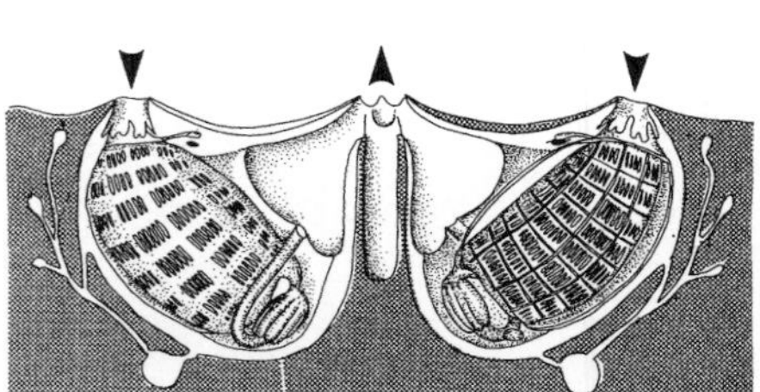

Aplidium

Botryllus

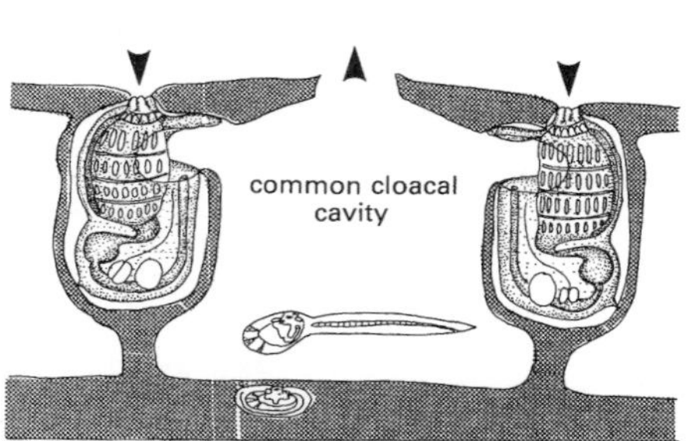

Didemnum

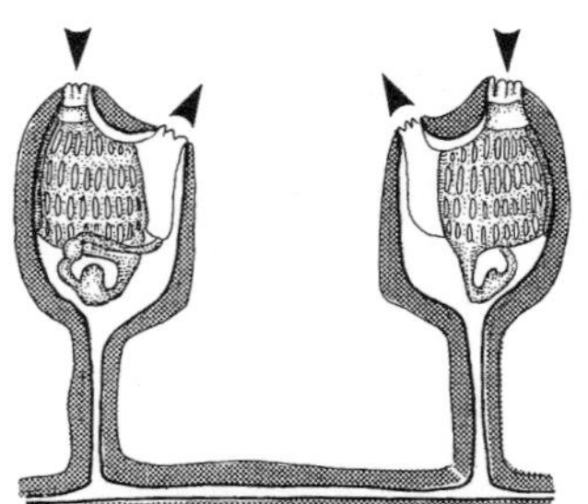

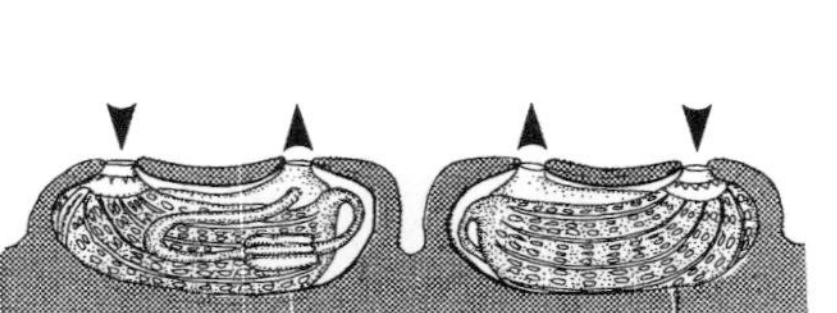

Distomus

Perophora

TYPES OF COLONIAL ORGANIZATION

Fig. 13.3

use of fine needles, needle knives, or sharp, small-bladed scalpels, and watch-maker's forceps; spring-bow scissors are also a help. The minute, close-packed zooids of didemnids are best recovered from a few vertical slices rather than by individual dissection. Didemnid colonies may contain densely aggregated granules: some should be saved for high-power microscopy but the colony may be cleared without the generation of gas bubbles using EDTA (ethylene diamine tetra acetic acid). This is prepared as a solution of 250 g in 1750 ml distilled water buffered to pH 7 by addition of sodium hydroxide pellets.

There are two orders of Ascidiacea: Enterogona and Pleurogona. The former has two suborders.

Order 1 ENTEROGONA

The larva has left and right atrial openings which meet and fuse medially during metamorphosis to form the atrial siphon. Gonads are associated with the intestinal loop, or are postabdominal.

Suborder Aplousobranchia

Branchial sac without internal longitudinal vessels (*haploos*, simple), except in families Cionidae and Diazonidae. Body divided into thorax, abdomen, and, sometimes, postabdomen. Gut abdominal; gonads abdominal or postabdominal. Colonial except Cionidae.

Suborder Phlebobranchia

Branchial sac with internal longitudinal vessels (*phleps*, vein). Body not divided into thorax and abdomen. Gut loop and gonads beside the branchial sac, left or right. Unitary except for Perophoridae. *Perophora* has stolonate colonies and rather small zooids; rows of papillae project from the transverse vessels into the lumen of the pharynx. In other genera the papillae bifurcate and fuse with their neighbours fore and aft, to form the inner longitudinal vessels. These vessels are evenly distributed around the walls of the pharynx, which are not longitudinally folded.

Order 2 PLEUROGONA

Suborder Stolidobranchia

The atrial opening forms as a single, median invagination. Gut loop alongside the pharynx; gonads also lateral, developing in the mantle wall, usually on both sides. Branchial sac (except botryllines) with longitudinal folds (*stolidotos*, hanging in folds) and with inner longitudinal vessels usually distributed unevenly, mostly along the lateral infoldings. Unitary or colonial.

KEY TO SPECIES

1. Unitary ascidians occurring singly or in dense aggregations **2**

   Colonial ascidians without systems (i.e. each zooid having two externally opening apertures); zooids joined at the base by test or stolons **22**

   Colonial ascidians with systems (i.e. oral apertures opening to the surface but atria leading to an internal cavity opening by one or more common cloacal apertures); zooids completely embedded in common test **26**

2. Ascidian attached to a firm substratum (e.g. rock, seaweed) **3**

   Ascidian in or on sand or other soft bottom **19**

3. Ascidian with soft, translucent test; loop of gut below the long branchial sac ***Ciona intestinalis*** (p. 697)

   Test firm, commonly rather opaque; loop of gut alongside the branchial sac **4**

4. Ascidian somewhat rectangular in outline, laterally flattened; test transparent; gut loop to the right of the branchial sac ***Corella parallelogramma*** (p. 705)

   Shape and test not as above; gut loop to the left of the branchial sac **5**

5. Ascidian elongate, tapering below to a long, slender attachment stalk; test leathery; southern Britain **Styela clava** (p. 707)

   Ascidian not as above **6**

6. Oral siphon with 6–8 lobes; atrial siphon six-lobed; test often thick, cartilaginous; branchial sac without folds; gonads in the gut loop; oviparous (Phlebobranchia) **7**

   Oral siphon with 4–6 lobes; atrial siphon four-lobed; test leathery or thin; branchial sac generally with folds; gonads in the body wall; oviparous or larviparous (Stolidobranchia) **12**

7. Surface of test with rounded swellings; southern coasts only **Phallusia mammillata** (p. 707)

   Surface of test smooth, rough, with irregular swellings, or with attached particles **8**

8. Siphons close together; ascidian upright **9**

   Siphons well separated, the atrial siphon at least one-third way down body; ascidian often recumbent **10**

9. Test cartilaginous, opaque, white to pink; branchial test vessels papillate at intersections **Ascidia virginea** (p. 705)

   Test semi-transparent, often showing red pigmentation; branchial sac vessels without papillae at intersections **Ascidiella scabra** (p. 705)

10. Test translucent, greyish, rough (finely papillate); branchial sac vessels without papillae at intersections **Ascidiella aspersa** (p. 705)

    Test not as above; branchial sac papillate at intersections **11**

11. Test thick, smooth surfaced, often pink **Ascidia mentula** (p. 705)

    Test thin, greenish; smooth, rough, or covered with particles **Ascidia conchilega** (p. 705)

12. Both siphons four-lobed; oral tentacles simple; folds in branchial sac four or fewer each side; stigmata straight **13**

    Oral siphon six-lobed; oral tentacles branched; 5–7 folds each side of branchial sac; stigmata in spirals **16**

13. Both siphons four-lobed; stigmata straight **14**

    Oral siphon six-lobed; stigmata in spirals **16**

14. Ascidian flat (depressed), its test tesselate (with mosaic-like markings) **Pyura tessellata** (p. 709)

    Ascidian flask-shaped or, if flat, its test not tesselate **15**

15. Ascidian flat or flask-shaped; siphons usually disposed more or less symmetrically; gonad extensive, on the right side only, with large (almost 0.5 mm) red eggs **Dendrodoa grossularia** (p. 707)

    Ascidian flask-shaped; siphons usually disposed asymmetrically; gonads subdivided into 'tear-drops', in both walls; eggs small **17**

16. Surface of test tough, leathery, wrinkled; not predominantly red **Polycarpa pomaria** (p. 707)

Test smooth, reddish; southern and western coasts only
Polycarpa rustica (p. 707)

---

17. Lobes of siphons fringed; ovaries (on each body wall) with short oviduct directed anteriorly (toward the six-lobed oral siphon)
Molgula complanata (p. 709)

Lobes of siphons entire; oviduct directed posteriorly (toward the four-lobed atrial siphon) **18**

---

18. Branchial sac with six folds each side; oviparous *Molgula tubifera* (p. 711)

Branchial sac with seven folds each side; larviparous
Molgula citrina (p. 709)

---

19. Ascidian elongate, club-shaped
Pelonaia corrugata (p. 707)

Ascidian rounded or flask-shaped **20**

---

20. Body up to 80 mm; siphons well separated; test covered with particles except between the siphons
Molgula oculata (p. 711)

Body not exceeding 30 mm; siphons rather close **21**

---

21. Oviduct (on the body wall) directed posteriorly (toward the four-lobed atrial siphon); test fibrils with attached particles; may occur in aggregations
Molgula tubifera (p. 711)

Oviduct directed anteriorly (toward thesix-lobed oral siphon); test completely covered with particles
***Molgula occulta* (p. 710)**

---

22. Zoooids minute (<5 mm high), lentil-shaped, shortly stalked, well separated, linked by stolons; test translucent; gut loop beside the branchial sac *Perophora listeri* (p. 705)

Zooids taller than wide, joined basally, translucent; gut below the branchial sac **23**

Zooids upright or flattened, joined by stolons or lateral expansions of test; test opaque, yellow to reddish **25**

---

23. Zooids up to 20 mm high, united only at the very base by short stolons
Clavelina lepadiformis (p. 697)

Zooids arising from, and partially immersed in, an encrusting base; free portion <20 mm **24**

---

24. Basal mass thick, colonies often >10 cm across; oviparous
Diazona violacea (p. 697)

Basal incrustation up to about 5 mm thick, colonies small; larviparous
Pycnoclavella aurilucens (p. 697)

---

25. Colony encrusting; zooids closely packed, usually depressed
Distomus variolosus (p. 709)

Colony composed of more or less separated, upright zooids budded from stolons *Stolonica socialis* (p. 709)

---

26. Colony cushion-, mound-, or club-shaped; zooids long and slender (the key now uses characters of zooids, which should be dissected out with cuts through the test parallel to them) **27**

Colony flat, encrusting (rarely a lobe); zooids short **36**

---

27. Colony a flat cushion or mound, *not* pear-shaped with a narrow base or a distinct stem **28**

Colony pear- or club-shaped, narrowed at the base or with a distinct stem; or of lobes that are taller than wide **33**

---

28. Zooids comprising thorax and abdomen; embryos brooded in an embryo sac (diverticulum of the atrium); oral siphon with six shallow lobes ***Distaplia rosea*** (p. 698)

    Zooids comprising thorax, abdomen, and postabdomen, with gonads in the postabdomen; embryos brooded in the atrium; oral siphon with six or eight triangular lobes **29**

---

29. Oral siphon with eight lobes ***Sidnyum elegans*** (p. 699)

    Oral siphon with six lobes **30**

---

30. Stomach surface smooth ***Polyclinum aurantium*** (p. 699)

    Stomach surface tuberculate ***Synoicum pulmonaria*** (p. 702)

    Stomach with linear folds **31**

---

31. Colonies to 60 mm diameter, with three or four well-defined systems of zooids, each surrounding a common cloacal aperture ***Aplidium nordmanni*** (p. 699)

    Colonies not >40 (usually not >20) mm across, each cushion comprising a single system of zooids **32**

---

32. Colony heavily encrusted with sand, occurring on rock faces; atrial opening of zooid overhung by a triangular flap (languet) ***Aplidium densum*** (p. 698)

    Colonies mostly free of sand, generally on algae; atrial opening without languet ***Aplidium pallidum*** (p. 699)

---

33. Oral siphon with six lobes **34**

    Oral siphon with eight lobes **35**

---

34. Stomach surface smooth ***Polyclinum aurantium*** (p. 699)

    Stomach with six linear folds ***Aplidium punctum*** (p. 699)

    Stomach with about 30 vertical columns of tubercles ***Aplidium proliferum*** (p. 699)

---

35. Colony of reddish capitate lobes up to at least 40 mm in length; stomach with prominent tubercles ***Morchellium argus*** (p. 699)

    Colony a cluster of short lobes, each not >20 mm when intertidal; stomach with linear folds, these continuous or broken into short lengths ***Sidnyum turbinatum*** (p. 699)

---

36. Colony soft or gelatinous, not containing small, spherical, calcareous granules; the zooids often visible internally or of contrasting colour to the test (beware of some forms of *Trididemnum tenerum* which are mauve-purple and almost devoid of granules) **37**

    Colony containing small, spherical, calcareous granules, often so densely packed that the colony is uniformly or streakily white **40**

---

37. Colony greyish, translucent, often soft **38**

    Colony firm, the zooids often vividly contrasting in colour to the test **39**

---

38. Colony thin (<2 mm), transparent; intertidal and just below, often on algae ***Diplosoma listerianum*** (p. 702)

Colony thicker (4 mm), opaque; subtidal, especially on rock surfaces ***Diplosoma spongiforme*** (p. 702)

39. Zooids arranged in small, star-shaped clusters ***Botryllus schlosseri*** (p. 709)

Zooids arranged in sinuous parallel lines ***Botrylloides leachii*** (p. 709)

40. Colonies white to violet, densely packed with granules; oral openings of zooids not very apparent; zooids with four rows of stigmata; larvae with two fixatory papillae (the indistinguishable *D. coriaceum* has larvae with three fixatory papillae) ***Didemnum maculosum*** (p. 702)

Colonies white, densely packed with granules; oral openings of zooids conspicuous at the surface; zooids with four rows of stigmata; larvae with three fixatory papillae ***Lissoclinum perforatum*** (p. 703)

Colonies variable in colour (depending on the abundance of granules), whitish through mauve to black, with granules generally few (when abundant the colony resembles *Didemnum maculosum*); oral openings of zooids not very apparent; zooids with three rows of stigmata; larvae with three fixatory papillae ***Trididemnum cereum*** (p. 704)

(The species included in this triplet can be difficult to separate.)

Order Enterogona

Suborder Aplousobranchia

1 CIONIDAE

Unitary. Pharynx with inner longitudinal vessels and many rows of stigmata.

Ciona intestinalis (Linnaeus) Fig. 13.2

Test soft and translucent; body reaching 100–150 mm, nearly cylindrical, pale yellow-green, gelatinous and contractile, with five longitudinal muscle bands clearly visible in the mantle on each side. Apertures may have bright yellow margins and small red pigment spots between the lobes.

On rocks and piles, LWST and shallow water; common in harbours; all coasts.

2 DIAZONIDAE

Colonial. Pharynx with inner longitudinal vessels and many rows of stigmata.

Diazona violacea Savigny Fig. 13.4

Colonies 100–300 mm across. Zooids soft, translucent, pale green, like a bunch of small *Ciona* with their proximal ends embedded in a common basal mass. Oviparous.

Sublittoral and mostly in deep water; Outer Hebrides and western Ireland to Adriatic.

3 CLAVELINIDAE

Zooids arising freely from a basal stolon or encrustation. Thorax transparent but the ciliated tracts marked by lines of bright opaque pigment. Pharynx without inner longitudinal vessels. Apertures on short siphons, with unlobed margins. Larvae brooded in atrium.

Clavelina lepadiformis (Müller) Fig. 13.4

Conspicuous white or yellow lines mark the endostyle, peripharyngeal band, and dorsal lamina. Stomach brown with cream-white markings. Eggs and larvae reddish in the atrial cavity.

On rocks, etc. from LW pools to about 50 m; all coasts; Norway to Adriatic.

Pycnoclavella aurilucens Garstang Fig. 13.4

White, yellow, or orange pigments mark the pharyngeal feeding tracts. Basal test greenish-brown, sometimes sand encrusted.

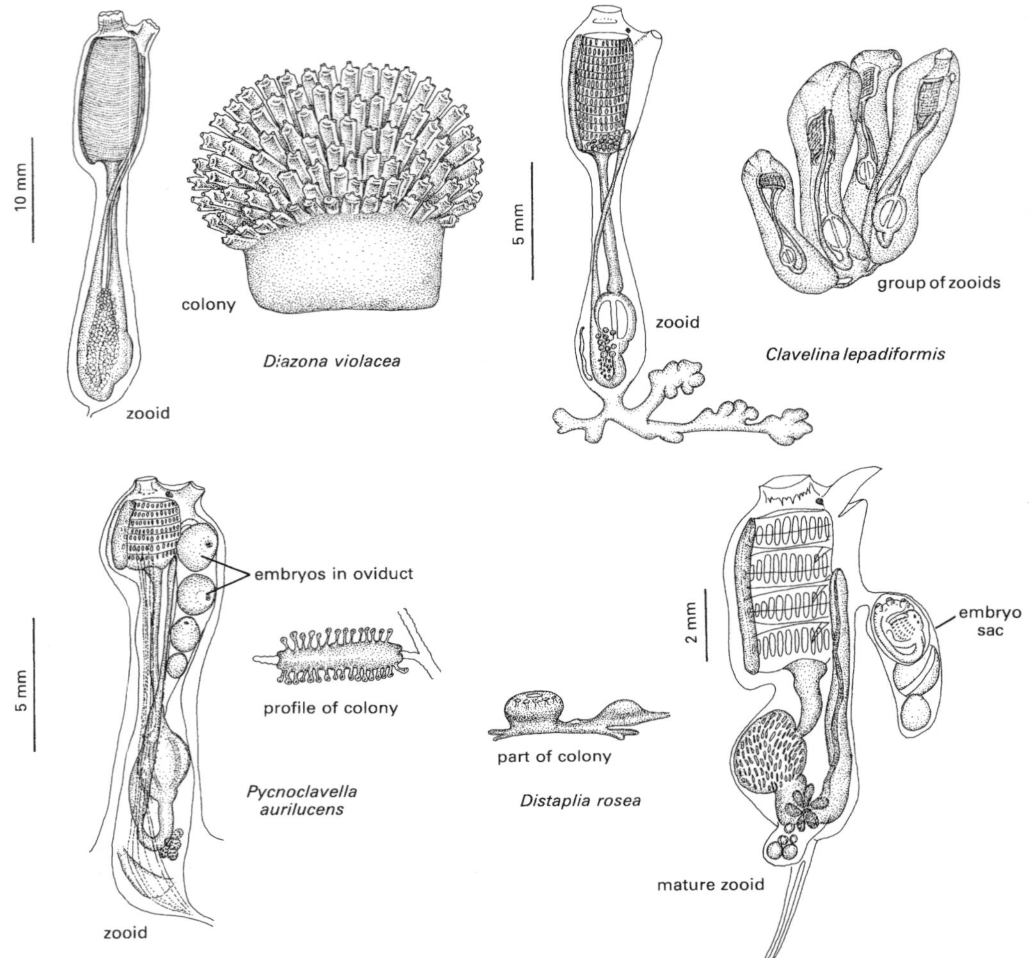

Fig. 13.4

On current-swept rocks at 10–30 m from Plymouth, south-west Britain, and the west of Ireland to Galway Bay; Brittany and perhaps south to Mediterranean.

4 HOLOZOIDAE

Fleshy colonies with zooids embedded in groups, each group surrounding a common atriopore. Gonads alongside gut loop.

Distaplia rosea Della Valle — Fig. 13.4

Colonies in form of rounded, dome-shaped, or lobate masses, 10–30 mm across, arising from basal plates or strands. Zooids with long atrial languet.

On stones and shells, 0–100 m, from Scotland to the Mediterranean.

5 POLYCLINIDAE

Colonies more or less mound-shaped or capitate, with the zooids wholly immersed and forming systems. Zooids elongate, divisible into thorax, abdomen, and postabdomen; branchial sac with numerous rows of stigmata; heart and gonads in postabdomen. Development viviparous. The British species are classified into five genera.

Aplidium densum (Giard) — Fig. 13.6

Colony composed of sessile cushions, each with a single common cloacal opening; sides and upper surface heavily encrusted with sand; 6–10 mm across. Oral siphon six-lobed; atrial opening with a languet. Stomach with 8–10 vertical folds.

Rock faces, etc., *Laminaria* zone down to 50 m; western coasts including Irish Sea up to Northern Ireland; southwards to Mediterranean.

Aplidium nordmanni
(Milne Edwards) Fig. 13.6
Colonies sessile, flat, up to 60 mm across, about 6 mm (or more) high. Zooids in regular, circular, or linear systems, each with a common cloaca. Oral siphons six-lobed, white, contrasting with the pink upper surface. Stomach with almost 30 simple, longitudinal folds. Larvae large, trunk about 1 mm long (cf. *A. proliferum*).

Under boulders on protected shores and from shallow water. South-west England and Brittany, to the Mediterranean.

Aplidium pallidum (Verrill) Figs 13.5, 13.6
Colonies small, 15–20(40) mm across, flattish, attached by a broad base; yellowish-buff; mostly free of sand. Systems with 6–8 zooids; small colonies comprising a single system. Zooids small, <4 mm long. Oral siphon six-lobed: atrial siphon without languet; stomach with 10–12 undivided folds.

Lower shore to 200 m, especially on algae and *Zostera*; common on south-western coasts.

Aplidium proliferum
(Milne Edwards) Figs 13.5, 13.6
Colony capitate with red or yellowish zooids in a transparent test; colonies to 50 mm high. Oral siphon six-lobed; atrial languet present, simple; stomach with transversely divided longitudinal folds. Larval trunk about 0.7 mm long.

Lower shore and in shallow water; on and under rocks, etc., also on shells and sponges. On all except North Sea coasts, commonest in the south-west; western Norway to the Mediterranean.

Aplidium punctum (Giard) Fig. 13.6
Distinctive clavate or capitate colonies, 6–12(–40) mm long; arising from a creeping, sand-encrusted base. Test clear; zooids whitish, not obviously in systems; up to 40 or more in each lobe. Oral siphon six-lobed; a red spot over the anterior end of the endostyle; atrial opening with a trifid languet; stomach with six thick folds.

On algae or in rock crevices; lower shore and shallow water; western English Channel.

Morchellium argus
(Milne Edwards) Figs 13.5, 13.6
Colony of reddish capitate lobes up to about 40 mm in length (longer when hanging below rock faces); stalks sand-encrusted, heads translucent, flocculent creamy with numerous fine red spots. Zooids long; oral siphon eight-lobed; branchial sac with four bright carmine spots anteriorly; atrial languet present, simple. Stomach with numerous, tubercle-like swellings.

Lower shore and shallow water, under boulders and overhangs, sometimes abundant. West and south-west Britain to France (Brittany).

Polyclinum aurantium
Milne Edwards Figs. 13.5, 13.6
Colony of round, pear-shaped, or flat-topped heads up to 30 mm across; yellow-grey or yellow-brown, sand encrusted. Each head with zooids in one or few systems. Zooids distinctive: postabdomen with slender attachment to abdomen. Oral siphon six-lobed; stomach wall smooth, the intestinal loop below the stomach twisted.

On stones, rock, and algae; lower shore to about 100 m; mainly southern and western coasts; from Norway to the Mediterranean.

Sidnyum elegans (Giard) Figs 13.5, 13.6
Colonies cushion-shaped, narrowed below but rarely capitate; 5–10(20) mm high, up to 60 mm across; pinkish-white in colour. Zooids in conspicuous irregular systems, arranged along canals, with somewhat indeterminate common openings; systems oval in small colonies. Oral siphon eight-lobed; branchial sac rose-coloured, siphons white; atrial languet with short lateral lobes. Stomach with 18–22 longitudinal, rarely broken folds.

Low on shore and in shallow water; mainly English Channel.

Sidnyum turbinatum Fleming Figs 13.5, 13.6
Colony of several lobes rising from encrusting stolons. Intertidally the lobes contain 2–4 systems and reach a size of 20 × 10–15 mm. Subtidal specimens reach 40 mm in height. Test semi-transparent, yellow, grey, and white. Oral siphon eight-lobed; three, or sometimes four red pigment spots; atrial opening with languet. Stomach with

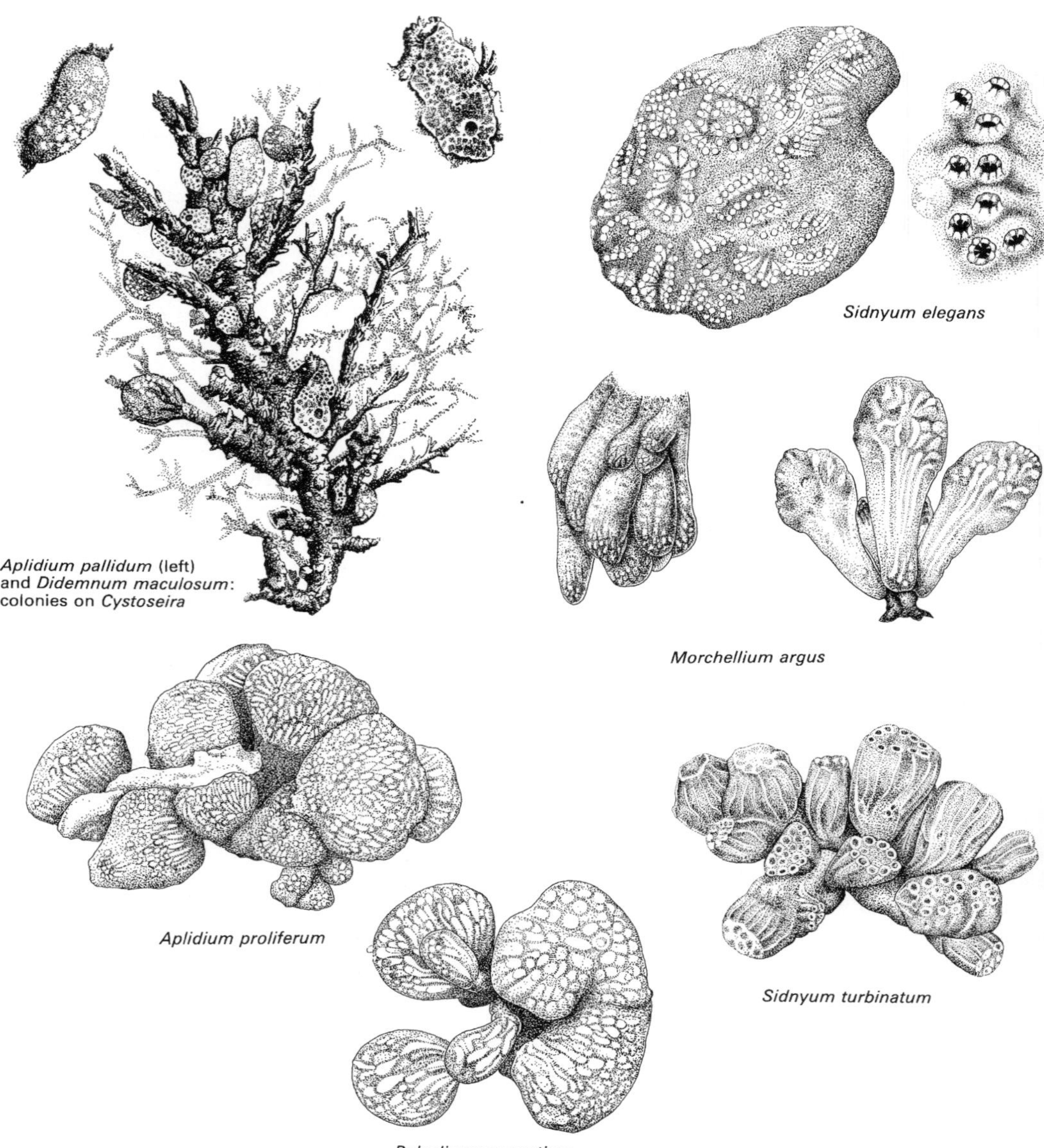
Sidnyum elegans
Aplidium pallidum (left)
and *Didemnum maculosum*:
colonies on *Cystoseira*
Morchellium argus
Aplidium proliferum
Sidnyum turbinatum
Polyclinum aurantium

Fig. 13.5

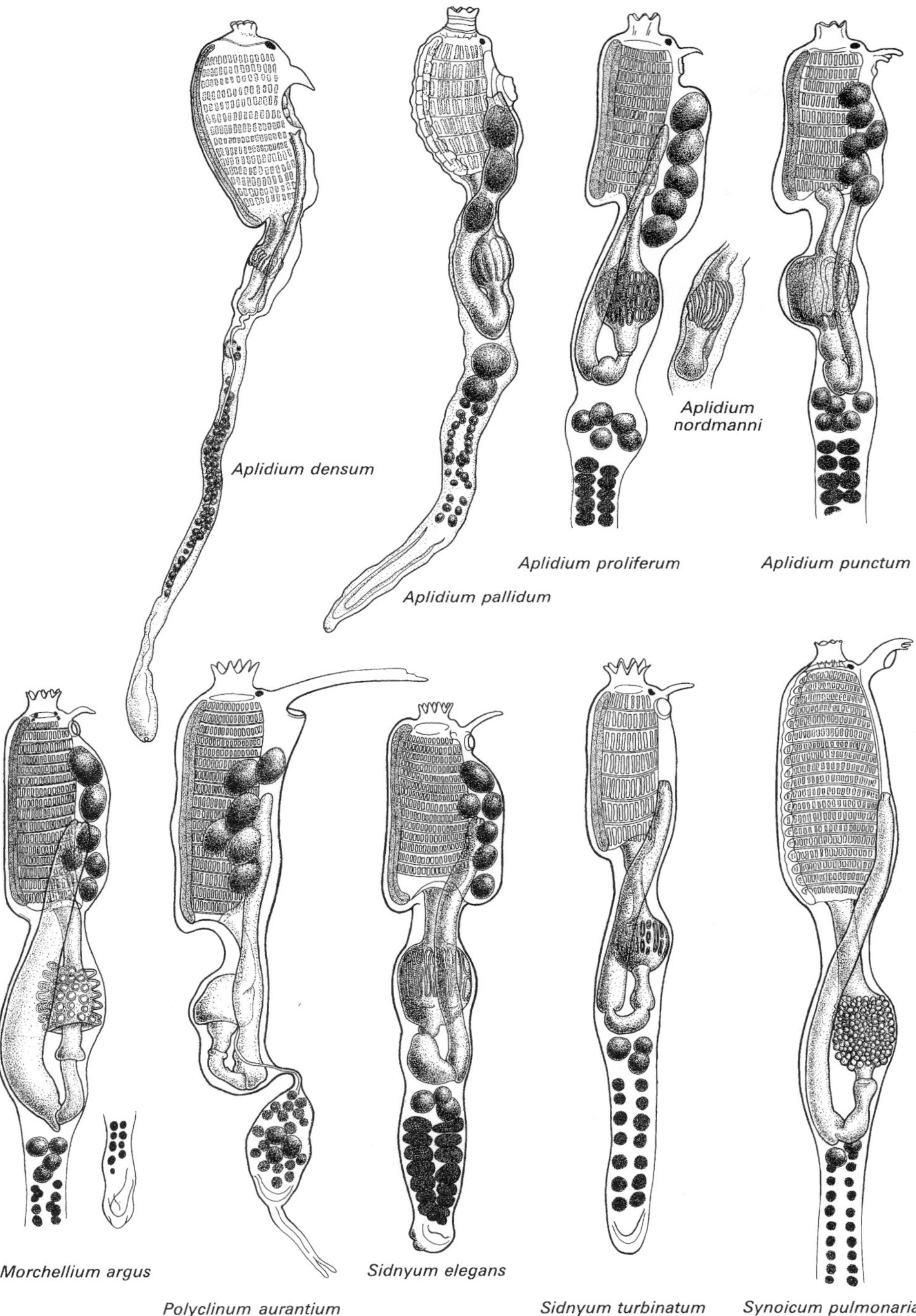
Aplidium densum
Aplidium nordmanni
Aplidium proliferum
Aplidium punctum
Aplidium pallidum
Morchellium argus
Sidnyum elegans
Polyclinum aurantium
Sidnyum turbinatum
Synoicum pulmonaria

Fig. 13.6

10–15 folds, continuous or variously broken into short lengths.

On stones, rocks, and algae; lower shore to 200 m; all British coasts, rarely North Sea; Norway to Mediterranean.

Synoicum pulmonaria
(Ellis and Solander) Fig. 13.6
Colonies pear-shaped or rounded, to about 100 mm in diameter; semi-transparent, yellowish-grey, with or without sand. Many circular systems. Oral siphon six-lobed; atrial siphon short, with a broad languet terminating in three processes. Stomach wall with rounded markings or low papillae; intestine twisted.

Sublittoral; 20–60 m on hard sandy bottoms; Arctic–Boreal.

6 DIDEMNIDAE

Colonies thin and encrusting, the base coextensive with the surface; gelatinous or packed with fine calcareous, stellate *granules*. Zooids very small, grouped around common cloacal openings, constricted below the pharynx and abdomen; no postabdomen. Branchial aperture six-lobed; atrial aperture variable. Three or four rows of stigmata. Saucer-shaped *lateral thoracic organs* (LTOs), one each side of the thorax, produce the calcareous granules. Gonads in the intestinal loop. Eggs large, developing one at a time. Testis single or subdivided; sperm duct straight or coiled spirally (anticlockwise from the apex) around the testis; when the latter, the rectum crosses to the left of the oesophagus. Tadpole larvae with two or, more usually, three adhesive papillae and with anterior ampullae; developing in the common test toward the base of the colony. Colonial budding oesophageal: two buds grow from the waist of the zooid and, as the zooid itself divides, the anterior bud forms a new abdomen for the thorax and the posterior bud forms a new thorax for the abdomen.

Features of very transparent didemnids may be observed direct. Spicules can be removed by treatment with ethylene diamine tetra acetic acid (EDTA), but usually it is necessary to cut thin vertical slices through a previously narcotized and fixed colony.

Didemnum coriaceum (Von Drasche) Fig. 13.7
Colonies thin, whitish to violet; rigid from densely packed granules (and indistinguishable from *D. maculosum*). Larvae large, 0.5 mm in length, and with three fixatory papillae (cf. *D. maculosum*); present May–September (France).

On rocks; in Britain recorded only from Plymouth; elsewhere from Brittany to western Norway and the Faroes.

Didemnum maculosum
(Milne Edwards) Fig. 13.7
Colony encrusting, smooth or rugose, hard, packed with granules; white, grey, violet, or whitish marked with violet. Granules generally with >15 rays, summits pointed (45°). Zooids about 1 mm but contracting strongly on preservation; four rows of stigmata; atrial opening simple, large, exposing some of the stigmata. Three of the six oral lobes are elongate in var. *dentata*. LTOs vertical, on the edge of the atrial opening, between the third and fourth rows of stigmata. Gonads to the left; testis undivided, sperm duct spiral, with 7–10 turns. Larvae small, 0.4 mm, with two fixatory papillae and 4–6 pairs of ampullae.

On rocks, stones, *Laminaria* holdfasts, *Cystoseira*, *Delesseria*, etc. from the lower shore and sublittoral; southern and western coasts, with an extensive further distribution. Commoner than *D. coriaceum* (3 : 1 at Plymouth).

Diplosoma listerianum
(Milne Edwards) Fig. 13.7
Colonies thin (<2 mm), moderately extensive, transparent, sometimes with brown or grey pigmented cells. Common cloacal apertures slightly raised in life, scarcely visible from above. Zooids about 2 mm, with four rows of stigmata; atrial opening simple, large, almost completely exposing the stigmata: no languet. No LTOs. Gonads to the left of the gut. Testis usually bilobed; sperm duct not proximally coiled, aligned with the rectum. Ovary containing several eggs. Larval trunk 0.4–0.6 mm, with three fixatory papillae.

Intertidal and in shallow water, especially on algal fronds, *Zostera* leaves, etc. Apparently widely distributed but in Britain not distinguished from the two following species.

Diplosoma spongiforme (Giard) Fig. 13.7
Colonies thick (commonly about 6 mm), soft but tough, pigmented; forming extensive, loosely attached crusts (up to 0.25 m^2) on rock faces. Greyish, containing white, non-calcareous bodies; zooids sometimes with black speckles. The surface and basal test joined by strands which branch

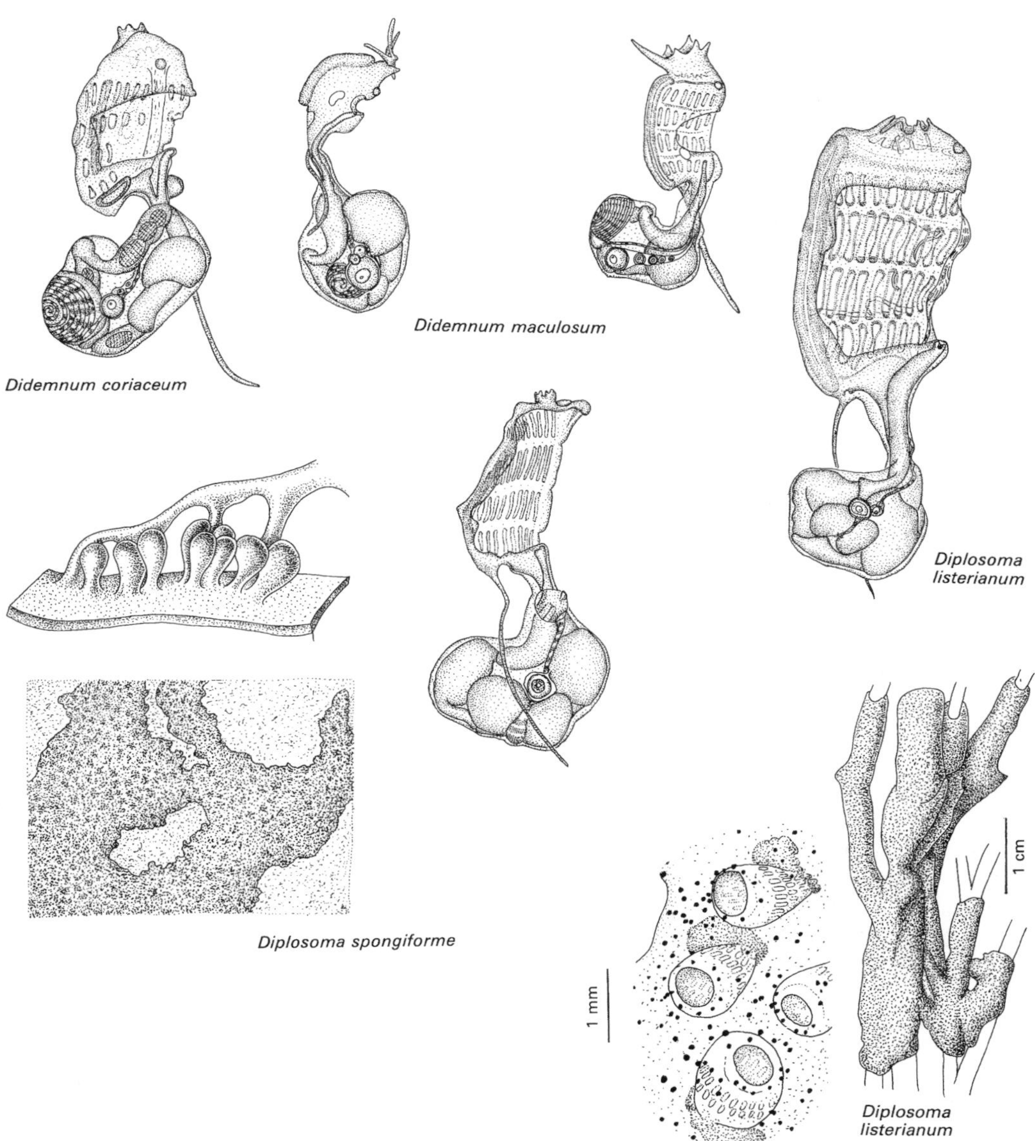

Fig. 13.7

toward the surface and enclose zooids in clusters of 4–6 around the generator zooid. Zooids about 1.5 mm. Atrial opening large. No LTOs. Gonads to the left of the gut. Testis bilobed; common sperm duct straight, aligned with the rectum. Ovary containing only one oocyte at a time. Larvae large, the trunk 0.6–0.8 mm in length, with 3–8 fixatory papillae and 2–7 pairs of ampullae.

Sublittoral, on rock faces (rarely algae) in the *Laminaria* zone, down to 40 m; western coasts, Brittany; also Mediterranean.

Lissoclinum perforatum (Giard) Fig. 13.8
Colonies rarely exceeding 30 mm across, about 2 mm thick; white or creamy with dark spots marking the oral openings and larger slits opening

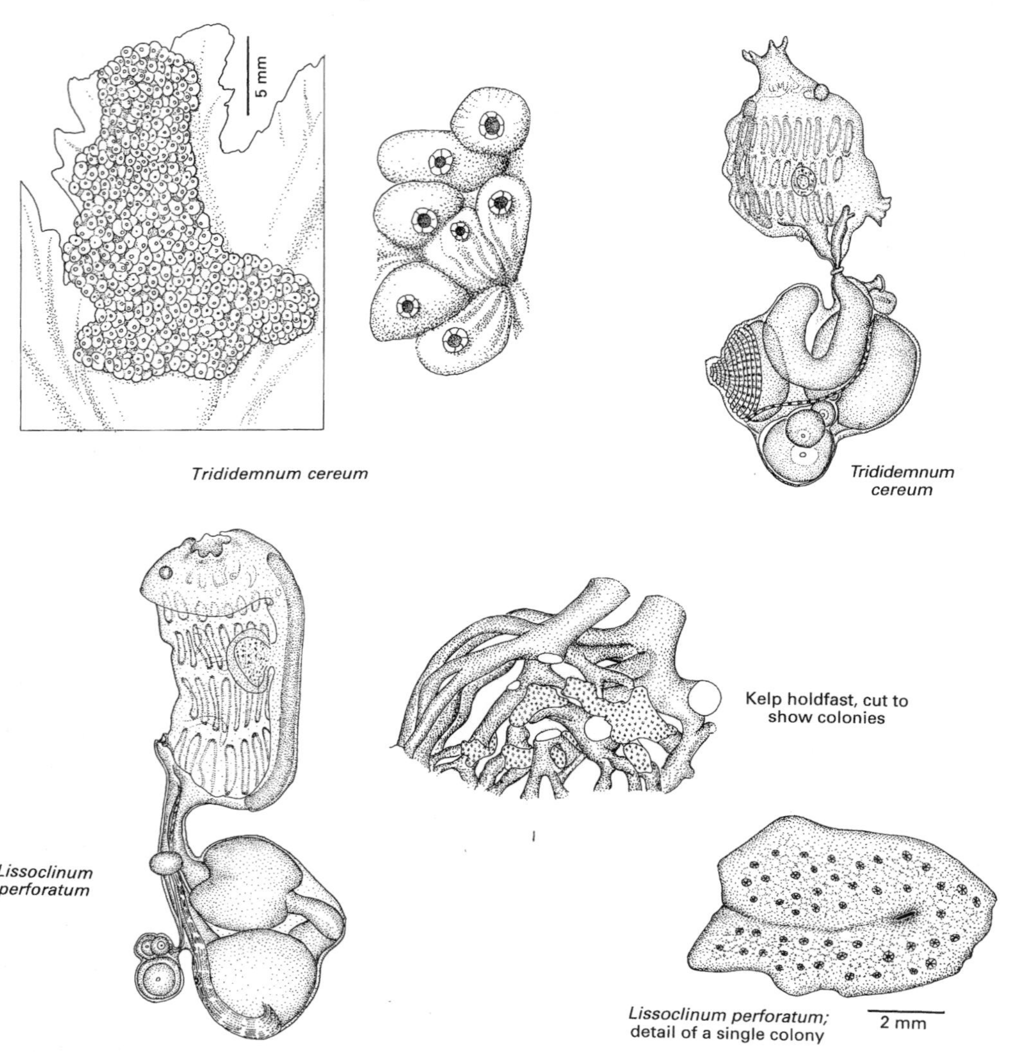

Fig. 13.8

from the extensive common cloaca. Zooids not clearly in systems. Colony border distinct, without zooids. Granules densely packed: rays <15, blunt ended; about 0.03 mm diameter. Zooids about 2 mm; a large atrial opening, no languet. LTOs wing-like, vertical, on the edge of the atrial opening between the second and third rows of stigmata. Gonads on right of gut; testis single, sperm duct straight, aligned with the rectum. Larval trunk about 0.6 mm long; three fixatory papillae in vertical row, four pairs of ampullae.

On *Laminaria* in the shallow sublittoral and on rocks, especially in strong tidal flow, to 30 m; southern and western coasts; Brittany.

Trididemnum cereum (Giard) Fig. 13.8

Colonies encrusting, up to 5 mm thick, of variable extent; semi-transparent, whitish, yellowish, mauve or brownish-black, with black pigment spots. Granules generally few, between the basal and surface test; when abundant in the surface test the colony resembles *Didemnum maculosum*. Granules very variable, the rays acicular, linear, or tapering. Zooids with three rows of stigmata, and an atrial siphon level with the lowest row of stigmata; often a dark spot anteriorly over the endostyle. LTOs conspicuous, mid-lateral, between the second and third rows of stigmata. Gonads on the left; testis undivided; sperm duct proximally spiral, with 7–12 turns. Larval trunk 0.6 mm long; three papillae in vertical row and four pairs of anterior ampullae.

On rocks, algae (e.g. haptera of *Laminaria*), and other ascidians; lower shore down to about 200 m; south and west coasts of Britain; Norway to Mediterranean.

Suborder Phlebobranchia

7 PEROPHORIDAE

Colonial, with zooids united by basal stolons. Branchial sac with or without longitudinal vessels. Gut to left of branchial sac; gonads in the intestinal loop.

Perophora listeri Forbes — Fig. 13.9

Colonies of well-separated zooids arising from creeping stolons. Zooids shortly stalked, to about 4 mm high, somewhat laterally compressed; transparent. Siphons short; four rows of stigmata; no inner longitudinal vessels. Larvae develop in atrial cavity.

On stones, hydroids, algae, etc., lower shore to about 30 m; Irish Sea, south and west coasts; southwards to Mediterranean.

8 CORELLIDAE

Unitary. Gut to right of branchial sac. Gonads in or around the intestinal loop. Oviparous.

Corella parallelogramma (Müller) — Fig. 13.9

Body subquadrate and laterally compressed, up to about 50 mm high, attached by a small part of the base. Test hard, thin, smooth, and transparent; mantle with yellow, white, and red markings. Stigmata in subquadrate spirals.

On stones, shells, *Laminaria*, etc. in clear water down to about 200 m; all coasts; Norway south to the Mediterranean.

9 ASCIDIIDAE

Unitary. Oral siphon with 6–8 lobes, atrial with six. Gut to left of branchial sac. Stigmata straight. Gonads within the loop of, or spreading over, the intestine.

Ascidia conchilega Müller — Fig. 13.9

Body elliptical oblong, attached along the left side; oral siphon terminal, atrial siphon two-thirds down body; test thin, translucent, greenish, usually with small rough projections; naked or covered with shell fragments; up to 60 mm in length. In branchial sac, the right face of the dorsal lamina with small irregular papillae.

On stones, and shells; lower shore to deep water; all coasts; from the Faroe Islands and Norway to the Mediterranean.

Ascidia mentula Müller — Fig. 13.9

Body elongate, usually 50–180 mm; test thick, cartilaginous, translucent, usually pink, raised into irregular swellings; attached mostly by the left side; oral siphon terminal, atrial siphon one-half to two-thirds down body. In branchial sac the right face of the dorsal lamina without papillae.

On rock, stones, and shells; from the lower shore to about 2300 m; all coasts; from Norway to the Mediterranean.

Ascidia virginea Müller — Fig. 13.9

Body subrectangular, attached by part or all of one side; usually 30–80 mm. Test smooth, cartilaginous, translucent, white or pink. Oral siphon terminal, atrial siphon a short distance from it. Rectum long, extending beyond the intestinal loop.

On stones and shells; 30 to about 400 m; all except North Sea coasts; Norway to the Mediterranean.

Ascidiella aspersa (Müller) — Fig. 13.9

Body ovoid, usually attached by the left side; 50–130 mm. Test cartilaginous, greyish, surface rough (finly papillate). Oral siphon terminal, atrial siphon about one-third down body. Dorsal tubercle with elliptical opening, both horns incurved. Oral tentacles 30–40, well separated, fewer than the inner longitudinal vessels. Ovary on left side of intestinal loop, testis on both sides.

On stones, algae, and man-made structures, often aggregated in masses. Common on all British coast; distributed Norway to the Mediterranean.

Ascidiella scabra (Müller) — Fig. 13.9

Body ovoid, usually attached by the left side; 20–50 mm. Test cartilaginous, smooth or wrinkled, with small red papillae around the siphons; semi-transparent, with red pigment on the upper mantle often

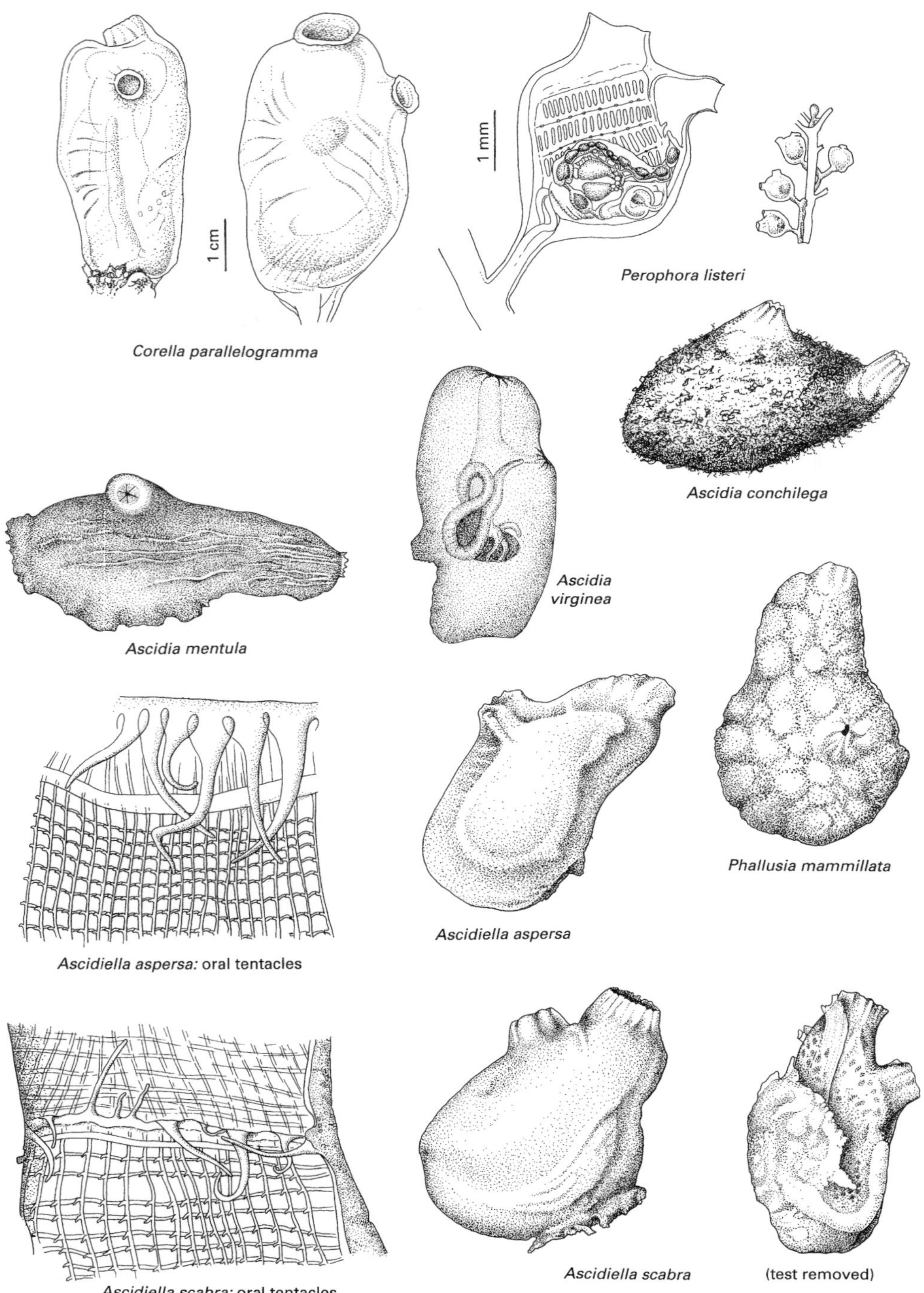
1 cm
1 mm
Perophora listeri
Corella parallelogramma
Ascidia conchilega
Ascidia virginea
Ascidia mentula
Phallusia mammillata
Ascidiella aspersa
Ascidiella aspersa: oral tentacles
Ascidiella scabra
(test removed)
Ascidiella scabra: oral tentacles

Fig. 13.9

visible. Oral siphon terminal, atrial siphon close to it (about one-quarter down body). Dorsal tubercle U-shaped, with the horns variously coiled. Oral tentacles 30–110, always more numerous and closer spaced than the inner longitudinal vessels; no papillae at the intersections of the longitudinal and transverse vessel.

On algae, stones and shells; lower shore to about 300 m; all coasts; Norway to the Mediterranean.

Phallusia mammillata (Cuvier) Fig. 13.9

Body ovoid, broader at the base, up to 140 mm in length. Test thick, cartilaginous, smooth, raised into rounded mammillae; whitish to brown. Oral siphon terminal, atrial siphon one-third to one-half down body. Posterior part of branchial sac recurved upon itself.

Attached to stones, often in silty situations; lower shore and shallow water down to 180 m; south coast of Britain and south-west Ireland (possibly further north); southwards to the Mediterranean.

Order Pleurogona

Suborder Stolidobranchia

10 STYELIDAE

Unitary or colonial. Siphons smooth edged or four-lobed; oral tentacles simple. Branchial sac with never more than four folds each side; stigmata straight; dorsal lamina smooth-edged. Stomach with ridges and pyloric caecum, but no 'liver'. Gonads in the body wall, alongside the branchial sac; variously developed, single, paired, or many; those in the form of multiple small bodies, embedded in the body wall or projecting from it, being termed *polycarps*.

The family is divided into three subfamilies:

A. STYELINAE. Unitary but often aggregated. Both siphons four-lobed. Branchial sac usually with four folds per side. Gonads on one or both sides, usually multiple.

B. POLYZOINAE. Forming loose colonies or clonal aggregations by budding from the mantle wall; without systems. Branchial sac with or without folds.

C. BOTRYLLINAE. Colonial, with systems. No branchial folds.

10A STYELINAE

Dendrodoa grossularia (Van Beneden) Fig. 13.10

Body depressed dome-shaped on a spreading base, or upright; aggregated or solitary, reddish-brown; up to about 20 mm in length, not more than 15 mm in diameter. Branchial folds rudimentary, their position marked by groups of inner longitudinal bars. Gut on the left. One gonad only, on the right, cylindrical, parallel with the endostyle; ovarian follicle surrounded by testis; viviparous, the tadpoles developing in the atrium.

On rock, stones, shell, etc., from the lower shore into deep water; all coasts, in aggregations, particularly in the south and west; locally common; Arctic south to Isles of Scilly and Brittany peninsula.

Polycarpa pomaria (Savigny) Fig. 13.10

Shape variable, ovoid, subglobular or conical, up to 70 mm long. Test leathery, wrinkled and mammillated; brownish; not particle-encrusted. Branchial sac with four low folds each side. Gut on the left, stomach horizontal. Polycarps drop-shaped, numerous, scattered each side of the body. Viviparous (at least sometimes).

Generally attached to stones and shells, frequently clustered, shallow water down to about 450 m; all coasts; locally common; Arctic to Mediterranean.

Polycarpa rustica (Linnaeus) Fig. 13.10

Body ovoid-cylindrical, to about 20 mm. Oral siphon terminal, atrial siphon slightly dorsal to it. Test smooth, red or pink. Branchial sac with four folds each side. Gut on the left. Polycarps numerous, on both sides of the body (cf. the superficially similar *Dendrodoa grossularia*).

On stones, etc., LWST to shallow water; Irish Sea, English Channel, Brittany.

Styela clava Herdman Fig. 13.10

Body large, inverted club-shaped, up to 120 mm. Siphons terminal, close together. Test leathery, wrinkled.

On stones, pilings, etc. South-west from Portsmouth to Swansea. Introduced from Pacific.

Pelonaia corrugata Forbes and Goodsir Fig. 13.10

Body large, up to 140 mm long; club-shaped, with the siphons terminal and close together at the narrow end. Test transversely wrinkled, with fibrils towards the lower end, particle-encrusted. Branchial sac without folds. Gut on the left, stomach vertical. Gonads linear, the testis follicles marginal to the ovary, U- or J-shaped, the arm to the oviduct being the longer. Oviparous; development without tadpole.

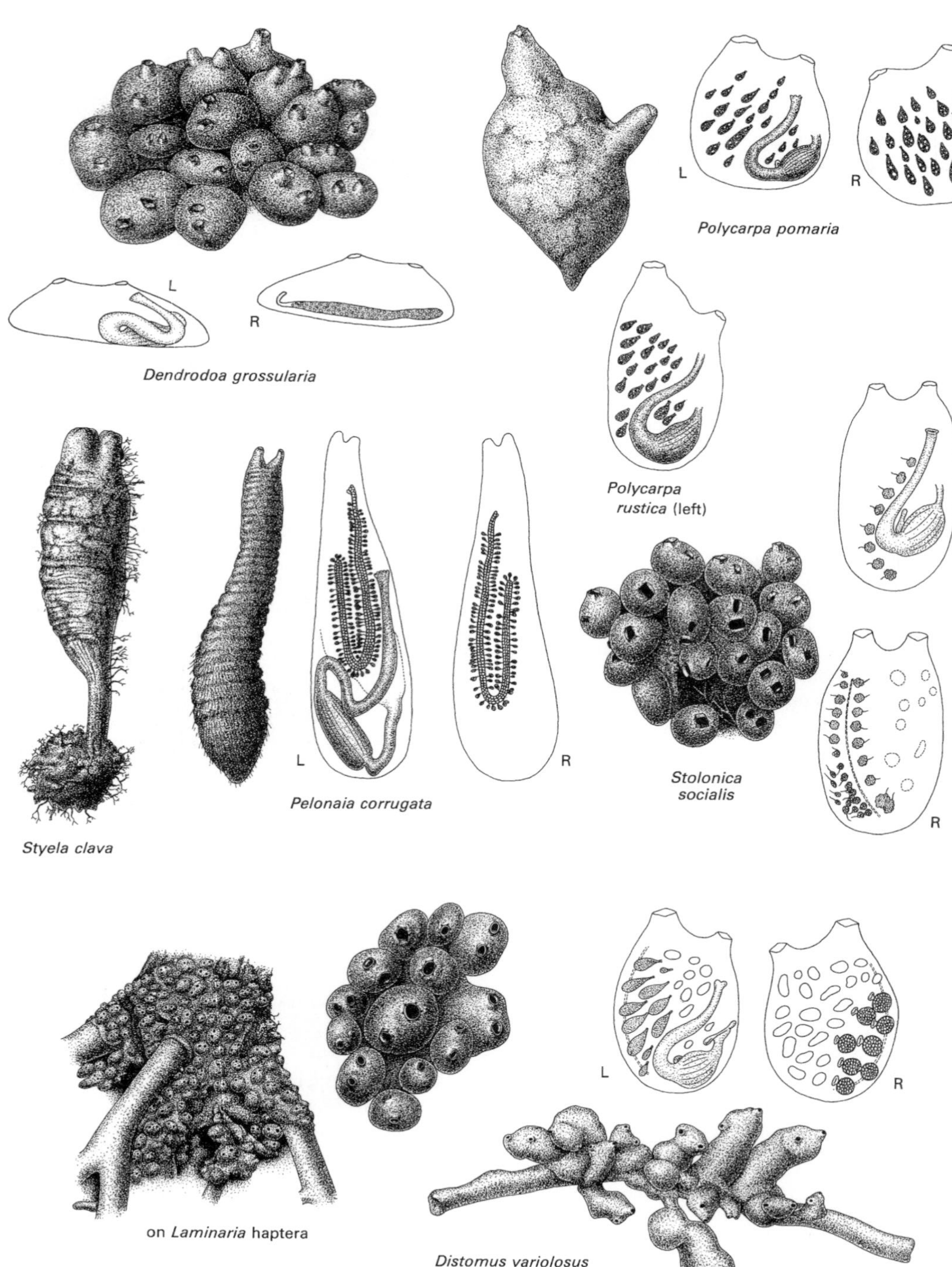
L
R
Polycarpa pomaria
L
R
Dendrodoa grossularia
Polycarpa
rustica (left)
L
R
Pelonaia corrugata
Stolonica
socialis
R
Styela clava
L
R
on Laminaria haptera
Distomus variolosus

Fig. 13.10

Lives buried in sand or mud; northern waters only (Clyde to north-west England); North Sea to the Arctic.

10B POLYZOINAE

Distomus variolosus Gaertner Fig. 13.10

Colony a crust of zooids united basally and, to some extent, laterally by test; ovoid, depressed, or sometimes taller with the common base intermittent; about 10 mm long; reddish; siphons terminal, opening independently. Ovaries 6–10, each with a single large egg and short oviduct, restricted to the right side; testes about 12, ovoid-elongate, only on the left. Viviparous; larval trunk 0.7 mm. Budded zooids develop at the end of atrial stolons (outgrowths from the thoracic body wall), which subsequently atrophy.

Confusion is possible between *D. variolosus* and densely aggregated *Dendrodoa grossularia*. If doubtful, *Distomus* has the ovary on the right side and several horizontal testicular follicles on the left; *Dendrodoa* has an elongate ovotestis on the right side only.

On rocks, etc., but especially around the haptera and up the stipe of *Laminaria hyperborea*, often washed up; local, from south Devon, Lundy, to Galway Bay; south to Portugal.

Stolonica socialis Hartmeyer Fig. 13.10

Zooids in clusters, upright with terminal siphons, to about 20 mm, united basally by stolons; yellowish or orange. Branchial sac with two or three folds each side. Gonads in three series, two ventral paralleling the endostyle, the third intestinal.

Subtidal rocks, 5–35 m; from south Devon, Lundy, Skomer, to Donegal; also Channel Islands and Brittany.

10C BOTRYLLINAE

Botrylloides leachi (Savigny) Fig. 13.11

Colony flat, encrusting, usually grey, orange, or red-brown. Zooids in two parallel rows with common cloaca between them.

On algae, stones, ascidian tests, etc., on the lower shore and in shallow water; all coasts; from northern Norway to the Mediterranean.

Botryllus schlosseri (Pallas) Fig. 13.11

Colony flat, encrusting, or lobate; test of almost any colour, though blue is commonest. Zooids contrasting in colour with the test, in stellate systems of 3–12.

On algae, ascidian tests, and inert surfaces; lower shore and shallow water; all coasts; from northern Norway to the Mediterranean.

11 PYURIDAE

Unitary, test leathery. Siphons both four-lobed; oral tentacles branched. Branchial sac with 5–8 folds on each side; dorsal lamina continuous or broken into languets. Stigmata small and straight. Gut on left; stomach bearing hepatic tubules, the 'liver'. Gonads mixed, on both sides of the body; oviparous.

Pyura tessellata (Forbes) Fig. 13.11

Body flat, ovate in plan, the siphons far apart; to about 10 mm in length. Test strongly tessellated (i.e. a mosaic), divided into oblong or hexagonal, concentrically marked, platelets.

On stones, shells, other ascidians, etc. from the lower shore to about 300 m; all except North Sea coasts; from Faroes and Norway to the Mediterranean.

12 MOLGULIDAE

Unitary, often subglobular and dwelling in soft substrata. Test fairly thin, usually bearing fibrils (helping to anchor the animal in particulate substrata). Oral siphon six-lobed; oral tentacles branched; atrial siphon four-lobed. Stigmata in spirals. Gut on the left; stomach bearing hepatic tubules, the 'liver'. A large, sausage-shaped, renal organ attached to the body wall on the right side. Gonads mixed, on both sides of the body; oviparous or viviparous (the larvae sometimes anuran).

Molgula citrina Alder and Hancock Fig. 13.11

Body ovoid, subglobular or slightly depressed, to about 18 mm, basally attached. Test hard, often greenish, naked or with some adhering sand grains. Siphons short, well separated. Branchial sac with seven folds each side. One gonad each side, that on the left lying in the upper curve of the intestine; the duct as long as the gonad. Viviparous.

On rock, stones, pier piles, other ascidians, etc., occurring in clusters; lower shore and shallow (generally <50 m) water; all coasts; Arctic to Western Approaches.

Molgula complanata Alder and Hancock Fig. 13.11

Body subglobular or somewhat compressed, to about 20 mm. Test with fibrils, to which are attached sand and shell particles. Siphons well sep-

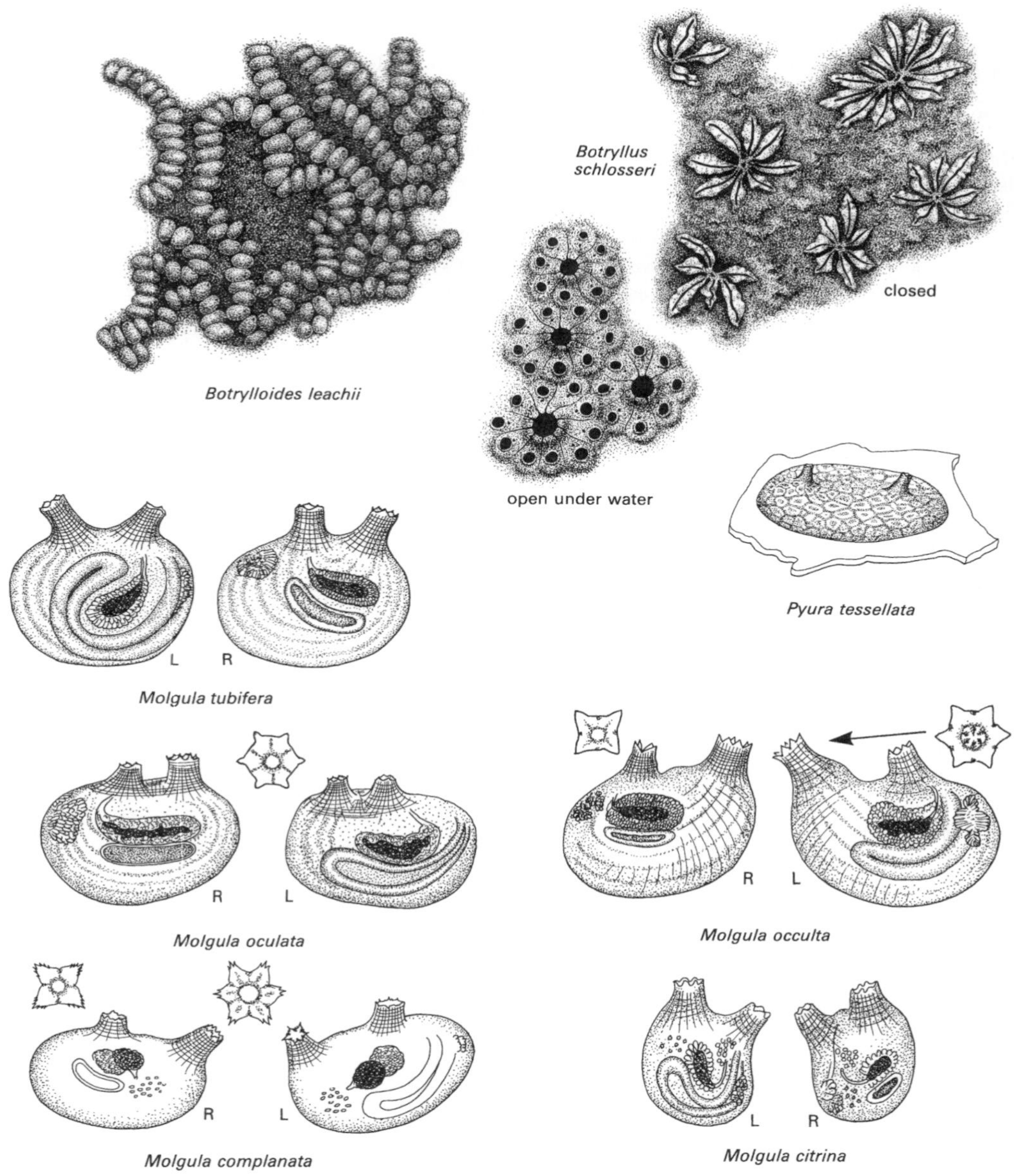

Fig. 13.11

arated; siphonal lobes fringed. Branchial sac with 6–7 small folds each side. Gonads on both sides; the short ducts leading anteriorly. Viviparous.

In crevices and depressions on stones and shells on a rocky or sandy bottom, probably widely overlooked. Lower shore and shallow water; all coasts; Arctic to Brittany.

Molgula occulta Kupfer Fig. 13.11

Body ovoid to globular, to 30 mm long; test (apart from the siphons) completely covered with particles; siphons short. Branchial sac with seven folds each side; 5–6 longitudinal bars on each fold; stigmata spiral on the folds, almost straight between them. The gut almost horizontal; left gonad immediately above it, testis surrounding the ovary; duct rather short, leading straight upwards toward the atrial siphon. When living, the renal sac contains a yellowish concretion lying in brownish fluid. Oviparous.

Unattached, on mud, sand, and shell gravel; from just below LWST to about 100 m; south and west coasts; from Norway to the Mediterranean.

Molgula oculata Forbes Fig. 13.11
Body ovoid to subglobular; to 80 mm; test heavily coated with sand or shell particles, except for the siphons and the space between them. Branchial sac with seven folds, each with 10–11 longitudinal bars; stigmata spiral on the folds. Gut almost horizontal; left gonad above it, the ovary enveloped by testis; duct moderately short, leading straight upwards towards the atrial siphon. When living, the renal sac contains a purplish fluid. Oviparous.

Unattached, on soft substrata; LWST to about 80 m; widely distributed from Shetland, north-east England to the south-west and also Biscay.

Molgula tubifera Oersted Figs 13.2, 13.11
Body rounded, 10–30 mm, grey or greenish, covered with fibrils to which particles may be attached. Siphons basally close together, diverging. Branchial sac with six folds each side. Gut sharply looped, disposed in a 'C'. Gonads club-shaped, the left one almost encircled by the gut; the right one above the renal organ; ducts shortly directed to the base of the atrial siphon. Oviparous.

On stones, pilings, algae, etc., or lying loose in sand; often in aggregations; lower shore and shallow water, including harbours and estuaries, to about 90 m; all coasts; from northern Norway to Portugal. To be regarded as different from the American *M. manhattensis*.

14 FISH

Fish (Subphylum Euchordata (vertebrata): Pisces) are sufficiently familiar to need little general introduction. It should be realized, however, that 'Pisces' does not represent a natural group but, corresponding to the word 'fish', embraces the jawless cyclostomes, the cartilaginous sharks and their allies (Chondrichthyes), and the bony fishes (Osteichthyes); each of these groupings constituting a separate and distinctive vertebrate class. Many of the technical terms refer primarily to bony fishes, but the names of fins and parts of the body apply equally to cartilaginous fishes.

The gross external features of a fish's body are indicated in Fig. 14.1. The generalized shape is termed *fusiform*, but the body may be variously elongated or flattened laterally (*compressed*) or dorso-ventrally (*depressed*). Fins in the mid-line are the *dorsal* (D), the *anal* (A), and the *caudal* (C); while paired fins supported by the pectoral and pelvic girdles are referred to as the *pectorals* (P) and the *pelvics* or *ventrals* (V), respectively. In cartilaginous and primitive bony fishes the pelvic fins are *abdominal* and situated near the vent; but in advanced bony fishes, in which the pectorals are placed higher on the flanks, the pelvics have moved forward to below or in front of the pectorals, being then respectively *thoracic* or *jugular*. The fins are supported by rays of various types, horny in Chondrichthyes but bony in Osteichthyes.

The morphology of the caudal fin varies considerably between the classes. In Chondrichthyes it is basically unequal or *heterocercal*, with the vertebral column entering and supporting the upper fin lobe. In most Osteichthyes the tail is outwardly symmetrical or *homocercal*. Most of the rays articulate with the modified haemal arches (called *hypurals*) of the last few vertebrae.

In primitive bony fishes the finrays are soft, being flexible, articulated and, in general, branched; in advanced fishes the anterior rays of the dorsal and anal fins, and the outer ray of the pelvics, are unjointed and unbranched. Such *spiny rays* may be thin and flexible but are more often stiff and pointed. When summarizing the number of finrays for convenience in formula form, spiny rays are denoted by roman and soft rays by arabic numerals: thus, DV/10–11 indicates a dorsal fin composed of five spiny rays succeeded by 10 or 11 soft rays. Where there is more than one dorsal fin, each is indicated by a preceding numeral, i.e. 1D, 2D.

Also important in identifying fishes is squamation, or the structure and arrangement of scales. In sharks the scales are *placoid*, or tooth-like, but in modern bony fishes the scales are thin flakes of bone. *Cycloid* scales (Fig. 14.1), found in most soft-rayed bony fishes are disciform, with the exposed rear margin entire; *ctenoid* scales, characteristic of many spiny-rayed fishes, have a toothed rear margin. Scale counts are helpful, and the number of scales along the lateral line is used most frequently (e.g., lat. sc. 50–60), counting from the free edge of the gill cover to the origin of the caudal fin, ignoring any scales that lie actually on the base of the fin itself. If the number of scale rows is required, count downwards and backwards from the commencement of the (first) dorsal fin, and upwards and forwards from the commencement of the anal fin, reaching—but not including—the scale overlying the lateral line. Thus, 5/11 indicates five longitudinal rows of scales above the lateral line and 11 below.

The external gill openings of cartilaginous fishes are separate slits; in bony fishes the gills lie together in a common cavity, overlain by the gill cover, with a single exterior opening on each side. Less conspicuous are the olfactory openings. Cyclostomes possess a single nostril on top of the head. In sharks and rays there is, on each side, a single, ventral nostril. The olfactory organs in bony fishes have (usually) two dorsal nostrils on each side, placed most often just in front of the eyes.

The skull of a bony fish is complex, and the form and disposition of bones has been modified considerably during the course of evolution. The few bones commonly referred to in the descriptions of species are labelled in Fig. 14.1. In addition to the cranium, notice the jaws and the flat bones constituting the gill cover, especially the *preoperculum* (PreOp) and *operculum* (Op). Below these and supporting the jugular continuation of the opercular membrane, is a series of slender *branchiostegal rays* (br.) attached to the bones of the hyoid arch. The number of rays may be helpful in taxonomy. In fairly primitive bony fishes, such as salmonids, skull features include the delimitation of the upper jaw by two tooth-bearing bones, *premaxilla* and *maxilla*, and the presence of a series of small *circumorbital bones* around the eye. In more highly evolved fishes, such as gadids, the maxilla lies behind the premaxilla; the maxilla provides the upper jaw pivot, but the premaxilla—which in some fish orders is protrusible—alone is toothed and defines the upper side of the gape. The maxilla, however, delimits the upper jaw in external view, and jaw measurements are made to its posterior end. The first circumorbital bone (*preorbital* or *lachrymal*) may extend forwards as far as the premaxilla, while in scorpaeniforms the third circumorbital or *suborbital* is elongated and articulates with the preoperculum. Teeth may be present not only on the jaws but in the roof of the mouth, on the *palatines* and/or *vomer*, and in the floor of the mouth, on the terminal bones of the hyoid arch.

The branchial skeleton in bony fishes consists of four gill-bearing arches. Typically each arch comprises four paired elements, of which the middle two bear gill filaments, and an unpaired basal element lying in the midline. The number of *gill rakers*, the bony knobs or bristles opposing the gills in each arch, is frequently of systematic importance. The first gill arch is utilized and should be removed in its entirety prior to counting. Pharyngeal or gill-teeth may also be present and taxonomically helpful, as in wrasses. *Hyperpharyngeal teeth* are situated on the upper elements of the gill arches; *hypopharyngeal teeth* are borne on bones representing the lower elements of the fifth branchial arch, the components usually being fused into a single structure.

The pectoral girdle of bony fishes is intimately associated with the skull, and is joined to it dorsally by the *post-temporal* bone (Fig. 14.1). The largest constituent bone on each side is the *cleithrum*. In fishes with thoracic or jugular ventral fins the pelvic girdle is anchored to the cleithra.

The nomenclature employed in this book characterizes orders with the ending -formes. The classification is as follows.

Superclass Agnatha

Craniates without jaws.

Class I Cyclostomata (Marsipobranchii)

Notochord persistent. Gills in the form of pouches supported by a complex cartilaginous skeleton. Elongated body; no paired fins.

Order 1 Petromyzoniformes

Adults with a terminal sucker surrounding the mouth. Scaleless body. Series of gill openings on each side. Lampreys.

Subphylum Gnathostomata

Craniates with jaws.

Class I Chondrichthyes (Elasmobranchii, Selachii)

Skull and skeleton cartilaginous. Paired fins present. Gill filaments attached throughout their length to the septa; five or more branchial openings each side. Skin usually rough, with placoid scales ('dermal denticles'). No air bladder. Intestinal spiral valve present. Males with copulatory organs ('claspers').

Subclass Euselachii

Sharks and rays.

Order 1 Lamniformes

Two spineless dorsal fins. Anal fin present.

Order 2 Squaliformes

Two dorsal fins, each preceded by a spine. No anal fin.

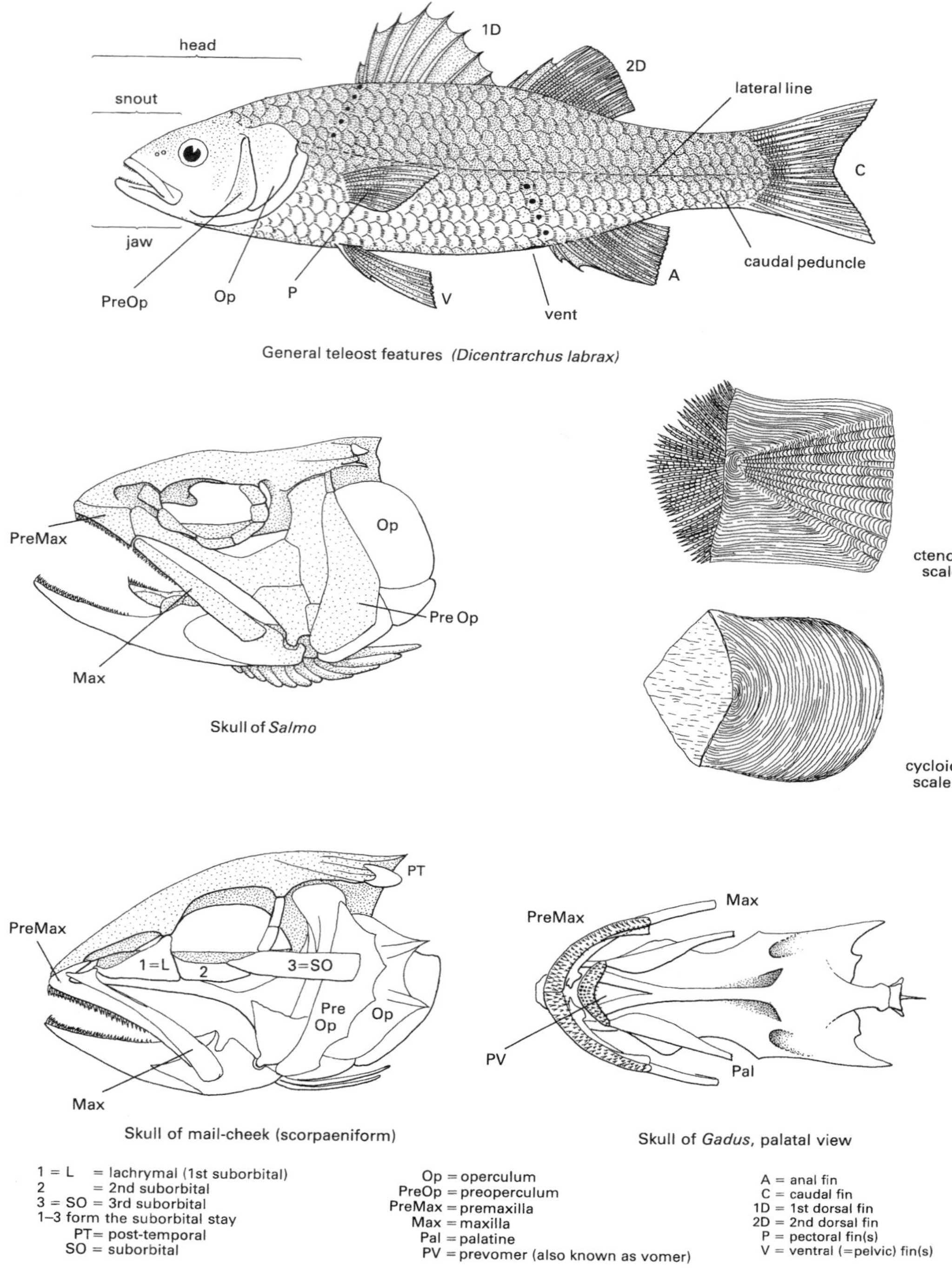
head
snout
jaw
1D
2D
lateral line
C
caudal peduncle
A
vent
V
P
Op
PreOp
General teleost features (Dicentrarchus labrax)
PreMax
Op
Pre Op
Max
Skull of Salmo
ctenoid scale
cycloid scale
PT
PreMax
1=L
2
3=SO
Pre Op
Op
Max
Skull of mail-cheek (scorpaeniform)
Max
PreMax
PV
Pal
Skull of Gadus, palatal view
1 = L = lachrymal (1st suborbital)
2 = 2nd suborbital
3 = SO = 3rd suborbital
1–3 form the suborbital stay
PT= post-temporal
SO = suborbital
Op = operculum
PreOp = preoperculum
PreMax = premaxilla
Max = maxilla
Pal = palatine
PV = prevomer (also known as vomer)
A = anal fin
C = caudal fin
1D = 1st dorsal fin
2D = 2nd dorsal fin
P = pectoral fin(s)
V = ventral (=pelvic) fin(s)

Fig. 14.1

Order 3 Rajiformes

Body depressed; gill slits opening ventrally. Skates and rays.

Order 4 Torpediniformes

Electric rays.

Class II Osteichthyes

Skeleton bony. Paired fins present. Gills filamentous; a single, operculate gill opening. Air bladder or lung usually present. Scales bony.

Subclass Actinopterygii

Ray-finned fishes.

Infraclass Teleostei

The most advanced and diverse bony fishes, with symmetrical (homocercal) tails and thin scales.

Superorder Elopomorpha

Primitive teleosts with toothed maxillae forming part of gape and no spiny finrays; generally of eel-like form and having a leptocephalus larva.

Order 1 Anguilliformes

Eels. Larva a 'leptocephalus' (flat, leaflike, transparent and planktonic).

Superorder Clupeomorpha

Primitive teleosts with compressed silvery bodies and caducous scales. Toothed maxillae form part of gape. Pectoral fins low on the flank, pelvics abdominal; no spiny finrays. Diverticula of the swimbladder extend into the head and end in a characteristic arrangement of bullae around the ears. No lateral line pores on trunk.

Order 2 Clupeiformes

Characters of the superorder.

Superorder Protacanthopterygii

Primitive teleosts. Maxilla included in gape. Pectoral fins low on the flank, pelvics abdominal; no spiny finrays. A rayless (adipose) fin frequently present.

Order 3 Salmoniformes

Adipose fin present. Salmon-like fishes.

Superorder Paracanthopterygii

Advanced marine teleosts, with or without spiny finrays, characterized (and distinguished from Acanthopterygii) by, for example, non-protrusible jaws, the presence of a particular jaw muscle (m. levator maxillae superioris), and the structure of the caudal skeleton. Pelvic fins thoracic or jugular.

Order 4 Gobiesociformes

Pelvic fins modified into a complex sucker. Scaleless. No spiny rays in the median fins. No swimbladder. Clingfishes.

Order 5 Lophiiformes

First dorsal finray modified as a lure. The pectoral fins modified to facilitate crawling. Scaleless. Angler fishes.

Order 6 Gadiformes

Nektonic fishes. Scales cycloid. Fins without spiny rays. Swimbladder well developed.

Superorder Atherinomorpha

A small superorder of osteologically distinct fishes with protrusible jaws of a characteristic type.

Order 7 Atheriniformes

Characters of the superorder.

Superorder Acanthopterygii

The largest superorder, comprising advanced teleosts of varied form and habit, characterized by: protrusible jaws having a distinctive mechanism, the absence of the m. levator maxillae superioris, the exclusion of the maxilla from the gape, the structure of the caudal skeleton, ctenoid scales, presence of spiny finrays, and the thoracic or jugular position of the pelvic fins.

Order 8 Gasterosteiformes

Dorsal fin preceded by separate spines. Leading ray of pelvics spiny. Sticklebacks.

Order 9 Syngnathiformes

Tubular-mouthed fishes with pelvic fins abdominal (or absent). First dorsal fin with spiny rays, but no free spines preceding it. Pipefishes.

Order 10 Scorpaeniformes

Differ from Perciformes by the presence of an enlarged suborbital bone which extends posteriorly to articulate with the preoperculum (Fig. 14.1).

Order 11 Perciformes

A vast order of spiny-finned teleosts, central to the Acanthopterygii. Leading ray(s) of dorsal, anal, and pelvic fins usually spiny.

Order 12 Pleuronectiformes

Benthic, strongly compressed teleosts (flatfishes), with asymmetric head having both eyes placed on the same side of the body.

The species described are those found typically in coastal waters. They include shore fish, those that can be trawled, and those caught by rod and line with the omission of the more oceanic sport fishes. On rocky shores fish can be found under boulders (rocklings, blennies, butterfish, and *Nerophis lumbriciformis*) and in pools: netting in the fringing algae may produce *Crenilabrus melops* and *Gobiusculus flavescens*, while *Lipophrys pholis* and *Gobius paganellus* are more likely to be hiding among stones on the pool bottom or in crevices. At extreme low water the holdfasts of *Saccorhiza polyschides* may conceal clingfishes and *Liparis montagui*. On sandy shores push-netting at lowtide collects young flatfish, *Syngnathus rostellatus* and *Pomatoschistus minutus*; if *Zostera marina* is present *Spinachia spinachia*, pipefishes, wrasses, and many others will be found. Different species occur when the water is brackish.

Specimens may be preserved in formalin or 70% ethanol, or be deep frozen, but colour patterns should be recorded before storage in preservative. The measurements employed in fish identification should be largely self-explanatory. Fishery biologists measure length as from the tip of the snout to the mid-point of the caudal fin (*fork length*), but the *standard length* utilized by systematists is the length excluding the caudal fin. *Head length* is measured from the snout tip to the rear edge of the operculum, and the snout is taken to end at the anterior edge of the orbit.

KEY TO FAMILIES

1. More than one gill opening each side — 2

   A single gill opening each side (Osteichthyes: Teleostei) — 8

2. Seven gill openings each side. Paired fins absent (Cyclostomata) — **1. Petromyzonidae**

   Five gill openings each side. Paired fins present (Chondrichthyes: Euselachii) — 3

3. Gill openings ventral. Body depressed — 4

   Gill openings lateral. Body elongated or fusiform — 6

4. Mouth terminal — **5. Squatinidae**

   Mouth ventral — 5

5. Rounded anteriorly; caudal fin well developed (electric rays) — **7. Torpedinidae**

Pointed anteriorly; caudal fin absent or rudimentary (skates and rays) **6. Rajidae**

---

6. Anal fin absent; both dorsals preceded by spines **4. Squalidae**

Anal fin present; dorsals without spines **7**

---

7. Anterior dorsal above or behind pelvics **2. Scyliorhinidae**

Anterior dorsal above gap between pectorals and pelvics **3. Carcharinidae**

---

8. Body long and slender, covered by a rigid exoskeleton in the form of bony rings. Head drawn out into a long tubular snout (pipefishes) **19. Syngnathidae**

Body not covered by a rigid exoskeleton of bony rings. Head without tubular snout **9**

---

9. Body eel-like, supple. Pelvics absent. Scales hidden or absent. No barbel (eels) **10**

The above characters not present in combination **11**

---

10. Dorsal fin commencing well behind pectorals. Lower jaw more prominent **8. Anguillidae**

Dorsal fin commencing above free end of pectorals. Upper jaw more prominent **9. Congridae**

---

11. Pelvic fins abdominal **12**

Pelvic fins thoracic, jugular, or absent **13**

---

12. Single dorsal fin present **10. Clupeidae**

Adipose fin present behind the rayed dorsal fin **11. Salmonidae**

---

13. Fish lying on their sides, with asymmetrical head and both eyes on the same side (flatfishes) **14**

Symmetrical fish, not lying on their sides **16**

---

14. Eyes placed on the left side **35. Bothidae**

Eyes placed on the right side **15**

---

15. Preopercular bone with a free edge. Mouth terminal **36. Pleuronectidae**

Edge of preopercular bone covered or hidden by skin. Mouth subterminal **37. Soleidae**

---

16. Anterior dorsal finray situated on the head and transformed into a 'fishing line' (angler fish) **13. Lophiidae**

Anterior dorsal finray not situated on the head and not transformed into a fishing line **17**

---

17. Spiny finrays generally absent. Pelvics with 5–17 soft rays. One or two anal fins **18**

Some spiny rays present. Pelvic fins with a maximum of five soft rays (or sometimes transformed into a ventral sucker). Only one anal fin **19**

---

18. Two dorsal fins. No barbel on lower lip **15. Merlucciidae**

Two or three dorsal fins. Usually (always when there are only two dorsal fins) a barbel on the lower lip **14. Gadidae**

---

19. First dorsal fin represented by a series of separated spines **18. Gasterosteidae**

First dorsal fin continuous, not represented by separated spines **20**

---

20. Pelvic fins subjugular, widely spaced, separated by an adhesive disc. Dorsal fin lacking spiny rays **12. Gobiesocidae**

Adhesive disc (if present) formed from the pelvic fins. Anterior dorsal fin (or part of fin) composed of spiny rays **21**

---

21. Rear part of dorsal and anal fins broken up into a series of separate finlets **34. Scombridae**

These fins not broken up into finlets **22**

---

22. Pelvic fins not recognizable, fully transformed into an adhesive organ **23. Liparidae**

Pelvic fins not transformed into an adhesive organ **23**

---

23. Head armoured with bony plates **24**

Head unarmoured **26**

---

24. Lower pectoral finrays separate; pelvics with five soft rays **20. Triglidae**

Lower pectoral finrays not separate; pelvics with fewer than five soft rays **25**

25. Body to a large extent naked **21. Cottidae**

Body covered by a series of bony plates **22. Agonidae**

26. Pelvic fins united, forming a disc **33. Gobiidae**

Pelvic fins independent, sometimes reduced or absent **27**

27. Dorsal fins widely separated. Pelvic fins behind the point of insertion of the pectorals **28**

Dorsal fins feebly separated or contiguous at the base or fused. Pelvic fins thoracic or jugular **29**

28. Anterior dorsal fin with four spiny rays; anal fin with three spines **25. Mugilidae**

Anterior dorsal fin with 5–9 spiny rays; anal fin with one spine **17. Atherinidae**

29. First dorsal fin very short, second dorsal separate and much longer **30**

Dorsal fin(s) not of this form **31**

30. Head somewhat laterally compressed. Anterior dorsal fin with 5–7 spiny rays, posterior dorsal with 21–32 rays **27. Trachinidae**

Head strongly depressed. Anterior dorsal fin with four spiny rays, posterior dorsal with nine or ten rays **32. Callionymidae**

31. Pelvic fins thoracic **32**

Pelvic fins jugular or absent **33**

32. Lips normal. Palatal teeth present; pharyngeal teeth absent. Scales ctenoid **24. Moronidae**

Lips relatively thick. Teeth on the jaws but not on the palate; pharyngeal teeth present. Scales cycloid **26. Labridae**

33. Caudal fin forked; pelvics absent **31. Ammodytidae**

Caudal fin rounded or tapering, pelvics small but usually present **34**

---

34. Caudal fin completely fused with dorsal and anal; dorsal fin with a depressed region posteriorly **16. Zoarcidae**

Caudal fin distinct **35**

---

35. Dorsal fin more or less distinctly divided into two regions **28. Blenniidae**

Dorsal fin uniform, made up of spiny rays only **36**

---

36. Pelvic fins rudimentary; dorsal marked with a series of black spots **30. Pholidae**

Pelvic fins jugular; dorsal not spotted **29. Lumpenidae**

Class Cyclostomata

Order Petromyzoniformes

1 PETROMYZONIDAE

Eel-like body; scaleless. Mouth jawless, rounded, suctorial. Single dorsal nasal opening. Eyes present. Seven branchial openings each side. Distinctive larva ('ammocoetes').

Lampetra fluviatilis (Linnaeus).
Lampern, lamprey Fig. 14.2
Body eel-like, without paired fins. Oral disc subcircular, provided with sharp, horny teeth. Eyes rather small, followed by seven rounded gill openings. Dark blue above with silvery flanks or grey above and creamy below. To 50 cm.

Parasitic on other fish, but also taken in pelagic townets; coastal; entering rivers during September–October prior to spring spawning; all coasts except the extreme north.

Class Chondrichthyes

Order Lamniformes

2 SCYLIORHINIDAE

Small, bottom-living sharks. Mouth ventral. Spiracle present; five gill slits. Dorsal fins spineless, the first placed above or behind the pelvics; anal fin present; caudal fin with shallow ventral lobe. Oviparous.

---

1. Markings consisting of large brown spots on a light background. Nasal flaps separated by a wide gap ***Scyliorhinus stellaris***

Markings consisting of small brown spots on a light background. Nasal flaps continuous across the front of the mouth ***Scyliorhinus canicula***

Scyliorhinus canicula (Linnaeus).
Dogfish Fig. 14.2
Origin of 1D behind pelvics; origin of 2D level with hind end of A. Nasal flaps hardly separated; almost joined in front of mouth. Colour sandy, with small brown spots; pale below. To 75 cm. Egg capsules not more than 6–7 cm long, with a twisted tendril at each corner.

In shallow water, generally near the bottom. Off all coasts; common.

Scyliorhinus stellaris (Linnaeus).
Nursehound Fig. 14.2
Origin of 1D level with hind edge of pelvics; origin of 2D in front of hind end of A. Nasal flaps separated by a wide gap, not joined. Colour sandy, with large brown spots; pale below. To 150 cm. Egg capsules at least 10 cm long.

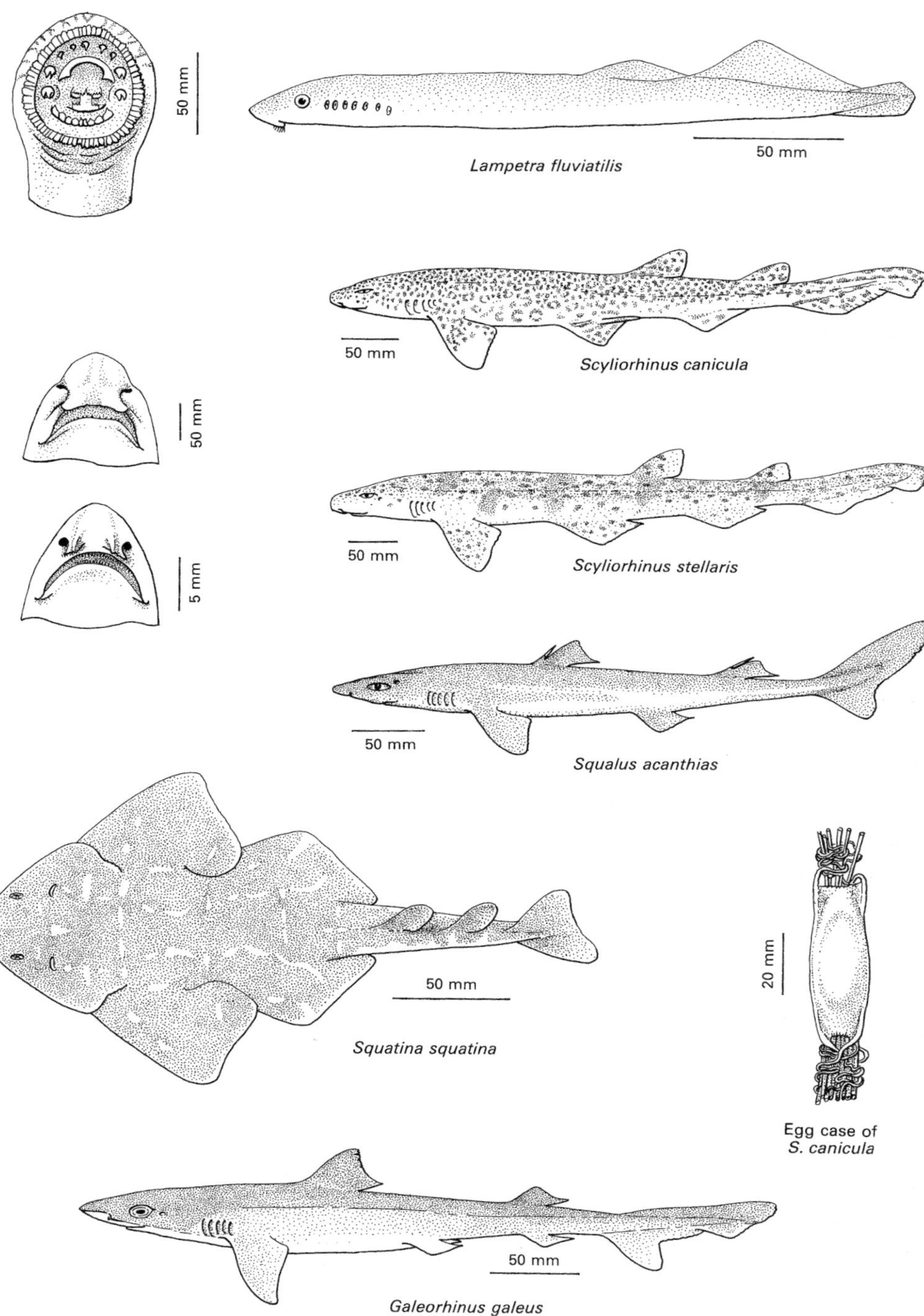
50 mm
50 mm
Lampetra fluviatilis
50 mm
Scyliorhinus canicula
50 mm
50 mm
Scyliorhinus stellaris
5 mm
50 mm
Squalus acanthias
50 mm
Squatina squatina
20 mm
Egg case of
S. canicula
50 mm
Galeorhinus galeus

Fig. 14.2

Inshore and in deeper water, over rough ground. Off all coasts, but most frequent in the south and west.

3 CARCHARINIDAE

Sharks. Snout elongated; mouth crescentic, ventral. Teeth triangular, with serrated edges. First dorsal fin spineless, placed above the interspace between pectorals and pelvics; anal fin present.

Galeorhinus galeus (Linnaeus). Tope Fig. 14.2
Shark-like form, no lateral keel on the caudal peduncle; skin rather rough. Pectorals pointed. Lower lobe of tail moderately developed. Spiracle present. Teeth unicuspid, serrated on the inner side, in 3–4 rows. Grey above, white below. To 150 cm.

Bottom-living in shallow water over sand, coming inshore during summer; off all coasts.

The smooth hounds (*Mustelus* spp.) lack the pointed pectorals and prominent lower tail lobe of the tope.

Order Squaliformes

4 SQUALIDAE

Spiny sharks. Snout elongated, mouth ventral, with a deep groove on either side. Spiracles present. Two dorsal fins, each preceded by a spine. No anal fin.

Squalus acanthias Linnaeus.
Spurdog Fig. 14.2
Body long and slender; origin of 1D not in front of rear edge of pectorals, 2D placed well behind pelvics; caudal with well-developed lower lobe. Teeth unicuspid. Grey above, with a row of white spots along the flank; pale below. To more than 1 m.

Pelagic, forming shoals; may occur inshore. Off all coasts; common.

5 SQUATINIDAE

Body depressed; pectoral fins large and overlapping pelvics, not attached to the head. Mouth terminal, flanked by nostrils which have valvular flaps. Spiracles large. Two dorsal fins; no anal.

Squatina squatina (Linnaeus).
Monkfish Fig. 14.2
Distinctive body shape. Colour sandy above, with darker and lighter markings; white below. To about 2 m.

Bottom living on sandy and gravelly seabeds; inshore or in fairly shallow water. Off all coasts, but especially in the south and west; common during summer.

Order Rajiformes

6 RAJIDAE

Skates and rays. Body depressed, comprising a rhombic or circular disc and a slender tail. Pectoral fins ('wings') greatly expanded, joined to the snout. Two dorsal fins; caudal fin rudimentary or absent. Oviparous; the egg case with a curved, pointed horn arising from each corner. The long-snouted species are customarily termed skates, and the short-snouted species, rays. Specific identification may be difficult. Infraspecific variation is frequently extensive, with morphometry, spinulation, and colour changing during growth. In general the spinules on the wings develop with age, while the median spines may become less obvious. Males and females may differ (i.e. apart from the presence or absence of claspers).

---

1. Snout long; under-surface greyish (skates) ***Raja batis***

   Snout short; under-surface white (rays) **2**

---

2. Wings rounded laterally **3**

   Wings more or less pointed laterally, their front and rear edges meeting at an angle of roughly 90° **5**

---

3. A prominent black and yellow ocellus (round marking) approximately in the centre of the upperside of each wing ***Raja naevus***

   No ocelli **4**

---

4. Upper surface marked with stripes, with or without spots **5**

   Upper surface spotted or of uniform colour, not striped **6**

---

5. Stripes lighter than the background colour ***Raja microocellata***

   Stripes darker than the background colour ***Raja clavata***

6. Spots dark (often sparsely developed in juveniles), and not extending to the margins of the wings; skin smooth and slimy except for the spines or spinules on the tail, around the snout, and on the wing tips ***Raja montagui***

   Spots dark or light, sometimes extending to the margins of the wings; skin rough **7**

7. Skin very rough and (in adults) bearing large spines with swollen bases; upper surface usually grey with dark and/or light spots ***Raja clavata***

   Skin with texture of fine sandpaper, spines small and without swollen bases; upper surface light brown covered overall with small dark spots ***Raja brachyura***

Raja batis Linnaeus. Skate Fig. 14.3
Snout long; leading edge of the wing concave; tips obtusely pointed. Median tail spines present in adults, including one or two between the dorsal fins. Dark or brownish-grey above, with lighter spots; grey below. Ventral mucus pores black. To 2 m.

The most common of the deep-water skates. Off all coasts except the south-east.

Raja brachyura Lafont. Blond ray Fig. 14.3
Short-snouted; leading edge of the wings fairly straight; tips roughly right-angled, obtusely pointed. Upper surface with the texture of fine sandpaper, spinulose in adults. Jaws with 60–90 rows of teeth. Fawn-coloured above, with many small dark spots extending all over the disc and tail; no ocellus; white below. To over 1 m.

Prefers water less than 100 m deep; young fish coming inshore. Common off south and west coasts.

Raja clavata Linnaeus.
Thornback ray Fig. 14.3
Short-snouted; leading edge of the wings fairly straight; the tips roughly right-angled, pointed. Large, broad-based spines on the back, which is rough and coarsely spinulose in all stages. Jaws with 36–44 rows of teeth. Dorsal ground colour grey, often with darker stripes and dark and/or light spots; a mottled ocellus may be present; white below. To 1 m.

In shallow water, often close inshore. Off all coasts; common.

Raja microocellata Montagu
Painted ray Fig. 14.3
Short-snouted; leading edge of the wings fairly straight; tips nearly right-angled, obtusely pointed. Spinules present on the front of the disc, the remaining area smooth. Jaws with 44–52 rows of teeth. Light brown above with large white spots and white stripes parallel with the edges of the wings; no ocellus; white below. To 80 cm.

On sandy grounds in shallow water, often close inshore. Confined to south-west Britain and to the south and west of Ireland; locally common.

Raja montagui Fowler. Spotted ray Fig. 14.3
Short-snouted; leading edge of the wings fairly straight; tips nearly right-angled; obtusely pointed. Jaws with 38–60 rows of teeth. Light brown above, with small black spots sometimes grouped as an ocellus but not extending to the lateral margins of the disc; white below. To 75 cm.

In shallow to moderately deep water, especially on sandy grounds. South and west coasts; common.

Raja naevus Müller and Henle.
Cuckoo ray Fig. 14.4
Short-snouted; leading edge of the wings fairly straight, outer angles broadly rounded. Fawn above, with a conspicuous black and yellow ocellus on each wing; white below. To 70 cm.

In shallow water and from moderate depths; all coasts; moderately common.

Order Torpediniformes

7 TORPEDINIDAE

Electric rays. Disc broad, anteriorly rounded, flabby, smooth skinned. Two dorsal fins; caudal fin

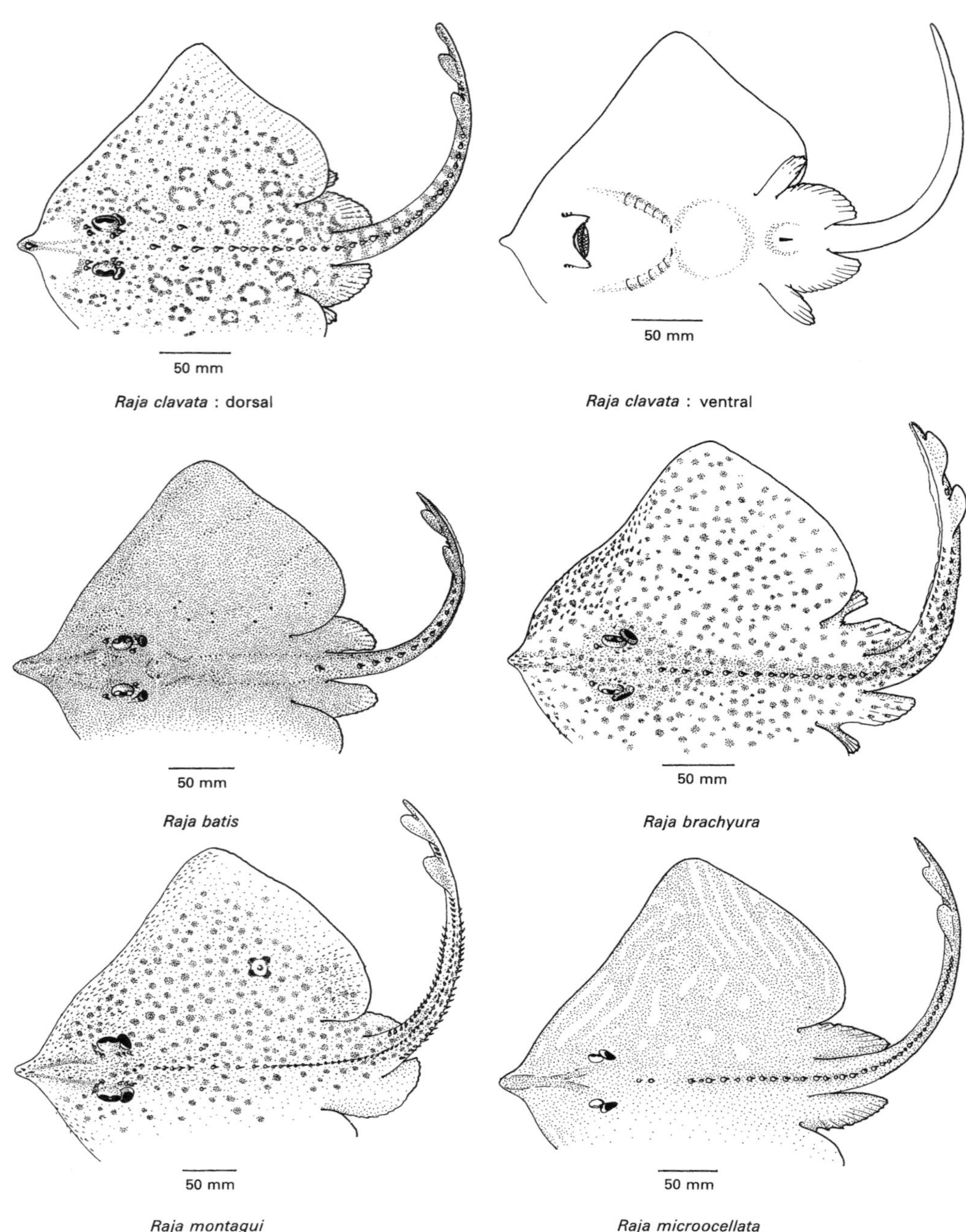

Raja clavata : dorsal

Raja clavata : ventral

Raja batis

Raja brachyura

Raja montagui

Raja microocellata

Fig. 14.3

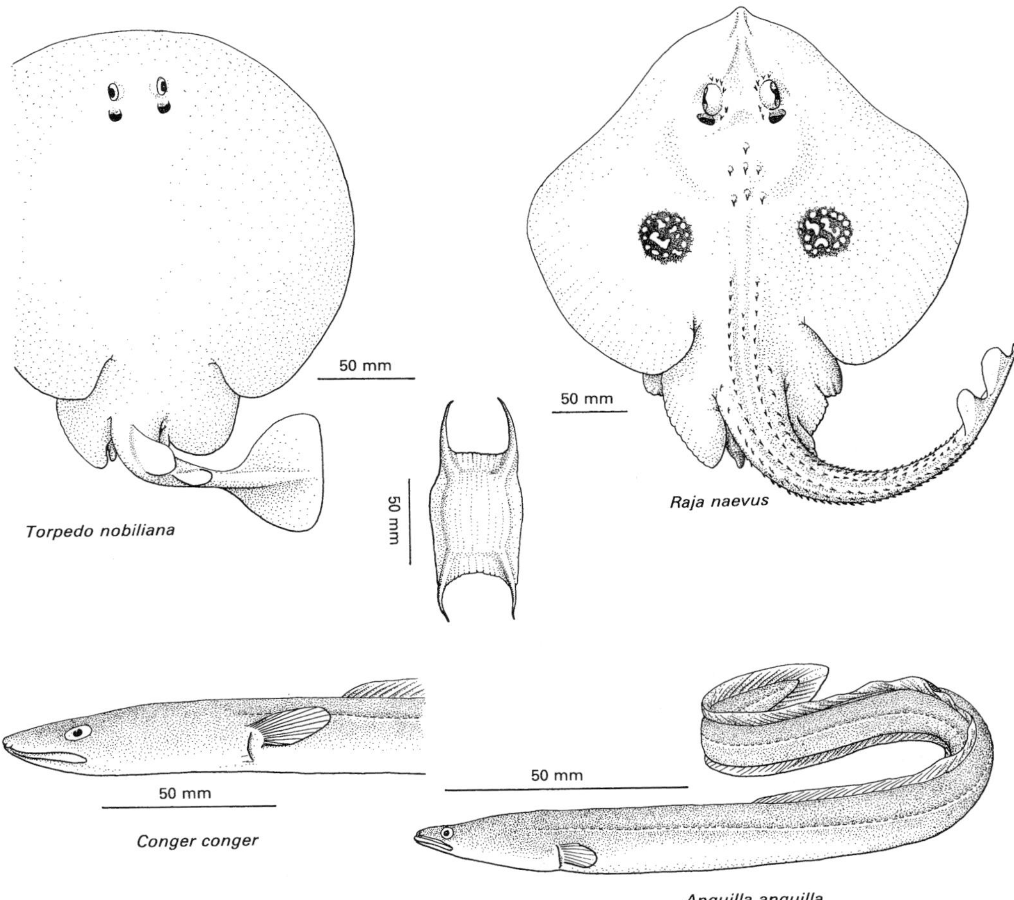

Fig. 14.4

well developed. Electric organ (used primarily for stunning prey) filling most of each wing.

Torpedo nobiliana Bonaparte.
Electric ray Fig. 14.4
First dorsal fin larger than second. Spiracles smooth. Dark brown or grey above; white below. To 180 cm. Large, freshly caught rays can discharge powerful electric shocks.

In shallow to moderately deep water; off southern and western coasts; frequent.

Class Osteichthyes

Subclass Actinopterygii

Infraclass Teleostei

Superorder Elopomorpha

Order Anguilliformes

8 ANGUILLIDAE

Snake-like body; mouth terminal; dorsal and anal fins continuous; caudal fin rudimentary. Pectoral fins present; pelvics absent. Scales minute. Lateral line well developed. Leptocephalus larva.

Anguilla anguilla (Linnaeus). Eel. Fig. 14.4
Lower jaw longer than upper. Dorsal fin originating far behind pectorals. D 245–275; A 205–235; P 17–18. Dark or black above, yellowish below (silvery when approaching sexual maturity). To 60 cm.

Catadromous. Leptocephali arrive in British waters in autumn and metamorphose. Elvers (7 cm) ascend rivers in winter or early spring. Adults live in fresh water or in estuaries. Maturing 'silver' eels may be caught at river mouths at the start of their oceanic migration. All coasts; common.

9 CONGRIDAE

Snake-like body; mouth terminal; dorsal and anal fins continuous; caudal fin rudimentary. Pectoral fins present; pelvics absent. Scaleless. Lateral line well developed. Leptocephalus larva.

Conger conger (Linnaeus). Conger Fig. 14.4
Upper jaw longer than the lower. Dorsal fin originating at the level of the pectoral fin tip. D 270–300; A 205–230; P 16–19. Brown above, paler or white below. Larger than eel; to 2 m or more.

Found intertidally on rocky shores, hiding in crevices, and offshore; principally on the south and west coasts. Breeds in the eastern Atlantic, reaching British waters as leptocephali.

Superorder Clupeomorpha

Order Clupeiformes

10 CLUPEIDAE

Pelagic fishes frequently forming shoals. One dorsal fin; pectorals low on the flank, pelvics abdominal. Bodies completely covered with caducous cycloid scales, except on the head. No lateral line. No spiny finrays. Herring-like fishes.

---

1. Upper jaw notched in the mid-line, the lower jaw fitting inside it **_Alosa fallax_**

   Upper jaw without notch **2**

---

2. Operculum with radiating ridges **_Sardina pilchardus_**

   Operculum without radiating ridges **3**

---

3. Origin of D behind that of pelvics; 21–23 scales between the throat and the pelvic origin; pelvic fins with seven or eight rays **_Sprattus sprattus_**

   Origin of D behind or slightly in front of that of pelvics; 27–30 scales between the throat and the pelvic origin; pelvic fins with 8–10 rays **_Clupea harengus_**

Alosa fallax (Lacepède).
Twaite shad Fig. 14.5
Upper jaw notched, the lower jaw fitting inside it. Belly keeled and strongly serrated. 40–60 rakers on the first gill arch (cf. 80–130 in the similar but less common Allis shad, *A. alosa* (Linn.)). Br. viii; D 18–21; A 19–25; P 15–16; V 9; lat. sc. 60–70; vert. 55–59. Steel-blue above, silvery-white below; a large dark blotch on the shoulder, usually succeeded by a further five or six; golden yellow on the side of the head. To 50 cm.

Mainly on southern and western coasts; anadromous, entering rivers for spawning in May–June; quite common.

Clupea harengus Linnaeus. Herring Fig. 14.5
Lower jaw longer than upper. Belly weakly keeled, not strongly serrated. Origin of D in front of base of pelvics, 27–30 scales on the belly between the throat and the pelvic fins; 13–15 between the pelvics and the vent. 18–23 pyloric caeca. Br. viii; D 17–20; A 16–18; P 17; V 8–10; lat. sc. 53–60; vert. 51–58. Dark-blue above, silver-white below. To 40 cm.

Off all coasts; formerly in vast offshore shoals, now less common.

Sardina pilchardus (Walbaum)
Pilchard, sardine Fig. 14.5
Jaws of equal length; the gape not extending past the eye. Belly rounded; operculum marked with radiating ridges. Origin of D in front of base of pelvics; two posterior rays of A longer than the others. Br. vi–viii; D 17–18; A 17–18; P 16–17; V 6–8; lat. sc. 28–30; vert. 50–53. Greenish-olive above, silver-white below. To 25 cm.

Pelagic and shoaling; common off south-western coasts, rarer northwards.

Sprattus sprattus (Linnaeus). Sprat Fig. 14.5
Lower jaw longer than the upper. Belly sharply keeled and strongly serrated. Origin of D behind base of pelvics. Anal fin with more rays than in herring, pelvics fewer. 21–23 scales on the belly

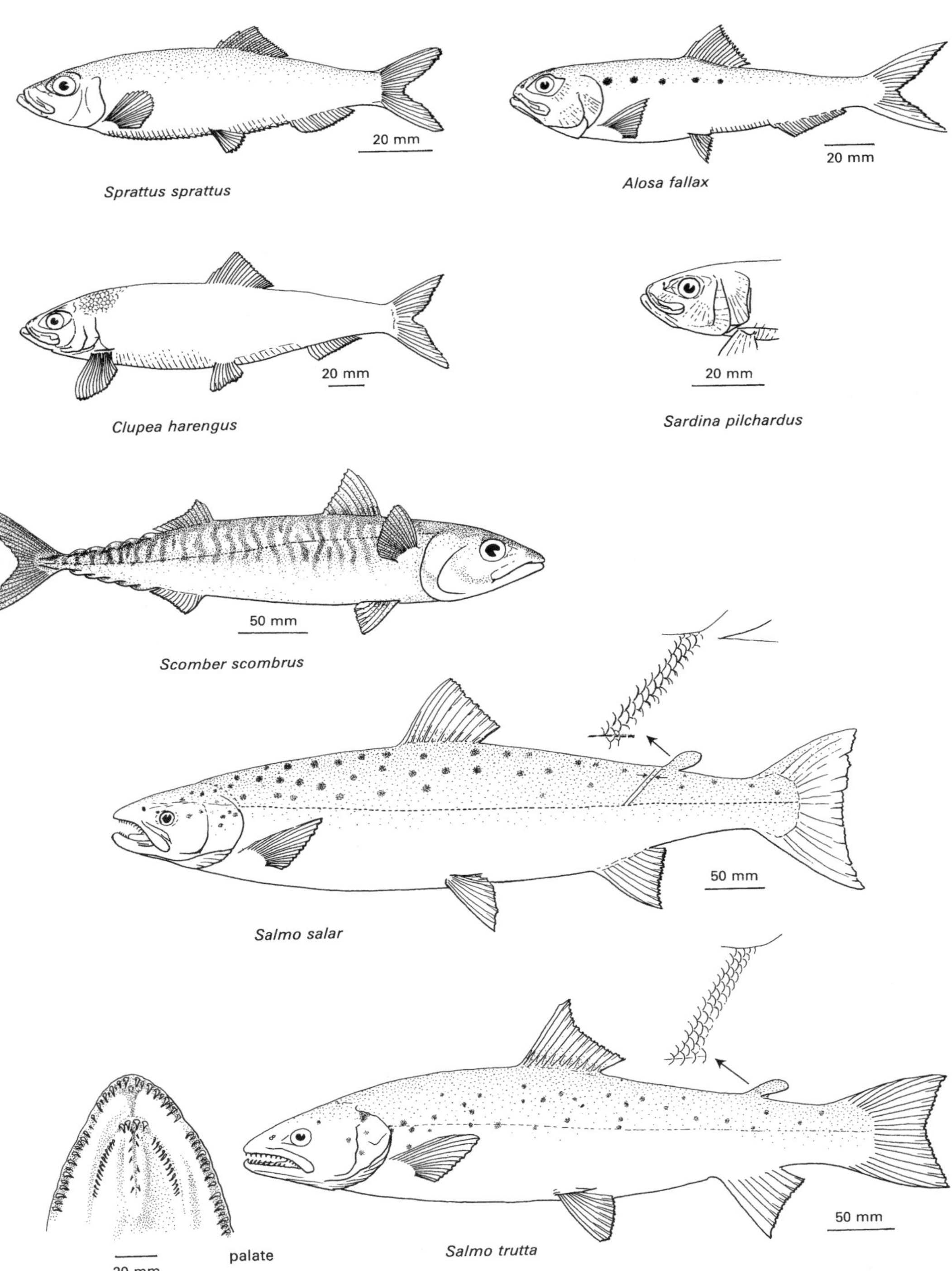
20 mm
Sprattus sprattus
20 mm
Alosa fallax
20 mm
Clupea harengus
20 mm
Sardina pilchardus
50 mm
Scomber scombrus
50 mm
Salmo salar
20 mm
palate
50 mm
Salmo trutta

Fig. 14.5

between the throat and the pelvic fins. 11–12 between the pelvics and the vent. Seven pyloric caeca. Br. vi-vii; D 16–19; A 18–20; P 16–17; V 7–8; lat. sc. 47–50; vert. 46–49. Blue or blue-green above, silver-white below. To 15 cm.

Pelagic, forming large shoals close inshore; all coasts; common.

Superorder Protacanthopterygii

Order Salmoniformes

11 SALMONIDAE

Two dorsal fins, the second *adipose* (rayless); pectorals low on the flank, pelvics abdominal. Body, but not the head, covered with cycloid scales. Jaws toothed. Tail square or slightly forked.

---

1. Ten to 13 rows of scales between rear edge of the adipose fin and the lateral line (count downwards and forwards) ***Salmo salar***

   Thirteen to 16 rows of scales between the rear edge of the adipose fin and the lateral line ***Salmo trutta***

Salmo salar Linnaeus. Salmon Fig. 14.5
Maxilla not extending greatly behind rear of the eye. 17–24 rakers on the first gill arch. No teeth on the head of the vomer. 10–13 rows of scales from the rear edge of the adipose fin to the lateral line. Br. xi–xii; D 12–14; A 9–11; P 14; V 9; lat. sc. 120–125; vert. 57–60. Colour in the sea: blue-green above, flanks silvery, belly white, with small black spots on the back and flanks. To 150 cm.

Anadromous. Returns from the ocean to coastal waters during the spring and summer, entering estuaries and ascending rivers prior to spawning during the winter. All coasts, particularly the north and west.

Salmo trutta Linnaeus. Trout Fig. 14.5
Maxilla extending behind rear of the eye. 13–18 rakers on the first gill arch. Three or four teeth on the head of the vomer. 13–17 rows of scales from the rear edge of the adipose fin to the lateral line. Br. x–xii; D 12–14; A 10–12; P 13–14; V 9; lat. sc. 115–130; vert. 56–61. Colour in the sea: brownish-grey above, flanks silvery, belly white; with reddish and small black spots on the back and flanks. To 70 cm.

Anadromous. Inshore waters and estuaries, ascending rivers for spawning; all coasts.

Superorder Paracanthopterygii

Order Gobiesociformes

12 GOBIESOCIDAE

Clingfishes or suckers. Small, somewhat depressed fishes, having a complex ventral sucker (modified pelvic fins) below the pectorals; V I/4. Scaleless.

---

1. Dorsal fin with 15–20 rays, anal with 9–12 rays **2**

   Dorsal and anal fins with 4–7 rays **3**

---

2. Dorsal, caudal, and anal fins continuous ***Lepadogaster lepadogaster***

   Dorsal and anal fins not continuous with the caudal fin ***Lepadogaster candollei***

---

3. A small dark (not red) spot at the origin of the dorsal and anal fins (♀) or a large dark spot on each (♂); first anal finray below or in front of the second dorsal finray ***Apletodon dentatus***

   Dorsal and anal fins lacking a dark spot (sometimes with a red one); first anal finray below or behind the third dorsal finray ***Diplecogaster bimaculata***

Apletodon dentatus (Facciola).
Small-headed clingfish Fig. 14.6
A small clingfish with D and A short and similarly sized, both clearly separated from the caudal; first anal finray roughly below the first dorsal finray. Premaxillary teeth of two kinds, small and rounded in front, with one to three larger teeth at the sides. D 5–6; A 5–7; P 20–24; vert. 29–32. Brownish, with a small (female) or large (male) dark spot on the dorsal and anal fins. To 5 cm.

Lower shore, often in the hollow bases of *Saccorhiza*. On Atlantic coasts, but distribution uncertain.

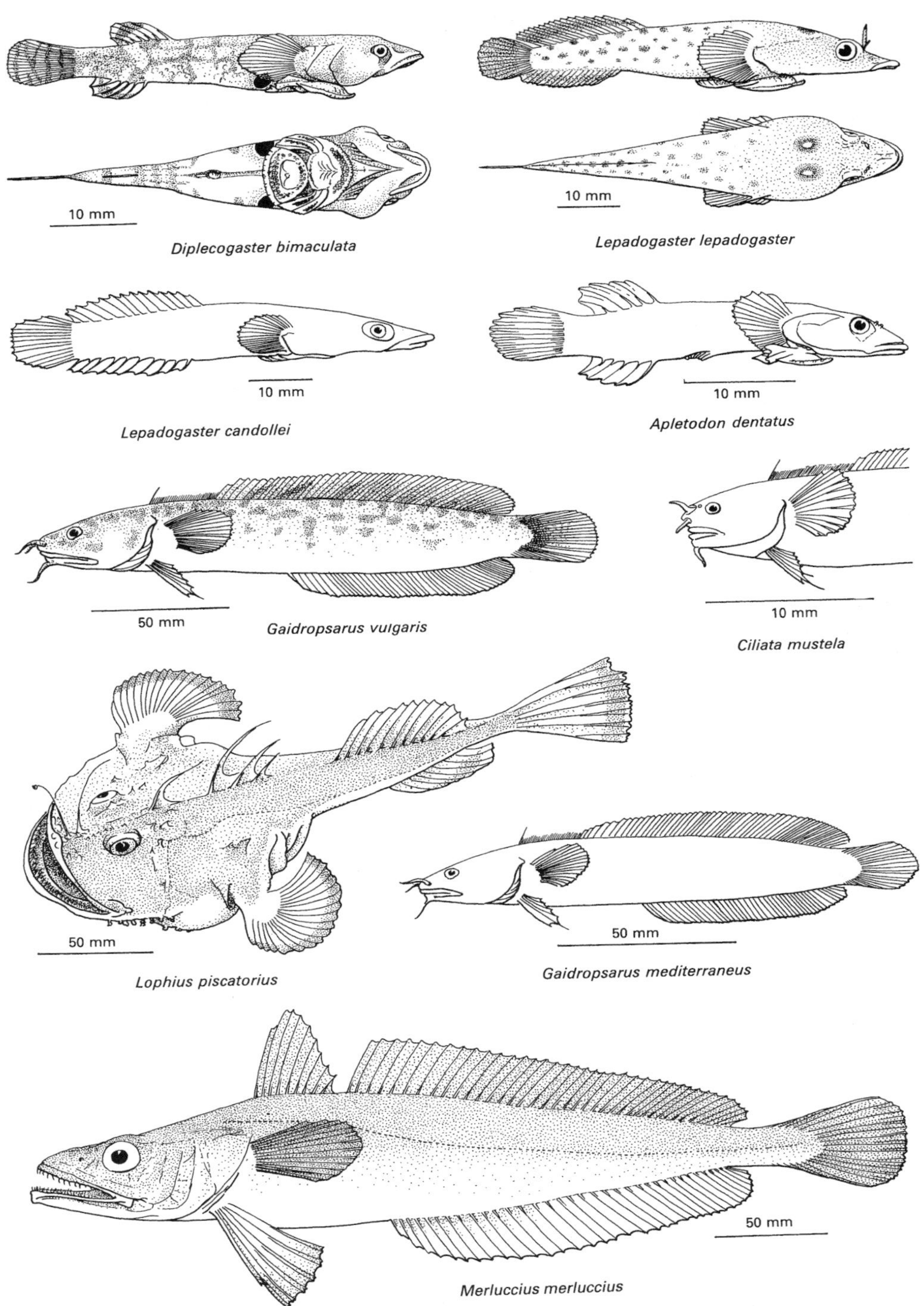
10 mm
Diplecogaster bimaculata
10 mm
Lepadogaster lepadogaster
10 mm
Lepadogaster candollei
10 mm
Apletodon dentatus
50 mm
Gaidropsarus vulgaris
10 mm
Ciliata mustela
50 mm
Lophius piscatorius
50 mm
Gaidropsarus mediterraneus
50 mm
Merluccius merluccius

Fig. 14.6

Diplecogaster bimaculata (Bonnaterre)
Two-spotted clingfish — Fig. 14.6
A small clingfish with D and A short and unequal, both clearly separated from the caudal; first anal finray behind the second dorsal finray. Teeth of one kind only. D 5–7; A 4–6; P 20–26; vert. 30–31. Reddish, often spotted; males with a large pectoral spot, purple outlined with yellow, partly hidden by the fins. To 5 cm.

Offshore, on shelly or gravelly grounds; western coasts; locally quite common.

Lepadogaster candollei (Risso).
Connemara clingfish — Fig. 14.6
The largest British clingfish. Head depressed, of tapering profile. D and A moderately long, separated from the caudal. Front nostril bearing a very small tentacle. D 13–16; A 9–11; P 22–25. Colour variable, males reddish and spotted above, females greenish above; pale below. To 10 cm.

At or below low water mark on rocky shores. South and west Ireland, Devon, Cornwall, and Pembrokeshire; local.

Lepadogaster lepadogaster (Bonnaterre).
Cornish clingfish — Fig. 14.6
Head depressed, of tapering profile. D longer than A, both continuous with the caudal. Front nostril bearing a conspicuous tentacle. Br. vi; D 16–20; A 9–11; P 20–25; vert. 32–33. Reddish, with a pair of conspicuous spots, blue outlined with red, on the back.

Between tidemarks on rocky, weed-covered shores; south and west; often common. The western European populations are referred to subspecies *purpurea*.

Order Lophiiformes

13 LOPHIIDAE

Angler-fish. Body depressed; head large, jaws wide; the lower jaw projecting anteriorly. Jaws and vomers toothed. Gill slit opening behind pectoral fin. First dorsal fin spinous, preceded by three separate rays, the first of them modified as a lure.

Lophius piscatorius Linnaeus
Angler, Monk — Fig. 14.6
Broad, depressed body with huge mouth. Loose, scaleless skin, the whole fish flabby; with a lateral fringe of small flaps. Anteriormost dorsal finray long and slender, with a terminal flap. Br. vi; 1D II + I + III; 2D 10–13; A 9–11; C 8; P 23–28; V I/5; vert. 28–31. Brownish-grey (changeable) above, white below. To nearly 2 m.

Occasionally found on the shore, but extends to the continental slope; all coasts; quite common.

Order Gadiformes

14 GADIDAE

Cods and rocklings. Body elongate, with small cycloid scales. Fins without spiny rays. Pectoral fins on the flank, pelvics thoracic or jugular. Lower jaw usually with a barbel. Swimbladder present (reduced in rocklings). Branchiostegals seven.

---

1. Two dorsal fins (1D sometimes reduced); one anal fin. Rocklings — **2**

   Three dorsal fins (1D, 2D, 3D); two anal fins (1A, 2A) — **4**

---

2. Five barbels present — ***Ciliata mustela***

   Three barbels present — **3**

---

3. Uniform brownish colour; on the shore and in shallow water — ***Gaidropsarus mediterraneus***

   Pinkish with dark-brown markings; sublittoral — ***Gaidropsarus vulgaris***

---

4. A barbel present on the lower jaw — **5**

   Barbel absent or minute — **9**

---

5. Lower jaw projecting — ***Pollachius virens***

   Upper jaw projecting — **6**

---

6. 1A long, its origin below the middle of 1D — ***Trisopterus luscus***

   1A shorter, its origin not, or hardly, in front of 2D — **7**

---

7. Barbel short, less than half eye diameter. A black patch below 1D; lateral line black **Melanogrammus aeglefinus**

Barbel longer, three-quarters or more of eye diameter. No black patch below 1D; lateral line not black **8**

8. Pectoral fins reaching the origins of 2D and 1A; lateral line darker than the ground colour **Trisopterus minutus**

Pectoral fins not reaching the origins of 2D and 1A; lateral line lighter than the ground colour **Gadus morhua**

9. Upper jaw projecting **10**

Lower jaw projecting **11**

10. A black patch present above the pectoral fins; lateral line black **Melanogrammus aeglefinus**

A black spot present at the base of the pectoral fins; lateral line brown **Merlangius merlangus**

11. Origin of 1A below the middle of 1D; pectorals passing the origin of 1A; lateral line curved above origin of 1A **Pollachius pollachius**

Origin of 1A not or hardly in advance of 2D; pectorals not reaching the origin of 1A; lateral line straight **Pollachius virens**

Ciliata mustela (Linnaeus).
Five-bearded rockling Fig. 14.6
Body elongate; 1D modified, with a single long ray anteriorly followed by a series of short hair-like rays. Five barbels: one on the lower jaw, two on the upper jaw and one on each anterior nostril. Head about one-fifth of body. 1D 1 + hairs; 2D 50–55; A 40–41; P 15–17; V 7; vert. 47. Dark brown above, white below. To 30 cm.

Littoral; under boulders and among weed on rocky shores; all coasts; common.

A second species *C. septentrionalis* (Collett) has been found off western and northern coasts. Its head is large, about one-quarter of body.

Gadus morhua Linnaeus. Cod Fig. 14.7
Three dorsal and two anal fins, 1A originating under the front of 2D. Pectoral fins not reaching 2D. Lower jaw shorter than upper; barbel long. 1D 14–15; 2D 18–22; 3D 17–20; 1A 19–23; 2A 17–19; P 16–21; V 6; vert. 49–53. Mottled golden brown above, lateral line white, belly white. To 150 cm.

Inshore and in shallow water; all coasts; common.

Gaidropsarus mediterraneus (Linnaeus).
Shore rockling Fig. 14.6
Body elongate; 1D modified, with a single long ray anteriorly followed by a series of short hair-like rays. Three barbels: one on the lower jaw and one on each anterior nostril. 1D 1 + hairs; 2D 53–63; A 43–53; P 15–17; V 5–7; vert. 47–49. Uniform dark-brown colour above, paler below. To 25 cm.

Littoral; under boulders and among weed on rocky shores; western and southern coasts; common.

Gaidropsarus vulgaris (Cloquet).
Three-bearded rockling Fig. 14.6
Body elongate; like *G. mediterraneus* but larger and with more rays in the pectoral fins. 1D 1 + hairs; 2D 56–64; A 46–52; P 20–22; V 8; vert. 46–49. Ground colour pink with dark-brown markings, paler below. To 50 cm.

Sublittoral; inshore, usually on hard grounds; off all coasts.

Melanogrammus aeglefinus (Linnaeus).
Haddock Fig. 14.7
Three dorsal and two anal fins. 1A originating under the first rays of 2D. Lower jaw shorter than upper; barbel small. 1D 14–16; 2D 19–24; 3D 19–22; 1A 21–25; 2A 20–23; P 19–20; V 6; vert. 52–57. Dark brown above, with a black oval patch above the pectoral fin; lateral line black; belly white. To 80 cm.

Inshore and offshore; off all coasts except in the English Channel.

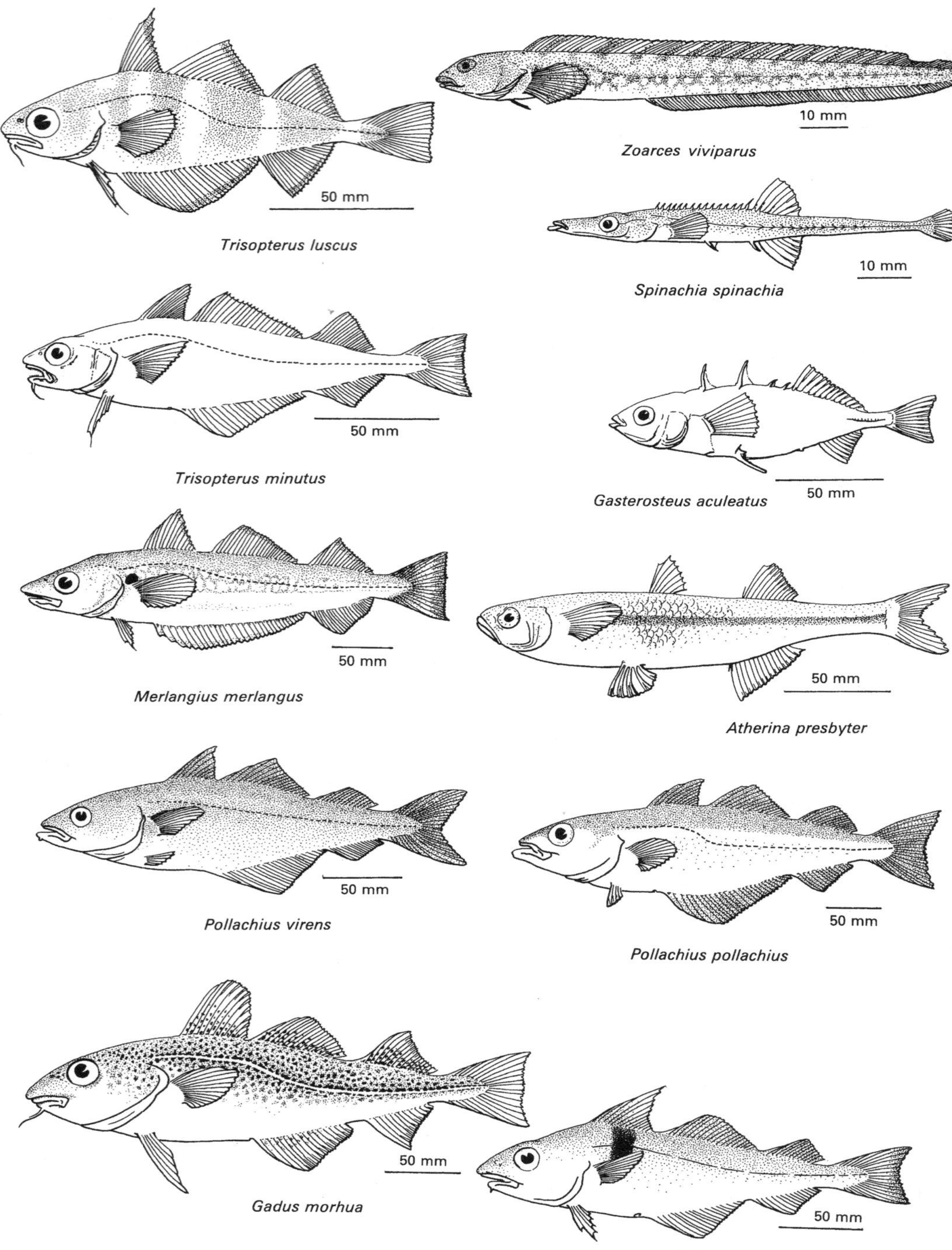
50 mm
Trisopterus luscus
10 mm
Zoarces viviparus
10 mm
Spinachia spinachia
50 mm
Trisopterus minutus
50 mm
Gasterosteus aculeatus
50 mm
Merlangius merlangus
50 mm
Atherina presbyter
50 mm
Pollachius virens
50 mm
Pollachius pollachius
50 mm
Gadus morhua
50 mm
Melanogrammus aeglefinus

Fig. 14.7

Merlangius merlangus (Linnaeus)
Whiting Fig. 14.7
Three dorsal and two anal fins, 1A originating under the central rays of 2D. Lower jaw slightly shorter than the upper; barbel rudimentary or absent. 1D 12–15; 2D 18–25; 3D 19–22; 1A 30–35; 2A 21–23; P 19–20; V 6; vert. 53–57. Brown or blue-green above, with a dark spot at the pectoral finbase; lateral line dark; silvery-white below. To 70 cm.

In shallow water, sometimes inshore; off all coasts; abundant.

Pollachius pollachius (Linnaeus).
Pollack Fig. 14.7
Three dorsal and two anal fins. 1A originating under the central rays of 1D. Lower jaw longer than the upper; no barbel. Lateral line dark, with a sharp curve below the first dorsal interspace. 1D 12–14; 2D 17–21; 3D 16–20; 1A 24–34; 2A 17–21; P 17–19; V 6; vert. 52–55. Dark olive-brown above, lighter mottled on the flanks; belly white. To 80 cm.

Mainly inshore, over hard ground; sometimes in small shoals. Off all coasts, particularly the south and west.

Pollachius virens (Linnaeus).
Saithe, coalfish, coley Fig. 14.7
Three dorsal and two anal fins. 1A originating below the rear end of 1D. Lower jaw slightly longer than the upper; barbel rudimentary. Lateral line light, almost straight (without a sharp curve below the first dorsal interspace). 1D 13–15; 2D 19–24; 3D 19–24; 1A 25–30; 2A 17–24; P 19–20; V 6; vert. 53–56. Dark greenish or blackish above, flanks paler, belly white. To 1 m or more.

Inshore and offshore, forming small shoals. Common along all coasts except in the south.

Trisopterus luscus (Linnaeus).
Bib, pout Fig. 14.7
Body rather deep, with three dorsal and two anal fins. 1A originating below the middle of 1D. Pectoral and pelvic fins extending well past the vent. Lower jaw shorter than the upper, with a long barbel. Lateral line gently curved anteriorly. 1D 11–14; 2D 20–24; 3D 18–20; 1A 30–34; 2A 19–22; P 17; V 6; vert. 47–49. Bronze above, with several broad vertical bands of darker colour down the flanks (visible only in living or freshly caught fish); a black spot at the pectoral finbase. To 40 cm.

Inshore and coastal waters all round the British Isles; very common.

Trisopterus minutus (Linnaeus).
Poor-cod Fig. 14.7
Small, fusiform, with three dorsal and two anal fins, 1A originating below the first dorsal interspace. Pectoral and pelvic fins short, the latter scarcely reaching the vent. Lower jaw shorter than the upper, with a long barbel. Lateral line gently curved anteriorly. 1D 13; 2D 23–26; 3D 22–24; 1A 28–29; 2A 23–25; P 13–16; V 6; vert. 47–51. Brownish above, paler below; a small dark spot at the pectoral finbase. To 23 cm.

In shallow water, mainly offshore; common except off south-east coasts.

15 MERLUCCIIDAE

Similar to Gadidae in most respects. No barbel on lower lip. Two dorsal fins; single anal fin. Br. vii.

Merluccius merluccius (Linnaeus).
Hake Fig. 14.6
Body elongate. Mouth large; two rows of teeth on lower jaw, one row on upper. Lateral line straight. 1D 9–10; 2D 36–40; A 36–40; P 12–14; V 7; lat. sc. 135–150; vert. 50–52. Dark blue-grey above, paler below; buccal cavity black. To 1 m.

Midwater or demersal, inshore during summer. Mainly in the south-west and west; absent from the southern North Sea.

16 ZOARCIDAE

Elongate to eel-shaped fishes. Scales cycloid or absent. Dorsal and anal fins long, merging with caudal. Pelvic fins jugular or absent. Jaws toothed. No swimbladder.

Zoarces viviparus (Linnaeus).
Eel pout, viviparous blenny Fig. 14.7
Body long. D long, with a depressed area near the tail where short, weak spines replace soft rays; A long, both it and D continuous with caudal. Pectoral fins well developed, pelvics jugular. Br. vi; D 72–85/V–XI/16–24; A 80–95; C 10–13; P 16–21;

V 3; vert. 108–118. Brownish above, paler below; a row of dark spots along the back extending on to the dorsal fin, and irregular blotches on the flank. To 35 cm.

Intertidal (though moving offshore during summer) to 10 m, and also in brackish waters; mainly north-eastern.

Superorder Atherinomorpha

Order Atheriniformes

17 ATHERINIDAE

Slender, fusiform fishes, with large cycloid scales on the body. Jaws oblique. Two dorsal fins, the first spinous. Pectorals high on flank, anterior to pelvics.

Atherina presbyter Cuvier.
Sand smelt Fig. 14.7
Small, slender fish with two dorsal fins and an emarginate tail. 27–30 rakers on the first gill arch. 1D VII/VIII; 2D II/10–13; A II/13–16; C 17; P 13–15; V I/5; lat. sc. 57–62; vert. 50–52. Green above; a broad intense silvery band on the flank, white below. To 15 cm.

Inshore, including estuaries, in shoals; south and west coasts. Eggs less than 2.0 mm in diameter, with long filaments; attached to coastal algae during summer.

The very local *A. boyeri* Risso is similar but has 21–26 gill rakers, 50–58 lateral line scales, and 43-47 vertebrae.

Order Gasterosteiformes

18 GASTEROSTEIDAE

Spiny finned fishes with elongate bodies. Second dorsal fin preceded by a series of isolated spines, in shape and position symmetrical with the anal. Pectoral fins high on the flank, pelvics represented mainly by a single spine behind the pectorals. Scales absent, but flanks bearing bony scutes. Br. iii; V I/1. Males construct nests and guard the eggs.

---

1. Body long and thin; 15 isolated spines in front of the second dorsal fin; marine **_Spinachia spinachia_**

   Body fusiform; three spines in front of the second dorsal fin; generally in brackish water **_Gasterosteus aculeatus_**

Gasterosteus aculeatus Linnaeus.
Three-spined stickleback Fig. 14.7
Body hardly elongate but with slender caudal peduncle. 2D preceded by three spines, two long and one short. D III/8–14; A I/6–11; P 10–11; vert. 29–33. Back green or dark, flanks silver, belly white (underside of head and belly red in males during spring). To 10 cm.

In fresh and brackish water, occasionally on the shore; throughout and all round the British Isles.

Spinachia spinachia (Linnaeus).
Fifteen-spined stickleback Fig. 14.7
Body elongate, the caudal peduncle long and slender. Dorsal fin preceded by 14–16 isolated spines. D XIV-XVI/6–7; A I/6–7; P 9–10; vert. 40–41. Brown or olive above, sometimes with dark bars on the flanks; paler below. To 15 cm.

Marine, but extending into brackish water; found inshore amongst seaweed and *Zostera*; all coasts, but uncommon in the south-east.

Order Syngnathiformes

19 SYNGNATHIDAE

Elongate fishes, long and thin (pipefishes) or with compressed, somewhat elevated trunk and prehensile tail (sea horses). Mouth small, at the end of a slender snout. Body covered by bony rings ('cristae'). Pelvic fins absent. Swim mainly by undulations of the dorsal fin. Eggs incubated by the males, adhering to the abdomen or within a caudal brood pouch. Live amongst weed.

---

1. Caudal fin absent or rudimentary; pectorals absent (except in young stages) **2**

   Caudal fin well developed; pectorals present in all stages **3**

---

2. Dorsal fin with 24–26(–28) rays. (A common species under rocks on the shore) **_Nerophis lumbriciformis_**

   Dorsal fin with 34–40 rays **_Nerophis ophidion_**

---

3. Snout laterally compressed, nearly as deep as the postorbital portion of the head **_Syngnathus typhle_**

Snout rounded, less deep than the postorbital portion of the head **4**

---

4. 13–16 pre-anal body rings, 37–42 postanal; snout (tip of snout to anterior margin of orbit) about the same length as rest of head (anterior margin of orbit to posterior edge of operculum). A common fish of sandy bays, up to 17 cm long ***Syngnathus rostellatus***

   17–21 pre-anal rings, 43–46 postanal; snout 1.3–2× as long as the rest of the head. Usually sublittoral, up to 45 cm long ***Syngnathus acus***

Nerophis lumbriciformis (Jenyns)
Worm pipefish Fig. 14.8
Body smooth. Snout short, upturned. Vent in front of length centre point. 17–18 pre-anal rings, 46–54 postanal. D placed over the vent; anal, caudal, and pectoral fins absent. D 24–26. Eggs carried in a shallow abdominal groove. Dark brown with white markings on the head and throat. To 15 cm.

Intertidal; found under boulders or amongst weed on rocky shores; all coasts; common where the habitat is suitable.

Nerophis ophidion Linnaeus.
Straight-nosed pipefish Fig. 14.8
Body smooth. Snout about equal in length to the rest of the head (and longer than in *N. lumbriciformis*), straight above. Vent at about the length centre point. 28–32 pre-anal rings, 68–82 postanal. D placed over vent; anal, caudal, and pectoral fins absent. D 33–34. Eggs carried in a shallow abdominal groove. Green, darker dorsally, sometimes with pale blue markings on head and underside. To 30 cm.

Coastal; amongst seaweed or *Zostera*; widespread but local.

The less-coastal snake pipefish, *Entelurus aequoreus* (L.) (Fig. 14.8) has a rudimentary caudal fin and D is largely in front of the vent.

Syngnathus acus Linnaeus.
Greater pipefish Fig. 14.8
Body rings prominent. Snout long, greater in length than the rest of the head; round in section. 17–22 pre-anal rings; 39–43 postanal. D 36–45; A 3; C 10; P 11–13. Eggs carried in a postanal brood pouch. Brown above, paler below, with dark markings or bands. To 46 cm.

Coastal; over hard or soft bottoms, amongst seaweed or *Zostera*; all coasts; fairly common.

Syngnathus rostellatus Nilsson.
Lesser pipefish Fig. 14.8
Body rings prominent. Snout about the same length as the rest of the head; round in section. 13–17 pre-anal rings, 37–42 postanal. D 36–45; A 2–3; C 10; P 10–13. Eggs carried in a postanal brood pouch. Brown above, paler below. To 17 cm.

Inshore; a common fish of sandy bays in southern Britain and Ireland. Distribution in Scotland unknown.

Syngnathus typhle Linnaeus.
Deep-nosed pipefish Fig. 14.8
Body rings distinct. Snout long, greater in length than the rest of the head, deep and compressed. 16–18 pre-anal rings, 33-39 postanal. D 28–41; A 2–4; C 9–11; P 12–16. Eggs carried in a postanal brood pouch. Brown or green, the head region marked with fine brown spots. To 30 cm.

Coastal; amongst seaweed or *Zostera*; widespread but local.

Order Scorpaeniformes

20 TRIGLIDAE

Body with scales. Head covered with rough bony plates; the preorbitals (lachrymals) extending to the tip of the snout. Spines may be present on the shoulder bone (post-temporal), on the preoperculum and operculum, and on the pectoral girdle (scapula or coracoid). Small teeth present on the jaws. Two dorsal fins, the first spinous. Pectoral fins large, the three lowest rays separated as slender tactile processes. Pelvic fins below pectorals. Br. vii

---

1. Pectoral fins not reaching the vent; anal fin with 18–20 rays; lateral line spiny ***Eutrigla gurnardus***

   Pectoral fins reaching or extending beyond the vent; anal fin with 14–17 rays; lateral line usually not spiny **2**

---

2. Lateral line covered by soft, narrow, transversely elongated, smooth-edged scutes ***Aspitrigla cuculus***

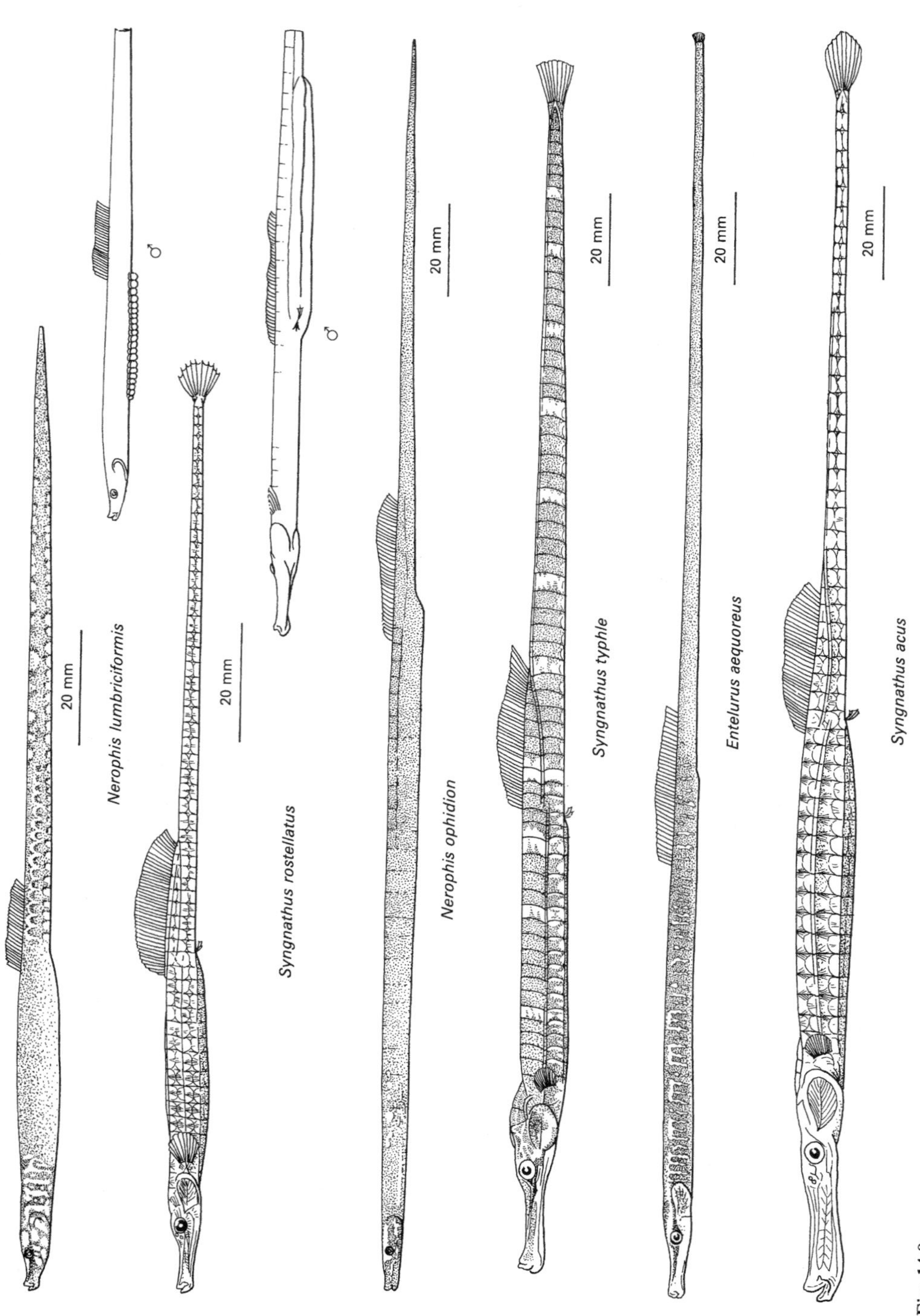
Nerophis lumbriciformis
♂
20 mm
Syngnathus rostellatus
♂
20 mm
Nerophis ophidion
20 mm
Syngnathus typhle
20 mm
Entelurus aequoreus
20 mm
Syngnathus acus
20 mm

Fig. 14.8

Lateral line not covered by transversely elongated scutes **3**

---

3. Snout flattened and anteriorly produced; lateral line scales the same size as the remainder, not spiny; pectoral fins coloured intense blue on the hinder surface ***Trigla lucerna***

Snout not flattened, not anteriorly produced, lateral line scales somewhat larger than the remainder, weakly spinous posteriorly; body transversely ridged ***Trigloporus lastoviza***

Aspitrigla cuculus (Linnaeus).
Red gurnard Fig. 14.9
Head 0.2–0.25 of total length; snout length 1.2–1.5× eye diameter, its upper profile slightly concave, descending rather abruptly; preorbital with acute angle and short pointed teeth anteriorly. Angle of preoperculum shortly spinous, with two small projections below the spine; operculum with a well-developed spine; and a spine on the coracoid. Pectoral fin about as long as the head, reaching the anal fin. Lateral-line scales unarmed but forming deep, linear plates, looking like short folds of skin. D IX-X/17–18; A 16–17; C 13; P 10 + iii; V I/5; lat. sc. 70–73; vert. 37; pyloric caeca 8. Rose-coloured, paler below. To 30 cm.

In shallow water; off all coasts, but uncommon in the North Sea.

Eutrigla gurnardus (Linnaeus).
Grey gurnard Fig. 14.9
Head 0.25–0.35 of total length; snout length about 2× eye diameter, its upper profile nearly straight; preorbital with acutely pointed angle, toothed anteriorly. Angle of preoperculum with two or three teeth; operculum with a large spine and a short prominence above it; a strong spine on the coracoid. Pectoral fin much shorter than the head, not reaching the anal fin. Lateral line scales shortly spinous. D VII–IX/18-19; A 17–19; C 15; P 10 + iii; V I/5; lat. sc. 72–77; vert. 38; pyloric caeca 7–9. Grey, brownish, or (rarely) dull red, whitish below; back and flank often white-spotted; frequently a large black blotch on 1D. To 45 cm.

Inshore and shallow waters, off all coasts; common.

Trigla lucerna Linnaeus.
Tub, Sapphirine gurnard Fig. 14.9
Head about 0.25 of total length; snout length about 2× eye diameter, its upper profile slightly concave; preorbital with acutely pointed angle, toothed anteriorly. Angle of preoperculum acute; operculum with a spine in juveniles; a ridge and a spine on the coracoid. Pectoral fin as long or longer than the head, overlapping the anal fin. Lateral line smooth. D IX–X/15–17; A 14–16; C 12; P 10 + iii; V I/5; lat. sc. 70; vert. 34; pyloric caeca 10. Brownish-red above, white below; pectoral fins reddish externally with bluish bands and/or margins, bright or dark blue internally with light spots or darker bands. To 60 cm.

Inshore and into moderate depths; common around the southern half of the British Isles, rare off northern coasts.

Trigloporus lastoviza (Brunnich).
Streaked gurnard Fig. 14.9
Head 0.2–0.25 of total length; snout length 1–1.5× eye diameter, its upper profile straight, descending abruptly, in the adult almost vertically; preorbital with obtuse angle, rough anteriorly. Angle of preoperculum blunt; operculum shortly spined; a strong spine on the scapula. Pectoral fin as long or longer than the head, reaching the anal fin. Lateral-line scales weakly spined, with vertical ridges spreading from them dorsally and ventrally. D IX–XI/16-17; A 14–16; C 12; P 10–11 + iii; V I/5; lat. sc. 66–71; vert. 35; pyloric caeca 10. Red above, white below; pectorals dark, with deep-blue spots almost forming transverse bands. To 36 cm.

In shallowish waters, mainly off the south and south-west (less often west) coasts.

21 COTTIDAE

Head broad and depressed; scaleless, but with covering bony or scale-like plates; spines on the head. Two dorsal fins, the first spinous; anal with spiny rays.

---

1. Branchiostegal membranes from each side joining ventrally, forming a continuous free posterior margin; lateral line without bony plates ***Myoxocephalus scorpius***

Branchiostegal membranes from each side not joining ventrally, but broadly

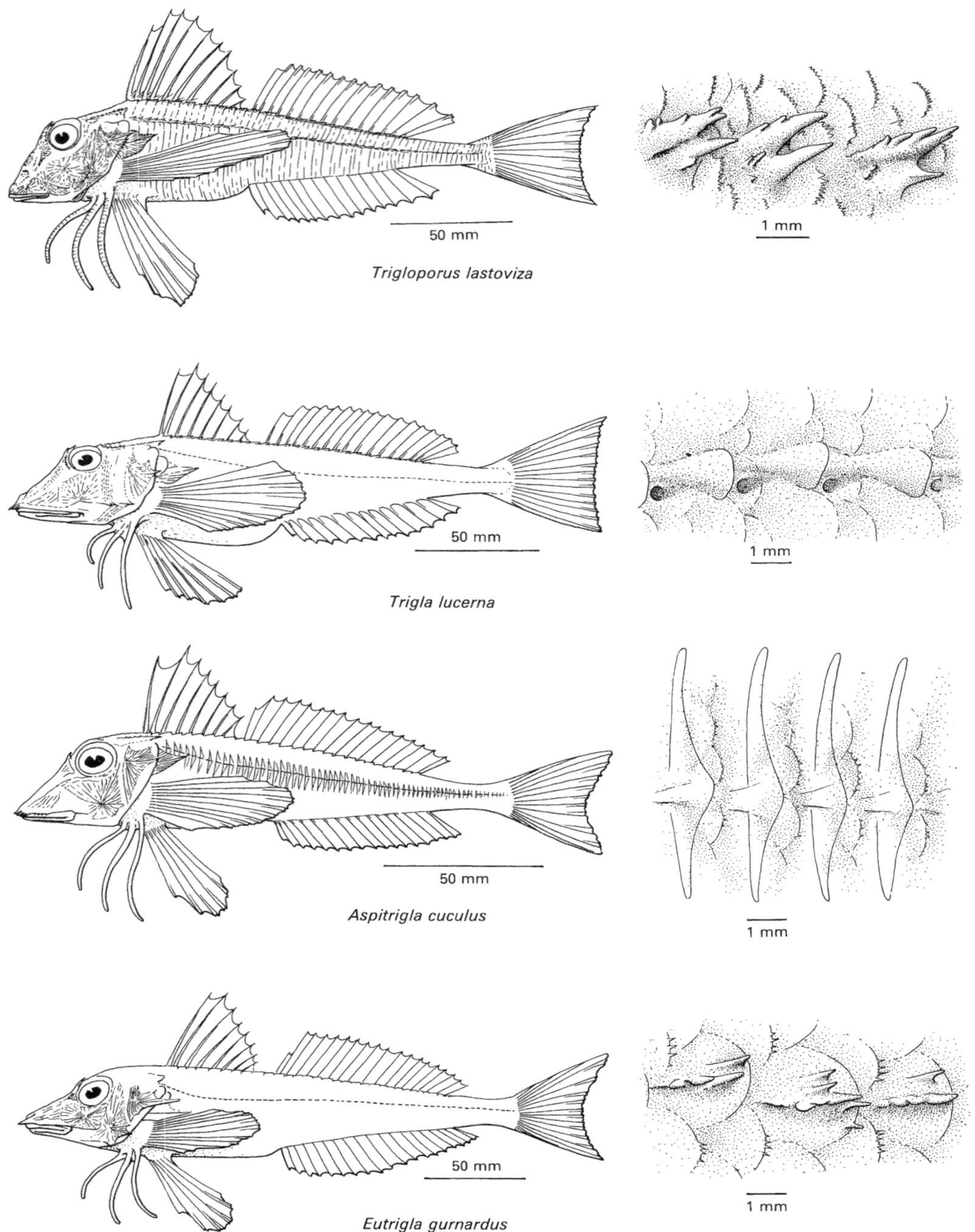
50 mm
1 mm
Trigloporus lastoviza
50 mm
1 mm
Trigla lucerna
50 mm
1 mm
Aspitrigla cuculus
50 mm
1 mm
Eutrigla gurnardus

Fig. 14.9

separated; lateral line marked by a row of small spinous scutes

Taurulus bubalis

Myoxocephalus scorpius (Linnaeus).
Bull rout, father lasher Fig. 14.10
Preoperculum with two posterior spines, the upper longer but reaching only to the base of the opercular spine; one small anterior spine. Skin smooth; lateral line unarmoured, but usually with small spiny plates (best developed in males) above and below it. Branchiostegal membranes joined ventrally across the 'isthmus', so that a single flap crosses the entire ventral surface. Br. vi; 1D VII–IX; 2D 12–17; A 10–14; P 14–18; V I/3; vert. 34–35. Light brown with dark bands and blotches; belly pale. To 30 cm (males smaller).

Inshore and in shallow water; off all coasts; not (at least in southern Britain) found intertidally.

Taurulus bubalis (Euphrasen).
Sea scorpion Fig. 14.10
Preoperculum with four spines, the uppermost the longest, almost reaching to the end of the opercular spine. Skin smooth; lateral line armed with spiny plates. Branchiostegal membranes from each side not joined ventrally across the 'isthmus'. Br. vi; 1D VII–IX; 2D 10–13; A 8–10; P 14–16; V I/3; lat. sc. 32–35; vert. 29–30. Light brown with dark bands and blotches; pale below. To about 15 cm.

Intertidal on rocky shores and inshore on hard grounds; all coasts; common.

22 AGONIDAE

Body covered with bony plates, usually in eight rows. Two dorsal fins; finrays unbranched in all fins; no spines in the anal fin.

Agonus cataphractus (Linnaeus).
Pogge Fig. 14.10
Head wide, depressed, wider than long, two pairs of massive spines on the snout; mouth inferior. Two barbels below the snout, and a divided barbel at each corner of the mouth; entire underside of head and branchiostegal membranes covered with barbels. Br. vi; 1D V–VI; 2D 6–8; A 6–7; C 12; P 16–17; V I/2; dorsal scutes 31–34; vert. 36. Dull brown above, white below. To 15 cm.

Inshore and in shallow water; all coasts; common. English Channel represents southern limit of range.

23 LIPARIDAE

Head and trunk thick, tail compressed. Skin loose, with minute spinules; no scales or bony tubercles. Single long dorsal fin; anal long. Pelvic fins transformed into a sucking disc. Pectoral fins extending ventrally.

Liparis montagui (Donovan).
Montagu's sea snail Fig. 14.10
Slimy and gelatinous, tadpole-shaped. D low anteriorly, not reaching the caudal; A reaching the caudal, but its membrane not continuous with it. Posterior nostrils in adult covered by skin. Br. vi; D 28–30; A 22–25; C 14; P 27–30; V I/5; vert. 33–36. Brownish, very variable. To 6 cm.

Littoral and inshore, on rocky grounds; associated with algae, and may occur in *Saccorhiza* holdfasts. All coasts, except in the south-east.

In the more offshore *Liparis liparis* (L.) the last dorsal and, particularly, anal finrays overlap the base of the caudal fin.

Order Perciformes

24 MORONIDAE

Body fusiform, covered with small ctenoid scales which extend on to the head. Two dorsal fins, or the spinous fin continuous with the second; anal fin with three spines. Pelvic fins thoracic, with one spine and five rays. Mouth protrusible.

Dicentrarchus labrax (Linnaeus).
Bass Fig. 14.1
Spiny- and soft-rayed dorsal fins separate; caudal emarginate. Posterior edge of preoperculum serrated, the teeth forward-pointing; two weak spines on the operculum. Br. vii; 1D VIII–IX; 2D I/12–13; A III/10–12; P 15–16; V I/5; lat. sc. 66–74; vert. 25-26. Grey or blue above, flanks silvery, belly white. To 1 m.

Inshore during the summer months, common off southern and western coasts, uncommon further north; migrating offshore during winter.

25 MUGILIDAE

Grey mullets. Body fusiform, covered with cycloid or faintly ctenoid scales which extend on to the head. Mouth terminal, broad; teeth minute or absent. Two widely separated dorsal fins, the first short, consisting of four strong spines. Pelvic fins abdominal, but anterior to 1D. Lateral line absent.

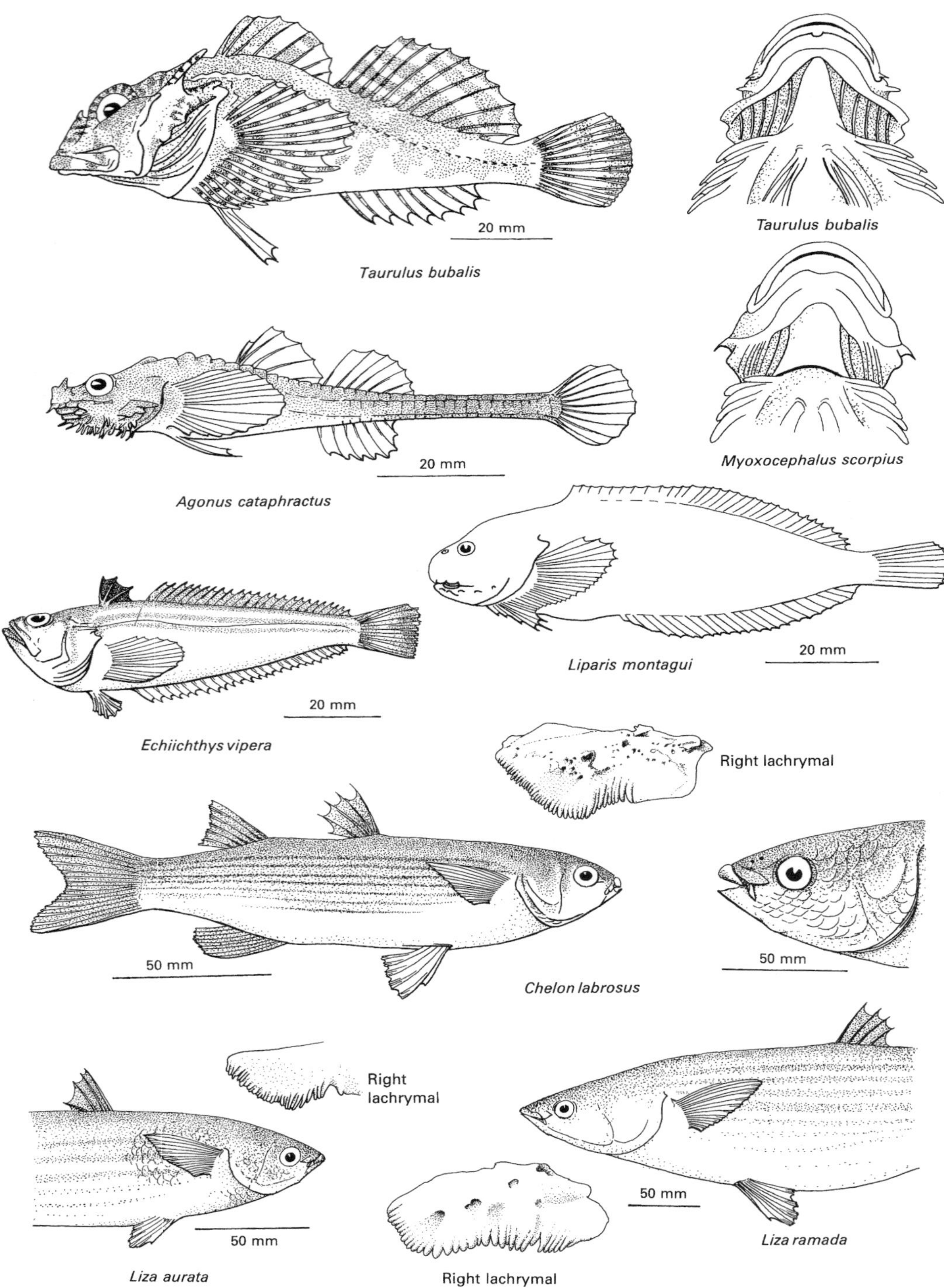
20 mm
Taurulus bubalis
Taurulus bubalis
20 mm
Agonus cataphractus
Myoxocephalus scorpius
20 mm
Liparis montagui
20 mm
Echiichthys vipera
Right lachrymal
50 mm
50 mm
Chelon labrosus
Right lachrymal
50 mm
50 mm
Liza aurata
Right lachrymal
Liza ramada

Fig. 14.10

Stomach divided into two parts. The three British species are morphologically very alike. Small fish are particularly difficult to identify but melanophore patterns on the lower jaw of juveniles (25–50 mm) are diagnostic. *Chelon labrosus* has large or dense melanophores anterior to the eye on the mandibular, ventro-opercular, and gular regions; in *Liza aurata* they are absent from the gular region; in *L. ramada* they are present on gular but absent from the ventro-opercular region

---

1. Upper lip broad (its greatest depth equal to or exceeding eye diameter), with 1–4 rows of papillae in fish of standard length greater than 90 mm. Six pyloric caeca **_Chelon labrosus_**

   Upper lip thin (its greatest depth less than half eye diameter), smooth. More than six pyloric caeca **2**

---

2. Posterior end of lachrymal bone tapered; pectoral fin 17–23% of standard length, reaching the orbit when turned forward; a well defined golden spot on operculum, no marked black spot at the pectoral axil; eight pyloric caeca, the dorsal ones longest **_Liza aurata_**

   Posterior end of lachrymal bone truncate; pectoral fin less than 20% of standard length (not reaching the orbit when turned forward); a vague, diffuse golden spot on operculum, and a black spot at the pectoral axil; seven pyloric caeca of uniform length **_Liza ramada_**

Chelon labrosus (Risso).
Thick-lipped grey mullet Fig. 14.10
Upper lip very thick, its greatest depth equal to or exceeding half the eye diameter, bearing 3–5 rows of small papillae (which increase in number and size as the fish grows); a line of fine teeth also present. Lachrymal (preorbital) bone coming to an obtuse point posteriorly, its slightly convex lower margin finely serrated; end of maxilla hardly obscured by lachrymal; jugular space scarcely visible. Pectoral length exceeding the distance from its base to the rear of the orbit. Br. vi; 1D IV; 2D 9–10; A III/8–9; P 17–18; V I/5; lat. sc. 45–46; vert. 24; pyloric caeca six. Dark grey above, flanks with longitudinal grey stripes, silvery below. To 66 cm.

The commonest grey mullet in the British Isles; all coasts; probably offshore during winter, coming into coastal and estuarine waters during spring and summer; forms shoals.

Liza aurata (Risso).
Golden grey mullet Fig. 14.10
Depth of upper lip not exceeding one-half of the eye diameter; without papillae. Labial teeth large. Lachrymal (preorbital) bone coming to an obtuse point posteriorly, serrated on its convex lower margin; maxilla covered by lachrymal; jugular space oval. Pectoral length exceeding the distance from its base to the rear of the orbit. Br. vi; 1D IV; 2D III/7–9; A III/9; P 17; V I/5; lat. sc. 40–46; vert. 25; pyloric caeca eight. Blue-grey above, flanks somewhat yellowish with longitudinal grey stripes; golden spots on the preoperculum and operculum. To 44 cm.

Possibly on all coasts; uncommon or overlooked on account of similarity to other grey mullets; perhaps not entering brackish water.

Liza ramada (Risso).
Thin-lipped grey mullet Fig. 14.10
Depth of upper lip not exceeding one-third of the eye diameter; without papillae; a line of fine teeth present. Lachrymal (preorbital) bone rounded posteriorly, with well developed teeth around the angle and on the lower margin; only the end of the maxilla visible below the lachrymal; jugular space oval. Pectoral fin length less than (or equalling in the young) the distance from its base to the rear of the orbit. Br. vi; 1D IV; 2D 8–9; A III/8–9; P 17; V I/5; lat. sc. 42–46; vert. 24; pyloric caeca seven. Dark grey above, flanks with longitudinal grey stripes, silvery below. To 50 cm.

May occasionally be found in the same habitats as *C. labrosus* on south and west coasts.

26 LABRIDAE

Wrasses. Body compressed, covered with cycloid scales which sometimes extend onto the head; the number of scale rows is important. Dorsal fin single, with a long, spinous anterior portion; anal with three or more spines. Lips thick, with powerful teeth; no palatal teeth. Hypopharyngeal bones fused, bearing large grinding teeth. No pyloric caeca. Closed swim-bladder. Often brightly coloured.

---

1. Anal fin with five spiny rays **Centrolabrus exoletus**

   Anal fin with three or four spiny rays **2**

---

2. 42–47 scales along the lateral line, and at least six rows of scales above it **3**

   32–38 scales along the lateral line, not more than four rows of scales above it **4**

---

3. Dorsal fin with 19–20 spiny rays and 9–11 soft rays; pectorals with 15 rays; mouth small **Labrus bergylta**

   Dorsal fin with 16–18 spiny rays and 11–14 soft rays; pectorals with 16–17 rays; large mouth **Labrus mixtus**

---

4. A dark spot at the base of the central rays of the caudal fin; 32–36 scales along the lateral line; teeth in a single row on each jaw **Crenilabrus melops**

   A dorsally situated dark spot on the caudal peduncle; 36–38 scales along the lateral line; teeth in several rows on each jaw **Ctenolabrus rupestris**

Centrolabrus exoletus (Linnaeus).
Rock cook Fig. 14.11

Body oval-elongate, head profile rounded; mouth very small, the maxilla not reaching the level of the nostrils. Five spines in the anal fin. Br. v; D XVIII-XX/5-7; A IV–VI/6–8; P 14; V I/5; lat. sc. 32–35; sc. rows 3–4/10–11; vert. 32–34. Greenish or brownish, paler below; caudal fin crossed by a broad, curved, dark band. To 15 cm.

Inshore over rocky grounds; south-west England and the west and north of the British Isles.

Crenilabrus melops (Linnaeus).
Corkwing Fig. 14.11

Body oval; mouth small, the maxilla not reaching the level of the nostrils. Br. v; D XV–XVIII/8–10; A III/8–10; P 14–16; V I/5; lat. sc. 32–36; sc. rows 4/12; vert. 32–34. Colour variable, bluish to yellowish-green, with an oval black spot below the lateral line on the caudal peduncle and a kidney-shaped mark behind the eye. To 15 cm.

Intertidal, in weedy shore-pools, or inshore; all coasts, but most common in the south and west.

Crenilabrus bailloni Valenciennes has been recorded rarely; it differs in lacking the dark mark behind the eye and, in life, by having a blue crescent at the base of the pectoral fin rays.

Ctenolabrus rupestris (Linnaeus).
Goldsinny Fig. 14.11

Body elongate, head-profile rounded; mouth small, the maxilla about reaching the level of the nostrils. Br. v; D XVI–XVIII/8–10; A III–IV/7–8; P 14; V I/5; lat. sc. 36–39; sc. rows 3–4/14–15; vert. 32–35. Brownish-red with a dark blotch on the commencement of the dorsal fin and another on the upper side of the caudal peduncle; paler below. To 18 cm.

Labrus bergylta Ascanius.
Ballan wrasse Fig. 14.11

Body elongate, head-profile rounded; mouth small, the maxilla scarcely reaching the level of the nostrils. Br. v; D XIX–XX/9–11; A III/8–10; P 14–15; V I/5; lat. sc. 41–47; sc. rows 6–7/12–15; vert. 37–38. Colour greenish or brownish, but very variable; each scale paler in the centre; flanks spotted or striped. To 50 cm.

Inshore, near rocks and over hard ground; common off southern and western coasts.

Labrus mixtus Linnaeus
Cuckoo wrasse Fig. 14.11

Body elongate, head-profile tapering, snout elongated; mouth large, maxilla passing the level of the nostrils. Br. v; D XVI–XVIII/11–14; A III/9–11; P 17; V I/5; lat. sc. 50–60; sc. rows 5–7/17–21; vert. 38–39. Sexually dichromatic: males with blue heads and orange bodies boldly striped with blue; females orange with two dark brown spots at the base of the soft dorsal finrays and a third on the upper side of the caudal peduncle. To 35 cm.

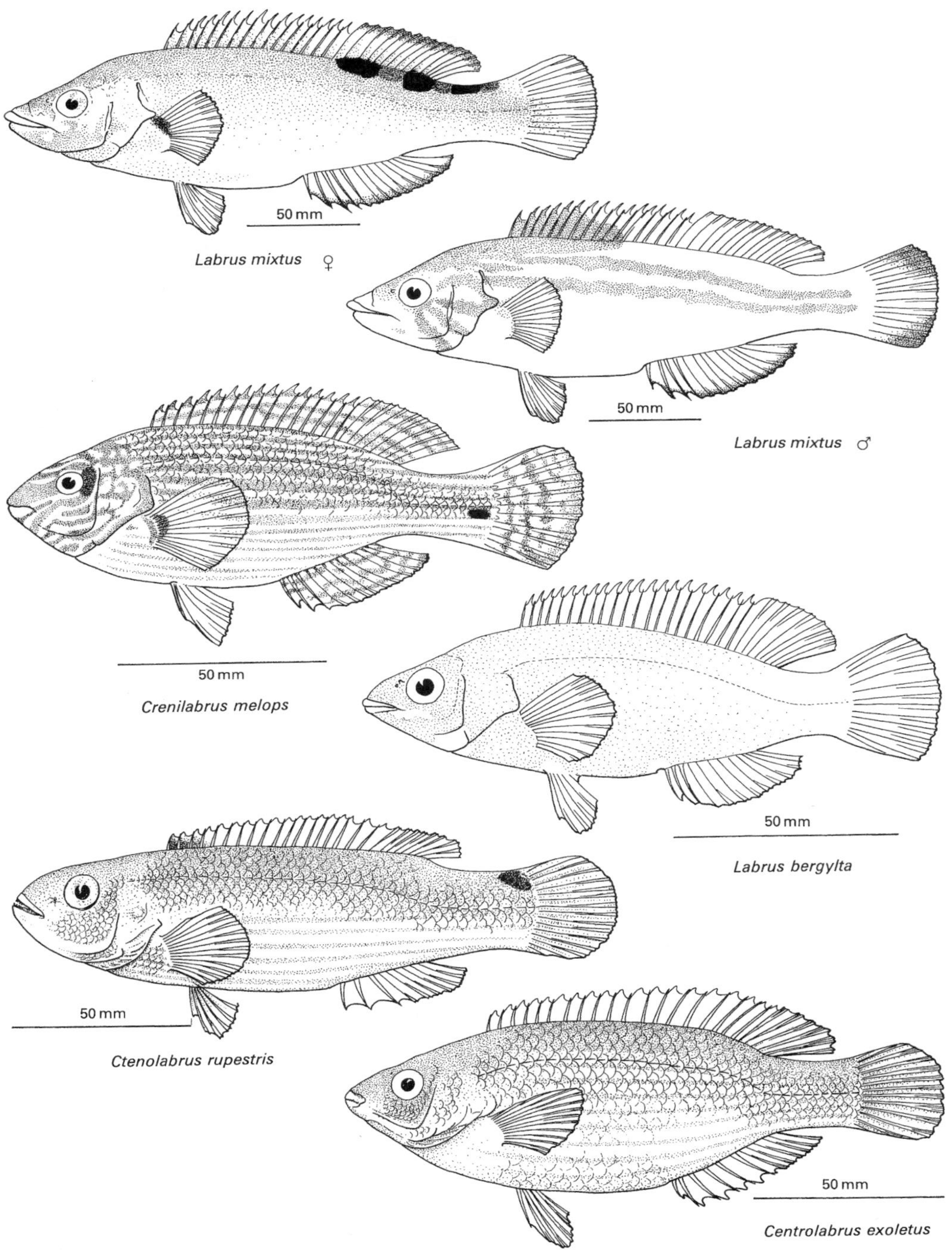
50 mm
Labrus mixtus ♀
50 mm
Labrus mixtus ♂
50 mm
Crenilabrus melops
50 mm
Labrus bergylta
50 mm
Ctenolabrus rupestris
50 mm
Centrolabrus exoletus

Fig. 14.11

Over rocks and hard ground, not on the shore or in very shallow water; mainly off southern and western coasts.

27 TRACHINIDAE

Benthic fishes with elongate, compressed bodies, dorsally placed eyes and an oblique mouth. The spinous dorsal fin is short, with a black membrane. Second dorsal and anal fins long; pelvics jugular. Operculum spined. Scales small, cycloid, arranged in distinctive oblique rows. Lateral line high on the flank, downcurved on the caudal peduncle. Poison glands situated at the base of the fin spines and in the vicinity of the opercular spine. Wounds inflicted by the poison spine are painful and can be dangerous.

Echiichthys vipera Cuvier.
Lesser weever Fig. 14.10
Body rather short, compressed; mouth large, steeply sloping; eyes on top of the head. No preorbital spines. Margin of pectoral fin evenly rounded. Br. vi; 1D V–VI; 2D 21–24; A I/24–26; P 14; V I/5; lat. sc. 60; vert. 34–36. Yellowish-brown above, paler below; first dorsal fin black. To 14 cm.

From about low water mark on sandy beaches and shallow water, living buried in sand with only the top of the head and the back uncovered; all coasts; common in suitable areas.

The rarer, offshore greater weever *Trachinus draco* L. is larger and has a less oblique mouth; small preorbital spines are present; and the pectoral fin is slightly emarginate.

28 BLENNIIDAE

Elongate fishes with a long, subdivided dorsal fin; pelvics jugular; caudal fin distinct. Skin scaleless. Each jaw with a single row of teeth.

---

1. Median cephalic tentacle and crest present **_Coryphoblennius galerita_**

   Median cephalic tentacle and crest absent **2**

---

2. Orbital tentacles present **_Parablennius gattorugine_**

   Orbital tentacles absent **_Lipophrys pholis_**

Coryphoblennius galerita (Linnaeus).
Montagu's blenny Fig. 14.12
Body smooth, elongate, with D divided by a shallow notch into subequal parts. A crest along the top of the head commences as a tall, wide, fringed, erectile flap, and continues as a series of short filiform tentacles. No orbital tentacles. Br. vi; D XII–XIII/15–18; A 17–18; P 12; V 2. Brownish with darker bands, and with bluish-white spots along the flank. To 8 cm.

Intertidal on rocky coasts, in mid-shore pools; very local, Cornwall, Pembrokeshire, and south-west Ireland north to Connemara.

Lipophrys pholis (Linnaeus).
Shanny Fig. 14.12
Body smooth, elongate, with D divided by a small notch into subequal parts. No orbital tentacles. Br. vi; D XI–XIII/18–20; A I/17–19; P 13; V 2. Dark brown and blotched, or blackish, with a dark spot behind the first dorsal finray. To 16 cm.

Intertidal; on any shore with rock and weed, or in pools; all coasts; common.

Parablennius gattorugine (Linnaeus).
Tompot blenny Fig. 14.12
Body smooth, elongate, with D divided by a dip into two parts of roughly equal height. Branched orbital tentacles present. Br. vi; D XII–XIV/17–20; A I/20–21; P 14; V I/2–3; vert. 39. Brownish with dark bands. To 30 cm.

Mainly inshore, occasionally at extreme low-water mark on rocky shores; south and west coasts.

The butterfly blenny, *Blennius ocellaris* L., has the anterior part of the dorsal fin tall and bearing a white-edged black ocellus; it may be caught off south-western coasts.

29 LUMPENIDAE

Elongate fishes with a long dorsal fin consisting of thin spiny rays only; pelvic fins under, or slightly in advance of, the pectorals, sometimes with a spiny ray. Scales small.

Lumpenus lampretaeformis (Walbaum).
Snake blenny Fig. 14.12
Body very long and slender. D long, of spiny rays only; depressed anteriorly, with the first few rays not united by a membrane. D LXVIII-LXXVI; A I/47–53; P 14–15; V I/3; vert. 80–85. Brownish-yellow, in places tinged blue; brown spots on the

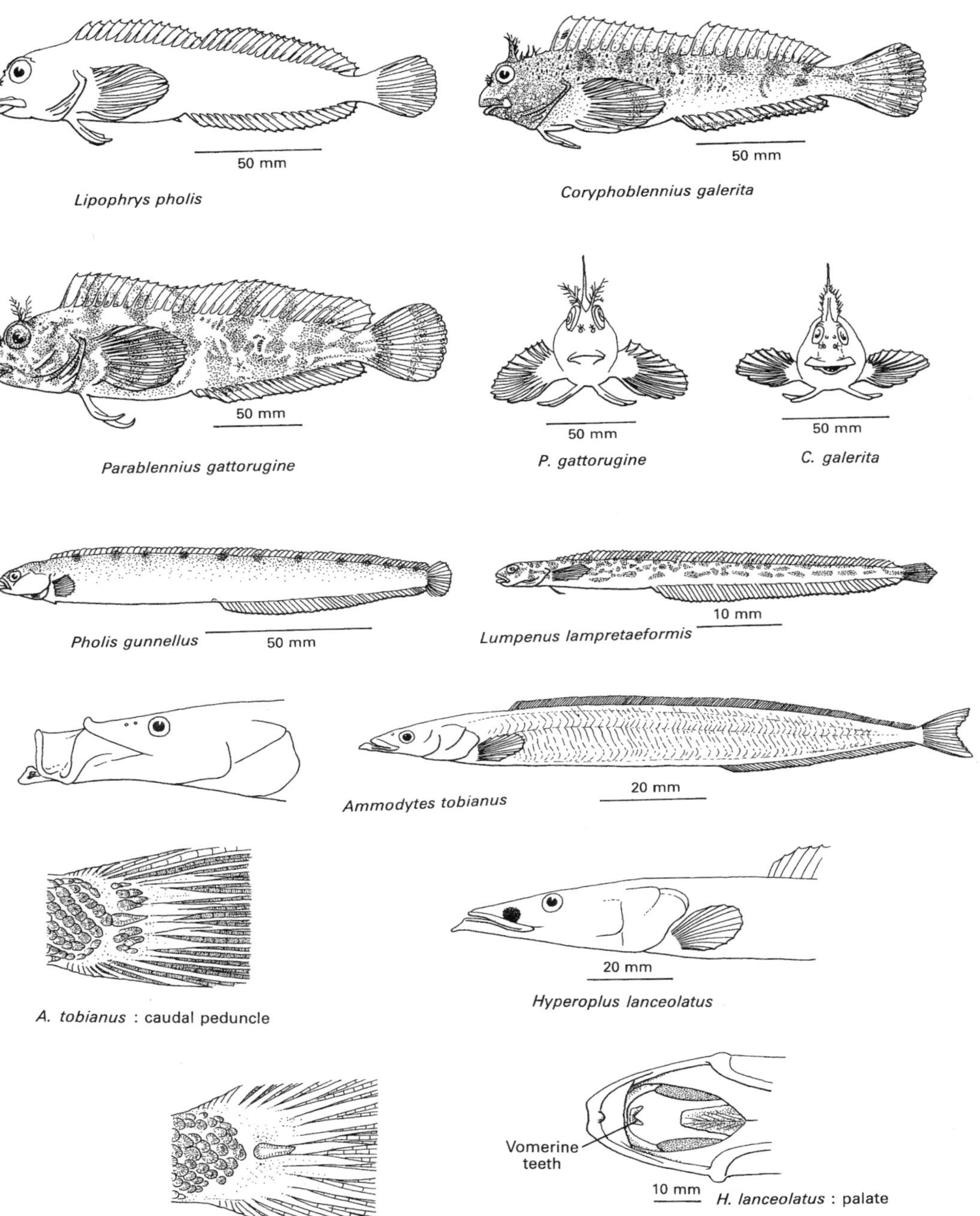
50 mm
Lipophrys pholis
50 mm
Coryphoblennius galerita
50 mm
Parablennius gattorugine
50 mm
P. gattorugine
50 mm
C. galerita
Pholis gunnellus
50 mm
10 mm
Lumpenus lampretaeformis
Ammodytes tobianus
20 mm
A. tobianus : caudal peduncle
20 mm
Hyperoplus lanceolatus
A. marinus: caudal peduncle
Vomerine teeth
10 mm
H. lanceolatus : palate

Fig. 14.12

flank; usually 8–10 larger spots along the dorsal fin. To 40 cm.

Inshore and in shallow water, over muddy bottoms; off Scottish coasts.

30 PHOLIDAE

Elongate, laterally compressed fishes with a long dorsal fin consisting of short spiny rays only. Pelvic fins under the pectorals, reduced; or absent.

Pholis gunnellus (Linnaeus).
Butterfish Fig. 14.12
Eel-like, compressed body; D of short, spiny rays only; pelvics small. Scales minute, embedded; skin slippery. Br. v; D LXXV–LXXXII; A II/39–45; P 10–12; V I/1; vert. 82–85. Yellowish-brown, with nine to 15 white-edged black spots along the base of the dorsal fin. To 25 cm.

Intertidal on rocky shores; also offshore; all coasts; common.

31 AMMODYTIDAE

Sand eels. Body elongate, covered with small cycloid scales; or scales absent. Head pointed, lower jaw projecting. Upper jaw bordered by premaxillary, extensible and sometimes protrusible; teeth on jaws absent or reduced. Median fins without spiny rays; a single dorsal fin; caudal forked; pelvics usually absent. Pelagic or found buried in sand.

---

1. Two 'teeth' on the vomer (in the roof of the mouth, Fig. 14.12). Dorsal fin commencing just behind the tip of the pectorals. Premaxillary not protractile
Hyperoplus lanceolatus

Vomer lacking 'teeth'. Dorsal fin commencing above pectorals. Premaxillary protractile
Ammodytes tobianus

Ammodytes tobianus Linnaeus.
Lesser sand eel Fig. 14.12
Eel-like body covered with small cycloid scales situated along numerous oblique dermal folds. Premaxilla protractile; anterior end of maxillae not fused, joined by a tendon. Vomer toothless. D commencing in front of the free end of the pectorals. Back in advance of the dorsal fin completely covered with well-developed scales; scales on the belly arranged in chevrons; dorsal and ventral muscle pads of the caudal fin (which are separated by a single elongate scale) bearing scales. Br. vii; D 50–56; A 25–31; P 10–13; vert. 60–66. Back greenish, belly and sides silvery. To 20 cm.

Inshore and offshore; may be dug out of sandy beaches below MTL; all coasts very common.

The similar but offshore *A. marinus* Raitt has no scales, apart from the central one, on the muscle pads at the base of the caudal fin (Fig. 14.12).

Offshore in the west and south-west.

Hyperoplus lanceolatus (Lesauvage).
Greater sand eel Fig. 14.12
Eel-like body covered with minute cycloid scales not lying along oblique dermal folds. Premaxillae not protractile; anterior ends of maxillae broadened and united at the tip of the snout. Vomer bearing two strong tooth-like processes. Dorsal fin commencing behind the pectorals. D 52–61; A 27–33; P 12–14; vert. 65–69. Back greenish, belly and sides silvery; a distinct dark spot on the side of the snout (which is absent in the offshore, south-western *H. immaculatus* (Corbin)). To 32 cm.

Inshore and offshore; may sometimes be dug out of sandy beaches; off all coasts; common.

32 CALLIONYMIDAE

Dragonets. Body elongate, somewhat depressed. Two dorsal fins, the first spinous; pelvic fins in front of pectorals. Skin scaleless. Jaws toothed. Eyes situated dorsally, interorbital space very narrow. Angle of preoperculum bearing a variously armoured spine. No swimbladder. Sexually dimorphic.

---

1. Base of anal fin (A) at least 70% of the distance between its commencement and the tip of the snout; caudal fin at least 80% of the distance between the commencement of 1D and the tip of the snout. Two rows of dark spots laterally on the body, and four parallel rows of dark spots on 2D of the adult male. Lateral line displaced dorsally at the caudal peduncle (seen in preserved specimens) ***Callionymus maculatus***

Base of A not above 60% of the distance between its commencement and the tip of the snout; caudal fin less than 80% of the distance between the commencement of 1D and the tip of the snout. Body variously patterned but without dark spots laterally or on 2D. Lateral line not displaced dorsally at the caudal peduncle (seen in preserved specimens) ***Callionymus lyra***

Callionymus lyra Linnaeus.
Common dragonet Fig. 14.13

Body elongate, head broad, somewhat depressed. Angle of the preoperculum tapering into a strong spine bearing three terminal processes, two directed upwards and one backwards; shortly in front of this a fourth, forward-directed process. 1D short, the rays very long in males; triangular in females and immatures; rays of 2D long in males. Br. vi; 1D IV; 2D 9 (the last ray divided from its base); A 9; P 20; V I/5; vert. 21. Body of male brownish, with blue spots dorsally and a bluish band along the flank; the second dorsal fin yellow with bright blue longitudinal stripes. Body of female and immature brown, paler below, with a series of dark spots along the flank and, from 1D, three ill-defined brown bands across the back; posterior two-thirds of first dorsal fin bluish-black. Males to 30 cm, females to 20 cm.

In shallow and coastal waters; small specimens even in sandy shore pools; all coasts; common.

Callionymus maculatus Rafinesque.
Spotted dragonet Fig. 14.13

Body form as in preceding species. Br. vi; 1D IV; 2D 9, rarely 10 (the last ray divided from its base); A 9; C 9–10; P 16; V I/5. Body greyish-brown, paler below, with two rows of dark spots and some blue spots on the flank and, from 1D, four irregular, softly outlined brown patches across the back. Males with 2D light-coloured, marked with four parallel rows of pearly-edged dark spots; females and immatures with indistinct spots on fin. Males to 14 cm, females to 11 cm.

In shallow and moderately deep water, less coastal and less common than *C. lyra*; apparently absent from the east coast.

A third species, *C. reticulatus* Valenciennes, is present in the south and west (Fig. 14.13). Males have oblique rows of dark spots on 2D. In both sexes 2D has 10 rays (the last one branched) rather than nine.

33 GOBIIDAE

Small fishes, generally of benthic habit, with elongate bodies of rounded section; head large, blunt, eyes dorso-lateral and close together. Two dorsal fins, the first short and with spiny rays. Pelvic fins joined, forming a disc (Fig. 14.14); the fin membrane extending across the disc as an anterior fold; V I/5. Frequently sexually dimorphic or dichromatic.

Lines of minute sensory papillae ramify over the head of gobies, and their arrangement is important in taxonomy. To reveal these papillae the fish should first be fixed in formalin (4% formaldehyde in sea water) for 45 minutes, washed well, and then transferred to chromic acid (2% aqueous solution of chromium trioxide) for 24 hours. Treated fish should be thoroughly washed and stored in 70% ethanol.

The rarer British gobies are still poorly known, and are not described here; but one species, *Lesueurigobius friesii* (Malm), found on muddy grounds in deepish water, deserves mention because it lives in association with the crustacean *Nephrops norvegicus*; it is grey-brown with a row of large yellow spots on the flank, and has 24–29 scales in the lateral series.

---

1. Pelagic fishes with translucent, unpigmented bodies ***Aphia minuta***

   Bottom-living fishes with opaque and generally pigmented bodies **2**

---

2. First dorsal fin (1D) with 7–8 spiny rays; a conspicuous black spot at the base of the tail ***Gobiusculus flavescens***

   1D with 6–7 spiny rays; base of tail lacking a conspicuous spot **3**

---

3. Gobies often exceeding 8 cm in length, with superior rays of pectoral fins separate and filamentous distally or with flanks prominently spotted with bright orange; four or more vertical rows of papillae on the side of the head

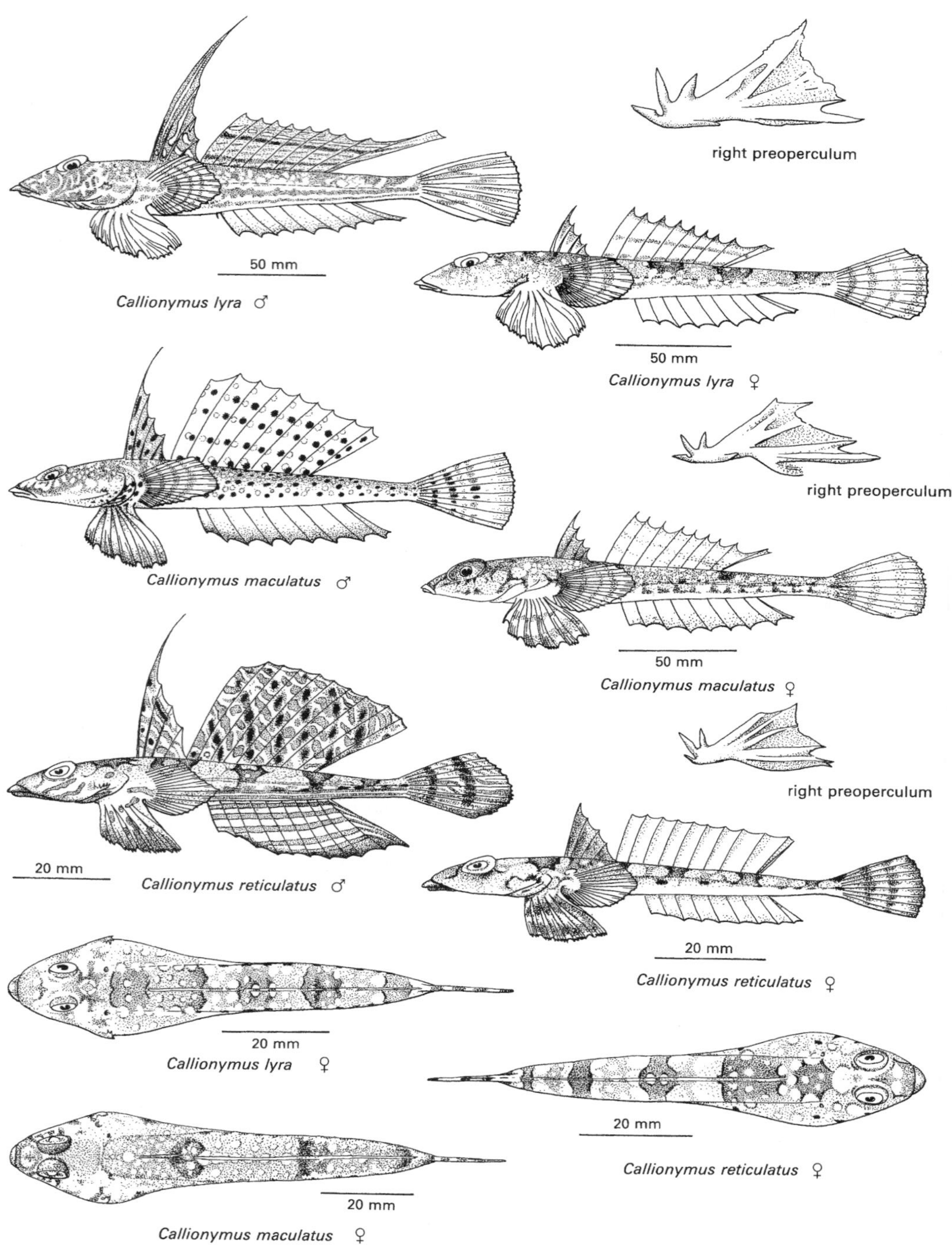
right preoperculum
50 mm
Callionymus lyra ♂
50 mm
Callionymus lyra ♀
right preoperculum
Callionymus maculatus ♂
50 mm
Callionymus maculatus ♀
right preoperculum
20 mm
Callionymus reticulatus ♂
20 mm
Callionymus reticulatus ♀
20 mm
Callionymus lyra ♀
20 mm
Callionymus reticulatus ♀
20 mm
Callionymus maculatus ♀

Fig. 14.13

below the orbit may be fairly conspicuous **4**

Small gobies generally less than 8 cm in length, without filamentous rays to the pectoral fins and not prominently spotted with orange; suborbital papillae inconspicuous **6**

---

4. Ground-colour pale, bearing orange spots; hinder part of 1D black (34-38 scales in a lateral series) ***Thorogobius ephippiatus***

   Colour pale or dark, but without orange spots **5**

---

5. Fewer than 50 scales in a lateral series; usually a dark marking on the upper anterior corner of each dorsal fin ***Gobius niger***

   Over 50 scales in a lateral series; 1D with a pale superior band ***Gobius paganellus***

---

6. 55–70 scales in a lateral series, without a marked increase in size below 2D ***Pomatoschistus minutus***

   Not more than 50 scales in a lateral series, showing a marked increase in size below 2D **7**

---

7. 35–40 scales in a lateral series; dorsal fins with one or two rows of black spots; three small, double, dark marks along the lateral mid-line below three pale, dorsal saddles ***Pomatoschistus pictus***

   40–50 scales in a lateral series; dorsal fins without rows of black spots; often in brackish water ***Pomatoschistus microps***

Aphia minuta (Risso).
Transparent goby Fig. 14.14
Body compressed, transparent. Pelvics united. Br. v; 1D V; 2D I/10–12; A I/11–13; P 16. To 6 cm.

Pelagic in shoals; inshore waters; all coasts; common.

A rarer transparent goby, *Crystallogobius linearis* (Von Düben), differs in having 1D small (two-rayed) or rudimentary, and 2D and A long and low.

Gobius niger Linnaeus. Black goby Fig. 14.14
Thick-set goby; head about 0.25 of total length; caudal peduncle about half of head length. 1D with central rays the longest, especially in the male; fin membrane of 1D confluent with the start of 2D; upper two or three finrays of pectoral free at their tips (for one-quarter of their length at most); margin of pelvic disc rounded, the anterior membrane without lateral lobes. Rim of anterior nostril with a simple flap. 1D V–VII; 2D I/11–13; A I/10–13; P 15–20; lat. sc. 32–42; vert. 27–28. Dark grey or brown with lighter markings; each dorsal fin with a small black mark on the anterior upper corner. To 17 cm.

In shallow coastal or estuarine waters, rarely on the shore, amongst *Zostera* or seaweed on soft substrata; all coasts; frequent or common.

Two other rather similar gobies are present in south-western waters. *Gobius couchi* Miller and El-Tawil and *G. gasteveni* Miller lack the elongated rays in 1D and the dark anterior spots on both dorsals. *Gobius couchi*, 1D V–VII; 2D I/12–14; A I/11–13; P 15–18; lat. sc. 35–45, is grey-brown with several lateral blotches, a higher-than-wide spot on the pectorals, and three gular spots under the head. *Gobius gasteveni*, 1D V–VI; 2D I/13–15; A I/12–14; P 18–22; lat. sc. 37–45, has similar though fewer lateral markings, a wider-than-high blotch on the pectorals, and no gular spots.

Gobius paganellus Linnaeus.
Rock goby Fig. 14.14
Thick-set goby; head 0.2–0.25 of total length; caudal peduncle about 0.65 of head length. 1D with the anterior ray longest; fin membrane of 1D extending to base of 2D; upper four or five finrays of pectoral free for much of their length. Margin of pelvic disc rounded, the anterior membrane with feebly developed lobes. Anterior nostril with a digitate tentacle. Br. v; 1D VI; 2D I/11–13; A I/10–12; P 20; lat. sc. 50–58; vert. 28–29. Uniform dark brown or nearly black; first dorsal fin with a pale yellow or orange marginal band. To 12 cm.

Intertidal on rocky shores, particularly in tide pools, and on hard grounds in shallow water; common on south-west coasts, absent from the east.

In Cornwall the giant goby, *G. cobitis* Pallas, with lat. sc. 60-68, the membrane of the pelvic disc lobed, and the first dorsal fin uniformly coloured, occurs in high-level shore pools. *Gobius cruentatus* Gmelin, with the lips and cheeks streaked with red, the rows of sensory papillae black, and lat. sc. 52–58, has been reported from southern Ireland.

Gobiusculus flavescens (Fabricius).
Two-spot goby Fig. 14.14
Slender body; interorbital space wide, at least half the eye diameter. Head about 0.2 of total length; caudal peduncle long, about equal to the head length. 1D usually with one ray more than in other gobies. Margin of pelvic anterior membrane entire. 1D VII (rarely VI or VIII); 2D I/9–11; A I/10–11; P 19; lat. sc. 34–40; vert. 31. Reddish-brown, darker above, with a series of lighter bands across the back; a distinct and conspicuous black spot at the base of the caudal fin, and a smaller dark spot just behind the pectoral origin; dorsal fins banded with red and off-white, and caudal fin transversely banded.

Inshore and in intertidal pools, amongst weed; does not rest on the bottom as much as other gobies; all coasts; common in the south and west.

Pomatoschistus microps (Krøyer).
Common goby Fig. 14.14
Small goby; head 0.2–0.25 of total length; caudal peduncle about equal to the head length. 1D and 2D separated. Margin of pelvic anterior membrane entire. Size of the scales increasing quite markedly at about the commencement of 2D. Top of head, nape and throat scaleless. 1D V–VI; 2D I/8–10; A I/8–10; P 18; lat. sc. 43–50. Grey or sandy-coloured, often with darker spots or bars; dorsal fins with spots or bands, males with a black spot on the posterior edge of 1D; upper part of pectoral fin base dark. To 7 cm.

In tide-pools, estuaries, saltmarshes, and brackish water; all coasts; abundant.

Pomatoschistus minutus (Pallas) *sensu lato*.
Sand goby Fig. 14.14
Small goby; head about 0.25 of total length; caudal peduncle almost equal to head length. 1D and 2D slightly separated. Margin of pelvic anterior membrane fringed with villi. Scales small and numerous, reaching forward on to the nape and on to the throat in front of pelvic disc. Cleft between branchiostegal membranes reaching forward 0.2–0.3 of head length. 1D VI–VII; 2D I/9–12; A I/9–12; P 18–21; lat. sc. 55–75; vert. 30–34. Sandy-brown above, with darker speckles. First dorsal fin often with a dark spot on its posterior margin. To 11 cm.

Inshore on soft bottoms and near MLW on sandy shores; all coasts; common.

Pomatoschistus minutus s. lato comprises two similar coastal species. *Pomatoschistus minutus s. stricto* tends to inhabit estuarine waters. It has a mean vertebral count of 33 and reaches 11 cm. There is always a blue-black spot, which does not reach the edge of the membrane, on 1D. *Pomatoschistus lozanoi* (de Buen) is coastal, has a mean vertebral count of 32 and reaches only 8 cm. The dark spot on 1D is present on males but only in spring; it reaches the edge of the membrane. The two species can best be discriminated by the arrangement of sensory papillae on the head (Whitehead *et al.* 1984–86).

Further offshore both species are replaced by *Pomatoschistus norvegicus* (Collett), in which the nape and throat are scaleless; the spot on 1D is present in maturing fish of both sexes and reaches to the edge of the membrane; and the cleft between the branchiostegal membranes reaches forward for about one-half of the head length. 2D I/8–10; A I/8–10; P 16–18; lat. sc. 55–60; vert. 32. To 6.5 cm.

Pomatoschistus pictus (Malm).
Painted goby Fig. 14.14
Small goby; head 0.2–0.25 of total length; caudal peduncle about 0.75 of head length. Dorsal fins slightly separated. Margin of pelvic anterior membrane entire. The size of the scales increases under 2D. No scales on top of the head or on the throat. 1D VI; 2D I/8–9; A I/8–9; P 18; lat. sc. 35–41; vert. 28. Brown; a line of four or five diffuse spots along the flank; four light bands across the back; upper part of pectoral finbase dark; dorsal fin with one or two rows of small black spots along the finbase, one (or two) between each ray, and longitudinal reddish bands (especially developed in breeding males). To 5.5 cm.

Inshore on rough, shelly or gravelly grounds, occasionally found intertidally; south and west coasts; frequent.

Thorogobius ephippiatus (Lowe).
Leopard-spotted goby Fig. 14.14
Head about 0.2 of total length; caudal peduncle about 0.8 of head length. Scales on body not

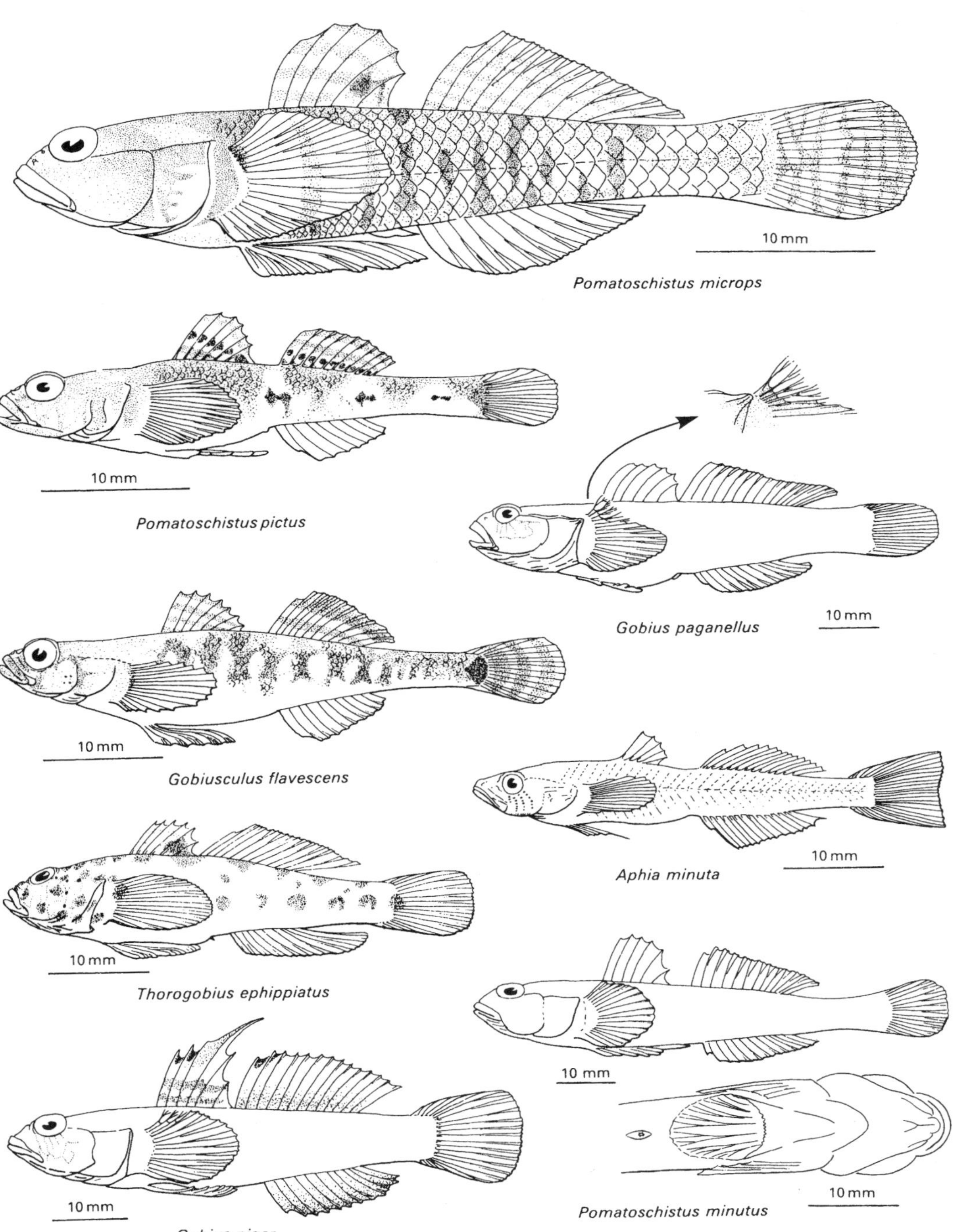
10 mm
Pomatoschistus microps
10 mm
Pomatoschistus pictus
Gobius paganellus
10 mm
10 mm
Gobiusculus flavescens
Aphia minuta
10 mm
10 mm
Thorogobius ephippiatus
10 mm
10 mm
10 mm
Gobius niger
Pomatoschistus minutus

Fig. 14.14

extending forward on to the top of the head. 1D and 2D separated by a small gap; upper finrays of pectoral not separate. 1D V–VII; 2D I/10–12; A I/10; P 17–20; lat. sc. 33–42; vert. 28. Pale brown with orange spots and blotches on the head, flanks, and dorsal fins; posterior part of first dorsal fin black. To 14 cm.

Inshore, on rocky grounds; shelters in crevices but seen frequently by scuba-divers; western coasts.

34 SCOMBRIDAE

Mackerel. Body fusiform, rounded in section, tapering to a thin caudal peduncle; caudal fin strongly forked. First dorsal fin short, with spiny rays, widely separated from the second dorsal; both the second dorsal and the anal succeeded by a series of finlets extending on to the caudal peduncle. Pectoral fins high on the flank; pelvics below pectorals.

Scomber scombrus Linnaeus.
Mackerel Fig. 14.5
Br. vii; 1D XI–XIII; 2D 11–13 + 5(–6); A I/11–13 + 5; P 17; A I/5; vert. 30–31. Deep greeny-blue above, crossed by narrow, black curving bands; silvery on the flanks, white below. To 50 cm.

Pelagic, forming shoals; off all coasts, except in the south-east, especially during the summer months.

Order Pleuronectiformes

35 BOTHIDAE

Flatfish with eyes on the left side of the head.

---

1. Fish oval in outline (width about half of the length) **2**

   Fish rounded in outline (width about three-quarters of length) **3**

---

2. Mouth small (scarcely overlapping the left eye); pelvic fin on the blind side shorter based than that on the eyed side; scales very readily detached ***Arnoglossus laterna***

   Mouth large (extending well past anterior of left eye); pelvic fin bases equal ***Phrynorhombus norvegicus***

---

3. Skin smooth; distal half of first dorsal finrays separated into filaments; dorsal and anal fins not reaching caudal; pelvics free from anal ***Scophthalmus rhombus***

   Skin on the eyed side bearing tubercles (except in small fish); only extreme tips of dorsal finrays separated; dorsal and anal fins not reaching caudal; pelvics free from anal ***Scophthalmus maximus***

   Skin smooth; dorsal finrays not separated; dorsal and anal fins with posterior lobes extending under the caudal fin; pelvics united with anal; conspicuous 'eye-brows' ***Zeugopterus punctatus***

Arnoglossus laterna (Walbaum).
Scaldfish Fig. 14.15
Elongate-oval outline; head about 0.22 of total length; mouth fairly small, the maxilla scarcely passing the front of the left eye. Finbase of pelvic shorter on the blind side. First dorsal finrays free for part of their length. Scales large but caducous. Lateral line arched above the pectoral fin. Br. vii; D 87–93; A 65–74; P 10–11; V 6; lat. sc. 48–56; vert. 37–42. Brownish above, white below. To 19 cm.

Inshore waters on sandy grounds around southern Britain; quite common; rarer northwards.

Phrynorhombus norvegicus (Günther).
Norwegian topknot Fig. 14.15
Oval flatfish with a distinct gap between anterior end of dorsal fin and upper jaw; a distinct caudal peduncle. Pelvic fins of equal size. Eyed side rough; scales not caducous. D 76–84; A 58–68; P 9–10; V 6; lat. sc. 46–52. To 12 cm. Brown with darker markings.

On rocky grounds, 10–50 m, more often seen by scuba-divers than caught in trawls. Brittany to Iceland and north Norway; absent on most of east coasts of England and Scotland.

Scophthalmus maximus (Linnaeus).
Turbot Fig. 14.15
Rhomboid outline, width about 0.7 of length; head large, almost 0.3 of total length. Dorsal and anal

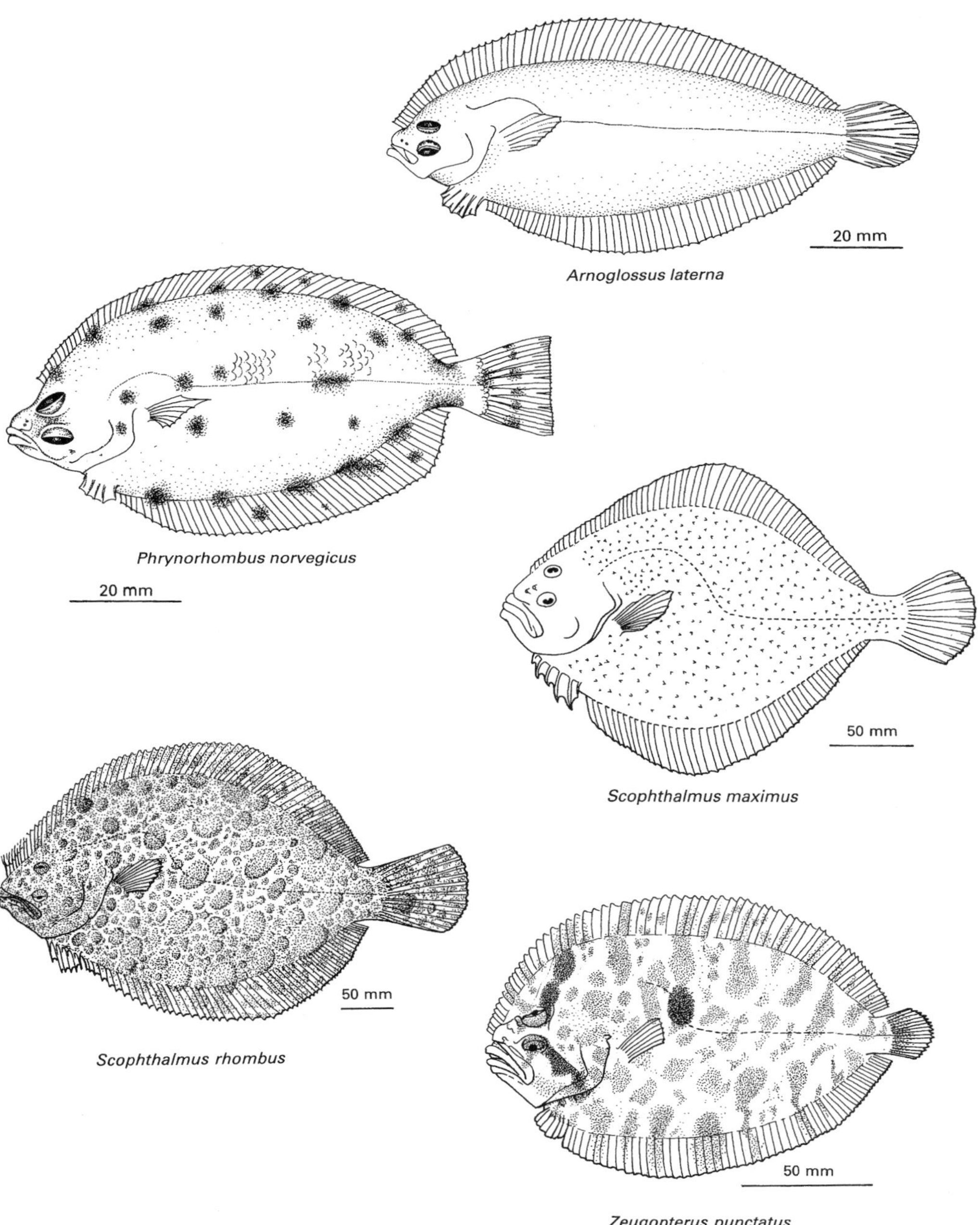
20 mm
Arnoglossus laterna
Phrynorhombus norvegicus
20 mm
50 mm
Scophthalmus maximus
50 mm
Scophthalmus rhombus
50 mm
Zeugopterus punctatus

Fig. 14.15

fins not reaching the caudal; tips of the first dorsal finrays free and branched. No scales but prominent tubercles scattered irregularly over the body on the eyed side. Lateral line strongly arched above the pectoral fin. Br. vii; D 57–71; A 43–52; P 11–12; V 6; vert. 30–31. Colour variable, often greyish-brown with darker speckles; white below. To 80 cm.

Mainly inshore on sandy or gravelly grounds; off all coasts but uncommon around Scotland.

Scophthalmus rhombus (Linnaeus).
Brill Fig. 14.15
Oval outline, width about 0.6 of length; head 0.22–0.25 of total length. Dorsal fin reaching caudal, but not extending underneath it. First dorsal finrays branched and free for about half their length. Scales present; tubercles absent. Lateral line strongly arched above the pectoral fin. Br. vii; D 73–83; A 56–62; P 11–12; A 6; lat. sc. 150; vert. 34–37. Greyish-brown with darker speckles, white below. To 61 cm.

Mainly inshore on sandy grounds; off all coasts but uncommon around Scotland.

Zeugopterus punctatus (Bloch).
Topknot Fig. 14.15
Flatfish with distinctive oblong outline. Dorsal fin originating close to the upper jaw; dorsal and anal fins reaching and extending under the caudal fin. Pelvics joined to the anal fin by membrane. D 88–10; A 67–76. To 25 cm. Mottled brown with dark 'eye-brows' and a dark blotch behind the pectoral fin.

On rocky grounds, to about 25 m. Biscay to western Norway; absent from eastern England.

36 PLEURONECTIDAE

Flatfish with eyes on the right side of the head (reversals occur); snout pointed and the mouth terminal.

---

1. Mouth large, reaching or passing the middle of the eye **_Hippoglossoides platessoides_**

   Mouth small, extending only to the front of the eye **2**

---

2. Eyed surface covered with ctenoid scales, rough to the touch **_Limanda limanda_**

   Eyed surface smooth to the touch (except where prickles are present) **3**

---

3. Head small (about one-sixth of total length); body outline oval; dorsal and anal fins almost reaching the caudal fin. Lateral line curved over pectoral fin; body oval, width more than half of length **_Microstomus kitt_**

   Head large (about one-quarter of total length); body outline rhombic; dorsal and anal fins distant from the caudal fin **4**

---

4. Short spines present along the base of the dorsal and anal fins, and in the region of the lateral line; eyed side typically very dark, blind side opaque white **_Pleuronectes flesus_**

   No spines along the base of the dorsal and anal fins or in the region of the lateral line; eyed side typically brown with orange spots, blind side pearly white **_Pleuronectes platessa_**

Hippoglossoides platessoides (Fabricius).
Long rough dab Fig. 14.16
Outline slender-oval; head about 0.2 of total length. Mouth large, the maxilla reaching the middle of the right eye. Skin on the eyed side rough from the ctenoid scales. Lateral line slightly curved above pectoral fin. Caudal fin obtusely wedge-shaped. Br. viii; D 78–98; A 60–79; P 10; V 6; lat. sc. 85–95; vert. 42–46. Brown above, white below. To 30 cm.

Offshore on fine grounds; common around Scotland, absent from south and south-east England.

Limanda limanda (Linnaeus). Dab Fig. 14.16
Outline oval; head about 0.2 of total length. Mouth small, the maxilla just reaching the right eye. Skin

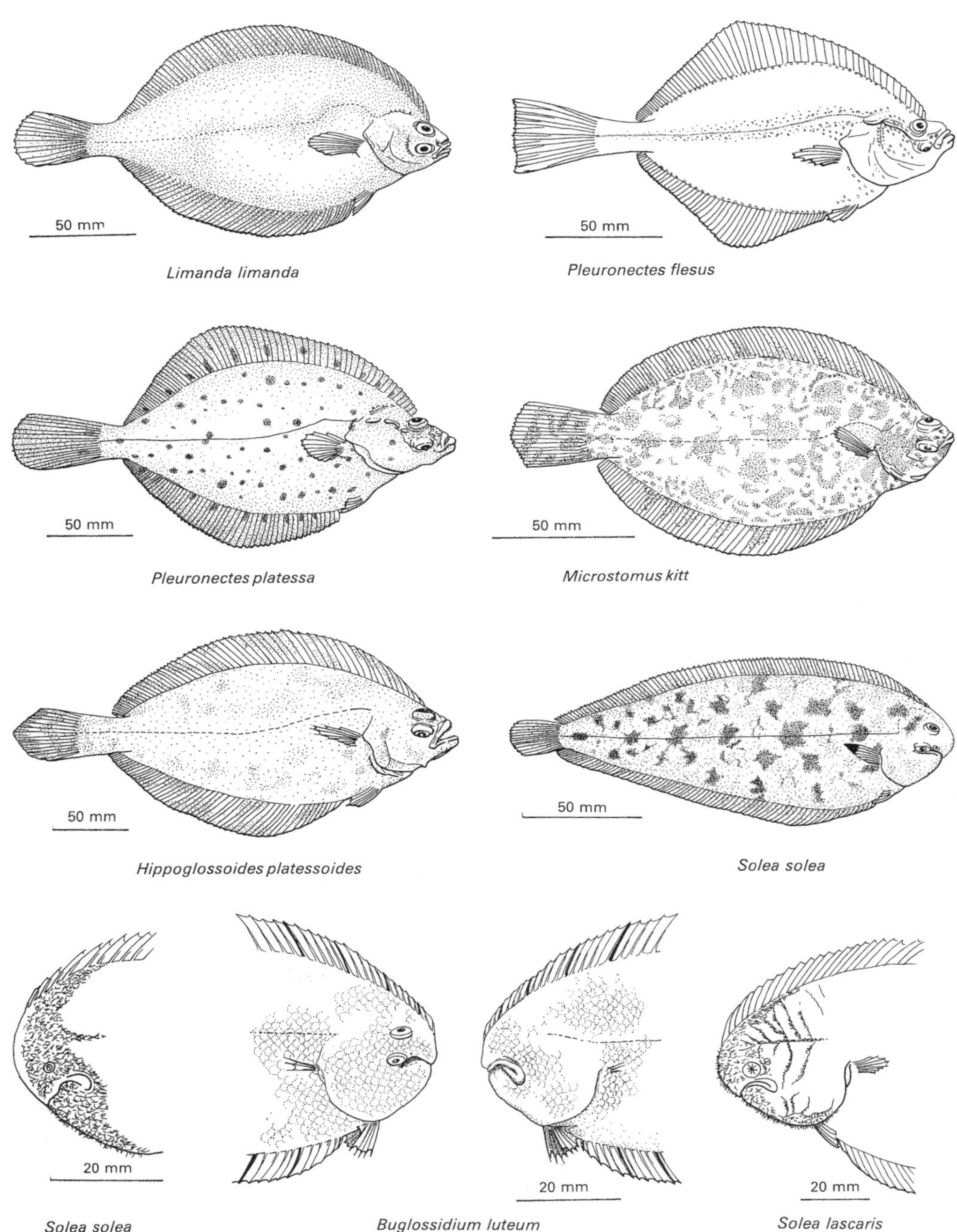
50 mm
Limanda limanda
50 mm
Pleuronectes flesus
50 mm
Pleuronectes platessa
50 mm
Microstomus kitt
50 mm
Hippoglossoides platessoides
50 mm
Solea solea
20 mm
Solea solea
20 mm
Buglossidium luteum
20 mm
Solea lascaris

Fig. 14.16

on the eyed side rough from the ctenoid scales. Lateral line strongly arched above the pectoral fin. Br. viii; D 65–81; A 50–64; P 10–11; V 6; lat. sc. 73–90; vert. 40–41. Sandy brown above, white below. To 25 cm.

Inshore and in shallow offshore waters, on sandy grounds; off all coasts; abundant.

Microstomus kitt (Walbaum).
Lemon sole — Fig. 14.16

Outline oval, greatest width about one-half of length; head small, less than 0.17 of total length. Mouth small, the maxilla extending to the front of the right eye. Right eye in advance of the left. Scales cycloid, the skin smooth and slimy. Lateral line with a shallow arch above the pectoral fin. Caudal peduncle very short. Br. vii; D 85–97; A 69–76; P 10; v 5–6; lat. sc. 130; vert. 48. Light brown with darker and lighter irregular blotches. To 66 cm.

Offshore on sandy or shelly grounds; off all coasts but commoner in the north.

Pleuronectes flesus (Linnaeus).
Flounder, fluke — Fig. 14.16

Outline oval; head large, 0.25 of total length. Mouth small; maxilla not extending beneath the right eye. Skin smooth except for a band of prickles along the lateral line and a row of tubercles along the base of the dorsal and anal fins. Lateral line slightly curved above the pectoral fin. Anal fin preceded by a spine. Caudal peduncle long, half the length of the fin. Br. v; D 52–67; A 35–46; P 10; V 6; vert. 36. Colour variable above, but often dark greeny-brown, occasionally with pale orange blotches; dense, opaque white below. To 50 cm.

Inshore, particularly in estuaries; all coasts; common.

Pleuronectes platessa Linnaeus.
Plaice — Fig. 14.16

Outline oval; head slightly less than 0.25 of total length. Mouth fairly small, the maxilla reaching to just below the right eye. Skin smooth, without prickles. Lateral line slightly curved above the pectoral fin; a line of tubercles extending between the lateral line and the interorbital ridge. Anal fin preceded by a spine. Caudal peduncle about one-third the length of the fin. Br. vii; D 65–79; A 48–59; P 10–11; V 6; vert. 42–43. Dark brown above, brightly blotched with orange; pearly (not opaque) white below. To 50 cm.

Shelf waters on sandy grounds, the young fish in particular coming right inshore; off all coasts; common.

37 SOLEIDAE

Elongate-oval flatfishes with eyes on the right side; head margin rounded, the mouth not terminal. Margin of preoperculum covered by skin. Scales small, ctenoid; the skin rough.

---

1. Pectoral fin on the eyed side unspotted, that on the blind side vestigial; every 5th or 6th dorsal and anal finray black. (A small fish not exceeding 13 cm in length) — ***Buglossidium luteum***

   Pectoral fin on the eyed side with a black spot; the blind side fin hardly smaller than that on the eyed side; dorsal and anal finrays uniformly coloured — **2**

---

2. Anterior nostril on the blind side tubular, but not enlarged; black spot extending to the free end of the pectoral fin — ***Solea solea***

   Anterior nostril on the blind side enlarged, rosette-like; black spot not extending to the free end of the pectoral fin; south and west — ***Solea lascaris***

Buglossidium luteum (Risso).
Solenette — Fig. 14.16

Pectoral fins small, unequal; that on the eyed side shorter than the pelvics, its uppermost ray much longer than the others; blind side pectoral vestigial. Dorsal and anal fins terminating just short of the caudal. Br. viii; D 71–78; A 55–56; RP 5, LP 3; V 5; lat. sc. 70–75; vert. 36. Light sandy brown, speckled darker; pectoral fin with a dark central spot; dorsal and anal fins pale with every fifth or sixth (occasionally fourth or seventh) ray black. To 13 cm. On sandy grounds in shallow water, inshore and offshore. Common on all coasts.

Solea lascaris (Risso). Sand sole Fig. 14.16
Pectoral fins approximately equal, much larger than pelvics. Dorsal and anal fin membranes continuous with the caudal. Anterior blind side nostril dilated, encircled by short branched filaments. Br. vii; D 71–90; A 55–75; P 10; V 5; lat. sc. 96–140; vert. 46–67. Light brown above, speckled and spotted with dark brown; a black spot on the right pectoral does not reach the fin margin. To 35 cm.

On sandy grounds in coastal waters, the young coming right inshore; south-west Britain only; quite common.

Solea solea (Linnaeus). Sole Fig. 14.16
Pectoral fins approximately equal, longer than the pelvics. Dorsal and anal fin membranes continuous with the caudal. Anterior blind side nostril small and tubular, not fringed. Br. vii; D 75–93; A 59–79; P 7; V 5–6; lat. sc. 140–165; vert. 49–52. Brown above, with darker blotches, a black spot on the right pectoral extends to the fin margin. To 50 cm.

On sandy and fine grounds in coastal waters, including estuaries. Off all coasts, but rare towards the north of Scotland.

REFERENCES AND FURTHER READING

CHAPTER 1 INTRODUCTION

Barnes, R. S. K. and Hughes, R. N. (1982). *An introduction to marine ecology*. Blackwell Scientific Publications, Oxford.

Boaden, P. J. S. and Seed, R. (1985). *An introduction to coastal ecology*. Blackie, Glasgow.

Briggs, J. C. (1974). *Marine zoogeography*. McGraw Hill, New York.

Earll, R. and Erwin, D.G. (1983). *Sublittoral ecology*. Clarendon Press, Oxford.

Forbes, E. and Godwin-Austen, R. (1859). *The natural history of the European Seas*. Van Voorst, London.

Gosse, E. W. (1907). *Father and son; A study of two temperaments*. William Heinemann, London.

Jones, W. E. and Demetropoulos, A. (1968). Exposure to wave action: measurements of an important ecological parameter on rocky shores on Anglesey. *Journal of Experimental Marine Biology and Ecology*, **2**, (1), 46–63.

King, C. A. M. (1975). *Introduction to physical and biological oceanography*. Edward Arnold, London.

Kinne, O. (1970). *Marine ecology*, Vol. 1, *Environmental factors, Part 1*. John Wiley, London.

Kinne, O. (1971). *Marine ecology*, Vol. 1, *Environmental factors, Part 2*. John Wiley, Chichester.

Kinne, O. (1975). *Marine ecology*, Vol. 2, *Physiological mechanisms, Part 1*. John Wiley, Chichester.

Kinne, O. (1975). *Marine ecology*, Vol. 2, *Physiological mechanisms, Part 2*. John Wiley, Chichester.

Kinne, O. (1976). *Marine ecology*, Vol. 3, *Cultivation, Part 1*. John Wiley, Chichester.

Kinne, O. (1977). *Marine ecology*, Vol. 3, *Cultivation, Part 2*. John Wiley, Chichester.

Kinne, O. (1977). *Marine ecology*, Vol. 3, *Cultivation, Part 3*. John Wiley, Chichester.

Kinne, O. (1978). *Marine ecology*, Vol. 4, *Dynamics*. John Wiley, Chichester.

Kinne, O. (1982). *Marine ecology*, Vol. 5, *Ocean management, Part 1*. John Wiley, Chichester.

Kinne, O. (1983). *Marine ecology*, Vol. 5, *Ocean management, Part 2*. John Wiley, Chichester.

Kinne, O. (1984). *Marine ecology*, Vol. 5, *Ocean management, Part 3*. John Wiley, Chichester.

Kinne, O. (1984). *Marine ecology*, Vol. 5, *Ocean management, Part 4*. John Wiley, Chichester.

Lee, A. J. and Ramster, J. W. (ed.) (1981). *Atlas of the seas around the British Isles*. Ministry of Agriculture, Fisheries and Food, London.

Lewis, J. R. (1964). *The ecology of rocky shores*. English Universities Press, London.

Moore, P. G. and Seed, R. (1985). *The ecology of rocky coasts: Essays presented to J. R. Lewis*. Hodder and Stoughton, London.

Price, J. H., Irvine, D. E. G., and Farnham, W. F. (ed.) (1980). *The shore environment*, 2 volumes. Academic Press, London.

Stephenson, T. A. and Stephenson, A. (1972). *Life between tidemarks on rocky shores*. W. H. Freeman, San Francisco.

Yonge, C. M. (1949). *The sea shore*. Collins, London.

CHAPTER 2 USING THIS BOOK

Erwin, D. and Picton, B. (1990). *Guide to inshore marine life*, (2nd edn). Immel Publishing, London.

Fish, J. D. and Fish, S. (1989). *A student's guide to the seashore*. Unwin Hyman, London.

Hayward, P. J. and Ryland, J. S. (ed.) (1990). *The marine fauna of the British Isles and north-west Europe*, 2 volumes. Clarendon Press, Oxford.

Higgins, R. P. and Thiel, H. (1988). *Introduction to the study of meiofauna*. Smithsonian Institution, Washington.

Howson, C. M. (ed.) (1987). *Directory of the British marine fauna and flora*. Marine Conservation Society, Ross-on-Wye.

Platt, H. M. and Warwick, R. M. (1983). Free-living marine nematodes, Part I. British Enoplids. *Linnean Society Synopses of the British Fauna (ns)*, **28**, 1–307.

Sims, R. W. (1980). *Animal identification: A reference guide. 1. Marine and brackish water animals*. British Museum (Natural History) and John Wiley, London.

Smaldon, G. and Lee, E. W. (1979). *A synopsis of methods for the narcotization of marine invertebrates*, Information series, Natural History, No. 6. Royal Scottish Museum, Edinburgh.

CHAPTER 3 PORIFERA

Ackers, R. G., Moss, D., Picton, B. E., and Stone, S. M. K. (1985). *Sponges of the British Isles*, (4th edn). Marine Conservation Society, Ross-on-Wye.

Arndt, W. (1935). Porifera. *Tierwelt der Nord- und Ostsee*, **3a**, 1–140.

Bergquist, P. R. (1978). *Sponges*. Hutchinson, London.

Burton, M. (1963). *A revision of the classification of the calcareous sponges*. British Museum (Natural History), London.

Hartmann, W. D. (1982). *Porifera*. In *Synopsis and classification of living organisms*, vol. 1, (ed. S. P. Parker), pp. 641–66. McGraw-Hill, New York.

Vacelet, J. and Boury-Esnault, N. (ed.) (1987). *Taxonomy of Porifera*. Springer-Verlag, Berlin.

CHAPTER 4 CNIDARIA AND CTENOPHORA

Allman, G. J. (1871, 1872). *A monograph of the gymnoblastic or tubularian hydroids*. Ray Society, London. Vol. I, part (1871), pp. 1–154; Vol. I, remainder (1872), pp. 155–231; Vol. II (1872), pp. 232–450.

Bouillon, J., Boero, F., Cicogna, F., and Cornelius P. F. S. (1987). *Modern trends in the systematics, ecology and evolution of hydroids and hydromedusae*. Clarendon Press, Oxford.

Broch, H. (1928). Hydrozoa, I. *Tierwelt der Nord- und Ostsee*, **36**, 1–100.

Corbin, P. G. (1979). The seasonal abundance of four species of Stauromedusae (Coelenterata: Scyphomedusae) at Plymouth. *Journal of the Marine Biological Association of the UK*, **59**, 385–91.

Cornelius, P. F. S. (1975). The hydroid species of *Obelia* (Coelenterata, Hydrozoa: Campanulariidae), with notes on the medusa stage. *Bulletin of the British Museum (Natural History), Zoology*, **28**, 249–93.

Cornelius, P. F. S. (1975). A revision of the species of Lafoeidae and Haleciidae (Coelenterata: Hydroida) recorded from Britain and nearby seas. *Bulletin of the British Museum (Natural History), Zoology*, **28**, 373–426.

Cornelius, P. F. S. (1982). Hydroids and medusae of the family Campanulariidae recorded from the eastern North Atlantic, with a world synopsis of genera. *Bulletin of the British Museum (Natural History), Zoology*, **42**, 37–148.

Hincks, T. (1868). *A history of the British hydroid zoophytes*, 2 volumes. Van Voorst, London.

Kirkpatrick, P. A. and Pugh, P. R. (1984). Siphonophores and Velellids. *Linnean Society Synopses of the British Fauna (ns)*, **29**, 1–154.

Kramp, P. L. (1961). Synopsis of the medusae of the world. *Journal of the Marine Biological Association of the UK*, **40**, 292–303.

Manuel, R. L. (1983). *The Anthozoa of the British Isles – a colour guide*, (2nd edn). Marine Conservation Society, Ross-on-Wye.

Manuel, R. L. (1988). British Anthozoa. *Linnean Society Synopses of the British Fauna (ns)*, **18**, (revised edition).

Millard, N. A. H. (1975). Monograph of the Hydroida of Southern Africa. *Annals of the South African Museum*, **68**, 1–513.

Naumov, D. V. (1961). Scyphomedusae of the seas of the USSR. *Opred Faune SSSR*, **75**, 1–98. (in Russian).

Naumov, D. V. (1969). Hydroids and hydromedusae of the USSR. *Fauna USSR*, **70**, 1–660 (Israel Program for Scientific Translation, Cat. no. 5108).

Russell, F. S. (1953). *The medusae of the British Isles. Anthomedusae, Leptomedusae, Limnomedusae, Trachymedusae and Narcomedusae*. Cambridge University Press, Cambridge.

Russell, F. S. (1970). *The medusae of the British Isles. II. Pelagic Scyphozoa, with a supplement to the first volume on hydromedusae*. Cambridge University Press, Cambridge.

Shick, J. M. (1991). *A functional biology of sea anemones*. Chapman & Hall, London.

Stephenson, T. A. (1928). *The British sea anemones*, Vol. 1. The Ray Society, London.

Stephenson, T. A. (1935). *The British sea anemones*, Vol. 2. The Ray Society, London.

Svoboda, A. and Cornelius, P. F. S. 1991. The European and Mediterranean species of *Aglaophenia* (Cnidaria: Hydrozoa). *Zoologische Verhandelingen*, **274**, 1–72.

CHAPTER 5 PLATYHELMINTHES AND NEMERTEA

Ball, I. R. and Reynoldson, T. B. (1981). British planarians. *Linnean Society Synopses of the British Fauna (ns)*, **19**, 1–141.

Brunberg, L. (1964). On the nemertean fauna of Danish waters. *Ophelia*, **1**, 77–111.

de Beauchamp, P. (1961). *Turbellaria*. In *Traité de Zoologie*, (ed. P. Grassé) pp. 1–212, Vol. 4, Part 1. Masson, Paris.

Douglas, A. E. (1985). Growth and reproduction of *Convoluta roscoffensis* containing different naturally occurring algal symbionts. *Journal of the Marine Biological Association of the UK*, **65**, 871–9.

Gamble, F.W. (1893). Contributions to a knowledge of British marine Turbellaria. *Quarterly Journal of Microscopical Science*, **34**, 433–528.

Gibson, R. (1982). British Nemerteans. *Linnean Society Synopses of the British Fauna (ns)*, **24**, 1–212.

Graff, L. von (1913). Turbellarien II, Rhabdocoela. *Das Tierreich*, **35**, 1–484.

Hallez, P. (1909). Biologie, organization, histologieet embryologiedun, *Paravortex cardii* n. sp. Rhabdocoele parasite du *Cardium edule L. Archives de Zoologie Experimentale et generale*, Ser. 4, **9**, 429–544.

Hyman, L. B. (1951). *The invertebrates*, Vol. II, *Platyhelminthes and Rhynchocoela*. McGraw-Hill. New York.

Meixner, J. (1938). Turbellaria (Strudelwurmes). *Tierwelt der Nord- und Ostsee*, **33**, (4b), 1–146.

Prudhoe, S. (1982). British polyclad turbellarians. *Linnean Society Synopses of the British Fauna*, **26**, 1–77.

Westblad, E. (1953). Marine Macrostomida (Turbellaria) from Scandinavia and England. *Arkiv för Zoologi*, **4**, 391–408.

CHAPTER 6 ANNELIDA

Brinkhurst, R. O. (1982). British and other marine and estuarine oligochaetes. *Linnean Society Synopses of the British Fauna (ns)*, **21**, 1–127.

Chambers, S. (1985). *Polychaetes from Scottish Waters, 2. Aphroditidae, Sigalionidae and Polydontidae*. Royal Scottish Museum, Edinburgh.

Chambers, S. J. and Garwood, P. R. (1992). *Polychaetes from Scottish waters, 3. Nereidae*. Royal Scottish Museum, Edinburgh.

Garwood, P. R. (1981). The marine fauna of the Cullercoats district. No. 9, Polychaeta, Errantia. *Report of the Dove marine Laboratory* ser. 3, **22**, 1–192.

Garwood, P. R. (1982). The marine fauna of the Cullercoats district. No. 10, Polychaeta, Sedentaria, incl. Archiannelida. *Report of the Dove marine Laboratory* ser. 3, **23**, 1–270.

George, J. D. and Hartmann-Schröder, G. (1985). Polychaetes: British Amphinomida, Spintherida and Eunicida. *Linnean Society Synopses of the British Fauna (ns)*, **32**, 1–221.

Giere, O. and Pfannkuche, O. (1982). Biology and ecology of marine Oligochaeta, a review. *Oceanography and Marine Biology Annual Review*, **20**, 173–308.

Holthe, T. (1986). Polychaeta Terebellomorpha. In *Marine invertebrates of Scandinavia*, Vol. 7. Norwegian University Press, Oslo.

Knight-Jones, P. (1983). Contributions to the taxonomy of Sabellidae (Polychaeta). *Zoological Journal of the Linnean Society*, **79**, 245–95.

Knight-Jones, P. and Knight-Jones, E. W. (1977). Taxonomy and ecology of British Spirorbidae (Polychaeta). *Journal of the Marine Biological Association of the UK*, **57**, 453–99.

Nelson-Smith, A. (1967). *Catalogue of main marine fouling organisms*, Vol. 3, *Serpulids*. OECD, Paris.

Pettibone, M. H. (1982). *Annelida*. In *Synopsis and classification of living organisms*, (ed. S. P. Parker), Vol. 2, 1–61. McGraw-Hill, New York.

Pjeijel, F. and Dales, R. P. (1991). Polychaetes: British Phyllodocoideans, Typhloscolecoideans and Tomopteroideans. *Linnean Society Synopses of the British Fauna (ns)*, **45**, 1–202.

Tebble, N. and Chambers, S. (1982). *Polychaetes from Scottish waters, 1. Polynoidae*. Royal Scottish Museum, Edinburgh.

Thorp, C. H., Knight-Jones, P., and Knight-Jones, E. W. (1986). New records of tubeworms established in British harbours. *Journal of the Marine Biological Association of the UK*, **66**, 881–8.

Westheide, W. (1990). Polychaetes: Interstitial families. *Linnean Society Synopses of the British Fauna (ns)*, **44**, 1–152.

CHAPTER 7 PRIAPULIDA, SIPUNCULIDA, ECHIURIDA, AND ENTOPROCTA

Gibbs, P. E. (1977). British Sipunculans. *Linnean Society Synopses of the British Fauna (ns)*, **12**, 1–35.

Gibbs, P. E. and Cutler, E. B. (1987). A classification of the phylum Sipuncula. *Bulletin of the British Museum (Natural History), Zoology*, **52**, 43–58.

Nielsen, C. (1964). Studies on Danish Entoprocta. *Ophelia*, **1**, 1–76.

Nielsen, C. (1989). Entoprocts. *Linnean Society Synopses of the British Fauna (ns)*, **41**, 1–131.

Stephen, A. C. and Edmonds, S. J. (1972). *The phyla Sipuncula and Echiura*. British Museum (Natural History), London.

CHAPTER 8 CRUSTACEA

Abele, L. G. (1982). *The biology of Crustacea. 1, Systematics, the fossil record, and biogeography*. Academic Press, New York.

Abele, L. G. (1982). *The biology of Crustacea. 2, Embryology, morphology and genetics*. Academic Press, New York.

Athersuch, J., Horne, D. J., and Whittaker, J. E. (1990). Marine and brackish water ostracods. *Linnean Society Synopses of the British Fauna (ns)*, **43**, 1–343.

Atwood, H. L. and Sandeman, D. C. (1982). *The biology of Crustacea. 3, Neurobiology: structure and function*. Academic Press, New York.

British Micropalaeontological Society (1977–1989). *A stereo atlas of Ostracod shells*, Vols 4–15. British Micropalaeontological Society, London.

Christiansen, M. E. (1969). Crustacea, Decapoda, Brachyura. In *Marine invertebrates of Scandinavia*, Vol. 2. Universitetsforlaget, Oslo.

Crothers, J. and Crothers, M. (1983). A key to the crabs and crab-like animals of inshore British waters. *Field Studies*, **5**, 753–806.

Edwards, E. (1978). *The Edible Crab and its fishery in British waters.* Fishing News Books, Surrey.

Fincham, A. A. and Wickins, J. F. (1976). *Identification of commercial prawns and shrimps.* British Museum (Natural History), London.

Gotto, V. (1993). Commensal and parasitic Copepods associated with marine invertebrates (and whales). *Linnean Society Synopses of the British Fauna (ns)*, **46**, 1–264.

Hoeg, J. T. and Lutzen, J. (1985). Crustacea: Rhizocephala. In *Marine invertebrates of Scandinavia*, Vol. **6**. Universitetsforlaget, Oslo.

Holdich, D. M. and Bird, G. J. (1986). Tanaidacea (Crustacea) from sublittoral waters off West Scotland, including descriptions of two new genera. *Journal of Natural History*, **20**, 79–100.

Holdich, D. M. and Jones, J. A. (1983). Tanaids. *Linnean Society Synopses of the British Fauna (ns)*, **27**, 1–98.

Ingle, R. W. 1980. *British Crabs.* Oxford University Press and British Museum (Natural History), Oxford and London.

Ingle, R. W. (1983). Shallow-water crabs. *Linnean Society Synopses of the British Fauna (ns)*, **25**, 1–206.

Ingle, R. W. (1992). *Hermit crabs of the north eastern Atlantic and the Mediterranean Sea.* Chapman & Hall and the Natural History Museum, London.

Jacobs, B. J. M. (1987). A taxonomic revision of the European, Mediterranean and north-west African species generally placed in *Sphaeroma* Bosc, 1802 (Isopoda: Flabellifera: Sphaeromatidae). *Zoologische Verhandelingen*, **238**, 3–71.

Jones, N. S. (1976). British cumaceans. *Linnean Society Synopses of the British Fauna (ns)*, **7**, 1–66.

Kabata, Z. (1979). *Parasitic copepods of British fishes.* Ray Society, London.

Kabata, Z. (1992). Copepods parasitic on fishes. *Linnean Society Synopses of the British Fauna (ns)*, **47**, 1–264.

Lincoln, R. J. (1979). *British gammaridean amphipods.* British Museum (Natural History), London.

Makings, P. (1977). A guide to the British coastal Mysidacea. *Field Studies*, **4**, 575–95.

Mantel, L. H. (1983). *The biology of Crustacea. 5, Internal anatomy and physiological regulation.* Academic Press, New York.

Mauchline, J. (1984). Euphausiid, Stomatopod and Leptostracan Crustaceans. *Linnean Society Synopses of the British Fauna (ns)*, **30**, 1–91.

Naylor, E. (1972). British marine isopods. *Linnean Society Synopses of the British Fauna (ns)*, **3**, 1–86.

Nillson-Cantell, C. A. (1978). Cirripedia Thoracica and Acrothoracica. In *Marine invertebrates of Scandinavia*, Vol. 5, pp. 1–133. Universitetsforlaget, Oslo.

Provenzano, A. J. (1983). *The biology of Crustacea. 6, Pathobiology.* Academic Press, New York.

Provenzano, A. J. (1985). *The biology of Crustacea. 10, Economic aspects: Fisheries and culture.* Academic Press, New York.

Sandeman, D. C. and Atwood, H. L. (1982). *The biology of Crustacea. 4, Neural integration and behaviour.* Academic Press, New York.

Smaldon, G. (1993). British coastal shrimps and prawns. *Linnean Society Synopsis of the British Fauna (n.s)*, **15**, (2nd edition, revised and enlarged by L. B. Holthuis and C. H. J. M. Fransen), 1–142.

Southward, A. J. (1963). Barnacles. *Catalogue of main marine fouling organisms*, Vol. 1. OECD, Paris.

Southward, A. J. (1976). On the taxonomic status and distribution of *Chthamalus stellatus* (Cirripedia), in the north-east Atlantic region: with a key to the common intertidal barnacles of Britain. *Journal of the Marine Biological Association of the UK*, **56**, 1007–28.

Tattersall, W. M. and Tattersall, O. S. (1951). *The British Mysidacea.* The Ray Society, London.

University of Leicester, Department of Geology. (1973–75). *A stereo atlas of ostracod shells*, Vols 1–2. University of Leicester.

University of Leicester, Department of Geology, 1976. *A stereo atlas of ostracod shells*, Vol. 3. Broadwater Press, Welwyn.

Vernberg, F. J. and Vernberg, W. B. (1983). *The biology of Crustacea*, Vol. 7, *Behaviour and ecology.* Academic Press, New York.

Vernberg, F. J. and Vernberg, W. B. (1983). *The biology of Crustacea*, Vol. 8, *Environmental adaptations.* Academic Press, New York.

Warner, G. F. (1977). *The biology of crabs.* Elek Science, London.

Yamaguti, S. (1963). *Parasitic Copepoda and Branchiura of fishes.* Interscience, New York.

CHAPTER 9 OTHER ARTHROPODS

Blower, J. G. (1985). Millipedes. *Linnean Society Synopses of the British Fauna (ns)*, **35**, 1–242.

Cheng, L. (1976). *Marine insects.* Elsevier-North Holland, Amsterdam.

Chinery, M. (1986). *Collins guide to the insects of Britain and western Europe.* Collins, London.

Evans, G. O. and Browning, E. (1955). Techniques for the preparation of mites for study. *Annals and Magazine of Natural History* Ser. 12, **8**, 631–5.

Evans, G. O. and Till, W. M. (1979). Mesostigmatic mites of Britain and Ireland (Chelicerata: Acari-Parasitiformes). An introduction to their external morphology and classification. *Transactions of the Zoological Society, London*, **35**, 139–270.

Evans, G. O., Griffiths, D. A., Macfarlane, D., Murphy, P. W. and Till, W. M. (1989).

International course in acarology, practical manual. 5 volumes. Department of Pure and Applied Zoology, University of Reading.

Gough, J. J. (1979). A key for the identification of the families of Collembola recorded from the British Isles. *Entomologist's Monthly Magazine*, **113**, 193–7.

Green, J. and Macquitty, M. (1987). Halacarid Mites. *Linnean Society Synopses of the British Fauna, (ns)*, **36**, 1–178.

King, P. E. (1974). British sea spiders. *Linnean Society Synopses of the British Fauna (ns)*, 5, 1–68.

King, P. E. (1986). A revised key to the adults of littoral pycnogonids in the British Isles. *Field Studies*, **6**, 493–516.

Legg, G. and Jones, R. E. (1988). Pseudoscorpiones (Arthropoda: Arachnida). *Linnean Society Synopses of the British Fauna (ns)*, **40**, 1–159.

Lockett, G. H., Millidge, A. F., and Merrett, P. (1974). *British spiders*, Vol. 3. Ray Society, London.

Lockett, G. H., Millidge, A. F. and Merrett, P. (1981). *British spiders*, Vols 1, 2. Ray Society, London.

Unwin, D. M. (1984). A key to the families of British Coleoptera. *Field Studies*, **6**, 149–97.

CHAPTER 10 MOLLUSCA

Clarke, M. R. (1966). A review of the systematics and ecology of oceanic squids. *Advances in Marine Biology*, **4**, 91–300.

Clarke, M. R. (ed.) (1986). *A handbook for the identification of cephalopod beaks*. Clarendon Press, Oxford.

Dell'Angelo, B. and Tursi, A. (1978). Guida bibliografica ai chitoni (Polyplacophora): 1970–1978. *Oebalia*, **4**, 79–151.

Dell'Angelo, B. and Tursi, A. (1978). Guida bibliografica ai chitoni (Polyplacophora): 1970–1979 primo supplemento. *Oebalia*, **7**, 43–133.

Fretter, V. and Graham, A. (1962). *British prosobranch molluscs*. Ray Society, London.

Fretter, V. and Graham, A. (1976). The prosobranch molluscs of Britain and Denmark. Part 1, Pleurotomariacea, Fissurellacea and Patellacea. *Journal of Molluscan Studies*, (Suppl. 1), 1–38.

Fretter, V. and Graham, A. (1977). The prosobranch molluscs of Britain and Denmark. Part 2, Trochacea. *Journal of Molluscan Studies*, (Suppl. 3), 39–100.

Fretter, V. and Graham, A. (1978). The prosobranch molluscs of Britain and Denmark. Part 3, Neritacea, Viviparacea, Valvatacea, terrestrial and freshwater Littorinacea and Rissoacea. *Journal of Molluscan Studies*, (Suppl. 5), 101–52.

Fretter, V. and Graham, A. (1978). The prosobranch molluscs of Britain and Denmark. Part 4, Marine Rissoacea. *Journal of Molluscan Studies*, (Suppl. 6), 153–241.

Fretter, V. and Graham, A. (1980). The prosobranch molluscs of Britain and Denmark. Part 5, Marine Littorinacea. *Journal of Molluscan Studies* (Suppl. 7), 243–54.

Fretter, V. and Graham, A. (1981). The prosobranch molluscs of Britain and Denmark. Part 6, Cerithiacea, Strombacea, Hipponiacea, Calyptracea, Lamellariacea, Cypraeacea, Naticacea, Tonnacea, Heteropoda. *Journal of Molluscan Studies*, (Suppl. 9), 285–363.

Fretter, V. and Graham, A. (1982). The prosobranch molluscs of Britain and Denmark. Part 7, Heterogastropoda (Cerithiopsacea, Triforacea, Epitoniacea, Eulimacea). *Journal of Molluscan Studies*, (Suppl. 11), 363–434.

Fretter, V. and Graham, A. (1985). The prosobranch molluscs of Britain and Denmark. Part 8, Neogastropoda. *Journal of Molluscan Studies*, (Suppl. 15), 435–556.

Graham, A. (1988). British prosobranch and other operculate gastropod molluscs. *Linnean Society Synopses of the British Fauna (ns)*, **2** (2nd ed), 1–662.

Hochachka, P. W. (1983). *The Mollusca. 1. Metabolic biochemistry and molecular biomechanics*. Academic Press, London and New York.

Hochachka, P. W. (1983). *The Mollusca*. 2. *Environmental biochemistry and physiology*. Academic Press, London and New York.

Hyman, L. H. (1967). *The invertebrates*, Vol. 6, *Mollusca 1*. McGraw-Hill, New York and London.

Jones, A. M. and Baxter, J. M. (1987). Mollusca: Caudofoveata, Solenogastres, Polyplacophora and Scaphopoda. *Linnean Society Synopses of the British Fauna (ns)*, **37**, 1–123.

Just, H. and Edmunds, M. (1985). North Atlantic nudibranchs (Mollusca) as seen by Henning Lemche. *Ophelia*, (Suppl. 2), 1–170.

Morton, J. E. 1967. *Molluscs*, (4th edn). Hutchinson, London.

Nixon, N. and Messenger, J. B. (eds) (1977). Biology of cephalopods. *Symposia of the Zoological Society of London*, **38**, pp. xviii + 615.

Purchon, R. D. (1977). *The biology of the Mollusca*, (2nd edn). Pergamon Press, Oxford.

Roper, C. F. E., Sweeny, M. J., and Nauen, C. E. (1984). *FAO species catalogue*, Vol. 3, *Cephalopods of the world*, FAO fisheries Synopsis, no. **125**.

Russell-Hunter, W. D. (1983). *The mollusca*, Vol. 6, *Ecology*. Academic Press, London and New York.

Saleuddin, A. S. M. and Wilbur, K. M. (1983). *The Mollusca*, Vol. 4, *Physiology*, *Part 1*. Academic Press, London and New York.

Saleuddin, A. S. M. and Wilbur, K. M. (1983). *The Mollusca* Vol, 5, *Physiology, Part 2*. Academic Press, London and New York.

Tebble, N. (1976). *British bivalve seashells*, (2nd edn). HMSO, Edinburgh.

Thompson, T. E. (1976). *Biology of opisthobranch molluscs*, Vol. 1. The Ray Society, London.

Thompson, T. E. (1988). Molluscs: benthic opisthobranchs. *Linnean Society Synopses of the British Fauna (ns)*, **8**. (2nd edn), 1–356.

Thompson, T. E. and Brown, G. H. (1984). *Biology of opisthobranch molluscs*, Vol. 2. The Ray Society, London.

Tompa, A. S., Verdonk, N. H., and van den Biggelaar, J. A. M. (1984). *The Mollusca*, Vol, 7, *Reproduction*. Academic Press, London and New York.

Verdonk, N. H., van den Biggelaar, J. A. M. and Tompa, A. S. (1983). *The Mollusca*, Vol. 3, *Development*. Academic Press, London and New York.

Yonge, C. M. and Thompson, T. E. (1976). *Living marine molluscs*. Collins, London.

CHAPTER 11 BRYOZOA

Hayward, P. J. (1985). Ctenostome bryozoans. *Linnean Society Synopses of the British Fauna (ns)*, **33**, 1–169.

Hayward, P. J. and Ryland, J. S. (1979). British ascophoran bryozoans. *Linnean Society Synopses of the British Fauna (ns)*, **14**, 1–312.

Hayward, P. J. and Ryland, J. S. (1985). Cyclostome bryozoans. *Linnean Society Synopses of the British Fauna (ns)*, **35**, 1–147.

Ryland, J. S. (1965). Polyzoa. *Catalogue of main marine fouling organisms*, Vol. 2. OECD, Paris.

Ryland, J. S. (1970). *Bryozoans*. Hutchinson University Library, London.

Ryland, J. S. (1982). Bryozoa. In *Synopsis and classification of living organisms*, 2, (ed. S. P. Parker), pp. 743–63. McGraw-Hill, New York.

Ryland, J. S. and Hayward, P. J. (1977). British anascan bryozoans. *Linnean Society Synopses of the British Fauna (ns)*, **10**, 1–188.

CHAPTER 12 ECHINODERMATA

Clark, A. M. (1977). *Starfishes and related echinoderms*. British Museum (Natural History), London.

Clark, A. M. and Downey, M. E. (1992). *Starfishes of the Atlantic*. Chapman & Hall, London.

Mortensen, T. (1927). *Handbook of the echinoderms of the British Isles*. Oxford University Press, Oxford.

Nichols, D. (1969). *Echinoderms*. Hutchinson University Library, London.

Picton, B. E. (1993). *A field guide to the shallow-water echinoderms of the British Isles*. Immel Publishing, London.

CHAPTER 13 HEMICHORDATA AND UROCHORDATA

Berrill, N. J. (1950). *The Tunicata, with an account of the British species*. The Ray Society, London.

Burdon-Jones, C. and Patil, A. M. (1960). A revision of the genus *Saccoglossus* (Enteropneusta) in British waters. *Proceedings of the Zoological Society, London*, **134**, 635–45.

Lafargue, F. and Wahl, M. (1987). The didemnid ascidian fauna of France. *Annales de l'Institute Océanographique, Monaco*, **63**, 1–46.

Millar, R. H. (1966). Tunicata Ascidiacea. In *Marine invertebrates of Scandinavia*, Vol. 1. Universitetsforlaget, Oslo.

Millar, R. H. (1970). British ascidians. *Linnean Society Synopses of the British Fauna (ns)*, **1**, 1–92.

Millar R. H. (1971). The biology of ascidians. *Advances in Marine Biology*, **9**, 1–100.

Picton, B. E. (1985). *Ascidians of the British Isles. A colour guide*. Marine Conservation Society, Ross-on-Wye.

Stebbing, A. R. D. (1970). The status and ecology of *Rhabdopleura compacta* Hincks (Hemichordata) from Plymouth. *Journal of the Marine Biological Association of the UK*, **50**, 209–21.

Stebbing, A. R. D. (1970). Aspects of the reproduction and life cycle of *Rhabdopleura compacta* (Hemichordata). *Marine Biology* **5**, 205–12.

CHAPTER 14 EUCHORDATA (VERTEBRATA) PISCES

Bone, Q. and Marshall, N. B. (1982). *Biology of fishes*. Blackie, Glasgow.

Lythgoe, J. and Lythgoe, G. (1991). *Fishes of the sea*. Blandford, London.

Norman, J. R. and Greenwood, P. H. (1974). *A history of fishes*, (3rd. edn). Benn, London.

Wheeler, A. (1969). *The fishes of the British Isles and north-west Europe*. Macmillan, London.

Wheeler, A. (1978). *Key to the fishes of northern Europe*. Warne, London.

Wheeler, A. (1992). A list of the common and scientific names of fishes of the British Isles. *Journal of Fish Biology*, **41**, (Suppl. A).

Whitehead, P. J. P., Bauchot, M.-L., Hureau, J.-C., Nielsen, J., and Tortonese, E. (1984–86). *Fishes of the north-eastern Atlantic and the Mediterranean*. Vols. 1–3. Unesco, Paris.

TAXONOMIC INDEX

SUBJECT INDEX